Renewable Energy

Renewable Energy
Sources for Fuels and Electricity

Edited by

Thomas B. Johansson
Henry Kelly
Amulya K. N. Reddy
Robert H. Williams

Executive Editor

Laurie Burnham

ISLAND PRESS

Washington, D.C. ❏ Covelo, California

About Island Press

Island Press, a nonprofit organization, publishes, markets, and distributes the most advanced thinking on the conservation of our natural resources—books about soil, land, water, forests, wildlife, and hazardous and toxic wastes. These books are practical tools used by public officials, business and industry leaders, natural resource managers, and concerned citizens working to solve both local and global resource problems.

Founded in 1978, Island Press reorganized in 1984 to meet the increasing demand for substantive books on all resource-related issues. Island Press publishes and distributes under its own imprint and offers these services to other nonprofit organizations.

Support for Island Press is provided by the Geraldine R. Dodge Foundation, The Energy Foundation, The Charles Engelhard Foundation, The Ford Foundation, Glen Eagles Foundation, The George Gund Foundation, William and Flora Hewlett Foundation, The John D. and Catherine T. MacArthur Foundation, The Andrew W. Mellon Foundation, The Joyce Mertz-Gilmore Foundation, The New-Land Foundation, The J. N. Pew, Jr. Charitable Trust, Alida Rockefeller, The Rockefeller Brothers Fund, The Rockefeller Foundation, The Tides Foundation, and individual donors.

© 1993 Island Press

Library of Congress Cataloging-in-Publication Data

Renewable energy: sources for fuels and electricity / edited by Thomas B.
    Johansson ... [et al.].
      p. cm.
    Includes bibliographical references and index.
    ISBN 1–55963–139–2 (cloth : acid-free paper)
    ISBN 1–55963–138–4 (pbk.)
    1. Renewable energy sources. I. Johansson, Thomas B.
621.042—dc20   92–14194
                                                    CIP

Printed on recycled, acid-free paper

Manufactured in the United States of America

10 9 8 7 6 5 4 3 2 1

Island Press wishes to thank the Energy Foundation for its funding to support the publication of this book. Additional funding was provided by the Geraldine R. Dodge Foundation and the Joyce Mertz-Gilmore Foundation.

The editors of *Renewable Energy* thank the governments of Sweden, Norway, and The Netherlands for financing the preparation and production of this camera-ready volume.

Renewable Energy was commissioned by the United Nations Solar Energy Group on Environment and Development as input to the 1992 United Nations Conference on Environment and Development in Rio de Janeiro.

CONTENTS

1 RENEWABLE FUELS AND ELECTRICITY FOR A GROWING WORLD ECONOMY: DEFINING AND ACHIEVING THE POTENTIAL
Thomas B. Johansson, Henry Kelly, Amulya K. N. Reddy and
Robert H. Williams   1

2 HYDROPOWER AND ITS CONSTRAINTS
José Roberto Moreira and Alan Douglas Poole   73

3 WIND ENERGY: TECHNOLOGY AND ECONOMICS
Alfred J. Cavallo, Susan M. Hock and Don R. Smith   121

4 WIND ENERGY: RESOURCES, SYSTEMS AND REGIONAL STRATEGIES
Michael J. Grubb and Niels I. Meyer   157

5 SOLAR-THERMAL ELECTRIC TECHNOLOGY
Pascal De Laquil III, David Kearney, Michael Geyer and Richard Diver   213

6 INTRODUCTION TO PHOTOVOLTAIC TECHNOLOGY
Henry Kelly   297

7 CRYSTALLINE- AND POLYCRYSTALLINE-SILICON SOLAR CELLS
Martin A. Green   337

8 PHOTOVOLTAIC CONCENTRATOR TECHNOLOGY
Eldon C. Boes and Antonio Luque   361

9 AMORPHOUS SILICON PHOTOVOLTAIC SYSTEMS
David E. Carlson and Sigurd Wagner   403

10 POLYCRYSTALLINE THIN-FILM PHOTOVOLTAICS
Ken Zweibel and Allen M. Barnett   437

11 UTILITY FIELD EXPERIENCE WITH PHOTOVOLTAIC SYSTEMS
Kay Firor, Roberto Vigotti and Joseph J. Iannucci   483

12 OCEAN ENERGY SYSTEMS
James E. Cavanagh, John H. Clarke and Roger Price   513

13 GEOTHERMAL ENERGY
Civis G. Palmerini   549

14 BIOMASS FOR ENERGY: SUPPLY PROSPECTS
David O. Hall, Frank Rosillo-Calle, Robert H. Williams and Jeremy Woods   593

15 BIOENERGY: DIRECT APPLICATIONS IN COOKING
Gautam S. Dutt and N. H. Ravindranath   653

16 OPEN-TOP WOOD GASIFIERS
H. S. Mukunda, S. Dasappa and U. Shrinivasa   699

17 ADVANCED GASIFICATION-BASED BIOMASS POWER GENERATION
Robert H. Williams and Eric D. Larson   729

18 BIOGAS ELECTRICITY—THE PURA VILLAGE CASE STUDY
P. Rajabapaiah, S. Jayakumar and Amulya K. N. Reddy   787

19 ANAEROBIC DIGESTION FOR ENERGY PRODUCTION AND ENVIRONMENTAL PROTECTION
Gatze Lettinga and Adriaan C. van Haandel   817

20 THE BRAZILIAN FUEL-ALCOHOL PROGRAM
José Goldemberg, Lourival C. Monaco and Isaias C. Macedo   841

21 ETHANOL AND METHANOL FROM CELLULOSIC BIOMASS
Charles E. Wyman, Richard L. Bain, Norman D. Hinman and Don J. Stevens 865

22 SOLAR HYDROGEN
Joan M. Ogden and Joachim Nitsch   925

23 UTILITY STRATEGIES FOR USING RENEWABLES
Henry Kelly and Carl J. Weinberg   1011

**A RENEWABLES-INTENSIVE GLOBAL ENERGY SCENARIO
(APPENDIX TO CHAPTER 1)**
Thomas B. Johansson, Henry Kelly, Amulya K. N. Reddy and
Robert H. Williams   1071

(Authors' affiliations are listed at the end of this volume.)

PREFACE

The United Nations Conference on Environment and Development (UNCED) held in Rio de Janeiro, Brazil, in June 1992, addressed the challenges of achieving worldwide sustainable development. The goal of sustainable development cannot be realized without major changes in the world's energy system. Accordingly, *Agenda 21*, which was adopted by UNCED, called for "new policies or programs, as appropriate, to increase the contribution of environmentally safe and sound and cost-effective energy systems, particularly new and renewable ones, through less polluting and more efficient energy production, transmission, distribution, and use."

Renewable Energy: Sources for Fuels and Electricity was prepared as an input to the UNCED process. It was commissioned by the United Nations Solar Energy Group for Environment and Development (UNSEGED), a high-level group of experts convened by the United Nations under the mandate of the General Assembly Resolution A/45/208 of 21 December 1990. That resolution requested that the UNSEGED prepare a comprehensive and analytical study on new and renewable sources of energy aimed at providing a significant input to UNCED.

In response to this mandate, UNSEGED prepared the report *Solar Energy: A Strategy in Support of Environment and Development* (published as document A/AC.218/1992/Rev.1.) on the value and role of renewable sources of energy in promoting economic and social development in environmentally sustainable ways and commissioned a state-of-the-art assessment of renewable energy.

Renewable Energy: Sources for Fuels and Electricity, the complement to the UNSEGED report, assesses the technical and economic prospects for making fuels and electricity from renewable energy sources. The preparation of this book was an international effort carried out by some of the world's leading specialists on renewable energy and was supported by the governments of Sweden, Norway, and The Netherlands.

We are grateful to the book's executive editor, Dr. Laurie Burnham, for her dedication to the project, to the many reviewers of the individual chapters, whose detailed and constructive comments were helpful in ensuring the technical integrity of the analyses, to the book's production managers, Samantha Kanaga and John Shimwell, for the long hours they spent designing and producing the volume, to Michele Brown, for coordinating the production process, to Donna Riley, Claudia Stoy, and Ruth Williams for their assistance in production, and to Ryan Katofsky, Gregory Terzian, and Elinor Williams for their help with proofreading the final drafts. Finally, we are grateful to the staff of Island Press for their patience and understanding throughout the book's gestation. In particular, we thank Charles Savitt, the publisher, for his commitment to the book's goals, Joe Ingram, editor-in-chief, for his unfailing support, and Beth Beisel, production manager, for her advice and assistance with the book's production.

The Editors   July 1992

1
RENEWABLE FUELS AND ELECTRICITY FOR A GROWING WORLD ECONOMY
Defining and Achieving the Potential

THOMAS B. JOHANSSON
HENRY KELLY
AMULYA K.N. REDDY
ROBERT H. WILLIAMS

MAJOR FINDINGS

If the world economy expands to meet the aspirations of countries around the globe, energy demand is likely to increase even if strenuous efforts are made to increase the efficiency of energy use. Given adequate support, renewable energy technologies can meet much of the growing demand at prices lower than those usually forecast for conventional energy. By the middle of the 21st century, renewable sources of energy could account for three-fifths of the world's electricity market (see figure 1) and two-fifths of the market for fuels used directly (see figure 2).[1] Moreover, making a transition to a renewables-intensive energy economy would provide environmental and other benefits not measured in standard economic accounts (see Box A). For example, by 2050 global carbon dioxide (CO_2) emissions would be reduced to 75 percent of their 1985 levels provided that energy efficiency and renewables are both pursued aggressively (see figure 3a and 3 b). And because renewable energy is expected to be competitive with conventional energy, such benefits could be achieved at no additional cost.

This auspicious outlook for renewables reflects impressive technical gains made during the past decade. Renewable energy systems have benefited from developments in electronics, biotechnology, materials sciences, and in other energy areas. For example, advances in jet engines for military and civilian aircraft applications, and in coal gasification for reducing air pollution from coal combustion, have made it possible to produce electricity competitively using gas turbines de-

---

1. This renewables-intensive energy scenario would satisfy energy demands associated with an eight-fold increase in economic output for the world by the middle of the 21st century, as projected by the Response Strategies Working Group of the Intergovernmental Panel on Climate Change. World energy demand continues to grow in the scenario considered in spite of a rapid increase in energy efficiency; demand for electricity increases 265 percent between 1985 and 2050 in this scenario.

rived from jet engines and fired with gasified biomass.[2] And fuel cells developed originally for the space program have opened the door to the use of hydrogen as a non-polluting fuel for transportation. Indeed, many of the most promising options described in this book are the result of advances made in areas not directly related to renewable energy, and were scarcely considered a decade ago.

Moreover, because the size of most renewable energy equipment is small, renewable energy technologies can advance at a faster pace than conventional tech-

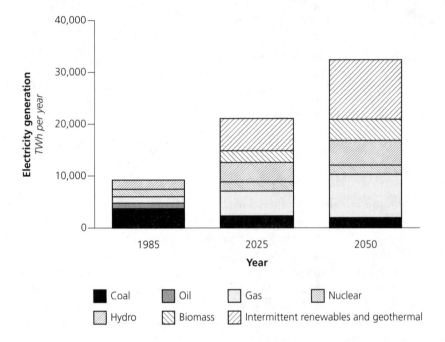

FIGURE 1: ELECTRICITY GENERATION FOR THE RENEWABLES-INTENSIVE GLOBAL ENERGY SCENARIO

Renewables can play major roles in the global energy economy in the decades ahead. In the global energy demand scenario adopted for this study, global electricity production would more than double by 2025, and more than triple by 2050. The share of renewable energy generation would increase from 20 percent in 1985 (mostly hydroelectric power) to about 60 percent in 2025, with roughly comparable contributions from hydropower, intermittent renewables (wind and direct solar power), and biomass. The contribution of intermittent renewables could be as high as 30 percent by the middle of the next century.

A high rate of penetration by intermittent renewables without electrical storage would be facilitated by emphasis on advanced natural gas–fired gas turbine power generating systems. Such power generating systems—characterized by low capital cost, high thermodynamic efficiency, and the flexibility to vary electrical output quickly in response to changes in the output of intermittent power-generating systems—would make it possible to "back up" the intermittent renewables at low cost, with little, if any, need for electrical storage. For the scenario developed here, the share of natural gas in power generation nearly doubles by 2025, from its 12 percent share in 1985.

---

2. In this document, the term "biomass" refers to any plant matter used directly as fuel or converted into fluid fuels or electricity. Sources of biomass are diverse and include the wastes of agricultural and forest-product operations as well as wood, sugarcane, and other plants grown specifically as energy crops.

nologies. While large energy facilities require extensive construction in the field, where labor is costly and productivity gains difficult to achieve, most renewable energy equipment can be constructed in factories, where it is easier to apply modern manufacturing techniques that facilitate cost reduction. The small scale of the equipment also makes the time required from initial design to operation short, so that needed improvements can be identified by field testing and quickly incorporated into modified designs. In this way, many generations of technology can be introduced in short periods.

Key elements of a renewables-intensive energy future

An energy future making intensive use of renewable resources is likely to have the following key characteristics:

▼ There would be a diversity of energy sources, the relative abundance of which would vary from region to region. Electricity could be provided by various combinations of hydroelectric power, intermittent renewable power sources (wind,

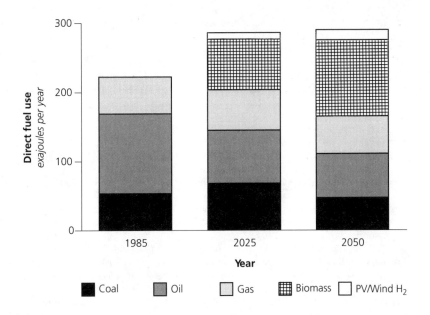

FIGURE 2: DIRECT FUEL-USE FOR THE RENEWABLES-INTENSIVE GLOBAL ENERGY SCENARIO

    In the global energy demand scenario adopted for this study, the use of fuels for purposes other than electricity generation will grow by less than one-third, much less than the generation of electricity (compare this figure with figure 1). The renewables contribution to fuels used directly could reach nearly one-fourth by 2025 and two-fifths by 2050, with most of the contribution coming from biomass-derived fuels—methanol, ethanol, hydrogen, and biogas. Methanol and hydrogen may well prove to be the biofuels of choice, because they are the energy carriers most easily used in the fuel cells that would be used for transportation.

solar-thermal electric, and photovoltaic power), biomass power, and geothermal power. Fuels could be provided by methanol, ethanol, hydrogen, and methane (biogas) derived from biomass, supplemented by hydrogen derived electrolytically from intermittent renewables.

▼ Emphasis would be given to the efficient use of both renewable and conventional energy supplies, in all sectors. Emphasis on efficient energy use facilitates the introduction of energy carriers such as methanol and hydrogen. It also makes it possible to extract more useful energy from such renewable resources as hydropower and biomass, which are limited by environmental or land use constraints.

Box A
Benefits of Renewable Energy not Captured in Standard Economic Accounts

Social and economic development: Production of renewable energy, particularly biomass, can provide economic development and employment opportunities, especially in rural areas, that otherwise have limited opportunities for economic growth. Renewable energy can thus help reduce poverty in rural areas and reduce pressures for urban migration.

Land restoration: Growing biomass for energy on degraded lands can provide the incentives and financing needed to restore lands rendered nearly useless by previous agricultural or forestry practices. Although lands farmed for energy would not be restored to their original condition, the recovery of these lands for biomass plantations would support rural development, prevent erosion, and provide a better habitat for wildlife than at present.

Reduced air pollution: Renewable energy technologies, such as methanol or hydrogen for fuel-cell vehicles, produce virtually none of the emissions associated with urban air pollution and acid deposition, without the need for costly additional controls.

Abatement of global warming: Renewable energy use does not produce carbon dioxide and other greenhouse emissions that contribute to global warming. Even the use of biomass fuels will not contribute to global warming: the carbon dioxide released when biomass is burned equals the amount absorbed from the atmosphere by plants as they are grown for biomass fuel.

Fuel supply diversity: There would be substantial interregional energy trade in a renewables-intensive energy future, involving a diversity of energy carriers and suppliers. Energy importers would be able to choose from among more producers and fuel types than they do today and thus would be less vulnerable to monopoly price manipulation or unexpected disruptions of supplies. Such competition would make wide swings in energy prices less likely, leading eventually to stabilization of the world oil price. The growth in world energy trade would also provide new opportunities for energy suppliers. Especially promising are the prospects for trade in alcohol fuels such as methanol derived from biomass, natural gas (not a renewable fuel but an important complement to renewables), and, later, hydrogen.

Reducing the risks of nuclear weapons proliferation: Competitive renewable resources could reduce incentives to build a large world infrastructure in support of nuclear energy, thus avoiding major increases in the production, transportation, and storage of plutonium and other nuclear materials that could be diverted to nuclear weapons production.

▼   Biomass would be widely used. Biomass would be grown sustainably and converted efficiently to electricity and liquid and gaseous fuels using modern technology, in contrast to the present situation, where biomass is used inefficiently and sometimes contributes to deforestation.

▼   Intermittent renewables would provide as much as one third of total electricity requirements cost-effectively in most regions, without the need for new electrical storage technologies.

▼   Natural gas would play a major role in supporting the growth of a renewable energy industry. Natural gas-fired turbines, which have low capital costs and can quickly adjust their electrical output, can provide excellent back-up for intermittent renewables on electric power grids. Natural gas would also help launch a biomass-based methanol industry; methanol might well be introduced using natural gas feedstocks before the shift to methanol derived from biomass occurs.

▼   A renewables-intensive energy future would introduce new choices and competition in energy markets. Growing trade in renewable fuels and natural gas

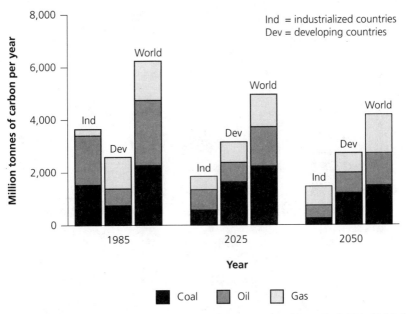

FIGURE 3a: EMISSIONS OF CO_2 FOR THE RENEWABLES-INTENSIVE GLOBAL ENERGY SCENARIO, BY WORLD REGION

   Global CO_2 emissions from the burning of fossil fuels associated with the renewables-intensive global energy scenario would be reduced 12 percent by 2025 and 26 percent by 2050.

   During this period, the CO_2 emissions from the industrialized countries (including former centrally planned Europe) would be reduced nearly in half by 2025 and nearly two-thirds by 2050. The industrialized country share of total worldwide emissions would decline from about three-fourths in 1985 to about two-fifths in 2025 to about one-third in 2050.

would diversify the mix of suppliers and the products traded (see figure 4), which would increase competition and reduce the likelihood of rapid price fluctuations and supply disruptions. It could also lead eventually to a stabilization of world energy prices. In addition, new opportunities for energy suppliers would be created. Especially promising are prospects for trade in alcohol fuels, such as methanol derived from biomass. Land-rich countries in sub-Saharan Africa and Latin America could become major alcohol fuel exporters.

▼ Most electricity produced from renewable sources would be fed into large electrical grids and marketed by electric utilities.

▼ Liquid and gaseous fuels would be marketed much as oil and natural gas are today. Large oil companies could become the principal marketers; some might also become producers, perhaps in joint ventures with agricultural or forest-product industry firms.

▼ The levels of renewable energy development indicated by this scenario represent a tiny fraction of the technical potential for renewable energy. Higher levels

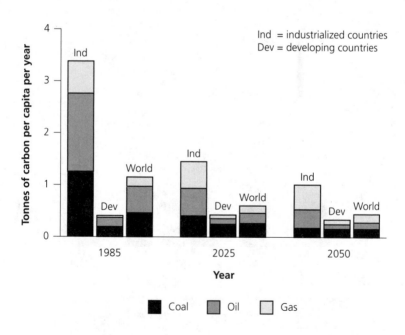

FIGURE 3b: PER CAPITA EMISSIONS OF CO_2 FOR THE RENEWABLES-INTENSIVE GLOBAL ENERGY SCENARIO, BY WORLD REGION
 Global CO_2 emissions per capita associated with the renewables-intensive global energy scenario would be reduced nearly in half by 2025 and by more than three-fifths by 2050.
 Despite the rising relative contribution of developing countries to total global CO_2 emissions (see figure 3a), per capita emissions of developing countries in 2050 would still be only one-third of those for industrialized countries.

might be pursued, for example, if society should seek greater reductions in CO_2 emissions.

Public policy issues

A renewables-intensive global energy future is technically feasible, and the prospects are excellent that a wide range of new renewable energy technologies will become fully competitive with conventional sources of energy during the next several decades. Yet the transition to renewables will not occur at the pace envisaged if existing market conditions remain unchanged. Private companies are un-

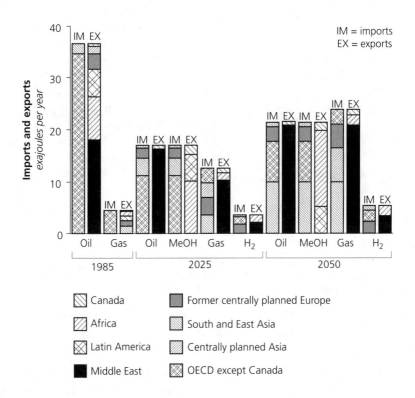

FIGURE 4: INTERREGIONAL FLOWS OF FUELS FOR THE RENEWABLES-INTENSIVE GLOBAL ENERGY SCENARIO

The importance of world energy commerce for the renewables-intensive global energy scenario is illustrated here. This figure shows that in the second quarter of the next century there would be comparable interregional flows of oil, natural gas, and methanol, and that hydrogen derived from renewable energy sources begins to play a role in energy commerce. This diversified supply mix is in sharp contrast to the situation today, where oil dominates international commerce in liquid and gaseous fuels.

Most methanol exports would originate in sub-Saharan Africa and in Latin America, where there are vast degraded areas suitable for revegetation that will not be needed for cropland. Growing biomass on such lands for methanol or hydrogen production would provide a powerful economic driver for restoring these lands. Solar-electric hydrogen exports would come from regions in North Africa and the Middle East that have good insolation.

likely to make the investments necessary to develop renewable technologies because the benefits are distant and not easily captured by individual firms. Moreover, private firms will not invest in large volumes of commercially available renewable energy technologies because renewable energy costs will usually not be significantly lower than the costs of conventional energy. And finally, the private sector will not invest in commercially available technologies to the extent justified by the external benefits (e.g., a stabilized world oil price or reduced greenhouse-gas emissions) that would arise from their widespread deployment. If these problems are not addressed, renewable energy will enter the market relatively slowly.

Fortunately, the policies needed to achieve the twin goals of increasing efficiency and expanding markets for renewable energy are fully consistent with programs needed to encourage innovation and productivity growth throughout the economy. Given the right policy environment, energy industries will adopt innovations, driven by the same competitive pressures that have revitalized other major manufacturing businesses around the world. Electric utilities will have to shift from being protected monopolies enjoying economies-of-scale in large generating plants to being competitive managers of investment portfolios that combine a diverse set of technologies, ranging from advanced generation, transmission, distribution, and storage equipment to efficient energy-using devices on customers' premises. Automobile and truck manufacturers, and the businesses that supply fuels for these vehicles, will need to develop entirely new products. A range of new fuel and vehicle types, including fuel-cell vehicles powered by alcohol or hydrogen, are likely to play major roles in transportation in the next century.

Capturing the potential for renewables requires new policy initiatives. The following policy initiatives are proposed to encourage innovation and investment in renewable technologies:

▼ Subsidies that artificially reduce the price of fuels that compete with renewables should be removed; if existing subsidies cannot be removed for political reasons, renewable energy technologies should be given equivalent incentives.

▼ Taxes, regulations, and other policy instruments should ensure that consumer decisions are based on the full cost of energy, including environmental and other external costs not reflected in market prices.

▼ Government support for research on and development and demonstration of renewable energy technologies should be increased to reflect the critical roles renewable energy technologies can play in meeting energy, developmental, and environmental objectives. This should be carried out in close cooperation with the private sector.

▼ Government regulations of electric utilities should be carefully reviewed to ensure that investments in new generating equipment are consistent with a renewables-intensive future and that utilities are involved in programs to demonstrate new renewable energy technologies in their service territories.

▼ Policies designed to encourage the development of a biofuels industry must be closely coordinated with both national agricultural development programs and efforts to restore degraded lands.

▼ National institutions should be created or strengthened to implement renewable energy programs.

▼ International development funds available for the energy sector should be directed increasingly to renewables.

▼ A strong international institution should be created to assist and coordinate national and regional programs for increased use of renewables, to support the assessment of energy options, and to support centers of excellence in specialized areas of renewable energy research.

There are many ways such policies could be implemented. The preferred policy instruments will vary with the level of the initiative (local, national, or international) and with the region. On a regional level, the preferred options will reflect differences in endowments of renewable resources, stages of economic development, and cultural characteristics.

The integrating theme for all such initiatives, however, should be an energy policy aimed at promoting sustainable development. It will not be possible to provide the energy needed to bring a decent standard of living to the world's poor or to sustain the economic well-being of the industrialized countries in environmentally acceptable ways, if the present energy course continues. The path to a sustainable society requires more efficient energy use and a shift to a variety of renewable energy sources.

While not all renewables are inherently clean, there is such a diversity of choices that a shift to renewables carried out in the context of sustainable development could provide a far cleaner energy system than would be feasible by tightening controls on conventional energy.

The central challenge to policymakers in the decades ahead is to frame economic policies that simultaneously satisfy both socio-economic developmental and environmental challenges. The analysis in this book demonstrates the enormous contribution that renewable energy can make in addressing this challenge. It provides a strong case that carefully crafted policies can provide a powerful impetus to the development and widespread use of renewable energy technologies and can lead ultimately to a world that meets critical socio-economic developmental and environmental objectives.

CONSTRUCTING A RENEWABLES-INTENSIVE GLOBAL ENERGY SCENARIO

The cornerstone of this chapter is a renewables-intensive global energy scenario developed to identify the potential markets for renewable technologies in the years 2025 and 2050, assuming that market barriers to these technologies are removed by comprehensive national policies (see "An agenda for action," below).

Some global features of the scenario are presented in figures 1 to 4. Separate scenarios were constructed for the 11 world regions listed in Box B. Details are presented in the appendix at the back of the book. Only the salient aspects of the analysis are presented here.

In constructing the scenario it was assumed that renewable energy technologies will capture markets whenever 1) a plausible case can be made that renewable energy is no more expensive on a life-cycle cost basis than conventional alternatives,[3] and 2) the use of renewable technologies at the levels indicated will not create significant environmental, land use, or other problems. The economic analysis did not take into account any credits for the external benefits of renewables listed in Box A.

Energy demand

The market for renewable energy depends in part on the future demand for energy services: heating and cooling, lighting, transportation, and so on. This demand, in turn, depends on economic and population growth and on the efficiency of energy use. Future energy supply requirements can be estimated by taking such considerations into account. For the construction of the renewables-intensive energy scenario, future levels of demand for electricity and for solid, liquid, and gaseous fuels were assumed to be the same as those projected in a scenario

---

Box B
World Regions for the Renewables-intensive Global Energy Scenario Analysis[a]

Africa
Latin America
South and East Asia
Centrally Planned Asia
Japan
Australia/New Zealand
United States
Canada
OECD Europe[b]
Former Centrally Planned Europe
Middle East

a. The regions chosen for analysis are the nine regions used in the scenario analysis carried out by the Response Strategies Working Group of the Intergovernmental Panel on Climate Change, disaggregated into the above 11 regions. The disaggregation involved separating Canada/OECD Europe into Canada and OECD Europe and OECD Pacific into Japan and Australia/New Zealand.

b. The industrialized market-oriented countries are members of the Organization for Economic Cooperation and Development (OECD). OECD Europe in particular consists of all the countries in Western Europe.

---

3. Assumptions about the cost and performance of future renewable energy equipment are based on the analyses presented in later chapters.

by the Response Strategies Working Group of the Intergovernmental Panel on Climate Change.[4]

The Working Group developed several projections of energy demand. The one adopted for the renewables-intensive scenario is characterized by "high economic growth" and "accelerated policies" (see figure 5). The accelerated policies case was designed to demonstrate the effect of policies that would stimulate the adoption of energy-efficient technologies, without restricting economic growth. Because renewable technologies are unlikely to succeed unless they are a part of a program designed to minimize the overall cost of providing energy services, the energy-efficiency assumptions underlying the accelerated policies scenario are consistent with the objectives of the renewables-intensive scenario.

The high economic growth, accelerated-policies scenario projects a doubling of world population and an eight-fold increase in gross world economic product between 1985 and 2050. Economic growth rates are assumed to be higher for developing countries than for those already industrialized. Energy demand grows more slowly than economic output, because of the accelerated adoption of energy-efficient technologies, but demand growth outpaces efficiency improvements—especially in rapidly growing developing countries. World demand for fuel (excluding fuel for generating electricity) is projected to increase 30 percent between 1985 and 2050 and demand for electricity 265 percent (see figure 5).

The Working Group's assumptions about energy efficiency gains are ambitious; nonetheless cost-effective efficiency improvements greater than those in the

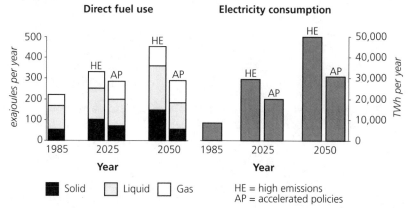

FIGURE 5: ALTERNATIVE GLOBAL ENERGY SCENARIOS DEVELOPED BY THE INTERGOVERNMENTAL PANEL ON CLIMATE CHANGE

The two alternative scenarios for the direct use of fuels (left) and electricity consumption (right) shown here were developed by the Response Strategies Working Group (RSWG) of the Intergovernmental Panel on Climate Change (IPCC), as a contribution to that group's assessment of strategies for responding to the prospect of climatic change arising from the buildup of greenhouse gases in the atmosphere. Both scenarios are characterized by high economic growth. The lower energy demand scenario provides the basis for developing the renewables-intensive global energy supply scenario in the present study. For details see the appendix to chapter 1 at the back of the book.

---

4. The Response Strategies Working Group projections were made for nine world regions. Simple scaling rules were used to convert this into an 11-region projection. The energy demand projections were otherwise not modified.

scenario are technically feasible, and new policies can help speed their adoption. Structural shifts to less energy-intensive economic activities may also reduce the energy needs of modern economies below those projected.[5]

Energy resources

Construction of a global energy supply scenario must be consistent with energy resource endowments and various practical constraints on the recovery of these resources.

Renewable energy

In the renewables-intensive energy scenario, global consumption of renewable resources reaches a level equivalent to 318 exajoules per year of fossil fuels by 2050—a rate comparable to total present world energy consumption.[6] Though large, this rate of production involves using less than 0.01 percent of the 3.8 million exajoules of solar energy reaching the earth's surface each year. The total electric energy produced from intermittent renewable sources (some 34 exajoules per year) would be less than 0.003 percent of the sunlight that falls on land and less than 0.1 percent of the energy available in the winds. Moreover, the electric energy that would be recovered from hydropower resources, some 17 exajoules per year by 2050, is small relative to the 130 to 160 exajoules per year that are theoretically recoverable (see chapter 2: *Hydropower and its Constraints*). The amount of energy targeted for recovery from biomass, 206 exajoules per year by 2050, is also small compared with the rate (3,800 exajoules per year) at which plants convert solar energy to biomass (see chapter 14: *Biomass for Energy: Supply Prospects*).

The production levels considered are therefore not likely to be constrained by resource availability. A number of other practical considerations, however, do limit the renewable resources that can be used. The scenario was constructed subject to the following restrictions:

▼ Biomass must be produced sustainably (see Box C), with none harvested from virgin forests. Some 62 percent of the biomass supply would come from plantations established on degraded lands or, in industrialized countries, on excess

---

5. For example, per capita energy use in OECD Europe is presently about 20 percent less than in Eastern Europe and the former Soviet Union. In the accelerated-policies scenario, per capita energy demand declines in OECD Europe and increases 60 percent by 2050 in Eastern Europe and the former Soviet Union. In light of the rapid economic and political changes now under way, it is doubtful that these two regions will take such divergent paths.

6. Most results for the renewables-intensive global scenario are reported in terms of exajoules (EJ) of direct fuel use or terawatt-hours (TWh) of electricity replaced by renewable resources. If all electricity is produced from fossil or other fuels, it is straightforward to convert electricity production in TWh into the fuel quantities (in EJ) required to produce the electricity (the "primary energy" requirements for electricity generation). But the conversion is ambiguous if electricity is produced from hydroelectric or intermittent renewable sources. If it is assumed that non-fuel-fired electricity generators substitute for generators converting fuel to electricity at the worldwide average efficiency for fuel-based power generation in a given year, the renewables-intensive global scenario results for the year 2050 are that renewables provide 318 EJ of primary energy equivalent (including the losses associated with producing synthetic liquid and gaseous fuels) out of a total world primary energy demand of 561 EJ. The renewables contribution is made up of 206 EJ from biomass, 64 EJ from intermittent renewables, 32 EJ from hydropower, 14 EJ from electrolytic hydrogen, and 1 EJ from geothermal energy. [Primary commercial energy demand was 323 EJ in 1985. In addition, non-commercial biomass energy is consumed at a rate of about 50 EJ per year (see chapter 14).]

agricultural lands. Another 32 percent would come from residues of agricultural or forestry operations. Some residues must be left behind to maintain soil quality or for economic reasons; three fourths of the energy in urban refuse and lumber and pulpwood residues, one half of residues from ongoing logging operations, one fourth of the dung produced by livestock, one fourth of the residues from cereals, and about two thirds of the residues from sugar cane are recovered in the scenario. The remaining 6 percent of the biomass supply would come from forests

---

Box C

Toward Sustainable Biomass Production for Energy

The renewables-intensive global energy scenario calls for some 400 million hectares of biomass plantations by the second quarter of the 21st century. Three questions are raised by the scenario. First, are net energy balances sufficiently favorable to justify the effort? Second, can high biomass yields be sustained over wide areas and long periods? And third, would such plantations be environmentally acceptable?

Achieving high plantation yields requires energy inputs—especially for fertilizers and for harvesting and hauling the biomass. The energy content of harvested biomass, however, is typically 10 to 15 times greater than the energy inputs.

But can high yields be achieved year after year? The question is critical because essential nutrients are removed from a site at harvest; if these nutrients are not replenished, soil fertility and yields will decline over time. Fortunately, replenishment is feasible with good management. Twigs and leaves, the parts of the plant in which nutrients tend to concentrate, should be left at the plantation site at harvest, and the mineral nutrients recovered as ash at energy conversion facilities should be returned to the plantation soils. Nitrogen losses can be restored through the application of chemical fertilizers; makeup requirements can be kept low by choosing species that are especially efficient in the use of nutrients. Alternatively, plantations can be made nitrogen self-sufficient by growing nitrogen-fixing species, perhaps intermixed with other species. In the future, it will be possible to reduce nutrient inputs by matching nutrient applications to a plant's cyclic needs.

Intensive planting and harvesting activities can also increase erosion, leading to productivity declines. Erosion risks for annual energy crops would be similar to those for annual food crops and so the cultivation of such crops should be avoided on erodible lands. For crops such as trees and perennial grasses, average erosion rates are low because planting is so infrequent—typically once every 10 to 20 years.

An environmental drawback of plantations is that they support far fewer species than natural forests. Accordingly, it is proposed here that plantations be established not on areas now occupied by natural forests but instead 1) on deforested and otherwise degraded lands in developing countries and 2) on excess agricultural lands in industrialized countries. Moreover, a certain percentage of land should be maintained in a natural state as sanctuary for birds and other fauna to help control pest populations. In short, plantations would actually improve the status quo with regard to biological diversity.

For a more detailed discussion, see chapter 14: *Biomass for Energy: Supply Prospects.*

that are now routinely harvested for lumber, paper, or fuel wood. Production from these forests can be made fully sustainable—although some of these forests are not well managed today.

▼ Although wind resources are enormous, the use of wind equipment will be substantially constrained in some regions by land-use restrictions—particularly where population densities are high. In the scenario, substantial development of wind power takes place in the Great Plains of the United States(where most of the country's wind resources are found), while in Europe the level of development is limited because of "severe land-use constraints" (see chapter 4: *Wind Energy: Resources, Systems, and Regional Strategies*).

▼ The amounts of wind, solar-thermal, and photovoltaic power that can be economically integrated into electric generating systems are very sensitive to patterns of electricity demand as well as weather conditions. The marginal value of these so-called intermittent electricity sources typically declines as their share of the total electric market increases. Analysis of these interactions suggests that intermittent electric generators can provide 25 to 35 percent of the total electricity supply in most parts of the world (see chapter 23: *Utility Strategies for Using Renewables*). Some regions would emphasize wind, while others would find photovoltaic or solar-thermal-electric systems more attractive. On average, Europe is a comparatively poor location for intermittent power generation, so that the penetration of intermittent renewables there is limited to 14 percent in 2025 and 18 percent in 2050.

▼ Although the exploitable hydroelectric potential is large, especially in developing countries (see chapter 2), and hydropower is an excellent complement to intermittent electric sources, the development of hydropower will be constrained by environmental and social concerns—particularly for projects that would flood large areas. Because of these constraints, it is assumed that only a fraction of potential sites would be exploited, with most growth occurring in developing countries. Worldwide, only one-fourth of the technical potential as estimated by the World Energy Conference, would be exploited in the scenario by 2050. Total hydroelectric production in the United States, Canada, and OECD Europe would increase by only one third between 1985 and 2050, and some of the increase would result from efficiency gains achieved by retrofitting existing installations.

Conventional fuels
By making efficient use of energy and expanding the use of renewable technologies, the world can expect to have adequate supplies of fossil fuels well into the 21st century. However, in some instances regional declines in fossil fuel production can be expected because of resource constraints.

Oil production outside the Middle East would decline slowly under the renewables-intensive scenario, so that one third of the estimated ultimately recoverable conventional resources will remain in the ground in 2050. As a result, non-Middle Eastern oil production would drop from 103 exajoules per year in 1985 to 31 exajoules per year in 2050. To meet the demand for liquid fuels that cannot be met by renewables (see below), oil production is assumed to increase in the Middle East, from 24 exajoules per year in 1985 to 34 exajoules per year in 2050. Total world conventional oil resources would decline from about 9,900 exajoules in 1988 to 4,300 exajoules in 2050.

Although remaining conventional natural-gas resources are comparable to those for conventional oil, gas is presently produced globally at just half the rate for oil. With adequate investment in pipelines and other infrastructure components, gas could be a major energy source for many years. In the decades ahead, substantial increases in gas production are feasible for all regions of the world except for the United States and OECD Europe. For the United States and OECD Europe, where resources are more limited, production would decline slowly, so that one-third of these regions' gas resources will remain in 2050. In aggregate, gas production outside the Middle East would increase slowly, from 62 exajoules per year in 1985 to 75 exajoules in 2050. But in the Middle East, where gas resources are enormous and largely unexploited, production would expand more than 12-fold, to 33 exajoules per year in 2050. Globally, about half the conventional gas resources would remain in 2050.

The renewables-intensive scenario was developed for future fuel prices that are significantly lower than those used in most long-term energy forecasts. It is expected that in the decades ahead the world oil price would rise only modestly and the price of natural gas would approach the oil price (which implies that the gas price paid by electric utilities would roughly double). There are two primary reasons for expecting relatively modest energy price increases: first, overall demand for fuels would grow comparatively slowly between 1985 and 2050 because of assumed increases in the efficiency of energy use; and second, renewable fuels could probably be produced at costs that would make them competitive with petroleum at oil prices not much higher than at present (see below).

Electricity generation strategies

Electricity appears to be the energy carrier of choice for modern economies since growth in electricity has outpaced growth in the demand for fuels. This trend is projected to continue in the IPCC accelerated-policies scenario assumed for the present analysis (see figure 5). In this scenario, average per capita demand for electricity is projected to increase by 70 percent between 1985 and 2050, while per capita direct use of fuels would decline by more than 30 percent. Demand is expected to rise especially sharply in developing countries, which are projected to experience almost a five-fold increase in per capita electricity consumption. Overall, world demand for electricity would increase some 265 percent by 2050 in the accelerated policies scenario—equivalent to adding more than 50 large [1,000 megawatts-electric (MW$_e$)] plants each year.

The context

Most of the world's electricity is produced and marketed essentially as a commodity by heavily regulated or state-owned utility companies. Such companies operate under rules based on the assumption that a natural monopoly exists in the production and delivery of electricity. But the rules that have governed utility operation are being undermined by 1) concerns that these rules have encouraged utilities to build new power plants when investments in energy efficiency would have been more cost-effective, 2) greater scrutiny being given to the economic inefficiencies of large generating facilities and large, hierarchical transmission and distribution systems, and 3) the development of attractive new generating technologies that can operate efficiently in small sizes at distributed sites.

In some countries, utility regulations have been changed to expose regulated utilities to competition, and investors have been forced to bear more of the risks of unwise investment. This competition, in turn, is forcing utilities, like many other contemporary businesses, to rethink the way they produce and market their products. Competition in most modern markets requires sensitivity to the needs of individual clients, an ability to react quickly to their changing demands, and a capacity to introduce new products and production technologies efficiently.

Some utilities are moving away from their traditional role of simply being purveyors of electricity and are becoming concerned with all aspects of their customers' electric service needs. A number of utilities, for example, have begun to invest in equipment at customers' sites that enables customers to make more efficient use of electricity. Some are also attempting to increase the efficiency of managing transmission and distribution networks, which account for nearly half a typical utility's capital investment.

Competitive pressures, growing concerns about environmental issues, siting difficulties, and other forces have also stimulated innovation in electric-generating technologies. Technical developments in renewables and in a number of other areas have combined to greatly enrich the set of available electric-generating equipment. Advanced gas turbines, originally developed for aircraft engines and now being applied for power generation, are a prominent example. With natural-gas firing, such turbines have achieved system efficiencies of 45 percent for commercial units (compared with efficiencies of 35 percent or less for conventional steam-electric power units), and efficiencies greater than 50 percent will likely be realized by 2000. Thereafter, advanced gas-turbine and fuel-cell technologies powered with natural gas are likely to achieve efficiencies ranging from 55 to 60 percent.

Evaluating technical alternatives

The changes under way in electric utilities, reinforced by growing regulatory pressures to reduce emissions, create a favorable investment climate for renewable electric technologies, in light of the rapidly declining costs for various promising renewable technologies, as shown in other chapters of this book.

The potential role for a renewable technology in an electric utility system depends on the cost of producing electricity with the technology and on the value of this electricity. The value, in turn, depends on the characteristics of both the renewable electric technology and the utility system in which it will operate (see chapter 23).

The ongoing innovation in electric generation from conventional fuels means that the question "How much is renewable energy worth?" must constantly be accompanied by the question "Compared with what?" Renewables that will be introduced commercially after 2000 will be competing not with today's conventional fuel-generating technologies but with more advanced technologies.

The value of electricity from renewable energy will vary depending on the technology. Because biomass generating plants, for example, can be dispatched much like conventional coal-fired plants, the value of the electricity they produce will simply be the cost of not having to provide the electricity with coal plants.

The determination of value is not as straightforward for wind, photovoltaic, and solar-thermal electric generating systems (intermittent generators), whose output is not governed by decisions of utility dispatchers. The value of intermittent renewables depends on the characteristics of all the other generating units on the system and on the variation in the demand for electricity, by time of day and time of year. While the fluctuations in intermittent electricity production pose new problems for the management of generating systems, these challenges are not fundamentally different from those involved in managing fluctuations in the electricity demands of buildings (often representing more than half of electricity demand), which are also heavily influenced by the weather.

The value of hydroelectric power should also be determined by examining the way it affects overall utility system costs. Hydroelectric output, for example, can be varied quickly to meet sharp increases in demand that might otherwise have to be met inefficiently with thermal power plants.

Electricity storage systems can reduce the cost of meeting fluctuating demand if they can be charged during periods when electricity is relatively cheap and discharged when utility power is comparatively expensive (e.g., during periods of peak demand). The extent to which renewable energy systems increase or decrease the value of storage depends on local details. For instance, photovoltaic systems that produce most of their output during periods of peak demand may actually decrease the value of energy storage. Storage systems are likely to be expensive and thus limited to specialized markets for the foreseeable future. In the present analysis, no electrical storage is considered except for the storage capabilities inherent in hydroelectric equipment.

Ultimately, the most important measures of the worth of renewable power sources are the roles they can play in an investment portfolio designed to minimize the cost of energy services, and the extent to which their adoption would provide environmental and other non-market benefits (see Box A).

Hydroelectric. Hydroelectric power is an attractive source of comparatively low-cost electricity, which presently provides about 20 percent of the world's electricity. Hydroelectric systems with large reservoirs are particularly valuable to electricity grids that have large amounts of intermittent renewables because their reservoirs can buffer the variable output of the intermittents. However, as noted, only a modest fraction of the hydroelectric potential will be exploited because of environmental and social concerns about hydroelectric projects.

Small-scale hydropower facilities can play an important role in local energy economies. This resource is already extensively developed in China, where it accounts for about one third of hydroelectric power generation. The global potential for small-scale hydropower, at installed capacities up to 2 megawatts, is equivalent to about 30 percent of world hydroelectric power generation in 1985. One drawback of small-scale hydropower is that it almost always involves run-of-river plants that lack reservoir capacity to store water (see chapter 2).

Biomass. Biomass, mainly in the form of industrial and agricultural residues, has provided electricity for many years with conventional steam-turbine power-generators. The United States currently has more than 8,000 MW_e of generating capacity fueled with such feedstocks. Existing steam-turbine conversion technologies are cost competitive in regions where low-cost biomass fuels are available, in spite of the fact that these technologies are comparatively inefficient at the small sizes required for biomass electricity production.

The performance of biomass electric systems can be improved dramatically by adapting to biomass advanced-gasification technologies originally developed for coal (see chapter 17: *Advanced Gasification-based Biomass Power Generation*). Biomass is a more attractive feedstock for gasification than coal because it is easier to gasify and has a very low sulfur content, so expensive sulfur removal equipment is not needed. Biomass integrated gasifier/gas turbine power systems with efficiencies of 40 percent or more will be demonstrated in the mid 1990s and will probably be commercially available by 2000. These systems offer high efficiencies and low unit capital costs for baseload power generation at relatively modest scales of 100 MW_e or less and will compete with coal-fired power plants under many circumstances—even when fueled with relatively costly biomass feedstocks (see figure 6). By 2025, conversion efficiencies as high as 57 percent may be feasible using fuel-cell technologies being developed for coal.

Wind. Total wind generating capacity is now about 2,000 MW_e—most of it in California and Denmark. The investments made to achieve this level of development have led to a steady accumulation of field experience and organizational learning. Taken together, many small engineering improvements, better operation and maintenance practices, improved wind prospecting, and a variety of other incremental improvements have led to steady cost reductions.

Technological advances promise continued cost reductions (see figure 7). For example, the falling cost of electronic controls has made it possible to replace mechanical frequency controls with electronic systems. In addition, modern computer technology has made it possible to substantially improve the design of blades and other components. At good wind sites, the cost of electricity using new technology for which prototypes were field tested in 1992 will be less than the cost of electricity from new fossil-fuel plants (see chapter 3: *Wind Energy: Technology and Economics*).

The value of wind electricity depends on the characteristics of the utility system into which it is integrated, as well as on regional wind conditions. Some areas, particularly warm coastal areas, have winds with seasonal and daily patterns that correlate with demand, whereas others have winds that do not. Analyses conducted in the United Kingdom, Denmark, and the Netherlands make it clear that

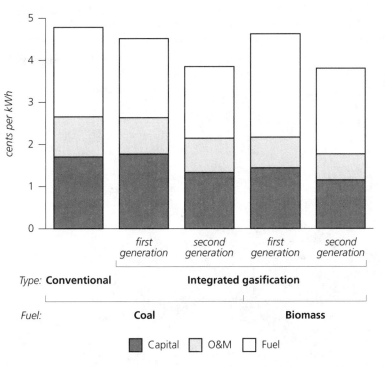

FIGURE 6: ESTIMATED COST OF ELECTRICITY FROM BIOMASS AND COAL
    Estimated electricity costs are shown at the right for two biomass-integrated gasifier/gas turbine (BIG/GT) power systems: a 109 MW$_e$ first-generation system that could be commercialized in the 1990s, based on the use of a steam-injected gas turbine (STIG) that is commercially available for firing with natural gas, and a 111 MW$_e$ second-generation system based on an intercooled steam-injected gas turbine (ISTIG) that could become available for natural gas-fired turbine systems in the 1990s, with applications to biomass in the year 2000+ time period. For comparison, the cost of electricity is also shown for a new 1,000 MW$_e$ coal-fired, steam-electric plant with flue-gas desulfurization; for a first-generation 800 MW$_e$ coal integrated gasification/combined cycle plant (CIG/CC); and for a second-generation coal integrated gasification/intercooled steam-injected gas turbine (CIG/ISTIG) plant. The figure shows that electricity generated from biomass using gasification is expected to be competitive with electricity from coal. For a more detailed discussion of these costs, see Endnotes for Figures.

wind systems have greater value if numerous generating sites are connected because it is likely that wind power fluctuations from a system of turbines installed at many widely separated sites will be less than at any individual site (see chapter 4).

Solar-thermal. Solar-thermal electric generation systems use sunlight to heat fluids that drive turbines. Such systems typically concentrate sunlight with mirrors that follow the sun. Limited dispatching is possible because the heated fluids can be stored comparatively inexpensively. And because the systems use conventional power turbines, backup power can be provided simply by burning natural gas to heat the fluids when sunlight is not available.

The capacity of commercial systems now connected to utility grids totals 350 MW_e. Most of these systems are in California and were constructed in response to financial incentives offered in the 1980s. As in the case of wind, continued engineering refinements have led to steady efficiency improvements and declining costs.

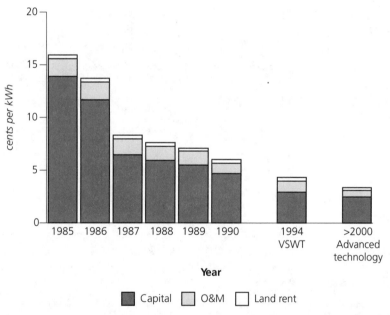

FIGURE 7: TREND IN COST OF ELECTRICITY FROM WIND IN CALIFORNIA
   Wind power costs for the period 1985–1990 are based on actual experience with wind farms in California. The cost shown for 1994 is for a new variable-speed wind turbine (VSWT) expected to go into commercial service in 1994, and for the time beyond 2000 the cost reflects expectations about improvements that could be realized over the next decade.
   The cost of electricity from units expected to be in commercial service by 1994 would be about the same as for a modern coal plant (see figure 6). For a more detailed discussion of these costs, see Endnotes for Figures.

A state-of-the-art solar-thermal electric system would produce electricity for about 9.3 cents per kWh.[7] This cost is significantly higher than the cost of base-load electricity in most regions. However, the value of solar-thermal electric power is enhanced by the fact that output varies directly with sunlight and thus often correlates well with peak electricity demands in warm areas with heavy air-conditioning loads.

Progress in reducing costs was interrupted in 1992 when the world's leading solar-thermal electric corporation, Luz International, went bankrupt, largely as a result of sudden changes in tax laws and regulations that all but eliminated financial incentives for solar-thermal electric technology. There is good reason to believe, however, that the cost of solar-thermal electric systems will continue to decline if adequate investments are made in research and field demonstrations (see chapter 5: *Solar-thermal Electric Technology*). With technological improvements, annual average solar-to-electric efficiencies could increase from the current 13 percent to 17 percent or more. Higher efficiencies and cost reductions for tracking mirror systems and other expensive components may reduce overall costs per kilowatt-electric (kW_e) by as much as 40 percent. With advanced technologies, electricity could be provided for a cost of the order of 6 cents per kWh (see chapter 5).

Photovoltaics.   Photovoltaic modules are solid-state devices that convert sunlight directly into electricity with no rotating equipment. Photovoltaic systems can be built in virtually any size, are highly reliable, and require little maintenance.

Worldwide photovoltaic sales are about 50 MW_e annually. The major problem limiting the widespread use of photovoltaics is the high cost of manufacturing the sheets of semiconductor materials needed for power systems. Photovoltaic electricity now costs 25 to 35 cents per kWh.

Despite their high cost, photovoltaic systems are cost-effective in many areas remote from utility grids, where alternative sources of power are impractical or costly (see chapter 11: *Utility Field Experience with Photovoltaic Systems*). For grid-connected distributed systems, the value of photovoltaic electricity can also be high because the electricity is produced during periods of peak demand, thereby reducing the need for costly conventional peaking capacity, and close to the sites where it is consumed, thereby reducing transmission and distribution expenses and increasing system reliability. Small, grid-connected photovoltaic systems may be competitive today where distributed generation has particularly high value (see chapter 23). Many such niche applications will provide early markets for photovoltaic systems. As photovoltaic prices fall, markets will expand rapidly.

Photovoltaic prices have fallen sharply since the mid 1970s, although prices have actually increased a few percent in the last three years because demand has grown faster than supplies. There is, however, good reason to be confident that prices will fall substantially by the turn of the century. There are several alternative

---

7. Calculated for the same financial assumptions used to construct figure 7.

paths by which costs can be reduced. Systems based on thin films of materials such as amorphous silicon alloys, cadmium telluride, or copper indium diselenide are particularly promising, both because they are well suited for the application of mass-production techniques and because the amounts of active materials required are small. Another strategy calls for concentrating sunlight on small, highly efficient photovoltaic cells using inexpensive lenses or mirrors. Early in the 21st century photovoltaic systems should be able to provide electricity at a cost of 4 to 6 cents per kWh (see chapter 6: *Introduction to Photovoltaic Technology;* chapter 7: *Crystalline- and Polycrystalline-silicon Solar Cells;* chapter 8: *Photovoltaic Concentrator Technology;* chapter 9: *Amorphous Silicon Photovoltaic Systems;* and chapter 10: *Polycrystalline Thin-film Photovoltaics).*

Other renewable resources.   Heat escaping from the earth's interior (geothermal energy), heat available in the temperature difference between the surface and deep waters of the ocean, and energy in the tides and waves can all be harnessed to produce electricity.

Geothermal power based on current hydrothermal technology can be locally significant in those parts of the world where there are favorable resources. About 6 gigawatts-electric (GW_e) of geothermal power is being produced today, and 15 GW_e may be added during the next decade. If hot dry-rock geothermal technology is successfully developed, the global geothermal potential would be large (see chapter 13: *Geothermal Energy).*

The various forms of ocean energy are abundant but often available far from consumer sites. Progress in reducing costs has been slow (see chapter 12: *Ocean Energy Systems).*

Natural gas.   Natural gas–fired generating technologies, especially those based on gas turbines, are well suited for power systems that incorporate large amounts of intermittent renewables. The technologies are economical in comparatively small (50 to 100 MW_e) sizes. Large gas turbine/steam turbine combined-cycle plants are also commercially available.

Because gas turbines have low capital costs, they have long been used by utilities for "peaking service"—that is, to meet infrequent demand peaks, such as those that occur only a few hours each year. But because they have been relatively inefficient, gas turbines have not been operated for extended periods. This situation is changing rapidly, however, with the emergence of a new generation of highly efficient gas turbines. Power systems based on these new turbines and fueled with natural gas can compete with coal-fired power plants in many areas even if the gas price doubles. Moreover, because these systems can be designed to have relatively low emissions, their deployment can help abate air pollution problems. Not surprisingly, utility interest in these systems is growing rapidly.

Competitively priced advanced gas turbine systems are available at relatively small scales. These smaller units are attractive because they can be installed quickly and located at dispersed sites—reducing the risks inherent in building large centralized plants when demand is uncertain. In addition, they can be started rapidly to meet fluctuating demand. This feature makes the advanced gas turbine power system an attractive complement to large numbers of intermittent electric generators on a utility system.

Coal. Coal is the dominant fuel for electricity production in many parts of the world. Technical advances during the past four decades have led to sophisticated large-scale steam-electric plants that are now typically 35 percent efficient. But the performance of such systems may be approaching practical limits, and costs have increased as a result of a increasingly stringent environmental regulations. New technologies are needed to achieve further efficiency improvements and cost reductions. The most promising new technology is an integrated gasifier/gas turbine power system. In this system coal is gasified at high temperatures, and the hot gas is used directly as fuel for a gas turbine power-generating cycle. This technology, which is being demonstrated at commercial scales, creates much less air pollution than steam-electric technology and is more energy-efficient. Advanced versions, which may be available by 2000 may achieve efficiencies of 42 percent and have lower unit capital costs than steam-electric plants (see chapter 17).

By 2025, coal plants may reach efficiencies as high as 57 percent, using integrated gasifier/fuel-cell systems. Fuel cells, which convert fuel directly into electricity without first burning it, are much more energy-efficient and cleaner than technologies based on combustion processes.

Nuclear. It is difficult to know whether coal or nuclear power will be the less costly during the next few decades. Technical advances are expected in both areas, and the cost of both will be heavily influenced by regulations. No effort was made to resolve this uncertainty in the present study, and it is assumed that the cost of nuclear power will not be much different from the cost of coal power.

Electric utility investment portfolios

Utilities in different parts of the world will select different sets of technologies depending on patterns of local electric demand, access to natural gas, coal or other resources, and the extent and mix of local renewable resources (biomass, wind, sunlight, and hydroelectric power).

One virtue of renewables is the large number of technological choices, although such variety also frustrates attempts to make simple, universal statements about their worth. However, a number of basic principles can be illustrated by examining the economics of alternative combinations of technologies that might be installed in a region with good, but not remarkable, access to renewable resources in the period after 2000 (see chapter 23).

A case study of the average cost of electricity with renewables. Northern California was chosen as a representative area because its wind- and solar-energy resources are close to world averages and its use of hydroelectric power approximates the world average of 20 percent. The region has already successfully integrated renewables into the utility grid, and good data have been collected on the actual field performance of large operational systems. Correlations between sunlight, wind, and utility loads do not appear to be unusual.[8]

Investment alternatives for this utility system were analyzed using the following assumptions:

▼   No economic credit was given for environmental or other external benefits of renewable electric generating technologies. Only the cost of meeting present environmental laws was taken into account in the analysis.

▼   No energy storage was used except for the storage inherent in hydroelectric reservoirs and in biomass fuel.

▼   The amount of wind, solar-thermal electric, and photovoltaic capacity was varied parametrically; no attempt was made to optimize the mix.

▼   Investments in a mix of biomass and conventional generating equipment were made to minimize production costs for each assumed level of intermittent electric and hydroelectric power generation, assuming all new equipment. (Real utilities, of course, would need to consider the sunk costs of existing equipment.)

▼   Any intermittent renewable output in excess of the load at any given hour is wasted.[9]

Using these assumptions, the fuel price assumptions for the renewables-intensive energy scenario, and capital and other electric generating cost estimates derived from other chapters in this book and elsewhere (see endnotes for figure 8 for details), the average cost of electricity generated with conventional equipment

---

8. The results of this case study are qualitatively similar to analyses of wind energy equipment operating in the United Kingdom and the Netherlands (see chapter 4).

9. This is not a realistic assumption since some market would undoubtedly be found for this electricity—however unreliable—since its cost would be very low. In the real world it would be reasonable to expect that electricity could be sold to neighboring utilities or used to produce a storable commodity.

FIGURE 8: COMPARING ELECTRIC UTILITY INVESTMENT PORTFOLIOS
   The average cost of meeting the annual electricity needs of a large utility in northern California (top) and the fraction of electricity generated by each energy source (bottom) are displayed for a variety of investment alternatives. The clear segments at the tops of the six bars on the right represent the value of distributed photovoltaic power to the utility system. The net cost is given by the level at the tops of the shaded bars—the gross cost less the distributed photovoltaic benefit. The figures in brackets on top of the bars (top) represent the amount of CO_2 released into the atmosphere relative to the conventional fossil case (100). The costs are the result of an hour-by-hour simulation of the utility that considered electricity demand, the variable output of intermittent renewable equipment, the load-leveling capabilities of hydroelectric facilities, and the dispatching of coal, natural gas, and biomass-fueled electric-generating plants. The selection of coal, biomass, and natural gas-burning plants was done to minimize the cost of serving loads not covered by other equipment within the constraints specified.
   The figure shows that a large fraction (30 percent) of electricity generation could come from intermittent sources without increasing the average cost of electricity, and that this utility could operate almost entirely on renewable sources of energy (case 9). The CO_2 emissions in this case would be 97 percent less than for the conventional case (case 1). See Endnotes for Figures for more details.

was found to be 4.9 cents per kWh. Costs would be reduced to 4.6 cents per kWh using the best performing gas turbines and coal gasification systems that are now coming onto the market and to 4 cents per kWh using advanced fossil-fuel generating equipment (3.8 cents per kWh if low-cost hydroelectric capacity is also available) (see figure 8).

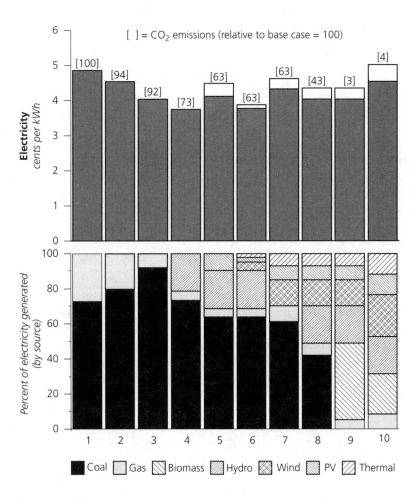

1 = conventional fossil

2 = best new fossil

3 = advanced fossil

4 = advanced fossil with 21 percent hydro

5 = advanced fossil with 10 percent PV and 21 percent hydro

6 = advanced fossil with 10 percent intermittents and 21 percent hydro

7 = advanced fossil with 30 percent mixed intermittents (three wind sites)

8 = advanced fossil with 30 percent mixed intermittents (three wind sites) and 21 percent hydro

9 = advanced biomass and gas with 30 percent mixed intermittents (three wind sites) and 21 percent hydro

10 = advanced biomass and gas with 50 percent mixed intermittents (three wind sites) and 21 percent hydro

Costs were also computed for utility portfolios involving various combinations and levels of renewable technologies. Systems in which 10, 30, and 50 percent of the utility's electricity was generated from intermittents were considered. When a mixture of intermittent renewable generators was considered, roughly half of the energy was assumed to come from wind, with the remaining divided equally between solar-thermal electric and photovoltaic systems. Systems with and without hydropower and with and without biomass were considered. Because the biomass systems could compete directly with coal in all of the portfolios considered (see figure 6), biomass could completely displace coal without increasing costs if it were available.

A striking finding of the analysis is that, with one exception, all the renewable portfolios considered could meet the system's load at a cost lower than for a system with typical new coal and gas using equipment (see figure 8). Also, in many of these cases, from 90 to 95 percent of all electricity comes from renewables. The high-renewables cases generate only 3 to 4 percent as much CO_2 as the advanced fossil system. Yet these large reductions in CO_2 emissions are achieved at costs between 0 and 1 cent per kWh—that is, the cost of electricity is only 0 to 1 cent per kWh higher than in the least costly case. For comparison, the average electric price for consumers today is 6.3 cents per kWh (a delivered price that includes transmission, distribution, and management costs, as well as the generation costs displayed in figure 8).

Although utility portfolios that involved intermittents for more than 10 percent of the electricity generated were somewhat more expensive than utility portfolios that made maximum use of advanced, highly efficient fossil fuel generators, even for the 50 percent intermittents case, the renewable system cost only 1.5 cents per kWh more than the advanced fossil system.

The reduced costs arising from buffering demand on a utility system with hydroelectric equipment are apparent for both conventional and renewables-intensive cases. But hydroelectric power is particularly attractive when used to buffer the large fluctuations of output that arise at high penetrations of intermittent renewables (see figure 8).[10]

An important parameter characterizing small-scale photovoltaic systems located at sites dispersed throughout the utility system is the credit these systems are due because of their value in reducing transmission and distribution costs and increasing system reliability. This so-called "distributed credit" can reduce the net cost of photovoltaic systems considerably, as indicated by the impact on the average cost of power generation in figure 8.

In the analysis of photovoltaic systems, a range of costs was considered because of the large uncertainties about future prices. In addition, because the value of the distributed benefit for photovoltaic systems is site-specific, a conservative

---

10. The cost advantages of buffering with hydropower would be greater than shown in figure 8 if conventional instead of advanced gas turbines were used to provide peaking power.

estimate of this value was developed based on the average cost of expanding transmission and distribution systems planned in northern California. Moreover, the credit was assumed to be zero for the least costly photovoltaic systems considered.

Reoptimizing the choice and use of thermal-electric equipment. Determining the value of the renewables on a utility system requires two steps: estimating the total capacity and mix of thermal-electric generating equipment needed to meet load at the lowest cost and dispatching the equipment thus selected. This second step may require assessing dispatching choices and costs for each hour of the year (see chapter 23). This exercise, which was carried out for the northern California case study, shows that the introduction of substantial quantities of renewables can profoundly affect the choice and use of thermal-electric generating equipment. The impact intermittent equipment will have on the loads that must be met with dispatchable plants can be illustrated using a load-duration curve, which shows the number of hours per year that loads exceed a specified level (see base case in figure 9). This curve can be used to determine the set of dispatchable thermal-electric plants that will minimize the cost of meeting this particular distribution of loads. With intermittent renewables and hydropower added to the system, the net load that must be met with thermal-electric equipment can be determined by subtracting the contributions made by various combinations of these renewables from the total load (see other cases in figure 9).

Given the comparatively poor match between wind output and electrical demand in most regions, the wind system illustrated in the plot at the upper right in figure 9 does not reduce the net utility peak load significantly but does provide large amounts of electrical energy. Adding wind power to the system would also shift the optimal mix of thermal-electric generating capacity from baseload (advanced coal) to less costly gas turbine-based equipment (simple-cycle and combined-cycle plants for peaking and load-following purposes, respectively). If large quantities of wind are integrated into the system, wind power output could exceed demand. This is illustrated in the plot at the upper right of figure 9, which shows that wind output exceeds demand for nearly 1,000 hours per year (indicated as zero net load) for the case where 30 percent of total electricity is provided by wind power.

The operational advantage of combining different intermittent renewables can be seen by comparing the plots on the lower left and upper right in figure 9. Because photovoltaic and solar-thermal electric systems are much more likely to operate during periods of peak demand, the net peak load would tend to be reduced if these technologies are included in the mix of intermittent renewables. Also, the net load for the utility would be less erratic for a diverse set of intermittent renewables than for a wind-only system at the same total level of penetration. (The impact of wind power on the utility system could also be made less erratic by diversifying the wind sources—that is, by combining the outputs from wind farms at different locations).

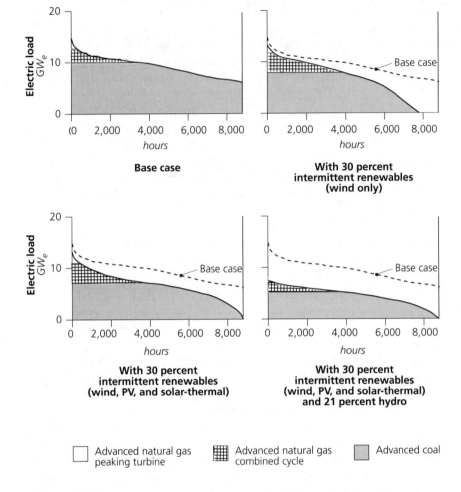

FIGURE 9: EFFECTS OF INTERMITTENT SOURCES ON THE THERMAL GENERATING SYSTEM

The solid curves in these four figures illustrate the loads served by thermal power plants (coal, gas, and biomass fueled) for a large utility in Northern California. These curves were constructed by taking the load at each of the 8,760 hours in a year and sorting these loads so that the peak annual load is at the left of the curve and the minimum load is on the right. Each point on a curve represents the number of hours each year that the indicated load is exceeded. The "base case" is the load actually served by the utility. The other solid curves indicate the load remaining after subtracting 30 percent of intermittent renewables sources (in different combinations of wind, solar thermal, photovoltaic) and a 21 percent contribution from hydroelectric power, from the load for the base case. Also shown are the outputs of advanced coal and gas turbine equipment at installed capacities that would minimize the cost of meeting the remaining loads, given the fuel prices assumed in the renewables-intensive global energy scenario. See Endnotes for Figures for more details.

The ability of hydroelectric power to be dispatched to buffer the fluctuating output of intermittent renewables and thereby reduce the need for peaking turbines and the variations in loads that must be met by thermal-electric plants is illustrated in figure 9, lower right.

Economies of scale. The new electric generating technologies just described raise fundamental questions about how utilities can and should operate in the future. In particular, they undermine long-standing assumptions about scale economies, especially about the value of scale for individual generating units. This critical issue must be reviewed by any utility considering significant expansion.

The new technologies affect the scale of generating equipment in a variety of ways. One is that most attractive modern generating systems come in small modules. Biomass systems that can compete with coal are far smaller (100 MW_e or less) than conventional coal systems, and advanced natural-gas combined-cycle systems achieve 50 percent conversion efficiency at similar sizes. Highly efficient fuel-cell systems, which may be available in the first quarter of the next century, could be economical at even smaller sizes. And there may be no real scale-economies in photovoltaic systems; indeed rooftop systems may actually be less expensive to install than larger systems installed in the field (although the issue is controversial).

Large generating systems also incur various operational and planning costs that have traditionally been offset by the low cost of the electricity they produce. But as the new competing technologies reduce the production cost advantages of large plants, these operational and planning costs must be considered more carefully. Whereas the economics of larger plants depends heavily on long-term forecasts, which have proven notoriously unreliable, small plants can be added quickly as they are needed and even disassembled and moved if loads decline. The uncertainty inherent in any demand forecasting is increased by new programs to encourage investment in energy efficiency and non-utility generation—programs whose outcomes are difficult to predict with precision. A utility must consider whether demand will increase by simply scaling up the base-case profile shown in figure 9 or whether the net loads that will be served by thermal-electric plants will change to resemble more closely one of the other net loads plotted in this figure. For these other net loads, many fewer large conventional plants are required. Large plants are also less flexible in meeting fluctuating loads and are costly to start and stop. If utilities make heavy use of intermittent renewables, demand for conventional power may be very low—even zero, for significant periods. Large plants would fare poorly in such environments.

Finally, reliability can probably be increased, rather than decreased, by installing a large number of small generators instead of a few large ones. A utility must maintain reserves sufficient to cover an unexpected failure of its largest operating plant. The reserve capacity required to meet the one-day-in-10-years loss-of-load reliability criterion (the reliability standard for power generation used by many utilities) is higher for a set of large plants than for a larger number of smaller plants with identical failure rates per plant.

Inexpensive generators that can be built in small sizes also change the logic of investments in electric transmission and distribution. Traditionally, investments in transmission and distribution networks have been justified for three primary reasons: 1) to aggregate loads to take advantage of large-scale generation (and its presumed lower cost power), 2) to create load diversity so that generators need not follow the sharp variations in demand for individual customers, and 3) to increase reliability by interconnecting many different thermal-electric plants and connecting remote hydroelectric generators to the system.

In some instances, investments in distributed renewables would be preferable to increased investments in transmission and distribution. A recent study of the value of distributed photovoltaic systems in northern California suggests that any system capable of reducing the demand on electrical transformers and associated transmission and distribution equipment at a utility substation can postpone the need for investment to expand the capacity of this load-bearing equipment. Such distributed siting will also reduce line losses and improve a utility's ability to control the quality of its power. In addition, distributed generators provide more options for routing power around faults in the distribution system and so improve the system's reliability. This reliability gain is important since the loss-of-load actually experienced by utility customers is typically several times as large as the one-day-in-10-years loss-of-loadcriterion.

Under some circumstances, renewables can increase the need for electric transmission. It is much less likely, for example, that a series of connected wind-farms will have zero output than it is that a single wind machine will have zero output during a period of peak utility demand. Grids can also help dispatch intermittent energy from areas with temporary over-supply to regions lacking power because of wind or sunlight conditions. In the future it may even be useful to transmit direct solar power east to provide power in evenings.

These critical issues need much more attention.

Constructing the electricity scenario

Some of the electricity strategies developed in the above case study can be easily extrapolated to other regions while others must be adjusted to reflect local constraints and resources in constructing the electricity supply mixes for the renewables-intensive global energy scenario. (See the appendix at the back of the book for a detailed discussion of the scenarios for each of the 11 world regions.)

The extent to which the world electricity supply in 2025 will be renewables-intensive depends in part on utility investment decisions made over the next decade, because power plants last 30 years or more. In this regard, it is fortuitous that many utilities around the world are adopting natural gas–fired, gas turbine–based power systems as the thermal-electric technology of choice. The low cost and operational advantages of these turbine systems make them attractive in serving the needs of many utility planners today, as well as in the future when intermittent renewables could play large roles. Emphasis on gas turbines in the near term makes it possible to avoid commitments to large conventional power plants while alternatives, including renewables, are explored. And the dispatch of many small-scale gas turbines distributed around a utility's service territory will provide utilities with valuable experience that can eventually be transferred to dispersed renewable systems.

Because of these advantages, the role of natural gas is assumed to increase rapidly in the global scenario, with its share in power generation increasing from 12 percent in 1985 to about 25 percent in the period 2025 to 2050.

In many situations the cost of electricity from biomass will be about the same as that from coal (see figure 6). Since the scenario was designed to explore intensive investment in renewables, it was assumed that biomass is preferred over coal whenever coal shows no obvious economic advantage. Under this assumption, the biomass share in power generation would be about 18 percent in the period 2025 to 2050, and coal would be used for power generation primarily in regions where the availability of inexpensive renewable resources is limited. As a result the amount of electricity produced from coal at the global level would be reduced by 46 percent between 1985 and 2050; because of growing efficiency of coal use (with all coal power provided by fuel cells in 2050), the amount of coal needed for power generation would be reduced more—by 70 percent.

In most regions it was assumed that nuclear power generation between 2025 and 2050 would be at the level of the output of plants that are now either operating or under construction. Accordingly, the amount of nuclear power generated worldwide in 2025 and beyond would be 31 percent more than in 1985.

Only modest further reductions in CO_2 emissions could be achieved by complementing the renewables deployed in the renewables-intensive scenario with more nuclear power substituted for fossil fuels in power generation, because the power sector accounts for such a small fraction of the CO_2 generated in this scenario (see figure 10).

Although hydroelectric power is economically attractive, environmental concerns and other restrictions will limit its growth. As a result of considering such restrictions on a region-by-region basis, the global share of power generation accounted for by hydroelectricity would decline from the present 20 percent to 15 percent by 2050. However, its relative contribution to peak power might not decline, despite the relative decline in generation, if more generating turbines are added at each installation to improve the capability to follow load and the variable output of intermittent renewables.

The potential contribution of intermittents depends on regional weather and land-use restrictions. Taking the analysis of an "average" renewable region described in the case study as a base, judgments were made about the potential contribution of intermittents in each region. Worldwide, intermittents provide 22 percent of electricity in 2025 and 30 percent in 2050.

Strategies for fuels

Fuels used for purposes other than electric generation (called "direct fuel use") accounted for 65 percent of the primary commercial energy used in the world in 1985. Direct fuel use includes the coal, natural gas, and petroleum that are used for transportation, industrial-process heat, space heat, and other purposes. The demand for fuels increases only 31 percent between 1985 and 2050, compared with a 265 percent increase for electricity, given the assumptions of the IPCC accelerated policies scenario (see figure 5). However, the share of primary energy accounted for by the direct use of fuels in 2050 is 61 percent, only slightly less than at present, owing to both the expected higher efficiency for generating electricity and the losses incurred in manufacturing synthetic liquid and gaseous fuels.[11]

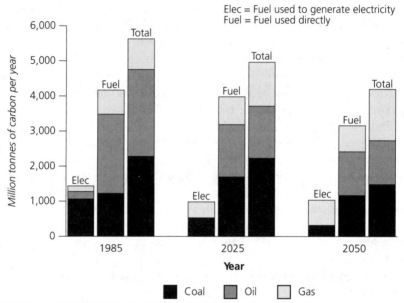

FIGURE 10: EMISSIONS OF CO_2 FOR THE RENEWABLES-INTENSIVE GLOBAL ENERGY SCENARIO, BY ENERGY CARRIER

Despite the expected growing share of energy going to power generation, the power sector's share of global CO_2 emissions would not increase in the period 1985–2050 from the present relatively low level (about 1/4 of total emissions), in a renewables-intensive global energy future. Accordingly, if society should choose to reduce emissions further than is indicated for the global energy scenario developed here, only modest further reductions could be achieved by substituting additional carbon-free power-generating sources for fossil fuel power-generating sources. Much greater reductions could be achieved by substituting renewable fuels for fossil fuels used directly. The least costly renewable fuel options for such substitutions would be biomass fuels (e.g. methanol and hydrogen). Much more land could be committed to biomass production than was assumed for the scenario illustrated here—especially in Africa and Latin America. In addition, vast quantities of hydrogen could be produced from wind and photovoltaic sources without encountering land or water availability constraints.

---

11. In the renewables-intensive energy scenario, 345 EJ of primary energy is required in 2050 to produce the 289 EJ of direct fuel use called for in the IPCC accelerated policies scenario. See footnote 6.

In constructing the renewables-intensive scenario, particular attention was paid to liquid and gaseous fuels. The demand for liquid fuels used directly, which accounted for more than half of fuel use in 1985, is projected to increase about 11 percent, driven largely by growth in the transportation sector, which accounted for 60 percent of all petroleum consumption in 1985. The demand for gaseous fuels used directly, which accounted for about a quarter of direct fuel use in 1985, is projected to double by 2050.

The principal candidates for transportation fuels from renewable sources are ethanol, methanol, hydrogen manufactured from biomass, and hydrogen manufactured by electrolysis of water using renewable sources of electricity. Measured in energy terms (dollars per gigajoule) all these fuels will probably cost more to produce than petroleum-based transport fuels in the early years of the 21st century. However, as in the case of renewable electric technologies, the economic viability of each of these fuels will be determined in the context of the system in which it will be used. When an alternative fuel is considered as part of a system designed and optimized for that fuel, the economic performance will often be much better than for the case where the alternative fuel is simply substituted for, say, gasoline in an internal-combustion-engine vehicle optimized to run on gasoline.

New transportation systems are likely to emerge that involve both new fuels and new vehicle types. The economic and environmental merits of alternative systems will be judged on the basis of the costs and environmental impacts of providing transportation services with alternative combinations of fuels and vehicle types.

As will be shown, all the candidate biomass-derived fuels have realistic chances of being competitive in providing transportation services early in the 21st century. While the cost of these services would be slightly higher with electrolytic hydrogen, this option offers an abundant resource providing pollution-free transportation services.

The context

Transportation systems have been dominated by internal combustion engines and gasoline for three generations. In fact, the internal combustion engine has been so successful that until recently prospects for radical alternatives have been considered dim, and little research has been directed to the search for alternatives. Not only would a shift to new kinds of vehicles and new fuels affect the business practices of vehicle manufacturers and petroleum companies, two of the world's largest economic sectors, but also a shift to a new fuel would require massive investments in new infrastructures, as filling stations, pipelines, and fuel storage facilities are replaced or repurposed.

Such obstacles should not be viewed as insurmountable, however. Starting in the late 1970s, Brazil, for example, has shifted about half of its automotive fleet to ethanol. And recent volatility in transportation markets increases the likelihood that renewable fuels will find major markets in other countries as well.

In addition, environmental reform is likely to transform worldwide markets for new fuels and propulsion systems. Already, acute air quality problems are creating markets for innovative transportation systems in some urban areas. Planners in these areas have come to realize that the solution is not as simple as mandating further incremental restrictions on tailpipe emissions and so are beginning to explore more radical changes. Research and development are now underway on zero emission propulsion alternatives—both battery-powered electric vehicles and fuel-cell vehicles. Also under investigation are fuel alternatives such as reformulated gasoline, compressed natural gas, alcohol fuels (ethanol and methanol), electricity, and hydrogen.

In response to air quality concerns, California has mandated that by 2003, 10 percent of the vehicles sold in the state must have zero emissions. Other U. S. states are considering similar requirements.

Although the search for alternative transportation strategies is motivated largely by environmental concerns, technologies that are superior to existing internal-combustion-engine vehicles in terms of cost, reliability, driving performance, and noise generation could emerge. The challenge for policymakers is to construct policies that facilitate the introduction of new transportation systems that offer consumers multiple benefits while meeting the environmental goals of society.

New vehicle technologies

To date, the battery-powered electric vehicle is the zero-emission vehicle that has received the most attention. But its market share is likely to be small unless there is a breakthrough in battery storage technology, because its range is limited and its batteries require several hours for recharging. Moreover, if the electricity for these vehicles comes from conventional fossil fuel-powered generators, air quality problems are not eliminated, only shifted from one site to another.

Fuel-cell-powered electric vehicles are at an earlier stage of development but are likely to be attractive alternatives to battery-powered vehicles. Fuel cells convert hydrogen directly into electricity without first burning it to produce heat. Hydrogen fuel-cell vehicles are perhaps three times as energy-efficient as comparable vehicles powered by gasoline-burning internal-combustion engines. The electricity produced by the fuel cell drives electric motors that provide power to the wheels. It is likely that a fuel-cell vehicle will use a battery (or perhaps a capacitor) to store electricity for starts and to provide extra power for passing and long uphill climbs. A battery can also store energy that would otherwise be lost in braking. The battery in a fuel-cell vehicle would be larger than the ones in standard vehicles but much smaller than the ones in battery-powered electric vehicles.

Fuel-cell vehicles are refueled quickly, either with hydrogen or with a hydrogen carrier such as methanol that is converted into hydrogen on board. Fuel-cell vehicles fueled with methanol are about 2½ times as energy-efficient as comparable gasoline internal-combustion-engine vehicles.

While several types are under development, the fuel cell attracting the most attention for transportation applications is the proton-exchange-membrane fuel cell, so-called because it employs a thin membrane to separate the hydrogen ions from electrons in its operation.

Fuel-cell vehicles have superb environmental characteristics. Hydrogen-powered vehicles emit only water vapor and methanol vehicles only carbon dioxide and tiny amounts of local air pollutants, along with water vapor. If the methanol is produced from biomass grown on a sustainable basis, this carbon dioxide would be absorbed by growing biomass, so that net carbon dioxide emissions would be zero.

With an aggressive development effort driven largely by environmental concerns, fuel-cell vehicles could be demonstrated in prototype vehicles during the 1990s and be ready for various captive fleet applications during the first decade of the 21st century.

Alternative fuels

During the next few decades, several fuels will be competing for markets now dominated by gasoline. In the near term, the most important renewable transportation fuels are likely to be ethanol and methanol used in internal-combustion-engine vehicles. In the longer term methanol and hydrogen used in fuel-cell vehicles may be preferred.

Ethanol is already widely used in Brazil, where it is made from sugarcane, and to a lesser extent in the United States, where it made from maize (corn). Although neither technology is economical at current oil prices, prospects are good for making cane-derived ethanol competitive in tropical countries where sugarcane can be grown (see chapters 17 and 20: *The Brazilian Fuel-alcohol Program*). It is much less likely that maize-derived ethanol can be made competitive (see chapter 21: *Ethanol and Methanol from Cellulosic Biomass*).

Ongoing advances in biochemical technology could make it possible to produce ethanol by 2000 at competitive costs from comparatively inexpensive wood chips or other cellulosic materials using enzymatic hydrolysis. Methanol can be made from these same feedstocks using a thermochemical rather than a biological conversion process. However, if production cost goals for enzymatic hydrolysis systems are realized, ethanol from cellulose would probably be less expensive per unit of contained energy than methanol. Thus, ethanol would be preferred to methanol for internal-combustion engine vehicles.

The alcohols are excellent fuels for use in internal combustion engines. When optimized for operation on alcohol fuels, these engines are about 20 percent more energy-efficient than when operated on gasoline. The use of alcohol fuels in internal-combustion-engine vehicles, however, may eventually be limited by air pollution regulations. Although vehicles fueled by alcohols are cleaner than when fueled by ordinary gasoline, they are not zero-emission vehicles. In fact, when run on alcohols they produce only marginally less air pollution than when run on reformulated gasolines.

Methanol fuel-cell vehicles, however, produce almost no pollution. Overall, the cost per kilometer for driving a highly energy-efficient methanol fuel-cell vehicle could be less than for an ethanol internal-combustion-engine vehicle, even if the methanol is more expensive to produce.

For fuel-cell applications, methanol serves as a convenient carrier for the needed hydrogen. Hydrogen can be generated from methanol with low-temperature heat in the presence of an inexpensive catalyst. This conversion takes place in a device called a reformer, which can be installed along with the fuel cell under the hood of the car. Heat at much higher temperatures and more costly catalysts are required to reform ethanol, making impractical the use of ethanol in motor vehicles with proton-exchange-membrane fuel cells, which generate only low-temperature waste heat. Thus, methanol will probably be preferred as a transport fuel for the next couple of decades if fuel-cell vehicles become well established.[12]

Because the production of petroleum is expected to decline in most regions outside the Middle East in the decade ahead, renewable fuels will be competing against fuels manufactured from coal or natural gas. Biomass-derived fuels are likely to be less expensive and involve fewer investment risks than fuels from coal, though they will be unable to compete with methanol manufactured from natural gas at today's prices. This situation will change over the next couple of decades. As worldwide gas demand grows and natural gas trade intensifies, natural gas prices throughout the world are likely to rise to at least $3.5 per gigajoule ($21 per barrel of oil equivalent), a price at which biomass-derived methanol would be competitive.

Because growing biomass requires considerable land and water resources, there are limits to the amount of biomass-based methanol that can be produced in regions where population densities are high (Europe and South and East Asia) or water is scarce (North Africa) or both. And a global commitment to major reductions in greenhouse-gas emissions might create demand for low-carbon fuels that could not be met with biomass fuels alone. The additional demand could be satisfied by generating electrolytic hydrogen from wind or direct solar sources for use in fuel-cell vehicles. The potential for producing hydrogen in this way is vast. For example, hydrogen produced from photovoltaic power sources on 1 percent of the world's desert areas and used in fuel cells could displace the equivalent of the present rate of world fossil fuel use.

While the technology for electrolyzing water to produce hydrogen is well known, techniques for storing hydrogen must be improved. Although hydrogen can be stored directly as a compressed gas or in a metal hydride, an alternative and somewhat radical new approach, which calls for the use of powdered iron as a hydrogen carrier, may turn out to be less expensive. The powdered iron would be oxidized on board the car to produce hydrogen with steam generated by the fuel cell (see chapter 22: *Solar Hydrogen*).

---

12. This situation could change in the longer term if advanced fuel cells that operate at high temperatures (such as the solid-oxide fuel cell) can be successfully developed for transportation. A wide range of alternative fuels could be reformed using the high-quality waste heat generated by such fuel cells.

Ethanol. Brazil launched the transition to an energy regime based on fuels from renewables in 1975 with its fuel-ethanol program. Under this program, ethanol from sugarcane (via the fermentation of sugar juice) has become a major fuel for Brazilian light-duty vehicles (automobiles and light commercial trucks). In 1989, Brazil's entire fleet of light-duty vehicles consisted of 4.2 million fueled with hydrated (neat) alcohol and 5 million fueled with gasohol (a gasoline/ethanol blend that is 22 percent ethanol).

A large-scale fuel-ethanol industry has since been established, with ethanol production increasing from 0.6 billion liters in 1976 to 12 billion liters in 1989. Substantial improvements in production technology took place as the industry expanded. Between 1977 and 1985 alcohol yield (in liters per tonne of cane) increased 23 percent and sugarcane yield (in tonnes of cane per hectare per year) increased 16 percent. Such productivity gains were reflected in a 4 percent annual rate of decline in the cost of ethanol production between 1979 and 1988.

Despite Brazil's success in launching an ethanol industry, the cost of producing ethanol today, which is some 23 cents per liter,[13] is higher than the value of neat ethanol as a gasoline substitute, which is about 20 cents per liter at the 1991 world oil price of $19 per barrel. However, Brazilian analyses indicate that the prospects for reducing the ethanol production cost another 23 percent over the next several years are good (see chapter 20). A reduction of this magnitude would make Brazilian alcohol competitive at the 1991 world oil price.

Brazilian alcohol could also be made competitive using sugarcane residues as fuel to run biomass-integrated gasifier/gas turbine cogeneration systems. Such systems used in energy-efficient distilleries could provide both the steam and electricity required for alcohol production plus excess electricity for sale to the electric utility. Both bagasse, the residue left after crushing the cane to extract the sugar juice, and the tops and leaves of the sugar cane plant could be used as fuel. Although the bagasse now provides energy for the distillery, it is used very inefficiently; and the tops and leaves of the sugar plant are burned off the cane fields just before harvest in many parts of the world. If alcohol and electricity were coproduced this way, the alcohol could be sold at a price competitive with gasoline at the 1991 oil price and the by-product electricity sold at a price competitive with electricity from new hydroelectric power plants (see chapter 17).

The prospects are thus auspicious for making alcohol competitively from sugarcane with technological improvements that are within reach. These advanced alcohol technologies could be adopted by other tropical countries where the conditions are suitable for growing sugarcane.

The United States has a fuel-ethanol program based mainly on maize. At present there are 50 fuel-ethanol facilities producing 3 billion liters of ethanol a year. The ethanol is used mainly in gasohol (a 10 percent blend with gasoline). In some years, maize-based ethanol is competitive, but because the prices for maize

---

13. All cost estimates for the production of fuels for transportation presented in this chapter are based on a 12 percent real discount rate. Taxes are not included.

and the coproducts of ethanol fluctuate widely from year to year, ethanol is profitable, on balance, only because of substantial subsidies.

A promising alternative involves shifting from maize to low-cost cellulosic feedstocks, such as wood chips, which can be converted to ethanol via an enzymatic hydrolysis process now under development. The production cost has fallen sharply, from about 95 cents per liter a decade ago, to 28 cents per liter today. Although this is still too high to compete with gasoline at the present world oil price, analysts at the U.S. National Renewable Energy Laboratory project that, with anticipated technological improvements, ethanol produced this way could be competitive by 2000 with gasoline based on oil selling for less than $25 a barrel (see chapter 21). Progress has been and is expected to be rapid—not only because research on the technology is progressing well but also because rapid progress is being made in a variety of related biotechnology fields.

Methanol. Methanol is produced from biomass via a thermochemical process that begins with the gasification of biomass at high temperature. The products of gasification, which include carbon monoxide, hydrogen, and methane, are then converted to methanol via well-established industrial processes developed originally for making methanol from natural gas and coal (see chapter 21). While most gasification efforts have focused on coal, biomass is far more reactive than coal and thus easier to gasify; moreover, it contains very little sulfur. These advantages mean lower production costs. A biomass-to-methanol plant costing $200 million and producing 0.5 billion liters per year could produce methanol at a lower cost per liter than a coal-to-methanol plant costing some $1.3 billion and producing more than two billion liters per year, even if the biomass feedstock cost nearly twice as much as the coal ($2.5 vs $1.4 per gigajoule) (see chapter 21). In addition, the biomass facilities would present fewer financial risks since they could be built in comparatively small units.

Although methanol from biomass may cost less than methanol from coal, it would probably be more costly as an internal-combustion-engine fuel than ethanol from the same biomass. But biomass-derived methanol would be an attractive fuel for fuel-cell vehicles. On a lifecycle-cost basis (cents per kilometer) a methanol fuel-cell vehicle in mass production may be competitive with a gasoline internal-combustion-engine vehicle at a world oil price below the present level (see figure 11)—despite the fact that the price paid for methanol at the pump may be 50 percent higher than for gasoline, and that a methanol fuel-cell vehicle may cost 50 percent more than a gasoline internal-combustion-engine vehicle with comparable performance. This remarkable result arises because the methanol fuel-cell vehicle is likely to be 2½ times as energy-efficient as the gasoline internal-combustion-engine vehicle and is expected to last much longer and have lower maintenance costs. The fuel-cell vehicle would also be less costly than the battery-powered electric vehicle (see figure 11) and would not have the drawback of long battery-recharging times.

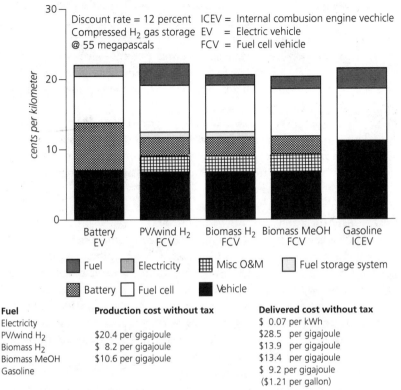

Fuel	Production cost without tax	Delivered cost without tax
Electricity		$ 0.07 per kWh
PV/wind H₂	$20.4 per gigajoule	$28.5 per gigajoule
Biomass H₂	$ 8.2 per gigajoule	$13.9 per gigajoule
Biomass MeOH	$10.6 per gigajoule	$13.4 per gigajoule
Gasoline		$ 9.2 per gigajoule
		($1.21 per gallon)

FIGURE 11: LIFECYCLE COST FOR AUTOMOBILES

The cost of owning and operating an automobile is shown here on a lifecycle cost basis (in U.S. cents per kilometer) for cars powered by alternative energy systems. The cost for the reference case— an internal-combustion-engine vehicle (ICEV) fired with gasoline at the price projected for the United States in the year 2000—is indicated by the bar on the right.

The three bars in the middle are for fuel-cell vehicles (FCVs) using alternative renewable fuels: hydrogen derived electrolytically from PV or wind sources, hydrogen derived thermochemically from biomass, and methanol derived thermochemically from biomass. The onboard storage mechanism assumed here for hydrogen involves the use of compressed gas canisters (even though more attractive storage systems are likely to become available in the coming decades). The hydrogen FCV creates no air pollution, and the methanol FCV very little.

Even though the price for the biomass fuels at the refueling station is about 50 percent higher than the gasoline price, the biomass-derived methanol and hydrogen FCVs would be competitive on a lifecycle cost basis with the gasoline-powered ICEV because the FCV would be much more energy-efficient, and it is expected to last longer and have lower maintenance costs.

The lifecycle cost for the FCV powered with hydrogen derived via electrolysis from photovoltaic (PV) or wind electricity is about the same as for the battery-powered electric vehicle (EV), shown on the left. Unlike the battery-powered EV, which requires several hours for recharging, the hydrogen FCV can be refueled in 2–3 minutes.

Though hydrogen derived from PV/wind sources would be much more costly than hydrogen derived thermochemically from biomass, its production would not be limited by land use or water availability constraints, as would eventually be the case for biomass-derived hydrogen or methanol. The PV/wind hydrogen FCV would be competitive with the gasoline-powered ICEV at a gasoline price (without retail taxes) of $0.39 per liter ($1.49 per gallon), corresponding to a crude oil price of $34 per barrel.

For a more detailed discussion of these costs, see Endnotes for Figures.

Uncertainties associated with these cost estimates can be resolved only with more fuel-cell development, the construction of prototype vehicles, and experience with test fleets. A key assumption is that fuel-cell vehicles will last longer than internal-combustion-engine vehicles, essentially because electric drive trains are more durable than mechanical drive trains. More experience with fuel-cell vehicles is needed to ascertain whether they do indeed have longer useful lives.

Hydrogen.   Hydrogen produced electrolytically from wind or direct solar-power sources and used in fuel-cell vehicles can provide zero-emission transportation. As for any fuel, appropriate safety procedures must be followed. Although the hazards of hydrogen are different from those of the various hydrocarbon fuels now in use, they are not greater (see chapter 22).

Electrolytic hydrogen may be attractive in regions such as Europe, South and East Asia, North Africa, and the southwest United States, where prospects for biomass-derived fuels are limited—either because of high population density or lack of water. Land requirements are small for both wind and direct-solar sources, compared to those for biomass fuels. Moreover, as with wind electricity, producing hydrogen from wind would be compatible with the simultaneous use of the land for other purposes such as ranching or farming. Siting in desert regions, where land is cheap and insolation good, may be favored for photovoltaic-hydrogen systems, because little water is needed for electrolysis: the equivalent of two to three centimeters per year of rain on the collectors, which represents a small fraction of total precipitation, even for arid regions.

Electrolytic hydrogen will probably not be cheap. If cost goals for wind and photovoltaic electricity for the period shortly after 2000 are met, the corresponding cost of pressurized electrolytic hydrogen to the consumer would be about twice that for methanol derived from biomass; moreover, a hydrogen fuel-cell car would cost more than a methanol fuel-cell car, because of the added cost for the hydrogen storage system. Despite these extra expenses, however, the lifecycle cost for a hydrogen fuel-cell car, in cents per kilometer, would be only 2 or 3 percent higher than for a gasoline internal-combustion-engine car—about the same as for a battery-powered electric vehicle (see figure 11).

The transition to an energy economy in which hydrogen plays a major role could be launched with hydrogen derived from biomass (see chapter 22). Hydrogen can be produced thermochemically from biomass using the same gasifier technology that would be used for methanol production. Although the downstream gas-processing technologies would differ from those used for methanol production, in each case the process technologies are well-established. Thus, from a technological perspective, making hydrogen from biomass is no more difficult than making methanol. In fact, the process would be more energy-efficient, and the production cost would be less. Biomass-derived hydrogen delivered to users in the transport sector would typically cost only half as much as hydrogen produced electrolytically from wind or photovoltaic sources; and, as for a fuel-cell car operated on biomass-derived methanol, a fuel-cell car operated on biomass-derived hydrogen would probably compete with the gasoline internal-combustion-engine car at a world oil price lower than at present (see figure 11).

While the adoption of hydrogen derived from biomass would facilitate a shift later to electrolytic hydrogen, the transition from a liquid fuel (gasoline) to a gaseous fuel (hydrogen) would be more difficult than switching from gasoline to methanol. Indeed, the transition to hydrogen would require an entirely new infrastructure, necessitating large capital investments.

It may be feasible to use hydrogen energy on a large scale without having to transport or store hydrogen—thus obviating the need for a gaseous fuel infrastructure. One way would be to employ powdered iron as a hydrogen carrier (see chapter 22).

In an iron-powered fuel-cell car, powdered iron would be oxidized under the hood with steam generated by the fuel cell, producing hydrogen and iron oxide (rust) in the process. When a tank of iron is fully oxidized, the tankful of rust would be exchanged for a new tankful of powdered iron at a refueling station. The rust in turn would be reduced to iron at recycling centers with reducing gases such as hydrogen or carbon monoxide. While the details of how hydrogen would be generated in the car and how the rust would be recycled are yet to be worked out, powdered iron could prove to be a relatively low-cost hydrogen carrier. Recycling centers might employ biomass gasifiers to generate reducing agents (hydrogen and carbon monoxide) or use electrolytic hydrogen if biomass supplies are constrained.

Powdered iron and methanol are competing hydrogen carriers for fuel-cell vehicles. The iron-powered fuel-cell car would be preferred, should biomass fuel supplies be constrained.

Constructing the fuels scenario

The potential contributions of renewable fuels to the renewables-intensive global energy scenario were estimated for 11 regions (see Box B), taking into account the economic analysis of alternative fuels presented above, as well as regional endowments of various biomass feedstocks and other renewable energy sources and fossil fuels. (See the appendix at the back of the book for details.)

Total global biomass supplies available for energy purposes (for both electricity and fuels production) are estimated to be the following:

▼ Biomass from plantations grown mainly on degraded lands in developing countries and on excess agricultural lands in industrialized countries amounts to 80 exajoules in 2025 and 128 exajoules in 2050.

▼ Residues from agricultural and forest-product industry activities and from urban refuse amount to 55 exajoules in 2025 and 68 exajoules in 2050.

▼ Biomass from forests that are already routinely harvested for fuelwood amounts to 10 exajoules in 2025 and 2050.

The contributions of biomass feedstocks and other renewable sources to fuels production are as follows:

▼   In sugarcane-producing countries it is assumed that cane grows in proportion to the population and that ethanol is produced with one-third the cane, via fermentation of sugar juice. Under these assumptions, ethanol from cane meets less than 1 percent of global liquid-fuel requirements from 2025 to 2050.

▼   Additional alcohol is produced from cellulosic feedstocks. Methanol is identified as the alcohol of choice for these feedstocks—largely to reflect the favorable prospects for methanol fuel-cell vehicles. (It may be feasible to overcome the problems of matching ethanol to fuel cells, but the scenario was constructed with the conservative assumption that methanol would be preferred. The energy balances would change little if ethanol derived from enzymatic hydrolysis of cellulosic feedstocks or a mix of alcohols were emphasized instead.)

▼   Liquid-fuel requirements in excess of domestic oil and alcohol production are met by imports. Half of imports are oil and half are methanol from biomass. This mix was chosen in light of the expectation that lifecycle costs would be about the same for gasoline internal-combustion-engine vehicles and methanol fuel-cell vehicles (see figure 11). (As in the case of domestic production, imported alcohol could also be ethanol or a mix of alcohols.)

▼   The regions with the greatest promise for exporting methanol are Latin America and sub-Saharan Africa, where there are large amounts of degraded lands suitable for reforestation that would not be needed for food production. Total methanol exports from these regions amount to 15 exajoules in 2025 and 20 exajoules in 2050 (see figure 4).

▼   Overall, by 2025 about one third of global liquid-fuel requirements for the renewables-intensive scenario are met with methanol; by 2050, one half.

▼   Biogas (methane) produced by the anaerobic digestion of recoverable dung contributes from 8 to 9 percent of global gaseous fuel requirements from 2025 to 2050.

▼   Biomass-derived hydrogen contributes 12 percent of global gaseous fuel requirements in 2025 and 16 percent in 2050.

▼   Wherever domestic natural gas, biogas, and biomass-derived hydrogen are inadequate to satisfy gaseous fuel demand, the remaining demand can be met by a mix of imported natural gas and electrolytic hydrogen derived from intermittent renewable sources, produced either domestically or imported. Hydrogen from intermittent sources provides from 8 to 9 percent of global gaseous fuel requirements from 2025 to 2050.

▼   The regions with the greatest promise for producing electrolytic hydrogen for export markets are desert areas in the Middle East and North Africa. Exports of electrolytic hydrogen from these regions total more than 3 exajoules in 2025 and more than 5 exajoules in 2050 (see figure 4).

AN AGENDA FOR ACTION

By the middle of the 21st century, renewable energy sources can play a central role in world energy markets. They can do this even if world energy prices increase very slowly and without subsidies or credits to reflect external benefits not tracked in standard economic accounting (see Box A).

The potential for renewables will not be realized, however, unless the regulations and incentives operating in existing energy markets are changed. The comparatively low level of world investment in renewable energy technologies, the inertia of existing energy systems, and the difficulty of maintaining investment interest during periods when markets are small and uncertain all act to slow the rate at which renewable energy technologies will move into the market.

Few private investors have the patience or the resources for the sustained research and demonstration programs needed to make renewable energy technologies profitable. Moreover, the merits of renewable energy technologies are often masked by national policies that subsidize fossil or nuclear energy technologies without providing corresponding support for renewables. Another difficulty is that world oil prices might be manipulated to undercut the prices for renewable energy. Indeed, this risk could be a major barrier to private investment in renewables. Moreover, even if the performance and cost projections for the various renewable energy technologies are realized, there are few markets where renewable energy equipment will be able to produce fuels or electricity at prices dramatically lower than those for fuels and electricity from conventional sources. Unlike such innovations as integrated circuits or automobiles, renewable energy technologies will not be so clearly superior on economic grounds that the earlier generations of technologies will be swept aside by the new. Under these circumstances, renewables will penetrate established markets very slowly.

Renewable energy technologies are not likely to be rapidly and widely adopted without a global commitment to support the research, development, and demonstrations needed to stimulate early market interest in renewable energy technologies, to eliminate or neutralize market biases embedded in current policies, to provide incentives to promote the risk-taking involved in launching new industries, and to ensure that renewable energy technologies are rapidly transferred to developing countries.

The industrialized countries have major responsibilities in the required global effort. These countries have the technical and financial resources to accelerate the development of renewable technologies and to convert technical concepts into practical products. Indeed, unless the industrialized countries make renewables a central part of their domestic energy programs and back this commitment with resources and policy reforms, a world commitment to renewable energy would have little credibility.

Both developing and industrialized countries can provide attractive markets for renewable energy systems. In fact, there may be occasions when developing areas can leapfrog industrialized countries because they are not constrained by existing energy infrastructures.

It is important to take steps to ensure that technologies are rapidly transferred from industrialized to developing countries. Joint ventures involving firms from developing and industrialized countries offer one useful approach for transferring technology and know-how to the firms in the developing countries most likely to be able to profit from the transfer. Such endeavors must be designed to benefit all participants; firms from developing countries must be closely involved in designing and installing equipment and in project management.

Outline of a strategy for increased utilization of renewable energy
If renewables are to achieve the prominence in world energy indicated by the renewables-intensive global energy scenario, coordinated national and international programs are needed to promote major institutional and technological innovations. Among the most important are those that shape energy markets through regulations and subsidies; foster research, development, and demonstration efforts; coordinate renewable energy development with agriculture and forestry; set utility policy; and encourage multinational planning, research, and economic assistance efforts. All the programs proposed here would support broader national programs for encouraging innovation and productivity growth.

Innovation is required not only in the energy sector but also in many other sectors of the economy (e.g., agriculture, transportation, and the chemical process industries). These changes cannot occur unless a country has a technology policy conducive to risk-taking and new investment. Likewise, national environmental goals, which can often be well served by introducing renewable energy technologies, can be met much more efficiently, and at far lower costs, if a country has an effective policy for encouraging adaptation, change, and innovation throughout its economy. Technology policy is set by macroeconomic and fiscal policies, education and training policies, and other regulations beyond the scope of the present analysis.

Innovation will be required in the structures of the industries that will produce and manage renewable energy. The nature of renewable energy resources and technologies implies that the required industries will have some features quite different from those of today's energy industries. However, present industries have unique capabilities that can help renewable energy reach its full potential.

Electric utilities, for example, have much to offer in facilitating the introduction of new sources of electricity. However, if utilities are to have major roles in renewable energy they must be suitably transformed. Historically, utilities have been operated as public or private monopolies in most countries, producing most of their power in large centralized plants. Utility systems that make full use of the potential for renewables will have broad portfolios of investments: a wide variety of renewable energy sources operated at many different scales—including those not connected to the grid, advanced conventional energy technologies, advanced controls for distribution systems, energy-efficient equipment located on customers' premises, and some storage equipment. Given the right incentives, electric utilities can create efficient markets for this complex set of investment opportunities. By acting as skillful brokers, utility managers can ensure that new renewable equipment is well designed and efficiently integrated into regional electricity markets. Regulatory policies should be changed so as to facilitate the transition to this new regime for electric utilities that is already underway in some parts of the world.

As in the case of electric utilities, deep structural changes will be required of petroleum companies if they are to make major contributions in renewable fuels; they would have to produce, distribute, and market fuels from diverse and decentralized production facilities. Large potential markets for biomass fuels provide opportunities for new relationships among multinational petroleum companies and agricultural and forestry producers. If the proposal to establish biomass plantations on large areas of deforested and otherwise degraded lands in Africa and Latin America were implemented to the extent that these regions became major exporters of biofuels, the forging of such relationships would undoubtedly be required. In such joint ventures, petroleum companies could offer important chemical processing skills as well as much needed capital resources. Also potential markets for electrolytic hydrogen could promote new relationships between petroleum companies and electric utilities. Petroleum companies are already feeling pressure to provide cleaner fuels in response to growing urban air quality concerns. Moreover, the prospect of declining oil production in most parts of the world outside the Middle East will require many oil companies to diversify away from their traditional business activities in order to grow. Renewable fuels provide an opportunity for continued growth.

The widespread use of renewable fuels requires that vehicles be designed to make efficient use of the unique properties of alcohols and hydrogen. Fuel-cell vehicles will be especially important. To produce such vehicles in substantial quantities by early in the 21st century will require major changes in the business practices of many large vehicle producers. Most companies will have to substantially increase investments in research and development and make unprecedented commitments to technical innovation.

Innovation is also called for in agriculture to promote diversification of production to include biomass for energy. Such a shift in agricultural policy would serve various pressing needs, in both industrialized and developing countries. This change could provide attractive cash crops in areas needing economic diversification, such as those regions of the industrialized world that have excess agricultural productive capacity. In developing countries, the establishment of biomass plantations on degraded lands could also both provide a mechanism for financing the restoration of these lands while providing modern energy carriers at competitive prices for rural development.

New programs should be crafted carefully. Specific programs that would have the best chances of meeting policy goals are those that: 1) lead efficiently to a clear objective, 2) have low transaction costs and are administratively simple, 3) follow a schedule that is predictable and clearly understood, and 4) work in an integrated manner with programs having other purposes (e.g., economic development, land management and transportation).

a. Creating rational energy markets

The accumulation of decades of incentives and regulations designed to promote favored energy technologies have badly distorted energy markets in many countries. Many of these policies discourage investments in energy efficiency and renewable energy technologies. These past interventions should be reviewed and revised as necessary to ensure that government intervention in energy pricing is clearly linked to current environmental, development, or security goals that markets do not address.

b. Research, development, and demonstration

Large increases in public and private support for research, development, as well as for technology and market demonstrations, are needed to support an accelerated transition to renewable energy.

c. Agricultural and forestry policies

Major growth in the world production of biomass for energy will require the development of techniques for producing biomass profitably and sustainably on both excess cropland and degraded lands and the formulation of policies that would facilitate the creation of industries that would produce biomass for energy on these lands. In most areas, these steps will not be taken without revised agricultural and forestry policies.

d. International programs

International organizations, with the active support of national governments, need to play expanded roles by coordinating national programs, facilitating technology transfer, and formulating agreements on environmental and resource accounting and other topics.

While much of the following discussion focuses on specific government interventions, some of the most important opportunities for governments are not programmatic. Indeed, governments can find a variety of creative ways to encourage development of renewables. They can raise expectations and provide social support for private investments by serving as deal-makers, by bringing technical information to the attention to interested firms, by widely disseminating success stories, and by helping to forge research and demonstration consortia. In addition, they can facilitate the aggregation of markets for new products and encourage innovative contracting.

Creating rational energy markets

Energy markets in many countries are heavily manipulated by government policies. Policies put in place over several generations to encourage exploitation of domestic resources or to encourage specific technologies often have the inadvertent effect of discouraging investment in renewable energy technologies for which these incentives do not apply. Many of these policies have outlived their usefulness and need to be reexamined in the context of current national energy and environmental policies.

The discussion that follows suggests a two-step process for rationalizing energy markets. First, it is important to understand the extent and implications of the distortions already in place. Second, it is necessary to articulate the objectives for intervention in energy markets and revise policies so that the net signal sent to the market is the one desired. It is reasonable, for example, to force markets to consider environmental and other external costs. It may no longer be reasonable to subsidize fossil fuel production.

Reviewing existing subsidies

Subsidies for fossil and nuclear energy production usually bias investors against renewable resources. Common biases include:

▼ Specialized tax benefits for production of coal, oil, and gas (including special allowances for the depletion of nonrenewable resources, exemption from minimum taxation requirements, etc.).

▼ Direct subsidies for kerosene and diesel fuels.

▼ Public financing for the construction and modernization of nuclear plants.

▼ Regulatory policies that permit utilities to receive a return on plants that are still under construction (designed to encourage investment in large coal and nuclear plants, which require many years to build).

▼ Disproportionate amount of publicly supported research and development devoted to conventional energy sources.

▼ Public indemnification of nuclear facilities and public management of nuclear materials production and waste disposal.

▼ Agricultural policies that discourage crop diversification to energy crops. (Price supports and other subsidies provided for food crops are withheld from energy crops in most countries.)

Although direct and indirect subsidies have the effect of distorting energy markets in ways that are often difficult to understand, their combined effect is powerful. Therefore, each country should carry out a comprehensive assessment of the direct and indirect impacts of all subsidies and regulations affecting competition between energy sources.

Rationalizing market interventions

The review of subsidies provides information that can be used in creating a set of government interventions in energy markets clearly linked to societal goals. The revision process should include:

▼ A statement of national energy objectives (addressing issues of energy security, environmental quality, economic development, etc.). Each objective should be associated with measurable indicators of success.

▼ Adoption of policies that link the net governmental impact on energy markets to measurable objectives. Inappropriate subsidies should be removed, but if this proves politically difficult, a second-best alternative would be to introduce offsetting subsidies for renewables.

Only government action can force private markets to consider environmental and security costs. While there may be disagreement about the magnitude of these costs and the most appropriate way to make markets reflect these costs, most would agree that the costs are not zero. International cooperation in this area is critical because countries are reluctant to impose taxes on domestic producers that might reduce the competitiveness of domestic businesses.

The most straightforward way to force markets to reflect external costs is to impose a tax equal to those costs. Taxes have comparatively small transaction costs and allow the market the freedom to select investments consistent with the new signals. Yet few countries have energy taxes specifically linked to environmental goals. Many countries are finding it difficult to impose energy taxes for political reasons. It is particularly difficult to impose taxes large enough to have a major impact on investor behavior.

If a program of taxation based on externality costs cannot be implemented or proves ineffective, other solutions must be considered. There are several options. Regulation of electric and gas utilities provides a powerful tool for controlling emissions. While countries have different methods of regulatory control, most can influence the choice of power sources that are connected to transmission and distribution grids. Regulations can be designed to ensure that utilities generate or purchase electricity or promote investments in end-use efficiency so as to minimize the cost of providing reliable electric services—including environmental and security costs. Under such regulations a utility may be required to compare generating options only after an environmental fee has been added to the cost of electricity from fossil or nuclear energy sources. The size of the fee would be set on the basis of the proposed plant's environmental or security costs. The fee would appear as an increase in the price charged for electricity.

Tax policy can also be used to encourage environmentally benign technologies by providing offsetting subsidies to non-renewable energy supply technologies in situations where it is politically difficult to remove existing subsidies.

Tax policies can also be used to provide temporary subsidies to help launch new industries. A variety of investment tax credits have been used for such purposes over the past few years. A particularly attractive option is also one of the simplest—providing tax credits for the amount of energy actually produced. For example, an electric utility might be given a tax credit equal to two cents per kilowatt-hour of renewable electricity it sells to customers. Such an incentive could be either transitional, designed to be phased out over a specified period as the industry matured, or it could be retained indefinitely as a measure of externality costs. An advantage of this production incentive over traditional investment tax credits is that investors cannot enjoy a significant subsidy unless they install equipment capable of working efficiently over a period of many years. A disadvantage is that investors must wait years to receive the full subsidy.

Lessons learned
Two ambitious efforts to introduce renewable technologies are the Brazilian ethanol program and programs to encourage renewable electricity production in California. With the wisdom of hindsight, both efforts could have been more efficient; nonetheless, both have resulted in major new domestic energy industries that, with continuing technological innovation, will grow and prosper. And both resulted in the construction of many renewable energy facilities that continue to produce useful output. The following discussion reviews these case studies, as well as alternative fuels programs in the United States and the Danish wind energy program, and the lessons learned from them.

Fuels. In the 1970s, Brazil launched a program to shift to ethanol as an automotive fuel, a conversion driven primarily by tax policy and the regulation of fuel and vehicles, as follows (see chapter 20):

▼   The relative prices of alcohol and gasoline were adjusted through Petrobrás, the state-owned petroleum company.

▼   The government guaranteed that Petrobrás would purchase a specified quantity of alcohol.

▼   Low-interest loans were given to alcohol producers as a transitional incentive.

▼   Sales taxes on alcohol cars were set below those for gasoline cars.

▼   It was required that all gasoline be sold as gasohol, a mixture containing 22 percent alcohol.

In 1981 the price of alcohol was set 26 percent below that of gasoline (based on a kilometers-driven equivalence); in 1991 the price difference had fallen to six percent.

The Brazilian program created a large and technically sophisticated domestic alcohol industry that the Brazilians estimate has created 700,000 jobs in rural areas. About 60 percent of all cars in Brazil today use 100 percent ethanol fuel. The sharp decline in the world oil price in the 1980s, however, means that ethanol fuels are not economic in today's market, and the World Bank has begun to press Brazil to curtail the program. Sales of pure ethanol vehicles have fallen while sales of vehicles designed for gasoline/alcohol blends are increasing. However, alcohol produced from sugarcane can be made competitive at the present world oil price even without taking credit for nonmarket benefits, if modern production technologies, such as those discussed earlier, are adopted. Because it has developed the needed infrastructure, Brazil will be readily able to exploit these technological opportunities.

The United States has also created a market for grain-derived fuel ethanol by providing a very high price subsidy—some 13 cents per liter (49 cents per gallon). The program has been criticized because of its high cost and because 80 percent of the alcohol is produced by a single firm.

The 1991 U.S. Clean Air Act established specific goals for the use of alternative fuels in vehicle fleets. It stipulated that owners of fleets (10 or more vehicles) located in regions that have not met specified environmental quality goals must include a fixed percentage of "clean-fuel" vehicles in all new vehicle purchases–by 2000, 70 percent of cars and light trucks and 50 percent of heavy-duty trucks. Fuels that qualify include hydrogen, alcohols and alcohol mixtures (85 percent or more alcohol by volume), electricity, and natural gas. Federal fleets (other than those owned by the Defense Department) are included in the provision.[14] The Clean Air Act also allows a California experiment that specifies goals for zero-emission vehicles (10 percent of all vehicles sold in the state in 2003);[15] as written, this all but requires electric vehicles; alcohol vehicles would not qualify.

Electricity.   California now has nearly 1,400 MW_e of wind capacity and over 600 MW_e of solar-thermal-electric capacity operating or under construction. This capacity was encouraged by federal and state tax credits and by national utility regulations[16] that required utilities to purchase power from independent power producers at attractive prices. During most of the 1980s, renewable electric systems were offered standard contracts for power at prices set by the expected future costs for oil and natural gas-fired electric power generation, under the assumption of a rapidly increasing oil price.

The California program created a market with many competing firms and ultimately resulted in a wind industry that shows promise of expanding into unsubsidized markets in California and elsewhere. Both United States and Danish firms benefited from this program, the costs of which were borne by taxpayers and utility customers. Although the program was not efficiently administered and in some instances led to the construction of poorly designed machines, its cost was low compared with other national programs to promote new energy technologies. One obvious lesson from the California experience is that programs must send

---

14. U.S. Public Law 101-549 (1990), Sections 241–250.

15. California Air Resources Board, Proposed Regulations for Low-Emission Vehicles and Clean Fuels Staff Report, 13 August 1990.

16. Federal Energy Regulatory Commission regulations [45 FR 12212, 25 February 1980; 45 FR 17959, 20 March 1980] implementing Sections 210 and 201 of the Public Utility Regulatory Policies Act of 1978 [P.L. 95-617, 9 November 1978].

consistent signals to investors. If subsidies are only temporary, they should be reduced according to a clear and predictable schedule. Consider the Luz Corporation, which built California's solar thermal-electric capacity; the company went bankrupt in 1991 when the subsidies available to them were removed abruptly.[17]

Danish experience in promoting wind power development provides a sharp contrast to that in California. Under a government program that began in 1979 independent wind power producers received a subsidy amounting to 30 percent of a wind turbine's purchase price plus part of the installation cost. The subsidy was designed to be gradually phased out by 1989. Denmark was able to avoid subsidizing faulty designs by a requirement that only turbines tested and approved at the Risø National Laboratory were eligible for the subsidy. The Danish program not only helped nurture the development of the wind power industry but also demonstrated that it is possible to craft efficient policies for promoting renewables.

Research, development, and demonstration

Government investment in research, development, and technological demonstration is justified in cases where it is difficult for individual firms to capture the benefits of new technologies or where the benefits are greater than can be measured in competitive private markets. Direct government sponsorship of these activities for renewable energy technologies meets these criteria.

Two major challenges for policymaking are: 1) establishing rational research priorities based on merit, and 2) managing research support so that the technology developed is compatible with the needs of practical markets and can be quickly converted to commercial products. Accomplishing the latter may often require joint public and private financing of research, development, and demonstration projects.

Setting priorities

The priority given renewable energy in national research budgets is low. Overall investment in energy research in member countries of the International Energy Agency (IEA) has fallen 30 percent since 1981, and funding for renewable energy has fallen even faster. More than 50 percent now goes to nuclear energy and 15 percent to fossil fuels, while renewable energy technologies receive less than 7 percent of IEA country research and development funding, down from 14 percent in 1981. Yet growing awareness of the need for an environmentally sustainable energy program should foster growth rather than reductions in research support for renewables.

Among the technologies that are presently supported, electric-generating technologies receive the vast majority of funding, while promising opportunities

17. Lodker, M. 1991. Barriers to commercialization of large-scale solar electricity: lessons learned from the LUZ experience, SAND91-7014.

for new fuels, particularly fuels from renewable resources, receive only a small fraction of the total.

Government research should not be limited to hardware development. Support is needed for gathering information about renewable resources and for developing analytical tools for assessing the costs and benefits provided by renewable energy technologies. New modeling tools are needed, because intermittent renewables, fuel-cell vehicles, alternative fuels for vehicles, and other technologies are difficult to evaluate using models designed primarily for conventional energy systems.

Distributed renewable power sources, for example, offer advantages to a utility that are not measured in conventional models. The ability to locate small generators close to consumer sites, for example, can increase system reliability and decrease transmission and distribution costs. The fact that renewables are often economical in small sizes means that renewable generating capacity can be added to the system quickly in response to changes in demand; larger units that require years to plan and build are vulnerable to forecasting errors. The development of renewables would be facilitated if utilities had planning tools that could demonstrate the value of such characteristics of renewables.

The renewables-intensive energy scenario developed in this study highlights key renewable energy technologies and helps provide a basis for setting priorities. Some research and development priorities suggested by this analysis are indicated in Box D.

Links to industry

An important lesson from the experience of the last decade is that research and development funding can be squandered without careful attention to the links between the laboratory and commercial markets. It seldom makes sense to support research and development with public funds to the point of commercial readiness and then attempt to transfer the technology to a commercial developer. The organizational learning involved in research and development is difficult to transfer. It is far better to develop technologies in collaborative efforts involving research institutions, private equipment developers, and potential users, thereby promoting learning simultaneously in the laboratory, factory, and field.

The challenge of converting a product of research and development into a marketable product is often more difficult and expensive than the challenge of solving basic technical and engineering problems. Carefully selected and well-designed demonstration projects are needed to bridge this gap. Such demonstrations can reveal design flaws and force attention to system integration, marketing, and other issues that may not have been obvious during the research and development phases. An advantage to involving private businesses early in the process, of course, is to ensure that such difficulties are recognized as early as possible.

Because demonstrations are much more costly than research, usually only a small fraction of promising research products are demonstrated. In this regard, renewable energy technologies offer a significant advantage over most competing energy sources. Demonstrating a new nuclear technology or a coal-based synthetic fuel facility can cost a billion dollars or more and an advanced coal gasification-based power plant can cost hundreds of millions of dollars. In contrast, demonstrations of advanced biomass gasification-based power systems, production-scale wind machines, or new photovoltaic manufacturing systems typically cost only tens of millions of dollars. Because of the relatively low costs for renewable energy technology demonstrations, many more innovative concepts can be tested than is feasible for conventional energy technologies.

Two stages of demonstrations are needed: 1)technical demonstrations of prototypes at scales large enough to demonstrate the practicality of the technology in realistic settings and 2)market demonstrations involving significant numbers of the systems deployed in a variety of applications. Public sector support is often required at both stages of demonstration. A list of some high-priority projects for demonstration is presented in Box E.

One way to transfer technology from industrialized to developing countries and to demonstrate new technologies in developing countries is through joint ventures between multinational corporations and firms in developing countries. Such joint ventures should be organized on terms acceptable to both the develop-

---

Box D

Some Research and Development Priorities Emerging from the Renewables-intensive Global Energy Scenario

Planning tools and assessment methodologies to help utilities, other businesses, and government organizations assess the merits of renewable technologies in specific regions

Detailed wind and solar resource assessments for promising areas worldwide

Advanced gasifiers for biomass (including atmospheric and pressurized systems) for fluid fuels production and electricity generation

Improved yields in enzymatic processes for ethanol production

Development of a wide range of productive biomass energy crop species tailored for different regions

Fuel-cell related:
◆ Proton-exchange-membrane, solid-oxide, and molten-carbonate fuel cells
◆ Advanced technologies for reforming ethanol and methanol to hydrogen
◆ Advanced systems for storing hydrogen on vehicles, including hydrogen carriers such as powdered iron
◆ Biomass integrated gasifier/fuel cell systems for stationary power generation (e.g., molten carbonate fuel cells)

Low-cost photovoltaic modules capable of efficiencies greater than 12 percent

Key components for high-temperature, solar-thermal-electric conversion systems, including high-temperature thermal storage systems and conversion systems that employ Brayton cycles

ing country and the foreign industry and designed to ensure that all parties profit from the relationship. Multiple mutual benefits can be generated in these joint ventures as companies share responsibility for design, field construction, local marketing, and legal and regulatory issues specific to each site. Firms in developing countries can gain access to technical know-how in its most useful forms. In moving from the demonstration phase to the domestic market development phase, local firms can develop maintenance and operational skills, and gradually they can become preferred suppliers of components. Joint ventures in the production of electronic components evolved in this manner in Southeast Asia. Joint ventures could similarly be created for various renewable energy technologies, such as the production and operation of photovoltaic equipment, wind turbines, and biomass gasifiers. However, capturing the potential for technology transfer in this way requires careful planning by domestic firms.

A recently announced demonstration project for a biomass integrated gasifier/gas turbine electric power-generating technology in the northeast of Brazil provides a good model for a renewable energy joint venture project that can benefit both developing and industrialized countries. The $50 million project is being funded in part by the Global Environment Facility (GEF) of the World Bank, the United Nations Development Program (UNDP), and the United Nations Environment Program (UNEP). An open design competition is being conducted to attract bids from international consortia involving Brazilian, European, and U.S. firms. The Brazilian firms are actively involved in project design and will play major roles in installing and operating the equipment.

Box E
Some Demonstration Project Priorities Emerging from the Renewables-intensive Global Energy Scenario

New generations of wind turbines

Advanced solar thermal-electric power systems

Innovative approaches for achieving low balance-of-system (exclusive of photovoltaic module) costs for photovoltaic power systems.

Integrated biomass gasifier/gas-turbine systems for electric power generation

Advanced gas-turbine cycles for power generation

Techniques for efficient biomass harvesting, transportation, and storage (including drying, chipping, and harvesting equipment)

Techniques for restoring degraded lands in different regions

Integrated fuel-cell vehicle designs

Low-head hydroelectric installations

Once a concept's technical feasibility has been demonstrated, the establishment of a commercial industry for producing the technology can be encouraged by identifying markets where the technology could be competitive in the near future. National programs can help identify and aggregate these markets through several mechanisms:

Direct government purchases through competitive bidding. For example, government purchases for its vehicular fleets could provide early markets for renewable fuels and for vehicles designed to use these fuels, and purchases of distributed photovoltaic and other generating systems for government buildings could provide early markets for renewable-electric technologies.

Incentives for introducing new technologies in regions that face extreme local air-quality problems. For example, incentives might be provided to introduce fleets of fuel-cell vehicles in areas where new vehicle types and alternative fuels are needed to meet air-quality goals.

Information programs. Such programs, which would include the development and dissemination of analytical design tools, could help potential purchasers understand the advantages of renewable energy equipment—for example, by providing utilities assistance in identifying niche markets for distributed photovoltaic systems.

Cooperation with international development assistance organizations. Governments could provide such organizations with financial support to help promote the transfer of renewable energy technologies to developing countries.

Agricultural and forestry policies

Policies to encourage development of a biofuels industry must be closely coordinated with national agricultural programs and with efforts to restore degraded lands. Indeed, energy production may become a major theme in worldwide efforts to promote rural economic development and environmental reclamation.

About 55 percent of the biomass in the renewables-intensive energy scenario in 2025 is provided by energy crops grown on about 370 million hectares of land, mainly on excess cropland in the industrialized countries (90 million hectares) and on deforested or otherwise degraded lands in the developing countries (280 million hectares). The total is approximately equal to the amount of land committed to rice and wheat production at present. (The total land area in food production is about 1,500 million hectares.)

The production of energy from biomass on such a scale may require a unique combination of business interests. Because large amounts of biomass would be used to produce liquid fuels for transportation, some of today's oil companies might become producers of fuels derived from biomass. They might be motivated to do so in part because of the pressures on them to provide cleaner transport fuels and in part by the prospect that a shift to biofuels would enable them to grow in the face of the prospect of declining oil production in most regions outside the Middle East.

Efforts to establish biomass energy plantations will confront all the challenges associated with producing agricultural food crops. In fact, biomass energy production will be carried out much like the production of food, with technologies from agriculture adapted to energy crops as appropriate. Agricultural businesses are therefore also likely to be involved in biomass energy production.

Processing centers will be relatively small and distributed throughout a region rather than concentrated at central sites. Facilities for processing biomass are best located in rural areas close to production sites, to avoid high biomass transport costs. Biomass growers may well work in joint ventures with the energy companies that would process the biomass feedstocks.

Large amounts of degraded lands that are suitable for plantations and not likely to be required for food production can be found in Latin America and sub-Saharan Africa (see chapter 14). Accordingly, some 161 million hectares in Latin America and 95 million hectares in Africa have been targeted for biomass plantations by 2025 in the renewables-intensive energy scenario. Together they represent one-eighth of degraded lands and three-fifths of degraded lands identified as being suitable for reforestation in these regions (see table 1). Establishing biomass energy plantations on these lands would require cooperative efforts involving biomass growers, energy companies, and governmental and other groups seeking ways for restoring these lands.

Despite the opportunities offered by biomass plantations, there are many hurdles that must be overcome to establish energy crop production as a routine agro-industrial activity.

One problem generic to all regions is that while energy crop production shares many common features with agricultural crop production, the technological base for energy crops is far weaker. The accelerated development of a wide variety of high-productivity energy crops tailored to the soil and climate conditions of different regions should be encouraged. A sustained commitment to research is needed, with continual monitoring of the results, so that priorities can be reassessed periodically.

In addition, all regions face a transitional problem: biomass growers will not make investments unless they are confident that there will be markets for their products. Thus policies to establish plantations must be coordinated with policies aimed at launching bioenergy industries. Once the bioenergy industries are established, financing for biomass plantations could be provided by the firms that will ultimately purchase the crops. In the interim, government financing might be necessary.

Plantations on excess agricultural lands in industrialized countries
The amount of cropland needed for food production is declining in the industrialized countries. In the European Community, more than 15 million hectares of

Table 1: Present and prospective land use patterns relating to biomass plantations in developing regions
millions of hectares

Region	Present cropland[a]	Potential cropland[b]	Reforestable degraded land[c]	Land committed to biomass plantations in Renewables-intensive Global Energy Scenario[d] 2025	2050
Africa	179	753	101 (+ 148)	95	106
Latin America	179	890	156 (+ 32)	161	165
Asia (ex. China)	348	413	169 (+ 150)[e]	–	–

a. Arable land plus land in permanent crops as of 1984. The Intergovernmental Panel On Climate Change (IPCC) has projected that the land area under cultivation in developing countries will increase at a rate of 1.2 percent per year through 2025, which would result in a 50 percent increase relative to the current amount by that time [Intergovernmental Panel on Climate Change. 1991. *Climate Change: the IPCC Response Strategies*, Island Press, Washingon DC].

b. Land classified by the Food and Agriculture Organization of the United Nations as potential agricultural land [Food and Agriculture Organization. 1991. *Land and water inventory*, Bruinsma, Rome], distributed by rainfall classification category, as follows:

Rainfall classification category category	Percent of potential cropland
low rainfall	8
uncertain rainfall	11
good rainfall	19
problem land	47
naturally flooded areas	13
desert	1

c. Alan Grainger of Resources for the Future in Washington DC has estimated, by region, the amount of tropical degraded land suitable for reforestation in the categories of logged forests in the humid tropics, forest fallow (intermittent use by shifting cultivators), deforested watersheds, and desertified drylands. He estimated that nearly all the land in the first three categories might be reforested, plus, overall, about one fifth of the desertified drylands. His estimates for land potentially suitable for reforestation in the desertified drylands category is presented here in parentheses. [A. Grainger. 1988. Estimating areas of degraded tropical lands requiring replenishment of forest cover, *International Treecrops Journal* 5:31-6].

d. See the appendix at the back of the book. The plantation land areas targeted for plantations in 2050 amount to 13 percent of the total degraded land area in Africa and 18 percent of the total degraded land area in Latin America.

e. Granger's estimate of degraded land suitable for reforestation is for all of Asia, including China.

land will have to be taken out of farming by 2000 if the surpluses and subsidies associated with the Common Agricultural Policy are to be brought under control; in the future, excess cropland in the Community could increase to as much as 50 million hectares, if crop productivities continue to increase and produce unmanageable surpluses. In the United States, more than one fifth of total cropland, some 33 million hectares, was idled in 1990, either to keep food prices up or to control erosion, and, as in the European Community, the amount of idle cropland is expected to grow. The conversion of excess cropland in the industrialized countries to biomass plantations for energy offers a major opportunity for making productive use of these lands.

Shifting cropland to biomass production for energy purposes is not easily accomplished under present policies. In many countries, farmers are deterred by a subsidy system that specifies what crops the farmer can produce in order to qualify for a subsidy, and energy crops are not allowed. While conversion to profitable biomass energy production will make it possible eventually to phase out agricultural subsidies, this will not be accomplished overnight because of the economic dislocations that would result; today these subsidies total about $300 billion per year for North America, Europe, and Japan.

As long as a system of subsidies is in place, the bias against energy crops should be removed. It might even be desirable to augment the subsidy for energy crop production initially, to provide an impetus for establishing this as a new agro-industrial activity, as Sweden is doing. Sweden has recently implemented a policy promoting the conversion of excess cropland to plantations of willow as an energy crop. By 1991, 4,000 hectares had been planted under this program, and the planted area is expected to increase to 10,000 hectares by 1993. The government supports this cropland conversion by providing substantial temporary subsidies to farmers for making the conversion. It is expected that after having gained a few years of experience, farmers will plant trees for energy without subsidies.

Plantations on degraded lands in developing countries
Worldwide interest in restoring tropical degraded lands is growing, as indicated by the ambitious goal of a global net afforestation rate of 12 million hectares per year by 2000, which was set in the 1989 Noordwijk Declaration.[18]

This is comparable to the biomass energy plantation establishment rate required for centrally planned Asia, Latin America, and Africa during the first quarter of the 21st century in the renewables-intensive scenario. Thus, the joint goals of establishing biomass plantations for energy and restoring degraded lands could be served simultaneously by planting biomass energy crops on degraded lands. Biofuels also offer farmers a new option for cash crops, thereby supporting rural development.

With skillful management, the energy industries involved may be able to provide the capital needed to finance land restoration activities. In principle, the industries would have an incentive to restore these lands in sustainable ways. Potential biofuels production in Latin America and sub-Saharan Africa would be large enough to serve large export markets; the value of the biofuels traded in world commerce could be comparable to that for oil (see figure 4). But in order to exploit this potential, the energy industries would require secure supplies of biomass feedstocks throughout the lifetimes of their capital-intensive energy conversion facilities—some 20 years or more. Such supply security could be assured only if the plantations were managed on a sustainable basis.

---

18. The Noordwijk Declaration on Atmospheric Pollution and Climatic Change, Ministerial Conference on Atmospheric Pollution and Climatic Change, Ministerial Conference on Atmospheric Pollution and Climatic Change, Noordwijk, The Netherlands, November, 1989.

The challenge of restoration is to find a sequence of plantings that can restore ground temperatures, organic and nutrient content, moisture levels, and other soil conditions to a point where crop yields are high and sustainable. Successful restoration strategies typically begin by establishing a hardy species with the aid of commercial fertilizers or local compost. Once erosion is stabilized and ground temperatures lowered, organic material can accumulate, microbes can return, and moisture and nutrient properties can be steadily improved. This can lead to a self-regenerating cycle of increasing soil fertility.[19]

Other difficulties that must be surmounted reflect general conditions in developing regions. Attention must be paid to a broad range of technical, socioeconomic, political, and cultural issues in developing regions, in addition to issues that are common to all regions. Many regions lack roads to transport biomass to processing facilities and also the means to move the biofuels to markets. Growers in poor areas cannot wait the 5 to 8 years that is typically required for cash returns on short-rotation tree crops. Also, land ownership may be complex and disputed. Indigenous peoples who have a tradition of grazing and measure their wealth in cattle may not wish to convert land to other uses.

Despite the difficulties involved, growing energy crops on degraded lands is feasible, as indicated by the fact that many of the most successful biomass plantations in developing countries have been established on degraded lands (see chapter 14).

It will be necessary to integrate plantation development into more general rural development programs, and these efforts must be sensitive to the short- and long-term economic interests of people living in the area.

Programs for establishing large-scale plantations on degraded lands should include the development of region-specific restoration plans that take into account local soil conditions, weather, and social conditions. Successful restoration activities conducted by both outside experts and local farmers should be investigated. Also, restoration plans that result in commercial energy crops should be demonstrated. Such demonstrations might be conducted as joint ventures among local agricultural producers and equipment supply firms, local and multinational energy companies, and local and international organizations interested in land restoration.

International programs

The benefits of making greater use of renewable energy resources and the costs of failing to exploit them are international in scope. World management or mismanagement of these resources would have a profound effect on socio-economic development, the global environment, and energy security. Accordingly, there is a strong need for international cooperation in developing renewable energy resources.

19. Office of Technology Assessment of the U.S. Congress. 1992. *Technologies to Sustain Tropical Forest Resources and Biological Diversity,* OTA-F-515, U.S. Government Printing Office, Washington, DC.

Centers of excellence for developing and transferring renewable energy technologies
The Castel Gandolfo Colloquium[20] outlined a specific proposal for "Centers of Excellence on New and Renewable Energy Sources." Such centers could specialize in regional energy crops, distributed photovoltaic systems, biomass gasification, or groups of technologies appropriate for specific climates. Located primarily in developing countries, the centers could conduct research and sponsor joint ventures to demonstrate advanced-design concepts in practical applications. The centers could also support local resource planning, education, training, and visiting scholar programs.

One possible model is the International Rice Research Institute in the Philippines, which has an excellent history as an international research center. This and similar centers specializing in other grains combine sophisticated research with an effective mechanism for disseminating research results to practitioners around the world. If this model is adopted, care must be taken to ensure that the renewable energy centers do not weaken effective research programs already underway in universities.

The energy research centers could also provide a venue for transferring technology from industrialized countries to developing regions. Research or demonstration projects undertaken at such centers could combine the skills of scientists and engineers from around the world in pursuit of common objectives. Technology is most effectively transferred when highly skilled people can work together in a supportive environment and are in a position to understand which technologies are relevant to practical problems. The effectiveness of such activity is undermined, however, if the support for research and demonstrations is not sustained and continuous.

Supporting renewable energy development with a carbon tax
Concerns about global climate change and other environmental issues motivate much of the interest in accelerating the development and utilization of renewables. Conditions for technological development of renewables would be much improved if the international community could agree on targets for reducing greenhouse gas emissions.

If agreement could be reached, various policy instruments could be used to reduce emissions, including carbon taxes and tradeable permits (the issuance of which would fix the levels of emissions allowable by different countries). An evaluation of the merits of the alternatives is beyond the scope of the present study.

However, the revenues from even a small carbon tax could be valuable if used to support research, development, and demonstration endeavors involving renewable energy, including activities at the centers of excellence just described. Consider, for example, a tax of $1 per tonne of carbon (equivalent to 12 cents per barrel of petroleum) on all fossil fuels. While such a tax would have little direct

---

20. Committee on the Development and Utilization of New and Renewable Sources of Energy, United Nations General Assembly.1988. *Note by the General-Secretary: colloquium of high-level experts on new and renewable sources of energy,* doc. A/AC.218/14, New York.

impact in reducing fossil fuel demand, it would generate nearly $6 billion per year. Industrialized countries, which produce three-fourths of CO_2 emissions from fossil fuel burning, would bear a corresponding fraction of the expense.

Energy plans compatible with sustainable development

An international consensus on energy paths that are compatible with sustainable development must grow from a shared vision of how such paths can be identified and managed, based on each country's assessment of the possibilities. National and international planning would benefit from the development of national energy programs designed around a consistent set of topics. The topics might include:

▼ Assessment of opportunities for minimizing the cost of providing energy services in a manner consistent with environmental constraints. The assessment would include options for making more efficient use of energy as well as energy supplies.

▼ Detailed review of renewable energy resources, including wind, hydropower, direct solar and biological resources.

▼ An assessment of the environmental impacts of large-scale use of renewable energy--particularly large-scale biomass production. This would include establishing specific, measurable criteria relating to biodiversity, soil conservation, water quality, etc.

▼ Regional and local plans for restoring degraded lands that would reflect opportunities for energy crops.

Donor organizations

The priorities of international donor agencies must be revised to reflect the opportunities renewable sources of energy present both for social and economic development and for protecting the global environment. Renewable energy sources presently receive too small a share of energy-related economic assistance.

Donor organizations should move to:

▼ Develop clear criteria relating to economic development and environmental goals for energy research, development, and demonstration programs, and energy-sector infrastructure-building. Adopting such criteria would lead to greatly expanded support for renewables projects.

▼ Support energy programs compatible with sustainable development in countries with limited financial resources. This will require extensive training of domestic analytical staff and providing analytical tools and other forms of technical assistance. Support should also be included for renewable energy resource assessments.

▼ Ensure adequate transfer of technology to developing countries. Joint ventures should be encouraged to promote technology transfer, but other mechanisms should also be employed to ensure that developing countries are aware of renewable energy technologies relevant to their needs, that the technology transferred is adapted to local conditions and needs, that the individuals and organizations most likely to use the technology are targeted recipients, and that indigenous technological capabilities are advanced through the transfer. As the benefits of environmentally sound technologies are enjoyed worldwide, mechanisms should be developed for making renewable energy technologies available to developing countries on favorable and concessionary terms.

World donor organizations are uniquely suited to accelerate the development of renewable energy technology and to facilitate its transfer to developing countries. Such organizations can help aggregate markets for photovoltaic arrays and other devices so as to encourage investment for early market development. Skillful use of large purchases can accelerate investments in the production of renewable energy equipment in both industrialized and developing countries.

The international community should establish a permanent institution to ensure coordination of international activities in renewable energy and to provide continuity in management and oversight. The organization should review and compare national renewable energy programs. And it should establish mechanisms for sharing research results and analytical tools. The organization should also serve as a clearinghouse for information on renewable energy and on the companies, research organizations, and individuals that can provide needed equipment or design assistance. Its work should be closely coordinated with the research centers proposed in the Castel Gandolfo Colloquium.

Renewables, the environment, and economic growth policies

There is no hiding from the fact that programs for promoting national economic growth must reflect environmental and security concerns. The question is not whether these issues will affect national investment, but when and how. While poorly designed environmental regulatory programs will dampen productivity and economic growth, environmental policies can also be designed to promote growth. Programs forcing attention on environmental issues can also focus on inefficient production practices, stimulate thinking about new approaches, and encourage investment in more productive equipment. Environmental rules have often had the effect of forcing innovation on a sluggish industrial sector. Such innovation has proved critical for international competitiveness and should have been pursued even in the absence of regulatory pressure. In short, challenging but rationally conceived environmental regulations can stimulate investment that can enhance productivity, competitiveness, and job creation, while ensuring that national environmental and security goals are met efficiently.

The renewable energy technologies described in this book exemplify the range of technologies coming onto the market that can increase productivity in the broadest sense–by making more efficient use of labor, capital, materials, energy, land, and other resources–while minimizing production of noxious wastes.

A policy aimed at accelerating the development of fuels and electricity from renewable sources could be a critical element in a broader policy designed to promote economic growth and environmental quality in both industrialized and developing countries.

ACKNOWLEDGMENTS

The authors thank Gregory Terzian for his assistance with the scenario analysis, the preparation of the graphics, and the finalizing of the manuscript. The research carried out by one of the authors (RHW) that provided the basis for his contribution to this chapter was supported by the Bioenergy Systems and Technology Program of the Winrock International Institute for Agricultural Development, the Office of Energy and Infrastructure of the U.S. Agency for International Development, the Office of Policy Analysis of the U.S. Environmental Protection Agency, and the Geraldine R. Dodge, W. Alton Jones, Merck, and New Land foundations.

ENDNOTES FOR FIGURES

Notes for figure 6

The following is a detailed description of the costs shown in figure 6 for alternative coal and biomass power plants:

Electricity generation costs for alternative coal and biomass technologies
cents per kWh

| | Coal conventional | Coal gasification | | Biomass gasification | |
	Steam-electric[b]	1st gen.[c]	2nd gen.[d]	1st gen.[e]	2nd gen.[f]
Capital[a]	1.71	1.76	1.32	1.44	1.15
Fuel	2.12	1.88	1.71	2.46	2.04
O&M	0.94	0.88	0.83	0.72	0.61
Total	**4.77**	**4.52**	**3.86**	**4.62**	**3.80**

a.   The annual capital charge rate (0.07765) is the sum of the capital recovery factor (0.07265) for a 6 percent discount rate (appropriate for utility investments in industrialized countries) and a 30-year plant life plus an insurance charge (0.005); all taxes are neglected. In all cases the annual average capacity factor is assumed to be 75 percent.

b.   A subcritical coal-fired steam-electric plant (two 500 MW$_e$ units) with flue gas desulfurization as characterized by the Electric Power Research Institute (EPRI) in the US [EPRI. 1986. *Technical Assessment Guide (TAG)*, Palo Alto, California]. It has a higher heating value (HHV) efficiency of 33.9 percent at full load, an installed capital cost of $1450 per kW$_e$, a variable O&M cost of 0.59 cents per kWh, and a fixed O&M cost of $23.1 per kW-year. The coal price is assumed to be $2 per gigajoule.

c.   A coal integrated gasification/combined cycle (CIG/CC) plant (one 800 MW$_e$ unit) based on oxygen-blown gasification in an entrained flow gasifier, with commercial service starting in 1994 [EPRI. 1989. *TAG*]. It has HHV efficiency of 38.2 percent at full load, an installed capital cost of $1489 per kW$_e$, a variable O&M cost of 0.36 cents per kWh, and a fixed O&M cost of $34.2 per kW-year. The coal price is assumed to be $2 per gigajoule.

d.   A coal integrated gasification/intercooled steam-injected gas turbine (CIG/ISTIG) plant (one 109 MW$_e$ unit) based on air-blown gasification in a fixed-bed gasifier (see chapter 17). The fixed-bed gasifier is commercially available. The intercooled steam-injected gas turbine could be commercialized for natural gas service before the turn of the century and could be readily modified for use with coal. The major unproven feature of this technology involves the requirement to clean the gas of sulfur while it is still hot and before it is burned in the gas turbine combustor. If this so-called "hot-gas cleanup" technology could be demonstrated at a commercial scale, this technology could be commercially available in the period near the turn of the century. The technology would be simpler, less costly, and more energy-efficient than first-generation coal gasification technology (note c). It is expected that the HHV efficiency at full load would be 42.1 percent, the installed capital cost $1122 per kW$_e$, the variable O&M cost 0.10 cents per kWh, and the fixed O&M cost $48.0 per kW-year. The coal price is assumed to be $2 per gigajoule.

e.   A biomass integrated gasification/steam-injected gas turbine (BIG/STIG) plant (two 51.5 MW$_e$ units) based on air-blown gasification in a fixed-bed gasifier (see chapter 17). The steam-injected gas turbine is commercially available for natural gas applications and could be modified for biomass applications with little effort. Hot gas sulfur cleanup, needed for coal, is not needed for biomass, because biomass generally contains little sulfur. The technology could be commerically available by the turn of the century. A BIG/STIG plant operated on a pre-dried biomass feedstock would have a HHV efficiency of 35.6 percent at full load, an installed capital cost of $1121 per kW$_e$, a variable O&M cost of 0.10 cents per kWh, and a fixed O&M cost of $40.8 per kW-year. A variant being developed by Imatran Voima in Finland involves drying the biomass in superheated steam in such a way that the moisture driven out of the fuel is recovered as steam for the power system. Such a drying system would add perhaps $100 per kW$_e$ to the capital cost but would enable the system to use less costly wet fuel with no energy efficiency penalty; this variant is assumed here (see chapter 17). Dried plantation biomass chips in the US are estimated to cost $3 per gigajoule delivered to the power plant, based on present technology; without drying the cost would be $2.43 per gigajoule (see chapter 17); the latter cost is assumed here.

f.   A biomass integrated gasification/intercooled steam-injected gas turbine (BIG/ISTIG) plant (one 111 MW$_e$ unit) would differ from a BIG/STIG plant only in that the steam-injected gas turbine would be replaced by a more energy-efficient and less capital-intensive intercooled steam-injected gas turbine (see chapter 17). After the latter is commmercialized for natural gas (which could be achieved before the turn of the century) it could be modified for use in BIG/ISTIG systems. A BIG/ISTIG plant operated on pre-dried biomass would have a HHV efficiency of 42.9 percent at full load, an installed capital cost of $874 per kW$_e$, a variable O&M cost of 0.09 cents per kWh, and a fixed O&M cost of $34.2 per kW-year. It is assumed that this is used in connection with an Imatran Voima biomass drier costing an additional $100 per kW$_e$. It is assumed that the biomass feedstock costs $2.43 per gigajoule.

Notes for figure 7

The following is a detailed description of the costs shown in figure 7 for wind turbine electricity (for details see chapter 3):

Wind electricity production costs
cents per kWh

	Data for new California wind farms[a]						VSWT[b]	Advanced technology[c]
	1985	1986	1987	1988	1989	1990	1994	> 2000
Capital[d]	13.9	11.7	6.5	5.9	5.5	4.7	2.9	2.5
O&M	1.7	1.7	1.5	1.4	1.3	1.0	1.1	0.6
Land rent	0.3	0.3	0.3	0.3	0.3	0.3	0.3	0.3
Total	**15.9**	**13.7**	**8.3**	**7.6**	**7.1**	**6.0**	**4.3**	**3.4**

Wind unit capital cost $ per kW$_e$

1,900	1,600	1,100	1,050	1,050	1,000	760	800

Wind farm capacity factor *percent*

13	13	16	17	18	20	25	30

a.   The performance and cost values for the period 1985-1990 are based on actual experience in California.

b.   The performance and cost values for 1994 are based on public filings with the California Energy Commission for a new variable-speed wind turbine that is expected to go into commercial service in 1994. This variable speed turbine has been developed by US Windpower, with contributions from the Pacific Gas and Electric Company, the Niagara Mohawk Power Company, and the Electric Power Research Institute.

c.   The performance and cost values for the post-2000 period reflect expectations about improvements that could be realized with R&D over the next decade. These values provide the basis for the calculations involving wind power in figure 8.

d.   The annual capital charge rate (0.0832) is the sum of the capital recovery factor (0.0782) for a 6 percent discount rate (appropriate for utility investments in industrialized countries) and a 25-year plant life plus an insurance charge (0.005); all taxes are neglected.

Notes for figure 8

The estimated costs of meeting the electrical load for the utility system shown in figure 8 are derived for electrical load data provided by the Pacific Gas and Electric Company for 1989 using the following procedure (see chapter 23 for details):

1.   The output of the wind, solar thermal, and photovoltaic equipment was computed and subtracted from the demand for electricity in each hour of the year. The output of photovoltaic and solar thermal equipment was derived by scaling actual hourly output to reflect installed capacity. The wind output was derived from a randomized Weibull wind speed distribution and an equation that models the output of a variable speed wind turbine for this wind speed distribution. As the installed capacity of wind turbines was increased, the average wind speed was progressively decreased, from 8 m/s to 7.5 m/s, to 7.0 m/s, to reflect a range of available wind resources. For each 5 percent of total system load met by wind turbines a different uncorrelated Weibull wind distribution was used to simulate the effects of geographical diversity. For further details, see note on wind costs following the table below on cost parameters for alternative power-generating technologies.

2.   Hydroelectric output was dispatched to reduce the peaks of the remaining load, subject to the following constraints: Output was allowed to range from a minimum of 1 GW_e to a maximum of 4 GW_e, but the system output was forced to average 2 GW_e for each week of the year (i.e. reservoirs ended the week at the same level they began). These assumptions were chosen for illustrative purposes and are roughly consistent with Northern California's peaks and averages. They are not intended to reflect the complex operating characteristics of systems actually operating in California and the Columbia River system which exports power to the region.

3.   A set of coal, gas, and biomass plants was selected that minimizes the cost of meeting the electric load that remains after wind, photovoltaic, solar thermal, and hydroelectric power has been subtracted. An analysis of the reliability of the system was also conducted to ensure that the plants selected can meet the standard utility reliability criterion that the system will fail to meet some part of the load for not more than 24 hours in 10 years. The set of plants selected is sensitive to fuel costs and the cost and operating characteristics of the equipment available for selection.

4.   The fossil and biomass equipment were dispatched in every hour so as to minimize the cost of meeting the remaining loads, considering the penalties associated with operating plants at part load and the cost of starting and stopping large plants.

5.   The total cost of meeting the utility's electrical load, considering both capital carrying costs, fixed and variable operating costs, and fuel costs, was computed.

Generation costs were computed assuming the following cost characteristics for the alternative technologies involved (see chapter 23 for assumptions concerning variable heat rates, outage rates, startup times, and other more detailed characteristics):

Generation costs

Plant type	Fixed costs[a] $/kW-year	Variable costs[b] $/kWh
Natural gas		
Conventional simple cycle[c]	31	0.064
Best new simple cycle[d]	28	0.060
Advanced simple cycle[e]	59	0.040
Conventional combined cycle[f]	44	0.041
Best new combined cycle[g]	42	0.038
Advanced combined cycle[h]	63	0.033
Coal		
Conventional steam turbine[i]	136	0.027
Best new coal[j]	142	0.022
Advanced coal[j]	135	0.018
Advanced biomass[j]	110	0.021
Hydroelectric[k]	277	0
Wind[l]	67	0.01
Solar thermal[m]	126	0.001
Photovoltaic[n]	70–140	0.002

a. Annual capital charges plus annual fixed operation and maintenance charges. Capital charges were calculated for a real discount rate of 6 percent and a 0.5 percent annual insurance cost; taxes were neglected. Interest charges during construction are included. All systems were assumed to have 30-year lifetimes unless otherwise noted.

b. (Annualized fuel costs)*(heat rate at 100 percent load) + variable operating costs. Coal prices were assumed to remain constant at $2 per gigajoule and biomass prices at $2.43 per gigajoule during the period of analysis. Natural gas prices were assumed to increase linearly from $4 per gigajoule at plant startup to $5 per gigajoule after 30 years. The fuel prices were levelized.

c. An 80 MW_e simple cycle gas turbine costing $368 per kW_e (overnight construction cost), with a peak HHV efficiency of 28 percent [EPRI. 1989. TAG].

d. A 140 MW_e simple cycle gas turbine costing $355 per kW_e, with a peak HHV efficiency of 29.5 percent [EPRI. 1989. TAG].

e. An 80 MW_e intercooled GE LM-6000 costing $440 per kW_e, with a peak HHV efficiency of 41 percent (based on data provided by the General Electric Company and Bechtel Power Corporation).

f. A 120 MW_e unit costing $503 per kW_e, with a HHV efficiency of 43 percent [EPRI. 1989. TAG].

g. A 210 MW_e unit costing $473 per kW_e, with a HHV efficiency of 45 percent [EPRI. 1989. TAG].

h. A 100 MW_e combined cycle using an intercooled version of the GE LM-6000 costing $600 per kW_e with a HHV efficiency of 50 percent (based on data provided by the General Electric Company and Bechtel Power Corporation).

i. A 1000 MW_e sub-critical coal plant with a HHV efficiency of 34 percent costing $1290 per kW_e [EPRI. 1986. TAG].

j. See notes for figure 6.

k. The units were assumed to have an installed cost of $1800 per kW_e (based on the peak capacity). This is in the upper part of the range of estimated costs for the Pacific Northwest of the US (Northwest Power Planning Council. 1991. Northwest Conservation and Electric Power Plan, Volume II, Part II, Portland, Oregon).

l. Advanced variable-speed wind turbines with a capital cost of $800 per kW_e. The output of the wind systems was computed as indicated in note 1 above. The capacity factors corresponding to average wind speeds of 8 m/s, 7.5 m/s, and 7 m/s were calculated to be 28.5 percent, 25.5 percent, and 21.5 percent, respectively.

m. A 200 MW$_e$ central receiver solar thermal system installed for $1625 per kW$_e$. The output of the system was scaled from the energy available on a fully tracking system in 1989 (see chapter 5).

n. The installed cost of photovoltaic systems is assumed to be in the range $900 to $1800 per kW$_e$. An annual fixed charge of $140 per kW-year corresponds to an installed cost of $1800 per kW$_e$. An annual fixed charge of $70 per kW-year corresponds either to an installed cost of $900 per kW$_e$ or to an installed cost of $1800 per kW$_e$ together with a distributed credit of $70 per kW-year that would reflect the reduced transmission costs and increased reliability benefits offered by distributed photovoltaic systems. Distributed credits were based on PG&E estimates of the value of deferring upgrades to transmission and distribution systems (taking into account the expected availability of photovoltaic power at the time of the utility's peak demand) and an estimate of the value offered by distributed systems in increasing system reliability. A conservative estimate of these distributed credits is $56 per kW-year for avoided transmission and distribution capital costs, $4 per kW-year for avoided transmission and distribution maintenance, and $10 to $20 per kW-year for reliability (5-10 percent of the credit estimated for the single PG&E site for which a detailed analysis is available).

Notes for figure 9

The graph labeled "base case" shown in figure 9 is the actual power delivered by the utility. The curve was computed by taking the power delivered in each hour of the 8760 hours in a year and sorting so that the hours with the highest load are on the left and those with the lowest are on the right of the figure. The indicated maximum load served by different kinds of generating capacity reflects the capacity of each plant type that would minimize the cost of meeting the utility's load, given assumptions about fixed and variable operating costs described in the notes for figure 8. The actual installed capacity purchased to meet load differs from these maximum load values for two reasons: (i) extra capacity must be purchased to ensure that reserves are available when equipment is being maintained, and (ii) plants can only be purchased in discrete units. The latter issue, a problem primarily for larger plants, can be illustrated by an example: a cost-minimizing mix might require 300 MW$_e$ of a given type of plant, whereas units of the required type might be available only in increments of 200 MW$_e$.

The other curves were constructed by subtracting the wind, photovoltaic, and solar thermal energy produced in each hour from the utility "base case" load curve just described. The hydroelectric equipment was then dispatched in a way that minimized peak loads while ensuring (i) an average output of 2 GW$_e$ for each week of the year, and (ii) hydroelectric output was in the range 1 to 4 GW$_e$. The first assumption has the effect of ensuring that reservoirs are not drawn down. After the optimum dispatch schedule for hydroelectric was determined for each hour of the year, the hourly hydroelectric power output was subtracted from the net load (hourly load for the base case minus intermittent renewable hourly output). The remaining loads were then sorted. As in the base case, the mix of thermal power plants displayed represents the mix that minimizes production costs.

Notes for figure 11

The following is a summary of salient vehicular characteristics for the alternative vehicles shown in figure 11:

Characteristics of alternative vehicles[a]

	Units	Gasoline ICEV	MeOH FCV[b]	Hydrogen FCV[b]	Battery EV[c]
Power-to-weight ratio[d]	*kW /tonne*	71.8	56.7	57.3	56.2
Gasoline-equivalent fuel economy	*liters/100 km*	9.08	3.59	3.05	2.18
	mpg	25.9	64.8	76.5	108
Curb weight	*tonnes*	1.37	1.27	1.24	1.44
Initial price	*10^3 dollars*	17.3	26.3	27.4	29.6
Lifetime range	*10^3 km*	193	289	289	289
Annual maintenance cost	*dollars*	516	416	401	358

a.   For a more detailed discussion of the technology and economics of alternative automobile energy systems, see chapter 22.

b.   It is assumed that the fuel-cell vehicles (FCVs) are powered by proton-exchange-membrane (PEM) fuel cells. The fuel cell is a device that converts fuel energy directly into electricity without first burning it to produce heat.   A fuel cell is inherently much more energy-efficient than fuel-burning engines. The PEM fuel cell in particular is well suited for automotive use because of its low operating temperature (25 to 120 °C) and high power density (1.33 kW per kg and 1.20 kW per liter, compared to 0.12 kW per kg and 0.16 kW per liter for the commercially available phosphoric acid fuel cell). The PEM fuel cell can be fueled either directly with hydrogen or with methanol (which would then be "reformed" onboard the car with steam--i.e., the steam reacts with the methanol to form the needed hydrogen fuel plus carbon dioxide). The hydrogen fuel-cell system considered here involves the use of onboard compressed hydrogen storage [at a pressure of 55 MPa (8,000 psia)].

c.   The battery considered for the battery-powered electric vehicle (EV) is a promising advanced battery—the bipolar lithium alloy/iron disulfide battery.

d.   The power listed here is the power delivered to the wheels. The power-to-weight ratio is one measure of performance. For the designs considered here the gasoline internal-combustion-engine vehicle (ICEV) would have better peak acceleration than the alternatives. However, because lags are inherently less with electric than with mechanical drive trains, the FCVs and the EV would have quicker responses than the ICEV under most driving conditions.

The following is a summary of the renewable fuel costs assumed for the automobile lifecycle cost analysis presented in figure 11:

Production and consumer costs for alternative renewable fuels[a]
$ per gigajoule

	Alternative sources of hydrogen[b]		Methanol from biomass
	Wind/PV	Biomass	
Production cost	20.4[c]	8.2 [d]	10.6[d]
Compression	1.4[e]	– [e]	–
Storage	1.1[f]	– [f]	–
Local distribution	0.5	0.5	–
Refueling station	5.2[g]	5.2 [g]	2.8[h]
Cost to consumer	28.5	13.9	13.4

a.   All fuel production costs (as well as the costs of compression and storage) were calculated assuming a 12 percent real (inflation-corrected) cost of money (discount rate); taxes are not included.

b.   For a detailed discussion of the technology and economics of alternative hydrogen production options, see chapter 22.

c.   This hydrogen production cost is a mid-range value for wind and photovoltaic (PV) power sources expected in the 2000+ time period, for the assumed discount rate. In the case of wind power, it is estimated that hydrogen can be produced at a cost in the range $16.2 to $25.0 per gigajoule, for projected ac wind electricity costs in the range $0.033 to $0.052 per kWh. In the case of photovoltaic power, it is estimated that hydrogen can be produced at a cost of $15.3 to $25.5 per gigajoule, for DC PV electricity costs in the range $0.030 to $0.061 per kWh. In both cases it is assumed that hydrogen is produced at atmospheric pressure using unipolar electrolyzers.

d.   Both methanol and hydrogen can be produced from biomass via thermochemical gasification. The present analysis is based on the use of biomass feedstocks costing $3 per gigajoule. The gasifier involved is the Battelle Columbus Laboratory (BCL) gasifier, which is under development and not yet commercially available. In the gasification process, the biomass is heated and converted into a gaseous mixture consisting mainly of methane, carbon monoxide, and hydrogen. This gaseous fuel mixture can then be converted into methanol or hydrogen, using well-established industrial technologies [E.D. Larson and R.E. Katofsky. 1992. Production of hydrogen and methanol via bimass gasification, in *Advances in thermochemical biomass conversion*, Elsevier Applied Science, London (forthcoming)]. The production of methanol based on the BCL gasifier is described in chapter 21.

e.   For the present analysis it is assumed that hydrogen produced from intermittent renewable sources is generated using electrolyzers operated at atmospheric pressure, so that the hydrogen must be compressed before it enters the pipeline. The output of the hydrogen-from-biomass plant would be hydrogen pressurized to 7.5 MPa (1,098 psia), so that in this case no further pressurization is needed before the hydrogen is put into the pipeline.

f.   Hydrogen storage is needed with intermittent renewable sources in order to keep the distribution lines operating at near capacity. In the case of hydrogen derived from biomass these storage costs would be small and are neglected here, as the biomass hydrogen plant would be producing hydrogen continuously.

g.   The refueling station cost is high because of the costs associated with pressurizing hydrogen for the FCV storage canisters.

h.   The cost for local distribution plus the cost for refueling operations is estimated to be $0.050 per liter ($0.19 per gallon of methanol), or $2.77 per gigajoule.

The following table presents the estimated lifecycle costs shown in figure 11, disaggregated by component (for details, see chapter 22):

Lifecycle costs for alternative motor vehicle systems
cents per kilometer

	EV battery	Fuel cell vehicles		PV/wind	ICEV gasoline
		Biomass			
		MeOH	H$_2$[a]	H$_2$[a]	
Vehicle	7.09	6.73	6.72	6.72	11.17
Fuel cell	–	2.60	2.25	2.25	–
Battery	6.71	2.52	2.67	2.67	–
Fuel storage	–	0.02	0.83	0.83	–
Home recharging system	0.05	–	–	–	–
Miscellaneous O&M	6.65	6.69	6.67	6.67	7.45
Fuel[b]	–	1.69	1.49	3.05	2.89
Purchased electricity[b]	1.47	–	–	–	–
Total	**21.97**	**20.25**	**20.63**	**22.19**	**21.51**

Breakeven gasoline price[c]					
$/liter	0.371	0.179	0.221	0.393	
$/gallon	1.40	0.68	0.84	1.49	

a. The costs for hydrogen fuel-cell vehicles include a $4000 cost to the consumer for compressed hydrogen gas canisters. Moreover, the cost of compressing hydrogen to the high pressure needed for these canisters makes a substantial contribution ($2.2 per gigajoule) to the price paid by the consumer for hydrogen fuel. With advanced hydrogen storage or carrier systems, costs may well be much less. For example, if direct reduced iron were used as a hydrogen carrier, with hydrogen generated onboard the car using steam produced by the fuel cell, as discussed in the text, the costs for the hydrogen canisters and for compressing the hydrogen at the refueling stration would be eliminated, and the extra costs incurred with the "iron-powered fuel-cell car" would probably be small.

b. For fuel pump prices (HHV basis, without retail taxes) of $9.18 per gigajoule ($1.21 per gallon) for gasoline, $13.4 per gigajoule for methanol, $13.9 per gigajoule for hydrogen derived from biomass, and $28.5 per gigajoule for hydrogen derived from photovoltaic or wind power sources; and for an electricity price of $0.07 per kWh ($19.4 per gigajoule).

c. This is the gasoline price (without retail taxes) at which the alternative fuel/vehicle systems would be able to compete on a lifecycle cost basis with a gasoline-fueled car having an internal combustion engine.

2
HYDROPOWER AND ITS CONSTRAINTS

JOSÉ ROBERTO MOREIRA
ALAN DOUGLAS POOLE

Hydropower, although an ancient technology, will emerge as a strategically critical source of new electricity in the decades ahead, especially for developing countries and the former Soviet Union, where most of the world's undeveloped potential exists. Although the global potential for hydropower is enormous, numerous factors may limit its exploitation. Nonetheless, much uncertainty surrounds existing estimates of potential. A number of key technologies, including pumped storage, refurbishment of old plants, and compensatory steps (such as reservoir redesign) to reduce the variable nature of river flow, have contributed to the successful exploitation of both large- and small-scale hydrologic resources.

Despite such technological advances, social and environmental impacts are argued to be the principal constraints and sources of uncertainty affecting hydropower's development. Large storage reservoirs exacerbate many problems, particularly in tropical regions. The impacts of hydropower can be categorized either as direct: those that arise during construction of the plant, filling of the reservoir, and changing of the river flow; or as indirect: those that affect the health and well-being of the community. Many of these problems, however, can be mitigated by improved planning that includes public participation at an early stage. Taking such externalities into account may prolong the planning process and increase overall costs, which are already high owing to the capital intensity of hydropower. Indeed, finance has become a major constraint affecting the development of hydropower in regions of greatest potential. Nonetheless, because of its ubiquity and low impact on the environment (relative to fossil-fuel plants), hydropower will remain a critical supply of new electricity in the developing countries and the former Soviet Union and therefore needs to be better understood.

INTRODUCTION

Hydroelectricity, which depends ultimately on the natural evaporation of water by solar energy, is the only renewable resource used on a large scale today for electricity generation. In 1986 it contributed 14.5 percent of the world's generated electricity. While basic conversion principles are well understood, hydroelectricity faces considerable constraints and challenges, for it is a resource whose exploitation interferes with natural hydrology, takes place in a variety of geographical situations, and can involve diverse objectives. Because hydroelectric generation is so site- and system-specific, it is difficult to cover the subject in a single chapter. Although we address factors that are likely to affect the future of hydropower, our perspective is colored by our experience with the system we know best: Brazil's.

Converting work to electricity

Electric power is generated in a hydroelectric plant when water, stored by dams, is released to turn water turbines that are coupled to generators. Transmission facilities then send the electricity to the market terminal and ultimately to the end user.

Dams, which are built across flowing bodies of water, generally have two purposes. One is to raise the water level, thus increasing its potential energy or hydraulic head; the other is to create a water reserve to compensate for fluctuations in river flow or power demand. The importance of both functions varies from site to site. Some dams have virtually no reserve storage capacity. Auxiliary works at a dam may include spillways, gates, or valves to control the discharge of surplus water; an intake structure to carry water into the power station; mechanisms to remove silt from the reservoir; and methods to allow ships or fish through the dam. Dams may also serve other purposes, such as providing water for human consumption, industrial operations, and irrigation; facilitating flood control; increasing a stream's depth for navigation; and creating recreational amenities [1-3]. In fact, most existing dams do not generate electricity, although many could be retrofitted to do so [4].

Hydraulic turbines convert the energy of an elevated water supply or flowing stream into mechanical energy on a rotating shaft. Whereas most old-style waterwheels used water weight directly, modern hydraulic turbines operate according to the impulse or reaction principle, which converts pressure and kinetic energy to rotational kinetic energy. The most common reaction turbines are known as the Francis, Kaplan, Bulb, Tube, and Straflo models. The main type of impulse turbine is the Pelton Turbine [2, 5]. Kinetic turbines, which would exploit only the kinetic energy of flowing water, without any hydraulic needs, are now being studied [6].

The turbine's rotating shaft drives an electric generator, which transforms mechanical power to electric power. Three types of generators are available for use with hydroelectric plants: for lower-capacity plants, the generator may be either an alternating-current induction type or a direct-current type; at higher capacities, a conventional synchronous type is used[5].

Transmission facilities consist of a step-up, or sending, substation at the electric generating plants; a step-down, or receiving, substation at the market terminal; interconnecting transmission lines; and, in some cases, intermediate switching stations [5]. Long distance transmission of electricity is often necessary to justify the development of larger sites.

Table 1: Annual world water balance[a]

Region	Surface area $10^6\ km^2$	Precipitation mm	$10^3\ km^3$	Evaporation mm	$10^3\ km^3$	Runoff[b] mm	$10^3\ km^3$
Europe	10.5	790	8.3	507	5.3	283	3.0
Asia	43.5	740	32.2	416	18.1	324	14.1
Africa	30.1	740	22.3	587	17.7	153	4.6
North America	24.2	756	18.3	418	10.1	339	8.2
South America	17.8	1,600	28.4	910	16.2	685	12.2
Australia and Oceania	8.9	791	7.1	511	4.6	280	2.5
Antarctica	14.0	165	2.3	0	0	165	2.3
Total land area	**149.0**	**800**	**119**	**485**	**72**	**315**	**47.0**
Pacific Ocean	178.7	1,460	260.0	1,510	269.7	–83	–14.8
Atlantic Ocean	91.7	1,010	92.7	1,360	124.4	–226	–20.8
Indian Ocean	76.2	1,320	100.4	1,420	108.0	–81	– 6.1
Arctic Ocean	14.7	361	5.3	220	8.2	–355	– 5.2
Total ocean area	**361.0**	**1,270**	**458.0**	**1,400**	**505.0**	**–130**	**–47.0**
Globe	**510**	**1,130**	**577**	**1,130**	**577**	**0**	**0**

a. [7].
b. Outflow of water from continents into ocean.

Global water availability and hydroelectric potential

On a global basis, the total amount of water precipitated over time equals the amount returned to the atmosphere by evapotranspiration (see table 1). On land, precipitation exceeds evapotranspiration, but the difference is eliminated by the runoff of rivers (and groundwater) to the ocean. In the oceans, an opposite process occurs [7]. Average precipitation, evaporation, and run-off in South America exceeds that of other continents by a factor of two; Asia accounts for the largest total run-off.

The energy potential of hydropower is determined by the volume of run-off water and by the distance it falls before reaching the ocean. Because run-off is not evenly distributed over continents, average altitudes (which range from 300 meters in Europe to 950 meters in Asia and 2,040 meters in Antarctica) are not adequate for calculating the theoretical potential, even on a gross basis. Seasonal variations in run-off also influence the theoretical potential. Estimates of the theoretical annual potential of world hydroelectricity range from 36,000 to 44,000 terawatt-hours (TWh) [1, 8, 9] (see table 2).

This gross theoretical potential is much larger than the technical usable potential (or theoretical exploitable potential), which in turn is substantially larger

Table 2: Annual hydroelectric potentials[a]

Region	Theoretical potential TWh	Technical potential TWh
Africa	10,118	3,140
North America	6,150	3,120
Latin America	5,670	3,780
Asia (excluding former USSR)	16,486	5,340
Australasia	1,500	390
Europe	4,360	1,430
Former USSR	3,940	2,190
World	**44,280**	**19,390**

a. [9].

than the economically exploitable potential. The uncertainty of the economic potential increases when such factors as geological constraints and social and environmental factors are considered. Despite the simplicity of the basic physics, hydropower is a surprisingly subtle resource whose evaluation requires a sequence of increasingly detailed studies. This type of medium- and long-term analysis has been a low priority in most developing countries.

Technical potential is evaluated on the basis of simplified engineering and economic criteria, with few if any environmental restrictions. The criteria may differ substantially from country to country and are relatively unsophisticated in many developing countries, which may inflate estimates of technical potential. Such estimates show a wide range of values (see tables 2 and 3). Most significant are the differences in regional values, especially with respect to Africa, North America, and the former Soviet Union.[1]

The economically exploitable potential, after detailed environmental, geological, and other economic constraints are considered, is substantially smaller than the technical potential. Given the uncertainties surrounding technical potential, any estimate of economic potential is clearly speculative, especially in the regions of greatest untapped potential. The historic experience of industrialized countries, where hydropower is most intensely developed, suggests that the economic potential of hydropower lies somewhere between 50 and 65 percent of technical potential. A more realistic range is likely to be 40 to 60 percent (see table 3). Thus, we believe the world's long-term economic hydroelectric potential may

1. The estimate of technical potential for developing countries (see table 3) is slightly higher than a somewhat earlier estimate by the World Bank [10]. Some of the difference may be attributed to the exclusion of certain countries from the World Bank estimate (Venezuela is the most significant).

Table 3: The world's technically exploitable hydropower potential and existing development by region[a]

Region	A Technically exploitable potential *TWh/year*	B Total hydro installed capacity *GW*	C Hydro generation in 1988 *TWh*	C/A *percent*
Africa	1,150	15.84	36	3
South Asia and Middle East	2,280	45.44	171	8
China	1,920	32.69	109	6
Former Soviet Union	3,830	62.20	220	6
Japan	130	20.26	87	67
North America	970	129.09	536	55
South America	3,190	75.98	335	11
Central America	350	10.71	32	9
Eastern Europe	160	16.56	49	31
Western Europe	910	128.44	436	48
Australasia	200	12.00	37	19
World	**15,090**	**549.2**	**2,040**	**14**

a. [11].

be on the order of 6,000 to 9,000 TWh per year, of which only about 2,000 TWh have been developed so far.[2]

Most of the undeveloped potential lies in the former Soviet Union and in the developing countries, whose annual technically exploitable potential is about 3,800 and 8,900 TWh respectively. To date, both regions have developed only a small fraction of their potential. Assuming, for example, that the former Soviet Union has a 40 to 60 percent range of economically viable and environmentally acceptable technical potential, only 220 TWh are currently used out of a total 1,500 to 2,300 TWh of economic potential. In the developing countries only 680 out of 3,600 to 5,300 TWh of economic potential are used.

Hydroelectric plants across a wide range of sizes contribute a significant share of the world's hydropower potential. In figure 1 the size profiles for southern, southeastern, and midwestern Brazil are shown. The estimates are probably biased against the relative potential of the smallest sites and illustrate a clear tendency to develop larger-scale potential first. Although the relative emphasis on large hydropower projects may have been exaggerated in Brazil and in some other developing countries, large-scale sites remain important, nevertheless. To simply eliminate large hydropower plants (those that generate, say, more than 1,000 megawatts) as

2. By comparison, total world electricity production in 1986 was 10,000 TWh.

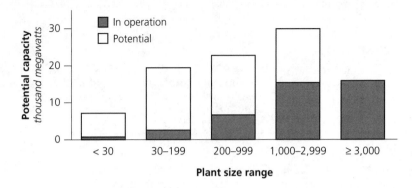

FIGURE 1: *Profile of scale of hydroelectric potential in Brazil, 1986: southern, southeastern, and midwestern regions. These regions have been selected because a more complete analysis is available for them than for the nation as a whole (especially Amazônia). Future potential is estimated to be competitive with nuclear power in* Plano 2010 [12]. *Capacity of plants in operation is defined in terms of their final capacity and not the capacity actually operating by the end of 1986.*

some small-hydropower enthusiasts advocate, would certainly eliminate a large fraction of the sector's economic potential.

Analysis of hydroelectric potential needs higher priority, and the quality of estimates must be drastically improved, especially with respect to socioeconomic and environmental factors. This vital step will probably require a serious international program to define medium- and long-term resources in key countries.

HISTORY OF HYDROELECTRIC POWER

Plants built up to 1930

Use of water power dates back to ancient times: primitive wheels, actuated by river current, once raised water for irrigation purposes, grinding, and myriad other applications. The devices were extremely inefficient, however, and used but a small part of a stream's available power. The development of undershot, breast, and overshot wheels represented great advances: water was confined to a channel, brought to the wheel, and used under a head or waterfall. Such technology was already in use in the Roman Empire, where the largest milling complex discovered so far had a power output equivalent to about 15 kilowatts [13]. Efficiencies grew from perhaps 30 percent for the undershot to about 80 percent or more for overshot wheels. About the middle of the 19th century, the hydraulic turbine superseded the overshot wheel [14].[3]

Hydropower technology is used at dams where flowing water can be regulated and stored. The first hydro facilities, known as run-of-the river plants, are un-

---

3. An overshot wheel, while efficient, has a large diameter, low turning speed, and is limited to heads no greater than 12 meters high [14].

able to generate power during the dry season when streams and rivers have low flow. Large dams, in contrast, are designed to generate continuous power, but require reservoirs of inundated land, which adds significantly to their construction costs.

Of the many turbines operating today, two were developed during the mid-19th century: the Francis turbine, which was designed in 1849 and powered the first hydroelectric plant in the United States, and the Pelton turbine, which appeared in 1890. In 1913, the Kaplan turbine was developed, followed in 1936 by the bulb turbine. In 1880, the development of hydroelectric power was launched when a small direct-current (DC) generating plant was built in the U.S. state of Wisconsin. The first transmission lines, built in 1890, further stimulated the growth of water power. Installed capacity in the United States grew from 2.5 gigawatts in 1890 to 10.4 gigawatts in 1930, with proportional increases in the rest of the industrialized world [14]. This was made possible by an increase in the range of developable head—from 30 meters, around 1900, to 240 meters or more by 1930, when energy conversion efficiencies reached 85 percent (with the Pelton turbine) and 90 percent (with the Francis and Kaplan turbines). During the same period, transmission voltages grew from 40 to 300 kilovolts, and plants having capacities in excess of 100 megawatts were installed.

1930 to the present
From 1930 until the mid 1960s, a concerted effort was made to increase the scale of hydroelectric generation. The creation of the U.S. Tennessee Valley Authority in the early 1930s, with its broad powers for finance, expropriation, and development, represented a critical institutional step [15]. In 1942, the Bonneville Power Authority began operating the Grand Coulee Dam, which (including additions) is still the third largest installed capacity plant in the world (see table 4).

Today more than 150 dams are classified as "large"[4] and account for 40 percent of existing world hydroelectric capacity. The largest represent some of the biggest engineering projects in the world. The Itaipu dam on the Brazil–Paraguay frontier, for example, is 8 kilometers long and 150 meters high. Although it is the largest power complex in the world (equivalent to 10 of the largest nuclear reactors), it is by no means the world's biggest dam. The reservoirs that accompany dams also rank as major bodies of fresh water: Ghana's Lake Volta, for example, covers 8,500 square kilometers, an area almost the size of Lebanon.

Most of the largest dams are now being built in the developing world, where between 1980 and 1986 the absolute increase in hydro generation was almost twice that of the industrialized world (172 versus 96 TWh). One reason why dam construction is on the wane in industrialized countries is that most of the promising sites have either already been developed or are excluded because of their natural beauty or other environmental concerns. In the United States, for example,

---

4. According to the International Commission of Large Dams, a large dam should have either an installed capacity of 1,000 megawatts, a reservoir capacity of 25 cubic kilometers (km^3), a volume of 15 million cubic meters (m^3), or a height of at least 150 meters.

Table 4: The world's largest dams and hydro plants in operation in 1989[a]

LARGEST GENERATING CAPACITY	Year of initial operation	Country	Rated capacity now megawatts	Rated capacity planned megawatts
1 Itaipu	1983	Brazil–Paraguay	10,500	12,600
2 Guri	1986	Venezuela	10,300	10,300
3 Grand Coulee	1942	United States	9,780	10,830
4 Krasnoyarsk	1968	Russia	6,000	6,000
5 La Grande 2	1979	Canada	5,328	5,328

HIGHEST EMBANKMENT DAMS	Year completed	Country	Type[b]	Height meters
1 Nurek	1980	Tajikistan	E	300
2 Chicoasén	1980	Mexico	E/R	261
3 Guavio	1989	Colombia	E/R	243
4 Mica	1972	Canada	E/R	242
5 Chivor	1975	Colombia	E/R	237

HIGHEST CONCRETE DAMS	Year completed	Country	Type[b]	Height meters
1 Grande Dixence	1961	Switzerland	G	285
2 Inguri	1980	Georgia	A	272
3 Vajont	1961	Italy	A	262
4 Mauvoisin	1957	Switzerland	A	237
5 El Cajón	1985	Honduras	A	234

LARGEST VOLUME DAMS Construction material	Year completed	Country	Type[b]	Volume $10^6 \, m^3$
1 Tarbela	1976	Pakistan	E/R	148.5
2 Fort Peck	1937	United States	E	96.0
3 Tucuruí	1984	Brazil	E/G/R	85.2
4 Guri	1986	Venezuela	E/G/R	78.0
5 Oahe	1958	United States	E	70.3

LARGEST CAPACITY RESERVOIRS	Year completed	Country	Type[b]	Reservoir volume $10^9 \, m^3$
1 Owen Falls[c]	1954	Uganda	G	2,700.0
2 Bratsk	1964	Russia	E/G	169.0
3 Aswan High	1970	Egypt	E/R	162.0
4 Kariba	1959	Zimbabwe–Zambia	A	160.3
5 Akosombo (Volta)	1965	Ghana	R	148.0

a. [16, 17].

b. Dam type: A = arch; E = earthfill; G = gravity; R = rockfill.

c. Major part of lake volume (Lake Victoria) is natural.

not a single major dam was approved for federal funding between 1976 and 1990. Interest in hydropower in these countries is now focused more on smaller sites, on the refurbishing and upgrading of existing hydroplants, and on retrofitting dams constructed for other purposes. In the United States, retrofitting non-power dams is estimated to have a potential of 26,000 megawatts [4]. Large- and intermediate-scale dams, however, will continue to be very important in developing countries, in the former Soviet Union and in some industrialized nations, such as Canada.

HYDROPOWER TECHNOLOGIES

Dam architecture

Hydroelectric power plants have typical elements that are modified to suit the particular requirements of each site (see figure 2). The dam blocks the downstream end of the reservoir. The spillway, or flood discharge structure, controls maximum water level and permits the passage of floodwater. The intake structure operates at designed water levels and is protected by a screen. The tunnels and penstock carry water from the intake structure to the turbines. The surge tank is located between the tunnel and penstock, or between the hydraulic machine and the tailwater tunnel; it is designed to dampen fluctuations in pressure and water level. The powerhouse contains turbines, generators and auxiliary equipment. The tailwater section connects the turbine outlet to the river [1].

Dams fall into several distinct classes according to their profile and building material, which is typically concrete, earth or rock. Concrete dams may be divided into three types: solid gravity, hollow gravity, and arch dams [3]. The gravity dam relies on its own weight to resist hydraulic thrust. The hollow gravity dam contains only about 35 to 40 percent of the concrete required for a solid dam, but is more expensive per unit volume. The arch dam, which is designed for narrow valleys with steep slopes, distributes the hydraulic thrust to its abutments. Earth dams fall into two categories: those that have a central impervious core and those that do not. The core usually consists of clay or concrete. A good core lowers the water's saturation line within the downstream toe, or base of the dam. Rock dams are constructed from loose rocks that are first dumped and then compacted to form an embankment. A watertight apron of concrete or wood is usually constructed on the upstream slope and sealed to the foundation [1, 14, 18].

Generating technologies

Modern hydraulic turbines are known as reaction and impulse wheels. When water enters a reaction wheel, it flows around the periphery, completely filling the wheel's passages and acting on the blades at a pressure greater than atmospheric, causing them to turn. The wheel is designed to use both pressure and kinetic energies. The main types of reaction wheels are the Francis turbine, which has a radial intake of water (water enters the turbine perpendicular to the shaft) and the

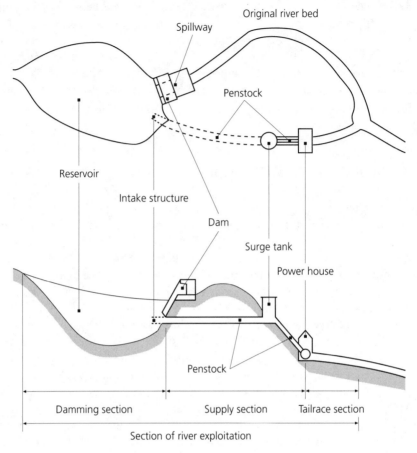

FIGURE 2: *The characteristic components of a river-diversion hydroelectric plant. [1].*

Kaplan turbine, which has an axial intake of water. The Kaplan turbine is pre-ferred when the hydraulic head is small (see figure 3). In contrast, water entering an impulse wheel does so at only a few points around the periphery; it does not completely fill the passages, and turns the wheel blades under atmospheric pres-sure. Moreover, the energy applied to an impulse wheel is entirely kinetic. The best known impulse wheel is the Pelton wheel, which is preferred in situations where the head is relatively large. Other types of turbines include the axial-flow hydraulic turbine, the tubular turbine, the straflo turbine, and the bulb turbine [5, 14].

Refurbishing existing plants

In the industrialized countries, an increasingly large fraction of operating hydro-plants are more than 50 years old. While the civil works of the plant (the dam, tunnels and spillway) may have an economic life ranging from 50 to 100 years or

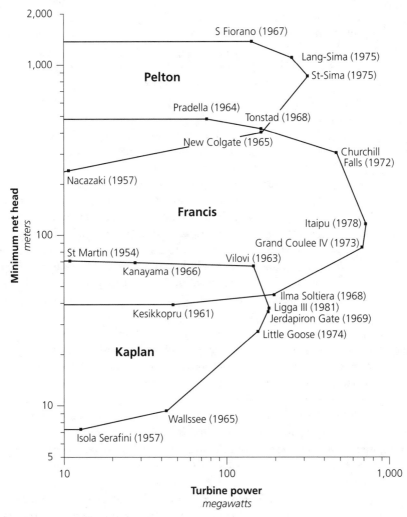

FIGURE 3: *Hydraulic turbines—domains of head and scale in the engineering practice of Pelton, Francis, and Kaplan turbines.*

more, a plant's mechanical and electrical components may last for only 60 years (as is the case for turbines; less in the case of some auxiliary equipment). Because civil works typically account for some 60 percent of a plant's initial investment, rehabilitating the mechanical and electrical equipment therefore represents an effective way to double the life of an old station at relatively low cost. Upgrading and refurbishing old plants not only extends their useful life, but enables them to be retrofitted according to modern operating regimes and brought to acceptable standards with respect to safety, the environment, efficiency, and reliability. In some cases, extra capacity can be added for relatively low cost [19–21]. In other cases, building a new plant is a better choice than refurbishing an old one [22], although the second alternative provides some clear advantages: feasibility studies are cheaper, the lead time is shorter, environmental concerns are less, and statutory approvals are more easily obtained. Moreover, significant cost and time overruns are rare.

Upgrading and refurbishing may call for installation of a new control system, modernization of the switching station, new insulation for the generators, and replacement of the turbine's runner blades with better designed high-strength steel components. The face of the dam may also need refurbishing with geosynthetic materials, or its height may be profitably increased [20, 23] (see "Technological advances," below). In addition, better understanding of local hydrology may call for redesign of the discharge outlet.

Refurbishing may be undertaken for a variety of reasons. A plant in New Zealand wanted to increase its energy production [21], whereas one in India needed increased peaking potential [24]. In India, more than 20 percent of operating hydropower plants were damaged by excessive siltation and, of those, 55 percent needed to be refurbished. Yet by investing a modest U.S.$212 million to renovate the plants (extending their useful life and improving their performance and reliability) a bonus of 528 megawatts of installed capacity and 396 gigawatt-hours (GWh) per year were gained [24].

Technological advances

Hydroelectric generation is usually regarded as a mature technology that is unlikely to advance. While some elements of hydropower, such as conventional turbine efficiency and costs, have reached a plateau, the same cannot be said for the system in its entirety. Construction techniques are changing, as is underground rock excavation. New turbine designs for low-head sites are still being developed, and transmission technology, which is of fundamental importance to hydropower development in isolated areas, continues to advance. Moreover, environmental concerns are driving changes in the design, construction, and operation of hydroelectric plants.

One newly developed technology has significantly reduced the overall costs of construction. Roller compacted concrete (RCC), which uses a large fraction of Pozzolani cement and is compacted with conventional earth-moving equipment (thus avoiding the need for joints and molds), requires about 40 percent less water and 30 percent less cement than ordinary concrete. The speed with which RCC dams can be constructed contributes to cost savings on the order of 66 percent compared with ordinary concrete and is one of the dam's main advantages. Moreover, RCC has great durability, high compressive strength, and low permeability. Not surprisingly, RCC dams are beginning to replace traditional immersion-vibrated concrete gravity dams, and are also being built at sites where rock dams were once the technology of choice. Indeed, the number of RCC dams rose from zero in 1980 to 47 in 13 countries by the end of 1989 [25, 26].

Geosynthetic materials represent another technological advance. Made of polymers such as polyamide and polypropylene, geosynthetics have been extensively used in other applications such as lining waste facilities or toxic storage structures, but have only recently been used in dam construction. A category of geosynthetics called geotextiles, which form a fiber matrix to hold particles and sand grains (thus preventing erosion yet allowing the passage of water), can be found in the filters and drains of dams, and geomembranes, which form impermeable sheets, are being tested in leaking and/or weakened dams [27, 28].

Another relatively new development is the use of inflatable weirs for regulating water level and/or for raising a dam's crest. The weirs are tube-shaped and made of flexible rubber-based fabric. They can be inflated either by air or water and are anchored to a concrete foundation. By varying pressure within the tube, the height of the weir is adjusted. If necessary, the weir can be easily and quickly deflated, thus allowing for the unrestricted passage of floodwater [29, 30]. Weirs are most appropriate for small-scale plants.

Transmission technology is evolving across a broad range of fronts [31]. The goal is to reduce the capital costs and space requirements of transmission lines, while increasing their reliability and efficiency. The specific requirements of hydropower are especially relevant to improving both long distance bulk transmission and interconnecting systems that have different frequencies (50 versus 60 hertz) or are asynchronous (they have the same frequency, but are calibrated to a different clock). Advances in high-voltage DC systems, especially those with multiple terminals, are particularly important. Direct-current applications for long-distance transmission were limited for a time to "bipoles" with a terminal at each end, but multiple terminal transmission technology has since prompted research on such improved ultrahigh-voltage alternating current systems as higher phase-order and half-wavelength transmission. In addition, line-insulating materials and new transformer cores are being developed to reduce transmission losses; improved conductivity remains a promising possibility.

Of potential use in both ocean energy systems and rivers are kinetic turbines that harness the energy of flowing water [2]. The technology, which was once widespread during the Middle Ages, consisted of waterwheels that were affixed to barges anchored in rivers. Although the technology faded away once dams became common, they are now seen as a relatively inexpensive and less damaging alternative to dams, as well as a way to tap otherwise inaccessible water flows. Several countries are trying to modernize the technique and to assess its costs [15]. The viability of floating waterwheels, however, is likely to be sensitive to the current's velocity. Thus, a huge, slow-flowing river like the Amazon has a theoretical maximum potential of only 650 megawatts. A proposal has been made in China to exploit some of the potential of the Three Gorges site on the Yangtze (16,800 megawatts), although building a dam at the site could dislocate a population in excess of one million.

Small hydropower plants
No consensus has been reached on the definition of small, mini, and micro hydropower plants (see table 5). In the United States, micro hydropower schemes

Table 5: Definitions of small, mini, and micro hydro plants found in the literature

Country	Micro *kilowatts*	Mini *kilowatts*	Small *megawatts*	Reference
United States	<100	100–1,000	1–30	32
United States	<100	100–1,000	–	33
China	–	<500	0.5–25	34
USSR	<100	–	0.1–30	35
France	5–5,000	–	–	36
India	<100	101–1,000	1–15	37
Brazil	<100	100–1,000	1–30	12
various	<100	<1,000	<10	38

are sometimes described as those having capacities below 100 kilowatts, mini hydropower plants are those ranging from 100 to 1,000 kilowatts, and small hydropower plants are those that produce from 1 to 30 megawatts.

Recent international surveys on small hydropower facilities (with capacities below 10 megawatts)[5] reveal that: 1) small hydropower plants are currently under construction or are being planned in 100 countries; 2) total installed capacity of small hydropower facilities in 1989 was 23.5 gigawatts, 38 percent of which was generated in China; 3) small hydropower potential (up to 2 megawatts) is on the order of 570 TWh per year, or 3.8 percent of the world's total technical hydropower potential; and 4) more than 130 companies now manufacture equipment specifically for plants having capacities ranging from 0.01 kilowatts to more than 10 megawatts [11, 39, 40].

In 1989, about 70 percent of the 205 small turbines ordered worldwide were of the same basic type as those used for larger projects: Francis turbines (27 percent), Kaplan turbines (24 percent), and Pelton turbines (17 percent); the balance consisted of crossflow, bulb, tubular, turbo, and other types. Of these, 15 percent were specifically for micro hydropower plants, 57 percent for mini plants, and 28 percent for small plants [40].

Several international agencies have provided technical and/or financial assistance for small hydropower-plant construction in developing countries. The Swiss Center for Appropriate Technology, for example, has worked with Nepal; the German Agency for Technical Cooperation has assisted Pakistan and the Philippines; and the Norwegian Agency for International Development has recently helped construct two mini hydropower plants in Mozambique. Each project considers the economic benefits, social acceptance, and management and operation

---

5. No differentiation among small, mini, and micro hydropower facilities was made.

of the plant in the context of the local culture. Experience indicates that instead of isolated projects, wider reaching programs should be encouraged to allow for the build-up of experience and to permit greater standardization of key components [41].

In 1950, China embarked on the most ambitious small hydropower program in the world. The technical potential of small hydropower facilities in China is estimated to be 220 TWh per year, or 70 gigawatts of installed capacity. By 1989, China had already developed 36 TWh per year (12.6 gigawatts of installed capacity), which corresponds to 37 percent of the country's total hydropower generation. By 2000, capacity is expected to reach 23.2 gigawatts [34]. China's small hydropower program offers several advantages. It is based on relatively simple technology. Equipment, construction materials, and labor are obtained locally. The work is standardized and serialized for different sites, and construction time and transmission lines are often short [34].

Small hydropower has one drawback: it is almost always obtained from run-of-river plants that lack the reservoir capacity to store water. Consequently, severe seasonal variations in power output may occur, depending on a site's hydrology. One source claims, for example, that many of the Chinese plants operate only part of each year and have a capacity factor of only 8 percent [42]. Thus the long-term viability of small hydropower facilities may depend on back-up electricity that is supplied either locally or via the grid.

Pumped-storage plants

Pumped-storage plants allow for large-scale storage of electric energy: they store hydro energy during off-peak hours by pumping water from a lower reservoir to an upper reservoir, and then produce electricity during peak hours, when electricity is most costly to produce, by turbining water from the upper to the lower reservoir. In addition, many storage plants also function as conventional hydroplants or as pumps to raise water for irrigation or public use [4, 43, 44].

Although similar to conventional hydropower facilities, pumped-storage plants are a mature technology and rely on specially designed reversible pump-turbines. Their overall efficiency has grown from 40 percent, when they were first introduced at the beginning of the 20th century, to 70 or 80 percent in 1991 [43, 45]. Moreover, they generally respond quickly to large changes in load, reducing the need for high-cost spinning-reserve and have relatively little environmental impact. In the future, the technology may be expanded to include underground reservoirs and to accommodate solar, wind, and tidal energy generation [43].

In 1990, 283 pumped-storage plants, representing 74 gigawatts of installed capacity, existed worldwide, and another 25 gigawatts were under construction. Europe has 40 percent of the total; the United States and Japan, 25 percent each [44].

Table 6: Seasonal flow variation of major rivers in Asia, Africa, and South America[a]

	Three months of highest flow		Three months of lowest flow		Ratio of average high to low flow
	Months	*Percent of annual river discharge*	*Months*	*Percent of annual river discharge*	
ASIA					
Yenisei	May–Jul	63	Feb–Apr	6	10.5
Lena	Jul–Aug	72	Feb–Apr	2	36.0
Huangho	Jul–Sep	49	Jan–Mar	10	4.9
Yangtze-Kiang	Jul–Sep	42	Dec–Feb	10	4.2
Mekong	Aug–Oct	61	Feb–Apr	4	15.3
Brahmaputra	Jul–Sep	52	Jan–Mar	6	8.7
Ganges	Jul–Sep	70	Mar–May	4	17.5
Indus	Jul–Sep	68	Jan–Mar	3	22.7
Tigris	Mar–May	54	Aug–Oct	6	9.0
AFRICA					
Congo	Nov–Jan	32	Jun–Aug	20	1.6
Nile	Aug–Oct	63	Mar–May	7	9.0
Volta	Sep–Nov	63	Mar–May	5	12.6
Niger	Dec–Feb	42	May–Jul	5	8.4
Zambesi	Mar–May	51	Sep–Nov	7	7.3
Limpopo	Jan–Mar	85	Aug–Oct	1	85.0
SOUTH AMERICA					
Paraná	Jan–Mar	38	Aug–Oct	17	2.2
Uruguay	Aug–Oct	37	Jan–Mar	16	2.3
São Francisco	Jan–Mar	44	Aug–Oct	10	4.3
Tocantins	Feb–Apr	52	Aug–Oct	6	8.4
Xingu	Mar–May	59	Aug–Oct	4	15.4

a. Sources: Africa and Asia, [46]; Paraná, São Francisco, Tocatins, Uruguay, and Xingu, Eletrobrás.

HYDROPOWER IN ELECTRICAL SYSTEMS

Natural river flow is highly variable, a characteristic that has important implications for the design of hydroelectric plants and their incorporation into the electrical generating system. Most rivers exhibit pronounced seasonal variation in their flow (see table 6). In some cases, the average three-month high flow may be more than 10 times greater than the average three-month low flow, while in others, it is less than double. In addition to seasonal fluctuations, there may be long-term variations in run-off. Some years are drier than average, and multi-year dry spells can occur. Such long-term variations are not only less predictable than seasonal patterns, but can influence the latter significantly, depending on the river basin. The Paraná river in Brazil, for example, is greatly affected by long-term changes, whereas the Xingu and Tocantins rivers are mostly influenced by seasonal changes.

Variability poses obvious problems. Electricity demand, after all, does not correspond to river flow. Indeed, seasons of peak electricity demand may coincide perversely with seasons of minimal river flow. Such is the case at high latitudes where electricity demand peaks in the winter, a time when there is low river flow because of ice. There are three broad approaches to this problem. The first is to regulate a river by building storage reservoirs: water is retained in the reservoirs when the river flow is high and discharged from them when the flow is low.

Storage capacity may also offer irrigation and flood control benefits, which in some cases are more important than energy. But storage does not come without its costs, which may be economic, environmental, and social. Because these costs are of growing concern, the tendency is to reduce reservoir size wherever possible when planning new projects. Providing the large storage volumes that are needed to even out natural flow variation to the long-term average becomes prohibitively expensive, more so in rivers with marked seasonal or long-term flow variability. On the Xingu river, for example, the additional regulation of flow achieved with added storage volume rapidly decreases above a regularization level of 70 percent, at which point economic and/or environmental marginal costs become critical (see figure 4). Moreover, to achieve a regulation of 75 percent means that, without other measures, 25 percent of the total long-term water resource is lost for generation purposes (see figure 4).

Such limits can be partially overcome by adopting complementation strategies. Hydroelectric complementation—the second approach to solving the variability problem—is based on the notion that different river basins can be connected, thus letting a higher flow in one basin compensate for low flow in the other. Under such an arrangement, more turbines can be installed on each river than if the basins were isolated. Thus in energy terms, transmission lines and extra turbines substitute for reservoir storage capacity. The success of the strategy clearly depends on the hydrologic diversity between basins. Unfortunately, neighboring basins are often similar hydrologically. For example, the Huangho, Yangtze-Kiang, Mekong, Brahmaputra, and Ganges rivers in Asia have roughly coincident seasons of high flow, as do the Paraná, São Francisco, and Tocantins rivers in Brazil (see table 6).

Hydroelectric complementation can be increased by introducing advanced transmission technologies to key areas of hydrologic diversity, most notably around the equator, where seasons tend to reverse from one hemisphere to the other. In South America, for example, high hydrologic diversity exists between the key headwater regions of rivers north of the equator (see areas labeled A and B on figure 5), which experience high flow from May to August, and those south of the equator in Argentina, Bolivia, Brazil, Paraguay, and Peru, where high flows

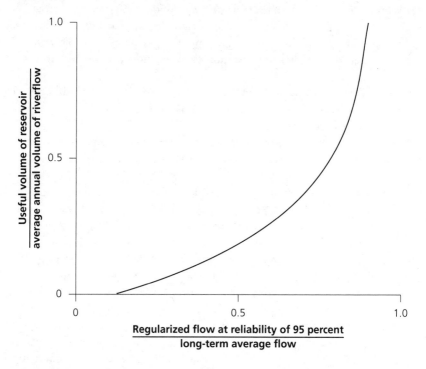

FIGURE 4: *Useful reservoir volume and regularization of river flow for the Xingu River hydroelectric plant at Altamira, Brazil. The average annual volume of riverflow at this site is 275 billion m^3 [47].*

occur from January to March (see area labeled C on figure 5). Interest in complementation has been severely hampered by the region's limited hydroelectric development until now, an effect that is likely to become increasingly relevant as electrical supply systems evolve.

The prospect, however, of Brazil developing large hydroelectric capacity (11,000 megawatts, as planned) on the "Great Bend" of the Xingu river raises the question of "interhemispheric interconnection." The parameters of this major project should be defined in the next few years, and would be greatly influenced by a decision regarding this interconnection. The northern pole for this hypothetical interconnection already exists. It is the Guri scheme on the Caroní river, (a tributary of the Orinoco in Venezuela), which represents one of the world's largest hydropower projects (10,300 megawatts projected for its final expansion). Two mega dams with such hydrologic diversity should justify the scale of transmission capacity (probably about 3,000 megawatts) needed to make the leap of some 2,500 kilometers pole to pole. This interconnection, with its possible sequels, is perhaps the most dramatic example of hydroelectric complementation. A similar "interhemispheric interconnection" may eventually be feasible in Africa (see table 6).

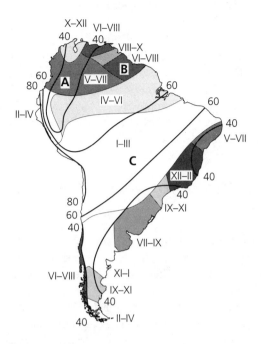

FIGURE 5: *Three-month periods of highest average river flow in tributaries and small basins of 1,000 to 30,000 km². Three-month periods, identified by roman numerals, are defined by areas of different shading. Lines with Arabic numerals indicate percentage of total annual runoff during the three months of highest flow [46].*

The third option for handling variability is called thermal complementation. The strategy substitutes standby thermal generating capacity for reservoir storage capacity. The thermal plant is activated when the water supply drops, either because river flow is low or because the reservoir is depleted. In systems that are predominantly hydroelectric, the average lifetime capacity factor of a thermal generating plant may be quite low; in Brazil, for example, a thermal plant operating in complementation with the (96 percent) hydropower system may have a lifetime average capacity factor of only 15 percent. The effective contribution to the system is substantially greater because the existence of thermal back-up permits greater hydroelectric generation when water is abundant. Despite its low capacity factor, a complementation thermal plant has very different characteristics than a more familiar peak-load plant. To begin with, the thermal plant must be designed to operate occasionally for many months at base load; the ability to adjust quickly to load changes is less important. Regular, dry-season complementation can also be quite economical when low-cost seasonal fuel, such as sugarcane bagasse, is available. Although the prospects for complementation between hydropower and solar power have not been carefully studied, insolation may often be greater in seasons and years when river flow is lower. Even with limited water-storage capacity, hydropower can complement diurnal variations in solar radiation. Thus, solar/hydro complementation may prove promising in many tropical regions.

This section has focused on hydropower's contribution to "firm power," which is the system's average continuous power requirement (ignoring peaks); much hydroelectric capacity in this case is operated as base-load. In mixed hydro-thermal systems, however, hydropower frequently provides peak power, a fact that is attributable in part to the excellent load-following characteristics of both conventional and pumped-storage hydro facilities.

ENVIRONMENTAL AND SOCIAL ISSUES

Overview

Hydropower-plant construction often has significant social and environmental impact. This fact, which was downplayed during the oil crises of the 1970s, has become more visible in recent years. Public resistance to, or questioning of, dam construction is crystallizing in many countries, and new constraints are being imposed that limit electricity generation in favor of other water-use objectives. In Sweden, for example, four northern rivers are legally protected from hydroelectric development, and in Norway legislation is pending that would prohibit hydro development on rivers that together account for 18 percent of the country's technical potential. In Czechoslovakia and Hungary, a major project on the Danube was halted in midconstruction by public opposition. In the United States, the Wild and Scenic Rivers Act excludes 32 gigawatts of potential power from development, and the Electric Consumer Protection Act of 1986 may make the development of another 22 gigawatts uneconomical [4]. Dams on the Columbia River may have to reduce their output by 10 percent for environmental reasons. And in Brazil, construction at two sites has been paralyzed by local opposition. Thus, social and environmental concerns have become key factors in determining how much of the remaining hydro potential is, in fact, viable.

Although the various impacts can be categorized generically, their significance and causality vary greatly from site to site. Any generalizations are heavily conditioned by such site-specific differences as topography, river flow, climate, ecology, land use, and so on, as well as by the scale and design of a particular hydroplant. With regard to the latter, the size of the dam, the characteristics of the reservoir, and the disruption of river flow may vary enormously. Some dams, for example, are virtually run-of-river, and if topography permits, inundate small areas relative to the amount of electricity generated. Other dams create relatively large reservoirs, either to store water against fluctuations in river flow or to create hydraulic head in regions of poor topography. These artificial lakes may be very different in terms of their hydraulic retention time or the biological oxygen demand that is created when large acres of biomass are flooded.

Most hydroelectric potential is located in regions where population densities are relatively low. The indirect impact of a major civil engineering project,[6] which may involve a large influx of migratory workers, can be considerable and is now recognized as a major understudied problem in tropical forest regions [48–50]. In more densely settled areas, hydroplant construction can dislocate large numbers of people, which requires high-level management attention, even when relatively small populations are involved (as in more remote areas). Reactions to such outside-induced changes vary, as does a people's relation to their river.

The frequent lack of competence in handling the social and environmental aspects of hydroelectricity, especially in large-scale projects, has contributed to an exaggerated pessimism in many quarters regarding the intrinsic negative impacts of hydropower. Such a situation can be blamed in part on the traditional ethos of the electric utilities, which are dominated by engineering and financial priorities and are insensitive to many other issues. While utilities in industrialized countries are becoming more enterprising, the same cannot be said for utilities in developing countries and in the eastern bloc countries where most hydroelectric potential exists. The problem is exacerbated by the relative weakness of environmental regulation in these countries, where utilities are often treated as direct extensions of the state.

New guidelines concerning the environmental and social impact of development projects undoubtedly add to the cost of hydroelectricity. Such guidelines not only affect new installations, especially in countries where standards have been less rigorous [51], but they influence the viability of older plants in the industrialized countries [4]. Hydroelectricity is not alone in suffering cost pressures for these reasons, and among conventional electricity resources today, it perhaps suffers the least. In addition to cost pressures, the development of some potential sites is simply prohibited by legislation, and such legislation is expected to multiply. Eventually, technical and institutional developments may help mitigate the impact of these restrictions.

Environmental concerns have fueled interest in smaller hydroelectric plants, which for many years have been undervalued. Whereas some environmental effects are unrelated to the scale of generating capacity, many others, including the indirect impacts of construction and induced seismicity (see page 27) are scale-related. Each site must be assessed individually, however. Although in many cases small may be better, big is not always bad,[7] especially when considering the impact per kWh of output [43]. If a certain potential can be obtained only by constructing a large-scale project, then the relative impacts of non-hydroelectric alternatives must be considered.

---

6. Even a 100-megawatt dam is considered a large engineering project in such regions.

7. This chapter does not systematically compare small- versus large-scale projects, a subject that deserves a chapter of its own. If our text is weighted toward large- and intermediate-scale plants (those that generate more than 30 megawatts), it is because most plants in the countries that have greatest potential fall into this size range.

The impact of construction
The construction phase of a hydroelectric plant brings social as well as direct and indirect environmental consequences. The social consequences can powerfully influence the indirect environmental impacts, which may often be more important. Direct environmental impacts are usually relatively minor. They include water diversion, drilling, and slope alteration, as well as reservoir preparation and the creation of an infrastructure for the workforce, which can be very large. In Brazil, for example, peak employment at Tucuruí reached about 35,000.

The social consequences of a large workforce are often amplified in developing countries by the large number of workers attracted by low-skill employment opportunities. In Brazil, for example, five or more persons are attracted to a construction site for every employed worker. As a result, housing for workers is often swamped and shanty towns emerge, creating additional social and environmental problems. Public-health problems can be severely aggravated unless adequate precautions are taken, as is rarely the case. Because hydro-plants are not generally isolated undertakings, local transportation networks and related industries often experience parallel growth. The additional infrastructure stimulates regional economic development and can represent a major change in economic dynamics, especially in more isolated regions. Hydropower is widely viewed as a regional development tool.

But its impact can also be negative. The construction of the Carajás mining and metallurgical complex (including the Tucuruí dam) in Brazil's eastern Amazônia is a recent example. The program provoked explosive migration, created pockets of urban and rural poverty, destroyed the character of the local communities, and was accompanied by intense deforestation. Such indirect impacts may be the biggest preoccupation for large hydropower facilities in areas of tropical rain forest. A large part of the world's remaining hydropower potential is in tropical forest areas; in South America perhaps as much as 80 percent of the potential is in rain forests. Unfortunately, the environmental evaluation process in these countries tends to define problems too narrowly [48]. It should be noted that deforestation and social disarticulation in the humid tropics are more likely to be related to cattle raising, logging, and mining than to dam construction. Indeed, the construction of hydroelectric facilities may be part of a strategy to consolidate more sustainable regional development [49].

Land inundation and reservoir filling
Most hydro-plants require a reservoir, which can significantly affect the terrestrial ecosystem, the human population, and the river itself. The terrestrial ecosystem flooded by the reservoir is often high in both biological productivity and diversity. This is especially true of the river floodplain, which is the first land area affected by a reservoir. Floodplains provide a crucial interface between riverine and terrestrial ecosystems, playing a crucial role in the reproduction and nutrition of aquatic species. At the same time, the alluvium and water supplied by the river are critical to many terrestrial animal species, including human beings. Because

Table 7: Ratio of power output to inundated area: selected plants in Brazil[a]

Hydroplant *existing and planned*	Scale *MW installed*	Power/area ratio *kW/hectare*
Jaguari	24	2,400.0
Sapucaia	300	714.0
Xingó	5,000	588.2
Segredo	1,260	152.7
Itá	1,620	116.7
Itaipu	12,600	93.6
Belo Monte	11,000	89.8
Machadinho	1,200	45.8
Garabi	1,800	22.5
Itaparica	1,500	18.0
Tucuruí	3,900	13.9
Três Irmãos	640	9.0
Porto Primavera	1,800	8.4
Serra da Mesa	1,200	6.7
Camargos	45	6.1
Manso	210	5.4
Samuel	217	3.3
Sobradinho	1,050	2.5
Balbina	250	1.1
Historical average	–	*21.7*

a. [12, 52].

floodplains have rich alluvial soils, they offer relatively high and stable crop yields and thus play a disproportionately key role in human agriculture. Even in areas lacking intensive agriculture activity, the river's edge concentrates human activity by providing fishing opportunities, transportation, and a way to discharge waste from homes and industry. Thus in both urban as well as rural areas, populations tend to be disproportionately represented along rivers.

The area inundated by hydroelectric plants is therefore generally of higher value than the surrounding region as a whole, a point that is often forgotten, but must be considered when comparing hydropower to other renewable energy options. Land exploited for the latter should be less valuable, in both economic and ecological terms. Calculating the ratio of a hydro-plant's installed capacity to the inundated area is one rough measure of environmental impact; many hydro-plants vary enormously by this measure (see table 7). Those with ratios of less than about 5 kilowatts per hectare require more land area than most competing

Table 8: Major examples of populations displaced by reservoirs[a]

Hydroplant	Country	Population displaced
Danjiangkou	China	383,000
Aswan	Egypt	120,000
Volta	Ghana	78,000
Narmada Sardar Sarovar	India	70,000[b]
Shuikou	China	62,500[b]
Sobradinho	Brazil	50,000–60,000
Kariba	Zambia	57,000
Itaparica	Brazil	45,000
Yacyreta	Argentina–Paraguay	45,000[b]
Tucuruí	Brazil	17,000

a. [53, 55].

b. Dam is in construction or at an advanced level of planning.

renewable technologies,[8] without considering land value itself.

A low kilowatt-to-hectare ratio is a danger signal, although it does not necessarily preclude construction. Much depends on the local context and the dam's function. Large reservoirs at Sobradinho and Serra da Mesa, for example, are used not only for electricity production but to regulate river flow for electricity production downstream (see table 7). Conversely, a relatively favorable ratio does not mean that inundation is simple. In Brazil, two plants with above-average ratios, Itá and Machadinho, are paralyzed by conflict with the local populations.

Disappropriation and population resettlement

Disappropriating inundated land and compensating relocated populations represents a major political and management challenge. Historically, such problems have been underemphasized and inadequately addressed. Indeed, a World Bank review concluded that "throughout the world, involuntary resettlement has probably been the most unsatisfactory component associated with dam construction" [53]. Large numbers of people are often affected (see table 8). If the Three Gorges project in China is constructed, for example, 1,400,000 persons may be displaced [54]. The importance and complexity of the dislocation process depends not only on the numbers involved, but also on the severity of its impact on the local population [53]. Resettlement can be highly disruptive, especially in rural areas [55], effectively dismantling a way of life. Those affected may be excluded from the de-

---

8. A conservative calculation assumes a biomass yield of 15 dry tonnes per hectare per year, with 30 percent overall conversion efficiency, and an average capacity factor of 50 percent. Sugarcane cogeneration systems may eventually produce more than 10 kilowatts per hectare at a 50 percent capacity factor.

velopment benefits of the project and permanently impoverished. Therefore "all resettlement programs must be development programs as well" [53].

The costs of disappropriating land (especially high-quality farmland) and resettling a population are significant, especially if the preparatory planning is poor. At Itaparica in Brazil, for example, resettlement problems are estimated to have increased the project's cost by $250 per kilowatt, or roughly 10 percent of the total project cost. At Porto Primavera (see table 7), costs may be as high as $500 per kilowatt. In the past, such externalities were often inadequately counted. Because attempts are now being made to incorporate these costs more fully, reservoir construction is relatively more expensive. This in turn may foster projects that deemphasize river regulation and increase complementation, as discussed in the section "Hydropower in electrical systems" above. In some cases, development of certain sites may simply not be viable.

Climatic and seismic effects

The presence of a large body of water may affect local climate. As a general rule, the greater a lake's surface area, the greater its climatic influence. Measurements of meteorological elements at the standard 2 meter height indicate that a lake's influence decreases exponentially inland away from the shore. In the tropics, there is some evidence that the introduction of man-made lakes decreases convective activity, thus reducing cloud cover. In temperate regions, fog typically forms over the lake and along the shore when the temperature drops close to freezing [56]. As a result of evaporation, humidity is slightly higher in the area immediately surrounding a man-made lake than it is downwind.

It is known that big reservoirs can influence tectonic activities (see table 9), but the precise relation between the two is still under debate. In the 1960s, earthquakes were linked to either the filling of the reservoir itself or to the new lake. Based on experience with reservoirs such as the Vouglans in France, that view has now been extended to include the postformation period. The Vouglans was first filled from April 1968 to November 1969 and then partially emptied from December 1970 to March 1971. Thereafter it was rapidly filled, reaching maximum capacity in July 1971. On July 21 the first earthquake (4.5 on the Richter Scale) occurred and was soon followed by another 20 earthquakes, epicentered 5 kilometers southeast of the reservoir. No earthquake had been registered before in the region.

In 1971 the 11 million m^3 Alengani reservoir in Corsica was closed; on September 19 an earthquake measuring 2.9 on the Richter Scale occurred. When the reservoir was refilled in April 1978, seismic activity returned, remaining until the end of 1980. In 1978, a strong earthquake occurred, preceded and followed by another 150 quakes. The activity is thought to stem from the lake's emptiness for many months and subsequent fast filling [57]. In yet another case, a highly destructive earthquake coincided with the building of the Koyna dam in 1967 in India [58]. In areas of low tectonic activity, the risk of an earthquake either does not exist or is minimal. In geologically unstable areas, however, large reservoirs may not be advisable.

The issue of seismicity leads naturally to the question of dam safety. The failure of a large reservoir's dam, such as the Aswan High Dam, would be a major disaster. But these dams tend to be well monitored, and thus are more vulnerable

Table 9: Reservoir-induced changes in seismicity[a]

Dam name	Location	Height of dam *meters*	Volume of reservoir *$10^6 m^3$*	Year of impound-ing	Year of largest earth-quake	Magnitude or intensity
MAJOR INDUCED EARTHQUAKES						
Koyna	India	103	2,780	1964	1967	6.5
Kremasta	Greece	165	4,750	1965	1966	6.3
Hsinfengkiang	China	105	10,500	1959	1962	6.1
Oroville	United States	236	4,295	1968	1975	5.9
Kariba	Zimbabwe–Zambia	128	160,368	1959	1963	5.8
Hoover	United States	221	36,703	1936	1939	5.0
Marathon	Greece	63	41	1930	1938	5.0
MINOR INDUCED EARTHQUAKES						
Benmore	New Zealand	118	2,100	1965	1966	5.0
Monteynard	France	155	240	1962	1963	4.9
Kurobe	Japan	186	199	1960	1961	4.9
Bejina-Basta	Yugoslavia	89	340	1966	1967	4.5–5.0
Nurek	USSR	300	10,400	1969	1972	4.5
Clark Hill	United States	67	2,500	1952	1974	4.3
Talbingo	Australia	162	921	1971	1972	3.5
Keban	Turkey	207	31,000	1973	1974	3.5
Jocassee	United States	133	1,430	1972	1975	3.2
Vajont	Italy	261	61	1961	1963	
Grandval	France	88	292	1959	1963	V
Canalles	Spain	150	678	1960	1962	V
TRANSIENT CHANGES IN SEISMICITY						
Oued Fodda	Algeria	101	228	1932		
Camarilles	Spain	44	40	1960	1961	3.5
Piasta	Italy	93	13	1965	1966	VI–VII
Vouglans	France	130	605	1968	1971	4.5
Contra	Switzerland	220	86	1965	1965	
DECREASED ACTIVITY						
Tarbela	Pakistan	143	13,687	1974		
Flaming Gorge	United States	153	4,647	1964		
Glen Canyon	United States	216	33,305	1964		
Anderson	United States	72	110	1950		

a. [54].

to military attack (which itself is unlikely) than to sudden structural failure. Rather, it is the world's smaller dams, which do not receive such close monitoring, that are much more likely to fail. Thus, periodic relicensing of civil structures is essential.

Earth and rockfill dams are perhaps the safest man-made structures in earthquake zones because they have great inertia, flexibility, and high damping that help to absorb an earthquake's energy. The world's highest dams—Nurek (which is 300 meters high), Rogun (335 meters; under construction), and Theri (262 meters; under construction)—are in areas of considerable seismic intensity, but are not endangered. And though an earthquake struck Mexico in September 1985, the dams Villita (60 meters high) and Infiermillo (146 meters), which were close to the epicenter and subjected to peak acceleration, remained safe.

Impact on flora and fauna

The need to protect biological diversity in an inundated area cannot be overlooked. Concern is greatest at the level of the ecological zones of the river basin, although its most practical manifestation is at the project level. The intensity of concern depends on the degree to which a habitat and its component species are at risk. Not only is biodiversity threatened in regions where anthropic change is already significant, but it is also threatened in regions, including much of Amazônia, where human impact is still minimal. The black-lion tamarin, *Leontopithecus chrysopygus*, which was once endemic to large areas in Brazil's São Paulo state, exemplifies the risk to wildlife. The species is now restricted to a few small areas, including one recently flooded by the Rosana dam.

The utility has attempted to minimize species loss in Rosana by transferring most of the vertebrates to new areas. Indeed, rescuing larger animals from the rising flood is one of the most dramatic aspects of reservoir filling, and naturally captures much media attention. Unlike the mythic Noah's Ark, however, most rescue exercises are relatively futile. Animals are too often thrown into habitats that are more or less at equilibrium and so they are unable to compete with existing populations, and usually experience high rates of mortality. Better planning is needed if the program is to succeed. Mitigating such losses requires that the species and individuals to be transferred be carefully selected, and that their destinations be carefully researched.

At the same time, gradual removal of vegetation in the area to be inundated can be beneficial because it encourages the spontaneous migration of some species [58]. Much is said about creating ecological reserves in compensation for the area lost to inundation. Such a policy is correct. It must be recognized, however, that the reserves are created only in the juridical sense since they already existed. The real reason for creating compensatory reserves is to protect the future against the accumulating effects of development. The same can be said for policies to replant reservoir margins with local endemic species. But such reserves may only approximate the original habitat because edaphic conditions and the rhythm of water rise and fall along the reservoir may be quite different from those that existed previously.

With respect to the flora, transferring species may not be practical, although saving their germplasm could be an important strategy for threatened species. Saving species reemphasizes the need to ensure that the necessary minimum habitat area is preserved. In the end, the question of animal and plant habitats cannot be separated since each depends to a considerable extent on the other.

An important issue to be addressed is the fate of the vegetation to be inundated, a process that can have major implications for water quality (see page 31). The gradual clearing and logging of forests appears preferable to sudden flooding; it also enables animal populations to adjust, as already observed. In the humid tropics the use of fire during the dry season can be an effective means to partially reduce some of the biomass (leaves, grass, vines, and small stems, which are of special importance to water quality) immediately before reservoir filling. Together they represent roughly 10 percent of the total biomass [59]. The resistant ligno-cellulosic trunks and main stems can be subsequently recovered, perhaps even after inundation.

Underwater pneumatic power saws are now being tested at Tucuruí where the electric utility failed to extract the high-value timber before filling the reservoir. The technique may prove to be competitive with conventional clearing, which can be quite expensive, especially for lower-value wood. Clearing Tucuruí, for example, would have added an estimated $440 million, or 9 percent, to the project's cost [60]. Yet clearing the forest, especially in shallower areas, is necessary to reduce breeding areas for disease-carrying insects, which are now a serious problem at Tucuruí. In deeper water, wood decomposes very slowly, contributing little to water quality problems and can retain its commercial value for years.

Changing the river: water quality, sedimentation, and aquatic ecosystems

Rivers play a key role in maintaining the global ecosystem by circulating fresh water and water-borne nutrients. Yet the impact of a hydroelectric plant on a river's ecological characteristics may be considerable. Even small dams can interfere with fish migration, and large storage reservoirs in humid tropical environments can convert a well-oxygenated river into an anoxic lake, hostile to most species. Moreover, dams tend to interrupt the natural flow of the suspended particles that carry important nutrients to downstream ecosystems. The slower the flow, the greater the interruption to the ecosystem. The result can be deleterious both to the hydroplant itself and to downstream ecosystems, especially when high sediment loads are involved.

Water quality

Water quality, a problem that has always concerned hydroelectric planners, is accentuated in humid tropical regions. There are several reasons for this. First, the amount of vegetation per hectare, especially leafy biomass that degrades rapidly, tends to be greater than in temperate regions. Second, relatively high ambient

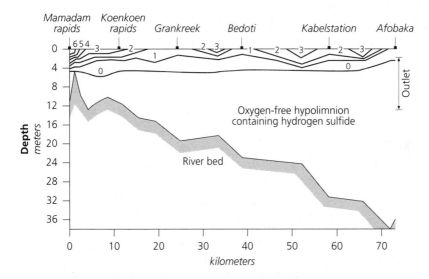

FIGURE 6: Oxygen isopleths in a length profile of the Brokopondo reservoir along the former Suriname River (during the dry season), 30 September–3 December 1965. The numbers on the figure refer to milligrams of oxygen per liter. The position of the intake gates of the hydroelectric power station was almost completely in the hypolimnion zone. Water passing through the turbines contained very little or no oxygen and caused fish mortality and oxygen deficiency over a long distance downstream [59].

temperatures favor rapid decomposition, thus increasing biological oxygen demand in the reservoir. Third, relatively constant air temperatures inhibit the thermally induced mixing that occurs in temperate latitudes. As a result, more or less permanent thermal stratification can develop, preventing oxygen exchange between the surface and deeper layers where biological decomposition takes place [59] (see figures 6 and 7 for an example of this phenomenon).

In the deeper, increasingly anaerobic layers, reduction of nitrates to ammonia and the release of inorganic phosphorus from sediments and organic matter occurs. Sulfates are reduced to foul-smelling sulfides under more strongly reducing conditions (when the redox potential is –120 to –150 millivolts), and methanogenesis occurs at –500 millivolts. Under these oxygen-free conditions, carbon is converted to methane, which is a potent greenhouse gas and therefore may have implications for global warming.[9] Phosphorus is also liberated, stimulating the surface growth of algae and other plant species such as the water hyacinth, *Eichornia crassipes*. Subsequent decay of these plants further depletes the lake of oxygen.

---

9. Hydroelectricity is relatively benign with respect to greenhouse-gas emissions. Construction of the plant releases roughly 1 percent of the CO_2 emitted per GWh by a coal-fired plant [4]. Decaying biomass in inundated areas may be a significant source of greenhouse gases, however, especially in tropical rainforests where the crop of standing biomass is so large. If the biomass were cleared and burned, CO_2 emissions might increase to 5 or 6 percent of those emitted by a coal-fired plant. If biomass were allowed to decay, methane gas, which has a greater greenhouse effect than CO_2, would be released, albeit slowly. Though the rate of release may be too slow to be significant, it must be investigated.

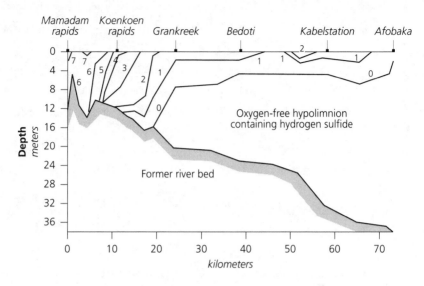

FIGURE 7: Reconstruction of oxygen isopleths in a length profile of the Brokopondo lake along the former Suriname River during the rainy season from 1–5 March 1966. The numbers on the figure at low water depths refer to milligrams of oxygen per liter [59].

The overall process, known as eutrophication, kills bottom-feeding animals, which need oxygen, although it may cause some surface-dwelling species to thrive. In addition, the growth of surface vegetation creates an ideal habitat for such human disease vectors as mosquitoes. Reducing conditions at the bottom may also release heavy metals such as mercury, which has occurred in some rivers of Amazônia, where gold mining creates heavy mercury contamination.

The most severe water quality effects may be felt downstream of the reservoir. Water passing through the turbines is generally taken from the deeper water layers and thus is highly deficient in oxygen. Its water may remain oxygen-poor for many kilometers downstream, killing fish and creating an unattractive odor [59]. The problem can be reduced by releasing oxygenated surface water over the spillway, but this may impact negatively on energy production during dry periods, exactly when environmental problems are the worst. It therefore becomes apparent that environmental issues are not limited to the design and construction phases of a hydro-plant but include its subsequent operation.

Fortunately, many of these environmental problems are not permanent. A reservoir tends to recuperate over time as the rate of biodegradation diminishes from its peak soon after reservoir filling, and as limiting nutrients, such as soluble phosphorus, are flushed downstream. Nonetheless, the hydroelectric plant may suffer from operating constraints during this period. The intensity of eutrophication that occurs tends to reflect the ratio of reservoir volume to river flow [59]. A larger relative volume implies longer turnover time, with a lower rate of renewal by incoming oxygenated water, and a greater volume of biomass relative to river flow. In contrast, reservoirs that have a relatively smaller volume-to-river flow ratio benefit from greater wet-season flushing effects and may experience less severe water quality problems.

Water quality must be considered when storage reservoirs that are large in relation to the river's flow are built in the humid tropics. Water quality may also pose problems for dams that are built in sequence. Mitigating measures, such as the clearing of biomass (especially its leafy components), the installation of multi-level water intake pipes (see figure 2), reaeration of deeper waters [59], and the use of aspirating turbines [4] exist in principle, but further investigation of these measures is needed.

Sedimentation

Dams artificially prevent the free flow of water. Although reservoirs are meant to store water, they also store the sediments carried by the river that flows into them. As flow velocity decreases, the rate of settling of suspended sediments increases. Initially, sedimentation occurs mostly in the reservoir volume below the dam's intake level, which does not impair the plant's functions. Eventually, however, without adequate control measures, sedimentation can shut down a hydroelectric plant or drastically reduce a dam's contributions to irrigation. Although the degree of sedimentation varies from basin to basin, the problem is sufficiently acute in some areas to raise questions about hydropower's long-term renewability. The issue has not yet received the attention it deserves.

The amount of sediment carried by a river is determined by the watershed, surrounding land use patterns, and the velocity and turbulence of the flow. Consider the yearly average silt content of the following rivers [61]:

River	kg/m^3
Yellow River	37.6
Colorado	16.6
Amur	2.3
Nile	1.6
Yangtze	0.4
Paraná	0.1

Rivers with high silt loads can rapidly create serious problems for hydroelectric facilities. After only four years, the Sanmen Gorge dam on the Yellow River lost 41 percent of its water storage capacity and 75 percent of its 1,000 megawatt installed capacity to sedimentation [15]. In extreme cases such as this, it is questionable whether the dam should have been built at all and surprising that planners could be so insensitive to an obvious problem.

Methods for mitigating the impact of sedimentation exist, although they have not yet been widely implemented. One method calls for installing sluice gates at the base of the dam that can be opened and closed. The gates create currents along the reservoir floor that sweep accumulated sediments downstream.

Because power is lost when the gates are operating, they are generally activated during periods of high flow. The effects on downstream ecosystems and dams must be considered. An advantage to this approach is that the flushing effect may also help oxygenate the deeper reservoir layers, thus limiting eutrophication. Such a strategy may be quite attractive, especially at certain tropical sites.

An alternative method is to pump sediment slurries over the dam crest, a technique that is best adapted to retrofit situations. The method has worked well for Brazil's 120-megawatt Mascarenhas plant, whose operations were jeopardized by sedimentation. Because the sediments are passed down the river, any dam that is downstream will have to adopt similar measures. Removing sediment from the river by conventional dredging methods or by pumping out slurry, which is dumped elsewhere, are two other options [62]. Dredging technology, however, can be prohibitively expensive (at least $1.00 per m^3) [62], as is disposal of the dredged sediments.

It is clearly important that hydroelectric systems avoid increasing sediment loads in the river, or perhaps even reduce them. To do so requires erosion-abatement strategies, which call for adoption of land use and agricultural and forestry practices in the basin itself. In places such as Costa Rica's Arenal dam, where the river basin is small, the utility itself may be able to assume responsibility. In large basins, management responsibility transcends the scope of a single utility. Nevertheless, in developing countries, utilities frequently have organizational and financial capabilities far superior to those of most government agencies and can therefore contribute significantly to erosion abatement. Many, however, lack a clear sense that participation is in their economic interest and do not understand what role they might play. Most are financially hard-pressed and are therefore reluctant to assume new, potentially open-ended commitments. Basin management is further complicated by the fact that part of the basin may be located in another country that does not benefit from the hydroelectricity being generated. Conflicts of this nature have arisen between India and Nepal, and between Egypt and Ethiopia.

Dams on rivers with high sediment loads may also create problems downstream by blocking the natural flow and deposition of sediments and their associated nutrients. The most celebrated example is the Aswan High Dam on the Nile river [63], where interruption of sedimentation and flooding has increased costs for Egyptian agriculture and promoted salinization. In the Amazon basin, if damming of white-water rivers such as the Madeira were to occur, then ecological problems might be created downstream.

Aquatic ecosystems
Dams have a significant impact on migratory fish species by preventing their migration. Fish ladders have been devised to facilitate upstream migration, but they represent only a partial solution to the problem: there are limits to the height that fish can climb, a situation that is exacerbated when many dams are built in sequence, as is often the case. Dams also interfere with the downstream movement

of smelts, which pass through the turbines in large numbers and are killed by the pressure. The Columbia river in the United States, with its series of dams, highlights some of the problems. Many subspecies of salmon, which were once abundant in the river, are now virtually extinct; overall numbers have fallen by more than 90 percent.

Moreover, many fish species require flowing water for reproduction and cannot adapt to stagnant reservoirs, even when eutrophication has not occurred (see page 31). The problem can be overcome in part with fish farms that release newly hatched fish to unimpeded tributaries. Some specialists also recommend limiting dam construction to allow for substantial stretches of free-flowing river. In the Amazon, for example, it has been proposed that at least 50 percent of a river's total extension be left as free-flowing [64, 65]. After the reservoir has existed for some number of years, it is expected that different aquatic and semi-aquatic ecosystems will develop, offering alternative habitats for some animal species.

Disturbance to aquatic ecosystems also raises the possibility that pathogens and their intermediate hosts will proliferate, especially in the tropics, leading to an increase of such fatal diseases as malaria, schistosomiasis, filariasis, and yellow fever. A common example of this problem is the rapid proliferation of water hyacinths, which provide an attractive habitat for many species of mosquitoes and water snails that harbor disease. Although some would advocate widespread use of pesticides and herbicides to combat these problems, chemicals create their own environmental and public-health problems.

Public health concerns

The impact of a hydroplant on human health during its construction and operating phases cannot be ignored. Unfortunately, few detailed epidemiological studies on the subject have been carried out, especially in developing countries, where they are most needed. Because construction leads to a sudden surge in the population, especially in remote areas, the generally precarious nutritional and hygiene conditions in developing countries are exacerbated. The situation favors the spread of diseases, including those whose microorganisms are transmitted directly from person to person (such as infantile diarrheas, hepatitis, and cholera) and those that pass through animal vectors (such as trypanosomiasis). The occupation and disturbance of the natural habitat can also provoke an increase in certain diseases, such as leishmaniasis in tropical forest areas. Health problems can also result from the resettlement of populations during the filling of the reservoir as well as during the subsequent demobilization of the labor force, when unemployment is high.

Habitat changes induced by the reservoir's filling (and by the subsequent development of irrigation systems that are often associated with hydropower) also favor the propagation of disease. In developing countries, the most important diseases associated with water-resource development are: malaria (264 million people are afflicted), schistosomiasis (200 million), lymphatic filariasis (90 million),

source development can introduce diseases into areas where they previously did not exist, as occurred with malaria at Itaipu. Although the tendency for disease vectors to proliferate can often be predicted according to the nature of the ecosystem as well as the reservoir (thus making it possible to take preventive measures), the tendency is to adopt "firefighting" strategies once health problems arise, a posture that is more expensive and less efficacious than preventive measures. Another traditional posture is to create more sanitary conditions for higher-income personnel, an elitist approach to public health that reflects a prevailing bias in most developing countries.

Incorporating social-environmental dimensions into planning

Including the social and environmental dimensions of hydroelectric plant construction and operation in the planning process represents a major challenge. Considerable changes in the planning methodology and priorities of utilities must be made, especially in the developing and eastern bloc countries. Such dimensions represent the main uncertainty regarding the ultimately exploitable fraction of hydropower.

In many countries, the environmental evaluation process (EEP) consists mainly of an environmental impact statement (EIS), which is drawn up for each project. While the EIS is better than nothing, it is still deficient in many ways. To begin with, it comes too late in the planning cycle to allow for serious modifications to the project and thus dictates a go–no-go situation. In addition, methodological problems exist, especially in developing countries. The area of impact, for example, is usually poorly defined. Finally, the EIS, although supposedly a public document, is frequently not available to the public [48].

Analysis of EEPs reveals that many impacts, both positive and negative, are inherently subjective [48, 50] and that real conflicts of interest exist between communities and utilities, as well as within communities themselves. Reducing conflict requires public participation during the planning stages. Although the principle is widely applied in industrialized countries, it still is not fully practiced in developing countries. Because hydropower has a long gestation period, dialogue must be initiated at an earlier stage than is customary. Such dialogue must focus on decisions that encompass the entire river basin (not just a single project) and must recognize that the water needed for hydropower has multiple uses.

These characteristics distinguish hydropower from most other electricity-generating technologies. Hydropower also differs from other technologies in its strong linkage to regional development and thus in its relation to state agencies as well as to the public. Cooperatively planned regional development is needed to maximize hydropower's side benefits (which can include navigation, irrigation, and watershed management), as well as to minimize its negative impacts. Unfortunately, such cooperation is notoriously difficult to achieve, especially in developing countries. It is interesting that the multilateral banks, which have played a and onchocerciasis or "river blindness" (25 million)[10] [44]. Moreover, water-re-

---

10. The total number of cases cannot be attributed only to water-resource development.

major role in financing hydroelectric projects and stimulating the management of social and environmental issues, have been averse to packages that integrate hydropower financing with other construction projects. Yet such multi-sector loans could greatly enhance cooperative endeavors between utilities and government agencies. For example, a utility might agree to help build and operate a public hospital (for its own employees as well as the public) provided that the public-health department commits itself to, say, half the cost. Unfortunately, such agreements are often broken or fail when the other participant fails to honor its commitment, allegedly for lack of funds.

It is also important that more emphasis be placed on preliminary social and environmental analyses. Such analyses are particularly important during the detailed inventory and pre-feasibility phases,[11] when sites and parameters such as the height of the dam begin to crystallize [50] but are not yet finalized. Unfortunately, because most utilities in developing countries are experiencing financial crises, inventory, and pre-feasibility analyses, which have no immediate economic benefits, have suffered heavily.

Clearly, social and environmental concerns must be given higher priority, both by the utilities themselves and by society at large. The process is likely to be slow, however, not only because such concerns are costly but also because resources, especially in developing countries, are scarce. In addition, all utilities are impeded to a certain extent by inertia. Fortunately, awareness of the costs of not incorporating these concerns into the planning process is growing.

ECONOMIC AND FINANCIAL ISSUES

Distinctive features of hydropower

Hydroelectricity has several characteristics that, taken together, make it unique in terms of economics and financing. To begin with, the technology is capital intensive, meaning that its capital costs far exceed its operating costs. Indeed, the operating costs of a hydroelectric plant are smaller than for any other large-scale electricity source. Hydropower's sensitivity to financial variables (such as discount rates or construction delays) is increased by the relatively long lead times that especially affect larger projects. Financial charges accumulated during construction can grow significantly, often accounting for 30 percent or more of direct investment.

Construction times for World Bank-financed projects in developing countries have recently been analyzed (see figure 8) [51]. Although there is a relation between plant size and construction time, the correlation is not very strong. Average construction time for a 20 megawatt plant is 3.5 years; for a 2,000 megawatt plant, it is 7.3 years. But 50 megawatt plants can take as long to build as 600

---

11. The planning cycle usually consists of the following sequence: preliminary inventory, detailed inventory, pre-feasibility, feasibility, basic project design, and executive project design.

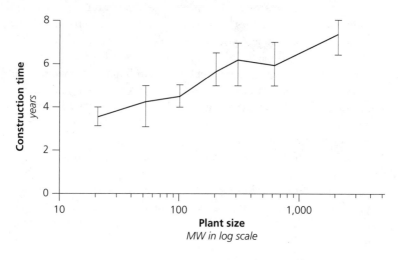

FIGURE 8: *Plant construction time in relation to scale. Points show average construction time of plants that are about the same size. Maximum and minimum times are also shown [51].*

megawatt plants. The lack of correlation reflects site variation, plant design, and the fact that different utilities are involved. A utility's financial health, for example, as well as the quality of its pre-construction planning and management, heavily influence construction time. If either is poor, serious slippage can occur in the construction schedule. The World Bank study also found that geological problems tend to be underestimated. As a general rule, most hydro projects experienced delays of 25 percent, a figure that is typical of large civil works projects. Environmental and social concerns also tend to increase licensing time. Planning problems are exacerbated by the relatively long lead time for hydro projects and the uncertainty of future electricity demand.

Another feature of hydropower is the large portion of investment costs (excluding financial costs) that go to civil engineering. Typically, 60 percent is allocated for this purpose, although the amount varies depending on the plant's design. Consequently, civil engineering contracts are commonly awarded to local firms, which often gain a major share of the project's direct investment. This is favorable from a balance-of-payments perspective, especially in developing and eastern bloc countries. Unfortunately, the amount of money involved can encourage the formation of powerful domestic lobbies to promote hydropower, sometimes inappropriately.

A further characteristic of hydropower is that the civil work of the generating plant typically has a long lifetime, measurable in centuries, when adequate maintenance occurs (unless sedimentation rates are excessive). Technical equipment can also be relatively long-lived. Therefore, once a hydro plant has been amortized, its operation contributes little to the costs of the electrical system as a whole. The savings can be significant within the context of a slow-growing electrical system, although conventional discount-rate analysis, which depreciates almost all the investment after 15 or 20 years, does not reflect the savings very well.

The final distinguishing feature of hydropower is the relatively high cost of transmitting its electricity to consumers. The location of hydro generation is fixed by local site characteristics. Initially, hydroelectric sites were developed near demand centers (contributing in some cases to their growth), but ever more distant sites, which are desirable from the perspective of low generation costs and usually the only available ones, are now being exploited. Thus a majority of the world's long-distance transmission projects are dedicated to the transmission of hydroelectricity (see table 10).

The costs of hydropower

Costs associated with hydropower are highly variable, depending on the site, conditions of financing, the degree to which environmental and social externalities are included, and the efficiency of administration (e.g. the optimization of construction, maintenance, and operational activities).

Hydropower costs are most commonly measured in terms of the cost per kilowatt installed. The measure is of limited usefulness, however, because it ignores the great differences (which may be twofold or more) in capacity factor between plants, especially those in regions or countries with distinct water-use needs and planning conventions. At some sites, dams serve multiple uses, and hydroelectricity may be a distinctly secondary function. Therefore, the most technically correct measure of a plant's cost is its contribution to reliable energy, although the concept has its difficulties. Not only is it rather abstract, but the amount of energy contributed can change substantially as the system evolves. As already observed (see the section "Hydropower in electrical systems"), a hydro-plant's reliable generation may depend on complementation with other elements of the system. Or it may benefit from the storage reservoirs of upstream dams. A plant that seems expensive may become considerably less so when such factors are weighed. Allocating benefits is clearly a tricky business, but may be resolved by considering all the hydroplants in a basin together, rather than individually. Unfortunately, most published analyses and discussions focus on individual plants.

A higher hydraulic head tends to decrease costs at individual sites, a relationship that is particularly sensitive when the head is less than 100 meters tall. Substantial cost differences arise, however, depending on whether the head is determined by the height of the dam or not. When the dam's height is almost insignificant relative to the head created (a common situation in mountainous areas), costs are cheapest. When the opposite is the case (valley dams, for example, may require extensive dikes if the topography is unfavorable), costs are increased.

Large and intermediate hydropower plants tend to be cheaper per kilowatt-installed than small ones, but the relation between size and cost is not simple [51]. The average cost of capacity in the Ivory Coast, for example, is lower than in Nigeria, despite the fact that the average unit size is one fifth as large (166 versus 983 megawatts) (see table 11). Relatively small plants can be among the cheapest, especially those that can exploit higher-head mountainous configurations. Still, much potential is found on rivers with relatively gradual fall. In order to exploit the economies of a higher head under these circumstances, a taller (and frequently larger) dam must be built. The same potential can sometimes be achieved with a series of smaller, lower-head dams, which need less total inundation. The trade-

Table 10: High-voltage transmission systems in service in the world[a]

System[b]		Total line distance *kilometers*	Rated voltage *kilovolts*	Generation source[c]
DIRECT CURRENT				
Pacific Intertie	United States	1,362	400	Complementation
Nelson River Bipole 1	Canada	890	450	Hydro
Caroba Bassa–Apolo	Mozambique–South Africa	1,414	533	Hydro
Square Butte	United States	749	250	Thermal
Nelson River Bipole 2	Canada	930	250	Hydro
CU Project	United States	710	400	Thermal
Inga Shaba	Zaire	1,780	500	Hydro
Itaipu	Brazil	783	600	Hydro
Pacific Incertie upgrade	United States	1,362	500	Complementation
Cross Channel 2	U.K.–France	72	400	Thermal
Quebec–New England	Canada–U.S.	1,000	450	Hydro
Intermountain	United States	764	500	Thermal
Rihand–Delhi	India	915	560	Thermal
ALTERNATING CURRENT				
James Bay–Churchill Falls	Canada	8,000	735	Hydro
AEP	United States	2,000	765	Thermal
USSR		3,500	750	Hydro
Itaipu	Brazil	890	750	Hydro
ESCOM	South Africa	2,200	800	?
EDELCA	Venezuela	1,237	765	Hydro
Siberia–Urals	Russia	900[d]	1,200	Hydro

a. [31].

b. Systems are listed in chronological order.

c. Refers to the principal generation source of electricity being transmitted. "Thermal" here can include nuclear (specifically in the Cross Channel system). "Complementation" refers to a two-way exchange of hydro and thermal (Pacific Intertie).

d. Total distance will be 2,300 kilometers.

Table 11: Unit costs for best potential hydroelectric projects in West Africa[a]

Region	Unit cost[b] 1986 $/kW	Average unit size megawatts	Number of projects	Capacity range megawatts
Benin	3,100	41	5	15–72
Burkina Faso	5,640	15	1	15
Ghana	1,680	169	5	51–450
Guinea	1,820	238	8	40–750
Ivory Coast	1,430	166	6	80–328
Liberia	2,740	146	5	74–214
Mali	2,600	78	8	17–300
Niger	2,550	111	2	72–150
Nigeria	1,610	983	3	400–1,950
Senegal	2,470	87	2	48–125
Sierra Leone	2,220	228	2	150–305
Togo	4,050	22	2	20–24
	1,920 *average*	*187* *average*	**49** **total**	**15–1,950** **total range**

a. [66].

b. Excludes transmission costs and interest during construction.

off between the economics of higher head and less inundation may favor smaller dams in areas where inundation is costly—depending on the nature of the sites that are available.

The distance of the site from its consuming center can greatly increase costs, as already noted, but there are strong economies of scale in transmission. Indeed, the economics of long-distance transmission explain in part why Brazil plans to export hydroelectric power from the Amazon to the South–Southeast interconnected grid (a distance in excess of 2,500 kilometers) and will emphasize the largest plants first.

There is surprisingly little systematic data available on the cost of individual plants. Recently built intermediate and large plants typically vary from $750 to $2,000 per installed kilowatt (total costs) or from $1,500 to $4,000 per reliable kilowatt.[12] The range reflects differences not only in the sites and their functions but also in the costs of alternatives in different regions. In parts of West Africa, for example, many hydro projects are expensive, but so are their alternatives.

According to the World Bank [51], real costs tended to increase at a rate of 3.5 to 4 percent per year from the late 1960s to the early 1980s. During much of this period, input costs (labor, construction equipment, and materials, etc.) in-

12. The ability to generate continuously 1 kilowatt with a given statistical reliability.

creased by 3 percent a year. Without improvements in the productivity of construction, which were not analyzed, this increase would be reflected in plant costs. In fact, until about 1980, increasing input costs closely tracked plant costs per kilowatt-installed. Thereafter, other factors became more prominent as the growth of input costs slowed. Among them was the increasing cost of mitigating social and environmental impacts.

Such externalities (see below) are expected to maintain the overall rate of cost increases over the coming years. Whether they fundamentally change hydropower's competitiveness with respect to alternative technologies remains to be seen. Certainly, external costs are expected to increase the cost of conventional thermal and nuclear technologies [51], especially in developing countries, where controls have historically been less strict. In contrast, new and renewable technologies should experience less cost pressure since they produce minimal carbon dioxide emissions. Another factor affecting future costs is that the least expensive sites tend to be exploited first. This process of site depletion is most apparent in the industrialized countries. Nonetheless, in some situations a system can be expanded to make large, relatively low-cost projects viable, as occurred at the Xingó plant on the São Francisco river in northeastern Brazil, the James Bay complex in Quebec, and the Guri dam in Venezuela.

Cost estimates for hydroelectric projects in regions of greatest potential give insight into the impact of site depletion, but such information is scarce, and what exists must be treated with great care. A recent expansion plan for Brazil [12] suggests, for example, that marginal costs will increase very slowly over the next 15 to 20 years (depending on the demand for hydropower) with a subsequent sharp rise in costs (see figure 9). More surprising, it suggests that the next 30 to 40 gigawatts of firm electricity (roughly equivalent to 60,000 to 80,000 megawatts installed) can be obtained almost as cheaply as existing capacity (which in theory averages $25 to 30 per megawatt-hour [MWh]). Although marginal costs may increase relatively slowly, the situation is not as optimistic as these numbers imply because there is a systematic tendency to underestimate costs. In Brazil, utilities compete for priority for their projects and suffer few consequences if there are significant cost overruns. In addition, many estimates are simplified, and problems that increase costs, such as geological conditions, are often not discovered until the detailed design stage [51].

A higher discount rate increases total investment in a plant, raising financial charges during construction and increasing the annual capital charge per unit of investment. Consider, for example, a plant with a direct investment (excluding finance charges during construction) of $1,200 per kilowatt, a construction time of six years, a 50-year lifespan and a capacity factor of 50 percent. With a discount rate of 6 percent, the plant's annual capital costs would be from ¢2.0 to 2.2 per kWh; at 12 percent, the costs would rise from ¢4.6 to 4.8 per kWh. By comparison, the operating costs of intermediate and large hydro plants would be on the order of only ¢0.2 to 0.4 per kWh [4].

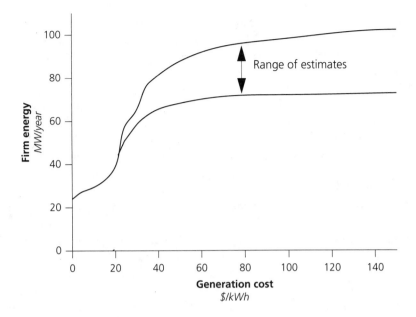

FIGURE 9: Unexploited hydroelectric potential as a function of generation cost in Brazil (1986 dollars). The potential used up to 1991 is shown at zero cost [12].

Externalities

It is widely accepted that the price of energy should include the externalities associated with its supply (see the section "Environmental and social issues," above). In many cases, the costs of minimizing hydroelectricity's negative social and environmental impacts can be clearly identified and entered into the overall costs of the project. Estimating these costs is relatively straightforward in principle but difficult in practice because the costs are often buried within general expenses. The costs also vary widely from project to project, both in absolute terms per kilowatt and as a percentage of total project costs. This makes it difficult to generalize about the total amount. A commonly mentioned figure is 10 percent, but based on historic precedent, this number should rise in the future. Measures to protect fish on the Columbia river, for example, may have reduced power output by 10 percent, and relocating Itaparica's human population added more than 15 percent to the cost of that project. When it became apparent that mitigating the social and environmental impacts of a medium-sized dam in Amazônia (Ji-Paraná) would increase costs by 20 to 30 percent, the project was postponed in favor of a natural-gas generating plant. Therefore, external costs are not only significant, they are increasing as societal standards become more stringent. Such increases affect both initial investment costs and operational costs.

Another class of negative externalities has emerged that may prove difficult to quantify. Taken together, the externalities reflect the "residual impact" that remains after the mitigating measures discussed above are included in the direct costs. How, for example, does one quantify the aesthetic loss caused by the disappearance of Glen Canyon in the United States or of Sete Quedas, the world's largest volume waterfall, which was lost when the Itaipu dam was built? Although some compensation can be made through royalty payments, these are usually not satisfactory and there is always the problem of whom to compensate. The situation is further aggravated because compensation costs are generally not included in the planning estimates.

Not all external costs are negative. Indeed, hydropower can have a positive impact on a region by stimulating the creation of a new infrastructure or by making greater use of water resources possible. Exploiting such benefits, however, requires the close cooperation of public and private agents outside the electric power sector, which presents a difficult institutional problem, especially in developing countries. Taking both negative and positive externalities (see page 36) into account thus leads not only to rising costs but renders the planning process more complex.

Financial constraints

Financial constraints have become a serious limiting factor to the growth of hydroelectricity and are likely to continue to be constrained for some years. Most of the countries with the greatest hydropower potential—the developing nations and the former Soviet Union—face severe economic problems that make investment capital scarce and/or prohibitively expensive. Given its capital intensity, hydropower is thus in a very unfavorable situation.

Most utilities will be forced to seek less capital-intensive solutions, even if in the long run some of those options may be more expensive. In addition, they will be forced to seek private non-utility financing, which is unlikely to stimulate hydropower (with the exception of small-scale projects in isolated areas) because private-sector discount rates are unfavorable to most hydropower projects. Even in the United States, it should be noted that most large hydropower facilities are government owned. The projected low price of petroleum (and fossil fuels in general) over the next few years is an additional impediment.

This financial crisis may force the electrical sector to think about supply solutions and demand management more innovatively. Until recently, almost all utilities (especially those in developing and eastern bloc countries) have been comfortable with a business-as-usual attitude. It is hoped they will reevaluate hydroelectricity and work creatively to reduce its costs, and so promote its development.

The fact that hydroelectric growth will be subdued in the near and medium term is not an indictment of its long-term potential. Although it is constrained by environmental and social concerns, many of these can be addressed and are actually small relative to those created by most other energy-supply technologies. Thus during the next 30 years, hydropower may possibly be the leading contributor to growth of the world's electricity supply. In order for that to happen, however, the structural and financial impediments to hydroelectric expansion need to be reduced. The industrialized countries must share responsibility for the process, especially in their capacity as co-protectors of the global environment. Although the process is multidimensional, attention must be directed in the short-term toward financing basic-planning initiatives that define potential, river-basin development, and individual projects. Increased attention should also be given to developing effective financing measures to minimize the negative and maximize the positive externalities of hydropower projects (despite the fact that doing so involves various government agencies, and so complicates coordination) (see page 36). Attention to the form of financing is also important not only to ensure efficient use of resources but to ensure successful negotiations with the public regarding the best use of a common good, water.

CONCLUSION

Hydroelectricity can clearly play a key role in expanding the world's electricity supply, although the magnitude of its role remains uncertain. The experience of some industrialized countries suggests that from 40 to 60 percent of the world's technical potential, or 6,000 to 9,000 TWh per year, may ultimately be exploited, representing a major increase over the 2,000 TWh presently developed. As we have emphasized, the basis for such a prognosis is fragile, and estimates may be sharply lowered by environmental or geological constraints. Eventually, electricity may have to be exchanged between countries, which will necessitate overcoming traditional political postures in many parts of the world.

It is indeed paradoxical that as concern for long-term energy strategies is growing, little is known about hydroelectric potential. The issues involved simply have not received the attention they deserve, perhaps because the scope for hydropower is limited in most industrialized nations. In the developing world and the former Soviet Union, however, the future of hydroelectricity is strategically critical not only now but for the coming decades.

ACKNOWLEDGMENTS

The authors wish to thank Maria Tereza Jorge Padua of FUNATURA, Francisco Correa of the University of São Paulo, and Jose Antonio Bulcao of Furnas for their contributions and comments during the execution of this work.

WORKS CITED

1. Raabe, I.J. 1985
   Hydro power—the design, use, and function of hydromechanical, hydraulic, and electrical equipment, VDI-Verlag, Düsseldorf, Germany.
2. McGraw-Hill. 1985
   McGraw-Hill encyclopedia of energy, Sybil P. Parker, ed., second edition, McGraw-Hill, New York.

3. Encyclopedia Britannica. 1976
   Dam, *Encyclopedia Britannica*, 15th edition. Encyclopedia Britannica Inc., Chicago.

4. Idaho National Engineering Laboratory, Oak Ridge National Laboratory, Sandia National Laboratories, and the Solar Energy Research Institute. 1990
   The potential of renewable energy—an interlaboratory white paper, SERI/TP-260-3674, DE 90000322, March.

5. U.S. Department of Energy. 1979
   Hydroelectric power evaluation, U.S. Department of Energy, DOE/FERC-0031, Washington DC, August.

6. Shea, C.P. 1988
   Renewable energy: today's contribution, tomorrow's promise, Worldwatch paper 81, Worldwatch Institute, Washington DC.

7. UNESCO. 1977
   Atlas of world water balance, explanatory text, UNESCO.

8. Boiteux, M. 1989
   Hydro: an ancient source of power for the future, *International Water Power & Dam Construction*, September 1989.

9. Bloss, W.H., Kappelmeyer, O., Koch, J., Meyer, J., and Schubert, E. 1980
   Survey of energy resources, prepared for eleventh world energy conference by the Federal Institute for Geosciences and Natural Sciences, Hanover, FRG, World Energy Conference, London.

10. World Bank. 1989
   The future role of hydroelectric power in developing countries, Energy Series paper 15, Washington DC, April.

11. International Water Power & Dam Construction. 1989
   The world's hydro resources, *International Water Power & Dam Construction*, September.

12. Eletrobrás. 1987
   Plano nacional de energia eletrica 1987/2010, Relatório Geral, Rio de Janeiro, Brazil.

13. Hodge, A.T. 1990
   A Roman factory, *Scientific American*, November.

14. Barrows, H.K. 1943
   Water power engineering, third edition, McGraw-Hill, New York.

15. Deudney, D. 1981
   Rivers of energy: the hydropower potential, Worldwatch paper 44, Worldwatch Institute, Washington DC, June 1981.

16. Mermel, T.W. 1990
   The world's major dams and hydro plants, *International Water Power & Dam Construction*, May.

17. International Commission on Large Dams. 1988
   World register of dams—1988 updating, International Commission on Large Dams, Paris.

18. Creager, W.P. and Justin, J.D. 1927
   Hydro-electric handbook, John Wiley & Sons, Inc., New York.

19. Bartle, A. 1987
   Uprating and refurbishment: our conference reviews the technology, *International Water Power & Dam Construction*, October.

20. International Water Power & Dam Construction. 1989
   The increasing importance of power plant refurbishment, *International Water Power & Dam Construction*, October.

21. Taylor, J. 1988
    Extending the operating life of hydro equipment, *International Water Power & Dam Construction*, October.

22. Bachofner, P. 1988
    The new Lebring scheme replaces the oldest plant on the Mur, *International Water Power & Dam Construction*, October.

23. Sins, G.P. 1989
    Reviewing the role of aging hydro-electric plants—a multidisciplinary approach, *International Water Power & Dam Construction*, October,.

24. Obero, B.R. and Naidy, B.S.K. 1988
    Uprating and refurbishing hydro plants in India, *International Water Power & Dam Construction*, October.

25. Dunstan, M.R.H. 1990
    A review of roller compacted concrete dams in the 1980's, *International Water Power & Dam Construction*, May.

26. Serafim, J.L. 1989
    Uma nota sobre a história dos materiais utilizados em barragens, Concreto massa no Brasil, Memória Técnica, 1. Registro Histórico, CBGB, Eletrobrás, IBRACON, Rio de Janeiro, Brazil.

27. Giroud, J.P. 1987
    A comment by the President of the International Geotextile Society, *International Water Power & Dam Construction*, 39:15.

28. Senbenelli, P. 1987
    Geosynthetics in dam construction: progress and prospects, *International Water Power & Dam Construction*, March.

29. Dumont, U. 1989
    The use of inflatable weirs for water level regulation, *International Water Power & Dam Construction*, October 1989.

30. Fujisawa, M. 1989
    Inflatable gates for raising the crest of dams, *International Water Power & Dam Construction*, October.

31. Frontin, S.O., Mosse, A., and Porangaba, H.D. 1988
    Advanced electric power transmission technologies, paper presented at ATAS VI: New Energy Technologies for Developing Countries, Suzdal, USSR, 17–21 October.

32. Fritz, J.J. 1989
    Small and mini hydropower systems—resource assessment and project feasibility, McGraw-Hill, New York.

33. Warnick, C.C. 1984
    Hydropower engineering, Prentice-Hall, Inc., Englewood Cliffs, New Jersey.

34. Xiong, S. 1990
    Small hydro development in China: achievements and prospects, *International Water Power & Dam Construction*, October.

35. Mikhailov, L., Feldman, B., and Linjuchev, V. 1990
    Small and micro hydro in the USSR, *International Water Power & Dam Construction*, October.

36. Monition, L., Lenir, M., and Roux, J. 1981
    Les microcentrales hydroelectriques, Masson S.A., Paris.

37. Control Electricity Authority. 1982
    Guidelines for small hydro, New Delhi, India.

38. Small hydropower survey. 1991
    Small hydropower survey, *International Water Power & Dam Construction*, May 1991.

39. Bartle, A. 1990
    Large potential for small hydro, *International Water Power & Dam Construction*, Small Power.

40. Small hydro power. 1990
    The world's small hydro power, *International Water Power & Dam Construction*, 1990.

41. International Water Power & Dam Construction. 1991
    Small hydro 1990—part one, *International Water Power & Dam Construction*, February.

42. Besant-Jones, J.
    The future of hydropower in developing countries, World Bank, Washington DC.

43. United Nations. 1981
    Report of the technical panel on hydropower on its second session, United Nations Conference on New and Renewable Sources of Energy, A/CONF.100/PC/30, United Nations, General Assembly, 21 January.

44. Douglas, T.H. 1990
    UK Conference reviews pumped-storage developments—comment, *International Water Power & Dam Construction*, April.

45. Considine, D.M. 1977
    Energy technology handbook, McGraw-Hill, New York.

46. UNESCO. 1980
    Balance hídrico mundial y recursos hidraulícos de la tierra, *Estudios e informes sobre hidrología* 25, Instituto de Hidrologia, UNESCO.

47. Consorcio Nacionel de Engenheiros Consultores/Eletronorte. 1980
    Inventário da Bacia do Rio Xingu, Brasília, Brazil.

48. Magrini, A., Teixeira, M.G., and Cerqueira, M.R. 1990
    Metodologia de avaliação de impactos ambientais, *Report analysis of the implantation of large energy projects—the case of the electric sector in Brazil*, part 5, Ford Foundation, Rio de Janeiro, Brazil.

49. Moreira, J.R. and Poole, A.D. 1990
    Alternativas energéticas e Amazônia, *Report analysis of the implantation of large energy projects—the case of the electric sector in Brazil*, part 4, Ford Foundation, Rio de Janeiro, Brazil.

50. SRL Projetos/Fundacao Instituto de Pesquisas Economicas. 1989
    Inserção regional de empreendimentos do setor eletrico—relatório final, for Eletrobrás, São Paulo, Brazil.

51. World Bank. 1990
    Understanding the cost and schedules of World Bank supported hydroelectric projects, Energy Series paper 31, World Bank, Washington DC, July.

52. Eletrobrás. 1990
    Plano diretor de meio ambiente do setor eletrico, Rio de Janeiro, Brazil.

53. Cernea, M.M. 1988
    Involuntary resettlement in development projects—policy guidelines in World Bank-financed projects, World Bank technical paper 80, Washington DC.

54. Goldsmith, E. and Hildeyard, N. 1984
    The social and environmental effects of large dams, Sierra Club, San Francisco.

55. Sigaud, L., et al. 1990
Avaliação de aspectos sociais da produção de energia hidrelétrica—os camponeses e as grandes barragens, *Report analysis of the implantation of large energy projects—the case of the electric sector in Brazil*, part 5, Ford Foundation, Rio de Janeiro, Brazil.

56. Landsberg, H.E. 1979
The effects of man's activities on climate, in M.K. Biswas, and A.K. Biswas, eds., *Food, Climate and Man*, 187–236, John Wiley and Sons, New York.

57. Rothé, S.P. 1973
Summary: geophysics report on man-made lakes in W.C. Achermann, et al., eds., *Manmade lakes*, American Geophysical Union, Washington DC.

58. Goodland, R. 1978
Environmental assessment of the Turucuí projects, Eletronorte, Brasília, Brazil.

59. Garzon, C.E. 1984
Water quality in hydroelectric projects—considerations for planning in tropical forest regions, World Bank technical paper 20, Washington DC.

60. Goldemberg, J. and Pádua, M.T. 1985
Impacto ecológico negativo das usinas hidrelétricas e usinas nucleares, *São Paulo Energia*, 2(16), CESP, São Paulo, Brazil.

61. Biswas, A.K. 1981.
Hydroelectric energy, in B. El-Hinnawi, and A.K. Biswas, eds., *Renewable sources of energy and the environment*, Tycooly International Publishing, Dublin.

62. Biswas, A.K. 1979
Water: a perspective on global issues and polities, *Journal of Water Resources*, planning and management division, American Society of Civil Engineers, proceedings paper 14815, 105:205–222.

63. Biswas, A.K. 1980
Environment and water development in the third world, *Journal of Water Resources*, planning and management division, American Society of Civil Engineers, proceedings paper 15295, 106:319–332.

64. Petrere, M. 1990
Alternativas para o desenvolvimento da Amazônia—a pesca e psicultura, proceedings of the meeting "Alternativas para o desenvolvimento da Amazônia," Brasília, Brazil, 19–20 September.

65. Petrere, M. 1990
As comunidades humanas ribeirinhas da Amazônia e suas transformações sociais, proceedings of the meeting "Populações humanas, rios e mares da Amazônia," Belém, Brazil, 6–9 July.

66. Lazenby, J.B.C. and Jones, P.M.S. 1987
Hydroelectricity in West Africa: its future role, *Energy Policy*, 441–455, October.

The figures in chapter 19 are incorrect. The correct figures are provided on these pages.

ERRATA

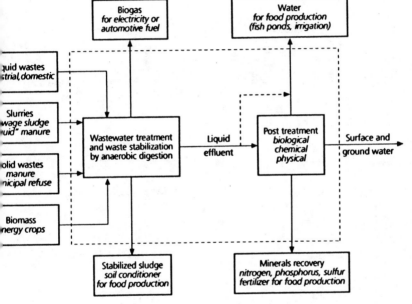

...ure 1: Integrated energy carrier, production, and environmental protection ...em based on anaerobic digestion and post treatment processes. End uses for ...products are also shown.

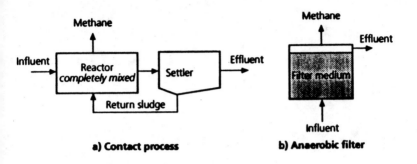

a) Contact process

b) Anaerobic filter

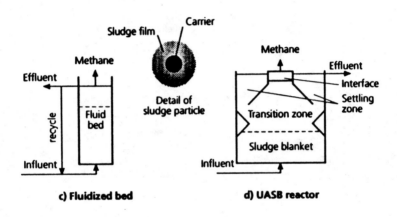

c) Fluidized bed

d) UASB reactor

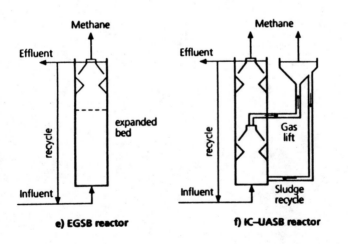

e) EGSB reactor

f) IC–UASB reactor

Figure 2: Schematic diagrams of some high-rate anaerobic wastewater treatment processes.

Alcohol	Bagasse	
	Used	Excess
38%	25%	25%

Current practice　　　　　　**a**

Energy production efficiency: 38 percent
Products per tonne of cane
Alcohol: 70 liters
Bagasse: 150 kilograms (20 percent solid)

Alcohol	Bagasse	
	Used	Excess
38%	25%	25%

With vinasse treatment　　　　**b**

Energy production efficiency: 48 percent
Products per tonne of cane
Alcohol: 70 liters
Bagasse: 150 kilograms (20 percent solid)
Methane: 7.7 kilograms
large increase in crop yield through vinasse use

Methane	Bagasse	
	Used	Excess
48%	10%	40%

Direct juice digestion　　　　**c**

Energy production efficiency: 48 percent
Products per tonne of cane
Bagasse: 240 kilograms (20 percent solid)
Methane: 35 kilograms
large increase in crop yield through vinasse use
no equipment for fermentation, centrifuge, or distillation

Methane	Bagasse
60%	38%

Direct cane digestion　　　　**d**

Energy production efficiency: 48 percent
Products per tonne of cane
Bagasse: 225 kilograms
Methane: 46 kilograms
large increase in crop yield through vinasse use
no equipment for fermentation, centrifuge, or distillation

re 4: Different options for energy conversion from energy crops. Percentage figures
t different energy content.

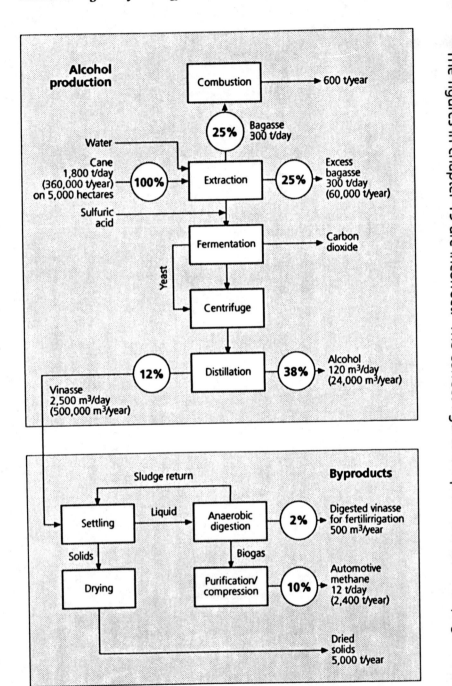

Figure 3: Flow chart of the production of alcohol at a medium-sized distillery in the northeast of Brazil. Percentage figures reflect energy content.

3

WIND ENERGY: TECHNOLOGY AND ECONOMICS

ALFRED J. CAVALLO
SUSAN M. HOCK
DON R. SMITH

Results from field operation of more than 17,800 wind turbines in Denmark and California during the past 10 years demonstrate that the present generation of wind turbine technology has been thoroughly tested and proven. Reliability is now satisfactory; in addition, wind farm operation and maintenance procedures have been mastered. Unit size has increased by a factor of 10 during the past decade: wind turbines rated at 0.5 megawatts are now available commercially from several manufacturers. Moreover, advances in wind turbine technology in the next 20 years, such as advanced materials for airfoils and transmissions, better controls and operating strategies, and improved high power-handling electronics, will substantially reduce capital costs as well as operation and maintenance costs. In areas with good wind resources [450 watts per square meter (m^2) wind power density at hub height], wind turbines now generate electricity at a cost of \$0.053 per kilowatt-hour (kWh) (6 percent interest, all taxes neglected). With a mature wind turbine technology, the cost is expected to decline to \$0.029 per kWh, rendering wind-generated electricity fully competitive with electricity from coal-fired generating stations. This development will have a profound impact on energy production industries all over the world.

INTRODUCTION

During the past 20 years outstanding progress has been made in the technology used to convert wind energy to electrical energy. More than 15,000 wind turbines in California and 2,800 in Denmark have been integrated into existing utility grids and are routinely operated in conjunction with conventional sources such as hydroelectric, fossil-fuel fired, and nuclear generating stations. Most important of all, in California the cost of wind-generated electricity has decreased substantially during the last several years (see figure 1), to the point where it is now an attractive contender among new generating options [1]. Installed capital costs have also dropped sharply, and wind turbine use has improved considerably (see figure 2), indicating that manufacturing techniques and wind farm operating methods are maturing rapidly. California and Denmark now produce 1.1 and 2.5

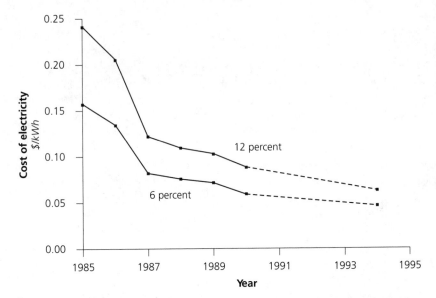

FIGURE 1: *Declining cost of electricity from wind turbines in California; data for 1994 are estimates [1, 2]. (a.) The average wind energy density in this region is > 450 watts per m², which is typical of the California mountain passes. (b.) Operation and maintenance costs are from [2]. (c.) 1994 estimated costs for the U.S. Windpower 33M-VS variable-speed turbine from the California Energy Commission, based on public filings by U.S. Windpower. Operation and maintenance plus land rental costs are quoted as $0.014 per kWh by U.S. Windpower. As field experience accumulates it is expected that operation and maintenance costs will drop below the present industry average of $0.01 per kWh. An important advantage of variable-speed operation is a reduction in stress and fatigue of turbine components (rotor, transmission, generator, and tower), which will lead to lower maintenance costs.*

percent, respectively, of their average electricity consumption from wind turbines, and Denmark has a stated goal that wind turbines should supply 10 percent of its electricity by the year 2005. Based on the extensive and encouraging experience of these two regions, it appears that wind-generated electricity is now the most economically competitive of the new solar-electric technologies.

Moreover, although the wind resources of California and Denmark are quite good, they are by no means unique. Similar or even higher quality resources are available in many other regions of the world (see chapter 4, *Wind energy: resources, systems, and regional strategies*). For example, a wind-generating electric potential of several times present world electricity consumption has been estimated [3] on the basis of a study for the World Meteorological Organization by Pacific Northwest Laboratories [4]. And in the United States, with reasonable assumptions and land use restrictions for wind turbine siting,[1] the wind electric potential of two

---

1. At sites where the windpower density is greater than 300 watts per m² at 50 meters elevation. Lands excluded from wind farm development include 100 percent of environmentally sensitive and urban lands, 50 percent of forest lands, 30 percent of agricultural lands, and 10 percent of rangelands.

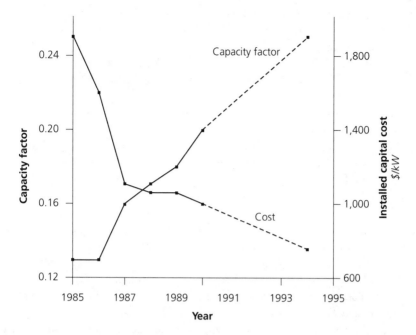

FIGURE 2: Wind turbine use, as indicated by the capacity factor, which is the ratio of the annual average power output to the rated power output (see the section "Power output" on page 10), and installed capital cost for wind turbines in California. The 1994 estimated cost and capacity factor for U.S. Windpower 33M-VS variable-speed turbine (see the section "Variable-speed wind turbines" on page 19) are based on public filings by U.S. Windpower. The capacity factor for 1990 is an average for all wind farms in California. Excluding obsolete machines in the Altamont Pass, the capacity factor for wind farms in this area is 0.24. The capacity factor for San Gorgonio Pass is 0.24 and for Tehachapi Pass is 0.19 [6].

states, North Dakota and South Dakota, is estimated to be 80 percent of present United States electricity consumption [5]. Clearly, wind-generated electricity will not be limited by the physical availability of wind resources.

Yet the potential of wind energy to contribute to a better standard of living in the modern world may come as somewhat of a surprise, perhaps due to its absence from contemporary industrialized societies. There is, of course, a long history of skillful exploitation of wind energy in many recent and ancient civilizations. For example, simple windmills were used in China several hundred years before the common era (BCE) to pump water, and as early as 200 BCE, vertical-axis windmills were employed to grind grain in Persia and the Middle East. Windmills were subsequently introduced to Europe in the 11th century by merchants and returning veterans of the Crusades [7]. There they were improved on first by the Dutch, and later by the English [8–10]. The maximum use of windmills in preindustrial Europe occurred in the 18th century, when there were more than 10,000 in the Netherlands alone; they were used to grind grain, pump water, and saw wood. Eventually, the mills, which were unable to compete with the low cost, convenience, and reliability of fossil fuels, were replaced by steam engines.

In the United States, windmills are best known for their role in the development of the American West at the end of the 1800s, and they were a vital asset [11] of early settlers on the Great Plains. Wind machines supplied water for the railroads and for cattle in areas remote from streams and springs, and were used by farmers to obtain water for small-scale irrigation and as stand-alone systems to generate electricity for their isolated homesteads [12].

These successful applications, however, offer little if any guidance to the role wind energy will have in the 21st century. In fact, the use of wind turbines as stand-alone systems may give a wrong impression of the way these machines will be used; independent wind machines are quite different from large arrays of wind turbines coupled to a utility grid that is also supplied by other types of generators (e.g., peaking, load, following, and baseload fossil-fuel power plants, and other intermittent renewable sources). In contrast to wind turbines that are isolated from the utility grid and thus require an auxiliary backup (rechargeable batteries in small systems, diesel engines in multikilowatt systems), which is both expensive and inconvenient, large numbers of wind turbines can be integrated into a utility grid with little or no need for storage. (Integrating intermittent renewable energy into utility grids creates other challenges—see chapter 23, *Utility strategies for using renewables.*)

There is no fundamental reason why wind turbines cannot be a major part of any utility grid where there are good wind resources, provided they can compete with more familiar systems on the basis of cost, reliability, and public acceptance. Although wind energy technology demands careful attention to scientific and engineering detail, it is well within the capabilities of most if not all countries and is becoming economically competitive in many regions around the world.

WIND ENERGY TECHNOLOGY

Basic considerations
The energy flux, or wind power density, of a stream of air of density ρ moving with velocity v is given by:[2]

$$P_w = \frac{\rho v^3}{2} \text{ watts per m}^2 \tag{1}$$

Not all of the wind power density is available for useful work; the maximum power [13] that can be extracted from a wind stream is $^{16}/_{27} \times P_w = 0.593 \times P_w$; this quantity is referred to as the Betz limit [14].

Because wind power density varies as the cube of the wind velocity, a wind

---

2. The kinetic energy, U, of a sample of air of volume, $A(\delta x)$, and density, ρ, moving with velocity, v, where

turbine must be able to function over very large variations in P_w to accommodate typical variations in wind speed. For example, if an area has a characteristic average wind velocity (v_{avg}) for wind speeds of $0.5 v_{avg}$ the available power density is only one-eighth that at v_{avg}, while at $2 v_{avg}$ the power density is eight times that at v_{avg}. In other words, wind velocities that are less than the average yield little useful power, while velocities much above the average can overstress turbine components. Thus, the technical challenge is to design a wind turbine that can function efficiently and reliably over the large variation in P_w, despite extremes of weather, with a minimum amount of maintenance for as low an initial capital cost as possible. Developing such a system has proved to be a demanding task.

Wind speed in any given region is not constant but varies over periods of seconds, hours (diurnal variation), days, and months (seasonal variation). (See chapter 23 for a discussion of the impact of these variations on the integration of wind-generated electricity to a utility grid.) Large changes in wind speed can be encountered by a wind turbine (see figure 3). Fluctuations in velocity that occur over seconds or minutes are referred to as turbulence and can cause fatigue and failure of wind turbine components (blades, transmissions, and generators).

To determine wind power density, wind velocity is usually averaged over a one-hour interval; the frequency at which different wind velocities occur is described by a wind speed frequency distribution, $f(v)$, which can vary diurnally and seasonally. If information on the frequency distribution is not available, a Rayleigh distribution is often assumed.[3]

---

2. (cont.) A is a unit area perpendicular to the wind stream and δx is parallel to the wind stream, is:

$$U = \frac{\rho A (\delta x) \, v^2}{2}$$

The energy flux, P_w, or wind energy density, is given by the time rate of change of U/A:

$$P_w = \frac{dU}{dt} \times \frac{1}{A} = \frac{\rho}{2} \left(\frac{\delta x}{\delta t} \right) v^2 = \frac{\rho v^3}{2}$$

The density of air must be calculated for the temperature, T, and pressure, P, at the location of the wind turbine as follows:

$$\rho = \frac{P}{R \, T}$$

where R is the gas constant. This correction can be substantial for summertime high altitude locations, relative to standard conditions (normally 15°C and 1 atm). For example, at Medicine Bow, Wyoming (altitude 2,000 meters), wind power density is 21 percent less than at sea level due to the higher elevation. At a temperature of 30°C, there is an additional 5 percent decrease in energy flux.

3. The Rayleigh probability density function (figure 4) has the form:

$$f(v) = \frac{\pi v}{2} (v_{avg})^{-2} \times \exp\left(-0.25\pi \left(\frac{v}{v_{avg}} \right)^2 \right)$$

For this function, $(v^3)_{avg} = 6/\pi \, (v_{avg})^3$, which illustrates the importance of high wind speeds in contributing to the average wind power density.

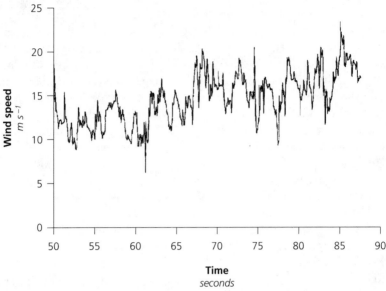

FIGURE 3: Wind speed as a function of time measured near the center of an array of 600 wind turbines in the San Gorgonio Pass, California. Measurements were made with a sonic anemometer with a 10-hertz frequency response. Turbines are spaced 2 rotor diameters (D) apart in rows 6D apart and perpendicular to the prevailing wind. Average wind velocity is 13.3 meters per second, and the standard deviation is 3.85 meters per second. Turbulence levels are characterized by the turbulence intensity, which is the ratio of the wind speed standard deviation to average wind speed; for this site the turbulence intensity is 0.29. A site with an intensity greater than 0.5 is considered too turbulent for a wind turbine [51].

Additionally, wind speed and frequency distribution may vary with elevation. Usually, wind measurements are made at a single elevation, often near 10 meters, which is very different from the possible hub height (25 to 50 meters) of modern wind turbines. To extrapolate these data to the required height, it is often assumed that wind speed increases as the one-seventh power of the elevation.[4] These rules were developed from a synthesis of many data sets and should be applied with this limitation in mind.

---

3. (cont.) This is a special case of the Weibull distribution, which fits a wide variety of wind speed data from many different locations. The Weibull function is a two-parameter probability density function of the form:

$$f(v) = \frac{b}{v_c} \cdot \left(\frac{v}{v_c}\right)^{b-1} \cdot \exp\left(\frac{-v}{v_c}\right)^b \quad (b > 1, v \geq 0, v_c > 0)$$

Here v_c is the scale parameter and b is the shape parameter. For b close to 1, the probability density as a function of velocity is relatively flat: this describes a wind regime that is quite variable. For $b > 2$, the probability density becomes more peaked and so describes a wind regime where the wind speed is relatively constant. Weibull density functions with different shape parameters can have the same v_{avg} but quite different $(v^3)_{avg}$ and wind power densities.

4. The increase of wind velocity with elevation, h, (above ground level) is usually termed wind shear; it is in general a function of surface roughness, wind speed, and atmospheric stability. Based on data from many

The above assumptions are critical to estimates of wind energy potential. The one-seventh power rule means that for a Rayleigh distribution, the wind power density at 50 meter elevations is twice that at 10 meter elevations. (Wind turbines mounted on tall towers will be able to take advantage of this favorable situation.) These assumptions form the basis, for example, for estimates [16] of the wind energy potential of the United States; they are used because detailed site-specific data are lacking.

In practice, wind speed distribution as well as the wind speed will differ at higher elevations from that at 10 meters: winds are usually steadier at higher elevations. Also, over large regions of the North American Great Plains the presence of strong night winds at higher elevations (a nocturnal jet [17]) substantially enhances the wind-electric potential of that region. Clearly, it is essential to measure $f(v)$ for at least one year at several different elevations at a given site in order to predict with confidence energy production and turbulence levels at that site.

Wind frequency distributions measured at the Altamont Pass in California and Bushland, Texas, are shown in figure 4. The average wind velocity at this Altamont Pass site is about 6.4 meters per second. Winds with three times this velocity, and thus 27 times the power density, are encountered one-hundredth as often as winds at the average velocity. The additional cost and weight of a turbine needed to capture all this energy at these very high wind speeds are not justified by the value of the extra energy obtained. To optimize energy capture as a function of turbine cost, turbines are designed to limit the energy captured above the rated wind velocity. (see the section "Wind turbine subsystems") At very high velocities, usually above 25 meters per second, the turbine must be stopped completely to protect it from damage.

The average energy flux at 10-meters elevation computed from the data shown in figure 4 is about 380 watts per m^2 in the Altamont Pass[5] (this is somewhat below average for this area [18]), while at Bushland it is about 330 watts per m^2. A wind power density of 330 watts per m^2 is typical of many areas of the world, including the northern coast of Europe, the United Kingdom, and Ireland (see chapter 4 for a more detailed discussion of world wind resources).

---

4. (cont.) locations, for areas of low surface roughness this is often approximated by:

$$v(h_2) \;=\; v(h_1) \times \left(\frac{h_2}{h_1}\right)^{1/7}$$

The European Wind Atlas [19] uses a number of methodologies to estimate wind shear, with results that are roughly equivalent to the one-seventh rule for seacoast and open-plain sites. However, there are important exceptions to these formulations. For example, the wind shear at the Altamont Pass is zero and negative at Solano County, California.

5. The wind power density can vary by a factor of three or more depending on the location in the Altamont Pass. The best sites are on the ridge lines, but there are many others that are not as well exposed to the wind. Unfortunately, data from only the best locations were used to estimate the average energy production when wind farms were first promoted in the early 1980s. The result was a shortfall by a factor of two in average annual energy output. This shortfall serves to underscore the need for detailed wind measurements above and below hub height at potential wind turbine sites.

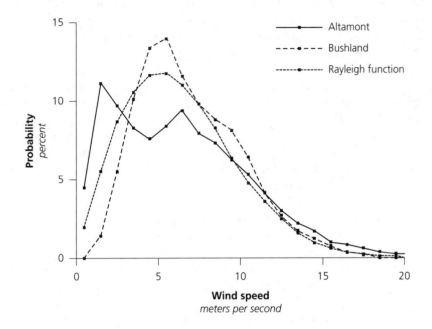

FIGURE 4: *The wind-speed probability density function at 10 meters elevation for Altamont, California, and Bushland, Texas (v_{avg} [Altamont] = 6.4 meters per second, v_{avg} [Bushland] = 7.0 meters per second). A Rayleigh probability density function for v_{avg} = 6.4 meters per second is shown for comparison [15]. The distribution at Bushland has a shape typical of that at a Great Plains site. Here the wind speed varies with elevation to the one-seventh power during the day and to the one-fourth power at night, indicating the presence of strong night winds at higher elevations (a nocturnal jet). The average wind power density is greater than 750 watts per m^2 at 50 meters elevation, indicating an excellent site. The distribution at Altamont has relative maxima at 1.5 meters per second and 6.5 meters per second, reflecting the unusual nature of these winds. They are driven by the temperature differential between the hot California Central Valley and the cool Pacific Ocean; the air in the Central Valley is heated and rises upward to be replaced by sea air flowing through an opening in the mountain range that separates central California from the coast. Wind speed is twice as high in the summer as in the winter and much higher in the early evening than in the late morning, following the temperature rise and fall in the Central Valley. These substantial periods of low wind give rise to the double maxima in the distribution. In addition, wind speed is independent of height in Altamont Pass, which is also a reflection of the driving mechanism.*

For comparison, a calculated Rayleigh probability density function for the case of v_{avg} = 6.4 meters per second is shown. The wind power density for this distribution is 305 watts per m^2, compared to 380 watts per m^2 computed from the measured Altamont data. The Rayleigh function is not a good fit to the Altamont data with the same average velocity, another indication of the necessity of making detailed wind speed measurements.

At this power density (350 watts per m^2) the area needed to intercept an average power of 2.1 gigawatts, which is the average thermal input for a typical 1–gigawatt-electric (GW$_e$) nuclear power plant[6], is 6 km^2; this would require wind turbines to be deployed over a land area of 2,100 km^2, which may seem large [20]. However, a significant advantage of wind energy systems is that at worst 5 percent, but typically less than 1 percent, of the land area on which wind turbines are deployed is needed for foundations, access roads, and electrical substations. Apart from the visual impact, the land itself is almost completely undisturbed and can continue to be used as rangeland, farmland, or for some other purpose, while royalties from energy production significantly enhance land values.

The emergence of wind electricity as the least costly solar-electric technology is due to two factors. The first is the higher average wind energy flux (450 watts per m^2 on average in the Altamont Pass) compared with the average daily solar-thermal flux (250 watts per m^2 in central California). The second is the higher conversion efficiency of wind turbines (25 percent) compared with solar-thermal electric power plants (15 percent projected, see chapter 5, *Solar-thermal electric technology*). The linear kinetic energy of the wind stream can be readily converted to the rotational kinetic energy needed to turn an electrical generator. The net result is that wind turbines can generate about three times as much electricity per m^2 of collector area as solar-thermal systems.

Aerodynamics

Modern wind turbines extract energy from the wind stream by transforming the wind's linear kinetic energy to the rotational motion needed to turn an electrical generator. This change is accomplished by a rotor, which has one, two, or three blades or airfoils attached to a hub; wind flowing over the surfaces of these airfoils generates the forces that cause the rotor to turn.

Air flowing smoothly over an airfoil (called laminar flow) produces two forces: lift, which acts perpendicular to the flow, and drag, which operates in the direction of the flow (see figure 5). If the flow becomes unattached, the lift is reduced and the airfoil is said to stall. Both lift and drag are proportional to the density of the air, the area of the airfoil, and the square of the wind speed for laminar flow, and are maximized at a single value of the angle of attack γ (the angle between the relative wind velocity and the chord line). (The vectorial addition of the wind velocity and the airfoil velocity is usually termed the relative velocity.) Since blade velocity increases with distance along the airfoil, the angle of attack must also change along the airfoil, i.e., the airfoil must be twisted to obtain maximum efficiency.

---

6. A thermal unit with a capacity of 1 GW$_e$ and an equivalent availability probability of 0.7 has an average electrical output of 0.7 GW$_e$ and a thermal input of 2.1 gigawatts-thermal (GW$_t$). For comparison, the installed rated capacity of wind farms in California in 1991 is about 1.5 GW$_e$, which, for a capacity factor of 0.24, has an average electrical output of 0.36 GW$_e$.

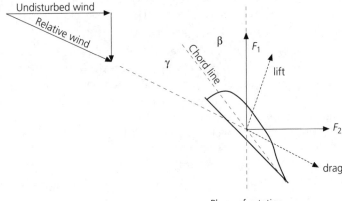

FIGURE 5: *Lift, drag, angle of attack, γ, and pitch angle, β, of a wind-turbine airfoil. The force* F_1 *is in the direction of motion of the airfoil; the force* F_2 *is referred to as the rotor thrust. The velocity of the end of the airfoil is typically 4 to 8 times the wind velocity and the angle of attack is typically less than 20°. Thus the diagram is indicative of a position far from the airfoil tip.*

Lift and drag forces are respectively perpendicular and parallel to the wind velocity as seen from the moving rotor and can be resolved into forces F_1 in the direction of airfoil translation and F_2 in the direction of the undisturbed wind. The force F_1 is available for useful work, while the tower and structural members of the wind turbine must be designed to withstand F_2 (which is termed the rotor thrust).

Power output

The fraction of power extracted from a wind stream by a wind turbine is denoted by C_p, which is the machine's coefficient of performance. The output power (P_{out}) of a wind turbine is:

$$P_{out} = C_p\,(v, \omega, \beta)\,\frac{A\rho\,v^3}{2} \qquad (2)$$

where C_p is a function of the wind velocity, the angular velocity of the rotor ω, and the pitch angle β (the angle between the airfoil chord line and the direction of translation of the airfoil; see figure 5), and also the airfoil shape and number of blades. Since an airfoil has optimum values of lift and drag for one angle of attack γ, or equivalently for one value of relative wind speed, C_p for a wind turbine with fixed blades and operating at a fixed angular velocity will also have a maximum value that decreases at higher or lower wind speeds. For existing airfoils, the maximum value occurs at a ratio of blade-tip velocity to wind velocity that is between four and eight. For example, a three-blade 100 kilowatt machine (USW-56-100, made by U.S. Wind Power) that is widely used in the Altamont

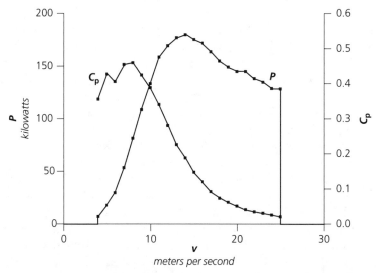

FIGURE 6: Power output P *and coefficient of performance* C_p *for a 150 kilowatt stall-regulated turbine [22]. Note the double relative maxima in the coefficient of performance, which result from an intelligent operations strategy. Since maximum airfoil efficiency is obtained at one value of γ, for maximum turbine efficiency the angular velocity of the rotor should change as the wind velocity changes. In practice such a change is difficult to achieve, and variable-speed turbines are only now being commercialized (see the section "Variable-speed wind turbines"). However, this can be approximated by two-speed operation of the rotor and the use of a generator that can be switched from 6 pole to 8 pole operation, or a high-speed and a low-speed generator. Doing so captures most of the efficiency advantages of variable-speed operation at a small increase in cost.*

Pass (v_{avg} = 7 meters per second) has a blade-tip velocity of 67 meters per second (150 mph).

There is a trade-off among several factors in choosing the number of blades: cost of the blades and transmission (the most expensive parts of a wind turbine), energy capture, and speed of rotation. A single-blade machine (with counter-weight) will have lower energy capture than a multiple-blade machine but it will rotate at high angular velocity. Because the rotor must turn a generator at 1,500 to 1,800 rpm to produce electrical power at 50 to 60 hertz, a higher rotor angular velocity permits the use of a transmission with a low gear ratio that is lighter, less expensive, and has lower losses than the transmission needed for a lower angular velocity rotor (see the subsection "Drive trains"). A three-blade machine has higher energy capture and better stability with respect to orientation in the wind stream but will have higher blade and transmission costs. Although most wind turbines now being manufactured do have three blades, it is not clear that this is the optimum choice for lowest overall cost of electricity.

The measured power output and coefficient of performance as a function of wind velocity of a commercially available wind turbine is shown in figure 6. This machine uses stall control (see the section "Wind turbine subsystems" on page 15) to limit the maximum power extracted from the wind and has a constant angular velocity (i.e., the rotation velocity of the blades is locked to the grid frequency of 50 or 60 hertz).

Power begins to be produced at 3 to 4 meters per second (the start-up velocity), and rated power[7], P_r, is produced at 10 meters per second. When wind velocities exceed 25 meters per second (the machine shutdown velocity) the turbine is stopped to protect it from damage. P_{out} increases by about a factor of 7 (from 17 to 130 kilowatts) as the wind velocity increases by a factor of 2 (from 5 to 10 meters per second), indicating a very efficient machine. This high efficiency is obtained by having two-speed rotor operation. At low wind-speed the rotor turns at lower angular velocity, while at high wind speed the angular velocity of the turbine is increased by about 50 percent, keeping the angle of attack, γ, and thus C_p of the airfoil approximately constant over this operating range. For wind speeds greater than 11 meters per second, C_p decreases rapidly due to the onset of stall (loss of lift), which limits the maximum power extracted from the wind stream.

The maximum coefficient of performance of 0.46 (attained at a wind speed of about 8 meters per second) is nearly 78 percent of the theoretical maximum coefficient of 0.593. Thus, turbine blades are already relatively efficient for a narrow range of operating conditions, although possibilities for significant improvement still exist (see the section "Future developments").

The average power output, P_{avg}, of a wind turbine for any time period of interest is determined by the power output at a given wind velocity multiplied by the probability of occurrence of that velocity summed over all possible wind velocities. The equation is written as:

$$P_{avg} = \int_v P_{out}(v) \times f(v) \, dv \qquad (3)$$

Furthermore, if P_{out} is written as $P_{out} = P_r \times g(v)$, then:

$$P_{avg} = P_r \times \int_v g(v) \times f(v) \, dv \qquad (4)$$

The integrated quantity is the ratio of the annual average power output to the rated power of the turbine. This is defined as the capacity factor (CF) and is an important parameter that will be used to calculate the cost of energy from wind turbines.

Capacity factor reduction
The ratio of achieved capacity factor to calculated capacity factor is often less than one; in 1986 the ratio for wind turbines at San Gorgonio, California, was 0.55

---

7. There is no generally accepted way of defining the rated power of a wind turbine; this term, therefore, is somewhat arbitrary. A more cumbersome but precise classification scheme is the specification of the expected energy production in a given wind regime from a given wind turbine.

for non–United States turbines and 0.3 for United States turbines [23] (the capacity factor has improved substantially since 1986, as mentioned previously).

Several effects can reduce the capacity factor, such as reduced turbine availability due to unforeseen maintenance problems, blade soiling (see the section "Future developments"), and wake effects.In a normal operating regime, maintenance should have minimal impact on turbine availability if scheduled during periods of low wind speed. High levels of turbulence during normal operation can induce high loading on turbine subsystems, including the rotor, gearbox, and tower. Such loading can cause premature component failure and high operation and maintenance costs. Turbulence at a given location can be determined by detailed measurements; sites where the turbulence is high should be avoided or use wind turbines with specially designed or strengthened components.

Reduced capacity may also be caused by the interaction between upwind and downwind turbines in a large array. Because a wind turbine extracts power from the wind stream, wind power density behind the turbine is decreased. Power density is gradually restored to its unperturbed level by the diffusion of energy downward from the wind stream above the array. The disturbed region behind the turbine is known as the wake; both the increased turbulence and reduced power density in the wake can degrade the performance of downwind turbines (i.e., turbines located at the center of an array relative to turbines on the array edge). For this reason, wind turbines cannot be close together.

In the California mountain passes [24], where winds are essentially unidirectional, machines are usually spaced 2.5 rotor diameters apart in rows perpendicular to the prevailing winds, with a row spacing of 8 rotor diameters. Wake effects have clearly caused problems at some wind farms, and there have been systematic field studies to measure these effects in detail [25]. Wake effects in large arrays have been studied theoretically [26] and with wind tunnel models [27]. Results indicate that for machines spaced 10 rotor diameters apart, wind velocity at the center of a large array is reduced by about a factor of 0.8 compared with the unperturbed velocity for wind turbines operating at the Betz limit. In practice, wind turbines will operate below the Betz limit, and the reduction in the average output of an array will be much smaller than would be expected from this result. Array losses are discussed further in chapter 4.

Modern wind turbines

Recent history of wind turbine development
There are two fundamentally different types of wind turbines (see figure 7). The first is the horizontal axis wind turbine (HAWT), which has the axis of rotation of its rotor parallel to the wind stream; the second is the vertical axis wind turbine (VAWT), which has the axis of rotation of its rotor perpendicular to the windstream.

Although the HAWT has long been used for small-scale applications, such as water pumping and nonutility electricity generation (see introduction), its devel-

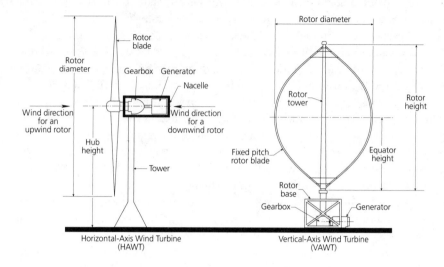

FIGURE 7: Two basic wind turbine configurations are shown: the horizontal axis wind turbine (HAWT) and the vertical axis wind turbine (VAWT). Upwind and downwind operation of the HAWT is indicated; most modern machines operate upwind to avoid shadowing of the blade by the tower, which can generate objectionable noise levels and increase stress on the blades. In practice, the hub height is approximately equal to the rotor diameter. The VAWT has its gearbox and generator at ground level, which simplifies routine maintenance, but cannot easily take advantage of greater wind speed and lower turbulence at higher elevation. It is supported by several guy wires fixed to the top of the rotor tower (not illustrated).

opment for large-scale power production began in the United States with the installation of the 1.25 megawatt Smith–Putnam [28] machine in Vermont in 1941. This machine was shut down in 1945 after several hundred hours of operation and was the last utility-scale wind development in the United States until 1975. In Europe, research continued after World War II in Denmark, France, the United Kingdom, and Germany. In Denmark, development of the Gedser Wind Turbine [29], a rugged and simple machine built to withstand the imposed wind loads, was continued until the early 1960s (by J. Juul) under the sponsorship of the Danish Utility Association [30]. In Germany, Ulrich Hutter [31] built a series of sophisticated machines that attempted to reduce component failures by shedding aerodynamic loads; these experiments ended in 1968. Results from both programs proved invaluable when wind turbine research was resurrected in the 1970s. Both design philosophies, load withstanding and load shedding, are reflected in the designs of present-day horizontal axis wind turbines.

The 1973 oil crisis focused the attention of governments in Europe and the United States on problems of energy supply security, and resulted in a large increase in funding for energy-related research. One of the many programs initiated at this time in the United States was a wind energy conversion research project. It was evidently decided, based on earlier work [32], that very large-capacity machines, of the order of 5 megawatts-electric (MW$_e$) or more, were needed if wind-generated electricity were to become competitive with fossil fuel power plants.

Beginning in 1975 [33], a series of successively larger machines, which attempted to implement advanced concepts of load shedding and variable speed (see the section "Variable-speed wind turbines" on page 19), were built. The first research machine was the MOD-0 (100 kilowatts); this was followed by the MOD-0A (200 kilowatts), the MOD-1 (2 megawatts), the MOD-2 (2.5 megawatts), and culminated with the MOD-5B, a 3.2 megawatt wind turbine (originally designed as a 7.2 megawatt machine). The last machine, which was built by the Westinghouse Electric Company on Oahu, Hawaii, was sold to Hawaiian Electric Renewable Systems in 1988. Because the cost of electricity from these large experimental machines [34] was significantly higher than from smaller systems, utilities and small power producers have not shown commercial interest in such muiltmegawatt wind turbines. It is now thought that the optimum machine size is between 0.2 and 0.5 megawatts, based on the simple argument [35] that energy capture increases with the square of rotor diameter, while the rotor mass, and thus cost, increases as the cube of the diameter for current designs. This judgment may change, however, as new designs and advanced materials become available.

In Denmark, development of smaller machines was based on the extensive experience obtained with the Gedser mill, and a wind turbine testing station was established at the Risø National Laboratory. Starting small has proved to be the best strategy for wind turbine development, in large part because field experience can be gained quickly, which in turn dictates rapid design improvements.

The vertical axis wind turbine was invented in the 1920s by the French engineer G.M. Darrieus [36]and is often called the Darrieus turbine. Major development work on this concept did not begin until the 1960s when the turbine was reinvented by two Canadian engineers. The VAWT has several advantages compared to the HAWT. First, it does not need a yaw system (see the section "Wind turbine subsystems") to turn it into the wind. In addition, its drivetrain, generator, and controls are located at ground level where they are accessible and can be easily maintained. Because the rotor blades operate under almost pure tension, relatively light and inexpensive extruded aluminum blades can be used. Finally, the VAWT is about as efficient as a horizontal axis turbine.

Darrieus wind-turbine development has been extensively supported by Sandia National Laboratories [37] for the U.S. Department of Energy (DOE), which built a 500 kilowatt, 34-meter-diameter VAWT at Bushland, Texas. The blades have both a variable chord length and a variable cross section to optimize performance. The use of a variable speed system in which the rotor speed can vary between 25 and 40 rpm to increase energy capture is currently being investigated.

The VAWT may prove to be cost effective in some applications, but it is limited because, unlike the HAWT, it cannot take advantage of the higher wind velocity and lower turbulence at higher elevations. The vast majority of wind turbines in use today are horizontal axis machines, and these will be the focus of this discussion.

Wind turbine subsystems

A modern HAWT is composed of six basic subsystems:

1. The rotor, which consists of one, two, or three blades mounted on a hub and may include aerodynamic braking systems and pitch controls.

2. The drive train, including gearbox or transmission, hydraulic systems, shafts, braking systems and nacelle, which encases the actual turbine.

3. The yaw system, which positions the rotor perpendicular to the wind stream.

4. Electrical and electronic systems, including the generator, relays, circuit breakers, droop cables, wiring, controls, and electronics and sensors.

5. The tower.

6. Balance-of-station systems including roads, ground-support equipment, and interconnection equipment.

The basic components (excluding balance-of-station systems) are illustrated in figure 8; the relative simplicity of a wind turbine assembly is remarkable. Initially, off-the-shelf industrial components (gearboxes, drive shafts, and generators) that had been developed for other applications were installed in wind turbines, a strategy that allowed the industry to grow very quickly. Now that a market has been established, components specifically adapted and designed for wind turbine applications are being built. These are expected to increase turbine efficiency and reduce maintenance costs.

Rotors. The rotor, which converts the wind's kinetic energy to kinetic energy of rotation, is a unique and critical part of a wind turbine. It is exposed to the full force of the elements and the full range of the variation in wind speed, direction, turbulence and shear (the change in the wind velocity with elevation). Because loads on the rotor are complex and difficult to model, they often cannot be simulated in a laboratory environment. For these reasons, this component represents the greatest engineering challenge in this field.

The rotor can be characterized as being rigid, with fixed (stall-controlled), or variable, pitch to limit maximum turbine output, or teetered, with fixed or variable pitch. In a teetered rotor, the rotor plane of rotation may vary a few degrees in a direction perpendicular to the average wind velocity. Such flexibility reduces stress on the drive train by uncoupling pitching moments at the hub; but it requires a design that can accommodate complicated dynamic loads and increased cost and complexity.

The rotor is also used to control the amount of energy extracted from the wind stream. Rotors with either variable-pitch blades or stall-controlled blades are commonly employed. With variable pitch, rotating the blade about an axis

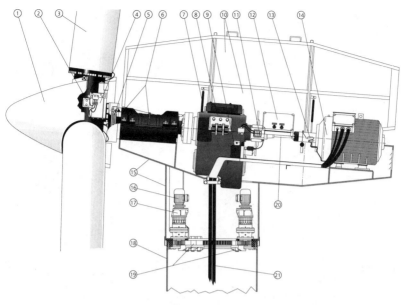

1. Nose cone
2. Hub
3. Blades
4. Hydraulics
5. Slew ring system
6. Main shaft
7. Shock absorber

8. Coaxial gearbox
9. Hydraulics
10. Nacelle
11. Brake
12. Controls
13. Vibration sensor
14. Generator

15. Bed plate
16. Yaw motor
17. Yaw gear
18. Tower
19. Yaw system
20. Transmission shaft
21. Power cables

FIGURE 8: A modern 150 kilowatt wind turbine, showing components [22].

along its length alters the pitch angle and thus the lift and drag forces on the blades. Variable pitch not only limits maximum energy capture but also reduces start-up speed and provides aerodynamic braking of the turbine. Such pitch control mechanisms, however, are subject to high loadings and must be carefully designed.

Stall-controlled blades, which limit the maximum energy capture of a turbine by loss of lift at high wind-speed, are attractive in their simplicity. At low wind velocity, airflow around the blade is laminar and the flow streamlines follow the blade's contour. At high wind velocities, the streamlines separate from the blade contour, causing the net force on the blade (F_1 in figure 5) to first level off and then decrease with increasing wind velocity. (This type of behavior is evident in figure 5, which shows power output of a stall-regulated wind turbine as a function of wind speed.) The velocity at which flow begins to separate from the blade is controlled by precisely shaping the blade contour.

Stall-controlled blades, although an elegant technique to limit energy capture, have the following limitations [38]:

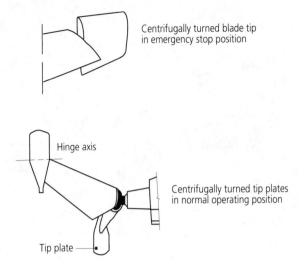

Centrifugally turned blade tip in emergency stop position

Hinge axis

Centrifugally turned tip plates in normal operating position

Tip plate

FIGURE 9: *An aerodynamic brake called the movable tip brake can stop the rotor on stall-controlled wind turbines [21].*

1.   Stall-induced turbulence may create additional structural loads.

2.   Wind-speed fluctuations about the stall speed can induce large fluctuating loads on the turbine.

3.   Rotor thrust (force F_2, figure 5) increases above the stall velocity, while it decreases with pitch control.

4.   Aerodynamic or mechanical brakes are needed to stop the rotor should loss of connection to the grid or transmission failure occur.

The most common type of braking system on stall-controlled machines is a movable tip brake (see figure 9), which deploys automatically when the rotation velocity exceeds some critical value. Because the tip brakes slow the rotor down by abruptly increasing drag, they increase loads on the other components of the turbine. Although early versions had reliability problems, modern systems have overcome the failings of their predecessors. Tip brakes can now bring the rotor to a complete rest, with the mechanical brake used only as a parking brake. Today's tip brakes are spring loaded, fail-safe, hydraulically controlled, and deploy simultaneously.

The blades for most existing wind turbines are based on aircraft airfoils and thus were designed to function in operating regimes other than those encountered by wind turbines. While such borrowed technology facilitated the rapid deployment of wind systems, the blades were later found to severely compromise wind turbine performance under some circumstances. Wind turbines with these blades have had the following problems:

1.   Generator failure resulting from excess energy capture in high winds.

2. Degraded energy capture as the result of blade soiling by dust or insect build-up.

3. Higher array losses due to the generation of higher levels of turbulence.

Advanced airfoils designed specifically for wind turbine applications have solved these problems (see the section "Future developments").

Drive trains.   The major components of the drive train are the low- and high-speed shafts, the mechanical braking system, bearings, couplings, gearbox, or transmission, and the nacelle. The gears of the drive train increase the angular velocity of the rotor, which is normally 0.5 to 2 hertz (30 to 120 rpm), to the output shaft rotational speed of 20 to 30 hertz (1,200 to 1,800 rpm), which is required by most generators to produce power at 50 to 60 hertz. For example, a 1,200 rpm generator requires a two- or three-stage gearbox with a ratio of 10:1 to 60:1 for typical rotor angular velocities. (The maximum ratio per stage is 6:1 and gearbox losses are 2 percent per stage at rated power [39].) Because existing wind turbines are locked to the grid frequency and thus operate at nearly constant angular velocity, the drive train must also partially dampen torque fluctuations caused by turbulence and shear. Loads are highly variable and cyclic, with peak loads as high as 10 times normal operating loads. Nonetheless, provided that a sufficient safety margin is allowed, commercially available units may be used.

Yaw-control systems.   Horizontal-axis wind turbines fall into two categories: upwind machines (the wind-stream encounters the rotor first) or downwind machines (the wind-stream encounters the tower first). Yaw systems are used to orient the plane of the rotor perpendicular to the wind-stream.

Downwind machines rely on a passive yaw-control system, which exploits the weather-vane action of the wind forces to position the rotor. This downwind system, while mechanically uncomplicated, has an intricate dynamic operating regime. These machines can oscillate about a stable position, which imposes high cyclic loads on the turbine's other components.

In contrast, upwind machines have an active yaw system. The yaw system drive-trains are subjected to very high loads because of wind turbulence. In older designs, all of the forces were taken up on one or two gear teeth, which led to fatigue-induced failures. Recent designs have eliminated this problem, however, by using yaw system brakes to hold the nacelle in position when the yaw drive is not activated.

Electrical systems.   Almost all modern wind turbines have induction generators, which consist of a stator, or stationary coils, and a generator rotor. The power output of this type of generator varies rapidly with the difference between the line frequency and the generator rotor angular velocity. Maximum power output is attained when this difference is a few percent above the line frequency; thus, the wind-turbine rotor angular velocity is locked to the line frequency.

The electrical equipment in a wind turbine, which includes electronic controls and sensors as well as the generator, must operate with minimum maintenance under a wide variety of harsh climatic conditions.

Operating experience

Millions of hours of operation have been accumulated on wind turbines built by manufacturers in the United States and Denmark. U.S. Windpower, Inc. has built and operates more than 3,400 wind turbines in Altamont Pass, California. About 50 percent of the approximately 15,000 wind turbines installed in California were built by such Danish firms as Micon, Bonus, Nordex, Vestas, and Danwin. The environments in which the machines function vary from the relatively mild conditions of the Danish countryside to the desert extremes of the southern California mountain passes, where sandstorms and temperature variations of −15°C to +35°C test the limits of reliability and durability.

Robert Lynette [2] made a detailed evaluation of the operation and maintenance of 4,500 wind turbines that were installed between 1981 and 1987 in California. The study documents many of the problems encountered with a relatively new technology: the average capacity factor at the time of Lynette's review (1987 data) was 0.13. Yet by 1990, the capacity factor had almost doubled to 0.24 [6], which was another indication of the remarkable progress that has been achieved in wind technology. Clearly, the technology associated with small- to medium-sized (50 to 250 kilowatt) machines has matured significantly. Many components or subsystems that had failed originally, such as mechanical brakes or tip brakes, were redesigned or upgraded. As a result, modern wind turbines will probably perform as specified almost anywhere in the world.

A selection of wind turbines presently available is given in table 1. The organizations listed have each manufactured more than 1,500 machines, all of which represent state-of-the-art equipment (1991) (see table 1). Although other manufacturers are also currently producing state-of-the-art equipment, they have not yet achieved such high production volumes.

There are now several manufacturers with in-depth experience. Moreover, machines of up to 0.5 megawatts are now offered commercially, which was not the case only one year ago. Thus, significant progress has clearly been made in both wind technology and manufacturing capability.

Variable-speed wind turbines

Currently installed wind turbines operate at a constant rotation frequency that is locked to the utility grid frequency. However, in decoupling the rotor angular velocity from the grid frequency, there is an increase in annual energy output, the quality of the power supplied to the grid is improved, and structural loads are reduced. The disadvantage is the increased cost associated with the power-handling electronics.

The increased energy output is a consequence of adjusting the rotor speed to increase the rotor efficiency at a given wind velocity. Rotor efficiency is highest

Table 1: A selection of presently available wind turbines (1991)
machines are upwind unless noted

Company	Model	Hub height *meters*	Rotor diameter *meters*	Rating *megawatts*
U.S. Windpower	USW 56-100[a]	18	17	0.1
	USW 33M-VS[b,d]	30	33	0.4
Nordtank	NTK-150[c]	32.5	24.6	0.15
	NTK 450/37[c]	35	37	0.45
Micon	M530-250[c]	30	26	0.25
Vestas	V27-225[d]	31.5	27	0.225
	V39-500[d]	40	39	0.5
Bonus	150 Mk III[c]	30	23.8	0.15
	450 Mk II[c]	35	35.8	0.45

a. Variable pitch, downwind

b. Variable speed.

c. Stall controlled

d. Variable pitch..

when the ratio of the rotor tip velocity to wind velocity is between 4 and 8. By varying the rotor angular velocity to optimize this ratio, annual energy output can be increased by about 10 percent [40]. In addition, the quality of the power supplied to the grid can be improved by using the power electronics to control the power factor[8] and to suppress harmonic currents. Finally, structural dynamic load reduction can be obtained by controlling the rotor angular velocity to avoid resonant interactions among wind turbine components, particularly the rotor and the tower. This latter advantage can be realized only if the interaction of the components is understood over the operating range of rotor angular velocity. Carefully designed control programs are therefore required to use this feature.

The most publicized variable-speed machine currently offered commercially is the U.S. Windpower 33M-VS (see table 1); the U.S. DOE and a number of European research groups are also pursuing variable-speed turbines. The 33M-VS is the result of a five year, $20 million project funded primarily by U.S. Windpower,

---

8. For resistive loads in AC circuits, the voltage and current flowing in the circuit are in phase, and the power consumed is the product of the voltage and the current. For circuits with inductive or capacitive elements, the voltage and current are not in phase and the power consumed is the product of the voltage and the current and the cosine of the difference in phase between them (cos θ). The latter is referred to as the power factor; utilities need to have this as close to 1 as possible. A power factor different from 1 indicates that currents are flowing simply to charge and discharge capacitive or inductive components. Because transmission losses are proportional to the current squared, this results in higher transmission losses. Wind turbine generators have large inductive elements (the coils in the generator) and may produce power with a power factor far from 1.

with contributions from the Electric Power Research Institute, the Pacific Gas & Electric Corporation, and the Niagara Mohawk Power Company. That this development effort is privately funded is an indication of the growing capabilities of the wind power industry.

Several variations [41] in the basic USW 33M-VS turbine design, such as a two-blade teetered rotor stall-regulated blades, were considered but rejected after detailed analysis. The present design incorporates a three-blade variable-pitch rotor, a parallel shaft transmission with dual generator output, and an active (upwind) yaw system. A power conversion module rectifies each generator output and converts it to power at the utility frequency. U.S. Windpower claims [42] that, when the 33M-VS is available in 1993, it will be able to generate electricity for less than $0.05 per kWh[9] in areas where the wind speed at hub height is 7.2 meters per second (16 mph, wind power density $\approx$ 450 watts per m^2).

Future developments

Wind turbine research and development programs are currently underway both in the United States and in the European Community. The European Community program [43] focuses mostly on turbines with a rated output of more than 750 kW$_e$. Such large machines may be well suited to offshore locations and to regions with a limited number of sites. In view of the high population densities in Europe, the development of larger machines is a reasonable strategy. Some of these large research machines are listed in table 2. Note they are much larger than the wind turbines currently available commercially and they incorporate advanced features such as variable speed and teetered rotors.

In the United States, DOE wind-program researchers work closely with turbine manufacturers to develop significant incremental improvements on the moderately sized machines now in use, while also investigating innovative technology that can be implemented by 2000. Advanced airfoil development and testing, structural-dynamics analysis, and modeling fatigue are major areas of activity. Two conceptual designs, the variable-speed wind turbine and stall-controlled wind turbine, have been analyzed for the U.S. DOE to estimate the potential benefits that could be achieved through incremental improvements. With both approaches, taller towers would be used to take advantage of greater wind speeds and lower turbulence levels at higher elevations.

The expected impact of advances in wind turbine technologies and siting strategies in the near term for these two development paths are significant (see table 3). The changes are given as a percentage, relative to the baseline 1990 design, for system costs, energy capture, and annual operation and maintenance (O&M) costs. For example, taller towers (40 versus 18 meters, baseline) that can

---

9. These assumptions are based on the accounting rules recommended by the Electric Power Research Institute for evaluating alternative power technologies [44]. With these accounting rules the assumed real discount rate is about 6 percent and both corporate and property taxes are taken into account. The resulting annual capital charge rate (10.2 percent) is about midway between the two values (8.3 percent and 13.3 percent) obtained here for the accounting rules used in the book: 6 and 12 percent real discount rates with all taxes neglected.

Table 2: Examples of European large-scale research wind turbines
1991

Country	Model	Hub height *meters*	Number of blades	Rating *megawatts*
Germany	Monopteros 50[a,b]	60	1	0.65–1.0
	WKA-60[a,c]	50	3	1.2
Italy	Gamma 60[a,b,c,d]	60	2	1.5
Netherlands	NEWCS 45[a,c]	60	2	1.0
Sweden	Nasudden-II[c,e]	≈80	2	3.0

a. Variable speed.
b. Teetered hub.
c. Variable pitch.
d. Yaw controlled power.
e. Carbon fiber/glass fiber composite blades.

take advantage of higher winds at higher elevations will increase system cost by 8 percent (a negative 8 percent improvement) although they will increase energy capture by 25 percent.

Most impressive are estimates that a variable-speed turbine on taller towers and other advances will increase energy capture by 56 percent and decrease operation and maintenance costs by $0.0051 per kWh. Use of better stall-controlled blades in conjunction with taller towers and other improvements will also increase energy capture and reduce maintenance by 49 percent and 0.0061 cents per kWh, respectively. Both approaches hold forth the promise of reducing the cost of electricity from typical sites in the Great Plains to less than $0.05 per kWh by the mid-1990s.[10]

From this analysis it is clear that significant reductions in cost are possible either using variable-speed technology or stall-controlled airfoils. Although U.S. Windpower is vigorously promoting its choice of a variable speed turbine, it is not clear at this time which of these approaches will prove superior.

It is not yet possible to formulate a detailed plan to improve wind turbines beyond the year 2000, nor is it clear which of the two competing technologies–the variable-speed or stall-controlled turbines–will prevail. The evolution of wind turbine technology depends to some extent on progress in fields such as materials science and high-power electronics. It is, however, possible to identify general areas that are most likely to reduce the cost of electricity from both large- and intermediate-sized wind turbines. Such developments include:

1. Advanced airfoil families designed specifically for wind turbines to reduce loading and increase energy capture.

10. See footnote 9.

Table 3: Estimated near-term percentage improvements
in performance and cost relative to 1990 baseline design,
for variable-speed wind turbines and stall-controlled turbines

Technical advances	Systems cost *percent improvement*	Energy capture *percent improvement*	O&M improvement *$ per kWh*
Codes			
Structural	5	–	–
Fatigue	5	–	–
Micro-siting	0	6	–
Variable speed			
Power electronics	–10	10	0.00
Control systems	–1	5	0.002
Advanced airfoils	0	10	0.001
Drive train	4	–	0.001
Tall tower *(40 meters)*	– 8	25	0.0001
Rotor hub	5	–	0.001
Total	**0**	**56**	**0.0051**
Stall controlled			
Aerodynamic controls	2	3	0.001
Control systems	–1	5	0.0015
Advanced airfoils	2	10	0.0015
Drive train	2	–	
Tall tower *(40 meters)*	–8	25	0.0001
Rotor hub	5	–	0.001
Total	**12**	**49**	**0.0061**

2.   Adaptive (neural network) types of controls that adjust system operating parameters based on wind characteristics.

3.   Incorporation of advanced materials and alloys into lighter, stronger components.

4.   Development of damage-tolerant rotors, by adapting aerospace techniques for manufacturing composite structures.

5.   Better understanding of micrositing effects on wind characteristics such as turbulence and wind shear, allowing for the optimum placement and height of individual turbines in large arrays or on complex terrain.

6.   Improved high-power handling solid-state devices.

As indicated in table 3 and the above list, one important opportunity to improve wind turbine performance is in the development of airfoils specifically adapted to wind turbine operating requirements. Present airfoils are based on designs used on aircraft and have marked drawbacks, as discussed previously. To eliminate the shortcomings of conventional blades, new families of so-called thick and thin airfoils have been designed [45] that have the performance characteristics required for stall-regulated wind turbines. The low-drag thin airfoil family is best suited to fiberglass rotors 10 to 20 meters in diameter. The thick airfoil family, having slightly more drag, can meet the more demanding structural requirements of fiberglass or wood composite rotors 23 to 30 meters in diameter. Both thin and thick airfoil families have performance characteristics that change from the blade tip (95 percent rotor radius) to blade root (30 percent rotor radius).

To control peak rotor power in high winds, the tip region of the blade must have a maximum lift coefficient (C_{lmax})[11] that is about 25 percent less than typical aircraft airfoils, while the root region of the blade must have a high C_{lmax} to aid rotor start-up and energy production at medium wind speeds. Unlike previous wind turbine blades, the new airfoils have a C_{lmax} that increases continuously from blade tip to blade root. In addition to controlling peak power, the new design permits the use of rotors that have a 15 percent greater sweep area for a given generator size, thus resulting in greater energy production.

In addition, the airfoil is designed to be less sensitive to surface roughness (caused by the accumulation of insects or dirt). The blade is shaped so that the airflow changes from laminar to turbulent on both the lower and upper surfaces of the blade as its maximum lift coefficient is approached.

Such calculated blade improvements have been verified [46] in side-by-side field tests; the new blades were found to produce from 10 to 30 percent more energy annually than conventional blades.

Further improvements [15] from better designed airfoils are possible (see figure 10). For example, by increasing the coefficient of performance of an airfoil to 0.5 ($C_p = 0.5$) over the operating wind speed, a significant increase in energy production can be achieved.

The mid-term goal of the United States Wind Energy Program is to reduce the levelized cost of electricity from areas with a wind speed of 5.8 meters per second at 10 meters elevation to $0.04 per kWh by 2000.[12] The targeted hub height of the turbines would be 40 meters; the wind speed at hub height is estimated to be 6.8 meters per second. Approximately 6 percent of the area of the continental United States has average wind speeds greater than or equal to this value, if the exclusions outlined in footnote 1 are applied.

---

11. The maximum lift coefficient $C_{lmax} = F_{lift}/(0.5\rho\,v^2 \cdot c\,B)$, where F_{lift} is the maximum lift force, $0.5\rho\,v^2$ is the kinetic energy per unit volume in the unperturbed wind stream, c is the distance between the front edge and the rear edge of the blade, and B is a unit length transverse to the flow (along the length of the blade).

12. See footnote 9.

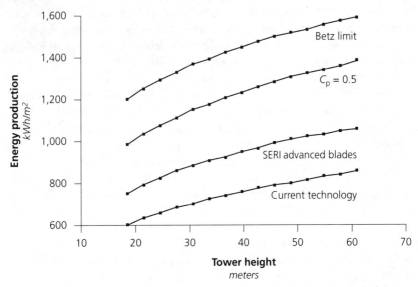

FIGURE 10: *Annual energy production as a function of tower height for stall-controlled turbines using the following blades types:*
(a.) current technology (aircraft-derived designs)
(b.) SERI/NREL advanced-technology blades
(c.) blades with a constant coefficient of performance of 0.5 over the entire wind speed operating range ($C_P = 0.5$)
(d.) ideal blades at the Betz limit.
The $C_P = 0.5$ blade assumes a constant energy extraction rate of 0.5 of the kinetic energy of the wind with no other losses, while calculations for the SERI/NREL advanced blade allow for 20 percent losses due to fouling, yaw errors, array effects, and machine inefficiencies. It is clear that substantial improvements in annual energy capture can be obtained from improved blade aerodynamic performance. For this site, a Rayleigh wind speed distribution is assumed for a wind speed of 5.8 meters per second at 10 meters.

ECONOMICS OF WIND ENERGY

The costs of electricity (COE) generated from wind are computed according to the rules outlined in chapter 1, *Renewable fuels and electricity for a growing world economy.* In addition to cost, factors such as tax policy, the attitude of the local utilities to which wind electricity must be sold, public acceptance of wind turbines, and government policy, are also of central importance. (These issues are discussed in chapter 4.)

The elements that contribute to the total cost of wind electricity are the installed capital cost, operation and maintenance, and other charges such as insurance and royalties or land rental.

Capital costs
The total installed capital cost of a wind farm not only includes the cost of the wind turbines but the balance-of-system costs, including roads, cables and con-

trols, and the utility grid substation. Balance-of-system costs represent about 20 percent of total costs for an on-shore wind farm in the United States and Europe (balance-of-system costs may be higher in other areas due to lack of skilled labor and remote locations). The capital costs are levelized, that is, the initial capital costs are spread over the assumed 25-year life of the generating unit. The average yearly contribution of the cost of capital (CC) to the total cost of electricity is then:

$$CC = \frac{ICC \times CRF}{P_r \times CF \times 8,766} \ \$ \text{ per kWh} \qquad (5)$$

Here ICC is the total installed capital cost in dollars, CRF is the capital recovery factor,[13] P_r (kilowatts) is the rated capacity of the wind turbine, CF is the average capacity factor, and 8,766 is the number of hours in one year.

Operation and maintenance

Lynette [47] estimates that the annual average O&M cost, including direct and indirect costs, will be about $0.008 per kWh for well-designed and built wind turbines of the type installed in the early to mid-1980s. If periodic major overhauls were to become necessary, the cost would increase to $0.013 per kWh. Lynette also indicates that turbines of higher output rating should have lower O&M costs, although in 1987 there was no evidence that this was true, perhaps because the larger turbines had less time in the field.

The Lynette report discusses the actual operational problems and maintenance costs, as well as actual versus projected power production, availability, capacity factor, and reliability and increase in reliability for a large number of wind farms in California. There was a significant decrease in O&M costs and an increase in reliability of the wind turbines during the period covered by the report (1981–1987), indicating a rapidly maturing industry.

Lynette's estimate of the average O&M cost is close to that for a recently installed wind farm of twenty-nine 225 kW$_e$ machines in an area with a wind speed of 7.4 meters per second at hub height (31.5 meters) in western Jutland, Denmark [48]. The estimated O&M costs were 1.75 percent of the installed cost of the turbine, or about 0.008 European Currency Unit (ECU) ($0.01) per kWh[14] (based on the extensive experience of the wind turbine manufacturer).

---

13. The capital recovery factor (CRF) is calculated as:

$$CRF = \frac{r}{1 - (1 + r)^{-n}}$$

where r is the interest rate and n the assumed life of the installation. For a 25-year lifetime, if r = 6 percent, CRF = 0.0782; for r = 12 percent, CRF = 0.1275. If the assumed lifetime is 15 years instead of 25 years, the CRF increases by 32 percent (r = 6 percent) and 15 percent (r = 12 percent).

14. One ECU = $1.27, December 1991.

Table 4: Labor and materials for maintenance of 150 kW$_e$ turbine[a]

Type	Interval	Labor/people	Materials
1	after first 100 hours	7 hours/2 people	Grease
2	twice yearly	3 hours/2 people	Grease, hydraulic fluid
3	yearly	7 hours/2 people	Oil filters, brake pads, grease, hydraulic fluid
4	every fifth year	16 hours/2 people	Oil, grease, oil pump, relief valves, wind gauge, paint

a. Source: [22].

Operating expenses include the cost of monitoring the power output and other parameters of the working turbines. Such monitoring is done routinely and remotely at all wind farms.

As with any machinery that must function reliably without intervention for long periods of time, it is essential that routine maintenance be done thoroughly at the intervals prescribed by the manufacturer. The labor and materials required for routine maintenance of a typical 150 kW$_e$ wind turbine (three blades, stall-regulated, 30- or 40-meter hub height, 27-meter rotor diameter) are not especially burdensome (see table 4).

Normal maintenance procedures, for example for the yearly maintenance interval (Type 3 in table 4), include, in abbreviated form, the following:

1. Tighten bolts with torque wrench to the specified tightness on the hub, generator suspension, and nacelle.

2. Inspect airfoils for cracks and report any immediately, noting the type and severity.

3. Check brake pads, replace if necessary.

4. Change oil filters, lubricate hub axle, yawing mechanism, generator, and transmission.

5. Check oil level in transmission, take oil sample and send it for analysis to registered laboratory.

6. Check starting current of generator.

7. Check overspeed safeguards (blade tip-aerodynamic brakes and mechanical brakes).

8. Check cables, sensors, generator cooling fan.

9. Clean area thoroughly.

The maintenance manual [49] lists 37 items in this particular routine.

Maintenance costs can be calculated, both for routine operations and for blade or generator replacement, directly from the information given by the manufacturer and a knowledge of the local labor rates.

Insurance and land rental

In the United States, insurance costs as well as land rental costs are included in levelized cost calculations. Insurance expenses account for approximately 0.5 percent of the capital cost per year. (The insurance charge is often added to the capital recovery factor as a component of the annual capital charge rate.) Land rental costs in the Altamont Pass are a fixed percentage of the price of electricity paid by the utility to the wind farm operator; at a typical royalty rate of 4 percent and a price of electricity of $0.08 per kWh, this is about $0.003 per kWh.

Costs and development history in Denmark and California

Because wind turbines are so different from the large centralized electric generating plants with which utilities are accustomed to deal, a major initiative from outside these organizations is required for this technology to be accepted as an alternative to fossil-fuel plants.

In the United States, acceptance of wind technology was accomplished with three separate yet coincidental federal governmental actions and one act passed by the state of California. These were the Public Utility Regulatory Policy Act (PURPA) of 1978, the Crude Oil Windfall Profits Act of 1980, the Economic Recovery Tax Act of 1981, and California state tax credits.

PURPA is the most important for it created a new class of electricity providers, the small power producers. According to PURPA, a utility was required to purchase the output of independent producers at the cost the utility could avoid by not providing this power itself. The new producers could generate up to 30 (now 80) MW_e and still be exempt from federal and state utility regulations and eligible for energy tax credits. This basically ended the monopoly power the utilities had previously enjoyed.

The Economic Recovery Tax Act (1981) reduced tax rates for the wealthy but allowed for an accelerated (five year) depreciation of the wind turbines. The Windfall Profits Act (1980) provided federal tax credits for producers using renewable energy, including wind; a 15 percent energy tax credit in addition to the normal 10 percent investment tax credit could be taken. These expired at the end of 1985. California allowed a 25 percent solar energy tax credit against state income taxes, which also ended in 1986.

In addition, utilities in California were encouraged to offer generous terms on long-term power purchase contracts for renewable energy. The standard offer number 4 (SO-4) guaranteed a rate of about $0.08 per kWh for 10 years and also had an escalation adjustment, but the program ceased after oil prices collapsed in 1985.

The net effect of these tax laws was that wind farms were sometimes operated as "tax farms" by a few unscrupulous promoters. Untested designs were rushed into production, and many wind farms had severe reliability problems with the wind turbines and produced less than 50 percent of the power promised in the initial promotional material. However, a few legitimate and technically competent companies used these incentives to lay the foundations for the successful industry that now exists in California.

Recent developments indicate that the cost of wind-generated electricity may decrease sharply after 1994. Based on public filings by U.S. Windpower [50], which assume the successful introduction and operation of the new USW 33M-VS (see table 1 and the section "Variable-speed wind turbines"), capital costs will be $760 per kilowatt for turnkey installations and the capacity factor will be 0.25. Operation and maintenance costs and land rental are conservatively assumed to be $0.014 per kWh, perhaps reflecting the uncertainty associated with a new machine. Because variable-speed operation is expected to decrease maintenance requirements, if these wind turbines perform according to specifications, the cost of electricity could drop much further, to less than $0.04 per kWh (assuming a 6 percent discount rate and neglecting taxes). It must be remembered, however, that the new machine represents a radical advance and thus may not achieve its published specifications under actual field conditions.

The Danish government has adopted a much more methodical and systematic approach to developing its wind resources [21]. As discussed previously, the country has had a long history of wind turbine use, research, and development. Beginning in 1979, private citizens who installed wind turbines were reimbursed 30 percent of a turbine's purchase price plus part of the installation costs. Only wind turbines tested and approved at the Risø National Laboratory were eligible for this subsidy, which has now been eliminated. This facility has been of prime importance in establishing the reliability of Danish wind turbines and a major factor in the success of Danish wind turbine manufacturers.

In addition, Danish utilities are obliged to purchase electricity generated by privately owned turbines at a rate of 0.071 to 0.078 ECU per kWh, which is a substantial fraction of the price charged to residential customers (0.116 ECU per kWh or $0.146 per kWh). The high price of electricity in Denmark, coupled with excellent wind resources and favorable price paid for wind-generated power, has created a very encouraging climate for wind energy development.

It is clear that the wind turbine is becoming a competitive electricity generating technology both in California and in Denmark (see table 5).Capital accounts for approximately 75 percent of the total cost of wind-generated electricity for an interest rate of 6 percent, and 83 percent for an interest rate of 12 percent (California data). Thus, reducing the capital cost of wind turbines will have a large impact on the cost of energy from wind.

Table 5: 1991 Cost of wind-generated electricity
6 and 12 percent interest rate, neglecting array interference effects

Location	P_w at hub height W/m²	Capacity factor	O&M $/kWh	Capital cost $/kW	COE $/kWh	
					6 percent	12 percent
California[a]	450	0.24	0.01	1,000	0.0526	0.076
California[b]	450	0.25	0.011	760	0.0429	0.0630
Denmark[c]	475	0.267	0.01	1,300	0.0552	0.084

a. The cost of electricity from wind farms in California's Altamont Pass. The installed capital cost and the capacity factor are based on data supplied to the California Energy Commission [6]. A 25-year turbine lifetime, insurance costs of 0.5 percent of installed capital costs, and land royalty payments at $0.003 per kWh are assumed.

b. USW 33M-vs variable-speed wind turbine.

c. The cost of wind-generated electricity in Denmark [52] is computed for a recently built wind farm of 29 machines with a maximum output of 225 kilowatts each at an installed cost of 1,028 ECU per kW$_e$. The O&M costs are estimated to be 1.75 percent of installed costs. Royalty payments are not included. The energy flux is estimated for a wind speed at hub height (31.5 meters) of 7.4 meters per second, assuming a Rayleigh wind speed distribution.

Other factors affecting the economics of wind-generated electricity

The above computations take into account only annual electricity production, but other wind characteristics strongly influence its economic value. The most important are the diurnal and seasonal patterns in wind-speed distribution relative to utility load characteristics (load/resource compatibility) and the distance of the resource from the consumer.

For example, the winds in the Altamont Pass are strongest in the summer, when the demand on the local utility (the Pacific Gas & Electric Company) is at a maximum. Although the daily load peak occurs around mid-afternoon (3 pm), winds do not reach maximum velocity until evening (about 8 pm). The electricity from wind at Altamont would be worth more if it were better matched to the utility's need for power during peak demand periods (see chapter 4).

The wind resources of the Great Plains in the United States, though quite substantial, are far from major demand centers. Thus, the cost of new transmission lines and the cost of and access to utility grids to "wheel" or transmit power long distances are of primary concern. An alternative to the long-distance transmission of electricity is the possible use of hydrogen as an energy carrier (see chapter 22, *Solar hydrogen*).

The load/resource compatibility is of central importance in the calculation of the capacity credit for wind energy: the amount of generating capacity that on average can be attributed to a given amount of installed wind turbine capacity. This can be treated in standard mathematical fashion for all types of generating units and is discussed in greater detail elsewhere (see chapters 4 and 23). One detailed study [53]of the Great Plains found a capacity credit of 30 to 50 percent

Table 6: Estimated future cost of wind-generated electricity[a]

	1990	1995	2000	2010	2020	2030
Hub height *in meters*	25	30	40	40	50	50
v_{avg} at hub height *m/s*	6.6	6.8	7.0	7.0	7.3	7.3
P_W at hub height W/m^2	333	360	408	408	450	450
System rating *kilowatts*	100	300	500	500	1,000	1,000
Installed cost *$ per kW*	1,100	1,000	950	850	800	750
Rotor diameter *meters*	18.3	33	40	40	51.7	51.7
Capacity factor	0.2	0.28	0.3	0.33	0.34	0.35
O&M *in $/kWh*	0.017	0.013	0.01	0.008	0.006	0.006
COE *in $/kWh*						
6 percent	0.072	0.05	0.043	0.036	0.0313	0.0294
12 percent	0.1032	0.07	0.061	0.050	0.446	0.0414

a. v_{avg} = 5.8 meters per second at 10 meters, Rayleigh wind-speed distribution; wind speed at 10 meters scaled to hub height using the 1/7 power law; includes insurance at 0.5 percent of installed cost, and land rental, $0.003 per kWh.

for a Kansas location, which is lower than the capacity credit for other technologies but is not at all zero as is often assumed [54].

Two important effects have been ignored in this economics discussion. The first is that of array interference. This is discussed in more detail in the next chapter; for example, for a 10 × 10 array of wind turbines spaced 9 rotor diameters apart, the array will have an efficiency of 87 percent compared with the same number of machines without interference. The second effect occurs when a larger fraction of the utility energy requirement is supplied by wind turbines (the level of penetration). Under these conditions, the value of wind-generated electricity decreases due to the intermittent nature of the resource. Both of these effects are taken into account in chapter 23 on the integration of intermittent resources onto utility grids.

Projected future costs

An estimate [55] of the future cost of wind-generated electricity for regions with wind speeds of 5.8 meters per second at 10 meters elevation is given in table 6. Cost reductions [56] in the near term (until 2000) will be achieved by both a reduction in capital cost and by increased energy output due to improved airfoils,

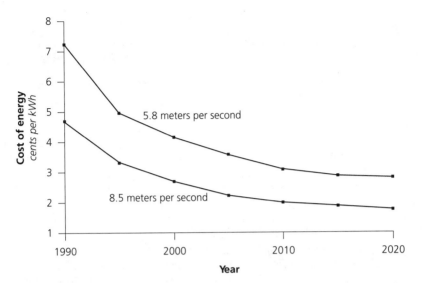

FIGURE 11: *The projected decrease in the cost of energy from wind turbines in regions with windspeeds at 10-meters elevation of between 5.8 meters per second and 8.5 meters per second through the year 2020 is shown. The discount rate is 6 percent, and the conditions listed in table 6 regarding hub height, machine rating, capacity factor, and operation and maintenance are assumed to apply.*

larger turbines, the introduction of variable speed turbines, and more efficient operating strategies. Improved designs and components will lead to the virtual elimination of replacement costs (above normal operation and maintenance costs). From 2000 to 2010, capital costs are expected to drop further as the result of innovation, including the use of lighter weight materials and designs leading to lower manufacturing costs. Continuing improvements in wind farm operations will also contribute to declining costs. Operation and maintenance costs are anticipated to decrease due to increased turbine output. From 2010 to 2020, further reductions in capital costs will be obtained from improved manufacturing procedures. By 2020 to 2030, it is anticipated that a stable, mature industry and market will be in place (see figure 11).

CONCLUSIONS

Advances in wind turbine technology and the associated reduction in cost of wind-generated electricity that have occurred during the past 20 years are dramatic. By 2000, wind-generated electricity in regions with good wind resources will be competitive with alternative electric generating technologies when wind energy provides low to moderate fractions of total utility electricity requirements. This development will have a profound impact on energy production industries around the world.

WORKS CITED

1. Rashkin, S. California Energy Commission, address to Windpower 1991 Conference, quoted in *Windpower Monthly,* p 15, November 1991. "The CEC's own analysis shows that wind has the lowest levelized cost of all renewable and fossil fuel technologies,... $0.048/kWh." This refers to the 1993–1994 period, and assumes the introduction of the USW 33M-VS variable speed wind turbine. See also the discussion in the sections "Modern wind turbines" on page 13 and "Costs and development history in Denmark and California" on page 28.

2. Lynette, R., Young, J., and Conover, K. 1989
   Experiences with commercial wind turbine design, EPRI Report GS-6245, EPRI, Box 50490, Palo Alto, California.

3. van Wijk, A.J.M., Coelingh, J.P., and Turkenberg,W.C. 1991
   Wind energy: status, constraints and opportunities, World Energy Council (to be published).

4. World Meteorological Organization. 1981
   Meteorological aspects of the utilization of wind as an energy source, technical note 175, Geneva, Switzerland.

5. Elliott, D.L., Wendell, L.L., and Gower, G.L. 1991
   An assessment of the available windy land area and wind energy potential in the contiguous United States, PNL 7789, UC-261, Pacific Northwest Laboratories, Richland, Washington.

6. Davidson, R. 1991
   Windpower Monthly, p. 15, November.

7. World Meteorological Organization. 1981
   Meteorological aspects of the utilization of wind as an energy source, technical note 175, p. 1, Geneva, Switzerland.

8. Kealey, E.J. 1987
   Harvesting the air: windmill pioneers in twelfth century England, University of California Press, Berkeley, California.

9. Kovarik, T., Popher, C., and Hurst, J. 1979
   Wind Energy, Domus Books, Northbrook, Illinois.

10. Torrey, V. 1976
    Wind catcher, p. 60, Stephen Green Press, Brattleboro, Vermont.

11. Webb,W.P. 1931
    The great plains, p. 348, Ginn & Co., Boston, Massachusetts.

12. Baker, T.L. 1985
    A field guide to American windmills, U. of Oklahoma Press, Norman, Oklahoma.

13. Dwinnell, J.H. 1949
    Principals of aerodynamics, McGraw-Hill, New York.

14. Betz, A. 1927
    Windmills in the light of modern research, *Die Naturwissenschafte,* 15(46):905–914.

15. Hock, S.M., Thresher, R.W., and Tu, P. 1991
    Potential for far term advanced wind turbines: performance and cost projections, SERI, Golden, Colorado.

16. Elliott, D.L. et al. 1991, op. cit.

17. Bonner,W.D. 1968
    Climatology of the low level jet, *Monthly Weather Review,* 96 833–850.

18. Smith, D.R. 1989
Altamont Wind Plant Evaluation Project: Report on 1988 Performance, Report 007.1-89.1, Pacific Gas and Electric Co., Research and Development, 3400 Crow Canyon Road, San Ramon, California.

19. Troen, I. and Petersen, L. 1989
European wind atlas, p. 69 and pp. 565–573, published for the EC Directorate-General for Science, Research and Development, Brussels, Belgium, by Risø National Laboratories, Roskilde, Denmark.

20. Elliott, D., Wendell, L., and Gower, G. 1991, op. cit., p. 44.

21. Smith, D.R. 1987
The wind farms of the Altamont Pass, *Annual Review of Energy* 12, p.163.

22. Nordex, A.S. 1991
DK-7323 Give, Denmark

23. Lynette, R., Conover, K., and Young, J. 1989
Experiences with commercial wind turbine design, GS-6245, pp. 2–15, EPRI, Palo Alto, California.

24. Lynette, R. et al., 1989, op. cit., pp. 2–7.

25. Elliott, D.L. 1991
Status of wake and array loss research, American Wind Energy Conference, Palm Springs, California, September.

26. Frandsen, S. 1991
On the wind speed reduction in the center of large clusters of wind turbines, *Proceedings of the EWEC*, pp. 375–380, Elsevier Science Publishers, Amsterdam, NY.

27. Builtjes, P.J.H. 1978
The interaction of windmill wakes, pp. B5-49-B5-58, *Proceedings of the International Symposium on Wind Energy Systems*, Amsterdam, October, BHRA, England.

28. Putnam, P.C. 1948
Power from the wind, Van Nostrand Reinhold, New York.

29. Lundsager, P.W. 1982
Experience with the Gedser windmill and small Danish windmills, Wind Energy Symposium, Energy Sources Technology Conference, New Orleans, Louisiana.

30. Rasmussen, B. and Oster, F. 1990
Danish Ministry of Energy, Copenhagen, Denmark.

31. Hutter, U. 1975
Review of past developments in West Germany, *Advanced wind energy systems,* workshop proceedings, Stockholm, STU/Vatentall.

32. Vargo, D.J. 1974
Wind energy development in the 20th century, NASA technical memorandum NASA TM X-71634, September.

33. Johnson, G.L. 1985, *Wind Energy Systems*. Prentice-Hall, Englewood Cliffs, N.J., p. 7–13.

34. Smith, D.R. and Ilyin, M.A. 1989
Solano MOD-2 wind turbine operating experience through 1988, GS-6567, Pacific Gas and Electric Co, Research and Development, San Ramon, California.

35. Smith, D.R. 1982
Optimum rotor diameter for horizontal axis wind turbines: the influence of wind shear assumptions, *Wind Engineering* 6(1):12–18.

36. Ramier, J.R. and Donovan, R.M. 1979
    Wind turbines for electric utilities: development status and economics, DOE/NASA/ 1028-79/23, NASA TM-79170, AIAA-79-0965, June.

37. Berg, D.E., Klimas, P.C., and Stephenson, W.A. 1990
    Aerodynamic design and initial performance measurements for the Sandia 34 Meter Diameter Vertical Axis Wind Turbine, *Proceedings of the Ninth ASME Wind Energy Symposium*, ASME, New Orleans, Louisiana, January.

38. Schmid, J. and Palz, W. 1986
    European wind energy technology, p. 51, D. Reidel Publishing Co., Dordrecht/Boston.

39 Johnson, G.J., 1985, p. 137, op.cit.

40. Wind Energy Technology Division. 1985
    Wind energy technology: generating power from the wind. *Five year research plan 1985– 1990*, US DOE, DOE/CE-T11.

41. Lucas, E.J., McNerney, G.M., DeMeo, E.A., and Steeley, W.J. 1989
    The EPRI-utility-USW advanced wind turbine program—status and plans, EPRI, 3412 Hillview Ave, Palo Alto, California.

42. Moore, T. 1990
    Excellent forecast for wind, *Electric Power Research Institute Journal*, June. This estimate used a fixed charge rate of 10.2 percent, as per EPRI TAG rules.

43. *International Energy Association Large Scale Wind Energy*. 1990
    Annual Report, National Energy Administration, S-11787 Stockholm, Sweden.

44. *Electric Power Research Institute Technical Assessment Guide*, December 1986, Volume 1, p. 4463.

45. Tangler, J., Smith, B., and Jager, D. 1991
    SERI advanced wind turbine blades, SERI, Golden, Colorado.

46. Tangler, J., Smith, B., Jager, D., McKenna, E., and Alread, J. 1989
    Atmospheric performance testing of the special purpose SERI thin airfoil family - preliminary results, *Windpower 89 Proceedings*, 115–120.

47. Lynette, R. 1989
    Assessment of wind power station performance and reliability, 8-3, EPRI Report GS-6256, EPRI, Box 50490, Palo Alto, California.

48. Nielsen, P. 1991
    Wind energy activities in Denmark, *Proceedings of the 1991 European Wind Energy Conference*, p. 177, Elsevier Science Publishing.

49. Nordex, A.S. 1991
    Maintenance manual for Nordex 250/150 kW Wind Turbine, DK-7323 Give, Denmark.

50. From a conversation with S. Rashkin, California Energy Commission, January 17, 1992.

51. Hock, S.M. 1992, National Renewable Energy Laboratory, personal communication

52. Nielsen, P. 1991, op. cit.

53. Marsh, W.D. 1979
    Requirements assessment of wind power plants in utility electric systems, EPRI Report ER-978, Palo Alto, California.

54. The potential of renewable energy: an interlaboratory white paper, p. F-2, SERI/TP-260-3674, March 1990.

55. The potential of renewable energy: an interlaboratory white paper, SERI/TP-260-3674; DE00000322, March 1990.

56. The potential of renewable energy: an interlaboratory white paper, p. F-5, SERI/TP-260-3674, March 1990.

4
WIND ENERGY: RESOURCES, SYSTEMS, AND REGIONAL STRATEGIES

MICHAEL J. GRUBB
NIELS I. MEYER

Wind power is already cost competitive with conventional modes of electricity generation under certain conditions and could, if widely exploited, meet 20 percent or more of the world's electricity needs within the next four to five decades. The greatest wind potential exists in North America, the former Soviet Union, Africa, and (to a lesser extent), South America, Australia, southern Asia, and parts of Europe. In all these areas, wind can make a significant contribution to the energy supply. In regions of the developing world and in island communities, wind can operate with storage and displace diesel fuel. In more developed areas, wind-generated electricity can be channeled directly into the grid, providing an environmentally benign alternative to fossil fuels. Indeed, wind power can contribute as much as 25 to 45 percent of a grid's energy supply before economic penalties become prohibitive; the presence of storage facilities or hydroelectric power would increase wind's share still further.

Despite a promising future, opportunities for wind power development are probably being missed because too little is known about either the resource or the technology. International efforts are badly needed to obtain better data and to disseminate technological information around the world. Even then, the extent to which wind is exploited will depend on public reaction and on the willingness of governments to embrace the technology. Action that governments might take to promote wind include providing strategic incentives to further its deployment, funding research on wind resources, taxing fossil fuels to reflect their social costs, and allowing independent wind generators adequate access to electricity systems.

INTRODUCTION

The previous chapter, *Wind Energy: Technology and Economics*, has demonstrated that wind power is already cost competitive with conventional sources of electricity generation under some conditions. This chapter shows that the total physical potential of wind power greatly exceeds present global electricity demand. Even when environmental, land-use, and systems constraints are taken into account, wind power can accommodate a substantial portion of global electricity demand, perhaps 20 percent. It is important therefore to consider the role of wind power in planning the future energy supply.

This chapter describes the practical potential for wind energy under different circumstances and analyzes the constraints and promising strategies for wind energy development and deployment. The broad themes of wind resources are considered, followed by a discussion of physical and environmental impacts and constraints on siting wind turbines, as well as the issues surrounding integration of wind energy into the energy supply. Detailed data are now available for a number of countries and regions, and these data are summarized. The global potential for wind power is estimated based on available data. Finally, policy issues and the prospects for the expansion of wind energy are discussed.

WIND ENERGY RESOURCES

General features

Wind energy is ultimately a solar resource, created primarily by temperature differences among the sea, land, and air, and by the overall temperature gradient that exists from the equator to the poles. About 0.25 percent of total solar radiation reaching the lower atmosphere is transformed to wind energy, which is then dissipated at a rate of about 30 times human energy consumption because of friction at the surface and within the atmosphere[1].

Only a small fraction of wind resources can be tapped in practice because of technical and social constraints. Limits imposed by technology include the conversion efficiency of wind turbines, their height, and the interference created between machines in wind parks, a phenomenon known as array loss.

The amount of energy available for capture by wind turbines is determined by a number of factors. To begin with, the cubic dependence of wind energy density (the energy flowing across a given vertical area) on windspeed means that the power output and economics of wind turbines are highly sensitive to windspeed: a 10 percent change in speed means about a 30 percent change in available energy. Windspeed is thus one of the most critical factors determining the amount of energy that can be extracted, and at what cost. Factors that influence windspeed include local factors such as surface roughness, the height of the air flow, scale of the terrain, altitude, and so forth.

Global wind systems

Large-scale wind systems are created by temperature differences between the earth's latitudes and by deflection caused by the earth's rotation. Dry air in the vicinity of 30 degrees north and 30 degrees south sinks and flows toward the equator, where it replaces rising hot air, a pattern known as the Hadley circulation (see figure 1). At mid-latitudes, between 30 and 70 degrees north and south, air flows toward the poles and is deflected westward, creating a wavelike pattern known as the Rossby circulation. Superimposed on the global system are many smaller circulations and weather systems reflecting regional variations in atmospheric temperature and pressure.

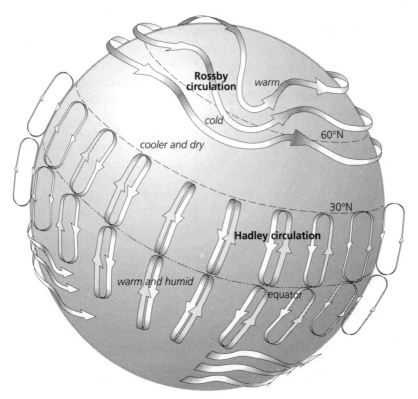

FIGURE 1: Global wind systems. [74]

Winds driven by large-scale atmospheric gradients are often (though imprecisely) known as synoptic winds. On a smaller scale, thermal winds may be created by local thermal effects. In hot regions, temperature differences between the land and sea may spawn strong winds. Valleys and mountains serve to channel and amplify winds. Thermal mountain winds of this type are notable at Altamont Pass in California, where cool air from the Pacific rushes through the mountains to replace hot air rising from desert regions to the east. Strong thermal winds may differ considerably from synoptic winds. The Altamont winds, for example, are relatively predictable, having a regular and strong daily cycle, and are strongest in summer.

Winds vary greatly in directionality. In some locations, strong prevailing winds are mostly unidirectional or bidirectional (as for example, California's coastal thermal winds); in other regions winds are broad-based and varied in direction.

Surface roughness and height above ground

In the atmospheric boundary layer, which on average extends roughly 1 kilometer above the earth's surface, air is slowed by surface friction. The extent to which it is slowed is influenced by the height and character of the surface. Near the ground, the terrain exerts considerable influence on windspeed. Other factors being equal, winds are strongest over grassland and open water, moderate over scrubland, and weakest over forested areas. Clearly, coastal areas and prairies or deserts with few wind breaks are favored regimes for wind energy (see table 7).

Whereas most wind data are collected at a height of about 10 meters, wind-turbine hubs may be as high as 75 meters (some experimental machines are taller). Thus, estimating how windspeed increases with height is an important part of resource assessment.

Traditionally, windspeeds have been extrapolated according to a power law, so that if a given windspeed is v_1 at height h_1, then the value v_2 at height h_2 can be derived as

$$v_2 = v_1 \left(\frac{h_2}{h_1} \right)^{\alpha} \tag{1}$$

where α is the height exponent. A value of $\alpha = \frac{1}{7}$ (0.143) has been widely used and is often appropriate for very smooth sites; values of 0.16 or higher are appropriate for inland sites, while values of 0.3 or even higher may apply in obstructed rural and urban sites.

Although the power law is a convenient approximation, it has no theoretical basis. Instead, aerodynamic theory suggests that when the atmosphere is thermally neutral,[1] which is generally the case on cloudy days, and when the wind is strong, air flow within the boundary layer should vary logarithmically with height [2], so that

$$v_2 = v_1 \frac{\ln (h_2/z_0)}{\ln (h_1/z_0)} \tag{2}$$

where z_0 is called the roughness length and is a parameter of surface roughness.[2] This approach, which is known as logarithmic extrapolation, is commonly used in Europe and is particularly useful when adequate data on surface roughness are available. Significant variations can exist between the two methods of extrapolation (see figure 3).

The logarithmic extrapolation is itself a great simplification, though often a

---

1. As a body of air rises in the atmosphere, it expands and cools because of the decreasing pressure. If there is no exchange of heat with the rest of the atmosphere, the rate at which it cools with height is the "adiabatic lapse rate" of about 9.8°C per kilometer. If the actual rate at which temperature declines with height at a given location roughly equals the adiabatic rate, air displaced up or down has the same temperature and density as surrounding air, and there is no net force up or down on it: the atmosphere is thermally neutral. This occurs when there is little heat input (e.g., thick cloud cover) and when strong winds mask thermal gradients.

2. The "roughness length" is a mathematical term that does not necessarily reflect the length-scale of obstacles in the landscape. Corresponding to the four roughness types illustrated in figure 2, the roughness lengths are Class 0 (Water)—0.0002 meters; Class 1 (Open)—0.03 meters; Class 2 (Farm land)—0.10 meters; Class 3 (Urban or obstructed rural)—0.40 meters.

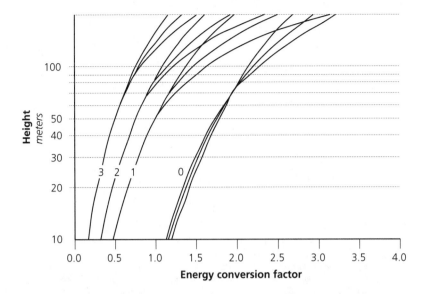

FIGURE 2: Variation of wind energy with height for four different roughness classes. These curves indicate the mean wind power density at heights between 10 and 200 meters over homogenous terrain for the four roughness classes of the European wind atlas [3], relative to the reference value at 50 meters and roughness class 1. The shading indicates the uncertainty due to climatic differences in Europe. The roughness classes are defined fully in the atlas and correspond roughly to
0 - Water areas
1 - Open areas with a few windbreaks
2 - Farm land with some windbreaks more than 1 kilometer apart
3 - Urban districts and farm land with many windbreaks

useful one. Local variations in surface roughness complicate the picture,[3] as do temperature variations, which can have a major impact on atmospheric characteristics, thus changing the vertical wind profile. In very "stable" atmospheric conditions,[4] the lower layers of the atmosphere are largely isolated from the winds above them, and ground-level winds may be quite still despite the presence of strong winds a few tens of meters above the ground. Under these conditions, the height exponent is quite large. For example, analysis carried out at a potential wind site in the Netherlands [4] showed that the estimated output of optimally rated wind turbines would increase by 20 and 48 percent at heights of 40 and 80

---

3. If the roughness upwind of a site varies significantly with distance, a simple logarithmic profile will not be correct, because there will in effect be various "internal boundary layers" at different heights.

4. If the temperature declines more rapidly with height than the adiabatic lapse rate (footnote 1), then a body of air displaced upward will be warmer than the surrounding air and will tend to rise further; the atmosphere is *unstable*, and there is greater mixing of different layers, reducing vertical windspeed gradients.

Conversely, if the temperature declines more slowly with height, air displaced upward is cooler (and hence denser) than the surrounding air, and tends to fall back; the atmosphere is then *stable*. The extreme case of this is an *inversion*, when warmer air overlies cooler air. A discussion of stability corrections to the simple logarithmic formula is given in reference [4].

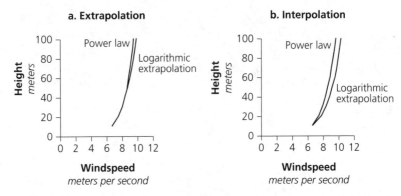

FIGURE 3: Comparison of power law and logarithmic height extrapolations. The graphs compare windspeeds extrapolated from measurements at two heights using the power law and logarithmic techniques. a) Shows windspeeds extrapolated on the basis of measurements at 10 meters (6.5 meters per second) and 30 meters (7.9 meters per second), b) shows windspeeds interpolated from 10 meters with an estimate of the constant "upper air" windspeed (13.5 meters per second at 950 meters).

For a fixed height exponent of ⅐, the calculated windspeed from 10 meters increases more slowly with height than any of the cases indicated in the figure. Noting that the power available depends upon the cube of windspeed, the variations can have a very significant impact on the estimated power available. For the conditions of figure (b), a ⅐ power law extrapolation would predict 25 percent less energy than the power law extrapolation.

meters, respectively, compared with the estimate obtained using extrapolation techniques, which ignored stability corrections. At the greater heights, there was also a marked change in speed distribution, with more constant winds and fewer periods of calm when stability effects were incorporated in the extrapolation.[5]

Such vertical effects may be particularly important in inland areas, where near-ground winds are generally lower than they are in coastal regions, but where vertical heat flux can be much greater.[6,7] The vertical profile also depends on general forcing conditions, such as the presence of synoptic winds or local thermal effects. At Altamont Pass, for example, the driving force is one of horizontal push rather than vertical shear, so there is much less increase of windspeed with height; in fact, there may be none depending on the terrain.

But for most flat sites with synoptic wind regimes, simple extrapolation techniques from 10-meter data may often err on the pessimistic side and overestimate

---

5. Reflected in an increase in the Weibull parameter of wind distribution (see the section "Windspeed distributions" on page 9) from 1.76 at 10 meters to 2.04 at 40 meters and 2.21 at 80 meters.

6. A study by Sisterton et al. in Illinois [5] found that extrapolation based on a ⅐ power law underestimated the likely wind energy by 40 percent, arising particularly from the night-time inversion, when light breezes at 10 meters were found to accompany strong winds at 45 meters (corresponding to a power law height exponent of about 0.5). See also the U.S. wind atlas discussed later.

7. "An unusually high wind pump (20 meters) can turn vigorously although trees up to 10 meters are still ...a balloon can drift straight up until reaching 15 to 20 meters, when it gets whipped sideways by strong winds" (Jerome Weingart, private communication). If this is a stability effect, it would tend to apply particularly in late afternoon and evening, when electricity demand is often at peak.

the frequency and strength of diurnal cycles. In view of the magnitude of the effects, more data on wind energy must be collected at the height of turbine hubs.

Terrain

Large-scale terrain variations, such as hills, have a major impact on wind patterns. Windspeed increases as air is squeezed around and over hills and ridges. Such locations therefore offer good sites for wind turbines, particularly if there are strong prevailing winds. Good hill sites will typically increase wind energy density by a factor of two to three at a height of 30 meters [6]. For winds accelerated by the terrain, the increase with height is greatly reduced and can even be reversed; studies of large hills show that windspeed can decrease for the first few tens of meters above the hilltop.

Computer models indicate the extent to which hills and ridges can increase the amount of exploitable energy compared with equivalent flat sites [7]. One such study, which had a resolution of 2 square kilometers (km^2) and was based on a hilly region in the United Kingdom, predicted that "...very large variations in output from individual turbines...would dramatically affect cost/resource estimates..." [8], yielding far more cost-effective sites and greater available energy than estimates that ignored terrain effects. Subsequent trial runs at 1-kilometer resolution yielded still higher results.

Although computer models illustrate the overall impact of terrain, they are no substitute for on-site measurements. As demonstrated at Altamont, changing the siting of a wind turbine by a few meters can have a substantial impact on energy output. Indeed, improved understanding and use of such micro-siting strategies is expected to increase the output of future windfarms by as much as 5 percent.

Height above sea level (ASL)

The height of the ground above sea level also affects wind strength. Several factors tend to increase windspeeds at higher elevations: first, the atmospheric boundary layer is thinner, which increases the direct driving force; second, there are often direct forcing effects from mountainous terrain, as noted above; and third, at higher elevations—above the timberline, for example—the terrain tends to be smoother and has fewer obstructions. Typically, there is a 5 to 10 percent increase in windspeed for each 100 meters ASL (see figure 4) [9]. Reduced air density at higher elevations offsets the increase in wind power to a small degree; to maintain the same power density, speed must increase by about 3 percent per 1,000 meters[8][10].

Because many factors influence windspeed and energy, wind resources must be assessed very carefully. Wind atlases (which summarize and standardize wind data for a region, often including relevant data such as land characteristics) pro-

---

8. Above a few thousand meters ASL this approximation becomes rather crude, but few areas at such heights are of practical interest for wind energy.

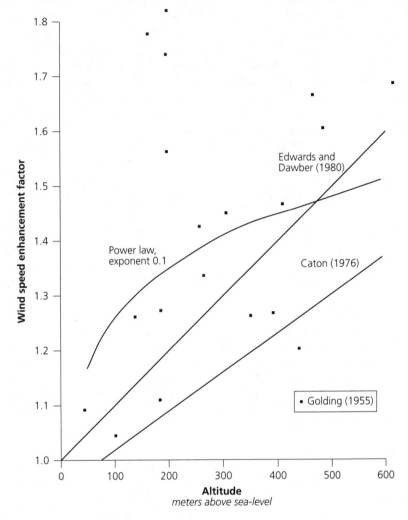

FIGURE 4: *Estimates of the effects of altitude on windspeed [9].*

vide information on the main patterns of wind regimes, but have limited applicability for assessing specific sites. Data on average wind statistics in the Danish wind atlas [11], for example, correlate well with production statistics from Danish wind turbines, although studies show that the output of individual turbines may differ by as much as 20 to 30 percent from atlas predictions [12]. Thus, while wind atlases may help estimate total resources and identify promising locations, they should not be used exclusively. It is imperative that individual site measurements be made before siting wind turbines, especially in regions where the terrain is complex or less data are available.

VARIATIONS IN WIND ENERGY

Windspeed distributions

Windspeeds are highly variable, and their distribution at any given site can be reasonably described in terms of the so-called Weibull distribution,[9] which is the proportion of time $t(v)$ for which the windspeed exceeds the value v:

$$t(v) = \exp\left[\left(\frac{-v}{v_c}\right)^b \right] \tag{3}$$

The parameter v_c is the characteristic windspeed, or Weibull scaling parameter; b is the Weibull shape parameter. Weibull statistics provide a convenient, if approximate, way of summarizing wind regimes for a period of time, typically one year. If the parameters v_c and b are known, then other information such as mean windspeed, power, variance, and so forth, can be calculated [13].

When b equals 2, the function is known as a *Rayleigh distribution*. Because synoptic wind regimes, such as those in many temperate regions, frequently yield b-values around 2, the Rayleigh distribution is often assumed when better data are unavailable. When $b = 2$, v_c typically exceeds the mean windspeed by about 10 percent. Sites near the equator have Weibull parameters ranging from 4 (which reflects the constant trade winds) to nearly 1 (which is indicative of highly variable, generally calm conditions, interrupted only by typhoon winds) [13]. Such variations exist in the Caribbean, where Weibull parameters range from 1 to 3. Because conditions differ so much from site to site, a generic wind turbine cannot be designed to suit all sites. Instead, different turbines can maximize output depending on site conditions.

Because wind is so variable, the mean energy density (proportional to $<v^3>$) at a site is much higher than the energy density at the mean windspeed $<v>$. For the Rayleigh distribution ($b = 2$):

$$<v^3> = \frac{6}{\pi}<v>^3 \tag{4}$$

In other words, mean energy density is nearly twice the energy density at the mean windspeed. The lower the b value, the greater the difference. Because energy density depends on the form of the Weibull distribution, the relation between mean windspeed and energy density can vary greatly. For example, at different sites where mean windspeed was measured as 6.3 meters per second, mean power densities ranged from 220 to 365 watts per square meter (m^2) [10]. As a result of

---

9. The formal density distribution given in the previous chapter is the derivative of this cumulative distribution.

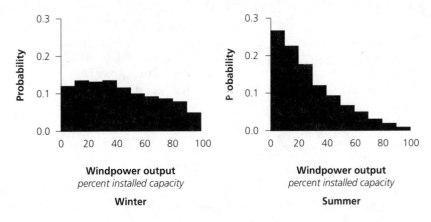

FIGURE 5: *Distribution of output from distributed wind energy in a temperate region (United Kingdom). The figure shows the probability of obtaining different levels of total wind power output (relative to installed capacity) if wind turbines were sited at many different locations in mainland Britain [14].*

such variations, site data are best expressed in terms of power density rather than mean windspeed[10].

In view of this relation between windspeed and density, an individual turbine (depending on its design and the wind regime), may spend a third of its time at full power, and a third not generating at all, with rapid fluctuations at intermediate levels as outlined above. However, by connecting many wind turbines at different locations to an integrated system, the characteristics of the output can be radically altered.

As an illustration of this, figure 5 shows the probability of obtaining different levels of total wind power (relative to the installed capacity) with wind turbines sited at many different locations around the United Kingdom). In summer, when the winds are weaker, the chances of obtaining maximum power (when some wind energy might have to be discarded) is small indeed, although there is a 25 percent chance that output will fall below 10 percent of the installed capacity. In winter, such low outputs would occur only about 10 percent of the time (or less, if variable-speed turbines are used), but the chances of obtaining output within 10 percent of maximum are still barely one in 20. Most of the time, the output would be at intermediate levels.[11]

---

10. For more detail on the wind regime, the Weibull distribution can be defined separately for different seasons, or even different months, to indicate differing characteristics in different periods. When divided in this way, large variations in *b* for a given site have been found over the year for example at Altamont Pass, California.

11. This study used a simplified wind-turbine characteristic that is relatively unresponsive to low windspeeds and used data collected at 10 meters, which for the reasons discussed above is more variable than the actual wind regime at the hub height of typical wind turbines. Thus, the actual probability of low or zero output would be still less than illustrated in figure 5.

Table 1: Length scales and windspeed variations for different timescales of wind fluctuations[a]

	Time scale	Coherence length scale *kilometers*	RMS[b] windspeed variation *percent*
Microscale	10 to 20 seconds	0.5	15
Mesoscale	30 minutes	14	13
Mesoscale	90 minutes	43	24

a. Source: [15]

b. RMS = root mean square.

Energy fluctuations

The output from individual wind turbines fluctuates on various timescales (see table 1), but the fluctuations tend to be correlated over a certain distance (the coherence-length scale). In stormy weather, variations of plus or minus 25 percent may be common. Microscale fluctuations, which may last from 10 to 20 seconds, cause power output from individual turbines to fluctuate, but such fluctuations are generally smoothed out within an array of turbines[12] [15]. The impact of fluctuations over longer timescales, which are imperfectly correlated between different sites, can also be estimated.[13]

---

12. If there are N sites sufficiently far apart for the fluctuations on a given timescale to occur independently, the variability of output (as percentage of the mean output) is reduced by $\sqrt{N}$. In practice, fluctuations are relatively independent at separations of more than the distance covered by the winds in the timescale of interest (e.g., microscale fluctuations, of a few tens of seconds, are largely independent between machines within a given array, while hourly fluctuations in an average 10 meters per second wind are more or less independent for sites of 30 to 40 kilometers apart (see table 1 above)).

   Thus to take an extreme example, even if 5,000 wind turbines of 2 megawatts capacity each were deployed in the long term and the average microscale power fluctuation from each machine was 15 percent of the capacity, the total variation would be

   Total average microscale variation = 5,000 × (15 percent × 2 megawatts)/$\sqrt{5,000}$ = 21 megawatts.

Such fluctuations—about 0.5 percent of the total wind capacity—would be negligible on large power systems. When the timescale stretches to 10 minute fluctuations, however (a timescale of particular interest, since it takes about 10 minutes to start up gas turbines for emergency supply), random fluctuations would only be independent between machines spaced more than 5 to 10 kilometer apart. In such cases the equations can be applied just as well to evaluate the effects of diversity between clusters of machines. If, for example, the same 5,000 turbines were arranged in 50 separate clusters of 100 machines each (many perhaps offshore), the root mean square (RMS) random fluctuations on timescales of 10 minutes to half an hour would be at most 200 megawatts—significant, but still small compared with demand variations on most systems that span such an area (e.g., compared with the 10 to 15 gigawatts diurnal demand cycle in the United Kingdom).

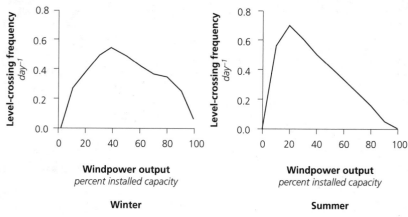

FIGURE 6: *Level-crossing frequency of output from distributed wind energy in temperate region (Britain).*

Most random fluctuations will have a negligible impact on the large-scale deployment of wind energy, especially when compared with demand: the gross output from wind energy usually varies much less than electricity demand. In Britain, for example, studies show that the (hourly averaged) output from dispersed wind energy will not increase across a given power level more than once a day on average (see figure 6).

Although extreme power fluctuations are important to system control, a Dutch study [16] showed that the probability declines rapidly for greater fluctuations (see figure 7). Data over a 10-year period showed a chance of only one in 10,000 that output would vary by 30 percent of installed capacity over the span of one hour, or by 60 percent over a four-hour period. For larger systems—or systems connected across a greater expanse—extreme fluctuations would be even less likely.

Fluctuations occurring on timescales of more than a few hours may not be smoothed out by diversity (that is, by having wind turbines dispersed over many different sites connected to the grid). Most wind sites do experience diurnal vari-

---

13. Diversity can be analyzed mathematically by calculating a "diversity factor" $D(t)$, which expresses the average variation in output from a group of wind turbines relative to the variations in one machine. This is defined as: (variation of total output/total capacity) = $D(t) \times$ (variation in single machine/machine capacity). The diversity factor for an array of machines can be found if the "coherence length" $L(t)$ of fluctuations (see text at table 1) is known.

For a square array of N machines spaced a distance d apart, it is given by

$$D(t) = \tanh[d/2L(t)]/\sqrt{N}$$

where $\tanh(x)$ is the hyperbolic tangent. In practice, $L(t)$ is approximately equal to the windspeed times the timescale t involved. For "microscale" variations, within tens of seconds, $L(t)$ will be less even than the spacing d between machines. The equation for $D(t)$ then approximates to $D(t) = 1/\sqrt{N}$, corresponding to the fact that the fluctuations between the machines are the independent.

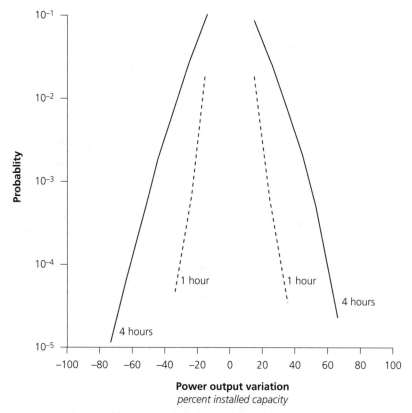

FIGURE 7: Probability of extreme output fluctuations for wind energy in the Netherlands. Even for the relatively small Dutch system, the probability of wind output increasing from zero to full power or vice versa, within four hours, is almost negligible [16].

ation. Thermally driven wind regimes, for example, reflect daily solar variations. Even in many synoptic regimes, an unstable atmosphere, which is created by sunshine, results in higher near-ground windspeeds. However, the nature and strength of the diurnal cycle varies greatly and declines rapidly with height, especially in synoptic wind systems.

For many wind regimes the seasonal cycle is most important. Frequently, seasonal change correlates well with electricity demand. In many hot regions, winds tend to be strongest during the summer when electricity demand is highest. In contrast, in cooler climates, electricity demand is usually greatest in winter, when the winds are usually stronger (see figure 8). This correlation between wind and load tends to increase the value of wind energy, although naturally this positive relationship is far from universal [14].

Predictability

Predicting variations in the energy output from wind turbines makes exploiting wind energy easier. By reducing the need for operational reserve, such predictions increase the amount that can be economically utilized. Adopting the simple algorithm that wind energy will persist at its present level can reduce the root mean square (RMS) prediction error in hourly forecasts by a factor of three, as compared with zero prediction [17]. Further improvements can be gained using relatively simple techniques that involve the manipulation of weather data (see figure 9).

Exploiting wind energy on a large scale—more than a few tens of megawatts—would involve placing wind turbines at diverse locations. This would greatly reduce output fluctuations and further reduce prediction errors. Large fluctuations would come only from major weather patterns, such as broad-based storm fronts. But with few exceptions, such major events are readily identified in satellite-based weather data, and predicted in weather forecasts. Thus, with modern weather forecasting, the output of large wind capacities is probably quite predictable on timescales up to about 24 hours. And as substantial investments are made in large-scale wind energy, prediction techniques will naturally improve. In time, wind forecasting will become part of weather forecasting, although predictions will be limited to only a few days because of the chaotic nature of weather formation.

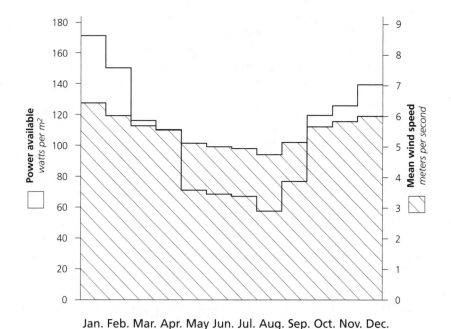

FIGURE 8: Seasonal variation of available wind energy in a temperate region (Britain). [14].

CONSTRAINTS ON ENERGY EXTRACTION

Array interference and siting densities

An operating wind turbine reduces the wind's speed for some distance downwind of the rotor. If the turbines are spaced too closely, they will interfere with one another, reducing the output of those downwind. But because the energy extracted by one turbine is steadily replaced by energy transferred from winds above, an equilibrium is eventually reached. The array efficiency is the actual output from clustered turbines compared with that which would be obtained without interference. The array efficiency depends on spacing between turbines and the nature of the wind regime—the forcing conditions, directionality, distribution of windspeeds, and flatness of the site.

Extensive theoretical and wind-tunnel studies [18] indicate that under typical conditions, interference increases quite rapidly when turbines are less than 10 rotor diameters ($10D$) apart. For an infinite number of turbines with $10D$ spacing, the limiting array efficiency is about 60 percent. But for a finite number, average losses are much lower, and closer siting is practical.[14] For example, clusters of 10 to 20 machines can be spaced about $5D$ apart. The approximate efficiency of square arrays (as estimated in a synoptic wind system) as a function of spacing and array size is presented in table 2.

Other factors also influence siting densities and related energy losses [19]. The incident wind angle, for example, can cause efficiency to vary considerably among turbines that are arranged in straight lines. Correspondingly, unidirectional or bidirectional winds allow for close crosswind spacings. At Palm Springs, Cal-

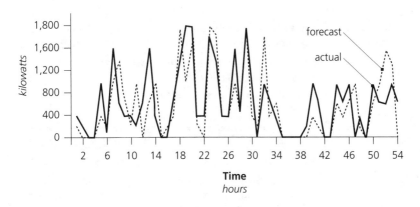

FIGURE 9: Windspeed prediction at Swedish site. Source: S. Engstrom, personal communication.

---

14. The rate at which individual turbine losses increase from the edge inward depends on the turbine spacing. At spacings of $6.5D$, conditions at the center of a square 10×10 machine array correspond closely to those in an infinite array. If $10D$ is used, a square array of 40 turbines has an overall array efficiency of 90 percent, decreasing to 80 percent for 200 turbines.

Table 2: Typical array efficiencies for different sizes and spacings of square arrays (non-directional synoptic wind regime)[a]

	Turbine spacing					
	4D	**5D**	**6D**	**7D**	**8D**	**9D**
Array size	Array efficiency (compared to equivalent number of machines with no interference) *percent*					
2 × 2	81	87	91	93	95	96
4 × 4	65	76	82	87	90	92
6 × 6	57	70	78	83	87	90
8 × 8	52	66	75	81	85	88
10 × 10	49	63	73	79	84	87

a. Derived from the empirical equations presented by David Milborrow and Philip Surman of the Central Electricity Generating Board [20]. These numbers are approximate but give a useful indication of the physical limits on siting densities and losses involved.

ifornia, wind turbines barely $1.5D$ apart face the wind, with rows spaced at about 5 to $7D$. For a given windspeed, this means that from two to four times as much power can be extracted from a given land area as compared with nondirectional winds. A few rows of wind turbines along a ridge can thus generate considerable power from a small area.

The rate at which the wind energy extracted by a turbine is replaced with energy from winds above depends on the vertical profile (wind shear) and on mixing between the different layers. Thus, in the stratified thermal winds of the Altamont Pass, where there is little increase of windspeed with height and little mixing between layers, disturbances carry far downwind, and interference between wind turbines is much greater than it was originally predicted. Because reduced windspeeds have been observed as much as $40D$ downwind of arrays, wind rights has become an important issue in wind-farm planning.

Array interference has other implications. Turbulence is greatly increased within an array. However, interference declines at higher windspeeds (especially above the rated windspeed (see chapter 3) because the wind turbines then extract a lower percentage of the energy). In other words, although individual turbine efficiency declines at higher windspeeds, array efficiency changes partially offset this and also can make the array output less variable than that from an individual turbine.

Land requirements

To collect large amounts of energy from the wind, turbines must be spread over a wide area and positioned so as to not interfere with one another. Spacing is particularly important to large windfarms where the turbines are typically separated by distances of five to 10 rotor diameters.

While such space requirements suggest that windfarms might consume large land areas, this is not the case. Overall, the entire machines occupy only a small fraction of the windfarm area. In Europe, for example, where wind turbines are often installed on costly land and spaced relatively far apart because the winds are not directional, as little as 1 percent of the land is actually occupied. On this basis, land consumption for a given energy output is comparable to that from conventional power sources. The bulk of a windfarm's land can be used for other purposes, such as farming and ranching. In fact, land values at Altamont Pass increased markedly after the windfarms were installed because of royalties they generated for the ranchers while allowing ranching to continue.

Noise

Many first-generation windmills were quite noisy. Although it is impossible to eliminate all noise (especially from the blades), modern turbines produce noise that is only slightly above general wind noise. And though blade noise increases with increasing windspeed, so does all other wind noise.

Few countries have made general noise rules for wind machines, but such regulations are needed to secure public acceptance of large-scale wind-power development and to ensure that manufacturers develop low-noise designs. In Denmark, the Ministry of Environment has set a maximum noise level for wind turbines of 45 decibels (45 dBA), measured at the nearest dwelling, representing a level that is "greater than in a bedroom at night, but less than in a house during the day," [21] and much less than most road noise. The requirement can be met by a single modern wind turbine of 300 kilowatts (operating at a windspeed of 8 meters per second) at a distance of about 200 meters, which in Denmark is the legal minimum distance from a wind turbine to a dwelling. A windfarm having some thirty 300-kilowatt machines would have to be at least 500 meters from the nearest dwelling in order to meet the 45 dBA requirement.

Bird strikes

The impact of windfarms on local bird populations has raised concerns that large numbers of birds might fly into the spinning rotor blades and be killed. Such fears, however, seem to be largely unjustified. Analysis of a 7.5 megawatt coastal windfarm in the Netherlands concluded that the wind turbines were far less detrimental to birds than a high-voltage electric power line and comparable to a 1-kilometer stretch of motorway [22]. And a study of nine windfarms in Denmark (ranging in size from two to 35 machines with rated power between 50 and 100 kilowatts) [23] found only two birds, presumably killed by collision, during the one-year investigation. Migrating wildfowl, especially swans and geese, often change course at a great distance from the wind turbines, and species that once fed in the vicinity of a windfarm apparently adapt rapidly to the presence of wind turbines and learn to avoid the rotors. However, in California, bird strikes are still an issue, where an average of about one identified bird of prey is apparently killed each month. [24]

Interference with telecommunications

Wind turbines can interfere with telecommunications (especially television) in some cases, but these problems should not seriously constrain the exploitation of wind power. Experience with windfarms in Europe shows that television interference is very localized and can be overcome by installing local amplification or cable connections to the small number of receivers that are affected. Nevertheless, the installation of large machines close to airports and other areas with sensitive electronics should be avoided.

Safety

To date, the wind industry has had a good safety record. However, like all industries involving moving machinery, hazardous working conditions can arise, especially if safety concerns are poorly managed and regulated. Some serious accidents have occurred during construction and maintenance.

Public safety is well protected. Modern turbines are equipped with redundant safety breaks that are automatically set off when blade-tip speeds exceed a specified value so as to prevent loss of a blade. Turbines also have monitoring systems that give early warnings of potential failures of sensitive components. In Denmark, the mandatory separation of 200 meters or more between a turbine and a dwelling can also be regarded as a safety distance, reducing risks of injury from potential equipment failures (such as blades breaking off). We can find no reported public injury from wind energy.

Visual impact and public acceptance

The visual impact of wind turbines on the countryside is one of their most contentious features. In some areas, conflicts may arise between the desire to preserve the land in its natural state (e.g., as wilderness areas) and the need to replace fossil fuels with energy sources that are more environmentally benign. While there is no simple solution to such conflicts, they can be reduced by excluding some areas for wind energy development and by ensuring that wind turbines are properly designed. Sites and positions of wind turbines must also be carefully assessed. In some cases, for example, larger numbers of small machines may be preferable to fewer numbers of big ones.

Public acceptance, which is critical to the large-scale development of wind power, depends on education and on participation in siting decisions. In general, the public must be provided with relevant information about general energy policies and the trade-offs associated with alternative sources of energy. A 1988 Danish opinion poll, for example, indicated that 77 percent of the Danish people are positive toward wind energy, and 48 percent are willing to pay extra for environmentally benign energy systems [25]. In general, a positive attitude toward renewable energy makes people less concerned about the visual impact of windfarms; for some, wind turbines are actually favored as visual evidence of clean energy. The ways in which the public is informed about a potential windfarm and allowed to participate in the decision process are also important; locally sponsored

Table 3: Estimated reduction factors due to environmental and land-use constraints[a]

Country	Population density *inhabitants per km²*	Reduction factors from exclusions	
		First order exclusions	Second order exclusions
Contiguous U.S.	3.14	1.6	4
Denmark	20	17	65
Netherlands	360	30	150

a. Estimates are derived from siting studies and experience in the United States, Denmark, and the Netherlands (see text). These are the only countries where land-siting surveys have been carried out in such detail to give meaningful results. First order exclusions are those that can be identified as infeasible or clearly unacceptable. Second order exclusions reflect more stringent exclusions, such as visual objections to siting.

projects tend to be more readily accepted than those introduced by utility companies without adequate interaction with the local community.

Economic spin-offs can also increase acceptability. Not only does the construction of windfarms stimulate the local economy, but the wind turbines can be an important source of additional revenue for those owning the land. In California, farmers and ranchers have benefited significantly from dual use of their land, which has raised land values considerably.

Siting constraints: experience and estimates

Not surprisingly, in sparsely populated regions, siting constraints are relatively few, but assume far greater importance as population densities rise (see table 3). Moreover, it is clear that in more sparsely populated regions, the wind potential (even after accommodating all siting constraints) may be substantial compared with electricity demand. For the relatively sparsely populated United States, a recent U.S. Department of Energy (DOE) report estimates that identifiable obstacles to siting wind turbines (e.g., cities, forests, etc.) reduce available wind energy by a factor of about 1.6. More severe land-use constraints, which exclude essentially all lands except for rangelands and barren lands in the U.S. West, reduce the available energy by a factor of four.[15] [26]

The Danish environmental authorities [27] assessed possible sites in Denmark for 1-megawatt and 2.5-megawatt wind turbines, with a minimum distance

---

15. The following definitions of the different levels of land use exclusions are used in the study by Elliott et al. [26]:

Environmental restrictions: excludes 100 percent of environmentally sensitive areas.

Moderate land use restrictions: excludes 100 percent of environmentally sensitive and urban areas plus 50 percent of forest areas plus 30 percent of agricultural areas plus 10 percent of range areas.

Severe land use restrictions: excludes 100 percent of environmentally sensitive, urban, forest, and agricultural areas plus 10 percent of range areas.

to dwellings of 200 meters. Conflicts of interest (visual impact, bird life, noise, etc.) were evaluated jointly with the local community. The initial screening identified about 200 areas that could accommodate around 6,000 wind turbines with a rated power of 2.5 megawatts each, generating about 36 terawatt-hours (TWh) per year or about ¹⁄₁₇th of the country's potential. After discussions with the local community, the number of wind turbines was subsequently reduced to about 550 with a yearly production of 3.5 TWh or about one-ninth of the country's present electricity production. Smaller wind turbines (from 200 to 450 kilowatts in size) would increase the number of potential sites, raising Denmark's total annual production to at least 10 TWh, which is about ¹⁄₆₀th of its wind energy potential[16] [28]. Studies in the densely populated Netherlands suggest that wind resources there (after all siting constraints are considered) may range from one to 16 TWh per year [29], compared with a gross potential of about 415 TWh per year.

SYSTEM INTEGRATION

Introducing the concept
Systems for supplying electricity range from individual home generators to diesel-based systems that power islands and isolated villages to large integrated power systems that supply the bulk of the world's electricity. In general, as a system's size increases, fuel and operating costs decrease, and the reliability, inertia, and robustness of the system increase. Two concerns are often raised about the utilization of wind-generated electricity. The first is whether or not wind's variable nature reduces its economic value compared with conventional power sources; the second is whether or not such variability makes it impossible (or highly uneconomic) to obtain more than a small percentage of system energy from wind. This section considers the issues as they relate to different system sizes.

Small-scale systems
Wind energy can be integrated into small-scale systems suitable for islands and other communities that are not connected to integrated power systems. In these markets, wind can reduce a community's dependence on high-cost imported diesel fuels for local power generation, and the potential benefits of doing so are well recognized. Many places, including islands and some communities in the developing world, not only need such power but have high wind potentials. In such cases it can be cheaper (and quicker) to build local wind systems rather than to expand the national grid.

Some communities in the developing world are without electricity. Others have diesel power, but must transport their diesel fuel over long distances, which can be expensive. Wind energy can be used alone with storage or, to save fuel, in

---

16. Just before going to press, a new evaluation from the Danish Environmental Agency was released, estimating a feasible resource closer to 5 TWh per year. The details of the assessment were not available at this time.

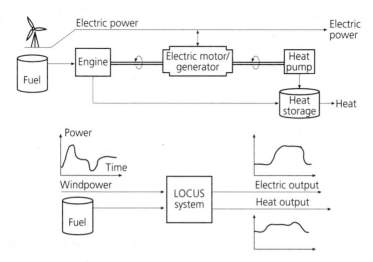

FIGURE 10: Schematics of a local cogeneration unit with heat pump and heat storage (LOCUS plant).

hybrid wind-diesel systems. Unfortunately, diesel plants can be slow to start or to respond to demand fluctuations, and they are often inefficient when run at low or varying output. And because wind energy for a small-scale system may involve only one turbine, it can be highly variable. Thus, unless sophisticated control strategies are adopted, many of the potential advantages offered by hybrid systems will be offset by the added burdens of a variable wind supply.

Control strategies vary widely. One hybrid wind-diesel system (on Fair Isle, between the Orkney Islands and the Shetland Islands off the northeast coast of Scotland) is designed to minimize the need for diesel operation by reducing loads (refrigerators, for example, are placed on circuits that can be interrupted should the wind temporarily abate) [30]. Other strategies rely on short-term energy storage from a flywheel or battery. Fuel savings as great as 80 percent can be obtained with appropriate control strategies [31]. Although optimizing wind-diesel systems is not simple, standardized packages for system design are now available [32].

On a larger scale, wind-diesel hybrid systems share several features with large grid-connected systems. Operators must adjust the scheduling of continuously operating diesel sets to account for variable wind input. Good control can yield considerable savings [33] but the high variability of wind input, combined with the poor operating characteristics of diesel sets, still creates difficulties. But such difficulties can be eased somewhat by installing several wind turbines, thus reducing fluctuations [34].

New systems have been designed that improve overall flexibility and efficiency. One such system is known as the LOCUS plant (see figure 10) [35]. In decentralized configurations, a LOCUS plant produces both electricity and heat using heat pumps, heat storage, and a district heating system. Because the relative output of electricity and heat can vary, these plants can accommodate variations in demand for heat and electricity as well as variable wind contributions. The benefits can be increased further by connecting several LOCUS units together [36].

Small-scale isolated wind systems are important because they provide electricity to communities that otherwise could not afford it or at lower cost than straight diesel systems. But such small-scale applications are limited by their often higher cost and the steady expansion of transmission networks. Thus, the overall potential for wind power seems to depend largely on its prospects for being integrated with grid-connected power systems.

Large-scale systems at low wind penetration

Integrating wind-generated electricity into grid-connected power systems presents no radically new problems [37] (see chapter 23, *Utility strategies*). Electricity demand may vary in its daily cycle by a factor of two, and intermittent sources such as wind act as a "negative load," simply reducing the demand on thermal units and thereby lowering operating costs. The key question that arises is, how does the added variability introduced by intermittent sources affect their value relative to other power plants?

Large systems generally create fewer problems for wind integration than small systems do. On one hand, if the installed wind capacity is small relative to the total demand, then wind fluctuations are simply lost among the fluctuations in electricity demand. On the other hand, if the installed wind capacity is large, then many wind turbines spread out among different sites will smooth the overall output (see the section "Wind-energy resources"). In addition, large systems have a greater natural reserve, with many thermal generating units connected at any time. Most large systems will also have sources, such as hydropower generators and gas turbines, that can respond rapidly to changing conditions. Consequently, wind energy can be exploited without the need for storage.

But in most areas of the world, wind energy cannot be relied on to meet peak demands, which raises concerns about the overall value of wind in displacing conventional power plants. Although wind power correlates to some extent with load—wind increases the thermal loss from buildings in winter, for example—the relation is not absolute. At Altamont Pass in California, peak demand and wind energy both increase in the summer, but they are frequently a few hours out of phase. In fact, in several very different locations, statistical analysis of hourly data indicates that wind availability and peak electricity demand are virtually independent [38–40].

In such cases, small amounts of wind energy still reduce the need for conventional plants. Because no power plant is completely reliable, a finite risk of failure always exists. To keep the risk low, plant capacity must always exceed the maximum expected demand by a large margin. Because wind power may be available at the critical moment when demand is high and other units have failed, it reduces a system's overall risk of failure and allows the conventional-plant margin to be reduced [41].

Therefore, when the statistical relation between wind availability and peak demand is weak, the capacity value of wind energy will equal that of a conventional thermal (coal, nuclear, or oil) power plant with the same energy output, depending somewhat on the season.[17] At locations where wind energy correlates well with peak demand—for example, in parts of California where the thermal winds are strongest when air-conditioning demand is at its peak—the capacity value of wind energy (relative to mean output) would be greater than that of a conventional plant. Conversely, in places where the wind rarely blows at times of peak demand, there is little or no capacity credit (see figure 11). Variations in capacity credit not only reflect differences in wind-load correlations but also wind-turbine capacity factors (that is, the average output compared with capacity).

Marginal capacity credit also declines steadily with increasing penetration, typically being cut in half when wind contributes from 5 to 8 percent of energy to these systems (see figure 11). The way in which capacity credit declines depends on the nature of the system and on the wind regime; studies of diverse wind energy in the United Kingdom [41, 42] suggest that capacity credit will not drop to half its initial value until wind supplies perhaps 10 percent of the system energy.[18] This relatively high figure in part reflects the benefits of so much available diversity, which makes periods without any wind in winter quite rare. Some capacity credit occurs at still higher penetrations, although excess thermal plant capacity above peak demand is always maintained for the rare occasions when there are both very high demand and little or no wind energy.

Another benefit of wind turbines is their ability to be installed rapidly, which reduces the "planning margin" requirement for installed capacity over maximum demand (currently dominated by load forecasting uncertainty) and so saves capital. The advantage has not yet been quantified, however. But despite the interest in capacity issues, the major savings come from the saving of fuel that is displaced by wind energy (the same is true for any baseload plant, such as a nuclear or hydroelectric plant). For the majority of regions where wind energy and electricity demand are not closely correlated, wind is equally likely to be available at any point in the demand cycle. Consequently, average fuel savings are the same as for

---

17. Note that the key factor is availability at times of system risk. For a conventional plant this exceeds the average availability (which takes into account both scheduled and unscheduled outages) because maintenance is scheduled for off-peak periods. For wind energy it depends on seasonal variations, but as discussed in the previous section, wind very often has a positive seasonal correlation with demand, as is the case for both examples cited there.

To a very close approximation the capacity value of wind energy at small penetrations in the United Kingdom equals that of a conventional plant with the same annual energy output. In the United States, the Pacific Gas & Electric Company includes a capacity credit term to windfarm owners, calculated on an hour-by-hour basis to reflect the value of wind power in reducing the loss-of-load probability. For 1986 it was calculated that Altamont windfarms reduced the chance of having lost load by 19 percent. The equivalent "firm power" of 66 megawatts almost exactly equaled the annual average output of the windfarms [40].

18. This would involve an installed wind capacity of 10 gigawatts (nearly 20 percent of thermal capacity, supplying nearly 10 percent of system energy), with a limiting capacity credit of about 5 gigawatts at high wind capacities (compared with a required system reserve margin of about 10 gigawatts without any wind capacity).

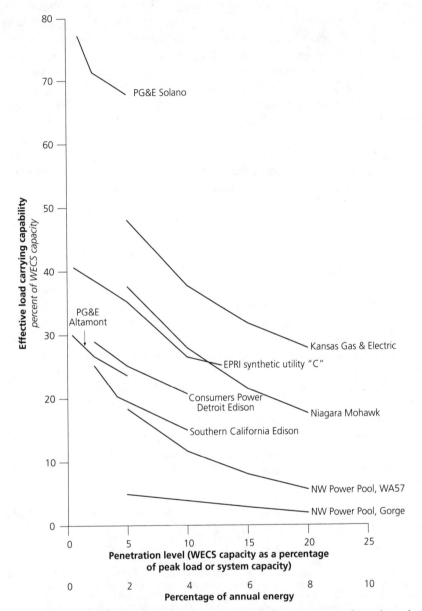

FIGURE 11: Capacity credit of wind energy on a range of U.S. systems. The figure shows the "effective load-carrying capability" (the capacity of a 100 percent reliable source that would be required to give the same contribution to system reliability) of wind energy, as a fraction of the installed capacity. Source: S. Hock, personal communication.

any thermal power source with the same energy output. As explained elsewhere [41], potential operational penalties arising from fluctuations in wind power[19] and uncertainties in wind prediction[20] do not become relevant until higher penetrations are reached (see also the appendix).

Exceptions arise when the correlation between wind energy and electricity demand is negative. In southern California, for example, the strongest winds often occur at times of low or moderate demand, and rarely at times of peak demand. Consequently, the energy derived from wind often displaces low-cost fuel for baseload plants, and thus has less value than if it were uncorrelated with demand.

But for the majority of systems in which wind energy and electricity demand are not closely correlated, the economic value of wind energy is directly comparable to that from conventional power sources: there are no penalties associated with wind's variability when its contribution is below a few percent of the total system energy. When a positive correlation exists, wind energy is more valuable than conventional sources. Only when the correlation is a negative one can wind be penalized because of its variability.

Large-scale systems at high penetration

As a system's wind capacity increases, this relatively simple picture breaks down. Various complex computer modeling studies have attempted to simulate the operation of power systems that incorporate larger capacities of wind energy; the most important of these are reviewed and compared in the appendix. This section focuses on the principles involved and on a study of British electricity supply sys-

---

19. Since the variations in wind input and demand usually add independently, the total variation can be represented as a "sum of squares" addition of the components—in a simplified form:

$$(\text{Total variability of load on thermal units})^2 =$$
$$(\text{total variability of electricity demand})^2 + (\text{total variability of wind energy})^2$$

Consequently, the marginal impact of wind energy fluctuations at low penetrations is practically zero. In practice, the fluctuations of wind energy are no more rapid than those of electricity demand itself, and very large capacities would have to be installed before fluctuations from wind energy become comparable to those already met.

20. Operating reserve is a complex topic, involving two important principles. First, operating reserve should be allocated to the system as a whole, not to back up any particular source; thus, however unpredictable the wind is, it needs to be considered in the context of other system uncertainties and reserve allocated to optimize the whole system.

Second, large power systems have many reserve options for preventing actual collapse of the system. Thus, managing reserve is a question of optimizing the mix between this "inherent" reserve, which is there anyway but may be expensive to use if load predictions are seriously in error, and "active" reserve, which is cheap to use in this event, but may be expensive to have continuously available. Thus active reserve levels are based on an economic trade-off, not an absolute security requirement; consequently the reserve costs are determined primarily by the average (RMS) prediction errors involved, not by maximum errors. Since the prediction errors in wind and load are usually independent, the combined error is again a sum-of-squares addition:

$$(\text{RMS prediction error for net load on thermal units})^2 =$$
$$(\text{RMS prediction error for electricity demand})^2 + (\text{RMS prediction error for wind energy})^2$$

Thus again, any errors in predicting wind energy when the penetration is small are lost among load fluctuations, with no associated penalty, and models that optimize system reserve levels confirm this. Reserve costs at higher penetrations are discussed in the appendix.

tem, which attempted to optimize the long-term operation and power supply of plants utilizing wind energy. As noted above, the capacity credit declines with increasing penetration because wind makes progressively smaller contributions to the system's reliability. In practice, however, capacity credit itself is less important than many assume. Old, inefficient plants may provide backup for newer plants, and a system that needs new plants just to maintain system reliability can have its needs met by building peaking plants—typically gas turbines—that have a capital cost often a third to a fifth of that of new baseload plants. Such units can be more expensive to run (although with the advent of highly efficient gas turbines this is less true than it used to be), but because (by definition) peaking plants are only used occasionally, these fuel costs are relatively unimportant.

At high levels of wind penetration, it is impossible to separate capacity issues and values from the fuel costs of other plants in the system. This can be illustrated by subtracting wind energy from electricity demand, and seeing how the distribution of the remaining load is affected. The residual distribution, which can be plotted as a net load duration curve, provides a look at the potential fuel savings attributable to wind (see figure 12). Clearly, as the wind capacity increases, wind energy will displace more and more baseload fuel, reducing both the value of the wind energy and the amount of time the baseload plant must run. In other words, for a given thermal plant mix, the fuel-saving value of wind energy declines steadily as capacity rises.

In the longer term, as the capacity of wind energy expands, thermal plants that have higher operating-fuel costs but are cheap to build will become more attractive because the reduced operating times will make fuel costs less important. Moreover, because the new thermal plants will be less capital-intensive, more capital will be available to invest in the rest of the system.

It is difficult to quantify the real value of reduced construction costs in the rest of the system because it depends on detailed assumptions about system development. But it is possible to calculate the long-run effects of redesigning the system to achieve given wind capacities for minimal cost (see figure 13). Studies show that the optimal capacity mix, total annuitized system costs, and long-run value of wind energy vary as a function of the installed wind capacity [14]. Even when the capacity of a baseload plant with low operating costs is unconstrained, wind energy is not rendered worthless, as is often assumed. Rather, savings accrue from the fact that less of the plant is economic when wind energy is present, which manifests itself as capital rather than as fuel savings (see figure 13b).

In this analysis, the impact of capacity credit at small penetrations is apparent: total thermal capacity can be reduced, and the electricity from wind turbines becomes as valuable initially as that from any competing thermal power source. In addition, even when wind energy requires backup, long-run savings can be substantial because of the "capacity remix" value. Indeed, total fuel costs are almost constant when the plant mix is optimized to accommodate the wind energy. In fact, in this case study, optimal wind capacity is nearly 30 gigawatts with the unconstrained cheap-operating baseload and more than 40 gigawatts if

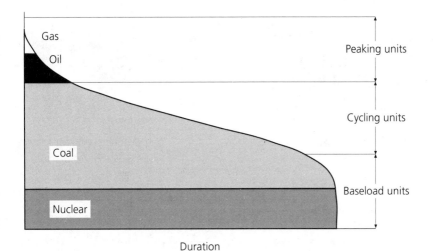

Peaking units

Cycling units

Baseload units

Duration

a. Total demand and merit order of loading

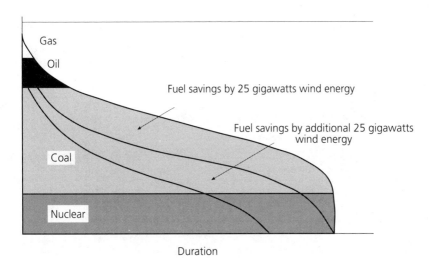

b. Impact of wind energy

FIGURE 12: *Load duration and fuel-saving impacts of wind energy at increasing penetrations are presented in the form of a so-called "load duration curve." The curve shows the duration for which load exceeds a given power level. Conventional plants are loaded under the curve in their "merit order" of operation, and the shaded areas represent energy supplied by each thermal plant type. By subtracting wind energy from the demand, and looking at the residual distribution (the "net load duration curve") it is easy to see the potential fuel savings from the wind energy, and the way in which it reduces the operating time of a thermal plant on the system.*

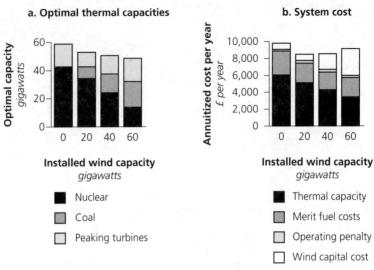

a. Optimal thermal capacities

Optimal capacity
gigawatts

Installed wind capacity
gigawatts

- ■ Nuclear
- ■ Coal
- ☐ Peaking turbines

b. System cost

Annuitized cost per year
£ per year

Installed wind capacity
gigawatts

- ■ Thermal capacity
- ■ Merit fuel costs
- ☐ Operating penalty
- ☐ Wind capital cost

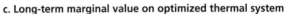

c. Long-term marginal value on optimized thermal system

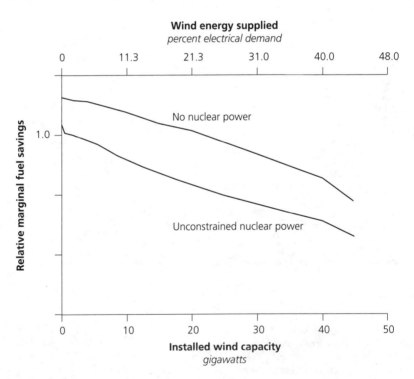

Wind energy supplied
percent electrical demand

| 0 | 11.3 | 21.3 | 31.0 | 40.0 | 48.0 |

No nuclear power

Relative marginal fuel savings

1.0

Unconstrained nuclear power

| 0 | 10 | 20 | 30 | 40 | 50 |

Installed wind capacity
gigawatts

FIGURE 13: Long-run integration of wind energy in a fully optimized system. A mix of three conventional plant types is shown: a baseload plant that is costly to build but cheap to run; a cycling plant that is cheaper to build, but more costly to run (and can vary its output without great difficulty); and peaking plants. The figure shows that the a) optimal capacity mix, b) total annuitized system costs, and c) long-run value of wind energy vary as a function of installed wind capacity, for a system with the demand and wind profiles appropriate to the mainland Britain[14].

such a cheap baseload is absent, compared with current thermal capacity of about 60 gigawatts.

Underlying these findings is a theme that must be grasped if the economic role of wind energy in large power systems is to be understood: the extreme conditions that can in principle occur with wind power have little economic relevance. For example, although some wind energy may be discarded as soon as the installed wind capacity (after availability and transmission losses) equals the minimum load (less a minimum thermal output—see the appendix), the amount is very small. For widely distributed wind systems, a combination of high wind output and low electricity demand occurs so infrequently as to be economically irrelevant. Wind power spillage is only significant if it occurs perhaps 5 to 10 percent of the time, which requires much higher wind penetrations. At the opposite end of the spectrum, combinations of low power and high demand can certainly occur, but these situations can be resolved at very low capital cost (say, by keeping old units in reserve); and if such events are sufficiently rare, the fuel costs incurred on such occasions are largely irrelevant.

The above results suggest that large amounts of wind energy can be accommodated even without storage.[21] But although there is no technical limit to wind penetration, there is a steady decline in its value as more is installed. The rate and form of this decline varies, but in most systems, wind should be able to contribute from 25 to 45 percent of the total electricity before operational losses become prohibitive, even in the absence of storage (see appendix). However, there are regions (especially sparsely populated ones with good resources, such as Canada and the United States), where exploitable resources may exceed those levels. In these areas obviously important issues of integration would arise, including the extensive use of long-distance transmission and/or storage methods, involving electricity directly or its conversion into fuels such as hydrogen (see chapter 22, *Solar hydrogen*).

---

21. As summarized in the appendix, a wide range of results have been reported from integration studies, claiming feasible contributions from wind energy ranging from a few percent to about 50 percent of overall system energy. The reasons for this are discussed in the appendix. They concern primarily differing modelling assumptions about the operation of the system, and the constraints imposed by operating limits placed on current power plants.

If large capacities of wind energy were placed on power systems structured and operated at present, with no significant measures taken either to make thermal units more flexible, or to predict wind energy better, then serious operational penalties could arise for wind contributions much above 10 to 15 percent of system energy. But in reality, such large wind capacities could only be built up over many years. The operating procedures for existing thermal units would be altered in this process to optimize their operation, and modifications might be made, as has already been demonstrated where the operating regime for units has altered for other reasons. New units would be constructed bearing in mind the need for greater operational flexibility, at little extra cost. As indicated in the chapter on utility strategies (chapter 23: Utility Strategies for using Renewables), many favored plants for new investments are anyway highly flexible. For systems relying heavily on combined heat and power production (CHP), their operation would be made more flexible by altering CHP techniques, or by using heat storage (which is far cheaper than electricity storage). Also efforts would be made to ensure good prediction of wind energy. Under such conditions, as explained in the appendix, operational penalties would be very small.

WIND ENERGY POTENTIAL

This section estimates wind energy potential in terms of its physical (theoretical) potential and its realistic potential (taking into account environmental, land-use, and systems constraints). Although the primary data are of varying quality and there are inherent uncertainties surrounding the various constraints, we believe the numbers derived give a reasonable indication of the magnitudes involved.

Technical assumptions

Today's typical wind turbine will have a hub height of about 30 meters above ground and a rotor diameter of about 30 meters, which corresponds to a rated capacity of from 250 to 400 kilowatts. The turbines' efficiency (the amount of incoming wind power converted into mechanical energy) typically ranges from 25 to 30 percent. As discussed in chapter 3, at present, wind resources can be exploited mainly in areas where wind power density is at least 400 watts per m^2 at 30 meters above the ground (or about 500 watts per m^2 at 50 meters), corresponding to U.S. wind-resource classes of 5 or higher.

Rapid advances in turbine technology have led to higher efficiencies, greater hub heights, and expanded unit capacities (see chapter 3). Thus, future wind potential should not be assessed on the basis of existing technology. For long-term resource estimates, we suggest the following technological assumptions:

▼   Hub height above ground = 50 meters.

▼   Rotor diameter D = 50 meters.

▼   Total conversion efficiency (turbine efficiency × array and system efficiency) = 26 percent (e.g., turbine efficiency of 35 percent; array and system losses of 25 percent).

▼   Average turbine spacing equivalent to $10D \times 5D$ (the $8D \times 6D$ spacing, which is more common in Europe, gives a similar packing density).

The losses are well within the limits suggested by array studies (see table 2). According to these assumptions, the density of energy production for a given land area is about 0.0041 of the vertical power density at 50 meters, so that the average annual production per km^2 of windfarm (in MWh per year) is 36 times the (vertical) wind power density (in watts per m^2) at 50 meters. These figures can reasonably be used to estimate average production from a given area after siting constraints are included.

Continuing technical advances will open new areas to development. Included will be regions where the annual average wind power density exceeds 250 to 300 watts per m^2 at 50 meters (300 watts per m^2 corresponds to class 3 or higher in the U.S. wind atlas notation). Such levels appear economically achievable (see chapter 3), especially as resource constraints and environmental concerns raise the price of fossil fuels.

Few regional evaluations have attempted to include offshore potential. This potential is of interest for densely populated countries with suitable coastlines, as is the case for a number of European countries. Estimates are given where available, but are not included in the main assessment.

Wind turbine systems will be a mixture of smaller and larger units installed in various configurations, from individual units to small clusters to large wind parks with tens or hundreds of turbines. Taking into account the lack of detailed data, the simplified assumptions described above are sufficient for approximate resource estimation.

Siting constraints are critically important, but to date there is no standardized way of presenting them. It is natural to start with the total (meteorological) wind energy resources as given by wind energy maps and then to derive the *gross electrical resource* using the above technical assumptions.

First-order exclusions, which reflect undisputable constraints from cities, forests, unreachable mountain areas, and the like, are subtracted, leaving the *first-order potential*. The most important reductions then come from social, environmental, and land-use constraints, including (and perhaps dominated by) visual impact, all of which depend on political and social judgements and traditions, and vary from country to country. We have estimated these *second-order* constraints, based in part on the surveys and field experience in the United States, Denmark, and the Netherlands, to produce a *second-order potential*. With these underlying assumptions, we now examine the potential for wind energy in different regions.

Wind energy potential in North America

In 1986, the U.S. wind atlas was published [10]. The book, which covers the contiguous United States, divides wind resources into seven categories (see table 4 and figure 14).

The most recent study of U.S. wind energy potential [26] draws on the atlas to estimate wind energy resources for each contiguous state, including available land areas and reduced potentials caused by various land-use constraints. The report assumes an average array spacing of $10D \times 5D$ and a total conversion efficiency (measured as turbine efficiency multiplied by array and system efficiencies) of 18.75 percent (that is, a turbine efficiency of 25 percent and array and system losses of 25 percent). These assumptions are consistent with currently available technology and are more conservative than the performance projected above, which indicates a factor of 1.4 greater potential output arising from reduced turbine and system losses.

The resulting electric potential for wind classes 5 and above (judged to be exploitable with today's technology) and for wind classes 3 and above (exploitable with advanced technology) and the corresponding land requirements are summarized in table 5. With the "moderate" land-use constraints, only about 0.6 percent of U.S. land area is exploitable today, but even so, the wind-electric potential on this land would correspond to about one-fourth total U.S. electricity generation

Table 4: Classes of wind power density in the U.S. wind atlas[a]

Wind power class	Wind power density at 10 meters *watts per m²*	Wind speed at 10 meters *m per sec*	Wind power density at 50 meters *watts per m²*	Wind speed at 50 meters *m per sec*
1	0–100	0–4.4	0–200	0–5.6
2	100–150	4.4–5.1	200–300	5.6–6.4
3	150–200	5.1–5.6	300–400	6.4–7.0
4	200–250	5.6–6.0	400–500	7.0–7.5
5	250–300	6.0–6.4	500–600	7.5–8.0
6	300–400	6.4–7.0	600–800	8.0–8.8
7	400–1,000	7.0–9.4	800–2,000	8.8–11.9

a. The categories are specified in terms of power densities at 50 meters height; the vertical extrapolation is based on a wind-speed power law with a $1/7$ height exponent, and mean windspeeds are estimated from the power density assuming a Rayleigh distribution and standard sea-level air density.

Table 5: U.S. wind energy resources[a]

	Percent of U.S. land area	Wind electric potential	
		TWh per year	*Percent of U.S. generation, 1990*
No land-use restrictions			
Wind classes ≥ 5	1.2	1,400	51
Wind classes ≥ 3	21.0	16,700	596
"Environmental" restrictions			
Wind classes ≥ 5	0.8	900	33
Wind classes ≥ 3	18.0	14,300	509
"Moderate" restrictions			
Wind classes ≥ 5	0.6	700	25
Wind classes ≥ 3	13.6	10,800	384
"Severe" restrictions			
Wind classes ≥ 5	0.4	500	17
Wind classes ≥ 3	5.7	4,600	165

a. See footnote 15, p. 19 for definitions of land-use restrictions. Source: Elliott, 1991 [10, 26].

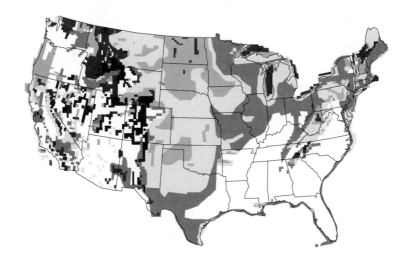

50 meters (164 feet)

	Power class	Wind power (W/m²)	Speed* (m/s)
☐	1	< 200	< 5.6
■	2	200–300	5.6–6.4
☐	3	300–400	6.4–7.0
▨	4	400–500	7.0–7.5
■	5	500–600	7.5–8.0
■	6	600–800	8.0–8.8
▨	7	> 800	> 8.8

* Equivalent wind speed at sea level for a Rayleigh distribution.

FIGURE 14: U.S. wind energy resources. Key results represent the average resource category of gridded cells of 0.333 degrees longitude by 0.25 degrees latitude.

at present. About 80 percent of this potential would come from three states: North Dakota, Wyoming, and Montana.

If wind technology advances as expected and land-use constraints are only "moderate," then more than 13 percent of the U.S. land area could prove economically viable for wind energy. The corresponding electricity potential would exceed by four times the total U.S. electricity generation today, and 90 percent of the potential would exist in 12 contiguous states in the Midwest.[22] Even with the "severe" land-use constraints, D. L. Elliott of Pacific Northwest Laboratories and his colleagues [26] estimate the potential to be 1.6 times the present U.S. electric-

---

22. In order of greatest potential they are North Dakota, Texas, Kansas, South Dakota, Montana, Nebraska, Wyoming, Oklahoma, Minnesota, Iowa, Colorado, and New Mexico.

Table 6: Canadian wind energy resources

Region	Area 1,000 km²	Percentage of area with wind power density Watts per m²			Gross electrical resource TWh/year[a]
		below 200	200–400	above 400	
Keewatin	591	0	4	96	12,500
Quebec	1,541	51	33	16	10,800
Manitoba	650	23	39	38	8,000
Newfoundland	405	25	21	54	5,600
Total					**36,900**

a. Assumptions as outlined on page 28, without exclusions from environmental, land-use, and systems constraints.

ity generation. Because the potential far exceeds local electrical needs, most of the wind energy would have to be exported to other regions of the country, or converted to hydrogen for use in fuel markets, both locally and in other regions (see chapter 22: *Solar Hydrogen*). Naturally there are uncertainties about the detailed figures and siting assumptions,[23] but the broad conclusions seem robust.

Canada has tremendous wind energy resources. Like the United States, it is a large and sparsely populated country. More than 90 percent of the country's wind potential is found in the Keewatin district, together with the provinces of Quebec, Manitoba, and Newfoundland, in order of decreasing potential (see table 6) [43]. The country's projected gross electrical wind resource of about 37,000 TWh per year exceeds its 1988 electricity consumption of 490 TWh by a factor of 70. And because Canada is so sparsely settled, environmental and land-use constraints are likely to be even less than in the United States.[24] In fact, using these resource data, with siting restrictions similar to the "severe" restrictions in the U.S. study, annual wind electricity potential would still exceed 9,200 TWh per year.

Existing hydropower facilities produce about 60 percent of Canada's electricity, which introduces great flexibility in accommodating wind energy, since the

23. There is some uncertainty about suitable assumptions for "severe" constraints. The Elliott et al. definition of 90 percent of rangelands, and nothing of other categories, accounts in part for the dominance of the Midwest states. It seems doubtful in practice whether covering 90 percent of such large areas with wind turbines at $5D \times 10D$ spacing would be acceptable, and clearly there would be substantial interference among arrays, perhaps involving interference with large-scale resource patterns. On the other hand, the assumption that no siting is allowed outside rangelands is clearly unduly pessimistic; equivalent assumptions for Denmark would exclude most of the wind turbines that have already been installed!

24. Although more than 45 percent of Canada is forested and thus unsuitable for wind energy, most of the higher windspeed areas reflected in the table are likely to have much lower forest density; otherwise they would be unlikely to have such strong winds. But even with much more severe siting assumptions the conclusions would be robust.

water can be stored and used for periods when there is little wind. The constraints are those of distance, and of need if much more hydro is exploitable; official interest in wind energy to date has focused upon its potential for remote supplies. However, an almost all-renewable Canadian electricity system, which could even export electricity to the United States, appears entirely feasible.

Overall, it appears that the long-term contribution of wind energy in North America could be very large and will ultimately be constrained not by resource limits, but by the high cost of exploiting weaker resources for conversion or transmission over long distances.

Wind potential in Europe

Detailed assessments of wind resources in Europe are summarized in the European Community wind atlas [3], which was completed in the late 1980s (see figure 14). Estimates of wind energy resources at a height of 50 meters above the ground are presented in the form of 15 maps for the 12 member countries of the European Community (EC). The maps depict the geographical distribution of five wind-energy classes, each of which is represented by a different color. Data

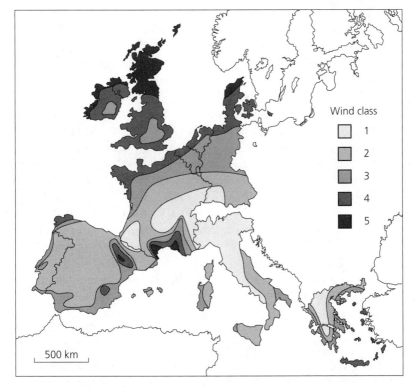

FIGURE 15: European Community wind energy resources (see table 7 for wind power densities in each wind class) [3].

Table 7: Wind resource classes and terrain types in the European wind atlas[a]

Terrain type	Sheltered terrain	Open plain	Sea coast	Open sea	Hills and ridges
Roughness class	3	1	1	0	2
Wind class	Corresponding wind power density at 50 m AGL[b] W/m^2				
1	< 50	< 100	< 150	< 200	< 400
2	50–100	100–200	150–250	200–400	400–700
3	100–150	200–300	250–400	400–600	700–1,200
4	150–250	300–500	400–700	600–800	1,200–1,800
5	> 250	> 500	> 700	> 800	> 1,800

a. The table shows for the five windspeed classes of the European wind atlas [3] the wind power density at 50 meters above ground level, each for five different topographic conditions. In the atlas the wind classes are indicated by different colors on wind maps; we have used a number classification to correspond as closely as possible to that used in the U.S. wind atlas.

b. AGL is above ground level

are also given for five different topographic conditions for each class (see table 7). Unfortunately, different classes have been chosen for the European and American atlases; correspondence between them is shown in tables 4 and 7. The atlas also provides conversion factors for assessing wind resources at heights and roughness classes other than the reference case (50 meters, roughness class 1) (see figure 2).

Wind resources are far from evenly distributed across Europe (see table 8 and figure 15). The United Kingdom, Ireland, and Greece are exceptional[25] and should offer substantial amounts of economic wind power under almost any conditions. The southern coast of France is also very windy. Many promising regions have been identified along a broad band of the northern European coastline, including most of Denmark, where average windspeeds of 7 to 8 meters per second at a height of 50 meters have been measured. Much of the rest of northern Europe and the Mediterranean coastal regions have average windspeeds ranging from 6 to 7 meters-per-second at 50 meters. Within each area, of course, local anomalies can create unusually good sites.

Resource and siting studies for the Netherlands, Europe's most densely populated country, show that siting constraints are likely to limit wind's onshore contribution there to at most 5 percent of consumption. Given such constraints, offshore siting could be important for some European countries. A Danish study, for example, examined the feasibility of installing 3-megawatt wind turbines at

25. The United Kingdom has high wind areas all along the west coast and in most of Scotland; the gross electric resource according to the assumptions on page 28 is about 2,600 TWh per year, compared with the EC's estimate of 1,760 TWh per year after first order exclusions (about one-third of the total EC-10— see footnote 26) estimated by the EC, and 760 TWh per year estimated by Grubb using more conservative technical and first-order siting constraints.

Table 8: Wind electric potentials in Europe

Country or region	Gross electrical potential[a] TWh/year	Population density people per km²	First-order potential TWh/year	Second-order potential TWh/year	1989 Electricity production TWh/year
Denmark	780	120	38 onshore	10 onshore 10 offshore	26
United Kingdom	2,600	235	760 onshore	20–150 onshore 200 offshore	285
Netherlands	420	360	16 onshore	2 onshore	67
EC[a,b]	8,400	140	490 onshore	130 onshore	1,600
Norway		13.1	32[c]	12[d]	109
Sweden	540[e]	19		30[f]	140
Finland		14.7	30[g]	10[g]	51

a. See the section on "Technical assumptions" (p.28).

b. Exclusion factors as for Denmark; see text.

c. For the whole Norwegian coast, including small island-cliffs [46].

d. Using only the best sitings along the coast [46].

e. Includes southern Sweden only, and only areas with mean annual wind power densities higher than 450 watts per m² at 100 meters height. Offshore sitings at 6 to 30 meters depth and more than 3 kilometers from land are also included [47].

f. About 7 TWh per year at land and 23 TWh per year offshore [48].

g. Including some offshore sitings [48].

sea depths between 6 and 10 meters. After accounting for possible conflicts (sailing, fishing, aesthetics, wildlife, offshore mining, telecommunications, air traffic, and military interests), it was determined that about 700 wind turbines, producing approximately 5 TWh per year, could be operated offshore [28]. Although this number may be reduced when actual negotiations with the interested parties (especially fishery organizations) begin, a number of smaller wind turbines could be accommodated in shallower waters (less than 6 meters deep). The result would be a total offshore resource for Denmark of around 10 TWh per year [28]. In the United Kingdom, the potential for wind turbines in 5 to 30 meters of water (*after* accounting for shipping, fishing, and military zones) is estimated to be about 200 TWh per year or 80 percent of current electricity demand [44].

Overall, the EC atlas data suggest that the total land area in the EC with average wind power densities above 250, 400, and 600 watts per m² covers 340,000,

240,000, and 90,000 km^2, respectively, yielding a gross electric resource of about 8,400 TWh per year. The average population density in Europe, which is about 145 persons per km^2, is close enough to that of Denmark so that average Danish exclusion factors may apply (see section "Siting constraints: experience and estimates"). Such factors would reduce the EC potential 17-fold, to about 490 TWh per year for first-order land-use exclusions, and 65-fold to 130 TWh per year for second-order exclusions,[26] or about 31 and 8 percent, respectively, of the total EC electricity consumption of 1,600 TWh per year. Offshore resources could add considerably to this, particularly for some countries.

Wind atlas methods have been used in both Sweden and Finland, but detailed wind maps are lacking for the rest of Europe. Yet promising wind resources exist along the coasts of Greenland, Iceland, Norway, Sweden, Finland, and the Baltic states, where population densities are relatively low. In particular, these countries are well positioned to exploit combined hydro–wind power systems, the advantages of which have been illustrated for a theoretical system linking wind turbines in Denmark with hydropower plants in Norway [49]. Such hybrid systems could be expanded to include all of Scandinavia, whose countries are already connected to a common grid, but doing so will require stronger interconnections. Although wind energy potential in Europe (unlike the United States) is limited by the availability of adequate sites, long-term onshore contributions of around 10 percent of current European electricity consumption, plus offshore resources, appear to be quite feasible.

Rest of the world

Although wind atlases for other regions of the world have been prepared, none yet matches the quality and consistency provided by the U.S. and European atlases. Global wind data from the World Meteorological Organization have been summarized by D.M. Simmons of the Noyes Data Corporation [50], and interpreted for wind energy applications [51]. The approximate and aggregated nature of such data needs to be emphasized, but the data still give a useful indication of probable resources in regions where more detailed evaluations are unavailable.

Africa

Africa straddles the tropical equatorial zones of the globe, and only in the extreme south does it touch the wind regime of the temperate westerlies. Simmons reports that the combination of subtropical anticyclones and strong sea-breeze effects create favorable sites along the coasts of Tunisia, and West Africa, in the Cape Verde and Canary Islands, and in Madagascar. According to his analysis, similar sites probably exist along the entire African coast from Egypt to francophone West Af-

---

26. An earlier estimate by the EC Commission of EC-10 resource *after* quantifiable first-order site exclusions is 4,000 TWh per year [45]. This was based on larger (100 meter) machines, included all windspeeds, and used a gridded approach in which sites were excluded only if there were specific identifiable obstacles. The total first-order resource estimated here is also much lower because of the much more severe estimate of constraints when extrapolated from the (much more detailed) Danish studies.

rica and along the western and southwestern coasts of South Africa. The wind regimes at these sites are reinforced to a great extent by sea-breeze effects caused by strong insolation, especially in the summer. Although inland areas of Africa have less wind potential, good sites may exist in regions such as Central East Africa and Sudan. Anecdotal evidence suggests that some inland areas may have greater resources at hub height than is generally acknowledged (see footnote 7).

One of the few areas for which detailed assessments have been published is Egypt, where recent estimates of wind resources have been made jointly by Egypt's New and Renewable Energy Authority (NREA) and institutions in the United States [52] and Denmark [53]. The mountainous coasts of Egypt on both sides of the Gulf of Suez channel large-scale flow along the coast. The flow is enhanced by large-scale sinking and modified by thermal effects, thus creating large wind potential along the gulf and the Red Sea. Much of the area is uninhabited desert with no constraints for wind turbine siting. One of the most promising areas, the flat coastal region from Ras Abu Darag to Hurghada, could alone generate as much as 30 TWh per year (see "Technical assumptions"). Depending on transmission constraints, another 30 TWh per year could potentially be obtained from other areas with high wind potential, notably those south of Hurghada and along the Mediterranean coast.

For comparison, present Egyptian electricity consumption is about 35 TWh per year. A substantial part of this is supplied by hydropower (about 6 TWh per year), which could help complement wind power. Substantial new transmission capacity would, however, be required to link these windy areas to main transmission routes along the Nile.

Asia

Wind flow on the continent of Asia, which extends from the equator to the polar seas, is very varied. Major features include a winter outflow from the Siberian anticyclone and a summer influx of monsoons across the equator into India and Southeast Asia. Simmons has noted that during the winter, persistent strong winds blow along the Pacific coast, including northeast trade winds around the subtropical Pacific anticyclone, and across the Philippine Islands and neighboring seas. During the summer, northerly winds persist along the Arctic coast, and southeasterly winds along the north Pacific coast blow around the Pacific anticyclone, which has shifted northward, following the movement of the sun. During seasonal transitions, most of the winds are light and variable.

The Chinese Academy of Meteorological Science has established a wind resources database [54]. Areas with highest wind potential, which are identified as having wind power densities greater than 200 watts per m^2 (height unspecified), are found in China's southeast coastal region and in northern Inner Mongolia. According to these wind maps, about 500,000 km^2 have wind power densities greater than 200 watts per m^2. If average annual wind power density in these regions is 400 watts per m^2 at a height of 50 meters, then total wind potential in China is on the order of 7,000 TWh per year, which compares favorably with an electricity consumption in 1989 of about 585 TWh (of this amount about 120 TWh was supplied by hydropower) [55].

Although substantial uncertainties exist in China's wind database (measurement heights and distribution are rarely cited), it should be possible to develop a hybrid hydro–wind system that could supply many times the country's present electricity consumption. One drawback to such a scheme, however, is that while the southeast coastal region proximates major population areas, the Inner Mongolian resource is thousands of kilometers away. Thus, transmission requirements over such large distances need to be closely analyzed.

Considerable interest in wind energy is emerging in India, where resource surveys point to a number of promising areas [56, 52]. Wind surveys of many countries in the Middle East have also been carried out, frequently indicating promising coastal or mountain wind resources [53].

The former USSR covers a vast expanse: its terrain includes steppes and major mountain ranges that would seemingly have great wind potential. In 1989, the USSR submitted a wind resource estimate of 2,000 TWh per year to the World Energy Conference, but the overall resource must be far higher. We have been unable, however, to obtain a more accurate estimate of the region's exploitable wind resources.

Australasia

A 1987 survey of wind resources in New Zealand [58] revealed that more than 11,000 km^2 of land lie within 2 kilometers of the coast, much of it smooth rounded hills with elevations of 50 to 500 meters. Average wind velocities across these lands exceed 8 meters per second at 50 meters above ground level, and thus the region is well suited for windfarms. The gross wind electric resource there (based on technical assumptions and on an average wind density of 700 watts per m^2) is about 280 TWh per year, compared with the country's total electricity demand of about 27 TWh in 1985. An estimate of the practical potential of 100 2.5-megawatt windfarms suggests they might meet 51 percent of electricity consumption. Such high penetration, however, would require close connection to New Zealand's hydropower system, which provides about 80 percent of the country's electricity in a normal year. The potential is thus very promising, limited more by economic requirements rather than by resource or system constraints.

According to Simmons, a strong sea breeze plays an important role in the wind regime along the coastal belt of Australia. Most wind potential is therefore likely to be in coastal regions, where its impact could be relatively large given the large expanse and low population density involved.

Latin America

Some of the best wind potential in Latin America exists along the coasts of Peru and Chile where strong southwesterly winds blow around the Pacific subtropical anticyclone. According to Simmons, the gradient is intensified during the summer because of the sharp temperature contrast that forms between the cold ocean

Table 9: World wind energy resource densities[a]

Region	Class 5–7		Class 4		Class 3	
	1,000km²	*percent*	*1,000km²*	*percent*	*1,000km²*	*percent*
Africa	200	1	3,350	11	3,750	12
Australia	550	5	400	4	850	8
North America	3,350	15	1,750	8	2,550	12
Latin America	950	5	850	5	1,400	8
Western Europe	371	22	416	10	345	8.6
Eastern Europe & former USSR	1,146	5	2,260	10	3,377	15
Rest of Asia	200	5	450	2	1,550	6
Total	**8,350**	**6**	**9,550**	**7**	**13,650**	**10**

a. Source: [62] and personal communication. The wind classes correspond to the notation used in the U.S. wind atlas [26] (see table 4). The areas corresponding to the different wind classes are given in thousands km² for the six continents.

and hotter inland areas. Good sites may also be found south of about 40 degrees latitude; Cape Horn, in fact, is one of the windiest places in the world.

Preliminary estimates suggest that wind resources are plentiful along the western coasts of Central and South America; a study of Costa Rica [59], for example, indicates that it has some of the best wind resources in the world, with mean windspeeds frequently higher than 10 meters per second. A preliminary set of wind atlases for Latin America was published in 1983 [60], but we could not obtain them in an accessible form.

Global wind-energy potential

Few estimates of exploitable global wind resources have been made. In 1981, the International Institute for Applied Systems Analysis (IIASA) estimated the worldwide wind potential to be 26,000 TWh per year [61], but this estimate included only coastal regions, apparently on the (incorrect) assumption that wind energy extracted in coastal areas is not replenished from the upper atmosphere. Consequently, the IIASA estimates are not relevant to a comprehensive assessment. More recently, World Meteorological Organization data have been used to estimate the global wind-energy potential [62] (see table 9).

These data, converted into estimates of wind potential using the technical assumptions, suggest a gross wind electrical potential globally of almost 500,000 TWh per year, with regional breakdowns as shown in table 10. We then estimate the impact of siting exclusions, noting their dependence on population density

Table 10: World wind electricity potentials[a]

Region	Resources class 3 and above		Population density	Estimated second-order potential
	Percent land area	*Gross electric potential, TWh/year*	*People per km²*	*TWh/year*
Africa	24	106,000	20	10,600
Australia	17	30,000	2	3,000
North America	35	139,000	15	14,000
Latin America	18	54,000	15	5,400
Western Europe	42	31,400	102	4,800
Eastern Europe and former USSR	29	106,000	13	10,600
Rest of Asia	9	32,000	100	4,900
World[b]	23	498,000		53,000

a. Numbers may not add due to rounding. For assumptions, see text.

b. Excluding Greenland, Antarctica, most islands, and offshore resources.

and drawing on the specific estimates discussed and the U.S. and European analyses presented above.[27]

With the exception of western Europe and Asia (excluding the former USSR), average population densities are less than those of the contiguous United States (see table 10). For the contiguous United States, Elliott and his colleagues estimated that "severe" siting constraints should have an exclusion factor of four, that is, windfarms could be sited on one-fourth of the total area, primarily in the central and midwest regions. However, noting that the United States may have unusually extensive rangelands, and noting that uncertainties exist concerning the extent to which the assumed array density could be maintained over large areas (footnote 23), we assume a more conservative average exclusion factor for low population-density continents of 10 (that is, one-tenth of the land area is available for windfarms). Combined with the fact that typically only from 25 to 30 percent of the land has class 3 or better winds, our assumptions correspond to windfarms sited on 1.5 to 3 percent of the total land area of the more sparsely populated continents.

For the more populous western Europe and Asia (excluding the former

27. Taking the average power density as the mid-point of the range for class 3 and class 4 winds and an average of 650 watts per m² for classes 5 to 7.

USSR), average population densities are slightly less than those of Denmark, and we assume siting densities corresponding to the Danish experience. Applying the corresponding reduction factors, this yields a global second-order potential of some 50,000 TWh per year. These results include land-based wind turbines only and compare with global electricity consumption in 1988 of about 10,600 TWh per year.

Most of the global wind potential comes from North America (including Canada and Alaska), the former USSR, Africa, and (to a lesser extent) South America and Australia. Unfortunately, much of the potential is in inhospitable areas far from regions of substantial electricity demand. Still, it appears that the global potential, *after* siting constraints, should exceed global electricity demand. However, this potential will be constrained by intraregional system limitations, for example, by the difficulty and costs of transporting surplus energy from resources in the central United States and Canada to coastal centers of demand. Without more information on regional wind distributions and long-term constraints, such issues cannot be adequately addressed. Nonetheless, the potential for exploiting wind energy as electricity is large and will become still larger if low-cost ways can be found for transporting wind energy from remote regions to demand centers, for example, as electrolytic hydrogen (see chapter 22).

POLICIES FOR DEVELOPING AND DEPLOYING WIND POWER

Introduction

Wind power is still a young and rapidly developing technology that must compete with such well-established sources of electricity as fossil fuels, hydropower, and nuclear power. Not only are traditional suppliers well entrenched and powerful, and energy planners and executives often suspicious of new and unfamiliar options, but economic analyses rarely include the external societal costs of energy production on environment, health, buildings, agriculture, and so forth. Thus, emerging technologies like wind power must overcome serious barriers before they can become a major source of energy (see chapter 1, *Renewable fuels and electricity for a growing world economy*).

In order to promote the penetration of wind power, government support programs have been introduced in a number of countries. These programs may include the following elements:

▼ Research and development of wind technology.

▼ Demonstration projects.

▼ Market stimulation through investment subsidies, tax rebates and favorable tariffs.

▼ Independent test stations that can certify the reliability of wind turbine designs.

▼ Resource evaluations.

▼ Buy-back regulations for private wind power sold to utilities.

▼ Regulations for private connections to the grid.

▼ National and utility targets for market penetration.

▼ Efficient planning and implementation procedures.

The most commonly implemented policy instrument has been government-funded research and development programs to promote megawatt-size wind turbines, which are too risky to be considered by small private companies. Many of these programs have run into technical problems (see chapter 3) and have yet to produce a machine that is economical, although promising developments are taking place. Most major developments in wind technology have resulted from government incentives to private companies, such as those incentives that have prevailed in California and Denmark.

California and Denmark

The rapid development of wind power in California in the early 1980s was a product of "sledgehammer" promotion. Generous federal tax credits, combined with state incentives, produced large subsidies for wind installations. In addition, the Public Utility Regulatory Policies Act (PURPA) mandated that utilities buy energy at its full avoided cost from independent generators, thus ensuring a market for wind power. Some contracts included 10-year price guarantees based on projected rising oil prices, and these "standard offer 4" contracts played a pivotal role in convincing wind investors that they would get adequate returns.

Entrepreneurs responded rapidly by constructing small-scale windfarms composed of 50- to 200-kilowatt turbines. Installation rates in California increased sixfold in one year, rising from 10 megawatts in 1981 to 60 megawatts in 1982, and by 1984 had reached 400 megawatts per year. By 1986, the cumulative investment in wind energy totaled about $2 billion, with the value of the energy generated put at $100 million per year [63].

Such rapid development had its drawbacks as well as its benefits. Many of the early machines were of poor quality and broke down during the first season of operation. As one manufacturer complained, some of the early companies knew more about tax minimization than they did about engineering. Machines were often sited carelessly, and some were sold on blatantly fraudulent promises. Such slipshod technology, as well as visions of thousands of motionless machines, threatened to destroy wind power in the United States. However, several companies, aided by favorable state-financing packages, invested heavily in wind energy technology and helped advance the field significantly. The total net cost of the program to the Californian economy (minus fuel savings) is estimated to be about $500 million [64], but without such a program, it is unlikely that wind energy would have reached its current stage.

A different strategy was pursued in Denmark, where interest in wind power has been long-standing, stimulated by a few innovative engineers. The first government research and development program for wind power was initiated in 1976. By 1989, support for the program totaled about 17 million European Currency Units (ECU) (or about $25 million in 1990 dollars). The program supported such endeavors as the construction of two experimental 600 kilowatt machines, preparation of the Danish wind atlas in 1980, and establishment of a test station for wind turbines at Risø National Laboratory, which later set standards for wind machines to ensure quality control. The first market stimulation program was initiated in 1979 when the government agreed to refund 30 percent of the capital cost of a certified wind turbine. As wind machines became cheaper and more profitable, this subsidy was gradually decreased and completely abolished in 1989, but overall it amounted to 35 million ECU (about $50 million).

Since 1979, the Danish government has regulated the price utilities pay for wind-generated electricity as well as the price paid by wind-machine owners for being connected to the grid. These regulations, which include energy-tax refunds, have provided private wind-machine owners with an after-tax return on their invested capital that ranges from 15 to 20 percent. In 1991, the Danish utilities sought to change these regulations to reduce payments to wind generators, but they have been unsuccessful so far.

The gradual increase in turbine size represents generational learning as well as more reliable machines. Danish wind turbines have experienced virtually no major technical setbacks, in part because the independent test station has helped ensure turbine reliability. California provided the first large market for Danish technology; by 1986, Danish turbines supplied half the Californian market, totaling export earnings of more than 640 million ECU (about $900 million). But the Danish wind industry suffered from severe financial difficulties in 1987 and 1988 with the collapse of the dollar and the withdrawal of wind subsidies in California. Today, the industry has been successfully reconstituted, bolstered by an upturn in the domestic market, which has subsequently expanded internationally. In 1991, approximately 45 percent of the world's wind power was supplied by wind turbines manufactured in Denmark.

Recently, the Danish Ministry of Energy has launched a special demonstration program to test the country's relatively large potential for offshore wind power. Completed in 1991, the first offshore windfarm consists of 11 turbines, each with a rated power of 450 kilowatts, and was built by the Elkraft utility with economic support from the EC. A second offshore windfarm is being planned by Elsam, another large Danish utility.

Other market stimulation programs exist in the Netherlands and in Germany, and more recently in the United Kingdom. Each kilowatt of installed wind capacity in the Netherlands is eligible for a direct subsidy of 330 ECU (with a specification for the maximum power per swept area); more recently, Germany has instituted subsidies of about 0.04 ECU per kWh of wind electricity with 10-year price guarantees. Wind farms that qualify under the United Kingdom's non-fossil-fuel obligation also receive a substantial credit for energy produced in the first few years; applications for systems producing more than 200 MW were accepted in 1991.

National policy instruments

This experience suggests six factors that are important to wind energy development:

Wind energy resource surveys.   Because wind resources are complex, and meteorological data on windspeeds are generally too inadequate to identify sites of greatest potential, national resource studies are crucial to wind energy development. At present, resource ignorance is a major constraint on wind development. Indeed, windspeed maps for California in the 1970s suggested there were no areas suitable for the development of wind energy. The later California Energy Commission surveys that identified strong winds near Altamont were essential to wind development in California. The more recent DOE appraisal of U.S. resources is similarly responsible for the upsurge of interest in the American Midwest.

Utility liberalization and buy-back regulation.   Electric utilities have not generally encouraged wind energy, having traditionally focused on large-scale conventional-power generating systems, and many utilities are poorly suited to handle dispersed, small-scale technologies, such as wind turbines. Although attitudes are slowly changing, at present the technology is driven largely by independent developers. Small-scale operations have several advantages: their owners are likely to be more knowledgeable than large utilities about local conditions, including land use. In addition, locally sponsored projects are less likely to encounter local resistance than large utility projects.

Regulations that allow independent generators to produce power and sell it to the grid are thus central issues. When oversight is lacking for the buy-back tariffs utilities must pay for independently produced power, independent generation has generally been squeezed out. In contrast, the California market was able to expand precisely because PURPA provided for regulation of such tariffs by utility commissions.[28] Similarly, Germany passed legislation in 1990 that requires utilities to pay 90 percent of the delivered electricity price to independent power producers; doing so has led to a boom in planned wind power developments.

Market stimulation incentives.   Additional incentives may be required to fully commercialize wind power. Such steps are needed because new energy sources face many obstacles, including simple barriers of scale and direct wielding of market power by established interests. Fossil fuels, for example, are not only subsidized in many countries, but most have high "external costs," which reflect their

---

28. The process of setting the appropriate tariffs can become quite contentious. While in principle the PURPA prescription that rates be equal to the utility's avoided cost is unambiguous, in practice there can be many plausible alternative interpretations of what the utility's avoided cost actually is, and, in the case of wind power, it is not an easy matter to determine the capacity value for wind power.

environmental impact and the implicit risks of dependence on foreign sources. Although taxing fossil fuels to reflect such external impacts makes economic sense, doing so in practice often proves difficult for political and welfare reasons. Incentives need to reward good machine performance with output credits, not just subsidize capital costs.

Utility and national targets. National deployment targets could substitute for subsidies if utilities were to lead wind development efforts. National targets can complement other measures and give companies an indication of how large the national market will be, which in turn helps justify their investment. Targets also provide a measure of progress and give a sense of coherence to wind energy prospects and development. Targets must, however, be flexible enough to accommodate new developments.

National certification. If government incentives for a fledgling industry, such as wind power, are too strong, they may attract investors seeking to make quick profits with inadequate technology, thus tarnishing the industry's reputation. Machine testing and certification programs, such as the one implemented in Denmark, have effectively avoided this problem.

Suitable planning processes. Inadequate planning can result in chaotic and unsightly developments and mounting public opposition. The converse can be equally damaging, however. If every windfarm has to go through lengthy and costly litigation such that a single objector can effectively block developments, then the industry will be moribund. Much has to be learned about appropriate planning procedures, but much of the responsibility must rest with wind developers, who should design sites with sensitivity and sympathy for local concerns.

A more general requirement is that government support be offered on a long-term basis. Any incentive scheme must recognize the capital-intensive nature of wind energy and hence provide a framework with a sufficiently long time span to permit commercial investment over the life of a wind turbine. In countries, such as Denmark, with strong support for renewable energy, investors may be willing to gamble, but in other countries, where the support is more tenuous, confidence will be greatly increased if legislation contains commitments or recommendations as to duration of support.

Market prospects

The California wind-power market, which boomed in the first half of the 1980s, faded after 1986 with the removal of state and federal tax credits and a decline in oil prices. The industry expanded again in the late 1980s, drawing on better machines, environmental concerns, and regained confidence in the technology. The expansion resumed with a broader international focus, initially in Europe but also with renewed interest in the United States and in a number of developing countries such as India, China, and Egypt. Most recently, Eastern Europe and the former Soviet Union have begun to assess the potential of renewable sources of energy, driven by an acute need for cleaner electricity supply systems.

Within the European Community alone, a tenfold expansion of wind power is expected during the 1990s, reaching total installed capacity of about 4,000 megawatts by around 2000 [65]. Similar, if not greater, patterns of expansion could occur in the United States, and perhaps elsewhere depending in part on the pace of technology transfer.

In 1990, global manufacturing capacity for wind turbines was close to 300 megawatts per year. However, if wind power is to increase substantially within the next three to four decades, manufacturing capacity will have to expand rapidly. Experience from other industrial areas (e.g., gas turbines) suggests that for electricity-producing technologies like wind, growth rates of perhaps 25 to 30 percent per year may be feasible, but they will be difficult to sustain for a long time [66]. For wind, if an average growth rate of 25 percent can be maintained until 2005, followed thereafter by a rate of 15 percent, by 2020 global installed wind capacity could be about 400 gigawatts. This capacity would produce close to 1,000 TWh per year, a figure that corresponds to nearly 10 percent of today's global electricity consumption.[29] Beyond such a level, wind deployment would start encountering more serious system and land constraints in the areas of most extensive exploitation. Further expansion might rely on greater development of power systems for bringing power from more remote and less populous or offshore regions. Growth would therefore be slowed, but a further doubling of capacity during the subsequent 10 to 20 years should be possible.

The rate and extent to which such expansion occurs will depend strongly on government policies. At present, low natural-gas prices in the United States and Europe make it hard for any source to compete with new gas plants, which are attractive to private investors because of their low capital cost. Nevertheless, most governments recognize the danger of overdependence on natural gas, and are supporting other energy options, including wind.

The expansion of wind energy in other countries will depend on such measures as wind resource surveys, utility reforms, and (for many developing and Eastern European countries) technology transfer, including reorientation of World Bank and bilateral energy assistance. Although the prospects for wind energy are very promising, it would be dangerous to take success for granted.

CONCLUSIONS

A fundamental reappraisal of the role of wind energy in the global energy economy is required. Under some conditions, modern wind turbines are already competitive with other supply options, notably coal, even when environmental,

---

29. For comparison, note that the Danish government expects to supply 10 percent of electricity consumption with wind power in 2005.

social, and resource-depletion costs are not taken into account. Internalizing these costs would render wind power competitive in many markets, especially with the advanced technologies just now entering the market.

Wind energy can be exploited in several ways. Wind turbines for island or rural communities in developing countries can displace diesel fuel and stimulate local economies. In developed areas, wind turbines can feed electricity directly into the grid. Most such grids could absorb from 25 to 45 percent of their energy from wind power before economic penalties become prohibitive.

Wind resources are very complex, and opportunities are probably being missed, and development slowed, because of inadequate knowledge of both resources and technical developments. International efforts must therefore be launched to develop adequate and standardized data and to disseminate information to all interested parties.

There is substantial regional variation in wind energy potential, with some regions limited by the availability of sites, while others are limited only by the capacity to absorb the energy and transmit it. Nevertheless, the global economic resource for wind energy (after allowing for siting constraints) is likely to be at least of the same order as current global electricity demand. Indeed, wind energy could supply as much as 10 percent of current global electricity demand within the next three to four decades, and double in the following 10 to 20 years, making a major and still expanding contribution to sustainable energy supplies. The extent to which the potential for wind energy is realized, however, will depend very much on the extent to which government policies provide strategic support for emerging energy industries, fund resource surveys, and adjust fuel prices to reflect their social costs.

Appendix
INTEGRATION STUDIES AND SYSTEM-OPERATING PENALTIES AT HIGH POWER SYSTEM PENETRATIONS

Many computer simulation studies of the integration of wind energy in power systems have been conducted, but there is no general agreement as to the potential for wind penetration on utility grids. Dutch utility studies reported that wind contributions would be limited to a few percent of demand [62]. Simulations of the Danish system [65] suggested that operating penalties would become prohibitive if wind were to supply more than 10 percent of demand. But L. Jarrass [67] studied the German system and concluded that no significant operating penalties would occur if input was less than 15 percent of demand; up to this level, savings were almost proportional to installed capacity. Two Australian studies suggested that losses become substantial if wind contributions exceed 20 to 30 percent of the system energy [68]. A number of U.S. studies reported feasible penetration ranging from 15 to 30 percent of demand [69].

For the former Central Electricity Generating Board (CEGB) system in England and Wales, operational penalties were not thought to be prohibitive for wind contributions below about 20 percent of demand [15]. Extensive simulation studies of this system by Reading University and the Rutherford Appleton Laboratories (Reading/RAL) suggested that contributions as great as 15 percent could be accommodated without difficulty, but that penalties would rise rapidly thereafter [67]. Application of the CEGB model to European Commission studies suggested that contributions of 20 percent, and perhaps higher, could be accommodated with little wind discarded [68]. Finally, extensive operational analysis of integration for a United Kingdom power system, optimized for long-term conditions, produced a range of very high wind penetrations [14, 70, 71]. Operating penalties were found to be fairly small, and economic wind contributions in the range of 25 to 45 percent of demand were found to be plausible. The studies did not include storage.

The different wind integration levels presented in the above studies reflect the nature of the various systems as well as a number of different operational assumptions, for example, start-up costs of steam-electric plants. But detailed comparison of the various studies demonstrates that different operational assumptions explain much of the variation in results.

Operational reserve protects the system against unexpected fluctuations. Reserve for fluctuations that last from 5 to 10 minutes can be provided by running steam-electric plants at partial load (so that their output can be rapidly increased to maximum power if required) and by providing hydropower or backup storage on standby. For slightly longer periods (more than 10 minutes), reserve can be provided with rapidly starting gas turbines (unlike steam-electric units, which may take an hour or more to start). But running plants at partial load is inefficient. Furthermore, there is usually a minimum partial-loading limit, below which plants become unstable. A high wind input may thus appear unacceptable, not because the wind input exceeds electricity demand, but because steam-electric plants need to be kept operating to provide operational reserve. Furthermore, if other power sources are assumed to be inflexible (i.e., they must generate at full power and cannot provide any reserve), then other units must keep generating over and above this fixed input solely to provide operating reserve.

Detailed analysis and comparison of the modeling studies cited above [72] emphasize the importance of these operational assumptions. Some of the studies cited use demonstrably incorrect methods for determining reserve levels. For example, the Dutch study cited assumed that reserve should be held equal to wind capacity, and others assume that reserve should be equal to the current wind power input. These assumptions inevitably result in large operational penalties. But operational reserve should be allocated to the system as a whole, and not used to back up one particular source. For example, allocating reserve just in case wind input suddenly falls to zero, without any consideration of the statistics involved, makes as little sense as allocating reserve to protect against the possibility that all consumers will suddenly switch on all their appliances. Such demand is theoretically possible, but so unlikely as to be irrelevant.

Because the short-term variations in wind and demand are usually statistically independent, the prediction errors and variations arising from a small amount of wind energy are drowned out by errors in predicting demand, so there is no operational penalty at low wind penetrations [41]. At higher levels, the uncertainties in wind prediction may dominate those of demand, at which point the maximum plausible variation in wind output, and its predictability, become important. As noted in the section "Energy fluctuations," the probability of wind output from dispensed machines fluctuating by more than 30 percent of wind capacity within one hour is less than one in 10,000, even across the relatively small area of the Netherlands. Moreover, systems can access one of several emergency options even if formal reserve levels are exceeded, so there is no need to hold spinning reserve to protect against unlikely occurrences (see chapter 23). Even if wind prediction is extremely poor, the reserve requirements are clearly less than those modeled in many of the more pessimistic integration studies; in fact, as discussed in the section "Predictability," wind predictability is likely to be quite good when large capacities are installed.

This is a simplified discussion. In reality, reserve control is a complex issue in which "active" reserve levels are optimized against use of "inherent" reserve if the active reserve is exhausted. Many modeling studies fail to optimize reserve levels adequately, resulting in unrealistically large penalties on wind energy.

Another major issue concerns the operational flexibility of baseload plants. Currently, such plants always run at full capacity when available, and simple rules of thumb indicate minimum accepted output levels (the "part-loading limit"). For example, the Reading/RAL studies assume that nuclear output is fixed; the Danish integration study assumes the output of combined heat and power (CHP) schemes to be wholly determined by (predetermined) heat demand, and many studies assume that baseload fossil-fuel plants cannot be partially loaded below output levels of 50 percent. But experience demonstrates that flexibility can usually be achieved when required. In France, even nuclear plants are partially loaded (sometimes to 30 percent of output) when the full nuclear capacity cannot be absorbed; many forms of CHP can alter the ratio of electric power to heat generation (and heat storage, which is far cheaper than electricity storage, can give greater operational flexibility in scheduling the CHP plants); and fossil-fuel plants can generally be adapted at little cost to run at loads well below the limits set in existing baseload duty [73].

Such seemingly detailed technical assumptions largely determine the results of integration studies. The comparison of modeling studies by Grubb [72] also examined the amount of wind energy that could be absorbed on the British power system under a range of different assumptions. Starting from the assumption used in a modeling study that suggested a maximum wind contribution of 15 to 20 percent, the analysis examined the impact of the following assumptions:

▼ Improved algorithm for optimizing operational reserve levels.

▼ Wind prediction errors to half the value arising from simply predicting persistence of current wind input.

▼ Nuclear partial load limit to 35 percent (from 100 percent).

▼ Fossil partial load limit to 35 percent (from 50 percent).

▼ Greater wind diversity.

In steady succession, these relatively modest changes raised the apparent limit on the penetration of wind energy[30] from about 20 percent to more than 45 percent of the electricity supplied.

Even the more sophisticated integration studies cited above generally reflect the impact of adding large capacities of wind power to power systems otherwise structured and operated as at present. In practice, power systems would evolve over decades to accommodate variable sources as the need arises; furthermore, new generating systems, such as combined-cycle plants, offer great flexibility with negligible operating penalties. Demand-side technologies also offer the prospect of greater flexibility in moderating demand according to supply conditions, and vice versa. Options for greater interconnection with other systems, such as hydropower, or for exploitation of wind diversity will also arise.

If the integration studies cited above were carried out with adequate optimization and if conditions were more appropriate to the probable evolution of power systems, operational penalties would be greatly reduced. Consequently, wind contributions of perhaps 25 to 45 percent of energy (depending on conditions) would be feasible before the economic penalties associated with wind's variability become prohibitive, even in the absence of hydro or storage. It seems most implausible that integration difficulties will significantly constrain the use of wind energy in most regions.

WORKS CITED

1. For a detailed account of global energy flows see Bent Sørensen, *1979 Solar Energy*, Academic Press.

2. Lumley, J.L. and Panofsky, H.A. 1964, The Structure of Atmospheric Turbulence, Interscience, London.

3. Troen, I. and Petersen, E.L. 1989
European Wind Atlas, Risø National Laboratories, Denmark.

4. van Wijk, A.J.M., Holtslag, A.A.M., and Turkenburg, W.C. 1984
EWEC'84, Hamburg.

5. Sisterton, D.L. et al. 1983, Difficulties in using Power Laws for Wind Energy Assessments, *Solar Energy* 31:2.

---

30. As emphasized in the text, there is no technical "penetration limit," but rather a steady decline in the value of wind energy as more is introduced. The "limit" is here taken as the percent contribution of wind to system energy at which the marginal value of fuel savings, taking into account all operational requirements on a fully re-optimized supply system, are reduced by about a third of the low-penetration value.

6. Allen, J. and Bird, R. 1977
   Energy Paper No. 21, HMSO, London.

7. Walmsley, J.L., Troen, I., Lalas, D.P., and Mason, P.J. 1990
   Boundary-Layer Meteorology 52, the Netherlands; Barnard, J.C, 1991. An evaluation of
   three models designed for siting wind turbines in areas of complex terrain, *Solar Energy*
   46:5, 283–294.

8. Newton, K. and Burch, S. 1983
   Energy Technology Support Unit, ETSU-R17, United Kingdom.

9. Holt, J.S., Milborrow, D.J., and Surman, P.L. 1989
   BWEA Conference, Glasgow.

10. Elliott, D.L., Holladay, C.G., Barchett, W.R., Foote, H.P., and Sandusky, W.F. 1986
    US Department of Energy, DOE/CH 100934, Washington DC.

11. Petersen, E.L., Troen, I., Frandsen, S., and Hedegaard, K. 1981
    Wind atlas for Denmark, a rational method for wind energy siting, Risø National Labo-
    ratories, Denmark.

12. Frandsen, S. and Christensen, C.J. 1990
    EWEC'90 Conference, Madrid.

13. Swift-Hook, D.T. 1979
    Wind Engineering, Multiscience, Vol. 3, No. 3.

14. Grubb, M.J. 1988
    The potential for wind energy in Britain, *Energy Policy* 12:1. For fuller details see Grubb,
    M.J. 1987, The integration and analysis of intermittent sources on electricity supply sys-
    tems, PhD thesis, Cavendish Laboratory, Cambridge.

15. Farmer, E.D. et al., 1980
    Economic and operational implications of a complex of wind generators on a power sys-
    tem, *IEE Proc. A,* Vol. 127, London; Bossanyii, E. et al. 1980, *Wind Engineering,* Vol. 4,
    No. 1.14.

16. van Wijk, A.J.M., Coelingh, J.P., and Turkenburg, W.C. 1990
    Modelling wind power production in the Netherlands, *Wind Engineering* 14:2. Similar
    results are reported for Germany in Beyer, H.G., Luther, J., and Stenberger-Willms, R.,
    1990. Fluctuations in the combined power output from geographically distributed grid
    couple wind energy conversion systems, *Wind Engineering* 14:3. For some statistics on
    U.K. wind variability see Palutikof, J.P. et al. 1990, Estimation of the wind resource at
    proposed wind turbine sites: the problems of spatial and temporal variability, *Wind Engi-
    neering* 14:1.

17. Bossanyi, E. et al. 1980
    Fluctuations in output from wind turbine clusters, *Wind Engineering,* Vol. 4, No. 1.

18. Builtjes, P.J.H. and Milborrow, D.J. 1980
    Modelling of wind turbine arrays, 3rd International Symposium on Wind Energy,
    Copenhagen.

19. Elliott, D.L. 1991
    Status of wake and array loss research, AWEA Windpower 91 Conference, Palm Springs,
    California.

20. Milborrow, D.J. and Surman, P.L. 1987
    CEGB wind energy strategy and future research, in J.Gault, ed., *Proceedings of the 9th
    BWEA Conference,* Glasgow.

21. European Wind Energy Association. 1991
    Wind Energy in Europe—Time for Action, EWEA, Oxford, October.

22. Winkelman, J.E. 1988
    Energie Spectrum 12, the Netherlands.

23. Ornis Consult. 1988
    Copenhagen, Denmark.

24. *Windpower Monthly* , June 1992 8(8); D. Smith, private communication.

25. Danish Ministry of Energy. 1985
    Renewable energy information, Copenhagen, Denmark.

26. Elliot, D.L., Windell, L.L., and Gower, G.L. 1991
    An assessment of the available windy land area and wind energy potential in the contiguous United States, Pacific Northwest Laboratory, PNL-7789, August.

27. Danish Energy Agency. 1986
    Large windmills in Denmark (in Danish), Copenhagen, Denmark.

28. Meyer, N.I., Pedersen, P.B., and Viegand, J. 1990
    Renewable Energy in Denmark (in Danish), Borgen Press, Copenhagen, Denmark.

29. Berkhuizen, J.C., de Vries, E.T., van den Doel, J.C., and Muis, H. 1986
    CEA, Rotterdam; Krekel, v.d.Woerd and Wouterse, 1987, KWW, Rotterdam.

30. Somerville, W.M. 1986
    In Anderson and Powles, eds., *Wind Energy Conversion 1986*, MEP, London.

31. Grauers, A. and Carlsson, O. 1991
    Wind-diesel system with a variable speed turbine, EWEC'91, Amsterdam.

32. Lundsager, P., Bindner, H., Infield, D., Scotney, A., Skarstein, O., Toftevaag, T., Uhlen, K., Pierik, J.T.G., Manninen, L., and Falchetta, M. 1991
    Progress with the European wind-diesel modelling software package, EWEC'91, Amsterdam.

33. Lipman, N.H. et al. 1989, Review of Wind Diesel Integration Studies,
    Rutherford Appleton Laboratory, Oxford.

34. Beyer, H.G., Degner, T., Gabler, H., and Waldl, H.P. 1991
    Wind farm diesel systems—an analysis of short term power fluctuations, EWEC'91, Amsterdam.

35. Illum, K. 1983
    Locus systems and plants, Aalborg University, Denmark.

36. Lund, H. and Rosager, F. 1985
    Analysis of electricity overflow in connection with wind power (in Danish), Aalborg University, Denmark.

37. Diesendorf, M. and Martin, B. 1981
    Wind Engineering, Vol. 4, No. 4.

38. Cook, H., Palutikoff, J., and Davies, T. 1988
    The effect of geographical dispersion on the variability of wind energy, *10th BWEA Wind Energy Conference,* London, March.

39. Grubb, M.J. 1988
    On capacity credits and wind-load correlations in Britain, *10th BWEA Wind Energy Conference,* London, March.

40. Smith, D.R. 1987
    The wind farms of the Altamont Pass area, *Ann. Rev. Energy* 12:145–83, 178.

41. Grubb, M.J. 1991
   The value of variable sources on power systems, *IEE Proc C,* 138 (2), London; Rocking-ham, A.P. and Taylor, R.H., System economic theory for WECS, *Proceedings of the 2nd BWEA Conference,* Cranfield; Swift-Hook, D.T. 1987, *Proceedings of the 9th BWEA Wind Energy Conference,* Edinburgh.

42. Bossanyi, E. 1983
   Wind Engineering 7:4.

43. Thompson, G. 1981
   The prospects for wind and wave power in North America, Report No. 117, Center for Energy and Environmental Studies, Princeton University, Princeton, New Jersey.

44. Milborrow, D. et al. 1992
   The UK offshore windpower resource, *Proceedings of the 4th International Symposium on Wind Energy Systems,* Stockholm, BHRA, Cranfield, March; Newton, K. 1983, *The UK wind energy resource,* ETSU-R20, Energy Technology Support Unit, Harwell, September.

45. Selzer, H. 1986
   Wind resource assessment, European Solar Energy R&D, Series G, Vol. 2.

46. Nygaard, T.A. and Tallhaug, L. 1990
   Wind power potential in Norway (in Norwegian), Norwegian Institute for Energy Technology, Kjeller, Norway.

47. Swedish wind power potential. 1988
   Swedish Government Report, SOU 1988:32, Stockholm, Sweden.

48. Lund, P.D. and Peltola, E.T. 1991
   Larger scale utilization of wind energy in Finland, an updated scenario, *Proceedings of the EWEC Conference,* Amsterdam.

49. Sørensen, B. 1981
   A combined wind and hydro power system, *Energy Policy.*

50. Simmons, D.M. 1985
   Wind Power, Noyes Data Corporation, Park Ridge, New Jersey.

51. Elliot, D.L., Aspliden, C.I., and Cherry, N.J. 1981
   World-wide wind resource assessment, Pacific Northwest Laboratory, World Meteorological Organization, USA.

52. New and Renewable Energy Authority (NREA). 1989
   Egypt Annual Average Wind Power Estimates, NREA Egypt and Battelle, USA.

53. Wind assessment at the Red Sea. 1991
   New and Renewable Energy Authority, Egypt and Risø National Laboratory, Denmark; Larsen, S.E. and Hansen, J.C., 1991, Risø National Laboratory, Denmark (unpublished).

54. Ruichao, Z. and Heng, X. 1990
   Evaluations of the wind energy resources and the potential of utilization in China, Academy of Meteorological Science, China.

55. Huifen, S. and Xuzhong, Y. 1991
   China's energy industry and its air pollution control, *Proceedings of Conference on Global Collaboration on a Sustainable Energy Development,* Snekkersten, Denmark.

56. Hosain, J. 1990
   An assessment of potential for installation of windfarms in India, PACER Conference on Role of Innovative Technologies, Tata Energy Research Institute, TERI/PFC, New Delhi, April.

57. *International Journal of Solar Energy* . 1988–1990
   Pergamon Press, various issues.

58. Cherry, N.J. 1987
    Wind Energy Resource Survey of New Zealand, New Zealand Energy Research and Development Committee, University of Auckland, New Zealand.

59. Electrowatt Engineering Services (Zürich). 1984
    Nonconventional energy sources, Vol. 1; Wind energy, for Instituto Costarricense Electricidad, San José.

60. Organizacion Latinoamericana de Energia. 1983
    Atlas Eolico Preliminar de America Latina y el Caribe, OLADE, Bogota.

61. Haefele, W. et al. 1981
    Energy in a finite world, International Institute of Applied Systems Analysis/Ballinger, Cambridge, Massachusetts.

62. van Wijk, A.J.M., Coelingh, J.P., and Turkenburg, W.C. 1991
    Proceedings of EWEC'91, Amsterdam; van Wijk, A.J.M., Coelingh, J.P., and Turkenburg, W.C., 1991, Wind energy, in Renewable Energy Report to the World Energy Conference (forthcoming 1992).

63. Gipe, P. 1986
    Maturation of the US wind industry, *Public Utilities Fortnightly,* USA, 20 February.

64. Cox, A., Blumstein, C., and Gilbert, R. 1989
    Wind power in California: a case study of targeted subsidies, Universitywide Energy Research Group report UER-191, University of California.

65. Grubb, N.J. 1991
    Wind energy in Britain and Europe: how much, how fast?, *Energy Exploration and Exploitation,* Vol. 12, January.

66. Grubb, M.J. and Walker, J., eds. 1992
    Emerging energy technologies: impacts and policy implications, RIIA/Dartmouth, Aldershot, United Kingdom, Chapter 1.

67. Jarrass, L. 1981
    Wind Engineering, Vol. 5, No. 3.

68. Halberg, N. 1983
    In Palz and Schnell, *Solar energy R&D in the European Community,* Series G, Vol. 1.

69. Hock, S. and Flaim, T. 1984
    Synthesis of value studies, SERI, Boulder, Colorado.

70. Bossanyi, E. and Halliday, J. 1983
    Proceedings of the 5th BWEA Wind Energy Conference ; Bossanyi, E. 1983, *Wind Engineering,* Vol. 7, No. 4.

71. Holt, D., Milborrow, D., and Thorpe, A. 1990
    Wind energy penetration study for the CEGB, Commission of the European Communities, DGXII, Brussels.

72. Grubb, M.J. 1988
    The economic value of wind energy at high power system penetrations: an analysis of models, sensitivities and assumptions, *Wind Engineering,* Vol. 12, No. 1.

73. Fenton, F.J. 1982
    Survey of cyclic load capabilities of fossil-steam generation units, IEEE transactions, PAS-101 (6).

74. World Meteorological Organization (WMO), 1981, Technical Note 175, Meteorological Aspects of the Utilization of Wind as an Energy Source, WMO, Geneva, Switzerland.

5
SOLAR-THERMAL ELECTRIC TECHNOLOGY

PASCAL DE LAQUIL III
DAVID KEARNEY
MICHAEL GEYER
RICHARD DIVER

Significant progress has been made in developing economically competitive solar-thermal electric technologies. During the early 1980s, several important pilot plants were constructed and successfully operated, thus establishing the technology's feasibility. Today more than 350 megawatts of electricity are generated by commercial solar-thermal plants in the United States, and the experience gained from those plants, in addition to research and development activities, has helped reduce the cost of solar-thermal systems to one-fifth that of the early pilot plants. Continued technological improvements are likely to reduce costs further, while enhancing performance levels. These advances, along with cost reductions made possible by large-scale production, construction of a succession of power plants, and scale-up to plant sizes of 100 to 200 megawatts of electricity, promise to make solar-thermal systems cost-competitive with fossil-fuel plants. Solar-thermal technologies are appropriate for a wide range of applications, including central-station power plants, where they can meet peak utility and intermediate load needs, and modular power plants in both remote and grid-distributed areas.

INTRODUCTION

The solar resource
In less than 40 minutes, the United States receives more energy in the form of sunlight than it does from the fossil fuels it burns in a year. Sunlight is the world's largest energy resource, and for thousands of years human civilization has effectively used nonconcentrated solar energy to produce light and heat and to grow food. Today, technologies are being developed that concentrate sunlight and utilize its energy for other applications, such as producing electricity, as well as steam and hot water for industrial processes.

Although the sun's direct and diffuse rays can be collected by solar arrays and used to heat a home or supply its hot-water needs, the unconcentrated rays are not nearly strong enough for efficient power generation. Solar-thermal electric technologies must, therefore, concentrate large amounts of sunlight onto a small area to permit the buildup of high-temperature heat, which in turn can be converted into electricity in a conventional heat engine. Solar-thermal power plants depend on direct sunlight, and so must be sited in regions having high direct solar radiation. The amount of solar radiation available at the earth's surface, at peak intensity, is approximately 1 kilowatt per square meter (m^2). Good solar power-plant sites typically have at least 2,500 kilowatt-hours (kWh) per m^2 of sunlight available annually, which corresponds to an average daily sunlight value of 6.8 kWh per m^2 [1, 2].

Background

Sunlight can be concentrated with mirrors or lenses to such an extent that it is theoretically possible to achieve temperatures approaching those of the sun's surface. In ancient Syracuse, Archimedes is said to have employed a large number of burnished shields as reflectors to ignite the sails of attacking Roman ships. About two centuries ago, the French scientist Antoine Lavoisier came close to melting platinum in a two-lens solar furnace. Today's solar-thermal technologies produce temperatures high enough to drive modern, efficient heat engines.

Three major solar-thermal technologies have been developed; each is characterized by the shape of the mirrored surface on which sunlight is collected and concentrated. Three major designs have been developed.[1] They are 1) the parabolic-trough system, which concentrates solar energy onto a receiver pipe located along the focal line of a trough collector; 2) the central-receiver system, which uses sun-tracking mirrors called heliostats to reflect solar energy onto a receiver/heat exchanger located on top of a tower; and 3) the parabolic-dish system, which uses a tracking dish reflector to concentrate sunlight onto either a receiver/engine or a receiver/heat exchanger mounted at the focal point of the dish.

Market potential

The size of the world's solar energy resource is truly enormous, and the amount that can be readily accessed with existing technology greatly exceeds the world's primary energy consumption. The size of the accessible resource in the United States, for example, exceeds 100 exajoules (1 exajoule = 10^{18} joules), which is significantly greater than the country's current primary energy consumption of 84 exajoules. Two recent studies, which assess the potential use of renewable energy technologies during the next 30 years, suggest that renewable resources can provide between 25 and 50 percent of the country's primary energy needs, depending on the degree of government and market support. According to the study, solar-thermal electric systems in 2020 are expected to produce from 2 to 3 exajoules per year. The amount is significant because at a typical solar-thermal power plant with a capacity factor of 35 percent, 1 exajoule per year represents a generating capacity of 30,000 megawatts of electricity!

---

1. One solar-thermal technology, the solar pond, does not concentrate sunlight, but absorbs both the direct and diffuse components of sunlight, employing a salt gradient to inhibit the natural convective action generated by the sun's heat. Although solar ponds do not achieve the high temperatures of concentrator systems, they can provide heat energy for desalting, process heat, and limited electricity generation.

But in order to fully exploit solar-thermal power, broader cooperation between governments, electric utilities, and private industry is needed. Government support for research, technology development, and demonstration projects, though necessary, is not sufficient. In addition, market incentives must be created to overcome barriers in the user and financial communities that exist because solar-thermal technologies are new, potentially risky, and viewed as a possible threat to existing technologies and interests. Finally, the major investments needed to properly develop and market solar-thermal electric technology must be supported by stable long-term regulatory policies [3, 4].

SOLAR-THERMAL TECHNOLOGIES

All solar-thermal power technologies rely on four basic systems: collector, receiver, transport-storage, and power conversion (see figure 1). The collector captures

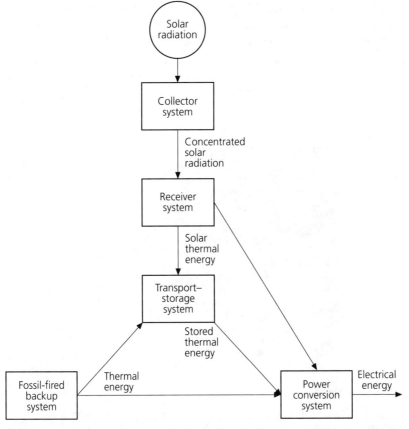

FIGURE 1: Solar-thermal electric technologies rely on collector, receiver, storage, and power conversion systems to convert sunlight into electricity. Some are also equipped with a backup fossil-fuel system. Annual solar-to-electric efficiency is a function of the reliability of these various systems.

and concentrates solar radiation, which is then delivered to the receiver. The receiver absorbs the concentrated sunlight, transferring its heat energy to a working fluid. The transport-storage system passes the fluid from the receiver to the power-conversion system; in some solar-thermal plants a portion of the thermal energy is stored for later use. The power-conversion system consists of a heat engine and related equipment for converting thermal into electrical energy. Some designs also include a secondary, fossil-fuel driven heat source that can either charge the storage system or drive the power-conversion system during periods of low sunlight.

The performance of a solar-thermal power plant reflects the efficiency and reliability of its four principal systems. A plant's solar-to-electric efficiency is measured as the ratio of its net electric output to the solar energy received by its collector field, measured annually (see figure 2). The amount of solar energy received is the product of the annual direct normal solar radiation, in kWh per m^2, multiplied by total collector aperture area. Because some solar input is lost when the plant cannot operate, the remaining sunlight is defined as the usable or available solar energy. A certain amount of energy is lost at each processing stage: sunlight is lost in the collector system as a result of sun-angle effects and optical losses; thermal energy is lost in the receiver and heat-transport systems; a significant loss occurs in the conversion of thermal energy to electricity; and a fraction of the electrical energy produced by the power-conversion system is consumed internally by the plant.

The efficiency of each component in a solar-thermal system can be calculated. The efficiency of the collector system, for example, depends on the reflectivity

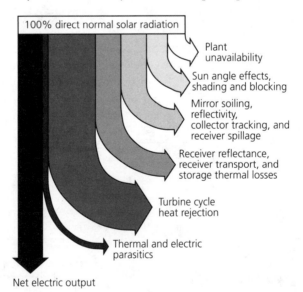

FIGURE 2: *Efficiency with which a solar-thermal system converts sunlight to electricity is charted. From 10 to 30 percent of direct sunlight reaching a system is converted into electricity.*

of its mirrors and the optical effectiveness of its geometry. Losses from the system result from sun-angle effects (also called cosine losses), soiling of the mirrors, shadowing and blocking of sunlight from the reflective surface by other components, and spillage of reflected sunlight outside the receiver boundaries. The collector's efficiency is the ratio of the energy incident on the receiver to the available direct normal sunlight on the aperture area.

The receiver's efficiency is the ratio of energy absorbed by the working fluid to the energy incident on the receiver. Losses result from sunlight reflecting off the receiver and from thermal energy that is radiated, convected, and conducted away from the hot receiver. Efficiency of the transport-storage system is determined by the ratio of thermal energy delivered to the power-conversion system to thermal energy absorbed in the receiver. Losses can usually be attributed to heat conduction through tanks, piping, valves, and other components. The efficiency of the power-conversion system is equivalent to the gross thermal-to-electric conversion efficiency of the power cycle. So-called parasitic losses refer to the amount of thermal and electrical energy required to run the plant during operating as well as nonoperating periods. Net electric output is gross turbine output minus parasitic losses.

Efficiency values are calculated for the plant's design operating conditions[2] and on an annual basis. Annual efficiency values measure the performance of the plant throughout a typical year and include the effects of equipment start-up and shutdown, overnight energy losses, part-load performance, and equipment availability. These values are either measured from actual plant data or they are simulated on a computer.

Although solar-thermal plants are capital-intensive, they are relatively cheap to operate because fuel costs (excluding supplemental fossil-firing) are almost nonexistent. Capital costs are typically dominated by the collector system, which effectively represents the plant's lifetime fuel supply. The next most expensive item is the power-conversion system, followed by the receiver and transport-storage systems. Additional costs for land, structures, and controls are small, but must be included. For example, solar-thermal plants use between 2 and 8 hectares per megawatt, but land costs normally represent less than 1 percent of a facility's total capital costs, in part because much of the land suitable for plant siting is unproductive desert.

Parabolic-trough systems

A parabolic trough is a linear solar collector with, as its name implies, a parabolic cross section. Its reflective surface concentrates sunlight onto a receiver tube located along the trough's focal line (see figure 3a). Fluid flowing in the tube is heated and then transported to a central point through a piping network designed to minimize heat loss. Parabolic troughs typically have a single horizontal focal line

---

2. The design point efficiency for a plant is typically calculated at noon on a solstice or equinox day at the design sunlight intensity. The design point efficiencies measure the instantaneous performance of the plant and its systems.

and therefore track the sun along only one axis, either north–south or east–west. A north–south orientation provides slightly more energy per year than an east–west orientation, but its winter output is low in mid-latitudes. In contrast, an east–west orientation provides a more constant output throughout the year.

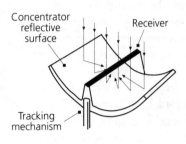

a) Parabolic trough

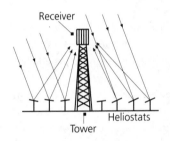

b) Central receiver

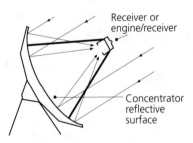

c) Parabolic dish

FIGURE 3: Three major solar-thermal technologies: the parabolic trough, the central receiver, and the parabolic dish are depicted. Parabolic-trough systems (a) concentrate solar energy onto a receiver tube that is positioned along the line focus of the trough collector. Central-receiver systems (b) have heliostats, or suntracking mirrors, that reflect solar energy to a receiver atop a tower. Parabolic-dish systems (c) use a parabolic two-axis tracking concentrator to focus the sun's rays onto a receiver mounted at the focal point of the dish.

Because sunlight is concentrated along a line rather than at a point, the concentration ratio of a parabolic trough typically ranges from 10 to 100, which is a much lower ratio than that of either the central receiver or the parabolic dish. Similarly, parabolic troughs operate at temperatures ranging from 100 to 400°C, which are significantly lower than those of other solar-thermal systems. But parabolic-trough systems are the most fully developed solar-thermal technology, and major installations for both process heat and electricity production already exist in the United States

Each parabolic-trough collector and its associated receiver tube, tracking device, and controls form a modular collector assembly that can be connected in various series- and parallel-flow circuits to achieve a range of performance characteristics. The plant in turn consists of a field of collector assemblies that transfer their collected thermal energy to a central power-conversion system, where the energy is converted into electricity. Commercial parabolic-trough electric plants use auxiliary gas-fired boilers or heaters to meet peak electricity demand during times when solar radiation intensity is low. Thermal energy storage for such systems has not yet proven to be economic.

Central-receiver systems

A central-receiver system consists of a field of heliostats, or sun-tracking mirrors, which reflect solar energy to a tower-mounted receiver (see figure 3b). The concentrated heat energy absorbed by the receiver is transferred to a circulating fluid that can be stored and later used to produce power. Central receivers have several key features: 1) they collect solar energy optically and transfer it to a single receiver, thus minimizing thermal-energy transport requirements; 2) they typically achieve concentration ratios of 300 to 1,500 and so are highly efficient both in collecting energy and in converting it to electricity; 3) they can conveniently store thermal energy; and 4) they are quite large (generally 10 megawatts or higher) and thus benefit from economies of scale. Central-receiver systems can operate at temperatures from 500 to 1,500°C.

Each heliostat at a central-receiver facility has from 50 to 150 m^2 of reflective surface. The heliostats collect and concentrate sunlight onto the receiver, which absorbs the concentrated sunlight, transferring its energy to a heat-transfer fluid. The heat-transport system, which consists primarily of pipes, pumps, and valves, directs the transfer fluid in a closed loop between the receiver, storage, and power-conversion systems. A thermal-storage system typically stores the collected energy as sensible heat for later delivery to the power-conversion system. The storage system also decouples the collection of solar energy from its conversion to electricity. The power-conversion system consists of a steam generator, turbine generator, and support equipment, which convert the thermal energy into electricity and supply it to the utility grid. A master control system coordinates the activities of the different systems [5].

There are three general configurations for the collector and receiver systems. In the first, heliostats completely surround the receiver tower, and the receiver, which is cylindrical, has an exterior heat-transfer surface. In the second, the heliostats are located north of the receiver tower (in the northern hemisphere), and the receiver has an enclosed heat-transfer surface. In the third, the heliostats are located north of the receiver tower, and the receiver, which is a vertical plane, has a north-facing heat-transfer surface.

In the final analysis, however, it is selection of the heat-transfer fluid, thermal-storage medium, and power-conversion cycle that defines a central-receiver plant. The heat-transfer fluid may either be water/steam, liquid sodium, or molten nitrate salt (sodium nitrate/potassium nitrate); the thermal-storage medium may be oil mixed with crushed rock, molten nitrate salt, or liquid sodium. All rely on steam-Rankine power-conversion systems, although a more advanced system has been proposed that would use air as the heat-transfer fluid, ceramic bricks for thermal storage, and either a steam-Rankine or open-cycle Brayton power-conversion system.

Parabolic-dish systems

A parabolic dish is a point-focus collector that tracks the sun in two axes, concentrating solar energy onto a receiver located at the focal point of the dish (see figure 3c). The receiver absorbs the radiant solar energy, converting it into thermal energy in a circulating fluid. The thermal energy can then either be converted into electricity using an engine-generator coupled directly to the receiver, or it can be transported through pipes to a central power-conversion system. Because the receivers are distributed throughout a collector field, parabolic dishes, like parabolic troughs, are often called distributed-receiver systems. Parabolic dishes have several important attributes: 1) because they are always pointed at the sun, they are the most efficient of all collector systems; 2) they typically have concentration ratios ranging from 600 to 2,000, and thus are highly efficient at thermal-energy absorption and power conversion; and 3) they have modular collector and receiver units that can either function independently or as part of a larger system of dishes. Parabolic-dish systems can achieve temperatures in excess of 1,500°C.

Parabolic-dish systems that generate electricity from a central power converter collect the absorbed sunlight from individual receivers and deliver it via a heat-transfer fluid to the thermal-storage and central power-conversion systems. The system is similar to a parabolic trough in that thermal energy, rather than reflected sunlight (as in the central-receiver system), is gathered from the collector field. The need to circulate heat-transfer fluid throughout the collector field raises design issues such as piping layout, pumping requirements, and thermal losses. To date, the central-generation parabolic-dish systems that have been built used either oil or water (steam) for their heat-transfer fluid.

Systems that employ small generators at the focal point of each dish gather energy in the form of electricity rather than as heated fluid. The engine-generators, which are close-coupled power-conversion systems, have several components: a receiver to absorb the concentrated sunlight and to heat the working fluid of the engine, which then converts the thermal energy into mechanical work; an alternator or generator attached to the engine to convert the work into electricity; a waste-heat exhaust system to vent excess heat to the atmosphere; and a control system to match the engine's operation to the available sunlight. This distributed parabolic-dish system lacks thermal-storage capabilities, but can be hybridized to run on fossil fuel during periods without sunshine.

Solar ponds

A solar pond does not concentrate solar radiation (see figure 4), but rather collects solar energy by absorbing both the direct and diffuse components of sunlight. Solar ponds contain salt in high concentrations near the bottom, with significantly diminishing concentrations near the surface. Such a disparity in concentration, known as a salt-density gradient, is obtained by preferentially dissolving heavy salt with increasing depth; this serves to suppress the natural tendency for heated fluid to rise to the surface and lose its heat to the atmosphere by evaporation, convection, and radiation. The density gradient permits heated fluid to remain in the bottom layers of the pond while the surface layers stay relatively cool. Temperature differences between the bottom and surface layers are sufficient to drive the vapor generator of an organic Rankine-cycle engine, to provide process heat and to desalt water. Temperatures of 90°C are commonly attained in the pond bottom. Solar ponds are sufficiently large to permit limited seasonal as well as diurnal storage of energy.

Hot salty fluid is withdrawn from the storage zone at the bottom of a solar pond, passed through an organic fluid evaporator or other heat exchanger, and returned to the pond bottom. During electricity generation, cool, less salty fluid can be drawn from the surface layers and used to cool the Rankine-cycle condenser; water is then returned to the surface.

The largest solar pond in the United States is a 0.3 hectare experimental facility in El Paso, Texas, which has operated reliably since its start-up in 1986. The pond runs a 70 kilowatt-electric (kW_e) organic Rankine-cycle turbine generator, and a 5,000 gallon per day desalting unit, while also providing process heat to an adjacent food processing company. The pond has reached and sustained temper-

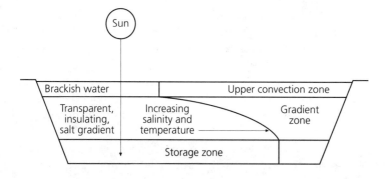

FIGURE 4: Solar pond with its characteristic salt gradient, is shown.

atures higher than 90°C in its heat-storage zone, generated more than 100 kilowatts of electricity during peak power output, and produced more than 80,000 gallons of potable water in a 24 hour period. During its five years of operation, it has produced more than 50,000 kWh of electricity [6]. In the early 1980s, a 20 hectare pond was built in Israel on the shore of the Dead Sea to power a 5 megawatt-electric (MW$_e$) Ormat turbine generator. Although the solar pond operated successfully for several years, in 1989 it was shut down for economic reasons [7].

Solar ponds are limited as a source for large-scale electricity by their need to consume vast amounts of water. Not only does the water that evaporates from the surface of the pond need to be replaced, but salt that diffuses from the bottom to the surface must be reclaimed by evaporating surface water and sending its dissolved salt back to the bottom. The amount of water consumed exceeds that of a conventional power plant by about 35 times. However, such hydrological demands do not limit the role of solar ponds in water-desalting applications, where electricity is needed primarily for internal pumping requirements. One such application has recently been proposed for California's Salton Sea [8–10].

The potential of solar ponds to provide freshwater, process heat, and electricity for island communities and coastal desert regions appears promising but has not been fully investigated. Although construction of the pond is the major capital investment, it can be performed locally. The turbine plant and the desalting equipment, which may need to be imported, are a small fraction of its total cost. In addition, the large storage capacity of a solar pond is well matched to the sunlight conditions of many island and desert coastal sites. Moreover, solar ponds function well at sizes suitable for villages and small cities.

PARABOLIC-TROUGH POWER PLANTS

In the 1880s, the Swedish American John Ericsson powered his hot-air engine with a parabolic trough, but it was not until 1912 that the trough was used in any significant way for power generation. At that time Frank Shumann and C.V. Boys of the United States constructed a 45 kilowatt steam-pumping plant in Meadi, Egypt, that covered an area of 1,200 m^2. Despite the plant's success, it was shut down in 1915 due to the onset of World War I and cheaper fuel prices.

Interest in the technology was not renewed until the 1970s and 1980s, at which time the U.S. Department of Energy (DOE) sponsored the development of several parabolic-trough process heat and water pumping systems. Two major programs were launched: a Solar Industrial Process Heat program [11], followed by a Modular Industrial Solar Retrofit program [12]. At about the same time, an International Energy Agency system for electric power production was tested in Spain [13].

To date, the most noteworthy privately financed, nonelectric parabolic-trough facility in the world is the successful 5,576 m^2 system in Chandler, Arizona, which has been operating since 1983. The system generates and stores thermal energy for copper-plating electrolyte tanks.

From 1984 to 1991, the Luz company dominated parabolic-trough technology, bringing it to a state of commercialization. Before operating their first 14-MW_e Solar Energy Generating Station (SEGS I), the company spent several years developing the components and systems at a test facility in Jerusalem, and was responsible for the construction and operation of two process heat facilities in Israel. Today, nine Luz-developed power plants, which generate a total of 354 megawatts of electricity, are operating in southern California.

Early pilot plants

Solar irrigation

In 1979, an irrigation pumping facility was constructed in Coolidge, Arizona. The objectives of the project, which ceased operation in 1982, were to test the system's performance, reliability, and operation/maintenance requirements. Parabolic troughs that covered an area of 2,140 m^2 furnished heat to a 150 kW_e organic-Rankine cycle to supply electricity to the local grid, which in turn supplied the pump motors.

The system achieved a solar-to-net electric efficiency of only 2.5 percent due to an inefficient power-conversion system and high receiver thermal losses. However, the solar-collector field achieved an availability of 98 percent.

SSPS/DCS

In 1981, a larger solar-electric parabolic-trough facility was constructed at the International Energy Agency test site at Tabernas, Spain [13]. Named the Small Solar Power Systems Project/Distributed Collector System (SSPS/DCS), the facility consisted of two solar fields with a total reflector area of 7,602 m^2, a steam generation system, and a conventional steam-turbine cycle power block connected to the local grid. Both single-axis and two-axis tracking collectors were evaluated, and testing lasted until 1986.

The SSPS/DCS provided considerable data over its lifetime with respect to the design and performance of parabolic troughs, thermal-storage systems, and power blocks. But the solar-to-electric efficiency of the system, with a design value approaching 9 percent reached only 2.5 percent on a daily basis. Solar-field heat losses, high thermal inertia, and an inefficient turbine cycle all account for the drop in efficiency. In general, the single-axis tracking design performed better than the two-axis design. While this is partly attributable to specific design features in the two collectors, the higher heat losses associated with the more extensive piping needed for the two-axis system were the primary factor [13].

Commercial projects

SEGS I and II

In 1984, the SEGS I plant was constructed and put on-line at Daggett, California, as a private investor-owned power plant supplying electricity to the Southern

California Edison Company's grid under the terms of a private power-purchase agreement. SEGS I, with an output of 13.8 megawatts, represented a major increase in field size from earlier Luz projects (see table 1). It was therefore a somewhat risky prospect that was accepted by investors because significant financial guarantees and other protections were offered.

The plant produces saturated steam at 35.3 bar, which is then superheated to 415°C by natural gas, and achieves a Rankine-cycle efficiency of 31.5 percent. A mineral heat-transfer fluid circulates through the collector fluid and returns to the power-conversion system to generate steam via shell-and-tube heat exchangers (see figure 5). Two large hot and cold oil tanks, each of which has a capacity of 850,000 gallons, provide enough storage to produce nearly three hours of full-load turbine operation. The system, which has a collector field aperture area of 82,960 m², was designed to generate 30,100 megawatt-hours (MWh) per year, given an average daily direct normal insolation of 7.46 kWh per m² and about 18 percent of its annual energy input from natural gas for superheating.

The 190,338 m² SEGS II plant, which immediately followed SEGS I, went on-line in 1985. The new plant differed from its predecessor in having greater turbine capacity (30 megawatts) and a natural gas–fired boiler (instead of thermal storage).[3] The independent natural gas firing improved the plant's ability to match output to the utility demands. However, to maintain eligibility as a qualifying facility under the Public Utilities Regulatory Policy Act, the Federal Energy Regulatory Commission requires that the natural gas input to a plant's total annual energy input be limited to 25 percent. The higher pressure and superheat

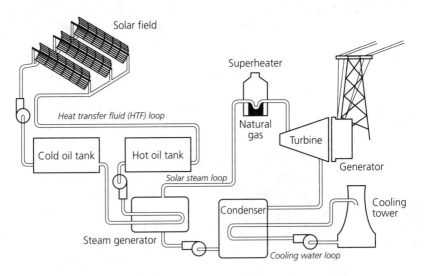

FIGURE 5: *Schematic flow diagram of the SEGS I power plant.*

---

3. Thermal storage was not economic in this or in subsequent SEGS plants because the higher solar-field operating temperatures required a more expensive synthetic heat-transfer fluid.

temperature of the steam produced in the gas-fired mode raises the turbine's efficiency sharply. The total annual design output of SEGS II is about 80,500 MWh of electricity.

SEGS I and II, which could be termed "pilot commercial plants" since they are the first large-scale parabolic-trough installations to generate electricity, are currently operating under a 30-year agreement with Southern California Edison. Experience with the SEGS plants has shown that one to two years of operation are normally required to bring a facility up to production capacity. Having solved early problems with the solar field and power-block systems, the reliability of these plants is now excellent. Although their annual performance levels are only about 85 percent of original projections, this performance is more a reflection of limited experience with the technology and inadequate performance projection methods. Revenues from 1991 to 1992 are considerably lower than projections, however, because low natural gas costs have led to low utility electricity rates, thus jeopardizing the continued economic operation of these facilities.

SEGS III–SEGS VII

Like SEGS I and II, a series of five 30 MW_e SEGS plants were subsequently developed as investor-owned facilities to supply power to Southern California Edison. The design, installation, and operation of each successive SEGS plant gained from the experience of previous plants in areas such as system configuration, field design, power-block component selection, construction techniques, and operating methods. Improvements to the solar field, for example, raised collector operating temperatures from about 321°C to 390°C, which in turn led to higher turbine inlet-steam pressure and a moderately higher superheat temperature. This gain, plus the introduction of reheat to the steam cycle, raised gross turbine-cycle efficiencies in the solar mode from 29.4 percent, which was achieved with SEGS II, to 37.5 percent. At the same time, optical efficiencies increased from 71 to 80 percent which, together with the improved-cycle efficiency, increased projected annual electrical output by 25 percent—from about 0.42 to 0.53 MWh per m^2. Projected annual solar-to-electric efficiencies are 11.5 percent for SEGS plants III–V; and 14.5 percent for plants VI to VIII. Design-point efficiencies are 23 and 26 percent, respectively.

SEGS III and IV began service in 1986 at Kramer Junction, California. Identical in design, they are both characterized by an increase in solar-mode steam conditions to 43.5 bar/327°C but have retained the same gas-fired steam conditions as SEGS II at 105 bar/510°C. SEGS V, which entered service in 1987, duplicates the design. Each plant consists of a solar field, power block, adjunct services (water and natural-gas supplies, power transmission lines), and a water-treatment system (see figure 6). The axes of the collector assemblies are oriented north–south. The power block, which contains all major mechanical and electrical subsystems for power production, sits near the center of the solar field. Together, the solar field, power block, and water-supply system cover about 2 hectares per MW_e net.

Table 1: Plant characteristics of SEGS I – SEGS IX

	Units	I	II	III	IV	V
Power block						
Turbine-generator output	gross MW$_e$	14.7	33	33	33	33
Output to utility	net MW$_e$	13.8	30	30	30	30
Turbine-generator set: Solar steam conditions						
Inlet pressure	bar	35.3	27.2	43.5	43.5	43.5
Reheat pressure	bar	0	0	0	0	0
Inlet temperature	°C	415	360	327	327	327
Reheat temperature	°C	na	na	na	na	na
Gas mode steam conditions[a]						
Inlet pressure	bar	0	105	105	105	105
Reheat pressure	bar	0	0	0	0	0
Inlet temperature	°C	0	510	510	510	510
Reheat temperature	°C	na	na	na	na	na
Electrical conversion efficiency						
Solar mode[b]	percent	31.5	29.4	30.6	30.6	30.6
Gas mode[c]	percent	0	37.3	37.3	37.3	37.3
Solar field						
Solar collector assemblies						
LS-1 (128 m^2)		560	536	0	0	0
LS-2 (235 m^2)		48	518	980	980	992
LS-3 (545 m^2)		0	0	0	0	32
Number of mirror segments		41,600	96,464	117,600	117,600	126,208
Field aperture area	m^2	82,960	190,338	230,300	230,300	250,560
Field inlet temperature	°C	240	231	248	248	248
Field outlet temperature	°C	307	321	349	349	349
Annual thermal efficiency	percent	35	43	43	43	43
Peak optical efficiency	percent	71	71	73	73	73
System thermal losses	percent of peak	17	12	14	14	14
Heat transfer fluid						
Type		Esso 500	VP-1	VP-1	VP-1	VP-1
Inventory	m^3	3,213	416	403	403	461
Thermal storage capacity	MWh$_t$	110	0	0	0	0
General						
Annual power outlet	net MWh/year	30,100	80,500	91,311	91,311	99,182
Annual gas power use	10^9 m^3/year	4.76	9.46	9.63	9.63	10.53

a. Gas superheating contributes 18 percent of turbine inlet energy

b. Generator gross electrical output divided by solar field thermal input

c. Generator gross electrical output divided by thermal input from gas-fired boiler or HTF heater

Table 1: (cont.)

Units		VI	VII	VIII	IX
Power block					
Turbine-generator output	*gross MW$_e$*	33	33	88	88
Output to utility	*net MW$_e$*	30	30	80	80
Turbine-generator set:					
Solar steam conditions					
Inlet pressure	*bar*	100	100	100	100
Reheat pressure	*bar*	17.2	17.2	17.2	17.2
Inlet temperature	*°C*	371	371	371	371
Reheat temperature	*°C*	371	371	371	371
Gas mode steam conditions[a]					
Inlet pressure	*bar*	100	100	100	100
Reheat pressure	*bar*	17.2	17.2	17.2	17.2
Inlet temperature	*°C*	510	510	371	371
Reheat temperature	*°C*	371	371	371	371
Electrical conversion efficiency					
Solar mode[b]	*percent*	37.5	37.5	37.6	37.6
Gas mode[c]	*percent*	39.5	39.5	37.6	37.6
Solar field					
Solar collector assemblies					
LS-1 (128 m^2)		0	0	0	0
LS-2 (235 m^2)		800	400	0	0
LS-3 (545 m^2)		0	184	852	888
Number of mirror segments		96,000	89,216	190,848	198,912
Field aperture area	*m^2*	188,000	194,280	464,340	483,960
Field inlet temperature	*°C*	293	293	293	293
Field outlet temperature	*°C*	390	390	390	390
Annual thermal efficiency	*percent*	42	43	53	50
Peak optical efficiency	*percent*	76	76	80	80
System thermal losses	*percent of peak*	15	15	15	15
Heat transfer fluid					
Type		VP-1	VP-1	VP-1	VP-1
Inventory	*m^3*	416	416	1,289	1,289
Thermal storage capacity	*MWh$_t$*	0	0	0	0
General					
Annual power outlet	*net MWh/ year*	90,850	92,646	252,842	256,125
Annual gas power use	*10^9 m^3/year*	8.1	8.1	24.8	25.2

a. Gas superheating contributes 18 percent of turbine inlet energy

b. Generator gross electrical output divided by solar field thermal input

c. Generator gross electrical output divided by thermal input from gas-fired boiler or HTF heater

The Luz solar field is built up from solar-collector assemblies, each of which consists of a row of individual parabolic-trough collectors driven by a single drive train (see figure 7). The mirrored parabolic troughs concentrate direct normal solar radiation onto a heat collection element, which is a selectively coated steel pipe surrounded by an evacuated glass annulus to reduce heat losses from the collectors. An advanced microprocessor, in conjunction with a sun sensor, tracks the sun, keeping the collectors focused during periods of sufficient insolation. The assemblies are arranged in parallel rows with three to four per row. The row-to-row spacing is optimized to minimize piping costs and shadowing during morning and evening hours.

Heat-transfer fluid passes from the collectors through main header piping to solar heat exchangers in the power block. Superheated steam generated by the heat-transfer fluid is then fed to the low-pressure casing of a conventional steam turbine. Spent steam is condensed into water, which returns to the heat exchangers, where it reverts back to steam. After passing through the heat exchangers, the cooled heat-transfer fluid circulates once again through the solar field, thus repeating the process.

A gas-fired auxiliary boiler supplies an alternate source of turbine inlet steam to the high-pressure casing, thus allowing electricity to be produced at night or on low-insolation days, if necessary, in accordance with the Federal Energy Regulatory Commission's constraints on natural gas use.

The conventional power block uses feedwater heaters to increase cycle efficiency at the modest inlet-steam pressure and temperature conditions that are generated by the solar field. Solar field control is provided by assembly-based microprocessors linked to a central microcomputer; the power block is controlled

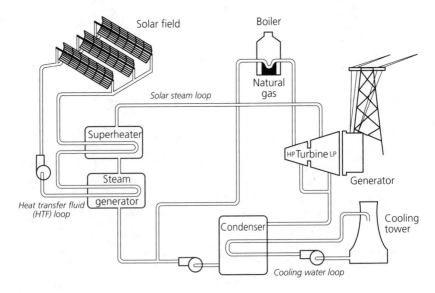

FIGURE 6: *Schematic flow diagram of the SEGS III–V plants is shown.*

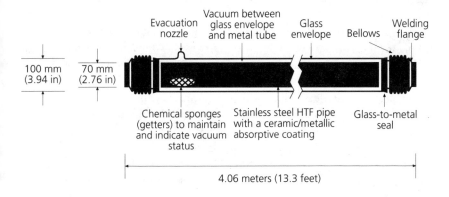

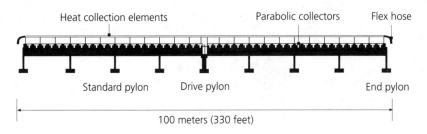

FIGURE 7: SEGS plant heat-collection element (top) and solar-collector assembly (bottom).

by a distributed process control system. Auxiliary services include water pumping, treatment and storage, natural gas transmission, and electric interconnection and transmission.

At SEGS III–V plants, heat-transfer fluid exiting the solar field reaches 349°C, which heats the turbine inlet steam to a superheated temperature of 327°C. The receivers are coated with a black chrome selective surface, and each assembly, which consists of Luz's second-generation trough collectors, has a mirror aperture area of about 235 m² [14].

SEGS VI and VII, which were installed in 1988, differ from SEGS III–V in the design of their power block, and thus in their efficiency. By raising steam conditions and changing to a reheat steam turbine, which consists of high- and low-pressure casings, the efficiency of the steam-Rankine cycle was increased by 20 percent (or about 7 percentage points). After the main steam expands through the high-pressure casing it passes through reheaters before being fed to the low-pressure casing. Reheat is supplied by the solar field during the solar mode and by the boiler during the gas-fired mode.

The increased efficiency of the SEGS VI and VII plants was made possible by changes in the solar field to give an outlet temperature of 390°C and, in turn, a steam pressure of 100 bar and superheat/reheat temperatures of 371°C. The receiver's black chrome was replaced with a ceramic/metallic selective surface, thus increasing absorptivity of the sun's rays and, most important, lowering emissivity to reduce reradiation losses.[4]

Performance of the SEGS plants can be viewed in several ways. As commercial plants, their primary objective is to meet revenue projections. In fact, owners were given a 10 year warranty of annual revenues, which was backed by the Luz company and financial instruments. Time-of-use electricity rates place high values on electricity produced during high, or on-peak, demand periods, which are weekday afternoons from 12 noon to 6 pm, from June through September. The summer mid-peak period (weekday mornings and evenings) has medium value, whereas the winter mid-peak period (weekdays from October through May), has lowest value. The contributions of SEGS III–VII to total projected annual revenues during summer on-peak, summer mid-peak, and winter mid-peak are 46, 15, and 29 percent, respectively.

The maturing process typically takes one or two years of adjusting and fine-tuning the power block and solar field, as well as learning to integrate the two operations most efficiently. The solar-to-electric efficiency of SEGS III–V has increased steadily, especially between 1989 and 1990 (see figure 8); today these plants function at a high performance level.

SEGS VI and VII are entering the final stage of their maturation period, and in 1990 they approached their warranty performance levels. During the critical on-peak period, the combined performance of SEGS III–VII actually exceeded 100 percent of their warranted output of 97 percent (see figure 9). Such strong on-peak performance produced good to excellent 1990 revenues, which ranged

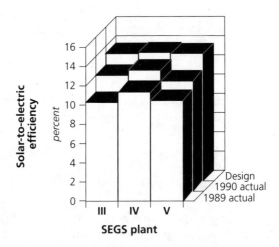

FIGURE 8: Predicted and actual annual solar-to-electric efficiency values from 1989 to 1990 have been calculated for SEGS III–V plants. The performance of SEGS V is slightly lower than expected because measurements were made after a number of third-generation collectors were added to the field but before they could be properly adjusted.

4. Note that in table 1 the turbine inlet pressure with gas-fired boiler steam is also 100 bar, but in this case the superheat temperature reaches 510°C and the cycle efficiency is slightly higher.

from a high of 108 percent of warranty for SEGS IV to a low of 90 percent for SEGS VII. Contributing to the plants' overall success was the fact that from 97 to 99 percent of all the solar collectors were available to track the sun, compared with a warranteed level of 97 percent [15].

SEGS VIII and IX

Federal regulations, which govern the size of solar power plants, were modified in 1989 to allow for an increase in SEGS plant size to 80 megawatts of electricity, thus increasing economies of scale in the power block. The first 80 MW$_e$ plant, SEGS VIII, entered service in 1989, and was followed in 1990 bySEGS IX (identical to its predecessor except for a slightly larger solar field). Both are located at Harper Lake, California [16].

The third generation of Luz parabolic-trough collectors, which were partially introduced in SEGS V and VII, are fully incorporated in SEGS VIII. The new collectors are larger (the size of a solar-collector assembly increased to 545 m^2), and so need fewer tracking systems per area of mirror surface. In addition, improvements were made to the collector's support structure, tracking system, and other components. The drive, for example, was changed from a gear-motor unit to a hydraulic system, markedly cutting costs without sacrificing performance or maintenance requirements.

The power block in the 80 megawatt plants is based on the reheat-turbine concept, which was first introduced in the SEGS VI and VII plants. But in these advanced plants, the boiler is replaced with a gas-fired heater that operates in parallel with the solar field (see figure 10). Because the turbine inlet steam has identical characteristics when the plant operates in either solar or gas-fired mode, the cycle efficiency is identical for each mode. In most other respects, however, (except size), SEGS VIII and IX are similar to SEGS VI and VII. When other SEGS

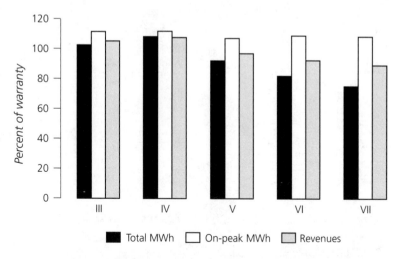

FIGURE 9: Performances of SEGS III–VII for 1990, compared with warranteed output.

plants are eventually built, centralization of the control systems, water pumping and treatment operations, fire-protection, and transmission switchyard will take place, further reducing costs and improving operations.[5]

From 1984 to 1990, the cost (in 1990 dollars) of constructing and operating SEGS plants fell considerably. Installation costs, which were $5,979 per kilowatt for SEGS I, dropped to $3,011 per kilowatt for SEGS VIII; in addition, performance levels rose. As a result, levelized electricity costs, as calculated by Luz for the financial assumptions of the SEGS plants, were 26.5 cents per kWh for SEGS I and dropped by almost 70 percent to 8.9 cents per kWh for SEGS VIII [15].

Future SEGS plants

Although Luz possessed contracts for four more plants having a total capacity of 300 megawatts, it filed for bankruptcy under U.S. law in late 1991 after failing to finance SEGS X. For the near future, the company's bankruptcy has rendered future commercial parabolic-trough developments at best uncertain. Several companies are exploring the option of continuing the development of SEGS technology and commercial trough projects.

Prior to its bankruptcy, the Luz company was actively improving the cost-effectiveness of its trough technology, with plans to achieve near cost competitiveness with conventional power plants. Unfortunately, the significant decline in fossil-energy prices in the 1980s, which occurred in parallel with marked reductions in regulatory and/or financial incentives (such as tax credits), created a dif-

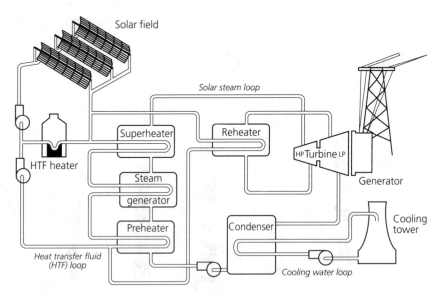

FIGURE 10: *Flow of heat-transfer fluid through the SEGS VIII and IX plants.*

---

5. Because SEGS VIII and IX began service in 1990, no significant measures of their performance are yet available.

ficult market environment in the United States for capital-intensive renewable technologies [17]. However, the situation may improve: burgeoning environmental concerns and a growing trend among public utility commissions to penalize power plants for their emissions promises to improve the market for renewable energy systems such as the SEGS plants.

Future SEGS plants may be configured quite differently from existing ones in order to meet base-load rather than just peak-load needs. In this case, the solar field would be integrated with a combined-cycle system[6] such that an oversized turbine could accept steam generated by both the solar field and by the combustion turbine's exhaust. Other plans called for an advanced collector, known as the LS-4, that would generate steam directly in the solar collector.

Direct-steam generation

Direct-steam generation (DSG), as its name implies, is a concept for steam generation in the solar field itself. By eliminating the need for heat-transfer fluid and centralized oil-heated steam generators, significant cost savings can be achieved (see figure 11). Performance gains are also possible with DSG. The gains derive from two mechanisms. First, by eliminating the heat-transfer fluid, the solar-field's operating temperature can be lowered without affecting the temperature of the steam to the turbine, and so will slightly reduce thermal losses. And second, pumping power requirements are lessened because the system will operate only boiler-feedwater pumps instead of both boiler-feedwater and heat-transfer fluid

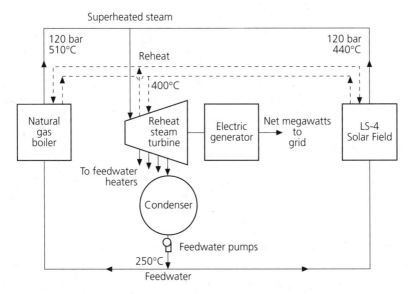

FIGURE 11: Process flow through a hybrid solar direct-steam generation (DSG) power plant.

---

6. A combined cycle is created when exhaust heat from a combustion turbine is used to generate steam, which then drives a conventional steam cycle.

pumps, provided that sophisticated controls successfully facilitate the two-phase flow of water and steam. Luz had planned to improve collector performance further by tilting the design, improving receiver tube absorptivity/emissivity characteristics and improving optical performance.

The DSG plant is designed to operate in either solar- or gas-fired mode, with steam supplied to the reheat turbine at 120 bar. Turbine-cycle efficiency is 40 percent in the solar mode, but 41 percent in the fossil mode because the boiler can superheat to a higher temperature.

Projections by Luz indicated that electrical performance can be increased by 4 percent, with a capital cost reduction of 13 percent and an overall 8 percent reduction in the levelized cost of electricity. Projections for a 200 MW_e DSG plant in the Mojave Desert gave an estimated installation price ranging from $2,100 to $2,300 per kW_e, a net annual production of 475,000 MWh, and a levelized electricity cost of about 8 cents per kWh.

DSG technology involves certain technical risks: tube overheating may occur in the boiling region, and flow instabilities may occur in parallel arrays. In order for the technology to be successfully commercialized, not only must such risks be reduced to an acceptable level, but performance must be field-tested. In 1989, Luz launched a major program to develop DSG technology and planned to collaborate with other interested groups to test plants of greater size and capacity, with an aim to commercialize the technology by 1996.

Combined-cycle with direct steam generation

A promising new configuration that combined SEGS parabolic-trough technology with a gas-turbine combined-cycle power plant was conceived to meet utility needs for continuous operation and peaking power with minimal environmental damage. Such a hybrid combined-cycle plant uses the solar field as the evaporation stage of an integrated system, with the gas-turbine exhaust being recycled for superheating and preheating. In other words, the solar field serves as the boiler in an otherwise conventional combined-cycle plant. This approach creates several synergies. First, the DSG system can take advantage of the steam turbine, generator, and other facilities of the combined-cycle plant at a modest increase in capital cost. Second, adding the DSG facility requires no additional operators or electrical interconnection equipment. And third, thermodynamic efficiencies are maximized because steam is evaporated *outside* the waste-heat recovery system; only the remaining thermal-heat exchange processes take place in the recovery heat exchanger. Thus, higher working-steam conditions can be achieved for the same degree of heat use, which increases overall cycle efficiency (see figures 12 and 13).

Because steam generation is moved from the heat-recovery boiler to the solar field, design of the heat exchanger is no longer constrained by the pinch point, thus giving the designer more latitude. In the solar field, where heat transfer in steam generation takes place from a radiation heat source, no equivalent to the pinch-point constraint exists. This new configuration is preferable from the perspective of the second law of thermodynamics because the solar field reduces the

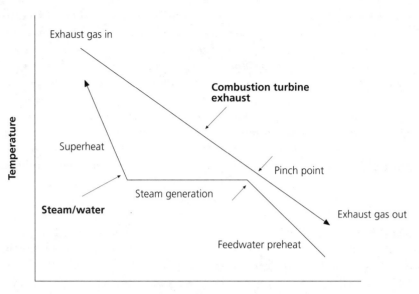

FIGURE 12: *Temperature paths for conventional evaporation are diagrammed. Temperature differences are plotted for hot exhaust gas and steam/water as they flow through a conventional heat recovery boiler. Most of the heat transfer occurs in the steam generation section; the temperature difference as the exhaust gases approach the saturation temperature, called the pinch point, strongly influences the amount of heat-transfer surface area needed.*

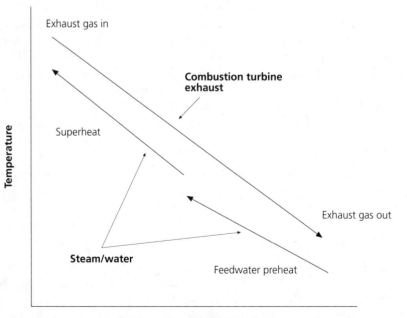

FIGURE 13: *Temperature paths for a solar hybrid combined-cycle plant are shown.*

production of entropy in the system. The hottest temperature source (the entering gas-turbine exhaust gas) is used for superheating, a moderate temperature source (the solar field) is used for steam generation, and a low temperature source (the exiting exhaust gas) is used for preheat. Consequently, the quality of energy used in the heat transfer process can be more appropriately matched to the temperature level required.

The solar hybrid combined-cycle system can achieve net cycle efficiencies of 53 to 55 percent. This compares favorably with conventional combined-cycle efficiencies of 44 to 47 percent for a single-pressure heat recovery boiler, 46 to 49 percent for a dual-pressure boiler, and 48 to 52 percent for a triple-pressure boiler.

Thermal storage concepts for parabolic-trough systems

Parabolic-trough plants have used both hot and cold tanks and single tank thermocline configurations for thermal storage. The two-tank storage system at SEGS I, for example, has a total storage capacity of 110 MWh and operates between 241°C and 307°C. In 1984, when the cost of heat-transfer oil was about $1.60 per gallon, specific storage investment costs amounted to $25 per kWh. Oil-storage systems based on mineral oils have been successfully tested, but are limited to a maximum operating temperature of about 300°C, which is too low for generating live steam for efficient turbine cycles [18, 19].

More recent SEGS parabolic-trough solar fields use an expensive synthetic oil that allows higher operating temperatures, but has a vapor pressure of about 7 bar, operating between 200°C and 400°C. Direct use of this fluid in a pressurized tank storage system would be prohibitively expensive in these systems.

Three thermal storage concepts have been identified as having potential for commercial systems with storage capacities of 1,000 MWh or more and a temperature range of 200°C to 450°C. All are suitable for thermal oil as well as for direct-steam generation, and promise to fall below the investment cost goal of $25 per kWh [20]. The first is a cascaded phase-change medium for passive storage of latent and sensible heat using different eutectic salts [21]; the second is an active two-tank storage system with eutectic molten salt as the intermediate heat-transfer and storage medium [20], and the third is a passive sensible heat dual-medium storage system in which the transport fluid passes through slabs of solid salt or concrete [22]. All three concepts, however, require significant development in order to reach technical readiness.

CENTRAL-RECEIVER SYSTEMS

The idea of using large numbers of heliostats or multiple-mirror facets to concentrate solar radiation has an ancient history, dating as far back perhaps as 212 B.C. when Archimedes is said to have used polished shields to focus sunlight on the Roman ships (a story believed to be more legend than fact by historians). In 1896, C.G. Barr was issued a patent for illuminating a solar engine atop a tower with sunlight reflected from an arrangement of parabolic mirrors on railroad cars sur-

Table 2: Central-receiver pilot plants have been constructed in several countries

Name	Location	Size MW_e	Receiver fluid	Start-up	Sponsors
Eurelios	Adrano, Sicily	1	Water/steam	1981	European Community
SSPS/CRS	Tabernas, Spain	0.5	Sodium	1981	Austria, Belgium, Italy, Greece, Spain, Sweden, Germany, Switzerland, United States
Sunshine	Nio, Japan	1	Water/steam	1981	Japan
Solar One	Daggett, California	10	Water/steam	1982	U.S. DOE, SCE[a], CEC[a], LADWP[a]
Themis	Targasonne, France	2.5	Hitec salt	1982	France
CESA-1	Tabernas, Spain	1	Water/steam	1983	Spain
MSEE	Albuquerque, New Mexico	0.75	Nitrate salt	1984	U.S. DOE, EPRI[a], U.S. Industry and Utilities
C3C-5	Crimea, former Soviet Union	5	Water/steam	1985	Former Soviet Union

a. SCE = Southern California Edison Company; CEC = California Energy Commission; LADWP = Los Angeles Department of Water and Power; EPRI = Electric Power Research Institute.

rounding the tower. In 1957 V.A. Baum, R.R. Aparasi, and B.A. Garf of the Energy Institute of Moscow designed a plant that would employ 19,000 m² of mirrors on 1,293 railroad flatcars, grouped into 23 trains on 23 separate concentric tracks. Although a 1:50 scale model of the plant was built, the project was eventually abandoned.

Central-receiver technology has been developed actively in the United States since the early 1970s when the Power Tower concept was first proposed by Alvin Hildebrandt and Lorin Vant-Hull of the University of Houston. Since then, Germany, Switzerland, Spain, France, Italy, the former Soviet Union, and Japan have launched programs to develop the technology. Several central-receiver pilot plants have been constructed around the world and several are still operated as test facilities (see table 2)[23, 24].

Water/steam pilot plants[7]

Solar One

The 10 MW_e Solar Central-Receiver pilot plant (Solar One) was constructed as a joint venture between the U.S. DOE, Southern California Edison, the Los

---

7. See figure 14 and table 3 for illustrations and general design parameters of water/steam systems.

Table 3: Key design parameters are listed for three major water/steam systems[a]

System Characteristic	Units	Solar One	CESA-1	Eurelios
Net plant rating	MW_e	10	1	1
Collector				
Collector area	m^2	71,084	11,880	6,216
Field configuration		Surround	North	North
Heliostat number		1,818	300	70/112
Heliostat size	m^2	39.1	39.6	52/ 23
Receiver				
Configuration		External	Cavity	Cavity
Tower height	m	55	60	77
Coolant		Water/steam	Water/steam	Water/steam
Boiler type		Once-through	Forced circulation	Natural circulation
Outlet temperature	°C	516	525	512
Outlet pressure	bar	105	108	62
Thermal storage				
Type		Single tank thermocline	Two tanks	Three tanks
Media		Oil/rocks/sand	Hitec salt	Hot water, Hitec salt
Capacity	MWh_t	135	16	0.36
Power conversion				
Type		Steam turbine	Steam turbine	Steam turbine
Inlet temperature/pressure				
from receiver	°C/bar	510/100	520/98	510/62
from storage	°C/bar	274/28	330/15	410/19
Heat rejection		Wet cooling tower	Dry cooling tower	Wet cooling tower

a. Although the three systems operated at essentially the same receiver outlet temperature, they differ in most other character-istics. At 10 MW$_e$, the Solar One plant is 1/10 the scale of a commercial plant, whereas the others are 1/100. Solar One employed an external receiver with a surround heliostat field while the others had cavity receivers in a north collector field. Moreover, each plant employed a different thermal-storage system. For all three, however, steam conditions generated by the thermal-storage system were significantly lower than those generated by the solar receiver.

FIGURE 14: Three major central-receiver pilot plants are shown: Solar One (top), CESA-1 (middle), and Themis (bottom).

Angeles Department of Water and Power, and the California Energy Commission. The plant, which was commissioned in 1982, was operated by Southern California Edison during its start-up, two-year test and evaluation period, and four-year power-production phase. After six years of successful operation, the plant had accomplished its intended objectives and was shut down. It established the technical feasibility of central-receiver systems; provided development, production, and operating and maintenance information to aid in the commercialization of similar plants; produced data on the environmental impact of solar-thermal plants; generated system operating and safety characteristics; promoted utility acceptance of the technology; and enhanced public familiarity with solar-thermal technology.

Solar One was a relatively straightforward design. Concentrated solar radiation from the collector field generated superheated steam at 510°C and 10.3 megapascals in a once-through receiver, and this superheated steam was routed directly to a steam turbine driving an electric generator (see figure 15). Some or all of the receiver steam could also be routed to a thermal-storage system. In the thermal-storage system, receiver steam heated heat-transfer oil, which was circulated through a tank filled with crushed rocks and sand, thus establishing a thermocline in the tank. The system was discharged by reverse circulating the oil through the tank to a steam generator. Because the oil had a maximum operating temperature of 315°C, the storage-generated steam reached only temperatures of 280°C at 2.8 megapascals. As a result, the turbine was equipped with a special admission port and gross cycle efficiency was reduced to 28 percent compared with 34 percent when operating at the rated receiver steam conditions.

Following a two-year test and evaluation period, Solar One entered its power production phase. Net electric output was maximized during this period, and operating procedures and equipment were modified to optimize plant performance (see figure 17). Heliostat availability was excellent and averaged 95 percent in the

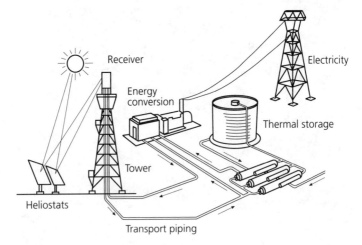

FIGURE 15: Solar One pilot plant, with its central receiver atop a tower, is shown.

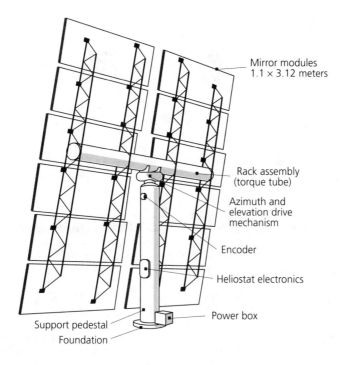

Mirror modules
1.1 × 3.12 meters

Rack assembly
(torque tube)

Azimuth and
elevation drive
mechanism

Encoder

Heliostat electronics

Power box

Support pedestal

Foundation

FIGURE 16: Solar One's heliostats consisted of glass mirrors with 39.1 m² of reflective area.

first year, 96.3 percent in the second, and 98.9 percent in the third. Individual heliostat availability of 99.6 percent was demonstrated. Overall plant availability was also excellent. During the first three years, the plant averaged 82 percent availability during daytime hours. During the fourth year, plant availability soared to 96 percent.

The Solar One receiver (see figure 18) is a once-through-to-superheat boiler that consists of 24 identical curved panels arranged as a cylinder. Initially, the receiver outlet temperature was reduced to 455°C as a result of thermal expansion problems. Although the problems were corrected and the receiver reached a 510°C outlet temperature, plant performance was identical at either operating temperature because reduced receiver losses at the lower outlet temperature offset reduced turbine efficiency. Thus, the plant was generally operated at the lower outlet temperature.

Although the thermal-storage system met capacity and efficiency criteria, it was used primarily to provide auxiliary steam to the system. Cost overruns during construction necessitated a 15 percent cutback in heliostats, and so there was never enough surplus energy to regularly charge the energy storage system.

Solar One was a showcase for modern digital-control technology. Operated automatically by a master control system, it was unique among electric-utility facilities at the time of its construction. The control system consisted of a computer

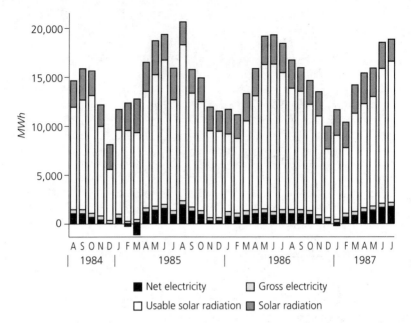

FIGURE 17: *Montly performance summaries for Solar One over its first three years of power production are shown. Of the total solar radiation, a large portion is above the threshold usable by the plant, and a fraction of that is converted to electricity. Net electricity is the amount remaining after subtracting the power consumed to run the plant and can be negative during extended shutdowns.*

to supervise the heliostat array controllers, three distributed process controllers to operate the plant's main flow loops, and five programmable logic controllers to control equipment and personnel protection circuits. The system reduced the need for operators, thus allowing employees to devote more time to improving plant reliability and efficiency.

The Solar One pilot plant was highly successful and achieved many technical milestones. It functioned well in all operating modes, and its water/steam receiver could operate continuously as long as available solar radiation remained above 300 watts per m². However, the plant's annual electric output fell short. Initial predictions of 13 percent annual efficiency were based on optimistic weather data and equipment performance parameters. A revised prediction based on 25 year average weather data and more realistic parameters indicated 8.2 percent annual efficiency and a net electric output of 15,000 MWh. Although the plant achieved only 70 percent of this value during its first three years, its performance continued to increase, and the power production rate during its fourth year, indicated that it would have reached its projected annual production level [25, 26].

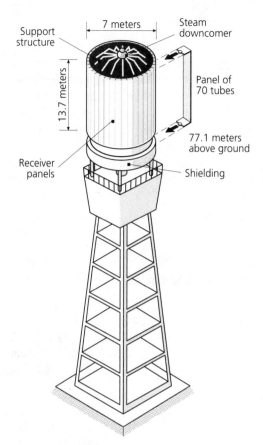

Support structure

7 meters

Steam downcomer

13.7 meters

Steam downcomer

Panel of 70 tubes

77.1 meters above ground

Receiver panels

Shielding

FIGURE 18: Solar One's external cylinder receiver is a once-through-to-superheat boiler in which feedwater is boiled and superheated in a single pass through a receiver panel.

Eurelios

Constructed near Adrano in Sicily, this 1 MW_e water/steam central-receiver plant was connected to the grid in 1981. The plant was sponsored by the Commission of the European Communities: its principal objective was to determine whether or not a solar power plant could be directly linked to the grid. The receiver, configured as a downward-tilted conical cavity, was a single once-through-to-superheat boiler. Short duration thermal-storage was provided by a water/steam accumulator and Hitec (sodium nitrate/sodium nitrite) salt for superheat.

Eurelios proved that a central-receiver power plant could produce grid-connected electricity and could be operated with standard procedures like those used for conventional thermal power plants. Its net power production, however, was disappointing, not only because sunlight at the site was poor, but because its solar receiver, storage, and heat transport systems were inadequately designed. The plant was shut down in 1984.

CESA-1

Located in Tabernas, Spain, this 1 MW_e water/steam central-receiver system operated from 1983 to 1984. The principal objective of the project, which was
sponsored by the Spanish Ministry of Industry and Energy, was to test the plant's
overall feasibility. The receiver consisted of a forced recirculation-type boiler, and
the thermal-storage system used was molten Hitec salt.

CESA-1's short period of operation provided limited, mostly qualitative information. The principal findings were that the time needed to start the receiver
was very long, because of the thermal inertia inherent in a recirculating steam
generator, and restarting the plant following a cloud passage was also delayed due
to inertia in the plant. In 1985, as part of a Spanish–German joint research
project, the facility became a testing site for developing and testing components
of an air Brayton-cycle solar power plant.

Lessons learned

Most of these experimental water/steam plants achieved their design performance
projections but fell short of the expected annual energy generation for several
reasons:

▼ The water/steam receivers, which were solar-driven boilers, required a long
start-up time; even then the turbine could only be started after the receiver was
producing a reasonable amount of steam. Start time (from sunrise until the turbine was on-line) ranged from two to three hours; much solar energy was lost during that time.

▼ Because the solar receiver and turbine generator were directly coupled, even
brief bouts of cloudiness caused the turbine to trip off-line; its restart and resyn-
chronization generally required a minimum of 45 minutes. Again, solar energy
was lost in this process. In addition, the turbine operated at part-load during the
mornings and afternoons when sunlight levels fell below the design (noontime)
level, thus lowering efficiency still further.

▼ The thermal-storage system, which transferred thermal energy from steam to
a storage medium and back to steam, lost a significant amount of thermodynamic
availability.

Advanced-technology pilot plants

Advanced central-receiver systems were developed to overcome the drawbacks of
water/steam systems. In advanced systems, which use sodium or molten salt for
the receiver and thermal-storage fluid, the receiver and turbine operate independently because they are separated by the thermal-storage system, which acts as a
buffer. The solar receiver can be started quickly because it is a single-phase-fluid
system. There is almost no thermodynamic loss in the thermal-storage system,
and the turbine need not run at part-load. In addition, a gas-fired heater can be
added to charge the storage system and to maintain plant operation during

Table 4: Key design parameters are listed for major advanced-concept system experiments

System Charactistic	Units	SSPS/CRS	Themis	MSEE
Net plant rating	MW_e	0.5	2.5	0.75
Collector				
Collector area	m^2	3,655	10,740	7,849
Field configuration		North	North	North
Heliostat number		93	201	211
Heliostat size	m^2	39.3	53.7	37.2
Receiver				
Configuration		Cavity and exposed panel	Cavity	Cavity
Tower height	*meters*	43	100	61
Coolant		Sodium	Hitec salt	Nitrate salt
Outlet temperature	°C	516	525	512
Thermal storage				
Type		Two tanks	Two tanks	Three tanks
Media		Sodium	Hitec salt	Nitrate salt
Capacity	MWh_t	5.5	40	7
Power conversion				
Type		Steam engine	Steam turbine	Steam turbine
Inlet temperature	°C	500	410	504
Inlet pressure	*bar*	100	40	72
Heat rejection		Wet cooling tower	Dry cooling tower	Wet cooling tower

cloudy periods, or if the solar plant is unavailable. Three major experiments on these systems have been conducted (see table 4).

Small Solar Power Systems Project/Central Receiver System(SSPS/CRS)
The Small Solar Power Systems Project/Central Receiver System (SSPS/CRS) began operations in 1982 under the aegis of the International Energy Agency. Located in Tabernas, Spain, the project involved a 500 kW$_e$ experimental system that used liquid sodium as the receiver and storage fluid. Nine countries (Austria, Belgium, Germany, Greece, Italy, Spain, Sweden, Switzerland, and the United States) helped design, construct, test, and operate the plant.

Initially, the plant experienced a series of equipment outages and operating complications that significantly limited the collection of long-term performance data. The most serious problems included leaks in the sodium equipment as well as frequent failures of both the electric trace heating system and the steam motor, which drove the generator. In addition, the system's high thermal inertia constrained operation and reduced efficiency.

In 1985, the plant was modified to test an advanced sodium receiver. Although the receiver performed very well at high solar concentrations (2.5 megawatts per m^2 peak flux), the entire facility suffered a setback in 1986, when a major sodium fire occurred following a valve maintenance operation. As a result of the fire, which destroyed a significant amount of equipment including the control room, the facility was converted into a test platform for advanced-receiver concepts [25, 13].

Themis

Built at Targasonne, France, and sponsored by Electricité de France, the Centre National de la Recherche Scientifique, and the Agence Française pour la Maîtrise de l'Energie (French Solar Energy Agency), this 2.5 MW$_e$ central-receiver pilot plant used a molten salt (Hitec) for its receiver and thermal-storage fluid. The project's objectives were to establish the technical feasibility of its overall design, as well as its components, and to evaluate its export potential.

The plant operated successfully from 1983 to 1986, demonstrating the advantages of decoupling solar-energy collection from power generation, and providing significant information for incorporation into future plants. However, its annual power production was substantially lower than predicted mostly because of inclement weather and design and construction deficiencies. The principal lessons learned were that the salt loops had to be simplified, and the design and reliability of the trace heating and salt pump had to be improved [25].

Molten-Salt Electric Experiment (MSEE)

Constructed at the Central-Receiver Test Facility (CRTF) in Albuquerque, New Mexico, this 750 kW$_e$ central-receiver system represented a cooperative project between the U.S. DOE, the Electric Power Research Institute, and a consortium of utilities and industries. Operated between 1984 and 1985, it used molten nitrate salt as the receiver and thermal-storage fluid and consisted of two previously tested molten-salt subsystems: a 5 megawatt-thermal (MW$_t$) receiver and a two-tank thermal storage system. A new molten-salt steam generator, a rebuilt turbine generator, heat transport equipment, and controls completed the system.

Testing and operating the MSEE underlined the technical feasibility as well as the flexibility of its molten-salt system. The plant could start up rapidly, operate through cloud transients, buffer the power-conversion equipment from solar transients, and shift power production using thermal storage. Furthermore, operation of the MSEE using the distributed-process control system was highly successful [25, 27]. But net power production during a 28-day test run was negative

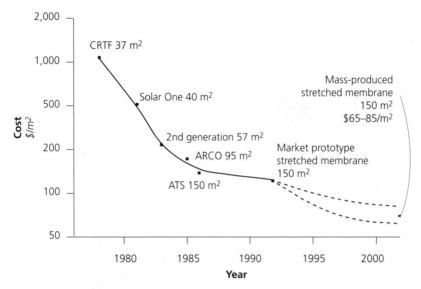

FIGURE 19: As heliostats have increased in size, their costs have dropped precipitously.

largely because the system's small size and use of nonoptimized components led to high electric parasitic loads. The experiment showed the need to further simplify the design of the salt loops and improve trace heating reliability.

Heliostat development and status

In a central-receiver power plant, the collector system is the largest single cost element. Therefore, significant effort has been devoted to lowering the cost of heliostats and to estimating their mass-production costs. All the test facilities and pilot plants have used glass/metal heliostats that consist of silvered-glass mirrors on steel support structures (see figure 16). By progressively increasing heliostat size, their cost has been reduced from about $1,000 per m^2 for the CRTF heliostats to less than $200 per m^2 today for a large onetime buy (see figure 19). These larger heliostats offer lower costs because, for a given size collector field, they have fewer drive assemblies, pedestals, foundations, controllers, and structural assemblies. In addition, they cost less to operate and maintain.

Glass/metal heliostats are well developed. The initial models, which were installed at the CRTF, Solar One, and other pilot plants, performed very reliably. The larger heliostats have also proved their reliability. In 1985, ARCO Solar produced almost 750 of its 95 m^2 and 43 of its 148 m^2 design as photovoltaic trackers for its 6.5 megawatt Carrisa Plains Photovoltaic Power Station. One 200 m^2 heliostat was produced by the Solar Power Engineering Company, but life-cycle cost studies indicate that optimum heliostat size is about 150 m^2. Larger heliostats, which are more susceptible to off-axis aberrations, tend to have reduced beam quality and also have higher unit costs due to the effect of wind loads on the structure and drive.

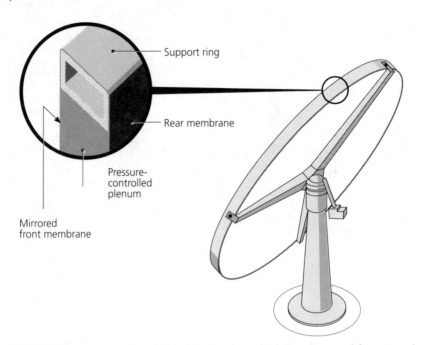

FIGURE 20: Stretched-membrane heliostat consists of two membranes: a mirrored front one and a supporting rear one, which are stretched across a support structure. Curvature of the heliostat is adjusted by changing the air pressure of the space between the membranes.

Recent research and development efforts have focused on polymer reflectors and stretched-membrane heliostats. A stretched-membrane heliostat (see figure 20) consists of a metal ring, across which two thin metal membranes are stretched. A focus control system adjusts the curvature of the front membrane, which is laminated with a silvered-polymer reflector, usually by adjusting the pressure (generally a very slight vacuum) in the plenum between the two membranes. Stretched-membrane heliostats are potentially much cheaper than glass/metal heliostats because they weigh less, use materials more efficiently, and have fewer parts.

Three generations of 50 m^2 stretched-membrane heliostats have been developed. Although the early models demonstrated excellent beam quality at low wind speeds, they had poor focus control in gusty winds. The second-generation models demonstrated excellent beam quality and focus control in gusty winds. A third-generation heliostat intended as a commercial prototype consists of two 50 m^2 mirrors mounted on a single structure (see figure 21). This heliostat, developed by Science Applications International Corporation (SAIC), is currently being evaluated at the Central-Receiver Test Facility.

To date, great uncertainty surrounds the durability and lifetime of the silvered-polymer reflector. Exposure tests suggest that the reflector has a lifespan of only five to 10 years, and issues such as removal and replacement of an old reflector remain relatively unexplored. In addition, current films are easily scratched

FIGURE 21: Commercial prototype of a 100 m² stretched-membrane heliostat has been installed at the Central-Receiver Test Facility (CRTF) by Science Applications International Corporation (SAIC).

and can be cleaned only by nonabrasive methods, which are not well developed and may be expensive.

Nitrate-salt receiver and storage technology

While the collector field represents the major cost element of a central-receiver power plant, the receiver is its greatest technical challenge. Not only must the receiver withstand many temperature cycles caused by daily start-up and shutdown and intermittent clouds, but tubed receivers must accommodate one-sided heating of the tubes. These factors place great stress on the receiver and result in thermal fatigue, which limits the amount of solar flux that can be concentrated on the receiver. In addition, high heat-transfer-fluid temperatures are needed to achieve high power-conversion efficiencies and thus reduce the size and cost of the collector field. Because the receiver represents the key risk element at a solar-thermal power plant, research and development efforts have been directed at improving its performance and reliability as well as lowering its cost. Therefore emphasis is

Table 5: Cost and performance estimates for the 100 and 200 megawatt plants developed by the Solar Central-Receiver Utility Studies[a]

Cost category	Units	200 megawatts fifth–tenth	100 megawatts first
Capital cost	*million 1987 $*	450	295
Annual O&M cost	*million 1987 $*	5.6	4.5
Annual net output	*GWh$_e$*	700	350
Annual capacity factor	*percent*	40	40
Levelized energy cost	*$/kWh$_e$*	0.075	0.10

a. These plant designs assumed Barstow, California, site parameters. The 100 megawatt plant represents the first commercial plant using existing equipment. The 200 megawatt plant represents the fifth to tenth commercial plant and assumes there have been modest improvements in equipment design.

placed on smaller receivers that can accommodate high solar-flux levels and have proportionally lower capital costs and reduced thermal losses.

Based on component development performed in the early 1980s, three attempts were made in the United States to develop demonstration power plants based on advanced receiver and storage designs. Southern California Edison and McDonnell Douglas attempted to finance and construct a 100 megawatt nitrate-salt central-receiver called Solar 100 [28]. Pacific Gas & Electric Company, ARCO Solar, Rockwell International, and Bechtel developed a 30 MW$_e$ sodium central-receiver through the detailed design phase. The Arizona Public Service Company and Black & Veatch hoped to repower the existing Saguaro plant with a 60 megawatt nitrate-salt central-receiver steam system. All these attempts failed to obtain adequate financing. Several factors explain why funding was so elusive: the U.S. DOE was unwilling to subsidize the projects, the tax credits were scheduled to expire at the end of 1985, there was a lack of equity commitment from key team members, and in California, the Public Utilities Commission wanted Southern California Edison and Pacific Gas & Electric to back a single central-receiver technology.

U.S. Utility Studies
In 1985, three U.S. utilities initiated a set of cost-shared cooperative studies co-sponsored by the DOE called the Solar Central-Receiver Utility Studies. The goal was to identify the preferred commercial plant configuration and to determine how the technology might be readied for commercialization. After a series of trade-off studies and conceptual design comparisons, the study concluded that nitrate-salt central-receiver designs can generate electric power more cheaply than sodium or water/steam central-receiver designs (see table 5). As a result, commercial plant designs based on an external cylindrical receiver and capable of gener-

ating either 100 or 200 megawatts of electricity were developed (see figure 22 and table 6) [29–33].

The annual electric output of these plant designs was calculated using the SOLERGY computer simulation program [34] with 15 minute sunlight and weather data for 1985 at Barstow, California, and the design-point system efficiencies given in table 7. The design-point component and system efficiencies reflect clear-day performance at solar noon. The annual energy calculation includes the effects of scheduled and forced plant outages, plant start-up, standby and shutdown thermal losses, and electric parasitic requirements. This calculation was later adjusted for Barstow 1977 weather data, because the 1977 annual insolation of 2,520 kWh per m² is very close to the 25 year average of 2,580 kWh per m². Annual energy and efficiency levels for the 100 MW$_e$ plant are shown in figure 23. Note that because of the losses mentioned above, annual efficiency values are significantly lower than design-point values. Capital cost estimates for the 100 MW$_e$ and 200 MW$_e$ commercial plant designs are given in table 8.

Table 6: Parameters for a commercial-size central-receiver power plant (based on the Solar Central-Receiver Utility Studies)

Item	Units	200 megawatts fifth–tenth	100 megawatts first
Land area	km^2	10.0	3.4
Maximum field radius	*meters*	17,782	1,314
Collector area	m^2	1,818,606	882,690
Number of heliostats		12,235	5,939
Receiver:			
Thermal rating	MW_t	936	468
Height	*meters*	28.4	21.1
Diameter	*meters*	22.7	192
Inlet/outlet temperature	°C	288/566	288/566
Tower height	*meters*	239	180
Thermal storage capacity	MWh_t	3,120	1,560
Salt storage tank sizes:			
Hot (H x D)	*meters*	13.0 × 28.7 (two)	13.0 × 28.7
Cold (H x D)	*meters*	12.2 × 40.5	12.2 × 28.7
Steam generator			
Rating	MW_t	520	260
Outlet temperature	°C	540	540
Outlet pressure	*bar*	125	125
Turbine gross rating	MW_e	220	110

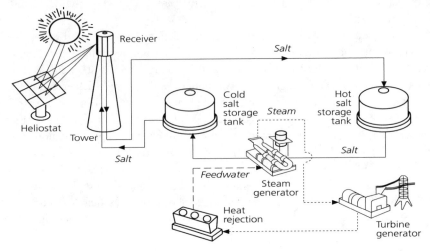

FIGURE 22: The preferred central-receiver plant configuration (based on Solar Central-Receiver Utility Studies).

Commercialization efforts

Although a 100 megawatt power plant is technically feasible, developers and investors are generally unwilling to risk building a first plant of this size. Thus, a 10 to 30 megawatt validation experiment must be undertaken to reduce the risk of building the first 100 megawatt plant. The final phase of the Utility Studies, which began in 1989, involved the design and evaluation of a plan to convert the 10 MW$_e$ Solar One pilot plant to advanced nitrate-salt technology. Called the Solar One Conversion Project, the study proposed removing the existing water/steam receiver, piping, thermal storage tank, charging and discharging heat ex-

Table 7: Central receiver design-point efficiencies

Plant system	200 megawatts fifth–tenth	100 megawatts first
Collector	0.643	0.661
Receiver	0.864	0.846
Heat transport	0.999	0.999
Thermal storage	0.999	0.999
Turbine	0.424	0.424
Power conversion	0.870	0.892
Total solar electric	0.205	0.211

Table 8: Commercial plant capital cost estimate summary[a, b]
thousands of dollars

	200 megawatts fifth–tenth	100 megawatts first
Land	2,351	1,140
Structures and improvements	4,652	3,161
Collector system	142,533	92,241
Receiver system	50,655	33,205
Thermal storage system	39,989	21,878
Steam generation system	23,639	14,951
Power generation system	88,142	53,587
Master control system	2,221	1,950
Total direct cost	**354,182**	**222,113**
Engineering, construction, and owner's costs	53,127	49,975
AFDC	45,619	26,991
Total capital cost	**452,928**	**299,079**

a. Capital cost estimates for the 100 and 200 MW$_e$ commercial plant designs are given in 1987 dollars. Major equipment costs are based on vendor quotations with contingency factors added as appropriate to the conceptual level of the design information. Capital cost estimates also include engineering and construction costs, owner's costs, an allowance for funds during construction (AFDC), and a 5 percent management reserve for the 100 MW$_e$ plant to reflect its first-of-a-kind nature.

b. Annual operating and maintenance costs for the 100 and 200 MW$_e$ plants were estimated to be $4.5 million and $5.6 million respectively. Estimates were based on staffing plans developed for utility ownership of the plants, typical quantities for spare parts, materials and supplies, service contracts for major equipment maintenance, and miscellaneous costs including safety, training, and office supplies.

changers, and oil transfer pumps. These were to be replaced with a new nitrate-salt receiver, piping, salt transfer pumps, thermal storage tanks, steam generator system, and gas-fired salt heater to provide a fully functioning hybrid solar/fossil central-receiver power plant. In addition, performance of the existing turbine plant was to be augmented with an additional stage of feedwater heating and an upgraded master control system (see figure 22) [35]. Net electric output of the converted Solar One plant was calculated to be 27,400 MWh per year, based on a number of factors, including measured Barstow sunlight and weather data for 1977, 25 percent fossil-fuel usage, and a modified version of SOLERGY, which accounts for hybrid solar/fossil operation.

The project's capital cost was estimated to be $30 million, including escalation to mid 1992. Private industry was expected to finance about 50 percent of the project's capital cost, with the remainder provided by a grant from the U.S. DOE, and revenues from the sale of electricity were supposed to repay the debt and provide a market rate of return to the equity investors. A two-year design and construction schedule, with commercial operation targeted for mid-1993, was developed, and a financial analysis was undertaken to identify the terms of a power-purchase agreement under which 50 percent of the construction costs could be provided through private sources. Two of these, a Standard Offer No. 4 (SO-4) utility power-purchase contract (like those used by Luz to develop SEGS III through VII) and a third-party sales contract, met the necessary criteria. However, neither approach was viable due to the unavailability of SO-4 power-purchase contracts and the complexities of developing a long-term power-sales contract for a demonstration project.

In 1991, facing new stringent emission-control regulations, Southern California Edison announced its intentions to sponsor the demonstration program to convert Solar One to a nitrate-salt system. The project, called Solar Two, will be funded as a cost-shared research and development program with the U.S. DOE; it has also attracted the participation of seven other major utilities in California, Arizona, Oregon, Idaho, and Utah, as well as EPRI, the California Energy Commis-

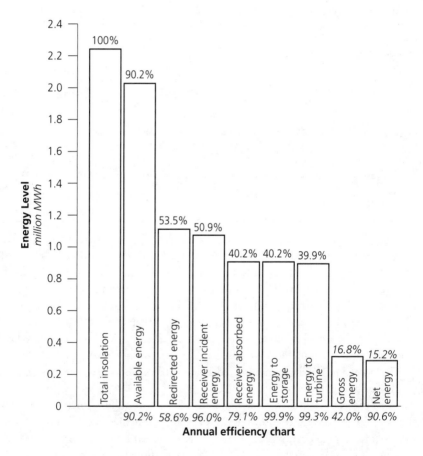

FIGURE 23: *Annual efficiency of a 100 MW$_e$ central receiver The efficiencies at the bottoms of the bars are component efficiencies; those a the top are cumulative efficiencies.*

sion, and Bechtel Corporation. The project participants hope that Solar Two will lead to the construction of an initial set of 100 megawatt power plants in the latter part of the 1990s.

Nitrate-salt direct-absorption receiver
Although the present generation of nitrate-salt receivers uses metal-tube heat exchangers, future receiver designs may eliminate the heat-exchange tubes. The direct-absorption receiver (DAR), as its name implies, absorbs concentrated solar radiation directly into a film of nitrate salt as it flows down a nearly vertical panel (see figure 24). The nitrate salt is blackened with a dopant to increase its absorptivity, thus enabling the DAR to tolerate significantly higher solar fluxes than tubed receivers. A DAR receiver promises to be smaller and lighter than a tubed receiver, and so should cost less and perform better [36]. A study by the Solar Energy Research Institute concluded that a DAR could reduce the cost of the receiver system by 50 percent. In addition, the study found that a DAR could increase a plant's annual energy output by 16 percent because of its higher absorptivity, reduced thermal losses, reduced pumping power, and quicker start-up and transient response [37]. A prototype DAR panel has been designed and built at the CRTF, and is currently undergoing salt-flow testing [38].

The DAR concept must overcome several major developmental challenges. The first concerns the need to maintain the integrity of the salt film given nonuniformities and eventual deformation in the receiver panel. Small hot spots could grow and lead to a burnout of the panel. However, initial tests indicate that the salt film is quite stable. The second concerns the choice and handling of an appropriate darkener. Because molten nitrate salt is essentially transparent to sunlight, it must be doped with a darkener, such as cobalt oxide particles, to achieve high absorptivity. Unfortunately, particulate darkeners quickly settle and tend to collect in the system's salt storage tanks. Furthermore, the particles may agglomerate, which decreases their absorptivity. To date, no suitable darkeners have been identified. The third issue concerns the loss of salt droplets when waves are inevitably created by the salt film as it flows down the vertical panel. Recent experi-

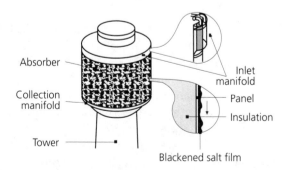

FIGURE 24: Direct-absorption receiver (DAR).

ments with a 6 meter test panel at the CRTF indicate that the waves begin to eject droplets after traveling about 5 meters down the panel. Two possible solutions have been proposed. One calls for the use of an intermediate salt manifold, located about halfway down the panel that would eliminate waves by catching the salt and reintroducing it onto the panel. (However, the manifold, which sits in a region of high solar flux would have to be cooled.) The other calls for an air curtain that would trap the ejected salt droplets and redirect them back onto the salt film [39].

Internal-film receiver
A modified version of the DAR, called the internal-film receiver (IFR), has been proposed that has many of the cost and performance advantages of the DAR. But because the salt film is contained on the inside surface of a cylindrical absorber, where it is heated convectively, a darkener is not required and salt drops are not lost as they are in the DAR. However, the IFR lacks the high solar flux capability provided by a darkened salt film, and so is likely to be larger in size [40, 41].

Air-receiver and storage technology
Whereas liquid coolants such as oil, water, sodium, and molten salts are limited to operating temperatures from 300 to 800°C, gaseous heat-transfer media such as air or helium allow for significantly higher receiver outlet temperatures [42]. Gases that can operate at outlet temperatures of 1,000°C and above make it possible to use such highly efficient equipment as lightweight gas turbines and a combined Brayton- and Rankine-cycle power plant for electricity generation. The gases would also make high-temperature processes possible, enabling concentrated solar energy to be coupled either directly or indirectly to chemical reactions such as synthesis gas and solar hydrogen production.

Because gas turbines require an operating pressure of 10 to 20 bar in an open-cycle configuration, and 20 to 60 bar in a closed-cycle configuration, efforts to design a high-pressure receiver focused first on metallic- and later on ceramic-tube receivers, both of which could tolerate pressures up to 10 bar. In 1985, a joint German–Spanish gas-cooled solar tower (GAST) technology program began testing components for a 20 MW$_e$ air-cooled solar plant with the CESA-1 heliostat field and a closed-loop gas supply system at the top of the CESA-1 tower. In 1986, operating conditions of 800°C at 9.3 bar outlet pressure were achieved with a metallic-tube receiver, and in 1987 a ceramic-tube receiver raised the outlet temperature to 1,000°C at the same pressure [43, 44].

Although gas-cooled ceramic-tube receivers can exceed temperatures of 1,000°C, they have significantly lower area-specific power output than do liquid-cooled tube receivers. Because their maximum allowable peak flux is approximately 100 kilowatts per m^2, they need five times as much area as a water/steam receiver and 10 to 25 times the absorbing area of an advanced salt or sodium receiver. Therefore, ceramic-tube receivers have higher heat losses as well as higher costs. To overcome this disadvantage, attempts are being made to distribute the heat-exchanging surface over a three-dimensional volume. Two such designs, known as volumetric and small-particle receivers, are under development.

Volumetric receivers

Volumetric receivers have either a porous-metallic or ceramic absorber, which allows both sunlight and air to penetrate to the interior of the receiver. Sunlight heats the elements of the absorber throughout its depth; atmospheric-pressure air is heated as it is drawn through the absorber. Volumetric receivers were developed concurrently in the United States and in Europe, and a number of metallic and ceramic designs have been investigated in 200 to 500 kW_t experiments [45–50].

The volumetric receiver offers a substantial increase in heat-exchanging surface area compared with a tubed receiver having the same aperture area. Indeed, because it can probably withstand heat fluxes up to 1,000 kilowatts per m^2, and its absorber could consist of potentially low-cost wire mesh, the volumetric receiver is a very attractive concept. Nonetheless, the receiver's advantages are balanced by the constraint that it must operate at atmospheric pressure. Until a transparent window is developed that can withstand the high operating temperatures, the stresses created by the 1,000 kilowatt per m^2 solar flux, and the pressure difference of 10 to 20 bar over an aperture surface of 30 to 50 m^2, the volumetric receiver concept will not allow for pressurized operation. At present such window material is not in sight; consequently, either compression of the heated air to the desired pressure level or an intermediate heat exchanger is necessary.

Typical high-temperature applications, like the Brayton cycle or steam reformation of methane, for example, need a minimum operating pressure of between 10 and 20 bar in order to keep the necessary balance of plant equipment within economic dimensions. However, atmospheric-pressure air at 700 to 800°C can be used to economically generate steam to drive a modern high-efficiency steam-turbine power plant.

PHOEBUS consortium

Volumetric air receivers have been most thoroughly developed by a European industry group called the PHOEBUS consortium. A study conducted by the group showed that central receivers using atmospheric air as the heat-transport medium to drive a steam-turbine power plant can generate electricity more cheaply than water/steam and sodium or nitrate-salt receivers. The air-receiver system was also judged to be simpler to design, operate, and maintain. Design simplicity is an important criterion for technicians and craftsmen in developing countries who must construct and operate the plants [51].

The PHOEBUS consortium identified a site in southwest Jordan (near Aqaba) considered suitable for installation of a demonstration plant, a 30 MW_e solar/fossil hybrid plant, which was hoped to become operational in 1995. According to the design (see figures 25 and 26), air at atmospheric pressure is drawn through the metallic wire-mesh volumetric receiver, heated to 700°C, and induced to flow down the hot gas duct and through the steam generator and/or the energy-storage system where it is cooled to about 200°C. Two variable-speed blowers control air flow, returning cooled air to the receiver aperture via the cold gas duct to recapture its available energy. Air at the inlet to the steam generator is heated with duct burners when fossil firing of the system is desired (see table 9 for system parameters) [52, 53].

The PHOEBUS air receiver consists of a north-facing absorber, an air-injection manifold, and a support structure (see figure 27). The planar absorber, which is elliptical, is tilted 15° forward to improve viewing angles to the heliostat field. The absorber consists of a large number of hexagonal modules, each of which contains a knitted and coiled metal wire-mesh absorbing element, supporting structure, and flow-control orifice. The air-injection manifold structure surrounding the absorber performs several functions. It contains the manifold for injecting warm air from the steam generator and thermal storage systems into the receiver; its inner surface acts as a secondary concentrator to reflect onto the absorber a portion of the incident flux that would normally fall outside the absorber; and the periphery of the structure contains a manifold and nozzles for an air curtain, which isolates the warm reinjected air from wind disturbances.

The thermal-storage system for PHOEBUS is based on ceramic regenerators, or Cowpers, which are structured stacks of ceramic bricks containing coolant passages. Studies show that regenerators can meet the requirements of a solar high-

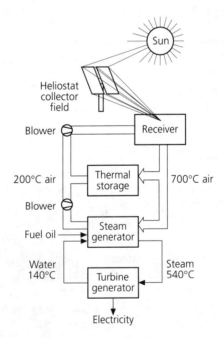

FIGURE 25: *Flow of energy through a PHOEBUS plant.*

FIGURE 26: Artist's rendering of a PHOEBUS plant is shown.

Table 9: Parameters for a 30 megawatt PHOEBUS plant

Receiver incident power	MW_t	46.5
Receiver absorbed power	MW_t	38.6
Salt inlet temperature	°C	288
Salt outlet temperature	°C	566
Salt flow rate	kg/second	91.5
Optical tower height	meters	75.44
Receiver diameter	meters	7.01
Receiver height	meters	7.32
Thermal storage capacity	MWh_t	78
	hours	2.25
Steam generator rating	MW_t	35
Salt heater rating	MW_t	17.5
Turbine plant efficiency	percent	36.1

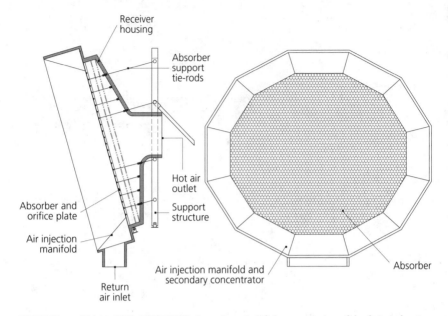

FIGURE 27: The 30 MW$_e$ PHOEBUS air receiver. At left is a cross section of the design, showing its knitted wire mesh absorber and orifice plate suspended from tie rods to a support structure at the rear of the receiver. The air injection manifold, surrounding the absorber, directs air returning from the system into the receiver absorber and also holds the manifold and nozzles for an air curtain (not shown). The receiver housing forms a manifold directing heated air into the hot air return duct.

At right, the front view shows how the absorber consists of over 4,000 hexagonal absorber elements, each constructed with a coil of knitted mesh. The entire absorber is backed by a plate containing flow-control orifices, and the absorber is organized into flow-control modules each consisting of seven absorber elements. The inlet air manifold surrounding the absorber also acts as a secondary concentrator to help flatten the flux distribution on the receiver.

temperature storage system: commercial modules exist that can store 100 MWh of usable thermal energy at 1,200°C with a 300°C temperature difference between charging and discharging [54, 55]. These regenerators have been adapted for the 700°C PHOEBUS thermal-storage system, with either a ceramic or metallic storage checker [56, 57]. However, to achieve cost-effective storage, bricks having greater surface area must be developed, and improved storage media, such as a ceramic containing phase-change salts, should be investigated [58].

It was estimated that the PHOEBUS plant would have an annual energy output of 98,000 MWh$_e$, a calculation based on SOLERGY and weather data for Barstow.[8] Energy calculations show that fossil fuel, if used only from one hour prior to sunset to three hours after sunset, would account for 28 percent of the plant's total output; the remaining 72 percent of the output would reflect a 14 percent annual solar-to-electric conversion efficiency. Total investment costs for the project were

---

8. Available sunlight data show that insolation levels at the Jordan site are equivalent to those at Barstow.

estimated at DM 330 million, including escalation to 1994. Annual operation and maintenance costs, including fuel, were estimated at DM 6.12 million.

Construction of the PHOEBUS project was not initiated for two reasons. First, the financing plan for the site, which was based on a combination of government grants, international development loans, and equity contributions from the PHOEBUS consortium, Jordan, and Arab development agencies, was rendered moot in 1991 by the Persian Gulf war. Second, the air-receiver system is in an early stage of development, and so a 2.5 MW_t prototype system experiment, currently planned for 1993 in Almería, Spain, is required to prove the receiver design.

A recent comparison of the latest designs for air- and salt-central-receiver systems for 100 MW_e plants, jointly performed by Sandia and the Deutsche Forschungsanstalt für Luft- und Raumfahrt (DLR) [59], shows a lower levelized cost of electricity for the salt system because it is more efficient and has a lower cost of thermal storage. The air receiver is less efficient largely because warm air is lost in the open receiver and the current ceramic brick thermal-storage design has a low effectiveness owing to its low surface to volume ratio. In addition, the turbine plant in an air receiver is less efficient because its feedwater temperature, which is determined by the temperature of the return air, is lower.

Small-particle receivers

Small-particle receivers use concentrated sunlight to heat an air stream that is darkened with a suspension of submicron-sized particles. Because the diameter of the particles is close to the wavelengths of visible light, they are highly absorptive and scatter little of the incoming light. In addition, their small size creates a relatively large surface area for heat transfer to the surrounding air; because they have low emissivity in the infrared region of the spectrum, they can produce high temperatures relatively efficiently. In theory, a small-particle receiver can attain temperatures greater than 2,000°C. Moreover, because heat absorption takes place at solar flux levels greater than 2 megawatts per m², the heat absorption area need not be very large, which in turn minimizes the area subject to reradiation and reflection losses. Furthermore, because heat absorption takes place in the air stream away from the receiver walls, these are relatively cool, which reduces radiative heat losses from the walls and simplifies the selection of wall materials.

The mass of particles required to effectively absorb highly concentrated sunlight and to heat the air is quite small, less than 0.2 weight percent. Absorber particles can either be reactive with air, such as pyrolyzed carbon, or nonreactive, such as silicon carbide or another ceramic material. To date, pyrolyzed carbon particles have been used because they have a high intrinsic absorptivity, can be readily generated at reasonable cost, and will eventually oxidize, leaving clean air at the receiver exit. They have one drawback, however. Once the particles completely oxidize, heat absorption stops, thus limiting the air temperatures that can be achieved with this receiver.

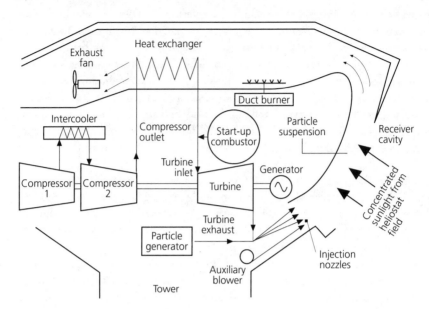

FIGURE 28: Particle injection receiver plant, which represents a solar/fossil hybrid Brayton-cycle configuration, is shown.

The small-particle concept was first proposed in 1978 by Arlon Hunt of Lawrence Berkeley Laboratory [60, 61], who constructed and tested a windowed forced-flow receiver in 1982 [62–64] that achieved an outlet air temperature of 750°C. Although 150 degrees less than predicted [62–64], the lower temperature probably reflected an increase in the particle oxidation rate created by a photolytic effect. Funding for the windowed receiver ceased thereafter because of shrinking R&D budgets and because its fabrication and operation could not be readily and economically adapted to large-scale applications.

In 1986, a windowless small-particle receiver was developed by the Bechtel Corporation. Called the Particle Injection Receiver (see figure 28), it operates much like a standard small-particle receiver, but does so at atmospheric pressure, and has an air curtain instead of a window to separate darkened from ambient air.

Gas-turbine exhaust, which is darkened with pyrolyzed carbon particles, is injected through a series of nozzles into the receiver cavity and the solar flux. The suspended particles, which are generated on-line from a hydrocarbon oil and are around 0.2 micron in diameter, absorb concentrated sunlight from a heliostat field with little scattering or reradiation. Because the particles are so small, they move at the same velocity as the air. Although there is no convective heat transfer, conduction to the surrounding air is extremely effective, and so the air closely tracks the particle temperature. The temperature of the particle/air suspension increases as it advances through the solar flux, and the particles oxidize at a temperature-dependent rate, disappearing as the air leaves the receiver cavity and flows to the heat exchanger. The heated atmospheric-pressure air stream is then induced through the heat exchanger by an exhaust fan. Because the heat exchanger is convectively heated, a relatively standard ceramic and metal design heats the pressurized air stream that drives the gas turbine.

Analysis of a proposed 40 MW$_e$ tower suggests that the outlet temperature of the receiver could reach 1,400°C [65]. According to a detailed two-dimensional heat transfer and fluid-flow model, receiver efficiency would reach 90 percent, and according to the SOLERGY code, annual solar-to-electric efficiency would approach 20 percent with an overall capacity factor of 26 percent.[9] The plant's near-term capital cost, which was dominated by the cost of heliostats and the air-to-air, ceramic/metal heat exchanger, was estimated at $2,900 per kilowatt. But by improving the cost and performance of the air-to-air heat exchanger, the capital cost potential for the plant could drop to an estimated $1,500 per kilowatt [66].

Regenerator systems

In theory, central-receiver systems delivering atmospheric pressure air could be adapted for applications requiring a pressurized-air working fluid by employing ceramic regenerators as air-to-air heat exchangers. Creating a large pressure difference between a heat source and its consumer is not a solar-specific problem. Ceramic regenerators are often used as heat exchangers, which are usually charged at ambient pressure by exhaust heat from a fossil heat source, and discharged at a significantly higher pressure. Such regenerators, if adapted for a central receiver, could link an atmospheric-pressure volumetric receiver to high-temperature thermal applications. Doing so could transform the pressure between the receiver and the thermal process, and extend operation up to 24 hours per day. The cost-effectiveness of such systems deserves investigation to determine if development efforts are justified.

Falling-sand receiver

This receiver—as its name suggests—absorbs solar energy directly into a falling curtain of millimeter-sized ceramic particles. Because temperatures ranging from 1,000 to 1,200°C can be achieved, the receiver can run a gas-turbine or combined-cycle power plant directly on solar energy. The technology is also attractive because the appropriate ceramic materials, either fused bauxite or zirconia, are relatively inexpensive, commercially available, and can provide cost-effective thermal storage. In addition, the particles are compatible with containment materials at high temperatures and can be used with modular, ceramic heat-exchangers developed for indirect-fired gas-turbine applications.

Nonetheless, several technical issues must be resolved before the receiver will be commercially attractive. These include the particles' tendency to agglomerate at high temperatures under moderate static pressure, the need to assess their durability and solar absorptance after repeated thermal cycling, the cost and com-

---

9. These figures are based on 1977 Barstow sunlight and weather data.

plexity of transporting relatively hot particles back up the tower, and the potential for particle and convective energy losses from the receiver aperture [67].

PARABOLIC-DISH SYSTEMS

Research efforts in the 1970s and 1980s produced parabolic-dish systems capable of providing thermal energy for central power generation or for driving an engine mounted at the focus of each dish. Among the engines considered for the latter purpose were organic Rankine-cycle engines, Brayton-cycle engines, sodium-heat engines, steam (Rankine) engines, and Stirling-cycle engines.

Organic Rankine-cycle engines, which use toluene as the working fluid and were developed for the Jet Propulsion Laboratory (JPL) by Ford Aerospace and the Barber-Nickels Company, began preliminary solar operation in the mid 1980s. But because the engine was relatively inefficient and expensive (large heat exchangers were required), it is no longer considered for solar applications.

Brayton-cycle engines have the potential for a long operating life, but have not yet been tested under field conditions. Attempts in the mid-1980s to develop a Brayton-based system depended on engines that were developed for other applications and were poorly suited for solar-thermal systems. An automotive gas turbine was modified for solar operation but never field-tested, in part because it performed poorly when operated on natural gas. A subatmospheric Brayton-cycle engine, which was originally developed for heat-pump applications, was integrated with a LaJet LEC-460 solar concentrator by Sanders Associates, but was not developed as a result of its disappointing performance as well as diminishing U.S. DOE budgets.

Interest in Brayton engines continues. James Kesseli of the Northern Research and Engineering Company has advanced the use of highly recuperated, Brayton engines built from turbocharger components for heat pumps as well as for solar applications. Such engines offer the potential for low cost, long life, and good efficiency. In addition, the infrastructure for manufacturing and maintaining turbo machinery already exists. Windowed ceramic-matrix receivers for Brayton engines, which were initially demonstrated on a testbed concentrator at the JPL Edwards test site in 1983, require further development and integration to Brayton engines to achieve complete dish-engine systems.

A relatively new concept, the sodium-heat engine, was investigated in the 1980s by researchers at Sandia National Laboratories because of its potential longevity. The engine, which lacks moving parts, relies on the flow of sodium ions through a solid electrolyte at high temperature. Driven by the pressure differential that exists between the high and low temperature zones of the engine, electrons pass through an external circuit where they are made to do useful work. However, because the cycle proved to be relatively inefficient and needed significant development, research on the technology was halted in 1987.

Steam-engine technology, although relatively mature, is also not especially well suited for distributed solar-energy systems. Dish-mounted steam engines must be relatively small, which prohibits the use of high-efficiency turbine expanders and so results in relatively low efficiency. Moreover, maintenance costs are high because of the need to maintain piston expanders and a complicated lubrication system. Consequently, there is relatively little interest in steam engines for solar applications.

Of all the engines, the Stirling is potentially the most efficient and has received the greatest attention. Although it has yet to be widely used, this externally heated piston engine (see the section "Stirling-engine systems" on page 61) is the focus of intense development worldwide. Having the potential for long life, reliable operation, and reasonable cost, the Stirling engine has emerged as the preeminent power-conversion module for dish-electric systems. Such advantages have also renewed interest in Stirling engines for other applications, including domestic heat pumps, automotive engines, refrigeration units, and space power-generators. Most of the research on Stirling engines in the United States is being supervised by the Lewis Research Center of the National Aeronautics and Space Agency (NASA) and by Sandia National Laboratories. Other countries, including Japan, Germany, and Sweden, are also actively developing the technology.

Pilot projects

Shenandoah

The U.S. Solar Total Energy Project at Shenandoah, Georgia, a central generation dish system, was designed to provide electricity, air conditioning, and process steam to a nearby industrial complex. The project, which began operations in 1982 and was jointly sponsored by the U.S. DOE and Georgia Power Company, consisted of 114 parabolic-dish collectors heating a synthetic oil from 260 to 363°C. The system included an oil heater to supplement energy from the solar collectors. The central power-conversion area consisted of a thermal-storage tank, steam generator, gas-fired heater, steam-turbine power plant, and absorption chiller (see figure 29).

Although all systems functioned, they failed to meet efficiency expectations. Performance data showed a significant mismatch between the output of the collector field and the system's thermal requirements. High thermal losses in the heat-transport system reduced thermal-collection efficiency to 45 percent, which was well below the design goal of 62 percent. Because the collectors provided only 20 to 25 percent of total thermal input to the system, the gas-fired oil heater dominated operations. However, the project, which was decommissioned in 1990, did demonstrate the solar total energy concept and provide several major lessons. To begin with, it was found that the closed-loop collector tracking system was inadequate; it was replaced with an open-loop system. In addition, the synthetic heat-transfer fluid needed to be reevaluated on the basis of economic and performance findings; and finally, it was determined that subsequent designs should consider parasitic loads, equipment availability, and integrated system performance, particularly at partial loads [25, 68].

White Cliffs

The White Cliffs Solar Power Station, which was supported by the New South Wales Government with technical assistance provided by the Australian National University, was designed to provide electricity and warm water for White Cliffs, a small mining town 1,100 kilometers west of Sydney, Australia. Engineering began in 1980, operation commenced in 1982, and design specifications were met by 1983.

The system, which is still operational, consists of 14 modular 5 meter-diameter parabolic-dish collectors. The parabolic substrate, which is fiberglass, is covered with 2,300 mirrored tiles. Steam is generated in semi-cavity receivers at temperatures as high as 550°C and at a pressure of 7 megapascals. Overall, the system produces 25 kilowatts of electricity and 100 kW$_t$ of low-quality heat at an insolation level of 1 kilowatt per m^2. The steam engine is a converted three-cylinder Lister diesel engine that required substantial modification of such components as valves, the system for extracting oil from the exhaust/steam condensate, and feedwater treatment before it would operate routinely. Although the system proved to be robust, reliably providing electricity and heat to a community in the remote Australian outback, the overall cost (about $5,000 per kilowatt of electric-

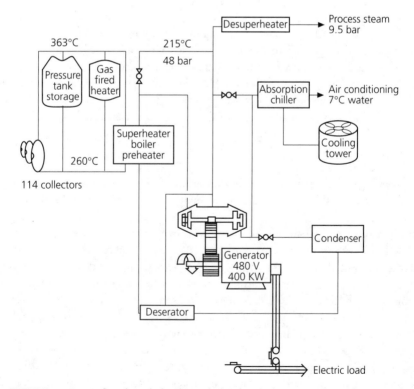

FIGURE 29: *Energy flow through the Shenandoah Solar Total Energy Project is diagrammed.*

ity, 1983 Australian $) is simply not competitive with conventional diesel engines.

Sulaibyah

The Sulaibyah Solar Power Station in Kuwait was designed to provide electricity and thermal energy for such agricultural purposes as greenhouse heating, desalination and irrigation. The system, which became operational in 1982, was jointly sponsored by the Kuwait Ministry for Electricity and Water and the German Federal Ministry for Research and Technology. A heat-transfer oil circulated through the collector field to the power-conversion system, which used an organic Rankine-cycle with toluene as the working fluid. The system also used thermal energy, both that from the collector field and that rejected from the power conversion system. Although additional thermal applications were added to the system in 1985, the system's performance remained at about half its predicted value. It became apparent that the high thermal inertia of the collector field led to large thermal losses and long start-up times.

Vanguard

Vanguard I was a 25 kW_e prototype parabolic-dish-Stirling engine module located at Rancho Mirage in California. Representing a cooperative effort by the U.S. DOE and the Advanco Corporation, the project was tested between 1984 and 1985. Its objectives were to identify and develop an early market for dish–Stirling systems, to determine the installation, operation, maintenance, and performance characteristics of the prototype module, and to prepare a plan for its commercialization.

The performance of the Vanguard module, which consisted of a 25 kW_e United Stirling 4-95 Mark II engine mounted on an Advanco parabolic dish, was outstanding. It achieved a maximum net solar-to-electric conversion efficiency of 29.4 percent and an average daily net efficiency of 22.7 percent. These record-setting performance levels can be attributed to the module's low thermal inertia (short start time and rapid response to cloud transients), high conversion efficiency, low parasitics, and good part-load performance. During its 18-month test period, the module achieved availability of 72 percent. The module also demonstrated fully automatic unattended operation from sunrise to sunset [25, 69].

Vanguard I also provided important information on the maintenance of dish–Stirling systems. The data were useful for identifying several problem areas. For example, the engine, which was designed for automotive use, was found to have a design life of 3,500 hours between overhauls, which is probably too short for electricity generation. In addition, testing uncovered problems related to the engine's initial solarization such as sensitivity of the directly illuminated heater heads to nonuniformities in the receiver's flux distribution and a complex control system.

Advanco planned to build a 30 megawatt commercial installation following the completion of its Vanguard I study, but was prevented from doing so by an exclusive agreement developed during the course of the project between the McDonnell Douglas Corporation and United Stirling AB (USAB) of Sweden.

King Abdul Aziz City

Between 1982 and 1985 three stretched-membrane concentrators with 17-meter diameters were designed and built in Saudi Arabia by Schlaich Bergermann & Partner of Germany. The project, which was jointly funded by the German government and the King Abdul Aziz City Center for Science and Technology, proved the viability of stretched-membrane technology for parabolic-dish concentrators. The two concentrators equipped with 50 kilowatt USAB Stirling engines (model 4-275) have proved successful both in tests and in continuous operation, operating up to 3,500 hours in grid-connected as well as stand-alone mode. Overall system efficiencies as high as 20 percent have been achieved.

McDonnell Douglas/United Stirling

During the Vanguard I project, the McDonnell Douglas Corporation and United Stirling agreed to jointly develop and commercialize a parabolic-dish-Stirling system based on the Mark II Stirling engine (model 4-95). Between 1983 and 1986, six prototype modules were built and enough hardware produced for two additional units (see figure 30). In 1984, three modules were tested by McDonnell Douglas at its solar test site in Huntington Beach, California. Three utility units were deployed by Southern California Edison, Georgia Power Company, and Nevada Power Company. However, the project was terminated when McDonnell Douglas decided that commercialization would be adversely affected by the decline in world oil prices and so decided to sell rights to the design to Southern California Edison. Prior to the sale, the Nevada Power unit (which was installed but never tested) was sold to Aisin Seiki Company, a Japanese developer of Stirling engines. In 1988, Southern California Edison decided to sell its rights to the Stirling dish to Hydrogen Engineering Associates, Inc., following a company decision to divest of its renewable energy interests.

Test data indicated that the system, like Vanguard I, had high performance capabilities and the potential for reliable operation [70]. Additional testing of three of the prototype systems, which was performed by Southern California Edison between 1986 and 1988, showed that system availability ranged from 50 to 87 percent, with the major causes of unavailability being circumstantial (loss of trained personnel, lack of spare parts, and special tests) rather than attributable to hardware failures (see table 10) [71].

Solarplant I

This was a privately owned and financed demonstration system located near Warner Springs, California. Built by LaJet Energy Company, the plant, which had a rating of 4.92 MW$_e$, was to provide power to the San Diego Gas & Electric

FIGURE 30: The McDonnell Douglas/United Stirling dish–Stirling module is shown.

Company. The plant employed 700 concentrators with water/steam as the receiver fluid; 600 of these were dedicated to generation of saturated steam; the remainder provided steam superheat. Each receiver was designed to absorb solar radiation using a nitrate-salt bath, which had sufficient storage to keep the receivers operating during brief cloud transients.

Solarplant I was started in mid-1984 but experienced significant equipment and operational problems that prevented it from ever reaching full rated operation. The major equipment problem was deterioration of the 1.52 meter diameter stretched-polymer reflector facets that comprise the parabolic-dish assemblies. Operationally, the system was plagued by prolonged start-up times resulting from overnight freezing of the nitrate salt and the low thermal mass of the system.

Collector technology
A dish collector is a two-axis tracking assembly that is functionally and physically similar to a heliostat except for two important differences: it must always point toward the sun during operation; and in order to concentrate energy, it must ap-

proximate a truncated paraboloid reflective surface. Dish concentrators have received less attention than heliostats, but benefit from many of the lessons learned during heliostat development. Most notable is the evolution of larger reflective areas and the transition from glass and metal to stretched-membrane designs (see figure 31).

The first generation of dish collectors, which were developed in the late 1970s and included those installed at Shenandoah, White Cliffs, and Sulaibyah, were relatively small and emphasized performance rather than cost. The next generation of concentrators, which was developed in the early to mid-1980s, reflected commercial cost goals and volume production. These designs are exemplified by the McDonnell Douglas/United Stirling system and the LaJet stretched-membrane concentrator used at Warner Springs. The latest generation of concentrators are large area designs that include innovative features intended to improve their cost or performance potential (see table 11) [25].

Stretched-membrane concentrators are currently the focus of considerable attention because they are most likely to achieve the goals of low production cost and adequate performance. Multifaceted designs, such as the LaJet system, as well as single-facet designs, such as the Schlaich Bergermann & Partner designs, are being pursued (see figure 32). Both Solar Kinetics Inc. of Dallas, Texas, and SAIC have produced stretched-membrane facets (3.5 meters in diameter) that exhibit excellent performance capabilities. Recently, Solar Kinetics has developed a 7-

Table 10: Efficiency of the McDonnell Douglas Corporation/United Stirling 25 kW$_e$ dish–Stirling system

Source	Efficiency percent	
	Component	Cumulative
Available direct sunlight	100.0	100.0
Reflectivity	91.1	91.1

Intercept
    Tracking
    Surface waviness
    Cant error
	96.7	88.1

Receiver
    Conduction
    Reflectivity
	90.0	79.3

Receiver temperature difference	99.5	78.9
PCU engine	39.8	31.4
Generator	94.8	29.8
Parasitics	95.5	28.4

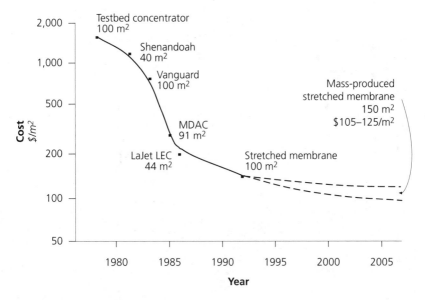

FIGURE 31: Dish collectors have become progressively larger and cheaper to produce.

meter single-facet dish, which demonstrated excellent performance in tests at Sandia National Laboratories [72]. In addition to a 17-meter diameter single-facet dish, Schlaich Bergermann & Partner has developed a 7-meter single-facet dish for use with the Stirling Power Systems V-160 Stirling engine, which achieved significantly better performance [73].

Stirling-engine systems

The greatest challenge facing distributed-dish systems is developing a power-conversion unit, which has low capital and maintenance costs, long life, high conversion efficiency, and the ability to operate automatically. Several different engines, such as gas turbines, reciprocating steam engines, and organic Rankine engines, have been explored, but in recent years, most attention has been focused on Stirling-cycle engines. These are externally heated piston engines in which heat is continuously added to a gas (normally hydrogen or helium at high pressure) that is contained in a closed system. The gas cycles between hot and cold spaces in the engine, as it does so, a regenerator stores and releases the heat that is added during expansion and rejected during compression.

In the United States, several industrial teams are collaborating with Sandia National Laboratories and NASA to develop the Stirling engine. Both kinematic and free-piston 25 kW$_e$ engines are being designed for utility applications, and at least one 5 kW$_e$ system is being developed for the remote power market. In a kinematic engine, the power pistons are connected to an output driveshaft that penetrates the engine casing, whereas in a free-piston engine, the power and dis-

Table 11: Design parameters for parabolic-dish concentrators

	MDC/USAB	LEC-460	PKI	SKI Advanced	Advanced membrane
Manufacturer	McDonnell Douglas	LaJet	Power Kinetics	Solar Kinetics	Schlaich Bergermann
Type	Parabolic dish, multiple facets	Stretched membranes multiple facets	Square dish secondary concentrator	Parabolic dish	Single stretched-membrane dish
Concentrator diameter	11 meters	9.5 meters	13.2 meters	14 meters	17 meters
Reflector area	91 m^2	43.7 m^2	135 m^2	154 m^2	227 m^2
Reflector material	Silvered glass	Aluminized Mylar (ECP-91)	Silvered glass	Silvered polymer (ECP-300X)	Silvered glass mirror tiles
Reflector assembly	82 curved glass mirrors bonded to stamped steel back structure	24 reflector membranes on aluminum frames, shaped by vacuum	360 mirror facets in venetian blind arrangement	30 curved aluminum parabolic Gore panels. Sandwich with corrugated core	Double steel membrane stretched over steel ring shaped by vacuum. Mirror tiles bonded to front
Structure/mounting	Trusses/beams/steel pedestal	Truss structure/ concrete pier	Space frame/secondary support system/ boxbeam track	Six radial arms, two rings, hub/tubular steel pedestal	Steel support girders on ring turntable base
Tracking axes	Elevation-azimuth	Polar-declination	Elevation: wheels and drag links azimuth: rotation on track	Elevation-azimuth	Elevation-azimuth: rotation on wheels
Concentration ratio	8,000	24–2,000 (changed by vacuum)	700	1,226	600
Number built	6	700	1	1 Gore	3

FIGURE 32: Dish concentrators that have been built and tested include an array at Solarplant 1 in Warner Springs, California, by the LaJet Energy Company (top) and one by Schlaich Bergermann & Partner of Germany (bottom).

placer pistons move freely within the engine and the casing is hermetically sealed. Power output occurs through a linear alternator or hydraulic coupling [74].

Kinematic Stirling engines

Stirling Thermal Motors, Inc., under contract to Sandia, is adapting the STM4-120 kinematic engine for parabolic-dish systems [75]. This 25 kW_e four-cylinder engine, which can run 50,000 hours between overhauls, has a pressurized crankcase, which increases the life of its rod seals, a variable-angle swashplate for simple and efficient power control, and a heat-pipe heat exchanger for near-isothermal heat input. Such features correct the shortcomings of previous kinematic engines.

During almost 200 hours of bench testing at Sandia, the STM4-120 has achieved an output of 14.5 kilowatts of electricity at 40 percent thermal efficiency. Additional tests at the full 25 kW_e are planned. In addition, a second STM4-120 engine is undergoing bench tests and will eventually be delivered to Sandia for tests on a parabolic dish. Testing will be conducted with a sodium-reflux solar receiver. The reflux receiver obviates the problems of local overheating experienced in previous tubed receivers [76].

A two-cylinder kinematic Stirling engine is being developed by the Stirling Power Systems Corporation, a former subsidiary of United Stirling. The engine, which is rated at 8 to 10 kW_e, is based on the USAB 4-95. Solarization of this gas-fired engine, which is called the V-160, has entailed modifying its lubrication, cooling, and control systems. In addition, the gas-fired heater tubes have been altered to absorb concentrated solar flux. To date, the V-160 has accumulated more than 350,000 running hours. Schlaich Bergermann & Partner is now developing a complete engine-concentrator system, and a test system, which was launched in 1989 at the University of Stuttgart, has achieved a maximum power output of 7.7 kW_e. Modifications to the receiver are now being made that should increase working gas temperature and raise power output. Three additional systems, which are to be built at the Plataforma de Solar Almería in Spain, will provide data on efficiency, reliability, and operating and maintenance costs [73]. A reflux heat-pipe receiver has also been developed by the University of Stuttgart and the Deutsche Forschungsanstalt für Luft- und Raumfahrt and successfully tested.

Free-piston engine

Sandia and the NASA Lewis Research Center are supporting development of 25 kW_e free-piston Stirling engines for terrestrial applications. It is thought that the free-piston engine, which offers a hermetically sealed configuration and few moving parts, might provide exceptional reliability, low maintenance, and a long life, although such attributes have yet to be demonstrated. Two versions of the free-piston engine have been studied by NASA and its contractors. The simplest employs a reciprocating piston to drive a permanent-magnet linear alternator, whereas the other uses the engine to drive a hydraulic pump/motor coupled to a rotating alternator.

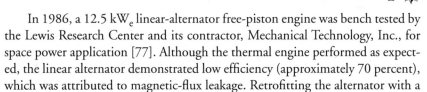

In 1986, a 12.5 kW$_e$ linear-alternator free-piston engine was bench tested by the Lewis Research Center and its contractor, Mechanical Technology, Inc., for space power application [77]. Although the thermal engine performed as expected, the linear alternator demonstrated low efficiency (approximately 70 percent), which was attributed to magnetic-flux leakage. Retrofitting the alternator with a new design yielded an efficiency greater than 90 percent.

Subsequently, the Cummins Engine Company of Indiana and the Stirling Technology Company (STC) have independently designed linear-alternator and hydraulic free-piston engines specifically for 25 kW$_e$ dish–Stirling applications [78, 79]. Design criteria demanded a longevity of 60,000 hours, with a major overhaul at 40,000 hours, and a cost of $450 per kW$_e$ for the complete conversion system including receiver, engine, and auxiliaries.

The STC Stirling design employs a single displacer piston and a helium working fluid, which is hermetically sealed from the two hydraulic power pistons nested-bellows seals. Solar input is through a liquid metal (sodium/potassium eutectic) reflux boiler; engine output is in the form of hydraulic power transmitted to a variable displacement pump/motor, which drives a rotary alternator. A comparison of the linear-alternator and hydraulic-output engines shows that the linear alternator is simpler, and the hydraulic engine is slightly less expensive to build and simpler to control [80].

The Cummins design uses a Thermacore sodium heat-pipe receiver and a linear alternator. In addition to its 25 kilowatt design, the company has been developing a 5 kW$_e$ dish–Stirling system for remote applications. A prototype engine/alternator design was solar tested in 1989. Although the system produced sufficient power for pumping water, it has been plagued with failures in the sodium heat-pipe absorber [81]. The receiver is undergoing continued improvement and endurance testing toward achieving its engine/alternator performance targets of 5 kilowatts net output and 32 percent engine/alternator efficiency [82].

In September 1991, Cummins and Sandia entered into a 3½ year joint venture program to commercialize 5 kW$_e$ dish–Stirling systems for remote power applications. Under this 50/50 government/industry cost-shared program, 16 systems, representing three generations of hardware will be developed and field-tested, leading to initial mass production in 1995.

Japanese Stirling engines

Ongoing Stirling-engine programs in Japan include a 3 kW$_e$ engine being developed by the Toshiba Corporation to power residential heat pumps, and a 2.4 kW$_e$ engine being developed by the Aisin Seiki company for testing in space [83, 84]. While the Japanese have devoted twice as much effort to Stirling engines as the Americans, none of their engines is being developed specifically for terrestrial solar-thermal systems.

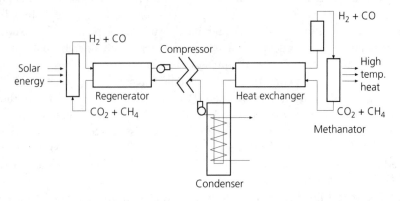

FIGURE 33: Simplified chemical heat-transport system shows carbon dioxide (CO_2) reforming methane (CH_4). In the receiver/reactor, high temperatures produced by concentrated solar energy drive the feed gases, CO_2 and CH_4, which react and form the product gas, a mixture of hydrogen (H_2) and carbon monoxide (CO) commonly called synthesis gas. The synthesis gas is then transported by pipelines to the point of use where a methanator recovers latent heat in the gas by converting it back to CO_2 and CH_4. High-temperature heat released during methanation can be used for power generation and other uses. CO_2 and CH_4 are returned by pipeline to the solar receiver/reactor, thus forming a closed-loop.

Thermochemical energy transport

A thermochemical-energy transport system transforms solar energy into chemical energy by way of a reversible endothermic reaction. The chemicals are then either stored or transported to an end-use site, which may be several hundred kilometers distant, where energy is released by reversing the chemical reaction [85, 86](see figure 33).

In recent years, the use of solar energy to power high-temperature chemical reactions has received international attention. Reforming reactions, particularly steam and carbon dioxide reformation of methane, are being investigated in at least five countries [87, 88]. Potential applications include the production of chemical feedstocks and transportation fuels, destruction of hazardous wastes, and thermochemical energy transport and storage.

Both central-receiver and parabolic-dish systems can generate sufficient solar flux (greater than 1,000 kilowatts per m^2), as well as the high temperatures (800 to 1,000°C) needed to drive these chemical reactions. But to date, most development work has focused on parabolic-dish systems because their size is best suited to the size of experimental receiver/reactors.

Several receiver/reactor systems have been tested; among them are two advanced designs that were developed at Sandia [89]. The first is a sodium-reflux heat pipe in which concentrated solar energy is absorbed on the vaporizer section of the heat pipe. Conventional catalyst-packed reactor tubes form the condenser section. Although the receiver was tested successfully at the solar furnace of the Weizmann Institute of Science, in Rehoval, Israel, the sodium-reflux heat-pipe receiver developed a small hole in a high flux region under simulated cloudy conditions, the result apparently of wick failure and improper refluxing of the device [90, 91]. The second design included a volumetric receiver that employed direct catalytic absorption using a porous ceramic absorber impregnated with a rhodium catalyst. The receiver, which was tested at Lampoldshausen, Germany, operated successfully during steady-state as well as during transient operating conditions, achieving a maximum methane conversion efficiency of 70 percent. Although the absorber performed satisfactorily in promoting reformation without carbon formation, several problems, including cracking and degradation of the porous matrix, nonuniform dispersion of the catalyst, and catalyst deactivation await resolution [92]. The Weizmann Institute has recently installed a methanator, which is specifically designed for solar applications and is the first component of a planned 450 kW$_t$ closed-loop test facility [93].

COST AND PERFORMANCE PROJECTIONS

While much has been accomplished in the field of solar-thermal technology during the past 15 years and several large-scale commercial power plants have been built, additional development is needed to render the technology truly cost-competitive. Each of the three major solar-thermal electric technologies has unique attributes, strengths, and weaknesses (see table 12) and for each a path exists by which costs can be lowered and performance increased without the need for technological breakthroughs [94].

Technology attributes

Parabolic troughs are utility-scale power plants that have been developed as solar–gas hybrid systems. Since 1984, nine fully commercial power plants have been financed and constructed in the Mojave Desert, and the technology has undergone three successive system-design improvements. Although these plants were built with mid-1980s power-purchase contracts, which are no longer available, future projects were planned on a competitive basis by Luz in the United States and Mexico, and by Flachglas Solartechnik in Brazil and Morocco. But in 1991 when Luz was forced to file for bankruptcy, its participation in future parabolic-trough endeavors became at best uncertain and probably unlikely. Prior to its demise, Luz expected that the next generation of SEGS plants would achieve near cost-competitiveness with conventional electric power plants. But a significant decline in fossil-fuel prices in the 1980s, coupled with marked reductions in regulatory and financial incentives, created a difficult market environment for capital-intensive technologies such as SEGS plants.[10]

Central-receiver plants are also utility-scale systems that have been developed because of their cost-effective thermal-energy storage capability. Although the technology has progressed through several pilot-scale experiments, to date no

---

10. Several companies, however, continue to pursue commercial projects and technology development.

commercial projects have been successfully developed, despite the fact that the systems have greater potential for low energy costs than trough systems, in part because of the uncertain cost and performance of large central receivers. The many pilot systems have performed less well than predicted, partly because of their relatively small size and partly because of their experimental nature. Central receivers are inherently large integrated systems, thus meaningful demonstration systems must also be large and costly. In contrast, trough and dish technologies are modular and demonstration plants can be smaller and less costly.

Parabolic-dish technology has accomplished its most significant achievements with distributed systems using Stirling engines. Such modular systems are suitable for small, remote sites as well as for utility-scale applications. However, these systems are the least mature of the three technologies and much basic development remains. Thus, dish–Stirling systems, which have the greatest potential for efficiency, also have the greatest uncertainty. Maintenance costs, for example, remain largely unknown in the face of inadequate data. In addition, recent devel-

Table 12: Comparison of the current strengths and weaknesses of the three solar-thermal electric technologies

	Attribute	Strengths	Weaknesses
Parabolic trough	Utility-scale system	Commercial	Limited performance potential
	Solar–gas hybrid	Proven good performance; growth by product investment; relatively simple system	Thermal storage not currently cost-effective
Central receiver	Utility-scale system	Potential for low cost and high performance	Cost/performance potential not confirmed
	High solar capacity factor	Cost-effective thermal storage	Demonstration size is large and costly; relatively complex system
Parabolic dish	Remote or utility applications	Very high performance	Engine and receiver require development
	Potential for gas hybridization	Modular system; development steps can be inexpensive	Engine development costs are high; receiver and engine lifetimes are uncertain; system maintenance costs are uncertain; no storage capability

opment efforts have emphasized advanced components that are just beginning to be integrated into dish–Stirling systems and so are unproven. Therefore, significant uncertainty remains with respect to the promised performance and reliability of these systems. The costs of deploying and testing modular dish systems is relatively small. However, the costs of developing a solarized Stirling engine that meets the necessary cost, availability, longevity, and maintenance parameters is not, and advances in dish–Stirling systems will probably depend on design modifications to Stirling engines that are made for other (nonsolar) purposes.

The SEGS plants have proven that solar-thermal electric power plants are a viable alternative to fossil-fuel plants. Luz's success during the 1980s can be attributed to a variety of factors. External conditions and market incentives supported the introduction of SEGS technology into the marketplace: oil prices were high, and federal and state tax credits offered a 40 percent capital-cost deduction on solar power projects. A five year accelerated depreciation plan reduced costs further, and both the Public Utility Regulatory Policy Act of 1978 and high fuel prices gave rise to attractive power-purchase contracts. Although these incentives were available to all solar-energy developers, Luz was especially well positioned to exploit them. Not only did the company possess long-term vision and a commitment to develop solar energy, it was also willing to accept risk and shelter the plant's investors from technical uncertainties; in doing so, it displayed both financial and technical acuity. Although Luz finally succumbed to the difficulties of the marketplace (low fuel prices and uncertain government support), the company can claim credit for the nine SEGS plants that today generate more than 350 megawatts of electricity for commercial use.

Levelized energy cost projections

Utility power plants are most commonly compared on the basis of their levelized electricity costs (in units of cents per kWh -- see Box A). The costs reflect initial capital investment and annual operating and maintenance costs over the life of the plant, but do not reflect the value of peak-demand power, which solar-thermal plants can provide through thermal-energy storage and/or fossil hybrid operation.

Predictions of levelized electricity costs for new technologies are subject to many uncertainties that significantly influence whether or not the component projections for capital cost, annual performance, and O&M cost are met. These uncertainties include system reliability, equipment efficiencies and lifetimes, organizational learning, manufacturing capability, and technological improvements (see table 13, which also characterizes the time frame for each technology).

Parabolic-trough power plants

The levelized electricity cost (LEC) for 80 MW$_e$ SEGS plants, which employ LS-3 collectors and oil heat-transfer fluid, ranges from 12 to 17 cents per kWh. The capital cost's upper boundary slightly exceeds the actual costs for SEGS IX, whereas its lower boundary falls slightly below the actual cost for SEGS VIII, sug-

Box A
THE RELATIVE ECONOMICS OF SOLAR-THERMAL POWER SYSTEMS

Levelized electricity cost (LEC) (in cents per kWh)
A standardized utility model and the following financial parameters were used to calculate the LEC.

$$LEC = \frac{CC \cdot FCR + O\&M + Fuel}{AkWh}$$

where:
   CC = capital cost
   FCR = fixed charge rate
   O&M = annual operating and maintenance expenditure including taxes and insurance
   Fuel = fuel cost adjusted for real escalation
AkWh = annual net electricity generation in kWh

Capital cost ($ per kilowatt)
Installed capital cost is projected as a function of technological improvements on a per-unit kilowatt basis. This measure, however, does not provide a useful comparison between the three solar-thermal technologies or between solar-thermal and other power-plant technologies, because the unit capital cost of some solar-thermal technologies reflects the capacity factor of the plant. Although solar-thermal plants with thermal-storage capabilities cost more, the additional costs are usually justified by increased energy production.

Fixed charge rate
Also called a levelized charge rate, the rate converts a plant's capital cost into an equivalent annual cost; it is calculated from the discount rate, depreciation schedules, and tax structures. For this analysis, two real discount rates, 6 and 12 percent, were used. Federal, state, and local property taxes were ignored, and 0.5 percent was added for insurance. These assumptions lead to fixed charge rates of 7.8 and 12.9 percent respectively.

Annual system efficiency
This term is the percentage of energy, calculated on an annual basis, that is extracted from the sun, converted to electricity, and delivered to the utility. It includes all operational and availability losses and is typically lower than the design-point system efficiency. Annual net electricity generation can be calculated by multiplying the annual system efficiency with the annual direct insolation at a site and the plant collector area.

Operation and maintenance cost (¢ per kWh)
This figure reflects the average amount spent yearly on labor, materials, and contracts to operate and maintain the plant. O&M costs are projected on a per-unit kWh basis. Fuel for hybrid plants is typically calculated separately from the O&M cost.

Capacity factor (percent)
This is the energy produced by the plant as a percentage of the total amount the plant would produce if operated at rated capacity during the entire year. By multiplying the capacity factor by the plant rating, annual net electricity generation can be determined. Because the solar resource is so variable, typical solar-only capacity factors are between 20 and 30 percent. But solar-thermal plants can increase their capacity factor either through fossil-hybridization or thermal storage.

Table 13: Levelized energy cost projections[a]

Timeframe	PARABOLIC TROUGH			CENTRAL RECEIVER				DISH–STIRLING		
	80 MWe LS-3	80 MWe LS-4	200 MWe LS-4	100 MWe first plant	200 MWe first plant	200 MWe baseload	200 MWe advanced receiver	3MWe/per year early remote market	30 MWe/per year early utility market	300 MWe/per year utility market
	Present	1995–2000	2000–2005	1995	2005	2005–2010	2005–2010	1995–2000	2000–2005	2005–2010
Capital cost range $/kWe	3,500–2,800	3,000–2,400	2,400–2,000	4,000–3,000	3,000–2,225	3,500–2,900	2,500–1,800	5,000–3,000	3,500–2,000	2,000–1,250
Collector system typical cost $/m2	250	200	150	175–120	120–75	75	75	500–300	300–200	200–150
Annual solar-to-electric range[b]	13–17 percent			8–15 percent	10–16 percent		12–18 percent	16–24 percent	18–26 percent	20–28 percent
Enhanced load matching method	25 percent natural gas	25 percent natural gas	25 percent natural gas	Thermal storage	Thermal storage	Thermal storage	Thermal storage	Solar only	Solar only	Solar only
Solar capacity factor range	22–25 percent	18–26 percent	22–27 percent	25–40 percent	30–40 percent	55–63 percent	32–43 percent	16–22 percent	20–26 percent	22–28 percent
Annual O&M cost range ¢/kWh	2.5–1.8	2.4–1.6	2.0–1.3	1.9–1.3	1.2–0.8	0.8–0.5	1.2–0.8	5.0–2.5	3.0–2.0	2.5–1.5
Solar LEC range[c] ¢/kWh	16.7–11.8	17.2–9.8	11.7–7.9	16.1–8.0	10.1–5.8	6.5– 4.6	8.2–4.5	32.8–14.6	18.6–8.8	10.6–5.5
Hybrid LEC range[c] ¢/kWh	13.0–9.3	13.5–7.9	9.3–6.5	–	–	–	–	–	–	–

a. The data in this table are compiled from several sources. The most comprehensive source is the U.S. DOE analysis performed for the National Energy Plan[3, 95–98]. The LEC calculations are based on a 6 percent real discount rate. They differ slightly from other LEC values given in this chapter that were calculated using different sets of economic assumptions.

b. Typical southwest U.S. site.

c. Fixed charge rate = 7.8 percent.

gesting there is room for improvement in LS-3 technology. The solar-capacity factor matches the performance of SEGS VIII; the upper boundary of the O&M cost range also approximates those of SEGS VIII.

The future of parabolic-trough technology depends on successful development of the LS-4 collector system, which uses direct steam generation, and on the construction of 200 MW$_e$ plants to take advantage of economies of scale. Luz predicted that the fifth such plant would have a lower boundary capital cost of $2,000 per kilowatt of electricity. The installed cost for solar hybrid combined-cycle systems is also thought to be $2,000 per kilowatt of electricity. The estimated upper boundary capital cost assumes only slight improvement over current systems. The LEC costs of the first 80 MW$_e$ LS-4 plant, which may be built during 1995 to 2000, are projected to range from 10 to 17 cents per hour, whereas those of subsequent 200 MW$_e$ LS-4 plants, which may be built during the 2000 to 2005 time frame, are estimated to be from 8 to 12 cents per kWh. The lower estimate of the capacity-factor range for the initial 80 MW$_e$ LS-4 plant reflects the current uncertainty in the performance of this developmental technology.

Trough power plants that employ direct steam generation are likely to continue hybrid operations because they do not yet have cost-effective storage systems. The LEC range for these solar–gas hybrid plants is based on a 1 percent real escalation in 1990 natural gas prices (see table 13), the value will increase, of course, if gas prices escalate at a higher rate.

The size of a parabolic-trough plant is largely limited by the solar field's pumping power requirements. The terrain at a particular site also influences field pumping requirements and strongly influences site preparation costs. The land can slope no more than 1 percent, for example, although terracing can be used effectively to overcome sloping. Typically, 2 hectares of land are needed per megawatt of electricity generated at a site.

Central-receiver power plants

The most economical sizes for central-receiver power plants are from 100 to 200 MW$_e$, with capacity factors ranging from 40 to 60 percent. Not only do large plants achieve significant economies of scale through their various components, but their operating and maintenance costs do not scale linearly with size.[11]

The U.S. Utility Studies provides the best basis for predicting the LEC range for near-term central-receiver power plants. The lower estimate of $3,000 per kW$_e$ for the first 100 MW$_e$ plant assumes salt-in-tube receiver technology and a collector system cost of $120 per m^2 (glass heliostat annual production level of 5,000 units and manufacturing tooling costs amortized over 10 years). The higher capital cost estimate of $4,000 refers to the same plant but with a $175 per m^2 collector system based on a one-time heliostat buy and a 25 percent increase in the cost of the rest of the plant. The plant capacity factor's lower boundary of 25

---

11. Solar–gas turbine systems, which are potentially economical in sizes as low as 10 MW$_e$, are an exception.

percent is based on the best performance (annual solar-to-electric efficiency) achieved by Solar One [94]. The lower boundary for o&m cost is the estimate found in the Utility Studies; the upper boundary is thought to be about 50 percent higher.

According to the Utility Studies, the fifth commercial central-receiver plant could be a 200 MW_e plant with a higher flux, salt-in-tube receiver and less expensive stretched-membrane heliostats. Furthermore, the system could achieve an LEC range of 6 to 10 cents per kWh by 2005. The lower capital cost estimate of $2,225 per kilowatt of electricity assumes capital costs of $75 per m^2 heliostats, and a 40 percent capacity factor. The higher estimate reflects a heliostat cost of $120 per m^2, and a 25 percent increase in the cost of the rest of the plant. Plant performance is estimated to result in a 30 to 40 percent capacity factor. The range of o&m costs are based on the Utility Studies and a 50 percent increase.

The LEC for a 200 MW_e central-receiver power plant may drop further with the development of high-capacity plants and advanced direct-absorption or internal-film receivers. Nitrate-salt central receivers employ a low-cost yet highly efficient thermal-storage system. Therefore, adding thermal storage improves the plant's capacity factor and reduces its LEC because the fixed costs in the power conversion system are amortized over more kWh. Thermal storage also enhances the plant's ability to generate power during peak-demand periods when electricity is most valuable. Advanced receivers, which are more efficient and have reduced pumping power needs, can potentially increase performance and lower cost. Both the high-capacity plants and advanced receiver concepts yield an LEC estimate that is less than 5 cents per kWh and is obtainable in the 2005 to 2010 time frame.

Central-receiver plants need between 2 and 4 hectares per megawatt of electricity, depending on their design capacity factor. Plant size appears to be limited by attenuation of the reflected sunlight from the furthest heliostats; for a plant with a receiver rating of 1,000 MW_t, the furthest heliostat is about 2 kilometers away from the receiver.

Parabolic-dish power plants

Modular dish–Stirling systems are being developed for remote power and utility applications. Remote power systems, which are designed to generate 5 to 10 kilowatts of electricity, are meant to compete with conventional diesel systems. In contrast, utility-based systems are expected to have 25 kW_e modules, with possibly larger ones in the future. The LEC projections for dish–Stirling systems begin with the production cost estimates of United Stirling, McDonnell Douglas, and Advanco for their glass/metal concentrator and the USAB 4-95 kinematic Stirling engine [94, 95]. This automotive derivative engine required a significant amount of maintenance. In the near future, advanced Stirling engines—both kinematic with swashplate control and free piston—are likely to be available in the 5 to 25 kW_e range. Because these advanced engines are so simple, both appear capable of achieving the low maintenance and longevity goals needed for cost-effective dish–Stirling systems. In addition, the simpler engines are expected to cost less than their predecessors. The current multifaceted concentrator designs, such as the LaJet concentrator, are also expected to drop concentrator costs below those for glass/metal designs. Given the developmental nature of the technology, however, cost and performance projections remain uncertain.

For the remote market, an initial annual production volume of three megawatts of electricity represents minimum use of production tooling. The lower capital cost estimate of $3,000 per kilowatt of electricity is based on $300 per m^2 in collector costs, a $1,000 per kW$_e$ receiver/engine/generator cost ($5,000 per unit), and a $750 per kW$_e$ balance-of-plant and installation cost ($3,750 per unit). The upper estimate represents $500 per m^2 in collector costs and a 50 percent increase in other values. The lower boundary of the capacity factor range is based on the demonstrated performance of the McDonnell Douglas/USAB system, and the upper boundary is based on a higher system availability. The lower O&M cost is estimated at $20 per m^2 [95]; the upper value is twice as high. Because the resulting LEC range of 15 to 33 cents per kWh is for a small remote system it should not be directly compared with LECs for utility systems.

Future improvements, such as single-facet stretched-membrane concentrators and higher receiver temperatures and engine efficiencies, will increase the cost-effectiveness of dish–Stirling systems. The predominant factor in lowering the LEC, however, will likely be mass-production economies for the collector and the receiver/engine/generator. In addition, as installed-system size increases, additional economies will be gained in the balance-of-plant cost. For the early utility market, at an annual production rate of 30 MW$_e$, the reduction in capital cost, improvement in capacity factor, and reduced O&M costs result in an LEC range of 9 to 19 cents per kWh in the 2000 to 2005 time frame.

Expanding production capacity to 300 megawatts of electricity per year can potentially cut system costs still further. In addition, improvements in engine efficiency and reliability may lead to increased capacity factors and reduced O&M costs. From 2005 to 2010, the projected LEC range is 5.5 to 11 cents per kWh.

Advanced systems

Because advanced solar-thermal systems are still at an early stage of development, their cost and performance potential is not easily quantified. Nonetheless, these systems have attractive features that are likely to spur their development.

Air central-receiver systems

Driving a steam-turbine power plant with a volumetric air receiver appears to be less cost-effective than using a nitrate-salt receiver. The air system, it seems, has higher thermal-storage costs as well as lower receiver and turbine efficiencies (see the section "Central-receiver systems" on page 24). But the receiver does have the potential to achieve significantly higher temperatures than a nitrate-salt system. Consequently, a volumetric receiver can drive a gas turbine or combined-cycle power plant, at least in theory. Because gas turbines are more cost-effective in smaller sizes than steam turbines, they are likely to be economical in the 10 to 100 MW$_e$ range.

Nonelectric applications
Several proven and potential applications for solar-thermal concentrators exist. Although these applications have not been discussed in this chapter, they represent a potential market that may have a positive and synergistic effect on the viability of solar-thermal electric systems.

Solar-thermal systems have been developed for a variety of industrial and commercial heating and cooling applications, but they have not yet succeeded in the marketplace. Other possibilities for the technology include destruction of hazardous organic chemicals and compounds using concentrated solar photons directly, detoxification of various substances at high temperatures, and creation of advanced materials, such as hardened ceramics or metal-clad ceramics.

SUMMARY

In addition to producing electricity, solar-thermal technologies produce hot water and steam for industrial applications; they can also provide hot air for industry and energy to photolytically or photocatalytically process fuels and chemicals and destroy hazardous materials.

Projected levelized electricity costs for the three solar-concentrator systems reflect the uncertainty inherent in the future cost, performance, and reliability of any developing technology (see figure 34).

Parabolic-trough systems are the most fully developed of the solar-thermal technologies, and major installations for both process heat and electric power production exist. Although trough systems typically operate at temperatures (100 to 400°C) that are significantly lower than those achieved by other concentrating systems, the trough design offers the advantage of flexibility and commercial experience. Because each module is connected to other modules via various series- and parallel-flow circuits, the system can be adjusted according to a range of performance characteristics. If necessary, a fossil-fueled auxiliary heat source can drive the power conversion system in parallel with or independent of the solar field.

Nine fully commercial parabolic-trough power plants are now operating in the Mojave Desert in the United States. These utility-scale power plants produce a total of 350 megawatts of electricity per year, and the two newest units generate 80 MW$_e$ each. They are solar–natural gas hybrid systems, because cost-effective thermal storage has not been developed for these plants. At present, the LECs for existing trough systems range from 12 to 17 cents per kWh.

Prospects for future plants, however, are uncertain following the recent financial failure of Luz International Ltd., the company that developed the technology as well as the nine existing plants.

Box B
SOLAR-THERMAL TEST FACILITIES

Advanced solar-thermal concepts are tested and evaluated at several facilities around the world. The principal features and capabilities of the major solar thermal facilities are provided below.

Odeillo
This 1 MW_t solar furnace, located at Odeillo, France, is the oldest major solar-thermal test facility. Constructed in 1970, this research station has been a center for investigations into materials behavior. The facility consists of a field of 63 heliostats that redirect sunlight to a large parabolic concentrator. The facility can achieve very high solar flux concentrations (as much as 3.5 megawatts per m^2) and test temperatures as high as 4,000°C [99].

Solar-Thermal Test Facility (STTF)
The Solar-Thermal Test Facility (STTF), which is operated for the U.S. DOE by Sandia National Laboratories, includes the Central-Receiver Test Facility (CRTF), the Distributed Receiver Test Facility (DRTF), and a Solar Furnace [100, 101]. The CRTF is a 5.5 MW_t facility consisting of 222 heliostats. A 61 meter concrete tower contains test platforms at several levels. The CRTF has tested water/steam, sodium, nitrate-salt, and air receivers, as well as advanced-glass and stretched-membrane heliostats. It was used for a 750 kW_e molten-salt electric experiment, and a full-scale nitrate-salt pump and valve experiment.

The DRTF consists of both dish and trough concentrators. Two 11 meter diameter, 75 kW_t parabolic-dish concentrators are used to test materials, receivers, and heat engines. Dishes, such as the LaJet concentrator, the Power Kinetics concentrator and the Solar Kinetics stretched-membrane concentrator, have also been evaluated. Trough collectors, originally built for the Modular Industrial Solar Retrofit (MISR) program [11], are currently being tested for a potential role in detoxifying contaminated groundwater. The 16 kilowatt-thermal solar furnace, which provides a peak solar flux of almost 4,000 suns, is used primarily for flux gauge calibration and materials testing.

Plataforma Solar de Almería (PSA)
Jointly operated by the Instituto de Energías Renovables of the Centro de Investigaciones Energéticas, Medioambientales y Tecnológicas in Madrid and the Deutsche Forschungsanstalt für Luft- und Raumfahrt in Cologne, the Plataforma Solar de Almería (PSA) in southern Spain is the largest solar-thermal test center in Europe today.

The PSA consists of three major test facilities: the central-receiver CESA-1 system, the SSPS/CRS tower plant, and the SSPS/DCS distributed power system. CESA-1, originally a full system experiment with a water/steam-cooled cavity receiver, contains an 11,880 m^2 heliostat field, which directs up to 7 megawatts of thermal power to the different experimental levels of the 80 meter tower and achieves a peak heat-flux density of 4 MW_t per m^2. In 1984, an experimental gas-cooled receiver loop, the Gas-cooled Solar Tower project, was installed on the tower's top level [41].

The SSPS/CRS tower plant, originally installed as a 500 kW_e full system experiment employing sodium as the receiver and storage fluid, was remodeled after a serious fire in 1985 into a multipurpose test facility, capable of providing up to 4 MW_t power for testing such central-receiver components as volumetric receivers, clustered Stirling engines, reconcentrators, and stretched-membrane heliostats.

The third major facility is the 7,600 m^2 distributed collector system SSPS/DCS, which underwent two phases of grid-connected operation. The collector fields were then dedicated to solar process heat experiments, including the connection of a 80 m^3 per day multiple effect desalination plant. Currently the implementation of a 11,500 m^2 test loop of advanced parabolic-trough collectors is being planned in order to test direct steam generation, thermal storage, and other advanced features of the next generation of commercial trough collectors.

Weizmann Institute of Science
The Weizmann Institute of Science operates two solar-thermal test facilities, a 7.3 meter diameter, 16 kW_t solar furnace capable of achieving a concentration ratio of 11,000, and a 2.9 MW_t central-receiver research facility. The latter contains 64 heliostats, each 54.25 m^2, and a 53 meter tower with three main experimental stations. The principal experiments performed at the facility involve solar-thermochemical heat transport and solar pumped lasers [99].

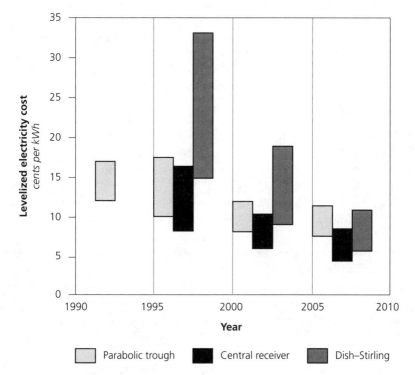

FIGURE 34: *Levelized energy cost projections for solar-only operations. (Upper and lower bounds reflect the best available data.)*

Nonetheless, a new type of trough, which generates steam directly in the field, has the potential to improve the system's overall cost-effectiveness. Its future development is currently under discussion by several companies. If successfully commercialized, this direct-steam trough (in combination with an increased plant size of 200 megawatts of electricity) will make an LEC range of 8 to 12 cents per kWh possible by 2000 to 2005.

Central receivers are considered the most promising among utility-scale thermal electric systems: they are inherently large, generally having a capacity of 10 megawatts or more, they achieve temperatures ranging from 500 to 1,500°C, and they collect solar energy optically, which means they have minimal thermal-energy transport needs. Moreover, the more recent designs include cost-effective thermal energy storage systems, which provide dispatchability and increase capacity.

At a central-receiver power plant, collected solar energy can either be converted directly into electricity or stored as sensible heat (in a thermal storage system) for subsequent delivery to the power conversion system. Because storage decouples the collection of solar energy from its conversion to electricity, it makes operation possible into the evening or during cloudy weather. The two types of central-receiver plants being developed use either molten nitrate salt as the receiver and thermal-storage fluids, or air as the heat-transfer fluid, with ceramic bricks for thermal storage.

Although central-receiver systems have been tested on a pilot basis, the technology has not yet been successfully commercialized, in part because its costs and performance characteristics are unproven. To be meaningful, demonstration systems must be large and costly; in contrast, the more modular trough and dish systems can be tested much more cheaply.

Central-receiver power plants that generate between 100 and 200 megawatts of electricity and have a capacity factor between 40 and 60 percent appear to be the most economical. In fact, the U.S. Utility Studies suggest that a near-term 100 MW_e central-receiver plant could generate electricity for as little as 8 to 16 cents per kWh. By 2005, the costs for an improved, 200 MW_e plant could drop to 6 to 10 cents a kWh. By 2010, as a result of further improvements in design and capacity, the LEC for 200 MW_e power plants may well reach less than 5 cents per kWh.

Parabolic-dish systems consist of modular collector and receiver units that can either function independently or as part of a larger system of dishes. Two distinct parabolic-dish power systems have been developed. One collects thermal energy from each dish and transports it to a centralized power conversion system; the other uses small engines coupled to generators to produce electricity at each dish. The centralized system, like the parabolic-trough system, gathers thermal energy rather than reflected sunlight.

The small engine-generator system gathers energy in the form of electricity rather than as heated fluid. Such systems operate as closely coupled power-conversion systems. Although the engine-generator parabolic-dish system contains no thermal storage, it can be hybridized to operate with fossil fuel during periods without sunshine.

A world record of 29.4 percent net solar-to-electric conversion efficiency was set by a 25 kW_e parabolic-dish-Stirling system. Although these modular dish–Stirling systems have the highest efficiency potential, they are the least developed of the three concentrating solar-thermal electric technologies. Even though the early systems demonstrated excellent efficiency and reasonable availability, maintenance costs remain highly uncertain given the lack of adequate data. In addition, the more recent designs emphasize advanced components that are mostly unproven. Therefore, significant uncertainties remain concerning their projected performance capabilities, lifespan, and reliability. Moreover, developing a solarized Stirling engine that meets the necessary cost, availability, lifetime, and maintenance parameters is a significant challenge.

The projected LEC range for a remote power system of 5 to 10 kilowatts, at an annual manufacturing production volume of 3 MW_e, is 15 to 33 cents per kWh in 1995.[12] Future improvements, such as single-facet stretched-membrane

---

12. Because these systems are intended for small remote applications, their LEC should not be directly compared to the trough and central-receiver LECs for large utility systems.

concentrators and higher receiver temperatures and engine efficiencies, will further increase the system's cost-effectiveness. But the predominant factor in lowering the LEC will be mass-production of the collector and the receiver and engine-generator systems. The LEC range for early utility systems at an annual production volume of 30 MW$_e$, is 9 to 19 cents per kWh in the 2000 to 2005 timeframe. By expanding production capacity to 300 MW$_e$ per year during the 2005 to 2010 timeframe, the LEC range is projected to drop to between 5.5 and 11 cents per kWh.

One solar-thermal technology—the solar pond—does not concentrate incoming solar radiation, but relies instead on a salt gradient to collect and store low-temperature heat. Temperatures at the bottom of the solar pond can reach 90°C, which is sufficient to drive the vapor generator of an organic Rankine-cycle engine, provide process heat, or desalt water. In addition, because solar ponds have a rather large storage capacity, they can store thermal energy seasonally as well as diurnally.

Solar ponds, however, are unlikely to generate large amounts of electricity because they are limited by their significant water consumption. Water must be continuously replenished to compensate for surface evaporation and to recycle salt, which is accomplished by evaporating surface water and sending its salt to the pond bottom. Nonetheless, solar ponds may be an excellent source of limited amounts of electricity, say, for desalting applications, where electricity is needed primarily to operate the internal pumps. Such desalting applications may be especially valuable in remote areas such as island and coastal desert sites in need of freshwater where the large energy storage capability of solar ponds matches the often diffuse insolation conditions.

Although the various solar-thermal technologies show great promise, they will not fully exploited until broader cooperation between governments, utilities, and private industry occurs. In addition, market incentives must be created to overcome barriers that exist, especially in the financial community, because solar-thermal systems are new (and thus a potentially risky investment) and may be viewed as competitive with existing technologies. Finally, solar-thermal systems must be supported by stable long-term regulatory policies.

WORKS CITED

1. Solar Thermal Power, Solar Technical Information Program, Solar Energy Research Institute, SERI/SP-273-3047, February 1987.

2. *Direct Normal Solar Radiation Manual. 1982*
   Solar Energy Research Institue, SERI/SP-281-1658, October.

3. The potential of renewable energy: an interlaboratory white paper, Idaho National Engineering Laboratory, Los Alamos National Laboratory, Oak Ridge National Laboratory, Sandia National Laboratories, Solar Energy Research Institute, SERI/TP-260-3674, March 1990.

4. Brower, M. 1990
   Cool energy: the renewable solution to global warming, Union of Concerned Scientists, Cambridge, Massachusetts.

5. Falcone, P.K. 1986
   A handbook for solar central receiver design, Sandia National Laboratories, SAND86-8009, December.

6. Smith, A.H.P. and Golding, P. 1991
   El Paso solar pond fact book, University of Texas at El Paso, presented to the US Bureau of Reclamation, November.

7. Telephone conversation with Eli Yaffe, Ormat Turbines Ltd., El Centro office, December 12, 1989.

8. Solar Salt Pond Generating Facility for California, California Energy Commission, P700-81-015, August 1981.

9. Preliminary planning alternatives for solving agricultural drainage and drainage-related problems in the San Joaquin Valley, US Department of the Interior and California Resources Agency, August 1989.

10. Efforts to save Salton Sea and enhance industrial, recreational and residential values in South-Central California, Salton Sea Task Force, September 1989.

11. Kutcher, C. et al. 1982
    Design approach for industrial process heat systems, SERI/TR-253-1356, August.

12. Alvis, R.L. 1980
    Modular Solar Industrial Retrofit Project (MISR), *Proceedings of the Line-Focus Solar Energy Technology Development Seminar for Industry*, Albuquerque, New Mexico, September, SAND80-1666, 93-113.

13. Grasse, W. 1985
    SSPS: results of test and operation 1981-1984, DFVLR, Report. No. SSPS SR 7, Cologne, Germany, May.

14. Jaffe, D., Friedlander, S., and Kearney, D. 1987
    The Luz solar electric generating systems in California, ISES Solar World Congress, Hamburg.

15. Kearney, D. et al. 1991
    Status of the SEGS plants, *Proceedings of the 1991 Solar World Congress of the International Solar Energy Society*, Denver, Colorado, August.

16. Application for Certification, SEGS VIII, Harper Lake, California, California Energy Commission, Docket No. 88-AFC-1, February 1988.

17. Lotkar, M. 1991
    Barriers to commercialization of large-scale solar electricity; lessons learned from the LUZ experience, contractor report SAND91-7014, Sandia National Laboratories, November.

18. Faas, S.E. et al.1986
    10 MWe solar thermal central receiver pilot plant: thermal storage subsystem evaluation, Sandia National Laboratories, SAND86-8212.

19. Geyer, M., Bitterlich, W., and Werner, K. 1987
    The dual medium storage tank at the IEA/SSPS project in Almería, Spain; Part 1: *Experimental validation of the thermodynamic design model, JSEE.*, Vol. 109, No. 3. 192–198, Washington DC, August.

20. Dinter, F., Geyer, M., and Tamme, R. 1991
    Thermal energy storage for commercial applications, a feasibility study on economic storage systems, Springer, Heidelberg.

21. Luz Industries Ltd., 1991
   Thermal energy storage for medium temperature solar electric power plants, final report of Luz, in F. Dinter, M. Geyer, and R. Tamme, eds., appendix to thermal energy storage for commercial applications, a feasibility study on economic storage systems, Springer, Heidelberg.

22. Beine, B. 1991
   Feasibility study for the concepts of a 200 MWh thermal energy storage with solid salt plates, concrete plates and phase change salts as storage media, final report of Siempelkamp, in F. Dinter, M. Geyer, and R. Tamme, eds., appendix to thermal energy storage for commercial applications, a feasibility study on economic storage systems, Springer, Heidelberg.

23. Baker, A.F. and Skinrood, A.C. 1983
   Characteristics of current solar central receiver projects, October 1982, Sandia National Laboratories, SAND83-8013, May.

24. Central receiver technology status and assessment, Solar Technical Information Program, Solar Energy Research Institute, SERI/SP-220-3314, September 1989.

25. Holl, R.J. 1989
   Status of solar-thermal electric technology, Electric Power Research Institute Report No. EPRI GS-6573, December.

26. Radosevich, L.G. 1988
   Final report on the power production phase of the 10 MWe solar central receiver pilot plant, Sandia National Laboratories, SAND87-8022, March.

27. Delameter, W.R. and Bergen, N.E. 1986
   Review of the molten salt electric experiment: a solar central rteceiver project, SAND86-8249, December.

28. Solar 100 conceptual study, Southern California Edison Company, McDonnell Douglas Corporation and Bechtel Power Corporation, August 3, 1982.

29. Hillesland, T., Jr. and De Laquil, P. 1988
   Results of the US solar central receiver utility studies, VDI Berichte 704.

30. Hillesland, T., Jr. 1988
   Solar central receiver technology advancement for electric utility applications, Phase I topical report, Pacific Gas & Electric Company, San Ramon, California, GM 633022-9, DOE Contract DE-FC04-86AL38740 and EPRI Contract RP 1478-1, August.

31. *Utility Solar Central Receiver Study,* Phase I, topical report, Arizona Public Service Company, Phoenix, Arizona, DOE Contract DE-FC04-86AL3874-1, November 1988.

32. *Utility Solar Central Receiver Study,* final report, Arizona Public Service Company, Phoenix, Arizona (for Alternate Utility Team), DOE Contract DE-FC04-86AL38741-2, September 1988.

33. Solar central receiver technology advancement for electric utility application, Phases IIA and IIB Topical Report, Pacific Gas & Electric Company and Arizona Public Service Company, DOE Contract DE-FC04-86AL38740, October 1989.

34. Stoddard, M.C. et al. 1987
   SOLERGY—a computer code for calculating the annual energy from central receiver power plants, SAND86-8060, Sandia National Laboratories, Livermore, California, May.

35. De Laquil, P. and Kelly, B.D. 1989
   Solar central receiver technology advancement for electric utility application, Phase IIC summary report, prepared for Pacific Gas & Electric Company by Bechtel National, Inc., DOE Contract DE-FC04-86AL38740, December.

36. Tyner, C.E. and Wu, S.F. 1988
Commercial direct absorption receiver design studies, in *Proceedings of the Fourth International Symposium on Research, Development and Applications of Solar Thermal Technology,* Santa Fe, New Mexico, June 13-17.

37 Anderson, J.V. et al. 1986
Direct absorption receiver system assessment, final draft, Solar Energy Research Institute, Golden, Colorado, November.

38. Chavez, J.M. et al. 1991
Design, construction, and testing of the direct absorption receiver panel research experiment, *Solar Engineering.*

39. Kolb, G.J. and Chavez, J.M. 1990
An economic analysis of a quad-panel direct absorption receiver for a commercial-scale central receiver power plant, 12th Annual ASME International Solar Energy Conference, Miami, Florida, *Solar Engineering,* April.

40. Tracy, T. 1992
Potential of modular internal film receiver in molten salt central receiver solar power system, to be published at 1992 ASME Solar Energy Conference, Maui, Hawaii, March.

41. Torre Cabezas, M. 1990
Receptor Avanzado de Pelicula Interna, Ciemat 652 DE91 741673, Madrid, ISSN 0214-087X.

42. Grasse, W. et al. 1987
Solar energy for high temperature technology and applications, DFVLR Brochure, Cologne.

43. Wehowsky, P. 1989
Goals, structure and main results of the technology program GAST, in M. Becker and M. Bohmer, eds., GAST: The Gas-Cooled Solar Tower Technology Program, *Proceedings of the Final Presentation,* 1–14, Springer-Verlag, Berlin.

44. Gas-cooled solar tower power plant GAST: Analysis of its potential (in German), Study prepared for the BMFT, Bonn, Germany, BMFT, 1985.

45. Karrais, B. 1986
Ultralight modular ceramic high flux receiver, in *Proceedings of the 3rd International Workshop on Solar Thermal Central Receiver Systems,* Constance, June 23-27.

46. Becker, M. and Böhmer, M. 1988
Achievements of high and low flux receiver development, in W. Bloss and F. Pfisterer, eds., *Proceedings of the ISES Solar World Congress,* Hamburg 1987,1744–1783, Pergamon Press, Oxford, United Kingdom.

47. Pritzkow, W. 1988
The volumetric ceramic receiver, second generation, in B. K. Gupta, ed., Solar thermal technology, *Proceedings of the 4th International Symposium,* Santa Fe, New Mexico, 635-643, Hemisphere Publishing Company, New York.

48. Chavez, C. and Chaza, C. 1990
Ceramic foam volumetric receiver, in *Proceedings of the 5th Symposium on Solar High Temperature Technologies,* Davos, Switzerland, August 27-31, Paul Scherrer Institute.

49. Böhmer, M. and Chaza, C. 1990
The ceramic foil volumetric receiver, in *Proceedings of the 5th Symposium on Solar High Temperature Technologies,* Davos, Switzerland, August 27-31, Paul Scherrer Institute.

50. Fricker, H.W. et al. 1988
Design and test results of the wire receiver experiment in Almería, in B. K. Gupta, ed., Solar thermal technology, *Proceedings of the 4th International Symposium,* Santa Fe, New Mexico, 265-277, Hemisphere Publishing Company, New York.

51. Fricker, H. W. and Meinecke, W. 1988
PHOEBUS results of the system comparison of the 30 MWe european feasibility study, in B. K. Gupta, ed., *Proceedings of the 4th International Symposium*, Santa Fe, New Mexico, 265-277, Hemisphere Publishing Company, New York.

52. De Laquil, P. et al. 1990
PHOEBUS Project 30 MWe solar central receiver plant conceptual desig, 12th Annual ASME International Solar Energy Conference, Miami, FL, 1-4 April.

53. PHOEBUS A 30 MWe solar tower power plant for Jordan, Phase IB, feasibility study, executive summary, March 1990.

54. Gintz, J.R.1976
Advanced thermal energy storage concept definition study for solar brayton power plants, Boeing Tech. Rep., Vol. 1, ERDA Contract No. EY-76-C-03-1300, SAN/1300-1.

55. Geyer, M. 1987
High temperature storage technology (in German), Springer, Berlin, Heidelberg, New York .

56. Kainer, H. et al. 1986
High temperature storage for gas cooled central receiver systems, solar thermal central receiver systems, *Proceedings of the 3rd Intrenational CRS Workshop*, Konstanz, Vol. 2, 897-916, Springer-Verlag, Heidelberg.

57. Kalfa, H. and Streuber, C. 1987
Layout of high temperature solid heat storages, solar thermal energy utilization, Vol. 2, 111-209, Springer-Verlag, Heidelberg.

58. Tamme, R., Allenspacher, P., and Geyer, M. A. 1986
High temperature thermal storage using salt/ceramic phase change materials, *Proceedings of the 21st IECEC*, San Diego, Vol. II, 846-849.

59. Second generation central receiver technologies: a status report, Sandia/DLR, to be published by DLR in 1992.

60. Hunt, A.J. 1978
Small particles heat exchangers, Lawrence Berkeley Laboratory Report 7841.

61. Hunt, A.J. 1979
A new solar thermal receiver using small particles, *Proceedings of the ISES*, Atlanta, Georgia.

62. Hunt, A.J. and Evans, D.B. 1981
The design and construction of a high temperature gas receiver utilizing small particle as the heat exchanger (SPHER), LBL-13755, Lawrence Berkeley Laboratory, University of California.

63. Hunt, A. J. and Brown, C.T. 1983a
Solar testing of the small particle heat exchanger receiver (SPHER), LBL-15807, Lawrence Berkeley Laboratory, University of California.

64. Hunt, A. J. and Brown, C. T. 1983b
Solar test results of an advanced direct absorption high temperature gas receiver, *Proceedings of the 1983 Solar World Congress*, Perth, Australia.

65. Schoenung, S.M., De Laquil, P., and Loyd, R.J. 1987
Particle suspension heat transfer in a solar central receiver, ASME-JSME-JSES Solar Energy Conference, Honolulu, HI., 22-27 March.

66. De Laquil, P. et al. 1988
A direct absorption solar central receiver gas-turbine power plant, ISES.

67. Martin, J. and Vitko, J., Jr. 1982
   ASCUAS: A solar central receiver utilizing a solid thermal carrier, Sandia National Laboratories, SAND82-8203, January.

68. Solar total energy project summary report, Georgia Power Co., SAND 87-7108, May 1988.

69. Washom, B.J. 1984
   Vanguard I Solar Parabolic Dish Stirling Engine Module, DOE/AL/16333-2, Advanco Corporation, September.

70. Coleman, G.C. and Raetz, J.E 1986
   Field performance of Dish–Stirling solar electric systems, *Proceedings of the 26th IECEC,* San Diego, California, August.

71. Lopez, C.W. and Stone, K.W. 1992
   Design and performance of the Southern California Edison Stirling Dish, *ASME International Solar Energy Conference,* Maui, Hawaii, March.

72. Grossman, J.W., Houser, R.M., and Erdman, W.W. 1992
   Testing of the Single-Element Stretched Membrane Dish, SAND91-2203, Sandia National Laboratories, Albuquerque, New Mexico, February.

73. Keck, T., Schiel, W., and Benz, Rainer 1990
   An innovative Dish/Stirling system, *Proceedings of the 25th Intersociety Energy Conversion Engineering Conference,* Reno, Nevada, August.

74. Diver, R. B. et al. 1990
   Trends in Dish Stirling solar receiver designs, *Proceedings of the 25th Intersociety Energy Conversion Conference,* Reno, Nevada, August.

75. Linker, K.L. 1989
   Testing of the STM4-120 Kinematic Stirling Engine for solar thermal electric systems, *Proceedings of the 24th IECEC Conference,* August.

76. Moreno, J.B. et al. 1990
   Test results from a full scale sodium reflux pool-boiler solar receiver, *12th Annual ASME International Solar Energy Conference,* Miami, Florida, April.

77. Slaby, J.G. and Alger, D.L. 1987
   Overview of free-piston stirling technology for space power application, *Proceedings of the 22nd IECEC Conference,* August.

78. Conceptual design of an Advanced Stirling conversion system for terrestrial power generation, DOE/NASA/0372-1, NASA CR-180890, Mechanical Technology, Inc., January 1988.

79. 25 kWe Solar Thermal Stirling Hydraulic System, DOE/NASA/0371-1, NASA CR-180889, 1988.

80. Shaltens, R.K. and Schreiber, J.G. 1989
   Comparison of conceptual designs for 25 kWe Advanced Stirling Conversion Systems for dish electric applications, *Proceedings of the 24th IECEC Conference,* August.

81. Davis. J. 1991
   Update on the CPG Dish/Stirling System development, SOLTEC 90, Cummins Power Generation, March.

82. Update on the CPG Dish/Stirling System Development, Cummins Power Generation, SOLTECH 91 Proceedings, Vol. II, Solar Energy Research Institute, Golden, Colorado, US Department of Energy, Washington, DC, Sandia National Laboratories, Albuquerque, New Mexico, March 1991.

83. Kagawa, N. et al. 1989
   Mechanical analysis and durability for a 3 kW Stirling Engine, *Proceedings of the 24th IECEC Conference,* August.

84. Nogawa, M. et al. 1989
    Development of solar stirling engine alternator for space experiments, *Proceedings of the 24th IECEC Conference*, August.

85. Fish, J.D. and Hawn, D.C. 1987
    Closed loop thermochemical energy transport based on CO2 reforming of methane: balancing the reaction system, *Journal of Solar Energy Engineering*, Vol. 109, No. 215.

86. Anikeev, V.I. etal. 1990
    Theoretical and experimental studies of solar catalytic power plants based on reversible reactions with participation of methane and synthesis gas, *International Journal of Hydrogen Energy*, Vol. 15, No. 4 275-286.

87. Levy, M. et al. 1990
    Storage and transport of solar energy by a thermochemical pipe, solar thermal technology-research development and applications, in B. P. Gupta, ed., *Proceedings of the 4th International Symposium*, 527-536.

88. Böhmer, M., Langnickel, U., and Sanchez, M. 1990
    Solar steam reforming of methane, *Proceedings of the 5th Symposium on Solar High Temperature Technologies*, August 27-31, Davos, Switzerland (to be published).

89. Diver, R.B. 1987
    Receiver/reactor concepts for thermochemical transport of solar energy, *Journal of Solar Energy Engineering*, Vol. 109, August,199-204.

90. Diver, R.B. and Ginn, W C. 1987
    Design of the Sandia-Israel 20-kW reflux heat-pipe solar receiver/reactor, SAND87-1533, Albuquerque, New Mexico, Sandia National Laboratories, September.

91. Diver, R.B. et al. 1988
    Solar test of an integrated sodium reflux heat-pipe receiver/reactor for thermochemical energy transport, solar thermal technology - research development and applications, *Proceedings of the 4th International Symposium*.

92. Muir, J.F. et al. 1991
    Solar reforming of methane in a direct absorption catalytic reactor on a parabolic dish: part I - test and analysis, *Solar Engineering*.

93. Spiewak, I., Epstein, M., and Meirovich, E. 1988
    Design of a methanator system for the Weizmann Institute solar thermochemical cycle, solar thermal technology - research development and applications, *Proceedings of the 4th International Symposium*.

94. Holl, R.J. and De Meo, E A. 1990
    The status of solar thermal electric technology, advances in solar energy, edited by K.W. Böer, American Solar Energy Society, Inc., Boulder, Colorado.

95. Williams, T. et al. 1987
    Characterization of solar thermal concepts for electricity generation, Pacific NW Laboratory Report PNL-6128, Richland, Washington.

96. Solar thermal electric technology rationale, August 1990, Solar Thermal and Biomass Power Division, Office of Solar Energy Conversion, US Department of Energy.

97. Renewable energy technology characterizations, prepared for National Energy Strategy, Science Applications International Corporation, March 1990.

98. Technology information module for solar thermal technology, prepared for Pacific Gas & Electric Company by Bechtel Corporation, September 1990.

99. Solar energy for high temperature technology and applications, DFVLR and the Energy Research Commission, ISBN 3-89100-012-X, Cologne, Germany, 1987.

100. Holmes, J.T. et al. 1980
Operating experience at the Central Receiver Test Facility (CRTF), Albuquerque, New Mexico, Sandia National Laboratories, SAND80-2504C.

101. Maxwell, C. and Holmes, J. 1987
Central Receiver Test Facility experiment manual, Albuquerque, New Mexico, Sandia National Laboratories, January, SAND86-1492.

6
INTRODUCTION TO PHOTOVOLTAIC TECHNOLOGY

HENRY KELLY

Photovoltaic devices convert sunlight directly into electricity using a method that differs fundamentally from the heat engines used in almost all other modes of electricity generation. Even wind power is in effect a heat engine, operated by differential solar heating of the atmosphere. Photovoltaic modules need have no moving parts, operate quietly without emissions, and are capable of long lifetimes with little or no maintenance. The major barrier to widespread adoption of photovoltaic equipment is its high cost.

There is ample room for optimism about cost reductions. The efficiency of commercial photovoltaic equipment is about half the efficiency of cells demonstrated in recent laboratory tests, and far below theoretical limits. And most photovoltaic equipment today is assembled in labor-intensive batch processes. Costs will fall sharply when mass production techniques are introduced. But the cost of photovoltaic electricity also depends heavily on reducing the high cost of such low-technology problems as weatherproofing cells and mounting them in the field.

A number of strategies for reducing photovoltaic system costs are being actively pursued around the world. The most promising seem to be methods for depositing thin (1 micron) layers of material on glass substrates, and methods using low-cost optical systems to concentrate sunlight on comparatively small, high-efficiency cells. One, or several, of these concepts is likely to produce power competitive with utility peaking power by the end of the century in average climates. Systems competitive with baseload utility power could be available shortly thereafter.

INTRODUCTION

Although the photovoltaic (PV) effect has been recognized since 1839 practical applications began in the early 1970s, when photovoltaic cells were adopted by the U.S. space program. The program spawned a number of innovations, and as prices fell commercial producers were able to sell PV systems for a growing number of land-based applications. In 1990, worldwide photovoltaic sales reached 48 megawatts [1] and included such small-scale markets as pocket calculators and other consumer goods (representing 22 percent of U.S. sales in 1989), as well as larger-scale markets such as communication systems (21 percent), buoys and other transportation-related systems (9 percent), and pumping stations (6 percent). Experimental grid-connected systems were responsible for 10 percent of U.S. sales in 1989; power arrays for isolated homes and other electric power applications accounted for 20 percent [2].

A decade of field experience now provides clear evidence that photovoltaic systems can be operated for several decades with minimal maintenance and high availability.

Unique features of photovoltaics

Because photovoltaic equipment differs in such fundamental ways from the fossil-fuel-powered equipment now used to provide much of the world's electricity, it is not surprising that photovoltaics offer a unique collection of benefits and problems (see appendix for a brief summary of the way photovoltaic cells work and the problems encountered in improving their performance).

Photovoltaic power is intermittent in that electricity can be generated only while the sun is shining. But such sunny periods are often when power is most valuable. The output of PV systems typically correlates with periods of high electricity demand, especially in southern regions where air conditioning systems create peak demands for electricity during hot summer days. Since a utility's costs are also high during systems peaks, PV power is often worth more than power from plants that must be run continuously or power from renewable systems such as wind, whose output may not correlate with demand.

Moreover, photovoltaic systems are inherently modular and can be located close to the sites where electricity is consumed. Generator systems near the end-user can reduce transmission and distribution costs and increase the reliability of electric services delivered.

The modular nature of photovoltaics also means that there is no shortage of appropriate sites for PV arrays; the only requirement is access to sunlight. Even in urban areas, where open land is at a premium, rooftops and parking lots are excellent locations for photovoltaic arrays. Moreover, the low profile of the units minimizes their visual impact. And their modular nature means that utilities can bring units on-line as they are needed, thereby reducing the risks inherent in long-term load forecasting.

There is no reason to believe that widespread use of PV equipment will lead to environmental or safety hazards if care is taken to anticipate possible problems. Some PV manufacturing methods use hazardous materials such as hydrogen selenide, and solvents similar to those used in the production of other semiconductor devices (see chapter 10: *Polycrystalline Thin-Film Photovoltaics*). The risks can be reduced to low levels if modern waste-minimization and recycling techniques are employed during manufacturing. Although the disposal of modules containing cadmium or other heavy metals could create environmental problems, discarded modules can be recycled economically, minimizing disposal problems.

System components

All photovoltaic systems consist of two basic elements: 1) the photovoltaic module, and 2) some form of mounting structure that provides support for the modules and, in some applications, follows the sun's apparent motion. A "module" consists of the photosensitive cells and encapsulating materials that protect the cells from the environment. Some modules also include rigid frames for attaching support structures. Modules can also include mirrors or lenses that focus light on the PV cells .

Since most applications require alternating current and not the direct current produced by PV cells, most PV systems also have inverters to make the conversion. In addition, PV systems not connected to a utility grid typically also use storage batteries to ensure continuity of power at night and during cloudy periods. In areas where electric demands peak during the day, however, the value of electric storage to utilities actually declines when PV systems are attached to the grid (see chapter 23: *Utility Strategies for Using Renewables*).

Design strategies

Two basic strategies are being pursued for reducing PV system costs. The first is to reduce the cost of photovoltaic modules by producing large areas of active materials at a low cost. These are called "flat-plate" systems. The second strategy avoids using large areas of photosensitive materials and instead uses lenses or other optical devices to focus sunlight on a small area of active material. These are called concentrator systems.

Virtually all PV systems operating in the world today are flat-plate systems. While some systems are rotated to follow the sun, most have fixed mounts and no moving parts. The central challenge is to find a way to mass-produce large amounts of active photovoltaic material, thus reducing costs.

Strategies for reducing flat-plate costs fall into two main categories: 1) reducing the cost of the cells manufactured (see chapter 7: *Crystalline and Polycrystalline Silicon Solar Cells*), and 2) perfecting the manufacture of thin-film cells, cells manufactured by applying small amounts (less than a micron) of PV material to glass or other substrates (see chapter 9: *Amorphous Silicon Photovoltaic Systems* and chapter 10). About 45 percent of modules sold in 1990 were made from traditional crystalline materials and 35 percent from thin films. Techniques for reducing crystalline material costs look much like strategies for reducing the cost of semiconductor devices such as simple silicon diodes (which are very similar to PV cells). Equipment designed to mass-produce PV thin-film is likely to resemble equipment now used to deposit thin coatings on glass.

The cost of flat-plate equipment is highly sensitive to the amount of energy that can be produced per square meter (m^2)of array, which is a measure of the array's efficiency. The glass and encapsulant used for weatherproofing will cost at least \$15 to 20 per m^2; mounting and site costs will add at least \$30 to 50 per m^2 even in mature markets. Such costs mean that a 10 percent efficient module would produce electricity at 2.5 and 3.5 cents per kilowatt hour (kWh) even if the active cell materials and cell fabrication cost nothing![1] It is unlikely that modules with efficiencies much lower than 9 to10 percent will ever move beyond limited niche markets.

There are two reasons to be optimistic about cost reductions for flat-plate systems: 1) efficiencies much higher than those exhibited in commercial systems have been demonstrated in test cells, which fall far short of theoretical efficiency limits (see table 1), and 2) most existing flat-plate modules are manufactured using labor-intensive, batch systems. Significant cost reductions will result if modern mass-production methods are employed.

The cost of concentrator systems also depends critically on system efficiency—in this case both the efficiency of the optical concentrators and the efficiency of the cell (see chapter 8: *Photovoltaic Concentrator Technology*). Concentrators intercept sunlight with inexpensive lens or mirror materials that focus light on a comparatively small area of cell material. Since cells cover one fifth to one hundredth of the area intercepted by the system, a comparatively high price can be paid per unit of cell area to achieve high conversion efficiency (see table 1).

Although concentrators have moving parts, they make only one rotation daily (11,000 rotations over 30 years) and can be quite reliable. Unlike flat-plate systems, concentrators cannot use diffuse sunlight—light scattered from clouds or humidity. This disadvantage, which is significant in areas with frequent cloud cover, is partially offset by the fact that most concentrator systems follow the sun and therefore receive more energy early and late in the day than fixed systems, which receive only oblique light during these hours. In most moderate to good climates, the energy available to concentrators is within 15 percent of the energy available to a stationary flat-plate system (see chapter 8, table 2).

The cost and value of photovoltaic electricity

The cost of PV electricity depends on the amount of sunlight available at the site, the cost of money, the cost of modules and other parts of the system (mounting, installation and so on), and the cost of operating the system and replacing parts [3].[2] If the equipment is mounted in a field, the cost of site development, fencing, and land rents must also be considered.

Most work on cost reduction has been focused on module costs. The average selling price of flat-plate photovoltaic modules (in constant dollars) fell from $20 per peak watt in 1976 to $7.1 per peak watt in 1984. The average 1990 price, $6.2 per peak watt, is actually slightly higher than the price in 1987 because demand has grown faster than supply. But module costs represent only 50 to 60 percent of the $10 to15 per peak watt typical of recently installed PV systems.

---

1. This assumes a constant dollar capital cost of 6 percent, insurance 0.5 percent, and use of an inverter costing 100 dollars per kilowatt.

2. The power of PV cells is reported in terms of their output in full sunlight (defined as the energy reaching the earth through a clear sky at sea level—approximately 1 kilowatt per m^2). Prices converted to 1989 dollars using the GNP deflator.

Table 1: Photovoltaic cell efficiency
percent

	Field experience (modules)[a]	Prototype modules	Experimental cells[b]	Theoretical limit
FLAT PLATE				
Crystalline Silicon	10–12[c]	17.8[d]	24.2[e]	30–33[f]
Polycrystalline Silicon	8–9[c]		18.2[g]	
Single-junction a-**Si**	3–5[c]	5[h]	6[b]	27–28[i]
Multi-junction a-**Si**	6[h]	8[h]	10[h]	
Mechanically stacked a-Si and CIS			15.6[h]	42[i]
CIS		11.1[j]	14.8[j]	23.5[k]
CdTe		10.0[j]	15.8[j]	27–28[k]
CONCENTRATORS				
GaAs		22[l]	28[l]	
GaAs on GaSb			34[m]	

a. Includes the effect of dust and other factors experienced in the field. The efficiencies are "stabilized efficiencies unless otherwise noted. (a-Si cells lose efficiency during the first year of operation).

b. Experimental cells are typically 1–4 cm^2. Modules in the field are less efficient than experimental cells because encapsulants reduce the light reaching cells

c. [15].

d. [36].

e. [37].

f. [38].

g. [39].

h. chapter 9, table 1. The efficiency for the mechanically stacked a-Si and CIS cell is not a stabilized efficiency.

i. [40].

j. [41].

k. [42].

l. chapter 8.

m. [43].

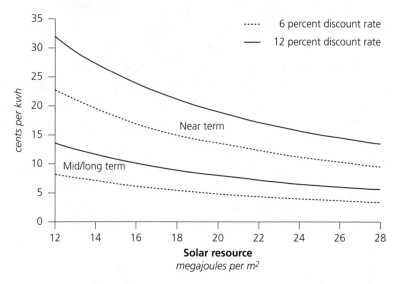

FIGURE 1: *The cost of electricity produced from* PV *systems financed with capital costs typical of regulated public utilities (6 percent real discount rate) and private investors (12 percent real discount rate). Near-term systems should be available by 2000 and mid/long-term systems shortly thereafter. The curves represent a range of alternative designs that may be available in the periods shown. For detailed assumptions about the systems considered, see table 13. Following the conventions of this document, costs are computed assuming that no taxes of any kind are paid. Care must be taken in using these costs since it is only appropriate to compare these "no tax" costs with a "no tax" cost estimate for a competing source of electricity. The costs of conventional electricity reflect approximate costs of baseload and peaking energy when taxes have been removed.*

Significant cost reductions are required for both modules and for associated equipment.

The value of PV electricity is highly sensitive to local conditions: the correlation between demand and available sunlight, the nature of the loads served, the cost and performance of other equipment available for power production (such as gas, coal, and hydroelectric equipment), and various other factors. These are examined in detail elsewhere (see chapter 23). Photovoltaic systems prove particularly valuable in regions such as the Southern United States where air-conditioning systems create peak electric demands on hot summer days when photovoltaic power is reliably available.

Both the value and the cost of PV systems depend on the system's size and location. Photovoltaic systems are inherently modular and can easily be located close to customers—even on rooftops. Systems close to customers can reduce the need for utility transmission and distribution equipment and increase the reliability of local electric service [4]. And using rooftops or other structures can reduce installation costs. Yet larger systems in remote areas may enjoy economies-of-scale in installation and also have lower overhead costs and markups. It is too early to predict how the market will sort out these factors and no clear statement can be made about the optimum size of photovoltaic installations.

At 1990 prices, PV systems are economical only where connection to a large utility grid is impractical. But large markets for off-grid systems exist even in the United States. The choice of building a new utility line or building an independent PV system depends both on the energy required at the site and the distance that a new utility line must run to provide grid service. The poles and transformers needed for a new line cannot be justified if the site needs only a small amount of power. At current prices it is cheaper to build an independent PV system to serve a site with a load smaller than 8 kWh per month rather than build a 65 meter utility line extension. Photovoltaic power for sites requiring 300 kWh per month would be less expensive than a 3,000 meter grid extension [5]. Potential markets are particularly large in developing nations where many sites are distant from well-developed utility grids.

However, utility-financed PV systems in regions with moderate to good sunlight should produce power for 10 to 15 cents per kWh before 2000 (see figure 1). At this price, PV equipment should be able to compete in a variety of niche markets even in sites already connected to the grid [4]. In the mid- to long-term, any of a variety of thin-film flat-plate and concentrator systems can be expected to deliver utility power for 3.5 to 7 cents per kWh depending on climate. If this happens, PV markets are likely to expand rapidly from remote to niche markets, particularly in areas with summer peaks.

By 2000, PV equipment could become a significant part of new utility investment. Analysis detailed in chapter 23 suggests that in a summer peaking utility, PV systems would compete directly with peaking and intermediate systems until the PV equipment provides from 10 to 15 percent of the utility's energy. At this point, the PV equipment begins to shift the utility's peak, and must compete with the less expensive baseload equipment. But in many areas, the photovoltaic equipment may be able to compete with power from baseload equipment meeting stringent clean air standards (4 to 6 cents per kWh). The cost of baseload power could decline if advanced equipment is introduced (see chapter 23).

FLAT-PLATE SYSTEMS

The vast majority of all PV systems installed to date are flat-plate systems. The costs of such systems reflect three basic elements: 1) the photovoltaic modules, 2) the cost of installing the modules in rigid mounts, and 3) the cost of power conversion, installation, maintenance, and various other factors. Present module costs represent 50 to 60 percent of total system costs and are expected to represent about the same fraction in the future.

The following section examines the three basic strategies for reducing flat-plate module costs: 1) improving the techniques that have been used since the 1970s to produce PV cells, which are based on crystalline silicon materials, 2) perfecting methods for mass producing thin-film modules, and 3) implementing several novel production concepts.

Table 2: Approximate production costs of polycrystalline photovoltaic modules
using conventional methods
$\$/m^2$

Wafer production	200
material	50
casting to ingot	50
saw into wafer	100
Cell fabrication	100
Laminate	100
Module assembly[a]	100
Total[b]	**500**

a. Assumes rigid frame (not required for rigid mounts).

b. $4 per watt at 12.5 percent efficiency.

Single-crystal and polycrystalline materials

Production of crystalline PV cells consists of four basic steps: 1) wafer production (the process by which silicon is cast and then sliced into wafers), 2) cell fabrication (when the slices are converted to PV cells), 3) encapsulation (typically laminating with weather-resistant materials), and 4) attachment of the frames and other devices needed for a self-supporting module. Approximate costs for contemporary production processes are shown in table 2.

Wafer production

Whereas single crystals are produced by slowly drawing an ingot from a high-temperature crucible of molten silicon, polycrystalline material is produced by slowly cooling a container filled with molten silicon (see chapter 7, figures 6 and 7). In both cases, the blocks of silicon are then cut into wafers, a time-consuming, and hence costly, batch process. It is also possible to grow polycrystalline materials in thin layers directly on a steel or ceramic substrate, thereby avoiding the need for crystal slicing (see chapter 10).

The first cells were made from single silicon crystals because polycrystalline materials have many crystal boundaries that can degrade efficiency. But polycrystalline materials can be made from lower-cost silicon materials and are less expensive to produce. Furthermore, modern cell designs have reduced the performance penalties resulting from use of polycrystalline material. In fact, the higher cost of producing single crystals often offsets the 1.5–2 percent efficiency gain that can be achieved by using single crystals.

The costs of raw silicon are high, and thick cells (350 to 450 microns) require a significant amount of material. Moreover, a nearly equal amount of material is lost as sawdust during the cutting process. Costs are somewhat contained, however, by purchasing silicon that is recycled from the production of semiconductors and sells for only about $22 per kilogram, or about half the price charged to the semiconductor industry. Nevertheless, the price is increasing as demand increases: PV production in 1990 consumed about 9 percent of the world's total semiconductor silicon material [6].

Significant reductions in the cost of semiconductor-grade silicon in the near future are unlikely. Instead, the greatest potential for near-term savings lies in improved sawing techniques. Such techniques make it possible to cut wafers that are only 200 microns thick, reducing the total amount of silicon needed to produce a finished wafer. However, these gains will be mitigated if demand exceeds the amount produced as scrap by the semiconductor industry. Because demand is increasing, the price of single-crystal wafers will most likely increase rather than decrease in the next few years. Indeed, by the mid-1990s, crystalline wafers could cost as much as $300 per m^2 and polycrystalline materials $200 per m^2 [7]. Such cost increases may be offset somewhat by applying mass production techniques to the sawing stage (see chapter 7).

Modern light-trapping designs can reduce material costs by greatly increasing the amount of light captured by comparatively thin layers of crystalline material. It may be possible to provide acceptable efficiencies from comparatively thin (20 micron) wafers (see chapter 7, figure 11).

A variety of methods for producing thin sheets of crystalline material are being explored. Polycrystalline material can be grown directly on a supporting material (see chapters 7 and 9). In the middle- to long-term, thin ribbons, or tubes, of silicon may be drawn directly from the melt (see chapter 7, figure 9). Such techniques can, in principle, reduce material requirements by factors of five or more. Film thickness would be reduced to only 50 microns, and losses from sawing would be eliminated. All these methods face difficulties: capital costs are high, production rates are slow, and defects in the materials limit efficiency. A number of firms have active research programs aimed at solving these difficulties: [8]

▼ Mobile Solar USA has devised a process whereby octagonal tubes 800 millimeters in circumference are drawn from a melt. With this method, production rates of 160 square centimeters (cm^2) per minute and efficiencies of 16 percent have been achieved.

▼ The Siemens company employs the so-called S-Web method to produce 100 millimeter-wide ribbons at rates that reach 1,000 centimeters per minute. Cells with 12 percent efficiency have been created using this method.

▼ The Westinghouse company uses the dendritic web method to pull 50 millimeter ribbons from a melt at rates as high as 10 cm^2 per minute. Efficiencies of 17 percent have been measured.

▼ The Astropower company has grown thin polycrystalline material directly on steel and ceramic substrates and then used the material to produce cells with efficiencies of 15 percent.

Table 3: Producing high-efficiency modules[a,b]

Module	Cost		Cell efficiency	Year
	$/W	*$/m²*		
1990 production	3.29	420	12.8	
Laser grooved, buried contact cells[c]	3.12	500	16.0	early 1990s
Laser grooved, buried contact cells[d]	2.06	330	16.1	1994
Transparent conductor and improved AR[e]	1.86	260	14.2	1994

a. Source: J.H. Wolgemuth, S. Narayanan, and R. Brenneman, "Cost Effectiveness of High Efficiency Cell Processes as Applied to Polycrystalline Silicon," *21st Annual IEEE Photovoltaic Specialists Conference*, 1990, p. 221.

b. All estimates assume use of inexpensive wafers and mass production technologies.

c. The Solarex modification of the process described in chapter 7.

d. Improved wafer production and mass production of cells and modules.

e. Uses a transparent conductive oxide on top surface of cell to improve collection of carriers and reduce shadowing by contacts on the front of the cell and use double layer of antireflective coating.

Cell fabrication

Converting a silicon wafer into an operating PV cell now costs about $100 per m². Reducing these costs will require improved processes as well as increased automation. Producers are experimenting with a variety of innovative strategies (see chapter 7), including some that have led to high-efficiency cell [9]. Studies show that the near-term production cost of a 17.8 percent efficient cell (or 15.9 percent efficient module) could fall to as low as $3 per watt [10], with production costs closer to $4 per watt if investor returns are included. However, a study by the Solarex company also suggests that the added cost of laser grooving and other techniques needed for improved efficiency may not be justified when comparatively straightforward extensions of current production technology are possible (see table 3) [11].

Encapsulation and module construction

Encapsulation and module construction may add another $200 per m² to module costs (see table 2), depending on the design. Significant savings can be realized by using less expensive encapsulation materials (the cost of the encapsulant and glass, for example, could be reduced to less than $20 per m²) (see chapter 9) and by integrating the designs of modules and support structures to minimize cost of framing materials.

Table 4a: Production cost of silicon modules (RTI method)[a]

		Czochralski crystal	Dendritic web	Amorphous multijunction
Plant output	*MW/year*	25	25	25
Module efficiency	*percent*	15	15	15
Cell thickness	*microns*	305	49.5	1.12
Plant capital cost	$\$\cdot10^6$	23.3	17.0	25.6
Selling price	$\$/m^2$	335	216	181
Direct materials		147	60	44
Direct labor		41	33	18
Overhead		81	71	63
Nonmanufacturing expense		12	12	11
Taxes		27	21	23
Net income		24	21	25
Total	**$\$/W$**	**2.22**	**1.44**	**1.21**

a. Source: Research Triangle Institute, "Photovoltaic Manufacturing Cost Analysis, EPRI (AP-4369), 1986. Summarized in R.A. Whisnat, P.T. Champagne, S.R. Wright, K.C. Brookshire, and G.J. Zukerman, "Comparison of Required Price for Amorphous Silicon, Dendritic Web, and Czochralski Flat Plate Modules, and a Concentrating Module," in IEEE Photovoltaic Specialists Conference, 1985 (IEEE: New York).

Notes: Silicon costs $47 per kilogram.
    Financing is 30 percent debt (14 percent interest) and 70 percent equity (18 percent return).
    Federal and state tax rate 50 percent.
    Prices converted from 1986 to 1989 dollars using a factor of 1.173 (GNP deflator).
    Manufacturing overhead = depreciation of production equipment and manufacturing floorspace, supervisory labor costs, and fringe benefits.
    Nonmanufacturing expense = salaries, office space, fringe benefits, personnel & accounting functions, support staffs.
    Net income = return on equity invested in manufacturing equipment and investment in floorspace, office equipment, and other needed capital investment.

Prospects

Overall, the potential for long-term cost reductions remains highly uncertain. The Electric Power Research Institute sponsored a project by the Research Triangle Institute to assess potential production costs in a 25 megawatt-per-year facility (see table 4a). The costs range from $2.20 per watt for modules drawn from single-crystal ingots (the Czochralski process) to $1.40 per watt for systems using silicon ribbons drawn with the dendritic web technique. Conservative assumptions were made about the risk premiums paid for initial plants (short depreciation times and high rates of investor return) but the technique is complex and difficult to reproduce.

To illustrate the significance of efficiency gains, costs were recalculated using a simplified financial model and rates of return closer to expectations for mature manufacturing facilities (see table 4b). Although cost reductions of 30 cents per watt are reasonable for the Czochralski and dendritic web processes, more work is needed to refine these estimates.

Table 4b: Production cost of silicon modules (alternative method)[a]

		Czochralski crystal	Dendritic web	Amorphous multijunction
Alternative estimate	$/m^2$	312	195	166
Direct materials		147	60	44
Direct labor		41	33	18
Indirect labor (@ 110 percent)		45	36	20
Capital (seven year depreciation)		49	36	54
Utilities and other		30	30	30
Price @ 10 percent	$/W		1.952	1.66
@ 15 percent		2.08	1.30	1.11
@ 20 percent		1.56	0.98	

a. Source: uses investment costs, materials, and direct labor costs from table 4a.

Notes: Total capital investment assumed to be 1.25 times production equipment (rough average of EPRI calculation in table 4a). The fixed charge rate (applied to the total capital investment—plant and equipment) is computed assuming financing is 30 percent debt at 10 percent, 70 percent equity at 14 percent. The marginal income tax rate is assumed to be 40 percent, property taxes 1 percent, and insurance 0.5 percent. Depreciation is assumed to be straight line. Using these assumptions, the fixed charge rate is 29 percent for a seven year depreciation period (assumed for "long-term" estimate) and 34 percent for five year depreciation ("near term"). These capital costs are for a mature facility. Clearly a risk premium (higher projected equity returns) would need to be paid for earlier plants.

A good approximation for the fixed charge rate (FCR) for capital is calculated as

$$FCR = PTI + F_d \cdot CRF(d,L) + (1 - F_d) \cdot (CRF(e,L) - T/L)/(1 - T), \text{ where}$$

$CRF(d,L)$ is the capital recovery factor of interest rate d for L years $= d(1 + d)^L/((1 + d)^L - 1)$
F_d = fraction financed from debt = 0.3
d = interest in debt = 0.1
e = return on equity expected after taxes = 0.14
T = marginal income tax rate = 0.4
PTI = property taxes and insurance = 0.015
(insurance = 0.005, property tax=0.01)

Thin films

A second strategy for producing flat plates, and one that could lead to module costs much lower than for crystalline cells, uses thin (1 micron) films of materials, such as amorphous (glassy) silicon (*a*-Si), copper indium diselenide (CIS) and cadmium telluride (CdTe) instead crystalline material. About one-third of the cells sold in 1990 were made from *a*-Si thin-film material. The share has grown rapidly in recent years.

There are two primary reasons for believing that thin films can eventually achieve lower costs than crystalline devices. First, thin films can be many times more efficient at absorbing sunlight than thick wafers. As a consequence, far less PV material is needed, which greatly decreases costs. And second, the techniques used to produce thin films are particularly well suited for mass-production. The costly batch processes of single-crystal production can be replaced by a continuous process in which active materials are sprayed or sputtered directly onto glass or metal. The central challenge is to develop cells with acceptable efficiency.

Competing approaches

To date, alloys of amorphous silicon have received the most attention. In recent years, however, interest in thin films based on copper-indium-diselenide (CIS), cadmium telluride (CdTe), and other materials has been growing. Amorphous silicon has the advantage of using a plentiful, benign material. Nevertheless, care must be taken to ensure occupational safety during the manufacturing process since some hazardous materials are used.

Commercial *a*-Si modules also suffer from the fact that performance degrades during the first year of operation. Table 1 reports only stabilized efficiency—i.e., steady-state efficiency reached after the initial performance decline. While steady progress is being made, the cells still remain far from their theoretical efficiency limits. Most commercial amorphous silicon thin-film systems operate with average efficiencies of 3.3 to 4.2 percent but modules with much higher efficiencies are now available (see table 1).

The CIS and CdTe thin films are not yet commercially available in large volumes but offer strong competition to *a*-Si because of their comparatively high stabilized efficiencies (see table 1). These new thin-films probably will not require the vacuum processes needed for *a*-Si and may therefore be less expensive to manufacture. They do, however, require somewhat more expensive materials, some of which could present environmental problems, unless they are appropriately recycled (see chapter 10).

High efficiency amorphous devices rely on multi-junction cell designs. These cells are made by building several cells on top of each other. Light not absorbed by upper layers is captured by deeper layers. Typically the cells are designed so that each layer is sensitive to a different color of incident light (i.e., one layer captures the red part of the spectrum and another the blue part). The layers can be made by using different alloys of silicon or they can be made by combining silicon cells with cells made from other materials. (chapter 9, table 1). The multijunction cells not only have access to more of the total light energy, but the thin layers of active materials seem to be much less susceptible than *a*-Si to degradation. Experimental *a*-Si multijunction cells have demonstrated stabilized efficiencies of 10 percent (initial efficiencies of 13.7 percent).

The competition between *a*-Si and competing thin-film materials will be interesting to watch during the next few years. Over the long term, however, the most promising strategy may involve both technologies in multijunction devices. Combined *a*-Si/CIS prototypes have been constructed by mechanically stacking *a*-Si cells on top of the CIS cells (they are not manufactured together as a single unit but are connected externally) and have demonstrated an efficiency of 15.6 percent [12].

Production costs

Relatively accurate estimates of capital and labor costs associated with mass pro-

Table 5a: Production cost of thin-film modules

Production rate	$10^3 \, m^2/year$	100	1,000	5,000
Depreciation time[a]	years	5	7	10
Yield[a]	percent	80	85	90
Equipment cost[b]	$m^2/year	100–120	70–100	24
Module area costs	$/m^2			
Capital[c]		43–51	25–35	7
Materials[d]		25–30	25–30	20–25
Direct labor[e]		10–30	5–15	5–10
Indirect labor (@110 percent)[f]		11–33	5–16	5–11
Utilities[a]		10	5	3
Yield losses[g]		20–32	10–16	4–6
Total		***119–186***	***75–117***	***44–62***
Module costs	$/W			
Module efficiency:[h]				
7 percent		1.70–2.66		
10 percent		1.19–1.86	0.75–1.17	0.44–0.62
15 percent.			0.50–0.78	0.29–0.41
18 percent				0.24–0.34

a. Follows assumptions in chapter 10.

b. Range of production costs for lower production rate volumes are taken from EPRI/RIT, and chapters 9 and 10. The cost of the 5 million m² per year facility is taken from chapter 10.

c. 1.25·CRF(18 percent, depreciation period)·investment in plant and equipment. For discussion see notes to table 4b.

d. See table 5a for lower production rates. It may not be possible to reduce material costs significantly for high-efficiency thin-film materials (such as multijunction a-Si). Improvements in manufacturing could increase use of materials and recycle wastes but these savings could be offset by the increased costs of multijunction materials. The low material utilization assumption for very high volume applies to single-junction CIS devices (see chapter 10).

e. Lower range from D. Carlson, "Low Cost Power from Thin-Film Photovoltaic," *Electricity*, T. Johansson, B.Bodlund, and R. Williams, eds. (Lund: Lund University Press, 1989). Upper range from chapter 10.

f. Indirect cost factors from Wolgemuth (1990). See table 3, note a.

g. (1-yield)·(capital + materials + labor + utilities).

h. It is assumed that the very high volume facilities would only be constructed for single-junction materials such as CIS. The capital cost savings would not be realized for systems requiring vacuum systems since large facilities would consist of multiple lines of identical equipment.

---

duced thin films can be based on similar production processes and on the experience of the fully-automated prototype line now operating at the Solarex Company. Table 5a suggests that production costs could range from $1.2 to 2.7 per peak watt in a facility that produced 100,000 m² of 10 percent efficient arrays each year. Costs could be reduced to $0.5 to $0.8 per peak watt in a facility producing 1 million m² per year of 15 percent efficient modules.

Still larger facilities could lead to even lower costs, but scale economies for *a*-Si are unlikely to be significant after 1 to 5 million m² are produced per year. Even a 1,000 m² per year plant would consist of a number of identical production lines

Table 5b: Material costs of thin-film modules
$/m^2$

	a-Silicon[a]	CIS[b]
Encapsulant	11.3	2.75
Glass	5.6	11.00[c]
Germane	5.0	
Silane	2.7	
Molybdenum		0.77
Indium		1.43
Selenium		0.58
Zinc oxide		0.71
Others		2.11
Frame		2.75
Aluminum	1.9	
Other materials	4.5	3.30
Total	**31.0**	**25.4**

a. Chapter 9.
b. Chapter 10.
c. Encapsulant.

for 30–60 centimeter wide module elements. Thin-film production techniques that do not require vacuum facilities can use larger sheets of material, and plants larger than 1 million m² per year might be easier to justify.

Table 5a also makes the advantage of high efficiency evident. Production costs change in direct proportion to efficiency since the steps required for high efficiency are unlikely to add significant material or processing costs.

The cost of producing higher efficiency, but more complex multijunction cells remains uncertain. Multijunction production is likely to be accomplished by adding additional stages to the production lines now being designed for single-junction *a*-Si devices. While this would add capital costs, the addition of a few steps need not add significantly to costs. And since the multijunction devices use such thin layers of active material, material costs need not rise much above those expected for single-junction devices (see chapter 9).

Active materials represent only about a quarter of the total materials costs of a photovoltaic module (see Table 5b). The bulk of the materials costs are attributable to glass, encapsulants, and frame materials. These costs begin to define the limits of what can be achieved in cost reduction. It is possible that savings somewhat beyond those shown in tables 5a and 5b could be achieved if glass manufacturing and array manufacturing operations were combined.

Novel designs

Most pv research focuses on crystalline and thin-film silicon, but a number of highly innovative strategies, including dye-sensitive films and silicon spheres, are now competing for attention.

Dye-sensitive films

Semiconductor material (titanium dioxide) can be covered with a thin layer of dye-sensitive film, or charge-transfer dye, which is an efficient light absorber and passes an energetic electron to the semiconductor material. Although the phenomenon has been known for many years, recent experiments have produced cells with surprisingly high efficiencies. The overall light to electric-energy yield is said to range from 7.1 to 7.9 percent under simulated solar light conditions and to be 12 percent in diffuse light [13].

Silicon spheres

Large sheets of photosensitive material can be fabricated from arrays of small, low-cost crystalline spheres. Concentric spheres built up from relatively inexpensive silicon raw materials can be used because the impurities remaining in the silicon migrate to the center of the spheres and away from the active regions. Material costs are further reduced because there are no silicon ingots to saw, and, thus, little is wasted.

The spheres are then embedded in a rigid sheet. The outer layer of the spheres is removed on one side of the sheet, exposing the interior of the spheres so that contact can be made. The key issue, of course, is whether the advantage of using low-cost silicon will offset the costs and lowered efficiency associated with large sheets [14].

Structures and other system costs related to area

Although lowering the cost of flat-plate modules represents the largest technological challenge for photovoltaics and has received the bulk of research attention, module costs represent only slightly more than half of the cost of an installed pv system. In recent years, system costs—exclusive of modules and power conditioners—have ranged from $400 to $500 per m^2. A major portion of those costs can be attributed to the structures that hold modules in place during strong winds, and to module connections, power cables (that bring DC power to the rectifier), and fencing. Other expenses include land acquisition, site preparation, and environmental impact assessment. Most of these costs are directly proportional to the area covered by photovoltaic modules. This area, in turn, is inversely proportional to the efficiency of the modules—a fact that further underscores the importance of high module efficiency.

Two strategies are being pursued for reducing area-related costs. The first focuses primarily on improving the engineering of support structures and making installation procedures more efficient. The second involves a variety of novel methods for mounting arrays on rooftops and other structures that can serve a dual function.

Table 6: Area-related costs in conceptual design studies[a]
1989 $/m²

Source	Year	Site prep	Support	Found-ation	Electrical	Total
Battelle	1981	11.8	26.4	12.6	18.4	69.2
JPL	1981	ns[b]	49.5	15	ns	65.5
Bechtel	1981	ns	30.7	17.4	ns	48.1
Martin Marietta	1982	4.4	80.5	8.9	c	89.4
Sandia	1984	1.1	20.8	8.4	23.6	53.9
EORI	1984	4.8	ns	ns	2.1	60.6
U.S. Dept. of Energy	1987	ns	ns	ns	(a)	54.9
Chronar	1989	0.4	ns	ns	8.9	39.9
Solar Energy Research Institute	1990	ns	ns	ns	9.0	44.0

a. Source: prepared by J. Ogden, Princeton University Center for Energy and Environmental Studies (1991).

b. ns = not specified.

c. Included in support estimates.

The cost of structures has fallen as engineering experience accumulates. Five of the eight experimental modules reported as a part of PVUSA (a cooperative government/electric utility venture) had installed foundation and structure costs of $110 per m², three had $125 per m², and one $260 per m² [15]. These systems range in size from 15.7 to 19.7 kilowatts. A 400 kilowatt "utility scale" system was installed with foundation and support structures that were approximately $35 per m². About one third of this cost was the foundation; the rest was spent on structures.

A number of conceptual studies suggest that costs could be reduced further (see table 6), especially in the structures themselves. However, cost reductions in other areas (such as site preparation, wiring, and installation) have proved stubbornly resistant to cost reduction.

Land costs are not likely to be significant except in dense urban areas, where large arrays are unlikely. Nontracking systems typically cover about half the surface area of a site, whereas tracking systems cover only about one-third of the surface to avoid shading. Land prices in the United States vary from $1 to $4 per m² in lightly populated suburban areas, to 1 to 5 cents per m² in rural areas.

The cost of mounting PV systems can be reduced significantly if modules are supported on rooftops or other structures that serve other purposes. In 1985, 30 residential units were installed in Garner, Massachusetts, with total costs (exclusive of modules and power-conditioning equipment) of $57 per m² (converted to 1989 dollars) [16]. Costs could have been reduced with greater experience. By the end of the installation, a four-man crew was installing four residential units a day.

Roof-mounted systems can also be installed on or near commercial buildings. Many parking lots in sunny areas have already mounted shading systems that would make excellent PV mounts. Low costs are possible if existing structures can be used. A recent bid to the New York Power Authority takes advantage of the fact that many modern roofing systems for commercial buildings use a heavy membrane covered with large concrete tiles (61 × 122 centimeters). The system proposes simply attaching PV arrays to these tiles and then using existing crews and equipment to lift them into place on a rooftop. Total costs (exclusive of the arrays and power conditioners) would be $13 per m² [17].

Photovoltaic arrays can also be made transparent and used as windows. Early designs had unattractive colors but modern designs use a laser to penetrate PV modules with an array of small dots. The units look much like tinted glass. At present, high costs have limited use of these systems to the sunroofs of luxury cars. But other applications are clearly possible if costs are reduced.

CONCENTRATOR SYSTEMS

Strategies for reducing PV costs face two contradictory goals: 1) covering the receiving area with a low-cost material, and 2) converting a high fraction of the energy intercepted into electricity. One way to resolve this riddle is to cover the area with an inexpensive mirror or lens surface and focus sunlight on a comparatively small, high-efficiency cell. Concentrator systems are often categorized by their concentration ratio, that is the ratio between the sun's intensity on the optical surface to the intensity incident on the cell. A concentration ratio of 10 means that the rate of energy reaching the cell is 10 times the energy that would reach the same surface if no optics were present. In general, systems that rotate about only one axis have lower concentration ratios than two-axis tracking systems. Practical one-axis systems are likely to have concentration ratios between 10 and 30 while concentration ratios of 1,000 to 5,000 are possible using two-axis tracking systems and advanced optics (see chapter 8, table 1).

Tracking systems

Most photovoltaic concentrators track the sun either on one axis (the tracker pivots around a north–south axis to follow the sun) or on two axes (the tracker pivots so that the system always directly faces the sun). Tracking systems have several attractive features. The demands placed on tracking units are comparatively modest and reliable systems are surprisingly inexpensive. Moreover, because the systems need only one rotation per day, wear on bearings and motors is minimized.

The cost and performance of concentrating systems depend both on the tracking units and on the modules, which are attached to the trackers and contain

Table 7: Projected costs of two-axis tracking structures[a]

	Cost	Cost/m²
Support structures	2,250	15.2
Drive	3,100	20.9
Foundation (hole, concrete, pedestal)	1,600	10.8
Tracking system	1,050	7.1
Field wiring for tracker	200	1.4
Installation and checkout	900	6.1
Total direct costs	***9,100***	***61.2***
Indirect costs @ 25 percent	2,275	15.3
Total cost	**11,375**	**76.5**

a. Source: The estimate was based on estimate prepared by Advanced Thermal Systems, Inc. (communication dated 26 September 1990). This firm controls the rights for the continuing use, sale, and further development of the heliostat and sun-tracking technology used for the ARCO Carissa Plains tracking photovoltaic plant. The firm has worked extensively with the Department of Energy on heliostats for central receiver facilities.

Notes: The estimates were made for heliostats produced at a rate of 5,000 units per year. Each unit is designed to hold 20 mirror modules each 1.2 × 6.1 meters (4 × 20 feet). Installation of each unit requires 30 person-hours. The original estimate was adjusted for purposes of this table by subtracting the cost of mirror materials from the direct cost and applying a 25 percent indirect-cost rate to the new total. This assumes that the weight and wind loading of concentrators would be approximately the same as that of the mirrors (a reasonable estimate given the glass mirrors assumed in the original analysis) and that the cost of attaching a photovoltaic module to the tracker would approximate the cost of attaching mirrors.

the system optics, the cells, and devices for dissipating heat. Most concentrators simply radiate heat into the air through metal fins, but active cooling is possible if there is use for low-temperature fluids in the immediate area (PV cogeneration systems). Active cooling keeps the cells at lower temperatures and improves electric generating efficiency. The cost of installing and maintaining the needed plumbing, however, may not be justified given the low quality of the heat produced.

Demonstration projects of solar-thermal and PV systems during the past decade have provided valuable experience for improving the design and maintenance of tracking structures [18]. Two-axis heliostats built in 1980 cost about $850 per m² [19] but much lower-cost systems have been built as parts of PV demonstrations in the mid-1980s. Flat-plate systems in Carissa Plains, California, and other locations were attached to tracking units because the modules were expensive and added cost in the structures could be justified in terms of the additional energy collected. The costs of the Carissa Plains facility have not been published, but a similar system, which has 148 m² of collecting surface mounted on a rotating pedestal, has installed costs in the range of $75 per m² (see table 7).

Single-axis tracking systems are even less expensive. These systems typically rotate around a north–south axis to follow the sun during the day, but cannot follow seasonal changes in the sun's elevation. The less than perfect tracking of one-axis systems means they capture only about 90 percent of the energy collected by

Table 8: Actual costs of a 300 kilowatt single-axis tracking system
built in Austin, Texas in 1986[a]
costs converted to 1989 $

	$/W	$/m²
PV modules (flat plate)	6.4	733
Power conditioner	0.88	100
Design/project management	0.74	85
Site preparation	0.34	40
Tracking and other structures	0.40	46
Electrical	0.41	47
Installation	1.12	128
Other (administrative, etc.)	0.60	69
Total	**11.92**	**1,365**

The experience suggested that $1.94 per watt ($214 per m²) could be
saved if the system were reproduced by reducing costs of electrical
conduits, installation, site preparation, and other work.

a. Source: J.H. Hoffner, "Construction Experience with a 300-Kilowatt Photovoltaic Plant in Austin, Texas," Proceedings of the 1987 Annual Meeting of the International Solar Energy Society, Portland, Oregon, 11–16 July 1987.

more precise tracking systems [20]. The 10 percent loss, however, may be justified by the comparatively low capital cost of the installations.

A one-axis tracking system installed in 1986 in Texas cost $46 per m², although installation and wiring costs added another $175 per m² (see table 8). It is difficult to determine whether the high installation costs result from the system's unique passive tracker, which is steered by the thermal expansion of tubes filled with a refrigerant. In contrast, most systems today are controlled by motor-drives and electronic sensors. In 1989, the Entech company installed a 300 kilowatt system with a two-axis tracking concentrator atop a parking lot at the 3M plant in Austin, Texas. Although clever designs may lead to large cost reductions in the near future, no low-cost system has yet proven itself. The SEA corporation projects costs as low as $17 to $33 per m² (depending on production volume) for lightweight structures that can be installed quickly on-site (see table 9).

While most work has concentrated on active tracking systems in which motors rotate the collectors, a variety of schemes have been demonstrated for passive trackers using fluids that expand and change the collector's position when sunlight reaches them. Low concentration ratios can even be achieved without tracking (see chapter 8).

Modules

Concentrating modules require an optical surface, a cell that functions well in concentrated sunlight, a system for dissipating heat from the cell, and structures

Table 9: Projected costs of a single-axis tracking structure[a]
$/m^2

	1 MW/year	10 MW/year	100 MW/year
Array structure	7.6	7.0	6.1
Tracker and drive	5.9	4.1	4.1
Assembly	16.1	10.4	6.6
Installation	4.1	1.8	0.8
Total	**33.7**	**23.3**	**17.4**

a. Source: SEA Corporation (San Jose, California) communication dated 17 June 1991.

Note: Assembly and installation include all overhead costs including return to investors for capital investment, taxes, shipping costs, and other expenses. Array and tracking costs are low because the modules are very light (weight dominated by plastic lens and heat sinks) and because the low concentration does not require high tracking precision (± 4 degrees is adequate). The system is a 10× concentrator using a linear fresnel lens. Each module is 2.67m × Installation is described by the producer as follows: "The arrays are shipped pre-assembled with the legs attached but rotated flat. As each array is unloaded, the tow legs are rotated vertically and the array is placed on a reasonably flat surface. Four pegs, sized by the local soil conditions, are then driven into the ground through holes in the array at the corners to secure the unit. Simple wire interconnections complete the installation. Up to four arrays [of 10 modules each] share a common tracking unit."

for holding the system together within reasonable alignment. Fresnel lenses provide an ideal optical surface because they are relatively cheap to manufacture, light in weight, and durable. Many systems also employ a secondary optical system to correct for slight tracking errors and misalignment of lenses. These secondary systems reflect or refract light that would otherwise miss the cell target.

Because the cell covers a small fraction of the total module area exposed to sunlight, it makes sense to invest in higher cost/higher efficiency cells. In fact, many of the high-efficiency single-crystal designs discussed earlier are good candidates for concentrator cells. Not only is their extra production cost justified in a concentrator, but many of these cells operate more efficiently in concentrated sunlight. Only 75 to 85 percent of the light falling on the lenses typically reaches the cells, however, because fresnel lenses are not perfectly transparent and are not perfect optical systems, thus, the overall module efficiency is from 75 to 85 percent less than the cells themselves.

Most modern modules are equipped with high-efficiency silicon cells, although cells using other materials are under development. Within the next few years, concentrator cells will probably achieve efficiencies ranging from 16 to 20 percent, lowering module costs to about $1 per watt for both low (10X) and high (100X) concentration ratios (see tables 10 and 11).

Several extremely high efficiency cells now in development are likely to be used primarily in concentrator systems, where comparatively high costs per unit of cell area are permissible (see table 1). Multijunction cells should be capable of efficiencies as high as 45 percent and it may even be possible to build cells that are 50 to 60 percent efficient using super lattice and other advanced technologies now being tested in the semiconductor industry. Experimental cells with multi-

Table 10: Projected cost of a 10X concentrating module[a]
$/m²

	1 MW/year	10 MW/year	100 MW/year
Cells	37	23	17
Optics	21	16	14
Heat sinks and ends	21	12	11
Other components	21	13	12
Assembly	9	5	4
Overhead	106	54	30
Total	**195**	**111**	**77**
Total less cells	**158**	**88**	**61**
		$/W	
16 percent cells @ 3.1 $/W *optics 75 percent efficient*	1.62	1.04	0.82
22 percent cells @ 12 $/W *optics 85 percent efficient*	2.04	1.67	1.53
22 percent cells @ 6 $/W *optics 75 percent efficient*	1.56	1.13	0.97

a. Source: SEA Corporation.

junctions have been built by mechanically stacking one device on top of another. The efficiency record for such devices stands at 34 percent.

SYSTEM DESIGN ISSUES

The cost of PV power depends primarily on two factors: the amount of sunlight available at a site and the cost of the module, but it also reflects other factors, including balance-of-system costs. The latter category includes the cost of support structures and their installation, power conditioning, and operation.

The solar resource

The energy available for a PV module depends on latitude and climate, and on whether the module is fixed or tracking. The best sites for PV systems obviously are desert areas, which receive about twice as much solar energy as the least attractive, cloudy northern sites. The difference in available resources translates directly into a difference in the cost of energy produced.

Systems that follow the sun capture significantly more energy than fixed systems. Moreover, tracking devices have the additional advantage of providing large

Table 11: Projected cost of a 100X concentrator module[a]

	2 MW/year (18 percent cells) 150 watts per module		24 MW/year (24 percent cells) 210 watts per module	
	$/m^2	$/W	$/m^2	$/W
Module (less cells)	196	1.90	128	0.89
Aluminum	31	0.30	27	0.18
Copper	33	0.32	26	0.18
Glass	27	0.26	5	0.04
Lens	34	0.33	17	0.11
Encapsulant	24	0.23	19	0.13
Wiring	28	0.27	23	0.16
Other	19	0.19	13	0.09
Cells	50	0.48	28	0.19
Module (with cells)	246	2.38	156	1.08

Module is 2.9 × 0.5 meters (145 m^2) using a flat fresnel lens for concentration and a secondary optical element to compensate for tracking and alignment errors. A copper heat dissipater is used.

a. Source: Alpha Solarco, July 1991.

amounts of power in the late afternoon when demand for electricity typically peaks. In desert areas, the amount of energy available to a tracking module is 50 percent greater than the energy available to a fixed module, whereas in cloudy areas, the difference is about 20 percent (see figures 2a and 2b). Tracking systems receive less energy than flat-plate systems in areas that are humid and comparatively rainy because significant amounts of energy in the form of diffuse sunlight are not available to concentrators. But even in these areas, the energy available to tracking systems is within 25 percent of the energy available to nontracking flat-plate systems.

While two-axis tracking may be needed for two-dimensional focusing, the energy available to a two-axis tracker is not appreciably larger than the energy available to a one-axis system. For example, a system in California with one north–south axis would collect 97 percent of the light energy available to a two-axis tracking system [21].

When only limited amounts of land are available, the tilt and spacing of arrays become important. This is particularly true for roof-mounted systems where tilt angles may be limited by a building's geometry or by wind loadings. For example, careful examination of a system in New Paltz, New York, indicated that a flat array tilted at 10 degrees received only 76 percent of the energy of an array

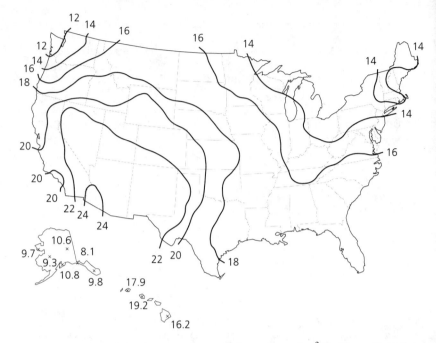

FIGURE 2a: *Average daily solar energy in megajoules reaching a 1 m² flat-plate collector facing south and tilted at the latitude angle. The energy includes sunlight received directly from the sun's disk and diffuse sunlight scattered from clouds and particles in the air.*

tilted at 30 degrees (close to the angle that maximizes annual energy capture). But overall, the lower tilt angle allowed 30 percent more array area on the roof without shading and produced more energy during winter months when total output was low [22].

Power conversion

For most applications, electricity produced by a PV array must be converted from direct current (DC) to alternating current (AC) in a power conditioner before it can be used. These systems are called inverters. If the system is to be connected to a utility grid, it must meet high standards for the quality of the AC delivered and must ensure safety for utility personnel working on lines connected to the PV systems.

The cost of the inverter and control systems required in a PV installation has been reduced significantly by recent advances in solid-state electronics. Although inverter and control systems now cost from about $400 to more than $1,000 per kilowatt, prices are expected to fall rapidly as markets develop. The insulated-gate bipolar transistors and MOS-controlled thyristors now being introduced into advanced motor control systems will also reduce the costs of power conditioners. Altogether, advanced power semiconductors, increased production volume, and improved circuit designs should reduce the costs of 0.05–5 megawatt power con-

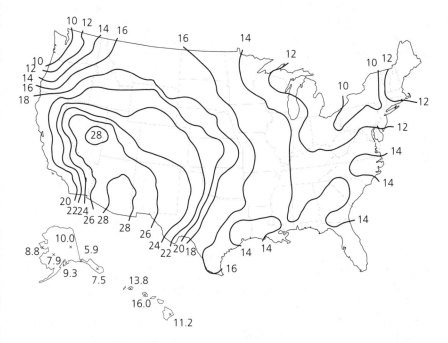

FIGURE 2b: Average daily solar energy in megajoules reaching a 1 m² surface directly from the sun's disk. It is assumed that the surface is rotated continuously through the day so that it is pointed directly at the sun (collector surface is always perpendicular to a line connecting the site with the sun). Sunlight reaching the surface from scattered sunlight is not included.

ditioners to $100 per kilowatt by the mid-1990s, if production reaches 50 megawatts per year [23]. The efficiency of power conditioners has also improved and is now on the order of 96 to 97 percent.

While power conditioners for systems smaller than 500 kilowatts are likely to cost more than for larger systems, mass production could reduce their costs. Mass-producing power conditioner boards small enough to be placed in the junction boxes of 30 to 50 watt modules would permit modules to be connected using standard 110 volt wiring [24].

Power conditioners used in experimental systems have proved to be the source of most system failures. But there is every reason to believe that the systems will become as reliable as other solid-state control systems given production experience. Overall photovoltaic system availabilities have averaged 95 percent in spite of difficulties with inverters [25].

Overhead costs

Overhead charges are associated with a number of items, including site-specific design work, environmental impact analysis, legal fees, licensing costs, planning, contingencies, and interest payments during construction. Typically, overhead charges for a mature, commercial process represent from 5 to 10 percent of total costs [26]. Such costs, however, can range from 70 to 100 percent or more for specially designed systems.

Utility overhead costs can be particularly high for large conventional central-station plants. But, for a number of reasons, most of these costs need not apply to PV systems. To begin with, PV systems built on a routine basis should require little site-specific design work. In addition, environmental permits, which are burdensome for conventional plants, should be straightforward for PV installations once a basic case has been established. Large conventional utility plants purchase expensive components as contingencies, but there is no need to make such purchases for PV systems. PV systems are inherently modular. A small number of modules and structures will be damaged in installation but these represent only a small fraction of system costs. And since modular PV systems will typically be constructed much more rapidly than large utility plants, interest during construction can be quite low. Considering all these factors, it is reasonable to assume that overhead costs will be 25 percent or less once PV installations become reasonably routine. In the absence of extensive field experience, however, this assumption can not be tested empirically.

The overhead for small PV systems can be much higher than for larger systems because of marketing, warehousing, transportation, retail, and other expenses. The retail price for PV modules purchased in small quantities (10 kilowatts or less), for example, is typically 50 to 75 percent above factory production costs. But these markups would not necessarily apply if PV systems are installed as a routine part of housing construction or by a large retrofit contractor associated with a utility.

Operations and maintenance

Enough experience has been gained with PV systems of many different sizes to be quite confident about their reliability and operating characteristics [27]. The operating costs of PV systems can be extremely low (see table 12). While some demonstration systems have been expensive, analyses suggest that most of the problems have straightforward remedies. Many of the costs resulted from design defects that have already been corrected. The data in table 12 suggest that operating costs should range from 11 to 15 cents per kWh for nontracking systems and 2 to 3 cents per kWh for mature tracking systems.

Efficiency issues

The output of installed PV modules is typically 80 percent of what is measured under standard laboratory test conditions.

Most modules are less efficient as the temperature increases. The effect of temperature on their efficiency depends on the type of cells used. Crystalline modules lose about 0.5 percentage points of efficiency per °C above the temperature at which the module was rated (typically 25°C). Modules are typically about 20°C warmer than the air temperature.[3] The operating efficiency of modules is therefore reduced about 10 percent because of temperature effects (0.5 percent per °C × 20°C).

Table 12: The operating experience of large photovoltaic systems[a]

	Power *megawatts*	System type	O&M costs *¢/kWh*			
			Observed		Potential	
			Tracker only	Total	Best parts	Double efficiency
Lovington, CA	0.10	FP/0D	0.00	0.39	0.13	0.11
Washington, DC	0.30	FP/0D	0.00	1.44	0.14	0.12
Sacramento, CA	2.00	FP/1D	0.02	0.61	0.15	0.13
Carissa Plains, CA	6.50	FP/2D	0.18	0.80	0.29	0.20
Lugo, CA	1.00	FP/2D	0.37	1.10	0.29	0.20
Phoenix, AZ	0.23	C/2D	1.78	4.81	0.53	0.30
Dallas/Fort Worth, TX	0.03	C/2D	0.82	6.97	0.73	0.35

a. Source: Electric Power Research Institute, "Photovoltaic Operation and Maintenance Evaluation" (EPRI GS-6625), December 1989.

Notes:
   FP = flat plate
   C = concentrator
   0D = no tracking
   1D = one-dimension tracking
   2D = two-dimensional tracking

Potential using "best parts" corrects known design defects and assumes use of parts with proven low O&M costs.

Potential using "double efficiency" assumes that best parts are used but module output is doubled by improved cell design (affects only some O&M).

Eighty percent of O&M in the Dallas/Fort Worth system resulted from problems with the power conditioner. More than half of the Sky Harbor costs result from moisture leakage into the arrays forcing extensive component replacement. The design defect has been corrected with improved seals.

Further performance reduction results from slight voltage mismatches between modules. Resistance losses caused by wiring, fouling with dust, snow, ice, and other effects can also reduce performance. The nature of these losses depends on the details of the system design and location, but losses of 10 percent are typical [28].

Figure 3 shows the measured performance of a variety of different modules that have been tested at the Pacific Gas & Electric facility in California. The crystalline systems operated at about 85 percent of the rated module efficiency. Obviously, the test procedures for establishing the rated efficiency of amorphous materials need improvement as standard methods are developed for including the

3. A tracking flat-plate system in Austin, Texas, had a module temperature $T_m = 18.41 + 1.14 T_a$ where T_a is the ambient temperature. A fixed system in Detroit, Michigan, had $T_m = 127.9 + 1.2 T_a$, and a similar system in Florida had $T_m = 20.8 + 1.1 T_a$ [34].

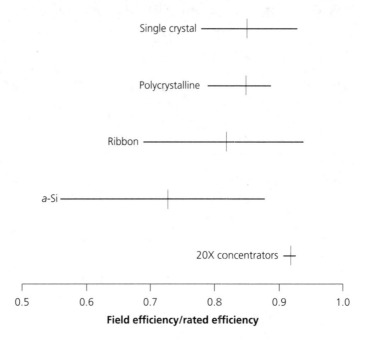

FIGURE 3: *Measured field performance of silicon modules showing the range of performance measured as well as the average performance point. The efficiency measured includes module efficiencies and losses due to field conditions (fouling of surfaces, high-temperature operation, etc.). Source: C. Jennings and Chuck Whitaker, "PV Module Performance Outdoors at PG&E," 21st PV Specialists Conference, IEEE, 1990.*

effect of performance degradation during the first few months of operation.

Rooftop units

Distributed units on residential or commercial buildings can have special problems. A rooftop PV system that feeds power into a transmission or distribution line may create a hazard for utility repair crews who think the unit is disconnected. Although inverters can be designed to disconnect automatically when utility power is lost, there is concern that if several PV units are feeding the same line, neighboring units will keep each other's inverters on. Modern inverters ensure that this will not happen [29] but utilities are only now working to establish uniform standards [30]. Islanding and other interconnection issues are being examined extensively at test facilities on Rokko Island in Japan [31] and in the United States by the Pacific Gas & Electric Company [32].

Residential units have been operated successfully for a number of years with surprisingly few difficulties [33]. Rooftop units mounted directly to shingles have had no reported leaks. When a homeowner wants to replace the shingles, installers can remove the PV system in two hours, replace the shingles, and reinstall the system in another two hours [33]. Nonetheless, routine maintenance presents a

Table 13: Assumptions used in PV system cost comparisons

	Efficiency percent		Cost $/m^2$		O&M ¢/kWh[a]	Electric cost ¢/kWh[a]		Source table
	module EM	system EPV·PPC	MOD	BOS		6 percent	12 percent	
Crystalline silicon								
Cz-Si	15	11.5	300	50	0.15	17	28	4b
Cast poly-Si	14.2	10.9	260	50	0.15	16	27	3
Poly/ribbon	15	11.5	200	50	0.15	12	21	4b
Thin film								
Near term	10	7.7	150	50	0.15	15	25	5b
Mid/long term	15	11.5	50	50	0.15	5.4	9.0	5b
Concentrators								
1D 1 MW/year	10	7.7	160	50	0.25	16	27	9, 10
1D 10 MW/year	10	7.7	60	50	0.25	9.1	15	9, 10
1D long term	20	15.4	60	50	0.25	4.9	8.0	9, 10
2D 1 MW/year	20	15.4	250	100	0.25	11	21	7, 11
2D 10 MW/year	20	15.4	150	100	0.25	8.5	16	7, 11
2D long term	35	27.0	150	100	0.25	5.7	9.4	7, 11

a. Assumes 1800 M J/m^2/year. Costs can be scaled to other sunlight regions (see Box A and Figures 2a and 2b).

problem for systems owned by homeowners who may not recognize system failure and may have difficulty finding repair services. Although automated meters may help detect failures, utility ownership appears to be the best long-term solution to maintenance.

THE COST OF PHOTOVOLTAIC ELECTRICITY

The bottom line in the maze of possible routes for reducing costs and improving system performance is the cost of the electricity finally produced. Electricity cost estimates require assumptions about such factors as balance-of-system (BOS) costs, overhead rates, and system lifetimes. Table 13 compares the cost of PV electricity derived from several possible flat-plate and concentrator systems operating in an average (1,800 kilowatt hours per m^2 per year) solar region. Two different costs are shown. One uses typical utility discount rates (6 percent real discount rate) and one uses typical industrial rates (12 percent real discount rate). Costs vary inversely with the sunlight available (see Box A and figure 1).

While the value of PV electricity is highly sensitive to the loads it is serving, to its location, and to the nature of the utility with which it is connected, the estimates summarized in table 13 strongly suggest that PV systems can provide power that is competitive with peaking power and even with baseload power by early in the next century. Large near-term markets can clearly be found in areas where utility electricity is not available. And near-term costs may also be low enough to capture significant numbers of niche markets in utility systems. Typically, these are areas where a PV system in an area with growing demand can delay investing in new transmission and distribution equipment. The value of a locally

---

Box A
Techniques for computing the cost of electricity from a photovoltaic system

The cost per kWh (PVCOST) is computed as follows:

$$\text{PVCOST} = \text{O\&M} + \frac{(\text{CRF}(d, L) + \text{INS}) \cdot (1 + \text{IND}) \cdot (\text{MOD} + \text{BOS} + \text{PC} \cdot \text{SP} \cdot \text{EPV})}{\text{SYR} \cdot \text{EPV} \cdot \text{EPC}}$$

Variable	Definition	Standard assumption
O&M	operating and maintenance costs per kWh	Table 13
CRF(*d,L*)	capital recovery factor of discount rate *d* for a period of *L* years	
d	discount rate	6 or 12 percent
L	lifetime of system in years	30
INS	insurance rate in $/year per dollar of capital invested	0.005
IND	indirect costs (fraction of direct capital costs)	0.25
MOD	module costs in $/m^2	Table 13
BOS	balance of area-related system costs (support structures, land, installation, wiring) in $/m^2	Table 13
PC	power conditioning costs in $/kW	100
SP	peak solar power incident on system in kW/m^2	1.1
EPV	efficiency of the PV system = EM·TC·ES	
	EM = efficiency of the module	Table 13
	TC = temperature correction	0.9
	ES = system efficiency	0.9
SYR	annual solar energy available to the system per m^2 in kWh/m^2	see maps
EPC	power conditioning efficiency	0.95

Source: Modification of methods recommended in the Five-Year Research Plan 1987–1991, National Photovoltaics Program, U.S. Department of Energy (DOE/CH10093-7), May 1987. Assumptions not discussed in the text are taken from this document.

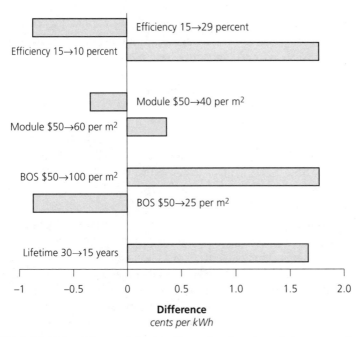

baseline = inexpensive thin film

FIGURE 4: Sensitivity of the cost of electricity from a flat-plate photovoltaic system to changes in important assumptions. The baseline system is an advanced thin-film system (see table 13). The figure explores the effect of changing efficiency from the baseline assumption of 15 percent to 10 percent and 19 percent, module costs from the baseline of $50 per square meter to $40–$60 per m², balance-of-system costs to a range of $25 to $100 per m², and the effect of cutting system lifetimes to 15 years.

sited system in such areas may be several cents per kWh higher than the value computed using avoided generation costs only [34].

Uncertainties inherent in estimating costs make it impossible to select a single best pv system. Although flat-plate systems using crystalline materials can capture significant niche markets in the near term, it appears that major cost reductions will be achieved over the long term only with thin-film materials or concentrators. The large improvements in efficiency and improvements in production costs are plausible, but as yet unproven. Many of the systems shown in table 13 can produce power at roughly comparable costs. In fact, the difference in prices falls well within the range of uncertainty about the input assumptions and competition among approaches is likely to be a part of the industry for the foreseeable future. The unique features of different systems will make them attractive to different market segments. Flat plates are much more likely to be built as a part of buildings, whereas tracking systems will be attractive primarily in larger field installations.

Figure 4 shows the sensitivity of cost estimates for a thin-film flat-plate system to changes in the more important assumptions. Electricity costs are obviously extremely sensitive to system efficiency. The importance of achieving long system lifetimes and low installation and mounting costs is also evident. Nearly 1 cent per kWh could be saved if flat-plate systems could be mounted on rooftops (the lower range of BOS estimates shown in figure 3) assuming that there were no other offsetting costs, such as much higher overhead costs.

While PV system costs as low as (and perhaps lower than) those estimated in table 13 are possible, cost reductions will not occur without significant effort. Investors willing to support major production facilities for PV must be found. Fortunately, investments in PV production facilities are 10 to 100 times less expensive than investments needed for most other new technologies (advanced reactors, synfuels from coal, fusion, etc.). Solutions to the practical and unglamorous problems that contribute substantially to system costs: site preparation, cable connections, site-specific design, etc. will only come about as the result of extensive field experience.

Photovoltaic technology is an area in which comparatively low investments in development could result in an extremely promising new energy system. Yet many promising areas of research in PV remain poorly funded. Increased public and private funding for research and development, and regulations crediting renewable electric generation for its environmental benefits, would lead to much more rapid progress. Under existing programs, PV markets will expand slowly from remote power markets to markets for grid-connected systems, and photovoltaic contributions to world energy and environment will fall far short of their potential.

Appendix
HOW PHOTOVOLTAIC CELLS WORK[4]

When a photon (a packet of light energy) enters a material it can free an electron from a stable position in the material's crystal structure and give it enough energy to move freely through the material. The minimum amount of energy required to free an electron from a fixed site is called the "band gap" of the material. Some electrons are also freed from fixed sites by atomic motion resulting from heat, but in semiconducting materials such as silicon, relatively few electrons are freed by room temperature heat.

The central trick of a photovoltaic device is to create a voltage inside a material that can direct electrons freed by photon collisions and produce a useful current. This is accomplished by combining semiconductor materials with different characteristics to form a junction. Junctions typically combine an "*n*-type" and a "*p*-type" material.

---

4. For a more detailed discussion, see [35].

In an "n", or negative-type material, electrons move freely at room temperature. The impurities added to create an *n*-type material create a new set of fixed sites for electrons. (See figure A-1.) These sites are filled with electrons when the temperature of the material is absolute zero, but at room temperature, vibrations in the material knock most of them loose and they move freely through the material.

In a "p" or positive-type material there are few free electrons but many "holes" that move freely at room temperature. Impurities added to make *p*-type materials also create a new set of fixed sites for electrons, but these sites are not occupied at absolute zero. At room temperature, however, a number of electrons in lower energy fixed sites are given enough energy to move into the new sites, leaving "holes" at the sites they formerly occupied (see figure A-1). Other electrons in fixed sites can move freely to occupy "hole sites" left vacant in neighboring atoms. In this way, holes move through the material, acting like positive charges.

The most common way to form *n*- and *p*-type materials is to dilute pure silicon with trace amounts of other elements. The addition of phosphorous to silicon, for example, creates an *n*-type material, whereas the addition of boron creates a *p*-type. Junctions can also be made from a variety of other materials, many of which are discussed in later chapters.

Forming a junction

When *n*- and *p*-type materials are brought together to form a junction, interesting things happen. Since the concentration of free electrons is much higher in the *n*-type material than in the *p*-type material, electrons drift across the junction from the *n*-type to the *p*-type. Holes drift across the junction in the opposite direction. This drift, or diffusion, creates a net electrical current (I_d) from the *p*-type material to the *n*-type. (See figure A-2.) (By convention, the direction of current is the direction of positive charge flow and opposite the direction of electron flow.)

Charges that cross the junction seldom get very far. Electrons entering the *p*-type material, for example, are trapped in holes near the junction (figure A-2). As charges move across the junction and stick, they create a voltage across the junction.

The voltage across the junction keeps increasing until the number of holes driven across the junction from the *n*-type material to the *p*-type material by the junction voltage (I_v) just equals the rate of flow of holes drifting across the junction from the *p*-type to the *n*-type materials junction because of the difference in concentrations (I_d). Precisely the same effect applies to the electrons.

Thermal energy in the material constantly creates pairs of free electrons and holes. Most quickly recombine but any holes created in the *n*-type material that drift near the junction are swept across to the *p*-type material by the junction voltage creating a current I_v. Free electrons that may drift near the junction in the *p*-type material are similarly driven towards the *n*-type material.

The number of charges that actually move across the junction to create the equilibrium situation is so small that the concentration of electrons and holes in

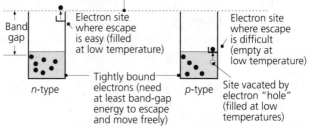

FIGURE A-1: n-*type and* p-*type materials insulated from each other.*

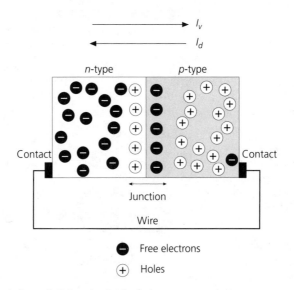

FIGURE A-2: *A photovoltaic junction in the dark.*

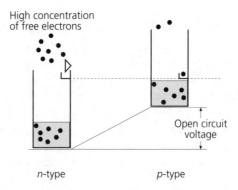

FIGURE A-3: n-*type and* p-*type materials connected at a junction.*

the bulk of the material is scarcely affected; no voltage can be measured across the bulk device. There is also no net current across the junction at equilibrium ($I_v + I_d = 0$).

Properties of a diode

The p–n junction just described operates as a "diode" or one-way valve for electrical current. If the positive terminal of a battery is placed on the p-side of the junction and the negative terminal to the n-side, the effect is to lower the voltage difference between the two materials. This means that a larger fraction of holes from the p-side can make it over the voltage barrier into the n-type material (I_d increases). Since the number of holes being driven across the junction from the n-type material (I_v) remains essentially the same as at equilibrium, the result is a positive current flow from p to n ($I_v + I_d > 0$) and electrons are flowing from the n-side to the p-side. The current can increase steadily with voltage. But if the battery terminals are reversed so that the voltage faced by diffusing charges increases, I_d can be driven to zero. For reasonable voltages, the largest achievable negative current is therefore I_v, which is comparatively small at room temperature.

The diode in light

The diode equilibrium is disrupted when photons strike the material. Photons with enough energy to free an electron from a fixed site in either an n- or p-type material create a new free electron and a new hole. This greatly increases the number of free electrons in the p-type material and the number of holes in the n-type material. If new holes spend much time in the n-type material, there is a high probability they will combine with an electron and be lost as charge carriers (the same is true for the new free electrons in p-type materials). In a well-designed cell, however, a large fraction of the holes created in n-type material, and free electrons in the p-type material, reach the junction (sometimes bouncing off the boundary of the material on the way) and are forced across by the voltage. The current I_v increases significantly as a result. The concentration of holes in the p-type material, however, is not changed significantly and I_d remains approximately the same.

If a low-resistance wire connects the n- and p-type materials (see figure A-2), a continuous net flow of current from the n- to the p-type material is created ($I_d + I_v < 0$) with no measurable external voltage. This is the "short circuit" current condition (see chapter 9, figure 14). Free electrons created in the p-type material could reach the n-type material by drifting to the electrical contact or by drifting to the junction. But holes cannot flow through the wire and the net flow of electrons through the wire is from the n- to the p-side.

If the wire connecting the n- and p-type materials is removed there can be no net current once equilibrium is reached. The distribution of the charge now creates a measurable open circuit voltage (see chapter 9, figure 4). The maximum voltage is limited by the band gap. As this voltage is approached, electrons trapped in the extra fixed sites in the p-type material can move to the extra sites in the n-type material with no change in energy (see figure A-3).

The current induced in the PV device when light is shining moves in the direction that would be blocked by the device when it operates as a diode (i.e., when it is in the dark). This means that if two PV cells are connected in series, and only one is exposed to light, little current can flow through the system. For this reason, shadows should never be allowed to fall on one or more modules that are connected in series.

Factors influencing system efficiency

Photovoltaic system costs are determined largely by the area of the collector and this area, in turn, depends on the efficiency with which sunlight is converted to electricity. This is true both for flat plates (where the area is largely active PV materials) and for concentrators (where the area is largely covered by lenses or mirrors).

The efficiency of the system depends on the fraction of solar power reaching the cell, which is converted to electrical power (current × voltage). Maximum efficiency is achieved when this product is at a maximum (obviously somewhere between the short-circuit and open-circuit points discussed earlier). Higher band gaps yield higher voltages, but lower currents, because only the most energetic photons create a new electron-hole pair, whereas lower band gaps yield more electrons at lower voltages. In most cells, the amount of voltage achieved in full sunlight is about one-half to two-thirds the band gap. If the sun's intensity is increased with a concentrator lens or other device, the voltage comes closer to the band gap.

Current can only be created if light photons have enough energy to dislodge a tightly bound electron. But photon energy that exceeds the amount needed to dislodge the bound electrons appears as random motion and is eventually lost as heat. Because excess photon energy is lost, cell efficiencies are highest for materials with band gaps that are closest to the average energy of the photons in sunlight.

Some losses are inevitable since solar photons reach the cell with varying amounts of energy (each color in the spectrum corresponds to a different energy). Solar energy peaks in the yellow–green region (photon energy of about 1.5 eV), which is close to the 1.45 eV band gap of cadmium telluride [33, 34]. The band gap of crystalline silicon is about 1.1 eV (see figure A-4).

In order to improve cell efficiencies, several limits to cell performance must be overcome. The problems and strategies for solving them are listed as follows:

1.   Light that is reflected, or lost, from the surface of the cell can be limited by using a variety of surface treatments (chapter 7).

2.   The loss of light that is reflected by electrical contacts on the front of the cells can be minimized by redesigning the electrodes or by using transparent electrical contacts.

3.   The amount of light that passes through the semiconductor material without colliding with an electron can be limited by selecting materials that are efficient light absorbers (i.e., have high absorptivity). Many of the thin films being consid-

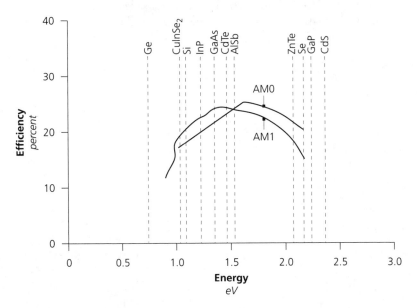

FIGURE A-4: Maximum efficiency calculated as a function of the energy needed to free an electron, assuming both the outer space solar spectrum (AM0) and the terrestrial spectrum (AM1); several semiconductors of interest are indicated. Source: P. Hersel and K. Zweibel, Basic Photovoltaic Principles and Methods, SERI, 1982, p.21.

ered, for example, are less then 1 micron thick, yet absorb 90 percent of sunlight. In contrast, single-crystal and polycrystalline silicon cells must be from 50 to 150 microns thick to have effective absorption; most are more than 300 microns thick.

4. Electrons and holes created by photons may recombine before reaching the junction and contributing current, if they reach impurities or flaws in the crystal structure. Hydrogen alloys are now being used to prevent electrons and holes from recombining in polycrystalline and noncrystalline (amorphous) silicon materials.

5. Electrical resistance within a semiconductor can reduce power, but such resistance can be minimized by careful attention to cell design.

6. The limits imposed by the choice of band gap can be overcome by stacking materials with different band gaps. The top junctions have large gaps and only capture the most energetic (blue light) photons, which allows photons of other colors to pass through to lower layers. The lower junctions have less tightly bound electrons and so capture less energetic (red light) photons. The record for photovoltaic efficiency, which is 35 percent, was achieved through stacking. Efficiencies as high as 60 percent or more are theoretically possible if large numbers of carefully tuned junctions are used, but practical systems will be difficult to manufac-

ture. There has been recent interest in using molecular beam epitaxy to create multi-layer super lattice systems. Such techniques would allow designers to optimize current and voltage generating characteristics separately [33].

WORKS CITED

1. Photovoltaic insider report, 1101 W. Colorado Blvd., Dallas, Texas, various issues.

2. US Department of Energy, Energy Information Administration. 1990
   Annual energy review 1990, 247 (DOE/EIA 0384(90)). 12.4% of US cells and modules were shipped to original equipment manufacturers and the uses not recorded in the EIA survey.

3. Maycock, P. 1991
   Photovoltaic technology performance, cost and market forecast to 2010, PV Energy Systems, Inc., PO Box 290, Casanova, Virginia.

4. Shugar, D. 1990
   Photovoltaics in the utility distribution system: the evaluation of system and distributed benefits, 21st IEEE Photovoltaic Specialists Conference, Orlando, May, 836-843.

5. Bigger, J. E., Kern, E. C., and Russel, M. C. 1991
   Cost-effective photovoltaic applications for electric utilities, presented at 22nd IEEE Photovoltaic Specialists Conference, Las Vegas, Nevada, October 7–11.

6. Herzer, H. 1991
   The development of the wafer cost and availability for the photovoltaic industry: ribbon/sheet approaches and its comparison with wafers; efficiency-cost trade off, presented at the Photovoltaic Solar Energy Conference, Spain. Prices converted using 1.8 DM = 1 $.

7. Herzer, op. cit.

8. Eyer, A., Raubner, A. and A., Goetzberger.
   Silicon sheet materials for solar cells, *Optoelectronics*, Vol..5, 239–257.

9. Mason, N.B. and Jordan, D. 1991
   A high efficiency silicon solar cell production technology, *Conference Record, 10th Photovoltaic Solar Energy Conference*, Lisbon, April, 250–253.

10. Mason et al., op. cit.

11. Wolgemuth, J.H., Narayanan, S., and Brenneman, R. 1990
    Cost effectiveness of high efficiency cell processes as applied to polycrystalline silicon, *Proceedings of the 21st Annual IEEE Photolytic Specialists Conference*, 221.

12. Morel, D.L., Blaker, K., Bottenberg, W., Fabick, L., Fedler, B., Gardenier, M., Kumamoto, D., Reinker, D., and Raifai, R. 1988
    Development and performance of 4-terminal thin film modules, *Proceedings of the 8th E.C. Photovoltaic Solar Energy Conference*, Florence, Italy, Kluwer Academic Publishers, Dordrecht, The Netherlands.

13. Oregon, B. and Gratzel, M. 1991
    A low-cost, high-efficiency solar cell based on dye-sensitized colloidal to$_2$ films, *Nature* 353, 737-740.

14. Texas Instruments and Southern California Edison. 1991
    Presentation at the 21nd IEEE Photovoltaic Specialists Conference, Las Vegas, Nevada, October 7-10.

15. Candelario, T. R., Hester, S.L., Townsend,T.U., Shipman, D.J. 1991
    pvUSA–performance, experience, and cost, presented at the 22nd IEEE Photovoltaic Specialists Conference, October 11.

16. Kern, E.C. 1986
    Residential solar photovoltaic systems: final report for the northeast residential experiment station, US Department of Energy Contract No. DE-AC02-76ET20279, June.

17. Kapner, M. 1990
    Power Authority of New York, private communication, December.

18. Alpert, D.J., Mancini, T.R., Houser R.M., and Grossman, J.W. 1990
    Solar concentrator development in the United States, presented at IEA's 5th Symposium on Solar High Temperature Technologies, August 27-31, Davos, Switzerland.

19. Mavis, C.L. 1989
    A description and assessment of heliostat technology, SAND87-8025, Sandia National Laboratory, Sandia, New Mexico.

20. Hoff, T. and Iannucci, J.J. 1990
    Maximizing the benefits derived from PV plants: selecting the best plant design and plant location, 21st Annual Photovoltaic Specialists Conference.

21. Hoff. T. and Iannuci, J.J. 1990, op. cit.

22. Kern, E.C. and Russell, M.C. 1992
    Rotating shadow band pyranometer irradiance monitoring for photovoltaic generation estimation, presented at the 23rd Annual Photovoltaic Specialists Conference.

23. Steitz, P., Goodman, F.R., and Key, T.S. 1990
    Photovoltaic power conditioning status and needs, 919, and Chaing, C.J., and Richards, E.H., A twenty percent efficient photovoltaic concentrator module, both presented at the 21st IEEE Photovoltaic Specialists Conference.

24. Kerns, E., private communication.

25. Risser, V. 1990
    Perspectives on PV systems operation and maintenance, CRIEPI presentation at 3rd EPRI/CRIEPI Workshop on Utility-Interconnected Photovoltaic Systems, Yountville, CA, March.

26. Electric Power Research Institute technical assessment guide: electricity supply 1989 (EPRI P-6587- L), 3–5.

27. Electric Power Research Institute photovoltaics system performance assessment for 1988 (EPRI GS-6696), January, 1990.

28. US Department of Energy, Five Year Research Plan 1987–1991 (DOE/CH10093-7), May 1987, 29.

29. Kobayashi, H., Takigawa, K., Hashimoto, E., Kitamura, A., and Matsuda, H. 1990
    Problems and countermeasures on safety of utility grid with a number of small scale PV systems, 850–855, and M. Russell and E. Kern. Lessons learned with residential photovoltaic systems, 898–902, both presented at the IEEE Photovoltaic Specialists Conference, Orlando, May.

30. Electric Power Research Institute. 1991
    Experiences and lessons learned with residential photovoltaic systems, EPRI GS7227, Project 1067-15, July.

31. Takigawa, K. 1990
    Performance test and evaluation facility of inverter for photovoltaic power systems, and Performance test and evaluation facility of storage batteries for PV applications, CRIEPI presentations at 3rd EPRI/CRIEPI Workshop on Utility Interconnected Photovoltaic Systems, Yountville, California, March.

32. Braun, G. 1989
    Evolution of the mini-utility concept, Integrated Electric Utility Workshop, Saipan, May.

33. Electric Power Research Institute 1991, op. cit.

34. Shugar, op. cit and the discussion in chapter 23.

35. Zweibel, K. 1990
    Harnessing Solar Power: The PV Challenge, (Plenum: New York).

36. Mason, M.B, Lord, B.E., Jordan, D., and Summers, J.G. 1991
    High-efficiency, low cost crystalline silicon production solar cells, Proceedings of the 10th
    E.E. Photovoltaic Solar Energy Conference, Lisbon, Portugal, April, Kluwer Academic
    Publishers, Dordrecht.

37. Zhao, J. Wang, A., and Green, M.A. 1990
    24% efficient PERL structure silicon solar cells", 21st IEEE Photovoltaic Specialists Con-
    ference, 333 (IEEE Publishing Services, New York).

38. Swanson, R. 1989
    Why we will have a 30 percent efficient silicon solar cell., 4th International Photovoltaic
    Science and Engineering Conference, Vol. 2, 573-580, Sidney, Australia.

39. Narayanan, S., Zolper J., Yun, F., Wenham, S.R., Sproul, A.B. Chong, C.M., and Green,
    M.A. 1990
    18% efficient polycrystalline silicon solar cells, 21st IEEE Photovoltaic Specialists Con-
    ference, 678 (IEEE Publishing Services, New York).

40. Fan, J.C.C., Tsaur, B.Y., and Palm, B.J. 1982
    Optimal design of high-efficiency tandem cells, Conf. Record, 16th IEEE Photovoltaic
    Specialists Conference, San Diego, CA, 692-701 (IEEE Publishing Services, New York).

41. Measurements taken by the National Renewable Energy Laboratory, 1992 (supplied by K.
    Zweibel, July 6, 1992

42. Sites, J.R. 1988
    Separation of voltage loss mechanisms in polycrystals attached to a utility grid. Batteries
    and other electric storage systems are useful only if line solar cells, 20th IEEE PV Special-
    ists Conference, Los Vegas, NV, Sept. 26.

43. Fraas, L.M., Avery, J.E., Sundaram, V.S., Nidh, V.T., Davenport, T.M., and Yerkes, J.W.
    1990
    Over 35% efficient GaAs/GaSb stacked concentrator cell assemblies for terrestrial appli-
    cations, 21st IEEE Photovoltaic Specialists Conference (IEEE Publishing Services, New
    York).

7
CRYSTALLINE- AND POLYCRYSTALLINE- SILICON SOLAR CELLS

MARTIN A. GREEN

Crystalline- and polycrystalline-silicon solar cells remain the "workhorse" for outdoor solar-power generation, despite significant advances with other photovoltaic (PV) devices. Recent improvements in the energy conversion efficiencies of silicon cells are likely to further extend their commercial life beyond the end of the current decade. Particularly intriguing are prospects for thin-film technology that is based on crystalline or polycrystalline silicon. Such a technology could maintain the high reliability and efficiency features of silicon cells while substantially reducing silicon material costs. Eventually, electricity generated by thin-film silicon may compete with that generated by fossil fuels, but would do so with much less damage to the environment.

INTRODUCTION

The photovoltaic (PV) effect was first demonstrated in 1839 by Edmond Becquerel while he was working in his father's laboratory in France [1]. Despite this early start, significant PV power generation did not become possible until the first efficient silicon solar cells were developed in 1954. Since then, silicon cells have dominated the commercial market for outdoor power PV modules. Other PV technologies have progressed rapidly (see chapters 6: *Introduction to Photovoltaic Technology*, 7: *Crystalline- and Polycrystalline-Silicon Solar Cells*, 8: *Photovoltaic Concentrator Technology*, 9: *Amorphous Silicon Photovoltaic Systems*, 10: *Polycrystalline Thin-film Photovoltaics*, and 11: *Utility Field Experience with Photovoltaic Systems*), but silicon technology has not stagnated. Although displaced from small-scale applications in consumer products by hydrogenated amorphous silicon alloys, silicon cells have undergone significant improvements in efficiency. They are thus likely to remain the "workhorse" for outdoor power applications at least until the end of this decade.

The overriding strengths of the technology lie in its high energy conversion efficiencies and in its well-documented reliability. To date, silicon technology is hampered only by its relatively high material requirements, which account for a substantial percentage of overall costs.

EVOLUTION OF CELL PERFORMANCE

Early history [2]

Interest in silicon as a PV material was sparked in the 1940s when silicon was found to perform well as a point-contact rectifier in applications such as early radar systems. While working with recrystallized melts of silicon, Russell Ohl of Bell Laboratories in the United States [3] found that rods cut from some recrystallized ingots had well-defined natural barriers, which gave rise to a good photovoltaic response.

One end of the rod developed a negative voltage when illuminated or heated. It also had to be biased negatively to create low resistance to current flow across the barrier. This led to the terminology of "negative-" or "*n*-type" silicon for material with these properties. The opposite type of material was termed "positive-" or "*p*-type" (see chapter 6). The role of donor and acceptor impurities in producing these properties was only subsequently shown. By carefully cutting cells from the cast ingot, Ohl was able to develop silicon PV devices that included the natural junction. Some were cut in such a way that the junction was parallel to the illuminated cell surface (see figure 1a), and some with the junction perpendicular.

Although Ohl's devices were comparable to the best photovoltaic cells then available, their laborious preparation precluded convenient manufacturing. By 1952, Kingsbury and Ohl had improved cell performance by using purer silicon, which prevented the formation of grown-junctions. Helium-ion bombardment produced a controlled rectifying junction along the surface (see figure 1b) [4].

Shortly thereafter, such developments were overshadowed by the emergence of technologies for silicon-crystal growth and for $p–n$ junction formation. These opened the door to the first modern high-performance silicon solar cell, reported in 1954 [5]. The new cells differed from their predecessors in having both contacts on the rear surface (see figure 1c) and an efficiency of 6 percent, which was about 15 times higher than that of earlier devices. Not surprisingly, they aroused considerable interest at the time, although it soon became apparent that such enthusiasm was premature [6–8]. Nonetheless, the cells proved suitable for use in spacecraft and were used primarily for that purpose until the early 1970s.

Solar cells in space

In 1958 the first solar cells to enter space were installed on Vanguard I, the second U.S. satellite launched into orbit. For space use, rapid changes in cell design took place in the late 1950s and early 1960s (see figure 2a); thereafter the design

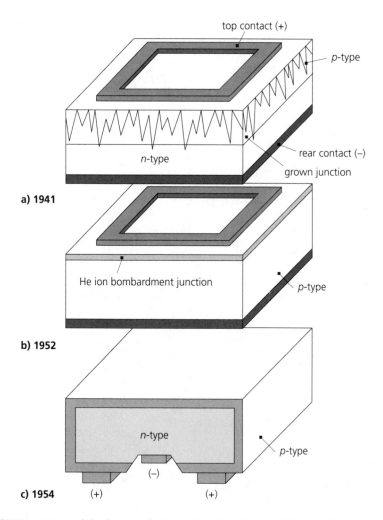

FIGURE 1: Historical development of silicon solar cells is illustrated. The first silicon cell was a "grown junction" device, which was created in 1941 (a); it was later followed by a helium-ion bombarded junction device (b) [4]; and a B-diffused junction cell (c), which was the first cell to achieve a useful energy conversion efficiency [5].

of solar cells remained essentially unchanged until the early 1970s. The resulting cells used a *p*-type substrate into which a thin *n*-type layer was introduced by high temperature diffusion of phosphorus impurities. Electrical contact was made to the *p*-type substrate by depositing a contact over the entire rear surface. The placement of a patterned finger-contact on the top *n*-type surface allowed light to enter the cell. The top surface was also coated with an antireflection layer, thus reducing reflection losses to tolerable levels. The cells converted only about 10 percent of solar radiation in space to electrical energy, but this conversion rate increased to

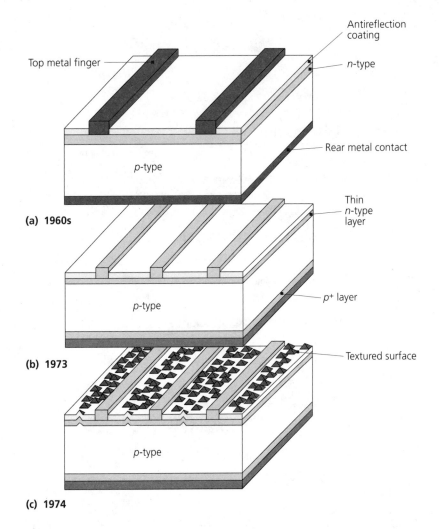

Antireflection coating

Top metal finger

n-type

Rear metal contact

p-type

(a) 1960s

Thin n-type layer

p-type

p+ layer

(b) 1973

Textured surface

p-type

(c) 1974

FIGURE 2: Silicon cells for use in space evolved considerably during the 1960s and 1970s. A conventional space cell was the standard from the early 1960s (a) but was outperformed by both the shallow-junction violet cell (b) [9] and the chemically textured nonreflecting black cell (c) [10]. (In practice, the pyramids on the black cell will fully cover its surface; partial coverage is shown here only for convenience in illustration.)

about 12 percent in terrestrial applications, where sunlight has less blue and shorter-wavelength light.

By the early 1970s, the benefits of alloying aluminum into the rear cell surface became apparent as the amount of cell output current and voltage increased. Although attributed initially to the gettering, or removal, of unwanted impurities from the cell [11, 12], the benefits were later shown to originate, at least partly, from a heavily doped region near the rear contact [13]. The doped region improved performance by creating a back-surface field. This shields regions of high carrier recombination (at the rear metal-to-silicon interface) from the rest of the cell by restricting movement of carriers to the rear. As a result of this design feature, which is widely used in modern cells, the efficiency of solar cells in space increased to 12 percent (corresponding to about 14 percent efficiency under terrestrial sunlight).

The conventional space cell responded poorly to blue and shorter-wavelength light, a deficiency attributable to the high concentration of phosphorus dopants used to convert the surface to an *n*-type layer. The high phosphorus concentrations degrade the electronic properties of the cell's surface region, where most of the blue light is absorbed.

The solution was simply to introduce less phosphorus into the surface. Doing so, however, increased resistance to current flow parallel to the top surface, and so had to be counteracted by setting the fingers more closely together (see figure 2a). Fortunately, photolithographic techniques developed for microelectronics made it possible to space the fingers close to one another without increasing shading (see figure 2b). In addition, the increased sensitivity of the cells to blue light could be exploited by coating them with new antireflection materials that were found to give better performance at blue wavelengths. Because the reduced optical thicknesses of these new coatings gave the cells a characteristic violet coloring, the cells became widely known as violet cells. All told, these improvements increased their efficiency in space to 13.5 percent [9], which corresponds to a terrestrial efficiency above 15 percent.

Another major improvement in cell design, known as surface texturing, occurred soon thereafter. Although the idea of reducing surface reflection by mechanically forming structures on the cell surface had been known for some time [14], forming such structures by directional chemical action was being newly explored by the microelectronics community. In 1974 Comsat Laboratories of Clarksburg, Maryland, reported that selective chemical etching could be used to form randomly located pyramids across the cell surface (see figure 2c) [10].

Such texturing affects solar-cell performance in two important ways. One is that light is reflected downward after it strikes the side of a pyramid and so has two chances to be coupled into the cell. The second is that, once light couples into the cell, it travels obliquely to its original direction. This means more light will be absorbed near the cell surface where the active junction region of the cell is located. Such cells, which have the appearance of black velvet after their surface has been coated with an antireflection layer, are referred to as black cells.

By 1974, as a result of such advances, efficiency for the best space cell had increased to 15.5 percent (to more than 17 percent for terrestrial cells). Although chemical texturing of crystalline cell surfaces subsequently became a standard feature in commercial silicon solar cells, it was almost a decade before any further efficiency improvement was demonstrated.

Recent improvements

A major breakthrough in cell performance came when techniques for reducing the electronic activity of cell surfaces and contact regions were developed [15, 16]. By sealing the cell surface with a layer of thermally grown oxide, a technique well known from microelectronics, additional design flexibility became possible. Flexibility was further increased by reducing contact recombination, for example, by introducing thin interfacial oxides between the contacts and silicon [16]. These features improved each cell's output voltage from 0.60 to more than 0.65 volts. Primarily due to this increased voltage, efficiency of terrestrial cells increased to more than 18 percent in 1983 and 19 percent in 1984 [17, 18].

In 1985, researchers at the University of New South Wales combined this flexibility with directional chemical texturing to produce the first silicon cell that exceeded 20 percent energy conversion efficiency [19] (see figure 3a). One difference between the new cell and earlier designs is the incorporation of a thin oxide layer along most of the cell surface to reduce electronic activity. Contact to the *n*-type region of the cell is made through a narrow stripe etched through the oxide, thus minimizing the metal-to-silicon contact area and total recombination at this interface.

The cell also demonstrated a reasonably high level of light trapping [20]. The rear interface, formed by aluminum alloying, although rough, was reasonably reflective. Weakly absorbed light reaches this interface where some is absorbed by the metal and thus wasted, and some is reflected in more or less random directions because of surface roughness. More than 90 percent of the light that is reflected back, however, strikes the cell's upper surface at an angle where it is internally reflected. This increases the effective optical path length or thickness of the cells by a factor of six to ten [20].

The next advance in cell design was made possible by an innovative cell structure developed primarily for PV concentrator systems [21]. The cell (see figure 3b) relies on two contacts on the rear of the cell but has no contacts on its illuminated surface. For the cell to function successfully, extremely high levels of surface and contact passivation are required, as are exceptionally high lifetimes for the charge carriers generated within the cell by the absorbed light (see chapter 8). The design, which was successfully implemented with state-of-the-art microelectronics processing, gave an exceptional boost to the performance of cells designed for concentrator systems [21]. Although its performance under normal, nonconcentrated sunlight was less marked, the design did achieve efficiencies greater than 22 percent under such conditions [22].

One strength of this design is that contact area to the cell is minimized. Contact occurs only at points that are scattered over the cell's rear surface. Recombination at each contact is limited by the heavy diffusion of phosphorus or boron, depending on the polarity of the contacts. The cell also has a higher level of light trapping than earlier designs because the rear contact arrangement permits much higher levels of reflection. Furthermore, the cells make extensive use of chlorine-based compounds, which are used to clean the furnaces

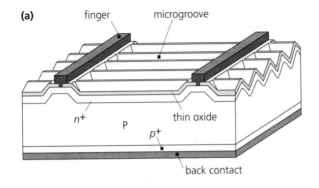

(a) finger · microgroove · n^+ · P · thin oxide · p^+ · back contact

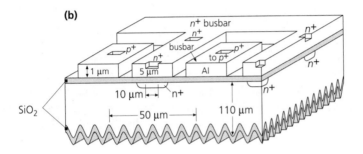

(b) n^+ busbar · n^+ busbar · p^+ · n^+ · p^+ · to p^+ · n^+ · 1 µm · 5 µm · Al · n^+ · 10 µm · n^+ · n^+ · 50 µm · 110 µm · SiO_2

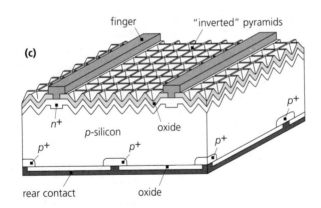

(c) finger · "inverted" pyramids · n^+ · p-silicon · oxide · p^+ · p^+ · p^+ · p^+ · rear contact · oxide

FIGURE 3: *Several high-efficiency silicon solar cells have been demonstrated recently. They include passivated-emitter cells (a), the first silicon cells to give 20 percent energy conversion efficiency by using oxide surface passivation, reduced contact area, and optimized junction diffusion [19] (the antireflection coating is not shown); point-contact cells (b) with microelectronics-quality oxide passivation, contacts on the nonilluminated surface and minimal diffused and contact areas [22] (the cell is inverted to show illumination striking from below); and (c) 23 to 24 percent efficient PERL (passivated emitter, rear locally diffused) cells that incorporate features of cells (a) and (b) [23].*

immediately prior to processing and during some of the high-temperature processing steps. This improves oxide quality and maintains high-carrier life-times during processing.

The next major improvement in silicon cell performance occurred when certain features of the point-contact cell were integrated into a more standard design called the passivated emitter, rear locally diffused (PERL) cell (see figure 3c). The result was a silicon substrate completely enshrouded by a microelec-tronics-quality oxide; top contact is made through narrow stripes in the oxide to regions where extra phosphorus dopants have been introduced. The extra phosphorus minimizes recombination at the contact. The PERL cell is much like the point-contact cell along its rear surface, where contact is made at iso-lated points that are heavily diffused with boron. Inverted pyramids are created along the top surface of the cell with directional chemical etching. The in-verted pyramids are advantageous because they allow for more effective light trapping than do the normal upright pyramids. By combining inverted pyra-mids with high reflectivity from the rear surface, an effective optical path-length boost (or effective increase in cell thickness) of between 25 to 30 times results. This type of cell has the highest efficiency of any silicon cell to date, with efficiencies in the range of 23 to 24 percent [23]. Ultimately, the PERL cell is expected to be more than 25 percent efficient, a rating that approaches the upper efficiency limit of 30 percent for silicon cells (see the section "Ultimate ef-ficiency and future prospects").

COMMERCIAL TERRESTRIAL CELLS

Overview

When designing a silicon cell, the most critical decision concerns the approach used to form metallic contact to the n- and p-type regions. Shortly after the 1973 oil embargo, several companies commenced production of silicon solar cells spe-cifically for the terrestrial market. By the mid-1970s, three separate techniques for forming metal contact to the cells were developed. From then until the mid-1980s, the Jet Propulsion Laboratory in Pasadena, California, managed a U.S. Government program to accelerate the development of silicon photovoltaics. During this time, a wealth of information on all aspects of silicon cell technology and economies was generated [24], including detailed information about these metallization approaches.

Metallization approaches

A metallization technique adopted by some manufacturers was simply a streamlined version of the one used for space cells. The technique involves evaporation of a metal that is heated to above its melting point in a vacuum [25], but unfortunately the process is not very cost-effective. The capital costs of the equipment are high, and a substantial amount of the metallization material is wasted. However, when the technique is combined with photolithographic definition of the top finger pattern, only minimal compromises in cell design have to be made, and so this approach results in higher performance than the other two approaches.

The second technique, which dates back to the beginning of the silicon-cell era [5], involved electroless plating of the metal contacts. Plating is accomplished by first defining the areas where the fingers are to be plated on the top surface; this is done by screen printing an appropriate pattern of masking material onto the surface. The plated layer, which is generally nickel, can then be built up either with additional layers of highly conductive metal, such as copper, or with a layer of molten solder.

The third technique calls for direct screen-printing of the metal onto the cell surface [26]. The metal, which is in paste form, is applied to the cell surface in defined patterns through a patterned screen. The most suitable metal pastes are silver-based, which are relatively expensive [24]. Overall, this deposition approach involves inexpensive equipment but high material costs.

Because both the nickel plating and screened silver techniques create much coarser finger patterns than does vacuum evaporation, the fingers must be liberally spaced to avoid excessive shading of the cell from incident sunlight. To keep the resistive losses arising from this increased spacing low, a high concentration of phosphorus dopant must be introduced into the cell surface. The drawback is that large amounts of phosphorus decrease the cell's response to blue light (see the section "Solar cells in space") and also restrict the cell's output voltage capability. (A further discussion of these issues and their partial remedies, is contained in a recent publication [27].)

Thus, to accommodate low-cost metallization approaches, sacrifices must be made in cell performance. Until recently, the performance of commercial cells has lagged far behind that of laboratory devices. The highest performing terrestrial cells using low-cost metallization are those made on crystalline silicon substrates. When chemically textured to reduce reflection losses, these cells show efficiencies as high as 14 percent.

Another common type of terrestrial cell is made on a large-grained polycrystalline silicon substrate formed by casting (see the section "Self-supporting sheets"). Because the polycrystalline grains are randomly oriented, surface texturing provides little performance advantage. The resulting reflection losses, combined with the detrimental electronic activity of the grain boundaries and the relatively poor electronic quality of the material between the grains, give lower efficiency values from 12 to 13 percent. The efficiency differences between crystalline and polycrystalline cells tend to be reduced, however, once the cells are encapsulated into modules where the square-shaped polycrystalline wafers allow for denser packing.

The efficiency of commercial cells could be increased further by changing the metallization process. One approach would be to write the silver paste directly onto the cell surface as opposed to applying it through a screen. Doing

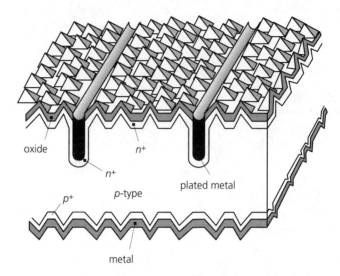

oxide   n+

n+

plated metal

p+   p-type

metal

FIGURE 4: The laser-grooved buried-contact solar cell, which offers high-energy conversion efficiency at low cost.

so would make it possible to reduce the width of the fingers and at the same time reduce wastage of the expensive paste. But heavy doping of the surface would still be required to give the cell low resistance contact. Although the procedure is being tested experimentally, it has not yet been commercialized.

A more dramatic efficiency improvement has recently been obtained with a design called the laser-grooved cell [28] (see figure 4), which is now being commercialized [29, 30]. The fine fingers that are essential for high efficiency are defined using a laser to form narrow grooves deep into the cell surface. The technique has the advantage of allowing separate heavily doped regions to be formed in the contact areas, which gives the cell low resistance contact. However, it maintains lightly diffused regions over most of the surface for good blue response and high output voltage. The deep narrow grooves are filled with metal by electroless plating, thus creating a large metal cross section without appreciable shading. The large cross section reduces resistance losses in the top surface compared to conventional commercial cells. Such improvements have produced pilot-production cells of 17.5 to 18.0 percent efficiency on commercial grade crystalline substrates [30]. Economic studies indicate that the costs per unit area for processing the laser-grooved cells are similar to, or lower than, those for standard screen-printed cells; in addition, power output for the grooved cells is 25 to 35 percent greater than for the standard cells [31]. Grooved-cell technology and its derivatives are expected to greatly improve the efficiency of commercial silicon solar cells during the next few years.

The first commercial cells produced by the laser-grooved method were tested during the 1990 World Solar Challenge, a 3,000 kilometer solar-car race from Darwin to Adelaide, Australia [32]. The winning car operated with a 1.3 kilowatt

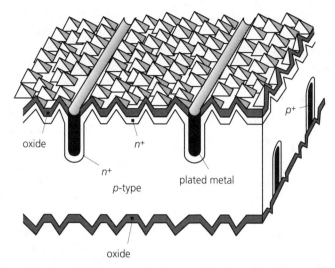

FIGURE 5: "Double-sided" laser-grooved cell may offer the highest commercially attainable efficiencies for silicon cells.

solar module, which had an energy conversion efficiency of 17 percent. The cells were largely responsible for the car's huge lead over its competitors, enabling it to finish the race a full day ahead of the others.

Recent developments such as demonstrated in the PERL cells (see the section "Recent improvements") have been incorporated into at least one variation of the laser-grooved approach (see figure 5). Laboratory cells fabricated in this way are already more efficient than those made via the standard laser-grooved process, and prospects for producing experimental cells above 22 percent efficiency with the new method are good. If successful, the process may well produce the first commercial silicon cells that exceed 20 percent efficiency. Likewise, the corresponding efficiency for polycrystalline cells could exceed 17 percent.

Extra processing steps beyond those used for crystalline silicon must be taken if equivalent performance levels from polycrystalline substrates are to be obtained [33]. Incorporating atomic hydrogen into polycrystalline silicon, for example, reduces the electronic activity of grain boundaries and defects within the grain and so boosts performance, increasing the cell's efficiency to more than 13 percent [33]. Structures with collecting junctions along both top and rear surfaces have also been suggested for polycrystalline silicon wafers [34]. The technique, in fact, has produced large-area cells (10 × 10 centimeters) with an efficiency of 16.8 percent [35] and a current density rivaling that of the highest quality material available [36].

Manufacturing costs

Ever since interest in the terrestrial use of silicon cells was renewed in the mid-1970s, the price of silicon modules has dropped dramatically. During the

1980s, module costs in real terms decreased by a factor of three to four, reaching an average price in volume of about $5.00 per peak rated watt in 1989 (1989 U.S. dollars). Further cost reductions are expected during the coming decade as a result of improved production processes, improved cell performance, and increased manufacturing volume.

The precise breakdown of manufacturing costs depends on many variables, including the source of the silicon wafers, the technology by which the wafers are processed into cells, the degree of automation both during the cell processing stage and the subsequent encapsulation of cells into modules, and the energy conversion efficiency of the final product.

An extensive independent analysis of manufacturing costs was conducted by the Jet Propulsion Laboratory in the 1980s as part of the Flat Plate Solar Array Project [24]. Various production processes were analyzed for hypothetical facilities with widely different production volumes, and levels of technology reflected start-up dates ranging from 1978 to 1988. Although production costs varied enormously, some commonalities emerged, thus providing insight into the structure of present and future manufacturing costs.

The study suggests that when silicon ingots are prepared using the relatively expensive Czochralski process (see the section "Polysilicon source material"), wafer costs account for 55 to 75 percent of the calculated manufacturing cost, regardless of facility size and start-up date. Processing the wafers into cells contributes from 10 to 25 percent of the total manufacturing cost, as does encapsulation. Thus, module costs are overwhelmed by the cost of the silicon wafer. According to the Jet Propulsion study, wafering alone contributes from 15 to 35 percent of the total manufacturing cost.

A more recent study [31] reports that lower costs can be achieved by combining high-efficiency laser-grooved techniques with 12.5 cm^2 polycrystalline wafers (see figure 4). Doing so lowers the estimated fabrication costs for solar modules to less than $3 per peak watt (1990 U.S. $) at the relatively small production volumes typical of commercial practice today (less than 10 megawatts per year throughput). When this process is used, polycrystalline wafers account for about 40 percent of the manufacturing cost, as opposed to 60 percent when they are obtained via the Czochralski process [30].

Improvements in solar-cell efficiency reduce costs in all areas, ranging from the cost of the silicon material in the cell to the cost of module encapsulation. Ultimately, however, it is the cost of slicing ingots into wafers that will determine the extent to which the costs of this general class of ingot technology can be reduced. For that reason, producing silicon directly in the form of sheets or ribbons, thus eliminating the cost of slicing ingots into wafers, offers great potential for cost reduction (see next section).

MATERIALS ISSUES

Overview

Recent developments suggest that highly efficient silicon cells can be obtained with low processing costs. The major challenge if silicon technology is to reach its full potential, is to reduce costs associated with the relatively large amounts of silicon that are required for the cells. These costs reflect the need to prepare high-purity silicon source material, generally in the form of fine-grained polycrystalline-silicon chunks; the cost of converting these chunks into large-grained or single-crystalline ingots of silicon; and finally, the cost of processing these ingots into silicon sheets. The last two operations can be bypassed by converting the large-grained material directly into sheets or ribbons.

Polysilicon source material

Pure silicon is required as the starting material for the production of cells. Although the silicon for cells need not be as pure as for microelectronics, the photovoltaics field has nonetheless benefited from the economies-of-scale created by the high-volume demand of the microelectronics industry. Moreover, because solar cells can use silicon that is "off specification" for microelectronics, further economies of scale are gained. However, the preparation of silicon source material for microelectronics involves several energy-intensive and inherently costly steps [25]. As a first step, silicon of metallurgical-grade purity must be extracted by the reduction of quartz with carbon. The silicon extract is then converted to a volatile liquid (trichlorosilane) and purified by fractional distillation. Finally, the purified trichlorosilane is reduced with hydrogen to produce extremely pure silicon.

Because the photovoltaics market can tolerate lower quality silicon, alternative processing methods have been investigated [24]. One such method, which is now at an advanced stage of development, calls for the formation of silane (SiH_4) as an intermediate product [24]. Another much cruder process begins with quartz material of higher purity than that normally used to produce metallurgical-grade silicon. The latter approach, when combined with relatively simple purification processes, could produce silicon at much lower cost than microelectronics-grade silicon, while also improving the energy efficiency of the extraction process.

Indeed, these low-cost approaches are likely to play an increasingly important role in the silicon solar-cell industry. The reason is that an expanding photovoltaics market will eventually demand more silicon than the microelectronics industry can provide (in 1991 photovoltaics consumed less than 10 percent of the high-purity silicon produced for microelectronics). To meet the demand, facilities for producing low-cost silicon can be expected to enter production before the end of the decade.

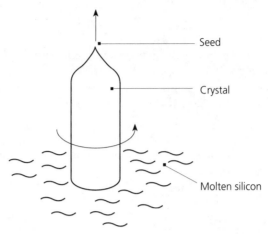

FIGURE 6: Czochralski crystal growth process for preparing crystalline silicon ingots is shown. A seed crystal is dipped into molten silicon and slowly removed, drawing out the cylindrical crystal

Crystalline-ingot technologies

For good cell performance, not only is reasonably pure source material required, but the silicon must have good crystallographic quality as well. Although the quality standards in this area are also not as severe as they are for microelectronics, the same techniques have been used to take advantage of economy-of-scale. The most common technique for producing ingots of good crystallographic quality is known as the Czochralski method (see figure 6). The technique involves drawing cylindrical crystal from a melt of silicon with an appropriate seed. Ingots that are 20 centimeters in diameter and more than 1 meter long are routinely produced by this method, although crystal diameters of 10 and 12.5 centimeters are more commonly used in photovoltaics. The grown crystals are then sliced into individual wafers using an inner-diameter slicing technique (see figure 7), which also was developed originally for the microelectronic industry. About half the silicon material is wasted during this cutting process.

Alternative cutting techniques use wires that are either impregnated with diamonds [24] or that serve to guide abrasive material through the ingot [37]. The latter methods are considered advantageous for photovoltaics because more wafers can be cut from a given length of silicon ingot. The thin wire not only reduces the wafer thickness but it also prevents excessive cutting loss. Wire-sawing produces wafers that are about half the thickness (0.17 millimeter) of the average wafer obtained by inner-diameter slicing [38].

The Czochralski process has recently been enhanced using magnetic fields to reduce circulation of the molten silicon during crystal growth [39]. Although creation of a magnetic field adds to the capital cost of the growth equipment, it greatly improves the photovoltaic potential of the crystals. In addition, the process offers better control, and higher rates of crystal growth are possible,

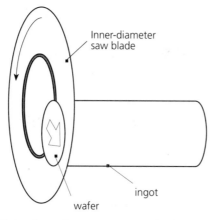

Inner-diameter
saw blade

ingot

wafer

FIGURE 7: Silicon wafers can be cut from grown ingots with an inner-diameter saw.

which offsets its higher capital costs. Moreover, costs are reduced by the improved cell efficiency that can be anticipated.

An alternative technique for preparing high-quality silicon is known as the float-zone process [39]. Although this technique also involves higher equipment capital costs than the Czochralski process, consumables such as the crucibles required to hold the melt are not needed, which reduces costs further. In fact, some researchers think the advantages of higher cell efficiency and lower consumable costs could make the float-zone process competitive with the Czochralski process [24, 40].

Polycrystalline ingot technologies

During the past 15 years, techniques for preparing large polycrystalline ingots specifically for photovoltaics have been developed [33]. The techniques generally involve the solidification of molten silicon at carefully controlled rates by cooling the silicon in a crucible (see figure 8a). A variety of materials has been used for the crucible or as crucible linings to reduce contamination of the melt and to counteract the detrimental effects of differential contraction during cooling. Several companies have developed processes for such ingot formation for their own use or for producing polycrystalline wafers sold to cell manufacturers. Diffusion of this technology has usually been by means of joint agreements between interested parties. Equipment that has recently been made available through normal commercial channels for this purpose [41] is likely to further accelerate the technology's adoption.

Large electronic differences can exist among polycrystalline ingots that are prepared by the various methods. Such differences reflect widely disparate levels of carbon and oxygen contamination, as well as the different activities of grain boundaries. Other issues needing resolution include throughput of the ingot preparation equipment. An interesting recent development is a casting

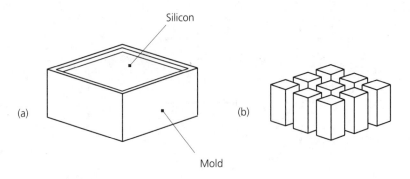

FIGURE 8: *(a) Polycrystalline silicon ingots are formed by controlled solidification in a crucible, or mold; the ingots are then cut into smaller sections (b) before being cut into wafers.*

process that provides magnetic support for the growing crystal, thus eliminating the need for crucibles and permitting continuous casting of ingots, at least in principle [42].

The polycrystalline ingots can be quite massive, as large as $40 \times 40 \times 40$ centimeters in volume and more than 100 kilograms in weight. These massive ingots are generally cut into more manageable blocks (see figure 8b), which are then sliced using the inner-diameter or wire-cutting method. The number of cell manufacturers using cast polycrystalline ingots in production has steadily increased in recent years, suggesting that this method, combined with wire-cutting, may become the main preparation technique for silicon wafers in the coming decade [38].

Self-supporting sheets

One drawback to silicon ingots is having to slice them into wafers. Not only are large amounts of silicon wasted in this way, but the slicing operation itself introduces additional costs. Such costs can be circumvented, however, by producing silicon directly in the form of sheets or ribbons of good crystallographic quality [43].

One of the earliest methods for producing silicon sheets is known as the dendritic web approach (see figure 9a). By controlling temperature of the melt, two dendrites, which are separated from each other by several centimeters, will form. As they are drawn from the melt, a thin sheet of molten silicon is trapped between them and subsequently solidifies into good crystallographic material. Unfortunately, the high quality of the silicon sheet is offset by the method's exacting temperature requirements and low throughput.

More rugged is the edge-defined film-fed growth (EFG) method (see figure 9b). Molten silicon moves by capillary action between two faces of a graphite die, and a sheet is drawn from the molten layer at the top of the die. In its most recent embodiment, the technique is used to form hollow octagonal tubes, each

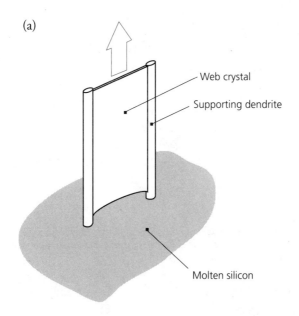

(a)

Web crystal

Supporting dendrite

Molten silicon

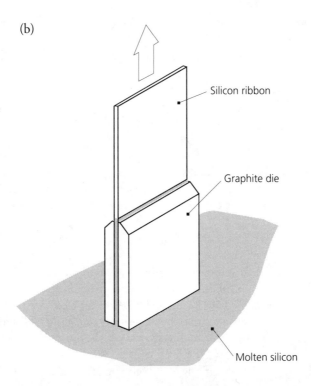

(b)

Silicon ribbon

Graphite die

Molten silicon

FIGURE 9: Growth of silicon ribbon is shown by the dendritic web process (a), and by the edge-defined film-fed growth (EFG) method (b).

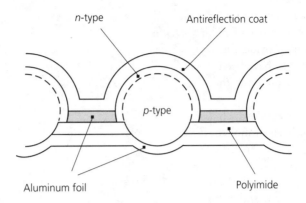

FIGURE 10: "Spheral" solar cells are created by a complex fabrication method that entails embedding 0.75-millimeter spheres of silicon in a perforated aluminium sheet, followed by additional processing to produce the structure shown.

face of which is 11 centimeters wide. This increases throughput and eliminates edge effects during growth.

Although the dendritic web and EFG methods are the two primary sheet-growth processes at present, various other approaches have been investigated [43]. These include the "S-web" approach, whereby a carbon web is coated with silicon when it is drawn through a silicon melt; the RAFT process, whereby silicon is solidified on reusable graphite substrates; and the powder method, whereby a layer of powdered silicon is recrystallized. A very different approach involves the "spin casting" of silicon wafers [43]. The wafers are created by spinning silicon between suitably coated graphite molds, which causes each wafer to individually crystallize. Producing silicon in sheets is potentially cheaper at higher volume than producing it in ingot form; nonetheless, the method will not be commercially viable unless high throughput and good-quality silicon can be guaranteed. If this challenge can be met, the sheet offers greater potential for lower cost and higher volume than the previous ingot approaches.

A relatively new approach to low-cost fabrication calls for embedding a matrix of small, spherical solar cells into perforated aluminum foil [44]. The spheres, which are 0.75 millimeters in diameter, are bonded to perforated aluminum foil that is 0.06 millimeters thick. The method offers two advantages: one is that relatively crude polysilicon can be used as the feedstock (see the section "Overview"), and the second is that the final sheet has considerable flexibility. Its disadvantages are that the steps required to process the tiny cells are fairly complicated (see figure 10) and this, combined with the relatively sparse packing of the spherical shapes, results in relatively low efficiencies. A pilot demonstration of the cost, repeatability, and throughput of the process is planned in the near future [44].

Supported sheets

The previous techniques generally produce silicon as self-supporting wafers or sheets. A number of methods for producing silicon on supporting sheets, which would provide the required structural rigidity and allow for thinner layers of silicon, have been investigated. The high melting point and reactivity of molten silicon limit the number of suitable candidates for the structural support layers.

One technique, which was partially developed in the 1970s, calls for preparing thin sheets of molten silicon on ceramic substrates [24,43]; another technique, also devised in the 1970s, involves chemically depositing silicon on a low-quality metallurgical-grade silicon substrate [24]. Neither technique, however, is being actively pursued.

A more recent and particularly promising development has been the deposition of silicon reportedly from solution onto a ceramic substrate [45]. By depositing silicon from solution, processing temperatures could be reduced, which increases flexibility in substrate choice. Energy conversion efficiencies greater than 15 percent for small-area cells and greater than 8 percent for larger cells have been demonstrated with this approach. Although the films producing the best performance are reasonably thick (more than 100 microns), thinner, high-performance cells are now under development.

ULTIMATE EFFICIENCY AND FUTURE PROSPECTS

Substantial improvements in the efficiency of commercial silicon cells are now being realized [29–31]. If silicon technology is to continue to dominate the market, its success will be closely linked to the level of efficiency that can be sustained in commercial production at low cost. Identifying the factors that limit the ultimate efficiency of a silicon cell and the directions that cell design must take to approach these limits are important to the commercial future of silicon cells.

Recent analyses show that cell thickness is an important parameter in determining efficiency [46, 47]. Decreasing cell thickness offers the potential for higher output voltage, but at the expense of decreasing the amount of light absorbed in the cell and thus the output current. It therefore follows that there is an optimum thickness for peak performance. Light trapping affects the thickness of silicon material required for high performance (see figure 11). Without light trapping, peak performance can be attained with cells that are about the thickness of present commercial cells (0.3 to 0.4 millimeters) or thicker. Cell performance falls off rapidly with decreasing thickness, which suggests that the cells must be at least 20 microns thick if reasonable efficiency is to be achieved. However, with light trapping, the optimum cell thickness decreases to about 100 microns. Significant, too, is the fact that efficiency does not decrease markedly with decreasing cell thickness in this case, even if the thickness

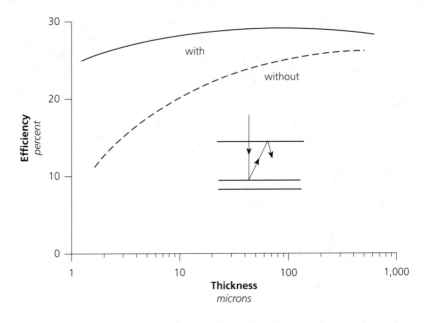

FIGURE 11: Efficiency limits are calculated for silicon cells under two different conditions. In one case (the dashed line), it is assumed that the light has only one pass across the cell; any light not absorbed in the cell material is assumed to be absorbed in the cell's rear contact. In the other case, light trapping effects are included in the analysis. (Here it is assumed that the cells respond well to light from any direction.)

is less than 10 microns. Thus, depositing thin-film silicon on a supporting layer would indeed seem to be a viable strategy.

Calculations (see figure 11) indicate that peak silicon-cell efficiency lies in the 29 to 30 percent range; to date (see the section "Recent improvements"), experimental cells with demonstrated efficiencies in the 23 to 24 percent range have been developed. It is likely that most of the difference between these figures will be bridged during the coming decade. During the same period, thin silicon cells may also achieve similar performance levels.

SUMMARY

Crystalline- and polycrystalline-silicon cells have served as the photovoltaic workhorse for outdoor power applications for the past two decades and are likely to continue in this role throughout the coming decade. Current consensus suggests that the technology may shift during the decade to thin polycrystalline wafers sliced from polycrystalline ingots. These wafers may be combined with improved cell-processing methods, such as the laser-grooved cell technique, which allows the inherent efficiency potential of such substrates to be realized.

Production of crystalline- and polycrystalline-silicon cells could well increase from levels of about 30 megawatts per year globally in 1990 to at least 10 times that amount by the end of the decade. However, because the cell approach is ultimately limited by its material intensiveness, it may well be superseded by a thin-film approach in the next century.

Particularly exciting is that silicon-cell technology may spawn a thin-film silicon alternative that maintains the high efficiency and reliability characteristics of the present cell technology. This would reduce photovoltaic costs to levels where they would be more than competitive with even large-scale conventional electricity generation sources and would provide opportunities for low-pollution electricity generation on a large scale in the next century.

WORKS CITED

1. Becquerel, A.E. 1839
   Recherches sur les effets de la radiation chimique de la lumière solaire au moyen des courants electriques, *Comptes rendus de L'Academie des Sciences* 9:145–149 (also *Annalen der Physick und Chemie* 54:18–34, 1841); Becquerel, A.E., Memoire sur les effects electriques produits sous l'influence des rayons solaires, *Comptes rendus de L'Academie des Sciences* 9:561–567, (also *Annalen der Physick und Chemie* 54:35–42, 1841).

2. Green, M.A. 1990
   Photovoltaics: coming of age, *Conference Record, 2lst IEEE Photovoltaic Specialists Conference,* Orlando, May, 1–8 (and references therein).

3. Ohl, R.S. 1941
   Light-sensitive electric device, U.S. Patent 2,402,622; Light-sensitive device including silicon, US Patent 2,443,542; both filed 27 May.

4. Kingsbury, E.F. and Ohl, R.S. 1952
   Photoelectric properties of ionically bombarded silicon, *Bell Systems Technical Journal* 31:802–815.

5. Chapin, D.M., Fuller, C.S., and Pearson, G.L. 1954
   A new silicon P–N junction photocell for converting solar radiation into electrical power, *Journal of Applied Physics* 25:676–677.

6. Bell Laboratories Record. 1954
   232–234; 436–437; 166.

7. Bell Laboratories Record. 1955
   241–246; 434.

8. Prince, M.B. 1990
   Early days of modern photovoltaics, *Optoelectronics* 5:326–328.

9. Lindmayer, J. and Allison, J. 1973
   The violet cell: an improved silicon solar cell, *COMSAT Technical Review* 3:1–22.

10. Haynos, J., Allison, J., Arndt, R., and Meulenberg, A. 1974
    The Comsat nonreflective silicon solar cell: a second generation improved cell, *International Conference on Photovoltaic Power Generation,* Hamburg, September, 487.

11. Iles, P.A. 1970
    Increased output from silicon solar cells, *Conference Record, 8th IEEE Photovoltaic Specialists Conference,* Seattle, 345–352.

12. Gereth, R., Fischer, H., Link, E., Mattes, S., and Pschunder, W. 1970
    Silicon solar technology of the seventies, *Conference Record, 8th IEEE Photovoltaic Specialists Conference,* Seattle, 353–359.

13. Mandelkorn, J. and Lamneck, J.H. 1973
    A new electric field effect in silicon solar cells, *Journal of Applied Physics* 44:4785–4787.

14. Baraona, C.R. and Brandhorst, H.W. 1975
    V-groove silicon solar cells, *Conference Record, 11th IEEE Photovoltaic Specialists Conference,* Scottsdale, May, 44–48.

15. Fossum, J.G. and Burgess, E.L. 1978
    High-efficiency p^+ and n^+ back-surface field silicon solar cells, *Applied Physics Letters* 33:238–240.

16. Godfrey, R.B. and Green, M.A. 1979
    655mV open circuit voltage, 17.6% efficient silicon MIS solar cells, *Applied Physics Letters* 34:790–793.

17. Green, M.A., Blakers, A.W., Shi Jiqun, Keller, E.M., and Wenham, S.R. 1984
    High efficiency silicon solar cells, *IEEE Transactions on Electron Devices* ED-31:671–678.

18. Green, M.A., Blakers, A.W., Shi Jiqun, Keller, E.M., and Wenham, S.R. 1984
    19.1% efficient silicon solar cell, *Applied Physics Letters* 44:1163–1165.

19. Blakers, A.W. and Green, M.A. 1986
    20% efficient silicon solar cell, *Applied Physics Letters* 48:215–217.

20. Green, M.A., Blakers, A.W., and Osterwald, C.R. 1985
    Characterization of high efficiency silicon solar cells, *Journal of Applied Physics* 58:4402–4408.

21. Sinton, R.A., Kwark, Y., Gan, J.Y., and Swanson, R.M. 1986
    27.5% efficient Si concentrator solar cells, *Electron Device Letters* EDL-7.

22. King, R.R., Sinton, R.A., and Swanson, R.M. 1988
    Front and back surface fields for point-contact solar cell, *Conference Record, 20th IEEE Photovoltaic Specialists Conference,* Las Vegas, September, 538–544.

23. Green, M.A. 1991
    Recent advances in silicon solar cell performance, *Proceedings of the 10th European Communities Photovoltaic Solar Energy Conference,* Lisbon, April, Kluwer Academic Publishers, Dordrecht, 250–253 (and references therein).

24. FPSAP. 1986
    Final Report, Flat Plate Solar Array Project, Volumes I–VIII, Jet Propulsion Laboratory Publication 86-31, 5101-5289, October.

25. Green, M.A. 1982
    Solar cells: operating principles, technology and system applications, Prentice-Hall, New Jersey.

26. Taylor, W.E. 1983
    Solar cell metallization historical perspective, *Proceedings of the Flat-Plate Solar Array Project Research Forum on Photovoltaic Metallization Systems,* report DOE/JPL-l 012-92, November, 9.

27. Szlufcik, J., Elgamel, H.E., Ghannam, M., Nijs, J., and Mertens, R. 1991
    Simple integral screenprinting process for selective emitter polycrystalline silicon solar cells, *Applied Physics Letters* 59:1583–1584.

28. Chong, C.M., Wenham, S.R., and Green, M.A. 1988
High efficiency laser grooved, buried contact solar cell, *Applied Physics Letters*, 52:107–109.

29. Boller, H.W. and Ebner, W. 1989
Transfer of the BCSC-concept into an industrial production line, *Proceedings of the 9th E.C. Photovoltaic Solar Energy Conference*, Freiburg, September, Kluwer Academic Publishers, Dordrecht, 411–413.

30. Mason, M.B., Lord, B.E., Jordan, D., and Summers, J.G. 1991
High-efficiency, low-cost crystalline silicon production solar cells, *Proceedings of the 10th E.C. Photovoltaic Solar Energy Conference*, Lisbon, April, Kluwer Academic Publishers, Dordrecht, 280–283.

31. Green, M.A., Wenham, S.R., Zhao, J., Bowden, S., Milne, A.M., Taouk, M., and Zhang, F. 1991
Present status of buried contact solar cells, *Proceedings of the 22nd IEEE Photovoltaics Specialists Conference*, Las Vegas, October.

32. Kyle, C.R. 1991
Racing with the sun: the 1990 World Solar Challenge, Society of Automotive Engineers, Inc., Warrendale, Pennsylvania.

33. Watanabe, H.,Hirawasa, K., Masuri, K., Okada, K., Takayama, M.,Fukui, K., and Yamashita, H. 1990
Technical progress of large area multicrystalline silicon solar cells and its applications, *Optoelectronics* 5:223–238.

34. Green, M.A. 1987
High efficiency silicon solar cells, Trans Tech Publications, Aedermannsdorf.

35. Saitoh, T. 1991
Recent progress of crystalline silicon solar cells in Japan, Joint U.S.–Japan Crystalline Silicon Solar Cell Meeting, Las Vegas, 7 October (unpublished).

36. Green, M.A., Wenham, S.R., Zhao, J., Wang, A., Yun, F., and Campbell, P. 1992
High efficiency silicon solar cells, *Proceedings of the 6th Photovoltaic Science and Engineering Conference*, New Delhi, February.

37. Anderson, J.R. 1980
Wire saw for low damage, low kerf loss wafering, *Conference Record, 14th IEEE Photovoltaic Specialists Conference*, San Diego, January, 309–311.

38. Claverie, A., Chabot, B., and Anglade, A. 1990
Photovoltaics in France, *Technical Digest, 5th International Photovoltaic Science and Engineering Conference*, Kyoto, November, 133–137.

39. McGuire, G.E., ed. 1988
Semiconductor materials and process technology handbook, Noyes, Park Ridge, Illinois.

40. Ludsteck, A. and. Fenzl, H.J. 1982
Solar grade floating-zone silicon, *Proceedings of 4th E.C. Photovoltaic Specialists Solar Energy Conference*, Stresa, May, 946–954.

41. Hukin, D.A. 1990
Polycrystalline photovoltaic silicon ingot production, *Technical Digest, 5th International Photovoltaic Science and Engineering Conference*, Kyoto, November, 303–306.

42. Kaneko, K., Misawa, T., and Tabata, K. 1990
100 cm^2 silicon square ingot made by electromagnetic casting, *Technical Digest, 5th International Photovoltaic Science and Engineering Conference*, Kyoto, November, 201–204.

43. Eyer, A., Rauber A., and Goetzberger, A. 1990
Silicon sheet materials for solar cells, *Optoelectronics* 5:239–258.

44. Levine, J.D., Hotchkiss, G.B., and Hammerbacher, M.D. 1991
Basic properties of the spheral solar cell, *Conference Record, 22nd IEEE Photovoltaic Specialists Conference,* Las Vegas, October.

45. Kendall, C.L., Checchi, J.C., Rock, M.L., Hall, R.B., and Barnett, A.M. 1990
10% efficient commercial-scale silicon-film solar cells, *Conference Record, 21st IEEE Photovoltaic Specialists Conference,* Orlando, May, 604–607 (and references therein).

46. Green, M.A. 1984
Limits on the open-circuit voltage and efficiency of silicon solar cells imposed by intrinsic auger processes, *IEEE Transactions on Electron Devices,* ED-31:671–678.

47. Tiedje, T., Yablonovich, E., Cody, D.G., and Brooks, B.G. 1984
Limiting efficiency of silicon solar cells, *IEEE Transactions on Electron Devices,* ED-31:711–716.

8
PHOTOVOLTAIC CONCENTRATOR TECHNOLOGY

ELDON C. BOES
ANTONIO LUQUE

Photovoltaic (PV) concentrator systems use lenses or mirrors to concentrate direct solar radiation onto smaller areas of solar cells. The advantage of this approach is that it circumvents the need for developing large areas of low-cost solar cells by replacing them with low-cost concentrators. Thus, a major emphasis in the development of PV concentrators is the design and development of high-efficiency "concentrator" solar cells. This explains why PV concentrator systems are generally more efficient than flat-plate PV power systems.

Several major PV concentrator power systems have been installed and are operational. A wide range of designs has been developed, including numerous designs using plastic Fresnel lenses. The amount of solar concentration employed has varied from as low a factor of two to as high a factor of 1,000. While most concentrators currently use silicon semiconductor cells, several groups are also developing compound semiconductor and multijunction concentrator cells, which have the potential for even higher efficiencies. Most PV concentrator designs require the use of sun-tracking collectors, and both tracking structures and controllers are under development.

Thus far, prices for PV concentrator systems have been comparable to prices for flat-plate PV systems. However, cost analyses indicate that costs for electricity from PV concentrators could drop below U.S.$0.15 per kilowatt-hour (kWh) in the immediate future with adequate markets, and that costs of approximately $0.05 per kWh are possible in 10 to 15 years. Because PV concentrators are more complex, applications for this technology are generally restricted to power levels of a few kilowatts or more. PV concentrators are also less effective in cloudier regions because they depend on direct solar radiation. Even with these limitations, however, substantial markets for PV concentrators for a variety of applications, such as centralized village power systems or distributed utility supplies, are expected to develop in the next few years. Some of these markets will result from the growing overall demand for PV power systems coupled with the difficulty that the flat-plate PV industry will encounter in trying to meet that demand. Over the long term, PV concentrators are likely to be a preferred choice for PV power technology for larger applications in sunny regions because of their high conversion efficiencies.

INTRODUCTION[1]

Photovoltaic (PV) technology has penetrated today's electricity market to only a limited degree, largely because PV power systems have such relatively high initial costs. One way to reduce costs is to replace expensive solar cells with lower-cost optical concentrators. Although more complex, the resulting collectors are expected to generate economical electricity through a combination of reduced solar-cell area and higher efficiency.

PHOTOVOLTAIC CONCENTRATORS: BASIC PRINCIPLES

Economics

As the name implies, photovoltaic concentrators use a lens or mirror to collect sunlight and then recast the concentrated light onto a solar cell of smaller size (see figures 1 and 2). Unlike flat-plate PV modules, which are equipped with large arrays of solar cells, PV concentrators first collect solar radiation with lower-cost lenses or mirrors and thus need substantially smaller areas of solar cells to convert the sun's radiation into electricity.

Because PV concentrator modules have a reduced solar-cell area, their costs can be significantly reduced, as can be demonstrated by plotting total module cost as a function of concentrator-cell cost and concentration ratio (see figure 3). If concentrator-cell costs range from $0.10 per square centimeter (cm^2) to $5.00 per cm^2, and assuming the cost of the lens or reflecting concentrator is $30 per square meter ($m^2$) (which is at least double the price of its material content) and the rest of the concentrator module is an additional $70 per m^2, then costs of less than $200 per m^2 for the entire module are possible. By comparison, the costs of today's flat-plate crystalline-silicon (*x*-Si) cells range from $0.02 to $0.04 per cm^2. The entire cost of a flat-plate *x*-Si module, including a profit margin, generally exceeds $500 per m^2. On a power basis, that translates into $4 to $7 per peak watt ($W_p$—the power output at a cell temperature of 25°C and at incident radiation of 1,000 watts per m^2).

Reducing module costs to less than $200 per m^2 appears to be necessary if photovoltaic technology is to achieve broad market penetration (see figure 4). A levelized energy cost (LEC) of $0.10 per kilowatt-hour (kWh) with flat-plate modules, assuming a peak efficiency of 15 percent, will be possible only if module costs drop to $100 to $225 per m^2, or $0.70 to $1.50 per W_p, depending on the assumed discount rate. The same levelized energy cost for a concentrator module, assuming a peak efficiency of 20 percent, calls for modules ranging from $100 to

---

1. This chapter cannot cover photovoltaic concentrator technology in its entirety. The authors therefore wish to call the reader's attention to the proceedings that result from three international conferences held approximately every 18 months. The conferences are the IEEE Photovoltaic Specialists Conference (PVSC), the European Community Photovoltaic Solar Energy Conference, and the Photovoltaic Science and Engineering Conference (PVSEC). The proceedings of these conferences provide an excellent source of information for all aspects of PV technology.

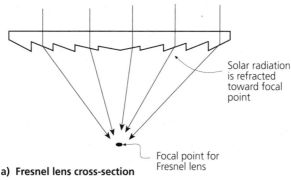

Solar radiation
is refracted
toward focal
point

Focal point for
Fresnel lens

a) Fresnel lens cross-section

Concentrator cells
mounted along focal
line

b) Linear Fresnel PV concentrator

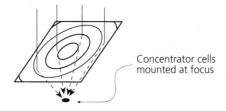

Concentrator cells
mounted at focus

c) Point-focus Fresnel PV concentrator

FIGURE 1: Fresnel concentrator design concept: a) cross section showing optical principle; b) linear Fresnel lens; and c) point-focus Fresnel lens.

$250, or $0.50 to $1.30 per W_p, again depending on the discount rate. Both represent challenging goals, which may be relaxed through higher efficiency and lower balance-of-system costs. Nevertheless, modules significantly cheaper than today's flat-plate *x*-Si modules must be developed in order to achieve an LEC approaching $0.10 per kWh. (See the section "Cost projections" for a discussion of lowering module costs with concentrator technology.)

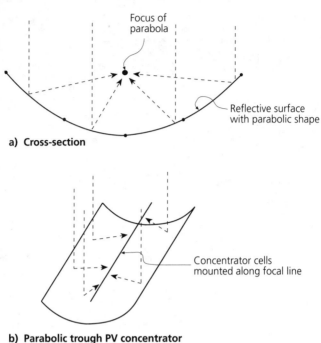

a) Cross-section

Focus of parabola

Reflective surface with parabolic shape

b) Parabolic trough PV concentrator

Concentrator cells mounted along focal line

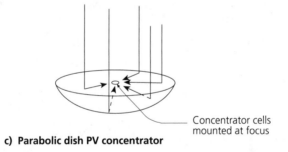

c) Parabolic dish PV concentrator

Concentrator cells mounted at focus

FIGURE 2: *Reflective parabolic-concentrator design concept: a) cross section showing optical principles; b) linear parabolic trough; and c) point-focus parabolic dish.*

In addition to offering cost advantages, PV concentrators provide another unique advantage: they are likely to be the only technology able to use the most efficient solar cells. There are two reasons for this. First, at the higher illumination levels that are typical of concentration systems, most solar cells exhibit higher conversion efficiencies (see "The development of concentrator cells"). And second, the most efficient cells may be too costly for flat-plate modules, yet be cost-effective in a high-efficiency concentrator system.

Indeed, increasing cell efficiency for PV concentrators has become the focus of major research and development efforts. Levelized energy costs are highly dependent on peak module efficiency (see figure 5). Because the cost con-

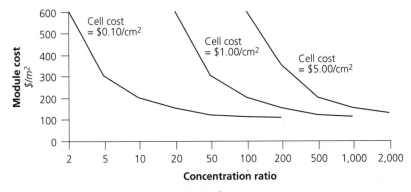

FIGURE 3: *Concentrator module costs in $ per m² are displayed as a function of concentration ratio for three different concentrator cell costs in $ per cm².*

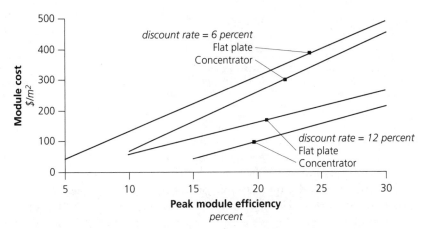

FIGURE 4: *Module cost versus peak module efficiency required to achieve a levelized energy cost of $0.10 per kWh. Assumes direct normal radiation of 2,600 kWh per m² per year, discount rates of 6 percent and 12 percent, and is based on the economic assumptions in chapter 6: Introduction to Photovoltaic Technology.*

BOX A

ILLUMINATION INTENSITY

The ratio of the concentrator area divided by the cell area is called the "geometric concentration ratio" and is frequently denoted by X. Thus, a PV concentrator that uses 100 cm² lenses and 1 cm² cells is called a 100X concentrator. The intensity of the illumination on a solar cell is measured in units called "suns," defined as the solar radiation incident on the cell divided by the solar radiation that would be incident on the cell under "one standard sun," that is, under bright sunlight with a total intensity of 100 milliwatts per cm² and a standard spectrum. Illumination intensity for a cell represents the "number of bright suns" incident on the cell. The illumination intensity and the geometric concentration ratio differ because of optical losses in a concentrator as well as the fact that incoming solar radiation is frequently of lower intensity than one standard sun.

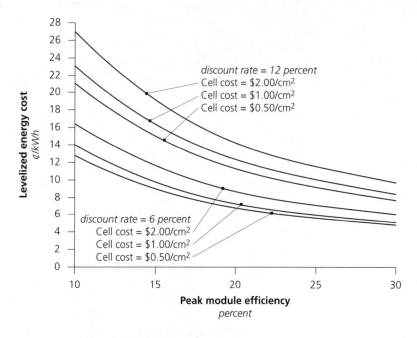

FIGURE 5: Dependence of levelized energy cost on concentrator module peak efficiency for three concentrator cell costs in $ per cm². Assumes a 200X concentration ratio, balance of module costs of $100 per m², and discount rates of 6 percent and 12 percent.

tribution of concentrator cells to total module cost can be controlled by the concentration ratio, relatively greater emphasis can be directed to increasing concentrator-cell efficiencies than to reducing their costs. Moreover, because module efficiency is determined more or less by optical efficiency multiplied by concentrator-cell efficiency (and optical efficiencies are generally about 80 percent), cell efficiency emerges as the key determinant of module efficiency. A module that contains 25 percent efficient concentrator cells, for example, will be approximately 20 percent efficient.[2]

Limits to concentration

In order to cast the incoming light collected by the concentrator surface area onto a smaller cell area, the angular spread of the light rays must be increased (see figures 1 and 2). Nearly parallel solar rays are redirected to assume a cone-like configuration incident on the concentrator cells. Mathematically, maximum concentration is achieved when the concentrator provides isotropic illumination (that is, light of equal intensity in all directions) on the cell, which is represented as $C = n^2/\sin^2\phi$, where ϕ is the half-angle of the cone of light being concentrated, and n is the refractive index of the transmitting me-

---

2. The average annual efficiency of the system will actually range from 15 to 17 percent as a result of the cell's operating temperature, dirt accumulation and other array field losses, and power conditioner losses.

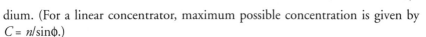

dium. (For a linear concentrator, maximum possible concentration is given by $C = n/\sin\phi$.)

In theory, the maximum concentration of solar radiation on the earth is about 47,000 times (assuming ϕ is about 0.27° and the refractive index of air is 1). However, the concentration ratios that are achievable in practice are much lower. The reason is that true isotropic illumination on the cell is impossible to achieve; in addition, optical losses result from such factors as material and surface shape imperfections in the concentrator. To offset these imperfections and to accommodate manufacturing and tracking errors, a concentrator must be designed to collect rays from a region larger than the sun.

Generally speaking, practical concentrators are designed to collect all rays within a half angle of about $\phi = 0.5°$ or 1°, which significantly decreases the concentration ratio. But the concentration ratio also depends on the type of concentrator. Linear concentrators, for example, are unidimensional (they only concentrate the incoming solar rays in one dimension onto a linear receiver), and thus have much lower maximum concentration ratios than do point-focus concentrators. In contrast, point-focus concentrators concentrate all the incoming rays from a two-dimensional area onto a "point" receiver, which is usually a single concentrator cell. (See figures 1 and 2 for the distinction between linear and point-focus concentrators.) Fresnel lenses are segmented (see figure 1a), a configuration that is preferred for PV concentrators because the more common convex lenses would be very thick and therefore too heavy and too expensive. The practical concentration ratio of Fresnel lenses is significantly reduced by chromatic aberration, which occurs because the refractive index of the lens material varies with wavelength—different wavelengths are bent at different angles by the lens.

By installing a secondary concentrator optical element just above the concentrator cell, however, the concentration ratio can be greatly increased because the secondary concentrator—if properly designed—makes it possible to focus isotropic illumination on the concentrator cell. Reflective or refractive secondaries are also used in some concentrator systems to increase tracking and alignment tolerances and to catch stray solar radiation that might otherwise narrowly miss its concentrator cell target [1]. The maximum theoretical and practical concentration ratios are listed for the principal types of solar concentrators in table 1.

Limits on concentration ratios are also imposed by the cells themselves. In an ideal solar cell, efficiency increases continuously as the concentration ratio increases. In a real solar cell, however, with series resistance R_s, maximum efficiency is achieved when $I_{sc}R_s = V_{th}$, where I_{sc} is the short-circuit current and V_{th} is the thermal voltage with a value of approximately 26 millivolts at 300 K [2]. To illustrate how this limits concentration, consider a silicon cell that has an area of 1 cm^2, a base resistivity of 0.1 ohms (Ω) per cm, and a thickness of 200 microns. Because its base series resistance is 2 milliohms, maximum efficiency occurs at 13 amps. Since current is directly proportional to incident radiation, which is 40

Table 1: Maximum theoretical and practical concentration ratios

Concentrator type	Maximum theoretical concentration $\phi = 0.27°$	Maximum practical concentration $\phi = 1.0°$
Circular reflective parabolic dish	12,000	820
Parabolic dish with secondary ($n = 1.49$ for secondary)	104,000	4,800
Square flat Fresnel lens, square cell	376	73
Square flat Fresnel lens with secondary	–	1,700
Linear flat lens	22	10
Linear arched lens	54	26
Linear parabolic reflector	108	29

milliamps for a good cell at one sun, concentration for maximum efficiency is limited to 325. (Both grid resistance and emitter resistance, which also contribute to series resistance, are ignored in this example.) Although series resistance is not the only factor that limits concentration, it is the most important one in most instances. In general, it is difficult to operate silicon (Si) concentrator cells above 300 suns with reasonable efficiency; a comparable practical concentration limit for gallium arsenide (GaAs) cells is about 1,000 suns.[3]

Another important topic concerns the concentration limits of cells fabricated for use in flat-plate modules. In general, today's one-sun cells are not efficient when operated above about 5X, because the cell processing methods result in relatively high series resistances. Significant contact resistance occurs between the grids and the semiconductor material itself, and emitter resistance is high because of widely spaced gridlines.

The desire to minimize the contribution of contact and emitter resistance to the cell's total series resistance leads to an important design trade-off. Although the resistance components can be decreased by making the contacts wider and more closely spaced, doing so blocks some incoming radiation, thus reducing cell current. The trade-off affects concentrator cells more than flat-plate cells because the higher currents in concentrator cells lead to larger losses from higher series resistance.

---

3. Considering optical losses and the fact that typical solar radiation intensities ate somewhat lower than 1,000 watts per m², these correspond to geometric concentration ratios of about 400X for Si cells and 1,300X for GaAs cells.

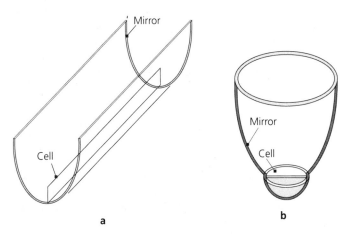

FIGURE 6: Static concentration concepts are shown: a) linear component parabolic concentrator (CPC) with a bifacial cell; and b) point focus CPC with a hemisphere for coupling the CPC to a bifacial cell.

Concentrating diffuse solar radiation

Diffuse sunlight is considered almost impossible to concentrate because the angular spread of incoming diffuse radiation, which spans the hemisphere of the sky from horizon to horizon, has an acceptance half-angle of $\phi = 90°$ that cannot be increased. Some tricks that allow for moderate concentration of diffuse light, however, exist. First, if the concentrator cell receiving the light is embedded in an optically dense medium of index n, the concentration level can be increased to $C = n^2/\sin^2 90° = n^2$. This is because the cone of rays entering the dense medium experiences a reduction of angular spread so that it once again increases where it impinges on the concentrator cell. Note that many transparent materials, such as glass and plastics, have indexes close to 1.5, rendering concentrations of 2.25 possible.

A second strategy calls for the use of bifacial cells [3], which are designed to accept radiation on both sides (see figure 6a). Because they accept light from any direction, the concentration limit in a medium of index n becomes $2n^2$, or 4.5 for many candidate materials.

Linear concentrators of the compound parabolic-concentrator (CPC) type (see figure 6a), are frequently used to collect both direct and, to a significant degree, diffuse radiation. The theoretical limit for full concentration of the diffuse light is $2n$. The point-focus CPC concentrator (see figure 6b) can approach a concentration of $2n^2$. Other concentrator designs, based on a variety of synthesis methods, are possible [4].

Linear CPCs are also attractive because their large acceptance angles enable them to collect the sun's rays without tracking. When designed to accept all direct radiation, they concentrate light by no more than a factor of six. The value can be increased by sacrificing some direct radiation in the summer, for example by not requiring that direct radiation occurring near sunrise and sunset be

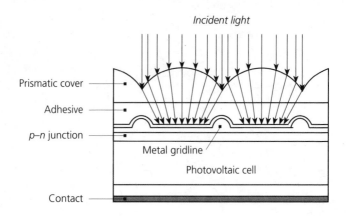

FIGURE 7: The prismatic cell cover uses a shaped transparent cover to direct incoming light away from concentrator cell gridlines.

collected. Certainly, such concentrators do not collect all diffuse radiation. Since diffuse radiation is not isotropic (it is not of equal intensity from all directions) but is generally more intense near the sun's position, these concentrators collect a significant amount of diffuse radiation. The spatial distribution of diffuse solar radiation under different sky conditions and with respect to the sun's position have been analyzed and modeled [5].

Finally, flat-plate collectors with bifacial cells offer another practical approach to static concentration. By painting the background behind and under the collectors white, energy output can be increased in some cases by 20 to 30 percent [6].

Novel cell–light coupling concepts

Several novel cell–light coupling schemes have emerged in the past five years. Prior to that time, the shadow produced by the cells' front grid was generally considered unavoidable, and so concentrator designers attempted to minimize the combination of shadow losses with resistance losses. Since then, several designs that greatly reduce or eliminate shadow losses have been proposed.

One design calls for a high-efficiency cell that lacks the front grid altogether, thus eliminating the shadowing problem [7]. It is not known how broadly applicable this approach will be because achievable costs for such cells are not yet known. However, it is likely that the cells will be expensive enough that concentrations of at least 100X will be required.

A second design is the ENTECH prismatic cover (see figure 7). This simple device deflects incoming solar rays from the metal fingers of the front grid, focusing them instead on the active cell area. The concept has been incorporated into several designs, including the 300 kilowatt ENTECH–3M system described below, and was instrumental in achieving concentrator cell efficiencies of 29 percent for GaAs cells and 20 percent for an experimental Si concentrator module (see the

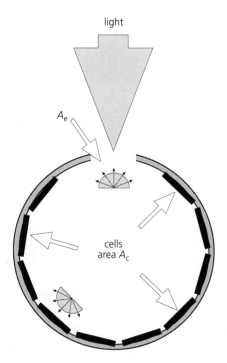

FIGURE 8: *Spherical cavity internally covered with solar cells. A_e is the area of the entrance aperture, A_c the cumulative area of the cells.*

section "Sandia's experimental and baseline modules"). The prismatic cover not only obscures the metal grid, but it allows for a denser grid coverage and therefore for a reduction of the cell's series resistance. Reducing resistance is especially important if one-sun cells are to be used successfully at 10X to 20X concentration. A related design reflects incident solar rays onto active cell regions using tilted metal-grid fingers, rather than a prismatic cover [9, 10].

Such grid obscuration schemes as the prismatic cover or tilted gridlines must be viewed as miniconcentrators, accepting rays over the whole cell area and redirecting them to the active cell regions. Doing so requires that the incoming rays conform to certain angular limitations, which limits the operation of the obscuration designs.

Another novel design, still in its infancy, involves light-confining cavities. Several cells are set into an integrating sphere or other cavity [11] (see figure 8), and incoming radiation is introduced through a small aperture, which prevents the light from escaping. Thus trapped, the rays have a high probability of being reflected onto an active region of another cell. In addition, cavity losses are reduced by making all noncell areas reflective, which allows reflected light to be absorbed elsewhere. Moreover, thin cells with reflective rear surfaces can be used because the photons that are not absorbed will reflect back into the

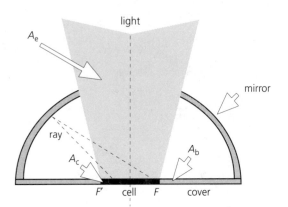

FIGURE 9: Ellipsoidal cavity with a solar cell in the bottom. F and F′ are the focal points of the half-ellipse. A_e is the area of the entrance aperture, A_c is the total cell area, and A_b the area of the back.

cavity and be absorbed by other cells. This reduces the base resistance of the cells and permits operation at higher concentrations of sunlight.

The main drawback to an integrating sphere or related device is the need for a small entry aperture, which suggests that sunlight is much more concentrated at the entrance to the cavity than it is on the inside. To relax this size limitation on the entry aperture, cavities based on the angular limitation of escaping light are being developed [12]. An ellipsoidal cavity, for example, from which only the rays joining the cell and the entry aperture can escape, reduces grid shadow by approximately the ratio of the entrance aperture area divided by the bottom cover area (see figure 9). Because an ellipsoidal cavity does not easily admit a secondary concentrator, the concentration of light on this cell is more limited than in noncavity designs. In the first experimental studies of this concept, an efficiency increase of 7 percent (from 21.0 to 22.5 percent) has been experimentally demonstrated for ellipsoidal cavities [13].

In addition, multiple-bandgap cells, with or without appropriate interference filters on the different cells, can be introduced into cavities to boost their efficiency. A prototype is currently being developed at the Polytechnic University of Madrid.

Applications

PV concentrators are best suited for larger, centralized power systems at sites with good solar resources. The primary advantage they offer over traditional flat-plate arrays is the relatively low cost of the module relative to power output. Such an advantage, however, is negated at small array sizes because the simple support structures for flat plates cost significantly less. In addition, concentrator systems require slightly more sophisticated operations and maintenance procedures for adequate alignment and functioning of their sun-tracking structures, rendering them less competitive on a small scale. Moreover, under cloudy conditions, when the amount of direct-beam solar radiation is zero, the output from a PV concentrator is zero, whereas flat-plate systems continue to produce a small amount of power from diffuse solar radiation.[4]

It is likely that concentrator systems will eventually generate at least 3 to 5 kilowatts in output power, and probably exceed 100 kilowatts. Although these power levels are small relative to global power generation, they are too high to enable concentrators to participate in the diverse and rapidly expanding markets for PV systems that are from 10 watts to a few kilowatts in size. Still, concentrators are suitable for all grid-connected applications and for larger, load-center applications such as centralized village power systems. Indeed, very large markets for PV concentrator systems are likely to emerge over the next five to 10 years as the attractive operating characteristics of PV power systems are more widely recognized.

To understand how solar resources influence the selection of flat-plate versus concentrator systems, the two forms of radiation—direct-beam and total radiation (direct and diffuse)—must be compared. The reason is that a PV power system's output generally matches the amount of solar radiation available. For concentrators, the output is proportional to the direct-beam radiation received; for flat-plates the amount is proportional to the total radiation. Of course, in either case shadowing losses from adjacent arrays or other objects may significantly reduce energy production and must also be considered.

As a general rule (see table 2), solar resources vary by about a factor of 2 or 2.5, depending on whether the location is sunny or cloudy. But the decrease from sunny to cloudy locations is somewhat larger for concentrator systems than for flat-plate collectors. Consider, for example, a comparison of radiation levels in Washington, DC, which is relatively cloudy, and Albuquerque, New Mexico, which has a relatively high number of sunny days. The amount of total radiation that can be exploited by a fixed latitude tilt–flat plate in Washington, DC, is 36 percent lower than it is in Albuquerque, but the amount of direct radiation exploitable by a concentrator is a full 50 percent lower. It is also interesting to note that the direct, normal resources are generally comparable to the total resources at a fixed latitude tilt (see figure 10).

EXPERIENCE AND PROGRESS WITH PV CONCENTRATORS

In this section we summarize the past 10 to 15 years of experience in developing concentrator designs and in installing and operating PV concentrator power systems.

---

4. Strictly speaking, a PV concentrator can also generate power from diffuse light, but its output is reduced by a factor approximately equal to the concentration ratio. Consequently the output under diffuse light is often 1 percent or less than the concentrator's output under bright sunlight, which is usually insufficient to operate the power system's controls.

Table 2: Average annual solar radiation resources available to
three types of collector[a]
kWh/m^2

Location	Total, two-axis tracking	Total, fixed at latitude	Direct, two-axis tracking
Albuquerque, New Mexico	3,450	2,530	2,630
Phoenix, Arizona	3,390	2,510	2,520
Almería, Spain	3,307	2,422	2,582
Zaragoza, Spain	3,293	2,437	2,552
Denver, Colorado	3,100	2,280	2,340
Sacramento, California	2,990	2,190	2,150
San Diego, California	2,720	2,110	1,860
Honolulu, Hawaii	2,580	2,000	1,610
Madrid, Spain	2,549	1,782	1,887
Austin, Texas	2,500	1,910	1,640
Omaha, Nebraska	2,490	1,850	1,680
Nice, France	2,405	1,745	1,790
Brasília, Brazil	2,397	1,877	1,649
Miami, Florida	2,380	1,870	1,420
Messina, Italy	2,354	1,742	1,706
Rome, Italy	2,288	1,677	1,664
Athens, Greece	2,268	1,678	1,622
Nashville, Tennessee	2,100	1,650	1,280
Pisa, Italy	2,099	1,547	1,492
Washington, DC	2,080	1,610	1,310
Boston, Massachusetts	1,920	1,470	1,170
Manaus, Brazil	1,776	1,430	1,128
Pittsburgh, Pennsylvania	1,760	1,390	990
Seattle, Washington	1,740	1,340	1,020
Stuttgart, Germany	1,729	1,276	1,167
Zürich, Switzerland	1,653	1,220	1,089
Hamburg, Germany	1,497	1,083	977

a. Source for U.S. locations, [14, 15].

Early development and design trends

Although PV concentrators were fabricated and tested before 1975, development of the technology did not begin in earnest until then. Thereafter, various concentrator designs were developed in both the United States and Europe and tested through the experimental stage. The various designs included reflective and refracting concentrators, including parabolic troughs and dishes, v-troughs, compound parabolic concentrators, heliostat/fixed receiver concepts, solid lenses, point-focus Fresnel lenses, and linear lens concepts (see figures 1 and 2).

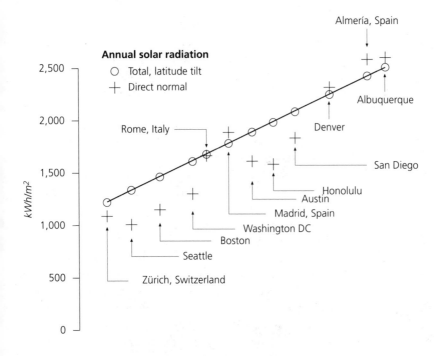

FIGURE 10: Comparison of annual average solar radiation available to a fixed flat-plate and a tracking concentrator.

The success of the designs varied substantially, thus leading to an interesting shakeout of the different technologies. The most important finding was that refractive concentrators equipped with Fresnel lenses generally performed better than reflective concentrators at both the prototype module and operating system levels. The performance difference, coupled with a decline in government funding for concentrator research, led to an almost complete cessation of reflective concentrator development for photovoltaics. As a result, most research activity today is directed at either point-focus or linear Fresnel systems. To our knowledge, only one firm, Sunpower, Inc., is developing a parabolic-dish PV concentrator and a PV central receiver[5] and no one is developing parabolic-trough concentrators. In contrast, reflective parabolic troughs have far outpaced other technologies in solar-thermal systems (see chapter 5: *Solar-thermal Electric Technology*), which also emphasize both reflector-based central receiver systems and parabolic-dish systems. Lens-based concentrators, however, have received little attention from the solar-thermal community.

There are several reasons why Fresnel lenses have been preferred in photovoltaic applications while mirrors are the preferred technology in solar-thermal ap-

---

5. The system uses an array of tracking reflective heliostats to focus direct radiation on a fixed receiver covered with concentrator cells.

plications. Fresnel-lens systems can be designed to provide a high-quality image on the target cells. Precision is important since the cells are very sensitive to non-uniform solar fluxes and elevated temperatures. Fresnel lenses allow more flexibility in optical design. Each facet can be individually shaped to ensure uniform flux on the cells and other desirable optical properties. Moreover, lens systems do not require as high tolerance in manufacturing as mirror-based systems. The angular error in a reflected beam is roughly twice the angular error in the mirror slope while in a Fresnel lens angular errors are partially self-correcting, because in the lens systems the error appears in both the front and back faces of the lens. In addition, with Fresnel systems the cells are behind the optical systems, and the finned aluminum supports or other devices designed to conduct heat away from the photovoltaic cells do not cast shadows on the optics. In a mirror system, the receiver is in front of the optics and may cast shadows.

But mirrors have the advantage of being inexpensive, both because mirrors are less expensive than lenses and because single-axis tracking mirror systems can be mounted horizontally without compromising the quality of the focused image. The focal position of light concentrated by Fresnel lenses changes when the sun is not perpendicular to the axis of rotation. This can create serious problems for photovoltaic systems, which are very sensitive to nonuniform solar fluxes and elevated temperatures. Most single-axis photovoltaic systems are mounted with the north end of the axis raised off the ground at the latitude angle in order to keep the sun angle as close to perpendicular as possible. Solar-thermal receivers can tolerate greater flux variations and typically use horizontal mounts. The horizontal tracking is particularly attractive for thermal systems since it reduces the cost of pipe interconnections.

Although most concentrators today rely on concentrating lenses, the lenses may be replaced in future designs with reflective mirrors, in part for cost reasons. In addition, a central receiver system eliminates the need for extensive field wiring to collect the output of distributed lens-based concentrators. Thus, it is quite possible that lens-based concentrators, which are considered superior to reflectors for photovoltaics today, will eventually be displaced.

The Fresnel lenses found in pv concentrators generally have a smooth (but not necessarily flat) upper surface with the Fresnel facets on the bottom side of the lens where they are protected from abrasion and are less likely to trap dust. Nearly all pv concentrator lenses are fabricated from acrylic, which has several preferred properties: optical clarity, moldability, outdoor stability, and relatively good abrasion resistance. Other lens materials have been tried, but so far none matches the performance of acrylic. Although glass is more weather resistant and stable than acrylic, Fresnel facets with adequate surface precision cannot easily be incorporated into glass lenses. Creating a composite lens by molding plastics or silicones to the glass has been tried, but the outcome has been lower efficiency than acrylic; in addition, delamination between the glass and lens material has occurred.

Most PV concentrators now being developed rely on passive cooling of the cells. Although active cooling of concentrators has been successfully demonstrated, the process is considered most feasible when there is a matching need for low-temperature thermal energy. But because most applications call only for the generation of electrical energy, and because passively cooled concentrators are much simpler, they have become the dominant type under development. The typical small lens size per receiver also facilitates passive cooling of the cells by reducing the amount of thermal energy to be dissipated from each concentrator cell.

Most PV concentrator modules under development are in general conformance with the optical principles presented in the section "Photovoltaic concentrators: basic principles" and belong to one of three broad categories: point-focus Fresnel modules using silicon concentrator cells, arched-linear Fresnel concentrators, or high-concentration designs.

Point-focus Fresnel modules using silicon concentrator cells generally operate in the 100X to 500X range of geometric concentration ratios. These modules frequently use secondary concentrators, which permit acceptance angles of about 1°. The cells in these modules are usually small (approximately one cm^2 in area), which facilitates cell cooling and effectively controls electrical resistance losses by limiting the magnitude of the cell currents.

Arched-linear Fresnel concentrators have concentration ratios that range from about 10X to 25X. Although these require relatively more concentrator cell area, the concentrator cells are operated at lower levels of incident radiation. Therefore, the cells are generally larger and thus can be fabricated more cheaply using processes that are similar to those for one-sun cells.

High concentration designs—those that have concentration ratios ranging from 500X to 1,000X—are also being developed by a number of research groups. These designs generally call for high-efficiency GaAs or multiple-junction cells. Because these cells are currently less developed and more expensive than silicon cells, they are not likely to be incorporated into commercially available designs for at least several more years.

In addition, several attempts have been made to develop low-concentration modules, those with concentration ratios ranging from 1.5X to 5X. In some cases, flat-plate modules have functioned as receivers, but to date these efforts have not been successful. The lack of success can be attributed to two problems: first, low concentration ratios do not provide much leverage to overcome solar-cell costs, which are the primary motivation for concentration. And second, when the concentration ratio reaches 2X or higher, the increased thermal energy generated within the cells must be dissipated by an active or passive cooling mechanism. Thus, even at 2X, the receiver must be more complex than traditional flat-plate modules.

FIGURE 11: *The 300 kW Soleras system built by Martin Marietta near Riyadh, Saudi Arabia, in 1981 under a joint U.S.–Saudi program.*

Field experience

Several major PV concentrator power systems have been installed around the world (see table 3; figures 11, 12, and 13). Two of the systems (Soleras in Saudi Arabia and Dallas–Fort Worth in Texas) are 10 years old and thus prove that durable and long-lived concentrator power systems are technically feasible. Two other systems (ENTECH–3M in Texas and Alpha Solarco in Nevada) were installed in 1989 and 1990, respectively. Both ENTECH–3M and Alpha Solarco represent major advances in module and array technology: they have simpler and more reliable tracking control systems and larger aperture areas for each tracker unit. The ENTECH–3M system [16], for example, requires only 12 arrays, each having an area of 167 m^2. The good long-term performances of both the Soleras [17] and the Dallas–Fort Worth systems [5] are supported by extensive data (see table 3 for a summary).

At present, several other concentrator systems are being planned or built. These include a 20 kilowatt system, which was installed by ENTECH in 1991 near Davis, California, as a part of the PVUSA Project[6] [18, 19], and a point-focus

---

6. PVUSA is an ongoing program to evaluate and demonstrate both established and emerging PV power-system technologies. The program is managed by the Pacific Gas & Electric Company and is funded by PG&E, the U.S. Department of Energy, the Electric Power Research Institute, the state of California, and several U.S. utility companies.

Table 3: Summary of major PV concentrator power systems

Project	Description	Performance	Status
Soleras 300 kilowatt system, Saudi Arabia (a joint U.S.–Saudi Arabia project)	160 pedestal mounted arrays, 33X concentration point-focus Fresnel modules, total array area of 3,806 m², built by Martin Marietta in 1981.	Array field has operated essentially continually since 1981. Average annual DC efficiencies of about 9 percent.	System converted from stand-alone to grid-connected in 1984. Currently used as a PV power source for the German–Saudi "Hysolar" hydrogen production project.
Sky Harbor 225 kilowatt system, Phoenix, Arizona	80 arrays, design as above. Total array area of 2,022 m². Array field built by Martin Marietta, system by Arizona Public Service in 1982.	System experienced numerous inverter, module, and array problems. Average annual DC efficiencies were only 6.5 percent.	System has been dismantled after several years of operation, as originally planned. Several arrays have been reinstalled at other sites.
DFW 25 kilowatt system, Dallas–Fort Worth, Texas	11 arrays of 10 linear Fresnel modules each; 25X concentration, PV and thermal; total aperture of 245 m². Built by E-Systems in 1982.	System operated very well from 1982 to 1987. Electrical and thermal efficiencies were 7 percent and 39 percent. No array degradation.	System has not operated for several years because of inverter problems. It will probably be dismantled in the near future.
ENTECH–3M Austin system Austin, Texas	12 arrays of 60 linear Fresnel modules each; 22X concentration. Cells use prismatic covers to eliminate grid shadowing losses. Total array area of 2,007 m². Built by ENTECH in 1989.	Long-term performance data not yet available. Average peak module efficiency 15 percent (cell at 25°C, 800 watts per m² insolation); array field operational efficiency is 13 percent (at 60°C cell temperature).	Operational.
Alpha Solarco 15 kilowatt system, Pahrump, Nevada	A single pedestal-mounted array of 500X point-focus Fresnel modules. Built in 1990 by Alpha Solarco.	This prototype array was built for evaluation purposes. Output from first modules was 10 kilowatts.	System is being retrofitted with new modules. Output projected to be 15 kilowatts.

FIGURE 12: *ENTECH's 300 kW linear Fresnel array at the 3M Company facility in Austin, Texas.*

Fresnel system currently being built by the Electric Power Research Institute (EPRI) at the same site. A number of other systems, including three based on parabolic-trough designs, have been built and tested. Performance for the parabolic-trough systems was not very good, however, with efficiencies of about 5 to 7 percent.

Although most PV concentrator systems have been developed in the United States, several have been built and tested in other countries. Japan has constructed and tested both a 50-kilowatt linear Fresnel system using silicon cells and a 10 kilowatt point-focus Fresnel system based on gallium arsenide cells. Developers in Spain, France, and Italy have also built and tested concentrators, but these have been small systems of less than 5 kilowatts and were operational for only a short time.

Module development

Most concentrator modules (like the power systems themselves) have been developed in the United States, where at least a dozen development efforts are currently underway (see table 4). Two designs—the Sandia 100X module and the Varian 1,000X gallium-arsenide module—have achieved peak efficiencies greater than 20 percent (measured at 25°C cell temperature and 800 watts per m^2 of direct normal radiation), values that include optical losses of about 10 to 20 percent and support the expectation that concentrator modules are more efficient than flat-

FIGURE 13: Alpha Solarco's array at Pahrump, Nevada.

plate modules. Overall module efficiency is determined primarily by the efficiency of the concentrator cells. For example, the silicon cells installed in Sandia's 100X experimental module (which were fabricated at the University of New South Wales) had peak efficiencies ranging from 23 to 25 percent when grid-obscuring prism covers were attached. The gallium arsenide cells in the Varian 1,000X module (which were fabricated by Varian) had peak efficiencies above 25 percent.

A few of the more representative and interesting designs are described in detail in the following subsections.

ENTECH *linear Fresnel modules*

ENTECH has developed, built, and tested a number of linear-focus Fresnel modules over the past several years that include actively cooled versions, suitable for combined PV-thermal applications, as well as passively cooled designs (see figure 14). Concentration ratios for the modules have ranged from 10X to 40X; the typical module measures about 3 meters long and 0.5 to 1.0 meter wide. The primary optical element in all the designs is an arched linear Fresnel lens, which has demonstrated excellent tolerance to slope errors, has eliminated most facet tip losses (losses caused by scattering when small amounts of incoming solar radiation hit a lens's facet tips), and has less chromatic aberration than flat lenses. Initially manufactured as flat components, the lenses are mechanically

Table 4: Summary of module development activities in the United States[a]

Developer	Module design description	Status
AESI	2 × 5 PFF, 0.25 m² aperture, plastic housing, TIR secondary, 350X.	Several prototypes have been fabricated.
Alpha Solarco	2 × 12 PFF, 1.25 m² aperture, aluminum housing, glass silo secondary, 492X.	100 first-generation modules installed at Pahrump, Nevada. Second-generation prototypes under test.
Black and Veatch	4 × 6 PFF, 0.7 m² aperture, metal housing, reflective secondary, 290X.	First prototype tested.
ENTECH	Several designs based on arched LFF lenses; module areas 1.5 to 3 m²; prismatic cell covers, 10X to 40X concentration.	Several hundred modules installed in three systems. Peak DC efficiencies as high as 17 percent.
EPRI	4 × 12 PFF module, metal housing, back-point contact Si cell, 500X.	Several prototypes tested, efficiency over 18 percent.
Isofoton (Spain)	111 × 27 cm² module with bifacial cells and CPC mirrors, 36 cells of 8 × 5 cm, no tracking.	In development; cell efficiency not yet satisfactory. Module power 25.6 W_p, module efficiency 8.5 percent.
Midway Labs	PFF, aluminum housing TIR secondary, 91X.	Several modules and an array built and under test.
SAIC/TFI	Reflective dish concentrator, PV receiver a dense array of cells, actively cooled.	Dish and first experimental receiver tested.
Sandia SBM3	2 × 12 PFF, 0.7 m² aperture, aluminum housing, reflective secondary, prismatic cell covers, 185X.	Prototype modules under test, best efficiency 16 to 17 percent.
Sandia Concept-90	PFF, 100 cm² lenses, completely encapsulated receiver, plastic housing, back-point contact cells.	Small prototype under test, peak efficiency of 19 percent.
Sandia 100X Experimental	2 × 5 PFF, 100X, designed for experimentation.	Used to establish Si module conversion efficiency of 17 percent in 1984 and 20 percent in 1989.
SEA Corp.	LFF design using extruded lens and housing sidewall, 10X.	Initial lenses and a minimodule have been made and tested. Lens efficiency above 70 percent confirmed.
SciTech	Low concentration designs using bifacial cells, tracking not required.	First prototypes being tested.
SKI	Design similar to SBM3 above.	First prototype being fabricated.
Varian	PFF, aluminum housing, TIR secondary, GaAs cells, 1,000X.	Minimodule tested with peak efficiency above 22 percent.
Wattsun	Thin profile module with 11 mm² lenses and 2.5 millimeter diameter Si cells.	Several prototype modules tested at 11 percent peak efficiency.

a. PFF = point-focus Fresnel optical concentrator; LFF = linear-focus Fresnel optical concentrator; TIR = totally internal reflecting optical element; approximate peak power ratings can be calculated as the product of the peak efficiency, the aperture area, and 1,000 watts.

arched to conform to the correct shape. Manufactured for ENTECH by the 3M Company, most of the arched lenses are formed by laminating a thin sheet of acrylic lens material to a thicker, flat acrylic superstrate. The arched lenses have efficiencies that typically range from 88 to 90 percent, which are about 5 to 10 percent higher than flat point-focus lenses.

One of the most interesting design innovations introduced by ENTECH in the past few years is the prismatic cover (see the section "Novel cell–light coupling concepts"), which is placed over the concentrator cells. Incorporation of the cover permits the 300 kilowatt system to use cells that have cell-grid coverage of 20 percent. In addition to reducing series resistance, such high grid coverage may lower manufacturing costs by permitting a lower cost metallization process to be used.

ENTECH is currently developing automated processes for concentrator cell assemblies and receivers, which represent the most difficult and costly parts of the manufacturing process. If successful, such automation will allow the company to manufacture and install systems at significantly lower prices than is possible today.

Sandia's experimental and baseline modules

In 1989 Sandia built an experimental silicon-based PV module that achieved a record 20 percent peak conversion efficiency. The module, which contains a

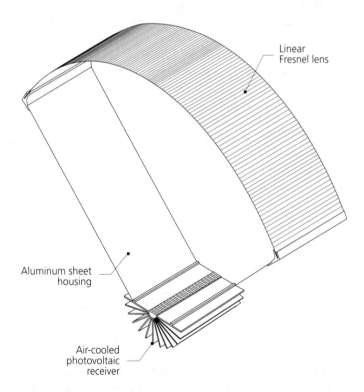

Linear
Fresnel lens

Aluminum sheet
housing

Air-cooled
photovoltaic
receiver

FIGURE 14: Schematic of the ENTECH 22.5X air-cooled linear Fresnel lens photovoltaic

2 × 6 matrix of point-focus Fresnel lenses and silicon concentrator cells, has a geometric concentration ratio of 100X. Its efficiency can be attributed largely to its silicon solar cells with prismatic cell covers, which have efficiencies of approximately 24 percent.

Because the experimental module is not especially complicated or advanced, it is expected that modules with similar efficiencies could become commercially available in the near future. The design's most critical breakthrough was achieving cell conversion efficiencies of 24 percent. It is expected that cell conversion efficiencies of around 25 percent will soon become routine and will not necessarily depend on the use of prism covers.

Currently undergoing parallel development is the Sandia baseline module III (SBM3) (see figure 15), which incorporates features from a number of earlier concentrators and is slated for commercial applications. Included in the design is the use of a metal housing without a separate heat sink (derived from Sandia's SBM2), a 2 × *N* arrangement of cells and lenses (adopted from the Martin Marietta and Varian companies), and a reflective secondary element (introduced by several development teams) [20].

Several other companies are developing modules similar to the SBM3. The Alpha Solarco module, for example, contains a 2 × 12 arrangement of cells and lenses and employs an aluminum housing without a separate heat sink. Rendering its design unique is a relatively high concentration ratio of nearly 500X and the use of a glass-silo secondary element, a concept that was first described by L.W. James of L.W. James and Associates [21]. The refractive glass silo offers considerably improved tracking and alignment tolerances and more uniform flux profiles for the concentrator cells than can be obtained with reflective secondaries.

Midway Laboratories is also developing a module that is encased in a metal housing, has no separate heat sink, and has refractive secondaries, but is unique in its projected ability to accommodate tracking errors as great as ±4°, which would enable it to operate with passive trackers. The concentration ratio for the module is 90X.

Solar Kinetics, a firm with considerable experience in the development and construction of solar-thermal systems, is also developing a concentrator module similar to the SBM3. The first prototype was fabricated in 1991 and is now being tested.

Varian's 1,000X minimodule

In 1989 the Varian company achieved a conversion efficiency breakthrough: it reported an efficiency of more than 22 percent in a two-lens, two-cell "minimodule" based on gallium arsenide [1]. Because the measurement was taken at an ambient temperature of 27°C, peak conversion efficiency, corresponding to a cell temperature of 25 °C, would have been significantly higher. The company is now incorporating the minimodule into a larger, 1,000X prototype module.

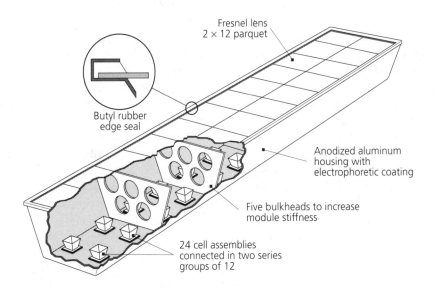

Fresnel lens
2 × 12 parquet

Butyl rubber
edge seal

Anodized aluminum
housing with
electrophoretic coating

Five bulkheads to increase
module stiffness

24 cell assemblies
connected in two series
groups of 12

FIGURE 15: Photograph and drawing of the Sandia SBM3 point-focus concentrator module.

SEA's 10X linear Fresnel module

The SEA Corporation is pursuing a linear Fresnel module [22] that is approximately 25 centimeters wide and designed to operate at a concentration ratio of approximately 10X. The most interesting aspect of the design is the proposed manufacturing method for the lens and sidewalls: it entails a single acrylic extrusion process, which is a well-known and low-cost manufacturing process that can effectively seal the lens and module sidewalls without the need for a separate joining step. The method has already led to lens efficiencies greater than 70 percent, which SEA believes is adequate for meeting the company's commercialization cost goals.

Sandia's Concept-90 module

Long-term durability and reliability are major issues confronting the developers of concentrator modules. Isolating the electrical circuit from other components, as well as from possible current leakage during wet conditions, is a critical consideration, for example. The obvious solution is to encapsulate the concentrator cells and circuit in a sealed package, a strategy that has worked successfully for ENTECH's linear Fresnel modules. Sandia National Laboratory's Concept-90 module [23] employs a similar strategy for point-focus modules by completely encapsulating the electric circuit in a flat, sealed package much like the one in a conventional flat-plate module. The concentrator cells are mounted on thin copper heat spreaders, which serve as the cells' rear contact and are nearly as large as the Fresnel lenses themselves. Top and bottom interconnects are isolated by a thin insulating material, such as a fiberglass mat. The entire solar cell and interconnect package is sealed in essentially the same way as flat-plate crystalline-silicon modules. The front side of the circuit/receiver sheet is potted in another plastic compound, which serves as the bottom of the acrylic module housing.

A small prototype containing six lenses and cells has been fabricated, and its peak efficiency tested at 19 percent. The design's success, however, depends on its performance and durability and on whether operating temperatures can be controlled within a completely encapsulated module. Performance and accelerated environmental testing is under way. Thermal analysis suggests that with relatively small 10 × 10 centimeter lenses and copper heat spreaders, cell temperatures should be equivalent to those for other PV concentrators, a projection that has been confirmed in module tests. The Concept-90 module therefore promises to have excellent conversion efficiency as well as the electrical durability and reliability characteristics of flat-plate modules.

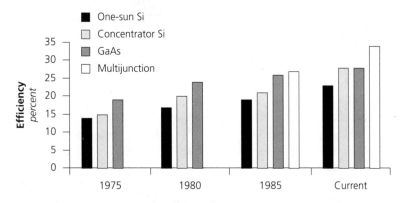

FIGURE 16: Efficiency progress for concentrator cells and one-sun silicon cells. The concentration ratios are those at which maximum efficiencies occurred, typically 100X to 200X for concentrator Si cells, 250X to 500X for GaAs cells, and 50X to 150X for multijunction cells.

COMPONENT DEVELOPMENT

Concentrator cells

Because concentrator cells are the critical determinants of PV concentrator efficiencies, a summary of their current status is warranted. (For more detailed information on concentrator-cell technology research and development see [24–26]). The fact that the conversion efficiencies of concentrator cells are generally higher than those of flat-plates (see Box B), along with the potential for low module cost, explains why interest in the technology has continued.

Three principal categories of concentrator cells have been developed: silicon (Si) concentrator cells, gallium arsenide (GaAs) concentrator cells, and multijunction (MJ) devices. The efficiency of each one has been plotted for the past 10 years (see figure 16) and compared with one-sun silicon cells. Silicon concentrator cells are the most developed of the three types and are found in virtually all operating concentrator systems. To date, gallium arsenide cells have been installed in only a few prototype modules and arrays; multijunction cells have been tested only in experimental modules.

BOX B
CONVERSION EFFICIENCY OF SOLAR CELLS
An important characteristic of PV cells is that conversion efficiency increases with light level (at least within certain limits). Put another way, a cell's output is the product of its short circuit current I_{sc}, open circuit voltage V_{oc}, and fill factor (FF). Its efficiency is the ratio of output divided by input: $EFF_c = (I_{sc} \cdot V_{oc} \cdot FF)/I$, where I is the intensity of the solar radiation incident on the cell. I_{sc} is a linear function that increases with intensity, whereas V_{oc} increases logarithmically. As long as the fill factor is not dominated by the series resistance, it will remain fairly constant. Efficiency, therefore, is approximately a logarithmically increasing function of intensity until the fill factor begins to increase (see the section "Photovoltaic concentrators: basic principles") [27].

Silicon concentrator cells

Most PV concentrator-cell research has been directed at silicon cells. As early as the 1970s, silicon concentrator-cell efficiencies (referenced to peak conditions of 25°C cell temperature and an airmass 1.5 solar spectrum) exceeded 15 percent; in 1980 the first cell reached 20 percent. A few years later, in 1985, more remarkable results were reported by R. Swanson and R. Sinton of Stanford University who had developed a 25-percent efficient, very thin silicon cell with both *p*-type and *n*-type junctions and contacts on the rear (thus the cell has no grid-shadowing losses) [28]. In 1987, M. Green of the University of New South Wales in Australia and P. Verlinden of the Catholic University of Louvain in Switzerland reported efficiencies in excess of 25 percent for silicon cells [29, 30]. Since then, Swanson and his colleagues have reported efficiencies of 28 percent; Green and his colleagues have reported 26 percent efficiency.

Such dramatic increases in conversion efficiencies were made possible by several advances in cell design, including the use of novel light-trapping schemes, which made thinner cells possible without sacrificing light absorption, and also patterned junctions that reduced the overall size of junction areas, permitting increased cell voltage. In addition, advances in cell processing contributed to efficiency gains. Improved surface passivation techniques, for example, reduced surface recombination losses (see chapter 7: *Crystalline- and Polycrystalline-Silicon Solar Cells*) [31].

In the past few years increasing emphasis has been placed on transferring these research results to production. To facilitate the development of industrial production of high-efficiency Si concentrator cells, several U.S. companies have joined forces with Sandia National Laboratories in a Concentrator Initiative program funded by the companies and the DOE. The program is expected to culminate in the manufacture within a few years of Si concentrator cells that are from 22 to 25 percent efficient. To date, the efficiency potential of Si concentrator cells is best demonstrated in the 300 kilowatt ENTECH–3M system, where peak module efficiency has reached 15 percent; and peak cell efficiencies average above 18 percent.

GaAs concentrator cells

It has been known for many years that solar cells made from III–V compound semiconductor materials (that is, compounds made of elements from groups III and V of the periodic table) have high conversion potential. Experimental GaAs concentrator cells and modules have been fabricated and tested for about 20 years; to date, the best module efficiency is the 22 percent obtained with the Varian minimodule. Varian has also reported an efficiency of 28 percent for a small GaAs cell with a prismatic cover operating at 200X, a cell that was designed for applications in space.

Gallium arsenide cells are used for some flat-plate applications in space, where output power is more important than cost. But the high cost of the cells renders them impractical for terrestrial flat-plate collectors. (GaAs cells were used, incidentally, for the collector panel on the General Motors Sunraycer car that so handily won the World Solar Challenge race across Australia in 1988.)

Multijunction concentrator cells
Because different solar cells vary in the efficiency with which they convert different parts of the solar spectrum to electricity, stacked multiple-junction cells have always been attractive. Such cells can provide higher overall efficiency by using each cell junction for the wavelength range for which it is best suited and combining their overall electrical output.

As early as 1980, Varian built a beam-splitter module that used dichroic filters (filters that reflect light below a certain energy level and transmit light above that level) to reflect the infrared portion of the spectrum to Si cells and to transmit the visible and ultraviolet portion to aluminum gallium arsenide (AlGaAs) cells [32]. Although the design is feasible, most researchers have pursued the simpler strategy of stacking two cells on top of one another. In this way, the top cell serves as a spectrum separator absorbing high-energy (ultraviolet and visible) radiation and transmitting lower-energy (infrared and some visible) wavelengths to the bottom cell. The two cells can be fabricated separately and mechanically stacked with a transparent adhesive, or they can be fabricated on a single starting substrate, forming a monolithic multiple-junction cell.

In spite of their theoretical advantages, the development of MJ cells has proved difficult. Not until 1988, when a mechanically stacked GaAs/Si cell achieved an efficiency of 31 percent (the first cell to exceed 30 percent), did the efficiency of an MJ cell substantially exceed that of single-junction cells [33]. Since then research on MJ cells has progressed rapidly. In 1990, L. Fraas of the Boeing Company achieved 34 percent efficiency in a mechanically stacked GaAs-on-gallium-antimony (GaSb) MJ cell [34].

But before mechanically stacked MJ cells can be used in a concentrator module, several engineering difficulties must be addressed. First, the metal contact-grid on the top and bottom of the upper cell, and on the top of the lower cell, must be properly aligned in order to maximize the transmission of long-wavelength light to the bottom cell. Second, the cells must be bonded with a thin yet durable transparent adhesive layer and mounted in such a way as to provide adequate heat dissipation, an especially difficult task for the top cell. And, finally, to maximize power, different series and parallel interconnection schemes are needed depending on the expected operating current and voltage characteristics of the cells. Such requirements, although challenging, are not insurmountable, and several interesting approaches have already been proposed [34, 33].

Monolithic MJ cells offer an attractive option for concentrator modules because they are inherently simpler than mechanically stacked cells. Like stacked cells, however, they are difficult to develop. Although efficiencies above 35 percent should be possible within the near future, the best efficiency to date is 31.8 percent for an indium phosphorus/gallium indium arsenide (InP–GaInAs) cell measured at 50X [35]. Increasing overall efficiency depends on the ability to grow two cells of appropriate bandgaps, each of which is highly efficient. In addition, most researchers are attempting to fabricate current-matched cells so that the top and bottom cells can be connected in series. Doing so is very attractive from the point of view of simplifying cell interconnects, but may have slowed the development of high-efficiency monolithic MJ cells.

Tracking structures and controllers

As indicated earlier, most pv concentrators exploit only direct solar radiation. To operate, they must track the sun's apparent motion, thus focusing the sun's rays on the concentrator cells. Nearly all pv concentrators under development use two-axis tracking, the two most common approaches being pedestals with azimuth and elevation tracking (similar to trackers for heliostats) and tilt-roll designs.

One of the greatest challenges is designing a tracking system capable of functioning under highly variable operating conditions such as occur daily and in different seasons and climates. Equally challenging is designing a tracker that can survive storm conditions, especially the onset of high winds. The solution to the latter problem has been to design an array that can be moved into a "stow position," thus exposing a relatively small surface area should strong horizontal winds develop. Most pv concentrator power plants are therefore equipped with sensors to detect increasing—and potentially dangerous—winds.

Another obvious concern is the need to keep costs low. There is general agreement that with proper tracking structure design, array-field balance-of-system (bos) costs of $125 per m^2 are possible for pv power systems. Studies also indicate that costs significantly less than $125 per m^2 are possible (see "Cost projections" below). The array-field bos cost again points out the importance of high conversion efficiencies. If a system has a relatively low efficiency of 12.5 percent, its bos cost of $125 per m^2 translates into a power-referenced cost of $1.00 per watt of power. If the efficiency is 20 percent, the array-field bos cost (excluding module and power conditioner) is only $0.63 per watt of power.

Whereas early tracking arrays generally relied on sun-sensors to provide signals to the tracking motors, most of today's trackers contain an open-loop, microprocessor-based controller. These controllers focus the arrays at the sun, according to preprogrammed time, date, and latitude information. Several designs, including one that has self-alignment and realignment capabilities [36, 37], are soon likely to be available at a cost of only about $100 each, which translates into less than $1 per m^2 for larger arrays.

COST PROJECTIONS

The major issue determining the ultimate role of pv technology in electric power markets is that of cost. Although concentrators are more complex than crystalline-silicon or thin-film flat-plate arrays, they achieve greater efficiency by using high-quality concentrator cells without the need and expense of large solar cell areas.

In 1988, J. Chamberlin of Sandia National Laboratories compiled an analysis of manufacturing costs based on data provided by several PV concentrator development teams [38]. The cost analyses, which included a return-on-investment markup, were provided for 1, 10, and 100 megawatt production levels (see table 5). Five major categories of module components were analyzed:

Cells	Concentrator cells (including prism covers, if applicable)
Cell assemblies	Cell mounts, interconnects, and electrical isolation
Optics	Concentrating lens or mirror and secondary optical elements
Housing	Basic module structural component that holds the optics and cell assemblies in position; includes heat dissipater
Other	Diodes, electrical feedthroughs, module assembly, testing, and other components

Manufacturing costs for the different components in a concentrator system vary according to the specific module design. The most costly components, and the elements that vary most in cost (see figure 17), are the concentrator cells and module housings. Concentrator cells usually represent a trade-off between cost and efficiency; frequently, more expensive cells are used to provide higher overall efficiency. Module housing costs, in contrast, are directly affected by basic mate-

Table 5: Average manufacturing costs for module components
$/m^2

Cost component	Annual production level		
	1 megawatt	10 megawatts	100 megawatts
Concentrator cells	104	63	53
Cell assemblies	43	27	21
Optics	42	32	19
Housing	72	54	42
Other	21	11	13
Total	**282**	**187**	**148**

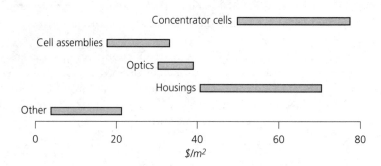

FIGURE 17: Projected cost ranges in $ per m² for concentrator module components at 10 megawatts per year production levels.

Table 6: Levelized electricity costs for PV concentrators[a]
cents per kWh

Module peak efficiency *percent*	$282/m² (1 MW/year)	$187/m² (10 MW/year)	$148/m² (100 MW/year)
10	20.6 (33.7)	15.9 (26.1)	14.0 (22.9)
15	13.9 (22.8)	10.8 (17.7)	9.6 (15.6)
20	10.6 (17.3)	8.3 (13.5)	7.3 (11.9)
25	8.6 (14.0)	6.8 (11.0)	6.0 (9.7)

a. Module manufacturing costs were obtained from table 5. The costs reflect an assumed discount rate of 6 percent (12 percent), array-field BOS costs of $125 per m², and direct normal radiation of 2,600 kWh per m² per year. Levelized energy costs resulting from module costs are provided for several values of module efficiency.

rial requirements that in turn are determined by the overall depth of the module. With the advent of newer module designs with shorter focal lengths, such as those by SEA, Sandia (the Concept-90 module), and Wattsun, significantly lower housing costs are possible.

Levelized electricity costs for several values of module efficiency have been calculated (see table 6). According to these estimates, electricity costs below $0.10 per kWh are likely with PV concentrator technology, even at relatively low production levels. Moreover, there is strong evidence to support the prediction that significantly lower costs are possible. Strategies for lowering electricity costs are offered in the following sections.

Concentrator-cell costs

Although PV concentrators require less cell area than other PV technologies, concentrator cells still represent the system's largest single cost component. Such high cost can be attributed in part to the cells' complexity and in part to today's small market for concentrators, a factor that is compounded by the lim-

Table 7: Hypothetical contribution of cells to total module costs
(for several concentration ratios)[a]

Case: market size and concentration level	Units	0.4 MW$_p$ flat plate	4.5 MW$_p$ 10X	54 MW$_p$ 100X	240 MW$_p$ 400X
Production of cells	cm^2/year	3·10^7	3·10^7	3·10^7	3·10^7
Si wafer cost	$/cm^2	0.020	0.020	0.04	0.04
Processing cost	$/cm^2	0.01	0.015	0.02	0.03
Wafer + processing cost	$/cm^2	0.03	0.035	0.06	0.07
Fabrication yield		0.95	0.90	0.70	0.25
Cell cost	$/cm^2	0.032	0.039	0.086	0.28
Price with markup	$/cm^2	0.042	0.051	0.112	0.546
Cell efficiency	percent	14	18	22	25
Module efficiency	percent	13	15	18	20
Cell cost contribution to module cost	$/cm^2	420	51	11.20	13.70

a. Mature fabrication corresponding to the market size indicated is assumed for all cases. Note that cell production levels in all cases are equal.

ited cell area required. As volumes increase, however, costs are expected to drop significantly (see table 7).

In the late 1970s, annual production of crystalline-silicon cells represented about 0.3 megawatts of power per year, which corresponded to a total cell area of about 4×10^7 cm^2 at a price of about $0.15 per cm^2. By 1990, however, output had increased by two orders of magnitude: annual production of crystalline-silicon cells represented about 30 megawatts of power worldwide, corresponding to a total cell area of roughly 2.5×10^9 cm^2, at a price of about $0.03 per cm^2. The price reduction largely reflects experience gained during production. In sum, it is not surprising that concentrator-cell costs are still high, not only because the present market remains well below the 10^6 cm^2 per year level, but because concentrator cells are at a more immature stage of industrialization than flat-plate crystalline-silicon cells were 10 years ago.

The two major cost components of silicon concentrator cells are the silicon wafer and the manufacturing process itself. Higher efficiency cells require float-zone (FZ) wafers, which are produced in limited quantity and thus are several times more expensive today than the Czochralski (CZ) or polycrystalline cells found in flat plates. However, if the market for concentrators grows to 10^7 cm^2 or more per year, manufacturers are confident that FZ costs will drop to within a factor of two of CZ costs.

Cell-processing costs should eventually approximate those for flat-plate cells, assuming an increase in production. But fabrication yields (the ratio of acceptable cells to cells entering the processing line) are likely to be lower than for flat-plate cells. The reason reflects the close link between yield and manufacturing experience as well as the complexity of the cells. Even a mature technology may have a low production yield if its components are complex. Therefore, the more complex concentrator cells are never likely to achieve the 95 percent production yields typical of flat-plate cells.

PV concentrator markets of 5 to 10 megawatts per year are likely in the near future with low concentration cells. Moreover, there is considerable similarity between flat-plate cells and 10X concentrator cells. At the higher concentrations of 100X and 400X, considerably larger markets are needed to justify increased cell production, although at such levels the contribution of concentrator cells to module cost is remarkably low. We believe, however, in view of the rapid growth of PV markets, that such goals are possible in the near term, especially if the indicated efficiencies (which are consistent with the current status of development for these cells), are achieved. We also think it likely that significant markets at intermediate production levels will be realized, although the cell costs will be significantly higher than those shown in table 7.

Although we have not addressed the potential costs of III–V compound semiconductor cells, the costs of such cells are similar in many ways to higher concentration silicon cells. Current manufacturing costs for gallium arsenide cells for space applications are thought to range from \$10 to \$15 per cm^2 but are expected to drop as low as \$2 to \$5 per cm^2 [39]. At a concentration ratio of 1,000X, such cells correspond to module cost contributions of \$20 to \$50 per m^2.

Lower array-field BOS costs

Although several studies indicate that array-field BOS costs of \$125 per m^2 are possible [40, 41], and these have formed the basis for many economic analyses, there is reason to believe that lower BOS costs can be achieved. An indication that lower costs are possible can be found in the cost projections for installed heliostat fields for central-receiver solar-thermal power systems. The two-axis tracking structures required for heliostats must meet more stringent technical parameters than most current PV concentrators.

A recent cost estimate for completely installed heliostat structures is shown in table 8. The "Field wiring (DC power collection)" category was added to this cost breakdown because it is necessary for PV concentrator systems but not for heliostat array-fields of solar-thermal central-receiver systems. Cost estimates are based on an annual production level of 5,000 heliostats. Each heliostat is designed to support 150 m^2 of concentrator aperture, which corresponds to a peak power level of 30 kilowatts at a peak module efficiency of 20 percent. Thus, the cost projection corresponds to an annual production of about 150 megawatts. These figures suggest that array-field BOS costs well below \$125 per m^2 may be possible with structures that provide excellent tracking accuracy.

Table 8: Cost estimates for recently installed heliostat structures [42]

Cost component	Cost $/m^2$
Attach modules	4.90
Support structures	15.20
Drive	20.90
Foundation (hole, concrete, pedestal)	10.80
Tracking system	7.10
Field wiring (drive power)	1.40
Field wiring (DC power collection)	5.00
Installation and checkout	6.10
Total direct costs	**71.40**

Cost potential of high-efficiency concentrators

Significantly higher cell conversion efficiencies are expected, especially for multijunction designs. We believe that concentrator-cell efficiencies above 40 percent are very likely in 10 to 15 years, and practical efficiencies of 45 percent are possible. If the optical efficiencies of concentrators reach 90 percent, then peak module efficiencies could reach 40 percent. Coupling such efficiencies with the module costs projected in the Chamberlin study, and with the heliostat tracking costs cited above, suggests that levelized energy costs of 3 to 5 cents per kWh at a 6 percent discount rate, and from 5 to 9 cents per kWh at a 12 percent discount rate, are entirely possible (see table 9).

Low-cost concentrator concepts

The 10X PV concentrator being developed by the SEA corporation appears to have significantly lower cost potential than the costs projected in the Chamberlin study. The design calls for manufacture of the concentrator lens and module housing sidewalls via a single, continuous extrusion process, an option that is attractive for several reasons: 1) plastics extrusion is very economical; 2) module assembly is simplified because two major module components are produced simultaneously; and 3) a secure, manufactured joint is formed between the lens and housing sidewalls. Moreover, the design is less costly because it calls for low concentration and less tracking accuracy. SEA cost projections confirm the company's belief that low costs are possible for both their modules and installed systems (see table 10) [43]. The levelized energy costs that can be achieved with a

Table 9: Levelized electricity costs resulting from higher-efficiency
modules and lower array-field BOS costs[a]
cents/kWh

Peak module efficiency *percent*	Module cost $200/m² Array BOS cost $100/m²	Module cost $150/m² Array BOS cost $75/m²
25	6.5 (10.6)	5.1 (8.2)
30	5.5 (9.0)	4.3 (7.0)
35	4.8 (7.8)	3.8 (6.1)
40	4.3 (7.0)	3.4 (5.4)
45	3.9 (6.3)	3.1 (4.9)

a. Assuming 6 percent (12 percent) discount rates and direct normal radiation of 2,600 kWh per m² per year.

Table 10: Manufacturing and field installation costs projected by the SEA
Corporation for a 10X linear concentrator system
$/m²

Cost component	1 MW/year	10 MW/year	100 MW/year
Concentrator cells	36.77	22.98	16.55
Interconnects, diodes, and wiring	0.77	0.65	0.56
Lens and module sidewalls (with reflective film)	21.28	16.09	14.03
Module ends, heatsink, miscellaneous	20.50	12.40	11.41
Module assembly	9.21	4.60	4.15
Total module costs	**88.53**	**56.72**	**46.70**
Total with markup	**194.77**	**110.61**	**77.06**
Array structure costs	13.47	11.03	10.12
Field installation costs	1.84	0.92	0.46
Total structure and installation costs, with markup	**33.68**	**23.30**	**17.45**
Installed system cost	**228.45**	**133.91**	**94.51**

10X linear concentrator system are projected to be as low as 4 to 6 cents per kWh assuming a 15 percent peak module efficiency at a production level of 100 megawatts per year (see table 11). The peak module efficiencies are assumed to range from only 10 to 20 percent because the developmental emphasis for the system is on low cost rather than on high efficiency.

SUMMARY AND CONCLUSIONS

Cost and performance studies make it clear that electricity costing less than 10 cents a kilowatt-hour can be generated using PV concentrator technology, and less than five cents a kilowatt-hour is quite possible. PV concentrator system costs and their corresponding levelized energy costs can be projected according to one of three scenarios: 1) the technology used is current, with production limited to two to five megawatts per year; 2) costs that are likely to be achieved in the near term at annual production levels of about 10 megawatts; and 3) costs that are potentially achievable in the longer term, 10 or more years from now, and at significantly higher production levels of 100 megawatts or more per year (see figure 18). These projections all assume 2,600 kWh per m^2 per year of direct normal solar radiation; electricity costs will be proportionally higher for locations with less direct normal radiation. Although many locations, such as central Europe, have significantly less radiation, there are also large areas (including southern Europe) with direct normal radiation in excess of 2,000 kWh per m^2 per year where levelized electricity costs would be within 25 percent of those projected above.

It should be remembered that the components and manufacturing processes for PV concentrators are generally similar to those for other mature technologies whose material and manufacturing costs are well known. Consequently, ac-

Table 11: Levelized energy costs derived from the SEA Corporation's
projected power system costs[a]
cents/kWh

Peak module efficiency *percent*	1 megawatt	10 megawatts	100 megawatts
10	11.7 (19.3)	17.1 (11.6)	5.2 (8.9)
15	8.0 (11.5)	4.9 (8.1)	3.7 (6.2)
20	6.2 (10.1)	3.9 (6.3)	2.9 (4.9)

a. The discount rate is 6 percent (12 percent) and direct normal radiation is 2,600 kWh per m^2 per year.

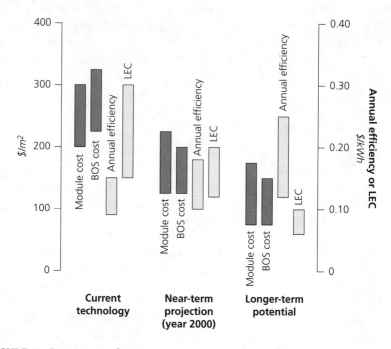

FIGURE 18: Cost projections for PV concentrator power systems. Power, conditioner costs of $500, $200, and $100 per peak kilowatt respectively, are included in the BOS costs for the three projections. Annual direct normal radiation is assumed to be 2,600 kWh per m². The LEC ranges reflect discount rates of 6 percent and 12 percent and realistic combinations of costs and efficiencies; they do not combine the lowest costs with the highest efficiencies.

curate cost projections for concentrators can be obtained by comparing PV plastic lenses with lighting-fixture covers, and module housings with formed metal or molded plastic products. For this reason, concentrators, despite being more complex than flat-plate arrays, offer a more certain path to competitive PV electricity.

If flat-plate PV technologies can attain a combination of high efficiency and low cost, they may obviate the need for continued development of more complex PV concentrators. But it is also likely that PV concentrators will always be preferred for larger power applications in sunny regions because of their efficiency and cost advantages over flat-plates.

ACKNOWLEDGMENTS

The authors gratefully acknowledge the assistance of the following individuals: A. Maish, J. Chamberlin, D. King, H. Post, and J. Gee of Sandia National Laboratories; E. Lorenzo, M.A. Egido, and M. Macagnan of the Polytechnic University of Madrid; N. Kaminar of SEA Corporation; E. Schmidt of Alpha Solarco; M. O'Neill of E-Systems; M. Imamura of WIP (Wirtschaft und Infrastruktur GmbH—Planungs—KG); and H. Kelly of the U.S. Office of Technology Assessment. Portions of this work were performed at Sandia National Energy Laboratories and the National Renewable Energy Laboratory supported by the U.S. Department of Energy under contracts DE-AC04-76DP00789 and DE-AC02-83CH10093.

WORKS CITED

1. Kuryla, M., Kaminar, N., MacMillan, H., Ladle Ristow, M., Virshup, G., Klausmeier Brown, M., and Partain, L. 1990
   22.7% efficient 1000X GaAs concentrator module, *Proceedings of the 21st IEEE Photovoltaic Specialists Conference*, Orlando, Florida, May.

2. Sanchez, E. and Araujo, G.L. 1984
   Mathematical analysis of the efficiency-concentration characteristic of a solar cell, *Solar Cells* 12:263–267.

3. Cuevas, A., Luque, A., Eguren, J., and del Alamo, J. 1981
   High efficiency bifacial back surface field solar cells, *Solar Cells* 3.

4. Minano, J.C. 1989
   Synthesis of concentrators in two-dimensional geometry; Design of concentrators; chapters 11 and 12 in Luque, A., ed., *Solar Cells and Optics for Photovoltaic Concentration*, Adam Hilger, Bristol.

5. Muneer, T. 1991
   A European solar radiation model, *Proceedings of the Biennial Congress of the International Solar Energy Society*, 1(part 2):939.

6. Luque, A., Lorenzo, E., Sala, G., and Lopez-Romero, S. 1984–85
   Diffusing reflectors for bifacial photovoltaic panels, *Solar Cells* 13.

7. Swanson, R. 1984
   Point contact silicon solar cells, *Proceedings of the 17th IEEE Photovoltaic Specialists Conference*.

8. O'Neill, M. 1987
   Measured performance of a 22.5X linear Fresnel lens solar concentrator using prismatically covered polycrystalline silicon cells, *Proceedings of the 19th IEEE Photovoltaic Specialists Conference*, May.

9. Blakers, A. and Green, M.A. 1986
   20% efficiency silicon solar cells, *Applied Physics Letters* 48:215–217.

10. Cuevas, A., Sinton, R., Midkiff, N., and Swanson, R. 1989
    Point junction concentration cell with a front metal grid, *Proceedings of the 9th EC Photovoltaic Specialists Conference*, 761–764.

11. Sinton, R. and Swanson, R. 1987
    Increased photogeneration in thin silicon concentrator solar cells, *IEEE Electron Device Letters*, EDL-8:547–549.

12. Luque, A. and Minano, J. 1991
    Optical aspects in photovoltaic energy conversion, *Solar Cells* 34:237–258.

13. Luque, A., Minano, J., P Davies, P., Terron, M.J., Tobias, I., Sala, G., Alonso, J., and Olivan, J. 1991
    Angle-limited cavities for silicon solar cells, *Proceedings of the 22nd IEEE Photovoltaic Specialists Conference*, Las Vegas, Nevada, October.

14. Menicucci, D. and Fernandez, J. 1986
    Estimates of available solar radiation and photovoltaic energy production for various tilted and tracking surfaces throughout the US based on PVFORM, a computerized performance model, Sandia Laboratories Report SAND85-2775, March.

15. Macagnan, M. 1991
    Private communication, M. Macagnan, Polytechnic University of Madrid.

16. O'Neill, M., Walters, R., Perry, J., McDanal, A., Jackson, M., and Hesse, W. 1990
Fabrication, installation and initial operation of the 2,000 sq.m. linear Fresnel lens photovoltaic concentrator system at 3M/Austin (Texas), *Proceedings of the 21st IEEE Photovoltaic Specialists Conference*, Orlando, Florida, May.

17. Salim, A. and Eugenio, N. 1990
A comprehensive report on the performance of the longest operating 350 kW concentrator photovoltaic power system, *Solar Cells* 29:1–24.

18. O'Neill, M., McDanal, A., Walters, R., and Perry, J. 1991
Recent developments in linear Fresnel lens concentrator technology, including the 300-kW 3M/Austin System, the 20-kW PVUSA System, and the concentrator initiative, *Proceedings of the 22nd IEEE Photovoltaic Specialists Conference*, Las Vegas, Nevada, October.

19. Candelario, T., Hester, S., Townsend, T., and Shipman, D.1991
PVUSA—performance, experience and cost, *Proceedings of the 22nd IEEE Photovoltaic Specialists Conference*, Las Vegas, Nevada, October.

20. Richards, E., Chiang, C., and Quintann, M. 1990
Performance testing and qualification of Sandia's third baseline photovoltaic concentrator module, *Proceedings of the 21st IEEE Photovoltaic Specialists Conference*, Orlando, Florida, May.

21. James, L. 1989
Use of imaging refractive secondaries in photovoltaic concentrators, Sandia National Laboratories report SAND89-7029, July.

22. Kaminar, N, McEntee, J., and Curchod, D. 1991
SEA 10X concentrator development progress, *Proceedings of the 22nd IEEE Photovoltaic Specialists Conference*, Las Vegas, Nevada, October.

23. Chiang, C. and Quintana, M. 1990
Sandia's Concept-90 photovoltaic concentrator module, *Proceedings of the 21st IEEE Photovoltaic Specialists Conference*, Orlando, Florida, May.

24. *Proceedings of the IEEE Photovoltaic Specialists Conference* 1991
Twenty-first IEEE Photovoltaic Specialists Conference—1990, 1.

25. Luque, A., Sala, G., Palz, W., Doa Santos, G., and Helm, P., eds., 1991
Proceedings of the Tenth EC Photovoltaic Solar Energy Conference, 8 – 12 April , Kluwer Academic Publishers, Dordrecht, the Netherlands.

26. *Photovoltaic Specialists Conference* 1990
Fifth International Photovoltaic Science and Engineering Conference, 26–30 November 1990, International. Technical Digest, Kyoto, Japan.

27. Solar Cells, 1982
Solar Cells. 6, special issue on concentrator cells.

28. Sinton, R., Kwark, Y., Gruenbaum P., and Swanson, R. 1985
Silicon point contact concentrator solar cells, *Proceedings of the 18th IEEE Photovoltaic Specialists Conference*, Las Vegas, October.

29. Green, M., Blakers, A., Wenham, S., Zhao, J., Chong, C., Taouk, M., Narayanan, S., and Willison, M. 1987
Improvements in flat-plate and concentrator silicon cell efficiency, *Proceedings of the 19th IEEE Photovoltaic Specialists Conference*, May.

30. Verlinden, P., Van de Wiele, F., Stehelin, G., Floret, F., and David, J. 1987
High efficiency interdigitated back contact silicon solar cells, *Proceedings of the 19th IEEE Photovoltaic Specialists Conference*, May.

31. Green, M., Wenham, S., and Blakers, A. 1987
Recent advances in high efficiency silicon solar cells, *Proceedings of the 19th IEEE Photovoltaic Specialists Conference*, May.

32. Borden, P., Gregory, P., Moore, O., James, L., and Vander Plas, H. 1981
A 10-unit dichroic filter spectral splitter module, *Proceedings of the 15th IEEE Photovoltaic Specialists Conference*, May.

33. Gee, J. 1987
Voltage-matched configurations for multijunction solar cells, *Proceedings of the 19th IEEE Photovoltaic Specialists Conference*, May.

34. Fraas, L., Avery, J., Sundaram, V., Dinh, V., Davenport, T., and Yerkes, J. 1990
Over 35%-efficient GaAs/GaSb stacked concentrator cell assemblies for terrestrial applications, *Proceedings of the 21st IEEE Photovoltaic Specialists Conference*, Orlando, Florida, May.

35. Wanlass, M., Ward, J., Emery, K., Gessert, T., Osterwald, C., and Coutts, T. 1991
High-performance concentrator tandem solar cells based on IR-sensitive bottom cells. *Solar Cells* 30:363.

36. Maish, A. 1990
Performance of a self-aligning solar array tracking controller, *Proceedings of the 21st IEEE Photovoltaic Specialists Conference*, Orlando, Florida, May,.

37. Gorman, D.
Advanced Thermal Systems, Englewood, Colorado.

38. Chamberlin, J. 1988
The costs of photovoltaic concentrator modules, *Proceedings of the 20th IEEE Photovoltaic Specialists Conference*, Las Vegas, Nevada, October.

39. Benner, J. Sopor, B., and Leboeuf, G. 1991
Can high efficiency photovoltaic technologies also be low cost? *Proceedings of the Biennial Congress of the International Solar Energy Society* 1:46, Pergamon, New York.

40. Post, H., Arvizu, D., and Thomas, M. 1985
A comparison of photovoltaic system options for today's and tomorrow's technologies, *Proceedings of the 18th IEEE Photovoltaic Specialists Conference*, Las Vegas, Nevada, October.

41. Five Year Research Plan. 1987–1991
National Photovoltaics Program of the U.S. Department of Energy, DOE/CH10093-7, May 1987.

42. Gorman, D. 1991
Private communication from David Gorman of Advanced Thermal Systems, Englewood, Colorado, to Henry Kelly, U.S. Office of Technology Assessment, June.

43. Kaminar, N. 1991
Private communication from Neil Kaminar, SEA Corporation to Eldon Boes, June.

44. O'Neill, M. 1988.
Five-year performance results for the Dallas-Fort Worth (DFW) airport solar total energy system, *ASHRAE Transactions*, 94, part 1.

9
AMORPHOUS SILICON PHOTOVOLTAIC SYSTEMS

DAVID E. CARLSON
SIGURD WAGNER

Amorphous silicon (*a*-Si) photovoltaic modules can now convert sunlight into electricity with initial efficiencies exceeding 10 percent. After a few months of operation, the modules stabilize at 85 to 90 percent of their initial efficiency. By the mid-1990s, manufacturing plants should be producing *a*-Si modules at a cost of about $1 per watt at a rate of about 10 megawatts per year. In the longer term, with increased plant size and higher efficiencies, module costs could fall below $0.50 per watt.

INTRODUCTION

Amorphous silicon (*a*-Si) is a glassy alloy of silicon and hydrogen (about 10 percent). Several properties make it an attractive material for thin-film solar cells:

1. Silicon is abundant and environmentally safe.

2. Amorphous silicon absorbs sunlight extremely well, so that only a very thin active solar cell layer is required (about 1 micron as compared with 100 microns or so for crystalline solar cells), greatly reducing solar-cell materials requirements.[1]

3. Thin films of *a*-Si can be deposited directly on inexpensive support materials such as glass, sheet steel, or plastic foil. Large-area photovoltaic (PV) modules can be manufactured using low-cost automated methods. Compared with those of crystalline solar cells, materials costs and other manufacturing costs (per unit area of solar cell) are much lower, although efficiencies are also lower.

4. Multijunction solar modules employing *a*-Si alloys (and possibly other thin-film materials) are likely to reach about 10 percent stabilized efficiency by the mid

---

1. One micron = 1×10^{-6} meters.

1990s and may eventually reach stabilized efficiencies of at least 18 percent. (As discussed in chapters 6: *Introduction to Photovoltaic Technology,* and 10: *Polycrystalline Thin-film Photovoltaics,* thin-film PV technologies could lead to low-cost PV power if conversion efficiencies of mass-produced PV modules can be increased to about 10 to 15 percent.)

Today amorphous silicon is a major commercial solar-cell material, with a worldwide market share in 1990 of about 32 percent of total sales [1]. First discovered in 1974 and introduced as a commercial product in 1980, *a*-Si photovoltaics have spanned an exceptionally wide range of applications, from initial markets powering calculators and small appliances to portable electricity generators to installations for the production of bulk power. (Amorphous silicon has proven to be a versatile thin-film semiconductor with numerous nonsolar applications, which have contributed vigorously to the rapidly expanding technology base of *a*-Si.)

Although rapid progress is being made in the materials science and manufacturing technology of *a*-Si modules, important challenges remain. The poor stability of early *a*-Si solar cells has been largely overcome, and the stabilized efficiency of the best laboratory multijunction cells is now about 10 percent. However, the stabilized efficiency of commercial *a*-Si modules in field tests is only about 4 percent, still too low for bulk power applications. Current *a*-Si research focuses on finding stable, more efficient materials and using more efficient solar-cell configurations (particularly multijunctions) and on the development of low-cost, large-scale manufacturing techniques.

HISTORICAL PERSPECTIVE

The discovery of *a*-Si solar cells

The unusual electronic properties of amorphous silicon deposited from a glow discharge in the silicon-bearing gas silane (SiH_4) were first noted in 1969 by R. Chittick and his colleagues at the Standard Telecommunications Laboratories in the United Kingdom [2]. Amorphous silicon solar cells were first investigated in 1974 by David Carlson (one of the authors of this chapter) and Christopher Wronski at RCA Laboratories in Princeton, New Jersey, in the course of experiments on the deposition of thin films of silicon for photovoltaic applications and reported in 1976 [3a, 3b]. (A typical single junction *a*-Si solar cell is shown in figure 1.) The potential of *a*-Si cells for solar conversion on a large scale was recognized immediately, and the development of *a*-Si technology was taken up worldwide.

Approaches to higher efficiency and better stability

The efficiency of the first *a*-Si solar cells in 1974 was only about 1 percent. Moreover, the efficiency of early *a*-Si cells quickly degraded when exposed to light, stabilizing at 25 to 50 percent of their initial efficiency. Much of the research on *a*-

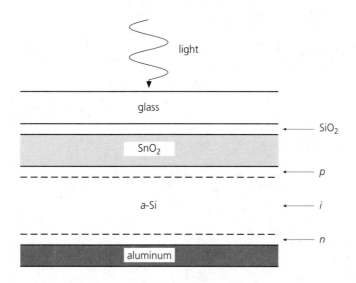

FIGURE 1: The device structure of a commercially available a-Si *photovoltaic module fabricated on a glass substrate.*

Si has centered on raising the conversion efficiency of the devices and on producing solar cells that are more resistent to light-induced degradation. In 1991, the best *a*-Si–based cells achieved stabilized conversion efficiencies of about 10 percent. As discussed in detail in the sections "Description of amorphous silicon as a solar cell material" and "Manufacturing technology", approaches to reaching higher efficiency include: 1) improving the solar-cell material by alloying to adjust the band gap (the energy barrier that conducting charges must "jump" to produce a current) or to enhance stability; 2) reducing light reflection at the cell surface and increasing absorption by light trapping within the cell; 3) improving the solar-cell structure; 4) reducing the light-induced degradation by using thinner solar-cell layers; and 5) enhancing efficiency and stability by using multijunction solar cells.

Multijunctions

The development of multijunction solar cells is a key step to increasing their efficiency and stabilizing them against light-induced degradation. Higher conversion efficiencies are obtained in multijunction devices by stacking cells of varying band gaps on top of each other. For example, in a two-junction or tandem-junction cell (see figure 2) the top cell has a wide band gap and will therefore only absorb high-energy photons (that is, short-wavelength light at the green and blue end of the spectrum). The rest of the spectrum (lower energy, red and infrared light) is transmitted to the bottom cell, which has a smaller band gap and can absorb most of the transmitted light.

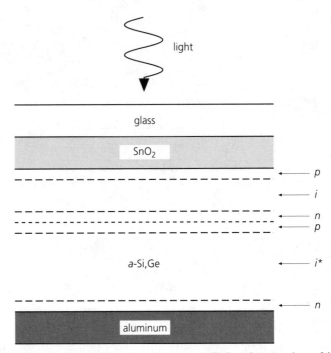

FIGURE 2: A schematic diagram of an a-Si *multijunction cell where the* i*-*type layer of the bottom cell is an amorphous silicon–germanium alloy with a narrow band gap.*

Amorphous silicon is well suited for use in multijunction solar cells because the band gap can be changed by alloying. Amorphous silicon can be alloyed with germanium to form a narrower band-gap material and with carbon to form a wider band-gap material. The data in figure 3 show that the optical properties of these amorphous silicon alloys can be readily varied by changing the alloy content. Presently, research efforts are focused on improving the electronic properties of these alloys for use in multijunction structures.

In a two-terminal multijunction device, each layer is deposited directly on top of the previous one using a continuous manufacturing line. The component cells are configured to produce equal currents under illumination and, because the cells are in series, the voltages add, giving a higher output power (higher efficiency) than either cell alone. In a four-terminal multijunction structure, the cells are fabricated separately, and the top cell is made with transparent contacts and optically coupled to (physically placed on top of) the bottom cell. Four-terminal cells would be more expensive to manufacture but would avoid matching problems at the interface between solar cells made of two very dissimilar materials.

Multijunction cells also tend to be more stable, because thinner layers can be used (which are inherently more stable) and because alloys of silicon and germanium can be made more intrinsically stable than *a*-Si. Although most commercial *a*-Si modules to date have been single-junction, some multijunction modules

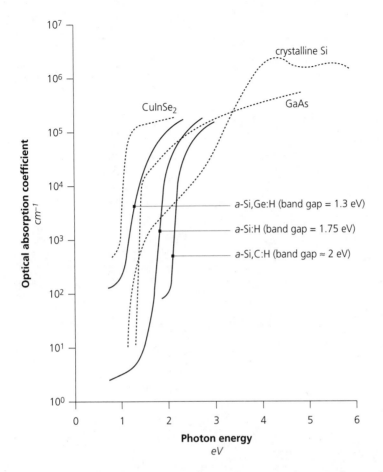

FIGURE 3: Optical absorption spectra for a-*Si, typical* a-*Si,Ge and* a-*Si,C alloys, and for crystalline Si, GaAs, and CuInSe₂.*

using different thicknesses of the same *a*-Si material were installed in 1991 as part of a project called PVUSA (see chapter 6). Most *a*-Si PV manufacturers are now concentrating on using alloyed multijunctions with cells of different band gaps as a means of attaining the efficiency and stability needed for power module applications.

Efficiency: best results, projections, theoretical limits

The best efficiency results (as of 1991) are shown in table 1 for single-junction *a*-Si cells, multijunction *a*-Si–based solar cells, and multijunction *a*-Si/CIS cells. (CIS stands for copper indium diselenide, $CuInSe_2$.) Efficiencies for small-area (about 1 cm^2) laboratory cells, larger-area (1,000 cm^2) laboratory modules, and large-area (more than 1,000 cm^2) commercial modules are given. To illustrate the effect of light-induced degradation, initial efficiency and stabilized efficiency are

Table 1: Efficiency of amorphous silicon–based solar cells
initial efficiency/stabilized efficiency

Material	Laboratory cell ~ 1 cm²	Laboratory module ~ 1,000 cm²	Commercial module > 1,000 cm²
Best results as of 1991 in percent			
a-Si single junction	12.5/6.0	10.1/5.0	7.0/5.0
a-Si–based multijunction	13.7/10.0[a]	10.0/8.0[a]	7.5/6.0[b]
a-Si/CIS four-terminal multijunction	15.6/na	12.3/na	–
Projections for practically achievable efficiencies			
a-Si single junction	15–20/na	–	–
a-Si-based multijunction	24/na		8–10/7–8 (early 1990s) 11/10 (mid-1990s)
a-Si/CIS multijunction or a-Si–based multijunction			na/18 (after 2005)
Theoretical efficiency limits for ideal materials			
Ideal single junction	28		
Ideal multijunction	42		

a. Three junctions each with different band gaps.

b. Two junctions with the same band gap produced by United Solar Systems Corporation.

shown. (It is important to know which efficiency is being quoted for *a*-Si.) For comparison, the theoretical limits for ideal materials are shown; it is 28 percent for single junctions and 42 percent for multijunctions. It is apparent that single-junction *a*-Si shows considerable light-induced degradation and that stabilized efficiencies are relatively low, only about 6 percent, even for laboratory cells that start at 12.5 percent. A better initial efficiency of about 13.7 percent is achieved by adding layers of *a*-Si,Ge alloy (silicon alloyed with germanium) to form a multijunction structure. The stabilized efficiency of these cells is about 80 percent of the initial efficiency. An even higher initial efficiency of 15.6 percent was obtained with optically coupled four-terminal multijunctions using *a*-Si and CIS.

Projections for practically achievable efficiencies indicate that small-area multijunction *a*-Si–based cells with initial efficiencies on the order of 24 percent might be possible. Commercial modules with stabilized efficiencies of 7 to 8 percent are projected over the next year or so, and by the mid-1990s, modules with efficiencies of 10 percent could become available. In the longer term (say, by the year 2005), it may be possible to reach an 18 percent stabilized efficiency with multijunction modules using *a*-Si alloys or other materials such as CIS.

Advances in manufacturing technology

The manufacturing technology of *a*-Si solar modules has advanced rapidly. When Sanyo Electric Co. introduced *a*-Si cells to power pocket calculators in 1980, the cell size was typically a few square centimeters. Today, the size of modules for power applications has reached 1 m^2 [4]. This growth in module size was made possible by the introduction of entirely new manufacturing technology, an area of continued rapid innovation.[2]

Current industrial development of *a*-Si photovoltaics is characterized by several other advances besides the impressive scaleup of module size. The efficiency gap between production modules and laboratory cells has been narrowed [7]. As discussed above, modules have been made more stable against light-induced degradation [8]; in the research laboratory, new, more efficient [9] and stable [10] cells have been demonstrated, and important advances have been made in the materials science of the thin films that make up *a*-Si cells. (Alloys of *a*-Si with germanium and carbon show particular promise.) Moreover, the description of the physical properties of these thin films has been made more quantitative, allowing for predictive modeling of cell performance.[3]

A critical challenge to further development of *a*-Si photovoltaics is the demonstration of the low-cost manufacturing anticipated from thin-film technology. A highly automated plant that produces profitably at least 10 megawatts per year of thin-film photovoltaics must be brought to operation to make this point. At present, annual worldwide production of amorphous silicon is about 14 megawatts, and most *a*-Si production plants produce about 1 megawatt per year. Several companies, including Solarex, Siemens, Phototronics, and Advanced Photovoltaic Systems, are planning to build automated 10 megawatt per year factories over the next few years. The cost of *a*-Si modules from near-term 10 megawatt per year plants is estimated to be about $0.6 to $1.2 per peak watt. In the longer term, with improvements in efficiency, further scaleup of plant size to 100

---

2. Because *a*-Si is an electrical insulator when not illuminated but becomes conducting in light, it is finding applications in electrophotography, including photocopiers, laser printers, and facsimile machines, and in optically addressed spatial light modulators for high-resolution projection displays, including high-definition television (HDTV) [5,6]. The electrical conductivity of *a*-Si can be altered by the controlled addition of impurities ("doping"). Conversely, *a*-Si can be made a permanent insulator by alloying with nitrogen. Combined with *a*-Si's other advantages, this adaptability of the electronic properties of *a*-Si has stimulated the development of *a*-Si thin-film electronics, with applications as diverse as driver circuits for flat-panel television screens and neural networks [7]. In this way, the results of what at first was exclusively *a*-Si solar cell development have spread to many other areas and greatly augmented worldwide interest in *a*-Si.

3. In contrast to single-crystal semiconductors, however, there is no comprehensive physical theory of *a*-Si. Therefore, any improvements to *a*-Si cells must be made on an empirical basis; the ultimate performance of *a*-Si cells cannot yet be predicted. Whether *a*-Si–based PV modules will just barely reach the 15 percent conversion efficiency benchmark set for large-scale utility applications, or whether they can exceed a 20 percent efficiency is an open question. The answer, for the time being, is expected to come from the experimenter rather than the theoretician.

Table 2: Amorphous silicon photovoltaic test programs

Organization	Array size kW$_p$	Manufacturer	Date installed
Pacific Gas & Electric	small	several	1985–1986
Philadelphia Electric	4	Solarex	1986
Arizona Public Service	small	several	1986
Alabama Power	55	Chronar	1986
Detroit Edison	4	Sovonics	1987
Florida Solar Energy Center	15	Arco Solar	1988
Wisconsin Power & Light	1.8	Sovonics	1988
PVUSA (Davis, California)	17.3	Sovonics	1989
PVUSA (Maui, Hawaii)	18.5	Sovonics	1989
PVUSA (Davis, California)	15.7	UPG	1989

to 1,000 megawatts per year, and with maturing manufacturing technology, it is likely that even lower costs could be achieved (see "Projected manufacturing costs for *a*-Si photovoltaic modules" and, in chapter 10, the sections "Manufacturing thin films" and "The pathway to low cost" for more detailed discussions of manufacturing costs).

Field experience with *a*-Si

As mentioned earlier, the first commercial application of amorphous silicon solar cells was in 1980, when Sanyo introduced a new line of solar-powered calculators. However, it was not until about 1985 that amorphous silicon solar cells began to be used in outdoor applications. Around 1986, test arrays of amorphous silicon photovoltaic modules were set up at several locations by U.S. utilities such as Alabama Power, Arizona Public Service, Philadelphia Electric, and Pacific Gas & Electric (see table 2). The modules were purchased from the major U.S. companies producing amorphous silicon solar cells at that time (Arco Solar, Chronar, Solarex, and Sovonics), and most modules showed initial conversion efficiencies in the range of 5 to 6 percent.

In general, all of these early modules showed significant degradation in the first several months of outdoor operation and then tended to stabilize at conversion efficiencies in the range of 3.3 to 4.2 percent [11], or about 56 to 88 percent of the nameplate rating.

More recently, Pacific Gas & Electric, in cooperation with the U.S. Department of Energy, the Electric Power Research Institute (EPRI), the California Energy Commission, and several other utilities, established the PVUSA program for testing and evaluating both established and emerging photovoltaic technologies. Two U.S. manufacturers, Sovonics and Utility Power Group (UPG), have delivered tandem-junction amorphous silicon modules that have been set up in three arrays ranging from 15.7 to 18.5 kilowatts with system efficiencies ranging from 3.3 to 3.7 percent [12]. Preliminary results indicate that these arrays are showing significantly less degradation than those set up in 1986. Solarex has recently started outdoor testing of triple-junction modules with stabilized efficiencies in the 7 to 8 percent range at the Philadelphia Electric/Solarex test site at Limerick, Pennsylvania. Testing of arrays of these modules should start in 1992.

DESCRIPTION OF AMORPHOUS SILICON AS A SOLAR CELL MATERIAL

Photovoltaic conversion proceeds via the absorption of light quanta that produce electron–hole pairs, followed by the production of electric current (see the appendix to chapter 6 for a description of the photovoltaic effect). Holes are positive-charge carriers, characterized by the absence of negative charges. The first step separates charge in energy and sets up a photovoltage that is proportional to the band gap. In the second step, the charge is separated in space by the field of a *p–n* junction diode. (In the "*p*" portion of a *p–n* junction diode, current is carried by holes, or positive carriers; in the "*n*" portion, by negative carriers.) The properties of *a*-Si relevant to solar cell action are: 1) its optical characteristics; 2) the transport and recombination of positive and negative charge carriers; and 3) the doping to adjust *p*- or *n*-type conductivity. (Doping is the controlled addition of small quantities of specific impurities.)

Optical properties of amorphous silicon

Optical absorption spectrum
The optical absorption spectrum of *a*-Si is shown in figure 3 along with those of crystalline silicon, gallium arsenide, copper indium diselenide, and selected amorphous silicon–germanium (symbolized by *a*-Si,Ge or *a*-Si,Ge:H, where H stands for hydrogen) and amorphous silicon–carbon alloys (*a*-Si,C or *a*-Si,C:H). (For comparison, figure 4 shows the solar spectrum, which corresponds well to absorption regions for these materials.) It turns out that *a*-Si absorbs the visible portion of sunlight so efficiently that *a*-Si solar cells can be made thinner than 1 micron. One advantage of *a*-Si is that its optical absorption spectrum can be changed slightly by varying its hydrogen content, and it can be changed greatly by alloying with carbon or germanium. Changing the absorption spectrum by alloying opens the opportunity for spectral splitting in high-efficiency multijunction cells.

Semiconductors absorb light with photon energies above their band gap ε_{opt} and are transparent to light of lower energy. Therefore, the photocurrent, which is proportional to the absorbed portion of the sunlight, increases as semiconductors with lower band gaps are chosen for the solar cell. The photovoltage, on the

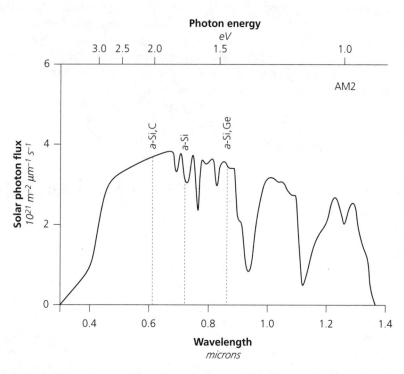

FIGURE 4: Solar spectrum striking the earth. A proposed combination of a-Si *materials for a triple-junction device is shown:* a-Si,C *with a gap of 2.0 eV for the top cell;* a-Si *(1.7eV) for the middle cell; and* a-Si,Ge *(1.45eV) for the bottom cell.*

other hand, is proportional to the magnitude of the band gap and decreases with lower band gaps. The cell efficiency is proportional to its power output—the product of the photocurrent and photovoltage. Because of the trade-off between current and voltage as a function of band-gap energy, cell efficiency peaks for semiconductors with gaps around 1.4 electron volts (eV). The band gap of *a*-Si that is suitable for devices is about 1.75 eV, which lies beyond the optimum value, and so the maximum attainable efficiency of *a*-Si is estimated at between 20 and 25 percent by single-crystal solar cell models, and at about 15 percent by empirical extrapolation from the known performance of *a*-Si cells. Higher efficiencies (theoretically up to 28 percent) can be expected from adjusting the band gap of *a*-Si to 1.4 eV by alloying with germanium.

Still higher efficiencies may be achieved with spectral splitting in multijunction cells; i.e., by using the light transmitted through a large-gap cell into a second cell with a lower band gap, and then possibly using the light transmitted through that cell into a third cell with a still lower gap (see figure 4). Multijunction devices that make use of this spectral splitting are built from individual cells with band gaps carefully adjusted by alloying (see figure 5). The efficiency of such multijunction cells could be raised approximately 50 percent above that of single-junction

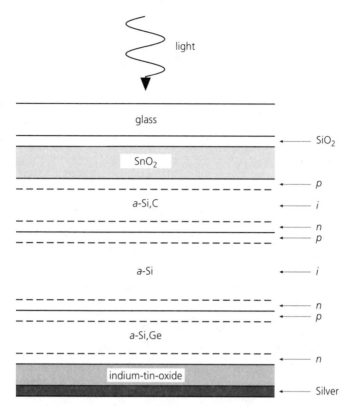

FIGURE 5: *The device structure of a high-performance* a-Si *multijunction module under development.*

cells. Single-crystal models predict efficiencies of about 42 percent for an ideal triple-junction cell [13] and more than 20 percent for empirical estimates based on best-observed properties of *a*-Si. To date, multijunction technology has brought the initial conversion efficiency from 12.5 percent in the best single-junction cell to 13.7 percent [14] and 15.6 percent [15] for two types of multijunction structures. Another advantage of multijunction devices is their inherently higher stability against light-induced efficiency loss, which we describe in a later section.

Light trapping
Light trapping is an important feature of the optical engineering of *a*-Si modules. When the *a*-Si structure is corrugated, or textured, light that enters it is deflected in such a way that the light is reflected internally, back into the cell. The optical path then becomes much longer than the cell thickness, and the optical absorption increases concomitantly. The theoretical maximum for the absorption enhancement is $4n^2$, where *n* is the refractive index of the semiconductor [16]. The *a*-Si structure is corrugated by depositing it on a transparent conductor (tin oxide) that is made to have a surface roughness on a submicrometer scale. The optical

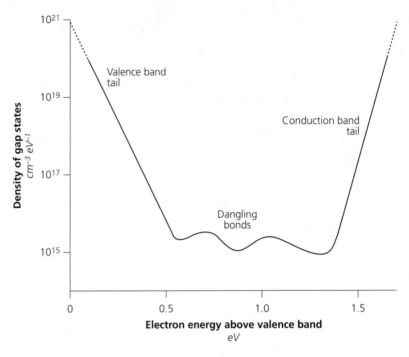

FIGURE 6: The distribution of electron states in the band gap of a-Si. The valence band tail and the dangling-bond states are the recombination centers that reduce the cell performance.

trapping is augmented by applying a rear electrical contact of aluminum, which reflects light back into the cell. When light trapping is used, the *a*-Si cell can be ultrathin (less than 0.4 microns) and hence particularly stable against degradation by light-induced defects.

Charge transport and defects
The electron energy levels in *a*-Si are distributed throughout the band gap. A typical distribution of the density of gap states in *a*-Si is shown in figure 6. In crystalline semiconductors, electronic levels located at energies within the band gap are not allowed by simple electron theory. When such levels occur, they are called defects because they degrade device performance. The continuous distribution of defects, or gap states, in the band gap of *a*-Si is a necessary consequence of the disordered arrangement of silicon atoms and the presence of dangling bonds (broken or incomplete chemical bonds). While this distribution can be detected in the optical absorption spectrum of *a*-Si, it does not affect the production of electron–hole pairs. On the other hand, the separation of the photogenerated charge and the photocurrent production can be strongly affected by gap states. The reason is that the continuous defect distribution promotes electron–hole recombination, which can reduce the photocurrent.

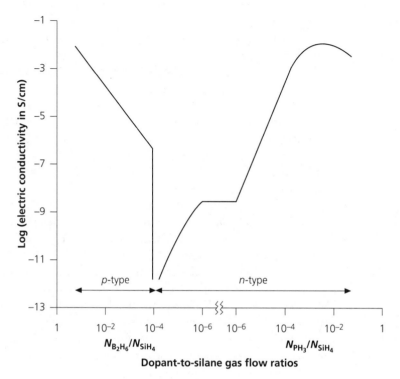

FIGURE 7: Electrical conductivity of a-Si *as set by the addition of dopant gases to the silane feedstock gas. N is the number of molecules per unit volume.*

The atomic arrangement of *a*-Si consists of silicon atoms in a local tetrahedral configuration, but the network becomes random at larger distances. Typically, *a*-Si contains about 10^{19} voids per cm^3: each void is the size of approximately 10 silicon atoms. The hydrogen is bound in Si–H groups. SiH_2 groups (two hydrogen atoms bound to one silicon atom) are considered a telltale sign of poor material. Approximately 5 atomic percent hydrogen is randomly distributed, and the remainder is clustered, in part at the surfaces of the voids.

Doping and alloys

The purpose of doping is to adjust, through the controlled addition of small amounts of specific impurities, the electrical conductivity type (*p* or *n*) and the magnitude of the conductivity. Alloying is the mixing of *a*-Si:H with large proportions of other elements, with the primary purpose of shifting the optical absorption spectrum of *a*-Si.

The principal doping elements used in *a*-Si are the same as in crystalline silicon: boron and phosphorus. (Figure 7 shows how conductivity varies with the concentration of dopant gas added to SiH_4.) The doping efficiency—the number of charge carriers produced per dopant atom—varies from 10^{-2} to 10^{-4}. This means that relatively large concentrations of dopant elements must be introduced if highly conducting material is desired. Doping *a*-Si raises its optical absorption and its density of recombination centers, so that the highly doped *p*- and *n*-type layers of a *p–i–n* cell absorb sunlight strongly but then destroy the photogenerated carriers by recombination. Therefore, the *p*-type layer of a conventional *p–i–n* cell is made with the smallest thickness allowed by device physics and is alloyed with carbon to raise the value of the band gap and restore optical transmission [17].

The doped *p*- and *n*-type layers in an *a*-Si *p–i–n* cell fulfill two functions. First, they set up the electric field in the *i*-type layer. This field increases with doping. The second function of these layers is to establish low-loss electrical contacts between the *a*-Si structure and a transparent conductor (tin oxide, SnO_2) on the *p*-side, and a metal (aluminum) on the *n*-side. In monolithic, or directly stacked, multijunction cells the *p*-type layer of one cell is in contact with the *n*-type layer of the cell on top of it. Photocurrent flows through such contacts in the blocking direction.[4] A very high recombination (or tunneling) current is desirable for minimizing power loss at such contacts, and this is achieved by doping as highly as possible or by interposing a special recombination layer.

We described in the section "Optical absorption spectrum" how alloys of *a*-Si with carbon or germanium are exploited to adjust the band gap to change the spectral response of the solar cell. Such alloys are used in both the doped and undoped layers of multijunction cells. As mentioned above, *a*-Si,C alloys also are used for widening the band gap of the *p*-type layer to make it more transparent. Furthermore, alloying with carbon is used to grade, or adjust, the band gap in the *i*-type layer for more efficient charge collection.

Light-induced degradation

The power output of *a*-Si modules is observed to degrade over time. Three groups of degradation mechanisms have been identified. One group is associated with the formation of particles in the deposition machines. These particles may become embedded in the solar cell structure and eventually lead to power loss. Such degradation mechanisms are being eliminated by improved deposition techniques. A second group of degradation modes is tied to the long-term exposure of the modules to the atmosphere and to temperature variations; these are discussed in the section "Reliability and stability". A third type of degradation is caused by the light-induced change of the *a*-Si itself. In essence, illumination of *a*-Si produces more recombination centers and so impairs photocurrent collection. This is the Staebler–Wronski effect [18].

The light-induced defects are dangling bonds that are produced by the trapping or recombination of photogenerated carriers. These defects are metastable and can be removed by annealing the *a*-Si for several minutes at about 200°C. The microscopic nature of the defect mechanism has not yet been identified, but a number of models have been proposed [19].

---

4. An unilluminated *p–n* junction in an electrical circuit acts as a diode, allowing current to flow in one direction but blocking it if the current direction is reversed.

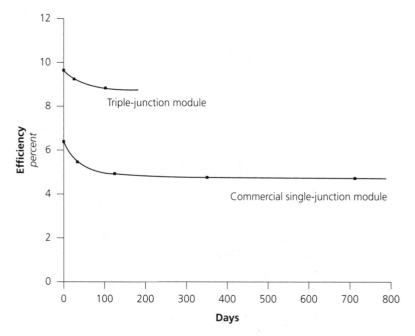

FIGURE 8: Variation of conversion efficiency with days of exposure to sunlight for a triple-junction module and a commercial single-junction module.

Two approaches have been taken to reduce the effect of light-induced degradation on cell efficiency. A powerful practical approach is a cell design that ensures a long carrier collection length even at a high defect density. Such designs use very thin cells, while keeping light absorption high by optical trapping or by stacking cells on top of each other. Tandem or triple cells for this purpose are designed for stability enhancement rather than efficiency enhancement. The second approach to reducing the susceptibility against the Staebler–Wronski effect is to improve the stability of the *a*-Si material itself.

The light-induced degradation of an amorphous silicon *p–i–n* cell can be minimized by making the *i*-type layer thin, as this ensures a strong internal electrical field that sweeps out photogenerated carriers before trapping or recombination can occur and create metastable centers [20]. The *i*-type layer of most commercial modules typically has a thickness of 250 to 350 nanometers, and, as shown in figure 8, the modules show a degradation in power output of about 20 to 25 percent in the first several months of outdoor operation before stabilizing. (This compares to a degradation of roughly 50 percent before stabilization in the first commercial modules, where the *i*-type layer thicknesses were about 500 nanometers.) The stabilized conversion efficiency of present modules is about 5 percent, corresponding to a power output of about 5 watts for a 1 ft^2 (900 cm^2) module.

Much higher stabilized conversion efficiencies can be attained with multi-junction cells. In fact, stabilized efficiencies of about 10 percent have recently been reported for two-junction small-area devices using the same band gap amorphous silicon in both junctions [21]. As figure 8 shows, 1 ft^2 multijunction modules are presently showing stabilized conversion efficiencies of about 8 to 9 percent.

The improved stability of multijunction modules is due to several factors. First, the *i*-type layer of the top junction is usually quite thin (60 to 100 nanometers), hence it is relatively stable. In addition, the other junctions absorb less light than a single-junction cell and thus they degrade more slowly. Slower degradation coupled with annealing leads to higher stabilized efficiencies. Furthermore, the amorphous silicon–germanium alloys that are commonly used in the bottom junction (see figure 5) are relatively stable when the germanium content is greater than about 30 percent [22], although there is not yet a good understanding of why this should be so.

At present, the best stabilized efficiencies are 10 percent in small-area cells, 8 percent in laboratory modules, and 5 percent in commercial modules. With projected advances in *a*-Si technology, it is likely that stabilized module efficiencies will reach 10 percent in mass-produced multijunction modules in the next few years and 18 percent or more in the longer term.

MANUFACTURING TECHNOLOGY

Methods of depositing amorphous silicon

Solar-cell-quality amorphous silicon films are deposited at temperatures between 200°C and 250°C. Approximately 10 atomic percent of hydrogen is incorporated at these temperatures, so that typical *a*-Si has the approximate chemical composition $Si_{0.9}H_{0.1}$ and is often referred to as *a*-Si:H. The source of silicon used for most deposition processes is the silicon-bearing gas, silane (SiH_4); *a*-Si films grow by attachment of reactive fragments (radicals) of silane to the surface of the growing film. Such radicals can be made by colliding electrons with silane molecules, by thermally dissociating or photodissociating silane or disilane (Si_2H_6), or by sputtering silicon atoms off the surface of a silicon electrode in the presence of hydrogen. In the search for *a*-Si with improved photovoltaic properties, many deposition techniques in a variety of geometries have been studied. Industrial and device laboratory practice is dominated by deposition from glow discharges.

In glow discharge deposition, the energy required to break up the gas molecules is provided by collisions with electrons. The electrons originate as secondary electrons in the collisions and receive their energy by acceleration in an electric field. To pick up enough energy for the fragmentation of silane molecules, the electrons must travel in the field for a sufficient distance. Therefore, the mean free path between collisions with silane molecules must be kept relatively large, which is accomplished by maintaining a low gas pressure. Some of the energy transferred to the silane in the electron collisions is radiated as visible light (from excited radicals), hence the name glow discharge. The process is similar to that in making a fluorescent lamp glow. Interestingly, the electric power density employed in *a*-Si

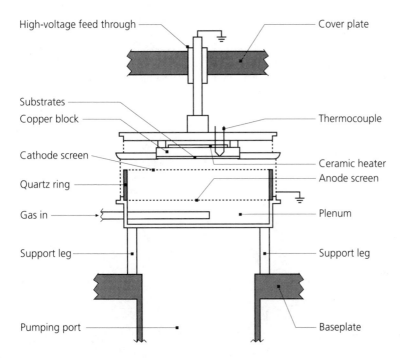

FIGURE 9: Cross section through the deposition region of a direct -current (DC) triode system. The discharge is maintained between the anode and the cathode screens. The substrate is grounded, so that energetic ions from the discharge are decelerated as they move from the cathode to the growing film.

deposition is comparable to the intensity of terrestrial sunlight, on the order of 1 kilowatt per m^2 of film surface.

The voltage setting up the electric field may be direct current (DC) or alternating (radio-frequency, or RF), as illustrated in figures 9 and 10. Of industrial importance are the number of electrodes per growth zone. In diodes, the glow discharge is maintained between two electrodes, one of which serves as the holder for the cell substrate. In triodes, the glow discharge is maintained between a solid electrode and a grid; the substrate on which the *a*-Si is grown is mounted on a third electrode, which is placed beyond the grid. This geometry moves the substrate away from short-lived highly energetic particles in the glow discharge, which could damage the growing film by bombardment.

The principal control parameters in glow discharge deposition of *a*-Si are the flow rates of the feedstock gases (silane, dopant gases, hydrogen), the pressure in the glow discharge, the electrical power fed into the discharge, and the temperature of the substrate on which the film grows. The deposition conditions may be varied over some range while maintaining the photovoltaic quality of the deposited *a*-Si. Scaleup to large substrates requires careful design to produce uniform thickness while ensuring efficient use of the source gases.

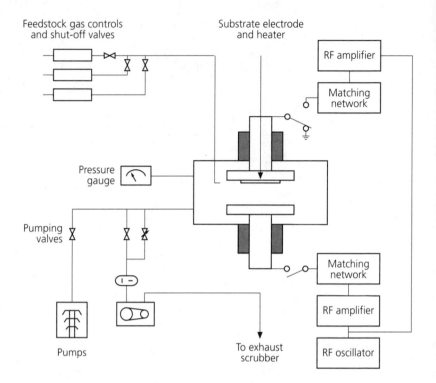

FIGURE 10: Layout of a radio-frequncy (RF) diode deposition system. Equipment controls the four deposition variables: gas flow on the upper left; pressure on the lower left; temperature at the top; power at the right.

Solar cell structures

While amorphous silicon solar cells can be fabricated in a variety of structures, the vast majority of commercial cells are made in the configuration shown infigure 1. This structure uses window glass (soda-lime-silicate glass) as a superstrate and has been used by companies such as Chronar (now Advanced Photovoltaic Systems, or APS), Sanyo, Siemens, and Solarex [23]. Generally, an electrode of transparent conductive oxide (TCO) (typically indium-tin-oxide (ITO), tin oxide, or zinc oxide) is first deposited onto the glass. In most commercial cells, the TCO electrode is deposited using atmospheric-pressure chemical vapor deposition (CVD) of fluorine-doped tin oxide. A thin silicon dioxide (SiO_2) buffer layer (about 50 nanometers) is often deposited on the glass before the tin oxide electrode to prevent diffusion of alkali atoms from the glass into the tin oxide and the amorphous silicon. The tin oxide layer is typically about 600 nanometers thick and is often grown with a submicron texture to enhance light trapping within the amorphous silicon.

Next, the amorphous silicon layers are usually deposited by means of plasma-enhanced chemical vapor deposition (glow discharge deposition) in an atmosphere of silane (SiH_4) or silane diluted in hydrogen (at a reduced pressure of roughly 0.001 atmospheres). Gases such as diborane (B_2H_6) or phosphine (PH_3)

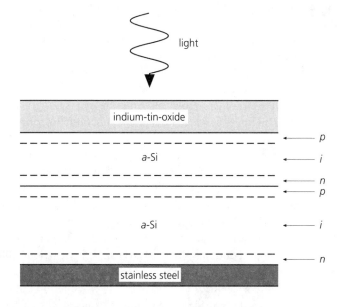

FIGURE 11: The device structure of a commercially available a-Si *double-junction photovoltaic module that is fabricated on a steel foil substrate.*

are added to the discharge atmosphere to obtain either *p*-type or *n*-type doped films, respectively. In most cases, the *p*-type layer is deposited first onto the transparent conductive oxide, and methane (CH_4) is also added to the discharge atmosphere to form a wide band gap ($\varepsilon_{opt} > 2.0$ eV) amorphous silicon–carbon *p*-type layer. The *p*-type layer is typically 10 nanometers thick, the *n*-type layer is about 30 nanometers thick, and the *i*-type layer is generally 250 to 350 nanometers thick.

The back contact of commercial *p–i–n* cells is usually formed by sputtering about 400 nanometers of aluminum onto the *n*-type layer. The cell is then protected from the environment by applying an encapsulation layer. Generally, the encapsulation is applied by either laminating materials such as ethyl vinyl acetate (EVA) and Tedlar to the aluminum, laminating a second sheet of glass to the back of the cell, or spray coating the back with an acrylic paint.

Another solar cell structure that has been used by companies such as Sharp and Sovonics is shown in figure 11 [24]. In this case, a multijunction amorphous silicon structure is deposited on a stainless steel foil and the top contact electrode is formed by the electron-beam evaporation of indium-tin-oxide (ITO). The amorphous silicon is deposited from a discharge atmosphere containing silane, silicon tetrafluoride, and hydrogen. The device consists of two series-connected *p–i–n* junctions (tandem structure), where the thickness of the top *i*-type layer is about 80 nanometers and the thickness of the bottom *i*-type layer is approximately 300 nanometers. The thicknesses are adjusted so that the photocurrents from each junction are comparable.

These multijunction cells are fabricated with heavily doped *p*- and *n*-type layers so that the connecting *p–n* junction is dominated by recombination and does not contribute an opposing voltage. The *p*-type layer in this kind of cell is usually boron-doped microcrystalline silicon. When the multijunction structure is scaled up to large areas, a metal grid is deposited on top of the indium-tin-oxide to assist in collecting the photocurrent.

Another type of multijunction solar cell structure that is not yet commercially available is shown in figure 5, where three *p–i–n* junctions are formed in series. In this case, the *i*-type layer of the top junction is fabricated from an amorphous silicon–carbon alloy (ε_{opt} = 1.9 to 2.0 eV), the *i*-type layer of the middle junction is amorphous silicon (ε_{opt} = 1.75 eV) and the *i*-type layer of the bottom junction consists of an amorphous silicon–germanium alloy (ε_{opt} = 1.4 to 1.5 eV). In this type of multijunction cell, short-wavelength light (about 400 to 500 nanometers) is converted into a photocurrent at a relatively high photovoltage in the top junction while intermediate-wavelength light (about 500 to 600 nanometers) is absorbed in the middle junction, and longer-wavelength light (about 600 to 800 nanometers) is absorbed in the bottom junction.

As shown in figure 5, the back contact of this triple-junction cell consists of a layer of indium-tin-oxide capped with a thin film of silver. The back contact is an effective mirror for the weakly absorbed infrared light, and therefore the combination of a textured tin oxide front contact and the highly reflective back contact increases the short-circuit photocurrent by increasing the optical path length of the weakly absorbed light within the cell [25].

The possibility of marrying *a*-Si–based solar cells with other materials such as CIS is being investigated, but monolithic solar cell structures have not yet been characterized.

Manufacturing methods: current state of the art

There are several manufacturing facilities in various parts of the world that are capable of producing on the order of 1 megawatt of amorphous silicon solar cells per year (e.g., facilities set up by Chronar in China, France, Taiwan, and Yugoslavia, the Canon–ECD plant in Michigan, and the Solarex facility in Pennsylvania). Advanced Photovoltaic Systems, the successor to Chronar, is planning to put a 10 megawatt per year plant into operation in California by 1992. Most of these plants use batch processes and are relatively labor intensive.

A new, totally automated pilot plant was brought on line in 1990 by Solarex. This automated manufacturing line can produce about 1 megawatt of 1 ft^2 single-junction solar cells (see figure 1) per year. The line produces encapsulated, fully functional, 5 watt modules without manual handling. The flow diagram for this automated facility is shown in figure 12.

Ordinary window glass is automatically loaded into a glass washer and then fed into an atmospheric CVD belt furnace. A thin layer (about 50 nanometers) of silicon dioxide is deposited onto the glass followed by about 600 nanometers of textured tin oxide. After the coated plates leave the furnace, the sheet resistance

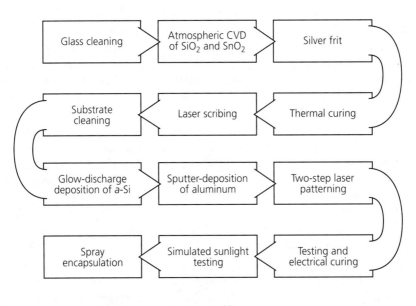

Automated line

FIGURE 12: The flow chart of a totally automated manufacturing line operating in Newtown, Pennsylvania (Solarex Thin Film Division).

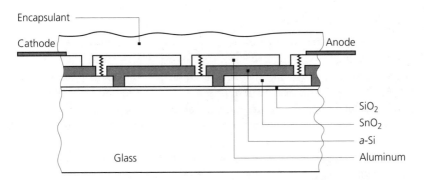

FIGURE 13: Diagram showing the interconnect region of a series-connected a-Si *photovoltaic module.*

of the tin oxide is measured in an automated test station. A silver frit is then applied to the tin oxide and cured in an annealing belt furnace to form conductive busbars. The tin oxide is then laser scribed to form the front contact segments of the individual cells (see figure 13).

The plates are cleaned and loaded into a multichamber vertical deposition system that can simultaneously coat 16 ft^2 (1.4 m^2) of substrates with amorphous silicon. The chambers are integrated with a vertical magnetron sputtering system that applies the aluminum back contact. The plates are fed into a laser system that uses a machine vision system to align the plates and scribe the aluminum back contact. Another laser system completes the patterning by fusing the aluminum to the tin oxide in the interconnect regions. A computerized system then tests each segment (cell) and applies a reverse bias to defective segments to burn out defects. The modules are tested under simulated sunlight conditions and in the final processing step are spray coated with an encapsulating paint.

The modules may undergo further processing depending on the application. For example, in the case of automobile battery maintainers, the square-foot plates are cut into three sections. The encapsulation is locally abraded in the region of the silver frit busbars so that current leads can be soldered to the conductive frit, and a sealant is subsequently applied in the vicinity of the soldered region. The modules are mounted in an injection-molded plastic frame and subjected to a final power test under simulated sunlight before they are shipped to customers.

Module performance

The performance of a solar cell is typically quantified by measuring the conversion efficiency under normal sunlight conditions (corresponding to a power density of 1 kilowatt per m^2) and by measuring the spectral response of the cell. Initial conversion efficiencies in the range of 12 percent have been measured in small-area single-junction amorphous silicon cells [25] where the device structure was somewhat similar to that shown in figure 1. However, to enhance performance, these cells are fabricated with a high-reflectivity back contact of ITO/silver (as in the case of the multijunction structure shown in figure 5), and the carbon content is graded in the vicinity of the *p–i* interface.

Somewhat higher conversion efficiencies have been measured in small-area triple-junction cells with the structure ITO/*p–i–n*/*p–i–n*/*p–i*–n*/metal, where the *i**-type layer is an amorphous silicon–germanium alloy with an optical band gap of about 1.5 eV. With this type of multijunction structure, initial conversion efficiencies as high as 13.7 percent have been obtained [14].

Even higher conversion efficiencies have been obtained in multijunction cells that combine amorphous silicon with other semiconducting materials. For example, a conversion efficiency of 16.8 percent has been obtained in a four-terminal structure consisting of an amorphous silicon *p–i–n* cell with transparent contacts that was optically coupled to a conventional *p–n* junction polycrystalline silicon cell [26]. An initial conversion efficiency of 15.6 percent has been obtained in another four-terminal multijunction structure that optically coupled an *a*-Si *p–i–n* cell to a copper-indium-diselenide cell [15]. (The *a*-Si cell contributed 10.3 percent and the CIS cell 5.3 percent to the total efficiency of 15.6 percent; the CIS cell alone was 12.4 percent efficient.) In general, the manufacturing cost of a four-terminal structure is expected to be higher than that of a two-terminal multijunction structure of comparable performance because two manufacturing lines are required in the former case. In theory, only marginal efficiency gains can be obtained in four-terminal devices, so that two-terminal multijunctions would be

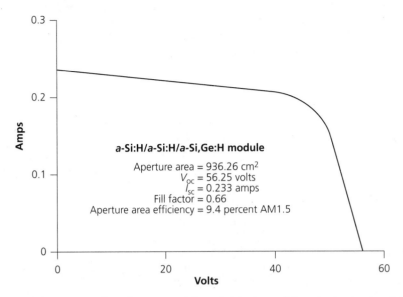

FIGURE 14: Current-voltage characteristics of a triple-junction module under simulated sunlight conditions.

preferable economically, once materials issues at the interface between the two cells are solved.

Good progress has been made in scaling up these laboratory cells into large-area photovoltaic modules. Modules exploiting the four-terminal, *a*-Si/CIS structure have shown initial efficiencies as high as 12.3 percent (aperture area = 843 cm^2) [27]. Initial conversion efficiencies as high as 10.0 percent have recently been observed at Solarex in two-terminal multijunction modules (aperture area = 935 cm^2) where the structure is similar to that shown in figure 5. The current–voltage characteristics of a 28 segment triple-junction module are depicted in figure 14, and the spectral response is shown in figure 15. Unlike a single-junction amorphous silicon module, this type of multijunction module is able to use near-infrared radiation at a wavelength of 800 nanometers due to the presence of an amorphous silicon–germanium alloy (ε_{opt} = 1.4eV) in the bottom junction.

Reliability and stability

The reliability and operational lifetime of amorphous silicon solar cells are largely determined by the encapsulation and packaging of the modules but can also be adversely affected by light-induced degradation of the amorphous silicon [18].

To estimate lifetime, modules are subjected to a battery of accelerated environmental tests in an attempt to simulate long-term outdoor conditions. These tests include: thermal cycling (250 cycles from −40°C to 90°C); humidity soak testing (4 weeks at 85°C, with 85 percent relative humidity); humidity-freeze cycling (10 cycles from −40°C to 85°C, 85 percent relative humidity); hail impact testing; torsional stress testing; wet high-voltage testing (< 50 microamps leakage

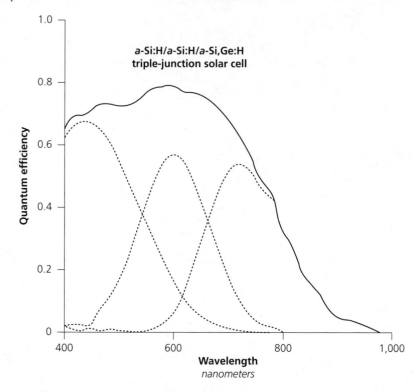

FIGURE 15: Spectral response of a triple-junction solar cell.

at > 1,500 V after spraying with water); hot water immersion testing (immersion in 40°C water for five days while illuminated); ultraviolet exposure testing; and salt-fog spray testing.

Most commercial amorphous silicon modules currently on the market do not pass the wet high-voltage test. This is not a problem for applications that involve battery charging, but the high-voltage test is required for high-voltage applications. It is typically needed for systems connected to a utility grid. Manufacturers have recently started producing amorphous silicon modules with an improved spray-coated encapsulation layer, and these modules pass all the tests mentioned above.

While some investigators have reported the diffusion of dopants at relatively low temperatures, properly fabricated cells show no evidence of diffusion of dopants or impurities that could limit the lifetime of modules under normal operating conditions [28]. Thus, the present state-of-the-art modules should last for at least 10 years outdoors and perhaps much longer before experiencing any environmental degradation. More field experience and testing are required before longer operational life can be ensured.

A number of failure mechanisms have been observed in the amorphous silicon modules at the various test sites, including delamination in those modules using plastic cover sheets, bad interconnects or junction boxes [12], cracks in the glass cover sheets or substrates, and corrosion effects due to moisture penetrating the edge seals [29]. In general, these problems have been found in only a small fraction (less than 4 percent) of the modules in the various test arrays.

In fact, the reliability of some amorphous silicon photovoltaic test arrays is quite good. A common measure of reliability is *availability*, which is defined as that percentage of time that an array actually operated compared with the total time that it could have operated (i.e., when there was sufficient sunlight to operate). For a 4 kilowatt array operated by Detroit Edison, the availability was on the order of 90 percent from 1987 to 1988 [30]. A 15 kilowatt array operated by the Florida Solar Energy Center has shown an availability of 99 percent over a 10 month period [29].

Another measure of array performance is the capacity factor, which is defined as the percentage of energy produced in a year compared with the energy that the array would produce if it operated at its maximum output continually for a year. The 15 kilowatt array at the Florida Solar Energy Center has shown a capacity factor of 21 percent, while a 17.3 kilowatt array at the Davis, California, PVUSA site has exhibited a capacity factor of 22 percent. In the case of the PVUSA site, the capacity factor of 22 percent is similar to that shown by an array using single-crystal silicon solar-cells.

In early field tests (see table 2), most *a*-Si modules showed light-induced degradation, stabilizing at a 3.3 to 4.2 percent efficiency, or only 56 to 88 percent of their nameplate rating. As discussed earlier (see "Light-induced degradation") the mechanism for the light-induced degradation of amorphous silicon solar cells has not been clearly identified. However, it can be minimized by a cell design that exploits thin layers and multijunctions. More recent tests show less degradation. At present, *a*-Si multijunction solar cells stabilize after a few months at 85 to 90 percent of their initial efficiency.

Projected manufacturing costs for *a*-Si photovoltaic modules

As mentioned earlier (see also chapter 6), photovoltaic modules are currently being used in a range of remote applications and may soon be used in some grid-supported applications where additional value can be assigned to the photovoltaic system. However, there must still be significant decreases in the manufacturing cost of photovoltaic modules before they find widespread use in most grid-connected applications. Here we examine how the cost of manufacturing *a*-Si PV modules is likely to evolve (see also chapters 6 and 10 for discussions of thin-film manufacturing costs).

Materials, labor, and equipment depreciation contribute to the overall manufacturing cost. Costs are estimated here for a highly automated near-term factory producing 10 megawatts per year of 10 percent efficient 2 × 4 foot frameless PV modules (10^5 m^2 per year). It is assumed that the plant operates three shifts per day, five days per week. The flow diagram for this process would be similar to that depicted in figure 12, where multiple *p–i–n* junctions are deposited using the

Table 3: Summary of PV module material costs for multijunction *a*-Si modules[a]

Material	Cost $/m^2$	Percent of total
Encapsulation	11.3	36.0
Glass[b]	5.6	18.8
Germanium	5.0	16.0
Silane	2.7	8.6
Aluminum (cathode)	1.9	6.2
All other materials	4.5	14.4
Total	**31.3**	**100.0**

a. Costs are given for a factory producing 10 MW_p per year of 10 percent efficient multijunction *a*-Si modules.

b. Normal window glass. Some modules have been fabricated using chemically strengthened glass that costs $14 per m^2.

glow discharge method. Estimates are also given for what might be possible in the long term in larger plants.

Materials costs

Table 3 lists the materials used in the manufacture of multijunction *a*-Si PV modules and also shows the cost per m^2, assuming a 90 percent yield. The estimates are based on actual procurement costs at Solarex in 1989. As shown in the table, the encapsulant and the glass are estimated to be the two largest material costs in future large-scale plants producing multijunction modules [31]. The total thickness of all the thin films in the structure is on the order of 1.5 microns, and the cost of the semiconductor feedstock ($7.7 per m^2) is less than one-fourth of the total materials cost of $31.3 per m^2.

Labor costs

Since the process is completely automated, the plant requires only a few people to monitor the equipment and assist in shipping and receiving. We assume that four operators and one supervisor are needed to run the plant and that other personnel include a quality control/process engineer, four maintenance people, and four people in administration, including a plant manager. At three shifts per day, five days per week, the total labor costs (including benefits and indirect labor costs) are about $6.44 per m^2.

Capital costs and other indirect costs

The equipment needed to process *a*-Si modules is listed in table 4. A rough estimate of the cost of the equipment is also given. Some of the equipment, such as in-line glass washers and CVD belt furnaces, is available commercially, while other

Table 4: Factory capital-equipment costs[a]

Equipment	Estimated cost
	millions $
Glass cleaner (4)	0.2
Chemical vapor deposition furnaces (2)	1.8
Laser systems (12)	1.8
a-Si systems (2)	2.0
Metallization (2)	1.4
Lead attachment	0.2
Encapsulation	0.6
Testing/curing	0.4
Computerization	1.0
Material handling	1.0
Total	**10.4**

a. Costs are given for a factory producing 10 MW_p per year of 10 percent efficient multijunction a-Si modules.

machines must be either constructed or modified to fit into an automated line. The total capital cost for a 10 megawatt per year facility is estimated to be $10.4 million. If we assume straight-line depreciation on the equipment over 10 years, then the average depreciation per year is $10.40 per m^2. (The cost here is given in constant dollars. The cash flow is not discounted. No profit or selling/marketing expenses are included. The effect of taxes and interest is not included in this estimate.)

Other indirect costs include rent, utilities, and insurance (see table 5). For a 10 megawatt per year plant, these costs add a total of about $8.30 per m^2.

Summary of costs
In table 6, we summarize the costs in dollars per m^2 and in terms of the percentage of the total cost. The total cost of 10 percent efficient modules from a 10 megawatt per year plant based on near-term manufacturing technology is about $56.5 per m^2 or $0.565 per peak watt, the cost of materials constituting more than half of the total manufacturing cost.

Since nonsemiconductor materials such as glass and encapsulation are a major factor in determining the cost of thin-film PV modules, the cost of the semiconductor material is secondary, and other factors such as performance, yields, equipment uptime, and equipment costs will play important roles in determining the lowest cost thin-film approach. While the transition from single-junction to multijunction *a*-Si modules adds material cost (about $0.05 per watt for germanium) and equipment cost (about $600,000 for additional deposition chambers),

Table 5: Other indirect manufacturing costs[a]

Item	Cost thousand $/year
Electricity	400
Rent	150
Repair of equipment	100
Janitorial services	50
Insurance	50
Waste disposal	10
Outside services	10
Telephone, water	10
Miscellaneous	50
Total	**830**

a. Costs are given for a factory producing 10 MW$_p$ per year of 10 percent efficient multijunction a-Si modules.

Table 6: Summary of PV module costs[a]

Item	Cost $/m^2$	Percent of total
Materials	31.3	55.4
Labor	6.5	11.5
Equipment depreciation	10.4	18.4
Other indirect costs	8.3	14.7
Total	**56.5**	**100.0**

a. Costs are given for a factory producing 10 MW$_p$ per year of 10 percent efficient multijunction a-Si modules. Depreciation is calculated assuming a straight-line 10 year depreciation of capital costs for equipment. No profit or inflation is included.

these additional costs are more than offset by the doubling of the stabilized conversion efficiency.

Several organizations have estimated the manufacturing costs for producing amorphous silicon solar cells in multimegawatt facilities, and the results of these analyses are shown in table 7. In general, the lower estimates were made assuming a greater degree of automation and also do not include taxes and nonmanufacturing expenses (such as marketing and sales support), which were included in the

Table 7: Estimates of PV module manufacturing costs[a]

Total cost $/W_p$	Plant size MW_p/year	Module efficiency percent	Module structure	Reference
0.57	10	10.0	multijunction	31
0.69	50	10.0	single junction	33
0.92	10	8.7	single junction	34
1.00	10	5.0	single junction	35
1.22	25	15.0	multijunction	32

a. All costs are in 1989 U.S. dollars.

high estimate of $1.22 per watt. The estimates for capital investment range from $10.4 million for a 10 megawatt per year facility [31] to $21.8 million for a 25 megawatt per year facility [32].

Cost estimates for larger plants
In the long term, further cost reductions might be realized in larger sized plants. Increasing the size of the plant to 1 million m^2 per year (100 megawatts) might reduce indirect costs by about a factor of two. The total manufacturing cost would then be about $0.47 per watt. Other cost reductions are possible. For example, the cost of glass could be reduced by building a float-glass plant at the front end of a large PV plant (> 10 million m^2 per year). The cost of silane could be reduced by producing it onsite (this would make economic sense at 1 million m^2 per year). Moreover, it might be possible to reduce the cost of the encapsulation material and use a less expensive source of germanium. Designing the plant to run larger plates of glass would also reduce the cost, provided the overall yield could be maintained at 90 percent. Further cost reductions would occur with yields greater than 90 percent, which should be possible with a well designed automated line. Finally, improving the efficiency of the module from 10 percent to 12 percent would reduce the cost from $0.57 to $0.47 per watt in a 10 megawatt plant. If efficiencies of 18 percent were achieved, costs would be even lower.

The optimum plant size may be as large as 1 gigawatt per year, because it may then be economical to integrate a float-glass plant as well as a silane production facility into the module manufacturing plant. While it may be economical to integrate a small float-glass plant into a 100 megawatt per year PV plant, the optimum size for a float-glass plant is about 10 million m^2 of glass per year.

Since manufacturing plants with capacities of about 10 megawatts per year should be producing modules at a total cost on the order of $1 per watt by the mid-1990s, it seems likely that photovoltaic systems using amorphous silicon multijunction modules will start to penetrate the utility grid system by the mid-

1990s, especially in those regions where the photovoltaic systems have additional value. By the late 1990s, costs would be about $0.6 per watt for 10 percent efficient multijunction modules from a 10 megawatt per year plant. In the longer term, with increased plant size and higher efficiency, module costs could fall below $0.50 per watt.

FUTURE PROSPECTS

The present amorphous silicon technology base should lead to the production of large-area multijunction modules with stabilized conversion efficiencies of about 10 percent by the mid-1990s. With manufacturing costs on the order of $1 per watt, these modules should find widespread use in remote applications and as peak power systems supporting feeders where investments in transmission and distribution equipment can be deferred. However, even higher conversion efficiencies may be required before thin-film photovoltaic modules can penetrate significant portions of the central-station generating market.

Efficiency is a very important factor in reaching low cost, and reaching high stabilized efficiencies is perhaps the biggest challenge facing *a*-Si technology. For *a*-Si, the key to higher conversion efficiencies is improvement in the electronic properties of the amorphous silicon-based alloys as well as further development of microcrystalline and polycrystalline thin-films (such as CIS) that can also be incorporated into multijunction structures. At present, both the narrow-band gap amorphous silicon–germanium alloys and the wide-band gap amorphous silicon–carbon alloys contain relatively high concentrations of defects associated with microvoids, germanium or carbon clusters, and hydrogen complexes. In particular, further improvement in the electronic properties of the amorphous silicon–carbon alloys could have a significant impact on the performance of triple-junction structures, such as the one shown in figure 5. In this type of structure, an improved amorphous silicon–carbon alloy could be used not only in the top *i*-type layer but also in all six of the doped layers.

Further improvements in performance would come from the elimination of the light-induced degradation shown by most of the amorphous silicon-based alloys. The conversion efficiency would also be increased if the transmission and conductivity of the transparent conductive oxide were increased and if the light-trapping behavior of the solar cell structure were enhanced.

Even further improvements in the performance of thin-film multijunction PV modules might come from incorporating polycrystalline thin-films into the device structure. For example, a triple-junction device with the structure *a*-Si,C/*a*-Si,Ge/CIS might eventually show conversion efficiencies in excess of 20 percent. In this case, the band gaps of the three materials would be about 2.0 eV, 1.5 eV, and 1.0 eV, respectively, in order to optimize the use of the solar spectrum.

With further improvements in performance and continuing reductions in manufacturing costs as well as balance-of-systems costs, photovoltaic power systems may be able to provide significant amounts of low cost power in the early part of the next century.

ACKNOWLEDGMENTS

The authors wish to thank Joan Ogden of the Center for Energy and Environmental Studies, Princeton University, for her help in reviewing and improving this chapter.

WORKS CITED

1. Maycock, P. 1991
   PV News 10(2), February.

2. Chittick, R.C., Alexander, J.H., and Sterling, H.F. 1969
   The preparation and properties of amorphous silicon, *Journal of the Electrochemical Society* 116:77–81.

3a. Carlson, D.E. 1977
   Semiconductor device having a body of amorphous silicon, U.S. patent 4,064,521.

3b. Carlson, D.E. and Wronski, C.R. 1976
   Amorphous silicon solar cell, *Applied Physics Letters* 28(11):671–773.

4. MacNeil, J., Delahoy, A.E., Kampas, F., Eser, E., Varvan, A., and Ellis, F. 1990
   A 10MW$_p$ *a*-Si:H module processing line, Conf. Record, IEEE Photovoltaic Specialists Conf., 21st, Kissimmee, Florida, pp. 1501–1505. IEEE, New York.

5. LeComber, P.G. 1989
   Present and future applications of amorphous silicon and its alloys, *Journal of Non-Crystalline Solids* 115:1–13.

6. Kanicki, J., ed. 1991
   Amorphous and microcrystalline semiconductor devices: optoelectronic devices, Artech House, Norwood, Massachusetts.

7. Stone, J.L. 1990
   Recent advances in thin-film solar cells, Tech. Digest Internat. PVSEC-5, Kyoto, pp. 227–232, Kyoto University, Kyoto.

8. Ichikawa, Y. 1990
   Recent progress in stability of *a*-Si solar cells, see Ref. 7, 617–621.

9. Hamakawa, Y. 1990
   Recent advances in solar photovoltaic technologies in Japan, see Ref. 7, 221–226.

10. Catalano, A. 1990
   Recent advances in *a*-Si:H alloy multijunction devices, see Ref. 7, 235–238.

11. Jennings, C. and Whitaker, R.C. 1990
   PV module performance outdoors at PG&E, see Ref. 4, 1023–1029.

12. Hester, S.L., Townsend, T.U., Clements, W.T., and Stolte, W.J. 1990
   PVUSA: lessons learned from start-up and early operation, see Ref. 4, 937–943.

13. Fan, J. C.C., Tsaur, B.-Y., and Palm, B.J. 1982
   Optimal design of high-efficiency tandem cells, conference record, 16th IEEE Photovoltaic Specialists Conference, San Diego, 692–701, IEEE, New York.

14. Yang, J., Ross, R., Glatfelter, T., Mohr, R., Hammond, G., Bernotaitis, C., Chen, E., Burdick, J., Hopson, M., and Guha, S. 1988
High efficiency multijunction solar cells using amorphous silicon and amorphous silicon–germanium alloys, conference record, 20th IEEE Photovoltaic Specialists Conference, Las Vegas, 241–246, IEEE, New York.

15. Morel, D.L., Blaker, K., Bottenberg, W., Fabick, L., Felder, B., Gardenier, M., Kumamoto, D., Reinker, D., and Rifai, R. 1988
Development and performance of 4-terminal thin film modules, *Proceedings of the European Community 8th Photovoltaic Solar Energy Conference*, Florence, Italy, 661–665, Kluwer Academic Publishers, Dordrecht, the Netherlands.

16. Tiedje, T., Abeles, B., Cebukla, J.M., and Pelz, J. 1983
Photoconductivity enhancement by light trapping in rough amorphous silicon, *Applied Physics Letters* 42:712–715.

17. Tawada, Y., Kondo, M., Okamoto, H., and Hamakawa, Y. 1982
Characterization of *a*-SiC:H as a window material for *p–i–n a*-Si solar cells, *Japanese Journal of Applied Physics* 21(2):291–297.

18. Staebler, D.L. and Wronski, C.R. 1977
Reversible conductivity changes in discharge-produced amorphous silicon, *Applied Physics Letters* 31(4):292–294.

19. Carlson, D.E. 1987
Hydrogen motion and the Staebler–Wronski effect in amorphous silicon, in eds., M. A. Kastner, G. A. Thomas, and S. R. Ovshinsky, *Disordered semiconductors*, 613–620, Plenum Press, New York.

20. Hanak, J.J. and Korsun, V. 1982
Optical stability studies of *a*-Si:H solar cells, see Ref. 11,. 1381–1383.

21. Ichikawa, Y., Fujikake, S., Yoshida, T., Hama, T., and Sakai, H. 1990
A stable 10 percent solar cell with *a*-Si/*a*-Si double-junction structure, See Ref. 4, 1475–1480.

22. Stutzmann, M., Street, R.A., Tsai, C.C., Boyce, J.B., and Ready, S.E. 1989
Structural, optical and spin properties of hydrogenated amorphous silicon–germanium films, *Journal of Applied Physics* 66(2):569–592.

23. Carlson, D.E. 1989
Amorphous silicon solar cells, *IEEE Transactions on Electron Devices* 38(12):2775–2780.

24. Morimoto, H., Hirobe, T., Katayama, M., Nanbu, H., Oka, H., Shiozaki, N., Izauri, H., Inui, M., and Takemoto, T. 1985
Mass production technology in a roll-to-roll amorphous silicon solar cell process, conference record, 18th IEEE Photovoltaic Specialists Conference, Las Vegas, 1523–1528. IEEE, New York.

25. Catalano, A., Arya, R.R., Fortmann, C., Morris, J., Newton, J., and O'Dowd, J.G. 1987
High performance, graded band gap *a*-Si:H solar cells, conference record, 19th IEEE Photovoltaic Specialists Conference, New Orleans, 1506–1507. IEEE, New York.

26. Matsumoto, Y., Miyagi, K., Takakura, H., Okamoto, H., and Hamakawa, Y. 1990
a-Si/poly-Si two- and four-terminal tandem type solar cells, see Ref. 4, 1420–1425.

27. Mitchell, K.W., Eberspacher, C., Ermer, J., Pier, D., and Mills, P. 1988
Copper indium diselenide photovoltaic technology, see Ref. 13, 1578–1582.

28. Carlson, D.E. 1982
Amorphous thin films for terrestrial solar cells, *Journal of Vacuum Science & Technology* 20(3):290–295.

29. Atmaram, G.H., Marion, B,. and Herig, C. 1990
Performance and reliability of a 15kW$_p$ amorphous silicon photovoltaic system, see Ref. 4, 821–830.

30. Pratt, R.G. and Burdick, J. 1988
Performance of a 4kW amorphous silicon alloy photovoltaic array at Oakland Community College, Auburn Hills, Michigan, see Ref. 12, 272–1277.

31. Carlson, D.E. 1989
Reducing the manufacturing cost of amorphous silicon solar cells to less than $0.50/W$_p$, *Proceedings of the 4th International Photovoltaic Science and Engineering Conference*, Sydney, Australia, 529–539, Institute of Radio and Electronics Engineering, Edgecliff, Australia.

32. Whisnant, R.A., Champagne, P.T., Wright, S.R., Brookshire, K.C., and Zuckerman, G.J. 1985
Comparison of required price for amorphous silicon, dendritic web and Czochrolski flat plate modules and a concentrating module, see Ref. 18, 1537–1544.

33. Firester, A.H. and Carlson, D.E. 1983
Harnessing the sun with amorphous silicon photovoltaics, *RCA Engineer* 28(3):40-44.

34. Barnett, A. 1982
Analysis of photovoltaic solar cell options, see Ref. 13, 1165–1171.

35. Sabisky, E. et al. 1989
Eureka—a 10 MWp *a*-Si:H module processing line, *Proceedings of the 9th European Community Photovoltaic Solar Energy Conference*, Freiburg, Germany, 1537–1544, Kluwer Academic Publishers, Dordrecht, the Netherlands.

10
POLYCRYSTALLINE THIN-FILM PHOTOVOLTAICS

KEN ZWEIBEL
ALLEN M. BARNETT

Flat-plate polycrystalline thin-film photovoltaic devices can be made from copper indium diselenide (CIS), cadmium telluride (CdTe), or thin-film crystalline silicon (thin x-Si). All are now either in a pre-commercial stage or about to reach the open market. CIS and CdTe are regarded as model thin-films that should reach low-cost goals of about U.S.$1 per watt by the end of the decade. Thin-film crystalline silicon is less advanced but appears just as promising for the longer term (past the year 2000).

Photovoltaic (pv) technology is expected to become economically viable during the 1990s, playing an emerging role in such multibillion dollar markets as peak power production, pv/diesel/battery hybrid systems and in providing power to remote villages in the less developed countries. By the first decade of the 21st century, pv thin-films should be fully developed as an energy alternative to conventional fuels and will become self-sustaining, generating electricity at a cost below $0.06 per kilowatt-hour.

INTRODUCTION

Exploiting the sun

Flat-plate photovoltaic (pv) technologies (which include polycrystalline thin-films) have an important edge over solar concentrators: they use total, or global, sunlight. Global sunlight consists of both direct sunlight, which is received straight from the sun's disk, and indirect sunlight, which is refracted or reflected once it enters the earth's atmosphere. In contrast, concentrators (pv systems that use lenses or mirrors to focus light on small cells) are unable to focus light that strikes them at different angles and thus can exploit only direct sunlight.

The use of global sunlight has two advantages: first, because a flat-plate pv array's output varies according to the amount of usable sunlight, arrays are more economical than technologies that depend entirely on direct sunlight (for example, pv or solar-thermal concentrators). Secondly, they are not limited to traditional solar sites, such as deserts or high plateaus. Indeed, the lack of geographical limitations imposed on flat-plates may be their most distinguishing feature.

The fact that flat-plates can exploit indirect light has enormous consequences. Global sunlight is from 25 to 75 percent more intense in most areas of the world than direct sunlight; even in deserts it is 20 percent more intense. In addition, global sunlight is much less variable than direct sunlight. In the United States, for example, direct sunlight varies by a ratio of about 2.5 to 1, whereas global sunlight varies by only 1.5 to 1. Thus flat-plate–generated electricity that cost U.S.$0.06 per kilowatt-hour (kWh) in Kansas City, Missouri, would cost a little less in desertlike Albuquerque, New Mexico, (less than $0.05 per kWh) but not much more in temperate locations such as New York (about $0.07 per kWh). In sum, flat-plate PV can effectively exploit a truly ubiquitous resource (total sunlight), which is neither localized geographically nor small in comparison to global needs.

System components

The cost of PV electricity depends on the application, which in turn determines the number of components needed for a system. The simplest system consists of a PV module, which is a set of electrically connected cells, and the wires that connect it to an application. Most applications demand additional components: a stand to hold the module and point it toward the sun; batteries (if storage or load-leveling is needed); controls and regulators for the electricity; and an inverter if alternating current (rather than direct current) is needed. Larger PV systems are sometimes equipped with single- or double-axis sun-trackers to maximize output, although often the added expense of the trackers and the inconvenience of their operations and maintenance outweigh their benefits.

Cost analysis

All economic projections in this chapter are based on average U.S. sunlight (about 1,800 kWh per square meter (m^2) per year for a fixed flat-plate, which is the simplest of PV support structures) rather than on optimal desert conditions. Consequently, our economic analysis is assumed to be widely applicable and not confined to best-case locations.

Multi-megawatt installations (those that generate 10 megawatts or more of electricity) provide a useful paradigm for future large-scale applications. A breakdown of component costs for such systems illuminates the basis for PV economics (see table 1). The module, which represents some 80 percent of a system's direct costs, for example, drives the overall costs of the entire system. Even though module costs are likely to drop, they will account for at least half of a PV system's direct costs for the foreseeable future. Consequently, development of lower-cost module technologies is regarded as the single most important strategy for lowering the cost of PV-generated electricity.

Yet the findings presented in table 1 are not entirely predictable. Flat-plate technologies may have efficiencies greater than 15 percent, but be more costly

Table 1: PV system component costs for central power[a]

Component	Today's cost $	Projected cost $
PV module	400 per m^2	50 per m^2
Fixed array	80 per m^2	50 per m^2
Power-conditioning	20 per m^2	10 per m^2
Land	4 per m^2	4 per m^2
Indirect	33 percent of direct	25 percent of direct
O&M	0.005 per kWh	0.001 per kWh
Cost of PV electricity	0.40 per kWh	0.06 per kWh

a. Today's costs assume 13 percent module efficiencies and 10 percent system efficiency due to losses from operating temperatures, wiring, power-conditioning, and dirt. Future costs assume 15 percent modules and 12 percent systems. Location assumes average U.S. sunlight of 1,800 kWh per m^2 per year for a fixed flat-plate.

to manufacture; or conversely, they may be less efficient, but cheaper. For instance, conventional crystalline silicon cells tend to be efficient (existing module prototypes exceed 15 percent efficiency) but costly (from $400 to $800 per m^2), whereas thin-film polycrystalline technologies are relatively inefficient (the best modules are in the 8 to 10 percent range) but less expensive (below $200 per m^2). Based on figure 1, the goals for polycrystalline thin-film technologies can be summarized in table 2. Module efficiencies are not likely to exceed 12 to 20 percent, and their costs may range from $150 per m^2 to $35 per m^2. In the worst case, flat-plate electricity will cost about $0.12 per kWh; at best, the price may drop to $0.045 per kWh, based on average U.S. sunlight.

Balance-of-system costs may also vary. Over the long term, reduced balance-of-system costs could drop the cost of PV-generated electricity in an average location to an astonishingly low $0.03 per kWh or provide the flexibility to meet a more conservative goal of $0.06 per kWh.[1]

In order to produce electricity at the price of $0.06 per kWh, flat-plate technologies must meet various requirements. First, they must have an outdoor life-span of at least 30 years; without such durability, costs rise substantially. Second, modules must represent a balance between low cost and high efficiency.

The cost of PV electricity could drop below $0.06 per kWh by the first decade of the 21st century if certain thin-film technologies are adopted. But reaching such cost goals will require improved module efficiencies and lower manufactur-

---

1. In subsequent discussions, we will use a more conservative analysis based on U.S. Department of Energy (DOE) and Electric Power Research Institute (EPRI) guidelines for our cost expectations, except we will assume an indirect, or overhead, rate of 25 percent. But the reader is cautioned that the ultimate costs of PV electricity could be lower than those we quote.

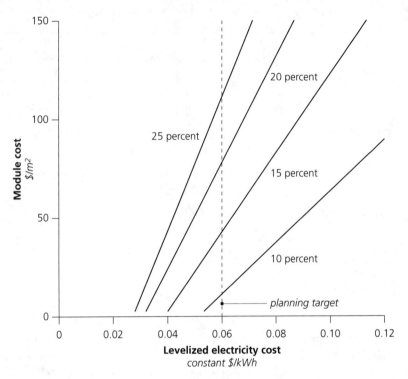

FIGURE 1: An analysis of the economics of fixed flat-plate modules, including module costs allowable at various system electricity costs and at various module efficiencies. All system losses are taken into account, and the location assumes average U.S. sunlight (about 1,800 kWh per m^2 per year). Balance-of-system assumptions are conservative (see table 1), but have a 25 percent indirect (overhead) rate.

Table 2: Module goals for reaching $0.06 per kWh: average U.S. location[a]

Technology	Efficiency goal *Efficiency now*	Cost goal *Cost now*
Thin-film silicon	**16 percent** *n/a*	**$50 per m^2** *n/a*
Copper indium diselenide	**15 percent** *9.7 percent*	**$45 per m^2** *$200 per m^2*
Cadmium telluride	**15 percent** *6.5 percent*	**$45 per m^2** *$150 per m^2*

a. The goals for polycrystalline thin-film flat-plate technologies, although approximate, are summarized. (All modules are assumed to be of relevant size for power production, i.e., at least 4 ft² [0.37 m²].) They reach universal cost-competitiveness at $0.06 per kWh in average U.S. locations.

ing costs. However, the economics of conventional energy options might change as a result of shortages or environmental concerns, thus rendering higher PV costs more competitive. Nonetheless, achieving a cost goal of $0.06 per kWh would clearly guarantee large-scale use of PV. Moreover, if PV electricity costs should drop to $0.03 to 0.05 per kWh (as more progressive assumptions would suggest), then PV electricity would be appropriate either for baseload electricity generation (with storage) or fuel production (see chapter 22: *Solar Hydrogen*), despite the increased costs of these more complex systems.

Although this chapter emphasizes long-term cost-reduction strategies, the path to low cost includes numerous interim markets that will hasten the growth of PV over the next two decades. Once PV reaches $0.10 per kWh (one-fourth of today's cost), multibillion-dollar markets such as peak power production, PV/diesel/battery hybrid systems, and power for developing nations will emerge. Thus, we emphasize the long-term potential of PV in order to present it as an important alternative to conventional energy, but do not want to create the impression that PV is insignificant in the short term. Indeed, PV will be viewed as an economically viable energy technology long before it is viewed as a replacement for conventional fuels.

THIN FILMS: AN OVERVIEW

Most thin films are made from semiconductors that absorb sunlight about 100 times more effectively than crystalline silicon. Whereas 50 to 100 microns[2] of crystalline silicon will absorb about 90 percent of global sunlight, the same amount can be absorbed by only 1 micron of thin-film semiconductor. This efficiency leads to lower manufacturing costs. Deposition of thin films (which may include electrodeposition, spraying, sputtering, chemical vapor deposition, chemical dipping, and glow-discharge) can also be much more rapid, much less energy intensive, and done on a larger scale than the manufacture of thick crystalline silicon. In addition, thin films require less handling to assemble into workable units because they are formed into modules (large-area devices) rather than individual cells.

Nonetheless, thin films have faced numerous technological hurdles. When first developed (between 1975 and 1980), thin-film materials and small-area cells had low (or no) efficiency; raising their efficiency was therefore the first requirement for success. Even today, module efficiency achievements of 10 to 15 percent remain a key goal of PV research and development efforts. Thin-film modules were also plagued by reliability problems: until recently, none was stable outdoors. Moreover, there were no processes for making them: considerable effort has been spent developing candidate processes (for candidate materials) and scaling these processes to make semiconductor films on large-area modules at high rates while maintaining premium material quality. Although

2. One micron is 1×10^{-6} meters.

these challenges have consumed more than a decade of research and development (R&D), they have led to several successful prototypes, which are now poised to make a major impact on the PV market.

Manufacturing thin films

The process of making thin-film modules consists of several steps:[3] 1) the acquisition or manufacture of the substrate; 2) cleaning and handling the substrate; 3) deposition of an electrical contact; 4) deposition of the active semiconductor layers; 5) deposition of another electrical contact; 6) monolithic integration (scribing, etc., which is usually intermingled with steps 1–6); 7) encapsulation and external electrical access; and 8) sundry procedures such as cleaning and handling that apply to all steps.

Glass is currently the preferred substrate for most thin films because it is one of the few relatively cheap materials that can withstand the high temperatures (up to 650°C) needed for thin-film fabrication. Alternative substrates, such as plastics, ceramics, or foils, are being investigated but are limited by stiff cost and durability requirements. Moreover, in some thin films, such as cadmium telluride (CdTe), the substrate forms the top of the device and so must be transparent enough to let light pass through to the cell's active layers. Few materials other than glass can meet this requirement. Moreover, because glass is relatively cheap (about \$4 to \$6 per m^2) and available in very large quantities, it is unlikely to be displaced in the near future.

Most thin films have a metal contact layer on the back of the module, which can be deposited in a number of ways. Sputtering, a process by which metal atoms are physically knocked off a target so that they strike and stick to a desired substrate, is the preferred method. Another method calls for depositing the metal contact by evaporation and condensation; this technique is especially effective with metals, such as tin, that have low boiling points.

In addition to metal back-contacts, most designs call for a transparent front contact, which is usually made from either tin oxide or zinc oxide. These highly transparent, highly conductive metal oxides can be sprayed, sputtered, or deposited by chemical or metal organic-chemical vapor transport. The efficacy of such processes depends on whether the layers are deposited early in the manufacturing process or near the end. If deposition occurs at an early stage, higher temperatures can be used and greater flexibility exists; but if it occurs toward the end, high temperatures can harm the existing layers and must be avoided.

The key layers in a thin-film module are the active semiconductors. Although most modules contain two semiconductors of opposite conductivity (called *n*- and *p*-type for negative and positive type conductivity), some designs call for three layers: *n*-type, *i*-type (for intrinsic-type conductivity) and *p*-type. All can be made by a variety of methods, which span the spectrum in terms of

---

3. The costs for equipment, labor, utilities, maintenance, waste treatment, and feedstock materials are included in each of these steps.

capital cost, deposition rates, temperature sensitivity, materials utilization, and complexity. Some require a vacuum, for example, whereas others do not. In general, high yield and reduced processing time and capital cost are desirable.

Unlike small crystalline cells, which must be wired together to form modules, thin films can be made as large-area modules. But, to improve efficiency, the large-area films must be subdivided into smaller cells. Each cell is electrically isolated from and then reconnected to its neighbor.[4] This procedure is known as monolithic integration. Dividing the film into cells requires a series of scribing, or separating, steps, which can be accomplished with lasers, chemicals or with a mechanical stylus or ablation technique. Contacts are deposited on each cell to provide a path for current flow between the cells. All of these steps can be easily mechanized and automated.

Once a large module is completed, it must be encapsulated to protect it from water, water vapor, and air. Water can corrode the metal contacts through which PV electricity flows, and water vapor can cause PV devices to degrade. Standard procedure is to use a transparent plastic, such as ethylene vinyl acetate, to seal the module surface to a protective piece of glass.

Thin-film manufacturing costs

Estimates of near-term PV costs vary according to the type of thin film, but copper indium diselenide ($CuInSe_2$ or CIS), cadmium telluride (CdTe), and thin-film amorphous silicon (thin a-Si) have comparable manufacturing costs (see [1–3] for CIS; [4] for amorphous silicon; and [5] for CdTe). Table 3 presents a summary of the ranges found in these projections. At the high end, thin-film modules are almost as expensive as crystalline silicon, and thus not competitive for power applications. At the low end, however, thin films will cost less than $1 per watt, one-third the cost of crystalline silicon. Moreover, if equipment costs, which are substantial, can be depreciated over 10 years instead of five, then the sensitivity of this capital cost is reduced by about half. Unfortunately, 10-year terms of depreciation are unlikely to occur in the near term; but they are likely to become available when thin-film technology matures, perhaps within the next decade.

Capital costs are sensitive to production rates, downtime, yields, and module efficiencies. Downtime and repair costs tend to scale with equipment costs. The more complex the equipment, the more costly it is to purchase and the more susceptible it is to breakage. When breakage does occur, the repair process is more tedious, requires greater skill, and may take significantly more time. Highly capital-intensive equipment, such as vacuum systems, thus increase the overall manufacturing risk. Therefore, in some cases inexpensive equipment combined with a backup system may be a better option.

---

4. Power losses in a module due to the module's resistivity increase with the square of the current in the module: interconnecting many small (low-current) cells limits the current in the module to that of one small cell while allowing each cell's voltage to add to the module's final output.

Fortunately, thin films can be made using less capital-intensive processes, such as non-vacuum systems. Spraying and electrodeposition are good, inexpensive, non-vacuum processes that reduce the capital cost of a unit of production.

More than 50 percent of materials costs (about $15 per m^2 or about $2 million for 10 megawatts of production) can be attributed to relatively conventional materials: the module's two glass sheets (front and back), encapsulation and framing materials, and the pieces that connect the module to outside circuits. Even these costs, which are comparable for all thin films, are likely to drop in the future as the result of technological developments.

Electrical materials (semiconductors and contacts) account for the remaining materials cost (about another $15 per m^2). Although semiconductors and feedstock gases are rather expensive, metal contacts are relatively cheap. Typical thin films (1 micron thick) have about 3 grams of material per square meter of coverage. However, the application process can be inefficient—as little as 10 percent of the feedstock may appear as final deposit. Still, because many contact metals cost less than $10 per kilogram, their inefficient use does not contribute significantly to overall costs; even at 10 percent efficiency (an extremely poor utilization rate), a layer costs only about $0.30 per m^2.

However, more expensive gases or rare materials may be used. Indium, for example, costs about $300 per kilogram. With perfect efficiency, 4 grams per m^2 of indium will yield 2 microns of CIS, at an acceptable cost of about $1.20 per m^2 (for comparison, the total cost of materials other than glass is about $15 per m^2). But if efficiency falls below 33 percent, then cost overruns become significant.

Table 3: Expense ranges for the manufacture of thin-film modules (near-term 10 megawatt plant)[a]

	$ millions	$ per m^2
Equipment ($5 to $40 million)	1–8	1–80
Direct labor	2–4	10–30
Materials	3–6	15–40
Utilities	0.5–1	5–20
Overhead	1–2	10–30
Total	**7.5–21**	**75–200**
Total, *$ per watt*	**$0.75–2.1**	

a. Cost projections for the manufacture of thin films within the next five years at the 10 megawatt annual capacity level are summarized. The data do not represent expectations about longer-term, large-scale, automated manufacturing methods (these will be less expensive). The data are based on the following assumptions: near-term thin films, five-year depreciation, efficiencies of 6 to 10 percent, yield about 80 percent, between one and three shifts, not fully automated; materials use varies but is assumed to be reasonably high (for example, the highest individual material cost for two sheets of glass is about $500,000 or $0.05 per watt at 10 percent efficiency).

Using gases as sources for layer deposition can create some problems. Although highly uniform layers can be deposited, utilization rates near 10 percent are typical, and thus costs are relatively high. One solution is to capture and re-use the gases on-site, which is theoretically easy, but not yet practiced, in part because the industry is still in its infancy.

The cost of modules (measured as dollars per watt produced by a 10 megawatt facility) vary for different capital costs per square meter of production (see figure 2).[5] In general, costs for thin-film production will be relatively high until all production processes are fully optimized and automated. Even then, module efficiency will be an important driver of final cost. Unless thin films cost less than $2 per watt, they are unlikely to ever have a role in power generation, yet costs will never fall below $2.5 per watt if module efficiencies stagnate at 4 percent. Even if module efficiency is raised to 6 percent, capital costs must fall below $25 per m^2 to reach the $2 per watt level. On the other hand, modules that are 8 percent efficient will always cost less than $2 per watt and sometimes as little as $1.25 per watt, if capital costs are low.

Crystalline thin-film production does not yet exist. But several 1 to 10 megawatt amorphous-silicon facilities have been built. These are prototypes and thus more costly than fully functional facilities, but they provide a basis for understanding thin-film costs. To date, capital-cost ratios (the total capital cost measured against annual production) for these first-generation amorphous-silicon plants run about $3 per watt, and module efficiencies average about 5 percent. At an annual production rate of about 2×10^5 m^2, a $30 million capital cost would

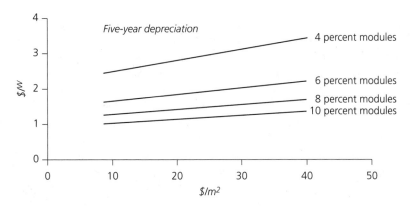

FIGURE 2: *The capital cost of making modules ($ per m^2) is defined as the total capital cost of the plant, divided by five years (for depreciation) and by the production volume (in m^2, including a yield correction) per year. The non-capital assumptions are similar to those for table 3: labor ($20 per m^2); materials ($30 per m^2); utilities ($10 per m^2); overhead ($10 per m^2); a five-year depreciation; 80 percent yield; three shifts and 260 workdays (including weekend maintenance); and relatively immature automation and process optimization.*

5. Note that capital cost per square meter is a more fundamental measure than capital cost per watt, which depends on the final power output of the modules.

be about $\$30 \times 10^6/(5 \text{ years} \times 2 \times 10^5 \text{ m}^2 \text{ per year}) = \$30$ per m^2. If the remaining costs (materials, labor, etc.) are comparable to other systems (see figure 2), then the final cost for amorphous silicon modules produced by these systems would be about $2.5 per watt, which is not economical. If the production capital costs could be lowered to about $15 per m^2 (representing $15 million in capital expenses for a 10 megawatt plant) and module efficiency raised to 5 percent, module costs would drop to about $2 per watt, which would be marginally acceptable. Later in this chapter (see "The pathway to low cost"), we will examine how costs will be affected as crystalline thin-films mature and large-scale production is achieved.

COPPER INDIUM DISELENIDE (CIS)

Historical perspective

In terms of easily measured achievements (efficiency and stability), copper indium diselenide (CuInSe$_2$ or CIS) emerged as the leading thin film during the 1980s. Yet because CIS was a fairly late starter among the thin films, it is strongly represented by only one company: Siemens Solar Industries (SSI), formerly ARCO Solar, based in Camarillo, California. Although other companies are developing CIS technology, they lag significantly behind SSI (for example, no one but SSI is making CIS modules).

CIS first attracted attention in the early 1970s. Pioneering work at Bell Laboratories in New Jersey, from 1973 to 1975 [6] culminated in the manufacture of single-crystal CIS cells with efficiencies greater than 12 percent [6]. Then during the mid to late 1970s, R. Mickelsen and W. Chen at the Boeing Corporation in Seattle, Washington, showed that efficient (about 9 percent) CIS cells could be made using a thin-film deposition method called vacuum evaporation [7–9]. This transition—from single-crystal devices to thin-film cells of reasonable efficiency—was a major stimulus for CIS development. CIS devices also attracted interest because they showed a stability more reminiscent of crystalline silicon than of existing thin-films (copper sulfide and amorphous silicon). As early as 1985, Boeing conducted a stability test on unencapsulated cells, exposing them to light for almost 8,000 hours, and showed they experienced no degradation.

The U.S. Department of Energy (DOE)–Solar Energy Research Institute (SERI)[6] program provided additional stimulus for work on CIS during the late 1970s, supporting most of the early work at Boeing and elsewhere. In 1980, in a parallel effort, ARCO Solar began work on CIS devices. More recently, from 1988 to 1990, SERI and ARCO Solar embarked on a powerful and successful collaboration. The result is CIS module performance that surpasses all other thin films and is now the benchmark against which others are measured.

---

6. SERI is now known as the National Renewable Energy Laboratory (NREL). For clarity, SERI will be used in this chapter.

CIS technology

The basic structure of CIS devices was defined by Boeing scientists from 1978 to 1981 (see figure 3). Although the original substrate on which the cells were deposited was alumina, it was subsequently replaced by glass, which is less costly. The back contact in almost all CIS devices is molybdenum, which is relatively nonreactive to the materials (copper, indium, and selenium) deposited on its surface. Molybdenum is usually deposited on the glass by sputtering, but does not always adhere properly; to enhance adhesion, a thin chromium layer is sometimes injected between the molybdenum and the glass. In addition, defects, or pinholes, must be minimized to prevent subsequent disruption to the CIS layer.

At Boeing, the original p-type CIS layer was deposited by vacuum evaporation of all three elements in two stages so that each of the elements could be individually heated [9]. Evaporation was controlled through a feedback mechanism within the vacuum chamber. It became apparent that the ratio of copper to indium was a key determinant of cell performance. The first layer of CIS contained evaporated copper and indium in almost equal amounts (there was a slight preponderance of copper). When the second layer was deposited (after the first was about 1.5 microns thick), the ratios were adjusted so that more indium than copper was deposited. Deposition temperatures during this stage were kept to 450°C, which resulted in a mixture of the two layers. The final, composite layer had a slight excess of copper over indium near the molybdenum, but an excess of indium over copper near the top of the CIS (at the junction). High efficiency could be achieved if the copper–indium atomic ratio near the junction ranged from 0.92 to 0.97. Below this ratio, CIS becomes n-type and does not make good junctions; above it, copper forms unstable and electronically undesirable compounds such as Cu_2Se, and in some cases copper nodules precipitate, shorting the device.

Subsequent to CIS deposition, an n-type layer, n-CdS, was deposited in order to induce an electric field with the p-CIS. A layer of n-CdS, about 3 to 4

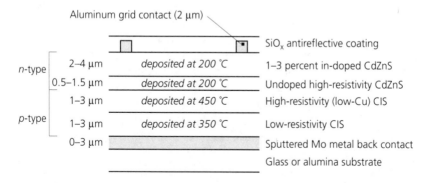

FIGURE 3: The early configuration of CIS cells was developed at Boeing. The n-*type layer was originally CdS; to improve efficiency, zinc was added to the CdS, forming CdZnS alloys.*

microns thick, was deposited by vacuum evaporation of pressed CdS powder. Like CIS, CdS was deposited in two layers. The layer closer to the CIS was relatively resistive in order to make a strong electric field; the other layer was doped with indium to induce high conductivity. The top half of the CdS had to be conductive to permit the lateral movement of electric charge, thus facilitating charge collection. To collect the current, aluminum grid fingers were evaporated on the CdS.

Almost without exception, vacuum-evaporated CIS devices were heat-treated to optimize their performance [10]. By heat-treating the cells in oxygen at about 200°C for half an hour, Boeing scientists were able to raise their efficiencies to 9 to 11 percent (from an untreated 1 to 8 percent). SERI investigators [11,12] subsequently demonstrated that evaporated CIS devices responded to oxidation/reduction treatments: the efficiency of the devices generally improved with oxidation, but suffered when they were subjected to reduction. It is believed that oxidation improves cell performance because it renders the CIS more *p*-type by tying up dangling bonds at the surface of selenium-deficient CIS grains. Reduction creates the reverse effect: it makes the film less *p*-type by increasing the number of dangling bonds. Although a defective *n*-type CIS layer cannot make an effective junction with *n*-CdS, it can be heat-treated to become *p*-type. Oxidation/reduction treatments of Boeing CIS devices are reversible, except that after several alternations of oxidation and reduction, a cell will stabilize at its highest efficiency, becoming immune to further alteration. (Oxidation/reduction treatments have little or no effect on the high-efficiency cells that are not made by vacuum evaporation.)

Although CIS cells reached 10 percent efficiency levels by the early 1980s, the design was seriously flawed because the thick CdS blocked too much light. Some of the sunlight, which should have penetrated the CdS layer, was absorbed instead, creating free charges that were too far from the cell's electric field to contribute to its current. The loss is significant because 95 percent or more of the sunlight that does penetrate to the CIS layer is absorbed within the electric field. Moreover, CdS absorbs all sunlight with energies above 2.4 electron volts (eV), which accounts for a substantial portion of the solar spectrum. In fact, if all photons in the spectrum above 2.4 eV could be transformed into electric current, they would account for about 6 milliamps per cm^2. A loss of 6 milliamps per cm^2 implies a 20 percent loss in efficiency for an average current of about 35 milliamps per cm^2. Boeing researchers attempted to solve the problem by adding zinc to the CdS, creating a CdZnS alloy with a higher band gap than CdS. But they found the CdZnS became too resistive if more than 20 percent zinc was added. Although the band gap rose slightly (to about 2.6 eV), only a small amount of current was gained.

In the mid 1980s, scientists at ARCO Solar [13, 14] devised a better solution: they replaced the thick CdZnS layer with two layers. One was a very thin (0.03 micron) layer of CdS; the other was a 1.5 micron layer of zinc oxide (ZnO) (see figure 4). Because the CdS layer was so thin, almost all the light above the 2.4 eV band gap passed through it; and because ZnO has a band gap greater than 3.0 eV, almost all photons passed through that layer, too. In addition, the conductivity of ZnO (like that of other metal oxides) can be improved by adding suitable impurities, such as aluminum and boron. ZnO therefore proved to be a better conductor than CdS; together the ZnO and CdS layers were far more transparent than one thick layer of CdS. As a result of this breakthrough, cell efficiencies improved by about 20 percent, from 10.5 to 12.5 percent [15]. Other CIS researchers have since adopted the same approach, making cells of similar efficiencies.

CIS technology faced another major challenge during the 1980s. Could product-sized modules be fabricated using scalable low-cost processes? In 1986, Boeing engineers began by constructing a record-setting 9.6 percent efficient, 100 cm² CIS submodule consisting of four cells connected in series. But the biggest advance occurred in the late 1980s, when ARCO Solar not only made larger CIS modules, but found a different way to manufacture CIS.

Today Arco Solar's successor, Siemens Solar Industries (SSI), deposits CIS by a two-step method called selenization [16–19]. After copper and indium are deposited, the copper–indium is selenized at elevated temperatures (about 400°C) using a selenium-bearing gas such as hydrogen selenide. The process produces CIS layers of very high quality: the grains are larger and better ordered than those formed by vacuum evaporation. Although the copper and indium can be deposited by several methods, sputtering is preferred. Sputtering equipment is as expensive as evaporation equipment, but because sputtering 1 micron of metal takes less than a minute (whereas evaporating the same thickness of CIS takes

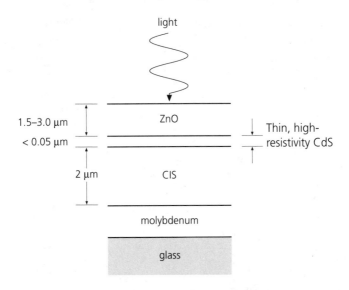

FIGURE 4: ARCO Solar increased the current of CIS devices by reducing the thickness of the CdS window layer enough to make it essentially transparent. Zinc oxide was added to provide increased lateral productivity.

about an hour) throughput is improved. Although the precursor copper–indium layers are usually sputtered separately, they mix during deposition, forming a complex alloy. As with the Boeing approach, the CIS copper–indium ratio determines efficiency, and ratios from 0.92 to 0.97 are preferred.

As a result of selenization, research at ARCO Solar progressed rapidly [20–23]. In 1985, scientists there made the first square-foot (0.0929 m^2) CIS module. By 1986, their CIS modules had obtained efficiencies of 5 percent, and by 1988, they had surpassed all others, reaching 11.1 percent efficiency (see figure 5). More remarkable is that such efficiency was achieved when the best active-area efficiency of a CIS cell was only about 12.5 percent. Thus, the best CIS module was 88 percent as efficient as the best cell, which suggests that CIS scales well to larger areas.

Once the ARCO modules were fabricated, their stability outdoors could be tested. In November 1988, ARCO Solar sent two prototype CIS modules to SERI for testing (see figure 6). Despite the fact that these modules are early prototypes, they have performed exceptionally well. After almost three years of testing, the drop in efficiency has been minimal, falling by no more than the 3 percent margin of error. To this day, these modules represent the only known independent outdoor test of CIS modules.

Since then, ARCO Solar and its successor SSI have continued to enlarge the scale and improve the efficiency of CIS technology. The efficiency of CIS devices is improved largely by increasing the quality of their materials. By reducing defects in CIS (and along its grain boundaries), voltages are raised. The adhesion of CIS to molybdenum, for example, can greatly affect a device's quality [3, 24–27]. When copper–indium layers are selenized into CIS (selenium is a larger atom than either copper or indium), a threefold expansion of the film occurs, creating

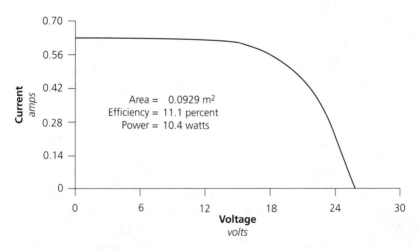

FIGURE 5: *The highest efficiency thin-film submodule was made by ARCO Solar in 1988 using CIS.*

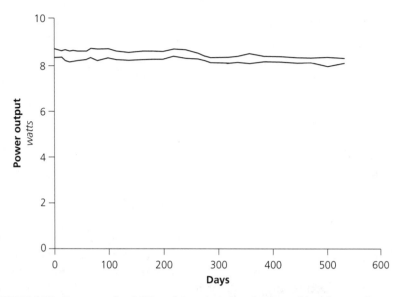

FIGURE 6: The first encapsulated CIS modules to be independently tested have been outdoors at SERI for over three years with little or no degradation (figure shows power output over a period of more than 500 days). The modules were made by ARCO Solar in 1988.

stress at the CIS–molybdenum interface that reduces both efficiency and yield. In extreme cases, the CIS film will actually peel off the molybdenum. Stability of such films outdoors is also affected.

International Solar Electric Technology (ISET), a small company in Inglewood, California, and SSI have both devised solutions to the adhesion problem. ISET puts a thin layer of tellurium between the molybdenum and CIS and is now fabricating high-efficiency CIS devices with much stronger adhesion [28–31]. SSI adds gallium to the back of the CIS layer to promote adhesion and CIS growth. The result is improved electronic properties and higher voltages.

Because both gallium and indium are group III elements, gallium can replace indium in the CIS crystal lattice without causing severe alterations. In fact, the use of gallium in CIS cells dates back to the mid 1980s, when Boeing began substituting gallium for indium to make $Cu(In,Ga)Se_2$ or CIGS cells [32]. Its rationale was that gallium would raise the band gap of the resulting material. CIS has a band gap near 1.0 eV, which places it below the best band gap (about 1.5 eV) for simple photovoltaic devices. Adding gallium can raise the band gap of the p-layer to as high as 1.7 eV, depending on how much gallium is used. By replacing 20 percent of the indium with gallium, Boeing found that the p-layer band gap increased to about 1.1 eV (almost the same as x-Si). Voltage, which naturally rises with increased band gap, was enhanced.

In late 1988, ARCO Solar reached 14.1 percent active-area efficiency on a 3.5 cm² cell. It is not known whether the cell was a gallium–CIS alloy or not. The cell voltage, which was 508 millivolts, was substantially higher than the 450 to 470 millivolt range seen in previous cells.

During the last two years, ssi has focused its efforts on scaling up CIS technology. In 1989, they produced an encapsulated 4 ft^2 (0.37 m^2) CIS module that was 8.7 percent efficient. In 1991 (after a corporate relocation that slowed operations for about a year), ssi produced a module that was 9.7 percent efficient, which represents by far the highest efficiency known for a thin-film module of this size.

Health and safety issues

CIS production and use raises several concerns. Hazardous feedstocks, including hydrogen selenide and cadmium, are often used during its production, and the sealed modules contain small amounts of solid cadmium in CdS thin-films and selenium in CIS devices [33–35].

Hydrogen selenide is a highly toxic gas [36]. Nonetheless, it can be used safely and economically, provided that well-known safety precautions (gas monitors, scrubbers, backups, and enclosure vessels) are adopted. From both a safety and economic perspective, the risks of hydrogen selenide can be further reduced by producing the gas on-site and recycling it after use. This minimizes the need to transport the gas and also ensures that only a small amount is on-site at any one time. In addition, several research organizations are searching for substitutes that might lead to elimination of the gas altogether.

Cadmium is another toxin, being both a poison and a possible carcinogen, but of less concern because it is used in such small quantities: 0.04 grams per m^2 of thin film. Within the workplace, cadmium feedstock can be safely contained, its possible entry into the air measured, and workers monitored for exposure [33]. Except in rare cases, cadmium is not an immediate threat to life, so monitoring it and ameliorating its impact are easier than for hydrogen selenide, which can kill quickly.

Within the CIS module, the cadmium (which is fixed in a stable semiconductor[7] and sandwiched between two sheets of glass), poses no measurable threat to the environment, either in terms of safety or toxic emissions. The same modules do, however, contain about 6 grams per m^2 of selenium, which might threaten groundwater if disposed of improperly. However, initial tests of CIS modules have been conducted using the U.S. Environmental Protection Agency's ep toxicity test and its more stringent tclp test [37]. Both call for grinding up a module, suspending the remains in various solutions, and then attempting to leach cadmium, selenium, and other materials from solution. CIS modules passed both of these tests by a factor of 10, and thus under present law cannot be considered hazardous waste. Nevertheless, we think that CIS manufacturers should recycle their modules to minimize waste and to reclaim valuable indium.

---

7. The semiconductor has a melting point greater than 1,000°C.

The indium supply

Indium availability will become an issue when CIS modules enter large-scale production [38]. Based on reasonable near-term assumptions (70 percent material utilization, 10 percent module efficiencies, and 2-micron-thick CIS devices), a gigawatt of CIS modules will need about 50 tonnes of indium. Although the world's current annual production of indium is only about 120 tonnes, indium is almost as abundant as silver (of which more than 8,000 tonnes are mined per year). This suggests that indium production can be increased to meet the growing demand for it. Usually found as a by-product of zinc production, indium can also be produced from lead, tin, and copper smelting. Nonetheless, the match between supply and demand could become an issue. If the production of indium lags behind CIS needs, its price could escalate, seriously impeding the growth of CIS production until the balance between supply and demand is regained. Fortunately, several companies have already expressed an interest in producing sufficient indium to meet CIS demand.

In the future, less indium may be needed per module. Theoretical limits of light absorption suggest that CIS layers as thin as 0.5 micron may be possible. If the CIS layer is 1 micron thick and has 70 percent utilization, then 25 tonnes of indium should be sufficient to produce a gigawatt of power. Another alternative is to replace about 25 percent of the indium with gallium, which would lower indium needs to less than 20 tonnes per gigawatt. Recycling CIS modules after 30 years would also ameliorate potential shortages.

Economics

Labor represents the largest expense of near-term CIS module production (see table 3). The breakdown of costs, by one estimate [3], are roughly as follows: equipment depreciation, $13 per m^2; direct materials,[8] $25 per m^2 (see table 4); labor and fringe benefits, $31 per m^2; and utilities and overhead, $21 per m^2. After factoring in an assumed yield of 80 percent, the total cost is $110 per m^2. The estimate is based on a 6 megawatt facility that employs one shift a day. In another study [2], materials costs were lower—only $16 per m^2—but the study was based on volume purchases, no module encapsulating frame,[9] and lower costs for such components as indium. For larger facilities, material costs of $16 to 20 per m^2 would be appropriate.

ISET overhead and equipment costs are probably unnecessarily high. They are based on prevailing labor rates in California (which are high relative to the U.S. average) and on minimal automation. If the facility were automated and operated three continuous shifts, then both capital and labor costs would drop significantly. More product would be produced and overhead, which would not be duplicated at night, would fall per unit of output. But using the more conser-

---

8. The process steps for this important cost category are: glass substrate, molybdenum sputtering, copper and indium sputtering, selenization, cadmium sulfide chemical dipping, zinc oxide sputtering, monolithic integration, and encapsulation.

9. Frames may not be needed for utility-scale applications.

Table 4: Materials costs and utilization rates for CIS modules in 1990

Material	Thickness	Utilization rate	Cost	
	microns	*percent*	*$ per g*	*$ per m*
Molybdenum	1.0	80	0.06	0.77
Copper	0.2	80	0.025	0.055
Indium	0.5	80	0.30	1.43
Selenium	1.0	50	0.06	0.58
CdS	0.05	30	0.22	0.18
Zinc II oxide	2.0	80	0.05	0.71
Miscellaneous				1.87
Total (active layers)				**5.60**
Glass (tempered)				7.00
Glass (substrate)				4.00
Encapsulant				2.75
Connectors, etc.				3.30
Other (frame, etc.)				2.75
Grand total (not corrected for 80 percent production yield)				**25.40**

Source: [3]

vative $110 per m^2, module costs would be $2.75 per watt at 4 percent efficiency, $1.83 per watt at 6 percent efficiency, $1.38 per watt at 8 percent efficiency and $1.10 per watt at 10 percent efficiency. At $13 per m^2 capital costs (with a $3.5 million capital investment), ISET is at the lower end of the module capital-cost curves (see figure 2). The company's materials utilization rates are good, but it buys small lots of material, which increases cost. It is clear, however, that glass, encapsulation, and module parts account for much of the materials costs; these vary little across the thin films.

CIS module production appears to fall within reasonable cost levels. As with other thin films, the efficiency of its modules determines its economics. Unlike other thin films, however, prototype CIS modules are now nearing efficiencies of 10 percent, which suggests that CIS will have the easiest time reaching high efficiencies in actual manufacture. Although such production has not yet occurred, and some yield and reliability issues remain to be solved, CIS is well positioned to take advantage of the natural cost reductions in overhead, equipment and labor that will accrue once production volumes increase. Thus, CIS can be regarded as a model thin-film, which may reach very low-cost goals by the end of the decade.

CADMIUM TELLURIDE (CdTe)

Historical perspective

In the late 1950s, cadmium telluride (CdTe) was one of the first thin films to be studied. At the time, its band gap of 1.5 eV was considered optimal for single-junction solar cells [39, 40]. Because CdTe represented the highest theoretical efficiency (about 28 percent), it became the focus of some developmental research. CdTe technology has undergone two periods of emphasis: in the late 1970s and early 1980s, when four companies (Kodak, Monosolar, Matsushita, and Ametek) attempted to commercialize it; and today, when Photon Energy, British Petroleum, Matsushita, and to a lesser degree, Solar Cells Inc. are about to commercialize it. Although earlier attempts to develop commercial CdTe modules failed, they sowed the seeds that are now bearing fruit; all the companies developing the technology today (with the exception of Photon Energy) have roots in the earlier companies [41].

In the late 1970s, it was found that CdTe cells of reasonable efficiencies (from 8 to 10 percent) could be fabricated by such relatively simple methods as electrodeposition [42,43]; closed-space sublimation [44, 45]; and screen printing [46], all of which are low in capital cost. Electrodeposition, for example, requires almost no complex equipment. It relies only on a potentiometer (a device to control the flow of electric current), vessels to hold the electrolyte, and two electrodes (one designed to hold the substrate). Megawatt production of CdTe layers can be achieved with an investment in the range of tens of thousands (as opposed to millions) of dollars.

The use of a postdeposition heat treatment to optimize CdTe devices also explains why high efficiencies can be achieved at relatively low cost [47]. Heat treatment is usually conducted at temperatures greater than 400°C and takes place in a medium that includes chemicals such as cadmium chloride. The treatment activates a cell's electric field at the n-CdS/p-CdTe interface and eliminates defects by causing the CdTe grains to regrow and become larger. Not only does less surface area mean fewer grain-boundary–related defects, but the defects themselves can be removed by chemically fixing them in bound states. Thus, the golden rule for making CdTe devices is to deposit cadmium and tellurium as inexpensively as possible and then to heat-treat the devices to activate them.

Although efforts to develop CdTe seemed to stall in the early 1980s, most of the technical work has been perpetuated [41]. Monosolar, a small U.S. company that developed many of the key electrodeposition processes [42] was bought in 1984 by SOHIO, which in turn was purchased by British Petroleum (BP) and renamed BP Solar. Today, Monosolar technology provides the foundation for BP Solar's work on CdTe. The company reports cell efficiencies of 12 percent and square-foot module efficiencies of 9.5 percent [48, 49].[10] In addition, it has reported good stability in outdoor tests of systems powered by CdTe modules. Thus, the Monosolar effort (which was partially supported through DOE/SERI subcontracts in the early 1980s) has not been lost.

Ametek also made substantial progress in CdTe electrodeposition [5, 43] during the 1980s. In the early 1980s the company attempted to commercialize a CdTe device based on a simple metal/n-CdTe structure, but the cells were less than 8 percent efficient and so proved uneconomic. During the mid 1980s, Ametek made significant progress when it switched to a n-CdS/p-CdTe device. By 1988, the company was producing cells with 11 percent efficiencies and had developed a new device based on n-CdS/CdTe/p-ZnTe [50]. Their small submodules were about 8 percent efficient. In 1989, the parent corporation restructured itself and eliminated its R&D division. Fortunately, Ametek's key technical investigator was hired by Solar Cells Inc., which has positioned itself to vigorously pursue CdTe.

Since the 1970s, the Matsushita company of Japan has also been committed to CdTe research [46] and was actually first in the early 1980s to commercialize CdTe cells. For almost a decade, the company has supplied Texas Instruments with small CdTe cells for its calculators. These CdTe devices, which are produced by a low-cost process called screen printing, are the only ones that can be purchased through retail outlets. Despite the inexpensive screen printing, the company's efforts to commercialize CdTe modules have not yet succeeded. Matsushita's modules have a serious flaw: the ratio of active to total surface area is only 60 percent. This means that a 10 percent small-area cell automatically loses 40 percent of its efficiency when encased in a module because of contact obscurations. Consequently, Matsushita reports that its best power output per square foot is only about 5 watts (about 50 W per m^2 or around 6 percent efficiency), an output that is far too low to justify commercialization.[11] The area problem, fortunately, does not apply to all CdTe modules; both BP Solar and Photon Energy have active to total surface-area ratios greater than 90 percent, which is comparable to other thin films. Thus, something unique to the Matsushita process has created a problem.

During the mid 1980s, although Monosolar and Kodak dropped their work on CdTe and Matsushita was having efficiency problems, research at both BP Solar and Ametek advanced significantly. But perhaps the most important progress in CdTe technology was occurring at a tiny Texas company called Photon Energy Inc [51].

---

10. In December 1991, BP Solar reported efficiencies of 14 percent for a small-area cell and 10 percent for a 1 ft^2 (0.0928 m^2) module.

11. In December 1991, Matsushita reported 80 percent area use and 8 percent for 1 ft^2 (0.0928 m^2) modules.

Photon Energy, Inc.

Photon Energy emerged in 1984 following the collapse of its predecessor company, Photon Power, which failed in its attempts to develop and commercialize copper sulfide thin-films. John Jordan of Photon Energy decided instead to use CdTe, which lacks the instabilities of copper sulfide. He and his colleagues subsequently embarked on a successful program to develop various low-cost wet-chemical methods of deposition [52]. Within two years, Photon Energy reported making cells that were 9 percent efficient [53]. By 1986, its 1 ft^2 (929 cm^2) modules had achieved efficiencies around 5 percent [54], and the company won a cost-shared subcontract from SERI and the U.S.DOE to accelerate its progress. The company achieved impressive results in 1988 and 1989 [55]: the efficiency of its small-area cells was 12.3 percent (0.31 cm^2 active area) and that of its modules was 7.3 percent (838 cm^2 aperture area). In 1991, Photon Energy won (jointly with SERI) a prestigious IR-100 award from *R&D Magazine* for its low-cost CdTe modules.[12]

Photon Energy's improvement of CdTe cell efficiencies was similar in many ways to ARCO Solar's experience with CIS. Like ARCO Solar, Photon Energy thinned the n-CdS layer, thus developing a means by which high-energy light could reach the cell's junction area. Currents of 25 milliamps per cm^2 were then obtained, which represented an increase of almost 20 percent, or 5 to 6 milliamps per cm^2, above previous cell designs (see figure 7).

In 1989, Photon Energy received an award from PVUSA[13] (as did ARCO Solar for CIS) to produce 20 kilowatts of modules by 1991. Like ARCO Solar, Photon Energy has brought the potential of its innovative CdTe thin-film nearer to fruition by developing low-cost production methods, demonstrating world-class cell and module efficiencies, and initiating prototype commercial production. Although module efficiencies of CdTe lag behind those of CIS, the potential of CdTe to meet and even surpass CIS appears to be quite strong. CdTe is easier to deposit (and so its low-cost potential is good), and its band gap (1.5 eV) is better matched to the solar spectrum in terms of its theoretical efficiency.[14] Although CIS is still leading CdTe as the thin-film of choice, its status may well be tested in the early 1990s.

Field stability

As we stated previously, one of the great strengths of polycrystalline thin-films is their lack of light-induced instability. Until recently, however, few independent tests had been carried out on the stability of these modules. But during the past two years, SERI has tested some of Photon Energy's CdTe modules. Although en-

---

12. In 1991 Photon Energy also raised its cell efficiency to 12.7 percent, its 1 ft^2 efficiency to 8.1 percent, and made the first 4 ft^2 (0.37 m^2) CdTe modules (6.5 percent efficiency).

13. Photovoltaics for Utility Scale Applications—a joint venture between the Pacific Gas and Electric Company (PG&E) and the U.S. Department of Energy.

14. The world record for CdTe cell efficiency was raised in November 1991 to 14.6 percent by T. Chu of the University of South Florida. This is now the highest efficiency of any thin-film cell.

gineering problems persist, the test results are surprisingly good (some modules left outdoors for almost 500 days showed no signs of degradation; see figure 8) and suggest that the stability issues associated with CdTe are those common to any module. Outdoor stability is key to the future of such thin films as CdTe and therefore will continue to be tested at SERI. Both Matsushita [56] and BP Solar [48, 49] have reported similar stability results for their outdoor modules.

Health and safety issues

As CdTe technology matures, related health and safety issues are becoming more relevant. Although the manufacture of most, if not all, PV materials can be potentially toxic, only CdTe and to a lesser extent CIS modules (which both contain cadmium) are associated with environmental concerns. Cadmium creates potential problems not only when CdTe devices are manufactured but also during field use [33, 57].

But safe manufacturing methods are fairly straightforward and thus easily implemented. In most cases, cadmium represents a chronic, not an acute, health threat. Although repeated exposure can lead to serious harm, familiar management options for monitoring equipment, maintaining worker hygiene, and biomonitoring personnel with urine tests provide reasonable assurances of safety. In fact, when compared with the toxic and explosive gases used to make most thin films,[15] cadmium may be far less dangerous, in part because it is handled more easily [58].

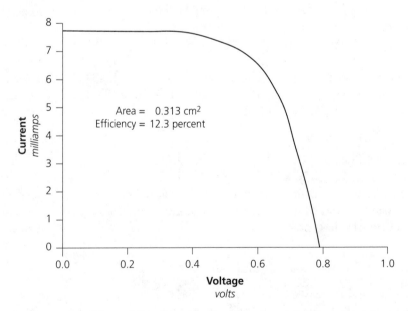

FIGURE 7: *A state of the art CdTe cell (12.3 percent efficient) made by Photon Energy Inc. and measured at SERI.*

15. Hydrogen selenide for CIS; arsine, phosphine, diborane, and germane for amorphous silicon.

The manufacturing process does, however, create waste. If materials utilization runs about 80 percent, and if 5 percent of all panels are rejected, then a 10-megawatt CdTe facility could produce about 250 kilograms of cadmium waste per year. The waste must either be landfilled or recycled. The latter option is more attractive because it minimizes waste and allows for reuse of cadmium and the more valuable tellurium.

Disposal assumes even greater importance for those who use CdTe modules. Whereas silicon-based modules are generally considered benign and can be discarded without health or safety concerns, CdTe disposal is not as straightforward. We have begun to investigate a cradle-to-grave strategy for handling CdTe and are being assisted in our efforts by ASARCO, a leading U.S. producer of cadmium and tellurium. Together, SERI and ASARCO intend to study ways to recycle cadmium waste produced during manufacturing and to recycle CdTe modules after their outdoor use. Preliminary findings suggest that recycling to avoid cadmium disposal and to regain tellurium is feasible; ASARCO smelters may be able to process CdTe waste and modules with minimal preprocessing. The same is true for CIS waste and modules.

Recycling, however, will work only for fairly large users such as central stations. Decentralized users might find recycling more difficult because they would have to devote considerable effort to creating a recycling infrastructure. Still, the 30-year life expectancy of modules provides a hefty time allowance during which these issues can be addressed.

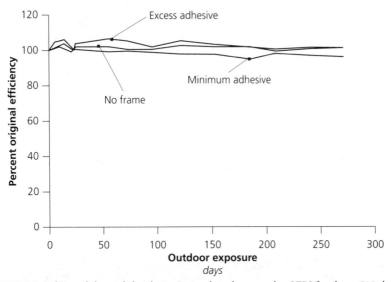

FIGURE 8: CdTe modules made by Photon Energy have been tested at SERI for almost 500 days without degradation. Figure shows power output over a period of almost 300 days.

Cadmium also creates a potential hazard during module use. Although it is sandwiched between layers of glass or metal encapsulants, and thus poses minimal danger, the module may break or be exposed to fire, releasing minute amounts of cadmium into the environment. Conservative assessments of volatilized cadmium suggest that fire can affect three types of installations [33]:

1.   Residential rooftop PV arrays. These contain such small amounts of cadmium that a person would have to be immersed in smoke (that is, she or he would have to be in danger of immediate death from either the fire or from smoke inhalation anyway) to be affected. Under these circumstances, cadmium adds little or no peril to overall fire hazards.

2.   Larger PV arrays (those greater than 100 kilowatts) on or near commercial buildings can create hazards downwind. If normal evacuation procedures are followed, however, such that evacuees avoid direct exposure to the smoke plume, the problems are eliminated.

3.   Large central-station PV arrays are in little danger of general conflagration. Except for grass fires, for which there are adequate safety procedures (such as keeping the grass cut and installing sprinklers), fires are not likely to spread beyond individual parts of the arrays. Centralized systems are also likely to have operating procedures to handle such emergencies and are not likely to be in populated areas.

For perspective, it must be mentioned that the amount of cadmium in CdTe modules is very small (on the order of 6 grams per m^2 in the CdTe modules; see table 5). Figure 9 shows the amount of cadmium feedstock needed to make CdTe layers of various thicknesses. The use of cadmium on a global scale is also quite small. For comparison, in 1986 the total amount of cadmium consumed was about 15,000 tonnes. Based on our assumptions (i.e., the need for about 5 to 10 grams per m^2 of cadmium feedstock), 1 gigawatt of CdTe module production requires some 50 to 100 tonnes, or 0.1 to 0.3 percent, of the annual world use of cadmium. Thus, even at a 1 gigawatt scale, the amount of cadmium needed is relatively insignificant.

Table 5: Amounts of cadmium in CdTe and CdS[a]

Material	Density	Amount of cadmium	
	g per cm^3	*percent weight*	*g per cm^3 or g/m^2/µm*
CdS	4.8	78	3.73
CdTe	5.9	47	2.75

a. Because polycrystalline material is slightly less dense than single-crystal material, the data provide an upper boundary for cadmium levels in CdTe modules. (Note that a 1 micron thick layer that is 1 m^2 in area has 1 cm^3 of volume, which accounts for the convenient equivalence expressed in the last column of table 5.)

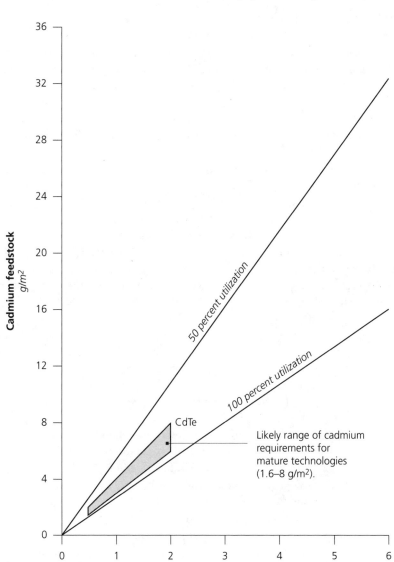

FIGURE 9: The amount of cadmium feedstock needed to make CdTe layers of various thicknesses. It also includes the amounts lost to less-than-100 percent utilization (the amount to be recycled).

Cadmium is also a factor in conventional energy production [59]: burning coal produces cadmium waste (about 1 kilogram per GWh of electricity). CdS/CdTe modules that contain 1 kilogram of cadmium (about 150 m² of module area) would produce about the same 1 GWh of electricity over their 30-year lifespan. But unlike the cadmium in coal, the cadmium in CdTe modules can be recycled (perhaps into another module), rather than released to the environment. Cadmium is also found in nickel–cadmium batteries. Each year about 1,000 tonnes of cadmium enters the U.S. waste stream in the form of discarded batteries [60]. By comparison, U.S. consumers would have to throw away 20 billion watts of PV modules (an absurd scenario) each year to produce a comparable amount of cadmium waste.

In addition, some cadmium will always enter the waste stream from the production of essential minerals, such as zinc and copper, of which it is a natural by-product. Cadmium is also present in phosphate fertilizers and thus can enter the human food chain directly—this represents one of the more serious sources of environmental cadmium [60]. Thus, removing cadmium from fertilizer, coal, and mineral tailings and sequestering it in PV modules would be a creative way to lessen the element's environmental impact. Moreover, because PV-generated electricity can substitute for conventionally generated electricity, its overall positive impact on the environment far outweighs its negative aspects, a fact that is relevant to any rational debate on the merits of CdTe modules. In short, CdTe modules can be manufactured and used safely if reasonable precautions are taken, especially if recycling of materials is practiced.

Economics

The economics of near-term CdTe production are similar to those for CIS (see section on the economics of CIS). If anything, equipment costs are lower while other costs are about the same. Thin-film CdTe is manufactured in much the same way as amorphous silicon (see chapter 9: *Amorphous Silicon Photovoltaic Systems*). Both start with glass substrates, on which a highly conductive transparent tin oxide is deposited. If the device is made of amorphous silicon (*a*-Si), then three *a*-Si layers are deposited on top of the tin oxide; for a CdTe device, layers of thin-CdS and CdTe are deposited. Both thin films are finished by adding a back contact, by scribing steps for monolithic integration and by encapsulation, perhaps with ethylene vinyl acetate and another sheet of glass.

But the costs of depositing CdS/CdTe or *a*-Si are quite different. Deposition of *a*-Si takes place via a slow vacuum process known as glow discharge. In contrast, CdS and CdTe are usually deposited by one of two wet-chemical processes: electrodeposition or spraying. Wet processes have almost negligible capital costs (tens of thousands instead of hundreds of thousands or millions of dollars). Thus CdTe production lacks the most expensive component of *a*-Si thin-film production (the glow-discharge chambers) and has capital costs from 50 to 80 percent below those of *a*-Si. Moreover, equipment maintenance, which is related to equipment cost and complexity, is much reduced for CdTe. Such low capital costs should facilitate the automation of CdTe technology, lowering labor costs without increasing capital costs significantly.

Materials costs are similar to other thin films, dominated as they are by glass costs (front and back) and encapsulation [5]. But because tellurium, the most expensive element, is one-third the cost of indium, CdTe materials costs are slightly less than those for CIS devices.

Photon Energy is now producing prototype CdTe modules. The company claims to have low costs, even with limited production capacity, because the cost of its capital equipment is so low (see figure 2). If their capital costs are estimated at about $10 per m^2, then 6 percent modules would cost about $1.6 per watt, and 8 percent modules about $1.2 per watt. At larger capacities and higher efficiencies, costs would drop much further.

CRYSTALLINE THIN-FILM SILICON (THIN *x*-Si)

Historical perspective

Thin-film silicon solar cells are defined as PV devices in which 1) the silicon layers are thin enough so that the materials costs associated with them are comparable to those for semiconductors in classic thin-films (that is, less than $10 per m^2); and 2) the silicon is grown as a film on an inexpensive supporting substrate. In an optimal configuration, silicon thin-films would be only about 40 microns thick and would be fabricated as large sheets, which would then be connected as monolithic modules. The efficiency and cost goals of thin-film silicon are similar to those for other polycrystalline thin-film technologies.

Semiconductors can be grown on foreign substrates using a number of techniques, including melt growth, vapor phase growth, solid state growth, and solution growth. Several investigators have successfully grown silicon films on graphite or graphite-coated ceramics [61]. The process provides relatively high growth rates, but requires exceptionally high temperatures (1,415°C) in order to melt the silicon.

Silicon thin-film can also be grown by depositing its vapor on a substrate [62, 63], a process that operates at much lower temperatures. Although this can result in poor quality small-grained silicon films, larger grains can be grown on a large-grain polycrystalline substrate [64]. The latter can be fabricated from inexpensive, less pure silicon. Fine-grain polycrystalline silicon, for example, has been grown first on molybdenum substrates and then recrystallized into larger grains, with good results [65].

Two hybrid solutions have also been reported. One calls for vacuum vapor deposition of silicon onto a heated aluminum substrate [66]. Unfortunately, the large grains of silicon that precipitated from the silicon–aluminum solution were so saturated with aluminum that solar-cell operations were unsatisfactory. The other technique calls for chemical vapor deposition (CVD) of silicon onto a heated tin-coated graphite substrate. This technique also causes large grains to grow from the tin solution [67], but impurities from the graphite result in unacceptable cell performance.

Efficiencies of 3.1 percent (not corrected for reflection) have been achieved for 11-micron-thick, small-grain polycrystalline silicon-solar cells grown on tita-

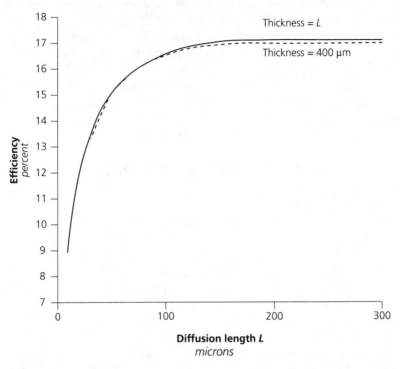

FIGURE 10: *The predicted efficiency of a thin-silicon solar cell is compared with that of a conventional device. The two curves represent devices of different thickness. The solid curve models a thin-silicon cell, where the thickness (H) is equal to the minority carrier diffusion length(L). The minority carrier diffusion length is a key parameter for silicon cells because it indicates the quality of the silicon: the longer the diffusion length, the higher the probability that an absorbed photon will contribute free carriers to the electric current. The curve predicting a slightly lower efficiency (dashed line) represents a 400 micron thick device. Efficiency improves with increasing diffusion length for both devices up to a diffusion length value of 200 microns. The model incorporates a minority carrier diffusion length dependent on doping. The curve predicting slightly higher efficiency (solid line) is modeled with a device thickness equal to the diffusion length. The back surface is modeled with a recombination velocity of 10^3 centimeters per second and no light trapping. The model shows that the thinner device structures generate slightly higher efficiency than the thicker devices modeled with the same material properties. The efficiency increase is due to reduced recombination in the base which leads to higher voltage.*

nium diboride (TiB$_2$)-coated alumina by physical vapor deposition [1]; efficiencies as high as 9.8 percent have been obtained for 25-micron-thick, large-grain polycrystalline silicon grown on graphite substrates [63]. A 12 percent efficiency has been reported for a 4 cm^2 solar cell grown by CVD on an upgraded metallurgical silicon substrate [64]. Silicon grown from a melt on ceramic is reported to have 10.5 percent efficiency [68]; and a microsize thin polycrystalline silicon (called silicon film) grown on a steel substrate achieved an efficiency of 9.5 percent (without an antireflection coating) [69]. These cells were the precursors of today's more efficient thin-film silicon cells.

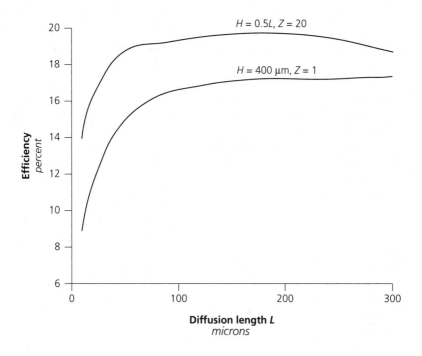

FIGURE 11: *Efficiency.increases when light-trapping effects are included. The top line predicts the efficiency for a device having a thickness equal to one half its diffusion length. Back-surface recombination velocity is modeled at 10^3 centimeters per second. The optical path length is modeled at 20 times the device thickness (Z=20). The light-trapping device is compared to a conventional 400 micron thick device without light trapping (lower line). The light-trapping device is shown to perform from 15 to 20 percent better than the conventional device. The difference in performance is greatest for minority carrier diffusion lengths of 40 to 80 microns.*

Cell design

Thin silicon cells can be as efficient as conventional thick cells, yet they use less material and are less sensitive to impurities (see figure 10). As long as the quality of the silicon layer is good, thinness, per se, does not defeat performance.

Designing silicon devices so that light is trapped within them (so-called light-trapping) increases their efficiency by 15 to 20 percent (see figure 11). Because thin devices are less dependent on diffusion length, they can be more conductive and achieve higher voltages. Such an option cannot be exploited by thick conventional devices, which lack the long diffusion lengths needed to collect electrons generated deep in the base layer. In thick cells, the long diffusion length requirement allows high efficiency to be obtained only with high quality, relatively expensive silicon wafers. Thus, thin devices that trap light and are made from modest quality material can achieve higher efficiencies than conventional devices made from high-quality material.

Light-trapping techniques

The light path in a thin silicon device can be extended many times through the use of textured surfaces. To compensate for partial reflections and interference effects, a model has been developed to evaluate different designs with respect to their light-confining ability [70]. The model has examined both front surface and back surface texturing (see figures 12 and 13), as well as random texturing and regularly spaced slat texturing. To minimize losses at the back of the silicon layer, an oxide of silicon, silicon dioxide (SiO_2), can be formed to reduce surface defects that would otherwise diminish current.

The model assumes that all generated carriers are collected, but the assumption holds true only when the thickness of the device is approximately half the length of the minority carrier diffusion value and when the back surface is passivated with SiO_2. The model also predicts the amount of current

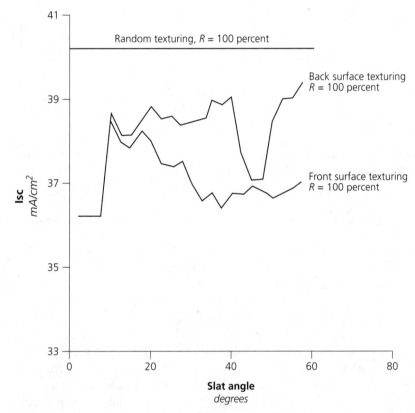

FIGURE 12: *Current densities (Isc) for thin-film silicon solar cells, with various front- and back-surface textures. Higher current densities indicate more efficient light trapping. The back surfaces of the textured devices have been modeled with wavelength independent reflectivities (R = 100 percent). The back surface of the 35 micron silicon layer is 100 percent reflective and has a double layer of antireflection coating.*

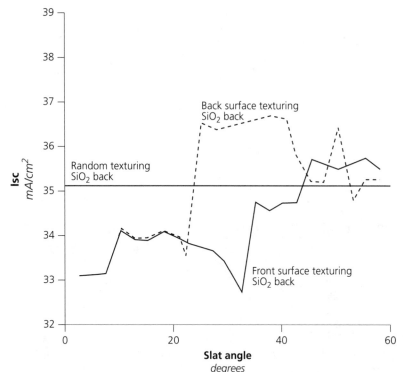

FIGURE 13: Current densities (Isc) for thin-film silicon solar cells with SiO_2 back-surface coatings and various front- and back-surface textures.

that can be obtained from the textured structure. High current levels, approximately 40 milliamps per cm^2, are predicted for 35-micron silicon layers that have 100 percent reflective and random back-surface texturing (see figure 12). Currents in the range of 39 milliamps per cm^2 can be obtained by either front or back surface texturing. Although front-surface texturing must meet a narrow range of design criteria with respect to slat angle and spacings, back-surface texturing can produce high levels of current over a much broader range of design criteria.

High current levels can be reduced by as much as 20 percent if significant recombination occurs at the back surface. As stated, recombination can be minimized, however, by placing SiO_2 coatings on the back surface to facilitate passivation. Such surfaces are highly reflective to confined light. According to the model, the highest levels of current are obtained for devices that have an SiO_2 coating and regularly spaced back-surface texturing, conditions under which total internal reflection levels are best controlled (see figure 13).

The extended light path achieved by light trapping can be modeled using the quantity Z, which describes the optical path length as Z times the device thickness. Z can be correlated to current. For a 35 micron device, $Z = 10$ corre-

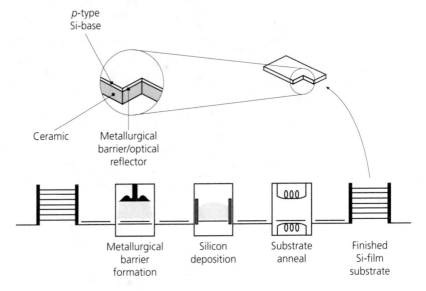

FIGURE 14: Silicon-film manufacturing steps.

lates to a current value of 40 milliamps per cm^2, while $Z = 5$ generates a current value of 38.9 milliamps per cm^2. For the devices modeled in figures 12 and 13, Z ranges from 5 to 10.

Crystalline thin-film technology

In the silicon-film manufacturing process, a metallurgical barrier/optical reflector is applied to a ceramic substrate, and the silicon (which was precipitated from solution) and the substrate on which it is deposited are annealed to reduce thermal stress (see figure 14). After the silicon/substrate complex is complete, solar cells are fabricated using standard silicon diffusion, metallization, and anti-reflection coating processes.

The silicon growth process

The silicon-film growth occurs from a saturated metal solution and is similar to the growth of silicon from a molten source material. This is called liquid-phase epitaxy.[16] Liquid-phase epitaxial devices are generally thought to be superior in performance to those grown by other methods [71]. Such devices include light-emitting diodes, semiconductor lasers, magnetic garnet bubble memories, and gallium arsenide solar cells. The improved performance of liquid-phase epitaxial devices (as opposed to vapor-phase or diffused devices) can be attributed to exact compositional control, longer diffusion lengths, fewer crystal defects, and the

---

16. Epitaxy is the growth on a crystalline substrate of a crystalline substance that mimics the orientation of the substrate.

tendency of impurities to remain in the liquid rather than enter the solidified silicon film. Liquid epitaxy also has a tendency to correct dislocations that may occur in the substrate, improving crystal structure.

Growing silicon on ceramic substrates in solution offers the advantages of low-growth temperatures, large-grain growth, and long diffusion lengths. To prevent crystalline overgrowth on a substrate, the growth process can be separated into the following sequential elements: wetting, nucleation, non-impinging crystal growth, fill-in crystal growth, and homoepitaxial film growth. The division allows for comprehensive theoretical and experimental analysis of the overall process.

Experimental technologies

Silicon film on steel
Silicon film grown on a low-cost coated steel substrate was first noted to have a photovoltaic effect in 1984 [69,72]. By 1985, the energy conversion efficiencies of these microsized devices had been increased to 9.6 percent, without antireflection coating. Although larger-area devices were plagued by fractures caused by thermal expansion differences between the steel and silicon, stress was reduced by adding a thick ceramic coating to the steel. The result in 1986 was a solar cell that measured 12 mm^2 and was 9.6 percent efficient (see table 6).

Table 6: Steel substrate

	Units	August 1984	April 1985	October 1985	November 1986[a]
V_{oc}	mV	35	484	515	535
I_{sc}	mA/cm^2		21.3[b]	24.0[b]	26
Fill factor	percent	25	67	77	68
Efficiency	percent		6.9[b]	9.6[b]	9.6
Area	cm^2		0.001	0.0023	0.12

a. Measured by SERI.

b. No antireflection coating.

Silicon film on ceramic
The need to buttress steel with a thick ceramic coating was sufficiently impractical that in 1987 researchers at Sandia National Laboratory replaced the steel with a ceramic material having the same thermal expansion properties as silicon. The result was a 1 cm^2, 10.2 percent efficient solar cell, which by 1988 had evolved at AstroPower in Newark, Delaware, into a 15.7 percent cell known as Product I of the Silicon-Film series [73]. These results were achieved with ac-

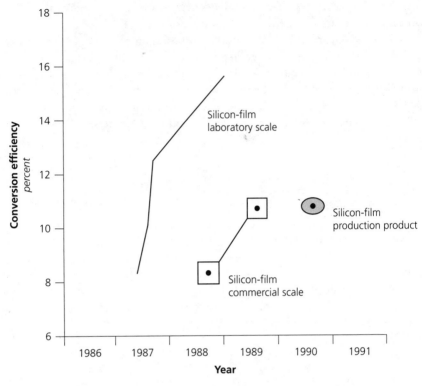

FIGURE 15: *Performance results for a 15.7 percent solar cell known as Product I of the Silicon-Film series are shown. Data are provided for both laboratory and commercial devices.*

tive silicon layers that were about 100 microns thick and did not involve light trapping (see figure 15 for a summary of Product I results).

Passivation and gettering

Reducing defects (passivation) and removing impurities (gettering) have both led to improved silicon solar-cell performance. When combined with other processing methods, such as deep diffusion and low-temperature annealing, the benefits of passivation and gettering are enhanced. These specialized processes are known to increase the overall energy conversion efficiency of a 1 cm^2 laboratory solar cell by 54 percent [74] (see table 7). Whereas the standard process calls for high-efficiency shallow diffusion; for contacts made from thallium, palladium, and silver; and for a double-layer antireflection coating, the specialized process also includes phosphorous gettering, hydrogen passivation, and low temperature annealing. Most of the improvement manifests itself as increased current, the rest as improved operating voltage, which is probably attributable to grain boundary passivation.

Table 7: Effects of passivation and gettering[a]

	Units	I	II
V_{OC}	*mV*	582	600
I_{sc}	*mA/cm²*	25.0	33.0
Fill factor	*percent*	70.2	79.2
Efficiency	*percent*	10.2	15.7
Area	*cm²*	0.85	1.02
L	*imicrons*	50	110

a. A silicon-film solar cell fabricated using a standard processing sequence (I) is compared with one fabricated using a specialized processing sequence (II) [75]. The standard process consists of high-efficiency shallow diffusion, Ti/Pd/Ag contacts and a double-layer refractive coating. The specialized process incorporates phosphorous gettering, hydrogen passivation, and low-temperature annealing.

Commercial-sized solar cells

To test the uniformity of the specialized process, 23 silicon-film solar cells were made at AstroPower in 1990 [75]. The resulting cells had efficiencies ranging from 6.8 to 10.9 percent, with 87 percent of them above 8 percent. The short-circuit current densities for commercial-scale cells are lower than those for laboratory-scale cells because of stress-induced defects during formation of the active silicon layer. Once stress is reduced during the production-scale process, improved short-circuit currents for the commercial-scale cells can be expected. At that time, energy conversion efficiencies should exceed 12 percent.

Field stability

Although data on the long-term stability of thin-film silicon solar cells do not exist, the cells are thought to have the same stability characteristics as single- and polycrystalline-silicon solar cells. These are known to be stable when well- encapsulated. Such high stability can be attributed in part to high processing temperatures, around 850°C, which are needed to create a photovoltaic junction. Accordingly, there is little reason to expect solar cell degradation at operating temperatures.

Health and safety issues

Thin-film silicon solar cells pose no apparent health or safety problems. Minor amounts of metals (silver, nickel, etc.) are used for grids; the use of solders (lead, for example) is minimized. Processing can be modified so that it does not depend on toxic materials.

Economics

The economics of silicon thin-film cells are similar to other thin films. Material costs are dominated by glass and by the module's backing material. Together, the semiconductor, silicon, and ceramic materials account for less than 10 percent of the cell's overall costs. Capital costs, which are low, account for

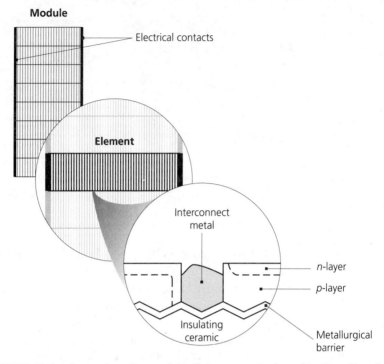

FIGURE 16: *AstroPower scheme for monolithically interconnecting long, thin silicon-film cells to make large-area thin-film modules.*

about 10 percent of the total. Unlike other thin films, however, the near-term costs for silicon cells include a large labor component because the individual cells must be fastened into arrays in order to form a module. Nonetheless, module costs of less than $2 per watt appear to be within reach.

Over the long term, improvements in module assembly, combined with improvements in solar cell performance, are expected to reduce module costs to less than $1 per watt. Further cost reductions may come when monolithic, interconnected arrays (similar to those utilized by other thin films) are developed. Such interconnected arrays can significantly reduce module costs by decreasing assembly costs. A key feature of a fully interconnected silicon module will be based on isolating the silicon-film wafers into segments (see figure 16). Because the wafers are only 20 to 50 microns thick and are supported by an insulating substrate, they can be readily isolated. A contacting scheme has been developed and metal interconnections will be on the edges or backs of the elements to reduce light losses.[17]

---

17. In January 1992 AstroPower created a 13.1 percent efficient silicon-film submodule made of four interconnected cells fabricated on a silicon substrate.

Remarks

Thin-film silicon solar cells may offer the cost and interconnection advantages of other thin-film cells, yet retain the high performance and stability of crystalline silicon. Moreover, thicker silicon film devices already display efficiencies as high as 15.7 percent without light trapping. Eventually, commercial thin silicon film products are expected to achieve efficiencies as high as 14.5 percent without light trapping, and greater than 17 percent with light trapping. At the same time, monolithic interconnection techniques will reduce module costs while retaining the high efficiencies associated with light trapping.

THE PATHWAY TO LOW COST

We have emphasized the near-term cost of thin films in this chapter because these technologies are unlikely to exceed 10 megawatt plant capacities during the next four or five years. Nevertheless, progress in production methods can be expected. We envision rapid progress, for example, in reducing production costs, modifying machinery to process several substrates simultaneously, automating the manufacturing line, and establishing an infrastructure to handle production, sales, health and safety issues, and management. When the scale of production reaches a reasonable level—perhaps 30 megawatts for the least capital-intensive processes and 100 megawatts for the rest—production costs should drop substantially (see table 8).

PV costs are inversely proportional to the amount of sunlight and are therefore lower in sunny locations than they are in average or cloudy regions. When PV costs are especially high, the location has a significant impact on overall costs, but when costs fall below $0.10 per kWh, the difference between sunny and cloudy regions diminishes. When PV costs drop to $0.06 per kWh, PV installations will become viable across 95 percent of the United States. In fact, at those prices, the price of conventional electricity exhibits more geographic variation than PV-generated electricity. For example, conventionally generated electricity varies from about $0.04 per kWh in the northwest to more than $0.12 per kWh on Long Island, New York or in San Diego, California.

It now appears that when production levels reach about 30 to 100 megawatts, PV costs will drop to a level (about $0.10 per kWh) appropriate for many utility-scale markets. At that point, the capital needed to lower costs still further (to about $0.06 per kWh) should be available. Reaching that point will depend on the following:

1. *Efficiency improvements.* As usual, everything depends heavily on efficiency. When efficiency goes up, production volume in m^2 per year drops proportionally; that is, a factory producing 10 percent modules needs to turn out only half as many modules as, say, a factory producing 5 percent modules. Thus the cost per watt drops sharply as the number of modules needed to produce a given amount

of megawatts goes down. PV-system costs are also heavily dependent on efficiency, since higher efficiencies reduce overall balance-of-system costs: if fewer modules are needed to produce a certain output, then the support structures and other associated components are proportionally less costly.

2. *Yield.* Yield of finished modules is expected to improve to 80 to 90 percent by 1995. Such an improvement seems reasonable given the anticipated pace of technological evolution. Large automated machines capable of repetitive processing, for example, are likely to improve overall yield.

3. *Depreciation period.* Early in a technology's development, depreciation periods are short because new technologies are likely to replace existing ones relatively quickly. But over time, as technological changes become incrementally smaller, equipment is kept for longer periods. We therefore assume that the depreciation period will increase from five years initially, to seven and then to 10

Table 8: Possible evolution of costs as plant size and technical maturity increase[a]

	Units	1.8×10^5 m² ~ 10 MW	4.2×10^5 m² ~ 30 MW	1.1×10^6 m² ~ 100 MW	3.7×10^6 m² ~ 500 MW
Efficiency	percent	7	9	11	15
Yield	percent	80	80	85	90
Depreciation period	years	5	5	7	10
Capital investment	$ millions	18	32	70	185
Investment/capacity	$/W	2	1	0.7	0.4
Equipment	$/m²	20	15	0.7	5
Labor and fringe	$/m²	20	17	13	10
Materials	$/m²	30	25	25	20
Utilities	$/m²	10	8	7	6
Overhead	$/m²	10	7	5	3
Cost	$/m²	113	90	70	49
	$/W	1.6	1.0	0.64	0.33
PV system cost	$/W$_p$	3	1.9	1.3	0.76
Sunlight					
below average	$/kWh	0.29	0.18	0.12	0.07
average	$/kWh	0.24	0.15	0.10	0.06
desert	$/kWh	0.18	0.11	0.075	0.045

a. Plant capacities in megawatts assume stated module efficiencies and production yields, which vary for assumed maturity. Three shifts and 260 workdays are assumed at all production levels. The equipment costs drop rapidly in the larger plants because longer depreciation periods are assumed. Final module costs include yield losses (they are the total of the component costs divided by the yield). PV system cost assumes balance-of-system costs in $ per m² of $100, $80, $70, and $65 for fixed flat-plates; a below-average location like Long Island, New York (1,500 kWh per m² of sunlight each year), an average location like Kansas City, Missouri (about 1,800 kWh per m²), and a desert location like Phoenix, Arizona (2,400 kWh per m²); and system efficiencies have been reduced 20 percent to account for losses expected under typical field conditions. See table 1 for breakdown of BOS costs.

years for a mature facility. Note in table 8 that as the equipment cost drops by a factor of four (from \$20 to \$5 per m^2), the depreciation period lengthens by a factor of two, so the actual drop in initial capital investment is only half (from \$20 to \$10 per m^2).

4. *Equipment costs.* It seems likely that equipment costs will decline by a factor of two as indicated above, given the immature state of existing manufacturing lines and the advantages that will accrue when larger machines replace smaller ones.

5. *Labor.* We also assume that labor costs will be cut by at least half, and perhaps more, since greater automation and bigger machines will reduce the labor force needed to produce PV modules.

6. *Materials.* Materials probably represent the most stubborn of all cost categories, and thus are expected to drop by only a third over the next five years. The drop will reflect the cost advantages of buying larger quantities from suppliers, improved materials utilization and the recycling of waste material, which is more practical at greater volumes. Several investigators, including John Jordan of Photon Energy, have proposed placing PV facilities at the end of glass manufacturing lines, thus greatly reducing the cost of glass substrates. Because glass is the most expensive component of thin-film modules, obtaining it directly from a glass manufacturer might cut costs by \$2 to \$5 per m^2 (or almost \$0.005 per kWh at the system level).

7. *Utilities.* A 40 percent drop in the amount of energy needed per unit of production is predicted. Some of the drop comes from larger production volumes per unit of building space, some from improved process design. For example, larger machines do not use proportionally more electricity than smaller machines because they perform some tasks (such as pumping down vacuum equipment and annealing parts) simultaneously.

8. *Overhead.* Overhead will naturally diminish as markets for PV improve and as PV prices fall far enough for large systems to become cost-competitive. For instance, the size of the sales staff does not increase with the size of the system being sold. Management and other categories of overhead would also fall proportionately.

An implicit factor in cost calculations is downtime. We have assumed 260 days of work per year (allowing weekends for repair and maintenance), but expect that downtime in excess of this buffer would be proportional to the complexity of the capital equipment. Thus for relatively expensive equipment, downtime may be greater, suggesting that in some cases simpler processes may be preferred over complex ones. For the most part, the throughput of thin-film manufacturing processes remains unproven. And so, like yield, downtime cannot be accurately predicted.

Reasonable assumptions about thin-film manufacturing capabilities suggest that low-cost PV might be produced before the end of the 1990s, provided that processes for rate, yield, reliability, and materials use are optimized; larger, more automated machines are designed; efficiencies continue to improve; and the infrastructure is expanded. Success in each area requires only that existing trends continue; there are no obvious showstoppers to impede progress toward low-cost thin-film PV.

It is beyond the scope of this chapter to discuss long-term prospects for reduced balance-of-system costs. Several of our colleagues predict that costs will be lower than we assume ($50 per m^2 or less, instead of our estimated $64 per m^2). In any case, it should be noted that a drop of $10 per m^2 (whether in reduced balance-of-system costs or in cheaper thin-film modules) corresponds to a drop of about $0.005 per kWh in the cost of PV electricity.

CONCLUSIONS

The development of conventional crystalline silicon photovoltaic materials for flat-plates has been steady and should continue to improve throughout the 1990s. Eventually, however, crystalline silicon will be replaced by various polycrystalline thin-films. As a result, the cost of PV modules is expected to drop to between $1 and $0.50 per watt by the end of the decade, depending on the speed with which manufacturing scale-up is achieved. Once those prices are reached, multibillion-dollar markets for PV will emerge, rendering the technology self-sustaining. Continued progress will assure that PV costs drop still further—to less than $0.06 per kWh in locations with average sunlight—thus rendering it fully competitive worldwide with existing conventional energy sources.

WORKS CITED

1. Russell, T.W.F., Baron, B.N., and Rocheleau, R.E. 1984
   Economics of processing thin film solar cells, *Journal of Vacuum Science and Technology* B2(4):840–844.

2. Jackson, B. 1985
   CdZnS/CuInSe$_2$ module design and cost assessment, SERI/TP-216-2633, Solar Energy Research Institute, Golden, Colorado.

3. Kapur, V.K. and Basol, B. 1990
   Key issues and cost estimates for the fabrication of CIS PV modules by the two-stage process, *Proceedings of the 21st IEEE Photovoltaic Specialists Conference*, 1:467–470.

4. Carlson, D. 1990
   Low cost power from thin film photovoltaics, in Johansson, T.B.; Bodlund, B. and Williams, R.H., eds., *Electricity: efficient end-use and new generation technologies, and their planning implications*, Lund University Press, Lund, Sweden.

5. Meyers, P.V. 1990
   Polycrystalline cadmium telluride n-i-p solar cells, SERI subcontract ZL-7-06031-2, final report, Solar Energy Research Institute, Golden, Colorado.

6. Shay, J.L., Wagner, S., Bachmann, K., Buehler, K., and Kasper, H.M. 1975
   Proceedings of the 11th IEEE Photovoltaic Specialists Conference, 503–507.

7. Chen, W.S. and Mickelsen, R.A. 1980
   Thin film CdS/CuInSe$_2$ heterojunction solar cell, *Proceedings of the Society of Photo-Optical Instrumentation Engineering* 248:62–69.

8. Chen, W.S. and Mickelsen, R.A. 1980
   High photocurrent polycrystalline thin film CdS/CuInSe$_2$ solar cell, *Applied Physics Letters* 36(5)371–373.

9. Chen, W.S. and Mickelsen, R.A. 1982
   Methods for forming thin film heterojunction solar cells from I-III-VI$_2$ chalcopyrite compounds, U.S. patent 4,335,266, assignee: Boeing Corporation.

10. Zweibel, K. 1985
    Post-deposition heat treatments of CdS/CuInSe$_2$ solar cells, SERI/PR-211-2468, Solar Energy Research Institute, Golden, Colorado.

11. Noufi, R., Souza, P., and Osterwald, C. 1985
    Effect of air anneal on CdS/CuInSe$_2$ thin film solar cells, *Solar Cells* 15(1):87–91.

12. Noufi, R., Powell, R.C., and Matson, R.J. 1987
    Low cost methods for the production of semiconductor films for CuInSe$_2$/CdS solar cells, *Solar Cells* 21:55–63.

13. Choudary, U.V., Shing, Y.H., Potter, R.R., Ermer, J.H., and Kapur, V.K. 1986
    CuInSe$_2$ thin film solar cell with thin CdS and transparent window layer, U.S. patent 4,611,091, assignee: Atlantic Richfield.

14. Vijayakumar, P.S., Blaker, A., Wieting, R.D., Wong, B., and Halami, A.T. 1988.
    Chemical vapor deposition of zinc oxide films and products, U.S. patent 4,751,149, assignee: Atlantic Richfield.

15. Potter, R.R., Eberspacher, C., and Fabick, L.B. 1985
    Device analysis of CuInSe$_2$/(Cd,Zn)S/ZnO solar cells, *Proceedings of the 18th IEEE Photovoltaic Specialists Conference*, 1659–1664.

16. Love, R.B. and Choudary, U.V. 1984
    Method for forming pv cells employing multinary semiconductor films, U.S. patent 4,465,575 (filed 28 February 1983), assignee: Atlantic Richfield Company.

17. Kapur, V.K., Choudary, U.V., and Chu, A.K.P. 1986
    Process of forming compound semiconductor material, U.S. patent 4,581,108, assignee: Atlantic Richfield.

18. Ermer, J.H. and Love, R.B. 1989
    Method for forming CuInSe$_2$ films, U.S. patent number 4,798,660 (filed 22 December 1986), assignee: Atlantic Richfield Company.

19. Eberspacher, C., Ermer, J., and Mitchell, K.W. 1989
    Process for making thin film solar cell, European patent application 0 318 315 A2 (filed 25 November 1988—pending), assignee: Atlantic Richfield Company.

20. Mitchell, K.W., Eberspacher, C., Ermer, J., and Pier, D. 1988
    Single and tandem junction CuInSe$_2$ cell and module technology, *Proceedings of the 20th IEEE Photovoltaic Specialists Conference*, 1384–1389.

21. Mitchell, K.W., Eberspacher, C., Ermer, J., Pauls, K., Pier, D., and Tanner, D. 1989
    Single and tandem junction CuInSe$_2$ technology, *4th International PV Science and Engineering Conference*, Sydney, Australia, 14–17 February 1989.

22. Pollock, G.A., Mitchell, K.W., and Ermer, J.H. 1989
    European patent application 89308108.3, Atlantic Richfield.

23. Ermer, J.H., Frederic, C., Pauls, K., Pier, D., Mitchell, K., and Eberspacher, C. 1989
Recent progress in large area CuInSe$_2$ L, *4th International PV Science and Engineering Conference*, Sydney, Australia, 475–480.

24. Kapur, V.K., Basol, B.M., and Tseng, E.S. 1985
Low cost thin film chalcopyrite solar cells, *Proceedings of the 18th IEEE Photovoltaic Specialists Conference*, 1429–1432.

25. Kapur, V.K., Basol, B.M., and Tseng, E.S. 1986
Preparation of thin films of chalcopyrites for photovoltaics, *Ternary & multinary compounds, Proceedings of the 7th International Conference*, Snowmass, Colorado, 219–224.

26. Kapur, V.K., Basol, B.M., and Tseng, E.S. 1987
Low cost methods for the production of semiconductor films for CuInSe$_2$/CdS solar cells, Solar Cells (Switzerland), 7th Photovoltaic Advanced Research and Development Project Review Meeting, 13 May 1986, Denver, Colorado, (21):65–73.

27. Chesarek, W., Mason, A., Mitchell, K., and Fabick, L. 1987
Spacial EBIC studies of ZnO/CdS/CuInSe$_2$ solar cells, *Proceedings of the 19th IEEE Photovoltaic Specialists Conference*, 791–794.

28. Kapur, V.K., Basol, B.M., Tseng, E.S., Kullberg, R.C., and Nguyen, N.L. 1988
High-efficiency, copper ternary, thin-film solar cells, SERI subcontractor annual report, SERI/STR-211-3226.

29. Basol, B.M. and Kapur, V.K. 1989
CuInSe$_2$ films and solar cells obtained by selenization of evaporated Cu–In layers, *Applied Physics Letters* 54:1918.

30. Basol, B.M. and Kapur, V.K. 1990
Deposition of CuInSe$_2$ films by a two-stage process utilizing E-beam evaporation, *IEEE Transactions Electron Devices* 37: 418.

31. Basol, B.M. and Kapur, V.K. 1990
CuInSe$_2$ thin films and high efficiency solar cells obtained by selenization of metallic layers, *Proceedings of the 20th IEEE Photovoltaic Specialists Conference*, 546–549.

32. Chen, W.S., Stewart, J.M., Stanbery, B.J., Devaney, W.E., and Mickelsen, R.A. 1987
Development of thin film polycrystalline CuInGaSe$_2$ solar cells, *Proceedings of the 19th IEEE Photovoltaic Specialists Conference*, 1445-1447.

33. Moskowitz, P.D., Zweibel, K., and Fthenakis, V.M. 1990
Health, safety and environmental issues relating to cadmium usage in PV energy systems, SERI/TR-211-3621 (DE90000310), Solar Energy Research Institute, Golden, Colorado.

34. Moskowitz, P.D. and Fthenakis, V.M. 1990
Toxic materials released from PV modules during fires: health risks, *Solar Cells* 29:63–71.

35. Zweibel, K. and Mitchell, R. 1989
CuInSe$_2$ and CdTe: scale-up for manufacturing, SERI/TR-211-3571, Solar Energy Research Institute, Golden, Colorado.

36. Moskowitz, P.D., Fowler, P.K., Dobryn, D.G., Lee, C.M., and Fthenakis, V.M. 1986
Control of toxic gas release during the production of copper indium diselenide PV cells, draft report, MIT/BNL-86-4).

37. Moskowitz, P.D. and Fthenakis, V.M. 1991
Environmental, health, and safety issues associated with the manufacture and use of II-VI PV devices, Brookhaven National Laboratory, Upton, New York (in press).

38. Zweibel, K., Jackson, B., and Hermann, A. 1986
Comments on "critical materials assessment program" (Smith and Watts, Battelle Memorial Institute), *Solar Cells* 16:631–634.

39. Rappaport, P. 1959
RCA Review 20:373.

40. Loferski, J.J. 1956
Journal of Applied Physics 27: 777.

41. Zweibel, K. ed. (invited), 1986
Special issue on CdTe, L.L. Kazmerski and T.J. Coutts, eds., *Solar Cells* 23(1–2).

42. Basol, B.M. 1988
Electrodeposited CdTe and HgCdTe solar cells, *Solar Cells* 23(1–2):69–88.

43. Meyers, P.V. 1988
Design of a thin film CdTe cell, *Solar Cells* 23(1–2):59–68.

44. Tyan, Y.S. 1988
Topics on thin film CdS/CdTe solar cells, *Solar Cells* 23(1–2):19–30.

45. Mitchell, K.W., Eberspacher, C., Cohen, F., Avery, J., Duran, G., and Bottenberg, W. 1988
Progress toward high efficiency thin film CdTe solar cells, *Solar Cells* 23(1–2):49–59.

46. Ikegami, S. 1988
CdS/CdTe solar cells by the screen-printing–sintering technique, *Solar Cells* 23(1–2):89–106.

47. Meyers, P.V., Liu, C.H., and Doty, M.E. 1989
Method of making pv cell with chloride dip, U.S. patent 4,873,198, assignee: Ametek, Inc.

48. Turner, A.K., Woodcock, J.M., Ozsan, M.E., Summers, J.G., et al. 1991
Stable, high efficiency thin film solar cells produced by electrodeposition of CdTe, *Proceedings of the 5th International PV Science and Engineering Conference.*

49. Ullal, H.S. 1990
European trip report (unpublished), Solar Energy Research Institute, Golden, Colorado.

50. Meyers, P.V., Liu, C.H., and Frey, T.J. 1987
Heterojunction *p-i-n* pv cell, U.S. patent 4,710,589, assignee: Ametek, Inc.

51. Jordan, J. and Albright, S. 1988
Large Area CdS/CTe pv Cells, *Solar Cells* 23(1–2):107–115.

52. Albright, S.P., Brown, D.K., and Jordan, J.F. 1987
Method and apparatus for forming a polycrystalline monolayer, European patent application EP 0 264 739 (A2).

53. Singh, V.P., Kenney, R.H., McClure, J.C., Albright, S.P., Ackerman, B., and Jordan, J.F. 1987
Proceedings of the 19th IEEE Photovoltaic Specialists Conference, 216–221.

54. Jordan, J. and Albright, S. 1988
Large area CdS/CdTe solar cells, *Solar Cells* 23(1–2):107–114.

55. Albright, S.P. and Ackerman, B. 1989
High efficiency large area CdTe modules, annual subcontract report 1 July 1988–30 June 1989, SERI/STR-211-3585.

56. Nakano, N. 1991
Reliability of *n*-CdS/*p*-CdTe solar modules in accelerated environmental tests and effect of oxygen, *Solar Cells* (in press).

57. Moskowitz, P.D. and Zweibel, K. 1990
Health and safety issues related to the production, use, and disposal of cadmium-based PV modules, *Proceedings of the 20th IEEE Photovoltaic Specialists Conference*, 1040–1042.

58. Doty, M. and Meyers, P.V. 1988
Safety advantages of a CdTe based PV module plant, in Luft, W., ed., *AIP Conference Proceedings 166, Photovoltaic Safety,* American Institute of Physics, New York, 10–18.

59. San Martin, R.L. 1989
Environmental emissions from energy technology systems: the total fuel cycle, U.S. Department of Energy, Washington DC.

60. Korzun, E.A. and Heck, H.H. 1990
Sources and fates of lead and cadmium in municipal solid waste, *Journal of the Air & Waste Management Association* 1220–1226.

61. Belouet, C., Texier-Hervo, C., Mautref, M., Belin, C., Paulin, J., and Schneider, J. 1983
Journal of Crystal Growth 61:615.

62. Feldman, C., Blum, N.A., and Satkiewicz, F.G. 1980
Conference Record of the 14th IEEE Photovoltaic Specialists Conference, 391.

63. Chu, T.L., Chu, S., Lin, C.L., and Abderrassoul, R. 1979
Journal of Applied Physics 50:919.

64. Khattak, C.P., Schmid, F., Robinson, P.H., and D'Aiello, R.V. 1982
Proceedings of the 16th IEEE Photovoltaic Specialists Conference, 128–132.

65. Lesk, I.A., Baghdadle, A., Gurtler, R.W., Ellis, R.J., Wise, J.A., and Coleman, M.G 1976
Proceedings of the 19th IEEE Photovoltaic Specialists Conference, 173.

66. Fang, P.H. and Ephrath, L. 1974
Applied Physics Letters 25(10):583.

67. Bloem, J., Gilin, L.J., Graef, M.W.M., and DeMoor, H.H.C. 1979
Proceedings of the 2nd European Community Photovoltaic Solar Energy Conference, D. Reidel, Berlin, 759.

68. Schuldt, S.B., Heaps, J.D., Schmit, F.M., Zook,J.D., and Grunt, B.L. 1984
Conference Record of the 15th IEEE Photovoltaic Specialists Conference, 934.

69. Barnett, A.M., Hall, R.B., Fardig, D.A., and Culik, J.S. 1986
Technical Digest of 2nd International PVSEC, Beijing, China, 167–170.

70. Rand, J.A., Hall, R.H., and Barnett, A.M. 1990
Light trapping in thin crystalline silicon solar cells, *Proceedings of the 20th IEEE Photovoltaic Specialists Conference*, 263–268.

71. Moon, R.L. 1980
Liquid phase epitaxy, in *Crystal growth,* Paplin, B., ed., Pergamon Press, New York.

72. Barnett, A.M., Hall, R.B., Fardig, D.A., and Culik, J.S. 1985
Conference Record of the 18th IEEE Photovoltaic Specialists Conference, 1094–1099.

73. Barnett, A.M., Domian, F.A., Ford, D.H., Kendall, C.L., Rand, J.A., Rock, M.L., and Hall, R.B. 1989
Thin silicon-film solar cells on ceramic substrates, *Technical Digest of the International PVSEC-4*, Sydney, Australia, 151.

74. Rock, M.L., Cunningham, D.W., Kendall, C.L., Hall, R.B., and Barnett, A.M. 1990
Process induced improvements in polycrystalline silicon-film solar cells, *Proceedings of the 20th IEEE Photovoltaic Specialists Conference*, 634–637.

75. Kendall, C.L, Checchi, J.C., Rock, M.L., Hall, R.B., and Barnett, A.M. 1990 10% efficient commercial-scale silicon-film solar cells, *Proceedings of the 20th IEEE Photovoltaic Specialists Conference*, 604–607.

11
UTILITY FIELD EXPERIENCE WITH PHOTOVOLTAIC SYSTEMS

KAY FIROR
ROBERTO VIGOTTI
JOSEPH J. IANNUCCI

Utilities are becoming increasingly aware of the potential of photovoltaic (PV) systems. Some utilities already employ such systems, and demonstrations of the production of megawatts (millions of watts) of power have been fielded at multiple sites in the industrialized world. As production, installation, and operation and maintenance costs fall, utilities throughout the world expect to rely on PV systems for substantial amounts of electricity.

INTRODUCTION

Electric utility companies have long been aware of the potential of photovoltaic (PV) systems for generating electricity. The early 1980s saw the more progressive utilities undertake PV research projects, and a much broader group was involved by the end of the decade. Utility involvement now includes demonstrations of the production of megawatts (millions of watts) of power. Sustained PV research budgets at individual private utilities are up to $3 million per year, and in the case of Italy's national utility, the budget is over $10 million per year.

Interest in PV systems centers on two broad areas. The first is the use of PV/battery or PV/battery/small generator systems in "stand alone" configurations to meet the power requirements of remote areas. The second is the connection of PV systems to power grids to meet peak power demands when the sun is shining; for example, on hot, sunny afternoons when air conditioners are in high use. In such grid-connected applications, the electricity would be produced to meet demand, avoiding the issue of storing electricity for later use.

Utilities expect to employ PV systems extensively in the future. In Europe, many utilities expect to be involved directly and through customer-owned systems, while all Japanese utilities expect to interconnect significant numbers of residential systems in the 1990s. In the United States, a cooperative effort among utility executives, regulatory officials, state and federal energy offices, PV research groups, and consumer advocate organizations has been formed to expedite the acceptance and use of PV systems within the utility industry. The coalescence of such a cooperative effort represents the beginnings of a major shift for the U.S. utility industry; as recently as 1988, only 9 percent of utility executives surveyed expected their own companies to use PV systems in the future [1]. Predictions of the extent of near-term use vary, with the most aggressive being the Japanese government's prediction of an annual market for 3,000 megawatts in Japan alone by the year 2000 [2].

Most electric utility involvement in PV systems is in industrialized countries, and this chapter will therefore focus on them. Perhaps somewhat paradoxically, however, the largest numbers of PV systems are in developing countries; in less-developed countries, many telecommunications and gas utilities have been employing PV power for over a decade. (A more detailed discussion of the applications of PV systems in developing countries will be found in the appendix.) The interest of developing countries in utility PV systems may increase as such systems become more standard in the developed world. In the meantime, private and government-supported PV programs are becoming more common in developing countries, primarily to supply small amounts of power to remote rural populations. Communications, water pumping and purification, vaccine refrigeration, and lighting are the most common applications of PV systems, as they fulfill the most urgent power needs of vast rural populations.

Photovoltaic research and development programs are actively supported in many national energy programs around the world. Many nations also have regulations, tax incentives, and other programs designed to encourage private investment in PV systems. National PV programs are reviewed in the sections that follow. The final section of this chapter summarizes the principal findings of the experiments conducted to date. Greater detail on national programs can be found in the appendix.

NATIONAL PROGRAMS TO ENCOURAGE PV SYSTEMS

Italy

A comprehensive, steadily increasing, and ambitious national PV development effort exists in Italy. The Italian National Energy Plan includes a growing effort to develop renewable energies, with particular emphasis on PV systems, which command a large share of the available financial resources. The currently approved five-year plan calls for the production of 25 megawatts of power by PV systems by 1995.

Italy's National Commission for Nuclear and Alternative Energy (ENEA) is the federal agency in charge of research and development and technology transfer to industry. The ENEA PV program has the following objectives:

▼ To improve on currently available industrial processes.

▼ To develop new processes, with emphasis on amorphous silicon thin-films.

▼ To optimize conventional components, improving performance and reliability.

▼ To standardize components and systems, in cooperation with Italian industry.

Photovoltaic demand in Italy is supported by the implementation of a series of ENEA-funded pilot plants for residential, agricultural, and grid power use. Additionally, incentives are in place to encourage energy conservation and the use of renewable energy. For example, Law no. 10 of 1991 provides a government contribution of up to 80 percent of the cost of installed PV systems used to supply electricity for buildings in residential, industrial, commercial, agricultural, or sports applications. The same law allows private and public organizations to build electricity-generating plants based on renewable sources. The electricity may be for on-site use, for distribution within a consortium of users, or for sale to the national utility, Ente Nazionale per L'Energia Elettrica (ENEL). The price paid by ENEL is set by a government body and is sufficiently high to encourage use of renewable sources. Presently, the price paid for PV-generated electricity is set at about 14 cents per kWh, a value close to the price paid by the final user.

Germany

Independent power generation is encouraged in Germany, with utilities required to pay competitive prices to renewable energy producers.

The German Federal Ministry for Research and Technology and the German federal states have announced a project to install grid-connected residential PV systems on 1,500 rooftops in Germany, with each unit producing from 1 to 5 kilowatts of power. Individual homeowners qualifying for the program will pay 25 to 50 percent of the cost of the system, depending on the state subsidy. The federal government will pay 50 percent of the installed cost. Through this project, utilities will gain an understanding of the effect of PV systems on peak demands, PV system interactions with the grid, and the reliability of small systems [3].

Austria

In Austria, nuclear power plants have been prohibited by popular referendum since 1978. Along with a large amount of hydro-power, 40 percent of Austria's electricity is generated by coal- or oil-fired thermal plants using imported fuel. In an attempt to reduce imports of oil and coal and to reduce emissions from these fuels, utilities are looking for environmentally acceptable alternatives. The national PV program is coordinated among the Austrian Federal Electricity Board, Verbundgesellschaft, nine provincial utilities, and various municipal utilities. All utility PV projects are monitored centrally by computer, and each site has on-line access to processed data from all sites.

United States

In the United States, the passage of the 1978 Public Utility Regulatory Policies Act (PURPA), under which all utilities are required to purchase power produced by qualifying nonutility generators, continues to affect the development of renewable energy technologies. Although PV systems are not yet competitive in central-station applications, as prices continue to decline PURPA will allow third-party investors to be among the first to install large-scale PV generation equipment in the United States.

Federal tax credits are also expected to be employed to encourage investment in PV. Although current federal tax credits are insufficient to make grid-connected PV systems economically competitive, they have proven effective in encouraging the development of wind and biomass industries. Lobbying efforts on behalf of the solar industries may result in a viable PV tax credit. To encourage extensive PV use, a tax credit might need to be as high as 10 cents per kWh initially and would then be phased out gradually.

Japan

Japanese law currently prohibits the "reverse flow" of electricity to the power grid, which is to say that any electricity not used cannot be returned to the grid. The only exception to the law is utility-owned central-station generating equipment. Residential-sized systems can be connected to the grid if they do not allow reverse power flow. The government is sponsoring extensive research on interconnection issues and on the development of approved inverters. (An inverter converts the direct-current, or DC, electricity produced by a PV cell to the alternating-current, or AC, electricity typically employed in power grids.) The Japanese government has announced a national goal of eventually meeting 2 percent of all electricity needs through the use of residential grid-connected PV systems. To support this goal, a 7 percent tax credit and low-interest loans are available for PV system purchasers. Also, for the first systems installed, $1 million has been allocated for direct government subsidies of 50 percent of PV system costs [4].

Switzerland

PV systems owned by individuals or companies are allowed grid-interconnection by Swiss law. Additionally, utilities are required to pay nearly peak load "high tariff" prices for PV-generated electricity fed into the grid. Prices paid are typically 9 to 12 cents per kWh. The Federal Inspectorate of electrical installations has agreed to allow residential PV systems to be hooked up to the grid without special permission. Also simplifying the use of residential PV systems in Switzerland, 10 large utilities have decided to allow a single reversible meter hookup for residential PV installations, thereby agreeing to pay the retail residential rate for PV-generated electricity.

RESEARCH, DEVELOPMENT, AND DEMONSTRATION

Government interest in the development of PV systems grew during the 1980s, with the notable exception of the United States, where federal PV research expenditures plummeted from a high of $150 million in 1981 to a sustained $35 million during the last several years of the decade. As the 1990s begin, the U.S. budget has increased once again, with a 1991 budget of $47 million. In 1990, the governments of Germany, Japan, and Italy had annual PV research budgets greater than that of the United States, with a high of $70 million in Germany.

Utility programs

In most cases, government research programs are closely coordinated, or at least interactive, with utilities. In Japan, the government's New Energy Development Organization (NEDO) plans and funds PV research, which is then carried out by utilities. Similar coordination takes place in Italy, where the national utility is essentially an arm of the government. In the United States, government and utility programs are more loosely coordinated.

Most utilities interested in the use of PV systems begin by building and operating a small (1 to 5 kilowatt) system. Several utilities have also sponsored medium-scale demonstration projects of at least several hundred kilowatts in size. Most such systems have been built in the United States. In the southwestern United States, utilities such as Arizona Public Service (APS) and the city of Austin, Texas, foresee central-station power plants as the most important future PV application. These utilities serve major urban areas surrounded by expanses of low-cost land, and they have many sunny days. Some larger systems have been installed in Japan, Germany, and Italy. Projects are also planned for Spain and Switzerland.

Test systems in the United States are usually connected to the grid. The grid-interaction allows the load to be powered from the PV system during sunny days and from the grid during nights or when PV output is insufficient. In Italy, emphasis has been placed on the development of stand-alone systems for bringing residential electricity to homes at a distance from the existing utility grid.

Japanese utilities expect the combination of a high proportion of urban areas, extremely high land costs, general affluence of the population, and high residential electricity rates to result in a major penetration of grid-connected residential rooftop systems. Although less constrained by land availability, some utilities in the eastern United States, such as New England Electric and Philadelphia Electric, also project residential rooftop systems to be the major PV application for many utilities. In Germany, Italy, Japan, Switzerland, and Austria, government policies are in place to encourage grid-connected residential rooftop systems. Stand-alone and diesel hybrid systems are the vision of the future of utilities serving large rural areas, and especially of utilities serving many islands.

Privately funded experiments

A number of firms other than utilities have also made a major contribution to the design and testing of PV systems. Perhaps the most important of these are the

ARCO Solar (now Siemens Solar Industries) facilities built during the first half of the 1980s. In Switzerland, the first hundred 3 kilowatt units of a planned 333 unit grid-connected residential PV system were scheduled for completion in early 1991. The units are available to homeowners from a Swiss company as kits at a price of $6 per watt, or as installed systems at $9 per watt. At least 10 large utilities throughout Switzerland are encouraging homeowner participation in the project by allowing grid hookup through a single reversible meter [5].

Three other significant projects have been built in the United States as part of housing development projects. In one of them, which is located in Gardner, Massachusetts, and sponsored by the New England Electric System, 2 kilowatt PV systems have been placed on 30 existing homes on the same distribution feeder. The John Long Homes project, built in Phoenix, Arizona, in 1985, included a 175 kilowatt grid-connected PV system to supply power for 24 individual homes. Utilities took note of this project as a prototype for possible future community power systems [6]. Finally, in San Diego, California, a subdivision of townhouses called Laguna Del Mar was built to incorporate 1 kilowatt PV systems on 36 of 112 townhouses. Each homeowner owns the associated PV system and sells electricity to the San Diego Gas & Electric (SDG&E) company. A 2 kilowatt inverter built into each system allows for future expansion of the PV array. SDG&E is studying the impact of the multiple PV systems on its grid, especially their effect on summer peak loads.

Test facilities

A variety of PV test facilities are owned by utilities, utility regulatory agencies, and utility-sponsored research organizations. Extensive PV module evaluations and comparisons are performed at utility test facilities in the United States, Italy, and Austria. Japan has the most ambitious test facility in the world for studying the impact of multiple inverters connected to a utility grid. Recognizing the complementary nature of much of the ongoing utility PV research, the Electric Power Research Institute (EPRI) in the United States and the Central Research Institute of the Electric Power Industry (CRIEPI) in Japan have set up a series of workshops for the purpose of sharing United States and Japanese operating experience. In a similar effort, the Italian Electric Authority (ENEL) in Italy and the Pacific Gas & Electric (PG&E) company in the United States have agreed to establish and maintain an information exchange on PV experience.

LESSONS LEARNED

The investment made in demonstrating PV systems over the past decade has resulted in a wealth of information that is being employed to set research priorities and improve engineering designs and installation procedures. Observations in a number of key areas are presented below.

Systems

After initial start-up problems are solved, utilities generally report high PV system availabilities, with most systems available 95 to 100 percent of the daylight hours [7]. Performance varies widely, being highly dependent on array configuration. Major accidents can also occur, such as lightning strikes or inverter breakdowns, that can render a plant's availability zero for substantial periods of time; in fact, a plant can be taken out of commission for weeks or months. In one case where mirrors are employed to enhance the output of a system, performance has fallen as the PV modules degrade due to overheating.

Problems in scheduling and installation

Poorly managed scheduling and component delivery dates can increase system costs. PVUSA, a cooperative government/utility project in the United States, found it imperative that primary contractors have the experience, resources, and organization to manage the entire scope of work. Schedule slips ranging from 1 to 18 months occurred at the PVUSA California site, primarily due to component availability and schedule control problems [8]. Fortunately, such problems are often "first of a kind" difficulties. The large Carrisa Plains plant in central California, for example, benefited from the kind of scheduling and management that are typical of more mature technologies.

Managing installation requires careful attention to local conditions. Mountain locations in Austria and flat, urban locations in the United States require different approaches to keep installation costs down. Mountain sites are often inaccessible during the winter, and a schedule slip can delay the completion date by many months. At the PVUSA test site in California, winter rains limit the access of heavy equipment to the array fields during several months of the year.

System safety

Personnel safety and (grid) system safety both rely on the automatic disconnection of inverters in the case of grid power outage. A major concern of utilities is the possibility that multiple inverters on a single distribution line could provide one another with a sufficient signal to remain on-line when grid power is lost. Such a "run-on" or "islanding" phenomenon could pose a significant threat to utility workers attempting to repair lines that are presumed dead, or devoid of power, and is the subject of extensive work in both Japan and the United States.

Another personnel safety issue is the allowable DC voltage of a residential PV system. The topic has not yet been fully discussed among utilities, although a suggestion has been made that the maximum allowable voltage be set at 50 volts. Any decision on this safety issue will have a severe impact on at least some inverter designs.

Yet another safety issue is interconnection. Interconnection requirements vary from one country to another, and often from one utility to another. Many utilities have not been faced with the interconnection of a PV system and have not had to think about what their requirements are or should be. The Salt River

Project in Arizona is one utility that has published detailed interface protection requirements for all sizes of PV systems interconnected to its grid [9]; the requirements are summarized in table 1.

Inverters

In Japan, concern about utility interconnection and safety has led to the development of two inverters. One, called the orthogonal core inverter, provides excellent protection against run-on, is simple, and requires little maintenance, but exhibits only a moderate level of performance. A higher level of performance is exhibited by the more advanced design of the so-called pulse-width-modulated inverter.

Most utilities testing inverters have reported at least some start-up problems. For example, at the PVUSA site, the four 25 kilowatt inverters now on-line required extensive adjustments before they functioned properly. Such experiences indicate a problem associated with the small market for grid-interactive inverters. But as inverter sales increase, problems will undoubtedly be worked out of designs, and manufacturers will be able to deliver high-reliability units that operate well at installation.

Predicting system power output

PG&E, APS, and the German utility company Rheinisch–Westfälisches Elektrizitätswerk have all published the results of extensive module testing programs [10, 11, 12]. Each group finds that inconsistencies among manufacturers and among different cell technologies make the module ratings provided by the manufacturers unsatisfactory for comparative analyses. Moreover, APS and PG&E found that few modules tested meet the manufacturers' claimed ratings. Test results from PG&E are shown in figure 1.

Because manufacturers' module ratings are not comparable to one another, a manufacturer's rating cannot serve as the basis for an algorithm that will predict system output. The PVUSA project has tried to get around this problem by insisting that bidders predict the actual array output under specified weather condi-

Table 1: Salt River Project grid interface protection requirements for PV systems[a]

System size	Requirements
0–10 kilowatts	No required isolation transformer; contactor; industrial-grade relay
10–50 kilowatts	Isolation transformer; contactor; industrial-grade relay
50–100 kilowatts	Isolation transformer; breaker; industrial-grade relay
> 100 kilowatts	Isolation transformer; breaker; utility-grade relay

a. From a presentation by Tom Lepley at IEA–ENEL Executive Conference on Photovoltaic Systems for Electric Utility Applications, Taormina, Italy, December 1990.

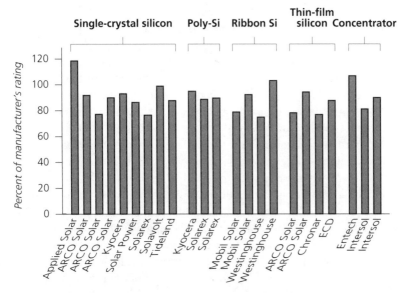

FIGURE 1: *Comparison of outdoor measurements made by PG&E and module ratings provided by the manufacturers. PG&E's number for each module's output is derived through statistical analysis of measurements made multiple times per day over a year or more and is corrected to Standard Test Conditions. The measurements were made between 1984 and 1986; modules were measured at 1,000 watts per m² insolation (flat plates), 850 watts per m² insolation (concentrators), and 25°C back-of-module temperature. (Data from* Models of photovoltaic module performance, *EPRI AP-6006, September 1988.)*

tions, and then requiring payment of a financial penalty if the prediction is not met. To date, predicted performance is still not usually met by PVUSA systems, which suggests that the financial penalties imposed are insufficient.

Generally, actual module power output is 5 to 10 percent lower than claimed, and inverter and transformer losses represent an additional 5 to 15 percent decrease in performance. As a rule of thumb, total system peak power output will be 12 to 25 percent lower than that calculated by multiplying the number of modules by the manufacturer's rating.

Operation and maintenance

In an EPRI-commissioned study of seven medium-scale U.S. projects, actual operation and maintenance costs were found to range from 0.4 to 7 cents per kWh, with an average of 2.3 cents per kWh. Three of the systems produced more than a megawatt of power, and they had substantially lower operation and maintenance costs than the four other systems, which were smaller. Unscheduled maintenance accounted for 73 to 100 percent of the costs, depending on the maintenance strategy at the site. Operation and maintenance costs for the same facilities in the future are estimated to range from 0.2 to 1.2 cents per kWh. The same study estimates that operation and maintenance costs for large-scale PV power plants will be under 0.35 cents per kWh for systems employing concentrated sunlight and lower than 0.11 cents per kWh for fixed flat-plate systems, with costs for tracking flat-plate systems falling in between [13].

While maintenance expenses varied by subsystem from one site to another, a consistent pattern emerged from the data. Inverters accounted for the highest maintenance expense at most sites. Fixed-array and tracking flat-plate systems were less expensive to maintain than concentrator systems. Trackers were generally a low-expense maintenance item; however, their malfunction can contribute to a costly loss of output.

An effective operation and maintenance strategy can stabilize costs. Researchers at the Southwest Technology Development Institute in the United States recommend preventive maintenance and periodic testing.

The interconnection of nonutility-owned equipment to the utility grid, especially in the case of residential systems, poses an important operation and maintenance question for utilities. Although a utility can require certain features in any inverter connected to its grid, in the case of systems belonging to customers, the utility could have a problem ensuring the correct operation of all safety features over time. The Philadelphia Electric company is proposing that the utility own and maintain all inverters; under this scenario, the utility would have to be willing to purchase DC power from customers having PV systems [14].

Remote systems present particularly challenging operation and maintenance issues. Austria's high alpine sites, for instance, which have exceptionally high PV power output during winter months, must operate under heavy snow loads and the constant danger of lightning strikes. Indeed, lightning is the most common cause of system outage at these sites, and the data system is always the first part of the system to be damaged. As a result, both data loggers and mechanical counters are installed at all the Austrian alpine sites for recording the intensity of incident sunlight as well as PV power output [15].

LOOKING TO THE FUTURE

The economic viability of stand-alone PV systems has been established under many circumstances, and prospects for grid-connected systems appear to be good in many parts of the world:

▼ PG&E has published an analysis suggesting that photovoltaic systems at carefully selected points in its distribution grid will be cost-effective by 1995 [16].

▼ EPRI and PG&E in the United States and the Center for Renewable Energy Sources in Greece recently estimated that the total installed cost of PV plants must reach $3,000 per kilowatt to "approach the threshold of significance," while $1,500 per kilowatt will be necessary before big markets for central-station plants will be open to PV. The authors suggest that the active participation of utilities in shaping the future of PV technology may be a prerequisite to the achievement of bulk-power PV [17].

▼ In Italy, the new ENEL program foresees a total installed capacity of 12 megawatts by 1995, including examples from all application areas. While small standalone systems are already clearly cost-effective, and small- to medium-scale systems can be cost-effective depending on local situations, multimegawatt plants are presumed by the Italians to become economically viable only in the long term. However, the Italians intend to continue development of large-scale systems for several reasons: to sustain and stimulate the Italian PV market, to acquire useful experience, to reduce balance-of-system costs, and to obtain relevant technical feedback to be used in other application areas.

▼ CRIEPI predicts energy costs for rooftop residential PV systems in Japan will be lower than residential electric rates by 2000. However, because of land costs, which would add an estimated $2,000 per kilowatt to the installed cost of a central-station system, the levelized energy cost of central-station PV plants is not expected to approach economic competitiveness in Japan until well beyond 2000 [18].

In reviewing the experiences of the past decade, PG&E and ENEL developed a proposal for a program of demonstrations that would proceed as follows:

1.   Utility uses of PV systems today, in cost-effective, stand-alone applications (typically 10 watts to 10 kilowatts per site) should be encouraged. Although the total capacity for these systems is relatively small, their significance should not be underestimated. Small-scale PV systems can pay for themselves in six months to two years. Additionally, small systems provide a low-risk means for PV systems in general to gain credibility in the eyes of utility engineers.

2.   PV systems should be installed to support local utility grid needs (100 kilowatts to 1 megawatt per site). The chief value of introducing PV systems to the utility grid is their potential to handle extra demand during heavy peak loads.

3.   PV systems should be constructed in remote villages and islands to provide stand-alone power (10 kilowatts to 1 megawatt per site). Village and island power systems have already been installed throughout the world. Often, PV systems represent the most appropriate and cost-effective technology for rural electrification. Traditionally, utilities have not become involved in serving remote power needs, having declined to serve in areas distant from the existing grid because of high costs. PV systems provide a new, low-cost approach to rural electrification that has proven feasible and effective in the United States and northern Europe as well as in developing countries.

4.   Rooftop PV systems should be connected to the grid for residential or commercial applications (1 to 10 kilowatts per site). A number of utilities are testing systems in the 2 to 25 kilowatt size range, to understand their performance, reliability, personnel safety, grid interactions, and other impacts. In some areas of industrialized countries, rooftop residential systems are expected to represent the first widespread use of PV systems. As discussed earlier in this chapter, Germany,

Italy, and Japan have existing or planned government programs to provide homeowners with the necessary incentives to install rooftop PV systems.

5. Scalable PV central-station power plant design and test activities should be initiated, and demonstration multi-megawatt plants built. Operating central-station PV power plants will generate invaluable utility experience and confidence regarding modular installations, short lead times, speed of installation, power quality, safety, operation and maintenance costs, and reliability. An iterative design approach to these plants will lead to improvements and cost reductions in components, engineering, and installation [19].

Two major keys to the growth of PV systems are the continued development of real markets (niche markets where PV systems are the low-cost option) and the fostering of carefully structured commercialization alliances. Given the extensive benefits to society that PV systems may provide, alliances involving the supply industries, utilities, the research and development community, and government agencies are needed, primarily to share risks, but also to optimize communications and resource applications.

Continued steady growth of the overall PV module market will allow for orderly business development to materialize in the supply industry, leading to increased mass production, lower module prices, and a convincingly profitable PV industry. Commercialization alliances can speed the introduction of PV systems into intermediate niche markets and enhance the cost-effectiveness of PV systems within those markets.

ACKNOWLEDGMENTS

The authors would like to thank the following people for their help in assembling the information presented in this chapter: Bob Hammond, at the time at the Southwest Technology Development Institute, now at Arizona State University; Mat Imamura of the German consulting firm WIP; García Martín of Hidroeléctrica Española; Paul Maycock of PV Energy Systems Inc.; Rudolph Minder of Electrowatt Engineering Services Ltd.; Armin Räuber of the Fraunhofer Institute; Markus Real of Alpha Real AG; and I.A.Di. Ritter and Heinrich Wilk of the Austrian utility company Oberösterreichische Kraftwerke AG.

The authors are also grateful for the review and suggestions provided by Bill Howley of Siemens Solar Industries, and by Paul Wormser, Dot Bergin, Eric Tornstrom, and Mark Farber of Mobil Solar Energy Corporation.

Appendix
ELECTRIC UTILITY PV PROJECTS

Austria
Austrian utilities operate PV test facilities, stand-alone systems, and grid-connected systems. Research emphasis is on alpine sites, because initial cost-effective use of PV systems is associated with mountainous areas that are inaccessible during the winter, but need reliable electric power. In some places, an existing grid distribu-

tion line is difficult to maintain during the winter, and a very reliable backup system is needed. Other sites are sufficiently remote so that no grid is nearby [20].

Test facilities

One utility company as well as the electric regulatory agency are involved in the PV test facilities in Austria:

▼ PV Solar Panel Test Station. The Austrian Federal Electricity Board, Verbundgesellschaft (VG), built a module test facility in Vienna, where 19 PV modules are being tested. Single-crystal, polycrystalline, and amorphous silicon cell technologies from various manufacturers are represented.

▼ Kanzelhohe Solar Center. A cooperative project of the University of Graz, Siemens of Austria, VG, and the utility company Österreichische Draukraftwerke AG (ODK), this facility uses heliostat structures to track a 1 kilowatt PV array, the output of which is then compared to that of a fixed 1 kilowatt array.

Medium-scale demonstrations

Oberösterreichische Kraftwerke AG (OKA), the electric utility of upper Austria, and VG, in cooperation with the European Community, plan a 100 kilowatt grid-connected plant. The project will be used to test tandem amorphous silicon module technology, optimum design of structures, and other system components, as well as economies of scale.

Other demonstrations

Austrian utilities have built and now operate six PV demonstration projects, ranging in size from 1 to 30 kilowatts:

▼ Loser. A 30 kilowatt plant was built on Mount Loser in the Austrian Alps (see figure 2).

▼ Spering Radio Link Station. VG operates a radio communication network that covers all of Austria, used to enhance reliability of the high-voltage grid. All radio link stations in this network are on mountain tops, and many are inaccessible during the winter.

▼ Kesselbach. A 1.5 kilowatt PV system operates the intake gate and drain valve of a hydrostorage plant in the Tyrolean Alps.

▼ Leonding Technical High School. OKA, in cooperation with ESG, the electric utility of Linz, has built a 1.4 kilowatt grid-connected amorphous silicon system at Leonding Technical High School, near Linz.

▼ Gmunden. OKA operates a 1.3 kilowatt grid-connected system, built in 1987, in Gmunden. This was the first grid-connected system in Austria, designed to test all aspects of grid-coupling and automatic operation.

FIGURE 2: The 30 kilowatt PV plant on Mount Loser in the Austrian Alps. (Photo courtesy of Oberösterreichische Kraftwerke AG.)

▼ Hochleckenhaus. OKA cooperated with the Austrian Alpine Club to supply 1 kilowatt of power to a three-season mountain refuge used by hikers and tourists.

Germany

The major German utility, Rheinisch–Westfälisches Elektrizitätswerk (RWE), has been involved in PV research since 1981, when a 3 kilowatt test system was placed on the rooftop of RWE headquarters in Essen. Emphasis in the German PV program is on the medium-scale demonstration of power plant technology [21]. Another large German utility company, Bayernwerk AG, jointly owns, with Siemens AG, the entire Siemens Solar Group, including Siemens Solar Industries in the United States and Siemens Solar GmbH in Germany.

RWE has built and is operating the first phase of a planned 1-megawatt plant, and is beginning the second phase. Begun in 1986, the goal is to design, construct, and operate three large PV plants having a combined capacity of 1 megawatt. Phase one, completed in October 1988, is a 340 kilowatt system in Kobern-Gondorf on the Mosel River near Koblenz. The major portion of the system has been in continuous operation since May 1989.

The Kobern-Gondorf plant was designed to test various PV module technologies, inverters, installation techniques, support structures, and architectural and ecological concepts for integrating the plant into a natural environment. The PV array is divided into three large (100 kilowatt) and five small (1.5 to 12 kilowatt) fields, all at a fixed tilt of 30°. In an unusual component-sizing test, inverters for the large arrays are rated at about 80 kilowatts, 25 percent lower than the rated output of the arrays. Researchers have found that the PV output exceeds inverter

FIGURE 3: *At Rheinisch–Westfälisches Elektrizitätswerk's 340 kilowatt Kobern-Gondorf plant, wide spacing of array fields and building on existing land contours makes the area a sanctuary for rare and endangered species. (Photo courtesy of Rheinisch–Westfälisches Elektrizitätswerk.)*

capabilities during only a few hours each year; at such times a controller limits the PV output. Both self- and line-commutated inverters from six different manufacturers are being tested.

Unique to the Kobern-Gondorf plant has been the desire to prove that land covered by large PV plants can also provide sanctuary for rare and endangered species [21]. As shown in figure 3, a 10 to 1 ratio of ecologically-intact surface to array area provides pleasing aesthetics as well as environmental protection.

RWE has also evaluated fixed-array rack designs. Designs at the Kobern-Gondorf plant have been modeled at ¹⁄₁₅th scale and subjected to extensive wind-tunnel testing. Conclusions from these tests and from measurements made at the site indicate that wind load assumptions for array design could be decreased by 20 percent relative to the applicable standard. Additionally, substantially lower-cost structures could be installed if 10 percent spacing were left between modules [21].

Italy
Italian PV research projects cover a range of applications, including medium-scale demonstrations of power plant technology, on-site generation for industrial purposes, such as water pumping and refrigeration, stand-alone residential systems for use in rural areas without electricity, and diesel hybrid systems for use in island electrification. Italy's federal agency in charge of research, development and technology transfer to industry, ENEA, has funded a series of PV demonstrations, including a 600 kilowatt system. Additionally, the Italian utility ENEL has supported a growing effort to develop and use PV systems since 1982, resulting in the installation of PV systems producing more than 100 kilowatts of power in 1990, includ-

FIGURE 4: The existing first phase of Italy's Delphos Plant is 300 kilowatts.

ing both grid-connected and stand-alone systems. The excellent results of initial experimental systems, all of which are still in operation, has led ENEL to increase its PV activities. In 1990, a new research and development project began, with a goal of contributing significantly to the objectives of the Italian National Energy Plan—25 megawatts of PV-produced power by 1995. ENEL's position is that PV systems can play an energy-significant role only through large and wide-spread diffusion into electric utility use [22].

▼ Adrano PV Test Field. The Adrano PV Test Field is a facility built and operated by ENEL to test innovative module technologies, assess grid-connection problems, and compare the performance of fixed versus tracking installations. ·

▼ Delphos Plant. The 300 kilowatt Delphos (Demonstration Electric Photo-voltaic System) Plant has long been the largest PV installation in Italy. Built by ENEA, Delphos is situated in southern Italy on a 7 hectare site at Monte Aquiline near Foggia. The site was chosen as representative of the geomorphology of marginal untilled land in Italy, as well as for its good annual insolation (amount of solar radiation received). The facility has served as an experimental station for the field testing and evaluation of PV arrays and other system components. Technical and economic data from Delphos will be used to design and market more advanced systems, both for grid-connection and stand-alone use [23]. Figure 4 shows the first phase of the Delphos Plant.

The next major medium-scale PV demonstration planned in Italy is to be a 3.3 megawatt plant near Naples in southern Italy. Initial design criteria have been established by ENEL, and plant construction is scheduled for completion by the end of 1993, with the first megawatt on-line in late 1992. The 3.3 megawatt plant will contribute to the growth of the PV module market, improve balance-of-system technologies, and enhance the field experience of ENEL and the utility industry [24].

Other demonstrations
ENEL has built and now operates two demonstration sites, as well as the Remote Homes project. ENEA is also now operating a series of demonstration projects [16].

▼ Vulcano Island. Built in 1984, the 80 kilowatt Vulcano Island PV system originally operated as a stand-alone power supply for 54 homes. In 1986, a new grid-connected inverter was added to the system, connecting it to a 5 kilometer network of 20 kilovolt line. System use thereby changed to that of a peaking plant in parallel with a diesel generator.

▼ Ancipa Project. Built with funding from the European Community, the Ancipa Project is a 1.5 kilowatt demonstration site designed to prove the reliability of PV systems at a large hydropower facility in Sicily. The system is considered to be the prototype for remote mountainous region applications in Italy.

▼ Remote Homes Project. This project consists of three phases. During the first phase, which took place from 1985 to 1986, nine small PV systems, ranging in size from 300 watts to 1.5 kilowatts, were built to test the feasibility of supplying power to remote homes. The nine systems powered isolated houses of nonprofit organizations. The second phase of the project, the Ginostra Project, was implemented from 1987 to 1988. Thirty 350 watt residential systems were built in the small town of Ginostra on the south side of Stromboli (Sicily). The purpose of this phase was to refine an optimum PV system for remote residential use throughout Europe. The third phase of the project encourages industrial production of small packaged systems for common use wherever grid connection would be costly. Approximately 350 small stand-alone systems for remote users and/or equipment will be installed in the near future, with partial funding from the European Community.

▼ Zambelli Plant. Built in Verona in 1984, the Zambelli Plant is a 70 kilowatt PV system that operates a 200 m^3-per-day water pumping station for the municipal water system.

▼ Giglio Island. On Giglio Island, in Tuscany, a 45 kilowatt PV-powered refrigeration system was built and began operation in 1984. As with many of Italy's demonstration projects, this one was built with the cooperation and funding of the European Community.

PV power is also used to light archaeological sites at Cetona, Sienna (20 kilowatts), Sovana, Grosseto (6 kilowatts), and airfield landing strips at the Lucca airport. PV systems are used as well in the electrification of two alpine huts as well as a small farm on the presidential estate near Rome. In a few places, cost-effective PV systems are already in use within ENEL's electric system, for powering, signalling, communications, and safety equipment [23].

Japan
Japan's research program emphasizes two future applications for PV systems. Primary PV use in Japan is expected to be in grid-connected rooftop systems, chiefly residential. A secondary use is planned for decreasing diesel operating costs in island electrification systems. Land prices in most of Japan prohibit ground-mounted PV systems, and high electricity costs are expected to encourage homeowners to invest in PV systems. Research concerns focus on grid-connection issues for small systems [25].

Test facilities
Japanese utilities operate two major PV test facilities [26]. The Rokko Island facility is the most extensive test center in the world for evaluating the effects of multiple grid-connected inverters on one another and on the utility's distribution system. At the Akagi test center, extensive testing of individual inverters and batteries is conducted. The Japanese have shown great caution in the use of grid-interactive inverters and have established these inverter test facilities to investigate all inverter operating modes and their possible effects on the utility grid.

▼   Rokko Island. Kansai Electric Power Co. and the Central Research Institute of the Electric Power Industry (CRIEPI) jointly designed and built the Rokko Island test site for alternate and renewable electric generation technologies. As shown in figure 5, 100 two kilowatt systems are in place, each with its own grid-connected inverter. The purpose of the Rokko island facility is to identify and analyze utility interface issues and to establish appropriate controls for interconnecting new energy technologies to the grid. Tests focus on islanding, reverse power flow, harmonics, voltage fluctuations, and system protection [27].

▼   Akagi Test Center. Stand-alone and grid-interactive inverters up to 10 kilowatts in size can be tested at this facility, which includes a PV array simulator and a simulated distribution line capable of 50 and 60 hertz operation, as well as various line disturbances and fault conditions. Also at Akagi is a facility for testing various storage batteries, including a battery developed by the New Energy Development Organization (NEDO). Tests typically include lifetime, cycle life, overcharge endurance, mechanical strength, storage capacity at various charge and discharge rates, efficiency, and self-discharge rates.

▼   Saijo PV Power Generation Test Plant. In 1986, NEDO funded a 1 megawatt demonstration plant and entrusted a team formed by Shikoku Electric Power Co., Shikoku Research Institute, Inc., and CRIEPI with its construction and the subsequent research to be performed. Built in Saijo City, the plant is the only grid-connected utility PV demonstration to include batteries. The Saijo plant's 1,800 kilowatts of battery storage add stability to the PV output [28].

FIGURE 5: Kansai Electric operates the Rokko Island test facility. One hundred 2 kilowatt PV systems are interconnected to a simulated distribution feeder. (Photo courtesy of Kansai Electric.)

Other demonstrations

Other Japanese utility demonstration projects include diesel hybrid systems and concentrator systems [29].

▼ Tokashi. An island electrification project uses 250 kilowatts of power produced by PV systems to supplement the diesel generators that supply electricity to five islands in the Kerame archipelago. Okinawa Electric Power Company, in partnership with Mitsubishi, built the system in 1987, with funding from NEDO. Battery storage and PV allow a single diesel generator to operate when needed, thus improving the load factor, and decreasing the need to adjust the number of active diesels.

▼ Kyushu Electric. Kyushu Electric Power Company built a PV–diesel hybrid system for island electrification in 1987. This system uses 100 kilowatts of PV power with a kVA diesel generator.

▼ Yamasaki Test Center. Built in 1984, the Yamasaki PV system is a 50 kilowatt linear Fresnel-lens concentrator system. With a 25X solar concentration ratio, the array is mounted for one-axis tracking at a 30° tilt. A three-phase line-commutated inverter connects the system to the local grid [30].

▼ Technical Research Center. The Technical Research Center in Amagasaki also has a concentrator project. In 1985, a 10 kilowatt gallium-arsenide array was built, using 218X point-focus Fresnel lenses. This test also includes a three-phase line-commutated inverter.

Spain

The Spanish utility company Hidroeléctrica Española (HE) operates seven PV demonstration plants, has a system- and component-level research and development program, and promotes PV technology training and information dissemination both within Spain and internationally [31]. HE is one of the largest utilities in Spain and is privately owned.

Test facility

At the Experimental Research Platform San Agustín de Guadalix in Madrid, HE conducts one- and two-axis tracking experiments with a 6 kilowatt array, amorphous-silicon degradation studies on a 2 kilowatt array, and advanced inverter testing using a 5 kilowatt array.

Medium-scale demonstrations

HE operates two 100 kilowatt plants and is involved in a feasibility study for a multimegawatt installation.

The grid-connected Photovoltaic Experimental Power Plant (CSFE) is in Madrid at the Experimental Research Platform. A second 100 kilowatt demonstration plant was built in collaboration with the European Community and several other companies, and is used for rural electrification in the village of Alicante on Tabarca Island.

Currently undergoing technical and economic feasibility studies is a proposed 2.5 megawatt plant. But because the Spanish government, HE, and the European Community are sufficiently serious about increased PV use, this plant may well become a reality during the early 1990s [32].

Other demonstrations

HE has three stand-alone PV demonstrations, which are designed to supply electricity to rural villages. At a rooftop laboratory at HE headquarters in Madrid, a 1 kilowatt experimental array has operated since 1983. The system is stand-alone, although it operates in a simulated grid-connected mode. System performance and hourly load data are connected and analyzed. Since 1982, HE has used a 0.5 kilowatt PV system to power a remote unmanned control system for a hydro flow-control valve near Cuenca.

Switzerland

Nine Swiss utilities and two industrial companies have formed a consortium, PHALK–Mont-Soleil, to construct and operate a 500 kilowatt demonstration project. Funded 50 percent by the consortium and 50 percent by the Swiss National Energy Research Fund and the canton of Bern, the PHALK (PHotovoltaics ALpines Kraftwerk) project will serve as a Swiss national PV research center, with part of the installation set aside for testing emerging module technologies and running special tests.

United States

In the United States, the diversity of climatic regions and land availabilities has led to widely divergent views among utilities as to the most reasonable and probable PV applications. Consequently, a broad spectrum of research and demonstration projects has emerged. Aggressive government support for PV research and demonstrations during the late 1970s and early 1980s, combined with the large number of electric utility companies in the United States, and increasing interest in the technology by many of those utilities have led to more than 100 utility PV demonstration and/or research projects in the United States.

Test facilities

U.S. utilities own, operate, and/or participate in funding an extremely wide variety of PV test facilities, the most important of which are described below:

▼ PVUSA. The Photovoltaics for Utility Scale Applications (PVUSA) project is jointly sponsored by the U.S. Department of Energy, Pacific Gas & Electric (PG&E), the California Energy Commission, the Electric Power Research Institute (EPRI), and nine other utilities and state governments, listed in table 2. The main test site for PVUSA is in Davis, California, where emerging module technologies are tested in 20 kilowatt systems, and innovative balance-of-system approaches

Table 2: Participants and cosponsors of the PVUSA project

Government agencies and research institutes

U.S. Department of Energy (DOE)
  Sandia National Laboratories
  National Renewable Energy Laboratory (formerly SERI)
  Jet Propulsion Laboratory

Electric Power Research Institute (EPRI)

California Energy Commission

U.S. Department of Defense Tri-Service PV Review Committee

Utilities and state research agencies

Pacific Gas & Electric Company (PG&E)

State of Hawaii/Maui Electric Company

City of Austin Electric Department

New York State R&D Authority

Virginia Power Company/Commonwealth of Virginia

Salt River Project

San Diego Gas & Electric Company

Niagara Mohawk Power Corporation

are tested in "utility-scalable" systems ranging from 200 to 400 kilowatts [33].

In addition to learning from the Davis site, utility sponsors are encouraged to build a 20 kilowatt system (similar to any one of the Davis site systems) at the utility's own site. As of late 1990, one off-site system had been built in Hawaii by the Maui Electric Company, one was planned by the Virginia Power Company and the commonwealth of Virginia, and one by the New York State Energy Research and Development Authority.

▼ VISTA. Virginia Power operates three 25 kilowatt systems at the Virginia Integrated Solar Test Arrays (VISTA) PV test facility [34]. The VISTA facility is designed to test and evaluate commercial/industrial-sized PV systems, anticipating initial PV penetration in such applications, while also serving as a model for future multi-megawatt systems.

▼ PG&E. Besides the PVUSA project, PG&E operates two PV test facilities at its research and development department in San Ramon, California. Since 1980, with co-funding from EPRI, PG&E has been testing individual PV modules. Performance and long-term operation are assessed; currently 38 modules from 17 manufacturers are being tested [35, 10].

In 1989, PG&E built a second test facility in San Ramon to investigate modular generation technologies in grid-connected, stand-alone, or hybrid configurations. A PV array simulator allows for testing of stand-alone or grid-connected inverters having capacities as great as 190 kilowatts.

▼ STAR Center. In 1988, the Arizona Public Service Company completed a PV test facility called the Solar Test and Research (STAR) Center, in Tempe, Arizona [11]. Seven two-axis tracking array structures allow side-by-side comparisons of flat-plate modules operating in fixed-, one-, and two-axis tracking configurations. Linear and point-focus Fresnel lens concentrators are also under evaluation. A nearby utility, the Salt River Project, has contributed operating funds to the STAR Facility, as has EPRI. The overall purpose of the facility is to evaluate PV technologies for use in central-station power plants in the desert southwest.

Medium-scale demonstrations

Several PV demonstrations larger than 100 kilowatts have been built in the United States. Those built and operated by electric utility companies are listed below.

▼ SMUD. In 1983, the Sacramento Municipal Utility District (SMUD) began a major PV demonstration project. The 1 megawatt system completed in 1984 was followed by a second 1 megawatt system in 1986.

▼ Austin. The municipal utility of the city of Austin, Texas, has built two 300 kilowatt PV systems. "PV300," which came on-line in late 1986, used single-crystal flat-plate modules with one-axis passive trackers, while the "3M/Austin" system uses linear Fresnel lens concentrator modules [36, 37].

▼ Sky Harbor. Arizona Public Service built a 200 kilowatt concentrator system that operated from 1982 to 1988. The system used point-focus Fresnel lens modules with a concentration ratio of 36X [13].

▼ PVUSA. The PVUSA project, plans to operate a series of medium-scale demonstrations. Two are now under construction: a 400 kilowatt fixed amorphous-silicon array and a 180 kilowatt one-axis tracking array of polycrystalline ribbon modules. Already completed and in initial test phases is a 175 kilowatt single-crystal array on passive one-axis trackers.

▼ Alabama Power. In 1986 Alabama Power, a subsidiary of the Southern Company, built the first medium-scale system to use amorphous silicon PV modules. Originally intended to be 100 kilowatts in size, module leakage and degradation problems required the system to be substantially derated [38]. Nonetheless, the demonstration served to identify strengths and weaknesses in the design of amorphous silicon modules.

Other demonstrations
Thirty utilities operate small grid-connected PV demonstration projects, and several others have existing or planned stand-alone demonstrations. Additionally, major customer-owned PV demonstrations impact utilities, sometimes to a large degree.

▼ Small utility-owned. Thirty utilities in the United States have small grid-connected PV demonstration projects [39]. A list of those utilities and projects appears in table 3. Although every utility demonstration project is worthy of discussion, in that every utility has its own philosophy, approach, and goals, the New England Electric Project is perhaps especially so. During 1986, with co-funding from the Department of Energy, New England Electric Systems chose a housing subdivision at the end of a distribution feeder in Gardner, Massachusetts, and with the cooperation of 30 homeowners, installed 30 two kilowatt PV arrays on residential roofs. Each home had an individual, self-commutated 2 kilowatt inverter, which was interconnected through a single, reversible meter. Data acquisition, in the form of a "sell back" meter that measured AC energy output to the utility distribution system and a separate meter monitoring total PV output, was placed on each system [40].

▼ Major customer-owned. No discussion of PV and utilities would be complete without mention of two major demonstration sites built by Atlantic Richfield Corporation's ARCO Solar (now Siemens Solar Industries) during the early 1980s. Although not utility sponsored, each of these PV systems was built with the cooperation of a major utility company and was afterwards closely monitored by those utilities. The first megawatt-scale PV demonstration was the 1 megawatt system at Southern California Edison's Lugo substation, near Hesperia [41]. At a site owned by PG&E, ARCO Solar built its second major demonstration, the 6.5 megawatt Carrisa Plains plant [42]. Both the Lugo and Carrisa plants were sold in early 1990

Table 3: Utility grid-connected PV projects in the United States[a]

Utility	Number of PV systems	Size kilowatts
Arizona Public Service Company	5	2
Austin City Electric Department	2	300
Austin City Electric Department	2	1
Bonneville Power Administration	1	10
Boston Edison Company	1	4
Central Maine Power	1	2
Corn Belt Power	1	4
Detroit Edison	1	4
Florida Power Corporation	1	15
Florida Power & Light	1	10
Georgia Power Company	1	4
Jacksonville Electric Authority	1	6
Lea County Electric Cooperative	2	50
Los Angeles Department of Water and Power	1	2
Narragansett Electric	1	3
Narragansett Electric	2	5
New England Electric System	32	2
New England Electric System	2	4
New England Electric System	1	7
Niagara Mohawk Power Corporation	1	15
Pacific Gas & Electric Company	4	20
Pacific Gas & Electric Company	1	5
Pennsylvania Power & Light	1	1
Philadelphia Electric Company	1	6
Philadelphia Electric Company	1	2
Public Service Electric &Gas	1	3
Platte River Power Authority	3	3
Puerto Rico Electric Power Authority	1	90
Sacramento Municipal Utility District	2	1,000
Salt River Project	1	1
San Diego Gas & Electric Company	1	2
Seattle Electric	1	< 1
Southern California Edison Company	3	2
Tennessee Valley Authority	1	10
Tennessee Valley Authority	3	2
Tennessee Valley Authority	2	4
Virginia Power Corporation	3	25
Wisconsin Electric	3	1
Wisconsin Power & Light	5	2

a. Blue Mountain Energy, 1991.

and are being dismantled for resale of the modules as remote power systems. The used PV modules are currently worth more for stand-alone applications than for the grid-connected application.

Cost-effective use

During 1989, researchers at PG&E discovered, somewhat to their surprise, that many operating divisions of the company were already using PV systems as the least-cost method of solving various problems [43]. Seventeen applications and over 400 installations were identified in 1989, as shown in figure 6. An estimated 700 installations and a combined installed capacity of 39 kilowatts were in place by the end of 1990. Each of these installations is a stand-alone PV/battery system. Systems range in size from 5 watts to 7.2 kilowatts.

Florida Power Corporation, Arizona Public Service, Georgia Power, KC Electric, Alabama Power, and 18 other U.S. utilities also have experience with cost-effective PV use. EPRI has sponsored a project to help utilities identify potential PV applications within their own operations and to provide utilities with the tools necessary to evaluate the cost-effectiveness of these applications [44].

Developing countries

In less-developed countries, many telecommunications and gas utilities have been using PV power for over a decade. Electric utility companies are increasingly involved in PV development or use programs. Private and government-supported PV programs are becoming common. Communications, water pumping and purification, vaccine refrigeration, and lighting are the most common applications of PV systems, as they fulfill the most urgent power needs of vast rural populations. Government interest in PV power is often expressed through the initiation of in-country module manufacturing or assembly capabilities. System kits are often imported, with assembly and/or component additions completed by a local value-added distribution infrastructure.

The most effective PV programs in the developing world encourage private and government use of PV systems, include in-country manufacturing or assembly, and do not discourage importation of PV modules and other system components by distributors. An example of such a program is that of Mexico, where PV use is growing fast enough to strain current world manufacturing capacity. More than 1,000 single-module home lighting kits assembled in Mexico have been installed by the Mexican utility CFE in association with the local Siemens Solar Industries distributor. CFE is involved in many aspects of PV use. Not only do they use PV arrays for internal purposes, but they sell power from PV systems to customers, finance entire systems sold to customers, and even in some cases give systems away to customers [45].

Other developing countries with government policies that contribute to growing PV use include Algeria, Brazil, China, India, and Venezuela, all of which manufacture modules; Colombia, where over 17,000 small systems are in use; South Africa, where an estimated 700 kilowatts of PV systems were installed during 1990; Thailand, where small entrepreneurial companies are rapidly making PV accessible to large rural populations; Honduras, Pakistan, and others.

Indonesia is an example of a country with a vast population for whom PV systems represent the least-cost power option. Indonesia comprises a huge archipelago of approximately 13,000 islands. Less than one third of Indonesia's 67,000 villages had electricity in 1990, but the government's goal is for the national util-

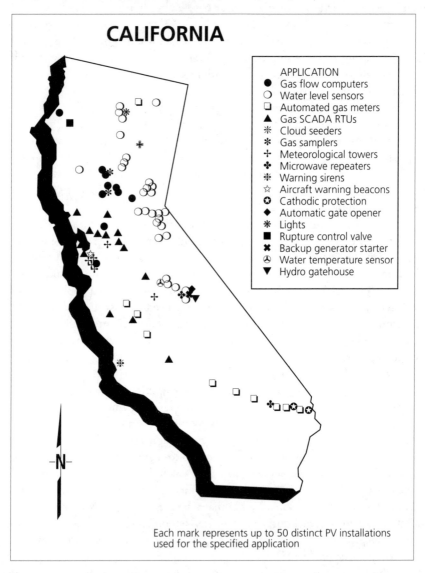

FIGURE 6: In 1989 PG&E owned and operated more than 400 cost-effective PV systems, sited throughout northern and central California. Since then, the number of systems has grown to 700, with a total installed capacity of 39 kilowatts.

ity, PLN, to electrify another 12,000 villages by 1993. The Dutch government, with cost sharing from the Indonesian government and from Royal Dutch Shell's subsidiary R&S Renewable Energy Systems BV, has successfully completed a PV village electrification project in a part of Indonesia with above-average rainfall and below-average insolation. The Indonesian government's Ministry of Mines and Energy, which is in charge of electricity development, has found PV to be more economical and more widely practical than diesel or hydro-power systems and now plans to install an additional 2,000 PV systems [46].

India has perhaps the most aggressive government-sponsored PV program in the developing world, with basic applied research, demonstrations, and manufacturing all supported heavily. India's approach is to develop in-country expertise and manufacturing capabilities, while essentially prohibiting imports of PV systems or components. One of the goals of India's program is to establish a new industry with dispersed manpower requirements. Although manufacturing is by necessity fairly centralized, PV use involves distribution and sales networks, as well as installation and maintenance specialists. PV use also increases markets for peripheral components such as electric pumps, storage batteries, and charge controllers.

Work under way in several industrialized countries is applicable to developing countries. The Italian utility, ENEL, has designed a mass-producible 300 watt remote power package intended for residential use [23]. Australian, Japanese, and United States utilities are involved in the development of stand-alone hybrid systems, sometimes called Remote Area Power Systems, or Mini-Utilities. All three countries have a number of inhabited areas that are remote from any central power grid and are looking into low-cost alternatives to extending the grid. In Australia and Japan, a number of inhabited islands have either no electricity or extremely high-cost diesel-generated electricity. Similarly, in the United States and Australia, rural or extremely remote sites often depend on small gas or diesel generators. Utilities are studying instances where the addition of PV systems can cut the cost of operating a generator dramatically or can provide a lower-cost alternative to the generator [29, 47, 48].

WORKS CITED

1. Rich, D., Baron, B.N., McDonnell, C., and Hajilambrinos, C. 1988
   Photovoltaics and electric utilities, *Project Report/EPP 88-04,* Institute of Energy Conversion, University of Delaware, December.

2. Electrotek Concepts, Inc. 1990
   EPRI Conference Memorandum, 3rd EPRI/CRIEPI Workshop on Utility-Interconnected Photovoltaic Systems, Yountville, California, March, p. 8–1.

3. Federal Government of Germany. 1990
   Translation of the legal document accompanying "The 1,000 Roof Program" brochure, Summary from the Federal Notice of September 22, 1990, September.

4. Strategies Unlimited. 1990
   Solar Flare, No. 90-6, 21 December, p. 23.

5. Maycock, P. 1990
   PV News, Vol. 9, No. 6, June.

6. Russell, M. and Kern, E. 1990
   Lessons learned with residential photovoltaic systems, *Proceedings of the 21st IEEE Photovoltaic Specialists Conference,* Orlando, May, pp. 898–902.

7. Rosenthal, A., Risser, V., Lane, C., and Bowling, D. 1990
   Photovoltaic system performance assessment for 1988, EPRI GS-6696, January.

8. Hester, S.L., Townsend, T.U., Clements, W.T., and Stolte, W.J. 1990
   PVUSA: lessons learned from startup and early operation, pp. 937–943.

9. Lepley, T. 1990
   Arizona public service and Salt River project testing experience, see reference 26.

10. Jennings, C. and Whitaker, C. M. 1990
    PV module performance outdoors at PG&E, see reference 6, pp. 1023–1029.

11. Lepley, T. 1990
    Results from Arizona Public Service Company's STAR Center, see reference 6, pp. 903–908.

12. Beyer, U., Dietrich, B., Pottbrock, R., and Lotfi, A. 1990
    Solar module testing at Kobern-Gondorf, *Modern Power Systems,* November.

13. Lynette, R. and Conover, K. 1989
    Photovoltaic operation and maintenance evaluation, EPRI GS 6625, December, p. 2.12.

14. D'Aiello, R.V., Twesme, E.N., and Fagnan, D.A. 1988
    Performance of Solarex/Philadelphia Electric Co. Amorphous Silicon PV Test Site, *Proceedings of the 20th IEEE Photovoltaic Specialists Conference,* Las Vegas, September.

15. Rockenbauer, F.
    On-line PV monitoring system of Austria, IEA - Solar Heating and Cooling Programme, Task XVI: Photovoltaics in Buildings.

16. Shugar, D. 1990
    Photovoltaics in the utility distribution system: the evaluation of system and distributed benefits, see reference 6, pp. 836–843.

17. DeMeo, E.A., Weinberg, C.J., and Tassiou, R. 1989
    Economic requirements for photovoltaic systems in electric utility applications, see reference 23.

18. Takigawa, K. 1990
    Investigation on the cost and perspectives of photovoltaic power applications, see reference 26.

19. Taschini, A. and Iannucci, J.J. 1989
    Potential of photovoltaic systems for present and future electric utility applications, see reference 23.

20. Nentwich, A., Schneeberger, M., Szeless, A., and Wilk, H. 1988
    Photovoltaic activities of Austrian electric utilities: projects and experiences, *Proceedings of the 8th European Photovoltaic Solar Energy Conference,* Florence, May.

21. Beyer, U. and Pottbrock, R. 1989
    RWE 1 MW photovoltaic project, phase 1: design construction and operation of a 340 kWp photovoltaic plant, *Proceedings of the 9th European Photovoltaic Solar Energy Conference,* Frieburg, September.

22. Taschini, A. and Iannucci, J.J. 1990
    Potential of photovoltaic systems for present and future electric utility applications, *Proceedings of the IEA-ENEL Executive Conference on Photovoltaic Systems for Electric Utility Applications,* Taormina, Italy, December.

23. Coiante, D. and Previ, A. 1989
    The Italian photovoltaic program, ENEL's Activity in Photovoltaics, Italian Electric Authority.

24. ENEL. 1990
    3 MW photovoltaic power station preliminary description, EC-US/DOE Workshop on Wind and PV Grid-Connected Systems, Madrid, September.

25. Electrotek Concepts, see reference 2, p. 7–1.

26. Takigawa, K. 1990
    Performance test and evaluation facility of inverter for photovoltaic power systems, and Performance test and evaluation facility of storage batteries for PV applications, CRIEPI presentations at 3rd EPRI/CRIEPI Workshop on Utility-Interconnected Photovoltaic Systems, Yountville, California, March.

27. Kobayashi, H., Takigawa, K., Hashimoto, E., Kitamura, A., and Matsuda, H. 1990
    Problems and countermeasures on safety of utility grid with a number of small-scale PV systems, see reference 6, pp. 850–855.

28. Kurokawa, M. 1990
    Saijo photovoltaic power generation test plant: operation results and a summary of the researches, see reference 26.

29. Miyagi, H. 1990
    Development of practical techniques for operating photovoltaic power generation systems: system for supplying power to isolated islands, and research and development of stand-alone dispersed photovoltaic power generation system: electric power supply for a small island, see reference 26.

30. Kansai Electric Power Co., Inc. 1990
    50-kW Concentrating Silicon Solar Photovoltaic System, see reference 26.

31. Hidroeléctrica Española, Department of Research and New Energies.
    Experimental Research Platform, Area: Photovoltaic Solar Energy, and H.E. Solar Photovoltaic Research Programme, Hidroelectrica Espanola, Madrid.

32. Hidroeléctrica Española, Department of Research and New Energies. 1990
    See reference 26; and personal communications with Senora García Martín, of Hidroeléctrica Española, March 1991.

33. Hester, S. 1988
    PVUSA: advancing the state-of-the-art in photovoltaics for utility scale applications, *Proceedings of the 20th IEEE Photovoltaic Specialists Conference,* Las Vegas, September, pp. 1068–1074.

34. Curry, R. 1988
    VISTA array efficiencies come close to rated module efficiencies, utility study shows, *Photovoltaic Insiders Report.*

35. Jennings, C. 1988
    PV module performance at PG&E, see reference 33, pp. 1225–1229.

36. Hoffner, J. and Panico, D. 1987
    Austin's PV300 plant specifications and cost, *Proceedings of the 1987 PV Annual Systems Symposium,* SAND 87-0097, February.

37. O'Neill, M., Walters, R., Perry, J., McDanal, A., Jackson, M., and Hesse, W. 1990
    Fabrication installation and initial operation of the 2,000 sq. m. linear fresnel lens photovoltaic concentrator system at 3M/Austin (Texas), see reference 6, pp. 1147–1152.

38. Personal communication with Herb Boyd of Alabama Power, April 1991

39. Smith, K. 1989
    Survey of US line-connected photovoltaic systems, EPRI GS-6306, March.

40. Kern E. 1987
Residential utility connected systems, *Proceedings of the 19th IEEE Photovoltaic Specialists Conference,* New Orleans, May, pp. 1007–1011.

41. Patapoff, N. 1985
Two years of interconnection experience with the 1 MW at Lugo, *Proceedings of the 18th IEEE Photovoltaic Specialists Conference,* Las Vegas, October, pp. 866–870.

42. Sumner, D. and Whitaker, C. 1990
Carrisa Plains Photovoltaic Power Plant: 1984–1987 performance, EPRI GS-6689, January.

43. Jennings, C. 1990
PG&E's cost-effective photovoltaic installations, see reference 6, pp. 914–918.

44. Kern, E., Russell, M., Slate, J., and Firor, K. 1991
Early applications of photovoltaics in the electric utility industry, EPRI (to be published).

45. Personal communication with Bill Howley of Siemens Solar Industries, March 1991.

46. Business Communications Company, Inc. 1990
Energy Conservation News, Vol. 12, No. 7, February.

47. Sheridan, N.R. and Gerken, K. 1988
Battery-inverter system for remote area power, *Proceedings of the 7th Conference on Electric Power Supply Industry,* Brisbane, Australia, October.

48. NEOS Corporation. 1991
Phase II Technical Assistance for K.C. Electric Association, Final Report, January.

12
OCEAN ENERGY SYSTEMS

JAMES E. CAVANAGH
JOHN H. CLARKE
ROGER PRICE

Energy is stored by nature in the tides, waves, and thermal and salinity gradients of the world's oceans. Although the total energy flux of each of these renewable resources is large, only a small fraction of their potential is likely to be exploited in the foreseeable future. There are two reasons for this. First, ocean energy is spread diffusely over a wide area, requiring large and expensive plants for its collection; and second, the energy is often available in areas remote from centers of consumption.

Tidal energy, which entails the use of estuarine barrages at sites having high tidal ranges, offers the best prospects in the short to medium term. Not only are its components commercially available, but many of the best sites for implementation have been identified. Indeed, on the basis of current field experience, tidal power may be regarded as a technically proven, dependable and long-lived source of electric power. The exploitation of wave energy, by comparison, is still in its infancy. Small shoreline and nearshore devices are likely to be developed first, but their applicability and potential is limited. More powerful, large-wave offshore energy plants are unlikely to be deployed for a few decades, although the bulk of ocean-energy potential is located offshore. Ocean thermal energy conversion (OTEC), which is currently in the prototype stage, is costly and largely restricted to tropical locations. Its applications are likely to be limited. Salt-gradient energy, once a focus of interest, is not expected to be exploited in the foreseeable future.

Overall, the pace and extent of commercial exploitation of ocean energy is likely to be affected by the rising environmental costs of fossil fuels and by the availability of construction capital at modest real interest rates. If the largest projects are to succeed, however, government support at the national level may be necessary.

INTRODUCTION

The oceans receive, store, and dissipate energy through various physical processes. Energy exists in the form of tides, waves, temperature differences, salt gradients, and marine biomass, each of which has been used or proposed for exploitation. Because these elements differ significantly from one another in terms of their physical processes, exploitation techniques, and state of development, they are discussed separately in this chapter.

Magnitude of the resource

Despite their differences, the four elements—tidal energy, wave energy, ocean-thermal energy, and salt-gradient energy—that comprise the ocean-energy resource base share certain characteristics. In each case, the total energy flux is large; about 2 terawatts for waves and tidal and salt gradients; and at least two orders of magnitude greater for thermal energy. The energy fluxes, however, are spread over a considerable geographical expanse, thus creating low-energy densities and raising the costs of collecting the energy. Moreover, because much of the technical potential exists in areas far from centers of consumption, it appears inevitable that only a small fraction of the global technical potential will be utilized in the foreseeable future.

Status of the technologies

Although ocean-energy systems have great potential, only tidal energy can be exploited using commercially available technology, and even then it is being exploited at only a handful of pilot sites. Until recently, serious interest in other ocean energy elements has been lacking and thus these technologies are relatively immature. The low cost of fossil fuels has also inhibited the development of ocean energy systems. Although wave energy, salt-gradient energy, and ocean thermal energy conversion (OTEC) systems could be greatly influenced by the pace and extent of technological improvement, it is difficult to predict—given the present immature state of these technologies—whether they will prove viable and competitive in the medium term.

Institutional constraints

The economic viability of ocean energy is affected by two institutional considerations: environmental costs and the prevailing financial climate. Under environmental costs, for example, ocean-energy technologies, which create little pollution, must be judged against conventional energy sources. But the latter are priced on the basis of internalized costs that do not include, or only partially include, the cost of their environmental impact.

The financial climate takes into account the sensitivity of capital-intensive projects to interest rates and other money market factors such as risk. The physical size and cost of machinery for extraction of tidal, wave, and ocean-thermal energy tends to be high for several reasons, including the intrinsically low-energy density of the resources, the low efficiency of wave-energy and OTEC plants, and the intermittent operation of tidal-energy systems. Because typical ocean-energy schemes have long construction times and are expensive, interest during construction can become a serious burden. Such schemes are particularly sensitive to the discount rate for capital employed, and therefore to the financing method. Overall costs must then be compared with those of a conventional thermal generation plant and also with other renewable energy plants, such as wind turbines or power systems utilizing landfill gas, both of which can be constructed on a smaller scale relatively quickly.

The cost of capital is also affected by risk, which is perceived by both lenders and investors to be high for all forms of ocean energy. The perception of risk can be attributed to the immature nature of the technology (in the case of wave and ocean-thermal systems), and to site-specific engineering problems and environmental uncertainty (in the case of tidal power). Until ocean-energy systems become common, or have at least been successfully demonstrated at full-scale or commercial operation, the cost of commercial capital for such ventures will remain high. In view of the financing required and risks involved, the very largest projects may well need support at the national level if they are to succeed.

TIDAL ENERGY

Origin of tides

Tides are created by the gravitational attraction of the moon and sun acting on the oceans of the rotating earth. The relative motions of these bodies cause the surface of the oceans to be raised and lowered periodically, according to a number of interacting cycles. These include the half-day cycle, which is created by the rotation of the earth within the gravitational field of the moon, resulting in a period of 12 hours 25 minutes between successive high waters; the 14 day cycle, which results when the gravitational fields of the sun and moon combine to give maxima and minima in the tides (called spring and neap tides, respectively); the half-year cycle, which is affected by the inclination of the moon's orbit and gives rise to a period of about 178 days between the highest spring tides in March and September. Other cycles, such as those that last for 19 and 1,600 years, arise from further complex interactions between the gravitational fields.

The range of a spring tide is commonly about twice that of a neap tide, whereas the longer period cycles impose smaller perturbations, such as ±11 percent due to the half-yearly cycle and ±4 percent due to the 19 year cycle.

Tides in the open ocean have a maximum amplitude of about 1 meter, whereas tides closer to shore, such as those that occur in estuaries, have substantially higher amplitudes, which are influenced by local effects such as shelving, funneling, reflection, and resonance (see figure 1). Consequently, the tidal range can vary substantially between different points on a coastline (see figure 2).

The resource

The amount of tidal energy available varies approximately with the square of the tidal range. The energy from a tidal power plant would therefore vary by a factor of four during a spring–neap cycle. However, unlike many other sources of renewable energy, tidal energy is highly predictable in both its timing and output.

Because ocean tides obtain their energy from the rotation of the earth, slowing it down in the process, they cannot be considered a renewable resource in the strict sense. Practically, however, the slowing of the earth is almost imperceptible over the span of human history and would not be exacerbated by the develop-

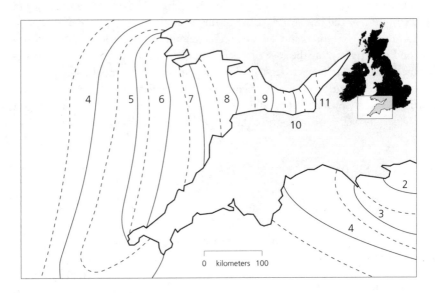

FIGURE 1: *Local effects combine nearshore to give a mean spring tidal range of more than 11 meters in the Severn estuary in the U.K. (tidal range shown in intervals of 0.5 meters).*

Table 1: Tidal-energy resources in Europe[a]

Country	Technically available tidal energy resource		European tidal resource
	GW	*TWh/year*	*percent*
United Kingdom	25.2	50.2	47.7
	(26.8)[b]	(49.5)[b]	–
France	22.8	44.4	42.1
Ireland	4.3	8.0	7.6
The Netherlands	1.0	1.8	1.8
Germany	0.4	0.8	0.7
Spain	0.07	0.13	0.1
Other European[c]	0	0	
Total Europe[c]	**53.8**	**105.4**	**100.0**

a. Estimates based on parametric modeling.

b. U.K. estimate based on more detailed barrage studies.

c. Excluding former USSR.

FIGURE 2: Tidal range map of northwestern Europe showing lines of equal mean spring range in meters.

ment of tidal power. The energy is dissipated by friction in shallow seas and along coastlines at an estimated rate of 1.7 terawatts.

Extraction of tidal energy is considered practical only when the tides are large and suitable sites for tidal plant construction can be found. Although such sites are not common, a number have been identified (see figure 3).

A parametric approach has been used recently to estimate the tidal energy potential for the European Community countries: France, Belgium, Germany, Denmark, the Netherlands, the United Kingdom, Ireland, Luxembourg, Spain, Greece, Portugal, and Italy (see figure 6). The results indicate that all reasonably exploitable sites (those with a mean tidal range exceeding 3 meters) could yield a total energy potential of about 105 terawatt-hours (TWh) per year. Most of the potential exists in the United Kingdom (50 TWh per year) and France (44 TWh per year), with relatively small contributions from Ireland, Holland, Germany, and Spain (see figure 4 and table 1). Scandinavia, the Baltic states, Portugal, Italy, Greece, and other countries surrounding the Mediterranean Sea lack significant potential because the tidal ranges of these countries are so low. On a worldwide basis, a total energy potential of perhaps 5 to 10 times the European potential (amounting to 500 to 1,000 TWh per year, or 3 to 7 percent of the total tidal energy being dissipated) might be expected. However, only a fraction of suitable sites (see figure 2) is likely to be economically exploitable.

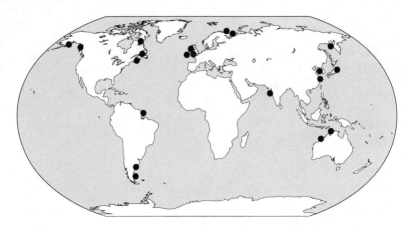

FIGURE 3: Principal sites for tidal power development (see also table 3).

The technology

Historical perspective

Tidal power is one of the older forms of energy exploited by human beings; records indicate that before AD 1100 tide mills were operating along the coasts of the United Kingdom, France, and Spain. Widely used for many centuries, they were gradually displaced by the cheaper and more convenient fuels and machines brought forth by the Industrial Revolution.

The old tide mills were remarkably simple: during the incoming (flood) tide, water would enter a storage pond through a sluice, where it would remain until the tide turned, and then, during the outgoing (ebb) tide it would flow back to the sea through a water wheel. Many other methods, including lift platforms, air compressors, and water pressurization, have been proposed, although none appears to offer a significant advantage over the old tide mill. Nonetheless, tidal energy has long enticed inventors, as evidenced by the hundreds of patents filed over the last 150 years.

Tidal barrage design and construction

A modern tidal energy scheme consists of a barrage, or dam, that is constructed across an estuary and (like the old-fashioned mill) is equipped with a series of gated sluices to permit entry of water to the basin. However, unlike its predecessor, the scheme extracts power using low-head axial turbines rather than a water wheel. If navigation to the upper part of the estuary is necessary, a ship lock may be installed (see figure 4).

The construction method usually proposed for the main barrage structure involves the use of caissons [2, 9, 11, 12, 14, 16]. These are large, prefabricated units of concrete or steel that are manufactured at shore-based construction yards and towed to the barrage site, where they are then sunk into position on prepared foundations. Caissons may house a group of turbine generators or sluices, or they

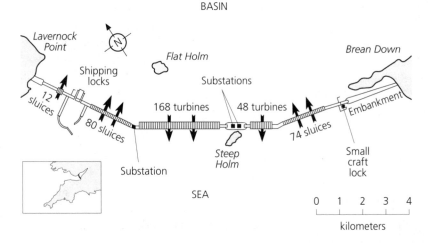

FIGURE 4: *Schematic representation of a possible 8.6 gigawatt barrage at Severn in the United Kingdom.*

may simply be blank to make up the remainder of the structure. In shallower waters, conventional embankment may serve as an alternative to caissons.

In the 1960s at La Rance, in Brittany, northern France, the French successfully constructed a barrage "in the dry" behind a temporary "cofferdam," which was later removed. However, this method is relatively costly and is generally considered to be too risky for larger estuaries. In addition, the construction process may create environmental problems. At La Rance, for example, the upper estuary was severed from the sea for more than two years, a situation that would probably be unacceptable today [3].

Another method calls for constructing diaphragm walls of reinforced concrete within a temporary sand island. But the approach offers no significant cost advantages over caissons and studies for the proposed Mersey Barrage in the United Kingdom indicate that the use of diaphragm walling could prolong construction time by about two years.

Electricity is generated at a tidal plant by large axial-flow turbines, having diameters as great as 9 meters. Because they are driven by a continuously varying head of water, the angles of the distributor, the turbine blades, or both, must be regulated for maximum efficiency. Also, if the turbine is to be used in both directions for electricity generation, or in reverse for pumping, double regulation is necessary. Such is the case at La Rance.

Two types of turbine generator exist: the conventional bulb turbine and the rim-generator turbine. The former contains the generator in a pod directly behind the turbine runner and normally has a directly driven multipole generator. This arrangement is generally preferred for larger machines, such as those at La Rance and those proposed for the Severn [12]. However, for smaller machines (such as those suggested for the Mersey and the Wyre in the United Kingdom), a geared bulb turbine in a pit configuration may be cheaper [15]. The alternative approach is to use a rim-generator (STRAFLO) turbine, in which the stator is outside the water passageway and the rotor is fixed to the periphery of the turbine runner. A large prototype has been successfully tested at Annapolis, Nova Scotia, in Canada and is favored for future tidal installations in Canada's Bay of Fundy [6]. Such large rim-generator turbines, however, lack double-regulation capabilities and have not been tested for reverse turbining or pumping. As a result, they may be restricted to simple ebb generation schemes.

Modes of operation
A tidal barrage can be operated in one of several ways. The simplest and commonly preferred method is known as ebb generation. During the flood tide, water enters the basin through sluices and is held there until the tide recedes sufficiently to create a suitable head. Water is then released through turbines, thus generating electricity. The release process is sustained until the tide turns and starts to rise, causing the head to fall below the minimum operating point. As the water rises, it once again enters the basin, thus repeating the cycle (see figure 5).

A second method, called flood generation, reverses the ebb cycle by generating electricity when tidal water flows into the basin. The technique is not especially efficient, however, because the sloping nature of the basin shores generally results in lower energy production. Moreover, reducing the water level in the basin may restrict navigation and be unappealing aesthetically. For these reasons, flood generation is rarely considered.

Another method, known as two-way generation, extracts energy from both ebb and flood tides. However, it does not usually yield more energy than simple ebb generation because any attempt to generate power during the flood tide will restrict the refilling of the basin and hence limit the amount of energy that can be generated during the ebb tide. In addition, two-way generation requires complex machinery and may impair navigation by lowering the maximum water level in the basin. Nonetheless, two-way generation may be advantageous in places where the electricity must be fed into a weak grid because the scheme can be operated over a longer period of the day. The La Rance barrage in France was originally operated in this way (see figure 5) [3].

Pumping is increasingly favored because of its ability to increase energy output. By operating the turbines in reverse to act as pumps, the water level in the basin (and hence the generating head) can be raised. Although energy for pumping must be imported, there is a net energy gain because water is released through the turbines at a greater head than when it enters the turbines toward high tide. Studies in the United Kingdom on a number of tidal-energy schemes indicate that the energy gain achieved by such pumping as compared with simple ebb generation is small but useful, typically ranging from 5 to 15 percent [12, 13, 15].

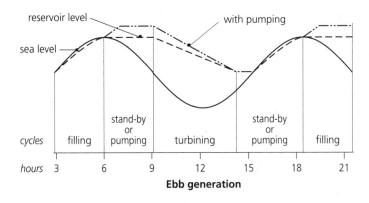

Ebb generation

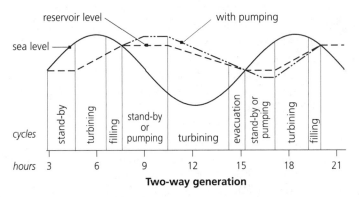

Two-way generation

FIGURE 5: Alternative operational modes at La Rance, France.

Other barrage configurations

In order to obtain greater continuity of supply and increased "firm capacity," a number of other barrage configurations and operating routines have been considered. But such complex schemes, including two or more linked or paired basins, are only realizable where the geography is favorable. Moreover, they have so far been found to be economically inferior to single basin schemes [11].

Tidal-energy plants have usually been designed to produce electrical energy only, without consideration for their firm capacity contribution. Single-basin schemes deliver one or two intermittent pulses of energy per tide, leading to low annual plant load factors ranging from 22 to 35 percent and providing little significant firm capacity. The output pulses recur at a period of 12 hours 25 minutes and thus move in and out of step with the rhythm of human activity. Such output can usually and without difficulty be injected directly into a strong electricity distribution network without the need for retiming, provided that the tidal generating capacity is a small percentage (say, less than 20 percent) of the total system capacity. When the barrage is generating, the output of fossil fuel–powered gen-

Table 2: Existing tidal energy plants

	Mean tidal range *meters*	Basin area *km²*	Installed capacity *megawatts*	Approximate output *GWh/year*	In service
La Rance, France	8.0	17	240.0	540	1966
Kislaya Guba, Russia	2.4	2	0.4	–	1968
Jiangxia, China	7.1	2	3.2	11	1980[a]
Annapolis, Canada	6.4	6	17.8	30	1984
Various sites, China	–	–	1.8	–	–

a. First unit in 1980, sixth in 1986

erating plants would be turned down and fossil fuel saved. In the United Kingdom, even the large (8.6 gigawatt) Severn Barrage could be accommodated in this way, subject to some grid strengthening [12].

Only if the tidal scheme represents a large proportion of total installed capacity on the network, or if transmission lines are weak, would storage be needed to smooth output. Studies undertaken on tidal energy schemes in the Bay of Fundy have tested the feasibility of three different methods of storage. The best one calls for construction of river hydropower plants with significant generating and storage capacity. The river reservoir can then be drawn down when the tidal plant is not operating and replenished when tidal output is available to displace normal hydro-electric generation. Although the method requires additional generating capacity, its throughput efficiency is close to 100 percent.

The other methods involve either conventional high-head pumped storage of water at an efficiency of around 75 percent or storage of compressed air in underground caverns at an efficiency of about 90 percent. Whereas the former is capital intensive, the latter requires gas-turbine fuel for air compression. These latter methods were both estimated to nearly double the delivered cost of electricity from the Bay of Fundy [6].

Existing tidal energy schemes
Relatively few tidal power plants have been constructed in the modern era. Of these, the first and largest is the 240 megawatt barrage at La Rance, which was built for commercial production in the 1960s and has now completed 25 years of successful operation. Others include the 18 megawatt plant at Annapolis, which was built to test a large diameter rim-generator turbine, the 0.4 megawatt experimental plant at Kislaya Guba in Russia, and the 3.2 megawatt Jiangxia station (as well as a number of small or multipurpose plants) in China (see table 2).

Operating experience at La Rance, Kislaya Guba, and Annapolis has generally been positive, although the stators of the La Rance generators and the field windings of the Annapolis generator had to be modified to overcome design weaknesses. But the plant has otherwise proved reliable, with availability in recent years of around 97 percent and only modest maintenance requirements [3,6]. Given the present level of technology, tidal power may be regarded as a technically proven, dependable and long-lived source of electric power.

Tidal stream devices
The high costs and possible environmental consequences of barrage construction have stimulated interest in extracting energy directly from tidal flows. A number of designs (which use turbines to extract energy from tidal streams and ocean currents), have been postulated and preliminary site studies have been undertaken in Florida in the United States, at Garolim in Korea, and at the Messina Strait in Italy. In the United States, turbines with large-diameter propellers have been studied; in both Canada and Russia, vertical-axis Darrieus turbines have been the focus of interest. The main drawback to obtaining energy directly from tidal streams is their low-energy density: the energy available from a turbine in a typical tidal stream with a flow of, say, 2 meters per second (4 knots), is one to two orders of magnitude lower than that available from a turbine of similar diameter in a suitably designed tidal barrage. This substantial reduction in energy output greatly outweighs the avoided civil-engineering costs of a barrage (which account for half to two thirds of the total barrage cost); hence, tidal-stream devices are unlikely to displace barrages in the near term, except perhaps in extreme scenarios or at unusual sites.

Development trends and possibilities
Despite tidal energy's long history, it was not until the 1960s that serious attempts were made to develop barrages for electricity generation. Since then, construction methods, machinery design, and operational techniques have improved substantially. Development efforts have been aimed largely at reducing capital costs and/or construction time, increasing electrical output or its value, and reducing the uncertainty of possible environmental effects.

Caissons, which were first adapted for tidal application at Kislaya Guba [2], are now standard in a wide range of marine applications. They may be constructed from either reinforced concrete or steel, depending on the local price of materials and labor, the availability of construction sites, foundation conditions, and the depth of water available for floating the units into place. Research on caissons made from composite materials (a steel/concrete/steel sandwich, for example) and hybrid construction (a concrete base with steel superstructure, for example) is continuing, especially in the United Kingdom. In addition, the potential for diaphragm-wall construction of either the whole dam structure or suitable elements of the dam structure, such as shiplocks, using the diaphragm-wall method, is being studied.

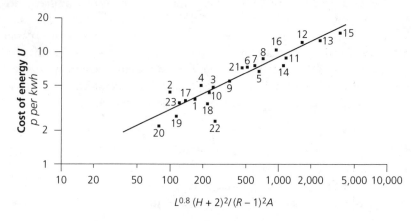

FIGURE 6: *Parametric assessment of potential tidal sites. [10]. A parametric approach has been developed in the United Kingdom to enable approximate analysis of potential schemes. The formula below relates the unit cost of energy generation to the tidal range and physical dimensions of the scheme and is calibrated using data from previous detailed barrage investigations:*

$$\log U = k \log \left[\frac{L^{0.8} (H + 2)^2}{A (R - 1)^2} \right]$$

where U = unit cost of electricity in p per kWh (1p = ¢56).
L = length of barrage in meters
H = maximum height of barrage above sea bed in meters
A = area of basin in km²
R = mean tidal range in meters.
k depends on financial assumption.

The double-regulated bulb turbine used at La Rance is still a favored technology, although pit and tube variants with geared generators have since been developed, albeit for limited power ratings (those less than 25 megawatts). The straight-flow, rim-generator turbine has been tested in ebb generation mode at Annapolis, but is not proven where pumping is required. Because machine efficiency and reliability are now high, the potential for further improvement is probably limited.

Design and feasibility studies undertaken in France, the United Kingdom, Canada, and Russia have led to techniques for quickly evaluating potential tidal sites on a parametric basis (see figure 6), [7, 17], for optimizing installed capacity [11], for maximizing electrical output or its value, and for integrating output into distribution networks.

In recent years, a number of small tidal-barrage schemes have been built in China as part of a broader, resource utilization plan involving aquaculture or navigational improvements. Such a plan may justify development when tidal power by itself is too costly. Studies in the United Kingdom have also emphasized the non-energy benefits of tidal barrages, including storm-surge prevention, road crossings, opportunities for watersport and marine activities, and increased land values.

The trend therefore has been toward a small decrease in the real cost of energy, a phenomenon that is attributable to marginal improvements in machinery, materials, and construction methods; optimization of barrage configuration and operation; and additional, non-energy benefits.

Tidal energy technology has clearly matured during the past three decades. As a result, future cost reductions, improvements in performance, and economic gains are likely to be relatively modest. Barring some radical technical break-through, further development is unlikely to drop the average cost of electricity from tidal barrages by more than 10 to 20 percent.

Economics and markets[1]

General considerations

On one hand, the cost of electricity generated from tidal power tends to be high, in part because construction times can be lengthy, amounting to several years for the larger projects. Operation is also intermittent, which results in a load factor of only 22 to 35 percent, thus increasing the overall cost per kilowatt of installed capacity. On the other hand, there are no fuel costs, and other annual operation and maintenance costs are very low (typically 0.5 percent of the initial capital cost). Moreover, a barrage can be expected to have great longevity: with reasonable maintenance the main structure should last at least 120 years and its associated machinery some 30 to 40 years.

Even so, the economics of tidal energy are highly site specific. Energy output varies according to the tidal range and geography of the enclosed basin, and costs reflect barrage height and length, as well as special needs such as shiplocks for navigation (see figure 6). Economy-of-scale seems not to be significant. Studies in the United Kingdom, for example, have found that sites ranging in capacity from 30 to 8,000 megawatts have broadly similar energy costs.

Social and environmental benefits should also be taken into account when assessing potential tidal schemes. Although such benefits may be substantial [12], they are often difficult to quantify in money terms. And, because they may not accrue to the developer, especially when the project is initiated by the private sector, they are often overlooked. Nonetheless, regional development opportunities can be created as the result of barrage construction. These include road crossings, ports and marinas, aquaculture farms, improved recreational sites, increased tourism, and higher land values.

The high capital costs and long construction times of large tidal barrages render tidal energy particularly sensitive to the discount rate for capital employed,

---

1. Unless otherwise indicated, all dollars and cents referred to in this chapter reflect 1989 values of U.S. currency with U.K. costs converted at the rate £1 = U.S.$1.8.

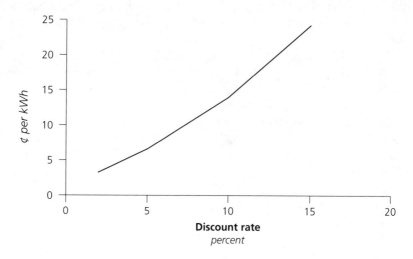

FIGURE 7: Cost of energy from the Severn Barrage showing sensitivity to discount rate.

and hence to the method of financing. Studies show, for example, that increasing the real discount rate from 5 to 10 percent would double the unit cost of electricity generated by the Severn Barrage, from U.S.7 cents to about 14 cents per kWh (see figure 7).

Estimation of economic potential

The parametric costing model (see figure 6) has recently been applied to most exploitable sites in western Europe that have a mean tidal range greater than 3 meters [18]. This in turn has led to the derivation of a resource cost curve (see figure 8), showing how much tidal energy can be extracted for a given unit cost of electricity as a function of the discount rate for capital employed. This cost curve makes it possible to estimate the economic resource for tidal energy in each country and also to identify priority sites for further investigation.

The cost curve indicates that about 64 gigawatts (105 TWh per year) can be obtained from reasonably exploitable sites in western Europe, excluding Russia. Most of the capacity, about 95 TWh per year, is available at a cost of less than 18 cents per kWh at a real annual discount rate of 5 percent (see table 3). Raising the discount rates to 10 and 15 percent would reduce the resource that is available below this cost threshold to 63 TWh per hour and just 3 TWh per year, respectively. At a lower threshold cost of 9 cents per kWh, the amount of electricity that could be generated would be 52 and 1.4 TWh per year at the respective discount rates of 5 and 8 percent. Tidal energy in Europe cannot effectively be exploited for less than 9 cents per kWh at discount rates above 10 percent or for less than 18 cents per kWh at discount rates above 15 percent. If this method were applied globally, the world economic potential for tidal energy might, on a basis of simple extrapolation, be roughly 5 to 10 times the above figures.

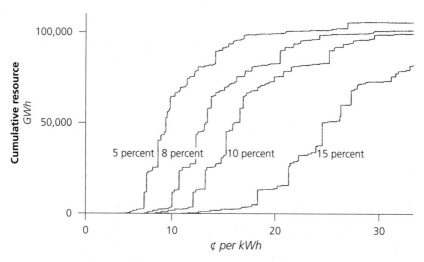

FIGURE 8: Cumulative tidal energy resource for western Europe against unit cost of generation for a range of discount rates. Notes: Generation cost at the barrage boundary are in 1989 U.S. dollars; costs are discounted over a project lifetime of 120 years.

National studies

In some countries, the potential market for tidal energy has been thoroughly investigated. In the United Kingdom, it is estimated that the best estuaries might provide about 28 TWh per year of electricity, while the total from all reasonably exploitable estuaries could exceed 50 TWh per year. Feasibility studies are presently underway at Severn (8,640 megawatts) [12–16], Mersey (700 megawatts), Wyre (63 megawatts), and Conwy (33 megawatts). For the Severn Barrage, capital costs are estimated to be about $1,800 per kilowatt; if capital were available at a discount rate of 6 percent per year, the resulting electricity would cost around 8 cents per kWh at the barrage boundary.

In Canada's Bay of Fundy, Cumberland Basin is not considered competitive at present[2] and Minas Basin, though economically more attractive with 30 percent lower unit costs, could adversely affect the coast of Maine in the United States [6]. In Russia, feasibility studies have been carried out in the White Sea and in the Sea of Okhotsk. In the latter area, the feasibility study of a major site at Tugur is presently under way. Provisional plans call for total installed capacity of 1,000 megawatts by 2005, 5,000 megawatts by 2010, and 10,000 megawatts by 2015. In Korea, a site at Garolim is being seriously considered. In India, some interest has been shown in the Gulf of Kutch, but the site is probably uneconomic given the present low price of fossil fuels. In Mexico, a site in the Colorado estuary is being investigated; in China, feasibility studies suggest that the southeast coast

---

2. The Cumberland Basin scheme would cost about $1,700 per kilowatt to build, plus $600 per kilowatt for transmission lines, and $1,200 per kilowatt for retiming using compressed air storage; at a 6 percent discount rate this would result in about 13 cents per kWh for retimed energy.

Table 3: Possible development sites (as of 1990)

	Mean tidal range meters	Basin area km²	Approximate installed capacity megawatts	Annual output TWh	Plant factor
Argentina					
San José Gulf	5.9	–	6,800	20.0	–
Australia					
Secure Bay 1	10.9	–	–	2.4	–
Secure Bay 2	10.9	–	–	5.4	–
Canada					
Cobequid	12.4	240	5,338	14.0	0.30
Cumberland	10.9	90	1,400	3.4	0.28
Shepody	10.0	115	1,800	4.8	0.30
India					
Gulf of Kutch	5.0	170	900	1.7	0.22
Gulf of Cambay	7.0	1,970	7,000	15.0	0.24
Korea					
Garolim	4.8	100	480	0.53	0.13
Cheonsu	4.5	–	–	1.2	–
Mexico					
Rio Colorado	6–7	–	–	5.4	–
Tiburón Island	–	–	–	–	–
United Kingdom					
Severn	7.0	520	8,640	17.0	0.22
Mersey	6.5	61	700	1.5	0.24
Wyre	6.0	5.8	47	0.09	0.22
Conwy	5.2	5.5	33	0.06	0.21
United States					
Passamaquoddy[a]	5.5	–	–	–	–
Knik Arm	7.5	–	2,900	7.4	0.29
Turnagain Arm	7.5	–	6,500	16.6	0.29
Russia					
Mezen	9.1	2,300	15,000	50.0	0.38
Tugur[b]	–	–	10,000	27.0	0.31
Penzhina	–	–	20,000	79.0	0.45

a. Borders the United States and Canada.

b. A 7,000 megawatt variant of the Tugur Barrage is also being investigated.

has considerable potential. In Australia, tidal-energy sites in the Secure Bay area are being reinvestigated, but the remoteness of potential markets is a major constraint.

Interest in sites that were once considered promising—Alaska and Passamaquoddy in North America, San José in Argentina, and the northeast coast of Brazil—lapsed when costs were found to be too high. Other areas with large energy potential that are no longer the focus of interest include Mont-Saint-Michel Bay in France and Ungava Bay in northern Canada. Financial constraints make the development of Mont-Saint-Michel unlikely; Ungava Bay is too remote.

Environmental effects

Tidal energy offers significant environmental benefits: it is nonpolluting and can displace coal and hydrocarbon fuels. By displacing coal, a tidal barrage can prevent about one million tonnes of carbon dioxide emissions per TWh generated (17 million tons (MT) per year for the Severn Barrage). A barrage also protects vulnerable coastlines against storm-surge tides. Still, tidal-energy projects may change the surrounding estuarine ecosystem; site-specific environmental impact assessments are therefore needed to identify these changes and determine their acceptability before construction begins. Elements of the estuarine ecosystem that must be characterized for an environmental statement include water quality, sediment type, and bird and fish populations. In addition, the likely impact a barrage will have on each of these must be addressed.

Construction of a barrage affects the hydrodynamic regime of the estuary, typically reducing the tidal range, currents, and intertidal area within the basin by about half. Such hydrodynamic changes can in turn influence both water quality (for example, the dilution and dispersion of effluents and turbidity) and the movement and composition of bed sediments. Any drop in turbidity that occurs may increase primary biological productivity, with consequent effects throughout the food chain, from phytoplankton to zooplankton, invertebrates, fish, and birds.

If a site is carefully selected, and a suitable barrage design and operating mode are chosen so that maintenance dredging is limited, then movement of silt need not be a problem. The reduction in intertidal mudflats for birds may, to some extent and in some cases, be compensated for by productivity changes that increase the availability of food. Some fish, especially migratory species, may suffer from increased mortality at the turbines or experience delayed passage, as observed at Annapolis. At La Rance, however, there is no evidence that any fish species has been significantly affected by the barrage. Although fish diversion systems can mitigate such impacts in principle, methods that combine low cost with efficiency have yet to be developed.

To date, no tidal-energy plant has been subjected to extensive environmental monitoring. Nonetheless, most operational experiences have been positive. Although some environmental uncertainties remain and require further investigation, especially at the site-specific level, no major factors have so far been identified that would inhibit the wider implementation of tidal energy, assuming proper attention to scheme design.

Remarks
The past 30 years have seen significant development of tidal-energy technology, particularly in France, the United Kingdom, Canada, and Russia. A number of demonstration plants have been constructed, including a full-scale (240 megawatt) power station at La Rance, and commercial projects are being considered by a number of countries. Future developments will focus on limiting construction costs, increasing output, and reducing the environmental uncertainties associated with barrages. Deployment will depend on the availability of capital at modest discount rates and, for the largest schemes, could require the involvement of national governments.

WAVE ENERGY

Ocean waves, created by the interaction of winds with the sea surface, contain both kinetic energy, which is described by the velocity of the water particles, and potential energy, which is a function of the amount of water displaced from the mean sea level. The energy transferred to the ocean from the wind depends on the latter's speed, the distance over which it interacts with the water, and its duration. The velocities of waves depend on their wavelength—the longer the wavelength, the faster the wave travels. This effect is seen in a hurricane where long waves travel faster than the generating storm and the hurricane is often preceded by heavy surf.

The power of a wave is quantified as the rate at which its energy is transferred across a 1 meter line at right angles to its direction, and is expressed in units of kilowatts per meter of wave front. The power in a wave train remains relatively constant in deep water, with small losses arising from the viscosity of the water and from interaction with the atmosphere (or turbulence). Thus, long smooth swells can persist for hundreds of kilometers, whereas shorter steeper seas decay rapidly. In water shallower than about half a wavelength, the motion of the water particles near the bottom is appreciable and energy is lost to friction with the seabed. The combination of loss mechanisms is complex. Once formed, waves continue to travel in the direction of their formation even after the wind dies down, which explains why a long swell can sometimes be observed in a calm sea, the residual effect perhaps of a distant storm that occurred days earlier.

A shelving seabed, where the water depth is progressively reduced toward the coastline, also causes reduction in wave speed and sometimes a change of wave direction if the wave fronts approach the seabed obliquely. This latter effect, called refraction, occurs as the wave front is progressively slowed down and thus becomes parallel to the beach—a phenomenon readily observed in nature. If the seabed contours are irregular, focusing and defocusing of waves can occur as the wave breaks up into components traveling in different directions. In addition,

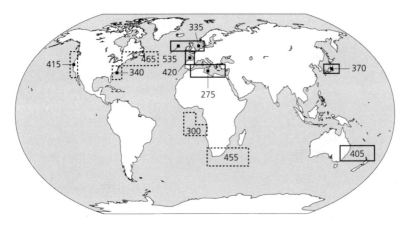

FIGURE 9: Annual wave energy in megwatt-hours (MWh) per meter from specific areas.

wave power at a specific coastal site may be reduced by land masses that obstruct the waves.

The resource

The highest concentration of wave energy occurs between the latitudes 40° and 60° in each hemisphere, which is where the winds blow most strongly. But latitudes of around 30°, where regular tradewinds prevail, may also be suitable for exploitation of wave energy. Winds blowing for long distances over the Atlantic and Pacific oceans can generate waves tens of meters high with more than 100 meters between crests and many tonnes of water displaced in each wave. The west coasts of Europe and the United States, and the coasts of New Zealand and Japan are particularly suitable for wave energy extraction [20] (see figure 9).

The technology

Large-scale offshore wave-energy devices

A wave-energy device extracts energy from the sea and changes it to another form—usually mechanical motion or fluid pressure. Converting this energy to electricity, however, is not simple because the low frequency of the waves (around 0.1 hertz) must be increased to the rotating speed of conventional mechanical and electrical power plants (around 1,500 rpm). Wave-energy devices can interact with waves in several ways in order to extract energy [29] (see figure 10). These include buoyant structures, called heaving floats, that are moored at or near the surface of the sea; hinged structures, called surface followers, that follow the contours of the waves; flexible-bag devices that inflate with air with the surge of the waves; structures with an enclosed oscillating water column (OWC), which acts like a piston to pump air (they can float or be fixed at, or below, the sea surface); and

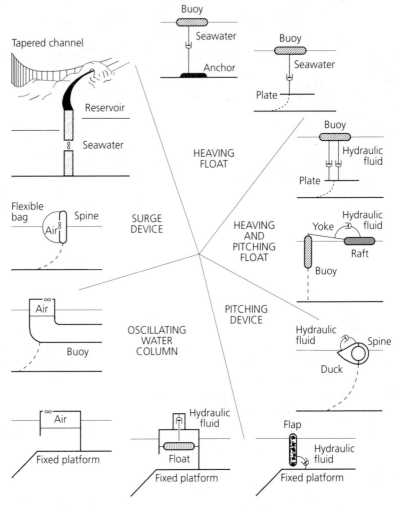

FIGURE 10: Symbolic representation of various types of wave-energy devices.

focusing devices that use shaped chambers to increase wave amplitude and hence drive pneumatic pumps or fill a reservoir on the shoreline.

All such devices must be constrained to resist the wave forces and to obtain a frame of reference against which the devices can react. Constraint can be achieved in one of three ways: by fixing or mooring the device to the seabed or shore, by mounting several devices on a large common frame, thus exploiting the relative motion between them to resist motion, or by using the gyroscopically induced inertia of a flywheel.

Terminators, attenuators, and point absorbers. The maximum available energy is intercepted when several devices are mounted on a linear spine oriented at 90° to the wave direction. The configuration is known as the terminator mode because energy is absorbed by terminating the waves. When energy is removed from a wave front having the same length as the device, the maximum capture rate is unity.

A device in this mode, constrained in heavy seas, experiences considerable forces. But these forces can be reduced by reorienting the device parallel to the wave direction, in the so-called attenuator mode. Doing so provides a smaller length of incident wave front so that the device rides the waves more like a ship. Although this can result in reduced energy capture, energy is extracted from waves at the head of the spine as well as from diffraction of the wave front into the sides of the spine. In addition, devices can be mounted on both sides of the spine, theoretically doubling the energy output. However, the capture ratio of an attenuator is only 62 percent that of a terminator of the same length [20].

A third group of devices, known as point absorbers, have linear dimensions that are less than the length of the waves. Although the point absorbers are capable of capturing energy equally from all directions, they are unable to capture energy from large waves because of their limited size.

Device sizes.   The optimum size of a device depends on the wave and sea characteristics in which it is intended to operate. The system must be tuned, for example, so that its natural oscillations match those of the most powerful and frequent waves. A simple calculation considers the volume of moving water in a typical wave and requires that for maximum energy capture an equivalent volume must be swept within or by the device. For a sea where the power-carrying waves have heights of 3 meters and wavelengths of 100 meters, calculation of swept volume gives linear-device dimensions of about 10 meters.

For a rigid spine with several devices, minimum length is determined by the requirement that in order to reduce pitching motion, the spine must always straddle at least two wave crests. Maximum length is determined by the strength of the spine and is limited to about twice the wavelength. A single spine unit could be 200 meters long and accommodate as many as 10 devices.

Arrays of devices.   Many devices are needed to generate appreciable quantities of electricity. Ideally, to minimize electrical collection and transmission costs, they should be close together. Yet if the devices are arranged in rows or are too close to one another, a number of hydrodynamic problems can arise. To begin with, shadowing by the front-row devices can block most of the energy to the rear row unless there is considerable fetch between rows. Also, one device might reflect oncoming waves, thus diminishing the power available to adjacent devices. And finally, efficiency losses can be created when the waves change direction because of the effect of adjacent devices; (the response of each device depends on the incident direction of the waves and, even with large gaps in each row to allow sufficient power to reach the rear, a loss of efficiency can occur). Consequently, a single row of devices is likely to be the most efficient. Moreover, if the distance between the devices can be correctly calculated, then each device will capture more energy, via a process called constructive interference, than if it operated alone [31, 32].

Small-scale shoreline and nearshore wave energy

The concept of large-scale (gigawatt) electricity generation by wave energy for distribution through a national grid is strictly associated with offshore devices. Shoreline devices, in contrast, because of restricted availability of suitable coastal sites and their lower wave potentials, are unlikely to provide large amounts of electricity for distribution to distant markets. They can, however, provide megawatt levels of power for local needs and may compete with diesel-generated supplies to island communities. Such devices include the Norwegian Tapchan, which has been developed for use in areas of low tidal range. Waves enter the Tapchan through a tapered channel and are thus focused, increasing their crest height. As the water exits the upper end of the channel it flows into a reservoir and returns to the sea via a turbine. A 350-kilowatt capacity Tapchan is currently operating near Bergen in Norway.

Other strategies call for combining natural and artificial gullies with an oscillating water column (OWC) and Wells turbine. (The Wells air turbine has the unique characteristic of rotating in the same direction irrespective of the direction of airflow; it is therefore particularly suited to this application in which the airflow reciprocates.) Once built, access to the sites should not be difficult, and operating and maintenance costs should be comparable with other shore-based generating plants, such as hydroelectric or diesel generators. Although shoreline wave-power levels may be an order of magnitude less than those offshore, waves are focused at certain locations by the nature of the coastline and the topography of the seabed, thus enhancing power output. However, sites with large tidal range or strong tidal currents are not suitable because these factors reduce efficiency.

Nearshore wave-energy stations, located at depths of 10 to 25 meters, offer greater potential than shoreline devices. Not only is the incident wave power greater farther from shore, but the sites themselves face fewer environmental and infrastructural restrictions. However, construction, operating, and maintenance costs are higher: tugs are needed to ferry crews and equipment to and from shore, and the stations are more vulnerable to weather conditions. As an example of this type of device, caisson breakwaters, which incorporate OWCs, have been developed on an experimental basis in Japan for electricity generation in combination with shoreline protection.

Development trends and possibilities

The main thrust of present development endeavors centers around different types of OWC and Wells-turbine devices for nearshore operations, whether they are incorporated into natural or artificial gullies, cliff-mounted columns or caisson breakwaters (whose primary function need not be electricity generation). The interest stems from increased confidence in the OWC and Wells turbines, favorable economics in remote locations where diesel power may be the only competition, the possibility of partnership with breakwater caisson projects, and the availability of land bases and infrastructure for construction and operation.

Because shoreline power stations will always be affected by environmental and infrastructure constraints, if wave energy is to make any sizable contribution, it will have to be obtained from offshore and nearshore waves. But solving the technical problems of nearshore and offshore devices, acquiring operating experience, and ultimately developing commercial confidence in the longevity and durability of such devices requires time. Nonetheless, a number of countries are supporting wave-energy research.

China has constructed a 3 kilowatt OWC shoreline device, which has an artificial gully and a Wells turbine, near Guangzhou on the Pearl River [21]. The country is also conducting research on wave-powered navigational vessels and marine buoys.

Denmark has undertaken research and development on a seabed-mounted hydraulic pump and electrical generator driven by a floating buoy. The first device was tested in an estimated maximum wave climate of 12 meters and suffered failure of some valves [22]. Further plans call for phase control on the float to provide short-term energy storage.

India has constructed an OWC caisson breakwater device, which has a 150-kilowatt Wells turbine, off the coast of Madras [23]. A further 20 modules may be added to the site to achieve a total capacity of more than 1 megawatt.

Japan has an active wave-energy program despite having annual average wave-power levels of only 10 kilowatts (kW) per meter, which is modest compared with the 50 kW per meter typical of the North Atlantic. Japanese interest in wave power has arisen for a number of reasons: the cost of importing fossil fuels is high, the country has many islands that provide low tidal range, and construction of nearshore breakwater devices could supply coastal protection, while at the same time facilitating the development of aquaculture, fish farming, and marine recreation.

In addition to pioneering work on wave-powered buoys, Japan has undertaken considerable research, including sea trials, on such devices as floating terminators, a floating OWC, a pendulum device, and an onshore OWC. A wave-power caisson breakwater with an OWC and a 40-kilowatt Wells turbine is currently being tested in the port of Sakata in the northern Sea of Japan [24].

Norway began an official wave-power research program in 1978 and to date has spent approximately $26 million on the program. In 1985, two onshore devices, the Tapchan and a shore-based OWC, were built after several years of theoretical development [33]. The OWC consists of a cliff-mounted hollow column whose lower entrance lies below the surface of the sea. Wave action causes the water in the cylinder to rise and fall, which in turn pumps air through a Wells turbine. A 0.5 megawatt prototype device was successfully operated at the same site for four years before being destroyed in a storm in December 1988.

Portugal plans to construct a 0.3-megawatt demonstration shoreline device at Porto Cachorro in the Azores where a population of about 1,500 people currently depends on diesel fuel [25]. The device, which will use an owc and Wells turbine, will be situated on a rocky coastline.

Sweden has developed and tested a floating buoy that drives a hose pump attached to a seabed anchor [26]. A shallow-water horizontal-rotor device is also under development.

The United Kingdom has conducted research and development on wave energy since 1974 at a total cost of about $64 million [8]. More than 300 wave-energy concepts have been examined, the most attractive of which have been tested at small scale in wave tanks; three have been tested at one tenth scale in sea conditions. Initial reference designs were developed for eight devices for a conceptual 2 gigawatt power station located off northwest Scotland. In total, the United Kingdom estimates that it has 6 gigawatts of technically extractable offshore wave-energy resources. The U.K. Department of Trade and Industry is currently reviewing the country's technical and economic prospects for wave energy (offshore, nearshore, and shoreline), with results expected in 1992. Some of the devices being reviewed [27] are the National Engineering Laboratory's (NEL) breakwater, the circular clam, the Salter duck, the solo duck, and the Queen's University of Belfast shoreline gully device. A prototype shoreline owc device with a nominal capacity of 75 kilowatts and an estimated annual energy delivery of 300 MWh is being tested on the Scottish island of Islay and is feeding electricity into the island's network [28].

Economics

The construction, assembly, and installation of a large offshore wave-power station, together with the necessary electrical connections would, if undertaken, represent a considerable investment amounting to several thousand million dollars. But, in view of its modular design, such a power station could be constructed in stages, thus allowing energy to be obtained before construction is complete. Nearshore power stations could be built on a smaller scale and towed into place for seabed mounting or mooring. Shoreline and breakwater caisson wave-energy stations have been constructed in the United Kingdom, Norway, Japan, and India and, despite their sometimes remote locations, present no greater technical difficulties than the construction of a small harbor or breakwater.

The difference in wave energy regimes and the considerable variety of devices, each of which varies in cost, performance, and availability, has resulted in a wide range of estimates for the cost of electricity from wave energy. A recent economic analysis of 11 wave-energy devices, based on evidence from hardware testing in wave tanks, in natural waves, and on commercial plant designs, indicates that the cost of electricity (COE) (in 1987 U.S. cents per kWh) [19, 29] is:

$$COE = 112.9/(\text{wave power})^{0.64}$$

where wave power is in kW per meter.

For the North Atlantic, when the wave power is 50 kW per meter, the cost of electricity is 9.23 U.S. cents. The data were assessed on a standard basis, with an assumed discount rate of 12.5 percent, and excluded submarine cable costs.

Many wave-power devices would need to be deployed offshore to produce considerable amounts of useful energy. It has been suggested that mass production might produce a decrease in the cost of each unit by 15 to 25 percent for each doubling of the cumulative production volume [34].

Environmental effects

Because wave energy can partly replace energy from fossil fuels, the technology would reduce greenhouse-gas emissions and atmospheric pollution. There may also be local environmental effects, depending on the choice of site. Benefits could include the creation of areas of sheltered water that are attractive to fish, sea-birds, seals, and seaweed. The possible effects on the local ecosystem of antifouling paints used at the site would need to be considered.

The coastal environment might also be affected by a wave-power station modifying the local wave climate. Although offshore floating devices of low free-board (the height above water) would probably have minimal effect on the coast-line, a station of seabed-mounted devices would reduce the wave energy reaching shores and shallow subtidal areas, and would in theory affect the density and species composition of resident organisms.

Wave-energy devices could create a potential hazard for ships; because of their low height above the water, the devices would be relatively invisible either by sight or by radar. Although the danger could be minimized by marking the devices with lights and transponders and by establishing navigation channels to avoid the arrays, some devices might drift as a result of mooring failure, thus creating hazardous conditions, not only for ships but also for coasts and harbors. Repair resources would therefore need to be deployed rapidly in order to retrieve the devices before they reached land. Doing so should not be difficult, however, since similar strategies have been successfully demonstrated in the offshore oil industry.

Remarks

Ocean waves represent a sizable renewable energy resource, yet wave technology is still at an early stage of development. Many small-scale experimental devices have been tested in wave tanks and at sea, and several prototype devices are now producing electricity for consumption. Constructed on the shoreline or as break-water caissons, these devices are thought to offer the best medium-term prospects for commercial development of wave-energy technology. Continued development of shoreline technology, however, will depend on whether the capital costs become low enough to compete with diesel-generated power or whether such devices can be linked to harbor-protection technologies.

Development of nearshore and offshore wave-energy technology is restricted by the inability to predict the longevity and operating and maintenance require-

Table 4: Key characteristics of OTEC plants

Unit size	100 kW to 1MW
Lifetime	30 years
Availability[a]	70–90 percent
Load factor[b]	70–90 percent
Efficiency	2.5 percent
Construction time	3–5 years
Planned maintenance time	2 percent

a. Availability describes the amount of time the plant is available to generate power when provided with an exploitable energy resource and is a measure of plant reliability. The load factor is the ratio between actual electrical units sent out during an average year and those that could theoretically be sent out if the plant operated at full capacity during that time.

b. The load factor is the ratio between actual electrical units sent out during an average year and those that could theoretically be sent out if the plant operated at full capacity during that time.

ments of these devices under severe sea conditions. Consequently, the prospects for economic exploitation of large-scale offshore energy are considered to be limited for at least the next decade. But if shoreline and nearshore wave energy technologies can be successfully developed and demonstrated over the next decade, then there may be the prospect of prototype offshore devices by early next century.

OCEAN THERMAL ENERGY CONVERSION

In the tropical and subtropical oceans of the world, a natural temperature difference exists between the surface waters and those at depth. The concept of ocean thermal energy conversion (OTEC) exploits this temperature difference to drive power plants to produce electricity. Because the surface waters are warmed by the sun, OTEC can be considered an indirect solar technology. Unlike other solar technologies, however, a reliable OTEC plant would be able to generate electricity continuously because the temperature difference lasts 24 hours a day.

The principles of OTEC have been known for more than 100 years, having first been proposed in 1881 by the French scientist D'Arsonval. In the 1930s his pupil Claude tested the concept experimentally and built the first generating station at Matanzas Bay in Cuba. The station produced 22 kilowatts of power for two weeks before it was partially destroyed in a hurricane and ceased operation. In the late 1950s, the French once more became interested in the concept and designed a number of plants, one of which was to be built off the Ivory Coast, but never went beyond the design stage. The technology was not resurrected until the 1960s when the Andersons, an American father-and-son team, formed a private

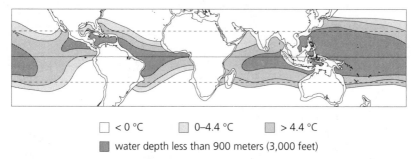

□ < 0 °C   ▨ 0–4.4 °C   ▩ > 4.4 °C
■ water depth less than 900 meters (3,000 feet)

FIGURE 11: The world's ocean thermal energy resource. The map shows average temperature differences between the surface and a depth of 900 meters. Source: Technology Review, U.S. Department of Energy.

company to exploit OTEC. Interest in the technology increased during the 1970s, driven by public concern for the environment and rising oil prices. Research and development programs commenced in France, Sweden, India, Japan, the Netherlands, the United Kingdom, and the United States. Although the research programs focused initially on the technological aspects, in more recent years they have addressed economic considerations as well. From these research and development efforts, a limited number of designs have emerged, some of which are likely to be tested through demonstration on a modest scale (up to 10 megawatts) over the next few years. The main thrust of the investigation will be questions of economics and reliability, which so far have limited the adoption of OTEC (see table 4). If tests prove successful and the economic and political climate remains favorable, OTEC might be commercially available between 2010 to 2020.

The resource

The oceans act as an enormous heat sink for solar radiation falling on the earth. About one-fourth of the 1.7×10^{17} watts of solar energy reaching the earth's atmosphere is absorbed by sea water. In tropical regions, the sun's heat can warm surface water temperatures to as much as 25°C, a sharp contrast to the 5°C temperatures that can exist at depths of about 1,000 meters. Because power cycles operating with such small temperature differences are low in efficiency, and therefore expensive, OTEC is only practical in regions where the temperature difference is 20°C or more. This restricts OTEC to tropical and subtropical regions (see figure 11). However, the area for possible exploitation is still very large, covering nearly 60 million km^2. Overall, OTEC's theoretical potential, if not restricted by economics or technical problems, is about two orders of magnitude greater than that of wind, wave, or tidal energy.

The technology

OTEC operates according to a simple principle, which has been described as nothing more than sophisticated plumbing [36]. Its low thermal conversion efficiency of approximately 2.5 percent requires enormous flows of warm surface water and

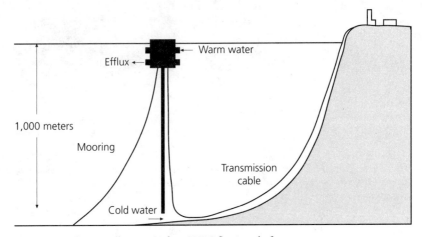

FIGURE 12: Schematic illustration of an OTEC floating platform.

cold deep water, thus necessitating an enormous operation. An OTEC station capable of generating about 100 megawatts of electricity, for example, must pump about 450 m^3 of both warm and cold water through its heat exchangers per second [37]. The engineering challenge, cost, and reliability of such huge components limit the prospects for OTEC.

OTEC plants may be based on land or installed on a floating platform or ship at sea. In either case, the essential and distinctive component of an OTEC plant is the large pipe required to bring cold water to the surface. For a 100-megawatt station, the pipe is expected to have a diameter of 20 meters and may range in length from 600 to 1,000 meters. Generating components vary according to the heat cycle used for power production. A 100 megawatt floating plant would have a displacement of around 200,000 to 300,000 tonnes (see figure 12), yet must be stable at sea and have suitable access for installation and maintenance of equipment. With these considerations in mind, large floating islands are not necessarily the best choice and therefore many platform designs have been proposed. Mooring in 1,000 meters of water creates further problems, but is necessary if electricity is to be transmitted to land. Some researchers have suggested that the platform be self-maneuvering, allowing it to move to areas where the temperature difference is greatest. The drawback is that electricity could not be transmitted by cable to land, but instead would have to be used at the station itself to produce an easily transportable energy-rich product, such as hydrogen produced by electrolysis (see chapter 22: *Solar Hydrogen*).

OTEC plants are designed to work with either a closed cycle or an open cycle (see figures 13a and 13b). In a closed cycle, warm surface water is pumpedthrough an evaporator in which a working fluid,[3] such as ammonia, is evaporated. The

---

3. The working fluid for the closed cycle must be a refrigerant capable of evaporation when exposed to warm water and condensation when exposed to cold. For cost and safety reasons, ammonia and Freon are currently the leading candidates.

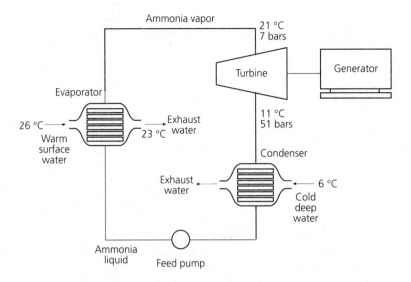

FIGURE 13a: OTEC plants operate in either a closed-cycle mode (above) or in an open-cycle mode.

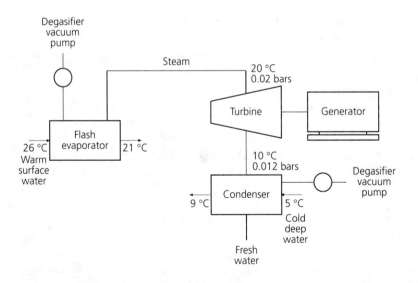

FIGURE 13b: OTEC plant in open-cycle mode.

vapor flows through a turbine to the condenser where it is cooled and condensed by cold water pumped from the ocean depths. The condensate is then collected and pumped back to the evaporator to close the cycle. In an open cycle, seawater serves as both the working fluid and the energy source. The warm seawater is evaporated at very low pressure (0.03 bars) in a flash evaporator. The resulting vapor then passes through a turbine and is condensed either by direct contact with cold seawater or by a surface condenser. If the second method is employed, freshwater is produced.

In both open and closed cycles, the condensation of the vapor causes a pressure difference across the turbine, which creates a flow of vapor sufficient to power a generator to produce electricity. Although a closed cycle has advantages over an open cycle because it uses a smaller turbine and does not require power-consuming degassing and vacuum pumps, the freshwater produced by an open-cycle system can be a valuable by-product in certain locations.

Development trends and possibilities

Research is currently underway on variations of the two basic operating cycles. One variation on the open cycle is called the "mist-lift" cycle, which entails the flash evaporation of warm surface water at the bottom of a tall vertical tube. The warm vapor rises to the top of the tube where it is trapped and condensed. The condensate is then used to power a water turbine on its way back to sea level. Another variation, which combines aspects of both the open and closed cycles, is called the "latent heat transfer closed cycle OTEC." In this system, warm surface water is flash evaporated and the resulting vapor condensed against ammonia, which then evaporates and drives a turbine:

Eight countries active in OTEC research and development have launched or are planning a number of interesting projects.

In the United Kingdom, scientists have designed a 10 megawatt closed-cycle plant to be installed in the Caribbean or Pacific. Also being designed is a 500-kilowatt closed-cycle onshore plant to be built in Hawaii that will combine freshwater production with mariculture. Because the cold water pumped from below has a higher nutrient content than surface water, an OTEC plant can enhance mariculture productivity.

In the Netherlands, feasibility studies for a 10-megawatt floating plant in the Dutch Antilles and a 100 kilowatt onshore plant for Bali have been completed, but funding for the projects has recently ceased.

A consortium involving interested parties from Sweden and Norway has agreed to develop a 1-megawatt unit in Jamaica.

France has decided to build a 5 megawatt onshore plant in Tahiti, but has recently postponed the commissioning date.

The government of Japan is supporting the design of a 10-megawatt floating plant and is also considering a land-based plant. In addition, it is sponsoring research on a combined OTEC–mariculture scheme involving a 5-kilowatt plant. Tokyo Electric plans to build a 20 megawatt plant on the island of Nauru following the success of its 100 kilowatt plant there.

In Taiwan, plans are being made for a 9-megawatt unit to use the cooling water discharged from a nuclear plant at Hon-Tsai. The plan not only provides a high temperature difference for the OTEC cycle but reduces the environmental impact of releasing the cooling water discharge directly into the ocean.

The United States is proceeding with a conceptual design for a 100-megawatt closed-cycle plant to be built in Puerto Rico.

The economics and potential market

The future of OTEC will be determined by economics. Will the technology be widely adopted throughout the tropics, or will it be suitable only for certain locations where the cost of alternative energy sources is high? Or will it be declared an unfortunate commercial failure altogether? Because only a few small plants have been built, it is difficult to accurately assess what the costs of a large commercial plant might be, especially since such a scheme is unlikely to be implemented for some years. At present, estimates based on existing OTEC designs give capital costs of the order of $10,000 per kilowatt-installed. An OTEC plant could conceivably be designed for unattended operation with operating costs as low as 0.8 cents per kilowatt-hour (kWh). Considering capital and operating costs together, the cost of electricity generation might fall within the range of 12 to 25 cents per kWh. In addition, costs could be further lowered by creating markets for freshwater or by developing a mariculture industry.

The thermal requirements of OTEC plants (that is, a temperature difference of approximately 20°C) restrict application of the technology to the area between the tropics of Cancer and Capricorn. The countries within this geographical band are mostly developing nations and many of them are islands. Therefore any commercial OTEC schemes are likely to be small-scale operations, probably less than 1 megawatt in size. The market for OTEC will be greatest wherever fossil fuels are most expensive; island communities that have to import oil at a premium price are an obvious target. But the vagaries of international oil prices make this market difficult to assess. Large grid-connected plants intended to meet the electricity needs of developed countries are not anticipated for at least a decade.

Environmental effects

The operation of an OTEC plant could have a number of possible environmental effects, most of which are difficult to assess. The large flows of hot and cold water might, conceivably, modify local or even global weather patterns, although the evidence is scarce. In addition, the carbon dioxide contained in deep ocean water could be released into the atmosphere when it is pumped up and heated in the condenser. The release of carbon dioxide is potentially more serious for open-cycle plants because the warm water must be deaerated before it is flash evaporated. However, even in the worst case scenario, the amount of carbon dioxide released would only be about one-fifteenth that released by oil or one twenty-fifth that released by coal in a plant of equivalent electrical output [19].

OTEC might also affect marine life. Fish, eggs, and larvae taken up by the plant could be harmed, and changes in temperature and salinity might adversely alter the local ecosystem. Concern has also been expressed that certain biocides, such as chlorine, required to control marine fouling may harm the ecosystem, although the required concentrations would be low.

One of the more effective working fluids for closed-cycle schemes is the chlorofluorocarbon Freon. Because its release has a damaging effect on the earth's stratospheric ozone layer, Freon would either have to be securely contained or its use avoided. Fortunately, alternatives to Freon are now becoming available.

Remarks

Overall, OTEC seems unlikely to play more than a minor role as a generator of electricity for the foreseeable future. Although proven at a small scale, the technology is hamstrung by poor economics. Its high costs can be attributed to the low efficiency of the heat-power cycle, which requires large-size components to handle the enormous water flows. Although cost reductions may be achieved by improving the design of the heat exchanger and coldwater pipe, these are unlikely to affect the technology's overall economics. The prospects for OTEC are therefore largely limited to locations where the cost of alternatives is high, and where power can be produced in association with another product, such as freshwater or mariculture.

SALINITY GRADIENTS

The resource

A great difference in osmotic pressure (equal to a 240 meter head) exists between freshwater and seawater. In theory, if this pressure could be utilized, every cubic meter of freshwater flowing into the ocean from a river could generate 0.65 kWh of electricity. A stream with a flow of 1 cubic meters (m^3) per second, for example, could produce a power output of 2,340 kilowatts. On the assumption that all the rivers of the world could be harnessed with devices of perfect efficiency, a total resource of 2.6 terawatts would result.

Even more dramatic is the potential of salt deposits. The osmotic pressure between freshwater and brine in the Dead Sea corresponds to a head of 5,000 meters, which is some 20 times greater than that for ocean water. The large deposits of dry salt in salt domes contain energy that is yet more compact.

The technology

A variety of techniques can be envisaged for reclaiming the energy in salinity gradients; in fact in the mid-1970s several investigatory studies on the technology were carried out.

Conceptually, the theoretical head could be developed by allowing freshwater to flow through a semipermeable membrane into a reservoir of saltwater. The pressure would raise the level of the reservoir to produce a head of water, which could then flow through a turbine. In practice, however, the flow of freshwater would dilute the seawater and so, in order to maintain the salinity gradient, more saltwater would have to be introduced into the reservoir. If the process were continuous, the surface level of the reservoir would rise to 240 meters above sea level and a great deal of energy would be needed to pump the seawater against such a head. Unfortunately, the costs of such an osmotic-pressure system are prohibitive: estimates indicate that prospective costs would average 10 to 14 cents per kWh in mid-1970s money (or 1989 22 to 30 cents in 1989 money).

An alternative method is reverse dialysis, which in effect involves creating a salt battery. Again, costs were found to be prohibitive, with capital costs in mid-1970s money reported at $50,000 per kilowatt (or $100,000 in 1989 money).

A third method, which is considered technically feasible, exploits the different vapor pressures of water and brine. The concept is a simple one: if freshwater were to be evaporated and condensed in seawater, the resulting vapor flow could drive a turbine. The process is attractive because the turbine conditions are similar to those in an open-cycle OTEC and the machinery would cost roughly the same. But the technology is severely hampered, both strategically and in terms of operating cost, because it calls for the consumption rather than the production of freshwater.

Development trends and possibilities
Despite the initial burst of interest in salinity gradients in the 1970s, research efforts in recent years have entirely disappeared and none of the techniques described above is still considered promising.

Remarks
The prospects for this technology are considered unpromising and do not warrant its further development at present.

WORKS CITED

Tidal energy
1. Bernshein, L.B. 1965
   Tidal power for electric power plants, Israel Program for Scientific Translations, Jerusalem.

2. Bernshein, L.B. 1986
   Tidal power engineering in the USSR, *Water Power & Dam Construction*, March, 37–41.

3. La Rance 20th anniversary colloquium papers. 1986

4. Tidal Power Corporation. 1982
   Fundy tidal power update, report of the Tidal Power Corporation of Halifax, Nova Scotia, Canada.

5. Department of Energy Mines and Resources. 1987
   Tidal power and Canada—a review, Department of Energy Mines and Resources, Ottawa, Canada, EMR-FPS-86-0003.

6. Fundy tidal power workshop papers. 1990

7. U.K. Institute of Civil Engineers. 1986
Tidal power, *Proceedings of 2nd Symposium on Tidal Power*, organized by the U.K. Institute of Civil Engineers.

8. U.K. Institute of Civil Engineers. 1989
Developments in tidal energy, *Proceedings of 3rd Conference on Tidal Power*, organized by the U.K. Institute of Civil Engineers.

9. Bay of Fundy Tidal Power Review Board. 1977
Reassessment of Fundy tidal power, report of the Bay of Fundy Tidal Power Review Board and Management Committee.

10. Baker, A.C. 1987
Tidal power, *U.K. IEE Proceedings*, 134:392–398.

11. U.K. energy paper. 1981
Tidal power from the Severn Barrage, U.K. energy paper 46.

12. U.K. energy paper. 1989
The Severn Barrage Project: general report, U.K. energy paper 57.

13. Energy Technology Support Unit (ETSU). 1990
The Severn Barrage project—detailed report, 5 volumes, ETSU TID 4060 P1–P5, Energy Technology Support Unit, U.K. Department of Energy, Abingdon, U.K.

14. ETSU. 1988
Tidal power from the Mersey: a feasibility study, stage 1, ETSU TID 4047.

15. ETSU. 1991
Mersey Barrage feasibility study, stage 2, ETSU TID 4071.

16. ETSU. 1990
Conwy estuary feasibility study of tidal power, ETSU TID 4075.

17. ETSU. 1989
The U.K. potential for tidal energy from small estuaries, ETSU TID 4048 P1.

18. *The potential for tidal energy within the European Community* (to be published).

19. World Energy Conference. 1992
Ocean energy, chapter 3 in *World Energy Conference Committee on renewable energy sources, opportunities and constraints 1990–2020*.

Wave energy

20. Leishman, J.M. and Scobie, G. 1976
U.K. National Engineering Laboratory report EAU M25.

21. Liang, X. et al. 1991
3rd Symposium on Ocean Wave Utilisation, Tokyo, organized by the Japan Marine Science and Technology Center (JAMSTEC), papers G-3 and G-4.

22. Nielsen, K. 1991
3rd Symposium on Ocean Wave Utilisation, Tokyo, organized by the Japan Marine Science and Technology Center (JAMSTEC), paper A-7.

23. Raju, V.S., Ravindra, M., and Koola, P.M. 1991
3rd Symposium on Ocean Wave Utilisation, Tokyo, organized by the Japan Marine Science and Technology Center (JAMSTEC), paper G-7.

24. Takahashi, S., Hotta, H., et al. 1991
3rd Symposium on Ocean Wave Utilisation, Tokyo, organized by the Japan Marine Science and Technology Center (JAMSTEC), papers D1–D8.

25. Falcão, A., Sarmento, A., Gato, L., and Pontes, M. 1991
    3rd Symposium on Ocean Wave Utilisation, Tokyo, organized by the Japan Marine Science and Technology Center (JAMSTEC), papers G5 and G6.

26. Berggren, L. and Bergdahl, L. 1991
    3rd Symposium on Ocean Wave Utilisation, Tokyo, organized by the Japan Marine Science and Technology Center (JAMSTEC), paper A-6.

27. ETSU. 1991
    Wave energy review—interim report, ETSU R60.

28. ETSU. 1991
    Islay shoreline gully device—phase 2, ETSU-WV-1680.

29. Hagerman, G. and Heller, T. 1988
    Proceedings of the International Renewable Energy Conference, Hawaii, 98–110.

30. Budal, K. and Falnes, J. 1975
    A resonant point absorber of ocean-wave power, *Nature* 256:478–479.

31. Budal, K. 1977
    Theory of absorption of wave power by a system of interacting bodies, *Journal of Ship Research* 21:248–253.

32. Budal, K. and Falnes, J. 1982
    Wave power conversion by point absorbers: A Norwegian project, *International Journal of Ambient Energy* 3:59–67.

33. Berge, H., ed. 1982
    Proceedings of the 2nd International Symposium on Wave Energy Utilisation, Trondheim, Norway.

34. Fisher, J.C. 1974
    The energy crisis in perspective, Wiley, New York.

35. Claeson, L. et al. 1987
    Energy from ocean waves, Allmanna Forlaget, Stockholm, Sweden.

Ocean Thermal Energy Conversion

36. Lavi, A. et al. 1973
    Plumbing the ocean depths: a new source of power, *IEEE Spectrum* October.

37. Griekspoor, W. and Van der Pot, B. 1979
    OTEC—principles, problems and progress, *Offshore Engineer* September.

38. Lennard, D. 1987
    Ocean thermal energy conversion—a base load system, *IEE energy options; the role of alternatives in the world energy scene*, conference publication 276.

39. Pergamon. 1980
    Ocean thermal energy conversion, Pergamon Press; ISSN 0360–5442.

40. International Conference on Energy Recovery (ICOER). 1989
    Ocean energy recovery, *Proceedings of the First International Conference on Energy Recovery ICOER '89*, American Society of Civil Engineers.

41. Solar Energy Research Institute (SERI). 1989
    Ocean thermal energy conversion: an overview, Solar Energy Research Institute, Golden, Colorado.

42. Green, H.J. and Guenther, P.R. 1989
    Carbon dioxide release from OTEC cycles, *Proceedings of the International Conference on Energy Recovery ICOER '89*, American Society of Civil Engineers.

13
GEOTHERMAL ENERGY

CIVIS G. PALMERINI

Geothermal energy, which has been used to generate electricity since the early 1900s, is a widespread resource found throughout the world. Its potential as an energy source is substantial: estimates show that if existing reserves are exploited, 12 billion tonnes of oil-equivalent (TOE) energy could be generated within the next 10 to 20 years. Although most prevailing geothermal systems are hydrothermal, technological advances now in the demonstration stage will pave the way for geopressured, magma, and hot dry rock systems.

Electricity and heat produced from geothermal sources are already economically competitive with conventional energy supplies, but commercialization of the process is hampered by the need for substantial capital investments. Nonetheless, in recent decades geothermal energy has achieved annual growth rates of 10 percent for direct uses such as heating and from 5 to 15 percent for power generation. The growth is expected to continue but at slightly lower rates.

According to recent estimates, installed geothermoelectric capacity, which was 6,000 MW_e (megawatts electric) at the end of 1989, will reach 15,000 MW_e by 2000, with a corresponding increase in production from 35 to 90 billion kilowatt-hours (kWh). Similarly, installed capacity for direct applications, which was some 11,500 MW_t (megawatts thermal) at the end of 1989, should rise to about 23,000 MW_t by 2000. Thus the energy equivalent for all geothermal uses should increase from 11 million TOE in 1989 to 26 million TOE by 2000. Some of the growth can be attributed to geothermal's high level of social acceptability, which in some cases renders it more advantageous than conventional sources of energy.

BACKGROUND

Geothermal resources: their nature, distribution, and global potential

In geological terms, geothermal energy is defined as the heat above the mean ambient temperature of the earth's solid core, which is about 8×10^{30} joules. The amount is enormous and represents 35 billion times the world's present total annual energy consumption. In reality, however, only a tiny fraction of natural heat can be extracted from the earth's crust, mainly for economic reasons, which limits exploitation to a maximum depth of 5 kilometers. To this depth, the temperature of the crust increases at an average rate of 30 to 35°C per kilometer (the so-called mean geothermal gradient).

The vertical thermal gradient of the earth's crust varies greatly (even over short distances), ranging anywhere from 50 percent of the mean value to twice or 10 times as much. Temperature distributions at 5 kilometers reflect that variation; in one zone they may be only 70 or 80°C; in another they may exceed 500°C. Because of the thermal gradient, natural heat rises by conduction (and by convection in some places), dissipating into the atmosphere when it reaches the earth's surface. Convection generally occurs when hot fluids (water, steam, and gas) flow to the surface, but will also occur when magma erupts from active volcanoes.

The conductive component of heat flow also varies from one point to another. In some places it is less than 30 milliwatts per square meter (m^2), while in others it exceeds 500 milliwatts per m^2. It averages 60 to 65 milliwatts per m^2, which is a few thousand times less than the mean solar radiation striking the earth's surface. Because of the sun's overwhelming dominance, human beings are not aware of the dissipation of terrestrial heat by conduction.

The presence of geothermal energy is apparent only in places where its heat dissipates into the atmosphere through a carrier fluid, such as the warm waters that flow from geysers and spas, or through the release of hot gases or volcanic eruption. The continuous dissipation of natural heat suggests the planet is slowly cooling, but whether it actually is or not is unclear. In fact, heat generated by the radioactive decay of unstable elements such as uranium, thorium, and potassium, which are abundant in the earth's crust, counters the heat lost to dissipation.

Terrestrial heat thus has two components: the first is linked to the planet's formation and consolidation some five billion years ago; the second is controlled by the isotopic decay of unstable elements and is thought to contribute some 40 percent to the conductive heat flow.

Geothermal energy, although nonuniformly distributed at shallow depths, is globally widespread. Nonetheless, the amount of heat that could theoretically be tapped within a depth of 5 kilometers (called the accessible resource base) [1] is on the order of 140×10^{24} joules (see table 1) [2]. In other words, the amount of accessible geothermal energy is about five orders of magnitude less than its global potential. Moreover, much of the in situ heat is too dispersed or too low in temperature to warrant exploitation. Thus, only a tiny fraction of the accessible resource base (probably four to five orders of magnitude less than total given above) is likely to constitute an economical source of energy within the next four to five decades. Smaller, by approximately another order of magnitude, are the reserves, which represent the portion of geothermal resources that might be exploited within the next one to two decades.

There are six regions where total heat flow and the concentration of geothermal energy are highest (see figure 1). They are the Circum-Pacific or Fire Belt, the Mid-Atlantic Ridge, the Alpine–Himalayan mountain chain, much of eastern Africa and the western Arabian peninsula, central Asia, and a few archipelagos in the central and south Pacific, including Hawaii, Western Samoa, and Fiji. All these

Table 1: Global geothermal potential

	Joules
Accessible resource base	$140 \cdot 10^{24}$
Resources	$5 \cdot 10^{21}$
Reserves	$500 \cdot 10^{18}$[a]

High-temperature resources and reserves (those at temperatures greater than 120 to 140°C) are mostly concentrated in zones with the highest thermal gradients and probably amount to no more than one third of the respective total values.

a. The value corresponds to the energy equivalent of about 12 billion TOE.

zones coincide with discontinuities in the earth's crust created by the extension, separation, shrinkage, or collision of independent crustal plates. The zones are therefore very active tectonically, experiencing frequent and often severe seismicities. In addition, they are characterized by numerous local lithologic and hydrogeological factors that contribute to their high conductive heat flow values. But simply because a site is located within one of these zones does not necessarily render it suitable for geothermal exploitation. Indeed, the converse may be true:

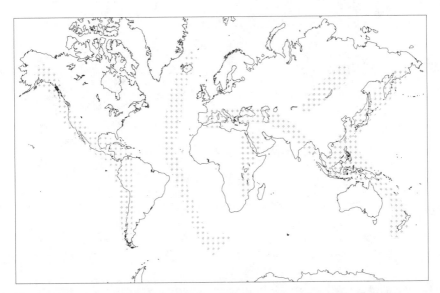

FIGURE 1: Sketch map showing the world's main geothermal zones (see text for explanation).

many areas that are being considered for geothermal development (especially for low-temperature projects) are in zones other than those listed above.

Classification and conceptual models

Geothermal systems are classified according to their thermal and hydrogeologic characteristics (see table 2). Numerous geologic and physical factors, including the structural setting of the site, its predominant rock type, porosity, and temperature distribution of the rock, combine to make each site and system unique (see figure 2).

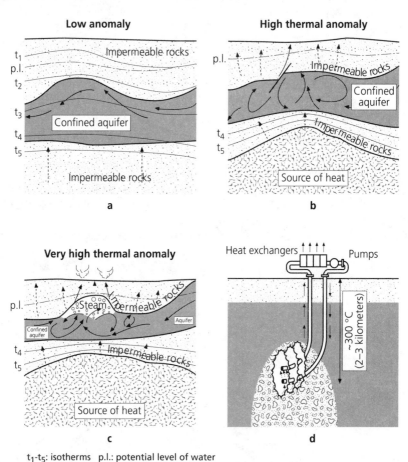

FIGURE 2: *Conceptual models of the three most common types of hydrothermal systems (a, b, and c) and of the hot dry rock (HDR) exploitation concept (d) (after [3]). a:Low temperature water-dominated system, with confined aquifer. b:High temperature water-dominated system, with confined aquifer. c:Steam-dominated system. d:Heat extraction from impermeable rocks by a closed-water loop.*

Table 2: Geothermal systems: classification and state of technology (after [3] with modifications and supplements)

System	Main characteristics	Synonym	State of utilization technology
HYDROTHERMAL	**Permeable formations with natural fluid circulation**		
Low to moderate temperature with unconfined aquifer	Temperature within 3 kilometers usually below 150°C Near-horizontal regional water table with local percolation of cold water	Low-enthalpy systems	Mature, but heat pumps are always needed
with confined aquifer	Pressurized water circulation with small concentration of energy in positive features (figure 2a)		Mature, but heat pumps are needed in some cases
High temperature	Temperature within 2 kilometers usually above 150°C	Water-dominated convective systems	
with unconfined aquifer	Open convective circuits with moderate concentration of energy above the heat source		Mature
with confined aquifer	Closed convective circuits with large concentration of energy in positive features (figure 2b)		Mature, with consolidation under way
Steam systems	Temperature within 1.5 kilometers usually above 200°C	Steam-dominated systems or high-enthalpy systems	
with confined aquifer	Closed convective circuits. Steam caps in trapping features bring about large accumulation of energy (figure 2c)		Fully mature and consolidated
DRY ROCK	**Impermeable rock formations with no natural circulation**	Hot dry rock or "artificial" or "man-made" systems	
Low to moderate temperature	Temperature within 3 kilometers usually below 150°C		Preliminary tests already started in a few cases
High temperature	Temperature within 3 kilometers usually above 250°C. Artificial water circulation between two wells is necessary for heat extraction (figure 2d)		Largely tested but further experimentation is needed
Other			
Geopressurized systems	Confined reservoirs with moderate temperatures (150 to 200°C) within 6 to 8 kilometers) and high pressure controlled by the weight of overlying sediments		Tests already carried out but further experimentation is needed
Brine systems	Temperature within 3 kilometers usually above 300°C. Fluid reservoir is a very concentrated saline solution (generally exceeding 100,000 ppm)		Tests already carried out but further experimentation is needed
Magma systems	Temperature within a few kilometers depth (usually above 500°C) controlled by magmatic intrusions		General studies carried out in a few cases and preliminary tests are about to start

The systems that are most attractive economically are those in which the above factors combine in a favorable way. An ideal site, for example, might be defined by the presence of an extensive heat source, a strong thermal anomaly (which creates high temperatures at shallow depths), an aquifer confined between two complexes of impermeable rocks (which prevents cooling of the geothermal fluid by fresh water) forming the so-called reservoir, an adequate water supply, high rock permeability (primary and/or secondary) to ensure that the production rate from the wells is adequate, and the presence of local convection circuits to keep temperatures elevated during recharge (when there is an inflow of cold water) and reinjection (see figures 2b and 2c).

HISTORY OF GEOTHERMAL EXPLOITATION

Terrestrial heat was first used in prehistoric epochs for bathing, or balneology. In ancient Rome, thermal waters were enormously popular (as they are today) and led to the proliferation of bath houses. The waters, which are rich in mineral salts, were widely viewed as therapeutic. In addition, the Romans found that the hot spring water was appropriate not only for bathing but could be used to heat their bath houses.

Thermal mineral salts have also been exploited since ancient times. The Etruscans almost certainly used boric acid, which they found in the Larderello region of Tuscany, to prepare the enamels with which they decorated their vases. Indeed, identification of boric acid in the geothermal fluids of this area by the German scientist Hubert Hoefer in 1777, and again by the Italian Paolo Mascagni in 1779, paved the way for the birth of the modern geothermal industry. The commercialization of boric acid can be traced back to 1818, when Francesco de Larderel, formerly of France, installed the first factory in Monte Cerboli, Italy, near present-day Larderello [4].

Beginning in 1827, boric-acid factories began exploiting geothermal steam to replace firewood as a fuel for concentrating the boric acid. In subsequent years, geothermal fluids (a mixture of water and steam at low temperature and pressure) were thus used not just to recover the boric salts contained in them, but also as a heat source for concentrating and drying boric brines and heating industrial establishments and housing facilities. Although some of these uses were abandoned in the mid 20th century, others such as space heating continue to the present day.

In the early 1900s, geothermal fluids provided energy for mechanical power and electricity. The first geothermal power plant, with a 250 kilowatt turbine generating unit, began operation in 1913. To avoid corrosion caused by mineral salts in the geothermal steam, and yet maintain a vacuum in the condenser,[1] an indirect cycle was used in which clean steam (at 0.15 megapascals) was generated in special boilers with the heat of the geothermal fluid [5, 6]. The scheme made it

---

1. A condensation cycle was necessary because steam pressure at that time was only 0.2 megapascals.

possible to dispose of noncondensable gases (mainly carbon dioxide) accompanying the steam and was retained almost unaltered for many years.

Installed geothermoelectric capacity in Italy increased rapidly, reaching 127 megawatts by 1944. The improved thermodynamic characteristics of steam coming from deeper wells boosted performance, permitting its direct use in exhausting-to-atmosphere turbines[2] and eventually in condensing power plants.[3] Direct steam cycles represented an alternative to indirect cycles, which were widely used at the time because they facilitated the integrated use (for both chemical and electric generation) of the geothermal fluid, from which boric acid, ammonia, and other products were recovered. Thus, the indirect cycle was abandoned only when power production became more important economically [7].

Impressed by Italy's success, other countries began to generate power from geothermal fluids. In 1923, the first wells were drilled at the Geysers, California, in the United States, followed by the installation of a 250 kilowatt generator. In 1925, Japan constructed a 1 kilowatt geothermal generator on the island of Kyushu [8]. In 1938, the first 300 kilowatt binary-cycle plant (see figure 4i) was installed near Naples. The questionable commercial success of these early generators, however, hindered further development.

Interest in geothermal energy was reborn in the 1950s. In Italy the plants at Larderello, which were destroyed during World War II, were reconstructed and enlarged. Japan installed a 30 kilowatt experimental unit in 1951, which was followed in 1966 by a 22 megawatt commercial plant. In 1952, a 275 kilowatt unit was built in Katanga, Zaire, and in 1959 a 3.5 megawatt unit was built in Mexico. At the Geysers in the United States an 11 megawatt system began operating in 1960. Between 1958 and 1963, the first large-scale plant fed by geothermal fluid from a water-dominated reservoir[4] was realized in New Zealand [8,9].

Throughout this period, geothermal energy was used for its heating properties. In 1930, for example, large-scale geothermal space-heating systems were built in Iceland, and later spread to France, Italy, New Zealand, the United States, and other countries. Since then geothermal heat has been exploited in a number of ways, most notably for horticulture, aquaculture, animal husbandry, soil heating, and industrial usage. (The first large-scale industrial application was the operation of paper mills in New Zealand.)

---

2. Turbines that discharge their outlet steam at atmospheric pressure.

3. The turbines in these power plants discharge steam at below atmospheric pressure (the condensers are kept under vacuum).

4. These reservoirs require an elaborate fluid-gathering system because the produced fluid is a mixture of steam and water.

USING GEOTHERMAL RESOURCES

Applications

The earth's internal heat supply can be used directly or indirectly by transforming its thermal energy into electricity, a more valuable and flexible form. Obviously, the higher the temperature of the heat-bearing fluid, the wider the range of practical applications. Applications that depend on high-temperature (above 100°C) fluids are usually referred to as high-enthalpy or high-temperature applications, whereas those for which moderate or low temperatures (below 100°C) are sufficient are known as low-enthalpy or low-temperature applications (see table 3).

High-temperature direct applications include water distillation and industrial-scale evaporation, which require thermal heat in the steam phase (saturated steam under pressure). Low-temperature direct applications encompass greenhouse heating and biodegradation, which require thermal heat only in the form of hot water. Above 140 to 150°C, and especially above 170 to 180°C, geothermal fluid lends itself to indirect applications, such as electric generation or industrial direct uses, such as the processing of wood pulp into paper. Residual heat discharged, say, downstream from a geothermal electric plant, can also be exploited for direct uses such as heating. In such cases, geothermal heat is said to have multiple or combined uses.

Exploration and drilling technologies

Discovering a geothermal field and exploiting the thermal fluid contained therein are similar to other aspects of the mining industry and as such always bring with them a certain degree of financial risk. To reduce the risk to an acceptable level, general surveys must precede more detailed prospecting. The first step entails fairly inexpensive regional geological and geochemical studies; in this way less interesting zones are quickly eliminated and only the most promising sites are subjected to detailed geological, geochemical, and geophysical analyses, the second step in exploration. The third step calls for a number of exploratory drillings. Such drillings involve substantial financial costs, but are necessary in order to reduce the risk of further drilling to economically acceptable levels.

Discovery and exploitation of a geothermal field are usually grouped into two main phases: the preinvestment phase and the initial investment phase. The first phase is subdivided into two stages, known as reconnaissance and prefeasibility studies, and includes all the surveys, prospecting, analyses, and surface studies that are needed to construct a geothermal model of a prospective site. The area's geothermal potential in terms of probable reserves and possible resources is also measured during this stage. Costs during the preinvestment phase are fairly modest (from $1 to $3 million), but provide a gauge of the benefit/cost ratio for the project. That ratio, in turn, determines whether or not the project should proceed into the more costly initial investment phase.

The initial investment phase, more commonly known as the feasibility phase, includes the drilling of wells (from a minimum of two or three to a maximum of

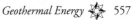

Table 3: Required temperature (°C) of geothermal fluids for different applications (modified after [10])

180	Evaporation of highly concentrated solutions Refrigeration by ammonia absorption Digestion of wood into paper pulp
170	Heavy water production via hydrogen sulfide process Drying of diatomaceous-enriched earth, which is used in filter systems
160	Drying of fish meal Drying of timber
150	Alumina production via Bayer process
140	Canning of food
130	Evaporation of water in sugar refining Extraction of salts by evaporation and crystallization
120	Production of freshwater by distillation Multiple effect evaporation and concentration of saline solutions
110	Drying and curing of light aggregate cement slabs
100	Drying of organic materials, seaweeds, grass, vegetables, etc. Washing and drying of wool
90	Drying of fish Intense de-icing operations
80	Space and water heating
70	Refrigeration (lower temperature limit)
60	Space heating of greenhouses
50	Mushroom growing Balneological baths
40	Soil warming
30	Swimming pools, biodegradation, fermentation Warm water for year-round mining in cold climates De-icing
20	Fish hatching and farming

six or seven) and related activities. The goal is to verify the existence and determine the physical characteristics of the geothermal reservoir; illuminate the physical and chemical characteristics of the geothermal fluid; permit experimentation under different production conditions; estimate probable production for the next 15 to 20 years; assess the sure reserves and estimate the probable resources; and begin environmental impact studies on the surrounding area. The cost of this second phase is about one order of magnitude

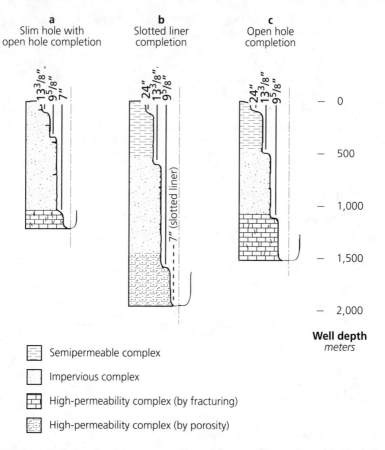

a
Slim hole with
open hole completion

b
Slotted liner
completion

c
Open hole
completion

13³/₈" 9⁵/₈" 7"

24" 13³/₈" 9⁵/₈"

7" (slotted liner)

24" 13³/₈" 9⁵/₈"

— 0

— 500

— 1,000

— 1,500

— 2,000

Well depth
meters

Semipermeable complex

Impervious complex

High-permeability complex (by fracturing)

High-permeability complex (by porosity)

FIGURE 3: Casing layouts for most common well completions are shown. After each hole is drilled, a cement casing is added.

a: Slim hole with open hole completion
   1. A 44 cm hole is drilled (down to a few tens of meters) with a 34 cm surface casing;
   2. A 31 cm hole is drilled (down to about 300 meters); casing is 24 cm;
   3. A 22 cm hole is drilled (down to about 1,100 meters); casing is 18 cm;
   4. A 15 cm open hole (i.e., without casing) is eventually drilled to the desired depth (about 1,300 meters).

b: Slotted liner completion
   1. A 76 cm hole is drilled (down to a few tens of meters); casing is 61 cm;
   2. A 44 cm hole is drilled (down to about 600 meters); casing is 34 cm;
   3. A 31 cm hole is drilled (down to about 1,100 meters); casing is 24 cm;
   4. A 22 cm hole is drilled to the desired depth (about 2,000 meters) and an 18 cm slotted liner (i.e., a casing with vertical slots to allow fluid passage) is hung at the bottom of the 24 cm casing.

c: Open hole completion
   1. A 76 cm hole is drilled (down to a few tens of meters); casing is 61 cm;
   2. A 44 cm hole is drilled (down to about 500 meters); casing is 34 cm;
   3. A 31 cm hole is drilled (down to about 1,100 meters) casing is 24 cm;
   4. A 22 cm hole is drilled to the desired depth (about 1,500 meters) and is left without casing.

Note: Different hole or casing diameters can be used; the depths to which each casing is sunk depend on the nature of the geological strata penetrated by the well.

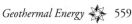

higher than the cost of the first phase, but the information provided by the feasibility phase is indispensable. Moreover, much of the initial investment is recovered once exploratory drilling begins because some (or most) of the wells can then be exploited for production or reinjection purposes.

The depth of exploratory drilling varies, depending on the depth of the reservoir and its intended use. Most low- or medium-temperature direct uses, for example, require shallow drillings (rarely beyond 2,000 meters), whereas projects aimed at geothermoelectric production almost always entail drilling to depths greater than 2,000 meters. In both cases, maximum depth is not generally determined by drilling technology but by economic constraints. (In rare instances, social and strategic reasons may override economic ones.)

Costs associated with the initial investment phase can be reduced by 30 to 40 percent by resorting to so-called slim holes: small-diameter exploratory drillings (see figure 3a). However, the flow rate of these wells is severely reduced compared with that of standard-diameter drillings. Therefore, if slim holes are to be used during field exploitation, exploratory drilling should be carried out with the same technical profile as production wells, with or without a slotted liner,[5] depending on the geologic nature of the reservoir (see figures 3b and 3c). Thus, the diameter of the production drill should measure from 9 to 13 inches (23 to 33 centimeters), which is large enough to provide total fluid flow rates ranging from a few tens to several hundred tonnes per hour.

By applying the same construction criteria to both exploratory and production drilling, two objectives (both equally important economically) can be achieved. First, knowledge of the reservoir and power generation can be obtained very quickly by means of small generating units placed at the wellhead (see figure 4a), resulting in the shortest possible payback period. Second, productive exploratory wells can produce much of the fluid needed to feed the first generating unit of an industrial capacity power plant, which shortens the investment period and ultimately reduces the capital cost of a geothermoelectric project.

Fluid utilization

Power generation from steam-dominated reservoirs

Fluid produced by steam-dominated geothermal reservoirs (those that lack a water phase) consists of superheated or dry-saturated steam with a highly variable content of noncondensable gases (from less than 1 percent to more than 10 percent). The fluid is available at pressures of 2 megapascals or more at wellhead, and can be piped directly to a power plant for use in steam turbines. Water-dominated

---

5. The deepest part of the well may be left unsupported (open-hole completion) (see figure 3c), or a slotted liner extending to the bottom and standing about 10 to 20 millimeters clear of the bore walls may be provided (see figure 3b). The slotted liner prevents collapse of the bore and damage by entrainment of large pieces of rock.

reservoirs, which always have a water phase present, are more common on a worldwide basis, but are less preferable than steam-dominated reservoirs because their exploitation requires complex equipment, including separators, flash-vessels, etc.

The noncondensable gas content of the steam determines whether the exhausting-to-atmosphere cycle (see figure 4a) or the condensing cycle (see figure 4b) is more practical. Although the condensing cycle has higher thermodynamic efficiency (until the gas content reaches 20 to 25 percent by weight) [11, 12], its components—the turbine, condenser (direct contact or surface type), cooling system, and noncondensable extraction system—are more expensive. The extraction system must be designed to optimize the vacuum in the condenser as well as to match the flow rate of the gas being extracted. For increasing amounts of gas, with a gradual increase in both cost and efficiency, the choice ranges from multistage ejector systems (with intercondensers) to mixed systems composed of ejectors followed by liquid-ring vacuum pumps or compressors, to multistage centrifugal compressors with intermediate cooling (powered by engines or mechanically connected to the turbine).[6]

The steam can also be channeled through an upstream reboiler, which is a heat exchanger/surface evaporator in which the steam condenses, and after its separation from the noncondensables is used to produce steam almost devoid of gas (see figure 4c). The scheme is practically identical to the one previously used at Larderello, but has mostly been applied in field tests to separate noncondensable gases as a means of treating them for hydrogen sulfide (H_2S) abatement [13, 14].

Variations on this scheme, which were developed specifically for steam with a high gas content, call for two surface reboilers in succession (see figure 4d) or for a single reboiler (a pressurized condenser) of the direct contact type (see figure 4e) [15]. Steam is produced at two pressures, in the first case by direct evaporation in the surface reboilers; in the second by means of two flash stages for water obtained in the direct contact reboiler. From a purely thermodynamic standpoint, the cycles with reboilers are less efficient than the condensing cycle (with a mechanical noncondensable extractor) when the gas content of the steam does not exceed 25 percent by weight [15].

Unlike the exhausting-to-atmosphere cycle, which emits all geothermal steam into the environment, the condensing cycle, in which condensed steam is

---

6. All gas extraction systems maintain a vacuum in the condenser by removing gas from the condenser and then compressing it to atmospheric pressure before venting. A multistage ejector system has two or three steam ejectors fed by motive steam (the same steam that feeds the power turbine). The first ejector, which is connected to the main steam (turbine) condenser, removes the noncondensable gas, which it compresses (by the action of the motive steam) to a pressure intermediate between the condenser and atmosphere.

An intercondenser then condenses the motive steam, now mixed with the noncondensable gas, and cools the gas before it is removed by the second ejector. At the outlet of the last ejector stage (which is the second or third, depending on the condenser), steam is condensed in an aftercondenser and the cooled noncondensable gas, which is now at atmospheric pressure, is vented.

In a mixed system, suction from the condenser and the first stage of compression are both carried out by an ejector, and are followed by a second stage of compression, which is performed by a liquid-ring vacuum pump powered by an electric motor. For large gas-flow rates, compression (subdivided in several stages) is performed by centrifugal compressors with gas coolers between adjacent stages.

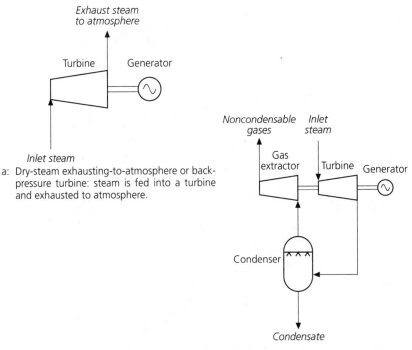

a: Dry-steam exhausting-to-atmosphere or back-pressure turbine: steam is fed into a turbine and exhausted to atmosphere.

b: Dry-steam condensing turbine with compressor for noncondensable gas removal: steam is fed to a turbine exhausting in a condenser under vacuum; noncondensable gas is extracted from the condenser.

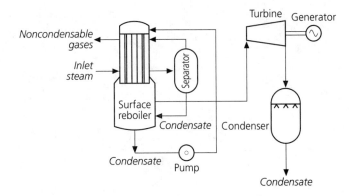

c: Dry steam turbine with upstream reboiler: steam almost free of non-condensable gas is generated in the reboiler and fed to the condensing turbine. Geothermal steam provides heat and water (condensate); noncondensable gases are vented to the atmosphere.

FIGURE 4: Conceptual schemes of the main types of geothermal-electric cycles. Cooling systems are not shown.

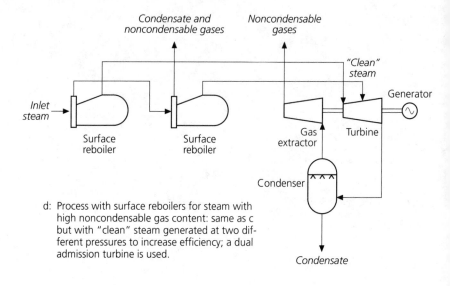

d: Process with surface reboilers for steam with high noncondensable gas content: same as c but with "clean" steam generated at two different pressures to increase efficiency; a dual admission turbine is used.

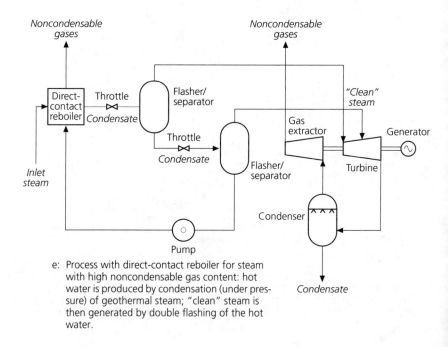

e: Process with direct-contact reboiler for steam with high noncondensable gas content: hot water is produced by condensation (under pressure) of geothermal steam; "clean" steam is then generated by double flashing of the hot water.

FIGURE 4 (continued)

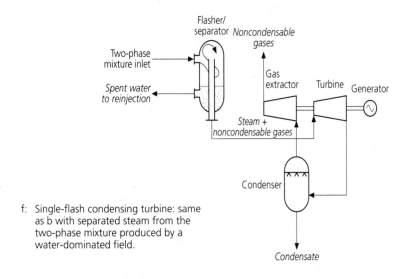

f: Single-flash condensing turbine: same as b with separated steam from the two-phase mixture produced by a water-dominated field.

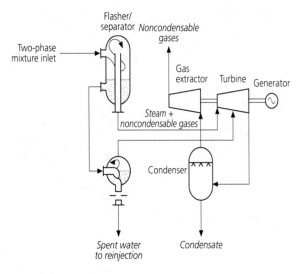

g: Double-flash condensing turbine: a flasher is added to cycle f to recover low-pressure flashed steam from the separated hot water before reinjection.

FIGURE 4 (continued)

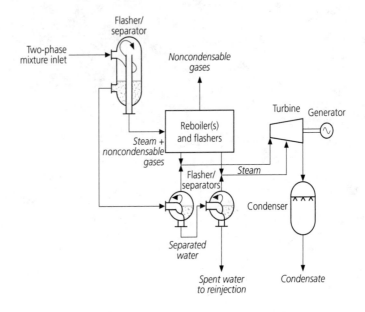

h: Double-flash process with preflash and reboiler(s) for water with high noncondensable gas content: one more flasher is added to cycle g; high-pressure separated steam (containing most of the noncondensable gas) is not fed directly to the turbine but used to generate "clean" steam in the same way as in cycles d and e.

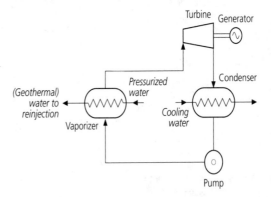

i: Binary power plant (organic Rankine cycle): geothermal hot water is used to vaporize the secondary working fluid that is fed to the turbine and then condensed before being sent to the vaporizer again.

FIGURE 4 (continued)

used in the cooling circuit, produces condensate amounting to some 20 to 30 percent of the steam flow rate. This excess condensate can be reinjected into the reservoir, thus partially recharging it, but the process must be done carefully to avoid the cooling of adjacent production wells, which would reduce their production and eventually kill them.

In practice, the exhausting-to-atmosphere cycle is limited (partly for environmental reasons) to small, less than 5-megawatt turbines, which are often installed at the wellhead to combine field tests with power production. Exceptions are made, however, when the steam from a back-pressure turbine can be exploited for heating.

In all cases, when the wells' back-pressure curves (the rate at which steam flow decreases with increasing pressure at wellhead) and the plant's steam rates (which decrease as turbine inlet pressure increases) are known, optimum production pressure must be chosen in order to maximize power production.

Steam rates for a geothermal plant vary from 10 to 20 kilograms per kWh for an exhausting-to-atmosphere cycle and from 6 to 9 kilograms per kWh for a condensing cycle [16]. Turbine sizes range from about 1 to 5 megawatts (for wellhead units) to 50 to 60 megawatts. Larger plants (those from 110 to 120 megawatts) operate two turbines in tandem.

The main impediment to plant operation is corrosion caused by chemicals present in the steam, particularly chloride and hydrogen sulfide (H_2S). Although corrosion affects mostly the turbines, it can also damage wells and steam pipelines. Strategies for combating its effects include injecting neutralizing reagents into the steam and selecting corrosion-resistant materials for the system; neither one is always fully successful.

Power generation from water-dominated reservoirs
The wells of water-dominated fields produce a mixture of water, steam, and non-condensable gases. The liquid phase can be obtained at wellhead only by using downhole pumps, which are centrifugal pumps set in the well below the flash level, where only the liquid phase is present. By maintaining the liquid phase under high pressure, the pumps prevent it from flashing—vaporizing and emitting gas and steam—in the well. In lineshaft pumps (the type widely used for high-temperature fluids), the motor is located outside the well, and the pump is operated by means of a long shaft. Other models of downhole pumps have an electric motor, which is connected to the surface via a cable from within the well.

Depending on reservoir temperature, either the flashed-steam/steam-turbine cycles or the binary cycle are used (though sometimes the two are combined). Steam for flashed-steam/steam-turbine cycles is obtained when a fraction of the liquid phase flashes. The process, which generally commences in the well, is completed at the surface, where the separation of flashed steam at one or more desired pressures can take place. The steam thus obtained then powers steam turbines in the same way that steam coming from steam-dominated systems does. Total-expansion cycles, which operate using special machines that can be fed either by a two-phase mixture or a pressurized liquid phase, are still in the experimental stage but are no longer the focus of much interest. The processes of fluid expansion, brought about by flashing, and energy conversion (when mechanical energy is produced from a decrease in the fluid's enthalpy) occur in the total expander.[7] From a thermodynamic standpoint, the process is equivalent to a flashed-steam cycle but is considered superior because it has an infinite number of stages. But existing expanders suffer from lower conversion efficiencies than do steam turbines and are of limited reliability and size (only one commercial device is available: the rotary separator turbine). Consequently, interest in them, which peaked during the 1980s, is now decreasing.

The steam in flash cycles is produced either at one pressure level or at several gradually decreasing levels by means of one or more subsequent expansion stages and fluid separation steps (the first expansion takes place inside the wells if they are not equipped with pumps). Separation is generally accomplished near the well site by centrifugal separators in various configurations (usually vertical but sometimes horizontal) or by other types of separators (gravity, impact, etc.). If separation occurs at the power plant, the geothermal fluid is piped there in two-phase flow (the form in which it emerges from the wellhead).

The number of stages and their relative operating temperatures and pressures provide opportunities for economic optimization. One criterion for optimization[8] calls for dividing the difference between the reservoir temperature and the condensing temperature in the power plant into equal intervals determined by the number of flash stages plus one [17]. Technical and economic considerations, however, limit the pressure of the last stage, which must not come too close to, or fall below, atmospheric pressure. Steam from each separation or flash stage is then used to feed steam turbines (which sometimes have multiple inlets)[9] in a manner similar to steam-dominated fields.

A cycle that has only one stage is called a single-flash cycle (see figure 4f) and those that have two stages are known as double-flash cycles (see figure 4g). Only rarely is a triple-flash cycle useful because the thermodynamic advantages of an additional flash stage gradually decline as the overall number of stages increases, while costs rise. Moreover, flash stages that operate at subatmospheric pressure are usually not advisable because the large specific volume of low-pressure steam requires large steam separators, pipelines, and so on. Moreover, air can contaminate subatmospheric steam, which creates additional load for the noncondensable gas extraction system.

---

7. An expander is a discharging-to-atmosphere turbine that may be used for steam with high noncondensable gas content in steam-dominated fields to obtain additional power production (see figures 4c, 4d, and 4e).

8. This method is valid as a first approximation for choosing the temperature of the flash stages once their number has been established.

9. A separate turbine may be used for each flash stage; or a multiple-inlet machine (that is, a dual-admission machine) may be used for a double-flash cycle.

In flash cycles, a "preflash" stage can be exploited[10] to obtain a moderate amount of high-pressure steam containing most of the undesirable gases, which can then be exhausted into the atmosphere, possibly by means of an expander. The preflash solution conveniently reduces the amount of noncondensable gas that must be extracted from fluids with high gas contents. In addition, overall efficiency can be increased by using the preflash steam in reboilers. In this case, the overall cycle with reboilers (see figure 4h) is more advantageous thermodynamically than in steam-dominated fields. For medium-temperature reservoirs, efficiency of the cycle becomes greater than a condensing flash cycle (without the preflash stage), when the gas contents exceed about 2 percent by weight of the total fluid [18].

Residual water from the last flash stage, which may represent close to 80 percent of the amount extracted, is usually reinjected into the reservoir. The extraction flow rate (for the same electric capacity) is much higher than that needed from a steam-dominated field owing to different specific enthalpies of the fluids and the failure to use the residual enthalpy of the water reinjected downstream from the last flash stage. The water can, however, be further exploited to feed binary cycles and/or thermal users in cascade.

Fluid requirements of single- and double-flash condensing plants, referenced to the total flow rate of the wells, are from 40 to 80 and from 30 to 60 kilograms of geothermal fluid per kWh, respectively, for reservoir temperatures around 200 to 250°C. Consumption increases significantly at lower temperature values (150°C), exceeding 150 and 120 kilograms per kWh for single- and double-flash cycles respectively (see table 4) [16]. Because great quantities of fluid are involved, the unit capacities of flash plants do not exceed 30 to 50 megawatts (compared with 110 to 120 megawatt power plants in steam-dominated fields).

Cycles that call for total expansion of the geothermal fluid offer considerable thermodynamic advantages since their potential efficiency theoretically equals that of a flash plant with an infinite number of stages [17]. The expanders needed to put this scheme into practice, however, have not reached commercial maturity, although a few prototypes, such as the 1-megawatt helical screw expander [19], have been developed. In addition, the rotary separator turbine, which enables electricity to be produced upstream from the steam turbine (in a configuration called "topping"), has been installed in one commercial flash plant [20].

In binary-cycle plants, the geothermal fluid, which is usually kept under pressure by downhole pumps, flows through heat exchangers, heating and vaporizing a secondary fluid, typically an aliphatic hydrocarbon or a fluorocarbon refrigerant, operating in a closed cycle of the Rankine type. When the refrigerant reaches the turbine, it expands, producing mechanical power.[11] The cooled geothermal fluid is then reinjected (see figure 4i) into the reservoir.

---

10. This should be done at the maximum pressure that is compatible with the characteristics of the wells.

11. The secondary fluid is then condensed, being cooled by air or water, and returned to the heat exchanger by a feed pump, thus closing the cycle.

Table 4: Comparison of cycles for electricity production from water-dominated plants (indicative data are relative to fluids with low noncondensable gas contents)

Cycle	Heat rate kg per kWh Reservoir temperature °C				Relative cost of investment[a] per unit mass flow rate extracted from wells (reservoir temperature ≈ 200°C)
	120	*150*	*200*	*250*	
Single flash					
exhausting to atmosphere	–	650	150	80	100
condensing. *figure 4f*	–	150	80	50	200
Double flash					
exhausting to atmosphere high-pressure turbine	–	140	75	40	175
condensing. *figure 4g*	–	130	60	30	260
Triple flash					
exhausting to atmosphere high-pressure turbine	–	–	65	35	225
with reboilers. *figure 4h*	–	–	60	30	250
condensing	–	–	55	25	300
Binary cycle *figure 4i*	400	140	70	–	300

a. Power plant and surface installations (gathering system, etc.).

The binary cycle is well suited for medium- and low-temperature fluids (those around 100°C), whose conversion efficiencies with flash cycles would be modest. Not only is the binary cycle more economic than the flash cycle at reservoir temperatures below 140 to 180°C, but it can use the residual water of flash cycles in a cascade configuration called "bottoming", and can also run on steam from flash processes or exhausting-to-atmosphere turbines.

The heat rates of binary-cycle plants depend on the range of hot and cold sources available to them. For example, when the geothermal fluid temperature ranges from 120 to 180°C, heat rates vary from 80 to 400 kilograms of geothermal fluid per kWh [20]. Most of today's power plants are composed of several modular units (as many as 20), each with a 1 megawatt capacity [21].

Water-dominated field exploitation cycles vary in their consumption of geothermal fluid per kWh produced and in their investment costs (see table 4). The problems posed by water-dominated fields are generally linked to the build-up of mineral deposits, or scale, in both the wells and fluid-handling equipment. The buildup reflects the composition of the geothermal fluid, which in turn is affected by reservoir temperature.

Whereas medium- and low-temperature fluids produce carbonate scale in the wells as well as in the upstream separating equipment, high-temperature fluids often give rise to sulfide, silica, and silicate deposits that accumulate mainly in the upstream separating equipment and in the downstream reinjection system. Methods of eliminating and controlling scale vary from site to site. In some places, typically at binary-cycle and some flash plants, scale inhibitors are added to the geothermal fluid; in others (for example, at Salton Sea in California) flash crystallizers containing suspended particles, which serve as nuclei for crystallization and deposition of scale, are preferred [22].

Direct uses of geothermal heat and its byproducts

Geothermal fluids can be used as a heat source in a wide variety of applications (see table 3). The applications, which are temperature dependent, range from balneology to aquaculture, from district heating and air conditioning to greenhouse farming and low-temperature industrial processes.

Direct use of geothermal heat is always advantageous from the standpoint of conversion efficiency, even in the case of fluids suitable for electric power generation, where conversion efficiencies are modest at best [23]. Indeed, efficiency rates, which represent the ratio between useful energy output and the thermal energy contained in the geofluid when it reaches ambient temperature and pressure, range from single-digit figures to about 20 percent for power production, and may exceed 90 percent for direct applications such as heating. Fossil fuels, in contrast, yield typical efficiency rates of 30 percent for power generation and 70 to 80 percent for heat production.

Direct applications are nevertheless limited by transportation difficulties and the low utilization coefficients typical of many applications. Therefore, when geothermal temperatures are sufficiently high, it is generally advantageous to combine power production with flash cycles or binary units and direct uses in cascade. In some cases, especially when steam is available, geothermal fluid can be used directly, but more often heat is transferred by heat exchangers to an intermediate carrier (usually water in a closed circuit), which then provides heat, sometimes in conjunction with supplementary heat pumps and/or boilers.

Geothermal fluids contain numerous byproducts, some of which, such as boric acid, have commercial value. Other byproducts include carbon dioxide, potassium salts, and silica. For some, commercial-scale exploitation is expected in the near future. In addition, highly saline geothermal fluids known as brines, which have a salt content of 30 percent or more, have been discovered at Salton Sea in the United States and at Cesano in Italy. Some of the solutes in the brines (potassium, lithium, cesium, rubidium, silver, and gold) are worth more commercially than the producible energy in the brine.

NEW TECHNOLOGIES

Present and short-term developments
Although hydrothermal power has been exploited for tens of years and is now technologically mature, research on risk reduction and cost lowering continues in the areas of exploration, drilling and energy use. In addition, exploitation of geopressured, hot dry rock (HDR), and magma systems is still experimental.

Geopressured geothermal systems
Geopressured systems are similar to hydrothermal systems, but are characterized by higher reservoir pressure (from 50 to more than 100 megapascals at depths of several thousand meters) and the presence of dissolved natural gas (3 to 12 grams per liter in the liquid phase). These systems can be exploited not only for their thermal energy content, which is typically available at temperatures ranging from 150 to 200°C, but for the hydraulic energy created by the pressure and for the calorific energy of the associated natural gas. Often discovered by crews drilling for oil, they are a normal phase of basin evolution and are found in many locations throughout the United States and the world.

The proposed exploitation cycles for geopressured systems (some of which have already been tested) call for a hydraulic turbine to be installed upstream from the separation between the liquid phase and the methane-containing gases; water is then channeled to binary units, possibly hybrid ones, where the residual heat of methane-fueled generators is used. Alternatively, the methane can be sold commercially [24].

Exploitation cycles are currently hindered by several problems. To begin with, reservoir potential can be difficult to assess. In addition, the high salinity (20 to 200 grams of dissolved salts per liter) of the geothermal fluids causes scaling and corrosion. Finally, reinjecting residual fluids into the original reservoirs is inconvenient, and thus reinjection wells must be drilled into other, shallower aquifers. Nonetheless, preliminary experiments conducted in the United States demonstrate that geopressured systems are technically feasible [24].

Electricity from such systems is not yet competitive with conventional energy sources, but may be by the latter half of the 1990s. Even then, geopressured systems are likely to play a larger role in direct applications, such as thermally enhanced oil recovery, than in power production. They are closest to commercial success in the United States, which is the only country currently providing research and development funds, albeit in modest amounts, for geopressured energy.

Hot dry rock (HDR) systems
HDR systems are high-temperature rock formations that lack fluid capable of carrying heat to the surface, In addition, they are not extensively fractured and so do not have the necessary surface area for heat exchange with artificially injected water. To exploit HDR systems, an artificial reservoir, which is connected to two or

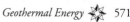

more production/reinjection wells, must be constructed. Heat contained in the rocks is extracted by circulating water through the reservoir (see figure 2d) and then passed through a binary cycle, which is appropriate when temperature levels are only modest.

The HDR concept originated in the United States in the early 1970s, and was first tested (by drilling, hydraulic fracturing, and flow tests) in the mid 1980s. The United Kingdom and Japan have also experimented with HDR systems. In the United Kingdom, a total of three wells were drilled and an artificial reservoir created at Rosemanowes, Cornwall. The reservoir, however, was considered to be about 100 times smaller than needed for commercial exploitation and its thermal performance was judged unsatisfactory. The production of energy from this site was never attempted, and the project eventually stopped.

In Japan, HDR wells have been established at three locations. The first site is that of an abandoned hydrothermal well near Hijori on the island of Honshu. High water losses have been a consistent problem during flow tests, but an extended flow test of a new HDR well is planned for 1993. At the second location, Iitate, shallow wells were drilled mainly for research purposes. The third site, near an area known as Ogachi, has focused mainly on drilling and fracturing technology. But multiple zone-fracturing experiments will be conducted there from 1991 to 1993, and in 1994 another production well will be drilled. In 1996, after flow testing experiments are completed, long-term testing of the reservoir will begin.

On the whole, HDR technology is still in the experimental stage. The main technological problems, which have been only partly solved, are linked to the creation of the reservoir, its durability, and to the recovery of the circulating fluid. Although HDR systems have been shown to be technically feasible in some cases, further studies are now underway or in the planning stages. One project near Soultz, in northeastern France, represents a collaborative effort between France, Germany, and the United Kingdom, with additional funding expected from the European Community (the site was considered more promising than Rosemanowes). In the United States, a long-term flow test, which is expected to last from one to two years, is under way at Fenton Hill, New Mexico. The project aims to evaluate reservoir potential at the site and thus to gauge the viability of its subsequent commercialization [25]; it is expected that HDR power production will be demonstrated first at Fenton Hill. Additional projects are planned elsewhere in the United States and Japan.

Economic studies of HDR technology, which are based on highly sophisticated models, show that costs are strongly influenced by the zone's geothermal gradient and hence by the depth of the wells that must be drilled. Nevertheless, these models are based on many assumptions, some of which have yet to be verified (including those based on reservoir geometry and fluid recovery), and must be viewed with some skepticism.

Current assessments show that HDR systems can be economically valid when the geothermal gradient is sufficiently high to permit fluid recovery at temperatures above 200°C and at depths no greater than 5 kilometers [26]. Ultimately, however, the economic feasibility of HDR will depend on the market trend of primary energy sources. Not only will fossil fuels determine the cost effectiveness of HDR, but their availability will indirectly influence R&D funding for HDR.

Magma systems

The objective in a magma system is to extract heat by tapping into the magma chamber. Although the potential of this resource has yet to be ascertained, the technology is currently being developed. Theoretically, fluid will be circulated in the top part of the magma chamber (which would be solidified and fractured by cooling around the well), thus creating a type of downhole heat exchanger. The heat will then power a closed Rankine-type cycle, generating as much as 25 to 45 megawatts for each well [27].

Field tests already conducted in the United States have been successful enough to justify follow-up economic studies. Further research on identifying magma systems by confirming and calibrating geophysical features of the surface is also planned. But more research must be undertaken—both from an economic and technical perspective—before magma can be considered a viable energy source.

Long-term developments

The energy potential of geopressured, HDR, and magma systems, which is considerable, justifies their continued development. Along the U.S. Gulf Coast, for example, the energy potential of geopressured systems alone amounts to 200 times the present annual energy consumption of the entire United States. HDR is much more widespread geographically than are magma and hydrothermal systems and thus may be important on a global scale. Magma, however, is a more concentrated energy source and therefore its exploitation may be more cost effective.

As mentioned earlier, the speed with which these technologies establish themselves depends on the cost trends of conventional energy sources. Rising petroleum costs, together with the modest environmental impact of geothermal energy, will likely accelerate these technologies during the next 10 to 20 years. Progress during a technology's initial stages is always slow; HDR, for example, has been under development for the past 15 to 20 years, yet still needs several more years to reach technical and economic maturity. Systems other than hydrothermal ones (geopressured, HDR, and magma systems) are likely to mature and become economically competitive only after 2000 (see "Present status and future prospects").

When maturation occurs (facilitated by R&D funding in the economically stronger countries), electricity production from these systems on a worldwide basis is unlikely to surpass 1,000 MW_e (megawatts electric) by 2010 and 10,000 MW_e by 2020. Even with such a scenario, which is a reasonably optimistic one, new geothermal technologies will account for only 20 to 30 percent of global geothermal electric production by 2020. Longer-term forecasts are even more tenuous because the continued development of other technologies, such as nuclear, will profoundly impact the overall energy market.

PRESENT ECONOMICS

The following treatment is mostly limited to a discussion of hydrothermal systems.

Power generation: total and share costs

Production costs of geothermoelectric power, which are highly variable, reflect the characteristics of specific geothermal fields and fluids. Technical activities that contribute to these costs include the following (see table 5):

1.   Surface exploration and research. The costs of these activities are viewed either as investments or losses depending on whether or not the field is productive, or they can be included in the overall operating and research budget (provided they are not borne by outside financing or institutions). Overall, their share amounts to 0.5 to 1.5 percent of the total money invested in the project.

2.   Drilling (production, reinjection, and dry wells) during exploration and field exploitation. Drilling costs are large and highly variable, depending on various production parameters, and may account for 30 to 60 percent of the cost of geothermal power production. Besides the expense of actual drilling, there are infrastructure costs (access roads, well sites, water supplies, etc.), as well as the cost of taking measurements and running tests during drilling.

3.   Construction of utilization plants (the gathering system and power plant). Water-dominated fields, which have a more complex exploitation cycle than steam-dominated ones, are more expensive, requiring greater outlays for both the gathering system and the power plant. Overall, these costs represent from 30 to 50 percent of the total costs.

4.   Major plant overhauls and maintenance. New wells (one for every 10 MW_e) must be drilled every eight to ten years. In addition, existing wells must be maintained in order to keep the flow rate to the units constant, and parts of the main machinery and the gathering system must be replaced. These expenses account for 3 to 5 percent of the cost of a kWh.

5.   Operation and planned maintenance. These expenditures vary depending on field characteristics and also on the discount rate; they may be relatively small for operations, but higher for maintenance. In some cases, costs are increased by environmental regulations that mandate hydrogen sulfide (H_2S) abatement and the proper disposal of solid sulfur produced by abatement.

The cost of producing a kWh can be calculated as follows (see figure 6):

1.   All costs associated with drilling and building the power plant must be capi-

Table 5: Costs of geothermoelectric power generation

Technical activities	Units	Steam-dominated reservoir	Water-dominated reservoir	Share and total costs			
				Discount rate 6%		Discount rate 12%	
				Steam	Water	Steam	Water
Surface exploration, including reconnaissance, prefeasibility and feasibility studies over a final area of 100–200 km²	$/sq. km	30,000–60,000	30,000–60,000	≈1%	≈1%	≈1%	≈1%
Drilling (1,000–3,000 meters, 9–13 inches) including related expenditures (access road, drilling site, water supply, move in/out)	$/m	1,000–1,650	1,000–1,650				
Success ratio K		0.50–0.75	0.50–0.75				
				36–55%	32–52%	40–58%	36–56%
Mean specific productivity Q	MW/well	2–5	3–8				
Well and field testing	$/well	45,000–70,000	80,000–150,000				
Reinjection of spent water: cost increase per drilled meter	$/m	70–170	350–850				
Gathering system	$/well	200,000–270,000	300,000–480,000				
Power plant							
Binary plant 1 MW	$/kW	1,700	1,700				
Back-pressure plant 5 MW	$/kW	1,000	1,000				
Back-pressure plant 15 MW	$/kW	600	600	29–40%	31–44%	32–44%	34–49%
Condensing plant 20 MW	$/kW	1,100	1,100				
Condensing plant 30 MW	$/kW	1,000	1,000				
Condensing plant 60 MW	$/kW	900	900				
Electric transmission line	$/km	50,000–100,000	50,000–100,000				
Field and plant O&M[a]				15–23%	16–23%	9–15%	9–14%
Generation cost	¢/kWh			3–8		4.5–12	

a. Including environmental control costs.

talized at the commencement of electric generation.

2. At the same time, major overhaul and maintenance costs estimated for the useful life of the installation must be discounted.

3. The equivalent annuity in relation to 1) and 2) must be calculated.

4. Annual operating and planned maintenance costs, and the annuity calculated in 3) must be combined so that the final cost of a kWh reflects the ratio of this sum to annual power production.

The cost of a kWh can be reduced by minimizing the time required to install production facilities, which can be accomplished, for example, by using modular power plants [28]. The cost, C, of a kWh varies in relation to the following production parameters: depth, D, success ratio, K, which is defined as the ratio between the number of productive wells and the total number of wells drilled, and specific productivity, Q, or the mean value of the capacity (in MW_e) that can be obtained with a standard generating cycle for each productive well. The value Q takes the hydraulic features of the geologic formation and well into account, as well as the chemical and physical characteristics of the geothermal fluid.

When K and Q are constant, C varies as a function of D, according to the exponential law: $C = A \times B^D$ (see figure 5), where A and B are constants that reflect fixed drilling costs and the technical profiles of the wells. Conversely, when D is constant, C varies according to the product $K \cdot Q$ (see figure 6) [29].

These parameters explain why the production costs of a geothermal kWh can range from 3 to 10 cents. Noteworthy is the fact that these are basic investment costs; unlike fossil-fuel plants, cost drifting will affect only the modest percentage represented by operating costs once the geothermal plant is operational. In countries where the power plant operator is not the field developer, the cost of the geothermal fluid is sometimes linked (by steam-sales contracts) to that of fossil fuels or to other factors, which leads to artificial variations in the cost of a geothermal kWh.

Future cost trends may depend on one of two conflicting possibilities: either the exploitation of deep reservoirs will increase drilling costs, or the refinement of surface exploration technology and the discovery of more economical drilling methods will reduce both the risk and cost of drilling new wells. If the latter occurs, cost increases should be limited to an acceptable 1 percent a year in constant dollars.

Of vital importance to the long-term success of geothermal energy is ample funding for research, especially in the areas of exploration and drilling. Such funds will economically offset the reduction in specific productivity, which is logically expected to rise as well-depth increases. Needed are techniques for locating deep-rock fractures and for directional drilling[12] into deep high-temperature reser-

---

12. Several wells or legs of the same well are drilled in different directions from one drilling site.

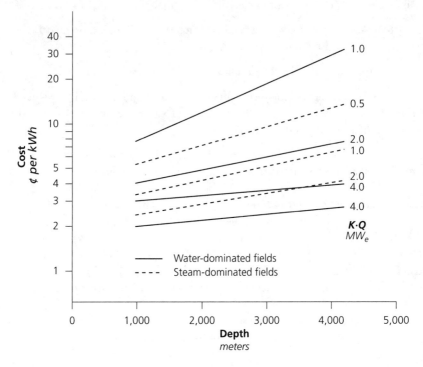

FIGURE 5: Cost of a kWh plotted against drilling depth as a function of K ·Q *(see text for explanation).*

voirs. Such research must be supported by national institutions in the framework of appropriate legislation.

Direct uses: total and share costs for different projects

Parameters that affect the cost of a kWh also apply to the cost of heat for direct uses, but in the case of direct heat they are limited to the equipment upstream from the distribution network. Substantial differences also exist in the following areas:

Low-temperature reservoirs are usually at relatively shallow depths (rarely deeper than 2,000 meters), and generally heat demand can be met with as few as one to four wells. But because the heat demand is often modest and uneven over the course of the year, the impact of a drilling failure or the high cost of fluid treatment to reduce corrosion or scaling may be economically fatal. In addition, the design of the heat installations cannot be completed until the geothermal fluid has been qualitatively and quantitatively characterized, which adds to the project's expense. Consequently, geothermal-heat systems often use fluids that have already been tapped or at least identified for other purposes. In some countries, the financial risks associated with fluid recovery are borne by the government, thus facili-

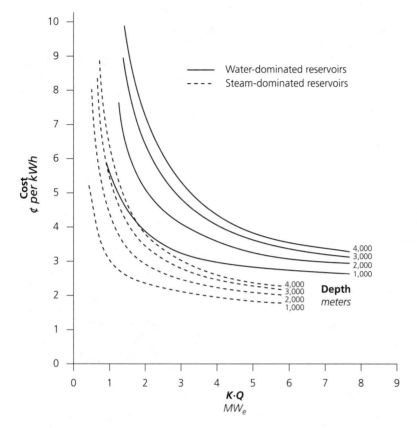

FIGURE 6: Cost of kWh plotteed against K · Q *as a function of drilling depth.*

tating the development of specific projects (see "Geothermal development: problems, constraints, and opportunities").

Optimizing the design of the utilization plants involves both the heat-transfer plant and its end users. Not only must the temperature of the fluid be optimally exploited, but installed capacity and energy consumption must be balanced by maximizing use,[13] which may mean relying on back-up fossil fuel during periods of peak load. Greater investments and higher operating costs are encountered when this goal is not met.

Because there are so many variables associated with direct applications, it is hard to generalize about economic parameters. Unit costs for investment and operation fluctuate widely, from high-investment projects that entail drilling and reinjection, to low-investment projects that rely on readily accessible low-salinity low-temperature fluids. In high-investment projects, about 70 to 90 percent of

---

13. The goal is to achieve maximum heat utilization so that the temperature of the geothermal fluid is reduced as much as possible before its disposal or reinjection.

Table 6: Costs of direct heat uses

Technical activities	Units	Unit costs	Share and total costs Discount rate 6%	Share and total costs Discount rate 12%
FLUID RECOVERY				
Drilling of production and reinjection wells				
depth < 1,000 m	$/m	300–800	24–78%[a]	26–80%[a]
1,000 m<depth<2,000 m	$/m	800–1,200		
Well testing and reservoir engineering	$/well	18,000–39,000		
THERMAL PLANT				
Possible production pump	$/well	40,000–130,000		
Wellhead equipment and possible scaling control equipment	$/well	120,000–180,000		
Pipelines, accessories, etc	$/kW$_t$	50–100	22–76%[a]	20–74%[a]
Heat exchangers	$/kW$_t$	140–270		
Possible reinjection pump	$/well	10,000–12,000		
Possible heat pump	$/kW$_t$	120–200		
Total investments	$/kW$_t$	320–1,500	20–89%[b]	33–94%[b]
Field and plant O&M			11–80%[b]	6–67%[b]
Heat cost	¢/MJ		0.15–1.5	0.2–2.5

a. Percent share of total investments.
b. Percent share of heat cost.

the heat cost can be attributed to the total investments (due mainly to drilling), whereas in low-investment projects, the percentage drops to 20 or 30 percent. The remainder in both cases consists of operating and maintenance costs.

The variability of the mean annual capacity factor, which is the ratio of the thermal energy used in a year to the energy obtainable by operating the plant year-round at maximum capacity, also influences heat costs. The capacity factor can range from 20 to nearly 100 percent, depending on how the heat is used (for district heating, agriculture, industrial usage, etc.), and on local climatic conditions. Operating and maintenance costs also vary as a result of the load factor, which influences the proportion of fixed to variable costs (see table 6).

Most of the heat cost (exploration expenses are extremely modest) reflects the following requirements: 1) Drilling and fluid characterization (an item that accounts for 25 to 75 percent of the investment cost, on average). 2) Construction of a heat-exchange plant, fluid-treatment plant (if necessary), and a temperature upgrading and backup system (also if necessary), which would contain heat pumps and/or boilers. These plants account for from 25 to 75 percent of the investment cost, on average. 3) Operation and maintenance activities, which account for 10 to 80 percent of overall costs. The total amount invested in 1) and 2) ranges from $320 to 1,500 per kilowatt-thermal and accounts for 20 to 90 percent of heat costs. The cost of geothermal heat can therefore range from ¢0.15 to 2.5 per megajoule. The cogeneration of electric and thermal energy will, however, reduce the cost of heat production.

GEOTHERMAL DEVELOPMENT: PROBLEMS, CONSTRAINTS, AND OPPORTUNITIES

The exploration and exploitation of geothermal resources is influenced by a number of factors, both in industrialized countries and to a greater extent in developing countries. Such factors vary from one country to the next, but include technical and scientific problems; economic and financial concerns; human resource availability; legal and institutional constraints; and environmental considerations.

The first category encompasses all the problems connected with exploration, fluid extraction, and the subsequent exploitation of geothermal fluids for power generation and/or direct heat. The subject is so vast, however, that specific treatment lies beyond the scope of this chapter. The second category takes the high cost of capital investment into consideration (especially for geothermoelectric projects) and addresses the issue of risk.

Together these two categories inhibit the start-up of many geothermal projects, especially in developing countries. Therefore, in order to attenuate the economic and financial difficulties presented during the initial stages (that is, during the reconnaissance, prefeasibility, and feasibility studies), geothermal projects may warrant national or international subsidies, or low-interest loans. Even so, such measures do not address the more serious problem of capital for the field development stage and for building utilization plants and facilities. In the long run, however, geothermal projects are characterized by their extended useful life and thus for the most part are economically attractive, especially when geared toward electricity production.

Still, from a strictly financial standpoint, the payback period is often long (sometimes five to 10 years), particularly in countries where inflation is high. As a consequence, private investors are often discouraged, and so shift the burden to government agencies or public institutions. In addition to being capital-poor, some developing countries also lack the necessary human expertise (geologists, engineers, technicians, etc.) to carry out and coordinate the basic organizational and operational activities needed to implement a geothermal project.

Legal constraints and institutional issues

Geothermal energy does not lend itself to long-distance transportation and must therefore be used in situ. That restriction may, however, be beneficial. By helping solve energy problems at the local or regional level, geothermal plants create a better balance-of-trade for oil-importing countries, allowing them to save currency that would otherwise buy petroleum. And oil-producing countries may be able to increase their oil exports by meeting more of their domestic needs with geothermal energy.

Geothermal energy should therefore be viewed as a resource of national interest: governments must look beyond a project's short-term financial picture and examine the overall benefits that it can provide. Some governments have already begun to recognize these added benefits and have initiated a series of legal and institutional measures to stimulate geothermal development. Such measures include: 1) establishing specific laws and regulations for the geothermal sector; 2) strengthening the infrastructure needed for geothermal development by funding basic research and training programs, providing resource inventories and technical assistance, facilitating information flow, etc.; 3) providing grants, soft loans, tax exemptions or reductions, and other incentives to encourage investment in geothermal energy; and 4) enacting energy policies that favor local sources over imported ones.

Among these measures, steps to reduce the risks of exploratory drilling are of special interest. From a legislative standpoint, regulation and classification of geothermal resources can also hasten the authorization process by recognizing the value of these resources to the public and by avoiding conflicts between decision-making bodies. Many countries have already adopted such legislation.

Geothermal projects are also aided by international initiatives: technical and financial assistance has been provided by the United Nations, financial assistance has been given by the World Bank, and funding for applied research and demonstration projects has been provided by the European Community. Other forms of support can be found in bilateral agreements between donor and recipient countries; such a pact was signed by Italy and Turkey in 1987. All these measures have stimulated geothermal projects, in often decisive ways, especially in developing countries.

Environmental considerations

Like most human activities, energy production involves interactions with the environment; geothermal resources are no exception to this rule. From a general point of view, however, use of geothermal energy, especially for heat, produces considerably fewer gaseous emissions than do fossil fuels. Although electricity generation has a greater impact on the environment than heat does, the impact of geothermal systems is still intermediate between fossil fuels and renewable energy sources such as solar, wind, and hydroelectric. When binary cycles are adopted, geothermal energy will equal other renewables in producing zero emissions.

Because geothermal energy must be used in situ, assessment of its environmental impact must be comprehensive and include all activities from the discovery and exploitation of a field to its final distribution network. Such impacts may hamper or even prohibit geothermal energy projects, especially in densely populated areas or areas of natural beauty in industrialized countries.

The environmental consequences of geothermal energy can be subdivided into two groups: temporary ones related to drilling and exploitation, and permanent ones that result from well maintenance, make-up drilling, and power plant operation. Plant operation involves such factors as land occupation and the aesthetic impact of the plants, including in some cases cooling-tower plumes; noise; the release of gaseous pollutants (H_2S, CO_2, radon, etc.) and toxic elements (mercury and arsenic) into the atmosphere; solid-waste and residual water disposal; and microearthquakes and subsidence.[14]

Each of these interactions, however, can be limited and/or controlled with appropriate countermeasures. Land use can be restricted, for example, by drilling several directional wells from one site. The aesthetic problem can be improved architecturally. Noise created largely by machinery and steam venting can be reduced during the design stage.

Air pollution is the greatest concern for most geothermal plants. For both technical and economic reasons, gases released during power production cannot be readily reinjected back into the reservoir.[15] Although abatement methods have been adopted in some cases for H_2S (at the Geysers, for example) they have disadvantages: the toxic chemicals used in abatement must be transported (increasing the risk of spillage) and solid sulfur must be disposed of properly.

Other gaseous pollutants normally do not constitute a significant problem. Radon, mercury, arsenic, and other compounds are present only in trace quantities and so do not significantly affect air quality. In fact, variations above normal background levels are barely detectable. Still, monitoring of these pollutants must continue. Emissions of CO_2 per kWh are so much less (23 times lower on average) than for thermoelectric plants [30], that geothermal energy may be promoted in the United States as a way to partially offset rising CO_2 levels.

Solid wastes are created when substances precipitate from geothermal fluids and from H_2S treatment plants. Although the wastes are sometimes toxic, they occur in modest amounts and thus can be reduced, if not eliminated, through careful management. Moreover, some wastes, such as sulfur, may actually have commercial value and be sold on the local market.

Residual waters are normally reinjected underground, generally into the same reservoir from which they were extracted. Doing so not only solves the disposal

---

14. Microearthquakes are generally produced when fluid is reinjected into a reservoir; subsidence occurs when fluid is extracted (without subsequent reinjection). Subsidence occasionally creates problems such as disruption to irrigation in flat areas.

15. Gas reinjection is known to occur only at the Coso geothermal field in California.

problem but partially recharges the reservoir. Because both production wells and reinjection drillings are isolated from shallow sedimentary layers, pollution of freshwater, including groundwater, is prevented.

Subsidence and increased microseismic activity (which is a natural characteristic of geothermal areas) are caused by the production and reinjection of geothermal fluids, respectively. Although reinjection eliminates or reduces subsidence, it can trigger microearthquakes, especially if large pressure differences develop vertically and/or laterally in the reservoir.

Although geothermal energy raises numerous environmental issues, many of those issues can now be addressed through appropriate control technologies, rendering geothermal energy compatible with the environment in which it is developed. Moreover, geothermal energy creates far less air and water pollution and hazardous waste, both in power production and to an even greater degree in direct applications, than do conventional forms of energy. The tendency to nationalize (or even internationalize) the environmental costs of power generation by levying a tax on CO_2 emissions, for example, will further boost geothermal development. For all these reasons, geothermal energy is more socially acceptable than conventional forms of primary energy in most regions of the world.

PRESENT STATUS AND FUTURE PROSPECTS

Achievements in power generation and direct uses up to 1989

Until the early 1960s, most geothermal production took place in Italy; since then geothermoelectric production has spread to an ever growing number of countries. The result has been an annual growth rate which to a great extent is influenced by the price of fossil fuel.

In 1950, installed geothermoelectric capacity was 145 megawatts (all in Italy); in 1960 it had increased to 385.7 megawatts (74 percent in Italy, the rest in New Zealand, the United States, and Mexico); by 1989 it had reached 6,000 megawatts in 18 countries (see table 7) [31, 32].

In the absence of good data, geothermal electricity production can be estimated by assuming that all plants have a mean annual capacity factor (the ratio between average load throughout the year and installed capacity) of 65 to 70 percent. According to those assumptions, world production in 1989 is estimated to have been about 35 billion kWh, which corresponds to about 8 million tonnes of oil–equivalent (TOE).[16]

The growth in capacity (see figure 7) [9], reflects an increase in the average annual development rate from 6 percent in 1969 to 11.3 percent in 1989. During the past decade the average annual development rate was 15.2 percent between

---

16. One tonne of oil–equivalent (TOE) equals 42 gigajoules, based on the unitary coefficient for fuel efficiency.

Table 7: Installed and planned geothermal power capacity (modified after [32], with updatings and roundings) *in megawatts*

Country	1985	1989	1995
United States	2,022	2,777	3,170
Philippines	894	894	2,164
Mexico	645	710	950
Italy	479	545	885
New Zealand	167	293	342
Japan	215	215	457
Indonesia	32	142	380
El Salvador	95	95	180
Nicaragua	35	70	100
Kenya	45	45	105
Iceland	39	45	110
China	11	21	50
Turkey	20	20	40
USSR	11	11	70
France	4	4	4
Portugal (Azores)	3	3	13
Greece	0	2	12
Thailand	0	0.3	3
Costa Rica	0	0	110
Guatemala	0	0	15
Dominica	0	0	10
St. Lucia	0	0	10
Total	**4,719**	**5,892**	**9,170**

1979 and 1985, but dropped to 5.7 percent between 1985 and 1989, a trend that reflects the oil shocks of 1973 and 1979, and the subsequent drop in oil prices.

Unfortunately, the data for direct geothermal applications are limited and nonhomogeneous. If only fluids above 35°C are considered, then total installed capacity is calculated to have been 3,100 MW_t (megawatts thermal) in 1975 [3]; 7,100 MW_t in 1984 and 11,500 MW_t in 1989 [33], corresponding to an average annual growth rate of 9.6 percent from 1975 to 1984 and 10.1 percent from 1984 to 1989. In 1989 (see table 8), the thermal energy (as compared with installed capacity) used in one year was 36,500 gigawatt-hours (GWh), which corresponds to an average load factor of 36 percent and about three million TOE per year.

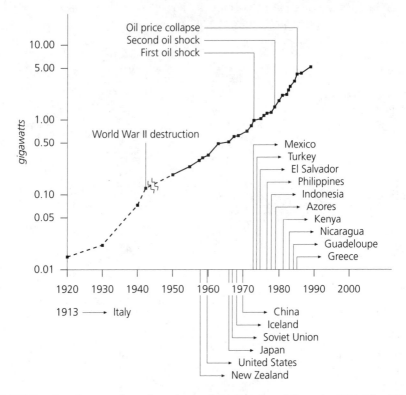

FIGURE 7: Growth pattern for geothermal power from 1920 through December 1989 (after [9]).

Prospects to the year 2000

From 1950 to 1989, the fluctuation in direct and indirect geothermal applications mostly reflected shifting oil prices, although after 1970, annual increases in capacity for indirect uses oscillated from 5 to 15 percent (with occasional 17 percent peaks) and for direct uses from 9 to 10 percent (with occasional 5 percent lows).

Projections for the next 10 years indicate that annual growth rates will approximate the maximum values given above. On that basis, if capacity in December 1989 was 6,000 MW_e and 11,500 MW_t for indirect and direct uses, respectively (see tables 7 and 8), then by the end of 2000, overall capacity could be expected to reach 23,000 MW_e for indirect applications and about 30,000 MW_t for direct ones [31].

Caution should be exercised in making medium- and long-term projections, however, not only because the world economic scenario faces considerable uncertainty, but because unforeseen operational difficulties always arise. Consequently, a more reasonable average annual increase might be 7 to 10 percent for indirect uses and 5 to 8 percent for direct uses. With regard to the latter, it must be pointed out that the balneotherapy market (which has grown at an average rate of about

Table 8: Direct uses of geothermal energy (modified after [33])

Country	Power MW_t		Energy GWh_t		Load percent	
	1984	1989	1984	1989	1984	1989
Algeria	–	13	–	117	–	100
Australia	11	11	29	29	30	30
Austria	4	4	10	10	29	29
Belgium	–	93	–	245	–	30
Bulgaria	–	293	–	770	–	30
Canada	2	2	5	5	29	29
China (+ Taiwan)	404	2,154	2,041	5,623	58	30
Colombia	12	12	32	32	30	30
Czechoslovakia	24	105	63	276	30	30
Denmark	1	1	8	8	91	91
Ethiopia	–	38	–	333	–	100
France	300	337	788	886	30	30
Germany	6	8	15	21	29	30
Greece	–	18	–	47	–	30
Guatemala	–	10	–	26	–	30
Hungary	1,001	1,276	2,615	3,354	30	30
Iceland	889	774	5,517	4,290	71	63
Italy	288	329	1,365	1,937	54	67
Japan	2,686	3,321	6,805	8,730	29	30
Mexico	28	–	74	–	30	–
New Zealand	215	258	1,484	1,763	79	78
Philippines	1	–	?	–	?	–
Poland	9	9	24	24	30	30
Romania	251	251	987	987	45	45
Switzerland	23	23	200	200	100	100
Tunisia	–	90	–	788	–	100
Turkey	166	246	423	625	29	29
United States	339	463	390	1,420	13	35
Former USSR	402	1,133	1,056	2,978	30	30
Yugoslavia	10	113	26	602	30	61
Others and roundings	28	115	43	374	–	–
Total	**7,100**	**11,500**	**24,000**	**36,500**	**39**	**36**

1 percent per year for the last 12 to 15 years and today accounts for more than half of all direct applications) appears to be nearly saturated [3]. Hence, the continued growth of direct applications chiefly reflects their expansion into the agribusiness, district heating, and industrial sectors (see figures 8a and 8b).

By 1995 (according to the same criteria cited on page 264), world geothermal power production will be about 60 billion kWh (or 13 million TOE); by 2000, it

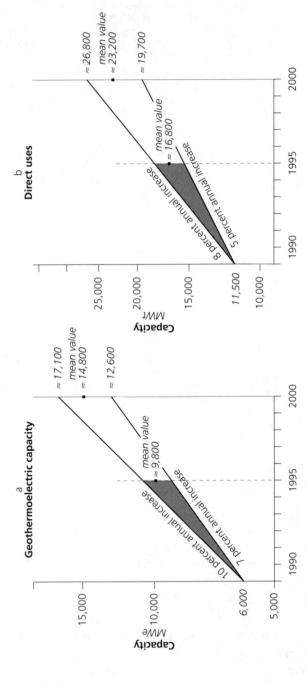

FIGURE 8: Foreseeable growth of geothermal energy to December 2000 a: Geothermoelectric capacity b: Direct uses.

will reach about 90 billion kWh (20 million TOE), in contrast to the 35 billion kWh produced during 1989.

Countries that are likely to initiate geothermal projects during the next 10 years include Argentina, Australia, Bolivia, Canada, Chile, Colombia, Costa Rica, Djibouti, Ecuador, Ethiopia, Guatemala, Romania, and St. Lucia. Others, which also have high-temperature potential and may pursue geothermal power, are Cape Verde, Dominica, Honduras, Pakistan, Panama, Peru, Iran, Tanzania, Sudan, Venezuela, and Zaire. Overall, some 40 countries (including the 22 listed in table 7) are likely to produce geothermal electric power by the end of this century. All of them fall, either wholly or partially, within the six belts of highest energy concentration (see figure 1).

These predictions refer only to the exploitation of hydrothermal reservoirs that have average depths of less than 2 kilometers (see figures 2b and 2c). Despite testing now underway, the promising new heat exploitation technologies (geopressured reservoirs, HDR, and magma systems) are unlikely to contribute significantly to the expansion of geothermal electric power by the end of the century.

Countries that are likely to develop their geothermal heat potential over the next decade (in addition to those listed in table 8) include Brazil, India, Israel, Madagascar, the Netherlands, South Korea, North Korea, Sweden, the United Kingdom, and Zambia. If countries that now use direct heat for balneotherapy shift to other uses by 2000, a total of 40 countries, not all of which are cited for geothermal electric generation, will be promoting geothermal energy.

Role of geothermal energy until 2000 and concluding remarks
The present and future role of geothermal energy within the framework of the global energy system can be predicted on the basis of data presented in the preceding sections (see also table 9). Two initial conclusions can be drawn. First, direct applications, including balneology, currently represent about one-fourth of geothermal use overall, and are likely to maintain that share for the next 10 years. If institutional and energy planning difficulties could be overcome, the contribution made by direct heat to the total world energy supply might increase from 0.04 to 0.07 percent by 2000 (see table 9). Given the current energy market, however, power production will remain the most important use of geothermal energy until 2000. Second, even if the use of geothermal energy increases over the next 10 years, it will nonetheless represent less than 1 percent of total world energy use and therefore remain a minor source of primary energy.

The share of total world electric capacity now occupied by geothermal power is modest (about 0.2 percent), but is expected to double by the end of the century (see table 10). Because geothermal plants have a higher mean annual capacity factor than other types of generating plants, geothermal power production represents a disproportionately large share of total world installed capacity; by 2000 its share will be about 0.6 percent.

However, the role of geothermal energy must be defined specifically for each country that chooses to harness it. In small countries, such as Djibouti and St.

Table 9: Geothermal versus total energy consumption in the world 1989-2000

	Energy consumption			
	1989		2000	
	$TOEx10^6$	percent[a]	$TOEx10^6$	percent[a]
Geothermal direct uses	3.1[b]	0.04	6.3	0.07
Geothermal electric generation	7.9[c]	0.10	19.8	0.22
Total geothermal uses	**11.0**	**0.14**	**26.1**	**0.29**
Total energy uses	**≈8·10³**	**100**	**≈9·10³**	**100**

a. Percentage over total energy uses.
b. 1 TOE = 42 gigajoules (fuel efficiency 100 percent). Only fluids with temperatures above 35°C are considered.
c. 1 kWh = 9.36 megajoules (1 TOE = 4,500 kWh_e).

Table 10: Geothermal versus total gross energy generation in the world 1989-2000

	1989				2000			
	Capacity		Production		Capacity		Production	
	GW	%[a]	TWh	%[b]	GW	%[a]	TWh	%[b]
Geothermal electric	5.9	0.22	35	0.30	14.8	0.41	89	0.58
Total electric	**2,700**	**100**	**11,500**	**100**	**3,600**	**100**	**15,200**	**100**

a. Percentage over total electricity capacity.
b. Percentage over total electricity generation.

Lucia for instance, where per capita energy consumption is low and a large increase in installed capacity is unlikely, geothermal electricity production may be sufficient to meet national demand. In countries that have relatively high geothermal potential (El Salvador, the Philippines, Kenya, Nicaragua, etc.) but require relatively little electric capacity (only about 10,000 MW_e), geothermal power can be expected to provide from 5 to 30 percent of their electricity needs by 2000. Among countries that are high energy consumers, such as Japan and Italy, where total installed capacity is tens of thousands of MW_e, and the United States, where capacity soars to hundreds of thousands of megawatts, geothermal energy is unlikely to ever contribute more than 2 percent to the total energy demand, even if

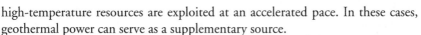

high-temperature resources are exploited at an accelerated pace. In these cases, geothermal power can serve as a supplementary source.

In conclusion, geothermal resources are not yet suitable for generalized worldwide development, and will not be before 2000. But because the number of countries at stake—30 or 40 at most—represents a manageable target, geothermal energy is likely to experience substantial growth within the next two or three decades. Not only can exploitation of the earth's geothermal resources be justified from an economic standpoint, but such exploitation can be justified from a strategic and social perspective. Geothermal development is therefore viewed as a global necessity.

WORKS CITED

1. Muffler, P. and Cataldi, R. 1978
   Methods for regional assessment of geothermal resources, *Proceedings of the ENEL–ERDA Workshop on Geothermal Resource Assessment and Reservoir Engineering, Larderello, 1977*, 131–207, ENEL, Rome.

2. Aldrich, M.J., Laughlin, A.W., and Gambill, D.T. 1981
   Geothermal resource base of the world: a revision of the EPRI's estimate, technical report LA 8531. University of California, Los Alamos National Laboratory, New Mexico.

3. Cataldi, R. and Sommaruga, C. 1986
   Background, present state and future prospects of geothermal development, *Geothermics* 15(3):359–383.

4. Barbier, E. 1984
   Eighty years of electricity from geothermal steam, *Geothermics* 13(4):389–401.

5. Luiggi, L. 1917
   La centrale termo-elettrica di Larderello (Larderello thermo-electric power plant), *Giornale del Genio Civile*, Rome, May 5–12.

6. Ginori Conti, P. 1917
   L'impianto di Larderello (Larderello power plant), *L'Elettrotecnica* 26–27:3–11, 15–25 September .

7. Mazzoni, A. 1954
   The steam vents of Tuscany and the Larderello power plant, second edition, 89–130, Arti Grafiche Calderini, Bologna.

8. DiPippo, R. 1980
   Geothermal energy as a source of electricity, 8–9, DOE/RA/28320-1, U.S. Department of Energy, Washington, DC.

9. DiPippo, R. 1988
   International developments in geothermal power production, *Geothermal Resources Council Bulletin* 17(5):8–19.

10. Lindal, G. 1973
   Industrial and other applications of geothermal energy, publication 135, UNESCO, Paris.

11. Palmerini, C.G. 1986
   La produzione di energia elettrica da fonte geotermica (Electricity production from geothermal resources), presented at Federelettrica national meeting, Padua, Italy.

12. Marconcini, R., Palamà, A., and Palmerini, C.G. 1982
    An energy equivalence criterion for geothermal fluids of differing composition and thermodynamic characteristics, *Proceedings of the International Conference on Geothermal Enery, Florence*, 2:217–225.

13. Coury and Associates, Inc. 1981
    Upstream H₂S removal from geothermal steam, EPRI final report AP-2100, Electric Power Research Institute, Palo Alto, California.

14. Instituto de Investigaciones Electricas. 1987
    Upstream hydrogen sulfide removal tests at the Cerro Prieto geothermal field, EPRI final report AP-5124. Electric Power Research Institute, Palo Alto, California.

15. Hankin, J.W., Cochrane, G.F., and Van Der Mast, V.C. 1984
    Geothermal power plant design for steam with high noncondensable gas, *Transactions of the Geothermal Resources Council* 8:65–70.

16. Armstead, H.C.H. 1983
    Geothermal energy, second edition, 162–222, E. & F.N. Spon, London/New York.

17. Eskesen, J.H, Whitehead, A., and Brunot, A.W. 1980
    Cycle thermodynamics, in Kestin, J., DiPippo, R., and Khalifa, H.E., eds., *Sourcebook on the production of electricity from geothermal energy*, 281–305. DOE/RA/28320-2, U.S. Department of Energy, Washington DC.

18. Allegrini, G, Sabatelli, F., and Cozzini, M. 1989
    Thermodynamical analysis of the optimum exploitation of a water-dominated field with high gas content, presented at UN Seminar on New Developments in Geothermal Energy, Ankara, Turkey.

19. McKay, R. 1982
    Helical screw expander evaluation project, final report DOE/ET/28329-1, U.S. Department of Energy, Washington DC.

20. Hudson, R.B. 1988
    Technical and economic overview of geothermal atmospheric exhaust and condensing turbines, binary cycle and biphase plant, *Geothermics* 17(1):51–74.

21. Bronicki, L.Y. 1988
    Electrical power from moderated temperature geothermal sources with modular mini-power plants, presented at the UN Seminar on New Developments in Geothermal Energy, Ankara, Turkey.

22. Featherstone, J.L. and Powell, D.R. 1981
    Stabilization of highly saline geothermal brines, *Journal of Petroleum Technology* 33(4):727–734.

23. Palmerini, C.G. and Paris, L. 1982
    Energia geotermica e territorio (Geothermal energy and territory), paper A.73, presented at the 83rd annual meeting of the Associazione Elettrotecnica Italiana, Bologna.

24. Campbell, R.G. and Hattar, M.M. 1990
    Operating results from a hybrid cycle power plant on a geopressured well, *Transactions of the Geothermal Resources Council*, 14(1):521–525.

25. Duchane, D.V., Brown, D.W., House, L., Robinson, B.R., and Ponden, R. 1990
    Progress in hot dry rock technology development, *Transactions of the Geothermal Resources Council* 14(1):555–559.

26. Tester, J.W. and Herzog, H.J. 1991
    The economics of heat mining: an analysis of design options and performance requirements of hot dry rock (HDR) geothermal power systems, *Proceedings of the 16th Workshop on Geothermal Reservoir Engineering, Stanford, California* (in press).

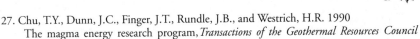

27. Chu, T.Y., Dunn, J.C., Finger, J.T., Rundle, J.B., and Westrich, H.R. 1990
    The magma energy research program, *Transactions of the Geothermal Resources Council* 14(1):567–577.

28. Allegrini, G, Giordano, G., Moscatelli, G., Palamà, A., Pollastri, G.G., and Tosi, G.P. 1985
    New trends in designing and constructing geothermal power plants in Italy, *Proceedings of the 1985 International Symposium on Geothermal Energy, Kailua Kona (HI)*, international volume, 279–288.

29. Allegrini, G. and Cappetti, G. 1990
    Economic analysis of geothermal projects, *Transactions of the Geothermal Resources Counci* 14(1):473–476.

30. Goddard, W.B., Goddard, C.B., and McClain, D.W. 1989
    Future air quality maintenance and improvements through the expanded use of geothermal energy, *Transactions of the Geothermal Resources Council* 13:27–34.

31. Baldi, P. and Cataldi, R. 1986
    Sviluppo e prospettive dell'energia geotermica in Italia e nel mondo (Development and prospects of geothermal energy in Italy and in the world), *Memorie della Società Geologica Italiana* 35:735–753.

32. Huttrer, G.W. 1990
    Geothermal electric power—a 1990 world status update, *Geothermal Resources Council Bulletin* 19(7):175–187.

33. Freeston, D.H. 1990
    Direct uses of geothermal energy in 1990, *Geothermal Resources Council Bulletin,* 19(7):188–198.

14
BIOMASS FOR ENERGY: SUPPLY PROSPECTS

DAVID O. HALL
FRANK ROSILLO-CALLE
ROBERT H. WILLIAMS
JEREMY WOODS

Biomass (plant matter) accounts for about 15 percent of world energy use and 38 percent of energy use in developing countries. But most biomass is used inefficiently, mainly for cooking and heating in rural areas of developing countries, and often in much the same way it has been for millennia. Biomass can also be converted into modern energy carriers such as gaseous and liquid fuels and electricity that can be widely used in more affluent societies. The prospects are good that these energy carriers can be produced from biomass at competitive costs under a wide range of circumstances. Moreover, the large-scale utilization of biomass for energy can provide a basis for rural development and employment in developing countries, thus helping curb urban migration. In addition, if biomass is grown sustainably, its production and use creates no net buildup of carbon dioxide (CO_2) in the atmosphere, because the CO_2 released during combustion is offset by the CO_2 extracted from the atmosphere during photosynthesis.

Biomass for energy can be obtained from residues of ongoing agricultural and forest-product industries, from harvesting forests, and from dedicated plantations. The harvesting of forests for biomass is likely to be limited by environmental concerns. Over the next couple of decades new bioenergy industries will be launched primarily using residues as feedstocks. Subsequently, the industrial base will shift to plantations, the largest potential source of biomass.

The most promising sites for plantations are deforested and otherwise degraded lands in developing countries and excess croplands in the industrialized countries. Revenues from the sale of biomass crops grown on plantations established on degraded lands can help finance the restoration of these lands. Establishing plantations on excess croplands can be a new livelihood to farmers who might otherwise abandon their land because of foodcrop overproduction. In either case, biomass plantations can, with careful planning, substantially improve these lands ecologically relative to their present uses. But a substantial and sustained research and development effort is needed to ensure the realization and sustainability of high yields under a wide range of growing conditions. Moreover, the establishment and maintenance of biomass plantations must be carried out in the framework of sustainable economic development in ways that are acceptable and beneficial to the local people.

Ultimately, land and water resource constraints will limit the contributions that biomass can make as an energy source in advanced societies. But biomass energy can nevertheless make major contributions to sustainable development before these limits are reached, if biomass is grown productively and sustainably and is efficiently converted to modern energy carriers that are used in energy-efficient end-use technologies.

INTRODUCTION

Photosynthesis offers a means of harnessing solar energy that deals effectively with two elusive features of sunlight—the high cost of collection and its intermittency. The collectors involved are simply the leaves of plants, which cost relatively little to grow compared with solar collectors for photovoltaic or solar thermal-electric power. And storage is conveniently provided by the plant matter, or biomass, produced.

Humans have used solar energy in the form of biomass for cooking and heating since the discovery of fire. Even today, the dominant use of fuelwood in the world is for cooking and heating in rural areas of developing countries. While biomass fuel is essential for survival in many places, its use is fraught with problems. Fuelwood stoves are typically very inefficient compared with modern stoves that burn gaseous fuels; they also generate considerable air pollution, causing health problems, especially for the women who do the cooking. The gathering of fuelwood is labor-intensive and in some areas contributes to deforestation. In rural areas, where fuelwood is scarce, crop residues or dung are used as fuel and are even less convenient to use than fuelwood. Such problems have led to biomass being widely regarded as "the poor man's oil" and therefore an energy source to be abandoned as economic development proceeds.

In most rural areas of the world, fuelwood is gathered outside the market system and is thus a non-commercial fuel. Because it is non-commercial and held in low esteem as an energy source, the role of biomass in the global energy economy is not well known. Official data on biomass energy use tabulated by the Food and Agricultural Organization (FAO) of the United Nations indicate that biomass is burned at a global rate of about 20 exajoules per year, which represents less than 6 percent of total energy use [1]. But detailed surveys in many countries indicate a much larger role. In Egypt, for example, surveys show that biomass accounts for 27 percent of total energy, while FAO data indicate only 2 percent; for China, the corresponding values are 27 percent and 7 percent [1].

Because such survey data are not available for many countries, and because existing data bases are often not maintained and updated, accurate values for the global use of biomass are not available. However, the best estimates [2] suggest that biomass is consumed globally at a rate of about 55 exajoules per year, which is some 15 percent of the world's energy-use (see figure 1). Biomass is the dominant source of energy (38 percent of total energy-use) in developing countries where nearly three-fourths of the world's people live. In some developing countries, biomass provides 90 percent of total energy. Biomass is also used for energy in some industrialized countries such as the United States (4 percent, equivalent in energy content to 1.4 million barrels of oil per day), Austria (10 percent), and Sweden (9 percent) [1].

Clearly biomass is a major energy source, one that is consumed globally at nearly the same rate as natural gas. Biomass may play an even larger role in the future if biomass is converted into modern energy carriers—mainly fluid (gaseous

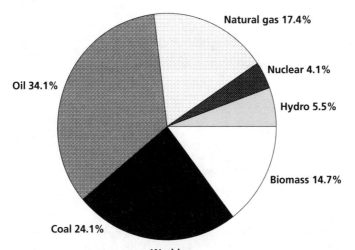

World
Total = 373 exajoules
Population = 4.87 billion
Energy use per capita = 77 gigajoules

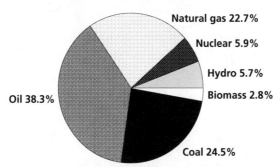

Industrialized countries
Total = 247 exajoules; 66 percent of world total
Population = 1.22 billion; 25 percent of world total
Energy use per capita = 202 gigajoules

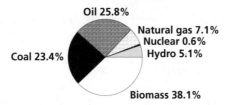

Developing countries
Total = 126 exajoules; 34 percent of world total
Population = 3.65 billion; 75 percent of world total
Energy use per capita = 35 gigajoules

FIGURE 1: Primary energy use for the world (top), industrialized countries (middle), and developing countries (bottom) in 1985. Biomass energy use estimates are from [2]. Other energy use data are from chapter 1: Renewable Fuels and Electricity for a Growing World Economy.

and liquid) fuels and electricity. There are several reasons for considering biomass in this context:

▼   Biomass is far more widely available than fossil fuels and, with good management practices, can be produced renewably.

▼   As development proceeds, consumers want modern energy carriers because they are cleaner and more convenient than traditional fuels.

▼   The prospects are good that modern energy carriers can be produced from biomass at competitive costs under a wide range of circumstances.

▼   If biomass is produced sustainably and then efficiently converted to modern energy carriers, biomass could emerge as a major player in the world's commercial energy economy.

▼   Modernized biomass energy can provide a basis for rural development and employment in developing countries, thereby helping curb urban migration.

▼   In developing countries, growing biomass for energy on deforested and otherwise degraded lands might provide a mechanism for financing the restoration of these lands.

▼   In industrialized countries, growing biomass for energy on excess croplands can provide a new livelihood for farmers who might otherwise abandon farming because of foodcrop overproduction.

▼   If biomass is grown sustainably, its production and use leads to no net build-up of carbon dioxide (CO_2) in the atmosphere, because the CO_2 released in combustion is offset by the CO_2 extracted from the atmosphere during photosynthesis.

SOME PHYSICAL AND CHEMICAL PROPERTIES OF BIOMASS

In the context of energy, biomass refers to all forms of plant-derived material that can be used for energy: wood, herbaceous plant matter, crop and forest residues, dung, etc. An appreciation of the physical and chemical properties of biomass fosters understanding of its potential roles as an energy source.

Because biomass is a solid fuel, comparisons with coal are instructive. On a dry-weight basis, heating values[1] range from about 17.5 gigajoules per tonne for various herbaceous biomass feedstocks (e.g., wheat straw, sugarcane bagasse, sudan grass) to about 20 gigajoules per tonne for woody feedstocks [3]. The corresponding values for bituminous coals and lignite range from 30 to 35 gigajoules

---

1. In this chapter the energy content of fuel is the higher heating value, which includes the latent heat of condensing the water vapor in combustion product gases. The lower heating value (which does not include this latent heat) is about 6 percent lower for most biomass feedstocks

per tonne and 23 to 26 gigajoules per tonne, respectively. At harvest, biomass contains considerable moisture, ranging from 8 to 20 percent for wheat straw, to 30 to 60 percent for woods, to 75 to 90 percent for animal manures, and to 95 percent for the water hyacinth;[2] (in contrast the moisture content of most bituminous coals ranges from 2 to 12 percent).[3] Thus, energy densities for biomass at the point of production are markedly less than those for coal.[4] The low energy density of harvested biomass and the dispersed nature of biomass production imply that the production of modern energy carriers (electricity, liquid, gaseous, and processed solid fuels) from biomass feedstocks should be carried out at dispersed installations that are relatively small, to avoid high transportation costs.

While the mass density of biomass makes it less attractive as a fuel than coal, its chemical attributes make it superior in many ways. The ash content of biomass is typically much lower than for coals, and the ash is generally free of the toxic metals and other trace contaminants that make difficult the disposal of coal ash in environmentally acceptable ways. Furthermore, the ash recovered at biomass conversion facilities has fertilizer value and can be dispersed over the biomass growing area to help replenish the nutrients removed from the site during harvesting. Also, sulfur (S), which is converted to sulfur dioxide (SO_2) during combustion and contributes to acid deposition, is a major environmental problem associated with combustion of coal, which contains typically 0.5 to 5 percent S by weight; for comparison, the S content of typical biomass feedstocks ranges from 0.01 to 0.1 percent.

---

2. Here moisture content, in percent, is measured on a wet basis as: 100 · (mass of contained water)/(mass of the wet biomass). An alternative convention measures the moisture content on a dry basis: 100 · (mass of contained water)/(mass of dry biomass).

3. Sub-bituminous coals often have moisture levels of 15 to 30 percent; the moisture content of lignites can exceed 35 percent.

4. Because the moisture content of biomass varies both among types of biomass and over time, biomass is often measured in volumetric terms. Care must be exercised in converting volumetric measures to mass measures, because different conventions are used in the literature, the convention used is often not stated, and even where it is stated, the density value can be ambiguous. The three mass density measures are true density, apparent density, and bulk density, defined as follows:

true density = (mass of wood)/(solid volume of wood)

apparent density = (mass of wood)/(solid volume of wood + pore volume)

bulk density = (mass of wood)/(solid volume of wood + pore volume + void volume)

True density is difficult to measure and is used primarily in scientific investigations. (However, its value for most woods is approximately 1.5 tonnes per m³.) Bulk density (sometimes called the "stacked volume" density) is easiest to measure but its use often leads to confusion because of uncertainty associated with the bulk density of biomass in various forms. For example, a cord of wood is defined as the amount contained in a stack 4 ft. × 4 feet × 8 feet (3.6 m³). If the wood pieces were perfect cylinders of equal diameter the apparent volume of wood would be 78.5 percent of the total volume. In practice, with rough pieces of unequal size, the apparent wood volume can be less than half the total volume. Apparent density is the most appropriate one for making conversions from volume to mass because it is relatively unambiguous and not as difficult to measure as true density; values for various dry woods range from about 0.4 tonnes per m³ for jack pine, yellow poplar, and hemlock, to 0.4-0.55 for *Eucalyptus grandis*, to 0.7 to 0.85 for *Acacia mearnsii*.

Biomass is also much more reactive than coal,[5] making it an especially attractive feedstock for thermochemical gasification for power generation (see chapter 17: *Advanced Gasification-Based Biomass Power Generation* and chapter 16: *Open-Top Wood Gasifiers*), for the production of methanol (see chapter 21: *Ethanol and Methanol from Cellulosic Biomass*), and for the production of hydrogen (see chapter 22: *Solar Hydrogen*). Exploiting this reactivity advantage in gasifier design makes it possible to carry out gasification and down-stream conversion processes cost-effectively at the modest scales needed for biomass conversion.

Another advantage is that biomass feedstocks, unlike coal feedstocks, are good candidates for a variety of biological conversion processes ranging from the anaerobic digestion of animal manures and other organic wastes (see chapter 18: *Biogas Electricity—the Pura Village Case Study,* and chapter 19: *Anaerobic Digestion for Energy Production and Environmental Protection*) to the production of ethanol from sugarcane (see chapter 20: *The Brazilian Fuel-Alcohol Program*) to the production of ethanol from lignocellulosic (woody) feedstocks (see chapter 21). These technologies have promise because the processes can be carried out under mild conditions, using biological agents designed to be highly specific in producing the desired products.

THE FUNDAMENTALS OF PHOTOSYNTHESIS

In photosynthesis, sunlight is absorbed by chlorophyll in the chloroplasts of green plant cells and utilized by the plant to produce carbohydrates from water (H_2O) and CO_2 taken from the atmosphere. The process can be presented in simplified form by the equation:

$$6CO_2 + 6H_2O \xrightarrow[\text{sunlight}]{} C_6H_{12}O_6 + 6O_2$$

where six molecules of H_2O and six molecules of CO_2 combine to produce one molecule of glucose (a carbohydrate) and six molecules of O_2.

Globally, the photosynthetic process produces an estimated 220 billion dry tonnes of biomass per year [4, 5], equivalent in energy value to about ten times global energy use. Humans already harvest significant fractions of this amount for food and forest products. Harvesting more biomass for energy purposes would involve both further use of natural photosynthetic production and augmented photosynthetic production through intensive management. An examination of some details of photosynthesis provides insights into the gains that might be achieved through intensive management.

---

5. The greater reactivity of biomass relates to its chemical structure. Neglecting minor chemical constituents, a typical biomass feedstock can be represented chemically as $CH_{1.45}O_{0.7}$, compared with $CH_{0.8}O_{0.08}$ for a typical coal. Thus, biomass has nearly twice as much hydrogen and nearly an order of magnitude more oxygen per carbon atom than coal.

Estimating maximum efficiencies for the field production of biomass

A key issue is the efficiency of converting incident solar energy to the chemical energy stored in carbohydrates in plants. Upper bounds on efficiency can be estimated from theoretical considerations. To begin with, plants use only light with wavelengths between 0.4 and 0.7 microns (visible light); this photosynthetically active radiation (PAR) makes up about 50 percent of the total energy in solar radiation that arrives at the earth's surface. About 80 percent of the intercepted PAR is captured by photosynthetically active compounds—the rest is lost by reflection, transmission, and absorption by non-photosynthesizing materials. The theoretical maximum energy efficiency of converting the effectively absorbed PAR to glucose is determined by the fact that a minimum of eight photons of PAR are required to produce glucose for each CO_2 molecule converted [6, 7]; the corresponding energy stored in the produced glucose is 28 percent of the light energy of these photons. And finally, about 40 percent of the energy stored in photosynthesis is consumed during dark respiration (the reverse of photosynthesis) to sustain a plant's metabolic processes. This sequence determines a maximum photosynthetic efficiency [2] of:

$$100 \times 0.50 \times 0.80 \times 0.28 \times 0.60 = 6.7 \text{ percent.}$$

The maximum applies primarily to C_4 plants (so-called because the first product of photosynthesis is a 4-carbon sugar), such as maize (corn), sorghum, and sugar cane, which grow best in relatively hot climates. For wheat, rice, soybeans, trees and other C_3 plants (for which the first product of photosynthesis is a 3-carbon sugar), which dominate temperate climates and account for 95 percent of global plant biomass, the maximum photosynthetic efficiency is lower. Additional losses occur because C_3 plants lose about 30 percent of the already fixed CO_2 during photorespiration (which competes with photosynthesis in the presence of light and does not occur in C_4 plants) and because the light-utilizing capacity of C_3 plants becomes light-saturated at lower light intensities than for C_4 plants, so that C_3 plants are unable to utilize perhaps 30 percent of the light absorbed by photosynthetically active compounds [8]; thus the maximum efficiency of energy conversion for C_3 plants is about $0.7 \times 0.7 \times 6.7 = 3.3$ percent.

These maximum efficiencies can be converted to biomass production rates (in dry tonnes per hectare per year) by noting that the heating value of typical herbaceous plants is about 17.5 gigajoules per dry tonne. Thus, for areas near Plymouth in the United Kingdom (50° N), where the average daily insolation is about 11.1 megajoules per square meter (m^2) (a relatively low level for temperate regions), the corresponding biomass production rate would be 156 tonnes per hectare per year for C_4 plants and half this value for C_3 plants; near Des Moines, Iowa (41° N), in the heart of the U.S. farm belt, where the average daily insolation is 14.9 megajoules per m^2, the biomass production rate would be 208 tonnes per hectare per year for C_4 plants and half this value for C_3 plants.

Such production rates could be achieved if a complete canopy were available throughout the year, if photosynthesis were not slowed by low temperatures in winter or by drought in summer, if adequate nutrients were available, and if there were no losses associated with pests and diseases.

Temperature can powerfully affect photosynthesis. Most C_3 plants photosynthesize at the maximum rate, provided the roots have access to adequate water and nutrients, when the temperature is between 20 and 30 °C, but cease photosynthesis altogether when the temperature falls to the range 0 to 5 °C. In contrast, many tropical plants, including members of the C_4 group, reach their maximum rates when the temperature is between 30 and 40 °C and cease photosynthesis between 10 and 15 °C [8].

While such temperature effects are not especially constraining in tropical regions, in many temperate climates they can limit the growing season to just a fraction of the year. For sites near Plymouth and Des Moines, this might be the 5-month period May through September, when the average daily insolation is 1.57 and 1.42 times the annual average, respectively. Accordingly, the temperature constraint on photosynthesis would reduce the maximum annual production of C_3 biomass plants to approximately $1.57 \times (5/12) \times 0.7 \times 0.7 \times 156 = 50$ tonnes per hectare per year near Plymouth, and to $1.42 \times (5/12) \times 0.7 \times 0.7 \times 208 = 60$ tonnes per hectare per year near Des Moines.

Not all produced biomass can be recovered for energy purposes. In the case of trees, for example, only the trunk and large branches would be utilized, not the roots, foliage, and twigs; with coppicing, where multiple rotations are obtained from a single planting, via regrowth from the stump after harvest, perhaps 58 percent of the total biomass produced could be recovered,[6] thereby reducing the recoverable amount to some 29 and 35 tonnes per hectare per year for sites near Plymouth and Des Moines, respectively.

Presently achievable yields are far less than these "practical maxima," largely because of inadequate water and nutrients and problems with pests and diseases. Yet even if these yields were routinely achievable, these calculations highlight a limiting aspect of biomass energy: photosynthesis is a relatively land-intensive way of using the sun's energy. The practical maximum yields for C_3 plants in temperate regions, for example, correspond to efficiencies of converting only about 1 percent of total sunlight received on an annual basis at the ground level into recoverable chemical energy. Even in warm climates, where low temperatures are not limiting and where C_4 plants can flourish, practical maximum efficiencies for recoverable plant matter will be no more than about 2 to 3 percent of incident sunlight. The losses associated with converting biomass into useful energy further reduce the overall efficiency of utilizing the sun's energy. Land requirements are

---

6. For well-nourished trees in temperate regions, it has been estimated that the root-to-shoot ratio would be 31/69 and that 65 percent of the shoot would be wood [9]. Thus, if there are three cuts for a single planting (i.e., two coppice rotations for each planting), with all growth taking place above ground after the first cut, the total recoverable wood per planting [assuming total biomass production is the same (= 100, arbitrary units) for each rotation] would be $0.65 \times (69 + 200)/300 = 0.58$.

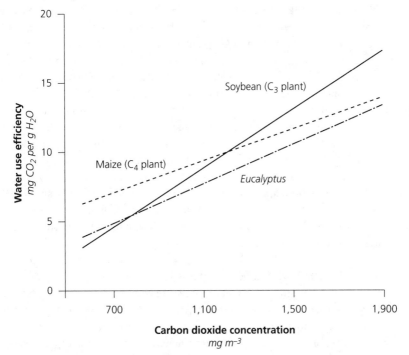

FIGURE 2: *Variation of water-use efficiency (WUE) with increasing concentration of atmospheric CO_2 for selected plant species [80]. The WUE is measured in milligrams of CO_2 (net) fixed in photosynthesis per gram of water transpired. The atmospheric concentration of CO_2 was 635 mg m^{-3} (353 ppm) in 1990.*

large, for example, compared with photovoltaic energy sources, for which efficiencies of 10 percent (sunlight to electricity) have already been achieved and efficiencies of 20 to 30 percent may be practical in the future.

Water and nutrient requirements for biomass production

Not only is biomass production land-intensive, but also it requires abundant water resources. Water is used to transport nutrients throughout the plant, and large amounts are lost via transpiration through the stomata, the openings in the leaves or needles that allow CO_2 to enter the plant.

Water requirements are inversely related to the water-use efficiency (WUE), measured in milligrams of CO_2 (net) fixed in photosynthesis per gram of water transpired (see figure 2); WUE is highly variable among plant types, but typically ranges from two to six milligrams CO_2 per gram H_2O. This range represents water requirements of 300 to 1,000 tonnes per tonne of dry biomass.

Producing biomass at a yield of 25 tonnes per hectare per year requires water for transpiration equivalent to 750 to 2,500 millimeters of annual rainfall. Thus, biomass production at high productivities will be limited to regions with abundant rainfall or alternative water resources.

Where water resources are limited, high productivities can be realized by selecting plant types better adapted to water limitations. There is a considerable range in WUEs for C_3 plants, and C_4 plants have higher WUEs than C_3 plants. This is illustrated in figure 2, which shows that at a CO_2 concentration of 353 parts per million (ppm) (the atmospheric concentration in 1990) the WUE for maize (a C_4 plant) would be 6 milligrams of CO_2 per gram of water, or equivalently, the water use requirement would be only 300 tonnes per tonne of biomass. C_4 plants use water much more efficiently than C_3 plants because under hot, dry conditions, when the stomata openings of leaves are kept small to conserve water, C_4 plants can continue to assimilate CO_2 while C_3 plants cannot.[7]

Nitrogen (N), phosphorus (P), potassium (K), and various trace nutrients are also needed to grow biomass. Many natural forests are deficient in N and P and, to a lesser extent, in K, S, Mg and other trace elements. When natural forests or plantations are harvested for biomass, nutrient balances must be restored, so that subsequent growth will be productive.

Biomass yields with adequate water and nutrients

The interactions between nutrients and moisture and the roles of root architecture and turnover of fine roots in agronomic efforts to increase yields at specific sites are not well understood [10]. In forests, the duration of the response to added nutrients is relatively brief, and the efficiency of trees in using fertilizers is relatively poor. The long-term correction of nutrient deficiencies in energy plantations requires additional strategies, such as nitrogen-fixing bacteria and slow-release fertilizers, coupled with efforts to minimize nutrient losses and moisture stress. These strategies will help sustain high productivities in ways that are both economical and environmentally acceptable; such attributes can be practically attained by selection and breeding of nutrient-responsive and nutrient-efficient genotypes [11].

In a 1977 paper [8], John Monteith of the University of Nottingham School of Agriculture in the United Kingdom showed that under conditions where crops are well fertilized and the supply of water is adequate for good growth, the average light-use efficiency is the same for various C_3 food crops (see figure 3). He found that the average light-use efficiency was 1.4 grams of dry plant matter per megajoule of total radiation intercepted by foliage throughout the growing season, with values for individual crops falling within $\pm$ 15 percent of this mean value. The mean value corresponds to a photosynthetic efficiency of 2.4 percent of radiation intercepted by the leaves during the growing season.

---

7. When the stomata close to conserve water, the concentration of CO_2 trapped in the leaf falls and the relative concentration of O_2 increases. The result in C_3 plants is that photorespiration increases at the expense of photosynthesis, and there is no net photosynthesis below CO_2 concentrations of about 50 ppm. In contrast, C_4 plants have an additional photosynthetic mechanism besides the Calvin cycle (characterized by competition between photosynthesis and photorespiration); they effectively pump up CO_2 concentrations inside the leaf, making it possible for C_4 plants to photosynthesize at CO_2 concentrations as low as 1 ppm.

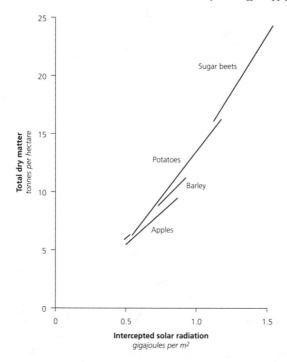

FIGURE 3: Relation between total plant matter at harvest and radiation intercepted by foliage throughout the growing season, for selected C₃ plants grown under conditions such that the crops are well fertilized and the supply of water is adequate for good growth [8].

It is now recognized that there is little difference in factors determining field potential between C_3 agricultural crops (e.g., wheat), various other herbaceous plants, and trees (all of which are of the C_3 type), even though the partitioning of the primary products of photosynthesis (i.e., sugars), into stored compounds such as starch and cellulose are different. The maximum net photosynthetic rates of individual broad-leaf tree leaves are similar to those for leaves of C_3 crops, while those for conifer needles are slightly lower. However, at the tree- or crop-leaf canopy level the CO_2 fixation rates of both conifer and broad-leaf trees are about the same as for agricultural crops. All tree species investigated behave similarly to C_3 crop plants, with about the same total biomass yield per unit of light intercepted by the leaves during the growing season (see table 1), for trees grown with nutrients and water supplies adequate for good growth [11].

The finding that the total biomass production is proportional to the amount of light intercepted during the growing season highlights the importance of (i) establishing a full leaf cover early in the growing season, and (ii) maintaining leaf cover as long as possible in the autumn, without risking frost damage, as a biomass production management goal.

These empirical data on light-use efficiency can be used to estimate maximum practical production rates for tree plantations. For example, the annual

Table 1: Some estimates of seasonal average light use efficiency for trees and C$_3$ agricultural crops that are well-supplied with water and mineral nutrients[a]

Species	Location	Light use efficiency[b]
		grams per megajoule
Salix viminalis	Scotland	1.4
Populus trichocarpa	Scotland	1.0
Pinus sylvestris	Sweden	0.9
Wheat and barley	England	1.2
Rape and fescue	France	1.0
Lucerne and beans	France	0.8
Soybeans	N. Carolina	1.3
Forage and grain legumes	N. Australia	1.1

a. [9].

b. Unlike the data in figure 3, which measure light use efficiency (LUE) in grams of total dry matter produced in a growing season per megajoule of intercepted total solar radiation, the LUE values presented here are measured in grams of total aboveground dry matter produced per MJ of solar radiation intercepted by the leaves in the growing season. Note that if the biomass partitioning for trees was 31 percent to root and 69 percent to shoot (see discussion in text), a LUE of 1.0 grams per MJ of above-ground wood would correspond to a total LUE of 1.4 grams per megajoule of total wood for trees.

harvestable wood yield (assuming a light use efficiency of 1.4 grams of total biomass production per megajoule and that 58 percent of total production is harvestable, see footnote 6) would be 22 and 26 tonnes per hectare per year at Plymouth and Des Moines, respectively, where the insolation, May to September, is 2,670 and 3,245 megajoules per m^2, respectively. (These empirically-based estimates of maximum practical yields are about 25 percent lower than the theoretical estimates presented earlier.) In tropical areas where trees grow year round, maximum yields would be much higher. For a tropical area where the annual average daily insolation is the same as for the period May to September in Des Moines (21.2 megajoules per m^2 per day), the maximum practical production rate would be $(12/5) \times 26 = 62$ tonnes per hectare per year of harvestable wood.

Not only can overall biomass production be increased when good nutrient balances are realized, but the trees thus produced tend to shift a percentage of their increased overall yield from roots to shoots (above-ground production). Generally, about 65 percent of above-ground growth is in wood (stem plus branches) and 35 percent in foliage. However, the partitioning between shoots and roots is much more variable and more susceptible to genetic manipulation and to nutrient and moisture optimization.[8] For example, increased nutrition of

8. For trees that can be coppiced, obtaining a high shoot-to-root ratio is not so important, since roots remain undisturbed from one rotation to the next in a physiologically and environmentally favorable system [13].

pines in a Swedish experiment produced trees with twice the total biomass production, with only 31 percent of the assimilated carbon going to the roots, compared with 59 percent in the unfertilized trees [9, 11, 12]. Experiments in the United States with hybrid cottonwood showed similar results.

Implications of rising CO_2 levels on biomass production

Biomass production, the atmospheric buildup of CO_2, and greenhouse warming are intimately related. In its 1990 scientific assessment of climate change, the Intergovernmental Panel on Climate Change (IPCC) estimated that the annual net release of CO_2 to the atmosphere from deforestation and related land use in the tropics is 1.6 ± 1.0 gigatonnes of carbon (GtC) per year (compared with 5.7 Gt C per year from the burning of fossil fuels), while at the same time regrowing forests in the northern hemisphere may be removing one to two GtC per year from the atmosphere [14].

Considerable attention has been given to massive reforestation efforts as a strategy for offsetting the release of CO_2 from the burning of fossil fuels by sequestering carbon in plants [14, 15], and more recently to the alternative strategy of growing biomass as a fossil fuel substitute [16, 17]. Much less attention has been given to the impact of rising CO_2 levels and associated climate change on the planet's flora in general, and on the managed growth of biomass in particular. As early as 1980, Roger Gifford of the Commonwealth Scientific and Industrial Research Organization in Australia estimated that, considering all changes in land use and bioproductivity, world biomass was already fixing CO_2 at a net rate of about one GtC more per year than at pre-industrial CO_2 levels [18]. There are many uncertainties associated with this issue, however. In its 1990 report, the IPCC summarized its assessment of the current scientific understanding as follows [14]: "At this time we have no evidence that elevated CO_2 has increased net carbon storage in natural ecosystems dominated by woody vegetation."

While little can be said with confidence about the impacts of increased CO_2 and associated climatic change on the general biosphere, the potential effects on biomass production under controlled circumstances is less uncertain.

The potential impacts of increasing CO_2 levels on biomass production include the following:

▼ Because the concentration of CO_2 would increase relative to O_2, photorespiration would be reduced in favor of photosynthesis; the effect on dark respiration is not so clear. CO_2 assimilation rates would probably increase, especially for C_3 plants. While C_4 plants are inherently more efficient at fixing CO_2 than C_3 plants (under moderate or warm conditions), experiments indicate that as CO_2 levels rise, the carbon fixation pathways of C_4 plants can rapidly become saturated while those of C_3 species (including all trees) will continue responding to the rising CO_2 levels [14]. In general, this CO_2 fertilization effect should be helpful for plantations of trees and C_3 herbaceous plants.

▼ Many studies indicate that as CO_2 levels increase, plants allocate proportionally more carbon below ground than above—thus increasing the root-to-shoot ratio [14]. This shift could offset, at least in part, the benefit for biomass energy crops of more rapid net CO_2 assimilation.

▼ Laboratory experiments indicate that the water-use efficiency of both C_3 and C_4 plants will increase with rising CO_2 levels (see figure 2). For example, if atmospheric CO_2 were to double, transpiration of water should decrease by 30 to 40 percent per unit leaf area for leaves of both plant types, while increasing the rate of photosynthesis approximately 30 percent for C_3 plants and having no major effect on the photosynthetic rate of C_4 plants. Considering both factors together, water-use efficiency should increase by about 35 percent for C_4 plants and 75 percent for C_3 plants [19]. The extent to which these findings hold for natural ecosystems or for cultivated crops is uncertain.

▼ A number of C_3 plants growing under nutrient-deficient conditions show increased growth when the CO_2 concentration is increased—one result being an increased carbon/nitrogen ratio. Various experiments also suggest that increased nitrogen fixation would be associated with increased atmospheric CO_2 levels [14]. Both effects could reduce fertilizer requirements for plantation biomass.

▼ With increased CO_2 levels, leaf canopies might be established earlier for annual plants, and there is evidence that delayed senescence might occur in some species. Although a prolonged growing season could increase crop yields, it might also expose plants to frost damage [14].

The effects on biomass production of changes in temperature and moisture induced by rising CO_2 emissions must also be taken into account. Unfortunately, even the direction of the effect is not known. One line of reasoning is that because respiration increases faster with temperature than gross photosynthesis, global warming could lead to a net reduction in carbon uptake by plants; alternatively, it has been argued that since for C_3 plants the optimum temperature for net photosynthesis increases with rising CO_2 levels, there would be more plant growth at higher temperatures [14]. These complex interactions clearly warrant closer scrutiny [20].

On balance, the evidence suggests that rising CO_2 levels may be beneficial to growing biomass for energy purposes. Many of the potential detrimental impacts of rising CO_2 levels on natural ecosystems relate to limiting factors such as nutrients and water, which could be adjusted for biomass plantations (except perhaps where even good management may be inadequate to offset adverse impacts, e.g., severe drought). However, there are as yet many uncertainties associated with this issue. Moreover, the potential benefits of increased CO_2 levels for biomass plantations, if any, would be small relative to the potential costs for the biosphere as a whole. If the climate changes rapidly, as is expected if current trends in energy production and use and deforestation persist, it may be difficult for many indi-

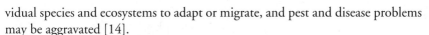

vidual species and ecosystems to adapt or migrate, and pest and disease problems may be aggravated [14].

But even with major changes in policy relating to energy and deforestation aimed at reducing global CO_2 emissions, the atmospheric concentration of CO_2 will continue to increase for decades. Accordingly, it is imperative to gain a better understanding of the potential effects of increased CO_2 levels on biomass production and to take this understanding into account in biomass energy planning.

ALTERNATE SOURCES OF BIOMASS

The most significant potential sources of biomass for energy are residues, wood resources from natural forests, and biomass from managed plantations.[9]

Biomass residues

Biomass residues are the organic byproducts of food, fiber, and forest production. The energy value of residues generated worldwide by the forest-products industry and in selected agricultural activities (three grains, sugarcane, dung) is estimated to be 111 exajoules per year (see table 2), equivalent to about one-third of total primary commercial energy use in 1985; for developing countries, the corresponding residue production rate is 69 exajoules per year, nearly equal to the rate of commercial energy consumption (see figure 1).

Not all residues generated can be utilized for energy. In some instances, it would not be economical to do so, such as when their wide dispersal or low bulk density makes recovery, transport, and storage too costly. Or residues may be more valuable if used for purposes other than energy. One such use would be recycling residues onto the land, to help restore nutrients or reduce erosion; or residues might be recovered for other domestic, industrial, or agricultural uses, such as for building materials, paper manufacturing, or animal fodder.

In parts of the developing world where fuelwood is scarce, such as China, the northern plains of India, Bangladesh and Pakistan, crop residues and/or dung are often the major cooking fuels for rural households. In villages, such residues can account for up to 90 percent of household energy. It has been estimated by the FAO that about 800 million people worldwide rely on crop residues and dung for energy.

In some industries, residues are routinely burned for fuel. For example, sugar factories and alcohol distilleries throughout the world use bagasse, the residue left after sugarcane is crushed to recover the sugar juice, to run their operations. In the United States the forest product and agricultural industries have some 8,000 megawatts-electric (MW_e) of installed electrical generating capacity fueled by

---

9. At present out-of-forest trees (including trees from agroforestry plots, on-farm trees, road-side trees, and home-garden trees) are an important biomass energy source for cooking and other local needs. Consideration of such resources is beyond the scope of this chapter, which focuses on biomass sources that could meet a substantial fraction of global energy needs.

Table 2: Energy content of selected biomass residues[a]
exajoules per year

Region	Maize[b]	Wheat[b]	Rice[b]	Sugar cane[c]	Dung[d]	Roundwood Industrial[e]	Fuelwood/ charcoal[f]
Industrialized							
US/Canada	2.95	1.93	0.13	0.19	3.08	7.66	0.92
Europe	0.61	2.39	0.04	0	4.22	4.12	0.41
Japan	0	0.02	0.24	0.01	0.30	0.41	0
Australia + NZ	0	0.29	0.02	0.19	1.36	0.35	0.02
Former USSR	0.23	1.97	0.04	0	3.58	3.92	0.60
Subtotals	*3.8*	*6.6*	*0.5*	*0.4*	*12.5*	*16.5*	*1.9*
Developing							
Latin America	0.71	0.38	0.29	3.58	7.21	1.47	2.12
Africa	0.48	0.25	0.20	0.54	5.38	0.75	3.31
China	1.23	1.75	3.43	0.48	4.81	1.27	1.34
Other Asia	0.51	1.88	5.29	2.70	10.91	2.31	4.62
Oceania	0	0	0	0.03	0.02	0.05	0.04
Subtotals	*2.9*	*4.3*	*9.2*	*7.3*	*28.3*	*5.8*	*11.4*
World	**6.7**	**10.9**	**9.7**	**7.7**	**40.8**	**22.3**	**13.3**

a. The energy content (higher heating value basis) of the total residues produced for the production rates for food crops and forest products indicated in Table 3, plus dung production rates by livestock, as described in note d.

b. The residue coefficients (in tonnes of residue per tonne of primary product) for air-dried residues are assumed to be 1.0, 1.3, and 1.4, for maize, wheat, and rice, respectively [82]. It is assumed that the residues contain 20 percent moisture. Higher heating values are 17.65, 17.51, and 16.28 gigajoules per dry tonne for residues of maize, wheat, and rice, respectively.

c. For sugar cane, approximately 150 kg of bagasse, 92 kg of attached tops and leaves, and 188 kg of detached leaves (dry weight basis) are produced along with each tonne of fresh cane (millable cane stems) [65]. Since the higher heating value of cane residues is 17.33 gigajoules per tonne, the energy content of these residues is 7.45 gigajoules per tonne of fresh cane.

d. Dung production rates are estimated from FAO and UN Population Division data on the numbers of animals for each commercial species and the following coefficients [65a]:

	Annual production	Heating Value
	tonnes, dry basis	*gigajoules per tonne*
Cattle	1.10	15.0
Sheep and goats	0.18	17.8
Pigs	0.22	17.0
Equines	0.55	14.9
Buffaloes and camels	1.46	14.9
Chickens	0.037	13.5

e. Industrial roundwood accounts on average for about 60% of the total biomass cut when trees are harvested. Milling and manufacturing wood wastes are approximately equal to the wood in the final forest product. Thus the total residue production rate is (40 + 30)/60 = 1.17 times roundwood production. The higher heating value of air-dry wood (assumed to be 20 percent moisture) is assumed to be 16 gigajoules per tonne, on average.

f. As in the case of industrial roundwood, it is assumed that the roundwood harvested for fuelwood and charcoal applications is 60 percent of the total biomass cut. Thus the residue production rate is 40/60 = 0.67 times the production rate of roundwood for fuelwood and charcoal applications.

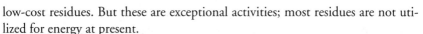

low-cost residues. But these are exceptional activities; most residues are not utilized for energy at present.

Even when residues are used for energy, they are usually burned inefficiently. Fuelwood cooking-stoves, for example, are very inefficient compared with stoves that use modern energy carriers (see chapter 15, *Bioenergy: Direct Applications in Cooking*). Most cane sugar factories are also extraordinarily inefficient in their use of biomass; although they obtain some electric power by burning bagasse, they are designed more as incinerators than efficient energy conversion facilities.[10] The inefficiency of residue use is not restricted to developing countries. Most biomass powerplants in the United States are about half as energy-efficient as central-station fossil fuel powerplants (see chapter 17).

The practical potential for utilizing residues for energy depends on the sustainability of the primary biomass production activity, its environmental impacts, and on the competition for the residues.

Crop residues

Crop residues can be collected either with conventional harvesting equipment, mostly by baling, after the primary crop has been harvested, or at the same time the crop is harvested. Although conventional forage harvesting equipment can be used to collect cereal straw, specialized collection systems are needed in most cases [22]. The appropriate technique depends on the crop, climate, topography, and on the relative costs of capital and labor.

A central question, however, is, how much residue should be left behind and recycled into the soil to help ensure that the sustainable production of the primary biomass product is not jeopardized? Recycling residues provides nutrients, helps prevent erosion, and provides a variety of other soil-quality benefits.

The adverse impacts of residue removal on soil nutrient balance can be serious for dryland agriculture, especially in the poorest areas of Asia, Africa, and Latin America, where chemical fertilizer inputs are low. Where the nutrient reserves are low to begin with, residue removal without compensating inputs will further degrade soil quality [23]. In contrast, organic matter from residues contributes little additional value as a source of nutrients in low-input wetland agriculture, for which farms receive substantial nutrient inputs from silt and the dissolved nutrients carried in irrigation water. In Bangladesh, for example, rice fields receive large doses of nutrient-rich silt carried down the rivers each monsoon season [23].

With high-input agriculture, chemical fertilizers provide most of the needed nutrients, so that residue recycling is not essential, as long as these chemical inputs are continuously available to compensate for nutrient losses from removals. How-

---

10. The energy content of cane residues is so large that with energy-efficient conversion technologies they could provide all onsite energy needs plus additional electricity up to 40 times onsite needs (see chapters 17 and 20 and [21]). Yet of the more than 2,000 sugar factories in more than 80 sugarcane-producing countries, only a small number in Brazil, Hawaii, Mauritius, Reunion, Thailand, and Zimbabwe produce more electricity than is needed onsite.

ever, residue recycling can help prevent the eventual depletion of various minor nutrients that are not routinely replenished via chemical inputs [23].

The amount of residues needed for erosion control varies markedly with site conditions. For example, erosion losses are generally much lower in flat areas than in hilly ones, unless terracing is maintained. Whereas water erosion and leaching hazards are most pronounced in tropical areas with heavy rainfall and in areas with intense snow melting, wind erosion is a hazard in drought-stricken areas before crops are well established. Clearly, fields with bare soil or a thin vegetation cover are the most prone to erosion [22].

For the U.S. corn belt, calculations have been made of the amount of residue recycling needed to bring erosion below the "soil-loss tolerance level," which is defined as the "maximum level of soil erosion that will permit a high level of crop productivity to be sustained economically and indefinitely." By this criterion, the fraction of residues that can be removed with conventional tillage practices averages 35 percent for the corn belt as a whole, with considerable variation from site to site [24].

The understanding of the extent that residues are needed for erosion control in tropical regions is not nearly as well developed as for temperate regions. Moreover, experience in temperate regions may not be especially relevant to tropical regions, where soil types and agricultural practices are different, topsoil is generally thinner, and rains tend to be more intense. Such differences imply, for example, that erosion can be a more pronounced problem [23].

Too much residue recycling can also be problematic. It can depress the growth of some crops (because mulched soil remains cooler than bare soil), inhibit nitrate formation, generate materials that are toxic to germination and growth, and promote the spread of plant diseases. As examples, cotton and tobacco stalks harbor pests and diseases and thus cannot be left to rot in the fields. It is also very difficult to recycle some residues, e.g. corn cobs, millet stalks, coffee prunings, and coconut shells.

In developing countries, where ploughing residues back into the soil by hand is difficult and the spread of pests and plant disease may be promoted by recycling, residues are often burned in the field. Rice residues are commonly disposed of this way in Burma, Indonesia, Malaysia, and Thailand [23]. In many sugarcane-producing regions, cane fields are burned just before harvest to facilitate harvesting; in so doing the tops and leaves are burned but the tough cane stem is undamaged. In cool, humid climates, where organic materials decompose slowly, straw is also often burned in the field. Because of growing air pollution concerns, efforts are underway in many parts of the world to curb the burning of residues in the field. Bans on this activity can free substantial amounts of biomass resources for energy purposes.[11]

---

11. Even if burning is curtailed, not all the residues should be recovered for energy purposes. Consider sugarcane. While there are no good empirical data indicating the amount of cane residues needed to sustain soil quality, one estimate is that about one-fourth of the tops and leaves should remain in the field for erosion control (private communication from L. Santo, an agronomist with the Hawaiian Sugar Planters' Association, 25 September 1991).

The United Kingdom, for example, produces about 14 million tonnes of straw annually, half of which is usually burned in the field. But because this practice will be banned in 1993, farmers and energy producers will be motivated to find ways to profit from this potential energy source.

The negative impacts of residue removal can be mitigated in various ways. If ash removed from the stack gases or fuel-processing equipment at a biomass conversion facility is returned to the fields, most mineral nutrients are restored. Of course, measures might have to be taken to restore nitrogen because all nitrogen is usually lost during energy conversion.

The adverse impacts of residue removal can also be reduced by changing agricultural production practices. A study of the U.S. corn belt showed, for example, that by shifting from conventional to conservation tillage the recoverable maize residues could be increased from 35 to 52 percent and that shifting to no-till agriculture could increase recoverable residues further, to 58 percent [24].

The amount of residues that can potentially be utilized for energy purposes also depends on recovery and storage technologies. In a 1980 study, the Office of Technology Assessment of the U.S. Congress estimated that on average for the United States, about one-fifth of residues could actually be utilized because only 60 percent of the residue that might be removed (subject to soil quality constraints) could be recovered with the technology then available and losses from storage would average about 15 percent [25]. But, of course, these parameters can change markedly over time, as technology improves.

Residues will never be exploited for energy purposes unless they can be recovered and delivered to energy conversion facilities at competitive costs. Costs will vary greatly, depending on the crop, climate, topography, costs of labor, and other inputs—as well as the opportunity costs associated with using the biomass for energy instead of for some other purpose. Some examples, however, suggest that residues will often be economically attractive.

In the case of sugarcane, trials involving 4,600 tonnes of cane tops and leaves were carried out in Thailand in 1989 and 1990, and estimates were made of the cost of the recovered residues, taking into account payments to farmers, the costs of harvesting, collecting, loading, and transporting the residues 10 kilometers to the factory, and the value of the nutrient losses (an opportunity cost). The resulting estimated cost was found to be about $43 per tonne (dry basis), or about $2.5 per gigajoule [26].

In the United Kingdom, the cost of straw recovered from the fields, baled and delivered to industrial customers ranges from $2.5 to $3.1 per gigajoule [27,28]. The significance of these costs depends on the conversion technologies involved.

For most commercially available equipment, which is typically inefficient, these costs are generally too high to justify residue recovery. However, if used in conjunction with the modern conversion technologies described elsewhere in this book, residues at such prices would be economically attractive (see, for example, chapter 17).

Dung residues

Animal dung is a potentially large biomass resource (see table 2): healthy animals produce dry dung that is equivalent to four or five times their body weight each year [23]. But dung is readily recoverable only for confined livestock or in settings where the labor costs associated with gathering dung are modest. Therefore, practically recoverable supplies are likely to be much less than the generation rates shown in table 2.

Because dung is an important source of fertilizer, concern is often expressed about its diversion to energy purposes. But in regions where animals feed primarily by grazing, as much as 80 percent of nitrogen in the dung is lost through ammonia volatilization [23].

If biogas (a mixture of methane and carbon dioxide) were produced from dung via anaerobic digestion, there would be no conflict between energy recovery and nutrient utilization, if the effluent from the digester were returned to the fields; in fact the effluent is a better fertilizer than the dung from which it is derived (see chapter 18).

Forest-product industry residues

Substantial biomass residues are generated by the forest-product industry, both in the forest at the time of harvest, and at mills and manufacturing facilities where the wood is converted into consumer products.

According to FAO statistics, about half of the world's roundwood (the recovered trunks and large branches of the harvested trees) is used for industrial products (wood, pulp and paper products) and half for fuelwood and charcoal production; industrialized countries account for three-fourths of the industrial roundwood, while developing countries account for five-sixths of the roundwood used for fuelwood and charcoal (see table 3).

Forest residues can be obtained by collecting branches, tops, and culled logs after the primary forest harvest, or they can be obtained via whole-tree harvesting, whereby a harvester cuts and transports the entire tree out of the area.

Although, in many parts of the world, mill and manufacturing residues are not yet utilized for their energy value, it would often be both economically and environmentally desirable to do so. Such resources are essentially free, and their use for energy purposes offers environmentally acceptable ways to dispose of unwanted and often polluting waste. The only environmental problems associated with this use are those associated with energy conversion, which can be minimized with modern conversion technologies. Moreover, returning ashes left after

Table 3: Production of selected foodcrops (1990) and roundwood (1989)[a]
million tonnes per year

Region	Maize	Wheat	Rice	Sugar cane	Roundwood[b] Industrial	Fuelwood/ charcoal
Industrialized						
US/Canada	209	106	7	25	410	86
Europe	43	131	2	0	221	38
Japan	0	1	13	2	22	0
Australia + NZ	0	16	1	26	19	2
Former USSR	16	108	2	0	210	56
Subtotals	*268*	*362*	*25*	*53*	*882*	*182*
Developing						
Latin America	50	21	16	481	79	198
Africa	34	14	11	73	40	309
China	87	96	188	64	68	125
Other Asia	36	103	290	362	124	431
Oceania	0	0	0	4	3	4
Subtotals	*207*	*234*	*505*	*984*	*314*	*1,067*
World	**475**	**597**	**532**	**1,037**	**1,197**	**1,249**

a. [83].

b. Original FAO data in cubic meters converted to air-dry tonnes of wood, assuming a mass density of 0.7 tonnes per cubic meter.

conversion to the forest can help restore nutrient losses associated with biomass removals.

The risks and benefits of residue removal from forests vary with the species, the region, and site conditions. One ecological concern is that residue removal can adversely affect detritivores that inhabit rotting logs. Residue removal can also adversely affect nutrient balances in the forest. In boreal conifer stands, for instance, where the soil is acidic, residue removal can increase acidification, by decreasing the neutralization effect of biomass decomposition. As the pH of the soil falls below 5, nutrients, especially calcium, potassium, and manganese, begin to leach out. Also, aluminum solubility increases when the pH drops below 4.5, releasing aluminum ions that are toxic to the root system [22].

But as with crop residues, residue removal can provide environmental and silvicultural benefits. For example, the U.S. Forest Service requires that loggers pile and burn logging residues on most of its land; many private land owners also require such practices to foster regeneration of the forest after logging; the recovery of logging residues for energy purposes would be preferable to burning this way. And greater removal of forest residues might help reduce the overload of nitrogen in the soil in areas where air pollutant deposition is significant, as is the case in southern Sweden [29].

On balance, some residues should be left in the forest, not only to contribute to soil fertility but also to help maintain biodiversity within the forest. Guidelines are needed for determining acceptable levels of residue removal. While some guidelines would vary from one situation to another, others would be generally applicable. Because about 70 percent of a tree's nutrients are concentrated in the foliage, twigs, and fine roots, it is desirable to leave these parts in the forest [30]. (Whole-tree harvesting technology makes it especially difficult to maintain adequate organic matter and nutrients in the soils.) Also, efforts should be made to minimize soil disturbances during harvesting. For example, in areas where the ground freezes, winter harvesting can minimize soil disturbance.

As with crop residues, forest residues will be recovered only if the costs are not prohibitive. The costs of forest residues delivered to the mill will often be in the range $2 to $3 per gigajoule [31], which is an attractive price, if modern energy conversion technologies are employed (see chapter 17).

Biomass from existing forests

Existing forests can often provide additional biomass for energy beyond that offered by logging residues. In many temperate forests, annual removals are much less than annual growth. For example, a study by the U.S. Office of Technology Assessment (OTA) [25] estimated that in the 1970s net annual growth in U.S. commercial forests was 400 to 800 million (dry) tonnes per year, while annual harvests of industrial roundwood totalled only 180 million tonnes per year. When harvests are much less than growth, forest yields tend to be lower than they otherwise would be. Much of the unharvested stock, which is often too low in quality for sale in traditional markets, is well-suited for energy applications. Removal of the low-quality woodstock for energy purposes can simultaneously lead to enhanced yields of high quality wood.

The OTA study [25] further estimated that, with full stocking, net annual growth of biomass in U.S commercial forests could be doubled, to 800 to 1,600 million tonnes per year (corresponding to an average productivity of four to eight tonnes per hectare per year). The energy value of the potential production in excess of current removals (220 million tonnes in 1986) is 12 to 28 exajoules per year. For comparison, total energy use in the United States in 1987 was 85 exajoules per year.

At the global level, the physical potential for utilizing wood from existing forests for energy is very uncertain. Mean annual increment data are not available in reliable forms for most countries. The determination of standing stocks on a

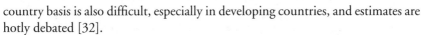

country basis is also difficult, especially in developing countries, and estimates are hotly debated [32].

Despite the difficulties, some attempts have been made to estimate global biomass production in forests. A 1975 estimate is that the total above-ground increment of wood in 1970 was 17.8 billion cubic meters on 3,800 million hectares of global forest lands [33]. A more recent estimate is that the annual yield from closed forests and open woodlands is 12.5 billion cubic meters (for an estimated growing stock of 417 billion cubic meters on 4,100 million hectares of global forests) [34].[12] For comparison, the estimated global average annual wood harvest in the period 1985 to 1987 was 3.3 billion cubic meters. Assuming an annual increment of 12.5 billion cubic meters, the unused increment is thus 9.2 billion cubic meters per year, representing an energy value of 105 exajoules per year.[13]

Some of the unused increment might be used for energy purposes. However, recovering wood from natural forests for energy purposes on a large scale is likely to run into strong opposition from environmental groups. Indeed, one of the major concerns voiced at the 1992 United Nations Conference on Environment and Development in Rio de Janeiro is the loss of biological diversity associated with the overuse of natural forests.

While the recovery of biomass for energy purposes from natural forests could be carried out sustainably and, in many instances, would increase forest productivity, the forests would be transformed into managed forests, with less biodiversity than natural forests.

The intensity of environmental debates on biological diversity varies considerably according to the type of forest in question. The greatest concerns are voiced for primary-growth forests: most notably tropical rainforests and old-growth conifer stands in areas like the Pacific Northwest of North America.

One strategy would be to preserve most of the remaining old-growth forests, compensating for their protection by more intensively managing regrowth forests. Harvesting low-quality wood for energy purposes would increase the productivity of high-quality wood in these regrowth forests, thereby helping alleviate the pressure to exploit original-growth forests.

Despite the appeal of such a strategy, its implementation requires societal consensus as to which forest lands will be preserved and which will be managed forests, and as to the degree of biological diversity desired in managed forests [29].

Until such consensus can be achieved, the use of existing forests for energy purposes will be associated largely with the recovery of residues from forest-product industry activities in accord with guidelines of sustainability [29].

---

12. This is for roundwood, not the total above-ground increment, so that this probably leads to an underestimate of the unused biomass.

13. Based on the FAO convention that one cubic meter contains 0.706 tonnes of air dry wood (assumed to be 20 percent moisture). There is considerable uncertainty as to whether this conversion factor is a good global average. For temperate hardwoods and softwoods typical densities are 0.53 and 0.44 tonnes per m^3, respectively [35]. Many tropical woods are much denser, but there are also many low-density woods in tropical regions.

Biomass plantations for energy

Biomass feedstocks for energy can be provided either by so-called short-rotation intensive-culture (SRIC) plantations of trees or plantations of herbaceous plants. The latter can be either thick-stemmed grasses, such as sorghum or sugarcane, or thin-stemmed grasses, such as switchgrass (a species endemic to the prairies of the midwestern United States).

Biomass energy crops would be intensively managed like agricultural crops. For annual row crops such as sorghum, planting, management, and harvesting would be virtually the same as for annual food crops. But SRIC crops would be harvested much less frequently—typically every three to eight years, with replanting every 15 to 30 years. (One or more additional rotations would usually be obtained by coppicing after the first cut.) For perennial grasses, harvesting would take place every six to 12 months, but replanting would be infrequent—perhaps only once a decade.

At present, there are about 100 million hectares of land in industrial tree plantations worldwide; most, however, are dedicated to relatively slow-growing trees for traditional forest-product markets. There are only about 6 million hectares of plantations with fast-growing non-coniferous (hardwood) trees, the type best suited for energy markets (see table 4). For comparison, the amounts of cropland and forest/woodland in the world are about 1,480 and 4,100 million hectares, respectively.

While the total amount of land in plantations is small, it is growing rapidly. According to the FAO, 60 percent of the reported plantation area in tropical regions was established between 1981 and 1990, at an average rate of 2.6 million hectares per year [36].

Review of experience with plantations

A review of plantation experience is helpful in understanding how plantations should be established and operated.

Jari. In 1968, the North American industrialist D.K. Ludwig launched an ambitious project to establish a large industrial complex at the 1.6 million hectare Jari estate in the rainforest on the north shore of the Amazon River near its mouth. (In 1982 Ludwig sold his interest to a consortium of Brazilian firms.) Tree plantations for pulpwood have been the central focus of Jari's activities. By 1988 a total of 90,000 hectares had been planted, mostly with *Gmelina arborea, Pinus caribaea,* and four species of *Eucalyptus.*

Jari has not been a financial success [37,38]. Aside from the adverse impacts of an unexpected depressed world pulp market, the effort to industrialize the tropical rainforests and establish plantations of exotic (non-native) trees there has been plagued by a multitude of unforeseen problems. It has been difficult to control weeds; pests and diseases have been devastating; some of the species planted have been poorly suited for the soils; and realized biomass productivities have been much lower than expected.

Table 4: Land area in plantations[a] (*million hectares*)

	Industrial plantations[b]			Industrial plantations of fast-growing trees[c]		
	Conif.	Non-conif.	Total	Conif.	Non-conif.	Total
North America						
United States				12.00[d]	0.50	12.50
Subtotal	*12.0*	*0.5*	*12.5*	*12.00*	*0.50*	*12.50*
Europe						
Spain				–	0.45	0.45
Portugal				–	0.40	0.40
Subtotal	*15.7*	*3.3*	*19.0*	*–*	*0.85*	*0.85*
Oceania						
New Zealand				1.18	0.02	1.20
Australia				0.90	0.06	0.96
Subtotal	*2.1*	*0.1*	*2.2*	*2.13*	*0.11*	*2.24*
Former USSR	16.9	–	16.9	–	–	–
Latin America						
Brazil				1.60	2.30	3.90
Chile				1.14	0.06	1.20
Argentina				0.46	0.18	0.64
Venezuela				0.18	0.02	0.20
Mexico				0.06	0.02	0.08
Other				0.08	0.35	0.43
Subtotal	*3.5*	*2.9*	*6.4*	*3.52*	*2.93*	*6.45*
Africa						
South Africa				0.50	0.80	1.30
Angola				0.02	0.05	0.07
Congo				–	0.04	0.04
Kenya				0.16	0.01	0.17
Zimbabwe				0.07	0.01	0.08
Other				0.55	0.33	0.88
Subtotal	*1.3*	*1.2*	*2.5*	*1.30*	*1.24*	*2.54*
Asia						
Indonesia				–	0.10	0.10
China		–		–	0.40	0.40
Other				–	0.17	0.10
Subtotal	*31.6*	*8.2*	*39.8*	*–*	*0.67*	*0.67*
Total	83.1	16.2	99.3	18.95	6.30	25.25

a. [84].

b. The majority of the temperate zone plantations tend to involve slow-growing trees, with increments of the order of 5 to 6 m^3 per hectare per year and rotation periods of up to 100 years.

c. Average yields for fast-growing industrial plantations are typically in the range 18 to 30 m^3 per hectare per year for *Eucalyptus* species (7 to 12 year rotations) and 15 to 25 m^3 per hectare per year for *Pinus* species (25 to 35 year rotations).

d. Borderline fast-growing (12 m^3 per hectare per year for pine plantations).

Major threats to its silvicultural operations in the short- and medium-terms include the spread of the *Ceratocystis* fungus in *G. arborea*, the most valuable commercial species, and increasing fire hazards from the agricultural activities of squatters adjacent to the plantations. Potential long-term problems include soil degradation from erosion and soil compaction, and a decline of soil fertility relative to newly cleared areas [39].

Jari's experience should be remembered when similar projects are planned. Replacing rainforests with managed plantations has proved to be undesirable. Besides posing substantial financial and technological risks, such projects face increasing opposition from environmental groups.

Dendrothermal Power.   One of the few large-scale plantation efforts dedicated to energy purposes is the Philippine Dendrothermal Power Program. Established in 1979 to help reduce dependence on imported oil, the program aimed to produce electricity from biomass, mainly with small-scale (3 MW_e each) steam-electric power plants, each fueled with an estimated 36,000 tonnes of green wood per year (and each requiring at least 1,200 hectares of plantations). Many of the plantations were established on grasslands, lands covered by brush, and secondary forest lands. Giant ipil-ipil (*Leucaena leucocephala*) was the most frequently planted species, with normal harvests 3 to 4 years after establishment and coppice crops every two or three years thereafter.

The program fell far short of expectations. Of 217 power plants that were to be completed by 1987, only nine were operational by the mid 1980s [40]. Aside from various factors that adversely affected the overall program [political crises, a low world oil price, and the inherently poor economics of producing electricity from plantation biomass in small steam-electric plants (see chapter 17)], a central problem for the plantation part of the program was that the cost of fuelwood production was typically higher than the price of wood purchased on the open market [40].

Wherever fuelwood from natural forests is abundantly available, it will be difficult to establish successful plantations, because plantation wood will be more costly. However, deforestation pressures or environmental restrictions on the harvesting of wood from natural forests will undoubtedly help support market development for plantation wood.

Eucalyptus plantations in Ethiopia.   *Eucalyptus globulus* was introduced into Ethiopia in 1895, as a response to a fuel shortage due to an overcutting of natural forests that threatened the very existence of Addis Ababa, the capital. The introduction was a success, and so early in the 20th century *Eucalyptus* plantations were initiated on a large scale. Today established plantations of *E. globulus* amount to about 100,000 hectares and the total plantation area (including other species)

is over 300,000 hectares. The Ethiopian government hopes to reforest 2.9 million hectares of the highlands.[14]

The introduction of *E. globulus* is judged a success because it is well adapted to the highland climate and soil conditions, coppices vigorously, is suitable for fuelwood and small poles for construction, and is unpalatable to livestock. Despite the fact that it was introduced as an exotic species, no serious ecological problems have been encountered since it was introduced almost a century ago [41].

Eucalyptus is sometimes described as ecologically undesirable, e.g., promoting erosion by preventing the growth of ground vegetation. But this argument does not hold under Ethiopian conditions, where the planting of *E. globulus* on barren highland sites has instead reduced erosion. Moreover, reestablishment of desirable native species is difficult; if seedlings are planted in an open area, they soon die in the sun; they usually regenerate only in the shade of a mature forest. In contrast, *E. globulus* is readily established on degraded lands and so may actually be the key to subsequently reintroducing desirable native species [41].

The most important lesson provided by the Ethiopian experience is that in some instances exotic species may be more successful than native species, especially in areas were the land is severely degraded. The most appropriate species for plantations are the ones best adapted to present site conditions and not to those in the distant past.

Eucalyptus in Brazil. Brazil has successfully established many plantations in the south and southeastern parts of the country, aided by its 1965 National Forestry Act, which provided considerable tax incentives for plantation establishment. A federal agency, the Brazilian Institute for Forest Development, which was established to oversee the program, immediately launched tree improvement research programs, including provenance trials for species of both *Eucalyptus* and *Pinus*. While many of the early plantations were initiated solely to gain tax benefits or to comply with reforestation laws and usually resulted in poor yields, Brazil now has a strong reforestation industry, supported by extensive research and development programs. The country has the largest concentration of SRIC plantations in the world, accounting for more than half of its total plantation area of over 6 million hectares.[15]

Brazil implemented a major reforestation policy in the 1960s, largely in response to deforestation in the south and southeast (by 1960 only 5 percent of the original forest remained) and to perceived needs for forest products for industrial growth. Besides the traditional products of the forest product industry (lumber, plywood, pulp and paper), Brazil also needed charcoal for iron- and steel-making.[16]

---

14. The forest cover in Ethiopia has been reduced from 40 percent in 1900 to 2.7 percent (3.5 million hectares) in 1985; the present rate of deforestation is about 100,000 hectares per year [41].

15. One estimate is that as of 1986 there were 6.2 million hectares of plantations in Brazil, of which 3.45 were *Eucalyptus*, 1.86 were *Pinus*, 1.12 were others [42]. Table 4 shows a lower estimate.

16. Brazil chose to evolve a steel industry based to a large degree on the use of charcoal because Brazilian coal resources are small and of low quality and because imported coke is costly.

As in Ethiopia, most of these plantations have been established on deforested and otherwise degraded lands. At well-managed plantations it is common practice to regenerate natural forests on a significant fraction of the land held by the plantation owner. For example, at Aracruz in Espirito Santo, where *Eucalyptus* is grown to make pulp, one hectare is held in natural reserves, on average, for each 2.4 hectares of plantations [43]. Maintaining these natural reserves is not just good public relations, it sustains a biologically diverse community of birds and other insectivores that helps control pests on the plantation, thereby increasing yields.

Improvements in soil preparation, planting, cultivation methods, species selection and vegetative propagation via clonal techniques,[17] and pest, disease, and fire control, have all contributed to significant increases in average yields per hectare over the past two decades. Large commercial plantations in Minas Gerais and Bahia have achieved average yields in the range 30 to 44 m^3 per hectare per year (see table 5). The favorable economics of plantation wood production in Brazil is indicated by the fact that Brazilian plantation-derived pulp and charcoal-derived steel are highly competitive in world markets.[18]

In the 1980s, the Hydroelectric Company of San Francisco (CHESF), the federally owned utility responsible for the generation and production of bulk electricity in the northeast, identified the production of electricity from biomass as a potentially major application of plantation wood. CHESF became interested in this strategy because its hydroelectric power supplies will be fully developed before the turn of the century. In searching for alternative electricity sources, it identified the biomass integrated gasifier/gas turbine power system (see chapter 17) as the most promising option. Interest in this potential power source led CHESF to carry out a major assessment of the prospects for large-scale production of biomass for energy in the northeast, including a major biogeoclimatic analysis of all the land in the northeast and its suitability for plantations [45]. For each of five major biogeoclimatic zones, detailed studies were carried out to estimate expected yields on lands that are not suitable for agriculture (mainly deforested or otherwise degraded lands), based on actual experience with commercial plantations (see table 5) and with experimental plots for these different zones. (No account was taken of future improvements in biomass production technology.)

The CHESF assessment showed that plantations could be established on up to 50 million hectares (one-third of the land area in the northeast) at an average productivity of about 27 m^3 (about 12.5 dry tonnes) of harvestable stem per hect-

---

17. In clonal propagation, trees are grown from identical genetic material, thus creating a uniform feedstock (especially important for pulpwood production) and making it possible to select for disease resistance, and other desirable traits that make trees well-adapted to local environmental conditions. A single clone might be propagated over an area of 10 to 25 hectares. Thus, many clones are used in a typical plantation, for diversity and optimum site specificity.

18. All of Brazil's pulp is from plantation wood. In 1991, about 32 percent of pig iron and 24 percent of steel were made with charcoal, 42 percent of which was derived from plantations—compared with 17 percent in 1984 [44]. The share from plantations is increasing as a result of regulations that discourage cutting indigenous forests for charcoal production.

Table 5: Data for some commercial *Eucalyptus* plantations in Brazil[a]

Company	Location	Planted area 10^3 ha	Average temp. °C	Altitude km	Annual precip. mm	Rotation length years	Yield[b] Peak solid m^3/year	Yield[b] Ave. solid m^3/year
CAF[c]	Bom Despacho Minas Gerais	30.0	22	0.70	1,375	4.4	50.3	44.0
CAF[c]	Carbonita Minas Gerais	25.4	21	0.73	1,025	6.0	24.5	21.0
FLORESA[c]	Vale do Jequitinhona Minas Gerais	100	21	0.73	1,025	4.4	37.6	31.8
COPENER[c]	Inhambupe Bahia	18.5	24	0.20	900	6.0	–	30.0
COPENER[c]	Alahoinhas Bahia	8.6	24	0.20	1,100	6.0	–	30.0
COPENER[c]	Entre Rios Bahia	13.9	24	0.20	1,100	6.0	–	40.0
Cimento Nassua	Barbalha Ceara	1.2	21	0.90	650	7.0	–	14.7

a. [45].

b. Stemwood only (no tops and branches).

c. CAF = Companhia Agricola Florestal Santa Barbara; FLORESA = Florestal Acesita; COPENER = Copene Energetica.

are per year (see table 6). The total potential biomass energy production from these plantations (some 12.6 exajoules per year) is equivalent to four times the total commercial energy consumption in Brazil in 1988. If this much biomass were used solely to generate electricity with advanced biomass integrated gasifier/gas turbine technologies, it could provide some 1,500 terawatt-hours (TWh) of electricity per year—more than 50 times the total electricity generated in the northeast or six times the total generated in all of Brazil in 1990 [45]. Thus, exploiting even a fraction of the biomass potential in the northeast could contribute significantly to Brazil's energy needs, as well as promote rural development in one of the country's poorest regions.

The average plantation yields estimated in the CHESF study are considerably less than yields being routinely achieved in the southeast, largely because biomass production is water-limited in most parts of the northeast.[19] Nevertheless,

---

19. This is indicated by a water deficit for all zones (see table 6). The water deficit (in meters of water) is defined as the difference between (i) the amount of water that could theoretically be transpired by plants and evaporated from the soil under the local environmental conditions with the incident solar radiant energy (so-called "potential evapotranspiration"), and (ii) the actual precipitation.

Table 6: Levelized cost of *Eucalytptus* wood for plantations in the northeast of Brazil, by bioclimatic region[a]

Levelized cost[b]	I	II	III	IV	V	Weighted Ave., NE
			dollars per gigajoule			
Plantation establishment						
Nursery production	0.03	0.04	0.05	0.09	0.24	0.05
Land	0.08	0.11	0.13	0.25	0.61	0.14
Planting	0.21	0.28	0.33	0.62	1.54	0.35
Administration	0.01	0.01	0.01	0.02	0.05	0.01
Subtotal	*0.33*	*0.33*	*0.52*	*0.98*	*2.44*	*0.55*
Plantation maintenance						
Management	0.01	0.01	0.01	0.02	0.04	0.01
Cultivation	0.05	0.07	0.09	0.16	0.40	0.09
Research	0.02	0.03	0.03	0.06	0.15	0.03
Harvest	0.35	0.35	0.35	0.35	0.35	0.35
Transport (85 km)	0.33	0.33	0.33	0.33	0.33	0.33
Subtotal	*0.76*	*0.79*	*0.81*	*0.92*	*1.27*	*0.81*
Total	**1.09**	**1.23**	**1.33**	**1.90**	**3.71**	**1.36**
Projected average yield[c] solid m^3/ha/yr	44	33	28	15	6	26.6
Potential plantation area[d] 10^6 ha	4.05	7.74	25.94	11.56	1.17	50.46
Precipitation m/yr	1.5–2.3	1.0–1.7	0.7–1.3	0.5–1.0	0.25–0.6	
Water deficit m/yr	0.0–0.1	0.05–0.3	0.2–0.6	0.5–1.0	0.8–1.3	
Ave. temperature °C	22–28	24–28	24–28	26–28	24–28	
Altitude meters	0–700	<900–1000	<700–1000	<700	<600	

a. [45].

b. Costs for delivered logs (33 percent moisture), based on average cost data provided by commercial plantation operators, assuming a mass density of 0.47 dry tonnes per solid m^3 and a higher heating value of 20 gigajoules per tonne, with three 6-year rotations per planting, and a 10 percent discount rate.

c. Potential yields (stemwood only) estimated at the Hydroelectric Company of San Francisco (CHESF), Recife, Brazil, by correlating yields from existing *Eucalyptus* plantations (commercial and experimental) with climate, soil and other characteristics of the plantations' bioclimatic regions. The potential yield for a bioclimatic region is taken to be the simple average of (i) the average yield for commercial plantations in that bioclimatic region (see table 5) and (ii) 0.75 times the average yield of experimental plots for that region. For bioclimatic region I the average yield for commercial plantations was assumed in this calculation to be the best yield for commercial plantations in region II (see table 5), as there are no commercial plantations in region I.

d. The potential plantation area is about 1/3 of the total area of the nine states in the northeast of Brazil—land that is generally unsuited for agriculture, primarily deforested land.

the analysis indicates that with present technology respectable yields can be widely achieved under much less than ideal conditions.

Plantation experience in temperate climates. There is much less commercial experience with plantations of fast-growing trees in temperate than in tropical regions. Most such plantations in temperate climates are concentrated in Portugal and Spain, where more than 800,000 hectares of Eucalyptus are grown for pulpwood. In northern parts of the European Community some 2,000 hectares of experimental plantations, based mainly on poplars and willows, have been established. It has been demonstrated in these experimental plots that yields of 10 to 12 tonnes (dry basis) can already be achieved on suitable sites and that there is considerable potential for improvement [46].

Sweden has recently begun to promote the conversion of excess cropland to SRIC willow (mainly *Salix viminalis* and *S. desyclados*) plantations for energy. Sweden's new policy, which is motivated by its excess capacity for food production and a need to find alternatives to food production on at least 500,000 hectares of agricultural land, provides substantial temporary subsidies to farmers making the conversion [47]. It is expected that after having gained a few years of experience farmers will plant trees for energy without subsidies. By 1991, 4,000 hectares had been planted under this program, and the planted area is expected to increase to 10,000 hectares by 1993.

The Swedish goal with current technology is a productivity of 12 dry tonnes per hectare per year [48]. In field trials by farmers with plots of the order of 100 hectares, total above-ground yields of 10 to 12 dry tonnes per hectare per year have been achieved [49]. If such yields can be realized on a large scale, it will show that good productivities are possible even in regions with long and cold winters. (All Swedish plantations are located between 55 and 60° north latitude.)

In the United States, commercial experience with SRIC plantations is mostly limited to some 24,000 hectares of hardwood plantations in the Pacific Northwest[20][50].

Despite the relative lack of commercial experience with biomass plantations, research is being conducted on plantations in various countries located in temperate climates. For example, after the oil embargo of 1973, the U.S. Department of Energy (DOE) launched an experimental program to investigate the prospects for producing wood for energy on SRIC plantations. More recently, the DOE has initiated research on various herbaceous crops. Significant yield increases have been obtained in experiments and field trials over the past two decades and continued gains are expected, along with substantial reductions in production costs [51].

---

20. The 12.0 million hectares of "fast-growing" conifers listed in table 4 for the United States are long-rotation pines grown in the southeast, not SRIC plantations.

The energy cost of growing biomass

If biomass energy is to be used as a fossil-fuel substitute, the energy provided should be greater than the fossil fuel energy needed to produce it.

Determining the overall energy balance requires taking into account both (i) the energy required to produce the biomass, and (ii) the energy required to convert the biomass feedstock into the energy carrier that will be marketed (e.g., electricity or a fluid fuel). The first part of this energy balance is examined here; the second, in other chapters concerned with energy conversion technologies.

Energy is needed to establish plantations, to make fertilizers and herbicides, and to harvest and transport the crop to the energy conversion facility. Although these energy inputs generally increase as the intensity of plantation management increases, so does the biomass yield. Thus, the relationship between energy output and input is key to understanding the energy implications of intensive management.

In a 1979 analysis, F. Thomas Ledig of the U.S. Forest Service examined this relationship for 11 situations involving biomass from trees, ranging from natural forests, to conifer plantations managed at average intensity, to intensively-managed SRIC hardwood plantations. He found that a linear relationship in which the net energy yield[21] equals 12 times the energy input provides a good fit to the data for this diverse set of conditions [52].

More recent estimates of the energy costs of plantation biomass (including both woody and herbaceous crops) grown under U.S. conditions have been made by analysts at the Oak Ridge National Laboratory. With near-term expected biomass yields (net of harvesting and storage losses) in the range 9 to 13 dry tonnes per hectare per year, net energy yields have been estimated to be in the range 10 to 15 times the energy inputs for these crops (see table 7). With projected higher future yields, these ratios would be somewhat higher.

Thus, the energy content of biomass grown on modern plantations will generally be much greater than the fossil fuel inputs required for its production.

The economic costs of growing biomass

The long-term prospects for plantation biomass as an energy source depend on its ability to compete with conventional energy. The cost of biomass production is one important indicator of the economic performance of plantations. Cost estimates are needed for production averaged over large areas involving biomass produced under a wide range of conditions.

The best available data for estimating costs come from countries such as Brazil, where biomass plantations are well-established commercially. Cost estimates based on data provided by commercial plantation operators in Brazil for plantations that might be established in the northeast of Brazil in each of five major biogeoclimatic zones are presented in table 6. The estimated cost averaged over the

---

21. The net energy yield is defined here as the energy content of the wood produced minus the energy input required for its production.

Table 7: Energy balances for biomass production on plantations[a]

	Hybrid poplar		Sorghum		Switchgrass	
	1990	2010	1990	2010	1990	2010
	gigajoules per hectare					
Energy input						
Establishment	0.14	0.14	1.29	1.29	0.39	0.39
Fertilizers	3.33	3.33	8.87	12.69	5.26	7.38
Herbicides	**0.41**	0.41	1.82	1.82	–	–
Equipment	0.17	0.17	–	–	–	–
Harvesting	7.31	11.69	3.72	8.24	5.47	8.41
Hauling[b]	2.40	3.07	3.81	6.90	2.79	3.60
Total	**13.76**	**18.82**	**19.51**	**30.94**	**13.91**	**19.79**
Energy output[c]	**223.74**	**366.30**	**232.75**	**528.50**	**157.50**	**252.00**
Net Energy Ratio[d]	**15.3**	**18.5**	**10.9**	**16.1**	**10.3**	**11.7**

a. [85].

b. The energy required to transport the biomass 40 kilometers to a biomass processing plant.

c. Yields net of harvesting and storage losses for present (future) production technology are assumed to be 11.3 (18.5) tonnes per hectare per year for hybrid poplar (with a heating value of 19.8 gigajoules per tonne), 13.3 (30.2) tonnes per hectare per year for sorghum (heating value of 17.5 gigajoules per tonnne), and 9.0 (14.4) tonnes per hectare per year for switchgrass (heating value of 17.5 megajoules per tonne).

d. The net energy ratio = (energy output - energy input)/energy input.

total potential plantation area (one-third of the land area of the northeast) is about $1.4 per gigajoule, and more than 99 percent of the total potential biomass could be produced at a cost no more than $1.9 per gigajoule.

That these are attractive costs is indicated by a comparison with international energy prices[22] in 1990 of $3.6 per gigajoule ($22 per barrel) for crude oil and $1.9 per gigajoule ($51 per tonne) for steam coal [53]. The low cost of plantation biomass in Brazil reflects both the country's extensive experience with biomass plantations and its low labor costs. Because biomass production is so labor-intensive, its cost will tend to be low in areas where labor costs are low. A preliminary assessment suggests even lower costs for plantation biomass in China.[23]

For temperate regions, where there is little commercial experience with biomass plantations, cost estimates cannot be made as reliably as for Brazil. Never-

22. These are average import prices for countries that are members of the International Energy Agency.

23. A cost estimate by analysts at the Oak Ridge National Laboratory [54] is 75 yuan per dry tonne ($16 per tonne or $0.8 per gigajoule) for wood grown on plantations in Yunan Province (at an average productivity of six dry tonnes per hectare per year) and delivered to a biomass conversion facility 85 km away (assumed here to be the same transport distance as for the Brazilian cost estimates in table 6).

Table 8: Estimated current and projected productivities and production costs for biomass grown on dedicated plantations in the United States[a]

	Annual yields dry tonnes per hectare per year				**Production costs** $ per gigajoule of net biomass[b]	
	1990		2010		1990	2010
	Gross[c]	Net[d]	Gross[c]	Net[d]		
Midwest						
Hybrid poplar	13.5	10.5	20.0	16.5	3.48	2.50
Switchgrass	13.0	9.0	20.0	14.4	3.86	2.73
Sorghum	22.4	18.3	35.0	29.3	2.73	1.87
Southeast						
Energy cane	22.6	18.5	35.0	29.3	2.97	1.86
Switchgrass	13.0	9.0	22.0	15.9	3.52	2.19

a. Source: [51]. Costs are calculated for a 6% discount rate and include the cost of transport 40 kilometers to the energy conversion center plus 6 months storage.

b. Based on higher heating value of the biomass—assumed to be 19.8 gigajoules per dry tonne for hybrid poplar and 17.5 gigajoules per dry tonne for the other (herbaceous) crops.

c. This is the standing yield at the time of harvest.

d. This is the yield net of the losses in harvesting and storage.

theless, rough but meaningful estimates can be made, based in part on extrapolations from experience with agriculture. Cost estimates for prospective U.S. plantation sites made by an analyst at the Oak Ridge National Laboratory are presented in table 8. These estimates are for plantations that might be established in the midwest or southeast, for one SRIC tree crop and for three herbaceous energy crops, and for yields and production technology estimated as achievable at present and in 20 years if research and development goals are met. Estimated present costs ($2.7 to $3.9 per gigajoule) are much higher than for Brazil. The cost targets for 2010 ($1.9 to $2.7 per gigajoule) are also higher. However, if these latter cost goals can be met, plantation biomass utilized with advanced conversion technologies would be competitive under a wide range of conditions.

Thus, plantation biomass energy is likely to be strongly competitive in many developing countries and prospectively competitive in temperate regions using advanced conversion technologies.

Toward productive and ecologically sustainable bioenergy plantations
If biomass plantations are to play major roles in the global energy economy, strategies are needed for achieving and sustaining high yields over large areas and long periods.

There is little long-term experience with high-yielding crops that can be drawn upon for guidance in formulating such strategies. Although the experience of sustaining high sugarcane yields for centuries in the Caribbean and elsewhere suggests that high sustained yields are feasible, research is needed to ascertain the optimal strategies for achieving high yields under a wide range of conditions.

Despite the embryonic state of understanding of what is required, some general guidelines can be provided, drawing on the limited experience with plantations and much broader experiences with agriculture and forestry. The practical issues that must be dealt with include site establishment, species selection, soil fertility, pests and diseases, erosion, water pollution, and the biological diversity of the plantation and its environs [55, 56].

Site establishment. Establishment issues are quite different for industrialized and developing country sites.

In industrialized countries, the most likely sites are reasonably good agricultural lands, for which establishment is not technically difficult. The problems involved in establishing annual energy crops, such as sorghum, are similar to those for establishing annual agricultural crops such as maize. As for agriculture, herbicides are needed to eradicate grasses and broadleaf weeds that would compete with young energy crops for water and available nutrients. And as for agriculture, biodegradable herbicides (e.g., glyphosphate) can be used to keep to low levels the environmental risks of herbicide use.

For many energy crops, planting is far less frequent than for food crops. For SRIC crops, replanting takes place every 15 to 20 years, if there are three cuts (one from a planted rotation and two from coppice rotations) per planting; for perennial grasses, replanting occurs perhaps once a decade. For such energy crops, herbicide applications would be much less frequent than for agricultural crops.

In most developing country situations, the leading candidate sites are deforested or otherwise degraded lands. The conversion of such lands into successful plantations is more challenging [57, 58]. However, the successes in establishing productive and profitable plantations on degraded lands in Brazil and elsewhere indicate that restoring these lands via plantations is feasible.

For degraded lands, the major technical challenge is to find a sequence of plantings which can restore the soil's organic and nutrient content, and improve moisture conditions, ground temperatures, and other soil conditions to a point where biological productivity is high and sustainable. Successful restoration strategies typically begin by establishing a hardy species with the aid of commercial fertilizers or local compost. Once erosion is stabilized and ground temperatures lowered, organic material can accumulate, microbiota can return, and moisture and nutrient properties can be steadily improved. This can lead to a self-regenerating cycle of increasing soil fertility [59]. While there is some experience with such approaches to land restoration, intensive efforts are needed to devise region-specific and site-specific strategies.

Species selection. Species selected for the plantations should be fast-growing[24] and well-matched to the plantation site with respect to water requirements and its seasonality, drought resistance, soil pH, nutrition, tolerance of saline soils, as well as susceptibility to herbivores, fire, disease, and pests. Consideration must also be given to the crop's potential impact on the surrounding habitat, with care taken to prevent aggressive exotic species from establishing themselves outside the plantation and displacing native flora [55].

Soil fertility. Attention must also be given to long-term soil fertility. Nutrients removed during harvesting must either be regenerated naturally or restored by adding fertilizers. Fertilizers are usually required when establishing plantations on degraded lands and may also be needed to realize high yields even on good sites.

Despite the favorable net energy yield for intensively managed biomass plantations (see table 7), it will often be desirable to reduce fertilizer inputs. There are several ways this can be accomplished.

Nutrient input requirements depend in part on the extent of biomass removal at harvest. For SRIC crops, it will usually be desirable to leave leaves and twigs in the field, as nutrients tend to concentrate in these parts of the plant. Also, the mineral nutrients recovered as ash at the energy conversion facilities should be returned to the site.

Another strategy for reducing fertilizer requirements is to select species that are especially efficient in their use of nutrients. As noted earlier, there is a wide range of nutrient-use efficiency among plants.

In addition, either selecting a nitrogen-fixing species for the biomass crop or intercropping the primary crop with a nitrogen-fixing species can make the plantation self-sufficient in nitrogen, even when there are large nitrogen losses in the harvested biomass [60]. The promise of intercropping strategies is suggested by 10-year trials in Hawaii, where yields of 25 dry tonnes per hectare per year have been achieved without nitrogen fertilizer when *Eucalyptus* is interplanted with nitrogen-fixing *Albizia* trees [61].

In the future it may be feasible to reduce fertilizer requirements through the use of techniques being developed for matching nutrient applications more precisely to the plant's time-varying need for nutrients [11, 62].

Pests and diseases. Plantations of only one or two species with plants of comparable age can be more vulnerable to attack by pests and pathogens than are natural forests or grasslands. Moreover, plantations in the tropics and subtropics tend to be more affected by disease and pest epidemics than those in temperate regions.

Despite the fact that the adoption of monocultural plantations has led to an increase in the number and severity of diseases and pests, much of this increase has been associated with land-clearing methods, poor matching of species to sites,

---

24. The measure of yield should be growth in mass (dry basis) rather than volume, because densities vary among species by a factor of three or more [55].

or the new ecological conditions arising from intensive management, rather than with the susceptibility of a single plant species to pathogens [55]. Thus care in species selection and good plantation design and management can be helpful in controlling pests and diseases, rendering the use of chemical pesticides unnecessary in all but extraordinary circumstances.

A good plantation design, for example, will include: (i) areas set aside for native flora and fauna to harbor natural predators for plantation pest control, and perhaps (ii) blocks of crops characterized by different clones and/or species. If a pest attack breaks out on one clone, a now common practice in well-managed plantations is to let the attack run its course and to let predators from the set-aside areas help halt the pest outbreak.

Careful monitoring is key to the control of pests and diseases; the compactness of intensively managed plantations facilitates this monitoring. Also very short rotations can reduce the risk that pests and diseases will spread, since the trees will often be harvested before the invading organisms reach epidemic proportions [55].

In some cases, exotic species grow extraordinarily well initially, because of the absence of pest species in the new environment. However, biological controls are also absent in the new environment, so that if a pest or disease is eventually introduced, the damage to the exotic species can be worse than in its original habitat.

Before establishing bioenergy plantation, there should be extensive tests of a variety of species and provenances for resistance to local pests and diseases. Exotic species should be chosen only when they are markedly superior to local species in disease resistance and hardiness. Because most C_3 plants grow at about the same rate per unit of light intercepted by the canopy during the growing season, with adequate water and nutrients, it may often be feasible to obtain high yields with indigenous species and clones by giving adequate attention to nutrient balances.

However, if plantations are established on degraded lands where regeneration of desirable native species is difficult or impossible, an introduced species may be the best option for restoring these lands—as evidenced by the successful introduction of *Eucalyptus* in Ethiopia [56].

Erosion. The potential for erosion control will be an important criterion in selecting the plantation species wherever erosion is a problem. Erosion tends to be significant during the year following planting. Thus, annual herbaceous crops like sorghum are no better in controlling erosion than annual agricultural row crops like maize, whereas either SRIC tree or perennial grass crops, for which planting is infrequent, can provide good erosion control. The effectiveness of perennial grasses and trees in controlling erosion is indicated by recent experience with the Conservation Reserve Program of the U.S Department of Agriculture. The erosion rate declined 92 percent on the 14 million hectares of highly erodible U.S. cropland taken out of annual production under this program and planted with perennial grasses and trees [63].

Water pollution.   Nutrient leaching from plantations can contaminate ground water and runoff, thereby degrading drinking water supplies and promoting algal blooms. Fortunately, the various strategies for limiting fertilizer input also reduce the potential for water pollution associated with nutrient leaching.

Biological diversity.   Biomass plantations are often criticized because the range of biological species they support is much narrower than for natural forests. While true, the criticism is not always relevant. It would be relevant if a biomass plantation replaced a virgin forest (as was the case with the Jari estate in Brazil). However, it would not be relevant if a plantation and associated natural reserves were established on degraded lands; in this instance, the restored lands would be able to support much greater biological diversity than was possible before restoration [29]. If biomass energy crops were to replace monocultural food crops, the effect on the local ecosystem would depend on the plantation crop species chosen, but in many cases the shift would be to a more biologically diverse landscape [29].

Achieving sustainable biomass production while maintaining biological diversity may ultimately require a shift to polycultural strategies (e.g. mixed species in various planting configurations) in many areas. Growing plants under nutrient-optimized conditions in particular could make it possible to achieve high yields with a mix of indigenous species and clones, thus facilitating the maintenance of a diverse landscape mosaic. A challenge posed by such strategies is that silviculturally compatible combinations of species are difficult to design and establish, because of the tendency of one species to dominate [55]. Also, monocultures tend to be favored for energy crops because management techniques now in use, borrowed largely from agriculture, were designed for monocultural systems. Such techniques are often necessary in serving agricultural markets, but they are not necessary for many biomass energy conversion systems, which can usually accommodate a variety of feedstocks. Polycultural establishment and management techniques therefore warrant high priority in research and development programs for energy crops.

As already noted, establishing and maintaining natural reserves at plantations can help control crop pests while enhancing the local ecosystem. However, preserving biodiversity on a regional basis will require land-use planning in which natural forest patches are connected via a network of undisturbed corridors (riparian buffer zones, shelterbelts, and hedgerows between fields), thus enabling species to migrate from one habitat to another [29].

While major expansions in research are needed to provide a sound analytical and empirical basis for achieving and sustaining high biomass yields in environmentally acceptable ways, there is time for such research as well as for extensive field trials, because major bioenergy industries can be launched with residues from the agricultural and forest products industries. If substantial commitments are made to biomass plantation research in the near term, plantation biomass could start to make contributions to energy supplies when residue supplies are no longer adequate to meet the needs of the growing biomass energy industry, perhaps at the turn of the century or shortly thereafter.

BIOMASS POTENTIAL FOR ENERGY APPLICATIONS

Ideally, estimates of the potential for utilizing biomass for energy should be made by first developing supply curves, which show how much biomass can be obtained at various costs from each source. The value of biomass for each application should then be determined from the technical and cost characteristics of the conversion technology and from energy prices in the markets where the biomass would compete. The potential supply is the quantity that is recoverable at costs up to this value. Unfortunately, biomass supply curves can be constructed only in isolated instances where good cost data are available. Despite this data constraint, rough estimates of potential biomass supplies can be obtained from rather general considerations.

Biomass residues

While the generation rate for biomass residues is large (see table 2), not all residues can be utilized for energy purposes. Some crop and logging residues should be left at the site to help ensure the sustainable production of the primary biomass product, and some recoverable residues would be better used for other purposes. Moreover, it will not be practical or cost-effective to recover all residues.

Rough estimates by region of the amount of residues recoverable in light of such considerations are presented in table 9. These estimates are consistent with the studies on residue recovery reviewed in the previous section. The residues considered here include the dung and forest-product industry residues listed in table 2 plus a more complete inventory of crop residues than was considered for table 2 (see note e, table 9).

With the exception of sugarcane, the recoverable fraction of crop residues is assumed to be one-fourth the generation rate. In the case of sugarcane, it is assumed that all the bagasse plus one-fourth of the tops and leaves are recoverable. In the case of dung, the recoverable fraction is assumed to be only one-eighth, because of the difficulties of recovering dung from grazing livestock. For forest-product residues, it is assumed that one-fourth of logging residues plus three-fourths of mill and manufacturing residues are recoverable.

Detailed studies of specific residues might show that higher recovery fractions are feasible. Also, as noted earlier, higher recovery fractions can often be realized by altering management of the primary biomass product (e.g., by shifting crop production to no-till agriculture). However, the recovery fractions indicated here are probably reasonable estimates of what could be recovered on average with present practice, consistent with the various constraints on residue recovery.

The global potential for recoverable residues under these assumptions is 31 exajoules per year. This is equivalent to about 10 percent of commercial energy-

Table 9: Commercial energy use and potential supplies of biomass for energy[a,b]
exajoules per year

Region	Commercial energy use[c]	Recoverable residues				Biomass plantations[d]
		Crop[e]	Forest[f]	Dung[g]	Total	
Industrialized						
US/Canada	87.9	1.7	3.8	0.4	5.9	34.8
Europe	79.8	1.3	2.0	0.5	3.8	11.4
Japan	16.6	0.1	0.2	-	0.3	0.9
Australia + NZ	3.6	0.3	0.2	0.2	0.6	17.9
Former USSR	56.9	0.9	2.0	0.4	3.3	46.5
Subtotals	*244.8*	*4.3*	*8.1*	*1.6*	*14.0*	*111.5*
Developing						
Latin America	17.4	2.4	1.2	0.9	4.5	51.4
Africa	9.2	0.7	1.2	0.7	2.6	52.9
China	23.0	1.9	0.9	0.6	3.4	16.3
Other Asia	27.7	3.2	2.2	1.4	6.8	33.4
Oceania	-	-	-	-	-	1.4
Subtotals	*77.3*	*8.2*	*5.5*	*3.6*	*17.2*	*155.4*
World	**322.1**	**12.5**	**13.6**	**5.2**	**31.2**	**266.9**

a. All energy values are expressed on a higher heating value basis.

b. Regional residue estimates are aggregates of country-by-country inventories compiled at the Information and Skills Center of the Bioenergy Users' Network, King's College, London [1].

c. Commercial energy use for 1985, as estimated by the U.S. Department of Energy [86], excluding biomass.

d. It is assumed that plantations having an average yield of 15 dry tonnes (with a heating value of 20 gigajoules per tonne) per hectare per year are established on 10 percent of the total amount of land now in forests/woodlands + cropland + permanent pasture—some 372 million hectares in industrialized countries and 518 million hectares in developing countries (see table 13).

e. Included are residues from cereals, vegetables and melons, roots and tubers, sugar beets, and sugar cane. Not included are residues from pulses, fruits and berries, oilcrops, tree nuts, coffee, cocoa and tea, tobacco, or fibre crops. Crop production data are from FAO sources [66, 83]. Except for sugar cane it is assumed that 1/4 of all residues generated are recoverable. For sugar cane it is assumed that all bagasse is recoverable and that 1/4 of the tops and leaves are recoverable (see note c, table 2). For crops other than sugar cane the following are the assumed residue generation rates per tonne of crop, and the assumed heating values of these residues:

	Residue generation rate	Residue heat content
	tonnes per tonne	*gigajoules per tonne, air-dry basis*
Cereals	1.3	13
Vegetables and melons	1.0	6
Roots and tubers	0.4	6
Sugar beets	0.3	6

f. It is assumed that three-fourths of the milling and manufacturing wood wastes and one-fourth of the forest residues are recoverable (see notes e and f, table 2).

g. It is assumed that one-eighth of the dung generated is recoverable (see note d, table 2).

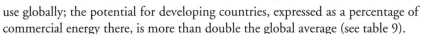

use globally; the potential for developing countries, expressed as a percentage of commercial energy there, is more than double the global average (see table 9).

While the potential residue recovery rate is about one-fourth the rate of global petroleum use or one-half the rate of natural gas use, the energy value of these residues would be less than these comparisons suggest if the biomass were used with currently available biomass conversion technologies, which are less efficient than fossil-fuel conversion technologies.

However, if biomass residues were used with advanced, energy-efficient conversion technologies, their value could be greater than these comparisons suggest. For example, recoverable residues would be sufficient to fuel a global fleet of light-duty vehicles (automobiles and light trucks) at the 1985 Western European level of light-duty vehicle use, if these vehicles were powered by fuel cells operated on hydrogen from biomass.[25] Alternatively, the same amount of biomass could provide one-sixth the electricity for a world having a per capita electricity production rate equal to that of Western Europe in 1985, if advanced biomass-integrated gasifier/gas turbine technologies were used to make electricity.[26] Thus advanced biomass-conversion and end-use technologies can convert plant residues, which are widely regarded as a marginal energy resource at best, into a significant global resource.

Plantation biomass
The potential for plantation biomass depends both on the yields that can be achieved and on the land areas that can be committed to plantations.

Potential yields
Estimates of average yields that can be sustained over large areas (millions of hectares) and over long periods (decades) are needed to assess the energy potential of plantation biomass. Estimating such yields is difficult because experience is limited. The concept of growing high-yield hardwood plantations for forest-product applications originated in the early 1960s and was extended to the production of energy crops only after the 1973 oil crisis. The notion of growing herbaceous crops for energy is even more embryonic. As noted, most biomass plantations were established after 1980 (see table 4).

Nevertheless, reasonable estimates can be made of potential yields—drawing on fundamental aspects of photosynthesis and laboratory experience as discussed earlier, experience with agricultural crops, field trials of plantation crops, and, to a limited degree, large-scale plantations.

---

25. For 0.36 vehicles per person, each driven 12,000 km per year [64], and a 1985 world population of 4.87 billion. For hydrogen fuel-cell vehicles with a gasoline-equivalent fuel economy of 3 liters per 100 km (1.06 kJ per km) and a 70 percent efficiency of converting biomass into hydrogen (see chapter 22).

26. At a conversion efficiency of 43 percent (see chapter 17), recoverable residues could provide 760 kWh per capita per year at the 1985 world population level. For comparison, per capita electricity generation in Western Europe was 4,520 kWh per year in 1985 (see appendix to chapter 1 at the back of the book).

Although production goals for agricultural crops are different from those for energy crops (with less emphasis in agriculture on maximizing yields and more on other crop qualities), agriculture provides a wealth of experience from which insights relevant to energy crops can be gleaned.

Among food crops, sugarcane is perhaps the most productive. Experience with sugarcane suggests what yields might be achievable with C_4 plants in tropical regions with good rainfall or irrigation. In 1987, the worldwide average yield of above-ground dry matter was 36 tonnes per hectare per year, averaged over 17 million hectares of cane plantations.[27] The highest country-wide yield was realized by Zambia—77 tonnes averaged over 10,000 hectares. The average for industrialized countries (mainly the United States, Australia, and South Africa) was 48 tonnes averaged over one million hectares of plantations [66].

Experience with grain crops in temperate climates is also instructive. In 1986 and 1987, the average annual maize yield on 28 million hectares in the United States was 7.5 tonnes, while in the United Kingdom, the average wheat yield realized on two million hectares was 7.0 tonnes. Taking into account residues as well (note b, table 2), the corresponding total above-ground yields in both cases were about 14 dry tonnes per hectare per year.

The history of agricultural yields also provides insights as to what might be accomplished with intensive management and improved technology. Average productivities of maize in the United States and wheat in the United Kingdom have more than tripled since the mid 1940s. Moreover, the agricultural community expects yields to continue increasing for major field crops. The U.S. Department of Agriculture projects that by 2030 annual maize and rice yields in the United States will reach 14 and 13 tonnes of grain per hectare—up 2.0-fold and 2.5-fold relative to 1982, respectively [67].

While large-scale field experience with plantation biomass is limited, extraordinary yields have been achieved in small-scale plots. For example, hybrid poplar coppice yields of up to 43 tonnes per hectare per year have been achieved in the U.S. Pacific Northwest [68]. Record yields of 113 m³ (65 dry tonnes) per hectare per year at seven years have been achieved with *Eucalyptus grandis* in Brazil [69].

In larger-scale plots, yields tend to be lower. In the United States, large-scale experimental plots have yields ranging from 15 to 22 tonnes per hectare per year [68]. In Brazil, yields of 70 m³ (about 40 tonnes per hectare per year) have been achieved for *Eucalyptus* grown over limited areas under favorable field conditions [70].

For commercial plantations, yields averaged over large areas (tens of thousands of hectares) are lower still. Average *Eucalyptus* yields of 10 to 20 tonnes per hectare per year are routinely achieved today in Brazil (see table 5). And, as noted earlier, it is believed (but remains to be demonstrated at large scales) that average yields ranging from 10 to 12 tonnes per hectare per year can be achieved on good

---

27. Yields are reported as the wet weight of the harvested stem. For each tonne of harvested stem the total above-ground dry biomass is 0.609 dry tonnes, consisting of 0.329 dry tonnes of stem plus 0.092 tonnes of tops and attached leaves plus 0.188 tonnes of detached leaves [65].

cropland with short-rotation intensive-culture plantations in the European Community and in Scandinavia.

The difference between yields that have been achieved at small scale and what can be realized today in large plantations might be taken as an indicator of the prospects of future improvement for commercial plantations with better management and improved technology. Before taking small-plot yields as a basis for projecting future possibilities, however, it is worth noting why there are such large differences between experimental and commercial yields. In essence, they arise because of the much higher degree of control that is possible with small experimental plots but not feasible with large-scale plantations. Specifically, the poorer yields resulting from shifts from small-scale to large-scale plots are due largely to inadequate weed control, poorer sites, and poor clone-site matching. In small experimental plots weeds are easily controlled by the experimenter. Also, the very best sites can be selected to achieve record yields. And many more clone-site combinations can be tried to identify the best clone for a small experimental plot than is possible with large field plantations [71]. In limiting attention to small plots one can also select examples where diseases or pests have not limited growth.

Good management or improved technology can overcome some of the problems associated with scale-up. It will probably be feasible to achieve a high degree of weed control in environmentally acceptable ways with advanced biodegradable herbicides. With adequate research and development, it may also be feasible to match disease-resistant clones to the various site conditions that exist in typical plantations, as has been done for agricultural crops. But most large plantations will have some average and some relatively poor sites, for which yields cannot be brought up to experimental plot values.

Such considerations suggest that today's small-plot yields represent upper bounds that can be approached but not realized in large-scale applications. Moreover, these yields can be approached only with intensive research and development aimed at identifying the appropriate site-matched, disease-resistant clones.

It does not follow, however, that the record yields achieved in small plots cannot be reached or exceeded, because record yields can be increased via genetic improvement of the crop. The potential gains to be made with genetic improvement are illustrated by recent experience in the Pacific Northwest of the United States, where 2,000 clones of cottonwood hybrids have been tested at 25 sites, yielding 25 to 30 tonnes per hectare per year averaged over a six-year growth cycle—some 1.5 to 2 times the yield of the parent stock [50]. Similarly, half the tripling in yield for maize and the doubling in yield for soybeans achieved between 1935 and 1975 was the result of genetic improvement [71]. But genetic advances along these lines require much more intensive research and development work on plantation biomass than is ongoing at present.

Making projections of biomass plantation yields that can be achieved routinely is fraught with uncertainty. The projections advanced here are neither optimistic not pessimistic. Rather they are extrapolated from experience and current understanding under the assumptions that: (i) society will support the research and development effort justified by the returns offered by "average" estimates of the success of such investments and (ii) hundreds of millions of hectares of plantations will be established worldwide over the next several decades.

In tropical regions, long growing seasons and favorable conditions for C_4 plants indicate good prospects for high average yields (averaged over C_3 and C_4 crops). Such favorable conditions are partially offset, however, by the expectation that plantations will be established primarily on degraded lands (see below). On balance, however, it is reasonable to expect that yields averaged over hundreds of millions of hectares could range from 15 to 20 dry tonnes per hectare per year during the first quarter of the 21st century (the upper half of the range of experience for well-managed plantations today) and 20 to 30 tonnes per hectare per year in the second quarter of the next century.

In temperate regions, average yields will be less than in the tropics because the growing season is shorter and most plantation crops will be C_3 plants. Offsetting these disadvantages is the prospect that most plantations will be established on relatively high-quality cropland (see below). In light of such considerations, average yields of 10 to 15 tonnes per hectare per year may be expected during the first quarter of the next century (the upper limit is slightly higher than projected yields for plantations in the near term in Europe) and 15 to 20 tonnes per hectare per year in the second quarter.[28]

Actual achievements could fall short of these goals if research and development and commercialization efforts are not significantly increased over present levels, while achievements could exceed these goals under more optimistic conditions.

Potential land areas for plantations

Land areas that can be considered for plantations are those that have climatic, topographical, and soil conditions that are suitable for supporting high-yielding plantation crops and are not better suited for other purposes. Areas with favorable growing conditions can be crudely identified as those where natural vegetation grows well. Various models have been developed to predict net primary productivity (NPP) patterns based on a range of limiting factors, of which the so-called annual moisture index—the ratio of precipitation (P) to potential evapotranspiration (PET)[29]—is generally dominant.[30] Tropical regions of high primary productivity tend to have annual moisture indices of one or more, although at high latitudes high annual moisture indices are not associated with high net primary productivities, because much of the precipitation occurs outside the growing sea-

---

28. The projected yields for temperate climates are modest compared to yields targeted in other studies. For example, researchers at the Weyerhauser Company have argued for targeted yields of 25 and 30 tonnes per hectare per year for plantations of Douglas fir in Washington and loblolly pine in North Carolina in the United States, respectively [72]. The more modest targets set forth here arise from the premise that considerations of long-term sustainability may limit maximum sustainable yields.

29. Potential evapotranspiration is defined in footnote 19.

30. An annual moisture index of less than one implies a water deficit, which will limit growth, while a ratio greater than one implies that there is excess water, so that runoff might occur.

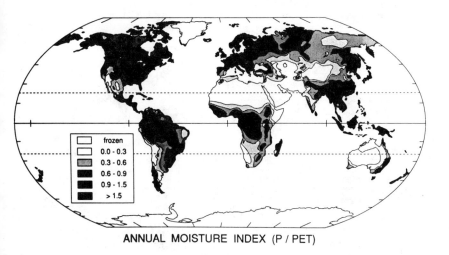

ANNUAL MOISTURE INDEX (P / PET)

FIGURE 4: World map of the annual moisture index [the ratio of the annual precipitation (P) to potential evapotranspiration (PET)], a measure of moisture availability for plant growth [81]. An annual moisture index of less than one implies a water deficit, which will limit plant growth, while a ratio greater than one implies that there is excess water, so that runoff might occur. Potential evapotranspiration is defined in footnote 19.

son, when temperatures are low (compare figures 4 and 5). Figure 5 shows that vegetation can grow very well in much of Latin America, sub-Saharan Africa, South Asia, and in the southeast United States, and that there are moderately good growing conditions in Europe, most of the eastern United States, coastal regions of Australia, and East Asia.

Food versus fuel.   A concern often voiced about biomass plantations is competition with food production. Because land is needed to grow food, but energy can be provided in many ways, food production should have priority. The key issues are the amount of land needed for food production and how this need compares with the arable land resource.

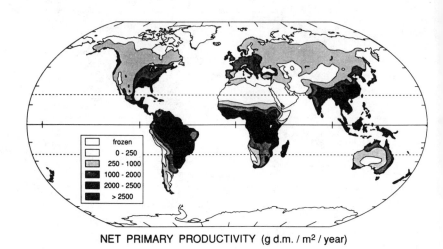

NET PRIMARY PRODUCTIVITY (g d.m. / m² / year)

FIGURE 5: World map of net primary productivity in grams of dry plant matter per square meter per year. (Note that one gram per square meter per year = 0.01 tonnes per hectare per year.) The net primary productivity values shown here were obtained by averaging values obtained from two models—the so-called "Miami" and "actual evapotranspiration-based sigmoid" models [81].

The amount of land needed for food production is declining in industrialized countries. In the United States, more than one-fifth of total cropland, some 33 million hectares, was idled in 1990, either to keep food prices high or to control erosion. Furthermore, the U.S. Department of Agriculture projects that the amount of idle cropland will probably increase to 52 million hectares by 2030 as a result of rising crop yields, despite an expected doubling of exports of maize, wheat, and soybeans in this period [67].

In the European Community, more than 15 million hectares of land will have to be taken out of farming by 2000 if surpluses and subsidies associated with the Common Agricultural Policy are brought under control [73]; in the future excess cropland in the Community could increase to as much as 50 million hectares if crop productivities continue to increase.

While the conversion of excess cropland in the industrialized countries to energy plantations presents an opportunity to make productive use of these lands, such a conversion would not be easily accomplished under present policies. In many countries, farmers are deterred by a subsidy system that specifies what crops the farmer can produce in order to qualify for a subsidy; and energy crops are not allowed. While conversion to profitable biomass-energy production will make it possible eventually to phase out agricultural subsidies, the phase-out will not be accomplished overnight, because of the economic dislocations that would result; today these subsidies total about $300 billion per year for North America, Europe, and Japan [74].

As long as a system of subsidies continues, the bias against energy crops should be removed. It might even be desirable to augment the subsidy for energy crop production initially, to provide an impetus for establishing this as a new agro-industrial activity, as Sweden is doing [47].

For developing countries the situation is quite different. Because of expected population growth and rising incomes, it is likely that more land will be needed for food production. The Response Strategies Working Group of the Intergovernmental Panel on Climate Change has projected that the land in food production in developing countries will increase 50 percent by 2025 [75]. The demand can be compared with potential supply—that is, land physically capable of supporting economic crop production, within soil and water constraints. Potential cropland was estimated for 91 developing countries in a 1991 FAO study [76]. For these 91 countries potential cropland was estimated to be about 2055 million hectares—nearly three times present cropland (see table 10).

Looking to 2025 and assuming that cropland requirements in developing countries will increase 50 percent by then, there would still be a substantial surplus potential cropland of nearly 1,000 million hectares in these countries (see table 11). There would be substantial regional differences, however, with surpluses totalling more than 1,100 million hectares in Latin America and Africa, and a 110 million hectare deficit in Asia. (China was not included in the FAO analysis.)

Thus it appears that substantial amounts of land suitable for energy plantations will be available in both Latin America and sub-Saharan Africa, even with major expansions of cropland to feed the growing population. But in Asia, with its high population density, conflicts with food production could become significant. The extent of the potential conflict, however, depends on future food crop productivities. It is possible that food/fuel interactions could prove to be synergistic, if some of the energy produced in biomass plantations were used to help make agriculture more productive. Increasing food production on the better agricultural lands, while growing trees or perennial grasses for energy on marginal lands, would generally be environmentally preferable to increasing agricultural output by bringing marginal lands into food production. Detailed assessments are needed, on a country-by-country basis, to understand the prospects for productivity

Table 10: Present[a] and potential[b] cropland for 91 developing countries
million hectares

	Present crop-land	Potential cropland						
		Low rainfall	Uncertain rainfall	Good rainfall	Natural flooded	Problem land	Desert	Total
Central America	37.6	2.2	13.3	18.5	5.7	31.4	3.5	**74.6**
South America	141.6	26.0	37.5	150.3	105.7	492.7	2.8	**815.0**
Africa	178.8	73.4	96.8	149.3	71.3	358.1	3.8	**752.7**
Asia (ex.China)	348.3	59.8	67.0	67.4	80.5	117.6	20.3	**412.5**
Total	**706.3**	**161.4**	**214.7**	**385.5**	**263.1**	**999.7**	**30.4**	**2,054.9**

a. [87].

b. As estimated by the Food and Agriculture Organization (FAO) in 1991 [76]. Potential cropland is defined by the FAO as all land that is physically capable of economic crop production, within soil and water constraints. It excludes land that is too steep or too dry or having unsuitable soils.

gains with more intensive agricultural management and the extent of food/fuel conflict if the agricultural sector is more intensively managed.

Unfortunately, the FAO study does not clarify where new cropland would come from. To be sure, some forestlands are involved. Clearly, it would not be desirable to cut down virgin forests in favor of intensively managed biomass plantations.

Establishing plantations on degraded lands. A land class that is emerging as a strong candidate for biomass plantations is deforested or otherwise degraded lands that are suitable for reforestation.

Interest in restoring tropical degraded lands is widespread, as indicated by the ambitious global net afforestation goal of 12 million hectares per year by 2000 set forth in the Noordwijk Declaration at the 1989 Ministerial Conference on Atmospheric and Climate Change in Noordwijk, The Netherlands [77].

Large land areas have been degraded. One estimate is that 2,077 million hectares of tropical lands are degraded, of which 758 million hectares have a theoretical potential for reforestation (see table 12). Another set of estimates is that (i) 500 million hectares of lands that once supported forests and are now unused for croplands or settlements could be available for reforestation in Africa, Asia, and Latin America; and (ii) an additional 365 million hectares of lands in the fallow cycle of shifting cultivation might also be targeted for reforestation [15].

In Asia, country by country assessments are needed to determine the extent to which its degraded lands will be needed for food production or other purposes. China, despite having a high population density, has a goal of increasing its forest

Table 11: The prospects for plantations in developing regions[a]
million hectares

	Cropland measures			Alternative measures of land areas potentially available for plantations		
	Present cropland[b]	Potential cropland[b]	Cropland required in 2025[c]	Excess potential cropland in 2025[d]	10% of cropland + perm. pasture + forests & wood-lands[e]	Degraded lands suitable for reforestation[f]
Latin America	179.2	889.6	269	621	171	156 (+ 32)
Africa	178.8	752.7	268	484	176	101 (+148)
Asia (ex.China)	348.3	412.5	522	–110	111	169 (+150)
Total	**706.3**	**2,054.9**	**1,059**	**995**	**458**	**426 (+330)**

a. Except for the last two columns on the right, the data here are for the 91 developing countries for which the FAO estimated potential cropland areas [76].

b. From table 10.

c. The Response Strategies Working Group of the Intergovernmental Panel on Climate Change has estimated that the area required for cropland in developing countries will increase 50 percent by 2025 [75].

d. This is the difference between the potential cropland and the cropland requirements in 2025.

e. From table 13, column V.

f. From table 12. The data refer to all countries on these continents, including China. The first entry is the sum of the land areas in logged forests, forest fallow, and deforested watersheds—all of which are estimated to be suitable for reforestation [88]. The number in parenthesis is 1/5 of the desertified drylands—the fraction of desertified drylands in developing countries estimated to be suitable for reforestation [88].

cover by 52 million hectares by 2000 (compared to the mid 1980s) and by an additional 93 million hectares over the longer term [78].

For Latin America and sub-Saharan Africa, the amount of degraded land suitable for reforestation is huge—some 156 million hectares and 101 million hectares, respectively (excluding degraded lands in the desertified drylands category) (see table 12). It is reasonable to assume that by 2025 an average productivity of at least 15 tonnes per hectare per year could be realized in plantations established on these lands. The resulting biomass energy production rates would be 47 and 30 exajoules per year for Latin America and Africa, respectively. Because these production rates are much greater than present energy consumption levels (see table 9), it may be possible for these regions to become major exporters of fuels derived from biomass (see chapter 1).

Despite the technical challenges involved, it is probably feasible to establish energy plantations on degraded lands—as evidenced by the many biomass plantations that have been established on such lands. Moreover, the favorable prospective economics for producing both electricity (see chapter 17) and fluid fuels (see

Table 12: Geographical distribution of tropical degraded lands and
potential areas for reforestation[a]
million hectares

	Logged forests	Forest fallow	Deforested watersheds	Desertified drylands	Total
Latin America	44.0	84.8	27.2	162.0	318.0
Africa	39.0	59.3	3.1	740.9	842.3
Asia	53.6	58.8	56.5	748.0	916.8
Total	**136.5**	**202.8**	**86.9**	**1,650.9**	**2,077.1**
Area suitable for reforestation	**137**	**203**	**87**	**331**	**758**

a. [88].

chapters 21 and 22) from plantation biomass suggests that the overall energy strategy is viable.

However, establishing large-scale energy plantations on degraded lands requires confronting a host of cultural, political, and developmental challenges. Land ownership may be complex and disputed. Indigenous peoples with a tradition of grazing and who measure their wealth in terms of cattle may not wish to convert land to other uses. Also, many of the regions have inadequate roads for transporting biomass to the processing facility and lack adequate transportation to move the produced biofuels to markets.

Therefore, plantation development efforts must be integrated into more general rural-development programs. Fortunately, prospects for simultaneously meeting local development needs, improving the environment, and providing profitable opportunities for biomass energy producers are good. In addition, the production of biofuel crops and associated conversion activities can provide substantial employment opportunities in rural areas. And with skillful management, energy industries may be able to finance land restoration activities.

A significant transitional problem, if energy crops do not mature in a single year, is that biomass producers in poor areas cannot afford to wait for cash returns. This problem can be dealt with in part by coproducing biofuels with annual food or fiber crops as plantations are being established. Also, financing for the biomass plantations can be provided by government or by the firms that would ultimately purchase the crops.

Plantation development might best be organized as joint ventures linking the interests of local agricultural producers and equipment suppliers, local and multinational energy companies, and local and international organizations interested in land restoration. Such joint ventures should be formed on terms acceptable to

Table 13: Present land use patterns[a]

| Region | Land use (million hectares) | | | | | Land use per capita[b] (hectares) | | |
| | I | II | III | IV | V | VI | VII | VIII |
	Total land	Forests & wood-lands	Crop-land	Perm. pasture	II+III+IV	Forests & wood-lands	Crop-land	Perm. pasture
Industrialized								
US/Canada	1,839	650	236	274	**1,160**	2.36	0.86	0.99
Europe	473	157	140	83	**380**	0.32	0.28	0.17
Japan	38	25	5	1	**31**	0.20	0.04	0.01
Australia + NZ	789	113	48	436	**597**	5.55	2.37	21.41
Former USSR	2,227	945	232	372	**1,549**	3.27	0.80	1.29
Subtotals	*5,366*	*1,890*	*661*	*1,166*	***3,717***	*1.57*	*0.55*	*0.97*
Developing								
Latin America	2,052	961	180	573	**1,714**	2.14	0.40	1.28
Africa	2,964	686	186	891	**1,763**	1.07	0.29	1.39
China	933	127	97	319	**543**	0.11	0.09	0.28
Other Asia	1,761	387	353	374	**1,114**	0.21	0.19	0.20
Oceania	54	44	1	1	**46**	7.07	0.16	0.16
Subtotals	*7,764*	*2,205*	*817*	*2,158*	***5,180***	*0.54*	*0.20*	*0.53*
World	**13,129**	**4,095**	**1,478**	**3,323**	**8,897**	**0.77**	**0.28**	**0.63**

a. [87].
b. At 1990 population levels.

both the developing country and the foreign firms and designed to ensure that all parties profit from the relationship.

The global potential for plantation biomass
Without detailed regional assessments, it is difficult to make meaningful projections of the global potential for plantation biomass. But by drawing on the general analysis presented here, a rough indication of the potential can be made.

By the middle of the 21st century, some 890 million hectares or 10 percent of the world's land area now in cropland, forests and woodland, and permanent pasture (7 percent of total world land area—see table 13) might be put into biomass production for energy. This is an ambitious but not implausible target. The developing country share of this total plantation land area—some 520 million hectares (see table 13), would require establishing plantations in the period 2000 to 2050 at about the rate targeted in the Noordwijk Declaration for the global net afforestation rate by 2000 [77]. Also, the Intergovernmental Panel on Climate Change estimated in 1992 that worldwide forests could be established on 250 to 850 million hectares as a strategy to sequester carbon, at relatively low cost [79].

If average yields of 15 tonnes per hectare per year were realized on 890 million hectares of plantations by 2050, the resulting global energy production rate would exceed 260 exajoules per year, or more than four-fifths of worldwide commercial energy use in 1985 (see table 9).

Many decades would be required to establish plantations on hundreds of millions of hectares around the world. The process would be lengthy in part because of the institutional challenges involved in launching such an effort and in part because of the technical issues that must be resolved. Much more research and development is needed on high-productivity energy crops that would be compatible with soil and climate conditions in many different regions and especially in areas where energy crops would be established on degraded lands. It is important that the research be sustained and constantly monitored so that priorities can be reviewed on a continuing basis.

But in light of the prospective favorable economics and the environmental and social benefits that can be derived from plantation biomass energy, intensive efforts to develop such plantations around the globe are warranted.

CONCLUSION

With good management, biomass for energy purposes can be produced sustainably and in large quantities in many parts of the world. Although potential biomass supplies include residues, low-quality wood recovered from existing forests, and biomass grown on plantations, most of the biomass that will be converted to modern energy carriers and used as commercial energy will probably come from residue and plantation sources because of growing environmental concerns about the overuse of natural forests.

By far the largest potential source of biomass for energy is plantation biomass. The establishment of plantations on deforested and otherwise degraded lands in developing countries and on excess cropland in industrialized countries offers major developmental and environmental benefits, as well as large and secure supplies of energy.

There is a rapidly growing body of knowledge in a number of countries around the world showing how biomass can be produced sustainably with high yields. However, a substantial and continuing commitment to research and development is needed to assure the realization and sustainability of high yields for many regions and under a wide range of growing conditions. Moreover, the establishment and maintenance of biomass plantations must be carried out in the framework of overall planning for economic development and environmental protection, and must involve the people living in the region.

Fortunately, efforts to modernize energy need not await resolution of the current uncertainties about plantation biomass. In the near term new bioenergy industries can be launched using residues as fuel. Although there are also uncertainties about how much of the residues can be recovered, given environmental and economic constraints, potential supplies even under severe constraints are sufficiently large that they can provide the basis for launching new bioenergy industries. If at the same time priority is given to research and development on the sustainable production of biomass, additional supplies of plantation biomass could meet the growing need for biomass feedstocks by the time recoverable residues are fully utilized.

The consensus among the world's leaders on the need for sustainable development, which emerged from the 1992 United Nations Conference on Environment and Development, provides a framework that is conducive to new thinking about biomass energy. If energy planners come to recognize the role bioenergy can play in achieving sustainable development and formulate new policies to support needed research and development and to help launch new industries, the world will move toward realization of the bioenergy potential.

ACKNOWLEDGMENTS

Two of the authors (DOH and FRC) thank Mr. Albert K. Senelwa, Moi University, Eldoret, Kenya, for his help in their efforts collecting data on biomass energy use and biomass residues for individual countries and the Bioenergy Users' Network for supporting that work. This comprehensive data base is being published separately by the Biomass Energy Users' Network Information Center, King's College, London [1]. The research carried out by one of the authors (RHW) that provided the basis for his contribution to this chapter was supported by the Bioenergy Systems and Technology Program of the Winrock International Institute for Agricultural Development, the Office of Energy and Infrastructure of the U.S. Agency for International Development, the Office of Policy Analysis of the U.S. Environmental Protection Agency, and the Geraldine R. Dodge, W. Alton Jones, Merck, and New Land Foundations.

WORKS CITED

1. Rosillo-Calle, F., Woods, J., and Hall, D.O. 1992
   Country-by-country survey of biomass use and potential for energy, Biomass Energy Users' Network Information Center, King's College, London.

2. Hall, D.O., Woods, J., and Scurlock, J.M.O. 1992
   Biomass production and data, appendix C, in D.O. Hall et al., eds., *Photosynthesis and production in a changing environment: a field and laboratory manual,* Chapman and Hall, London.

3. Jenkins, B.M. 1989
   Physical properties of biomass, in O. Kitani and C.W. Hall, eds., *Biomass handbook,* 860–891, Gordon and Breach, New York.

4. Hall, D.O. 1989
   Carbon flows in the biosphere: present and future, *J. Geol. Soc.* 146: 175–181.

5. Bolin, B. 1986
   How much CO_2 will remain in the atmosphere? in B. Bolin, B.R. Doos, J. Jager, and R.A. Warrick, eds., *The greenhouse effect, climate change, and ecosystems,* SCOPE 29, 93–155, Wiley, Chichester.

6. Bolton, J.R. and Hall, D.O. 1991
   The maximum efficiency of photosynthesis, *Photochemistry and Photobiology* 53: 545–548.

7. Walker, D.A. 1992
   Excited leaves; Tansley review no. 36, *New Phytol.* (forthcoming).

8. Monteith, J.L. 1977
   Climate and the efficiency of crop production in Britain, *Trans. R. Soc. Lond. B.* 281: 277–294.

9. Cannell, M.G.R. 1989
   Physiological basis of wood production: a review, *Scandinavian J. For. Res.* 4: 459–490.

10. Cole, D.W., Ford, E.D., and Turner, J. 1990
    Nutrients, moisture and productivity of established forests, *Forest Ecology and Management* 30: 283–299.

11. Linder, S. 1989
    Nutrient control of forest yield, in *Nutrition of trees,* Marcus Wallenberg Foundation Symposia 6: 62–87, Falun, Sweden.

12. Bevege, D.L. 1984
    Wood yield and quality in relation to tree nutrition, in G.D. Bowen and E.K.S. Nambiar, eds., *Nutrition of plantation forests,* 293–325, Academic Press, New York.

13. Mitchell, C.P. 1990
    Nutrients and growth in short-rotation forestry, *Biomass* 22: 91–105.

14. Houghton, J.T, Jenkins, G.J., and Ephraums, J.J., eds. 1990
    Climate change, the IPCC scientific assessment, Intergovernmental Panel on Climate Change, Cambridge University Press, Cambridge, United Kingdom.

15. Houghton, R.A. 1990
    The future role of tropical forests in affecting the CO_2 concentration in the atmosphere, *Ambio* 19: 204–209.

16. Hall, D.O., Mynick, H.E., and Williams, R.H. 1991
    Cooling the greenhouse with bioenergy, *Nature* 353: 11–12.

17. Hall, D.O., Mynick, H.E., and Williams, R.H. 1991
    Alternative roles for biomass in coping with greenhouse warming, *Science and Global Security* 2: 113–151.

18. Gifford, R.M. 1980
    Carbon storage by the biosphere, in E. Pearman, ed., *Carbon dioxide and climate,* 167–181, Australian Academy of Sciences, Canberra.

19. Nobel, P.S. 1991
    Physico-chemical and environmental plant physiology, Academic Press, London, 635 pp.

20. Long, S.P. 1991
    Modification of the response of photosynthetic productivity to rising temperature by atmospheric CO_2 concentrations: has its importance been underestimated? *Plant Cell and Environment* 14: 729–739.

21. Ogden, J.M., Williams, R.H., and Fulmer, M.E. 1991
    Cogeneration applications of biomass gasifier/gas turbine technologies in the cane sugar and alcohol industries, in J.W. Tester, D.O. Wood, and N.A. Ferrari, eds., *Energy and the environment in the 21st century,* 311–346, MIT Press, Cambridge, Massachusettes.

22. Stjernquist, I. 1990
    Modern woodfuels, in J. Pasztor and L.A. Kristoferson, eds., *Bioenergy and the environment,* 49–84, Westview Press, Boulder, Colorado.

23. Barnard, G.W. 1990
    Use of agricultural residues, in J. Pasztor and L. A. Kristoferson, eds., *Bioenergy and the environment,* 85–112, Westview Press, Boulder, Colorado.

24. Lindstrom M.J. et al. 1979
    Tillage and crop residues effects on soil erosion in the corn belt, *Journal of Soil and Water Conservation* March-April: 80–82.

25. Office of Technology Assessment of the United States Congress. 1980
    Energy from biological processes, Washington DC, September.

26. Jakeway, L.A.
    Cane energy recovery, *International Conference on Energy from Cane 1991,* Winrock International, Arlington, Virginia (forthcoming).

27. Energy Technology Support Unit. 1990
    Straw as a fuel—current developments in the U.K., UK Department of Energy, London.

28. Elliott P. and Booth, R.T. 1990
    Sustainable biomass energy, Selected papers, Shell International Petroleum Company, Shell Centre, London.

29. Beyea, J., Cook, J., Hall, D.O., Socolow, R.H., and Williams, R.H. 1992
    Toward ecological guidelines for large-scale biomass energy development, report of a workshop for engineers, ecologists, and policymakers convened by the National Audubon Society and Princeton University, May 6, 1991, National Audubon Society, New York.

30. Evans, J. and Hibberd, B.G. 1991
    Tree plantation review: study no. 8: plantation operations, Shell International Petroleum Company and Worldwide Fund for Nature, London, June.

31. Larson, E. 1992
    Biomass-gasifier/gas turbine cogeneration in the pulp and paper industry, *Journal of Engineering for Gas Turbines and Power* (forthcoming).

32. Brown, S., Gillespie, A.J.R., and Lugo, A.E. 1991
    Biomass in tropical forests of South and Southeast Asia, *Can. J. For. Research* 21: 111–117.

33. Earl, D.E. 1975
    Forest energy and economic development, Clarendon Press, Oxford, United Kingdom.

34. Openshaw, K. 1990
    Energy and the environment in Africa, Industry and Energy Department, The World Bank, Washington DC.

35. Office of Technology Assessment of the United States Congress. 1980
    Energy from biological process, volume III: appendices, part A: energy from wood, Washington DC, September.

36. Forest Resources Assessment 1990 Project. 1992
   The forest resources of the tropical zone by main ecological regions, Food and Agriculture Organization of the United Nations, June.

37. Fearnside, P.M. 1991
   Department of Ecology, INPA, Manaus, Brazil, personal communication.

38. Fearnside, P.M. 1988
   Jari at the age 19: lessons for Brazil's silvicultural plants at Carajas, *Interciencia* 13 (1):12–24.

39. Fearnside, P.M. and Rankin, J.M. 1989
   Jari revisitada: mudancas de perspectivas de sustentabilidade na Amazonia, *Brasil Forestal* 68:17–29.

40. Durst, P.B. 1986
   Wood-fired power plants in the Philippines: financial and economic assessment of wood-supply strategies, *Biomass* 11:115–133.

41. Pohjonen V. and Pukkala, T. 1990
   Eucalyptus globulus in Ethiopian forestry, *Forest Ecology and Management* 36: 19–31.

42. de Jesus, R.M. 1990
   The need for reforestation, paper presented at the US Environmental Protection Agency-sponsored workshop on Large-Scale Reforestation, Corvallis, Oregon.

43. Schmidheiny, S. 1992
   Changing course: a global business perspective on development and the environment, MIT Press, Cambridge.

44. ABRACAVE (Charcoal Producers Association). 1992
   Statistical yearbook 1992, Belo Horizonte, Brazil.

45. Carpentieri, A.E., Larson, E.D., and Woods, J. 1992
   Prospects for utility-scale, biomass-based electricity supply in northeast Brazil, CEES/PU Report No. 270, Center for Energy and Environmental Studies, Princeton University, Princeton, New Jersey, July.

46. Hummel, F.C., Palz, W., and Grassi, G., eds. 1988
   Biomass forestry in Europe: a strategy for the future, Elsevier Applied Science, London.

47. Ledin, S. 1990
   Plantations on farmland in Sweden, *IEA Bioenergy News* 2(3): 5, University of Aberdeen, Scotland.

48. Johansson, H.
   Energy forestry—opportunities, market, profitability, scale, environmental considerations, *Proceedings of the Symposium on New Crops for Europe,* King's College, Chapman and Hall, London (forthcoming).

49. Johansson, H. 1992
   Energiskog—energi some vaxer, newsletter from the Swedish University of Agricultural Sciences at Uppsala, *Mark/vaxter* 1, Uppsala, Sweden.

50. Abelson, P.H. 1991
   Improved yields of biomass, *Science* 252:1469.

51. Turhollow, A.H. 1991
   Economics of dedicated energy crop production, draft report, Oak Ridge National Laboratory, Oak Ridge, Tennessee.

52. Ledig, F.T. 1981
   Silvicultural systems for the energy efficient production of fuel biomass, in D.L. Klass, ed., *Biomass as a nonfossil fuel source,* ACS Symposium Series 144, American Chemical Society, Washington DC, 1981.

53. International Energy Agency. 1992
    Energy prices and taxes: third quarter 1991, Organization for Economic Cooperation and Development, Paris.

54. Perlack, R.D., Ranney, J.W., and Russell, M. 1991
    Biomass energy development in Yunnan Province, China: preliminary evaluation, a report prepared for the Committee on Renewable Energy Commerce and Trade and the Biofuels Division of the US Department of Energy, ORNL/TM-11791, Oak Ridge National Laboratory, Oak Ridge, Tennessee, June.

55. Davidson, J. 1987
    Bioenergy tree plantations in the tropics: ecological implications and impacts, Commission on Ecology Paper No. 12, International Union for the Conservation of Nature and Natural Resources, Gland, Switzerland.

56. Evans, J. 1992
    Plantation forestry in the tropics, 2nd ed., Oxford University Press, 400 pp.

57. Lovejoy, T.E. 1985
    Rehabilitation of degraded tropical forest lands, Commission on Ecology Occasional Paper No. 5, Gland, International Union for the Conservation of Nature and Natural Resources, Gland, Switzerland.

58. Sumitro, A. 1985
    Rehabilitation and utilization of forest fallows and degraded areas, GCPRAS/106/JPN, Field Document No. 7, Food and Agriculture Organization of the United Nations, Bangkok.

59. Office of Technology Assessment of the United States Congress. 1992
    Technologies to sustain tropical forest resources and biological diversity, OTA-F-515, Washington DC, May.

60. Borman, B.T. and Gordon, J.C. 1989
    Can intensively managed forest ecosytems be self-sufficient in nitrogen? *Forest Ecology and Management* 29: 95–103.

61. DeBell, D.S., Whitesell, C.D., and Schubert, T.H. 1989
    Using N_2-fixing *Albizia* to increase growth of *Eucalyptus* plantations in Hawaii, *Forest Science* 35(1): 64–75.

62. Kimmins, J.P. 1990
    Modelling the sustainability of forest production and yield for a changing and uncertain future, *The Forest Chronicle* 6: 271–280, June.

63. Council for Agricultural Science and Technology. 1990
    Ecological impacts of federal conservation and cropland reduction programs, Summary Report No. 117, Ames, Iowa, September.

64. Lashoff, D.A. and Tirpak, D.A., eds. 1991
    Policy options for stabilizing global climate, report to Congress, technical appendices, report prepared by the Office of Policy, Planning, and Evaluation, US Environmental Protection Agency, Washington DC.

65. Alexander, A.G. 1985
    The energy cane alternative, Sugar Series, Vol. 6, Elsevier, Amsterdam.

65a. Taylor, T.B., Taylor, R.P., and Weiss, S. 1982
    Worldwide data related to potentials for widescale use of renewable energy, PU/CEES Report No. 132, Center for Energy and Environmental Studies, Princeton University, Princeton, New Jersey.

66. Food and Agriculture Organization of the United Nations. 1990
    Production yearbook 1989, Rome.

67. US Department of Agriculture. 1990
The second RCA appraisal: soil, water, and related resources on nonfederal land in the United States: analysis of conditions and trends, Miscellaneous Publication No. 1482, Washington DC.

68. Betters, D.R., Wright, L.L., and Couto, L.
Short rotation woody crop plantations in Brazil and the United States, *Biomass and Bioenergy* (forthcoming).

69. Hillis, W.E. 1990
Fast-growing *Eucalyptus* and some of their characteristics, in D. Werner and P. Muller, eds., *Fast-growing trees and nitrogen-fixing trees,* 184–193, International Conference, Marburg, Germany, 8–12 October 1989, Gustav Fischer Verlag, Stuttgart.

70. Zobel, B., Campinhos, Jr., E., and Ikehemori, Y. 1983
Selecting and breeding for desirable wood, *Tappi Journal,* 70–74, January.

71. Hansen, E.A. 1991
SRIC yields: a look to the future, *Biomass and Bioenergy* 1:1–8.

72. Farnum, P., Timmins, R., and Kulp, J.L. 1983
Biotechnology of forest yield, *Science* 219: 694–702.

73. Hummel, F.C. 1988
Biomass forestry: implications for land-use policy in Europe, *Land Use Policy,* 375–384, October.

74. Organization for Economic Cooperation and Development. 1991
Agricultural subsidies, OECD, Paris.

75. Intergovernmental Panel on Climate Change. 1991
Climate change: the IPCC response strategies, Island Press, Washington DC.

76. Food and Agriculture Organization of the United Nations. 1991
Agricultural land use: inventory of agroecological zones studies, FAO, Rome.

77. The Noordwijk declaration on atmospheric pollution and climatic change, Ministerial Conference on Atmospheric and Climatic Change, Noordwijk, The Netherlands, November 1989.

78. Food and Agriculture Organization of the United Nations. 1982
Forestry in China, FAO Forestry Paper 35, Rome.

79. AFOS (Agriculture, Forestry, and Other Human Activities) subgroup. 1992
IPCC Report 1992, Working Group III, WMO/UNEP, Geneva, March.

80. Rogers, H.H., Bingham, G.E., Cure, J.D., Smith, J.M., and Surano, K.A. 1983
Responses of selected plant species to elevated CO_2 in the field, *Journal of Environmental Quality* 12: 569–574.

81. Box, E.O. and Meetemeyer, V. 1991
Geographic modeling and modern ecology, in G.Esser and D. Overdieck, eds., *Modern ecology: basic and applied aspects,* 773–804, Elsevier, Amsterdam.

82. Strehler, A. and Stutzle, W. 1987
Biomass residues, in D.O. Hall and R.P. Overend, eds., *Biomass: regenerable energy,* Wiley, Chichester.

83. Food and Agricultural Organization of the United Nations. 1991
Agrostat database, Rome.

84. Bazett, M.
The need for industrial wood plantations, Study No. 3 in the Tree Plantation Review, Non-Traditional Business Group, Shell International Petroleum Company, London (forthcoming).

85. Turhollow, A.H. and Perlack, R.D. 1991
Emissions of CO_2 from energy crop production, *Biomass and Bioenergy.* 1: 129–135.

86. Energy Information Administration. 1992
International Energy Annual, US Department of Energy, DOE/EIA-0219(90), Washington DC.

87. World Resources Institute. 1992
World resources 1992–93, a guide to the global environment, Oxford University Press.

88. Grainger, A. 1988
Estimating areas of degraded tropical lands requiring replenishment of forest cover, *International Tree Crops Journal* 5:31–61.

15
BIOENERGY: DIRECT APPLICATIONS IN COOKING

GAUTAM S. DUTT
N.H. RAVINDRANATH

Cooking stoves that burn traditional biofuels are used by half the world's population, yet many are inefficient and hazardous to the health of those who tend them. In recent years, however, a new generation of cookstoves needing less fuel and emitting fewer airborne particulates has emerged. Many of the new designs run on biomass that has been transformed into a liquid, gaseous, or improved solid-fuel form. Alternative cooking systems are compared, and data from cooking trials conducted by the authors in a south Indian village are provided.

INTRODUCTION

Biofuels, which include wood, charcoal, crop residues, and dung, provide cooking energy for half of the world's population, a situation that is unlikely to change in the foreseeable future. Although most households in the industrialized countries, and some in the less developed countries (LDCs), have shifted to petroleum fuels and electricity, these options are not available to most rural households in the LDCs. In those areas, cooking often represents 70 to 80 percent of total energy use, and biofuels make up most of the energy supply. In this chapter we review alternative technologies for cooking with biofuels. In a post-petroleum era some of these options may replace the needs now met by petroleum fuels.

EVALUATING COOKING SYSTEMS

A cooking system, which includes the stove, fuel, pots, and indeed the cook, can be evaluated from several perspectives [1]. The technical evaluation must include a stove's efficiency, cost, and its overall suitability for cooking and other needs. The environmental evaluation must consider the availability of the fuel source, the health impacts of the system, and the environmental consequences of a large-scale shift to a new cooking system. The social–cultural evaluation should include a system's impact on development and whether or not the technology is culturally acceptable. These evaluation criteria are discussed below.

Fuel availability

The availability and reliability of a particular fuel often limit the choice of cooking systems. In many rural areas and small towns in LDCs, modern cooking fuels such as kerosene, liquefied petroleum gas (LPG), and natural gas are available only on an intermittent basis, if at all. Moreover, many of these fuels, which must be transported to each household, require an infrastructure in the form of pipelines, usage meters, and billing systems that render them beyond the reach of poor communities. In other areas, deforestation may have severely limited the fuelwood supply, although lower quality wood in the form of twigs and shrubs may still be plentiful. Biogas, which is generated by the fermentation of organic matter such as dung, can be an effective cooking fuel, but its production requires sufficient amounts of raw material.

Stove suitability

All stoves must perform a variety of cooking tasks: grilling (broiling), baking, and preparing items on flat pans or griddles, as well as boiling. Whereas traditional cookstoves meet all cooking tasks, some improved fuelwood stoves will accommodate only round or flat-bottomed deep pots, and cannot be used for grilling or baking. Households that have this type of improved stove must therefore also use traditional cooking devices. The typical European-style gas stoves sold in Mexico, for instance, are not suitable for the preparation of tortillas, a principal food product throughout the country. Consequently, the few households in a village that do have access to LPG continue to use fuelwood for tortilla preparation. In urban areas, however, most households purchase their tortillas ready-made, and so find gas stoves satisfactory.

The suitability of a cooking system depends on a number of additional factors, including the speed with which it cooks, the ability of the cook to see the fire and control its heat output, the safety of the stove's design, and whether or not it is mobile, durable, and easy to maintain [2]. Does the stove, for example, coat the cooking pots with soot that must then be removed? Because many of these concerns have a cultural basis, and thus vary from region to region, a survey of potential users should precede the introduction of any alternative cooking system.

In rural areas, cookstoves may also perform a range of noncooking functions that affect human health and well-being. Heating water is one such function that is performed equally well by improved and traditional cookstoves. In colder regions, traditional cookstoves also provide space heat, although they require an inflow of (cold) air for combustion to occur, and thus are highly inefficient for heating purposes. An improved stove, especially when equipped with a chimney, will reduce the amount of available space heat; this reduction may be desirable in hot climates, but detrimental in colder regions. Yet chimneys remove smoke from the cooking area, thus reducing the health risks associated with smoke inhalation. In some areas smoke from a traditional open fire is used for insect control. Thus, space heating, insect control, and other functions may need to be met by other means when a modern cooking system is adopted. It is vital, therefore, that non-cooking functions be included in user surveys.

Fuel consumption

Alternative cooking systems usually offer improved energy efficiency and lower levels of indoor air pollution (see below). In the course of improving woodburning cookstoves, standardized procedures for measuring their fuel efficiency have been developed [3]. One such procedure, which was developed by Samuel Baldwin [4], can be adapted to any cooking fuel. Three types of tests are generally undertaken: water boiling tests (WBTs), controlled cooking tests (CCTs), and kitchen performance tests (KPTs).

In the WBT a measured quantity of water is boiled and simmered, while water temperature, time, and fuel use are recorded. WBT results are expressed as an efficiency, which may be defined in several ways. In the commonly accepted procedure, efficiency is the heat absorbed by the water in the pot (including the latent heat of the evaporated water) divided by the higher heating value (HHV) or gross calorific value of the fuel, which is adjusted for moisture content, an important factor in biomass fuels.

In the CCT, a meal typical of the region in which a new stove program may be implemented is prepared by a local cook. Different stoves are compared on the basis of fuel consumption, which is defined as the weight of fuel (in kilograms) needed to produce a given weight of cooked food (also in kilograms), or as the energy content of fuel needed per weight of cooked food (megajoules per kilogram), the latter being more useful in comparing stoves using different fuels. The drawback is that the test is conducted under controlled conditions and so may not reflect actual fuel consumption. Real-life users are less likely to operate the stove under optimal conditions and therefore less likely to obtain the best fuel economy. Moreover, once the stoves are constructed in large numbers they may fail to meet critical design specifications that affect their efficiency. In addition, the gains in fuel economy the stoves provide may be offset by the perception that fuel is no longer so scarce, and this may lead to increased cooking. In order to gauge the long-term effects of fuel economy, fuel use should be evaluated before and after a cooking system change.

Kitchen performance tests are fuel-use measurements carried out for a statistically meaningful number of households (often 100 or more), usually in conjunction with stove user surveys (see below). Because the measurements cannot be too intrusive, they are limited to the quantity (weight or volume) of fuel used per person per day, then converted to energy content and expressed in megajoules per person per day. For natural gas, LPG, and electricity, fuel use is generally determined by a meter. If LPG is used for cooking (and nothing else), fuel consumption can be determined from cylinder weight.

Smoke and air pollution

Biomass fuels generate a number of air pollutants, including particulates, carbon monoxide, and a variety of volatile hydrocarbons, all of which can produce adverse health effects. The health impact of biomass cookstoves most likely exceeds that of all other air pollutants combined. Other cookstove fuels, such as coal, generate some or all of these pollutants, as well as nitrogen and sulfur oxides. Nonetheless, cookstove pollution has been poorly characterized largely because the task is complex and requires sophisticated equipment. One method of evaluation calls for boiling water on the stove in a chamber where the ventilation rate and the concentration of two pollutants—carbon monoxide (CO) and total suspended particulates (TSP)—are monitored [5, 6]. The amount of fuel needed to heat the water is also measured. The data are then used to calculate (a) the emission factor, which is the weight of pollutants—CO and TSP—generated per unit weight of fuel burned; (b) the stove's efficiency; and (c) the emissions for each pollutant per standardized task. This last indicator measures the impact of fuel efficiency changes on pollutant emissions for a specific water heating task. The procedure, which is an extension of the WBT for measuring stove efficiency, also applies to stoves using fuels other than biomass.

Economics

Although not all features of alternative cooking systems are quantifiable, economic comparisons can be made on the basis of capital and operating expenses. Simple payback periods and internal rates of return are effective means of comparing two alternatives; when only one fuel type is saved or produced, the cost of saved energy is a popular index of energy conservation. An index that allows for a simple comparison among any number of alternatives is the annualized life-cycle cost (ALC). With this approach, capital investment in the cooking system (I) is annualized using a capital recovery factor (CRF), which depends on the discount rate, or the time value of money. To this is added annual fuel cost (FC) and other recurring operation and maintenance costs (O&M). The total is the annualized cost of owning and operating the cooking system: ALC = (I × CRF) + FC + O&M. The lowest value for ALC is the least-cost alternative for a given discount rate.

The ALC can be calculated from a user's perspective using market values for stove and fuel prices, or from a societal perspective by ignoring internal cash flows (taxes and subsidies) and including external costs (environmental impact and foreign-exchange requirements). The user's perspective, however, is meaningful only where stoves and fuels have a monetary cost, such as in urban areas. We have chosen a societal perspective based on reasonable stove and fuel costs, even though some stoves are self-built and fuel may be gathered for free. External costs have been omitted from our analysis.

Social and cultural factors

Cooking systems should be judged in terms of their social and cultural impacts as well as technical merit. Saving fuel, for example, may not be a priority among rural peoples faced with more pressing needs. In addition, changes in the cooking routine that alter social contact among household members may be undesirable. Yet for some a new fuel or change in stove design may be seen as a sign of progress and elevated status [1]. The participation of local women in the design and diffusion of alternative cooking systems is essential, for we do not necessarily want to see the adoption rate of an alternative system increase if that system does not benefit the people.

BIOMASS COOKING SYSTEMS

Three-stone open fire

In many parts of the world, cooking takes place over an open fire with the cooking vessel supported on three stones. There is abundant evidence to suggest that the three-stone fire was universal in ancient times; even today it carries a great deal of religious significance for many societies. Although the three-stone fire, which generally burns firewood and crop residues, has been reputed to have efficiency ratings as low as 5 percent [7, table 3], measurements now reveal that their efficiency may be as high as 36 percent [7, table 5]. Typical values are between 15 and 20 percent.

Traditional stoves

Many types of enclosed fires or stoves, which burn either fuelwood, crop residues, or dung, are found throughout the world. In Asia the stoves, which can be single-pot or multi-pot types, typically are made of mud and built by their users (see figure 1). In water boiling tests, traditional stoves fare little better than the open fire, achieving an efficiency rating no higher than 20 percent [7, table 5]. Another drawback is that traditional mud stoves, like the three-stone fire, lack chimneys and simply release their smoke into the kitchen.

Improved fuelwood stoves

Many attempts have been made to develop woodburning stoves that produce less smoke, burn less fuel, or do both. Indeed, the current wave of interest in improved fuelwood stoves may be traced back to the 1950s when a stove known as the Hyderabad chula was introduced in India, and to the 1960s when the Food and Agricultural Organization of the United Nations developed the Singer stove for use in Indonesia [8, 9]. Both are multi-pot designs that are known as "site-built stoves" because they are built into a kitchen. Since then two other types of biomass stoves have emerged. One is the portable stove, which can either be artisan built or mass produced; the other is a prefabricated artisan-made stove called an insert stove, which is assembled in the user's kitchen.

FIGURE 1: A three-pot traditional mud stove, used with fuelwood, crop residue, and cow dung (Karnataka, India).

Site-built stoves are made from a variety of materials depending on the regional availability of supplies. In China, for example, the stoves are made from bricks and cement, in Burkina Faso they are made from concrete, in India from mud (see figure 2), and in Central America from a mixture of volcanic ash and clay called lorena. Most have chimneys and are designed to accommodate two or three pots. On average, their efficiency appears to be slightly higher than that of three-stone fires: in one series of WBTs, site-built stoves were found to have efficiencies ranging from 15 to 22 percent, whereas the three-stone fire had an efficiency of 17 percent [4].

Still, an empirical study of three-pan mud stoves undertaken by the Indian Institute of Science indicated that much higher efficiencies are possible. By optimizing the dimensions of a stove called the ASTRA ole (ASTRA is the Centre for the Application of Science and Technology to Rural Areas), researchers at the institute were able to increase its efficiency from 15 to 45 percent [10]. Increasing the efficiency in this way, however, requires a trained stove builder who can build the stove according to precise dimensions. But because each stove is site built, quality control is difficult. In addition, many site-built stoves, including the ASTRA ole (see figure 2) are designed for use with cooking pots of a certain size; when pots of other sizes are used, the stove's overall efficiency may be greatly reduced.

Portable stoves, which are either mass-produced or artisan-built, are often cheaper to manufacture and more amenable to quality control than site-built stoves. The first energy-efficient sheet-metal stove was designed in 1983 for use in Burkina Faso [4]. Each stove is designed for a single pot size, and pots larger than the maximum size cannot be used (see figure 3). Although smaller pots can be used, they substantially reduce the stove's efficiency.

When operated correctly, portable stoves can be quite efficient. Baldwin recorded efficiencies of 29 percent for a sheet metal stove of simple design, 36 per-

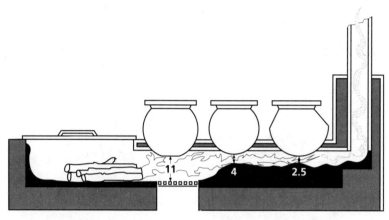

All dimensions in centimeters

FIGURE 2: ASTRA ole—a three-pot mud stove with chimney (India).

FIGURE 3: An improved metal stove (West Africa). Note how the pot sits inside the stove; stove and pot diameters must be matched for optimum performance.

cent for a ceramic stove, and 42 percent for an insulated metal stove. Other metal stoves, including Priyagni, Tara, and Swosthee models, have achieved efficiencies as high as 43 percent [11]. Moreover, because quality control can be rigorously enforced at a centralized manufacturing site, working stoves typically have efficiencies that approximate laboratory models. A sample of 200 Swosthee 1-kilo-

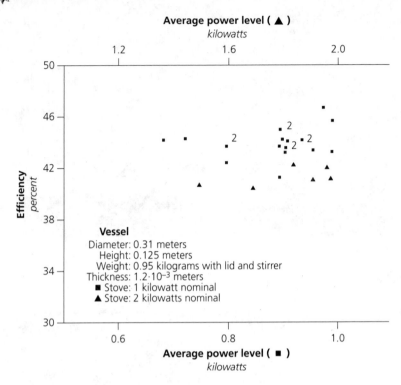

FIGURE 4: Efficiency of two metal stoves of 1 kilowatt and 2 kilowatts nominal power. The "2" refers to a second 1 kilowatt stove built to the same specifications but in a lot of 200. [11].

watt stoves (model S-1), for example, showed efficiencies indistinguishable from the prototype (see figure 4). A later prototype of Swosthee (model MS-4) is shown in figure 5.

In rural areas, where metal may be unavailable or expensive, portable stoves are often made from fired clay. Although ceramic stoves are cheaper to manufacture than metal stoves (a 1983 study in Burkina Faso showed that a fired-clay stove could be built for U.S.$0.57, as opposed to $2.25 for a metal stove [4]), the dimensions of a ceramic stove cannot be precisely maintained, and so its efficiencies are generally somewhat lower.

Insert stoves combine some of the advantages of an artisan-built stove with those of a site-built one. The stove's critical parts, usually ceramic, are artisan built and thus conform to precise specifications. The user purchases the parts and then installs them in the kitchen, cementing them together with mud to produce a highly durable stove. Insert stoves (with and without chimneys) have been promoted in Nepal [12], Sri Lanka [13], and elsewhere (see figure 6).

Charcoal stoves

Converting fuelwood to charcoal leads to improved fuel quality. Charcoal is relatively smokeless (although it generates a great deal of carbon monoxide) and was

FIGURE 5: *Prototype of the Swosthee MS-4 metal woodstove developed at the Indian Institute of Science, Bangalore. Unlike the West African metal stove of figure 3, the pot is not inserted into the stove; efficiency does decrease as pot diameter decreases.*

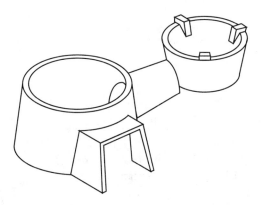

FIGURE 6: *An artisan-built pottery insert that is installed by the user in the kitchen to make a complete stove. Figure credit: Intermediate Technology Development Group.*

much favored by affluent households before petroleum fuels such as kerosene and LPG became available for cooking. Charcoal is still extensively used, particularly in the urban areas of many African countries.

Because charcoal is made from wood, the efficiency of charcoal production is as important as that of the charcoal cookstove itself. The efficiency with which wood is converted to charcoal depends on the wood's moisture content and the type of kiln used. According to studies carried out in Thailand [14], permanent (brick beehive) kilns are superior to both earth mounds and portable metal kilns,

Table 1: Conversion efficiencies and economics of charcoal kilns in Thailand[a]

Kiln type	Volume m³	Construction Cost[b] $	Yield[c] percent	Efficiency[d] percent	Production cost $/tonne	$/GJ
Earth mound	0.7	–	31.1	50.9	145	4.99
Mobile						
Tonga	0.2	13	22.7	36.0	403	13.91
Double drum	0.4	30	23.9	38.7	173	5.96
Permanent						
Brazilian modified	8.3	137	34.5	55.1	90	3.11
Hot tail	0.5	22	33.3	47.6	149	5.14
Mud beehive	7.2	38	32.0	44.8	101	3.49
Brick beehive 1	8.3	218	39.6	60.6	82	2.83
Brick beehive 2,3	2.0	105	37.5	62.5	95	3.26

a. Compiled from tables 4.1, 5.9, and 5.24 of [14. Not all charcoaling methods tested in Thailand are included.

b. All costs are in 1989 U.S.$, which were calculated using a conversion factor of 23 baht to 1 U.S.$ in 1983, and then inflated at 4 percent per year for six years.

c. Yield = $\dfrac{\textbf{weight of lump charcoal output (freshly recovered)}}{\textbf{oven dry weight of wood } - \textbf{ weight of brands}} \times \textbf{100}$

The brands result from the incomplete conversion of wood to charcoal.

d. Efficiency is the ratio of energy content in the charcoal output to the wood consumed.

producing high yields of good-quality charcoal at significantly reduced cost (see table 1). The study also found that traditional earth mounds, if properly tended, are more efficient than had been previously reported [14]. Because of conversion losses, and capital, labor, and transportation costs, charcoal is an expensive fuel (see table 2 for charcoal prices in several countries).

A typical traditional charcoal stove is illustrated in figure 7. Improved models have been introduced in a number of countries. Kenya's popular improved jiko (*jiko* means stove), for example, consists of an outer metal shell and a ceramic liner. Sandwiched between them is an insulating layer that reduces heat loss from the stove's sides and bottom (see figure 8).The United Nations International Children's Emergency Fund (UNICEF) developed an all-metal, double-walled stove called the Umeme that enables the user to control air flow and power (see figure 9). Ernst Sangen and Piet Visser of the Eindhoven University of Technology in the Netherlands [15] tested the performance of 12 charcoal stoves, using a boiling water test that calls for fuel weight to be kept roughly constant, which reduces the variability of test results [19]. According to this procedure, most of the stoves have efficiencies exceeding 40 percent in the high-power test (see table 3).

Table 2: Charcoal prices in selected urban areas

City	Year	Price $ per tonne[a]	Price 1989 $/GJ[b]	Reference
Khartoum	1987	$175–220	6.53–8.21	16
Nairobi	1985	$106–187	4.28–7.54	17
Raipur	1985	Rs.1,590	4.23	18
Hyderabad	1985	Rs.1,380	3.67	18
Bangalore	1985	Rs.1,170	3.11	18
Port au Prince	1989	$140	4.83	Destine, personal communication

a. Price per tonne corresponds to the year noted in the previous column as given in the source.

b. Dollar values have been inflated to 1989 at the rate of 4 percent per year; Indian rupees have been inflated at 7 percent per year and converted to dollars at Rs 17 per dollar. A higher heating value of 29 megajoules per kilogram for air-dried charcoal has been assumed.

FIGURE 7: A traditional charcoal stove, made of metal (West Africa).

The heat output of charcoal stoves, in contrast to fuelwood stoves, cannot be varied by changing the amount of fuel, but is instead controlled by adjusting airflow into the combustion area. Although four of the 12 stoves tested could not be controlled in this manner, the power output of four others could be reduced

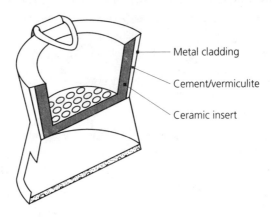

FIGURE 8: Cutaway of the Kenyan improved jiko, an improved charcoal stove in which a ceramic liner and an insulation layer placed within the metal shell reduce stove heat losses. Figure credit: Intermediate Technology Development Group.

FIGURE 9: The UNICEF jiko or Umeme—an all-metal charcoal stove developed by UNICEF in East Africa.

by a third or more. These stoves (the Thai bucket stove, the Haiti improved stove, the Kenyan improved jiko, the improved coalpot, and the UNICEF Umeme jiko) probably require the smallest amount of fuel in actual cooking; at least one of them (the Kenyan jiko) is already very popular.

Table 3: Performance characteristics of charcoal cookstoves[a]

Stove	Air control[b]	Power		Efficiency	CO/CO_2 ratio in combustion products		Time to boil
		kilowatts		percent			minutes
		high	low		high	low	
DUB 9 *Burundi*	yes	2.0	0.9	43.1	0.12	0.10	40
Haiti, improved	yes	2.0	0.6	45.6	0.12	0.15	35
Ethiopia, traditional	no	0.9	–	43.0	0.12	–	75
CEPPE *Ethiopia*	no	2.5	–	45.1	0.10	–	25
Sakkanal *Senegal*	yes	2.4	1.5	29.3	0.16	0.13	60
Feu malgache *Sahel*	no	1.8	–	29.2	0.12	–	65
Coalpot, improved[c]	yes	3.5	1.2	25.0	–	–	25
Sudan, traditional	no	1.7	–	41.5	0.14	–	55
Thai bucket, traditional	yes	3.6	2.0	45.0	0.10	0.06	25
Thai bucket, improved	yes	4.0	0.8	48.6	0.04	0.06	25
UNICEF jiko[d] *Kenya*	yes	3.5	1.1	37.1	0.05	0.09	30
KENGO jiko *Kenya*	yes	2.8	1.6	45.2	0.08	0.07	30

a. [15].

b. Stoves with air control were tested at both high and low power. The high-power test requires heating and boiling water for 30 minutes. Low power refers to water that was simmered for 60 minutes, with reduced air flow. Low power efficiency was recorded but is not presented here.

c. No gas analysis was done when this stove was tested.

d. Although this stove was also tested with an adapter for use with large pans, the data are not reported here.

Biomass-residue stoves

When crop residues and other biomass wastes do not serve a more valuable purpose, as in feed or fertilizer, they may be used for cooking fuel. Although plant stalks and straw burn in most traditional fuelwood stoves, other forms of biomass residue require modified stoves.

Powdery biomass, such as sawdust or rice husks, needs a specially designed cookstove. A popular design is the "kerosene can," a 16-liter rectangular container. A vertical cylindrical tube runs through the center of the can and is connected to a horizontal opening at the bottom for air entry. Biomass powder is stuffed into the cylinder and lit with a stick of wood. The stove requires little or no attention, but it has several disadvantages. Its heat output cannot be controlled, which in turn makes it about as efficient as an open fire. In addition, it lacks a chimney and thus is very smoky. According to S. Dasappa of the Indian Institute of Science (personal communication), a prototype is now being developed that has a secondary air intake, which helps combust volatiles, reducing smoke generation. Although its efficiency is greater, it remains below that of the most efficient wood stoves.

Densified biomass, powdery residues that are compressed and formed into briquettes, have several advantages over loose agricultural residues. The densified biomass has greater volumetric energy density than loose residues. Moreover, fuel briquettes have uniform properties and are an effective way of using loose agricultural residues that are unsuitable for fodder or fertilizer but are too difficult to transport and would otherwise simply be discarded.

Briquetting can be accomplished either with or without carbonization. In one method the biomass is charred to form a charcoal-like fuel; in the other no charring is involved. Although a small, manually operated press costs a few thousand dollars, large-scale briquetting operations require significant capital investment. An 800 tonne per year carbonizing plant in the Sudan, for example, cost $90,000 to build [16]. Moreover, packaging, storing, and transporting the briquettes add to the cost of the final product. Briquettes made from cotton stalk (which is densified without carbonization) and delivered to Addis Ababa, for instance, were priced at $114 a tonne in 1986 (equivalent to $128 in 1989, assuming 4 percent inflation per annum) [20]. Considering an energy content of 17.8 megajoules per kilogram [21, table 4.18], the 1989 price is equivalent to $7.19 per gigajoule. Briquette prices in other countries are comparable. Although lower-cost briquetting machines have been developed in India, all have had serious technical problems. Nevertheless, in 1990 biomass briquettes were available in India for only $40 to $60 per tonne [22]. H.S. Mukunda of the Indian Institute of Science estimates briquetting costs to be (1990) $18 per tonne, without counting the cost of raw material. If biomass residues can be sold for half the cost of plantation-grown wood, the total cost would be about $72 per tonne, substantially less than in Addis Ababa. Nonetheless, it must be noted that densified biomass briquettes are less efficient in traditional stoves or open fires than is wood

[23], and so briquettes must sell for less than wood in order to gain significant market penetration [24].

Although some energy is lost in conversion, carbonized briquettes represent a higher quality fuel than noncarbonized, densified briquettes. Partially pyrolyzed briquettes produce one-thirtieth as much carbon monoxide as densified briquettes [25]. Still, briquettes are more difficult to ignite than charcoal, generate more smoke, and in most places are more expensive (on a dollars per gigajoule basis). To counter these problems, the Intermediate Technology Development Group in Great Britain has devised cooking systems that combine briquetting machines with special stoves that have been promoted in the Gambia [25–27]. Lower-cost briquetting machines have also been commercialized in Thailand [28].

Biogas stoves

Biogas is a combustible gas produced by the anaerobic digestion of animal dung and other wastes. The gas, which is burned in specially designed stoves, produces little carbon monoxide and no smoke. An added environmental benefit is that slurry from the digester provides the soil with more fixed nitrogen than when dung is directly applied as a fertilizer.

In some countries, biogas digesters have been extensively promoted; China, for example, had some five million small biogas plants operating in 1986 [29]. By 1988, more than one million biogas plants had been installed in India [30], although many of them have had problems. Of 4,108 plants surveyed in the state of Maharashtra, only 35 percent were found to be working. In only 3 percent of the cases could the shutdown be attributed to technical defects; 29 percent were closed for lack of dung, 16 percent were unable to digest the dung correctly. For most cases (52 percent), however, plant closure resulted from "lethargy, lack of interest, lack of knowledge regarding importance of biogas, etc." [31].

Most of the Indian biogas plants are small units, intended for individual families. Greater economies of scale can be achieved by building community and institutional plants (see table 4 and chapter 18: *Biogas Electricity—The Pura Village Case Study*). By 1988, 430 large-scale digesters had been installed in India [32]. Although production costs drop, large-scale plants have additional expenses: they must distribute the biogas through pipelines, which in turn requires that billing meters be installed, two measures that considerably increase the amount of capital investment required. Instituting a billing and collecting system also adds to the overall cost of operating and maintaining the digester. Taking these factors into account, the estimated price of biogas delivered to households is expected to range from $11 to $12 per gigajoule (see table 5).

Biogas can also be produced from a number of other sources including sewage treatment plants (see chapter 19: *Anaerobic Digestion for Energy Production and Environmental Protection* and [36]), food-processing effluent, weeds, leaves, and nonedible starch, all of which are receiving considerable attention as alternatives to dung. Other advances in biogas generation are taking place at the level of

Table 4: Costs of biogas plants in India[a,b]

	Capacity *m³ per day*	Digester[c] *1990$*	O&M[d] *1900 $ per year*
Individual	2	369	9
Community	12	883	399
	24	1,322	548
	30	1,541	671
	36	1,760	745
	45	2,097	908
	54	2,418	1,113
	60	2,638	1,187

a. [33], [34] and [35], table 1. Plant life: 25 years.

b. Original data in 1985 and 1986 Rs.

c. Additional capital costs for a biogas plant include piping and connections of the digester, sand filters for the slurry ouput of the digester, and tools and accessories for maintenance.

d. For community-scale plants, the operation and maintenance costs include a fuel cost, equivalent to gathering dung and distributing the slurry produced by the digester.

the digester. Research is now under way to increase the mean residence time of a given digester volume (e.g., plug flow) to reduce plant cost or to alter the process so that residence time (and digester volume and cost) is reduced.

Burners that are designed specifically for biogas have achieved efficiencies around 60 percent, and thus are equivalent to natural gas or LPG burners [37]. But this may not be the case in practice; in one study of six Indian biogas stoves, efficiencies were found to vary from 40 to 65 percent [38].

Producer gas

A gas containing carbon monoxide and hydrogen as primary combustible components can be derived from raw biomass by thermochemical conversion. In many industrialized countries such a fuel was once obtained from coal for use in cooking, and before that for lighting, until natural gas became more widely available. In Sweden and elsewhere, charcoal-based producer gas served as a transport fuel during World War II, when petroleum availability was limited [39]. In Beijing and Shanghai, about half of all households still use coal gas for cooking [40]. In fact, cities in several developing countries still have coal-based producer gas distributed to houses. Recently, there has been renewed interest in biomass gasifiers as a means to produce fuel for petroleum-importing developing countries [41–43] (see chapter 16: *Open-top Wood Gasifiers* and chapter 17: *Advanced Gasification-based Biomass Power Generation*).

Producer gas has generally been proposed for rural areas for motive power (pumping, etc.) and for electricity generation (see chapter 16). Costs of Indian gasifiers from two sources are given in table 6. For such purposes, the gas must be

Table 5: System cost estimate for community biogas plant supplying gas to 40 households[a,b]

Capital investment	U.S. $	
Biogas digester, 45 m³/day	2,100	(rounded, from table 4)
Distribution piping	2,100	(estimated)
Meters, 40 @ $50	2,000	(estimated)
Total	**6,200**	

Annual operation	$/yr	
Operation and maintenance of digester	900	(rounded, from table 4)
Billing and administration	1,800	(estimated)
Total	**2,700**	

Life of biogas plant = 25 years

Total gas demand = 40 HH × 6 persons/HH × 0.15 m³/day × 22.3 MJ/m³
= 802.8 MJ/day or 293 GJ

	Discount rate	
	6 percent	12 percent
Annualized investment and operation *dollars*	3,185	3,491
Gas cost delivered *dollars per gigajoule*	11	12

a. We assume a six-person household, with per capita fuel consumption of 0.15 m³/day as measured in the Ungra experiments (see below, page 297). This gives an average gas demand of 36 m³ per day. A 45 m³/day capacity biogas digester is selected. The calorific value of biogas is 22.3 megajoules per m³.

b. A number of costs have been estimated and are indicative of the expected magnitude.

cleaned and cooled before it is fed to an internal combustion engine. In contrast, gas that is burned for cooking has no such requirements; it is simply blown through the gasifier and into a storage tank. Although cooking gas probably costs slightly less to produce than motive gas, the difference between them is not substantial. Overall costs of a producer-gas based cooking system must include the cost of building and maintaining an infrastructure (see table 7); as a consequence, the price of gas delivered to households is thought to range from $15 to $16 per gigajoule.

Table 6: Costs of wood gasifiers including cooling and cleaning
(for use in engines)

Source[a]	Capacity GJ/hour	Capital cost[b] $
Jain	0.0876	1,385
	0.1640	2,077
	0.3850	4,327
	1.4250	15,385
Baliga	0.0855	1,389
	1.4400	30,556

a. [44] and [45].
b. Original figures have been converted to 1990 $ using a conversion factor of Rs.(1985)13 = 1$ (1990).

Table 7: System cost estimates for a wood gasifier supplying gas to 40 households[a]

Capital investment	U.S. $	
Gasifier, 0.164 GJ/hr	2,100	(rounded, from table 6)
Distribution piping	2,100	(estimated, same as table 5)
Meters, 40 @ $50	2,000	(estimated, same as table 5)
Total	**6,200**	

Annual operation	$/yr	
Operation and maintenance	620	(10 percent of investment)
Fuelwood cost[b]	1,460	
Billing and administration	1,800	(estimated, same as table 5)
Total	**2,700**	

Life of gasifier   = 25,000 hours = about 10 years at 7 hours per day

Life, balance-of-system   = 25 years

			Discount rate	
			6 percent	12 percent
Annualized investment and operation	$		4,486	4,474
Gas cost delivered	$/GJ		15	16

a. Again we assume a six-person household with an annual cooking energy demand of 293 gigajoules, as in table 6. Although we assume the gasifier is 70 percent efficient, its actual efficiency is likely to be higher. We have not included the (primary) energy consumption or cost of the electric blower operation. Thus fuelwood input will be 419 gigajoules per year. For a 0.164 gigajoules per hour (input) gasifier this implies daily operation of seven hours. Increased demand could be met, provided wood was available, by operating the gasifier for longer periods.

b. For a fuelwood input of 419 gigajoules and a fuel price of $3.49 per gigajoule, we get $1,462.

Table 8: Estimates of the cost of ethanol production in Brazil[a]

Source	1990 U.S. ¢/liter	$/GJ
World Bank, 1984	22.5–24.3	10.2–11.1
Borges, 1984	20.1–27.4	9.1–12.5
CENAL, 1984	25.3–28.2	11.5–12.8
Mello and Pelin, 1984	49 –56	22 –25
Comissâo Nacional de Energia, 1987	21.3–23.1	9.7–10.5
Motta, 1987	28–44	12.7–20

a. [33], table 5.3, [48].

Ethanol

Kerosene is a liquid fuel widely used for cooking by the urban poor of many developing countries. Alcohol is an alternative liquid fuel that may be relevant to oil-importing developing countries. In a number of countries, alcohol is added to gasoline to enhance its octane number, and in Brazil provides fuel for automobiles (see chapter 20: *The Brazilian Fuel-alcohol Program*).

At least two types of alcohol can be produced from biomass: ethanol and methanol. Ethanol is made by the fermentation of such crops as sugarcane, sweet sorghum, cassava, sweet potato, maize, etc., but it can also be made from wood and other cellulosic materials (see chapter 21: *Ethanol and Methanol from Cellulosic Biomass*). Methanol, or wood alcohol, was once made from wood by a thermochemical process; today it is made primarily from natural gas.

The costs of producing ethanol in Brazil vary but typically fall around 1990 $11 per gigajoule (see table 8). A number of factors affect the price: in Brazil the most important one is the cost of delivered cane [33]. In Mauritius, where alcohol from sugarcane has been proposed for use in cooking, and the feasibility of alcohol cookstoves has been tested [46], production costs for a plant with an annual output of 18 million liters of low grade ethanol (85 percent alcohol) is estimated to be 20 (1990) U.S. cents per liter [46, table 2]. If the calorific value is 16.7 megajoules per liter (for 85 percent ethanol), the cost works out to $12 per gigajoule, which is slightly higher than the Brazilian average.

Production costs can be reduced by combining sugarcane processing for alcohol distillation with the production of electricity from the biomass residues of sugarcane [47]. Although the price of ethanol depends on the distillation technology, the duration of the milling season, and prices for sugarcane residues and electricity, its cost could be about $7.1 per gigajoule assuming advances in technology discussed elsewhere in this book (see chapters 17 and 20).

In many countries, land suitable for growing food crops cannot be diverted to grow sugarcane for fuel ethanol. Such a food–fuel conflict need not exist. For instance, ethanol can be fermented from the nonedible stem of sweet sorghum. Field tests have led India's Nimbkar Agricultural Research Institute (NARI) to propose that sweet sorghum be grown for three products: grain for human consumption, alcohol from the stems for fuel, and residues for fodder or combustion [49]. In 1979, 17.6 million hectares were devoted to sorghum cultivation in India. If this area were converted to sweet sorghum, the alcohol yield would be 18 million cubic meters (m^3) per year, or the energy equivalent of nine million tonnes of kerosene. Thus, sweet sorghum, which has a higher grain yield than normal sorghum, could significantly reduce the demand for imported kerosene, which may reach 18 million tonnes by 2000 [50]. One drawback, however, is that sweet sorghum lacks the taste of normal sorghum and thus may not be sufficiently appealing as a food item to permit a major shift in sorghum plantation.

Alcohol fuels are not commonly used for cooking, although a number of campstoves are designed to run on either ethanol or denatured alcohol. Some of the simpler, nonpressure models may be appropriate for use in LDCs. Efficiencies appear to be quite high. Jas Gill [51] conducted a series of WBTs on campstoves and obtained efficiency levels of about 60 percent. Baguant and Panray of Mauritius also found an efficiency of 60 percent in a stove they tested [46]. Yet another study found efficiencies comparable to kerosene stoves [52]. A study conducted for the World Bank by the Eindhoven University of Technology in the Netherlands pinpoints a number of other factors that affect the overall performance of alcohol stoves [53]. NARI has recently developed a wickless alcohol cooking stove that can run on alcohol concentrations as low as 45 percent [54].

COMPARATIVE ANALYSIS OF COOKING SYSTEMS

Cooking systems may be compared on the basis of user fuel preference, relative energy consumption and comparative costs.

User preference

Based on observed fuel choices, a fuel-preference ladder may be constructed for houses that have the option of more than one fuel (see table 9). People generally prefer an alternative higher up the ladder when it is available and affordable [55].

At its point of use, electricity is the cleanest "fuel," yet it is rarely used for cooking in LDCs because of its expense. Many people prefer natural gas to electricity; like electricity it is distributed directly to the house, but it is less costly. LPG is also popular but is delivered in gas cylinders that take up space and must be periodically replaced; delivery may be uncertain in rural areas (some houses have a fixed LPG tank that is refilled at regular intervals by a tanker truck). Kerosene, which is smelly and not as convenient as natural gas or LPG, is ranked lower than gaseous fuels but higher than any solid fuel. Among solid fuels, charcoal is ranked the highest and—until the recent arrival of petroleum fuels—was the pre-

Table 9: User preferences for cooking fuels[a]

Common fuels	Alternative fuels
Electricity	
Natural gas	
Liquefied petroleum gas	
	Biogas (centralized plants)
Kerosene	Producer gas
	Alcohol
Charcoal	
Coal and soft coke	
	Coal briquette
	Biogas (individual plants)
	Densified, carbonized biomass briquette
Fuelwood	
	Densified biomass briquette
Crop residue	
Dung	

a. Fuels listed in decreasing order of preference. The left column is based on observations in urban areas, where many of these fuels are available. An increase in income is associated with a rise up the fuel-preference ladder [55]. The preference for alternative fuels, relative to traditional fuels, is shown in the right-hand column.

mier cooking fuel. Stoves burning charcoal, coal, soft coke, and coal briquettes are harder to light and put out, and their heat output cannot be readily controlled.

Human beings seem to have a primordial attachment to firewood; even industrialized societies are reluctant to part with their fireplaces, although hearth-based fires offer little thermal benefit in centrally heated houses. In urban Java, even the relatively wealthy households use fuelwood for cooking if it is locally available [56]. In contrast, crop residues and dung tend to be fuels of last resort and are used principally in regions where fuelwood supplies are inadequate. In south and southeast Asia, residues of noncereal crops are used extensively for cooking. In many rural areas of China, straw serves as the principal cooking fuel [29, table 1].

A hierarchy of preferences also exists for alternative fuels (see table 9). Biogas, for example, like natural gas, can be conveniently piped to the kitchen, although its heating value is considerably less than that of LPG. Thus, piped biogas probably ranks intermediate between LPG and kerosene. Biomass-derived producer gas has not been widely adopted for cooking in part because it contains carbon monoxide, which can leak from fixtures, creating a health risk. For that reason, we consider producer gas to be on par with kerosene. Ethanol has a relatively low heating value and mixes well with water, which will dilute it; on that basis we consider it to have less value than kerosene. Moreover, in societies where alcohol production is illegal, ethanol is unlikely to be acceptable as a cooking fuel.

Energy use

Joy Dunkerley and her colleagues analyzed household energy use in several Indian cities [18, table 5]. They found that wealthier households cook more food and do so at a rate that increases rapidly with income. The total fuel consumed by each household, however, is not linked to income: the wealthy may cook more, but can afford more efficient fuels. A similar result was noted in urban Java [56, table 2]. In addition, they found fuel consumption varied from city to city, depending on fuel price and availability. In rural areas, where higher quality fuels are unavailable, higher income households consume more fuelwood.

The relative energy consumption in cooking with different fuels can be determined from fuel-substitution ratios, estimated by doing a regression analysis of energy use, adjusted for household income and size, cooking practices, fuel price, prices of competing fuels, index of fuel availability, and fuel choice. Fitzgerald and others used this method of analysis to compare energy use for cooking with fuelwood, kerosene, and LPG in urban Java [56]. Adjusting for the other factors, they found that households cooking with fuelwood and LPG consume 1.66 and 0.85 times as much energy, respectively, as those cooking with kerosene. Assuming that an average woodstove is 20 percent efficient, and a kerosene stove is 40 percent efficient, kerosene users consume more fuel than is expected from relative efficiencies alone. Similarly, LPG users consume more fuel relative to kerosene users. The conclusions are therefore somewhat surprising: after adjusting for socioeconomic factors, less energy is saved as households "upgrade" from fuelwood to kerosene to LPG than would be expected on the basis of technical efficiency. Such findings should be taken into account before initiating programs to promote upgraded cooking fuels.

Another consideration is the relation between fuel expenditures (brought about by efficiency improvements) and income. Less money spent on fuel means an increase in disposable income, which is clearly a desirable goal, but some of this would be spent on fuel. Some households, for example, will use their disposable incomes to buy more food, which requires additional fuel. The actual savings in fuelwood from an efficient stove therefore may be less than expected. In order to calculate actual savings, kitchen performance tests should be conducted before, and at various intervals following, the introduction of an improved stove. Such evaluations have been made for the ASTRA ole [57]. Donald Jones of the U.S. Oak Ridge National Laboratory [58] has also reported that fuel savings may be mitigated by efficiency: if consumer demand for a fuel drops, then the price of that fuel may decline, adding further to a user's disposable income.

Energy requirements

Some of the above conclusions [18, 56] are based on the relative efficiencies of different cooking systems. They must be interpreted cautiously, however, because

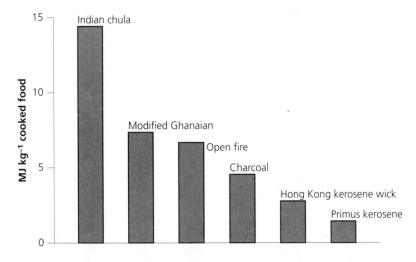

FIGURE 10: Specific fuel consumption (SFC) for various stove/fuel combinations in Fiji. SFC figures are averaged for five different meals typical of Chinese, Indian, and Fijian households [59].

in some cases efficiency figures for different stoves and cooking fuels were assumed [60, 61]. Moreover, water boiling tests carried out on the same stove show a wide range in efficiency values depending on test circumstances [7, 19].

The technical performance of different stoves can be compared by conducting controlled cooking tests (see page 277). One such comparison was conducted in Fiji [59], where two kerosene, one charcoal, and three fuelwood stoves were used to cook five typical meals (representative of the island's Fijian, Indian, and Chinese populations). The data obtained indicate a broad range of fuel consumption (energy input per kilogram of cooked food) (see figure 10). The least efficient stove, a four-burner cement fuelwood stove with a chimney called an Indian chula, used 11 times as much fuel as the most efficient stove, a kerosene pressure stove called a Swedish primus. The Indian chula consumes twice as much fuel as an open fire, which uses the least amount of fuelwood in cooking. The Hong Kong, or kerosene wick stove, burns twice as much fuel as the Swedish primus, but still uses about 2.5 times less energy than the open fire.

We conducted a similar comparison of energy use in cooking in December, 1990 in Ungra, a village in the Indian state of Karnataka. A typical meal for a six-person household (rice, millet, cowpeas, and vegetables) was cooked three times in each of 12 stove/fuel combinations. Eight fuels (firewood, charcoal, dung, sawdust, biogas, kerosene, LPG, and electricity) and 12 stoves (see table 10) were tested. A village cook familiar with several of the stoves did the cooking, except for the electric and biogas stoves, which for technical reasons could not be tested in the village. (See table 10 for the results of the study.)

Electricity has the lowest specific energy consumption (0.64 megajoules per kilogram) at its point of use. When the efficiencies of electricity generation (35 percent), transmission and distribution (optimistically 85 percent) are consid-

Table 10: Efficiency and fuel consumption based on water boiling and controlled cooking tests in Ungra, India

Fuel	Stove	Efficiency[a]	Specific fuel consumption[b]		Mean time to cook meal
		percent	Phys. units/ kg	MJ/kg	minutes
Biofuel stoves[c]					
Wood	Three-stone fire	15.6	217 g	3.44	101
	Traditional three-pan	14.2	271 g	4.31	62
	ASTRA ole three-pan	33.5	141 g	2.24	62
	Swosthee MS-4	17.2	183 g	2.91	111
Charcoal	Traditional metal	23.2	95 g	2.38	na
Dung cake	Traditional three-pan	11.1	304 g	4.00	na
Sawdust	IISc improved	30.4	253 g	4.02	na
Biogas	KVIC burner	45.1	0.05 m³	1.23	103
Other stoves					
Kerosene	Nutan	60.2	26.1 g	1.13	106
	Perfect	40.4	26.6 g	1.15	131
LPG	Superflame	60.4	20.1 g	0.91	76
Electric	Hot plate	71.3	0.17 kWh	0.64	99

a. Efficiency is the average of three water boiling tests, except for the three-stone fire and the ASTRA stove, which were tested twice.

b. Specific fuel consumption is based on 1 kilogram of cooked food, averaged over three meals. A typical meal consists of 1,102 grams of rice (+ 2,121 grams of water); 883 grams of ragi flour, a millet, (+ 1,554 grams of water); 256 grams of cowpeas, a legume, (+ 2,463 grams of water); and 150 grams of vegetables mixed with the cowpeas.

c. The stoves are described in the footnotes to table 11.

ered, primary energy consumption with electricity increases to 0.64 /(0.35 × 0.85) = 2.15 megajoules per kilogram, which is only marginally less than the most efficient fuelwood option (2.24 megajoules per kilogram). In contrast, the traditional woodstove consumes 3.76 as much and the LPG stove 0.79 times as much as the most efficient kerosene stove.

The Ungra cooking trials were accompanied by a series of standardized WBTs (see table 10). As expected, fuel consumption decreases as efficiency increases, but the relation is weak for fuelwood as well as kerosene stoves (see figure 11). The traditional three-pot woodstove and the three-stone fire, for instance, have virtually identical efficiencies, yet the traditional stove consumes 25 percent more fuel. In contrast, the two kerosene stoves have markedly different efficiencies (40 versus 60 percent) but burn virtually the same amount of fuel during cooking. Such discrepancies presumably arise from differences in low-power performance,

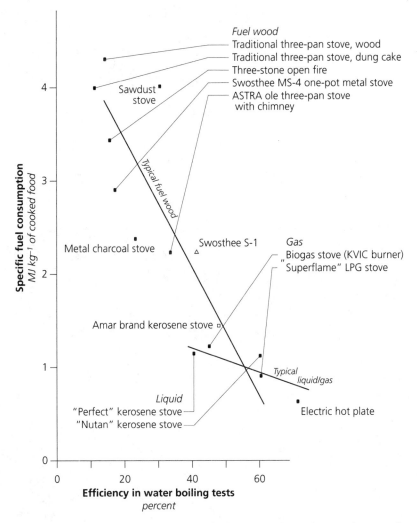

FIGURE 11: The relation between the amount of fuel a stove needs to cook a typical Ungra village meal and its efficiency at boiling water is shown for a number of stove/fuel combinations. The Swosthee S-1 and Amar brand kerosene stove are stoves tested in an earlier study [11].

which are not characterized in our simple measurement of water boiling efficiencies [53]. Average cooking time, which is an important consideration in stove choice, also varies from model to model (see table 10). The traditional and ASTRA three-pot stoves, for example, are the fastest, while a kerosene stove takes almost twice as long to cook the same item.

Economics

Cooking costs are calculated on the basis of annualized life-cycle cost, which includes the price paid for the stove and any auxiliary equipment (for example, a gas cylinder for LPG or a household biogas plant), as well as their respective life-

Table 11: Comparative economics of cooking stoves and fuel

Fuel	Stove	Stove cost[a]	Stove life[b]	Annual fuel use[c]	Fuel cost	Annualized life-cycle cost $ at real discount rates of			
		U.S. $	years	gigajoules	$/GJ	6 percent	12 percent	75 percent	150 percent
Part A: Based on cooking fuel use in Ungra village (from table 10)									
Wood	Three-stone fire	0.00	3	20.27	3.49[d]	71	71	71	71
	Traditional three-pan[e]	0.00	3	25.12	3.49	88	88	88	88
	ASTRA ole three-pan[e]	7.78	3	13.64	3.49	51	51	55	60
	Swosthee MS-4[f]	5.56	3	17.15	3.49	62	62	65	69
Charcoal	Traditional metal	(3.00)[g]	1	14.49	8.86	132	132	134	136
Dung	Traditional three-pan[h]	0.00	3	24.35	note[i]	–	–	–	–
Sawdust	IISc[j]	(5.00)	2	24.48	1.74	45	46	48	52
Biogas[k]	Individual	13.89[l]	7	7.33	0.00	38	56	286	563
	Community	13.89	7	7.33	(11.50)	87	87	95	105
Kerosene	Nutan[m]	6.94	3	6.35	5.30	36	37	40	45
	Perfect[n]	8.06	3	6.78	5.30	39	39	43	49
LPG	Superflame[o]	50.00	7	5.40	5.30	19[p]	53	107	180
Electric	Hot plate[q]	38.89	5	3.84	25.00	105	107	127	155
Part B: Estimates, not based on measurements									
Charcoal	Improved jiko[r]	5.49[g]	1	(11.00)	8.86	103	104	107	111
Briquette[r]		(5.56)	3	17.15	(7.19)	125	126	128	132
Producer gas[s]		(13.89)	7	7.33	(15.50)	116	117	124	134
Alcohol[t]	Brazil conventional	(8.00)	3	6.78	11.00[u]	76	76	81	87
	Advanced technology	(8.00)	3	6.78	(7.08)[v]	51	51	55	61

a. Stove costs are based on market prices in Bangalore; Swosthee is not commercially available—the price is based on an earlier production run [11]. A conversion rate of Indian Rs. 18 per U.S. dollar is used, corresponding to the official exchange rate in December 1990. Stove costs are annualized at the discount rates shown.

b. Kerosene stove life has been adjusted down, in part because the replacement cost of wicks is not considered. Stove maintenance costs are neglected for all models.

c. Annual fuel use is estimated by measured energy use per meal and is based on the assumption that a family of six has two identical meals a day, as is the custom in the Ungra region.

d. The price of fuelwood is set at 5.55¢ per kilogram, which is equivalent to the cost of plantation wood in India of about Rs 1 per kilogram. The price of charcoal is based on the fact that in Bangalore in the early 1980s it cost 2.54 times as much as fuelwood [18,62]. Using the ratio relative to plantation wood, we arrive at a cost of $8.86 GJ^{-1}. The price of sawdust, often a "waste" material, is set at half that of fuelwood. For biogas costs, see note k below. Costs of kerosene and LPG are assumed to be equivalent to $30 per barrel of petroleum (including refining and distribution); electricity is assumed to cost 9¢ per kWh, which corresponds to the marginal cost of supplying electricity from new power plants.

e. The traditional stove is a three-pan stove without chimney and is designed to be energy efficient (figure 2).

f. Swosthee is a single pan metal stove designed at the Indian Institute of Science in Bangalore (figure 1); the ASTRA stove is a three-pan stove with chimney and is designed to be energy efficient. Model MS-4 (figure 5) was tested.

g. The charcoal stove was a traditional metal stove, similar to the "Feu Malgache" of West Africa and the Kenyan jiko, which cost from 35 to 45 Kenyan shillings (KS) in 1985 [17]; we assume a cost of KS 40, equivalent to $2.92 in 1989. We use the Kenyan price to facilitate comparison with the Kenyan "improved jiko" shown in part B of the table. The parentheses here and elsewhere in the table indicate estimated values. The improved jiko (figure 8) sold for from 60 to 85 KS in Nairobi in mid-1985 and with a saturated market, a retail price of 55 to 75 KS was expected [17]. We set the price at KS 75, equivalent to $5.49 in 1989. Traditional stoves last about a year, while the various components of the improved stoves last from six months to three years [17]. We assume a one-year average lifetime for both. Energy consumption of an improved stove is expected to fall between an efficient woodstove and a kerosene stove.

h. Dung cake was burned in the traditional three-pan stove normally fueled by wood.

i. For dung, we should take the value of an equivalent amount of fuel as its opportunity cost, but this figure was not at hand.

j. The sawdust stove is similar to a traditional stove, which is modified to improve combustion by the provision of a secondary airflow. The prototype was developed at the Indian Institute of Science.

k. Two models of biogas digesters are considered: "Individual" is a 2 m^3/day family-sized digester (see table 5 for cost data). The digester is owned by the household and capitalized at its discount rate. The cost of family labor in animal husbandry and digester operation is not included; "community" is a community-scale biogas plant operated as a public utility that pipes and sells gas to households. Gas cost is an average for discount rates of 6 and 12 percent.

l. The biogas stove is a standard burner designed by the Khadi and Village Industries Commission, a biogas promotion agency in India.

m. The "Nutan" is a fuel-saving wick stove developed by the Indian Oil Corporation.

n. The "Perfect" is a popular make of wick stove.

o. The "Superflame" is a two-burner LPG stove.

p. Annualized cost for LPG includes a $50 gas cylinder with an expected life of 5 years.

q. For electricity, a single-burner hot plate was used.

r. The metal stove is considered equivalent to the Swosthee MS-4 with the same cost, lifetime, and fuel consumption rates as when it is used with fuelwood. We assume the fuel price estimated in the text.

s. The gasifier is assumed to be managed as a public utility that pipes the gaseous fuel to the household. The stove cost is the same as a single-burner biogas stove, has the same life, and consumes the same amount of fuel. We assume a unit cost of gas as estimated earlier in the text.

t. Not stove alternatives, but alcohol production technologies and costs. Ethanol-stove fuel consumption characteristics are assumed to be the same as for a kerosene stove.

u. The ethanol cost is typical of Brazilian distilleries (see text and chapter 20).

v. Advanced technology with bagasse gasification and gas turbine cogeneration. (See text and chapter 17.)

times. Capital costs are annualized according to two discount rates. The values of 0.06 and 0.12 (which represent real discount rates of 6 and 12 percent) reflect the cost of the capital to society. The discount rates implicit in purchasing decisions that involve a trade-off between initial capital cost and operating expenses are considerably higher for low-income households, reaching 150 percent for the poorest households [A.K.N. Reddy, personal communication, 1990]. Instead of actual fuel prices, which include the effects of taxes and subsidies, we have considered the costs to society. Thus, our calculations include international petroleum prices, marginal costs of electricity supply, and the cost of plantation firewood.

T. K. Das and his colleagues at Jadavpur University in Calcutta [63] show that if the costs of learning about a new technology, such as an improved stove, are included in an economic analysis, the overall cost shifts upward. Using such an analysis, some new technologies are not so cheap, and consumer reluctance to embrace them is indeed economically rational. Therefore, lower cost options should be promoted more effectively in order to reduce their information cost and render them more accessible.

An economic comparison of capital costs and actual fuel use in the Ungra cooking trials was made for the 12 stove/fuel combinations tested (see part A of table 11). For all discount rates, the kerosene stoves represent the least-cost option, whereas electric stoves, not surprisingly, are among the most expensive. But traditional charcoal stoves are even more expensive to run than electric stoves because the fuel is expensive and its consumption comparable to that for efficient woodstoves. Assuming that sawdust costs half as much as fuelwood, then the improved sawdust stove, which was developed at the Indian Institute of Science, would be the cheapest alternative biofuel stove. In most cases, however, the ASTRA stove is the least expensive option, especially when discount rates are high. If the discount rate is low—say, 6 percent—biogas obtained from a household digester will be slightly cheaper because labor costs are not counted. But as the discount rate increases, so does the cost of cooking with biogas from an individual digester.

A community biogas digester requires pipeline construction and operation, maintenance, and administration. Since it is likely to be built by the government or another organization with access to capital, we consider the social discount rate for plant cost to be low. Thus, the cost of cooking with biogas from a community plant increases only slightly with the discount rate. Community biogas digesters are nonetheless more expensive than most other options. At social discount rates of 6 and 12 percent, LPG and ASTRA stoves are virtually identical in cost; still, the high capital costs of the LPG stove, which also requires a gas storage cylinder, render it very expensive to low-income households in view of their high implicit discount rates.

We also estimated costs and energy use for several biofuel alternatives that were not tested at Ungra: improved charcoal stoves, biomass briquettes, producer gas, and alcohol (see part B of table 11). While these figures are somewhat tentative, several observations can be made. Among all the gas and liquid fuel options, kerosene still costs the least, whereas charcoal is expensive even when used in an improved stove. Briquettes are another expensive alternative, as is producer gas derived from plantation-grown wood. According to data obtained from Brazil (see chapter 20), alcohol would cost less than the more expensive alternatives (charcoal, electricity, producer gas, and briquettes) but still about twice that of kerosene.

Research into gas- and liquid-fueled stoves suggests that additional savings are possible through technological innovation. For instance, the water boiling efficiency of the biogas burner was only 45 percent, compared to 60 percent for an LPG burner. Even the efficiency of a conventional LPG stove could be increased to 65 or 70 percent by replacing its conventional burner with an infrared-impingement burner, which also emits significantly fewer nitrogen oxides [64]. Wick-type kerosene stoves could be made more efficient by giving them a wider range of power output, even without increasing their water boiling efficiency [53]. Because the efficiency of gas- and liquid-fueled stoves is less dependent on the operator, technical improvements are likely to reduce fuel use in actual cooking.

When costs are dominated by fuel consumption and cost, rather than by stove cost, the annualized life-cycle cost depends weakly on the discount rate (see table 11). Such is the case for all cooking systems except those that run on LPG, electricity, and biogas. These findings suggest that the more efficient stoves are likely to be more cost effective, despite their initial expense. For example, even if the cost of a kerosene stove is doubled to, say, $13.88, to achieve a fuel savings of only 10 percent, the annualized life-cycle cost will be lower (at the lower discount rates of 6 or 12 percent).

COOKING, RESOURCES, AND THE ENVIRONMENT

In the previous section we discussed fuel choice, energy use, and the economics of cooking systems without mentioning resource constraints or the environmental consequences of the various options. In this section we consider resource availability and the relation of cooking to deforestation, global climate change, and indoor air pollution.

Resources
Petroleum resources (and to some extent coal) are traded internationally. In contrast, biofuels are generally traded locally in part because (with the possible exception of ethanol) they have low energy density and are expensive to transport. Whether or not a given biofuel can meet existing demand depends on its availability and on competing demands. For instance, in addition to fueling a cookstove, biogas may be used to pump water, generate electricity, and provide light. Similarly, sugarcane can be processed for food or fermented to ethanol, which can then either be imbibed or combusted.

India's supply of biogas is currently derived principally from dung (see table 12). Cattle (including buffalo) theoretically can provide about half of India's

Table 12: Biogas resources for meeting India's cooking needs

	Peak season 5 months/year	Lean season 7 months/year
Livestock population (including buffaloes)	$259 \cdot 10^6$	
Human population, 1991	$844 \cdot 10^6$	
Dung yield/animal/day[a] *kilograms*	7.35	3.52
Dung production, all India/day *kilograms*	$1.90 \cdot 10^9$	$0.911 \cdot 10^9$
National biogas potential/day[b] *joules*	$1.48 \cdot 10^{15}$	$0.711 \cdot 10^{15}$
Population whose cooking needs can be met[c]	$443 \cdot 10^6$	$212 \cdot 10^6$
Percentage of total population	52	25

a. [65].

b. At 35 liters of gas per kilogram of fresh dung, and 22.5 megajoules per m^3 of biogas.

c. Assuming a consumption of 7.33 gigajoules per year per household of six, as in Ungra (see table 11).

cooking energy needs during peak dung-producing season and a quarter of its needs during off season, provided that dung is collected from all cattle and processed by community-based biogas plants. In reality the potential is much lower in part because household biogas digesters must be considered, and only those households with enough cattle to meet their cooking needs can be counted in the estimate. Moreover, excess gas produced by those having more than the minimum head of cattle may not be available to other households. Thus the country's cooking demand cannot be met with current levels of dung production, and biogas must be viewed as a supplementary cooking fuel whose distribution in most cases is not likely to be cost-effective (see table 12). Although the dung shortage can be solved, for instance, by improving cattle stock (and dung yield), or by supplementing dung in the digester with other forms of biomass, or even by mixing wood-based producer gas with the biogas [66], all of these alternatives require an additional commitment of land and resources.

Land requirements for wood-based fuels (fuelwood, charcoal, and producer gas) can be compared with one another and with those for ethanol (see table 13). By considering the yield of a typical tropical fuelwood plantation, we estimated the amount of land (in m^2) needed to provide cooking energy for a six-person household. The range is considerable. Among wood-based cooking systems, producer-gas stoves require the least amount of land (340 m^2 per household), whereas traditional charcoal stoves need almost four times as much. Even the best charcoal technology requires 75 percent more land than a producer-gas system; the best woodstove requires 40 percent more.

Table 13: Primary wood requirements for woodfuel-based cooking and land requirements for some biofuel alternatives

Fuel	Stove	End-use energy[a] GJ/year	Primary wood energy[b] GJ/year	Land requirements[c] m²/household
Fuelwood	Three-stone	20.27	20	660
	Traditional	25.12	25	840
	ASTRA ole	13.64	14	480
	Swosthee MS-4	17.15	17	580
Charcoal	Traditional	14.49[d]	40	1,340
	Traditional	14.49[e]	24	800
	Improved jiko	11.00[d]	31	1,040
	Improved jiko	11.00[e]	18	600
Producer gas		7.33	10[f]	340
Ethanol:				
Alcohol only from cane		6.78	–	730[g]
Alcohol plus electricity from residues using existing steam-electric power technology		6.78	–	300[h]
Alcohol plus electricity from residues using advanced gasifier/gas turbine power generating technology		6.78	–	160[h]

a. Numbers in this column are taken from table 11.
b. This column is rounded to integers to facilitate comparison.
c. Wood-yield rate is assumed to be 15 dry tonnes per hectare per year, typical of plantations in tropical countries.
d. Using a low-efficiency charcoal process, such as the Tonga drum, with an energy conversion efficiency of 36 percent (see table 1).
e. Using a brick beehive kiln, BB1, with an efficiency of 61 percent (see table 1).

f. Gasification efficiency of 70 percent, as assumed above. See [5].

g. Assuming a cane yield of 58.4 tonnes per hectare per year (the global average in 1987) and a hydrated (96 percent) alcohol yield of 70 liters per tonne of cane (typical for Brazilian distilleries), the alcohol production rate would be 93.2 gigajoules per hectare per year.

h. It is assumed that the land requirements are divided between alcohol and electricity production, with electricity's share determined by the amount of alcohol that would be required to produce that electricity in a modern (45 percent) efficient gas turbine–steam turbine combined cycle plant. With commercially available high-pressure steam-turbine–based cogeneration technology some 280 kWh of electricity can be produced per tonne of cane from cane residues (bagasse during the milling season and cane tops and leaves during the off season), or 4 kWh per liter of alcohol, in excess of on-site needs. With advanced biomass integrated gasifier/gas turbine–based power-generating technology that could be commercialized in the 1990s, 720 kWh per tonne of cane or 10 kWh per liter could be produced. See chapter17 for details on these alternative power-generating technologies.

Ethanol coproduced from sugarcane with electricity is the least land intensive of the cooking fuels shown in table 13. Biogas generated from dung (not shown) would be less land intensive, because the dung is produced by animals that serve primarily other sources (food and traction). In fact, because the fertilizer value of the digester output is at least as high as that of dung, land requirements for biogas may be considered to be negative.

Equity

The issue of equitable access to biomass resources is a complex one that should be included in development discussions. As a general rule, policymakers should ensure that the economic standing of the poorest segments of society will not be adversely affected by a program's implementation. Although subsidizing the purchase of energy-efficient stoves in rural areas is clearly beneficial to all, subsidies for individual biogas plants generally favor those who are relatively well-off—rural households with the requisite number of cattle and resources needed to complete the investment. Whether or not such selective subsidies actually have an adverse effect on the poor has been the subject of debate for a long time, at least in India. One side argues that people who once gathered dung for free may be denied access to the resource if biogas plants create a market for dung. The other side contends that the environmental benefits of gasifying dung, rather than burning it, are significant and thus outweigh any equity problems. They also argue that biogas plants reduce pressure on such cooking fuels as fuelwood and crop residues, making them more accessible to the poor. Evidence seems to support the latter group (V. Joshi, Tata Energy Research Institute, personal communication, 1990).

Deforestation concerns

Surveys of rural energy use indicate that the average person consumes approximately 1 kilogram of air-dried fuelwood equivalent per day [67]. Because about half the world's population cooks with biofuels, total consumption amounts to about 2.5 billion kilograms per day, or almost 1 billion tonnes a year. Because the demand for fuelwood exceeds the potential supply in some countries [68], a trend that is likely to grow worldwide [69], fuelwood has been declared the "other energy crisis," [70] and wood gathering and land clearing have been said to be the principal causes of deforestation [71].

Rural biomass usage, however, differs markedly from wood consumption in urban areas. Whereas the fuelwood consumed in cities comes almost entirely from whole trees and thus is likely to be a significant agent of deforestation [62], in rural areas such as Ungra, women and children collect mainly twigs, branches, and roots for cooking [72], and trees are not felled to meet local fuel needs [73].

Data from a rural energy survey [74] support these findings. According to the survey about 84 percent of total biofuel consumption in India is "in the form of crop residues, animal wastes or small branches and twigs, most of which...has fallen to the ground" and does not contribute to deforestation [75]. "Probably no more than 5 to 7 percent of total biofuel consumption involves 'permanent' tree loss," and a small part of overall tree loss. Similar conclusions have been drawn for other countries as well, leading the Energy Research Group of Canada's International Development Research Centre to conclude that "the use of firewood does not necessarily or frequently lead to deforestation" [76].

Yet in many parts of the world rural fuelwood users are the primary victims of deforestation, resulting from other activities such as hydroelectric dams that primarily benefit others. As a result, many households in China, India, and elsewhere have turned to crop residues or dung, fuels that are considered inferior to fuelwood (see table 9). Moreover, when dung is burned many nutrients that are useful for fertilization are lost.

Since biofuels used for cooking in urban areas can cause deforestation elsewhere, implementation of improved cookstove programs is easily justified. Moreover, because the distribution of fuelwood in cities is invariably commercialized, such programs are relatively easy to promote. Because stoves can be built much faster than biomass can be "produced" through afforestation, improved stoves can delay deforestation and buy time for other means to halt or slow down deforestation [77].

Indoor air pollution and health

The environmental consequences of cooking systems range from air pollution (both indoor and outdoor) to acid rain and global climate change. The combustion of cooking fuels generates air pollutants in the form of particulate matter and gases, including carbon monoxide, nitrogen oxides, sulfur oxides, and a variety of hydrocarbons. The quantity of each pollutant released is characterized by its emission factor, that is the number of grams of pollutant released per kilogram of fuel burned, a value dependent on fuel type, cookstove design, and operation. The actual airborne concentration of the pollutant, measured as milligrams per m^3, depends on a kitchen's volume and ventilation rate as well as on the emission factor. Human exposure to the pollutant is determined both by the concentration and by the time the cook and other people spend in the kitchen cooking. The actual dose received (by the lungs, for example), depends on the pollutant's specific characteristics; overall health effects depend on the dose and the toxicity of the pollutant, as well as the general health of the person.

Although respiratory diseases are the main cause of death in developing countries [78], there has been little systematic study of the relation between cookstoves and health. (For an excellent review of the health implications of biofuel cookstoves, see [67].) Nonetheless, concentrations of total suspended particulates, benzo-a-pyrenes and other pollutants in rural kitchens are known to exceed the ambient air quality standards for these pollutants by large margins in all countries.

Careful epidemiological studies conducted in Nepal in the 1980s clearly demonstrate a link between domestic smoke exposure and the incidence of chronic bronchitis, chronic cor pulmonale, and acute respiratory infection [79–81]. Studies in other countries are less conclusive, perhaps because of "different definitions of the biological endpoints, such as chronic bronchitis" [67]. If epidemiological studies on the health effects of rural cooking have been scarce,

Table 14: Emissions of carbon monoxide and total suspended particulates (TSP) in metal biomass and kerosene stoves[a,b]

Stove	Carbon monoxide grams per task			
	Fuelwood	Mustard stalk	Dung	Kerosene
Conventional metal	5	11	14	
Tara	5	16	–	
Priyagni	6	13	11	
Community Polytechnic	9	–	–	
Jwala	9	–	–	
Sahyog	9	20	–	
Thapoli	11	21	–	
Nutan				4
Kerosette				4

	Total suspended particulates grams per task			
Conventional metal	0.6	1.2	2.2	
Tara	0.3	1.7	–	
Priyagni	0.7	1.3	1.4	
Community Polytechnic	0.4	–	–	
Jwala	0.6	–	–	
Sahyog	0.5	1.4	–	
Thapoli	0.5	1.3	–	
Nutan				0.2
Kerosette				0.2

a. [83], tables 4.D.11, and 4.D.12.
b. A cooking task is simulated by raising the temperature of 3.5 liters of water to 60°C.

studies documenting the health impact of improved stoves have been entirely lacking.

The Tata Energy Research Institute (TERI) in New Delhi has measured laboratory emissions of carbon monoxide (CO) and total suspended particulates (TSP) for different stove/fuel combinations using the methodology described above [5, 82, 6, 83]. One of their tests [83] "confirmed that emission factors for biomass fuels increase with efficiency, but in half the stove-fuel combinations, the increase in emissions was offset by an increase in efficiency" (see table 14). Laboratory measurements indicate that crop residues and dung generate far more pollutants than fuelwood; surprisingly, the kerosene stoves produced almost as much CO and TSP as the cleanest fuelwood stoves. Similar tests carried out on two-pot mud and cement stoves demonstrated that simply venting the chimney outside the room will remove from 80 to 88 percent of the CO and from 62 to 75 percent of

Table 15: Carbon monoxide (CO) and total suspended particulates (TSP) in village kitchen: traditional versus improved stoves with chimneys[a]

Stove type	Number of stoves	Burn rate *kg/hour*	Mean CO concentration *mg/m³*	Mean TSP exposure *mg/m³*
Garhibazidpur village				
Traditional	34	1.4	7.0	3.3
Sahyog	22	1.4	6.0	2.9
Keelara and				
Pallerayanahalli villages				
Traditional	24	1.9	19.0	3.2
Traditional + hood	16	2.0	5.7	1.7
ASTRA ole	20	2.1	8.9	2.5

a. [84].

TSP. Laboratory measurements must now be extended to other stove/fuel combinations and pollution reduction goals verified through field measurements.

Jamuna Ramakrishna of the East–West Center in Hawaii [84] measured CO concentrations and TSP exposures (by having the cook wear a portable monitor) in houses with traditional open stoves and in those with improved stoves and chimneys. While the new stoves emitted somewhat fewer pollutants (see table 15), the difference was statistically significant ($p < 0.05$) in only two of three villages. Such a small reduction suggests the new stoves were not built or maintained as intended [85]. The study also indicated that substituting the wrong pots on pot-specific stoves and leaving unused pot holes open will increase the amount of smoke released into a kitchen. Pending further epidemiological studies, TERI has proposed that all new stoves be designed to emit fewer pollutants per task, in addition to meeting increased fuel efficiency standards and other desirable characteristics [83].

Biofuels and climate change

Carbon dioxide (CO_2) and methane are two of the most potent gases that contribute to global warming. Because CO_2 is emitted by burning biomass, and because methane (from a variety of sources) can be captured and burned off in cookstoves, cooking technology may have a direct impact on global warming [86]. While renewably-grown biomass leads to no net increase in atmospheric CO_2 (because the amount released during combustion is reabsorbed by growing trees), fossil fuels are major contributors to global warming, producing large amounts of CO_2. Our estimates of CO_2 emissions from a variety of stove/fuel combinations (see table 16) indicate that for fuelwood, charcoal, and producer gas, emissions range from zero (representing renewably grown wood) to an upper

Table 16: Carbon dioxide emissions from selected cooking systems

Fuel	Stove	Annual fuel use[a]	CO_2 emission factor[b]	Total emission of CO_2
Wood	Traditional	25.12	0–85	0–2,135
	ASTRA	13.64	0–85	0–1,159
Charcoal[c]	Traditional[d]	40.25	0–85	0–3,421
	Improved[e]	18.03	0-85	0–1,533
Producer gas[c]		10.47	0–85	0–890
Dung, biogas, crop residues, and alcohol: no net production of CO_2				
Kerosene	Nutan	6.35	77	489
LPG	Superflame	5.40	59	319
Natural gas[f]		(5.40)	51	275
Coal[g]		(14.49)	95	1,377

a. Fuel consumption data taken from table 11.
b. Carbon dioxide emission factors taken from [87].
c. For charcoal and producer gas, the primary fuelwood consumption is shown and used to estimate CO_2 emissions.
d. The charcoal was made in a 36-percent efficient kiln and burned in a traditional stove.
e. The charcoal was made in a 61-percent efficient kiln and burned in an improved jiko.
f. Energy consumption with natural gas is assumed to be the same as with an LPG stove.
g. Energy consumption with coal is assumed to be the same as with charcoal in a traditional stove.

limit if the wood comes entirely from deforestation, without replanting. But because biofuels used for cooking have not contributed significantly to deforestation in India and elsewhere (see page 306), their contribution to CO_2 emissions is relatively minor. In addition, it should be noted that many fuel crops (including sugarcane) are replanted and so produce no net CO_2; similarly, dung that is obtained from a stable cattle population produces no net CO_2.

Landfills are a significant source of methane gas (CH_4), producing the gas by a process similar to the one that occurs in a biogas plant. The estimated 2.7 million tonnes of CH_4 released by Indian landfills (see table 17) has an energy content of 135×10^{15} joules, which is enough to meet the cooking needs of some 150 million people. Landfills worldwide could meet the cooking needs of some 2.3 billion people, which coincidentally is roughly the number of people worldwide who depend on biofuels for cooking. Yet methane from landfills is rarely captured, a situation that must change. Not only does landfill methane have the ability to provide the world's people with a significant amount of energy, but burning it will significantly lessen its atmospheric buildup.

Table 17: Anthropogenic methane emissions in India and the world
million tonnes per year[a]

Source	Global	India
Animals	76	10.3–10.4
Rice	114	32.8–33.8
Biomass burning	65	4.3
Coal mines	35	1.3
Natural gas	47	0.1
Landfills	41	2.5–2.9
Total	378	51.3–52.8

a. [88].

SUMMARY AND DISCUSSION

A cooking system must satisfy multiple criteria. At the household level it must be easy to operate, relatively speedy, and produce minimal smoke. In addition, the initial and operating costs of the system must be relatively low and the fuel supply should be reliable and affordable. At the national level, other issues may be important, including a system's foreign exchange implications, the renewability of the energy source, its environmental soundness, and its role in meeting overall development objectives.

Studies indicate that users generally prefer systems that run on gaseous and liquid fuels, which also reduce energy use compared with solid fuels. Kerosene, which currently costs the equivalent of $30 per barrel of petroleum, is the least-cost alternative fuel from a societal perspective. From the perspective of the individual, fuelwood, which can be collected free from common lands, may be the least-cost alternative.

We believe that policies should reflect the societal optimum and so propose the following guidelines:

1. *Marginal cost pricing.* Price distortions in the form of subsidies or prices set below replacement cost for fuels and electricity should be removed gradually. Electricity prices should match long-term marginal costs; petroleum prices should match international levels with an appropriate import premium where applicable. Because the use of fuelwood and other wood products in urban areas has been implicated in deforestation, the nonrenewable felling of trees on public

lands should be internalized by charging a "stumpage" fee verified at urban entry points. Keith Openshaw and Charles Feinstein of the World Bank have devised a method for determining economically rational stumpage fees [89], a method that is being successfully adapted in the Sahelian countries (W. Floor, World Bank, personal communication, 1990).

2. *Financing of capital investments.* Social discount rates should be determined by interest rates for capital investment in development projects. These discount rates are much less than the implicit discount rates of low-income households. No subsidies are involved, and upper-income users would have the option of purchasing a new cooking system outright or with alternative financing.

3. *Subsidies for rural areas.* Because rural biomass users often depend on gathered wood, dung, or crop residues, they have little economic incentive to invest in energy-efficient cookstoves. Moreover, they may not have the necessary capital. Yet improved biofuel stoves and other alternative cooking systems may significantly improve village life, for instance, by reducing time spent gathering fuel and/or cooking, by increasing the amount of dung for fertilization, and by reducing indoor air pollution. If such advantages can be obtained, then alternative cooking systems should be subsidized, at least for lower-income households. Subsidy programs should also take into account the fact that woodstoves use more fuel and deteriorate over time and thus must be replaced after one or two years.

A number of key findings have emerged with respect to alternative cooking systems. To begin with, technological innovation promises to greatly improve the efficiency—both of biofuel production and of cookstoves—so that overall energy use per capita will decline. Because fuel is often the largest component of total cooking costs, substantial efforts to develop energy-efficient stoves for all types of fuel are therefore justified. Although fuel economy is best compared through controlled cooking tests, the overall impact of a cooking system must be evaluated under real-life conditions.

Resource requirements for biomass-based fuels should be considered before promoting new cooking systems. A resource that cannot meet all cooking energy needs and requires a significant infrastructure (for example, pipelines) may not be economical.

Indoor air pollution represents a significant health hazard, particularly in LDCs, and an alternative cooking system should not generate more air pollutants per cooking task than the system it is intended to replace. Several biofuel alternatives (dung, biogas, alcohol, and crop and fuelwood residues) generate no net carbon dioxide, nor does renewably grown fuelwood. In India only from 5 to 7 percent of the total biofuel consumed may contribute to permanent deforestation. And if the biogas and methane generated by landfills are redirected for cooking purposes, methane's contribution to global warming is mitigated.

Although this chapter focuses on the technical and economic evaluations of cooking fuel alternatives, social and cultural factors are also important, particularly in rural areas outside the monetary economy. Among indigenous people food preparation usually has ritual significance. Thus, the beliefs and customs of our planet's inhabitants should not be ignored when any kind of new development project is initiated.

ACKNOWLEDGMENTS

The information that forms the basis for this review has been collected from a number of places. The Intermediate Technology Development Group of Rugby, England, has been extremely helpful in providing reports as well as issues of its newsletter *Boiling Point*. Veena Joshi and others at the Tata Energy Research Institute, New Delhi, and Kirk Smith of East–West Center have provided us with a great deal of material, particularly in relation to air pollution from biomass stoves. H. Mukunda, Udipi Shrinivasa of the Department of Aerospace Engineering, Indian Institute of Science, Bombay, and Anil Date of the Indian Institute of Technology, Bombay, have all shared their research results over the years and provided many useful insights. Omar Masera of Lawrence Berkeley Laboratory, Sam Baldwin, of the Office of Technology Assessment of the U.S. Congress, and Krishna Prasad of the Eindhoven University of Technology have not only kept us informed on themes related to household energy, but also reviewed a draft of this chapter and gave many helpful criticisms. Beyond providing food for thought, A.K.N. Reddy has supported our efforts in this area from time out of mind, and Robert Williams convinced us to undertake this review. Our gratitude to these people and to the many others without whose help this review would not be possible. (GD)

We also greatly appreciate the help and patience of Mrs. Jayalaxmi of Ungra village in conducting the cooking tests and Mr. Mohan Nayak and Mr. Nagaraja in conducting the water boiling tests. (NHR and GD)

WORKS CITED

1. Masera, O. 1990
   Sustainable energy scenarios for rural Mexico: an integrated evaluation framework for cooking stoves, MS dissertation, Energy and Resources, University of California, Berkeley.

2. Joshi, V. 1988
   Rural energy demand and role of improved chulha, *Energy Policy Issues* 4:23–35, Tata Energy Research Institute, New Delhi.

3. Volunteers in Technical Assistance. 1982
   Testing the efficiency of woodburning cookstoves: provisional international standards, Arlington, Virginia.

4. Baldwin, S.F. 1987
   Biomass stoves: engineering design, development, and dissemination, Center for Energy and Environmental Studies, Princeton University, and Volunteers in Technical Assistance, Arlington, Virginia, 81–114.

5. Ahuja, D.R., Joshi, V., Smith, K.R., and Venkataraman, C. 1987
   Thermal performance and emission characteristics of unvented biomass-burning cookstoves: a proposed standard method for evaluation, *Biomass* 12:247–270.

6. Joshi, V., Venkataraman, C., and Ahuja, D.R. 1987
   Emission factors in biofuel burning metal cookstoves, Tata Energy Research Institute, New Delhi.

7. Gill, J. 1987
   Improved stoves in developing countries—a critique, *Energy Policy* 15:135–144.

8. Raju, S.P. 1953
   Smokeless kitchens for the millions, Christian Literature Society, Madras, India.

9. Singer, H. 1961
Improvements of fuelwood cooking stoves and economy in fuelwood consumption, report to the Food and Agricultural Organization, Rome.

10. Lokras, S.S., Babu, D.S.S., Bhogle, S., Jagadish, K.S., and Kumar, R. 1983
Development of an improved three-pan cookstove, *Proceedings of the ASTRA Seminar*, Indian Institute of Science, Bangalore, 13–17.

11. Mukunda, H.S., Shrinivasa, U., and Dasappa, S. 1988
Portable single-pan wood stoves of high efficiency for domestic use, *Sadhana*, Indian Academy of Sciences proceedings in engineering sciences, 13 (part 4):230–70.

12. Joseph, S., Shrestha, K.L., Pelinck, E., Campbell, J.G., and Bhattarai, T.N. 1984
Introduction of improved stoves for domestic cooking in Nepal, field document 10, His Majesty's Government (HMG), United Nations Development Program (UNDP), Food and Agricultural Organization (FAO) Community Forestry Development Project, Kathmandu, Nepal.

13. Amerasekera, R.M. and Sepalage, B. 1988
Sri Lankan stoves—past and present, *Boiling Point* (15):13–16, Intermediate Technology Development Group, Rugby, United Kingdom.

14. Royal Forest Department, Bangkok. 1985
Charcoal production improvement for rural development in Thailand, Forest Products Resources Division, Royal Forest Department, Ministry of Agriculture and Cooperatives, Bangkok, Thailand. Probable date of publication 1985.

15. Sangen, E. and Visser, P. 1986
A study of the performance of charcoal stoves, *Boiling Point* (10):15–18, Intermediate Technology Development Group, Rugby, United Kingdom.

16. Siemons, R.V., Ahmed, and Hood, H. 1989
Cotton stalk charcoal agglomeration in the Sudan, *Boiling Point*, special edition on briquetting, 10–16, Intermediate Technology Development Group, Rugby, United Kingdom.

17. Hyman, E.L. 1986
The economics of improved charcoal stoves in Kenya, *Energy Policy* 149–158, April 1989.

18. Dunkerley, J., Macauley, M., Naimuddin, M., and Agarwal, P.C. 1990
Consumption of fuelwood and other household cooking fuels in Indian cities, *Energy Policy* 92–99, January–February.

19. Bussmann, P.J.T. 1988
Woodstoves: theory and applications in developing countries, Ph.D. thesis, Eindhoven University of Technology, Eindhoven, The Netherlands, 11–34.

20. Leitman, J. 1987
Residue utilizing: a recent experience from Africa, *Boiling Point* (12):11–14.

21. Leach, G. and Gowan, M. 1987
Household energy handbook: an interim guide and reference manual, technical paper 67, World Bank, Washington DC .

22. *Energy Management*. 1990
Editorial opinion, *Energy Management*, Journal of the National Productivity Council, New Delhi, 14(2):4, April–June.

23. Louvel, R. 1987
Briquetting of vegetable residues, *Boiling Point* (14):15–19, Intermediate Technology Development Group, Rugby, United Kingdom.

24. *Boiling Point.* 1989
Boiling Point, special edition on briquetting, Intermediate Technology Development Group, Rugby, United Kingdom.

25. ITDG. 1985
Low energy briquetting, Project 508 report, Intermediate Technology Development Group, Rugby, United Kingdom, 8 December.

26. ITDG. 1984
Noflie I and *Noflie II: a portable wood/briquette stove,* Intermediate Technology Development Group, Rugby, United Kingdom.

27. Young, P. 1988
Promising performance from a new briquette burner, *Boiling Point* (16):8–9, Intermediate Technology Development Group, Rugby, United Kingdom.

28. V.S. Machine Factory
Brochure, V.S. Machine Factory (manufacturer), T & P Intertrade Corp. (exporter), Bangkok, Thailand.

29. Qiu, D., Shuhua, G., Liange, B., and Wang, G. 1990
Diffusion and innovation in the Chinese biogas program, *World Development* 18(4):555–563.

30. *Economic Survey 1989–90.* 1990
Government of India, New Delhi.

31. Dhawan, A.S., Mahajan, R.B., Ganure, C.K., Kulkarni, V.V., and Gote, A.G. 1990
Constraints in the functioning of biogas programme in Maharashtra State, *Energy Management* 34–37, July–September.

32. TEDDY. 1989
TERI Energy Data Directory and Yearbook, Tata Energy Research Institute, New Delhi.

33. Larson, E.D. 1991
A developing-country-oriented overview of technologies and costs for converting biomass feedstocks into gases, liquids and electricity, PU/CEES report number 266, Center for Energy and Environmental Studies, Princeton University, Princeton, New Jersey.

34. Reddy, A.K.N., Sumithra, G.D., Balachandra, P., and d'Sa, A. 1991
The comparative costs of electricity conservation and centralized and decentralized electricity generation, in R.K. Pachauri, L. Srivastava, and K. Thukral, eds., *Energy - Environment - Development, Proceedings of the 12th International Conference of International Associations for Energy Economics,* January 1990, New Delhi. Vikas Publishing House, New Delhi, 772–814.

35. Bhatia, R. 1990
Diffusion of renewable energy technologies in developing countries: a case study of biogas engines in India, *World Development* 18(4):575–590.

36. National Aeronautical Laboratory 1981
Utilization of sludge gas for operation of gas turbine engines, *Newletter of the National Aeronautical Laboratory,* 9 (11–12), Bangalore, India, November–December.

37. Jiang, Z., Zhang, C., Wu, N., et al. 1984
High primary air biogas range burner, *Proceedings of the 1984 International Gas Research Conference,* Government Institutes, Inc., Rockville, Maryland, 1043–1052.

38. Vasudevan, P. and Jain, S. 1989
Field level study on biogas burners, *Changing Villages, Journal of the Consortium on Rural Technology,* New Delhi, 8(2).

39. SERI. 1979
   Generator gas—the Swedish experience from 1939–45, Solar Energy Research Institute (now National Renewable Energy Laboratory), Golden, Colorado, NTIS/SP-33-140.

40. Zhu, H., Brambley, M.R., and Morgan, R.P. 1983
   Household energy consumption in the People's Republic of China, *Energy* 8(10):763–774.

41. Dasappa, S., Reddy, V., Mukunda, H.S., and Shrinivasa, U. 1985
   Experience with gasifiers for 3.7 kW engines, *Ambio* 14:275–279.

42. Kjellström, B. 1985
   Biomass gasifiers for energy supply to agriculture and small industry, *Ambio* 14:267–274.

43. Bernardo, F.P. and Kilayko, G.U. 1990
   Promoting rural energy technology: the case of gasifiers in the Philippines, *World Development* 18(4):565–574.

44. Jain, B.C. and Vyarawalla, F. 1985
   Biomass gasification—its relevance to India and development work at Jyoti, Inc. *Proceedings of the Second International Producer Gas Conference*, Bandung, Indonesia.

45. Baliga, B.N., Dasappa, S., Shrinivasa, U., and Mukunda, H.S. 1990
   Gasifier based power generation: technology and economics, ASTRA, Indian Institute of Science, Bangalore.

46. Baguant, J. and Panray, R. 1991
   Low-grade ethanol as a cooking fuel substitute for kerosene: a case study in Mauritius, in R.K. Pachauri, L. Srivastava, and K. Thukral, eds., *Energy - Environment - Development, Proceedings of the 12th International Conference of International Associations for Energy Economics*, January 1990, New Delhi. Vikas Publishing House, New Delhi, 929–937.

47. Ogden, J., Williams, R.H., and Fulmer, M. 1991
   Cogeneration applications of biomass gasifier/gas turbine technologies in the cane sugar and alcohol industries, in J.W. Tester, D.O. Wood and N.A. Ferrari, eds., *Energy and the Environment in the 21st Century*, MIT Press, Cambridge, Massachusetts.

48. Copersucar. 1989
   Proálcool: fundamentos e perspectivas, São Paulo, Brazil.

49. Rajvanshi, A.K. 1988
   Solarly distilled ethanol from sweet sorghum as cooking fuel, in T.K. Moulik, ed., *Food–energy nexus and ecosystem*, Oxford and IBH Publishers, New Delhi, 383–993.

50. ABE. 1985
   Towards a perspective on energy demand and supply in India in 2004/5, report of the Advisory Board on Energy, Government of India.

51. Gill, J. 1986
   A preliminary investigation of alcohol-fueled stoves, report of the Energy Research Group, Open University, Milton Keynes, United Kingdom.

52. Vidyarthi, A. and Vidyarthi, V. 1985
   A study on the prospects of using liquid fuels for cooking in developing countries, Intermediate Technology Development Group, London.

53. EUT. 1985
   Test results on kerosene and other stoves in developing countries, report of Eindhoven University of Technology, World Bank Energy Department paper 27, Washington DC.

54. Rajvanshi, A.K. 1984
   Distillation of ethyl alcohol from fermented sweet sorghum solution by solar energy, final project report to the Department of Non-Conventional Energy Sources (DNES), New Delhi.

55. Leach, G. and Mearns, R. 1988
    Beyond the woodfuel crisis, Earthscan, London.

56. Fitzgerald, K. B., Barnes, D., and McGranahan, G. 1989
    Interfuel substitution and changes in the way households use energy: the case of cooking and lighting behavior in urban Java, Industry and Energy Department report, World Bank, Washington DC.

57. Ravindranath, N.H., Shailaja, R., and Revankar, A. 1989
    Dissemination and evaluation of fuel efficient and smokeless ASTRA stove in Karnataka, *Energy Environment Monitor* 5(2):48–60.

58. Jones, D.W. 1988
    Some simple economics of improved cookstove programs in developing countries, *Resources and Energy* 10:247–264.

59. Siwatibau, S. 1981
    Rural energy in Fiji: a survey of domestic rural energy use and potential, IDRC-157e, International Development Research Centre, Ottawa, Canada.

60. Kishore, V.V.N. 1988
    Cooking energy systems—a comparative study, *Energy Policy Issues* 4:37–45, Tata Energy Research Institute, New Delhi.

61. Utria, B.E. 1989
    Evaluación del sector energético residencial en la Republica Dominicana, *Seminario Latinoamericano de planificación energética para el sector residencial*, San José, Costa Rica; available from the Industry and Energy Department, ESMAP/World Bank, Washington DC.

62. Reddy, A.K.N. and Reddy, B.S. 1983
    Energy in a stratified society: case study of firewood in Bangalore, *Economic and Political Weekly* New Delhi, 8 October.

63. Das, T.K., Roy, J., and Chakraborty, D. 1990
    New technological options in energy sector cost burden on rural households, *Integrated renewable energy for rural development, Proceedings of the Annual Congress of the Solar Energy Society of India*, Tata McGraw-Hill, New Delhi, 468–472.

64. Shukla, K.C. and Hurley, J.R. 1983
    Development of an efficient, low NO_x domestic gas range cook top, report by Thermo Electron Corp., Waltham, Massachusetts, for Gas Research Institute.

65. Ravindranath, N.H. and Chanakya, H.N. 1986
    Biomass-based energy system for a South Indian village, *Biomass* 9:215–233.

66. Reddy, A.K.N., Somasekhar, H.I., Rajabapaiah, P., and Jayakumar, S. undated
    Scenarios for the evolution of the Pura Energy Centre, report of ASTRA, Indian Institute of Science, Bangalore.

67. Smith, K.R. 1987
    Biofuels, air pollution, and health: a global review, Plenum Press, New York.

68. Forest Survey of India. 1987
    The state of forest report, Ministry of Environment and Forest, Dehra Dun.

69. De Montalembert, M.R. and Clement, J. 1983
    Fuelwood supplies in the developing countries, Food and Agricultural Organization, United Nations, Rome.

70. Eckholm, E. 1975
    The other energy crisis: firewood, Worldwatch Institute, Washington DC.

71. Eckholm, E. 1976
    Losing ground, Worldwatch Institute, Washington DC.

72. Centre for the Application of Science and Technology to Rural Areas (ASTRA). 1982
Rural energy consumption patterns—a field study, *Biomass* 2:255–280.

73. Ravindranath, N.H., Nayak, M.M., Hiriyur, R.S., and Dinesh, C.R. 1991
The status and use of tree biomass in a semi-arid village ecosystem, *Biomass and Bioenergy* (in press).

74. Natarajan, I. 1985
Domestic fuel survey with special reference to kerosene, National Council for Applied Economic Research, New Delhi.

75. Leach, G. 1989
Biomass energy, welfare and the CO_2 problem, in S. Gupta and R.K. Pachauri, eds., *Global warming and climate change: perspectives from developing countries*, 189–194.

76. ERG. 1986
Energy research: directions and issues for developing countries, IDRC-250e, Energy Research Group, International Development Research Centre, Ottawa.

77. Meier, P. and Munasinghe, M. 1987
Implementing a practical fuelwood conservation policy: the case of Sri Lanka, *Energy Policy* 125–134, April.

78. WHO. 1984
1983 Statistics Annual, World Health Organization, Geneva.

79. Pandey, M.R. 1984
Domestic smoke pollution and chronic bronchitis in a rural community in the hill region of Nepal, *Thorax* 39:337–339.

80. Pandey, M.R., Regmi, H.N., Neupane, R.P., et al. 1985
Domestic smoke pollution and respiratory function in rural Nepal, *Tokai Journal of Experimental and Clinical Medicine* 10:471–481.

81. Pandey, M.R., Boleij, J.S.M., Smith, K.R., and Wafula, E.M. 1989
Indoor air pollution and acute respiratory infection in children, *The Lancet*, 25 February; summarized in *Boiling Point* (18):15–18, Intermediate Technology Development Group, Rugby, United Kingdom.

82. Joshi, V., Venkataraman, C., and Ahuja, D.R. 1985
Performance of cookstoves: measurements of thermal efficiencies and emissions, *Proceedings of the Second World Congress on Engineering and the Environment*, New Delhi, Tata Energy Research Institute, New Delhi, November.

83. TERI. 1987
Stove efficiencies and harmful emissions, report of Tata Energy Research Institute, New Delhi.

84. Ramakrishna, J. 1988
Patterns of domestic air pollution in rural India, Ph.D dissertation (geography), University of Hawaii Resource Systems Institute, East–West Center.

85. Smith, K.R. 1987
Cookstove smoke and health, *Boiling Point* (13):2–9, Intermediate Technology Development Group, Rugby, United Kingdom.

86. Mintzer, I.M. 1987
A matter of degrees: the potential for controlling the greenhouse effect, World Resources Institute, Washington DC.

87. DeCicco, J., Cook, J., Bolze, D., and Beyea, J. 1990
CO_2 diet for a greenhouse planet: a citizen's guide for slowing global warming, National Audubon Society, New York.

88. Pachauri, R.K. and Suri, V. 1990
   Contribution to greenhouse gases through large-scale use of fossil fuels and biofuels, *Energy Environment Monitor* 6(1):11–16, citing D.R. Ahuja, Environmental Protection Agency, Washington DC.

89. Openshaw, K. and Feinstein, C. 1989
   Fuelwood stumpage: considerations for developing country energy planning, Industry and Energy Department Working Paper Energy Series paper 16, World Bank, Washington DC.

16
OPEN-TOP WOOD GASIFIERS

H.S. MUKUNDA
S. DASAPPA
U. SHRINIVASA

The technology and economics of a new class of open-top gasifiers for use with diesel engines in dual-fuel mode are described. The performances of systems that range in capacity from 3.7 to 100 kilowatts are discussed, with special emphasis placed on gasifiers at extreme ends of the capacity range. The essential differences and benefits of the new technology are compared with World War II closed-top models. Studies indicate that the open-top design achieves diesel replacement values greater than 80 percent and is less dependent on feedstock quality, moisture content, and density. The amount of diesel fuel saved per system among motivated users (mostly small farmers) exceeds 70 percent. A comparative analysis of two gasifier systems: a 5 kilowatt system that runs the village power station in Hosahalli, Karnataka (India), and a 100 kilowatt system that powers a sawmill on the remote island of Port Blair in the Andaman and Nicobar archipelago was undertaken. The cost of installing the larger system, including computerized data acquisition and control systems, was U.S.$625[1] (Rs 12,500) per kilowatt, with an energy cost of $0.074 (Rs 1.60) per kWh (the cost of energy subsidized by the state is Rs 1.25 per kWh).

INTRODUCTION

Solar energy captured by photosynthesis and stored in biomass can be converted by the process of gasification into a gaseous high-energy fuel (9,600 megajoules per m^3) that can be used in internal combustion engines for power generation. Gasification is a two-stage process: during the first stage, the biomass undergoes partial combustion to produce gas and charcoal; during the second, the charcoal reduces the product gases (chiefly carbon dioxide and water vapor) to form carbon monoxide (CO) and hydrogen (H_2). The process also generates methane and other higher hydrocarbons depending on the design and operating conditions of the gasifier.

---

1. 1 U.S.$ = 20 Indian rupees at the time of preparing this paper (April 1991).

The combustible gas so produced is composed of about 18 to 20 percent H_2, 18 to 20 percent CO, 2 to 3 percent methane (CH_4), 8 to 10 percent carbon dioxide (CO_2), with nitrogen making up the rest. The gas will fuel a spark-ignition engine, delivering about 60 percent of the power of gasoline, or it will run a compression-ignition engine in dual-fuel mode, eliminating the need for 75 to 85 percent of the diesel fuel. The latter application is attractive for developing countries, especially India, where large numbers of diesel engines are employed at various power levels.

Wood gasification has developed in spurts, the most intense activity taking place during World War II in response to petroleum shortages in both civilian and military sectors. Some of the most insightful studies of wood gasifiers during this period are well documented [1, 2]. Most subsequent work has been devoted to the replication of existing systems. Around 1980, T.B. Reed and his colleagues at the Solar Energy Research Institute (now the National Renewable Energy Laboratory) in the United States conducted systematic laboratory studies on an open-top reactor, a version of which had been previously and successfully adapted for rice-husk gasification in China [3]. Recent research and development efforts have produced a technology that can process wood at powers ranging from 5 to 100 kilowatts.

Problems with open-top reactors frequently cited in the literature include the buildup of tar and the lack of critical assessment of available designs. Although different designs are described, they often are not rated against one another in terms of their overall performance [1]. Consequently, many designs that differ from one another only slightly have been proposed. Attempts have been made to rate the various designs according to the amount of time needed for pyrolysis (the release of volatiles in the presence of heat) and the reaction time of charcoal with air [4]. Although some studies have produced relevant findings, the results have not found their way into prototype designs as expected [1, 2]. Our chapter reviews the above studies and presents an integrated picture of gasifiers designed to run strictly on woody biomass. Research we have initiated on novel approaches to the gasification of agricultural wastes is beyond the scope of the present chapter.

THE OPEN-TOP VERSUS THE CLOSED-TOP GASIFIER

The World War II vintage closed-top and the more modern open-top gasifiers differ significantly in their geometries (see figure 1). In the closed-top gasifier, the hopper region, into which the biomass is loaded, is relatively massive, its size decided by two values: the reactor's diameter divided by its throat diameter (d_r/d_t) and the height of the hopper divided by the throat diameter (h_3/d_t), which are determined by the amount of time (typically 2 or 3 hours) needed to run a single, uninterrupted cycle (see figure 1). The choice of d_t reflects the need to balance two factors: the higher the value of d_t, the greater the risk that tar-laden gases will escape the high-temperature combustion zone, yet the smaller the value of d_t the greater the velocity of the gases that sweep through the throat and the reduction

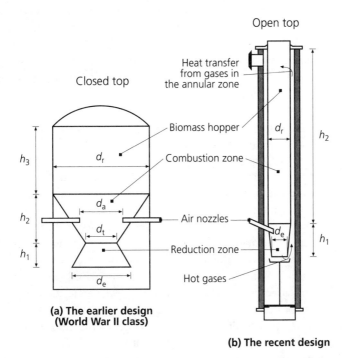

Open top

Closed top

Heat transfer from gases in the annular zone

Biomass hopper

Combustion zone

Air nozzles

Reduction zone

Hot gases

d_r

d_a

d_t

d_e

h_3

h_2

h_1

d_r

h_2

d_e

h_1

(a) The earlier design (World War II class)

(b) The recent design

FIGURE 1: The most important dimensions in the closed-top reactor are the throat diameter, d_t, the reactor diameter, d_r, the exit-plane diameter, d_e, and the relative heights of the reactor vessel, h_1, h_2, h_3 (see figure 1a). Correspondingly, the important dimensions of the open-top reactor are the lateral widths, d_r and d_e, and the heights, h_1 and h_2 (see figure 1b). Appropriate values for the ratios d_r/d_t, h_1/d_t, and h_2/d_t, are based on the "best" performance of some commercial designs.

zones, collecting fine dust and ash. Tar is reduced in the smaller open-top model principally because temperatures in the reactor are higher, which facilitates tar cracking (the reduction of complex hydrocarbons into simpler forms), as well as the completion of all reactions.

Another drawback to the closed-top gasifier is that the diameter of the hopper is so large that heat transfer from the high-temperature zone generally affects wood chips near the hopper's wall rather than in its center [5]. The problem is overcome in models, most notably the Imbert generator, that have an outer chamber around the hopper. Hot gases flowing along the outer wall transfer heat to the wood chips. In some cases, preheating the air is recommended [1], but this is generally ineffective because of the low heat capacity of air and the large area required for heat transfer.

Other closed-top designs, such as the monorator [1], include an outer zone next to the hopper for tar collection. Unless tar is reduced, the walls become laden with tar that is either encrusted and hard, or liquid and sticky. Tar in the latter state may form glue-like bridges, particularly when the gasifier is started, thus interfering with the flow of biomass. Another drawback to the closed-top gasifier is that generating combustible gas of reasonable quality is more difficult when wood with a high moisture content (20 to 30 percent) is used. This problem is one that cannot be entirely attributed to the moisture content of the wood and can be solved by installing a properly designed hopper. Because of these problems and the need to address them, the closed-top gasifier is considered less effective than the open-top model.

Tar can be reduced by distributing air nozzles around the periphery of the reactor. In this way fuel vapors are intercepted and combusted. During this process temperatures in the combustion zone also rise, which helps reduce tar [2]. The number of air nozzles is determined by the desired flow rate (and level of thermal and mechanical power). Air distribution around the nozzles, and in regions between the airflow zones where volatiles can escape (resulting in high-tar gas), can be mapped. Early gasifier programs in many countries encountered difficulties, in large part because inadequate attention was paid to tar problems. A key problem is that once tar escapes from the reactor, it cannot be readily eliminated because the cooling/spraying systems that are effective against dust do not trap tar-laden vapors. Thus, tar is best eliminated through thermodynamic and oxidative measures in the reactor itself.

The open-top gasifier, on the other hand, provides for more homogeneous airflow because the gases pass through a long porous bed of fuel in the vertical reactor [6]. Studies to determine flow distribution under cold conditions [6] indicate that the velocity distribution in a packed bed is homogeneous after a few particle depths. In addition, regenerative heating created by the transfer of heat from the gases (through the wall) to the biomass increases residence time in the high-temperature zone and thus leads to better tar cracking. Many of the configurations of the open-top design, including laboratory models [4], lack the air nozzles used in earlier designs. A commercial version developed by W.P. Walawender and his colleagues at Kansas State University [7] has a complex central air-nozzle that helps reduce tar, but does not facilitate start-up of the system. However, by combining the open-top with an air nozzle across the reactor, quick lighting with a simple wick flame is possible. In addition, the air nozzle stabilizes the combustion zone, preventing stratification, which occurs when the flame front moves in an opposite direction to the airflow. As a consequence, the high-temperature zone spreads above the air nozzle by airflow from the top, rather than by radiation, thermal conduction, and weak convection processes, as is the case with the closed-top gasifier (see figure 2).

Forced convection heat transfer from hot gases flowing in the annular gas passage (see figure 1) also contributes to upward flame propagation in the open-top. Such enhanced heat transfer makes it possible to gasify wood chips that have a moisture content as high as 25 percent because the added heat reduces the extra moisture. In general, air drawn through the top of the gasifier represents about 40 to 70 percent of the total inflow, depending on the size of the wood chips and rate of gas flow, which can reduce pressure along the path. Reliability is greatly en-

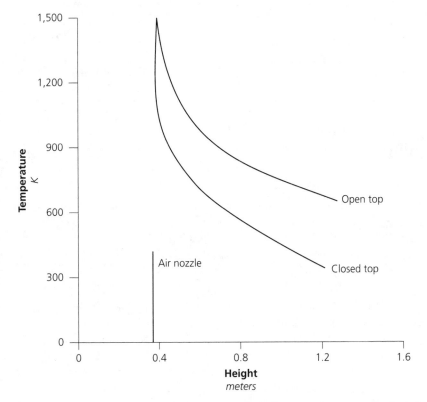

FIGURE 2: *The graph shows the temperature variation from the air nozzle region in the classical closed-top and the present open-top design. It is clear that the width of the high-temperature region, 600 K and above, is about 1 meter for the open-top design, whereas it is constant at about 0.4 meters for the earlier design. What is more, it can be controlled by decreasing the airflow through the air nozzle in the current design.*

hanced by the ability of the reactor to dry wood chips that contain variable amounts of moisture; insulation around the reactor helps maintain high temperatures in the reaction zone and thus also helps to reduce tar.

Wood-chip size also affects the reactor's operation. Studies indicate that to achieve power levels of 15 to 75 kilowatts, chips should be from 60 to 80 millimeters long and from 50 to 60 millimeters in diameter [2]. For open-top gasifiers, maximum size of the wood chip should be about one-sixth to one-seventh the diameter of the reactor in order to meet flow requirements, without prolonging gasification time [4] or creating excess porosity.

The actual conversion of the biomass to gas is a two-step process. The first step, called the flaming period, occurs when biomass burns in the presence of air drawn largely from the top, producing charcoal. The second step, known as charcoal conversion, occurs when charcoal reacts with carbon dioxide and water vapor to produce combustible gas. The duration of both stages has been determined by laboratory studies on cubes of wood [8]. The results, which are correctly expressed

Table 1: Reactor diameter and related parameters for various power levels

Power	Gas flow rate	Air flow rate	Air flow rate from top		Wood consumption rate	Diameter	Air flux
kilowatts	*g/second*	*g/second*	*g/second*	*percent*	*kg/hour*	*millimeters*	*kg/m²/s*
3.7	2.5	1.9	0.76	40	3.8	150	0.043
20.0	15.0	9.5	4.80	50	20.0	250	0.100
100.0	70.0	42.0	21.00	50	100.0	350	0.210

as volume-based diameters, apply to other geometries, such as cylinders and spheres, as well. Although the time needed to convert charcoal into CO and H_2 with CO_2 alone has been quantified kinetically, the effect of H_2O, which drives the conversion faster than CO_2 does, seems to have been ignored. The situation deserves further attention.

For a reactor of any diameter, the optimal height is determined by the properties of the woody biomass, namely, its density, specific heat, and heats of phase-change according to a simple model for heat balance. The distance, or height, traveled by the woody biomass must allow for a residence time at least equal to the sum of the flaming pyrolysis and charcoal conversion times. Thus, open-top gasifier designs can be based on a one-dimensionality consideration for which the ratio of height to diameter of the reactor should be large, typically a value from six to eight. Therefore, once height is determined, diameter can be calculated. Smaller diameters are generally preferred in order to minimize reactor size and maximize heat transfer to the entire cross section. Too small a diameter, however, necessitates the use of smaller wood chips and leads to unmanageable pressure drops at reasonably high flow rates. Such factors must be optimized, although at present there are no precise guidelines for doing so (see table 1).

The superficial mass flux (the flow rate of gas relative to area) of air drawn from the top of the reactor is shown in column 8 of table 1. As power increases (column 1), mass flux increases, a relation that reflects the reactor's more compact design. Increasing the diameter at higher power levels (for example, to maintain the same mass flux as for 3.7 kilowatt system) will not hinder performance, provided the reactor height is sufficient to allow for chemical reaction time. Diameters with a length-to-diameter ratio equal to 2 or 3, however, lead to the loss of one-dimensionality; consequently, the reactor's center may not receive as much heat as its perimeter, where heat transfer results in better flaming pyrolysis. The 3.7 and 20 kilowatt reactors are equipped with a single air nozzle; the 100 kilowatt system has six, which reflects its larger diameter. Thus, if the diameter size of the 20 kilowatt system is to be increased, additional air nozzles may be needed.

Height and volume requirements can be related mathematically as follows, with the velocity of movement of the wood chips v_w expressed in terms of the mass consumption rate of wood chips $\dot{m}_w$, the superficial density of wood chips, ρ_{sup}, and the cross-sectional area of the reactor A_r as

$$v_w = \frac{\dot{m}_w}{\rho_{sup} A_r} \tag{1}$$

If one assumes that the wood chips undergo heating and flaming pyrolysis during their passage from the top of the reactor to the air nozzle, this time t_f can be expressed as the diameter of the wood chips d_w (which can be a mean diameter, if based on volume, or simply a diameter if the wood chips are long and cylindrical) as

$$t_f = K_c d_w^{1.8} \text{ where } K_c = 3.5 \text{ s mm}^{-1.8} \tag{2}$$

where d_w is in millimeters and t_f is in seconds. The above equation is indicative of the diffusion limits of the flaming pyrolysis process; the exponent on d_w, which is derived from simple theory, works out to be 2. The value of 1.8 is a curve fit of laboratory data [9] that compensates for the influence of convection currents. The height of the reactor above the air nozzle h_2 is obtained as

$$h_2 = v_w t_f = \frac{4 K_c}{\pi} \cdot \frac{\dot{m}_w d_w^{1.8}}{\rho_{sup} d_r^2} \tag{3}$$

In the above equation K_c is about 3.5 seconds mm$^{-1.8}$ for d_w up to 25 millimeters. Although wood chips are about one-sixth to one-seventh the diameter of the reactor, they undergo a size reduction of 10 to 15 percent [9] during flaming. Moreover, those that are larger than 25 millimeters in diameter have a tendency to crack into smaller pieces. For much larger reactor sizes, the wood-chip size reaches an asymptotic limit caused by the availability of biomass sizes. In such cases, the height of the reactor decreases with the square of the reactor diameter. Height h_1 is governed by the reduction reactions of charcoal with CO_2 and H_2O.

As the above calculations indicate, reactor designs have a rational basis; nevertheless, it has not been possible to generate a design that will meet a target final composition under a given set of operating conditions. In order to do so, more work on basic kinetic parameters and modelling needs to be done.

THE MODERN OPEN-TOP GASIFIER

The reactor and gas-cooling and -cleaning systems are the most important components in an open-top gasifier (see figure 3); an additional feature is the auxiliary wood-processing system. A variety of designs, which were motivated by the need to adapt systems to individual site conditions, now exist. Some designs are retro-

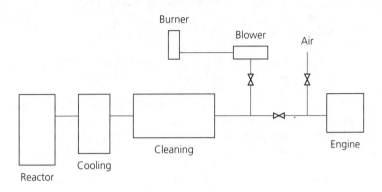

FIGURE 3: Block diagram of the elements of the gasifier system.

fits of existing systems; in others, the major components have been redesigned, as described below.

The reactor

We have proposed two reactor redesigns (see figure 4). In both, the annular stainless-steel reactor is wrapped with 75 millimeter light alumino-silicate insulation and encased in an outer aluminum sheet. The inner shell tapers to a cone at its bottom to create a small, intense combustion zone and to facilitate ash removal by high-velocity gases. In one design, the bottom is capped, which is useful where conditions do not allow for additional space in the bottom region. However, the maximum uninterrupted run for a bottom-capped 3.7 kilowatt system is 17 hours and about 10 hours for a 100 kilowatt system at 80 percent load. Continuous operation is possible with the alternative design, which is equipped with a water seal that enables the ash to be washed away by the water, thus obviating the need to interrupt the system for cleaning. Although water evaporation through the seal caused by radiant heat transfer from the reduction zone was a problem initially, a radiation shield has since been designed [10] that has enabled the seal to function satisfactorily at all power levels.

Cooling and cleaning systems

After a 10-year effort, a number of effective gasification cooling and cleaning systems have been designed, all of which had to first overcome a number of constraints: cooling had to make minimal demands on the local water supply, the dust extractor had to be available for a low price, and the time between overhauls had to be comparable to that for the engine itself. These constraints have now been met for dry and wet cooling systems, both of which can meet the requirements efficiently. Whereas wet cooling systems recirculate and spray water through a cooling tower (see figure 6), dry cooling systems dissipate the heat through large-area metal surfaces. Both depend on cleaning systems that use coir (coconut fiber) mats, cloth filters, and sandbed filters to cleanse the gas of partic-

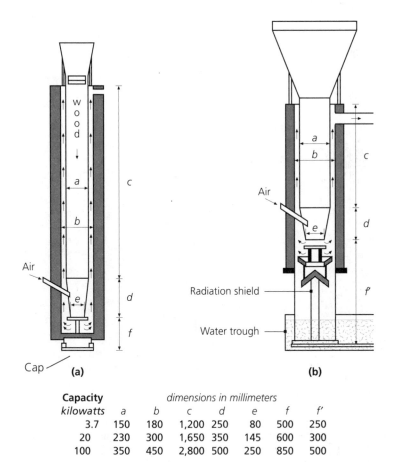

Capacity	dimensions in millimeters						
kilowatts	a	b	c	d	e	f	f'
3.7	150	180	1,200	250	80	500	250
20	230	300	1,650	350	145	600	300
100	350	450	2,800	500	250	850	500

FIGURE 4: Two reactor redesigns have been proposed. In one (a) the bottom is capped; in the other (b) the bottom is set in a trough of water.

ulates, and both reactors, fortunately, produce little tar; the small amount generated is carried into the engine without being deposited in the passages, and thus poses no operational difficulties [11].

The 3.7 kilowatt gasifier

After leaving the reactor, the gas enters the cleaning system, which in the 3.7 kilowatt gasifier is a high-efficiency cyclone: a centrifugal device that extracts fine particulate matter. Coarser particles are left behind in the annular region of the reactor itself (see figure 5). The gas then enters a long vertical tube tangentially so that it spirals down to a water trough at the bottom of the tube. There the gas skims across the water's surface and enters a second vertical tube, again tangentially. The tangential movement facilitates heat transfer from the gas to the wall. At the end of a typical six-hour run at a load of 3 kilowatts, the temperature of

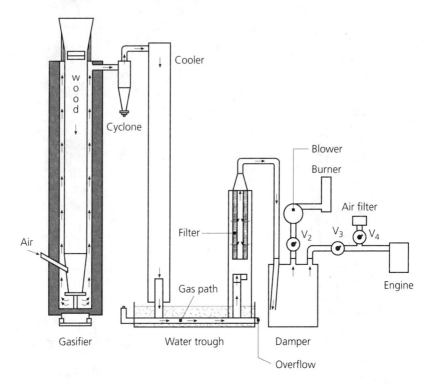

FIGURE 5: The 3.7 kilowatt system with dry cooling facility.

the gas entering the cyclone is about 600 K; when it leaves the cyclone it has reached 400 K; by the time the gas enters the vertical tube containing the filter, it is near-ambient temperature. The cooling raises the density of the gas and the mass flow of the charge ingested by the engine.

The choice of a water seal for the wet-cooling reactor (see figure 6) is somewhat incidental, since both developments took place around the same time. It is possible to combine the bottom-cap reactor with the water cooling system. The gas enters a long vertical tube at the top where a spray nozzle is mounted. Water, which enters the sprayer from the pump outlet and is controlled by either the size of the tube or by a valve, mixes with the incoming gas, cools it, and removes a major portion of the fine dust. The fine mist of dust and water vapor is eliminated when the gas passes through the cyclone. Measurements have shown that dust and tar levels at this stage are low, but an additional filter of sand and coir pith is nonetheless recommended. A pressure drop of about 400 pascals occurs in the reactor and another of 1,000 pascals occurs in the cooling and cleaning system, amounting to a total pressure drop of about 1,400 pascals; only when the pressure drops by more than 2,000 pascals will blockages affect performance by reducing the diesel replacement.

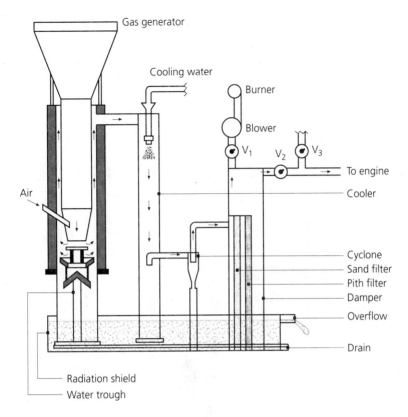

FIGURE 6: The 3.7 kilowatt system with water cooling facility.

The 100 kilowatt gasifier

The cooling and cleaning system of a 100 kilowatt gasifier differs from a 3.7 kilowatt system primarily because cooling occurs in two stages (see figures 7 and 8). The two stages reflect higher (about 30 kilowatt) cooling requirements, but are convenient to implement. Cleaning is performed by a 100 millimeter thick, 1.2 m^2 cross-section sandbed consisting of particles ranging in diameter from 250 to 600 microns. A 100 millimeter thick bed of coir pith absorbs residual moisture. Pressure drops by 600 pascals as the gas crosses the reactor and by 1,200 pascals as it crosses the cooling and cleaning system.

Blower, valves, and burner

Once the gas is cooled and cleaned, it passes through a damper either to a burner or to the engine air inlet. The damper isolates the engine from pressure fluctuations in the gasifier and helps to increase induction efficiency. One branch from the gas line passes through a valve to the blower, which is either hand cranked or electrically driven, and then goes to the burner, which is a simple cylinder with tangential entry for gas at its base (see figure 5). The design differs from earlier

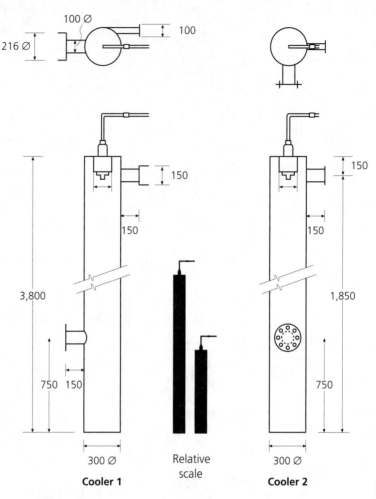

Cooler 1

Relative scale

Cooler 2

Scale 1:30
All dimensions in millimeters

FIGURE 7: The cooling and cleaning system for a 100 kilowatt system.

designs in the flares—tubes in which the gas is burned—and benefits from increased flame stability. Because the flame length is short, it is unaffected by wind currents. A second branch of the gas line is linked to air to permit the engine to draw the gas–air mixture.

The gasifier is started with a blower and runs until combustible gas is obtained, at which time the engine is cranked and made to run for a few minutes on diesel or gasoline before the gas line is opened. Because the gasifier is loaded with charcoal to some 300 millimeters above the air nozzle, it functions more like a simple charcoal gasifier in the early stages. Some users start the gasifier by running the engine directly, drawing air through the gasifier itself during the early stages.

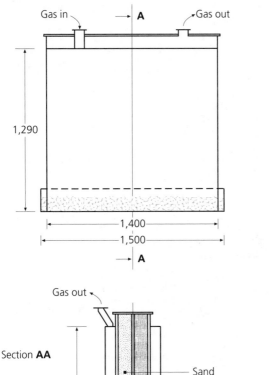

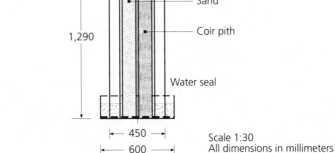

FIGURE 8: The cooling and cleaning system for a 100 kilowatt system. A frontal view is presented (top); as well as a cross-sectioanl view (bottom).

The air is then subsequently replaced by gas. Such a strategy renders use of the blower redundant in small power systems (those 3.7 kilowatts in size or less).

The wood-processing system

The woodstock available to users of 3.7 kilowatt systems varies. Some comes from the green and tender stems of mulberry plants after the leaves have been stripped to provide fodder for silkworms. The stems, which are from 10 to 15 millimeters in diameter, are easily chopped into pieces that are from 20 to 30 millimeters

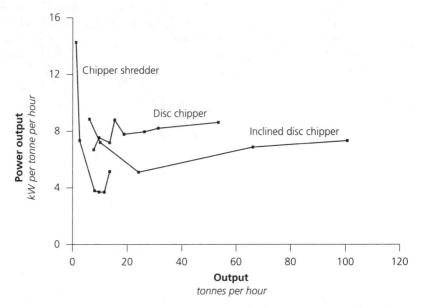

FIGURE 9: *The power (kW) per tonne per hour of wood chips relative to ouput of woodchipping machines.*

long. Other woodstock comes from dead or old trees that are from 50 to 70 millimeters in diameter and cut into pieces that are 1 or 2 meters long. In the latter case, the wood must be cut with a 0.75 kilowatt electrically driven circular saw in a separate cutting and sizing facility. Such a saw can process 40 kilograms of wood per hour, which is adequate for most applications below the 20 kilowatt level.

The 100 kilowatt system runs on wood chips that are from 25 to 30 millimeters in diameter and from 50 to 70 millimeters long. One such gasifier has been installed at the Chattam Island sawmill as part of a demonstration program run by India's Department of Nonconventional Energy Sources (DNES). Although some of the wood, which ranges from 100 to 200 millimeters in diameter and from 400 to 600 millimeters long, is light, most is dense and hard. The tree species processed at the sawmill include padauk (*Pterocarpus dalbergiodes*), mowa (*Maduca indica*), and garjan (*Dipterocarpus indicus*). Most wood wastes are of high density (900 kilograms per m^3), but some are of low density (450 kilograms per m^3). Although the mass consumption rate of both types of waste is the same, lower-density wood must be consumed volumetrically faster, and thus must be processed by a taller reactor.

The wood waste is transported manually to the gasifier plant from the mill and then fed through a chipping machine. In searching for appropriate chipping machines for the demonstration site, we found that most equipment, including inclined disc chippers, chipper shredders, and ordinary disc chippers is manufactured in Europe. Technical information on the power needed to produce a certain tonnage of chips each hour was put together from the brochures (see figure 9).

The power per unit output (1 tonne per hour) rises sharply, reaching 15 kilowatts for smaller loads of 500 kilograms per hour, but then asymptotes to about 8 kilowatts for loads larger than about 20 tonnes per hour. To produce about 250 kilograms per hour of chips requires a machine having about 5 kilowatts of power. At the Port Blair sawmill, a 20 tonne hydraulic power press with a 3.7 kilowatt capacity motor was modified to operate a hardened spring steel blade with its cutting edge angled to facilitate slicing. Trial runs show that the machine can easily cut about 200 kilograms per hour of wood. Chip preparation facilities, however, should be custom designed for each site; in some situations, for example, cutting machines that have a 3 kilowatt circular saw are adequate for sizing the wood.

The chip-loading system

Wood chips are loaded manually into both the 3.7 and 20 kilowatt systems. Every 30 and 45 minutes, loads of 2.5 and 10 kilograms, respectively, are emptied into the reactors' hoppers. In larger gasifiers, such as the 100 kilowatt system, a machine-driven loading system is necessary (see figure 10). When the chips drop below a certain level, the operator activates a conveyor belt that carries a new supply to the top of the gasifier. The arrangement, however, takes up considerable floor space, adding to the system's initial costs. Efforts are therefore being made to develop a more economical system. One possibility is split-level flooring: by constructing a platform about 1 meter below the top of the gasifier, a few bags of wood chips can be loaded into the reactor in about half an hour, thus eliminating the need for a conveyor belt.

Instrumentation and control

The diesel tank of the 3.7 kilowatt gasifier is equipped with a bypass valve and a measuring cylinder. By measuring diesel-flow rates in this way, the performance of the diesel engine can be compared in diesel-only and dual-fuel modes. Initially both gasifier systems were provided with a manometer to measure total pressure drop; if the drop was excessive, the system could be shut down for maintenance. Field observations, however, indicate that the manometer is almost never used; the systems run until a breakdown occurs and then they are dismantled for maintenance. During operation, diesel flow is optimized by reducing airflow to the engine, which increases gas flow and thus energy flow. If diesel flow becomes excessive, maintenance of the system becomes necessary. The diesel governor on the engine maintains steady operations until a decrease in airflow signals a fuel-rich condition in the dual-fuel mode and the engine cannot take the load. Airflow is then reduced only enough to support the load with a high diesel replacement. Typically, the stall region occurs when diesel replacement exceeds 90 percent; therefore, the system can be run in a stable manner at diesel replacements in excess of 85 percent, but below 90 percent.

The 100 kilowatt gasifier is fitted with a fair amount of instrumentation, some of which was specifically developed for the system. Voltage, current, frequency, and diesel flow rates are recorded. When the system operates in the dual-

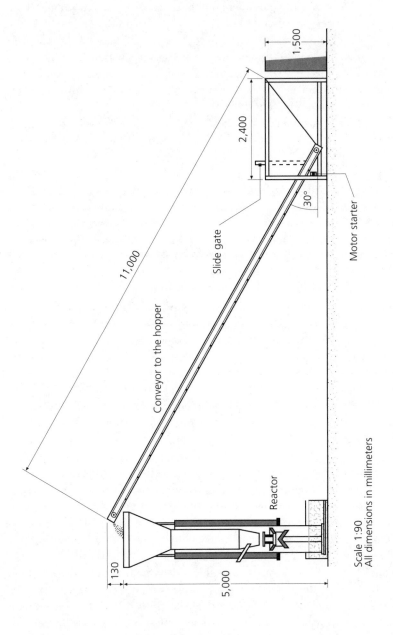

Scale 1:90
All dimensions in millimeters

FIGURE 10: *The wood chip loading system for a 100 kilowatt gasifier.*

Table 2: Tar and particulate levels

Flow rate *g/second*	Tar *ppm*	Particulates *ppm*
3 (3.7 kilowatt system)	30–40	50
40 (100 kilowatt system)	30–40	50

fuel mode, diesel consumption is reduced by decreasing airflow, thus increasing gas flow. Steady operating conditions are maintained by increasing resistance in the air line, which can be quickly reversed if the engine is unable to cope with the large demand near full-load conditions.

Frequency is measured electronically with a 0 to 10 millivolt sensor. The change in the wire's resistance resulting from the flow of diesel fuel is converted to an output signal with the help of appropriate electronics. After extensive calibration and the installation of high-performance electronic components, the measurement unit for the 100 kilowatt gasifier seems robust and accurate in its performance. Moreover, the system has now been linked to a personal computer, which continuously records the operating data and provides instructions to the control system for corrective action during the run.

Performance of the gasifier–engine system

Gasifier performance was measured initially by running it with a blower at various flow rates. Gas quality was assessed by measuring its composition, calorific value, and tar and particulate levels (see table 2). Engine performance in dual-fuel mode provided diesel replacement data at various loads. Apart from this, some monitoring of the engine's condition, including examination of its lubricating oil, valve seating, and engine head, were also made. Long-term monitoring of engine condition is on-going in Hosahalli, a small village in India, where a 4.4 kilowatt engine coupled to a 3.5 kVA alternator is being used for electrification. In addition, DNES had the system rigorously tested at a national testing center at the Indian Institute of Technology in Bombay before initiating its large-scale implementation. Most of the results obtained by the authors have been confirmed by tests carried out by P.P. Parikh and her colleagues at the Institute [11].

Every internal combustion engine can accept a certain amount of dust, but dust levels must not be excessive. Gasifiers produce dust in the form of fine carbon and tar, which if deposited at bends and valve seatings can cause engine seizure under extreme conditions. Therefore, the levels of both must be routinely measured. Tests on the relation between flow rate and the production of tar and particulate matter show that if the gas has a calorific value of 5 or 6 megajoules per cubic meter and a methane content of 1.5 to 3.0 percent, then tar and particulates will measure less than 40 and 50 ppm respectively. Shifting from diesel to dual-

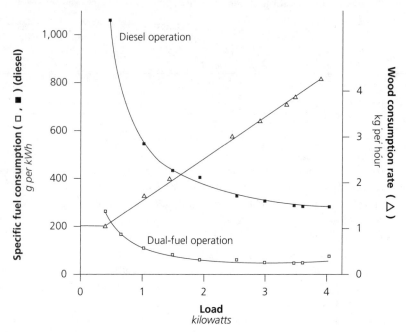

FIGURE 11: *Diesel consumption in diesel-only and dual-fuel run with load for a 3.7 kilowatt system in electrical mode.*

fuel mode is smooth; reducing the airflow raises the diesel replacement level to 85 percent or more at all loads. Fine tuning the air valve raises the replacement value still higher, to 93 percent. Minimum specific fuel consumption occurs at 80 percent load and is about 285 grams per kWh (see figure 11). In a normal cycle, with some variation in load, diesel fuel is consumed at a rate of 300 grams per kWh; maximum electrical output is 4 kilowatts. In the diesel-only mode, overall efficiency of the system is 28 percent. In the dual-fuel mode, efficiency drops to 15 percent, a situation that is related in part to the reduced combustion efficiency of the engine cylinder, which results from the lower flame speed of the wood gas–air mixture.

In the dual-fuel mode, lubrication oil needs to be changed about half to three-quarters as often as in the diesel mode. Also, the engine head is not affected until after about 1,200 hours of operation. But piston rings had to be replaced after about 1,600 hours, in part a result of overloading the engine.

DEMONSTRATION PROJECTS

The 3.7 kilowatt gasifier system has been turned over to private industry, and under a DNES program more than 300 such systems have been disseminated to beneficiaries, mostly in rural areas. Studies of dissemination, user motivation, and field performance have been carried out by the Karnataka State Council for Science and Technology, DNES, and the authors [10]. Two of the study's findings are particularly significant: first, users find the open-top system easier to operate than the classical closed-top model and second, mean time between maintenance and failure has increased [10].

The Hosahalli project

An open-top gasification system has been deployed since 1987 in Hosahalli, a village located about 110 kilometers southwest of Bangalore that has 43 houses and a population of about 270. The project, which is supported by various government agencies, was conceived as a way to provide rural villages with energy services derived from biomass. Hosahalli was chosen for a number of reasons. It was one of 13 villages in the state of Karnataka that had yet to be electrified, it had nearby land that was suitable for biomass generation, and it had a government willing to convert the land to biomass plantation and to construct a small power station.

Two and a half hectares were provided for biomass plantation and 60 m^2 of land were made available for the power station. The state electricity board helped draw electrical lines throughout the village. Each house was given one 40 watt fluorescent lamp and one 15 watt incandescent lamp; six 40 watt tube lights were provided for street lighting. An early version of the open-top gasifier with a dry cooling system and a single phase 3.5 kVA gasifier-based engine was installed to meet a total electrical load of 2.6 kilowatts.

The project was implemented in a series of stages. The first phase was initiated with the planting of such species as *Leucaena luecocephala*, *Acacia auriculiformis*, *Delbergia sissoo*, *Eucalyptus*, and others [12], and was completed when the plantation began to yield 10 tonnes per hectare of dry woody biomass in the first year and 7 tonnes per hectare in subsequent years. In May 1988, villagers received the first flow of electricity, which was provided for four hours every day during the first eight months and thereafter for six hours every day.

The second phase was initiated in response to a demand for drinking water, which at that time was obtained from a tank about half a kilometer from the village. The tank would dry up toward the end of December, creating significant water shortages during the four months before monsoon season. A decision was therefore made to pump deep-bore well water into two tanks inside the village. A 3 kilowatt pumpset was installed, which enabled the water to be piped over a distance of 500 meters. The pumping station, which is run by a second gasifier, has been operational since September 1990. Two village boys operate the gasifier, cutting tiny tree branches to size, drying them in the sun (or in a specially designed dryer that connects to the engine's exhaust pipe), and then loading them in the gasifier. When branches are unavailable the boys use a 0.5 kilowatt circular saw to split and cut logs. During monsoon months the dryer is necessary [10].

During the 15 month operating period from June 1988 to August 1989 [13], the amount of diesel fuel saved equaled 72 percent of the diesel-only mode and the 6.4 tonnes of wood burned (see figure 12) was far below the plantation yields

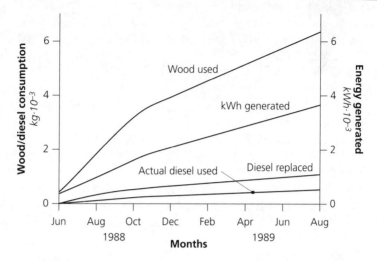

FIGURE 12: *Energy delivered measured in terms of diesel and wood consumption during 15 months of open-top gasifier operation at Hosahalli.*

of 25 and 15 tonnes. Electricity has reduced kerosene consumption for lighting purposes by an estimated 0.95 tonnes in one year [14]. The next stage calls for the construction of a flour mill with management of the system transferred to a village council.

The Port Blair sawmill project

A 100 kilowatt open-top system has been running a sawmill in Port Blair since January 1990. During daily use the engine delivers 50 kilowatts of lighting and from 10 to 27 kilowatts of power for welding and machining operations. The engine, which runs at diesel replacements of 76 to 85 percent, seems to have no difficulty handling the load. Minimum specific fuel consumption is 260 grams per kWh for loads ranging from 45 to 70 kilowatts. In diesel-only mode, overall efficiency increases to 34 percent, whereas in dual-fuel mode (with wood consumption at about 1.0 kilogram per kWh), overall efficiency drops to between 24 and 26 percent [15]. The gasification efficiency is about 85 percent, a value consistent with any well-designed gasifier.

Gasifier research and development continues. Recent efforts, for example, have resulted in a new ceramic reactor shell that can withstand the hostile, reactive, and thermal environment inside the reactor. The new shells, which are expected to significantly lengthen a reactor's life as well as improve its economics, are currently undergoing field trials.

OTHER FIELD TRIALS

Despite a lack of enthusiasm in the developed world for gasifiers as substitutes for petroleum fuels, research on them continues, motivated in large part by the desire to aid developing countries [16].

One open-top system, which was designed at Twente University in the Netherlands and installed in central Java, has a central air supply through the top, a refractory lined reactor and a rotating grate at the bottom. The system, which can generate 20 kilowatts of power, has run for more than 12,000 hours. Despite the fact that tar and dust in the treated gas reach levels as high as 2,000 and 130 parts per million (ppm) respectively, the system is said to have had no serious operational problems. No excessive wear on engine components was observed during overhauls that took place after 4,000 and 11,500 hours of operation. Studies indicate that although 70 percent diesel replacement was achieved, the annual average proved closer to 50 percent. Unfortunately, the annual diesel replacement level was assessed at greater than 70 percent in order for the gasifier to be economically viable.

One of the largest wood-gasifier plants based on the Imbert design is in Paraguay. The plant has three 420 kilowatt generators, which are powered by 1,000-rpm natural-gas engines and operate almost 7,400 hours a year. Of the power generated, about 56 percent goes for agro-industrial purposes; about 24 percent goes to households. The amount lost in transmission over an area of 240 km^2 is thought to be 20 percent. Reports do not indicate that performance of the control system is affected by a varying load situation, so it may be presumed that the control system was working satisfactorily.

The ferrocement open-top system [17] uses low-cost materials and fabrication techniques and so is relatively cheap to buy. Data show that the system cost is about $350 per installed kilowatt for a 5 kilowatt system. The system includes several large cylinders equipped with cloth filters, security filters, and a complex fabrication procedure for cleaning [17]. Despite the system's low cost, we believe it will have difficulty with air leakage, and thus may be unable to maintain diesel replacement levels of 75 to 85 percent. Although the design of the reactor itself may be of some value in terms of low cost, the total system is unlikely to succeed in the long term. It must be emphasized that in order to minimize leakage and render a field system reliable, the number of joints must be absolutely minimal. Another drawback to this design is that it runs on charcoal; the system's economy and user-friendliness would have been higher had it been designed for wood chips.

Chinese rice-husk gasifiers are also open-top designs [18]. The gasifiers produce gas that is high in tar because pyrolysis is the dominant process in these systems. Rice-husk char is far less reactive (by about one-tenth to one-twentieth) than wood charcoal, so char conversion and tar cracking normally do not occur in rice-husk gasifiers as they do in wood gasifiers.

Wood gasifiers have also been installed in Indonesia [19], but details of their performance have not been reported. Most of those systems are either pilot plants or demonstration projects. One pilot plant, which has a capacity of 80 kilowatts, produces gas that circulates upward through the biomass (in an updraft mode) and is reported to run 12 hours a day, generating electricity up to 65 kVA [19]. The system operates with a diesel engine, although the diesel replacement percentages have not been published. Power generation equivalent to 100 kilowatt systems is thought to be uneconomic.

In recent times, Imbert gasifiers with capacities that range from 50 to 500 kilowatts [19, 20] have been installed in several countries. In Guyana two gasifiers, with capacities of 60 and 125 kilowatts respectively, have run for approximately 12,000 hours in a sawmill; a third gasifier is expected to be added to them, increasing the system's total capacity to 4.8 megawatts. The investment costs of an Imbert plant is said to range from DM3,000 to DM4,000 ($900 to $1,100) per kilowatt of power output. Average wood consumption is about 1.25 kilograms per kWh, indicating an overall efficiency of about 18 percent.

SAFETY AND ENVIRONMENTAL CONSIDERATIONS

Exposure to carbon monoxide in the gas is of little or no concern because pressure in the system is below ambient levels—air can leak into the system, but gas cannot escape to the outside. At points where air leakage occurs and temperatures are high, burn-off at the reactor bottom may occur; in some instances of transient operation, pressure build-up from instantaneous combustion may create a potentially explosive situation. In early trials with closed-top gasifiers, such explosions did occur, but the problem has been eliminated in the current design, where pressure is released at one of the water seals with no untoward effect other than splashing of water.

In pumping applications, tar and dust, which are carried directly into the engine in a dry cooling system, are instead transported by the water of the wet cooling system and deposited on the ground, where they may pose an environmental hazard. Some countries, including India, regulate such discharges, quantifying them in terms of biological oxygen demand (BOD), chemical oxygen demand (COD), phenolic content, and other pollutants. Regulations are more stringent for discharge into inland surface waters than for isolated irrigation fields (see table 3).

The open-top gasifier produces BOD levels of 3.5 milligrams per liter, COD levels of 182 milligrams per liter and phenol levels of 12.0 milligrams per liter. The unacceptably high phenol levels should be reduced by constructing a filter bed, as has been done in China.

THE ECONOMICS OF GASIFIERS

For a technology to be economically viable, the capital cost of equipment as well as operating costs must be kept to an acceptable value. The open-top gasifier has been developed according to those criteria. An in-depth analysis by Reddy et al.

Table 3: Standards for treated industrial effluents in India

	Tolerance limits		
Feature	**into inland surface waters** *mg/liter*	**into irrigation lands** *mg/liter*	**into marine coastal areas** *mg/liter*
BOD	30	100	100
COD	250	–	250
Phenols	1	–	5

[21] indicates that decentralized power generation, such as can be obtained with renewable technologies, has a significant economic edge over conventional, centralized technologies.

In the present calculations, construction and installation times are considered to be negligible (less than six months) because the system is mostly made up of prefabricated items and site preparation is minimal. The payback period is determined by computing the extra costs incurred beyond those of a conventional diesel generator and the savings generated from the use of the new technology. Total operating costs include the initial capital outlay as well as subsequent running costs. Included under capital costs are equipment and construction costs (see table 4).

Costs are in dollars and Indian rupees (within parentheses) based on 1991 values. Values are nominal and based on actual numbers in most cases. The choice of 2,500 hours for the 4.4 kilowatt system is derived from the Hosahalli system, although an alternate scenario, which would raise the number of hours of operation per year to about 4,000, is also possible. Fixed costs, operating costs per kWh, and the total cost per kWh were obtained for diesel-only and dual-fuel modes, respectively. The total cost per year is the sum of fixed and operating costs; the total life of the project refers to the life of the building. Thus, replacement times for elements are based on the lifespan of the individual components. Present values are computed for all costs involved, with discount rates of 6 percent and 12 percent.

Maintenance costs

Maintenance costs for the diesel generator in both diesel-only and dual-fuel modes are the same, except for the cost of lubrication oil, which is higher in the dual-fuel mode. For 3.5 kilowatt systems in the dual-fuel mode, lubrication oil should be changed after every 250 hours of operation, as opposed to once every 500 hours in the diesel mode. For 100 kilowatt systems, the oil should be changed every 100 and 200 hours, respectively. From 3 to 6 percent of the total cost of a

Table 4: Cost parameters for 4.4 and 96 kilowatt systems[a]

		3.5 kW	80 kW
P	Load	3.5 kW	80 kW
N	Number of hours per year	2,500	5,000
R	Diesel replacement	70 percent	70 percent
C_{eg}	Cost of engine gen-set	170 (34,000)	15,000 (300,000)
C_w	Cost of wood per tonne	5 (100)	15 (250)
C_{lub}	Cost of lubricating oil per liter	3 (60)	3 (60)
C_d	Cost of diesel per liter	0.28 (5.60)	0.28 (5.60)
C_{be}	Cost of building (diesel mode)	500 (10,000)	1,500 (30,000)
C_{bg}	Cost of building (dual-fuel mode)	750 (15,000)	5,000 (100,000)
C_g	Cost of reactor	150 (3,000)	1,000 (20,000)
C_k	Cost of cooling and cleaning system	225 (4,500)	7,500 (150,000)
C_{cs}	Cost of control system, if any	–	10,000 (200,000)
$L_{dg,r}$	Life of engine gen-set, reactor hours	25,000	25,000
$L_{cc,b}$	Life of cooling and cleaning system building, years	10, 40	10, 40
I_{dg}	Interest rate on all elements	12/6 percent	12/6 percent
S	Salvage value	10 percent	10 percent
m_{dg}	Maintenance of diesel gen-set	10 percent	10 percent
m_{gs}	Maintenance of gasifier system	5 percent	5 percent
m_b	Maintenance of building	5 percent	5 percent
S_{fc}	Specific fuel consumption diesel	280 g/kWh	280 g/kWh
S_{lc}	Specific lubricating oil consumption	1.36 g/kWh	1.36 g/kWh
T_c	Lubricating oil tank capacity	5 liters	5 liters
S_{wc}	Specific wood consumption	1.3 kg/kWh	1.3 kg/kWh
LC_{dg}	Labor cost in diesel mode	Rs2/hour	Rs2/hour
LC_{gs}	Labor cost in dual-fuel mode	Rs4/hour	Rs4/hour

a. The costs are in 1991 U.S. $ (1 $ = Rs 20) and Indian rupees (in parentheses).

system generally goes for maintenance, depending on the number of hours of operation [20].

Operating costs
Experience reveals that a gasifier engine–generator system in the dual-fuel mode can be operated by one skilled person and one semiskilled helper. Since operation-

Table 5: Economics of diesel and dual-fuel modes at 6 percent interest rate[a]

Rated capacity *kilowatts*	4.4	96
	dual (diesel fuel)	dual (diesel fuel)
Load in kilowatts	3.5 (*3.5*)	80 (*80*)
Life of the plant in years	40 (*40*)	40 (*40*)
Capital cost per kilowatt engine gen-set + reactor	542.6 (*500*)	200 (*187.5*)
Present value of replacement of engine gen-set + reactor	566.51 (*522.15*)	514.35 (*232.2*)
Capital cost per kilowatt cooling, cleaning, and control system	64.3 (*0*)	218.75 (*0*)
Present value of replacement of cooling, cleaning, and control system	67.1 (*0*)	228 (*0*)
Total capital cost per kilowatt	**1,411.7 (*1,165*)**	**1,224.1 (*688.5*)**
Life cycle fuel cost per kilowatt	1,373.6 (*3,439.1*)	3,185.6 (*6,878.2*)
Life cycle operating and operating and maintenance cost per kilowatt	3,037.7 (*1,879*)	945 (*744.4*)
Fixed cost per kWh	0.033 (*0.0255*)	0.01905 (*0.00825*)
Operating cost per kWh	0.1275 (*0.152*)	0.0545 (*0.0735*)
Total cost per kWh	**0.1595 (*0.1775*)**	**0.0735 (*0.1135*)**

a. Italic figures denote the engine running on diesel alone. All cost figures are in U.S. dollars.

al costs do not increase during dual-fuel operation, the relative economics do not change. Any increase in operator costs, however, would lead to an increase in the cost of power. Consider, for example, a 3.7 kilowatt system run by two operators.

While use of a gasifier-based system raises the fixed costs of power generation considerably, reduced operating costs (made possible by the reduction in diesel fuel) more than offset fixed cost increases (see table 5). Indeed, the total cost of installation compares favorably with values quoted for coal-based thermal power generation systems because the economies of scale normally expected in megawatt-class power stations is obtained even at these power levels. The payback period depends both on the number of hours a system operates in a year, which is a key parameter, and the prevailing discount rates (see figure 13). If the number of hours of operation can be increased, the payback period will be lowered.

Larger systems (those with power outputs greater than 60 kilowatts), show a strong relation between payback, operating hours and the cost of wood chips (see figures 14 and 15). It is clear that as the interest rate increases from 6 to 12 per-

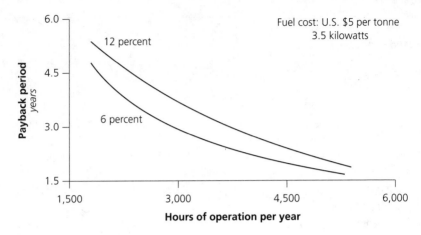

FIGURE 13: Payback period versus hours of annual operation for a 3.5 kilowatt gasifier system.

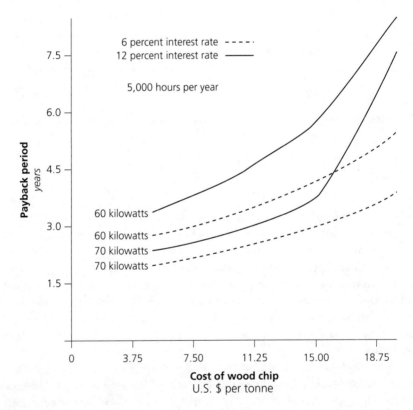

FIGURE 14: The variation of the payback period with the cost of wood chips for 5,000 hours of operation with 6 percent and 12 percent interest rates.

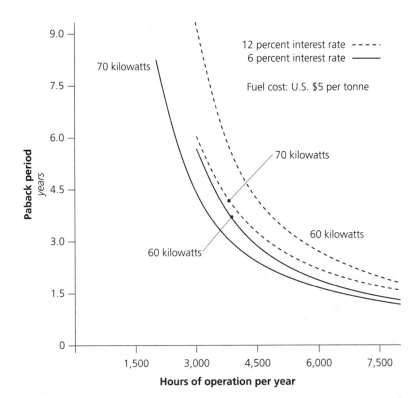

FIGURE 15: Payback period versus number of hours operated at 60 and 70 kilowatts power for a fuel cost of $5 per tonne of wood chips, at 6 and 12 percent interest rates.

cent, the payback period increases by 20 percent. Yet, increasing the number of operating hours, from 2,000 to 6,000 per year, reduces the payback period by a factor of about five. Although a fourfold increase in the cost of wood—from $5 (Rs 100) per tonne to $20 (Rs 400) per tonne—increases the payback period from 2.1 to 3.7 years, overall costs are affected much more by the number of operating hours than by the price of wood chips.

In 3.5 kilowatt systems, the difference in the cost per unit of electricity in diesel and dual-fuel mode is small, considering 2,500 hours of annual operating hours. Although operating costs are higher in the smaller system than they are in the larger systems, most of the extra cost can be attributed to the need for additional labor costs; trimming these costs would reduce the cost of unit electricity generated.

The value added to a system in the agricultural sector is very significant. Farmers in Karnataka benefited greatly from the 3.7 kilowatt water pumping system during the 1991 Persian Gulf crisis, during which diesel fuel was rationed. Thus, despite a doubling in labor costs, operating costs remained low. In addition, by doubling the operating hours from 2,500 to 5,000 per year, the payback period was reduced by about 40 percent. In such situations the economics of the 3.5 kilowatt system begins to match those of the larger system.

Results indicate that the cost of energy in the dual-fuel mode is substantially lower than in the diesel-only mode. The cost of energy provided by the state is around $0.065 (Rs 1.25) per kWh (which is generally subsidized). Thus the cost of energy from an engine–generator system in the dual-fuel mode is about 10 percent higher than what is charged by the state electricity board. The difference, however, is insignificant because the user is no longer dependent on the vagaries of a state-provided supply, especially when the value added to an industrial operation by the availability of energy is considerable.

CONCLUSIONS

Water pumping and electricity generation at various power levels are among the many applications that can be met with gasifier-based energy sources. Water-pumping demands are met more easily than the electric power demand because the load problems are less complex. But electricity can be supplied to an industry that has from 25 to 35 percent load variation by equipping a diesel engine in the dual-fuel mode with an "A" class governor, permitting strict control on frequency and load demands. Although no existing gasifier can generate more than a few hundred kilowatts of power, we see no particular difficulty in designing an efficient gasifier for megawatt power levels. The limiting factor, if one exists, will not be design-related but rather will be the availability of biomass, which must be harvested without undue environmental impact.

Economic considerations show that for the range of power levels now available, investing in gasifier-based power generation systems is a commercially attractive proposition, provided that the system is heavily used. Other beneficial aspects include the value of being able to install pumping or electricity generation systems in regions where the grid power supply is not reliable, which reduces investment in the system, lowering payback to a year or two for larger power systems. Thus, having looked at the system as a whole and subjected each element to scientific, technical, economic and aesthetic scrutiny, we see a bright future for the gasification technology around the world—if not immediately, then in the near future.

ACKNOWLEDGMENTS

The data presented in this chapter were obtained from wood gasifier projects administered and financed partly by the Karnataka State Council for Science and Technology in Bangalore and by the Department of Nonconventional Energy Sources in New Delhi. We thank them for their support. We would also like to thank the chapter's reviewers for their many useful comments. Finally, we are especially grateful to Amulya K.N. Reddy and P. Balachandra of the Indian Institute of Science for help with economic calculations.

WORKS CITED

1. Solar Energy Research Institute, Golden , Colorado, U.S. (SERI). 1979
   Generator gas—the Swedish experience from 1939–1945.

2. Kaupp, A. and Goss, J.R. 1984
   Small scale gas producer engine systems, German Appropriate Technology Exchange, Eschborn, Germany.

3. Coovattanachai, N., ed. 1986–1990
   Rural Energy, RAPA Bulletin, especially pages 12–51, 1990/1, FAO Office, Bangkok.

4. Reed, T.B. and Markson, M. 1982
   Biomass gasification reaction velocities, in R.P. Overend, T.A. Milne, and L.K. Mudge, eds., *Fundamentals of thermochemical biomass conversion,* 951–965, Elsevier Applied Science, New York.

5. Dasappa, S., Reddy, V., Shrinivasa, U., and Mukunda, H.S. 1985
   Experience with gasifiers for 3.7 kW engines, *Ambio* 14:275–279.

6. Groeneveld, M.J. 1980
   The co-current moving bed gasifier, Krips Repro Meppel, Twente University, Gröningen, the Netherlands.

7. Walawender, W.P., Chern, S.M., and Fan, L.T. 1985
   Wood chip gasification in a commercial downdraft gasifier, in R.P. Overend, T.A. Milne, and L.K. Mudge, eds., *Fundamentals of thermochemical biomass conversion,* 911–922, Elsevier Applied Science, New York.

8. Huff, E.R. 1982
   Effect of size, shape, density, moisture and furnace temperature on the burning times of wood pieces, in *Fundamentals of thermochemical biomass conversion: an international conference,* Solar Energy Research Institute, Golden, Colorado.

9. Mukunda, H.S., Paul, P.J., Shrinivasa, U., and Rajan, N.K.S. 1984
   Combustion of wooden spheres—experiments and model analysis, presented at the twentieth symposium (international) on combustion, The Combustion Institute, 1619–1628.

10. Dasappa, S., Shrinivasa, U., Baliga, B.N., and Mukunda, H.S. 1989
    Five-kilowatt wood gasifier technology: evolution and field experience, *Sadhana,* Indian Academy of Sciences proceedings in engineering sciences, 14:187–212.

11. Parikh, P.P., Bhave, A.G., Kapse, D.V., and Shashikantha 1989
    Study of thermal and emission performance of small gasifier dual fuel engine system, *Journal of Biomass,* 19:75–97.

12. Ravindranath, N.H., Dattaprasad, H.L., and Somashekar, H.I. 1990
    A rural energy system based on forest and wood gasifiers, *Current Science* 59:557–560.

13. Ravindranath, N.H., Dattaprasad, H.L., and Mukunda, H.S. 1989
    Village electrification using wood gasifiers, *Proceedings of the first meeting on recent advances in biomass gasification technology,* Bombay, 211–229.

14. Dasappa, S., Shrinivasa, U., Dattaprasad, H.L., and Mukunda, H.S. 1989
    Experience on running gasifier for village electrification, internal report, ASTRA, Indian Institute of Science, Bangalore.

15. Baliga, B.N., Dasappa, S., Shrinivasa, U., and Mukunda, H.S. 1991
    Gasifier based power generation: technology and economics, to appear in *Sadhana,* Indian Academy of Sciences proceedings in engineering sciences.

16. Bokalders, V. and Kjellstrom, B., eds. 1990
    Renewable energy for development, a Stockholm environment institute newsletter, 4:1–24.

17. Graf, U. 1990
Small charcoal gasifiers in projects of technical cooperation, reports 1–3, University of Bremen, Bremen, Germany.

18. Kaupp, A. 1984
Gasification of rice hulls: theory and praxis, German Appropriate Technology Exchange, Eschborn, Germany.

19. Kardona, 1988
The potential of biomass gasifier generator sets as an alternative power plant for rural electrification in Indonesia, Proceedings of the International Conference on Energy from Biomass and Wastes XII, New Orleans, Louisiana, Institute of Gas Technology, Chicago, U.S.

20. Zerbin, W.O. 1984
Generating electricity by gasification of biomass, in A.V. Bridgwater., ed., *Thermochemical processing of biomass,* Butterworths, London.

21. Reddy, A.K.N., Sumithra, G.D., Balachandra, P., and D' Sa, A. 1990
The comparative costs of energy conservation and generation from decentralized and centralized sources, *Economic and Political Weekly,* 2 June, 1201–1216.

17
ADVANCED GASIFICATION-BASED BIOMASS POWER GENERATION

ROBERT H. WILLIAMS
ERIC D. LARSON

A promising strategy for modernizing bioenergy is the production of electricity or the cogeneration of electricity and heat using gasified biomass with advanced conversion technologies.

Major advances that have been made in coal gasification technology, to marry the gas turbine to coal, are readily adaptable to biomass applications. Integrating biomass gasifiers with aeroderivative gas turbines in particular makes it possible to achieve high efficiencies and low unit capital costs at the modest scales required for bioenergy systems. Electricity produced with biomass-integrated gasifier/gas turbine (BIG/GT) power systems not only offers major environmental benefits but also would be competitive with electricity produced from fossil fuels and nuclear energy under a wide range of circumstances. Initial applications will be with biomass residues generated in the sugarcane, pulp and paper, and other agro- and forest-product industries. Eventually, biomass grown for energy purposes on dedicated energy farms will also be used to fuel these gas turbine systems.

Continuing improvements in jet engine and biomass gasification technologies will lead to further gains in the performance of BIG/GT systems over the next couple of decades. Fuel cells operated on gasified biomass offer the promise of even higher performance levels in the period beyond the turn of the century.

INTRODUCTION

Power generation is a route to the modernization of biomass for energy offering opportunities for substantial industrial development before the turn of the century. Already in the United States, installed biomass-electric generating capacity is of the order of 9,000 megawatts-electric (MW$_e$) [1] (see table 1). Much of this capacity was installed as a result of incentives provided by the Public Utility Regulatory Policies Act of 1978 (PURPA), which requires a utility to purchase electricity from cogenerators and other qualifying independent power producers at a price equal to the utility's avoided cost. There is not yet much biomass power generating capacity in the rest of the world, where PURPA-type incentives have not been available.

Table 1: Electricity generating plants burning biomass fuels in the United States as of 1989[a]

State	Number of facilities		Installed capacity MW_e		
	Stand-alone	Cogeneration	Stand-alone	Cogeneration	Total
Alabama	0	15	0	375	375
Arizona	2	0	45	0	45
Arkansas	1	4	2.4	10	12
California	64	30	736	255	991
Connecticut	4	3	155	14	169
Delaware	1	0	13	0	13
Florida	12	15	314	474	788
Georgia	0	5	0	36	36
Hawaii	2	13	70	129	199
Idaho	1	6	0.2	116	116
Illinois	0	1	0	2	2
Indiana	0	7	0	36	36
Iowa	2	1	11	2.2	13
Kentucky	1	1	1	1	2
Louisiana	1	12	11	300	311
Maine	4	22	88	704	792
Maryland	2	2	214	94	308
Massachusetts	2	9	38	252	290
Michigan	3	13	78	247	325
Minnesota	3	23	63	161	224
Mississippi	0	10	0	230	230
Missouri	0	2	0	60	60
Montana	2	17	18	340	358
New Hampshire	3	5	15	65	80
New Jersey	2	0	14	0	14
New York	11	17	154	425	579
North Carolina	3	27	60	351	411
Ohio	1	6	17	90	107
Oklahoma	2	1	8	17	25
Oregon	3	24	69	185	254
Pennsylvania	0	9	0	144	144
South Carolina	1	13	49	46	95
Tennessee	2	12	6	43	49
Texas	1	9	2	146	148
Utah	0	1	0	20	20
Vermont	5	3	80	218	298
Virginia	0	9	0	136	136
Washington	3	11	72	120	192
Wisconsin	5	9	55	117	172
Total	**149**	**367**	**2,459**	**5,962**	**8,421**

a. Based on [59]. The total here is an underestimate because the cited reference is incomplete.

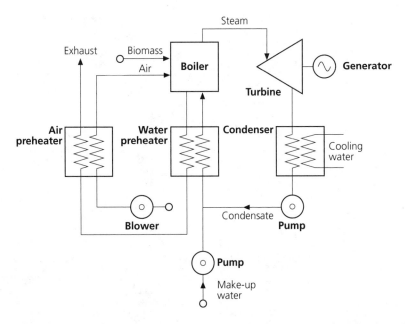

FIGURE 1: Biomass-fired condensing steam-turbine cycle for the production of power only.

Steam-turbine cycle technology

Essentially all biomass power plants today operate on a steam–Rankine cycle, a steam-turbine technology which was initially introduced into commercial use about 100 years ago. The biomass is burned in a boiler producing pressurized steam, and the steam is expanded through a turbine to produce electricity. In the production of power only, a fully condensing turbine is used (see figure 1), while in the production of electricity and heat (cogeneration), a condensing–extraction turbine (see figure 2) or a back-pressure turbine (see figure 3) is commonly used. Boiler systems in common use today include manually fed, brick-lined "Dutch ovens" in which the biomass burns in a pile (pile burners). More sophisticated units include automatic (stoker-fed) systems in which the biomass burns on a stationary or moving grate (grate burners), suspension burners in which the biomass burns while free-falling, and recently introduced fluidized-bed units in which the biomass burns as it is fluidized from below by a continuous jet of combustion air.

In general, the higher the peak temperature of the working fluid, the higher the thermodynamic efficiency of a power-generating cycle. In the case of steam turbine cycles, the higher the peak temperature and pressure of the steam working fluid, the more sophisticated and costly the equipment (e.g., at higher pressures and temperatures, higher quality steels are needed, water purity must be higher, etc.). Biomass-fueled steam turbine power units operate with steam conditions that are far more modest than those used in large, modern electric-utility coal-fired steam-electric systems. For example, the majority of the 94

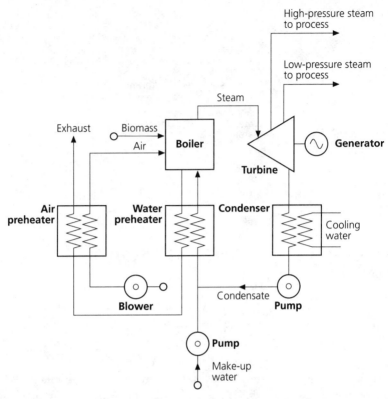

FIGURE 2: Biomass-fired condensing-extraction steam turbine (CEST) cycle with two extractions of steam for process use.

well-documented biomass plants in California (see table 1) operate with a steam pressure and temperature of about 6 megapascals and 480°C [2], compared to typical steam pressures of 10 to 24 megapascals and temperatures of 510 to 540°C in modern utility-scale coal plants [3]. Most biomass plants in developing countries are less sophisticated than those in California [4].

The modest steam conditions in biomass plants arise primarily because of the strong scale-dependence of the capital cost ($ per kW) of steam turbine systems—the main reason coal and nuclear steam-electric plants are built big. Biomass plants are usually of modest scale (less than about 100 MW$_e$) because of the dispersed nature of biomass supplies, which must be gathered from the countryside and transported to the power plant. If bio-electric plants were as large as coal or nuclear power stations (500 to 1,000 MW$_e$), the cost of delivering the fuel to the plant would often be prohibitive. To help minimize the dependence of unit cost on scale, vendors use lower grade steels in the boiler tubes of small-scale steam-electric plants and make other modifications that reduce capital cost, but also require more modest steam temperatures and pressures, thereby leading to reduced efficiency. Plants operating in California have effi-

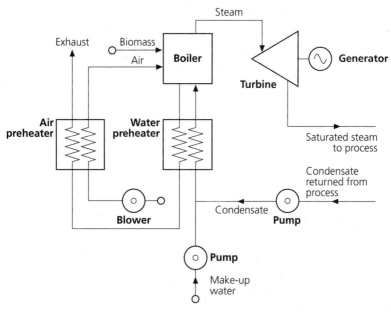

FIGURE 3: Biomass-fired back-pressure steam turbine for cogeneration of heat and electricity using biomass as fuel.

ciencies of 14 to 18 percent, compared with 35 percent for a modern coal plant.[1]

Such low efficiencies explain the reliance of the biomass power industry in the United States, as well as biomass power facilities elsewhere in the world, on low-, zero-, or negative-cost biomass feedstocks (primarily residues of agro- and forest product–industry operations and urban refuse). Largely as a result of the growth in biomass-based power generation, the supply of such feedstocks is dwindling in some parts of the United States today, although there are still significant unused or underused supplies of such feedstocks in much of the rest of the world. Once such low-cost feedstocks are fully used, however, continued biomass power expansion will require the use of higher cost feedstocks, such as residues that are hard to recover and biomass that is grown for energy on dedicated energy farms. In order to make higher cost biomass resources economically interesting for power generation, it is necessary to have technologies that offer higher efficiency and lower unit capital cost at modest scale.

One technological initiative in the United States aimed at improving the economics and efficiency of utility-scale steam cycle systems would use whole trees as fuel rather than more costly forms of biomass (e.g., woodchips). The whole-tree burner concept, developed by Energy Performance Systems, Inc., of Minne-

---

1. In this paper, efficiencies and fuel heating values are presented on a higher heating value basis.

sota, has yet to be commercially demonstrated, but an assessment of the technology was recently completed for the Electric Power Research Institute (EPRI) [5]. The EPRI report projects an efficiency and installed capital cost for a 100 MW$_e$ plant employing a reheat-steam cycle to be 34 percent and $1,365 per kilowatt-electric (kW$_e$).[2] The size was selected as a likely initial size for central station utility applications. The reheat-steam cycle is a more sophisticated cycle than is typically found in existing biomass power plants, nearly all of which are substantially smaller than 100 MW$_e$. Scale economy gains at the 100 MW$_e$ size may permit economical use of the more complex cycle.

In comparing the whole-tree technology to a conventional 100 MW$_e$ biomass reheat-steam system burning wet wood chips (assuming low-cost wood recovered from natural forests in the United States in both cases), the report estimated a 10 to 30 percent lower electricity production cost for the whole-tree case, because of the higher efficiency, lower capital cost, and lower fuel cost.

The higher efficiency of the whole-tree approach compared with conventional biomass-power technology would make the use of higher cost biomass feedstocks more economical at the 100 MW$_e$ scale. The biomass power market for installations at the relatively large scales needed for whole-tree burner technology may be quite limited, however. Moreover, the technology is not designed for use with the large quantities of biomass residues that are generated in relatively smaller concentrations at industrial sites today, e.g., waste wood in the forest products industry. And finally, the technology is not likely to improve much over time beyond what has been proposed, since the steam-turbine cycle is a mature technology for power generation. The efficiency of modern fossil fuel–fired, steam-electric power plants has not increased since the late 1950s (see figure 4), when peak steam temperatures of the order of 540°C were reached. While it is technically feasible to increase the peak steam temperatures further, doing so is probably not worthwhile because of the higher capital costs involved [6].

Gas-turbine cycle technology

A promising alternative to the steam-turbine cycle for biomass power generation is a set of biomass-integrated gasifier/gas turbine technologies that offer the potential for low unit capital cost and high thermodynamic efficiency at modest scales. This set of technologies involves marrying advanced Brayton cycle (gas turbine) power-generating or cogenerating cycles, which have already been developed for natural gas and clean liquid fuel applications, to closely coupled biomass gasifiers, which can be based to a large extent on gasifiers already developed for using coal in gas-turbine power cycles.

In contrast to steam-cycle technology, the unit capital costs of gas-turbine systems are relatively low and insensitive to scale. Thus, from a capital cost perspective, the gas turbine is an interesting candidate for biomass-based power generation.

---

2. In this paper, costs and prices are presented in 1989 U.S. dollars.

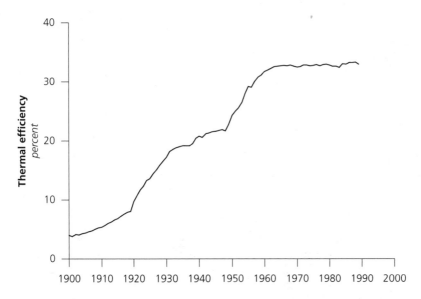

FIGURE 4: Historical trend in the average efficiency of electricity generation in central-station thermal power plants in the United States [6].

The gas turbine is also a good candidate for achieving higher thermodynamic efficiency because the peak cycle temperature of modern gas turbines (about 1,260°C for the best gas turbine for stationary power applications on the market) is far higher than that for steam turbines (about 540°C). This inherent advantage of the gas turbine is not exploited in the simple-cycle gas turbines used for electric utility peaking service (which are typically less efficient than steam-electric plants) because the hot turbine exhaust gases (425 to 540°C) from peaking units are discharged to the atmosphere, thus wasting large quantities of potentially useful energy. However, the inherent thermodynamic advantages of the gas turbine can be exploited by using its exhaust heat in various ways. The exhaust can be used, for example, to produce steam in a heat recovery steam generator (HRSG), which can then be used for industrial process needs in a cogeneration configuration (see figure 5) or to produce more power. As will be discussed below, with present technology, gas-turbine cycles with turbine exhaust heat recovery can achieve much higher thermodynamic efficiencies than are possible with steam turbine cycles.

Moreover, unlike the situation with steam turbines (see figure 4), gas turbines are being steadily improved. There have been continual advances in turbine blade materials and turbine cooling technologies for jet engines, and, as a result, there has been a 20°C average annual increase in the state-of-the-art turbine inlet temperature for jet engines and a continual increase in engine efficiency since the end of World War II, a trend that is expected to continue (see figure 6) [7, 8]. Such improvements are largely a result of U.S. Department of Defense support for re-

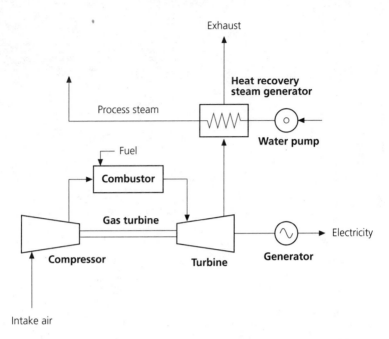

FIGURE 5: Simple-cycle gas turbine configured for cogeneration applications. Fuel burns in air pressurized by a compressor; hot combustion products drive a turbine; hot turbine exhaust gases are used to raise steam in a heat recovery steam generator (HRSG) for process applications.

search and development (R&D) on jet engines for military aircraft applications, which has averaged about $0.5 billion per year over the past decade [6].

There are two general classes of gas turbines that can be used for power generation: heavy-duty industrial turbines designed specifically for power generation and lightweight, compact, aeroderivative gas turbines. While eventually ongoing improvements in jet engine technology will be incorporated into industrial turbines, these advances are automatically incorporated into aeroderivative machines. Thus emphasis on aeroderivative turbines provides a direct, low-cost way to exploit advances in aircraft engines for stationary power applications.

Aeroderivative turbines also offer the advantages of high efficiency and low unit capital costs at the modest scales that will characterize much of the biomass energy market—for both cogeneration and power-only applications. Moreover, the compact, modular nature of aeroderivative turbines facilitates maintenance. When an aeroderivative engine fails, it can be replaced quickly by a spare trucked or flown in from a centralized lease-pool maintenance facility, at which the failed engine would be repaired. This feature should make aeroderivatives especially attractive for developing countries, where sophisticated onsite maintenance is often not readily available. The required maintenance infrastructure is largely already in place in those countries that have their own commercial airlines.

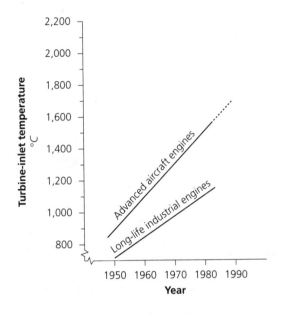

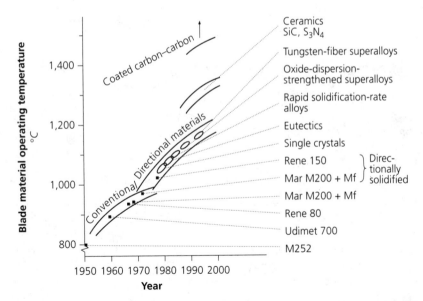

FIGURE 6: *The trend in turbine-inlet temperatures for advanced aircraft jet engines and long-life industrial turbines (top) and turbine blade material operating temperatures (bottom)[7]. Note: When an aircraft engine is modified for stationary applications, the rated turbine-inlet temperature is reduced about 110°C to promote long-life operation.*

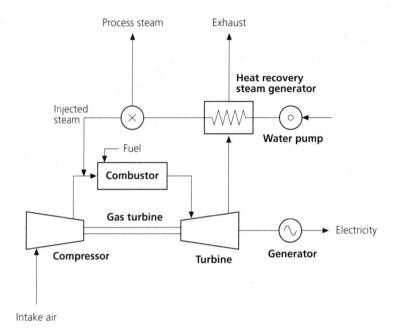

FIGURE 7: Steam-injected gas turbine (STIG). Similar to the simple-cycle gas turbine used for cogeneration, except that steam not needed for process use is injected into the combustor and at points further down the flow path for increased power output and higher electrical efficiency.

EFFICIENT GAS-TURBINE CYCLE OPTIONS

Although no gas-turbine cycle is commercially available for biomass applications, several cycles of interest are commercially available or are under development for applications involving high-quality fluid fuels.

Steam-injected gas turbine

One commercially available aeroderivative turbine cycle is the steam-injected gas turbine (STIG), (see figure 7). As in the simple-cycle gas turbine used for the co-generation of electricity and steam for process use (see figure 5), steam is produced in a STIG cycle from the gas turbine exhaust heat using an HRSG. But in this case steam not needed for process use is injected back into the gas turbine combustor (and at further points along the gas flow path), where it is heated to the turbine inlet temperature and then passed through the turbine. With steam injection, the gas turbine produces more power at higher electrical efficiency. STIG technology makes it possible (if the extra electricity produced when the steam load decreases can be used) to overcome the problem of poor part-load performance that has limited the use of simple-cycle gas turbines to cogeneration applications involving constant steam loads [9].

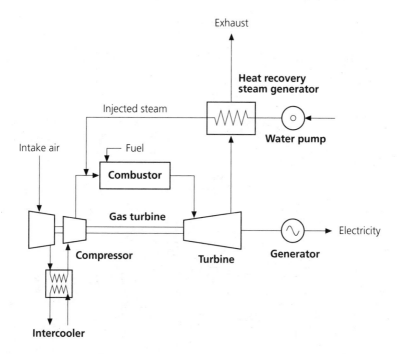

FIGURE 8: Intercooled steam-injected gas turbine (ISTIG). Similar to STIG, except that an intercooler between compressor stages leads to much higher efficiency and much larger electrical output, because less compressor work is required and the turbine can operate at a higher turbine-inlet temperature owing to improved cooling of the turbine blades with air bled from the compressor.

More than 30 STIG units in the U.S., which can burn high-quality gaseous and liquid fuels, are either operating, under construction, or on order [6]. The largest STIG unit commercially available is based on the General Electric (GE) LM-5000 turbine (derived from the jet engine used in the Boeing 747, the DC-10 Series 30, and the Airbus 300). As a simple cycle, it produces 33 MW$_e$ at 33 percent efficiency operated on natural gas fuel. With full steam injection, this engine produces 51 MW$_e$ at 40 percent efficiency. Turnkey STIG units of this capacity packaged on a skid and without a building are commercially offered for about $700 per kW$_e$ [10].

Intercooled steam-injected gas turbine

An advanced version of the STIG is the intercooled steam-injected gas turbine (ISTIG), (see figure 8). The installed cost of a 47 percent efficient, natural gas–fired 114 MW$_e$ ISTIG (essentially the 51 MW$_e$ STIG unit described above, but modified with compressor intercooling) would be about $500 per kW$_e$ [10]. Although ISTIG is not commercially available, it could be brought to market in 3 to 5 years, if there were sufficient commercial interest [11].

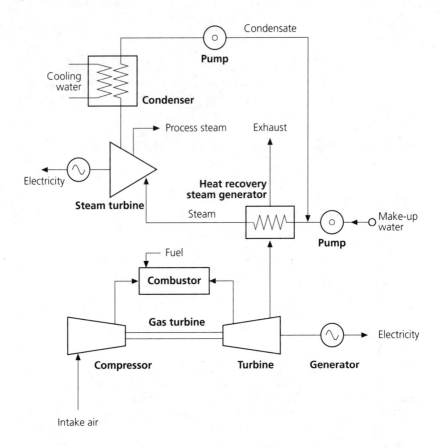

FIGURE 9: Gas turbine/steam turbine combined cycle (CC). Similar to the simple cycle gas turbine used for cogeneration, except that steam from the HRSG is used to produce extra power in a condensing steam turbine, from which some steam might be bled for process applications.

Combined cycle

An alternative way to use the steam recovered in the HRSG for power generation is in a separate steam turbine bottoming cycle. In this case the overall system is called a gas turbine/steam turbine combined-cycle (CC), (see figure 9). The combined-cycle is the most energy-efficient, power-generating cycle on the market today and is the generating technology of choice in many utility and independent power markets with natural gas firing.

Most combined cycles are based on heavy-duty industrial rather than on aeroderivative turbines. An important distinction between the two turbine types is that the combustors of the latter operate at much higher pressures (25 atmospheres or more, compared with 12 to 16 atmospheres for heavy-duty industrial turbines). High pressures are needed to optimize the performance of jet engines at today's high turbine inlet temperatures. Heavy-duty industrial turbines are usually designed instead for optimal performance in the combined-cycle mode.

For a given turbine inlet temperature, the turbine exhaust of heavy-duty industrial turbines is hotter and capable of producing more steam than is possible with aeroderivatives. Typically, the steam turbine bottoming cycle provides about one third of the total output of these combined cycles. In light of the strong scale economies of steam-turbine cycles, combined cycles based on heavy-duty industrial turbines are not the best candidate engines for applications at the modest scales needed for biomass.

Ordinarily, the relatively low turbine exhaust temperatures of the aeroderivatives makes them poor candidates for combined-cycle configurations, suggesting that the various steam-injected cycles would offer more favorable economics at the modest scales needed for biomass. This situation may change, however, with a new generation of aeroderivatives coming onto the market in the 1990s [12].

For example, the GE LM-6000, which will enter commercial service in 1992, will produce 42.4 MW$_e$ at a simple-cycle efficiency greater than 36 percent on natural gas and have an estimated equipment price of $250 per kW$_e$ [13],[3] which is much less than the $400 per kW$_e$ price for the most efficient (33 percent) aeroderivative on the market today [14]. Combined cycles based on the LM-6000 are expected to produce 53.3 MW$_e$ at an efficiency of 48 percent, making them as efficient as the most efficient combined cycle on the market [15]. The LM-6000 combined cycle might be economically competitive with much larger combined cycles based on heavy-duty industrial turbines, despite the scale-economy problem of the steam turbine bottoming cycle, not only because the cost of the gas turbine is so low, but also because the steam turbine accounts for such a modest fraction (one-fifth) of the total output.

The air bottoming cycle (ABC), (see figure 10) is an alternative promising way to recover exhaust heat. In the ABC the air working fluid is heated via a heat exchanger with exhaust heat from the gas turbine topping cycle. While not quite as efficient as a gas turbine/steam turbine combined cycle, a combined cycle involving the ABC would be much simpler (e.g., it requires no boiler) and more rugged, and it is expected to be much less costly to build, operate, and maintain [16]. Because it involves no advanced technology, the ABC could be developed and commercialized quickly.

Prospects for continuing improvements in gas turbine technology

The performance of gas turbines is expected to improve considerably in the decades ahead, both because of continuing improvements in jet engine technology (see figure 6) and because there are many untapped opportunities for improving the performance of stationary turbines that are not relevant for jet engines.

---

3. In converting a jet engine to an aeroderivative gas turbine, the fan is removed, and, ordinarily, the thrust-producing nozzle is replaced by a power turbine, so that the system produces net power instead of thrust. The extraordinarily low expected price for the LM-6000 arises because this engine is derived from a high-bypass ratio jet engine, in which the final expansion of combustion products is through a turbine instead of a thrust-generating nozzle. In the jet engine configuration, the output of the turbine drives the compressor and bypass fan. But when the fan is removed for stationary power applications, the unit produces net power without the addition of a costly power turbine.

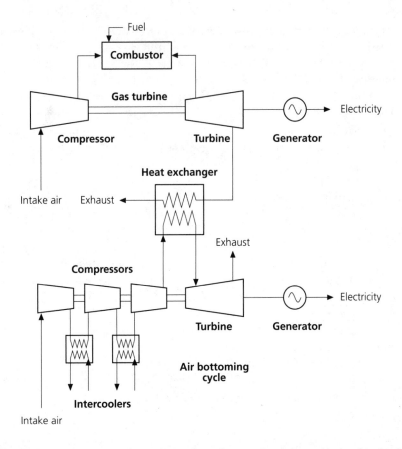

FIGURE 10: Air bottoming cycle (ABC). Similar to the gas turbine/steam turbine combined cycle, except that the exhaust heat from the gas turbine topping cycle is recovered not as steam but by heat transfer via a heat exchanger to the air working fluid of an indirectly heated air bottoming cycle.

Besides adapting aircraft engine blade cooling advances to stationary power applications, it would be desirable to replace air as the turbine blade coolant with steam, which offers several advantages. One is that steam, having a higher heat-carrying capacity per unit mass, is a more effective coolant than air. Also, steam provides the flexibility to choose higher pressures for cooling, thus making it feasible to achieve higher coolant velocities and therefore to provide more intensive heat removal from the components being cooled. In addition, the temperature of the steam used for cooling might be lower than that of air. And finally, when air is replaced by steam, the compression work requirements for the coolant become negligible. Despite such advantages, little progress has been made in steam cooling to date, in large part because steam cooling is not relevant to aircraft engines, because it is not practical to carry large quantities of water aboard airplanes.

Other possible cycle modifications could lead to improved performance over what can be achieved with combined cycles and ISTIG. One such modification would be to add a "reheat" combustor ahead of the final expansion stage in the turbine. Adding reheat increases not only turbine output but also efficiency, because with reheat there is an increase in the average temperature at which heat is added to the cycle through fuel combustion. Reheat is feasible because the high air-to-fuel ratio of modern gas turbines leaves plenty of oxygen in the primary combustor exhaust for reheat firing. The Pacific Gas & Electric Company has estimated that adding a reheat combustor to the ISTIG unit described above would increase output and efficiency on natural gas fuel from 114 MW_e and 47 percent to 185 MW_e and 49 percent [17].

FIRING GAS TURBINES WITH LOW-QUALITY FUELS

Gas turbines cannot be fired directly with biomass, because the biomass combustion products would damage the turbine blades. However, by first gasifying the biomass and cleaning the gas before combustion, it is feasible to operate gas turbines with biomass fuels.

Coal-integrated gasifier/gas turbine technologies

While little attention has been given to biomass use in gas turbines, many hundreds of millions of dollars of public and private-sector investment funds have been committed in the United States, Europe, and Japan to R&D efforts aimed at marrying the gas turbine to coal, through the use of coal-integrated gasifier/gas turbine (CIG/GT) systems. These efforts have been motivated in part by the large thermodynamic advantages offered by the gas turbine for power generation and the desire to exploit these advantages with coal (which is much more abundant than oil and natural gas), and in part by the prospect that the burning of coal can be accomplished with much less environmental damage through gasification than with alternative approaches.

One noteworthy development has been the successful construction and operation of a 94 MW_e combined-cycle power plant coupled to a Texaco oxygen-blown, entrained-flow coal gasifier with cold-gas cleanup (scrubber) at Cool Water, California. The Cool Water facility was operated from 1984 to 1989 by the Southern California Edison Company as part of a joint industrial effort involving the Electric Power Research Institute (EPRI), the Bechtel Corporation, the General Electric Company, Texaco, and a Japanese consortium, the Japan Cool Water Program. The Cool Water project demonstrated that it is technically feasible to operate a gas turbine on gasified coal with very low emissions [18]. Similar systems for marrying coal to the gas turbine through the use of oxygen-blown coal gasifiers have been developed by Dow and Shell. Dow is operating a 160 MW_e plant at its Louisiana Division in Plaquemine, Louisiana [19]. Shell, after operating a pilot plant at Deer Park, Texas, since mid-1987, is building a 250 MW_e commercial-scale plant in Buggenum, the Netherlands [20].

A commercial-scale (600 MW_e) plant using a Texaco gasifier and the best available combined-cycle would be cleaner than and roughly cost-competitive with a conventional, coal-fired, steam-electric power plant equipped with flue gas

Table 2: Estimated installed capital cost for IG/STIG and IG/ISTIG power plants fueled with coal and biomass
$ per kW_e

	CIG/STIG[a]	BIG/STIG[b]	CIG/ISTIG[a]	BIG/ISTIG[b]
I. Process capital cost				
Fuel handling	44.4	44.4	41.2	41.2
Blast air system	15.1	15.1	10.8	10.8
Gasification plant	180.5	180.5	93.3	93.3
Raw gas physical cleanup	9.9	9.9	8.6	8.6
Raw gas chemical cleanup	197.4	0.0	169.3	0.0
Gas turbine/HRSG	330.4	330.4	287.7	287.7
Balance of plant				
Mechanical	45.1	45.1	37.0	37.0
Electrical	72.9	72.9	54.3	54.3
Civil	73.5	73.5	68.1	68.1
Subtotal	*969.2*	*771.8*	*770.3*	*601.0*
II. Total plant cost				
Process plant cost	969.2	771.8	770.3	601.0
Engineering home office *10%*	96.9	77.2	77.0	60.1
Process contingency *6.2%*	60.1	47.9	47.8	37.3
Project contingency *17.4%*	168.6	134.3	134.0	104.6
Subtotal	*1,294.8*	*1,031.2*	*1,029.1*	*803.0*
III. Total capital requirement				
Total plant cost	1,294.8	1,031.2	1,029.1	803.0
AFDC[c] *2-year construction, 6% dr*	38.8	30.9	30.9	24.1
2-year construction, 12% dr	77.7	61.9	61.7	48.2
Preproduction costs *2.8%*	36.3	28.9	28.8	22.5
Inventory capital *2.8%*	36.3	28.9	28.8	22.5
Initial chemicals, catalysts	2.8	0.0	2.6	0.0
Land	1.5	1.5	1.5	1.5
Total *6 percent discount rate*	**1,410.5**	**1,121.4**	**1,121.7**	**873.6**
12 percent discount rate	**1,449.4**	**1,152.4**	**1,152.5**	**897.7**

desulfurization [6]. Because electricity produced with this technology would not be substantially less costly than that from conventional plants, this technology might not be widely adopted by utilities. However, it may be feasible to reduce costs with alternative, simpler CIG/GT systems.

A 1986 study for the U.S. Department of Energy (DOE) by the General Electric Corporate R&D Center analyzed various alternative CIG/GT systems to assess how the economics could be improved relative to oxygen-blown gasifier/combined cycle systems [21]. This study identified the following two promising gasification strategies for reducing capital costs and improving the conversion efficiency:

1. Replacing the oxygen-blown, entrained-flow gasifier with an airblown, dry-ash, fixed-bed gasifier, thus eliminating the need for the costly and scale-sensitive oxygen plant.

2. Employing hot-gas desulfurization instead of a scrubber (cold-gas cleanup) for sulfur removal, thus improving efficiency.

Moreover, the GE study found that by coupling these gasification strategies to the use of high performance aeroderivative gas turbines it would be feasible to achieve very attractive economics at modest scale. Specifically, the GE study found that the gasifier/gas turbine system offering the best overall performance is an airblown, coal-integrated gasifier/intercooled steam-injected gas turbine (CIG/ISTIG). It was estimated that a 109 MW$_e$ unit would have an overall coal-to-busbar efficiency of 42.1 percent at an installed cost of less than $1,200 per kW$_e$ (see table 2). This finding is consistent with other analyses by the U.S. DOE [22], which show that over a wide range of installed capacities, CIG/GT systems involving airblown gasifiers with hot-gas cleanup would be substantially less costly than those involving oxygen-blown gasifiers with cold-gas cleanup (see figure 11). The prospective performance of these airblown gasifier-based CIG/GT systems is expected to be markedly better than that of conventional coal-fired steam-electric plants with flue-gas desulfurization in the United States—34.6 percent efficiency for a plant having two 500 MW$_e$ units costing more than $1,400 per kW$_e$ (see note

---

Table 2 notes:

a. The CIG/STIG plant consists of two 50.5 MW$_e$ STIG units, each coupled to a Lurgi Mark IV dry-ash, air-blown, fixed-bed gasifier; the heat rate is 10.11 megajoules per kWh. The CIG/ISTIG plant consists of a single 109.1 MW$_e$ ISTIG unit, coupled to a single dry-ash, air-blown, Lurgi Mark IV fixed-bed gasifier; its heat rate is 8.55 megajoules per kWh. Costs were estimated in a study carried out at the GE Corporate R&D Center [21] using accounting rules set forth in the EPRI *Technical Assessment Guide* [60].

b. The output, performance, and costs of the biomass versions of these systems are estimated by starting with the coal systems and modifying them to account for the major differences arising from operation on biomass. The biomass gasification efficiency is assumed to be the same as the coal gasification efficiency. One difference is that only about 40 percent as much high pressure steam is needed to gasify 1 gigajoule of biomass as 1 gigajoule of coal, and the steam not needed for gasification can be injected into the turbine. However, this is not likely to have a significant effect on overall performance, since the injection of high-pressure steam into the combustor gives rise to approximately the same mass flow through the turbine and would require the same steam heating in the combustor as injection into the gasifier. An important difference, however, is that some low-pressure steam needed for the sulfur recovery unit with coal is not needed in the biomass systems. Here it is assumed that this low-pressure steam is injected into the turbine to increase power output and efficiency. As a result, the output and heat rate of the BIG/STIG are 103.0 MW$_e$ and 9.92 megajoules per kWh, while the corresponding quantities for BIG/ISTIG are 111.2 MW$_e$ and 8.39 megajoules per kWh, respectively. It is assumed that BIG/STIG (BIG/ISTIG) costs are the same as CIG/STIG (CIG/ISTIG) costs, except that the raw gas chemical cleanup phase required for coal would not be needed for biomass, because of its negligible sulfur content

c. AFDC = allowance for funds during construction (i.e., interest charges that accumulate during construction).

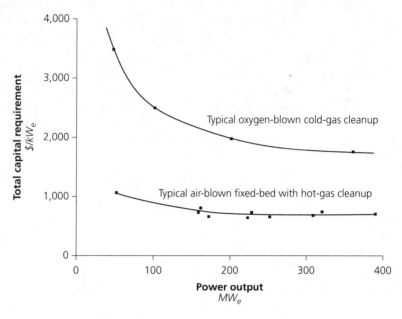

FIGURE 11: Estimated installed capital costs for CIG/GT power systems [22]. The upper curve is for typical systems involving oxygen-blown gasification with cold-gas cleanup, while the lower curve is for typical systems involving fixed-bed, air-blown gasifiers with hot-gas cleanup.

a, table 3). Not only does this alternative technology look interesting for coal, but because it offers the prospect of high performance and lower cost at modest scale, the combination of airblown gasifiers and aeroderivative gas turbines also warrants close attention for biomass tapplications.

Biomass-integrated gasifier/gas turbine technologies

Despite their inherent advantages, CIG/GT systems based on airblown gasifiers are not as well developed as systems involving oxygen-blown gasifiers, because the required technology for removing the sulfur from the hot gases exiting the gasifier is not proven at a commercial scale. But hot-gas sulfur cleanup technology would not usually be needed for biomass, because most biomass contains negligible sulfur (see table 4). Furthermore, the higher reactivity of biomass compared with coal makes it easier to gasify (see table 4 and figure 12). These considerations imply that it should be feasible to commercialize biomass versions of airblown gasifier/gas turbine systems more quickly and with less technological effort than the coal versions.

Figure 13 is a schematic representation of a biomass-integrated gasifier/gas turbine system based on the use of a steam-injected gas turbine, a leading first-generation candidate turbine for BIG/GT systems. The principal generic gasifier design options for BIG/GT applications are fixed-bed updraft and fluidized-bed gasifiers(see figure 14). (Fixed-bed downdraft units are less attractive for BIG/GT

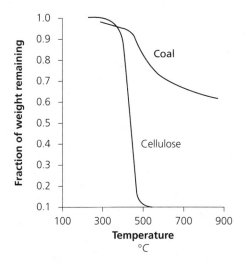

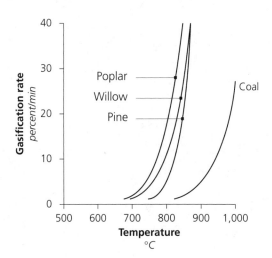

FIGURE 12: Comparison of the gasification characteristics of biomass and coal. The top figure [25] shows the rate at which Wyodak coal and cellulose (which accounts for typically half the weight of biomass) lose weight, or devolatilize, as they are pyrolyzed, i.e., heated in the absence of air. Pyrolysis is one of the main processes involved in converting solid fuels into combustible gases. Nearly complete devolatilization of cellulose occurs at under 500°C. In contrast, only about 40% of coal is devolatilized and only after heating to close to 900°C. The slower weight loss with coal reflects its inherently lower thermochemical reactivity. The much higher fraction of weight remaining even after heating to 900°C reflects the much lower content of volatile components in coal compared to cellulose.

The lower figure [26] shows the rate at which solid carbon that remains after pyrolysis (char) is converted into carbonaceous gases in the presence of steam. Char gasification is another of the major processes involved in converting solid fuels into combustible gases. Because of the higher reactivity of biomass chars, these gasify much more rapidly and at lower temperatures than coal chars. Thus, lower temperatures can be used in biomass gasifiers compared to coal gasifiers to achieve the same level of char conversion to gas.

Table 3: Busbar costs for alternative power technologies
cents per kWh

	CS[a]	BS[b]	CIG/ STIG[c]	BIG/ STIG[d]	CIG/ ISTIG[c]	BIG/ ISTIG[d]
Fuel[e]	$1.061 \cdot P_c$	$1.536 \cdot P_b$	$1.011 \cdot P_c$	$0.992 \cdot P_b$	$0.855 \cdot P_c$	$0.839 \cdot P_b$
Labor[f]			0.30	0.20	0.28	0.19
Maintenance[g]			0.42	0.32	0.33	0.24
Administration[h]			0.14	0.10	0.12	0.09
Total fixed O&M	**0.35**	**0.80**	**0.86**	**0.62**	**0.73**	**0.52**
Water requirements[i]			0.028	0.028	0.026	0.026
Catalysts/binder[j]			0.018	–	0.016	–
Solids disposal[k]			0.071	0.069	0.060	0.059
H_2SO_4 by-product credit[l]			–0.273	–	–0.231	–
Total variable O&M	**0.59**	**0.50**	**–0.16**	**0.10**	**–0.13**	**0.09**
Capital[m]						
6 percent discount rate	1.71	2.27	1.66	1.32	1.32	1.03
12 percent discount rate	3.19	3.99	2.85	2.26	2.26	1.76
Total	**$1.061 \cdot P_c$+**	**$1.536 \cdot P_b$+**	**$1.011 \cdot P_c$+**	**$0.992 \cdot P_b$+**	**$0.855 \cdot P_c$+**	**$0.839 \cdot P_b$+**
6 percent discount rate	**2.65**	**3.57**	**2.36**	**2.04**	**1.92**	**1.64**
12 percent discount rate	**4.13**	**5.29**	**3.55**	**2.98**	**2.86**	**2.37**

Examples (total busbar cost):

For P_c = $1.8 per GJ [n]						
6 percent discount rate	4.56		4.18		3.46	
12 percent discount rate	6.04		5.37		4.40	
For P_b = $3.0 per GJ [o]						
6 percent discount rate		8.18		5.02		4.16
12 percent discount rate		9.90		5.96		4.89

applications than updraft units, despite their lower levels of tar production, because of typically lower gasification efficiencies.)

Fixed-bed gasifiers

The fixed-bed, updraft gasifier (see figure 14, top) is a simple, efficient system suitable for biomass feedstocks having high bulk density (e.g., woodchips or densified biomass). The pressurized fixed-bed Lurgi Mark IV dry-ash gasifier, a mature system with extensive coal experience, appears to be a good candidate for biomass applications [23, 24]. Its adaptation to CIG/GT applications has been extensively evaluated [21]. While there has been no focused effort to develop this technology for biomass feedstocks, limited pilot-scale testing was carried out in 1991 by GE with wood chips and sugarcane bagasse pellets as fuel. A gas with adequate heating value was generated, but the tests indicate that development work is needed on pressurized feeding and on removal of contaminants from the product gas at elevated temperatures [27].

Hot-gas cleanup is perhaps the most important system-development issue for airblown gasification systems (for both coal and biomass feedstocks). A high degree of removal is required for alkali compounds (formed primarily from potassium and sodium in the feedstock) and particulates. Estimates of the tolerable concentration of alkali vapors in fuel gas for gas-turbine applications are very low—100 to 200 parts per billion at the gasifier exit [28, 29], with corresponding several-fold lower concentrations at the turbine inlet. The extent of alkali production and required removal from biomass gas is not well documented. Based on coal-related work, however, the gasifier exit temperature appears to be the most important controlling parameter. At sufficiently low temperatures the alkalis appear to condense on particulate matter and can be controlled by controlling particulates.

Table 3 notes:

a. CS = a subcritical, coal-fired steam-electric plant (two 500 MW_e units) with flue gas desulfurization, east or west central U.S. siting. EPRI estimates for heat rate (10.61 megajoules per kWh), overnight construction cost ($1,217 per kW_e), other capital ($78 per kW_e), O&M costs ($23.1 per kilowatt per year fixed; $0.0059 per kilowatt variable), and the idealized plant construction time (5 years) [60]. Including AFDC, the total capital cost amounts to $1,450 per kW_e ($1,624 per kW_e) for a 6 percent (12 percent) discount rate.

b. BS = a 27.6 MW_e biomass-fired steam-electric plant. Based on an EPRI design for a 24 MW_e condensing/extraction cogeneration plant producing 20,430 kilograms per hour of steam at 11.2 bar for process [60]. Here it is assumed that this steam is instead condensed, thus producing an additional 3.6 MW_e; the heat rate is 15.36 megajoules per kWh (corresponding to steam conditions of 86 bar and 510°C at the turbine inlet and a turbine efficiency of 80 percent). EPRI estimates for the overnight construction cost ($1,693 per kW_e), other capital ($127 per kW_e), and idealized construction period (3 years). Including AFDC, the total capital cost amounts to $1,924 per kW_e ($2,031 per kW_e) for a 6 percent (12 percent) discount rate.

c. CIG/STIG = a coal-integrated gasifier/steam-injected gas turbine and CIG/ISTIG = a coal-integrated gasifier/intercooled steam-injected gas turbine (see table 2).

d. BIG/STIG = a biomass-integrated gasifier/steam-injected gas turbine and BIG/ISTIG = a biomass-integrated gasifier/intercooled steam-injected gas turbine (see table 2).

e. P_c = coal price, and P_b = biomass price, in $ per gigajoule (HHV basis).

f. The coal-based systems require three operators for the gasification system, four for the hot-gas cleanup, and three for the power plant. At $22.55 per hour, operating labor costs for the coal systems are $1.977 million per year. Because hot-gas desulfurization is not needed for the biomass systems, it is assumed that seven operators are needed for the biomass systems—four fewer because hot-gas desulfurization is not needed and one more because of increased fuel handling requirements. Thus annual operating labor costs would be $1.384 million.

g. Annual maintenance costs (40 percent labor and 60 percent materials) are estimated to be $2.812 million for CIG/STIG (including $0.634 million for chemical hot-gas cleanup) and $2.342 million for CIG/ISTIG (including $0.591 million for chemical hot-gas cleanup). The corresponding values for BIG/STIG and BIG/ISTIG, without chemical hot-gas cleanup, are $2.178 million and $1.751 million, respectively.

h. Annual administrative costs, assumed to be 30 percent of O&M labor, are $0.930 million for CIG/STIG, $0.874 million for CIG/ISTIG, $0.677 million for BIG/STIG, and $0.625 million for BIG/ISTIG.

i. Raw water costs are $0.189 million per year for all systems.

j. Annual catalysts and binder costs are $0.121 million ($0.113 million) for CIG/STIG (CIG/ISTIG) and zero for BIG/GT systems.

k. Annual costs for solids disposal are $0.469 million ($0.428 million) for CIG/STIG (CIG/ISTIG) and are assumed to be the same for the corresponding BIG/GT systems.

l. Annual H_2SO_4 by-product credits are $1.815 million for CIG/STIG, $1.659 million for CIG/ISTIG; and zero for BIG/GT systems.

m. The capital charge rate includes the capital recovery factor for an assumed 30-year plant life [0.0726 (0.1241) for a 6 percent (12 percent) discount rate] plus an insurance charge rate of 0.5 percent of the initial capital cost per year. The capacity factor is assumed to be 75 percent.

n. The levelized price of coal, 2000–2030, delivered to utilities in the west/north central United States, as projected by the U.S. Department of Energy.

o. Delivered cost of wood chips from short-rotation, poplar crops in the United States, including the costs of 40 kilometers transport, drying, and 6-month storage, for a 6 percent discount rate (see table 6).

Table 4: Compositional data and heating values for biomass and coal (dry basis)

Feed stock	Proximate analysis percent by weight			Ultimate analysis percent by weight						HHV GJ/t	kg N/ GJ
	Volatile	Fixed Carbon	Ash	C	H	O	N	S	Ash		
Biomass[a]											
Red alder											
	87.10	12.50	0.40	49.55	6.06	43.78	0.13	0.07	0.41	19.30	0.067
Black locust											
	80.94	18.26	0.80	50.73	5.71	41.93	0.57	0.01	1.05	19.71	0.289
Poplar											
	82.32	16.35	1.33	48.45	5.85	43.69	0.47	0.01	1.53	19.38	0.243
Douglas fir											
	87.30	12.60	0.10	50.64	6.18	43.00	0.06	0.02	0.10	20.37	0.029
Casuarina											
	78.94	19.66	1.40	48.61	5.83	43.36	0.59	0.02	1.59	19.44	0.303
Eucalyptus grandis											
	82.55	16.93	0.52	48.33	5.89	45.13	0.15	0.01	0.49	19.35	0.078
Leucaena											
	80.94	17.53	1.53	49.20	6.05	42.74	0.47	0.03	1.51	19.07	0.246
Sugarcane bagasse											
	73.78	14.95	11.27	44.80	5.35	39.55	0.38	0.01	9.91	17.33	0.219
Coal[b]											
West Kentucky bituminous											
	33.12	48.18	18.70	65.78	4.62	4.86	1.26	4.74	18.74	27.81	0.453
Illinois No. 6 bituminous											
	37.50	43.40	18.18	65.34	4.20	6.59	1.02	4.55	18.30	26.67	0.382
Wyoming subbituminous											
	44.68	46.12	9.20	68.75	4.89	15.55	0.89	0.69	9.24	26.78	0.332
East Texas lignite											
	44.55	38.86	16.59	60.98	4.45	15.82	1.08	1.08	16.65	24.36	0.443

a. [61].

b. West Kentucky bituminous coal characteristics data from [62]; other coal characteristics data from [60].

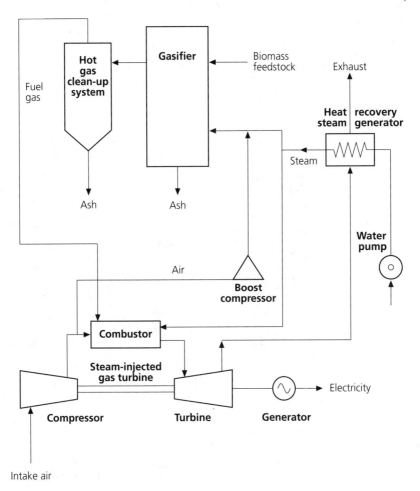

FIGURE 13: A BIG/GT cycle based on a STIG unit.

The fixed-bed gasifier may offer a simple mechanism for controlling alkalis. At the relatively low temperatures of the fuel gas exiting the fixed-bed gasifier (500 to 600°C), most of the alkalis appear to condense on particulates and can thus be controlled by controlling particulates. Data from coal gasification experiments (see figure 15) suggest that particulate cleanup with fixed-bed biomass gasification may be possible using cyclones.

While a high rate of tar formation has long been a concern for fixed-bed gasifiers in other applications, this should not be a concern for BIG/GT applications, because the temperatures of the fuel gas exiting the gasifier would be high enough that the tars would be in the vapor phase. By close-coupling the gasifier and gas turbine, the tars can be burned (without condensation problems) in the gas turbine combustor. Tars are desirable, in fact, to boost the heating value of the gas. Thus, for fixed-bed gasifiers there appears to be a practically realizable exit-

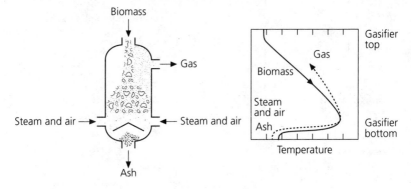

Fixed-bed gasifier (dry ash)

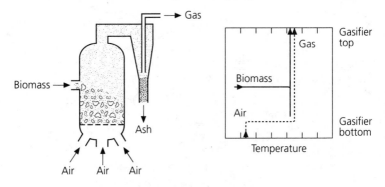

Bubbling fluidized-bed gasifier

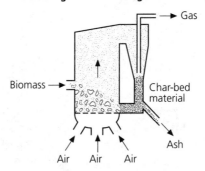

Circulating fluidized-bed gasifier

FIGURE 14: Operating principles and temperature profiles for fixed-bed (top) and bubbling fluidized-bed (center) gasifiers, and the operating principle for a circulating fluidized-bed gasifier (bottom).

gas temperature window of 500 to 600°C, within which problems with both condensed tars and vaporized alkalis may be avoided. This temperature window also coincides with material limits for valves that would be used to control gas flow to the turbine [21].

Fluidized-bed gasifiers

Fluidized-bed gasifiers have higher throughput capabilities and greater fuel flexibility than fixed-beds, including the ability to handle low-density feedstocks like undensified crop residues or sawdust [23]. Their ability to handle a wide range of biomass fuels with minimal preprocessing may ultimately make fluidized-bed gasifiers the technology of choice for many biomass applications, because of the diversity of biomass feedstocks.

For fluidized-bed gasifiers, gas quality control is more problematic than for fixed-bed gasifiers, for two reasons. First, at the high temperatures of the fuel gas exiting the gasifier (800 to 900°C), alkalis will be in the vapor phase. Dealing with this problem will probably require fuel gas cooling to condense the alkalis, with attendant efficiency and capital cost penalties. (Some gas cooling would be needed in any case to meet control-valve material constraints.) Second, there is much more particulate carryover with a fluidized-bed gasifier, so that control of particulates with cyclones would not be adequate. Ceramic or sintered-metal barrier filters, neither of which is fully proven commercially, are probably needed. A

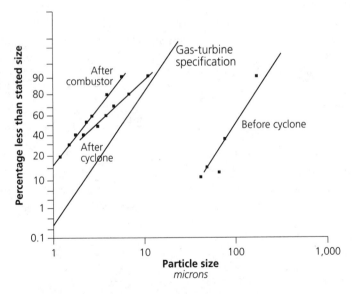

FIGURE 15: Measured particulate size distribution from tests with coal in a pilot-scale fixed-bed dry-ash gasifier [30]. The data are for the raw fuel gas exiting the gasifier, for the fuel gas after the cyclone, and for the combustion product gases after a simulated gas turbine combustor. Also shown are recommended gas turbine specifications, which were developed by GE in the late 1970s as part of a U.S. DOE–supported program in pressurized fluidized-bed coal combustion for gas turbine applications to locomotives.

number of such filters are being developed in Finland [31], the United States [22], and elsewhere.

Several candidate fluidized-bed systems for BIG/GT applications are in various states of development. The bubbling bed (see figure 14, center) was the first fluidized-bed design developed. The Rheinbraun–Uhde (Germany) HTW (High Temperature Winkler) is a commercially mature, pressurized bubbling fluidized-bed technology using coal. It has operated successfully with biomass, though not extensively. The world's only commercial pressurized gasifier operating on a feedstock other than coal is a peat-fired HTW unit in Finland. Additional testing is needed to fully demonstrate the HTW performance on biomass and its ability to meet gas turbine fuel gas specifications. Like all fluidized-beds discussed here, it has no actual or simulated operating experience coupled to a gas turbine.

The Institute of Gas Technology (IGT) Renugas system (United States) is another bubbling bed that has perhaps more experience operating at elevated pressure with biomass than any other gasifier. Its operation has only been at pilot scale, however, so a successful scaleup effort is needed before the technology can be considered mature. Its capability to meet gas turbine fuel gas specifications must also be demonstrated.

Circulating fluidized beds (CFBs), (see figure 14, bottom) allow for more complete carbon conversion and permit higher specific throughputs than bubbling beds. Lurgi (Germany), Ahlstrom (Finland), and Studsvik (Sweden) have commercially operating biomass-fired, atmospheric-pressure CFBs. There are no pressurized CFBs operating on any feedstock (including coal). However, there are some promising candidates. Ahlstrom is developing a pressurized CFB for biomass applications that would have gas cooling for alkali control and ceramic-filter particulate cleanup. Also, a hybrid HTW/Lurgi-CFB is under development for CIG/GT combined cycle applications.

System performance and cost

Initial applications of BIG/GT technologies are likely to be in industries where biomass residues of industrial activities can be used as fuel for the cogeneration of electricity and steam for on-site use. Cogeneration system performance parameters are shown in table 5, along with associated capital costs, 1) for various BIG/GT systems that involve coupling a fixed-bed gasifier to various steam-injected gas turbine systems,[4] and, for reference, 2) a double-extraction/condensing steam turbine (CEST) cogeneration system. From table 5 it is seen that in the maxi-

---

4. The BIG/STIG performance and cost estimates presented in table 5 can be compared to the estimated performance and cost for a 37 MW$_e$ BIG/combined cycle plant designed in a feasibility study carried out at the Shell International Petroleum Company. The Shell design consists of an Ahlstrom circulating fluidized-bed gasifier with ceramic-filter gas cleanup, feeding a Rolls Royce RB-211 aeroderivative gas turbine, which would provide 27 MW$_e$ of the plant's total output. Overall efficiency on biomass fuel with 15 percent moisture content was estimated to be 39 percent. The total installed cost was estimated to be $1,200 to $1,300 per kW$_e$ for commercial plants and $1,600 to $1,700 per kW$_e$ for the first demonstration plant [32]. (In the present analysis a 37 MW$_e$ BIG/STIG unit is estimated to have an efficiency of about 34 percent and an installed cost of about $1,200 per kW$_e$.)

Table 5: Performance and capital cost estimates of biomass cogeneration systems

	Cogeneration performance				Performance for maximum electric power		Installed capital cost[a]
	Electricity		Maximum process steam				
	MW_e	Effic.,%	kg/hour	Effic.,%	MW_e	Effic.,%	$/kW_e$
15 percent moisture content fuel[b]							
BIG/ISTIG							
LM-8000	97	37.9	76,200	25.4	111.2	42.9	870
BIG/STIG							
LM-5000	39	31.3	47,700	30.0	51.5	35.6	1,120
LM-1600	15	29.8	21,800	33.8	20	33.0	1,380
LM-38	4	29.1	5,700	32.4	5.4	33.1	1,840
50 percent moisture content fuel[b]							
BIG/STIG[c]	38.3	29.5	47,700	28.9	50.8	33.5	1,220[d]
CEST[e]	37	10.0	319,000	52.1	77	20.9	1,520

a. Unit installed costs for BIG/ISTIG, BIG/STIG, and CEST scale with capacity according to: ($ per kW_e)$_{ISTIG}$ = 2,463 $(MW_e)^{-0.22}$, $ per kW_e)$_{STIG}$ = 2,669 $(MW_e)^{-0.22}$, and ($ per kW_e)$_{CEST}$ = 6,100 $(MW_e)^{-0.32}$

b. The indicated fuel moisture content (mc) is the percentage of the wet weight of the biomass.

c. For the LM-5000. The lower efficiencies and slightly lower electricity production compared to the case with 15 percent mc fuel reflect the estimated energy use associated with drying the fuel to 15 to 20 percent mc.

d. Includes $84 per kW_e for a steam-based drier. Drying is accomplished using a commercial system (private communication from C. Muenter, Stork Friesland Scandinavia AB, Gothenberg, Sweden, September 1990) that condenses 15-bar saturated steam to provide heat for a dryer that evolves 2/3 as much 3-bar saturated steam from the wood chips during drying, for process use in the plant.

e. This is a double-extraction/condensing steam turbine, with assumed boiler efficiency 68 percent, feedwater temperature 182°C, turbine inlet steam conditions 6.2 megapascals, 400°C. Maximum process steam corresponds to operation with minimum flow to the condenser. Saturated process steam conditions are 12.9 bar (119 tonnes per hour) and 4.4 bar (300 tonnes per hour). Maximum electricity corresponds to minimum required extraction of 72 tonnes per hour of saturated steam at 4.4 bar.

mum steam-producing mode, all systems convert about 60 percent of the energy of the biomass fuel into steam and electricity, but the fraction of fuel energy converted to the more valuable electricity is three or more times as large for gas turbines as for the steam turbine. In the maximum electricity-producing mode, the BIG/GT systems produce from 1.6 to 2.1 times as much electricity per unit of biomass energy as the CEST system. BIG/GT systems are also expected to be less capital-intensive, and their capital costs less sensitive to scale, than CEST systems (see table 5 and figure 16).

In areas where biomass residues are in short supply, BIG/GT systems could be fired with biomass grown on plantations dedicated to biomass production. Such applications will involve the production of only electricity in central-station power plants, as well as cogeneration. Biomass fuel delivered from plantations will often be more costly than biomass residues. The current average delivered cost of dried wood chips grown on *Eucalyptus* plantations in Brazil is estimated to be from $2.2 to $2.4 per gigajoule (see table 6). Costs for delivered dry wood chips from short-rotation, intensive-culture poplar plantations on good quality agricultural land in the United States are projected to be from $3.0 to 3.8 per gigajoule (see table 6).

In both Brazil and the United States, plantation wood costs are higher than projected long-run costs of coal ($1.8 per gigajoule) delivered to coal-fired power plants in most regions of the world. However, because of their higher efficiency and lower unit capital cost, BIG/GT systems could compete with conventional coal-fired, steam-electric plants with biomass that is much more expensive than coal. Specifically, with coal priced at $1.8 per gigajoule, BIG/ISTIG plants could compete at biomass prices that are up to double the coal price. If CIG/ISTIG should eventually be commercialized, the competition from coal would be much tougher. However, BIG/ISTIG could still compete at biomass prices as much as 20 to 30 percent higher than coal prices owing to the lower capital cost of BIG/ISTIG (see table 2).

Commercialization prospects

At least six initiatives bearing on the commercial development of BIG/GT technology are underway or have recently been announced.

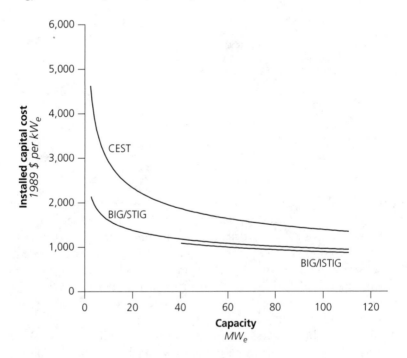

FIGURE 16: *Installed capital cost estimates for alternative biomass cogeneration technologies versus scale (table 5, note a).*

Table 6: Estimated costs of plantation wood: commercial *Eucalyptus* in Brazil today and year 2000 projection for short-rotation, intensive-culture poplar in the United States

$ per oven dry tonne

	Brazil[a]	United States[b]
Production cost[c] *(6 percent discount rate)*		
Establishment	5.17[d]	5.74[e]
Land rent[f]	0.99	7.95
Maintenance[g]		
Management	0.29	2.72
Land taxes	na	0.99
Research	0.58	na
Cultivation	1.55	
Insecticides/fungicides	na	0.93
Fertilizer	na	1.09
Subtotal	*8.58*	*19.42*
Harvesting[h,i]		
Harvester, tractor	na	4.58
Baler	na	3.87
Subtotal	*na*	*8.45*
Transport[h]		
Loader/unloader	na	4.46
Tractor/trailer[j]	na	5.15
Subtotal	*na*	*9.61*
Harvest/transport[k]		
Subtotal	*14.09*	*na*
Chipper/conveyor[h]		
Subtotal	*3.15*	*3.15*
Storage/drying[h]		
Storage[l]	6.77	6.77
Drying[m]	11.08	11.08
Subtotal	*17.85*	*17.85*
Total $ per oven dry tonne	**43.6**	**58.5**
$ per GJ	**2.2**[n]	**3.0**[o]
$ per GJ, 12 percent discount rate	**2.4**	**3.8**

In the fall of 1991, Sydkraft, the second largest electric utility in Sweden, began construction of a 6 MW$_e$ BIG/CC cogeneration demonstration plant in Varnamo, in southern Sweden (private communication from Owe Jonsson, Sydkraft Konsult, Malmö, Sweden, May 1992). Ahlstrom, a Finnish gasifier manufacturer, will provide a pressurized circulating fluidized-bed gasifier, a proprietary alkali removal system, and ceramic filters for particulates. Preliminary tests by Ahlstrom and Sydkraft indicate that wood fuels can be gasified at pressure and

---

Table 6 notes:

a. Average of high and low estimates given in [63] for eucalyptus plantations currently operated by various forest-products and charcoal industries. (na indicates not applicable)

b. For short-rotation poplars on good-quality agricultural land. Based on the use of a production model incorporating findings from the U.S. DOE Short-Rotation Woody-Crop Program [64]. (na indicates not applicable)

c. The levelized production cost is given by $[CRF(i,N) \times E + i \times L + M]/Y_L$, where

i = discount rate = 0.06
N = plantation life
$CRF(i,N)$ = capital recovery factor = $i/[1 - (1 + i)^{-N}]$
t = rotation period
L = land price
E = plantation establishment cost
M = annualized maintenance cost
Y_L = levelized yield = $CRF(i,N) \cdot \Sigma_t Y_t \cdot (1 + i)^{-t}$
Y_t = yield at each harvest

For Brazil, N = 18 years; t = 6 years, L = $208 per hectare; E = $707 per hectare; M = $30.5 per hectare. The total yield is estimated to be 97.3, 87.6, and 70.0 dry tonnes per hectare at the first, second, and third cuttings, respectively, corresponding to a levelized yield of 12.6 tonnes per hectare per year. This corresponds to an actual average yield of 14.1 tonnes per hectare per year. Carpentieri [63] estimates average yields to be 12.3 tonnes per hectare per year in the northeast of Brazil and 15.3 tonnes per hectare per year in the state of Minas Gerais, which contains most existing plantations in Brazil today. (In experimental plantations, yields as high as triple today's averages have been achieved in the northeast of Brazil, where soil and climate are not as well suited to tree production as in the southeast and south regions.)

For the United States, N = 12 years; t = 6 years; L = $1,800 per hectare; E = $654 per hectare; M = $78.5 per hectare. The total yield over two equal-yield cuttings is assumed to be 190 dry tonnes, corresponding to a levelized yield of 13.6 tonnes per hectare per year (15.8 tonnes per hectare per year actual average yield).

d. Includes nursery production of seedlings and planting.

e. Includes mowing/brushing, plowing, herbicides, liming, fertilization, planting.

f. The land rent is $(i \cdot L)$ (see note c). The Brazil land price of $208 per hectare is the average of high and low land prices as reported in [63]. The U.S. land price is typical for a good corn production site.

g. For Brazil, the maintenance costs include: 1) start-up management costs of $19.6 per hectare and $1.9 per hectare per year in years 1 to 18; 2) cultivation costs of $19.5 per hectare per year in years 1 to 18; and 3) research costs (to improve future productivity) of $7.3 per hectare per year in years 1 to 18.

For the United States, the maintenance costs include: 1) insecticides, fungicides applied every other year beginning in year 2 at a cost of $26 per hectare per application; 2) fertilizers applied every other year beginning in year 3 at a cost of $37 per application; 3) management at $37 per hectare per year; and 4) land taxes at 0.75 percent of the land price per year ($13.5 per hectare per year).

h. [65].

i. For a harvesting strategy in which trees are cut, crushed, field-dried, and baled before loading and transport to the storage/conversion site. (It has been found that for bolts of crushed wood averaging 10 centimeters in diameter, moisture contents (wet basis) have dropped from 50 percent to 20 to 30 percent after 6 days in the field [66]. Crushing tree-length stems with diameters up to 18 centimeters at a rate of 14 meters per minute requires only modest amounts of energy—some 0.88 kWh per tonne [67].)

j. Round-trip truck transport costs for a conversion facility located 40 kilometers from the harvesting site.

k. For Brazil, harvesting costs are $1,316 per hectare for the first rotation, $1,185 per hectare for the second rotation, and $950 per hectare for the final harvest. The harvested wood air dries to about 33 percent moisture before it reaches the user. Transport costs assume a 70 kilometer haul.

l. For 6 months of storage, with the wood covered by heavy polyethylene film.

m. Drying with unheated, forced-air system, based on a study by Frea [68].

n. Eucalyptus is assumed to have a higher heating value of 20 gigajoules per oven dry tonne.

o. Poplar has a heating value of 19.38 gigajoules per tonne (HHV basis).

the gas filtered at elevated temperature to specifications for gas turbines. Gas production at the demonstration site is scheduled to begin in March 1993.

Vattenfall, Sweden's largest utility, is also planning a BIG/GT combined cycle demonstration. Construction of a 60 MW$_e$ district heating cogeneration plant is scheduled to start in 1993. Vattenfall has selected Tampella, a Finnish company, to provide the gasifier for the project. Tampella has a license agreement covering the U-Gas fluidized-bed coal gasification system originally developed at the Institute of Gas Technology in Chicago. It plans to adopt the U-Gas system for use with biomass.

In Brazil, the Companhia Hidro Elétrica do São Francisco (CHESF), a major electric utility in the northeast, has an ongoing R&D program aimed at developing biomass from planted forests as a major fuel source for power generation, with conversion to electricity using BIG/GT units [33]. CHESF is leading the development of a BIG/GT demonstration project inBrazil, an effort that includes the participation of Eletrobrás (the Brazilian federal electric utility), Shell Brasil, Companhia Vale do Rio Doce (a major Brazilian forest products company), and Fundação de Ciência e Tecnologia (a Brazilian research institute). Potential equipment suppliers include GE, Studsvik, and Ahlstrom. Initial grant funding of $7 million has been provided by the Global Environment Facility (GEF) of the World Bank for a two year engineering design and development effort to be completed in 1994. The GEF will commit an additional $23 million at the end of a successful two year engineering effort. The project is seeking to build and run a demonstration combined cycle of 18 to 30 MW$_e$ output on plantation-derived wood chips in the northeast of Brazil, perhaps at the site of a forest-product manufacturing facility.

In the United States, the DOE announced in late 1990 a major new initiative to carry out R&D on BIG/GT technology [34]. The U.S. DOE also recently selected the IGT pressurized, bubbling fluidized-bed Renugas gasifier for a large-scale biomass gasification demonstration [35]. A pressurized pilot-scale Renugas unit has extensive operating experience on a variety of biomass feedstocks [36]. The scaled-up unit will be built in Hawaii and run initially on sugarcane bagasse (50 tonnes per day capacity). Start-up of the gasifier is anticipated in late 1992, with a gas turbine being added at a later time.

Also in the United States, the Vermont Department of Public Service, in cooperation with in-state electric utilities, is exploring possibilities for a commercial demonstration of BIG/GT technology fueled by wood chips derived from forest management operations (private communication from R. Sedano, Commissioner, Department of Public Service, State of Vermont, Montpelier, Vermont, February 1992). In preparation for this demonstration project, the U.S. DOE, the U.S. Environmental Protection Agency, and the U.S. Agency for International Development jointly supported gasification tests of wood chips and alternative biomass fuels at the pilot-scale fixed-bed gasifier at the General Electric Corporate Research and Development Center, Schenectady, New York (see earlier discussion).

The Finnish electric utility, Imatran Voima Oy (IVO), has begun development of a modified BIG/STIG cycle designed to take advantage of the moisture in wet feedstocks[37]. In the "IVOSDIG" cycle, wet fuel is first dried in a pressurized dryer, so that the moisture evaporated from the feedstock can be recovered as high-pressure steam. The recovered steam is then injected into the gas turbine, as in a conventional STIG, while the dried biomass is fed to the gasifier. IVO is targeting initial development of the IVOSDIG process for peat, which would have an initial moisture content of 60 to 75 percent before being dried to 10 to 30 percent for gasification. A 92 MW$_e$ IVOSDIG cycle is estimated to have an efficiency of 35 percent, starting with 70 percent moisture content peat. IVO is developing the fuel supply and drying systems, which they plan to couple to gasifier and gas turbine systems developed elsewhere. Commercialization of the cycle is targeted for the late 1990s.

Advanced BIG/GT technologies and beyond

BIG/ISTIG as described above does not represent the ultimate in performance for biomass power-generating technologies. Improvements in turbine blade materials and advances in blade cooling technology that will permit gas turbine operation at higher turbine inlet temperatures as well as cycle modifications such as the addition of reheat combustors will lead to overall system efficiency improvements in future BIG/GT systems. Also costs and performance will be improved with advances in biomass gasification and biomass feedstock preparation—e.g., as circulating fluidized-bed gasifiers and pressurized biomass drying technologies become well established.

In addition, in the time frame beyond the turn of the century it is very likely that the fuel cell will begin to make inroads in power-generating markets. In a fuel cell, the chemical energy in a hydrogen-rich fuel gas is converted to electricity by a process analogous to that in a battery. But unlike the situation for the battery, where the chemical energy is regenerated by recharging, chemical fuel for the fuel cell is supplied continuously [38] (see figure 17). Because it offers the potential for converting the chemical energy of fuels directly into electrical energy, its ultimate performance is not constrained by the Carnot thermodynamic limit that applies to heat engines. A variety of fuel cell designs are under development [39]. A strong candidate for power generation applications is the molten carbonate fuel cell, which, like the gas turbine, could ultimately be operated on solid fuels, such as coal and biomass, through the use of closely coupled gasifiers. The Electric Power Research Institute has projected that Advanced Coal-Integrated Gasifier/ Molten Carbonate Fuel Cells (ACIG/MCFCs) could reach coal-to-busbar efficiencies of more than 55 percent by 2020 [40]. Just as efficiency levels projected for CIG/GT technologies can be expected to be realized in BIG/GT systems, so also should the biomass versions of these fuel cells (ABIG/MCFCs) reach performance levels comparable to those expected for coal.

It is desirable to pursue such opportunities for future improvements in biomass power-generating technologies in R&D programs, because as the biomass power industry grows, it will be necessary to exploit higher cost biomass feed-stocks, as low-cost biomass supplies become exhausted and land use constraints on biomass production are approached. Such improvements in conversion technology would make it feasible to extend the roles for biomass in power generation.

ENVIRONMENTAL ISSUES

Environmental concerns will play major roles in shaping the course of energy development in the coming decades. For biomass energy, the outstanding issues concern local air pollution and global warming, and the challenges of producing biomass sustainably and preserving biological diversity.

In an environmentally constrained world, biomass energy, including power generation, has much to offer because of inherent characteristics of the biomass resource. A major challenge is to shape biomass energy development in ways that exploit these inherent advantages.

As a fuel, biomass is inherently cleaner than coal because it generally contains negligible sulfur, and it usually contains much less ash (see table 4). However, actual emissions depend on the conversion technology used. Air pollutant emissions from biomass combustion have often degraded local air quality and posed

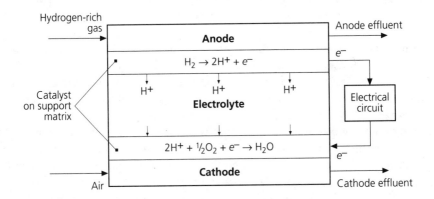

Net reaction:
$$2H^+ + \tfrac{1}{2}O_2 \rightarrow H_2O$$

FIGURE 17: The basic operating principle of a hydrogen fuel cell. A hydrogen fuel cell is a device for converting the chemical energy stored in hydrogen directly into electricity without first burning the hydrogen to produce heat. As hydrogen-rich fuel gas is passed over the anode of the fuel cell, the hydrogen is dissociated into two hydrogen ions and two electrons. The electrons contribute to an electric current that is delivered to an external circuit where it performs useful work, while the hydrogen ions pass through an electrolyte (or a hydrogen-permeable membrane) to the fuel cell's cathode. At the cathode, the hydrogen ions and the electrons returning from the external circuit combine with oxygen from the air to form water, the by-product of fuel cell operation.

human health problems. Biomass burning under a range of circumstances in tropical areas of developing countries also contributes to global climatic change, through emissions of greenhouse gases [especially carbon dioxide (CO_2) and methane (CH_4)], and to acid deposition, through emissions of organic acids (especially formic acid and ascetic acid) [41].

Using BIG/GT technologies for power generation with biomass would keep noxious emissions to very low levels. This prospect arises in large part from the inherent characteristics of the gas turbine. Unlike the situation in a steam-cycle boiler, combustion in a gas turbine takes place at a uniform high temperature at nearly 100 percent combustion efficiency—assuring complete burnup of potential organic pollutants and very low emissions of carbon monoxide. Moreover, very low emissions of particulates are assured by the need to protect the gas turbine blades from damage.

The local pollutant of greatest concern with BIG/GT systems is oxides of nitrogen (NO_x). Two major types of NO_x arise from different processes: thermal NO_x, which arises from oxidation of the nitrogen in combustion air at the high flame temperatures in the combustor, and the NO_x that is formed from nitrogen in the biomass. While thermal NO_x is the dominant concern in the combustion of natural gas, thermal NO_x emissions from BIG/GT systems would be very low, because the low heating value of the gas discharged from the gasifier (about 5 megajoules per normal m^3 (Nm^3), compared with 35 to 40 megajoules per Nm^3 for natural gas) gives rise to lower flame temperatures, at which little thermal NO_x forms.

NO_x from fuel-bound nitrogen (not present in natural gas) could be a concern for biomass, which, like coal, contains nitrogen (see table 4). Tests of fixed-bed gasifiers operated on coal give rise to high levels of ammonia (NH_3) in the fuel gas, some fraction of which will be converted to NO_x in the turbine combustor [42]. These NO_x emissions can be reduced with staged combustion; if emissions reduction achieved this way is not adequate, a selective catalytic reduction device could be placed in the turbine exhaust to reduce emissions further [42]. NO_x from fuel-bound nitrogen is likely to be much less a concern for fluidized-bed gasifiers, because the fuel gas from these gasifiers contains less ammonia and the higher gas temperature makes it feasible to use catalysts to decompose the ammonia to molecular nitrogen [42]. In both cases, measurements are needed to determine how serious the NO_x problem will be for biomass. This problem is getting considerable attention in ongoing coal gasification R&D programs, and solutions developed for coal will probably be readily transferable to biomass. Because biomass often contains less nitrogen than coal (see table 4), reducing NO_x emissions to target levels should be easier for biomass than for coal.[5]

Biomass energy systems also contribute essentially zero net CO_2 emissions if the biomass is produced sustainably (with the amount of biomass used for energy

5. When fertilizers are used, the resulting biomass may have greater nitrogen concentrations than biomass grown without artificial fertilization.

equal, on average, to the amount of biomass grown in a given period): the CO_2 released in combustion under these conditions is just equal to the CO_2 extracted from the atmosphere during photosynthesis. A minor net CO_2 release would result if fossil fuels were used in the biomass production process (petrochemical fertilizers, harvesting machinery, transport-vehicle fuel, etc.).

BIG/GT power generation in particular would be an especially cost-effective way to reduce carbon dioxide emissions. Since producing a kWh of electricity with coal-fired steam-electric plants emits some 2.6×10^{-4} tonnes of carbon (tC) as CO_2, the cost of reducing emissions of CO_2 (in $ per tC) by substituting a BIG/ISTIG unit for a coal-fired steam-electric unit would be, for the 6 percent discount rate case (see table 3):

$$\frac{(0.0164 + 0.00839 \times P_b) - (0.0265 + 0.01061 \times P_c) \text{ \$/kWh}}{2.6 \times 10^{-4} \text{ tC/kWh}}$$

where P_b and P_c are the costs of biomass and coal, respectively. For a typical coal price of P_c = $1.8 per gigajoule (see table 3), the cost of reducing CO_2 emissions by building a BIG/ISTIG unit instead of a coal-fired steam-electric plant would be negative up to a biomass price of $3.5 per gigajoule—a price level that is likely to be met or beaten in many plantation applications (see table 6).[6]

A concern often raised about using biomass for energy is that doing so would contribute to deforestation by creating new incentives for cutting down natural forests. But with capital-intensive energy conversion technologies like BIG/GT, there would instead be powerful built-in economic incentives for growing the biomass sustainably. When building a BIG/GT power system, the producer wants assurance that biomass supplies will be available throughout the expected lifecycle of the facility, some 30 years. As it is very costly to gather biomass supplies over a wide area and transport them to the power plant, affordable supplies must be available locally. For biomass to be locally avaliable on a continual basis, steps must be taken to ensure that biomass production is carried out sustainably. The power producer will typically find that it is worthwhile to take such steps,

---

6. If eventually CIG/ISTIG technology were commercialized, the economics of reducing CO_2 emissions would not be quite so favorable. In this case, the cost of reducing CO_2 emissions is given by (see table 3):

$$\frac{(0.0164 + 0.00839 \times P_b) - (0.0192 + 0.00855 \times P_c) \text{ \$/kWh}}{2.09 \times 10^{-4} \text{ tC/kWh}}$$

and, for coal and biomass prices of $1.8 and $3.5 per gigajoule, respectively, the cost of reducing CO_2 emissions would be a positive $53 per tC. One way of interpreting this result is that a carbon tax of this amount would have to be levied on coal to make BIG/ISTIG power competitive with CIG/ISTIG power. If that were done, the busbar cost of electricity from either BIG/ISTIG or CIG/ISTIG plants would still be lower than for a coal-fired steam-electric plant today (see table 3).

because the required investment would be relatively small compared with the investment required for the power plant. The latter is about $100 million for a 111 MW_e BIG/ISTIG plant, while the investment required to establish plantations to support such a power plant might be about $21 million.[7] Various approaches can be taken to promote sustainable biomass production (see chapter 14: *Biomass for Energy: Supply Prospects*).

One issue of increasing concern to environmentalists is the maintenance of biological diversity as biomass energy systems are developed [43]. Unlike the situation relating to sustainable biomass production, there are no strong, built-in safeguards in bioenergy systems to promote the maintenance of biological diversity. However, the impact of bioenergy production depends sensitively on how biomass is produced. If monoculture biomass plantations were to replace old-growth natural forests there would be substantial loss of biodiversity. But the status quo could be improved if plantations were established on deforested or degraded lands. Also, short-rotation tree crops and various perennial grasses would be an improvement on annual row-crop agriculture in this regard. As for the challenge of sustainable biomass production, there are various approaches that can be pursued to promote biological preservation goals (see chapter 14).

There are two ways in which the choice of downstream conversion technology can help meet biodiversity and other environmental goals relating to biomass production. First, emphasis on high efficiency conversion technologies makes it possible to extract more useful energy from a given amount of biomass, thereby making it easier to respect constraints on biomass production without severely limiting the role of biomass in the energy system. This would be the situation for power generation, for example, if biomass fuel cells become commercially available two or three decades from now—providing 2 to 4 times as much electricity from the same amount of biomass as biomass power plants now in operation. Second, if there are various other promising ways to provide the produced energy carrier besides from biomass, it would be easier to respect environmental constraints by curbing further expansion of biomass production as production constraints come into play, than if other production routes were not available. By the time biomass production levels are so high that such environmental or other constraints would limit further expansion of biomass production, a variety of other clean, renewable electric technologies are likely to be commercially viable. Because electricity is an energy carrier that can be provided by a rich diversity of primary supply sources, society should be able to choose the least problematic mix of electricity-generating options in meeting its electricity needs.

---

7. If the average plantation productivity were 14 tonnes per hectare per year, some 30,000 hectares would be needed to support a 111 MW_e BIG/ISTIG unit. The establishment cost of such plantations would be about $700 per hectare (roughly the U.S. and Brazilian establishment costs shown in table 6).

BIG/GT APPLICATIONS USING RESIDUES

Biomass residues from agricultural and forest product industries might be used as fuel for power generation or for cogeneration—providing heat and electricity for on-site needs and electricity for export to the local electric utility. The significance of these markets is suggested by the data in table which shows that global residue production in selected industries in 1988 was some 56 exajoules. This amount of residues, had it been used to produce power only in BIG/ISTIG units, would have been able to produce some 6,600 terawatt-hours (TWh), about as much electricity as was produced by all fossil fuel–fired power plants in the world in 1988 [44]. Of course, not all these residues could be recovered and used for power generation (see chapter 14), but this back-of-the-envelope calculation suggests that residue applications of BIG/GT technologies are worth exploring. The results of detailed analyses that have been carried out for sugarcane and kraft pulp industries are summarized below.

Sugarcane

Sugarcane is a well-established, photosynthetically efficient crop. Global sugarcane production totals nearly one billion tonnes per year. Most sugarcane is used to produce sugar, though in some areas (notably Brazil) the juice extracted from sugarcane is also fermented to ethyl alcohol (ethanol), which is used as a substitute for gasoline in transportation (see chapter 20: *The Brazilian Fuel-Alcohol Program*).

Reduced demand growth for sugar worldwide, a widely fluctuating world sugar price, and the inability of ethanol from cane to compete with gasoline at low world oil prices are stimulating interest in alternative products from cane, such as electricity. Two residues of the sugarcane plant in principle can be used as fuel in the production of by-product electricity: bagasse (the residue left after extracting the sugar juice from the cane stalk) and barbojo (the word used in Latin American countries for the tops and leaves of the sugarcane plant). Typically bagasse and barbojo are produced at rates of 150 kilograms and 279 kilograms (dry basis) per tonne of fresh cane.[8]

In most parts of the cane-producing world the barbojo is burned off just before harvest to facilitate harvesting the cane stalks, while the bagasse is used as fuel to provide the energy needs of sugar factories and alcohol distilleries. Small, bagasse-fired steam-turbine systems supplied with steam at 1.5 to 2.5 megapascals provide just enough steam and electricity to meet onsite factory needs, which are typically from about 350 to 500 kilograms of steam, and from 15 to 25 kWh of electricity per tonne of cane (tc) milled [45, 46]. Typically, factories are designed to be somewhat energy inefficient, consuming all the available bagasse while just

---

8. This estimate of barbojo production [48] is lower than one (330 kilograms per tc) published previously [46, 47]. The latter was determined from harvest trials in a Caribbean country where more of the cane top is removed as barbojo than is typical for most other regions of the world.

Table 7: Selected estimated global residue production rates
exajoules per year, 1988

Forest product industries[a]	
Kraft pulp[b]	
Hog fuel	0.7
Black liquor	2.7
Forest residues	0.8
Subtotal	*4.2*
Sawn wood and wood panels[c]	
Mill residues	3.6
Forest residues	6.2
Subtotal	*9.8*
Agricultural industries[a]	
Sugarcane[d]	7.2
Wheat	12.9
Rice[e]	10.6
Maize[e]	7.3
Barley[e]	3.8
Subtotal	*41.8*
Total	**55.8**

meeting factory energy demands, so that excess bagasse does not accumulate and become a disposal problem.

In a few sugar factories and alcohol distilleries, modern condensing-extraction steam turbine cogeneration systems (CEST) operated at turbine inlet pressures of 4.0 to 6.0 megapascals have been installed [46, 47]. With these systems, it is possible to produce enough steam to run a typical factory (350 to 500 kilograms per tc), plus 70 to 120 kWh per tc of electricity, or about 50 to 100 kWh per tc in excess of onsite needs. The extra electricity can be made available to other users by interconnecting the cogenerator with the utility grid. During the milling season (typically half the year) the CEST cogeneration system is fueled with 50 percent wet bagasse, as it comes from the mill. In the off-season, CEST units could, in principle, be operated in the full condensing mode producing power only, fired with barbojo that is harvested and stored for use in the off-season.

Much better thermodynamic and economic performance could be realized if BIG/GT units could be used instead of CEST units. As in the case of CEST technology, the BIG/GT systems would be fueled with bagasse during the milling season. In the off-season the system could be fueled with barbojo. Because of their

Table 7 notes:

a. Assuming higher heating values of 20 gigajoules and 15 gigajoules per dry tonne of woody and agricultural residues, respectively.

b. Assuming hog fuel, black liquor, and logging residues (which excludes roots, stumps, branches, needles, and leaves) of 7.0 gigajoules, 25.3 gigajoules, and 8.0 gigajoules per tonne of pulp, respectively (characteristic of the kraft pulp industry in the U.S. Southeast), for the 1988 global chemical pulpwood production of 105 million tonnes [51].

c. Assuming mill (note d) and forest (note e) residues of 0.30 tonnes per m³ and 0.52 tonnes per m³ of sawn-wood/wood-panel products, respectively (characteristic of the U.S. forest products industry in 1976), for the 1985 to 1987 world sawn-wood/wood-panels production rate of 600 million m³ [69].

　　Primary and secondary mill residues of the U.S. forest products industry not used by the pulp industry in 1976 amounted to 34.7 million dry tonnes [70], while U.S. sawn-wood and wood-panels production amounted to 115.4 million cubic meters [71]. Thus 34.7/115.4 = 0.30 tonnes of mill residues were produced for each m³ of sawn-wood and wood-panels produced.

　　U.S. forest residues totalled 76.4 million tonnes in 1976 [70]. Assuming each of the 40 million tonnes of pulp produced in the U.S. in 1976 [71] was associated with 0.42 tonnes of forest residues [51], the residues associated with sawn-wood/ wood-panels production in 1976 amounted to 59.6 million tonnes. Thus some 59.6/115.3 = 0.52 tonnes of forest residues were associated with each cubic meter of sawnwood and wood panels production.

d. Assuming bagasse amounting to 2.6 gigajoules and cane tops and leaves amounting to 4.8 gigajoules per t (wet) of harvested stem [48], for the 1987 cane production rate of 968 million tonnes worldwide [72].

e. Global grain production rates, 1986 [73] and associated residue production rates, assuming residue production coefficients characteristic of U.S. grain production in the period 1975 to 1977 were:

Grain	1986 production	Residue coefficient	Residue production
	million tonnes		*million tonnes*
Wheat	538	1.6	861
Rice	473	1.5	710
Maize	485	1.0	485
Barley	182	1.4	255

　　Selected U.S. grain production rates, 1975 to 1977 [74] and grain residue production rates [75], along with the corresponding residue coefficients, were:

Grain	Annual production	Residue production	Residue coefficient
	average, 1975 to 1977	*average, 1975 to 1977*	
	million tonnes	*million tonnes*	
Wheat	57.2	90.7	1.6
Rice	5.2	7.8	1.5
Maize	155.6	155.3	1.0
Barley	8.5	12.1	1.4

higher electrical efficiencies and lower unit capital costs (table 5), BIG/GT systems could provide more electricity per tc at lower costs than CEST systems.[9]

BIG/GT cogeneration systems, which are more efficient producers of electricity, produce less steam than CEST systems per tc processed. With BIG/STIG technology, steam can be produced at a maximum rate of about 275 kilograms (@ 2.0 megapascals, 316°C) per tc milled. With BIG/ISTIG, maximum steam production is about 210 kilograms per tc. Since a typical existing sugar factory or autonomous alcohol distillery requires about 350 to 500 kilograms per tc of process steam, factory steam economy improvement measures must be carried out before gas turbine cogeneration systems can be installed. By adapting to the cane sugar industry steam economy measures widely used in other process industries

---

9. For use in a fixed-bed gasifier designed originally for coal, the bagasse and barbojo would probably have to be densified (e.g., briquetted or pelletized). Densification would probably not be needed with a fluidized-bed gasifier. For both gasifiers the fuel moisture content should be in the range 15 to 25 percent.

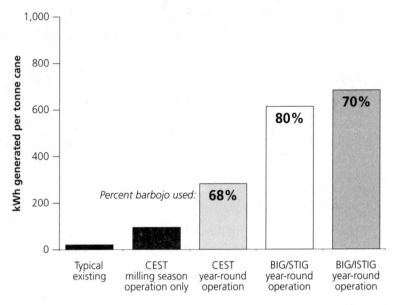

FIGURE 18: *The potential for cogeneration using sugarcane residues [46]. The left bar is for a typical existing situation at a sugar factory or alcohol distillery during the milling season. The next bar is for a CEST system operating during the milling season at a factory where steam-saving retrofits have been made [45]. The next three bars are for year-round operation of CEST, BIG/ STIG, and BIG/ISTIG systems, respectively, for a plant at which steam-saving retrofits have been made and where the milling season is 160 days. The number at the tops of the three bars to the right is the percentage of the barbojo (assuming 279 dry kilograms per tc total) used for power generation during the off-season.*

and by employing modern heat integration techniques to alcohol distilleries, it would be feasible to reduce steam requirements to the levels needed to accommodate BIG/GT cogeneration technologies [45–47].

Some alternative electricity production scenarios for sugar factories or alcohol distilleries are displayed in figure 18 and compared with a typical existing sugar factory or autonomous distillery. With CEST, BIG/STIG, and BIG/ISTIG systems operated year-round, annual electricity production would be 280 kWh per tc, 608 kWh per tc, and 677 kWh per tc, respectively, for a 160-day milling season (typical of Brazil), of which 12 to 13 kWh per tc would be needed for on-site use. For the scenarios with year-round operation, it is assumed that 20 to 30 percent of the barbojo would be left in the field (see figure 18).[10]

---

10. No studies have been done to determine the appropriate percentage of barbojo that should be left in the field. In tropical regions, the primary consideration is protection of the soil from raindrop impact and subsequent topsoil erosion. One estimate of the barbojo fraction that should remain in the field is 25 percent (private communication from L. Santo, Agronomist, Hawaiian Sugar Planters' Association, Honolulu, Hawaii, 25 September 1991). It is worth noting that any amount of barbojo left in the field would be an increase over current practice in most regions of the world, where barbojo is burned after harvesting, leaving essentially no soil cover.

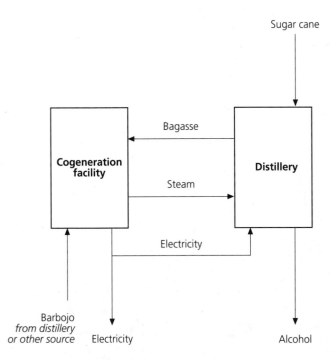

Sugar cane

Bagasse

Cogeneration facility

Distillery

Steam

Electricity

Barbojo
*from distillery
or other source*   Electricity

Alcohol

FIGURE 19: It is assumed that during the milling season the cogenerator buys bagasse fuel from the distiller and sells him steam and electricity. It is assumed that during the off-season, the cogenerator buys barbojo as fuel—either from the distiller (if the distiller owns the cane fields) or from independent cane producers.

For the situation where electricity is cogenerated at an autonomous alcohol distillery, the economics of coproducing electricity and ethanol have been analyzed for the setup indicated in figure 19 [46]: the distiller markets alcohol and sells cane residues to the cogenerator; and the cogenerator sells steam and electricity to the distiller and the excess electricity to the electric utility. It is assumed that during the milling season the distiller sells bagasse to the cogenerator and pays for process steam and electricity in return. For the off-season, it is assumed that the cogenerator uses barbojo as fuel, purchased either from the distiller (if the distiller owns the cane) or from independent cane growers.

Prospective total levelized costs for alcohol and cogenerated excess electricity are summarized in figure 20, for alcohol production of 73 liters per tc and for the alternative cogeneration options considered, assuming Brazilian cost conditions (table 8), a 160-day milling season, and a 12 percent discount rate. This figure indicates that when efficient BIG/GT technologies are used for cogeneration, using both bagasse and barbojo as fuel and taking credits from the sale of both against the cost of alcohol production, it would be feasible to produce both alcohol and electricity at costs that are competitive with the conventional alternatives in the Brazilian context, even at a low world oil price. (The conclusions are still stronger if a 6 instead of 12 percent discount rate were used.)

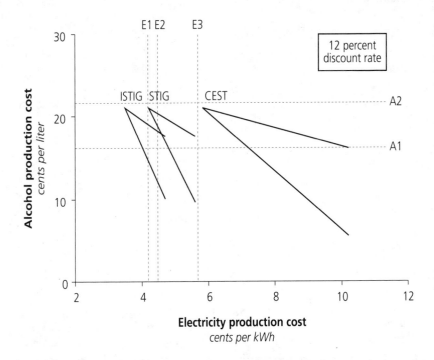

FIGURE 20: *The cost of ethanol and electricity using alternative cogeneration technologies, for Brazilian autonomous distilleries operated 160 days per year (based on reference 46). For cogeneration based on using bagasse as fuel during the milling season and barbojo during the off season. For each technology, a range of ethanol and electricity costs is shown, corresponding to different prices for the cane residues. As the residue price increases (moving from left to right along each line), the cost of electricity increases and the corresponding cost of ethanol decreases, because it is assumed that the distiller takes the increased revenues as a credit against the cost of producing alcohol. Two lines are shown for each technology. The top line represents the case where only bagasse revenues are credited against the alcohol cost. This would be the case if the barbojo were purchased from independent cane growers. The steeper line is for the case where the distiller owns the cane fields and sells barbojo as well as bagasse to the cogenerator. The point where these lines meet indicates the ethanol and electricity costs when no net payments are exchanged between the cogenerator and the distiller: bagasse is given to the cogenerator in exchange for the steam and electricity needed to run the distillery. The right endpoint of each line represents the point where the maximum residue cost to the cogenerator (purchase price plus the cost for processing the residues into a gasifiable form) is $3 per gigajoule, an estimate of the cost in Brazil of delivered, air-dried wood chips, for the case where the cogenerator has access to wood chips as an alternative biomass feedstock during the off-season.*

Also shown are: the prices at which ethanol would be competitive as a neat fuel (A1) with gasoline (for a world oil price of $24 per barrel), assuming a liter of alcohol is worth 0.84 liters of gasoline, and as an octane-enhancing additive (A2), assuming a liter of alcohol is worth 1.16 liters of gasoline; and the prices at which the cogenerated electricity would be equal to the operating cost of an oil-fired power plant at the present world oil price (E1) (assuming a heat rate of 13,120 kilojoules per kWh and a fuel oil price of $2.63 per gigajoule), the busbar cost of a new hydroelectric plant (E2) (assuming a capital cost of $1,500 per kW_e), and the busbar cost of a new coal plant (E3), assuming a capital cost of $1,400 per kW_e and a coal price of $1.7 per gigajoule.

Table 8: Estimated capital and operating costs for an autonomous distillery (excluding the costs for boilers and generating equipment), based on Brazilian experience[a]

Total installed capital cost *million $*	**18.083**
Fixed operating costs *thousand $ per year*	
Labor	560
Maintenance	362
Supply	36
Insurance	90
Total	***1,048***
Cane cost *$ per tc*	8.07
Other variable costs *$ per tc*	0.176

a. For a distillery processing 4,000 tc per day and producing 73 liters of ethanol per tc. Sources: [76] and private communication from G. Serra and J.R. Moreira, University of São Paulo, Brazil, 1989.

To give an indication of the potential role of electricity from cane residues in the global energy economy, consider a scenario (see table 9) for which it is assumed that cane production in the 80 cane-producing developing countries continues to grow for 40 years at the historical rate (about 3 percent per year), and that by the end of this period cogeneration systems having the performance of BIG/ISTIG units are fully deployed throughout the industry. Total cane electricity production in excess of onsite needs with this scenario would be about 2,100 TWh per year by 2027—some 28 percent more than total electricity generation from all sources in the 80 cane-producing developing countries in 1987.

Kraft pulp

Among several processes used to produce pulp from wood for papermaking, the kraft process dominates, accounting for three fourths of global pulp production [49]. About 90 percent of all kraft pulp is produced in industrialized countries today, but significant production growth is expected in developing countries [50], where feedstock production costs are often much lower.[11] Kraft pulp mills are significant producers of two forms of biomass energy (see table 7): hog fuel, which

---

11. The Food and Agriculture Organization of the United Nations has projected for the period 1980 to 2000 growth rates of 2.1 percent per year, 4.2 percent per year, and 6.8 percent per year, for chemical pulp production in the industrialized market economies, the former centrally planned economies, and the developing countries, respectively [50]. The high growth in developing countries is driven by high expected demand growth there. Developing countries consumed 8 kilograms of paper and paperboard products on average in 1984, compared to 133 kilograms per capita in all industrialized countries and 290 kilograms per capita in the U.S. (the world's highest consumption rate).

is primarily sawdust and bark produced during de-barking and chipping of logs, and black liquor, a lignin-rich by-product of cellulose extraction. Residues of pulp wood harvesting that are currently left in the forest are another potential biomass energy source (table 7).

Most kraft pulp mills today are designed to use all the biomass fuels available at the mill to meet their on-site steam and electricity needs with back-pressure steam-turbine cogeneration systems. Hog fuel is burned in conventional boilers to raise steam. Black liquor is consumed in Tomlinson recovery boilers, which raise steam and produce a sodium- and sulfur-laden chemical smelt that is recycled to the pulping process.

BIG/GT cogeneration systems could be considered for meeting mill energy needs and producing excess electricity for export. Hog fuel and forest residues brought to the mill could be gasified for this purpose using the fixed-bed or fluidized-bed technologies discussed earlier. Gasifiers for black liquor would also be needed. These have been under development for about the last 20 years [51], motivated by the high capital costs and smelt-water explosion risks of Tomlinson boilers and the large market potential for Tomlinson boiler retrofits over the next two decades [52]. Promising black-liquor gasifier development efforts are being carried out on an entrained-bed gasifier by Chemrec, a Swedish company [53], and on a fluidized-bed unit by MTCI, a U.S. company [54]. Atmospheric-pressure versions of these black-liquor gasifiers are likely to be commercially ready by the mid-1990s. Pressurized versions appear to be good candidates for gas turbine applications [55, 56].[12] Gas cooling and wet scrubbing would probably be needed in this application, in light of the high alkali content of black liquor.

Using hog fuel and black liquor available at typical modern kraft pulp mills (1,000 tonnes air-dry pulp (tp) per day capacity), BIG/GT cogeneration systems based on steam-injected or intercooled–steam injected gas turbines could generate three to four times more electricity per tonne of pulp produced than typical existing steam turbine cogeneration systems (figure 21). Steam use at the kraft mill would have to be reduced from a typical 16.2 gigajoules per tp by 40 to 50 percent for the BIG/GT systems to meet onsite steam demands. Such large reductions in steam use appear to be cost-effectively achievable, based on detailed assessments of Swedish mills [57, 51], which are considered to be the most energy-efficient in the world; onsite electricity use could probably also be reduced by as much as 25 percent. Still higher levels of BIG/GT electricity production (nearly five times current production) could be achieved by using currently unused and environmentally recoverable forest residues associated with pulp production (see figure 21).

---

12. While desirable, pressurization may not be necessary. For example, the MTCI technology produces, without using oxygen, a relatively high heating value gas that is cooled during wet scrubbing, so that the cost of compressing the gas to gas turbine combustor pressures might not be prohibitive. While producing a gas of comparable heating value using the Chemrec technology would require the use of oxygen, the oxygen could probably be provided at acceptable cost at the relatively large scales of most pulp mills. (In the future, many pulp mills will have oxygen plants on-site to supply oxygen delignification systems to reduce chlorine use in bleaching.)

Table 9: Net electricity production potential with BIG / ISTIG in the sugarcane industries of developing countries

| | Cane production million tc per year | | Electricity production TWh per year | | |
	1987[a] A	2027[b] B	Potential from cane in 2027[c] C	Actual from all sources in 1987[d] D	E = C/D
Africa					
South Africa	20.00	69.00	45.8	122.30[e]	0.37
Egypt	9.50	32.78	21.8	32.50	0.67
Mauritius	6.23	21.50	14.3	0.49	29.13
Sudan	5.00	17.25	11.5	1.06	10.81
Swaziland	4.00	13.80	9.2	–	–
Kenya	4.00	13.80	9.2	2.63	3.48
Zimbabwe	3.80	13.11	8.7	7.01	1.24
Reunion	2.11	7.29	4.8	–	–
Madagascar	1.80	6.21	4.1	0.50	8.25
Ivory Coast	1.75	6.04	4.0	2.20	1.82
Ethiopia	1.65	5.69	3.8	0.81	4.66
Malawi	1.60	5.52	3.7	0.58	6.32
Nigeria	1.55	5.35	3.6	9.91	0.36
Cameroon	1.29	4.45	3.0	2.39	1.24
Zambia	1.25	4.31	2.9	8.48	0.34
Zaire	1.09	3.75	2.5	5.30	0.47
Tanzania	1.08	3.71	2.5	0.87	2.83
Morocco	0.80	2.76	1.8	8.32	0.22
Senegal	0.70	2.42	1.6	0.75	2.14
Mozambique	0.67	2.31	1.5	0.50	3.07
Uganda	0.60	2.07	1.4	0.66	2.08
Congo	0.51	1.76	1.2	0.24	4.87
Somalia	0.37	1.28	0.8	0.26	3.27
Burkina Faso	0.33	1.14	0.8	0.13	5.82
Angola	0.32	1.10	0.7	0.81[f]	0.90
Chad	0.29	1.00	0.7	–	–
Mali	0.22	0.76	0.5	0.20	2.52
Guinea	0.20	0.69	0.5	0.50	0.92
Liberia	0.16	0.54	0.4	0.83	0.43
Gabon	0.14	0.48	0.3	0.88	0.36
Niger	0.11	0.38	0.3	0.16	1.58
Ghana	0.11	0.38	0.3	4.71	0.05
Sierra Leone	0.07	0.24	0.2	0.20	0.80
Rwanda	0.032	0.11	0.1	0.17	0.43
Subtotals	*73.3*	*253.0*	*168*	*216.4*	*0.78*

Table 9 (cont.)

	Cane production million tc per year		Electricity production TWh per year		
	1987[a] A	**2027**[b] B	**Potential from cane in 2027**[c] C	**Actual from all sources in 1987**[d] D	E = C/D
Oceania					
Fiji	3.49	12.05	8.0	0.43	18.61
Papua New Guinea	0.23	0.80	0.5	1.80	0.30
Subtotals	*3.7*	*12.9*	*8.5*	*2.2*	*3.83*
Central America					
Cuba	65.60	226.32	150.3	13.20[e]	11.38
Mexico	42.56	146.83	97.5	104.79	0.93
Dominican Republic	8.60	29.67	19.7	5.00	3.94
Guatemala	6.90	23.80	15.8	2.08	7.60
El Salvador	3.18	10.97	7.3	1.89	3.85
Honduras	3.00	10.35	6.9	1.81	3.80
Costa Rica	3.00	10.35	6.9	3.13	2.20
Haiti	3.00	10.35	6.9	0.45	15.27
Nicaragua	2.58	8.88	5.9	1.24	4.76
Jamaica	2.01	6.93	4.6	2.39	1.93
Panama	1.60	5.52	3.7	2.85	1.29
Trinidad and Tobago	1.24	4.26	2.8	3.30[e]	0.86
Belize	0.86	2.98	2.0	0.075	26.38
Guadaloupe	0.75	2.57	1.7	–	–
Barbados	0.73	2.52	1.7	0.43	3.89
St. Kitts and Nevis	0.25	0.86	0.6	–	–
Martinique	0.25	0.86	0.6	–	–
Bahamas	0.24	0.83	0.6	–	–
Subtotals	*146.4*	*504.9*	*335.2*	*142.6*	*2.35*
South America					
Brazil	273.86	944.79	627.3	202.29	3.10
Colombia	24.97	86.15	57.2	35.37	1.62
Argentina	14.00	48.30	32.1	52.17	0.61
Peru	5.95	20.53	13.6	14.20	0.96
Venezuela	7.00	24.15	16.0	50.21	0.32
Ecuador	5.20	17.94	11.9	5.67	2.10
Guyana	3.30	11.38	7.6	–	–
Paraguay	3.19	11.00	7.3	2.83	2.58
Bolivia	2.73	9.42	6.3	1.52	4.12
Uruguay	0.65	2.24	1.5	4.53	0.33
Suriname	0.11	0.38	0.3	–	–
Subtotal	*341.0*	*1,176.3*	*781.0*	*368.8*	*2.12*

	Cane production million tc per year		Electricity production TWh per year		
	1987[a] A	2027[b] B	Potential from cane in 2027[c] C	Actual from all sources in 1987[d] D	E = C/D
Asia					
India	182.48	629.55	418.0	217.50	1.92
China	52.55	181.30	120.4	497.30	0.24
Pakistan	31.70	109.37	72.6	28.40	2.56
Thailand	24.45	84.35	56.0	29.99	1.87
Indonesia	21.76	75.09	49.9	34.81	1.43
Philippines	13.33	45.97	30.5	23.85	1.28
Bangladesh	6.90	23.79	15.8	5.90	2.68
Vietnam	6.60	22.77	15.1	5.20[e]	2.91
Myanmar	3.28	11.32	7.5	2.28	3.30
Iran	1.15	3.97	2.6	36.80[e]	0.07
Malaysia	1.15	3.97	2.6	16.22	0.16
Sri Lanka	0.80	2.76	1.8	2.71	0.68
Nepal	0.62	2.13	1.4	0.54	2.62
Cambodia	0.21	0.71	0.5	–	–
Laos	0.11	0.36	0.2	0.88	0.27
Subtotal	*347.1*	*1,197.4*	*795.1*	*902.4*	*0.88*
Grand Total	**911.5**	**3,144.4**	**2,088**	**1,632.4**	**1.3**

a. [72].

b. Assuming cane production grows at an annual rate of 3.1 percent.

c. For sugar factories or alcohol distilleries operated 160 days per year, with 664 kWh of electricity generated per tc with BIG/ISTIG units in excess of on-site needs (13 kWh per tc).

d. Except where otherwise indicated, from [77].

e. For 1986, from [78].

f. Public electricity production in 1989, from [79].

The estimated costs of producing excess electricity at a hypothetical energy-efficient greenfield pulp mill using alternative cogeneration technologies are shown in table 10, assuming more energy-efficient end-use equipment needed at the mill is no more costly than conventional equipment.

Assuming utility financing with a 6 percent (12 percent) discount rate, and using only hog fuel and black liquor as fuel, the busbar costs are 7.8 (12.5), 3.2 (4.5), and 2.6 (3.6) cents per kWh, respectively, for the CEST, BIG/STIG and BIG/ISTIG. Adding the forest residues, the estimated costs with these alternative technologies, are, respectively, 6.6 to 7.6 (9.1 to 10.1), 3.9 to 4.3 (5.0 to 5.4), and 3.2 to 3.5 (4.1 to 4.4) cents per kWh. The busbar costs for the BIG/GT systems would be competitive with most alternative central-station power sources in most regions of the world (the major exception being electricity produced from low-cost natural gas using efficient gas turbine–based power systems).

Alternatively, with private ownership, the real internal rate of return before taxes and corrected for inflation would be less than 3 percent per year for CEST, 13 to 16 percent per year for BIG/STIG, and 21 to 23 percent per year for BIG/ISTIG, when excess electricity is sold for 5 cents per kWh (table 10). Revenues from electricity sales would be as much as $124 per tonne of pulp. For comparison, the production cost of one tonne of kraft pulp at modern mills in regions of the world with relatively lower feedstock costs is from $250 to $300 [58].

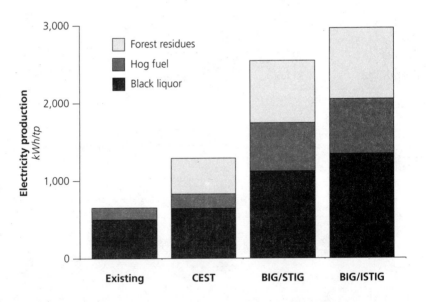

FIGURE 21: Electricity production at a hypothetical large (1,000 tonnes of pulp per day) modern bleached kraft pulp mill with alternative biomass cogeneration systems [51]. The quantities of available biomass fuel are based on southeastern US conditions. A typical existing mill has a steam demand of 16.3 gigajoules per tonne of pulp. For the CEST and BIG/STIG systems, the mill's steam demand is assumed to be 9.6 gigajoules per tp. For the BIG/ISTIG it is assumed to be 8.2 gigajoules per tp.

Table 10: Economics of electricity generation for export at a hypothetical 1,000 tonne per day kraft pulp mill[a]

Fuels and cogeneration technologies[b]	Utility ownership[c] busbar costs, *cents per kWh*						Private ownership[d] IRR, %	Electricity sales revenue[e]	
	Capital[f]		O&M[g]	Fuel[h]	Total			GWh/ year	$/tp
	6 %	12 %			6%	12%			
Black Liquor + Hog Fuel									
CEST	7.06	11.7	0.72	0.0	7.8	12.5	2.8	113	16
BIG/STIG	1.93	3.21	0.72	0.56	3.2	4.5	18.7	417	60
BIG/ISTIG	1.56	2.60	0.60	0.43	2.6	3.6	25.1	546	78
+ Forest Residues									
low-cost case									
CEST	3.83	6.37	0.72	2.00	6.6	9.1	2.7	275	39
BIG/STIG	1.63	2.71	0.72	1.53	3.9	5.0	16.1	683	97
BIG/ISTIG	1.37	2.28	0.60	1.20	3.2	4.1	23.0	866	124
high-cost case									
CEST	3.83	6.37	0.72	3.00	7.6	10.1	neg	275	39
BIG/STIG	1.63	2.71	0.72	1.92	4.3	5.4	13.5	683	97
BIG/ISTIG	1.37	2.28	0.60	1.51	3.5	4.4	20.6	866	124

a. [51].

b. Steam (electricity) needs met by cogeneration are 9.6 gigajoules per tp for CEST and BIG/STIG, and 8.2 gigajoules per tp for BIG/ISTIG (492 kWh per tp in all cases). Typical annual operating hours for a large pulp mill (8,400 hours per year) are assumed.

c. Assuming a 6 percent or 12 percent discount rate, a 30 year life, and an insurance charge equal to 0.5 percent of the initial capital cost per year.

d. Real internal rate of return, assuming a 25 year life, an insurance charge of 0.5 percent of the initial capital cost per year, and electricity sold for 5 cents per kWh.

e. Assuming electricity is sold for 5 cents per kWh.

f. Separate gasifier/gas-turbine units are assumed for black liquor and solid feedstocks. The capital cost for drying equipment needed with BIG/GT units is included in the fuel cost. A single steam turbine unit is assumed for the CEST system. In all cases a capital cost credit (equivalent to the cost of a Tomlinson recovery boiler) is assumed, since even the gasification system would also be serving the mill's chemical recovery requirements.

g. The O&M costs for the BIG/ISTIG and BIG/STIG are from table 2. The CEST O&M costs are assumed to be the same as for BIG/STIG.

h. Fuel costs for CEST are assumed zero for hog fuel and black liquor. For BIG/STIG and BIG/ISTIG, $1 per gigajoule is charged for drying hog fuel and for other pre-gasification handling. No charge for black liquor. The low and high forest-residue costs are assumed to be $3 per gigajoule and $4 per gigajoule for the gas turbine systems, and $2 per gigajoule and $3 per gigajoule for the CEST system, respectively. The lower costs for CEST arise because less pre-processing (e.g., drying) is needed.

Table 11: Global potential for electricity generation for export with BIG / ISTIG cogeneration technology in the kraft pulp industry

Region	1988			2020	
	Chemical pulp production[a] 10^6 t/year	Potential electricity from pulp[b] TWh/year	Utility fossil-fuel electricity generation[c] TWh/year	Projected pulp production[d] 10^6 t/year	Potential electricity from pulp[b] TWh/year
Industrialized	95.8	238	5,231	204.6	507
North America	57.8	143	2,106	105.1	262
CIS (former USSR)	7.1	18	1,181	34.7	86
Western Europe	19.5	48	954	30.4	75
Japan	7.9	19	470	19.0	48
Oceania	0.93	3	116	11.2	27
Eastern Europe	2.7	7	404	3.7	9
Developing	8.8	23	1,432	68.6	169
Latin America	5.6	14	216	35.4	88
Asia	2.5	7	1,021	24.0	59
Africa	0.67	2	194	9.2	22
World	**104.7**	**261**	**6,662**	**273.1**	**676**

a. [49].

b. Assuming 2,474 kWh per tp of electricity production in excess of process needs at efficient kraft pulp mills.

c. [44].

d. Assuming projected production growth rates 1980–2000 [50] persist till 2020. The annual growth rates are 1.9 percent, North America; 5.1 percent, CIS (USSR); 1.4 percent, Western Europe; 2.8 percent, Japan; 8.1 percent, Oceania; 1 percent, Eastern Europe; 5.9 percent, Latin America; 7.3 percent, Asia (excluding Japan); and 8.5 percent, Africa.

The worldwide potential for electricity exports from BIG/GT systems at kraft pulp mills could be significant. Assuming global chemical-pulp production grows to the year 2020 at regional rates projected for the period 1980 to 2000 by the Food and Agriculture Organization [50], up to some 676 TWh per year of exportable electricity could be produced by 2020, using black liquor, hog fuel and forest residues as fuel (see table 11) from some 100 gigawatts (GW) of installed BIG/ISTIG generating capacity. This electricity production is equivalent to 10 percent of the current global total from fossil fuels, or 14 percent of the total from coal.

PUBLIC POLICY ISSUES

Not only does biomass-based power, with proper management of biomass production, offer major local and global environmental benefits compared with fossil fuel–based power, but also it is likely that biomass power would be competitive with power from conventional sources in a wide range of circumstances, in both industrialized and developing countries.

The first priority in launching a major biomass-based power industry is to demonstrate BIG/GT systems based on present technology. It appears that this will happen over the course of the next several years.

But just demonstrating the technology will not be sufficient to launch a major biomass power industry. In addition, consideration must be given to the institutional reforms needed to create a hospitable environment for the industry. In light of the fact that BIG/GT power-generating units will generally be much smaller than conventional central-station power-generating units, a regulatory environment is needed that is conducive to power generation at modest scale— e.g., by independent power producers. In most parts of the world, utilities, whether privately owned or government owned, have tended to focus their investments on large, central-station systems and have discouraged decentralized power generation by independent power producers or otherwise. In the United States this problem has been successfully addressed with the passage of the PURPA legislation. PURPA-like reforms or alternative measures that would serve to encourage power generation at modest scales are needed in other parts of the world as well.

PURPA-like reforms should be complemented by regulatory or tax measures that would attract investment not just to biomass power but to biomass power based on new technologies like the BIG/GT. Incentives are needed to reward those investors willing to risk trying new technologies—incentives that would be operative for a few years and then phased out as the new technologies become well established.

Consideration should also be given to industrial structural issues that relate to the fact that biomass is an unusual fuel and often is not readily available for long-term contracts, as is the case for coal or natural gas. Accordingly, prospective biomass power producers may sometimes want to produce not just electricity but also biomass, in order to secure fuel supplies for the life of the plant investments. Or they may wish to form joint ventures with firms in the forest product or agricultural industries to increase biomass supply security. In some instances, institutional reforms might be needed to facilitate such joint ventures between utilities or independent power producers and various possible producers of biomass.

It is also important to support research and development on promising advanced biomass power-generating technologies, and to promote a continuing flow of innovations to market, so that as the biomass power industry develops and the use of biomass for power increases, there are improvements in the conversion technology that will enable biomass power to remain competitive as biomass costs rise.

Finally, achieving the high levels of biomass production that would be needed globally to support a major biomass power industry would be a significant undertaking. Great care will need to be taken to ensure than it is done in ecologically sound ways—paying close attention to considerations of sustainability, biological diversity, and other environmental issues.

ACKNOWLEDGMENTS

The research on which this chapter is based was supported by the Bioenergy Systems and Technology Project of the Winrock International Institute for Agricultural Development; the Office of Policy, Planning, and Evaluation and the Air and Energy Engineering Research Laboratory of the U.S. Environmental Protection Agency; the Office of Energy and Infrastructure of the U.S. Agency for International Development; and the Geraldine R. Dodge, W. Alton Jones, Merck, and New Land Foundations.

WORKS CITED

1. Office of Policy, Planning, and Analysis. 1990
   The potential of renewable energy: an interlaboratory white paper, SERI/TP-260-3674, US Department Energy, Washington DC.

2. Turnbull, J.H. 1991
   PG&E Biomass Qualifying Facilities Lessons Learned Scoping Study—Phase I, Pacific Gas & Electric Company, R&D Department, San Ramon, California.

3. Babcock and Wilcox Company. 1978
   Steam, its generation and use, Babcock and Wilcox Co., New York.

4. Mahin, D.B. 1991
   Industrial energy and electric power from wood residues, Winrock International Institute for Agricultural Development, Arlington, Virginia.

5. Research Triangle Institute (contractor), Energy Performance Systems (subcontractor). 1991
   Whole Tree Energy™ *: engineering and economic evaluation,* draft final report RP2612-15 to the Electric Power Research Institute, Electric Power Research Institute, Palo Alto, California.

6. Williams, R.H. and Larson, E.D. 1989
   Expanding roles for gas turbines in power generation, in T.B. Johansson, B. Bodlund, and R.H. Williams, eds., *Electricity: efficient end-use and new generation technologies and their planning implications,* 503–553, Lund University Press, Lund, Sweden.

7. Wilson, D.G. 1984
   The design of high-efficiency turbomachinery and gas turbines, MIT Press, Cambridge, Massachusetts.

8. Kano, K., Matsuzaki, H., Aoyama, K., Aoki, S., and Mandai, S. 1991
   Development study of 1,500°C class high temperature gas turbine, Paper 91-GT-297, American Society of Mechanical Engineers, New York, New York.

9. Larson, E.D. and Williams, R.H. 1987
   Steam-injected gas turbines, *Journal of Engineering for Gas Turbines and Power* 109(1):55–63.

10. Bemis, G.R., Soinski, A. J., Rashkin, S., Jenkins, A., and Johnson, R.L. 1989
    Technology characterizations: final report. Staff Issue Paper #7R, Energy Resources Conservation and Development Commission, Sacramento, California.

11. Horner, M.W. (Mar. and Ind. Eng. and Serv. Div. [Cincinnati, Ohio] of the General Electric Co.). 1988
    Position statement: intercooled steam-injected gas turbine, testimony presented at the Committee Hearing for the 1988 Electricity Report of the California Energy Commission, held at the Southern California Edison Co., 21–22 November.

12. Stambler, I. 1990
    New generation of industrialized aero engines coming for mid-1990 projects, *Gas Turbine World* 20(4):1922.

13. de Biasi, V. 1990
    LM 6000 dubbed the 40/40 machine due for full-load tests in late 1991, *Gas Turbine World* 20(3):1620.

14. Jersey Central Power & Light Co. and Sargent & Lundy Co. 1989
    A comparison of steam-injected gas turbine and combined-cycle power plants: technology assessment, GS-6415, Electric Power Research Institute, Palo Alto, California.

15. Macchi, E. 1990
    Power generation (including cogeneration), draft manuscript, Polytechnic University, Milan.

16. Anonymous. 1991
    Low-cost "air bottoming cycle" for gas turbines, *Gas Turbine World* 21(3):61.

17. de Candia, F. 1989
    ISTIG enhancement evaluation, vol. 1, Pacific Gas & Electric Co., R&D Department, San Ramon, California.

18. Cool Water Coal Gasification Program and Radian Corp. 1990
    Cool water coal gasification program: final report, GS-6806, Electric Power Research Institute, Palo Alto, California.

19. Roll, M.W. and Payonk, R.J. 1990
    Operation of the Dow coal gasification process during 1990, presented at the Ninth Annual Conference on Gasification Power Plants, Electric Power Research Institute, Palo Alto, California.

20. Wolde, D.G. 1990
    The 250 MW Netherlands IGCC project, presented at the Ninth Annual Conference on Gasification Power Plants, Electric Power Research Institute, Palo Alto, California.

21. Corman, J. C. 1986
    System analysis of simplified IGCC plants, report prepared for the US Department Energy by General Electric Co., Corporate R&D, Schenectady, New York.

22. Pitrolo, A.A. and Graham, L.E. 1990
    DOE activities supporting the IGCC technologies, chapter 1a, *Proceedings of the Conference on Integrated Gasification Combined Cycle Plants for Utility Applications,* Canadian Electrical Association, Montreal.

23. Larson, E.D., Svenningsson, S., and Bjerle, I. 1989
    Biomass gasification for gas turbine power generation, in T.B. Johansson, B. Bodlund, and R.H. Williams, eds., *Electricity: efficient end-use and new generation technologies and their planning implications*, 697–739, Lund University Press, Lund, Sweden.

24. Corman, J.C. 1987
    Integrated gasification–steam injected gas turbine (IG–STIG), presented at the Workshop on Biomass-Gasifier Steam-Injected Gas Turbines for the Cane Sugar Industry (Arlington, Virginia), organized by the Center for Energy and Environmental Studies, Princeton University, Princeton, New Jersey.

25. Antal, M.J. 1980
Thermochemical conversion of biomass: the scientific aspects, *Energy from biological processes,* Vol. IIIC, Office of Technology Assessment, Washington DC.

26. Waldheim, L. and Rensfelt, E. 1982
Methanol from wood and peat, 239–259, in T.B. Reed and M. Graboski, eds., *Proc. Biomass-to-Methanol Specialists Workshop,* SERI/CP-234-1590, Solar Energy Research Institute, Golden, Colorado.

27. General Electric Company. 1992
Biomass feedstock evaluations: Vermont program, Corporate R&D, Schenectady, New York, 11 March.

28. Horner, M. W. 1985
Simplified IGCC with hot fuel gas combustion, ASME Paper 85-JPGC-GT-13, American Society of Mechanical Engineers, New York.

29. Scandrett, L. A. and Clift, R. 1984
The thermodynamics of alkali removal from coal-derived gases, *Journal of the Institute of Energy* 57:391–397.

30. Corman, J.C. and Horner, M.W. 1986
Hot gas clean-up for a moving bed gasifier, chapter 14, AP-4680, *Proceedings of the 5th Annual Contractors Conference on Coal Gasification,* Electric Power Research Institute, Palo Alto, California.

31. Kurkela, E., Stahlberg, P., Laatikainen, J., and Nieminen, M. 1991
Removal of particulates and alkali metals from pressurized fluid-bed gasification of peat and biomass—gas cleanup for gas turbine applications, in D.L. Klass, ed., *Energy from biomass and wastes XV,* Institute of Gas Technology, Chicago.

32. Elliott, P. and Booth, R. 1990
Sustainable biomass energy, Selected Paper PAC/233, Shell International Petroleum Co., London.

33. Carpentieri, E. (Division Chief, Alternative Energy Sources, Hydroelectric Company of São Francisco, Brazil). 1990
Forest plantations for utility electricity in northeast Brazil, presentation at Winrock International Institute for Agricultural Development, Arlington, Virginia, 15 November.

34. San Martin, R. (Deputy Assistant Secretary, Office of Utility Technologies, Division of Conservation and Renewables). 1990
DOE research on biomass power production, presentation at the Conference on Biomass for Utility Applications, Tampa, Florida, 23–25 October.

35. Trenka, A.R., Kinoshita, C.M., Takahashi, P.K., Caldwell, C., Kwok, R., Onischak, M., and Babu, S.P. 1991
Demonstration plant for pressurized gasification of biomass feedstocks, in D.L. Klass, ed., *Energy from biomass and wastes XV,* Institute of Gas Technology, Chicago.

36. Evans, R.J., Knight, R.A., Onischak, M., and Babu, S.P. 1988
Development of biomass gasification to produce substitute fuels, PNL-6518, Battelle Pacific NW Laboratory, Richland, Washington.

37. Hulkkonen, S., Raiko, M., and Aijala, M. 1991
New power plant concept for moist fuels, IVOSDIG, Paper 91-GT-293, American Society of Mechanical Engineers, New York.

38. Blomen, L.J.M.J. 1989
Fuel cells: a review of fuel cell technology and its applications, in T.B. Johansson, B. Bodlund, and R.H. Williams, eds., *Electricity: efficient end use and new generation technologies, and their planning implications,* Lund University Press, Lund, Sweden.

39. George, T.J. and Mayfield, M.J. 1990
    Fuel cells: technology status report, US Department of Energy, Morgantown Energy Technology Center, Morgantown, West Virginia.

40. Douglas, J.1990
    Beyond steam: breaking through performance limits, *The EPRI Journal* 15(8):411.

41. Crutzen, P.J. and Andeae, M.O. 1990
    Biomass burning in the tropics: impact on atmospheric chemistry and biogeochemical cycles, *Science* 250:1669–1678.

42. Bajura, R.A. and Bechtel, T.F. 1990
    Update on DOE's IGCC Program. U.S. DOE, Morgantown Energy Technology Center, Morgantown, West Virginia.

43. Cook, J.H., Beyea, J., and Keeler, K.H. 1991
    Potential impacts of biomass production in the United States on biological diversity, *Annual Review of Energy* 16:401–431.

44. Energy Information Administration. 1989
    International Energy Annual 1988, DOE/EIA-0219(88), US Department of Energy, Washington DC.

45. Ogden, J.M., Hochgreb, S., and Hylton, M. 1990
    Steam economy and cogeneration in cane sugar factories, *International Sugar Journal* May/June.

46. Ogden, J.M., Williams, R.H., and Fulmer, M.E. 1991
    Cogeneration applications of biomass gasifier/gas turbine technologies in the cane sugar and alcohol industries: getting started with bioenergy strategies for reducing greenhouse gas emissions, in J. Tester, ed., *Proceedings of the Conference on Energy and Environment in the 21st Century*, MIT Press, Cambridge, Massachusetts.

47. Larson, E.D. and Williams, R.H. 1990
    Biomass-gasifier steam-injected gas turbine cogeneration, *J. Eng. for Gas Turbines and Power* 112:157–163.

48. Alexander, A.G. 1985
    The energy cane alternative, Elsevier, Amsterdam.

49. Food and Agriculture Organization. 1990
    1988 yearbook of forest products, United Nations, Rome.

50. Food and Agriculture Organization. 1982
    World forest products demand and supply 1990 and 2000, Forestry Paper 29, United Nations, Rome.

51. Larson, E.D. 1991
    Biomass-gasifier/gas turbine cogeneration in the pulp and paper industry, *J. Eng. for Gas Turbines and Power* (forthcoming).

52. Jaakko Poyry Oy. 1989
    Assessment of market prospects for the chemrec black liquor gasification process, Jaakko Poyry Oy, Helsinki, March.

53. Stigsson, L. 1989
    A new concept for kraft recovery, in *Proceedings of the 1989 International Chemical Recovery Conference*, Tappi, Atlanta, Georgia.

54. Manufacturing and Technology Conversion International, Inc. 1990
    Testing of an advanced thermochemical conversion reactor system, Battelle Pacific Northwest Laboratory, Richland, Washington.

55. Kelleher, E.G. 1985
Black liquor gasification and use of the product gas in combined-cycle cogeneration, phase II, *Tappi Journal* 106–110, November.

56. Kignell, J-E. 1990
Novel concept for chemicals and energy recovery from black liquor, in *Proceedings of the EUCEPA*, Swedish Pulp and Paper Association, Stockholm.

57. Alsefelt, P. 1989
Energy from the forest industry, Swedish Technology Development Board, Stockholm.

58. Shell Briefing Service. 1990
Focus on Forestry, Shell International Petroleum Co., London.

59. National Wood Energy Association. 1990
National biomass facilities directory, National Wood Energy Association, Arlington, Virginia.

60. Electric Power Research Institute. 1986
Technical assessment guide 1: electricity supply 1986, Electric Power Research Institute, Palo Alto, California.

61. Jenkins, B.M. 1989
Physical properties of biomass, 860–891, in O. Kitani and C.W. Hall, eds., *Biomass Handbook*, Gordon and Breach Science Publishers, New York.

62. Smelser, S.C. and Booras, G.S. 1990
An engineering and economic evaluation of CO_2 removal from fossil fuel-fired power plants, presented at the Ninth Annual Conference on Gasification Power Plants, Electric Power Research Institute, Palo Alto, California.

63. Carpentieri, E. 1991
An assessment of sustainable bioenergy in Brazil: national overview and case study for the northeast, PU/CEES Working Paper 119, Princeton, Center for Energy and Environmental Studies, Princeton University, New Jersey.

64. Strauss, C.H. and Wright, L.L. 1990
Woody biomass production costs in the United States: an economic summary of commercial populus plantation systems, *Solar Energy* 45 (2):105–110.

65. Strauss, C.H., Grado, S.C., Blankenhorn, P.R., and Bowersox, T.W. 1988
Economic valuations of multiple rotation SRIC biomass plantations, *Solar Energy* 41(2):207–214.

66. Barnett, P.E. 1985
Evaluation of roll splitting as an alternative to chipping woody biomass, presented at the Biomass Energy Research Conference, University of Florida, Gainesville, Florida, March 12–14.

67. Ashmore, C. 1985
Preliminary analysis of roll crushing of hybrid poplar using the FERIC roll crusher (unpublished).

68. Frea, W.J. 1984
Economic analysis of systems to pre-dry forest residues for industrial boiler fuel, in D.L. Klass, ed., *Energy from biomass and wastes VIII*, Institute of Gas Technology, Chicago.

69. World Resources Institute. 1990
World resources 1990–91, Oxford University Press, New York.

70. Office of Technology Assessment. 1980
Energy from biological processes, Vol. III, Appendices, part A: energy from wood, US Congress, Washington DC.

71. Food and Agriculture Organization of the United Nations. 1980
1978 yearbook of forest products, United Nations, Rome.

72. Food and Agriculture Organization of the United Nations. 1987
    FAO production yearbook, Vol. 41, United Nations, Rome.

73. U.S. Department of Commerce. 1990
    Statistical abstract of the United States 1990, US Government Printing Office, Washington
    DC.

74. U.S. Department of Agriculture. 1978
    Agricultural Statistics 1978, US Government Printing Office, Washington DC.

75. Office of Technology Assessment. 1980
    Energy from biological processes, Vol. II, technical and environmental analyses, US Con-
    gress, Washington DC.

76. Goldemberg, J., Moreira, J.R., Dos Santos, P.U.M., and Serra, G.E. 1985
    Ethanol fuel: a use of biomass energy in Brazil, *Ambio* 14 (4–5):293–297.

77. Escay, J.R. 1990
    *Summary data sheets of 1987 power and commercial energy statistics for 100 developing coun-
    tries,* Industry and Energy Department Working Paper, Energy Series Paper No. 23, World
    Bank, Washington DC.

78. U.S. Department of Commerce. 1990
    Statistical abstract of the United States 1989, US Government Printing Office, Washington
    DC.

79. Moore, E.A. (IENED) and Smith, G. 1990
    Capital expenditures for electric power in the developing countries in the 1990s, Industry and
    Energy Department Working Paper, Energy Series Paper No. 21, World Bank, Washing-
    ton DC.

18
BIOGAS ELECTRICITY— THE PURA VILLAGE CASE STUDY

P. RAJABAPAIAH
S. JAYAKUMAR
AMULYA K.N. REDDY

A potentially useful decentralized source of energy is biogas, which is an approximately 60:40 mixture of methane (CH_4) and carbon dioxide (CO_2), produced by the anaerobic fermentation of cellulosic biomass materials such as bovine wastes.

Since 1987, the traditional system of obtaining water, illumination, and fertilizer in Pura village in south India has been replaced with a community biogas plant electricity-generation system. The technical, managerial, and economical aspects of this system are the subject matter of the present paper. Various subsystems are described, and the problems of operation and maintenance under field conditions are also discussed.

A comparison of Pura's present community biogas system with its traditional means for obtaining water, illumination, and fertilizer shows that the households are winners on all counts, having obtained such benefits as improved hygiene and convenience at relatively low-cost.

The Pura community biogas plant is held together and sustained by the convergence of individual and collective interests. Noncooperation with the community biogas plant results in a heavy individual price (access to water and light being cut off by the village), which is too great a personal loss to compensate for the minor advantages of noncooperation and noncontribution to collective interests.

INTRODUCTION

Thus far, energy planning choices have been confined to *centralized* energy supply technologies—hydroelectric, oil-based or coal-based thermal, nuclear, and of late, natural gas–based power plants. But this restriction runs increasingly into two major difficulties: 1) shortages of capital and 2) opposition to such plants as a result of local and global environmental degradation. It is therefore essential to include *decentralized* sources of supply.[1]

---

1. In fact, planning must not be limited only to *supply* options but extended to include *energy efficiency improvement and other conservation options* (because saving energy means less need be generated). But energy saving is outside the scope of the present paper. It is, however, dealt with in [1].

One of the potentially useful decentralized sources of energy is biogas [2]—an approximately 60:40 mixture of methane (CH_4) and carbon dioxide (CO_2)—produced by the anaerobic fermentation of cellulosic biomass materials. Biogas has a calorific value of 23 megajoules (MJ) per cubic meter (m^3) and can fuel engines that in turn drive generators to produce electricity.

Many cellulosic biomass materials are available in the rural areas of developing countries. In particular, because of the huge bovine holdings in many countries such as India,[2] bovine wastes represent a cellulosic biomass source of considerable potential. Traditionally, these wastes are carefully collected in India and used as fertilizer, except in places where villagers are forced by the scarcity of firewood to burn dung cakes as cooking fuel. Insofar as biogas plants also yield fertilizer (as a sludge that performs better than farmyard manure), the generation of biogas fuel and/or electricity is a valuable bonus.

It is this bonus output that has motivated the large biogas programs in a number of developing countries—particularly in India and China [3–5]. Almost all biogas programs are based on family-sized plants rather than community biogas plants.[3] However, family-sized biogas plants lose significant economies of scale; also, their biogas output is suited more for cooking than for running an engine and generating electricity [6]. In addition, the low body weight of free-grazing bovine animals, particularly in drought-prone areas, can make bovine wastes inadequate to meet cooking energy needs, even though the bovine to human population ratio may seem satisfactory. In such situations, the use of community biogas plants for electricity generation is worth considering. Community biogas plants are more economical, but their main problems are social rather than technical [7]; they bring in their wake serious difficulties of organization and possibly issues of equity in the distribution of costs and benefits.

Biogas is an ideal fuel to run an engine for it can be converted into shaft power that can then drive a generator and generate electricity. Diesel engines are suitable for this purpose because:

▼ They have a high thermal efficiency compared with other types of engines and are well suited for the low-flame velocity of biogas [8].

▼ They are more extensively used in rural areas than other types of engines.

▼ Their normal working life (4 to 8 years) is more than that of other types of engines.

▼ They are reliable and simple to maintain.

---

2. The bovine to human ratio in India is 0.36, where bovine refers to cattle and buffaloes. For comparison, the cattle to human ratio is 0.27 and the livestock to human ratio is 0.57, where livestock includes sheep and goats and bovine animals.

3. David Stuckey has pointed out (personal communication) that there are perhaps no more than 20 community (as opposed to large-scale) plants in the world.

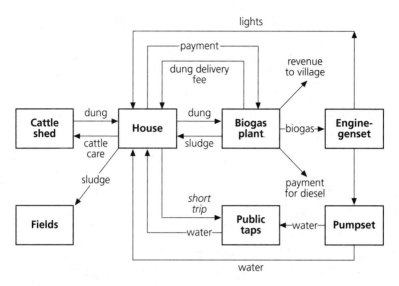

FIGURE 1: The existing community biogas plant system at Pura.

▼ They can be easily converted to the dual-fuel (biogas–diesel) mode, which is the most practical and efficient method of using biogas.

▼ They can switch smoothly without interruption to conventional diesel operation in case of a shortfall in biogas supply during an important operation.

Moreover, the use of biogas in dual-fuel engines is ideal for electricity generation in rural areas because

▼ It is a clean fuel, producing little or no pollution during combustion, unlike diesel fuel.

▼ It is a locally available and renewable source of energy.

▼ It can be produced cheaply with indigenous technology.

▼ It can provide employment to local people.

▼ It makes rural electricity systems self-reliant.

The decentralized biogas electricity system in Pura village

It was against this background that a decentralized biogas electricity system was established and demonstrated at Pura village (at Kunigal Taluk in the Tumkur district of Karnataka state in south India) as an alternative for providing rural electricity. In September 1987, the traditional system of obtaining water, illumination, and fertilizer in Pura was replaced with a community biogas-plant electricity-generation system, (see figure 1). The technical and economic aspects of this system are the subject matter of the present paper; the evolution and managerial aspects are dealt with elsewhere [9].

Table 1: Pura village: some statistics
April 1991

Population	463	
Cattle population	248	
Human/cattle ratio	1.9	
Households	87	
with grid electricity	39	(45 percent)
with grid + biogas electricity	24	(28 percent)
with biogas electricity	39	(45 percent)
with private water taps	29	(33 percent)
Water consumption before biogas system	17 liters per person per day	
Present water consumption	24 liters per person per day	

When the new system was introduced, Pura had already been connected to the Karnataka Electricity Board grid. In general, only 20 to 30 percent of the homes are electrified in electrified villages in India; in Pura, 45 percent of the homes were electrified (see table 1).

Even this step of limited electrification that took place in Pura may not be possible in all other villages in the near future because:

▼ Electricity has become scarce and expensive in India.

▼ Rural areas have been neglected in conventional electricity planning, apart from the recent energization of irrigation pumpsets. For example, in Karnataka only 20 percent of electricity produced flows to rural areas [1].

▼ There are enormous costs and losses involved in transmission and distribution (T&D) lines. For example, the T&D losses are about 21.5 percent in Karnataka.

▼ Electricity has become extremely unreliable in rural areas with regard to both duration (there is frequent load-shedding) and voltage.

▼ Even in electrified villages electricity is not accessible to most of the people.

THE PURA COMMUNITY BIOGAS SYSTEM

Technical subsystems
The community biogas plant system consists of the following subsystems:

▼ The biogas plant itself, in which bovine waste is anaerobically fermented to yield biogas.

▼ A sandbed filtration subsystem to filter the biogas plant slurry output and deliver filtered sludge with approximately the same moisture content as dung.

▼ The electricity generation subsystem.

▼ The electricity distribution subsystem for the electrical illumination of homes.

▼ The water supply subsystem.

Biogas plants

The two most popular conventional digesters for highly concentrated bovine and other animal dung, as well as agro-wastes in India, China, and other developing countries are the Indian floating-drum biogas plant[4] and the Chinese fixed-dome-type biogas plant.

In the Indian design, an inverted drum with a diameter very slightly less than that of the cylindrical digestion pit (usually but not necessarily below ground level) serves as a gas tank and provides an anaerobic seal while floating up and down depending upon the amount of biogas stored. Such a plant delivers gas at uniform pressure, provides a good seal against gas leakage, is highly reliable and robust, and has a proven performance for bovine dung digestion. Its drawback is that the gas holder is usually made of steel or ferrocement, and so is comparatively costly and requires regular maintenance.

The Chinese fixed-dome type biogas plant can be constructed locally with standard building materials such as cement and is cheaper because it is less materials-intensive. However, it is skill-intensive and is prone to gas leaks (notwithstanding epoxide coatings of mortar on the inside surface) if construction is not of high quality.

A plug-flow biogas reactor is useful for high volumetric gas production rates relative to typical fixed-dome and floating-drum plants. Its construction is similar to these two types of plants or a combination of both; however, to ensure true plug-flow conditions, length has to be considerably greater than width and depth. Though plug-flow biogas reactors may be appropriate to developing countries because of their low capital cost, they are still in the initial stages of dissemination in these countries.[5] Plug-flow reactors may not display special advantages in the case of bovine dung digestion, but they permit continuous gas production from biomass sources such as the water hyacinth and other aquatic weeds that tend to float.

---

4. The Indian design is also known as the Khadi and Village Industries Commission, or KVIC, design.

5. According to David Stuckey (personal communication), "there have been full-scale (120 m³) plug-flow reactors for almost 20 years (e.g., at Cornell University in the U.S.) and the small-scale ones developed by the Taiwanese have been operating for 15 years. They are easy to install and operate, and relatively economical although they have not diffused to a large extent probably because they take up more land area than below-ground units."

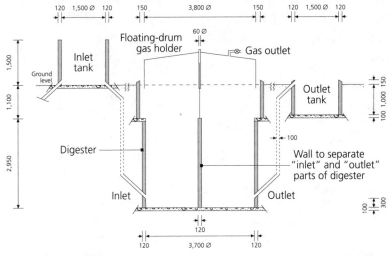

All dimensions in millimeters
Scale = 1:1,250

FIGURE 2: Sectional elevation of the biogas plant at Pura.

The biogas digesters in Pura are the Indian floating-drum type. The detailed dimensions were based on the cost minimization theory developed earlier [10] and on realistic residence times that had been observed under similar conditions elsewhere. Each digester in Pura is 4.1 meters in diameter and 4.2 meters deep. Also, low-cost construction techniques were used (see figure 2).[6]

The following are the salient features of the modified design:

▼ The volumetric ratio (gas produced per unit volume of the digester) is as high in the Pura floating-drum design as in plug-flow reactors, i.e., 0.5 compared with 0.2 to 0.3 in conventional fixed-dome and floating-drum plants.

▼ The plants have performed better than the original Khadi and Village Indisturies Commission (KVIC) plants: they produce 14 percent more biogas at ambient temperature in spite of the 40 percent reduction in digester volume.

▼ The plants are shallower and wider than conventional plants, thereby accelerating the rate of gas release from the production zone to the gas holder; hence, the modified plants are easier to construct wherever the groundwater table is high.

---

6. The low-cost techniques adopted by ASTRA (the Centre for the Application of Science and Technology to Rural Areas) included the following: 1) based on structural analysis, the minimum thickness used for the 4.2-meter-high digester wall is just 120 millimeters compared to the 360 millimeters of the conventional digesters; 2) ordinary plastering for the interior of the digester wall (because the dung slurry itself is a good sealant) in contrast to the multilayer plastering with a coating of leak-proofing compound; and 3) precise excavation to the size of the digester plus walls to enhance the strength of the wall as well as to minimize the refilling and thereby reduce the cost.

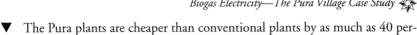

▼ The Pura plants are cheaper than conventional plants by as much as 40 percent.

In order to increase the reliability of the system, it was decided to construct two plants (each with half the rated gas production capacity), having a common inlet tank, instead of a single plant.

The maximum input to the biogas plants is 1.25 tonnes of cattle dung mixed with 1.25 m³ of water per day [10]. At this maximum loading, the influent slurry mixture contains 212 kilograms of dry matter (8.5 percent) having a volatile matter content of 177 kilograms (7 percent). The carbon content of the mixture is 57 kilograms (27 percent of the dry matter); the nitrogen content is about 3.6 kilograms (1.7 percent), and the carbon–nitrogen ratio is 16.

The plants can produce, at an average ambient temperature of 25 to 26°C, a maximum of 42.5 m³ biogas per day. In addition to the gas, the charging of the dung + water slurry would displace about 2.45 m³ per day of digested slurry, which yields about 1.2 tonnes per day of sludge (after the water has been removed by filtration). This slurry contains 164 kilograms (6.67 percent) of dry matter and 109 kilograms (4.45 percent) of volatile matter, 39 kilograms (24 percent of the dry matter) of carbon, and 3.6 kilograms (2.2 percent) of nitrogen—the same amount of nitrogen as in the input. The output carbon–nitrogen ratio is 11.

Human waste as an input

In China, human excrement is traditionally used directly on the fields as a fertilizer even though there is a risk of spreading intestinal parasites and other pathogens. Chinese biogas plants usually have a settling chamber below the digestor where the detention time is very long (about six months), leading to the destruction of more than 90 percent of the intestinal parasites and other pathogens. Thus, biogas plants perform an important sanitary function in China.

However, intestinal parasites, which are endemic in rural areas of India, are unlikely to be destroyed in the short detention times of Indian biogas plants. As a result, if biogas sludge is used as a fertilizer, it is likely to increase the spread of intestinal diseases. Moreover, human waste is not traditionally used as a fertilizer in India, and "contamination" of the sludge with human waste may create resistance to the acceptance of the sludge fertilizer. Hence, human excrement is *not* used as an input to the community biogas plant at Pura.

Sludge fertilizer from the biogas plants

Because nitrogen does not volatilize during anaerobic digestion, the effluent sludge from the biogas plant contains the same amount of nitrogen as the input slurry. However, the nitrogen content increases as a percentage of total solids (because the latter percentage decreases from 8.5 to 6.7 percent) and, furthermore, is converted into a form that is more readily usable by plants. Hence, biogas plants are often known as *biofertilizer plants*.

In fact, anaerobically digested biogas sludge has a higher nitrogen content than farmyard manure obtained by composting bovine dung. This is because bovine dung is traditionally placed in open-air compost pits before it is transfered to the fields. Due to the aerobic decomposition that takes place in open air, the nitrogen in farmyard manure decreases from 1.7 percent on a dry-weight basis to a constant value of 0.9 percent in about ten days. In contrast, the nitrogen content of biogas plant sludge decreases from 2.2 percent to a constant 1.9 percent in about three days in open air [12]. Thus, biogas sludge stabilizes with *double* the nitrogen content of farmyard manure. The greater nitrogen content of biogas sludge relative to manure implies a saving of energy that would otherwise have to be used to manufacture an equivalent amount of nitrogen in the form of artificial fertilizer.

Based on their nine years of experience, the farmers of Pura assert that weed growth is far less with biogas sludge fertilizer and, therefore, they use it for premium purposes, such as nurseries. Whereas farmyard manure sows the seeds of the weeds that are ingested by bovine animals and pass through their digestive systems into their dung, the biogas plant either destroys these seeds or makes them less fertile through anaerobic digestion.

The anaerobic process of digesting cattle dung also has an environmental advantage. Unlike cattle dung undergoing aerobic decomposition, biogas sludge neither smells nor attracts flies and mosquitoes. The people of Pura even say that biogas sludge repels termites—raw dung attracts them, and they harm the plants fertilized with manure. For these same reasons, the villagers prefer digested slurry to fresh dung for plastering their threshing yards.

Sandbed filters

Because of its multiple benefits, sludge fertilizer attracts special attention in community biogas systems. The households in Pura refused to sell the dung to the biogas plant; instead, they agreed to "loan" it to the plant so that it can be used for anaerobic digestion but insisted that the sludge should be returned to them on a pro-rata basis. However, because the dung, which is 18 percent solids, enters the plants after being mixed with an equal volume of water it becomes a diluted effluent with only about 6.5 percent solids. This watery effluent was unacceptable to the villagers because they could not transport it back to their homes.

Separating solids and liquid in the slurry effluent is not possible with sewage-type sludge sandbed dryers. It was necessary therefore to develop a special filtration system for biogas plant effluent. Such a system would make it easier to transport digested sludge from the biogas plant to the villagers' homes and compost pits. In addition, the system would enable the filtrate, which contains some anaerobic microorganisms,[7] to be mixed with the input dung, thereby marginally enhancing gas production and reducing the amount of water needed to charge the biogas plants.

---

7. Unpublished laboratory data of Stuckey (personal communication) suggests that "almost all the anaerobes are attached to the solid lignocellulosic particles, hence recycling liquid saves water, but would not increase cell concentration in the reactor." Chanakya and Ramaswamy (personal communication) have both found anaerobes suspended in the filtrate also, though most of them are attached to the solid particles.

The sandbed filtration system that was developed for digested slurry is simple but effective. The 11 filters constructed at Pura can together handle as much as 1.7 m^3 per day of slurry. Each filter measures 4 meters × 1 meter and consists of three layers: 5 centimeters of gravel at the bottom, then 3 centimeters of sand, and on top, wire mesh. Digested slurry effluent is poured to a height of 10 centimeters above the wire mesh.

The area of filter required is about 1 m^2 for every 100 liters of the digested slurry. The maximum residence time for filtration varies with the season and is about 72 hours in the rainy season and about 60 hours in the summer. Thus, for a steady-state operation, 3 m^2 area is required for 100 liters of slurry effluent. The maximum recovery of water from the filter is about 70 percent.

The sand beds last about a year, after which the sand has to be completely removed and relaid. The gravel and water outlet pipes also have to be cleaned and relaid after a year. The lining material, consisting of low-density polyethylene (LDPE) sheet, may have to be repaired or replaced.

The operation and maintenance of these filters is so easy that two village youths have, in a short period of a month, learned to do it, as well as the routine cleaning and upkeep of the filters. They even innovated a simple technique to prevent the dried sludge from clogging the wire mesh, distributing a thin film of wet sand over the wire mesh before spreading the slurry.

After sandbed filtration, the slurry displaced from the digester by the daily charging of dung mixed with water contains 17 percent total solids; i.e., the filtered sludge resembles cattle dung, which contains 18 percent total solids. At this stage, it would be possible to return filtered sludge to the villagers at the rate of 750 grams per kilogram of dung received. But because the villagers now understand the whole biogas process and have confidence in the distribution system, they do not withdraw the sludge after sandbed filtration. Instead, they use the biogas system as a "sludge bank" and allow considerable time to elapse before the sludge is withdrawn. During this period, there is further decrease in moisture content and an increase in total solids. Thus, it has become the accepted practice to return filtered and dried sludge to the villagers at the rate of 600 grams per kilogram of dung delivered to the biogas plant.

The long-term performance of Pura's biogas plants

Biogas plants are normally designed on the basis of either the minimum dung available or the maximum gas consumption that is required. Gas production depends on the amount of cattle dung and the ambient temperature. This temperature dependence is the reason for the universal complaint that biogas plants produce very little gas in winter and so other fuels are necessary to supplement biogas. But at Pura for the past nine years, gas production has been virtually uniform throughout the year. Paradoxically, reduction in output, if at all, occurs in summer and not in winter. This can be understood as follows.

The amount of dung available to the plant depends on the number of bovine animals and the fodder intake of these animals. In the case of free-grazing bovines, fodder intake depends on the grass cover (in the pastures), which in turn depends on the rainfall, which is seasonal. As a result, the dung yield varies by a factor of two between the seasons, which means that the loading rate (= digester volume × total solids concentration) also varies. The ambient temperature also has seasonal variations. It is interesting and fortunate that the shifts from minimum to maximum and vice versa in both dung yield and ambient temperature are gradual, and in summer the temperature is highest but the dung yield (loading rate) is lowest, while in winter the temperature is lowest but the dung yield is highest. Earlier findings [13] have emphasized that the response of biogas plants to these variations in loading rates is slow and gradual. And, the other important parameter—pH—is uniform throughout the year.

Dung loaded in winter stays in the digester for a long time, and because of lower summer loading rates, contributes to gas production even in the summer. Hence, the bigger the plant, the slower the response and the more uniform the gas production. The gas yield (gas per unit weight of input) also increases with the size and diameter and depth of the plant.[8] At locations where, despite the economies-of-scale in biogas plant costs, the cost of the plant is not as important as the availability of dung, long residence times of up to a year can be recommended. In south India, it has turned out, quite surprisingly, that the dung available for loading the biogas plant in the summer is the most important parameter for plant designers.

Electricity generation

A 7-horsepower (5.2 kilowatt) water-cooled biogas–diesel (dual-fuel) engine has been installed in an engine room (5.05 × 3.50 meters) next to the fields at the edge of the village. The engine has been mounted on antivibration footings and bolted firmly to the ground. The exhaust pipe, attached to a residential-type silencer, has been extended through the engine room wall to the open air in a direction toward the fields and away from the village. With such a construction, the engine is hardly audible in the village. Biogas from the biogas plant passes through a condensation trap and then enters the engine where it is mixed with diesel to provide the fuel. The engine is coupled to a 5 kVA 440 volt three-phase generator (to enable the operation of a three-phase submersible pump).

Electrical illumination of homes

The lighting system was energized on 2 October 1988.[9] It consists of 103 twenty watt fluorescent tube lights—97 in homes, two tube lights in a public temple, and four tube lights in the biogas plant complex. Forty-seven houses have one tube

---

8. This observation is applicable to digesters with a depth of greater than 0.5 meters, i.e., almost all conventional digesters.

9. October 2 was chosen to inaugurate the illumination of homes because it is celebrated in India as the birth anniversary of Mahatma Gandhi, who urged the country to "wipe every tear from every face." To an implementor of energy plans, Gandhi's call translates to illuminating homes that are an "area of darkness" (the title of a novel by V.S. Naipaul).

Table 2: Replacement of tube lights
99 tube lights installed on 2 October 1988

Month and year	Number replaced	Total hours of operation
August 1990	1	1,580
August 1990	7	1,600
November 1990	15	1,890
November 1990	5	1,893
December 1990	30	1,957

light and 25 have two tube lights. The load is distributed equally over the three phases, i.e., with 34 tube lights in each phase. The low power factor of 0.43 of the tube light system (consisting of the fluorescent lamp and the choke) has been improved to 0.72 by connecting each tube light with a 4 microfarad capacitor in parallel and, as a result, the power consumption has decreased from 31 watts to 27 watts. The life of the fluorescent lamps has been rated at 6,000 hours by manufacturers . However, it was decided to make an empirical study of this important parameter. From August to December 1990, 58 out of 99 tube lights were replaced after a life of between 1,580 and 1,957 hours of operation (see table 2).

Water supply
The water-supply system has been operating since September 1987. It consists of a three-phase, 3 horsepower (2.24 kilowatt), 6.75 m^3 per hour submersible pump fitted into a bore well. This pump lifts water from a 50 meter depth to an overhead tank. The water is then distributed by gravity through nine street taps in the village. The location of the taps was decided by the villagers. One of the taps is for livestock and one tap is in the biogas plant compound. In addition, as of September 1990, there were 29 private taps inside the households.

A 1977 study [14] of the traditional system of water collection for domestic purposes in Pura showed that, on average, a family made two trips per day spending 1.5 hours (45 minutes per trip) to cover 1.6 kilometers and transport 104 liters (4 potfuls) home for a per capita water consumption for domestic purposes of 17 liters per day. A survey in September 1987 showed that consumption had not changed significantly. However, between September 1987 and September 1988, when a 24-hour supply of piped water became available through public taps, consumption jumped immediately to 22 liters per capita and then slowly stabilized at 26 liters. After the villagers took over the management of the community biogas plant, they restricted access to the water supply (three times in a day) and the consumption came down to 22 liters between October 1988 and

August 1990. The initial 5 liter increase in consumption can be attributed to the fact that villagers allowed their bovines to drink the piped water.

But after providing private taps in 29 households, consumption again increased to 30 liters between September 1990 and April 1991. During this period, it was observed that some households were using water from their private taps for vegetable gardening within their house plots.[10] Apart from this, it is not yet known for what other purpose the water is being used. However, their habits, for example, the practice of bathing once in a week do not seem to have changed.

Operation of the system

The operation of the system consists of the following activities (see figure 1):

▼ Delivery of cattle dung by households owning bovines (24 percent by women, 27 percent by girls, 27 percent by boys, and 22 percent by men).

▼ Weighing the dung delivered by the households and recording the quantities in their passbooks and in the ledger books of the plant.

▼ Returning processed sludge to those who want any.

▼ Mixing the dung with water in a 1:1 ratio (by volume) and charging the plants with the dung + water mixture.

▼ Pouring the slurry displaced from the biogas plant by the dung and water mixture on sandbed filters.

▼ Releasing biogas from the plants and feeding it to the engine, adding the requisite amount of diesel, and starting the dual-fuel engine and the electrical generator.

▼ Supplying electricity either for running the submersible pump and pumping the borewell water to the overhead tank or for electrical illumination of homes.

▼ Keeping the biogas plants and their surroundings clean.

▼ Visiting households to receive payments for the electric lighting and to make payments for the delivery of dung to the plant.

▼ Maintaining the plant records and accounts.

Apart from delivery of dung to the plant and withdrawal of sludge, all the activities involving the operation of the biogas plants, the electricity generation and distribution subsystem, and the water supply subsystem are carried out by two village youths employed by the project.

---

10. Vegetable gardening with tap water in places other than their house plots was banned by the village development society long ago.

Maintenance of the biogas plants

The gas holders are painted regularly once in two years with chlorinated rubber black paint. At the time of fabrication, the material was made rust-free, primed with a noncorrosive primer followed by two coats of chlorinated rubber paint. In spite of these corrosion-prevention measures, after five years of operation, corrosion was observed at the joint where side sheet and top sheet are welded.

After four years of regular operation, when the plants were being renovated after a temporary suspension period, it was observed when the plants were emptied that about a 0.3 meter depth of sand and mud had settled at the bottom of the digester in spite of attempts to keep the charge free of sand. The sediments were removed.

Maintenance of the electricity generation subsystem

Maintenance of the electricity generation subsystem is carried out by the youths who are operating and maintaining the biogas plants and electricity and water subsystems. The salient features of the maintenance operations during the 44 months and 4,521 hours between September 1987 and April 1991 are as follows:

▼ The engine-genset has not required major repairs. The fuel injection nozzle of the engine has been cleaned once and replaced once, and the filter has been changed once; the rectifier, carbon brushes, and field coil of the generator have each been replaced once.

▼ The engine-genset repairs, which have been quite minor, have been mainly associated with the engine accessories: foundation bolts, radiator, silencer, etc.

▼ The daily operation and maintenance activities have been simplified by means of a flow chart (see figure 3) and a trouble-shooting chart (see figure 4).

▼ In addition, the operators carry out preventive engine maintenance and minor repairs after every 50, 200, 500, and 800 hours of operation.

▼ The training and skills acquired by the operators during the past 44 months of operation represent a significant contribution.

▼ The dual-fuel engine has proved reliable for biogas electricity generation systems.

Field performance

Between September 1987 and April 1991, the engine ran for about 4,521 hours [9] (see table 3), and an overall diesel replacement of 77 percent was achieved, with 4,483 liters of diesel replaced by 11,894 m³ of biogas. When abnormal values are excluded from the averaging, the diesel replacement turns out to be 80 percent (see tables 4 and 5; figures 5 and 6). Hence, the dual-fuel engine is efficient and can be switched over smoothly to diesel whenever the gas stock becomes zero, but the operators must avoid running on pure diesel if the aim is to get high diesel replacement values.

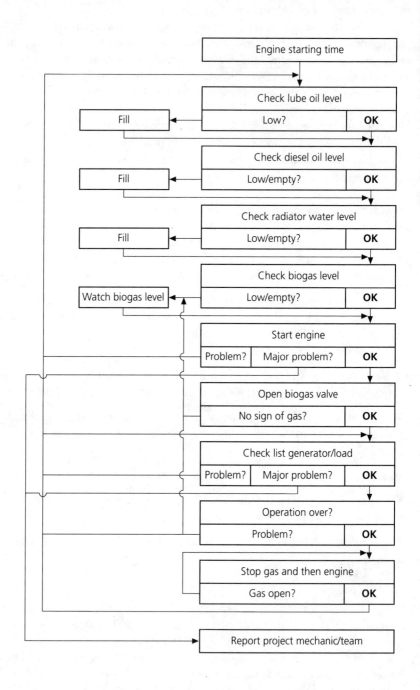

FIGURE 3: Flow chart for the operation of the dual-fuel engine at Pura.

Administration, organization, and institution-building

Community technologies require the proper operation of administrative arrangements, the establishment of suitable organizations, and the building of appropriate institutions. The crucial administrative step in Pura was establishing a scheme for dung collection and sludge return based on a delivery fee (of Rs. 0.02 or 0.1586 cents per kilogram), which goes primarily to the women. This ensures the involvement of the women, who are the principal beneficiaries of the water supply and the electric lights.

The essential institution had to be built at village level, and was designed to oversee the maintenance and operation of the rural energy center, the contribution of dung, the collection of payments for biogas outputs (electric lights and water) to the home, and the formulation and execution of plans for further development of the rural energy center. A *grama vikasa sabha* (village development society) was established, involving those villagers who lead traditional community

PROBLEMS	CAUSES	REMEDIES
1. Low speed 2. Engine cannot take load 3. Emits white or black smoke 4. Engine gets stopped	1. Engine overloaded 2. Biogas consumption high	1. Let the engine run idle 2. Put on load 3. Switch to biogas gradually
1. Low speed 2. Engine fails to take load	1. Improper positioning of the governor control (throttle handle)	1. Start with diesel 2. Put on load 3. Place throttle handle at correct position 4. Switch to biogas gradually
1. When switched to biogas, engine speed decreases or engine stalls	1. Intake of biogas is too much and too soon	1. Open biogas valve and increase slowly 2. When intermittent sound comes, close a little until the sound is silenced
1. Intermittent sound and knocking while the engine is on run	1. Sudden load reduction while the engine is on run 2. Engine takes too much biogas	1. Close the biogas valve gently (or) increase the load until the sound is silenced
1. Low efficiency 2. Low diesel replacement	1. Low biomass supply 2. Load imbalance 3. Short running durations 4. Operator less efficient	1. Check condensed water in pipes 2. Load steadily 3. Run for longer periods 4. Operator must be alert

FIGURE 4: Trouble-shooting chart for the operation of the biogas–diesel engine operation at Pura lists potential problems, causes and their remedies.

Table 3: Engine operation at biogas plant in Pura
September 1987 to April 1991

	Engine operation *hours*	Average daily operation time
Water	2,211	1 hr 40 min
Lighting	2,310	2 hr 29 min
Total	**4,521**	**4 hr 9 min**

Table 4: Field performance of biogas–diesel engine in Pura
average fuel consumption: January 1988 to April 1991

Diesel	Biogas		Total fuel	Diesel replacement
	Biogas	Diesel equivalent		
ml/hp/hour	*l/hp/hour*	*ml/hp/hour*	*ml/hp/hour*	*percent*
60.3	530.8	199.7	260.1	77

Table 5: Performance of biogas–electricity generation system in Pura
average input: January 1988 to April 1991

Diesel	Dung	Total fuel diesel equivalent	Diesel replacement
ml/kWh	*kg/kWh*	*ml/kWh*	*percent*
121	27	520	77

activities, for example, festivals and dramas. The village development society achieved a 93 percent collection of dues from 1988 to 1991, which is outstanding when compared with the dismal record of the large electric utilities in India.

The financial operation of the system is shown in the income-expenditure statement for September 1990 to April 1991 (see table 6). The average for this period shows that revenues from lighting and private water taps cover almost all (93 percent) of the expenditures except for the two workers' salaries, which are

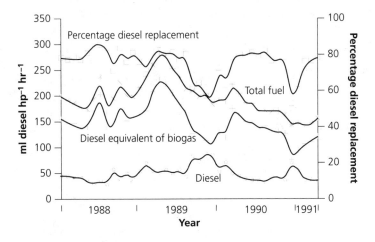

FIGURE 5: *Diesel fuel replaced by biogas.*

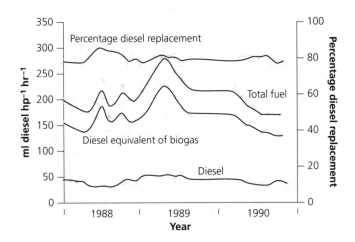

FIGURE 6: *Diesel fuel replaced by biogas at Pura excluding "abnormal" points.*

provided by the project sponsor, the Karnataka State Council for Science and Technology.

Many possibilities for increasing revenues and creating a surplus exist. These involve decreasing engine idle time and generating more electricity to produce more income. But doing so would require revenue-yielding loads such as industries and irrigation pumpsets, as well as more fuel for the engine generator set. The villagers are considering a dairy development scheme in the village with hybrid cattle to sell milk to the State Milk Marketing Board as well as more dung to the biogas plant. They are also establishing an energy forest to grow wood to run a

Table 6: Average monthly expenditure and income statement for Pura biogas plant
1989 US$ (1 dollar = 20 rupees) September 1990 to April 1991

Expenditure		Income	
Labor	20.98	Lights[a]	18.41
Diesel	10.19	Private water taps[b]	5.51
Dung	8.14	KSCST grant[c]	29.98
Repairs	2.08	–	–
Uncollectable[d]	1.52	–	–
Miscellaneous	1.67	–	–
Total	**53.90**	**Total**	**59.32**

a. The income from 81 working tube lights is 81 × Rs. 5 /connection /month = Rs. 405/month = 1989 $18.40/month.

b. The income from 25 working private water taps is 25 × Rs. 5 /connection /month = Rs. 125/month = 1989 $5.68/month.

c. Rs. 660/month ($33/month); i.e., the salaries of the two workers who run the community biogas plant is still being borne by the Karnataka State Council for Science and Technology, which has sponsored the project.

d. The income from the 97 tube lights and 29 private water taps that have been connected in the houses should be [(97 + 29) × Rs. 5 /connection /month] = Rs. 630/month = 1989 US$28.62/month. In fact, an average of Rs. 45.00/month (= $2.04/month) is not collected because about nine (7 percent) of the tube lights and/or taps are not functional for various reasons.

gasifier, similar to one demonstrated in a neighboring village, to fuel the same engine-generator set (see chapter 16: *Open-top Wood Gasifiers*).

The Pura biogas plant is held together and sustained by the convergence of individual and collective interests. It is customary to discuss the problem of individual gain versus community interests in terms of the famous "Tragedy of the Commons" [15], which states that the personal benefits gained by each individual or household promotes further destruction of the commons (that is, community resources) because such benefits are larger and more immediate than the personal loss from the slow and long-term destruction of the commons. Hence, each individual or household chooses to derive the immediate personal benefit rather than forgo it and save the commons.

Experience with the factors holding together and sustaining the Pura community biogas system appears to illustrate a converse principle that may be termed the "Blessing of the Commons" [16], which states that the price for not preserving the commons far outweighs whatever benefits might be obtained by ignoring the collective interest. In other words, the Blessing of the Commons is based on the coincidence of self-interest and collective interest. Thus, in Pura, noncooperation with the community biogas plant results in a heavy individual price (access to water and light being cut off by the village), which is too great a personal loss to compensate for the minor advantage of noncooperation and noncontribution to collective interests.

There must be many examples of the Blessing of the Commons that for centuries contributed to the survival of Indian villages in spite of the centripetal forc-

Table 7: Capital cost of a biogas-based electricity system

Item	Cost *1989 US$*
Biogas plant	2,554
Piping, etc.	166
Sand filters	83
7 horsepower diesel engine	754
5 KVA three-phase genset	1,563
Accessories, tools, etc.	415
Engine room	331
Total	**5,866**

es tearing them apart. Among those examples must have been maintenance of village water tanks, common lands, woodlots, etc. It is important to discover and use such examples for the design of rural development projects in general, and for rural energy centers in particular. Such *local* community control is a distinct alternative to the privatization (deregulation) option being offered as a solution to the defects of state control, operation, and regulation of the commons.

ECONOMICS OF THE COMMUNITY BIOGAS SYSTEM

The following considerations and data have been used to calculate the unit cost of energy (dollars per kilowatt-hour) and the unit cost of power (installed capacity in dollars per kilowatt) from biogas-based electricity generation (for various nominal discount rates).

Capital cost
The biogas-driven engine and generator are off-the-shelf items, and therefore the period of construction of a biogas-based electric power plant is determined by the time needed to construct the biogas reactor. Because a biogas reactor takes only about six months to build, it has been assumed that the entire expenditure is incurred at the commencement of the project.

The items involved in the capital expenditure are the biogas reactor, piping, sand filters, engine, generator, accessories and tools, and engine room. A breakdown is given in table 7, in which it is seen that the overnight construction cost on all items in 1990 U.S. dollars is $6,035, or $1,207 per kilowatt, including the first engine.

Matching the life of the biogas project with a 25-year central station plant

Decentralized biogas-based electricity generated at the village level represents an alternative approach to rural electrification, which is electricity generated from a central-station plant and transmitted via a grid to rural areas. The economics of decentralized biogas-based electricity generation must, therefore, be compared with the cost of central-station generation.

For this comparison to be on the same terms, it is essential that the lifetime of the biogas project be extended to match the longest-life central-station electricity project, which is about 25 years. This matching can be achieved by making repeated installations of the components of the biogas-based system that have shorter life spans until the standard 25 year central-station life is attained.

The longevity of the biogas reactor, generator, and all other items except the engine are assumed (with proper maintenance) to be 25 years, which means they are already matched to the central station. In contrast, the life of a biogas-driven engine is only 5,000 hours, after which its life can be extended by 5,000 hours per overhauling up to three overhaulings. To extend the life of the biogas project further, the engine must be replaced. Obviously, the number of replacement engines (including overhaulings) required to keep the system going for 25 years depends on the average number of hours that the system operates each day.

The unit cost of power and the unit cost of energy from the biogas-based electricity system also depend on the load factor, which is the average number of hours of operation per day. The results are demonstrated in figure 7, which shows that the unit cost of power increases linearly with the average number of working hours per day of the biogas-based system. In contrast, the unit cost of energy from biogas electricity (see figure 8) decreases with an increase in the number of working hours per day, though at a decreasing rate.

At present, the Pura system is operated for only about 4.2 hours per day, which corresponds to a dung input of 291 kilograms per day. This limited operation leads to a unit cost of energy of more than $0.25 per kWh.

But the economics of the system must not be judged on the basis of this high unit cost of energy for the following reasons. First, the plant has been sized and constructed to take an input of about 1,250 kilograms of dung per day, and to operate about 18 engine hours per day. Second, there are no technical obstacles to operating the plant for as long as 18 hours per day, provided it is supplied with suffiicient dung. Third, the daily dung input of 1,250 kilograms cited above corresponds to the output of the present population of 250 cattle yielding 5 kilograms of dung per day, which is an achievable target, given a mix of about 15 percent hybrid cattle (with dung outputs in excess of 10 kilograms per day) and 85 percent average Pura cattle (with dung outputs of about 4 kilograms per day).

It is well known that an industrial or power plant does not achieve full capacity as soon as it is started and that there is always a start-up or gestation period. In the case of conventional plants (for example, a thermal power plant), the gestation period is determined by technical factors such as temperature, etc. In the case of the Pura biogas plant, however, the gestation period is determined by so-

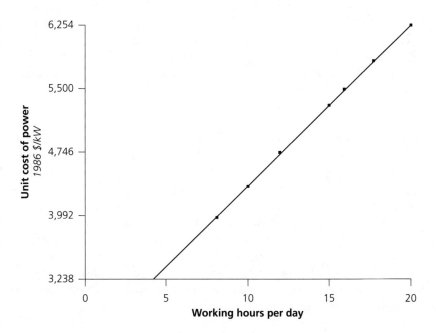

FIGURE 7: Variation of the unit cost (at 12 percent) of biogas electrical power measured against the capacity utilization of the plant (in working hours per day).

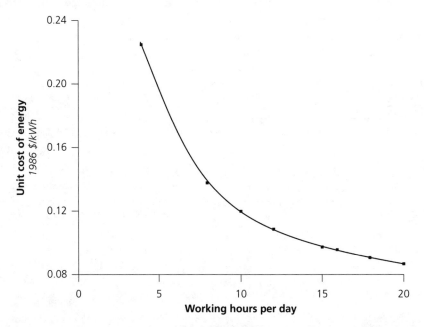

FIGURE 8: Variation of the unit cost (at 12 percent) of biogas electrical energy measured against the capacity utilization of the plant (in working hours per day).

ciological factors associated with the local production and supply of feedstock for the plant. Unlike an urban plant, the biogas plant in Pura does not purchase its feedstock from a global market; indeed the Pura plant cannot even get dung from the next village. Also, unlike an urban plant, the Pura plant does not meet a pre-existing demand; it must stimulate a demand that does not exist, a crucial development challenge.

As a rule, economic analysis of a plant (for example, a nuclear power plant) is never based on the load factor at start-up or before the gestation period is over; full-load performance is always assumed. Similarly, it would be quite incorrect to assess the economics of the Pura plant on the basis of its present low load factor—before the gestation period is over—of 17 percent, which corresponds to 4.2 hours per day.

For the purpose of comparing decentralized biogas-based electricity generation with the current approach to rural electrification, it is assumed that the gestation period for the Pura plant is over and that it now has the same load factor (62.8 percent) as a baseload central-station plant. On this basis, the number of working hours per day is 15.1.[11] Such an operation would require a daily input of 1,055 kilograms of dung, which is well within the 1,250 kilograms generated per day by the 250 cattle in the village.

Disaggregation of the costs of biogas electricity
By disaggregating the unit cost of energy for 15.1 hours of operation per day into its components, it can be seen (see figure 9) that, unlike centralized generation plants, a substantial percentage of the expenditures are incurred locally. This means that the system stimulates local prosperity.

---

11. The following are costs of engines, overhaulings, and fuel inputs (in 1989 $) for a system operating an average 15.1 hours per day:

Cost of engine	745
Cost of each overhauling	261
Cost of an engine plus three overhaulings discounted at 12 percent to the date of purchase of the engine	1,384
PV of replacement engines per kilowatt	497
PV of overhaulings per kilowatt	128
Total cost of replacement engines plus overhaulings at commissioning per kilowatt	625
PV of replacement engines plus overhaulings at the start of the project per kilowatt	590
Total capital cost (including $52/kW working capital) per kilowatt	1,807
Fee for collection of dung and delivery to plant @ 1986 Rs.0.02 per kilogram	0.001753
Dung consumption per engine hour	70 kilograms
Diesel consumption per engine hour	0.280 liters
Annual operation and maintenance (O&M) expenses for n shifts	$42.07 + 69.42n$
Annual O&M expenses for three shifts	250.34

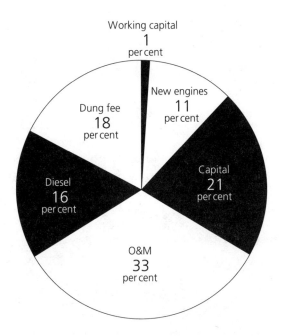

FIGURE 9: Components of the $0.12 per kWh cost of electricity from the Pura biogas electricity system.

Equivalent generation-end costs of biogas electricity

In order to compare the costs of electricity from a biogas-based system with those from a central-station power plant, it is necessary to ensure that the costs are computed at the same location. Doing so is necessary because biogas electricity is generated at the consumption end (in a village), whereas central-station electricity is generated far from the village and then transmitted through a grid. A comparison is made assuming the cost of the grid is a sunk cost.

To convert the unit cost of power of biogas-based electricity at the consumption end to the unit cost of power at the generation end, it is necessary to correct for 1) the grid's transmission and distribution (T&D) loss factor, 2) the ratio of the capacity factors of the biogas and central-station plants, and 3) the ratio of consumption by auxiliaries in the two plants. Karnataka's T&D losses of 21.5 percent mean that 1 kWh generated at the consumption end is equal to 1.27 kWh at the generation end or, equivalently, the unit cost of energy from village-level biogas-based electricity is actually 21.5 percent less when it is compared with central-station electricity. Since the capacity factors of the biogas and central-station plants are considered equal (62.8 percent) and the consumptions by auxiliary equipment within the plants are 1 and 10 percent, respectively, it turns out that 1 kilowatt generated by the biogas-based system at the consumption end is equivalent in Karnataka to the generation of 1.40 kilowatts at the generation end (see figures 10 and 11).

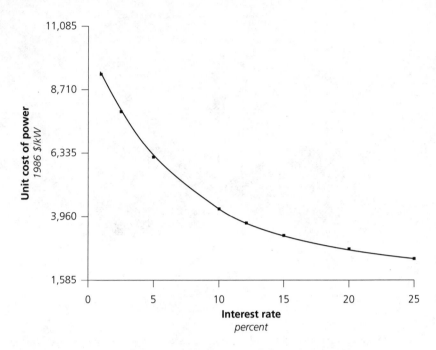

FIGURE 10: *Variation in the unit cost of biogas-generated electrical power with interest rate.*

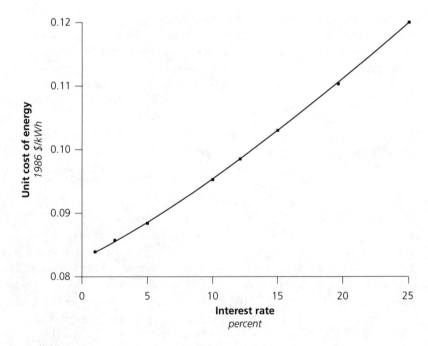

FIGURE 11: *Variation in the unit cost of biogas-generated electrical energy with interest rate.*

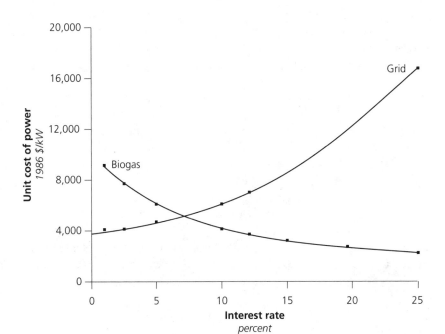

FIGURE 12: *Variation with interest rate of the unit cost of electric power from a biogas plant and from a central-station plant via the grid.*

Comparison of biogas and central-station electricity

When the unit costs of power from the biogas-based electricity system and from central-station power plants are plotted (see figure 12), two interesting results emerge:

▼   At interest rates greater than about 7.5 percent, the unit cost of power from the biogas-based system is less than that from the central-station power plant.

▼   The cost advantage of the biogas-based electricity system over the central-station plant increases with interest rate.

The interest rate reflects the scarcity of capital—the greater the scarcity of capital, the higher the interest rate. Hence, in capital-starved developing countries, such as India, high interest rates should be used for investment decisions. For such interest rates, the unit cost of power clearly favors biogas-based electricity systems rather than central-station power plants.

Economic viability of biogas electricity generation

In assessing the viability of electricity generated from village-level biogas plants, it is clear that the current limited-hours operation of only about 4.2 hours per day leads to a high unit cost of energy of more than $0.25 per kWh. In addition, current income from the biogas system (from lights and private water taps) can meet only about half of the recurring expenses of the plant while making no contribution to the capital investment charges (see table 6). Hence, a subsidy would be required to operate the biogas system at its present low-load factor of less than 20 percent. Though such a subsidy seems to be a pattern for urban water and electric utilities, it means that widespread replication of biogas electricity generation systems is unsustainable at such low load factors.

Fortunately, the limited-hours operation can be viewed as a transitional situation characteristic of the gestation period. As demand for electricity builds in the village and the dung supply increases to meet that demand, the income from the plant will also rise. Using the tariffs that the Pura households are currently paying for their lights and water taps, it can be shown that the deficit of income with respect to expenditure decreases as the number of hours of operation increases, and when the number becomes greater than about 6 hours per day, the system starts generating a surplus that can be used to return the capital investment. And at the same load factor as a central-station plant (15.1 hours per day), the unit cost of electrical energy and power is actually lower than that from a central-station plant. Hence, after a gestation period, the biogas system does not require subsidies and in fact is economically preferable to central-station plants supplying electricity to the village via a grid.

Decentralized biogas-based electricity systems

Thus, the above cost results argue in favor of decentralized electricity systems (based on biogas in the context of this paper but also on producer gas, small hydroelectric plants, etc., as shown by separate analyses) for rural electrification. But rural electrification via centralized generation plants (for instance, hydroelectric, coal-based thermal, and nuclear power plants) and grid transmission and distribution has already progressed almost to completion in many parts of India. Despite such developments, it might make economic sense to keep the grid as a standby and provide rural electricity through decentralized systems instead. In addition, quite apart from the economic aspects, there is no doubt that decentralized electricity systems strengthen self-reliance and, therefore, advance rural development efforts.

Advanced biogas plants of the future

The case for decentralized village-level biogas-based electricity generation systems should become even more attractive as the present generation of biogas plants gives way to plants based on advanced technology. A Pura-type traditional biogas plant has an energy conversion efficiency of around 27 percent, yet biogas technologies now exist that can achieve conversion efficiencies of around 60 percent.

The first step toward attaining efficiency increases is to operate the plant at optimum, instead of ambient, temperatures. Whereas average ambient temperatures in Pura are around 26°C, the optimum temperatures for the mesophilic bacteria responsible for biogas production are about 35°C, and operation at these temperatures can almost double the gas production rate.

Still greater improvements can be obtained with advanced biogas technologies that are now being used to process industrial wastes (see chatper 19: *Anaerobic Digestion for Energy Production and Environmental Protection*). These advanced technologies represent achievable goals and may well be adapted some day for handling animal, rather than industrial, wastes.

CONCLUSIONS

Traditionally, a family in Pura would make two trips per day taking 1.5 hours (45 minutes per trip) to cover 1.6 kilometers and transport 104 liters (4 potfuls) to obtain an average per capita water consumption for domestic purposes of 17 liters per day. Moreover, although Pura was electrified, 55 percent of the households were not electrified and had to rely on inefficient and expensive kerosene open-wick-lamps and chimney lanterns for lighting.

Comparing the present biogas plant system with this traditional system shows that the households are winners on all counts, having gained the following:

▼ Better water than what was available from the open tank.[12]

▼ Less effort expended to get this improved water.

▼ Better and more reliable illumination than the traditional kerosene lamps or unreliable low-voltage grid electricity.[13]

▼ Cheaper illumination of households compared with costlier kerosene lamps—a 20 watt fluorescent tube light costs $0.40 per month compared with about $0.80 per month for the approximately 3 liters per month of kerosene for kerosene lamps generating 0.02 less lumens.

▼ Improved biogas sludge fertilizer that has double the nitrogen content of farmyard manure and is less susceptible to the growth of weeds.

▼ A dung delivery fee to those (mainly women and children) who deliver the dung to the plant and take back the sludge.

In addition, the village (as a collective) through its *grama vikasa sabha* (village development society) has gained the following:

▼ Training and skill upgrading for two of its youths in the operation and maintenance of the biogas system.

▼ Revenue for the village when the total payment received for household electricity and water exceeds the expenses for diesel and dung delivery fees.

---

12. A morbidity and mortality study has not yet been carried out, but the women of Pura assert that the health status of their children with respect to intestinal problems has improved noticeably. Their observation is understandable because deep borewell water is far less contaminated than surface sources.

13. In the evening peak hours, the voltage of the grid supply, which is supposed to be 220 volts, goes down to as low as 150 volts so that the fluorescent tube lights do not start.

▼ A powerful mechanism to initiate and sustain village-scale cooperation.

▼ Distinct improvement in the quality of life with regard to water (and therefore health) and illumination.

▼ A small but significant advance in self-reliance, thanks to the realization that the current status and future development of the energy system can be decided and implemented by the village. In other words, the future in this matter is in their own hands.

ACKNOWLEDGMENTS

The authors would like to thank David Stuckey and Derek Lovejoy for their thorough and insightful reviewing of the first draft and for their very useful suggestions for improving the manuscript. Thanks are also due to Robert Williams for stimulating a careful analysis of the economics.

This is an appropriate context for placing on record the gratitude of the authors to the Executive Committee and the Secretariat of the Karnataka State Council for Science and Technology for steadfastly supporting the Pura community biogas plant project through all its vicissitudes. Of equal importance has been the faith of the people of Pura in the project team and their conviction that science and technology can improve the quality of their lives.

WORKS CITED

1. Reddy, A.K.N., Sumithra, G. D., Balachandra, P., and D'Sa, A. 1990
   Comparative costs of electricity conservation and centralized and decentralized electricity generation, *Economic and Political Weekly*, 1201–1216, 2 June.

2. Khandelwal, K.C. and Mahdi, S.S. 1986
   Biogas technology—a practical handbook, Tata–McGraw-Hill Publishing Company Limited, New Delhi.

3. Anonymous. 1986
   Biogas in Asia and the Pacific, report of the regional expert consultation on biogas network, 28 October to 1 November 1986, Bangkok, Thailand.

4. Anonymous. 1979
   A Chinese biogas manual, translated from the Chinese by Michael Crook and edited by Ariane van Buren, Intermediate Technology Publications Ltd., London.

5. Anonymous. 1982
   Diffusion of biomass energy technologies in developing countries, National Academy Press, Washington, DC.

6. Goldemberg, J., Johansson, T.B., Reddy, A.K.N., and Williams, R.H. 1988
   Energy for a sustainable world, Wiley-Eastern Limited, New Delhi.

7. Maniates, M. 1983
   Community biogas plants: social catalyst or technical fix? *Soft Energy Notes* 6(2).

8. Rajabapaiah, P. and Jayakumar, S. 1990
   Biogas-diesel engines for lighting and pumping drinking water in villages: Pura community biogas plant system experience, *Proceedings of the third international conference on small engines and their fuels for use in rural areas: development, operation and maintenance*, University of Reading, United Kingdom.

9. Rajabapaiah, P., Somasekhar, H.I., Jayakumar, S., and Reddy, A.K.N. 1991
   The saga of the Pura community biogas plant, *Economic and Political Weekly* (in press).

10. Subramanian, D.K., Rajabapaiah, P., and Reddy A.K.N. 1979
Studies in biogas technology. Part II: optimization of plant dimensions, *Proceedings of the Indian Academy of Sciences,* C2:365–376.

11. Reddy, A.K.N., Rajaraman, I., Subramanian, D.K., and Rajabapaiah, P. 1979
A community biogas plant system for Pura village—feasibility study and proposal, Karnataka State Council for Science and Technology, Indian Institute of Science Campus, Bangalore 560 012, India.

12. Rajabapaiah, P. 1979
Unpublished results referred to in P. Rajabapaiah, K.V. Ramanaiah, S.R. Mohan, and A.K.N. Reddy.

13. Rajabapaiah, P., Ramanaiah, K.V., Mohan, S.R., and Reddy, A.K.N. 1979
Studies in biogas technology. Part I. Performance of a conventional biogas plant, *Proceedings of the Indian Academy of Sciences,* C2:357–364.

14. Centre for the Application of Science and Technology to Rural Areas (ASTRA). 1981
Rural energy consumption patterns—a field study, *Biomass* 2(4):255–280.

15. Hardin, G. 1968
The Tragedy of the Commons, *Science* 162:1243–1248.

16. Reddy, A.K.N. 1991
Some reflections on rural energy issues (in press).

19
ANAEROBIC DIGESTION FOR ENERGY PRODUCTION AND ENVIRONMENTAL PROTECTION

GATZE LETTINGA
ADRIAAN C. VAN HAANDEL

Anaerobic digestion is the decomposition of complex molecules into simpler substances by micro-organisms in the absence of oxygen. Anaerobic digestion processes can be employed for resource conservation, for the production of biogas and other useful end products from biomass, and for environmental protection through waste and wastewater treatment. Modern high-rate anaerobic wastewater-treatment processes can effectively remove organic pollutants from wastewater at a cost far below that of conventional aerobic processes. These anaerobic wastewater treatment processes can also be profitably applied for the generation of biogas from energy crops such as sugarcane. In fact, these methods might even be an attractive alternative for the alcohol fermentation extensively employed in Brazil for the production of fuel alcohol from sugarcane. The potential of modern anaerobic processes for this purpose has not yet been widely recognized. This paper describes the principles and use of these processes and demonstrates their prospects for producing energy from sugarcane 1) by treating vinasse, the wastewater generated during the production of ethanol from sugarcane, and 2) as a direct method for producing biogas from sugarcane juice.

INTRODUCTION

Although the process of anaerobic digestion has been exploited by human beings for many centuries, the development of anaerobic wastewater treatment systems is relatively recent. Anaerobic digestion reactors are technologically simple and do not generally consume high-grade energy because the method does not depend on the supply of electricity or other energy sources. The end product of the process is biogas, a useful energy carrier. A wide range of wastewaters can be treated anaerobically, including domestic wastewater and sewage in tropical and subtropical regions. An important advantage is that the process can be applied almost anywhere and at any scale.

Complete removal of all pollutants from wastewater by anaerobic digestion is impossible; it is a mineralization process, and thus chemical compounds and ions such as ammonia, phosphate, and sulphide can remain in the effluent of an anaerobic treatment system. However, a number of effective physical-chemical and biological post-treatment processes have already been developed to recover the valuable mineralized bulk products, and others could be developed.

Anaerobic treatment can be part of integrated energy-carrier production and environmental protection systems, and the products from the digestion process can be profitably used. For example, clean treated water can be employed in fish ponds: the liquid effluent for irrigation and fertilization, the biogas as an energy source, and sludge for soil conditioning (see figure 1).

BASIC ASPECTS OF ANAEROBIC DIGESTION

The biochemistry and microbiology of anaerobic digestion are extremely complex and not completely understood. In the digestion process, a consortium of anaerobic bacteria[1] (very different from those found in aerobic processes) participates in the decomposition of a complex substrate. In the anaerobic degradation of complex wastes, the conversion of each group of organic compounds (proteins, carbohydrates, and lipids) requires its own characteristic group of organisms. The following main processes can be distinguished [1]:.

1. Hydrolysis or liquefaction of biodegradable biopolymers, produces soluble sugars, amino acids, and long-chain fatty acids.

2. Acidogenesis, the formation of hydrogen, short-chain volatile fatty acids (VFAS), and alcohols from monomeric compounds occurs.

3. Acetogenesis, the formation of acetate/acetic acid (ethanoate/ethanoic acid) and hydrogen from longer-chain VFAS and alcohols takes place.

4. Methanogenesis, the formation of methane (CH_4) and carbon dioxide (CO_2) from the methanogenic substrates acetate/acetic acid, hydrogen, carbon dioxide, and methanol occurs.

In the overall process, complex organic compounds are transformed into stable gases, mainly methane and carbon dioxide. Much of the chemical energy originally present in the organic material remains in the produced methane. Substantial removal of the biodegradable compounds occurs only during the methanogenic step. This last stage is possible only when the original organic compounds have been converted effectively into the methanogenic substrates in the preceding steps. For the digestion to proceed stably it is vital that the various biological conversions remain sufficiently coupled during the process, so that no serious accumulation of intermediate compounds occurs. In the steps prior to methanogenesis, a decrease in the chemical oxygen demand (COD),[2] which is a measure of the chemically oxidizable organic material in the waste, occurs only

---

1. Fungi such as yeast may also be used; for convenience we refer to bacteria here.
2. Chemical oxygen demand is the amount of oxygen in grams required to oxidize the organic material in 1 liter of the waste using dichromate as an oxidizing agent.

if hydrogen (H_2) is produced and released from the liquid phase (hydrogen has a stoichiometric chemical oxygen demand of 16 grams COD per gram of H_2).

Methane production potential

The amount of methane than can be produced from organic material is directly proportional to the material's content of convertible COD. In an anaerobic digester, oxidation by atmospheric oxygen cannot occur, hence the biodegradable COD present in the digested organic material is preserved in the end products, namely the methane and newly formed bacterial mass. Stoichiometrically, methane has a COD of 2 moles (64 grams of COD) of oxygen per mole (16 grams) of methane, so 1 gram of methane is equivalent to 4 grams of COD.

The production of carbon dioxide depends on the average oxidation state of the carbon atoms as can be seen from the general reaction equation for a compound with a structural formula $C_nH_aO_bN_d$:

$$C_nH_aO_bN_d + (n - \tfrac{1}{4}a - \tfrac{1}{2}b + \tfrac{3}{4}d)\, H_2O \rightarrow$$

$$(\tfrac{1}{2}n + \tfrac{1}{8}a - \tfrac{1}{4}b - \tfrac{3}{8}d)\, CH_4 + (\tfrac{1}{2}n - \tfrac{1}{8}a + \tfrac{1}{4}b + \tfrac{3}{8}d)\, CO_2 + d\, NH_3$$

For many naturally occurring compounds (such as carbohydrates) $a \approx 2b$ and $d \ll \tfrac{1}{2}n$, so that roughly equimolar amounts of carbon dioxide and methane are formed. However, generally the biogas evolving from the liquid phase during the digestion of carbohydrates contains less carbon dioxide than it does methane. This is due to the high solubility of carbon dioxide in water (relative to methane) and to the fact that it is bound chemically in the form of hydrogen carbonate (HCO_3^-) ions, which 1) neutralize the inorganic cations (such as Na^+) that were present in the original solution and remain after the degradation of the VFA ions (such as ethanoates), and 2) neutralize cations formed in the anaerobic digestion of nitrogen-containing organic compounds (i.e., the formation of NH_4^+).

Environmental factors affecting anaerobic digestion

Anaerobic digestion, like most biological processes, is strongly affected by environmental factors. Because it is beyond the context of this paper to go into great detail, we will discuss only the most important factors.

Temperature. Anaerobic digestion can occur at temperatures as low as 0°C, but the rate of methane production increases with increasing temperature until a relative maximum (mesophilic digestion) is reached at 35 to 37°C. At temperatures above this optimum, mesophilic organisms are replaced by thermophiles, and a maximum methanogenic activity is reached at about 55°C.[3]

---

3. Mesophilic digestion processes take place at an optimum temperature range of 20 to 40°C; thermophilic processes take place optimally between 45 and 60°C.

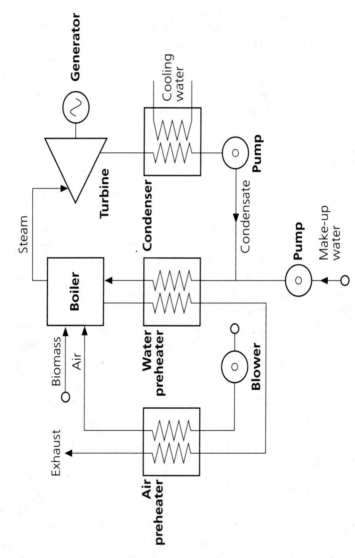

FIGURE 1: Integrated energy carrier, production, and environmental protection system based on anaerobic digestion and post treatment processes. End uses for the products are also shown.

pH and alkalinity. While the first steps of anaerobic digestion can occur over a wide range of *p*H values, methanogenesis only develops well when the *p*H is near its neutral value of 7. For *p*H values outside the range of about 6.5 to 7.5, the rate of methane production is distinctly lower. The presence of a sufficient amount of hydrogen carbonate (frequently denoted as bicarbonate alkalinity) in the solution is important because it will keep the *p*H in the optimal range for methanogenesis.

Toxicity. The toxic effect of several compounds on the agents of anaerobic digestion and notably on methanogenesis has been observed by many authors. Excessive concentrations of ammonia, cations such as Na^+, K^+, and Ca^{2+}, heavy metals, sulphide, and xenobiotics adversely affect methanogenesis. Several excellent reviews of toxicity in anaerobic digestion have been published [2].

Growth factors. Toxicity notwithstanding, the presence of certain substances is sometimes indispensable for the production and maintenance of anaerobic sludge. Along with macronutrients (nitrogen, phosphorus, potassium, sulfur, calcium, iron, and magnesium), several trace elements have been shown to be necessary, notably nickel, cobalt, and molybdenum [3]. Since sludge production is appreciably lower in anaerobic than in aerobic processes, the demand for nutrients is correspondingly smaller.

Principles of anaerobic wastewater treatment

Unlike those of aerobic wastewater treatment systems, the loading rates of anaerobic reactors are not limited by the supply of a reagent but by the processing capacity of the microorganisms and/or because a sufficiently large bacterial mass must be retained in the reactor. The following conditions are thus imperative for high-rate anaerobic reactors:

▼ A high concentration of anaerobic bacterial sludge must be retained under high organic (> 10 kilograms per cubic meter [m^3] per day) and high hydraulic (> 10 m^3 per m^3 per day) loading conditions (liquid detention times less than 8 hours).

▼ Maximum contact must occur between the incoming feedstock and the bacterial mass.

▼ There must be minimal transport problems with respect to substrate compounds, intermediate and end products.

During the past few decades several high-rate anaerobic treatment systems have been developed that differ in the way the above conditions are met. In essence, all high-rate processes have a mechanism either to retain the bacterial sludge mass in the reactor or to separate bacterial sludge from the effluent and return it to the reactor (see figure 2). The high-rate processes can be divided into two categories: 1) systems with fixed bacterial films on solid surfaces; and 2) systems with a sus-

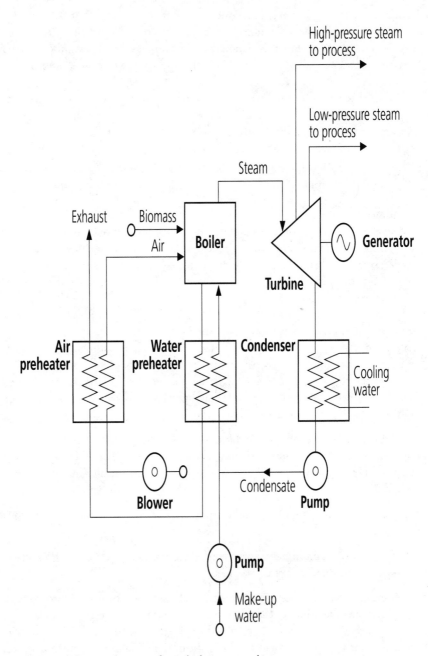

FIGURE 2: *Schematic diagrams of some high-rate anaerobic wastewater treatment processes.*

pended bacterial mass where retention is achieved through external or internal settling.

The most important high-rate anaerobic processes are the *contact process* (see figure 2a, [4]), the *anaerobic filter* (AF), operated in upflow (see figure 2b, [5]) and downflow versions [6], the *upflow anaerobic sludge blanket* (UASB) reactor (see figure 2c, [7, 23-28]), and the *fluidized bed* (FB) reactor (see figure 2d) developed by [8] and [9].

The contact process

The contact process is the anaerobic equivalent of conventional aerobic activated sludge processes for liquid wastewater treatment. The required sludge–wastewater contact is achieved by mechanical mixing and/or by gas recirculation, and high sludge retention is accomplished by separating solids from liquids in an external settling tank and returning the sludge to the reactor. The drawback of this system is that the anaerobic sludge settles with difficulty because of its buoyancy in the settler. The contact process is in fact not a high-rate system, because the applicable organic space loads do not exceed more than from 4 to 5 kilograms (kg) COD per m³ per day.

The anaerobic filter

In the anaerobic filter, high bacterial mass retention is accomplished 1) by bacterial attachment to a fixed support inside the reactor and 2) by the formation of bacterial aggregates, which are trapped in the voids between the support medium. The anaerobic filter was the first real high-rate process to demonstrate the tremendous potential of anaerobic systems. Conventional aerobic systems can generally handle up to 2 kg COD per m³ per day. However, the anaerobic filter can accommodate organic space loads of up to 10 to 15 kg COD per m³ per day for mainly liquid wastewaters . The main practical problem with upflow AF systems is clogging of the filter, which reduces contact between the retained sludge and wastewater. Clogging occurs less when the downflow version is used, but in this case sludge retention is more difficult. Another drawback of the AF system is the relatively high cost of the support medium. Due to these difficulties, the number of installed full-scale AF systems is still rather small (about 30 or 40) and likely will not increase significantly.

Fluidized-bed systems

Fluidized-bed systems retain sludge by coating inert mobile carrier particles with biofilm (see figure 2c). Contact with the feedstock is achieved by keeping these biofilm particles in a fluidized state. The central problem in the operation of fluidized biofilm-bed systems is to maintain biofilm particles of uniform shape, size, and density. As this is almost impossible, a stable fluidized bed cannot be guaranteed. Moreover, some of the recently introduced anaerobic FB systems [9, 10] require a separate preacidification reactor, and they generally also need to have their effluent recycled to fluidize the biomass-carrier particles. Consequently the energy requirements are higher, and additional pumps have to be installed. Thus prospects for these systems appear small.

The upflow anaerobic sludge blanket reactor

The UASB reactor relies on the formation of active sludge aggregates that settle without difficulty. The reactor consists of a digester compartment containing the sludge bed, a gas–solids separator in the upper part of the reactor, and an internal settler for sludge retention at the top of the reactor vessel. The UASB reactor is presently the most widely used high-rate anaerobic wastewater treatment system. More than 300 full-scale UASB reactors have been installed and the number of full-scale installations is still expanding rapidly. Further, a number of promising systems based on the UASB concept are either under development or have been introduced recently. This concept and some recent modifications will be discussed in more detail in the following section.

UPFLOW ANAEROBIC SLUDGE BLANKET (UASB) TECHNOLOGY [7, 23-28]

The UASB reactor concept is based on: 1) the formation of a "well-settling" granular or flocculent[4] type of anaerobic sludge, thereby eliminating the need for mechanical mixing in the reactor; 2) natural agitation caused by gas production and an even distribution of feed inlets, which provide the required contact. Complete fluidization is not needed; and 3) a well-designed gas–solids separator (GSS) system[5] in the upper part of the reactor, which will retain the well-settling sludge. A GSS is crucial for a UASB reactor, but this is also true for some FB systems [9, 10].

Operation of high-rate anaerobic wastewater-treatment reactors

One must distinguish between start-up and steady-state operation of anaerobic wastewater-treatment systems. The following discussion focuses on the UASB process.

Operation of UASB reactors during primary and secondary start-ups

Compared with aerobic bacteria, anaerobic bacteria grow very slowly; the cultivation of a sufficient amount of well-adapted anaerobic sludge from a poor-quality seed sludge may take several months. However, as anaerobic sludge can be stored unfed for very long periods (months or even years) without loss of its conversion capacity, this problem of primary start-up will disappear once a sufficient amount

---

4. Contains mechanically fragile sludge aggregates.

5. Some researchers and contractors propose to replace the GSS by a packed bed in the upper part of the reactor (e.g., [11–14]). Although there exist some full-scale installations of this hybrid UASB–AF system, in our opinion it cannot compete with the conventional UASB concept in retaining viable sludge, even when the sludge is granular, because the AF part of the reactor cannot sufficiently separate the sludge from the liquid, particularly at higher loading rates. In fact, the biogas evolved in the sludge bed beneath the AF will force most of the suspended sludge particles through the fixed AF bed.

of adapted sludge is available. In view of the large number of UASB reactors that are in operation or under construction, it is only a matter of time for this situation to become reality. Such sludge represents an almost ideal seed material for new reactors, even when the wastewater to be treated is very different from the one that generated the sludge. Experience thus far with this so-called secondary start-up has generally been very satisfactory.

Over the past decade, comprehensive studies on the primary start-up of UASB reactors have dealt in particular with the formation of granular types of sludge [15–18]. This granulation process can be summarized as follows:

1.   Bacterial growth in UASB reactors is delegated to a limited number of well-settling primary nuclei consisting of inert organic or inorganic bacterial carrier materials as well as bacterial aggregates.

2.   Washout of finely dispersed bacterial matter from the reactor occurs once higher loading conditions are imposed. As a result, film and aggregate formation on the primary growth nuclei are greatly enhanced.

3.   As both the size of the aggregates and the thickness of the biofilm are limited (i.e., dictated by internal binding forces and the degree of intertwinement), eventually a new, second generation of growth nuclei will evolve from detached films and fragments of disrupted granules. These secondary nuclei will in turn grow and form a third generation, and so on. Later generations of the granules will gradually grow denser, and the bacterial filaments contained in the granules will become shorter.

The mechanisms of immobilization of bacterial matter in UASB and FB reactor systems are similar. As in UASB reactors, granulation will proceed quite satisfactorily in FB systems with soluble types of wastewater, provided these systems are operated in a mode similar to UASB systems. This has led to the abandonment of the idea of "complete fluidization" [19, 20] and is currently advocated and practiced by companies marketing FB systems.

The difficulties experienced during initial start-up of a UASB or FB reactor should not be underestimated, but they should not be exaggerated either. Nonetheless, skilled operators must be available during this critical phase of the operation.

It should be emphasized that granular sludge is not a prerequisite for satisfactory performance of a UASB system. Excellent results have been obtained using a well-settling and sufficiently active flocculent sludge, although the loading potential of such flocculated sludge bed systems is lower.

Operation after completion of start-up
Few if any problems are generally encountered in the full-scale application of UASB systems, provided adverse conditions such as severe overloading, toxicity,

and sudden changes in temperature do not occur (this is true for all biological treatment systems). The degree of supervision and the required skill of the operators depends fundamentally on the characteristics of the wastewater. In the case of sewage, stability of the biological process is almost automatically guaranteed, and supervision may be limited mainly to the prevention of mechanical problems. However, skilled supervision is indispensable in industrial wastewater treatment plants, where large or sudden fluctuations can occur in composition, strength, temperature, *p*H, or bicarbonate alkalinity.

Applicability and present use of UASB processes

The anaerobic treatment process can be applied to practically any type of wastewater that contains organic material and in which toxic compounds do not occur at inhibitory concentrations. Although originally developed for treating mainly soluble and medium-strength (5 to 10 grams COD per liter) industrial wastewaters, the system is suitable for more complex (partially soluble, slightly and moderately toxic), as well as low-strength wastewaters such as sewage, even under suboptimal mesophilic temperatures. It is very likely that the concept will also work under psychrophilic[6] temperatures.

The experience obtained so far with full-scale installations that treat a wide range of industrial wastewaters as well domestic sewage has been good (see table 1). Very few cases exist with a less than satisfactory performance; those that do generally are poorly designed or operated.

UASB reactors are not appropriate for treating high-strength slurries, but they can be profitably employed for slurries with suspended-solid concentrations as high as 4 to 6 grams per liter. Most of the other high-rate systems are less suitable for that purpose. For the digestion of high-strength slurries, such as liquid manure and sewage sludges, conventional digester systems should be applied.

Design of anaerobic treatment systems

Because the full-scale operation of anaerobic wastewater-treatment systems is very recent, there are no standardized design parameters. On the other hand, sufficient information is available for a satisfactory design. As in the case of conventional aerobic systems, different companies have their own specific approaches.

Depending on their characteristics, suspended solids can adversely affect the anaerobic treatment system. If non- or poorly biodegradable suspended solids accumulate in the sludge bed, they may reduce the overall specific methanogenic activity of the sludge. In addition, the formation of granular sludge during start-up may be retarded or even completely inhibited. Apart from the characteristics of the suspended solids, the concentration of the suspended matter is also of great importance: beyond a certain concentration of suspended solids (approximately 4 to 6 grams per liter), a UASB reactor becomes less feasible or even nonapplicable. The above-mentioned problems can be overcome; for example, the solids re-

---

6. Between 5 and 15°C.

moved by placing a settler ahead of the anaerobic reactor can be digested in a separate sludge digester.

Sometimes using a separate liquefying/acidifying reactor before the methanogenic reactor has been recommended even for nonacidified soluble types of wastewaters. Particularly for these type of wastewaters, the separation of acidification and methane production is not an attractive option, because 1) it increases capital investment (two reactors instead of one) and operational costs (e.g., the use of chemicals, and need for more control); 2) it leads to a slower growth rate of granular sludge; and 3) it may lead in specific cases to problems of retained granular sludge in the methanogenic reactor, as can occur when the acidogenic sludge is insufficiently removed from the influent of the methanogenic reactor.

Future developments of UASB technology

An interesting variant of the UASB concept is the expanded granular sludge bed (EGSB) system (see figure 2e). These systems can be operated at superficial liquid velocities exceeding 6 meters per hour and are particularly suitable for removing soluble pollutants from wastewaters. As a result of the excellent contact between

Table 1: Types of wastewater treated in UASB reactors, numbers of installed reactors, and reactor volumes as of June 1990[a]

Wastewater	No.	Volume m^3	Wastewater	No.	Volume m^3
Alcohol	20	52,000	Fruit juice	3	4,600
Bakers' yeast	5	9,900	Fructose production	1	240
Bakery	2	347	Landfill leachate	6	2,500
Brewery	30	6,600	Paper and pulp	28	67,200
Candy	2	350	Pharmaceutical	2	1,720
Canneries	3	800	Potato processing	27	25,600
Chemical	2	600	Rubber production	1	650
Chocolate	1	285	Sludge liquor	1	1,000
Citric acid	2	6,700	Slaughterhouse	3	950
Coffee	2	1,300	Soft drinks	4	1,380
Dairy and cheese	6	2,300	Starch (barley, corn, wheat)	16	33,500
Distillery	8	24,000	Sugar processing	19	21,100
Domestic sewage	10	10,000	Vegetable and fruit	3	2,800
Fermentation	1	750	Yeast	4	8,550
Total				**205**	**287,722**

a. Information provided by Dutch contractors and consultants.

wastewater and the sludge, these systems can handle higher organic loading rates than conventional UASB reactors. They are, however, rather inefficient in removing suspended solids. Full-scale FB reactors at a yeast factory in Delft, the Netherlands, have recently been successfully converted to EGSB reactors.

Another modification of the UASB reactor is the so-called internal circulation (IC) UASB reactor (see figure 2f, [21, 22]). Like EGSB systems, these IC systems employ tall reactors (up to 20 meters) but, unlike EGSB systems, they are equipped with a built-in gas separator halfway up the height of the reactor. The lifting forces of the separated biogas are employed to recirculate granular sludge through the lower part of the reactor. In some cases such recirculation might be beneficial.

A very significant improvement in the performance of the UASB system can be achieved by building the reactor in modules and operating these modules (two or more) in series. One of the modules could consist of a EGSB system.

ROLE OF ANAEROBIC PROCESSES IN ENERGY GENERATION

The great potential of anaerobic digestion for environmental protection as well as energy production can best be illustrated with a practical example. For this we choose the Brazilian renewable energy program, Pro-álcool, which was instituted in 1975 by the Brazilian government. This program promotes the partial substitution of ethanol for petroleum-based fuels. Present annual production of ethanol is about 13 million m^3, which is almost entirely from the fermentation of sugarcane juice by yeast.

Along with alcohol, there are three by-products: 1) carbon dioxide formed during fermentation; 2) the solid cane residue, called bagasse, which remains after extraction of the juice; and 3) a high-strength wastewater called vinasse, which is the bottom fraction of the distillation process (see figure 3). The production rate and possibilities for energy generation from bagasse are discussed elsewhere in this book (see chapter 17: *Advanced Gasification-based Biomass Power Generation* and chapter 20: *The Brazilian Fuel-alcohol Program*).

The amount of vinasse depends strongly on local production methods and especially on yeast quality, which determines the maximum alcohol concentration in the fermented cane juice. A high-quality yeast allows a maximum alcohol concentration of 8 percent by volume, which sets a minimum alcohol/wastewater ratio of 1 liter of alcohol per 12 liters of wastewater. In practice, the maximum permissible alcohol concentration is generally lower because of the relatively low yeast quality. Moreover, cane and plant wash waters are usually added to the vinasse and so a more realistic ratio is 1:20. From our own experimental determinations using vinasse from four distilleries in the northeast of Brazil with alcohol/wastewater ratios between 1:15 and 1:25, we found an inversely proportional relationship between the pollutant strength of the vinasse and the amount of alcohol produced. Consequently the total amount of organic pollutants per m^3 of

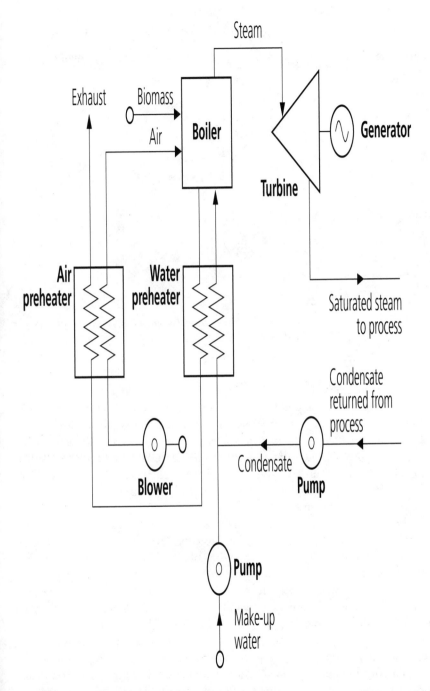

FIGURE 3: *Flow chart of the production of alcohol at a medium-sized distillery in the northeast of Brazil. Percentage figures reflect energy content.*

Table 2: Principal vinasse characteristics (raw vinasse from direct juice fermentation)

Parameter	Units	Value
Flow	*liters per liter of alcohol*	20.0
Biological oxygen demand[a]	*g/liter*	12.0
Chemical oxygen demand	*g/liter*	25.0
Settleable solids	*g/liter*	12.0
pH		3.5
Volatile fatty acids	*grams ethanoic acid/liter*	5.0
Total nitrogen		0.4
Phosphate		0.1
Potassium		0.8

a. Biological oxygen demand is the amount of oxygen required per liter of wastewater in the biological degradation (= aerobic conversion) of the polluting substances.

---

alcohol produced is independent of the alcohol/wastewater ratio. We found for the total organic load a value of approximately 500 kg COD per m³ of ethanol. The typical composition of vinasse for an alcohol/vinasse volumetric ratio of 1:20 is shown in table 2.

It is interesting to note that the actual useful energy production at a medium-sized distillery comprises only a fraction of the total energy production potential of sugarcane. As shown in figures 3 and 4a, only 38 percent of the energy content of the cane is actually converted into alcohol, while 12 percent is discharged as wastewater (vinasse) and 50 percent remains in the bagasse. In practice, about half of the bagasse (one-fourth of the energy content of cane) is burned at the distillery to satisfy its energy requirements. Consequently, the energy balance of most alcohol production plants in Brazil is not presently favorable.

Digested vinasse can be used for irrigation and fertilization of cane fields right after a harvest, but it is more often stored in "evaporation lagoons" or infiltrated into the soil, for which part of the cultivated land in a rotation scheme (about 5 percent) must be sacrificed. Another common practice is discharging vinasse into surface waters, which leads to a dramatic deterioration of water quality. Unfortunately, despite its high potential useful energy content, vinasse still has a negative economic value and a deleterious impact on the environment. From the point of view of sustainable development as well as economics, dumping vinasse into waterways should stop as soon as possible.

The organic pollution load of a medium-sized distillery can be estimated from the data in table 2. During the 200 day yearly harvest period, a medium-sized distillery produces about 24,000 m³ of alcohol and 500,000 m³ of vinasse at a total organic pollution load of 12,000 tonnes COD per year. A population of

300,000 would produce this much pollution annually, assuming one person produces 110 g COD per day of pollution. Moreover, this amount of waste is released over a six-month period. For the current production level of 13 million m^3 alcohol per year, the pollution power of this industrial activity is equivalent to 160 million inhabitants, which is more than the actual population of Brazil! Because the pollutants present in vinasse are highly biodegradable and include relatively large amounts of suspended proteins (up to 3 grams per liter, or 30 percent of the suspended matter), and because a large amount of vinasse is produced, methods should be developed and/or implemented as quickly as possible to exploit this potential energy resource and to combat pollution at the same time.

The following methods are available:

▼ Solid–liquid separation methods can produce a protein-rich animal fodder or possibly an organic fertilizer from the solids present in vinasse. Practical results have shown that applying the settled solids and the digested vinasse to cane fields results in a significant increase of productivity.

▼ High-rate anaerobic wastewater-treatment systems can convert soluble and poorly settleable organic pollutants into biogas and, to a lesser extent, into a settleable and stable bacterial sludge. The treated solution is virtually free from biodegradable material and can be used for unrestricted irrigation and, to some extent, for fertilization of the fields.

Protein-rich cattle fodder can be produced from settled vinasse slurries by spray-drying, and solid fertilizer by sun-drying of the settled solids. As explained above, the technical viability of anaerobic treatment has been well established, both for treating the soluble part of the vinasse, and—if economically attractive—the total vinasse wastewater. Currently more than a dozen UASB reactors have been installed at distilleries. These plants are operating quite satisfactorily at COD removal efficiencies exceeding 90 percent, at organic space loads of 20 kilograms (kg) COD per m^3 per day and more. Therefore, the production potential of the two by-products from vinasse in a medium-sized distillery amounts to 5,000 tonnes per year of protein-rich solids, and 2,700 tonnes of methane per year, based on 90 percent treatment efficiency and the stoichiometric relationship that 1 kilogram of methane is produced from the complete conversion of 4 kg COD.

Energy production from vinasse

From the above figures it follows that the proportion of vinasse methane per m^3 of alcohol amounts to 2,700 tonnes per 24,000 m^3 = 2,700 tonnes per 19,000 kilograms or 142 kilograms of methane per tonne of ethanol. As the higher heating values of methane and alcohol are 55.5 and 30.2 megajoules per kilogram respectively, the net amount of energy that can be produced from vinasse in the form of methane amounts to 26 percent of the energy content of the alcohol. Thus, the useful energy production from sugarcane juice can be increased by 26 percent with anaerobic wastewater treatment (see figure 4b).

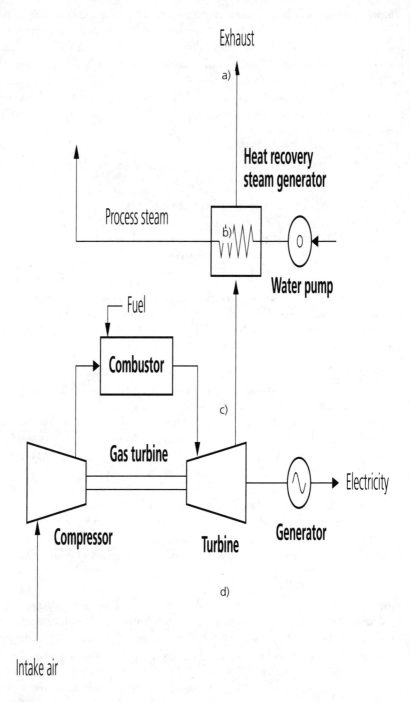

FIGURE 4: Different options for energy conversion from energy crops. Percentage figures reflect energy content.

Table 3: Comparison between aerobic and anaerobic treatment of vinasse for an assumed alcohol production of 120 m^3 per day (year-round treatment of 30,000 kg COD per day in 1,200 m^3 per day of vinasse)

Parameter	Units	Aerobic activated sludge	Anaerobic UASB
Load	kg COD/m^3/day	2	20
Volume	m^3	15,000	1,500
Removal efficiency	percent	> 90	> 90
Oxygen input	t O$_2$/day	15	–
Power requirements	kilowatts	750	–
Sludge production	t/day	8	2
Nitrogen demand[a]	t N$_2$/day	0.8	0.2
Phosphorus	t P/day	0.2	0.05

a. Present in vinasse: 0.5 t N$_2$/day and 0.1 t P/day.

Produced biogas can be used profitably in the alcohol production process as a substitute fuel for transporting the cane to the distillery and to generate power for the distillery. It is well established that automotive methane, produced from biogas by carbon dioxide absorption followed by compression, can provide as much as 80 percent of the normal fuel in diesel engines. For the substitution of 800,000 liters per year of diesel one would need 600 tonnes per year of methane, about one-fourth of the production potential. As a replacement for bagasse to generate steam and electricity for the distillery, vinasse would increase bagasse residues. After using methane as a fuel for the distillery's truck fleet, roughly 75 percent of the biogas produced from vinasse would still be available for steam and electricity production. The vinasse biogas could also be employed for the generation of electricity for export from the distillery, an option that is becoming increasingly important in Brazil. If the production of alcohol remains at 13 million m^3 per year, the total methane production potential from vinasse can be estimated at $110 \times 13 \times 10^6$ kilograms per year or 4,000 tonnes per day, which is equivalent to 40,000 barrels of oil per day. Assuming a (very conservative) conversion efficiency of 35 percent, useful electric power output can be estimated at 900 MW, which is the production capacity of a large power station. Of course, it is possible to couple electricity generation to steam production at the distillery and thus achieve a more efficient use of the produced methane.

Clearly, anaerobic wastewater treatment offers great potential for resource preservation and environmental protection (see figure 4). It is interesting to note that aerobic treatment (for example, in an activated sludge process) is not a realistic option for vinasse treatment. To illustrate this, the main characteristics for aerobic and anaerobic vinasse treatment are compared in table 3. It was assumed that an activated sludge process could be operated at a load of 2 kg COD per m^3 per day, which is rather high, but amounts to only one-tenth of the load commonly applied in UASB reactors. Therefore, the aerobic system would be 10 times larger than the UASB reactor. In addition, there would be significant operational costs associated with oxygenation and nutrients, whereas sludge from the aerobic system would be a mixed blessing. The differences between the produced active and UASB sludge are not only quantitative but also qualitative; the concentration of UASB sludge is significantly higher than that of activated sludge, even after thickening. Moreover, activated sludge has to be stabilized by digestion before it can be disposed. Even after stabilization, activated sludge is more difficult to dry than UASB sludge.

Direct production of energy from sugarcane

Modern high-rate anaerobic wastewater-treatment technology offers enormous potential for the direct production of energy as well as for environmental protection. Cane juice can be digested directly to produce methane, without the need for alcohol fermentation, centrifuging, and distillation, and the consumption of high-grade energy associated with these processes. Some 48 percent of the total energy in cane would be present in the converted methane, equal to the conversion efficiency of alcohol production with vinasse treatment, but the need for bagasse combustion would be much less, because energy would be needed only for the juice extraction process (see figure 4c).

A further reduction of investment and operational costs and an increase of energy output could be obtained by subjecting not only the juice but also the bagasse to anaerobic digestion (see figure 4d). In that case the expensive machinery for juice extraction would not be necessary. Much cheaper cutting machines would be substituted to produce small stalks for digestion. Assuming that 25 percent of the bagasse can be converted into methane, which is a very conservative figure for a low-lignin plant such as sugarcane, then the energy conversion efficiency would increase to 60 percent of the energy content of cane, compared with the 38 percent in a distillery with no anaerobic treatment system. If necessary, all the produced methane can be converted into automotive fuel by purification and compression.

Another important advantage of substituting anaerobic digestion for alcohol fermentation is that other energy crops can be used, such as starch-producing cassava and sweet potatoes. Anaerobic digestion can easily transform the starch into methane, whereas alcohol digestion can be applied only to some soluble sugars. Thus, in regions where agricultural activity is low (notably arid areas), it would be possible to produce these far less demanding raw materials for methane production. In this way good agricultural lands, presently used for energy production, could be used for other purposes such as human food production.

The energy produced by anaerobic digestion differs in one important aspect from fossil-fuel sources: the carbon dioxide released during combustion was previously extracted from the atmosphere by photosynthesis. Thus, energy produc-

Table 4: Rough estimates of methane production costs using UASB reactors in Europe and Brazil with 6 percent and 12 percent discount rates

		Europe				Brazil			
COD load $kg\ COD/m^3/day$		20		30		20		30	
Plant capacity m^3		**4,000**	**10,000**	**4,000**	**10,000**	**4,000**	**10,000**	**4,000**	**10,000**
Methane production rate[a] 10^6 GJ/year		0.308	0.770	0.462	1.155	0.308	0.770	0.462	1.155
Fixed capital[b] 10^6 $		1.6	3.5	1.6	3.5	1.0	2.0	1.0	2.0
Working capital[c] 10^6 $		0.08	0.18	0.08	0.18	0.05	0.10	0.05	0.10
Annual costs $10^5$$/year									
Fixed capital[d]									
6 percent		1.73	3.78	1.73	3.78	1.08	2.16	1.08	2.16
12 percent		2.43	5.32	2.43	5.32	1.52	3.04	1.52	3.04
Working capital[e]									
6 percent		0.05	0.11	0.05	0.11	0.03	0.06	0.03	0.06
12 percent		0.10	0.22	0.10	0.22	0.06	0.12	0.06	0.12
Labor and supervision		0.60	1.00	0.60	1.00	0.40	0.70	0.40	0.70
Maintenance and replacements[f]		0.64	1.40	0.64	1.40	0.40	0.80	0.40	0.80
Analysis and control		0.20	0.30	0.20	0.30	0.20	0.30	0.20	0.30
Total annual costs									
6 percent		3.22	6.59	3.22	6.59	2.11	4.02	2.11	4.02
12 percent		3.97	8.24	3.97	8.24	2.58	4.96	2.58	4.96
Unit production cost $/GJ									
6 percent		1.05	0.86	0.70	0.57	0.69	0.52	0.46	0.35
12 percent		1.29	1.07	0.86	0.71	0.84	0.64	0.56	0.43

a. Assuming a 95 percent COD reduction rate, a stoichiometric methane production rate of 0.25 kg per kg of COD, a methane yield that is 80 percent of the stoichiometric rate, and a methane higher heating value of 55.5 megajoules per kilogram.

b. The capital cost is estimated to be $400 per m^3 for a 4,000 m^3 digester and $350 per m^3 at 10,000 m^3 in Europe; for Brazil, the corresponding costs are $250 per m^3 for a 4,000 m^3 digester and $200 per m^3 at 10,000 m^3.

c. The working capital is assumed to be 5 percent of the fixed capital investment.

d. Assuming a 15 year plant life and an insurance cost of 0.5 percent of the installed capital cost per year, the annual capital charge rate for fixed capital is 0.108 for a 6 percent discount rate and 0.152 for a 12 percent discount rate.

e. The capital charge rate for working capital is 0.06 for a 6 percent discount rate and 0.12 for a 12 percent discount rate.

f. Assumed to be 4 percent of the fixed capital investment.

tion by anaerobic digestion of biomass does not lead to a net increase of carbon dioxide in the atmosphere and hence does not contribute to the greenhouse effect (see chapter 14: *Biomass for Energy: Supply Prospects*).

THE ECONOMICS OF ANAEROBIC TREATMENT

High-rate anaerobic wastewater-treatment can be accomplished in technically simple, low-cost reactor systems. The investment costs of a UASB reactor depend on the size of the reactor, land and labor prices, the characteristics of the liquid to be treated, the desired degree of sophistication, and auxiliary equipment (gas-storage and post-treatment facilities). The investment costs, therefore, depend strongly on the local situation. Table 4 provides rough guidelines for the investment costs of UASB systems for Europe and Brazil, based on practical experience with full-scale UASB reactors in both places [23]. Since energy consumption is nil, the nutrient requirements are generally negligible, and operation is not labor-intensive, total operating costs are extremely low. As a result, biogas can be generated in a medium-sized UASB plant (1,000 to 5,000 m^3 reactor volume) at a production cost of U.S.$0.03 to 0.05 per Nm^3 (1 m^3 of gas at 1 atmosphere and 0°C). For the huge UASB plants that produce biogas from sugarcane juice in Brazil, production costs would even be lower: about $0.02 per Nm^3.

CONCLUSIONS

High-rate anaerobic treatment using systems such as the UASB reactor can be regarded as fully mature. Considering its performance over a wide range of wastewaters and its extremely attractive technological and economic features, this technology should find a significantly wider application, especially for treatment of industrial and domestic wastewater. Anaerobic wastewater treatment is particularly attractive in countries with a tropical or subtropical climate, where the rate of methane production would be higher than in temperate areas because of the higher ambient temperatures. Almost all developing countries would benefit from the implementation of these systems. Not only would such systems help protect the environment at low cost and in a completely self-sufficient way, but anaerobic digestion would produce useful by-products with important applications in industry (methane) and agriculture (sludge).

Modern high-rate treatment methods are also very attractive for energy production. In terms of energy conversion into automotive fuel, anaerobic digestion is far more efficient than current alcohol fermentation technology. In addition, anaerobic digestion is significantly less expensive because it requires less equipment and is far simpler to use.

WORKS CITED

1. Gujer, W. and Zehnder, A. 1983
   Conversion processes in anaerobic digestion, *Water Science and Technology* 15:127–167, proceedings of an IAWPR specialized seminar *Anaerobic Treatment of Wastewater in Fixed Film Reactors*, M. Henze, ed., Copenhagen, 16–18 June 1982.

2. Henze, M. and Harramoes, P. 1983
   Anaerobic treatment of wastewater in fixed film reactors—a literature review, in *Water Science and Technology* 15:1–101, proceedings of an IAWPR specialized seminar "Anaerobic treatment of wastewater in fixed film reactors," M. Henze, ed., Copenhagen, 16–18 June 1982.

3. Speece, R.E. 1983
   Environmental requirements for anaerobic digestion of biomass. Report of the Environmental Studies Institute, Drexel University, Philadelphia.

4. Coulter, J., Soneda, E., and Ettinger, M. 1957
   Anaerobic contact processes for sewage disposal, *Sewage and Industrial Wastes*, 468–477.

5. Young, J. and McCarty, P. 1969
   The anaerobic filter for waste treatment, *Journal of Water Pollution Control* 4:R160–R173.

6. Van den Berg, L. and Lentz, C. P. 1971
   Comparison between up- and downflow fixed film reactors of varying surface-to-volume ratios for the treatment of bean blanching waste, *Proceedings of the 34th Industrial Waste Conference*, Purdue University, Ann Arbor, Michigan, 319–325.

7. Lettinga, G., Van Velsen, A., Hobma, S., de Zeeuw, W., and Klapwijk, A. 1980
   Use of the Upflow Sludge Blanket (UASB) reactor concept for biological wastewater treatment, *Biotechnology and Bioengineering* 22:699–734.

8. Jeris, J. 1982
   Industrial wastewater treatment using anaerobic fluidized bed reactors, *Water Science and Technology* 15:169–176, in M. Henze, ed., *Anaerobic treatment of wastewater in fixed film reactors*, Copenhagen, proceedings of an IAWPR specialized seminar, 16–18 June.

9. Heijnen, J.J., Mulder, A., Enger W., and Hoeks, F. 1986
   Review on the application of anaerobic fluidized bed reactors in waste-water treatment, *Proceedings of the Aquatech '86 conference—Anaerobic Treatment, a Grown-up Technology*, 15–19 September, Amsterdam, 159–174.

10. Heijnen, J.J. 1988
    Large scale anaerobic–aerobic treatment of complex industrial wastewater using immobilized biomass in fluid bed and air-lift suspension reactors, reprints Verfahrenstechnik Abwasserreiningung (Process technology of wastewater treatment), GVC-Diskussionstagung, 17–19 October, Baden Baden, 203–218.

11. Reynolds, P. and Colleran, E. 1986
    Comparison of the start-up and operation of anaerobic fixed bed and hybrid sludge-bed/fixed-bed reactors treating whey wastewater, *Proceedings of the Aquatech '86 Conference—Anaerobic Treatment, a Grown-up Technology*, 15–19 September, Amsterdam, 515–532.

12. Guiot, S., Kennedy K., and Van den Berg, L. 1986
    Comparison of the upflow anaerobic sludge blanket and sludge bed-filter concepts, *Proceedings of the Aquatech '86 Conference—Anaerobic Treatment, a Grown-up Technology*, 15–19 September, Amsterdam, 533–546.

13. Garavini, G., Mercuriali, L., Tilche, A., and Yang Xiushan, ENEA. 1988
    Performance characteristics of a thermophilic full scale hybrid reactor treating distillery slops, *Proceedings of the 5th International Conference on Anaerobic Digestion*, Bologna, 22–26 May, poster papers, A. Tiche and A. Rozzi, eds., 509–515.

14. Garutti, G., Bortone, G., Fagnocchi, G., Piccini, S., and Tilche, A. 1988
    Full-scale mesophilic sludge bed anaerobic filters treating distiller slops, *Proceedings of the 5th International Conference on Anaerobic Digestion*, Bologna, 22–26 May, poster papers, A. Tiche and A Rozzi, eds., 517.

15. de Zeeuw, W. 1987
    Granular sludge in UASB-reactors, in G. Lettinga, A. J. B. Zehnder, J.T.C. Grotenhuis, and L. W. Hulshoff Pol, eds., *Granular anaerobic sludge: microbiology and technology*, proceedings of the GASMAT workshop, 25–27 October, Lunteren, 132–146.

16. Hulshoff Pol, L.W. and Lettinga, G. 1986
    New technologies for anaerobic wastewater treatment, *Water Science and Technology* 18(12):41–53.

17. Hulshoff Pol, L.W., Heynekamp, K., and Lettinga, G. 1987
    The selection pressure as a driving force behind the granulation of granular sludge, in G. Lettinga, A. J. B. Zehnder, J.T.C. Grotenhuis, and L. W. Hulshoff Pol, eds., *Granular anaerobic sludge: microbiology and technology*, proceedings of the GASMAT workshop, 25–27 October, Lunteren, 146–153.

18. Hulshoff Pol, L.W. 1989
    The phenomenon of granulation, PhD Thesis, Department of Water Pollution Control, Agricultural University, Wageningen, the Netherlands.

19. Iza, J., Garcia, P.A., Sanz, I., and Fernandez-Polanco, F. 1987
    Granulation results in anaerobic fluidized bed reactors, in G.Lettinga, A. J. B. Zehnder, J.T.C. Grotenhuis, and L. W. Hulshoff Pol, eds., *Granular anaerobic sludge: microbiology and technology*, proceedings of the GASMAT workshop, 25–27 October, Lunteren, 146–153.

20. Iza, J. 1991
    Fluidized bed reactors for anaerobic treatment, *Water Science and Technology* 24(8)109–132), in M.S. Switzenbaum, ed., *Anaerobic treatment technology for municipal and industrial wastewaters*, proceedings of an international workshop held in Valladolid, 24–26 September 1990.

21. Vellinga, S., Hack, P., and Van der Vlugt, A. 1986
    New type "high rate" anaerobic reactor, *Proceedings of the Aquatech '86 Conference— Anaerobic Treatment, a Grown-up Technology*, 15–19 September, Amsterdam, 547–562.

22. Voorter, P. 1989
    New type anaerobic treatment, *Procestechniek* (10):52, October (in Dutch).

23. Lettinga, G., Hobma, S., Hulshoff Pol, L., de Zeeuw, W., de Jong P., and Roersma, R.1983
    Design operation and economy of anaerobic treatment, *Water Science and Technology* 15:177–195.

24. Lettinga, G., Van Velsen, A., de Zeeuw, W., and Hobma S. 1979
    Feasibility of the UASB-process, *Proceedings of the National Conference in Environmental Engineering*, 9–11 July, Sand Francisco, 35–45. American Society of Civil Engineers, New York.

25. Lettinga, G., Van Velsen, A., de Zeeuw, W., and Hobma, S. 1979
    The application of anaerobic digestion to industrial pollution treatment, in Stafford et al. eds., *Anaerobic digestion* 167–186, Applied Science Publishers, London.

26. Lettinga, G., Hulshoff Pol, L., Koster, I., Wiegant, W., de Zeeuw, W., Rinzema, A., Grin, P.; Roersma, R., and Hobma, S. 1984
    High-rate anaerobic wastewater treatment using the UASB-reactor under a wide range of temperature conditions, *Biotechnology and Engineering Reviews* 2:253–284.

27. Lettinga, G., and Hulshoff Pol, L.W. 1986
Advanced reactor design, operation and economy, *Water Science and Technology* 18(12):99–108.

28. Lettinga, G. and Hulshoff Pol, L.W. 1986
UASB-process design for various types of wastewaters, *Water Science and Technology,* 24(8)87–107), in M.S. Switzenbaum, ed., *Anaerobic treatment technology for municipal and industrial wastewaters,* proceedings of an international workshop held in Valladolid, 24–26 September 1990.

20
THE BRAZILIAN FUEL-ALCOHOL PROGRAM

JOSÉ GOLDEMBERG
LOURIVAL C. MONACO
ISAIAS C. MACEDO

The substitution of ethanol for gasoline in passenger cars and light vehicles in Brazil is one of the largest commercial biomass-to-energy programs in existence today. Engines that run strictly on gasoline are no longer available in the country, having been replaced by neat-ethanol engines and by gasohol engines that burn a mixture of 78 percent gasoline and 22 percent ethanol, by volume.

Technological advances, including more efficient production and processing of sugarcane, are responsible for the availability and low price of ethanol. The transition to ethanol fuel has reduced Brazil's dependence on foreign oil (thus lowering its import–export ratio), created significant employment opportunities, and greatly enhanced urban air quality. In addition, because sugarcane-derived ethanol is a renewable resource (the cane is replanted at the same rate it is harvested), the combustion of ethanol adds virtually no net carbon dioxide to the atmosphere and so helps reduce the threat of global warming.

BACKGROUND

History and original objectives

Ethanol derived from sugarcane has been used as an engine fuel in Brazil since 1903, when the 1º Congresso Nacional sobre Aplicações Industriais do Alcool (first national congress on industrial applications of alcohol) proposed that an infrastructure be established to promote alcohol production and use. During World War I, in fact, the use of alcohol was compulsory in many areas of the country. By 1923, production of ethanol had grown to 150 million liters per year; in 1927 it was blended with diethyl (ethylic) ether and castor oil. In 1931 a federal decree mandated that alcohol be added to gasoline to the extent of 5 percent of the mixture (by volume) and established guidelines for its transportation and commercialization. By 1941, ethanol production had reached 650 million liters. The growth in ethanol production can be attributed in part to the Brazilian government's desire to reduce the country's dependence on foreign oil.

The government's fears of overdependence on outside oil supplies proved to be well founded. When the first world oil crisis occurred in 1973, Brazil was importing 80 percent of its oil and thus suffered a severe setback in its balance of payments, resulting in continuously growing deficits. Representing a mere 11 percent of Brazil's total imports in 1973, oil increased to 57 percent by 1983 only 10 years later [1]. Today a large percentage of Brazil's foreign debt can be blamed on the sharp increases in oil prices and interest rates during the 1970s.

Faced with economic crisis, the country adopted a number of stringent measures, including a vow to increase and diversify export commodities and to enact an energy policy that emphasized conservation and oil substitution. From this policy was born the Brazilian fuel-ethanol (Pro-álcool) program in 1975, whose subsequent success can be attributed to several factors [1]. To begin with, the country was familiar with ethanol production and consumption and knew how to blend ethanol with gasoline. In addition, it had a plentiful sugarcane crop, abundant agricultural land, a climate suitable for rapid vegetative growth, and an adequate labor force. Furthermore, industrial facilities that could easily handle ethanol produced by small, annexed distilleries already existed. And finally, sugar prices in the international market were low and so diminished sugar's value as an export crop.

The Pro-álcool program also called for such ambitious objectives as a reduction in regional income differences, an increase in job opportunities for both skilled and unskilled workers, and an increase in industrial production (brought about in part by the increased demand for Brazilian-made manufacturing equipment). The outcome was the emergence of the sugarcane agro-industrial system [2].

Since then, the program has been adjusted many times, and its original goals [3] reevaluated in view of such factors as the growth in domestic oil production, fluctuating international prices, and other strategic considerations.

In short, Brazil's 15 year experience with the Pro-álcool program has proved beneficial on many fronts. The Pro-álcool program has served as a model for biomass-energy programs in other countries, provided invaluable insights about land use (for both food and energy), and led to the creation of quality jobs. Moreover, it has proved to be an effective tool for managing energy costs, fostering agro-industrial development and diversification, and learning to integrate new fuels into existing commercialized end-use systems. Some of these issues will be considered here.

Patterns of energy consumption

From 1973 to 1987, total energy consumption in Brazil increased from nearly 90 × 10^6 TOE (tons of oil equivalent) to 160 × 10^6 TOE [4]. Yet during the same period gasoline dropped from 12 percent of the energy market to only 4 percent, a clear indication of the success of the Pro-álcool program. Other effects of the oil-substitution program include an increase in ethanol production, which accounted for 18 percent of total fuel consumed by the transportation sector in 1987 and 4 percent of the total energy market. Similarly, bagasse (the woody fibers that remain after the juice is crushed from sugarcane) grew to 8 percent of the energy market (consumed mostly by sugarcane processors). Fuel oil consumption dropped from 13 to 6 percent. At the same time, production of domestic oil and natural gas rose from 24,000 to 83,000 TOE per day from 1974 to 1989, reducing the country's dependence on foreign oil from 78 to 47 percent [5].

Brazil's energy policy has had little effect on the consumption of diesel fuel, whose share of the total energy market rose from 9 percent in 1974 to 12 percent in 1987. To date no substitute for diesel has been actively pursued despite the fact that heavy goods are transported almost entirely by diesel trucks.

The switch from gasoline to ethanol

Since 1979, production of cars powered by pure gasoline has been halted. Today all cars and light vehicles sold in Brazil are required to run at least in part on ethanol. Ethanol fuel is available in two formulations: either anhydrous ethanol blended with gasoline (a mixture called gasohol that contains 22 percent ethanol) or hydrated ethanol for combustion by neat-alcohol engines. In 1989, the country's total fleet (passenger cars and light commercial vehicles) consisted of about 4.2 million neat-ethanol cars and 5 million gasohol cars. Since 1985, sales of neat-ethanol cars and light vehicles have exceeded those of gasohol vehicles (see figure 1), a trend that was reversed in 1990 owing to a shortage of ethanol.

Such growth was made possible by an increase in ethanol production, which rose from 0.6 billion liters per year in 1976 to 12 billion liters per year in 1989. Current projections call for fuels for Otto-cycle (spark-ignited reciprocating internal combustion) engines to be limited to the two formulations: anhydrous alcohol and hydrated alcohol. Neat-ethanol cars may account for 30

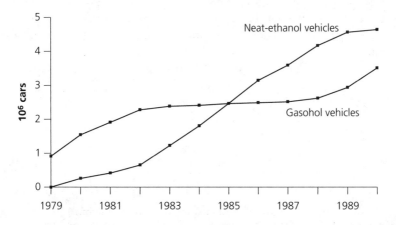

FIGURE 1: Cumulative sales of passenger cars and light-duty vehicles sold in Brazil beginning 1 January 1979. Data are from ANFAVEA, the Brazilian automakers' association.

to 40 percent of total vehicle sales by 1995, which in turn would demand an annual ethanol consumption of 16 billion liters. Existing industrial capacity is only 16.3 billion liters per year, and sugarcane production is expected to reach this capacity by the mid 1990s. Further increases, both in sugarcane and in ethanol production, will depend on the scope of the Brazil's energy policy.

Government incentives

Increasing the production and use of ethanol as a fuel was made possible by three government actions. First, the government agreed that the state-owned oil company, Petrobrás, would purchase a guaranteed amount of ethanol. Second, the government provided economic incentives for companies willing to produce ethanol, offering low interest rates from 1980 to 1985 and nearly U.S.$2.0 billion in loans, which represented 29 percent of the total investment needed to reach present installed capacity [1]. And third, the government made ethanol attractive to consumers by selling it at the pump for 59 percent of the price of gasoline. Today ethanol sells for 75 percent of the price of gas, which represents almost no subsidy because its technical equivalence to gasoline (based on engine output) is 80 percent. In the Center–South region of Brazil, economic incentives—either to producers or consumers—are no longer considered necessary, and thus are practically nonexistent today. In the Northeast region, incentives are still offered to producers within the context of the state's regional development policy.

Recent trends

The future of the Pro-álcool program has been so extensively discussed in Brazil that the general public is probably more familiar with its main issues (cost, environmental and social benefits and problems, and its strategic position) than with issues associated with oil and electricity production. Unlike oil and gas, ethanol is produced entirely by the private sector, although the price is controlled by the government. While low consumer prices were essential to the early success of the Pro-álcool program, the government from time to time has artificially depressed the price of ethanol to hold down inflation. But doing so has destabilized the industry, leading, for example, to ethanol shortages in 1989 and 1990, which in turn resulted in a drastic reduction in sales of ethanol cars. Such setbacks led the government to raise prices in 1991.

ETHANOL AS A FUEL

In Brazil, ethanol is used either as an octane enhancer in gasoline in the form of 22 percent anhydrous ethanol at 99.6 Gay Lussac (GL) (0.4 percent water) or in neat-ethanol engines in the form of hydrated ethanol at 95.5 GL. In other countries, gasohol blends typically contain only 10 percent ethanol. In the United States today, for example, nearly 7.5 percent of all automotive fuel sold contains 10 percent ethanol and 90 percent gasoline.

Ethanol makes an excellent motor fuel: it has a research octane number (RON) of 109 and a motor octane number (MON) of 90, both of which exceed those of gasoline, which has a RON of 91 to 98 and MON of 83 to 90.[1] Moreover, it is possible to design an ethanol engine that runs 30 percent more efficiently than a gasoline engine [6]. Ethanol also has a lower vapor pressure than gasoline (its Reid vapor pressure [the pressure at 38°C in a closed vessel with a vapor volume four times the volume of the liquid] is 16 kilopascals versus 71 for typical gasoline), which results in fewer evaporative emissions. Ethanol's flammability in air (1.3 to 7.6 percent, by volume) is also much lower than that of gasoline (3.5 to 19 percent, by volume), which reduces the number and severity of vehicle fires [6]. Anhydrous ethanol has lower and higher heating values of 21.2 megajoules per liter and 23.4 megajoules per liter, respectively; for gasoline the values are 30.1 and 34.9 megajoules per liter. The advantages that ethanol offers, including reduced emissions, however, will not be maximized until engines specifically designed for ethanol become available [6, 7].

NET ENERGY ANALYSIS

Sugarcane versus corn

The ethanol debate is sometimes confused by mixing data from the two primary sources of ethanol fuel: corn and sugarcane. This is unfortunate because the energy balances differ considerably for the two crops. The energy balance for ethanol from sugarcane under average and best conditions is presented in table 1 [8]. Sugarcane productivity, industrial conversion efficiency, and the presence of cogeneration (combined heat and power) facilities are quantified for 60 mills in the state of São Paulo. Under average conditions, the ratio of energy in ethanol/energy for fossil fuel spent (for both agriculture and industry) was 5.9; under best conditions the number rose to 8.2. (The energy content of surplus bagasse not needed as fuel is not included in these figures.) The net energy ratios for corn-derived ethanol are much lower [9] (see chapter 21: *Ethanol and Methanol from Cellulosic Biomass*) because almost all corn-processing operations need fossil fuels to produce ethanol, whereas sugarcane-derived ethanol can be produced using bagasse for fuel.

Use of bagasse and other residues

The amount of energy obtained from sugarcane can be increased through more efficient use of bagasse and other cane residues (the leaves and tops of the sugarcane plant). The energy potential can be summarized as follows:

▼ In 1990, the 98 million-tonne cane crop in Brazil's Center–South region consisted of 14.2 percent fiber on average, which corresponds to 283 kilograms

---

1. RON: as defined by the American Society for Testing and Materials (ASTM) in D908-47T; MON: as defined in ASTM D357-47.

Table 1: Energy flows in sugarcane and ethanol production

	Averages[a] MJ/tonne cane	Best values MJ/tonne cane
External energy inputs for sugarcane production[b]	**221.75**	**197.46**
Agricultural operations	36.95	26.33
Transportation	57.59	45.90
Fertilizers	69.18	64.74
Lime	6.77	6.77
Herbicides, insecticides	12.03	15.39
Seeds	8.11	7.23
Equipment	23.28	23.28
Labor	7.86	7.86
External energy inputs for ethanol production[c]	**70.10**	**40.59**
Electricity (purchased)	12.54	0.00
Chemicals and lubricants	6.56	6.56
Buildings	14.17	9.45
Heavy equipment	17.22	11.50
Light equipment	19.60	13.08

External energy flows,[d] **agriculture and industry**

	Input	Output	Input	Output
Agriculture	221.75	–	197.46	–
Industry	70.10	–	40.59	–
Ethanol	–	1,707.11	–	1,941.02
Bagasse surplus	–	175.14	–	328.55

a. 1985 sample of 60 annexed distilleries in the state of São Paulo. These distilleries produce one third of Brazil's ethanol; best values correspond to the best value for each item, that is, low energy input; high energy output [8].

b. Corresponding energy output is in sugar, bagasse, and cane residues. Most of the bagasse, produced at a rate of 283 kilograms (wet basis) per tonne cane and having an energy content of 2,690 megajoules per tonne cane (HHV) is used as fuel in the industrial process. The tops and leaves represent an untapped residue potential of 180 kg (dry matter) per tonne cane or 3,460 megajoules per tonne cane.

c. External energy only; does not include stored energy in the form of sugar and bagasse, or the electricity cogenerated with bagasse.

d. External energy used by the agro-industrial system comes mainly from fossil fuels (diesel or fuel oil), except for the small portion of electricity used in the manufacture of equipment, buildings, chemicals, etc. Even so, "totals" must be viewed carefully.

Table 2: Production of sugarcane tops and leaves

Country	Residues dry mass/Clean stalks fresh mass[a]	Remarks[b]
Brazil[c]	0.09 to 0.17	Three SP (Copersucar) varieties; yield: 80 tonnes cane per hectare
Brazil[d]	0.18	Six-year experiment; variety SP-71-1406 considered a "lowtrash" variety; yield: 80 tonnes cane per hectare
South Africa[e]	0.14	Variety NCO376; harvest at 19–20 months; average yield: 120 tonnes cane per hectare
Hawaii, U.S.[f]	0.09	Four Hawaiian varieties; harvest at 23.7 months (average); yield: 338 tonnes cane per hectare
Thailand[g]	0.16	Reported 11.1 tonnes per hectare of dry mass residues; we have assumed yields of 70 tonnes cane per hectare
Puerto Rico[h]	0.28	

a. Residues = green leaves, tops, and dry leaves of unburned cane. Clean stalks (fresh mass) = cane entering the mill.

b. Differences in yields are considered normal depending on whether the cane is harvested every 1 or 2 years, and whether or not it is irrigated.

c. [10].

d. Copersucar, private communication.

e. [11].

f. [12].

g. [13].

h. [14].

of bagasse with a moisture content of 50 percent and an energy content of 2.69 gigajoules per tonne of cane [on a higher heating value basis (HHV)] or 2.13 gigajoules per tonne [on a lower heating value basis (LHV)].

▼   The amount of additional residue in the form of tops and leaves as a fraction of clean stalks depends on the cane variety, time between harvests (which ranges from 1 to 2 years), the amount of fertilizer applied, and whether or not the cane is irrigated (see table 2). On average, however, about 150 kilograms of dry mass residue is produced per tonne of cane in Brazil, which represents an energy equivalent of 2.9 gigajoules per tonne (HHV).

ECONOMIC ANALYSIS

Recent findings

In recent years, numerous studies aimed at quantifying the true cost of ethanol production have been undertaken by both Brazilian and international organizations. A 1984 World Bank study [15] set the cost at $0.20 per liter (in 1984 dollars), which represents a weighted average taking into account different regions, the type of ethanol (anhydrous versus hydrated) and whether or not the distillery is annexed to a cane-processing plant or is autonomous. A correction factor of 0.83 over the official exchange rate was then applied to increase the U.S. dollar value with respect to the cruzeiro, thus making the figure more accurate. Outside Brazil's Northeast region (where cost estimates are less accurate), the cost averaged $0.185 per liter of alcohol, with most of the values falling between $0.17 and $0.23 [16, 17]. The average is equivalent to $7.9 per gigajoule.

Lifecycle costs

Yearly analyses have been made by the Copersucar company[2] using actual cost data from nearly 50 sugar and alcohol producing firms in the state of São Paulo. The data presented below are for an autonomous ethanol distillery system.

Production costs for anhydrous ethanol are shown in table 3, with sugarcane and capital investment costs itemized in tables 4 and 5. Analysis indicates that from 1979 to 1988 ethanol production costs dropped by 4 percent a year, which correlates with an increase in combined yield from agriculture and industry [3]. For example, ethanol production in 1977 was 2,663 liters per hectare; in 1985 it grew to 3,811 liters per hectare (an average annual increase of 4.3 percent). During the same period, agricultural yield increased 16 percent (measured in tonnes of cane per hectare) and industrial yield increased 23 percent (liters of ethanol per tonne of cane). By 1989, the average yield in the state of São Paulo was 4,700 liters of ethanol per hectare. Prospects for maintaining this trend are analyzed in a following section.

Value analysis

Ethanol was used initially as a gasoline extender. Consequently, early blending studies were based on the engines and gasoline commonly used in Brazil at the time. Gasoline typically had a low octane number, and tetraethyl lead was the additive used to raise the MON to 73 [19]. Today ethanol has displaced tetraethyl lead in gasoline and blends up to 20 percent ethanol are possible without modification of existing gasoline engines. The suitability of various blend ratios to engine characteristics (mainly compression ratio) are now well known.

---

2. Copersucar is a cooperative of sugarcane, ethanol, and sugar producers, responsible for one-third of Brazilian sugarcane production.

Table 3: Ethanol production costs in 1990[a]
1989 U.S.$/liter

	Average[b]	Higher average[b]
Direct costs		
Labor	0.006	0.007
Maintenance	0.004	0.006
Chemicals	0.002	0.002
Energy	0.002	0.003
Other	0.004	0.004
Interest on working capital and commercial costs	0.022	0.029
Cane[c]	0.127	0.134
Fixed costs		
Capital[d]		
6 percent	0.030	0.032
12 percent	0.051	0.053
Other	0.011	0.013
Total		
6 percent	0.208	0.230
12 percent	0.229	0.251

a. Based on a sample of 50 mills. Data are for anhydrous ethanol; hydrated ethanol is 7 to 10 percent lower in cost. [From Copersucar, DEAD: internal report.]

b. "Average" is the sample average; "Higher average" is the average for mills operating at above average cost.

c. See table 4: cost per tonne of cane. "Average" and "Higher average" correspond to 77.7 and 73.6 liters of ethanol per tonne of cane respectively.

d. See table 5; data are based on a 25 year lifecycle, with 10 percent residual value.

A 20 percent blending ratio brings the octane level to a minimum of 80 MON [20], and as a result, 1 liter of anhydrous ethanol, used as an octane enhancer, is considered equivalent in terms of automotive performance to 1.04 liters of gasoline [15]. In a 22 percent ethanol blend, ethanol not only acts as an octane enhancer but also as a gasoline extender (with a relatively lower value). Yet ethanol's value (even as only a gasoline extender) depends on the local cost of refining an equivalent high-octane gasoline, and so it must be evaluated for each location. Such a comprehensive location-by-location study is needed in Brazil; even disregarding environmental benefits there is some indication that its equivalence as an octane enhancer is higher than 1.04 [6].

Measurements of neat-ethanol engines indicate that 1.20 liters of hydrated ethanol are equivalent to 1 liter of gasohol [3]. This equivalence figure may change, however, as more efficient ways of combusting ethanol are developed. Studies of optimized methanol vehicles by the U.S. Environmental Protection

Table 4: Sugarcane production costs[a]

Item	Cost U.S.$ per tonne of cane
Variable costs[b]	5.78
Fixed costs (including land)	2.32
Social tax	0.28
Capital charges[c]	1.17
Total	**9.55**

a. Based on a yield of 75 tonnes of cane per hectare and 1990 analysis of 50 sugar producers in the state of São Paulo [from Copersucar, DEAD: unpublished reports]. Average pol. percent cane = 13.6 [18].

b. Total labor costs (including a fraction of fixed costs) are $1.72 per tonne of cane, or 18 percent of total cane cost.

c. At 12 percent annual rate.

Table 5: Capital investment required for a distillery with a capacity to produce 240,000 liters of ethanol per day[a]

Sector	Costs[b] 1989 1,000 U.S.$
Cane reception	559
Milling sector	5,306
Juice treatment	558
Boilers and energy	3,786
Fermentation	1,067
Distillery	2,947
Ethanol storage	1,067
Building	1,019
Other[c]	2,294
Total	**18,603**

a. Up to 4,000 tonnes of cane per day; processed 150 days a year.

b. Including assembly and engineering.

c. Piping, instrumentation, and electrical systems.

Agency, for example, suggest that a 30 percent efficiency improvement over the best gasoline engine can be achieved when using neat ethanol in a dedicated vehicle [6]. Under those conditions, the equivalence value is 1.15 liters of ethanol to 1 liter of gasoline. Further gains (measured against the best gasoline engine that can be produced using existing technology) are possible if vehicle optimization is achieved [6].

Cost estimates based on the equivalences of 1.20 liters of ethanol per liter of gasoline (neat-ethanol engine) and 1.04 liters of gasoline per liter of ethanol (gasohol engine) can be made. For example, Copersucar has estimated (based on refining data) that if crude oil sells for $20 a barrel, then ethanol and gasohol would be valued at $22.30 ($0.14 per liter) and $28 ($0.18 per liter) a barrel, respectively. If the average cost of anhydrous gasohol were $0.21 a liter, then an equivalent crude oil price would be $24 a barrel. Hydrated ethanol, which costs 7 percent less than anhydrous per liter, would compete with crude oil priced at $29 a barrel.

Most economic analyses of ethanol call for at least two price adjustments: one for the overvalued official exchange rate and the other for the reduction in land values (they should be lowered to reflect the existence of large uncultivated areas, not small local markets). Copersucar estimates that making such adjustments would lead to a 20 percent reduction in costs. Consideration is not given here to other factors that increase the economic value of ethanol, such as its role in environmental protection and its contribution to employment in agriculture and industry.

ENVIRONMENTAL ISSUES

Soil quality

Sugarcane is a monocultural crop, grown year after year on the same land and for that reason might be expected to produce diminishing yields over time. Yet the reverse is true: after decades of harvesting, Brazilian sugarcane yields have been steadily increasing. Much of the increased productivity can be attributed to better soil preparation, the development of superior cane varieties and the recycling of nutrients (stillage). Such techniques have greatly reduced erosion, and less topsoil is lost from sugarcane fields than from most other monocultures in the state of São Paulo[3] [21]. Nonetheless, further erosion abatement strategies are still needed.

Water contamination

Steps have been taken to prevent water contamination caused by the runoff of cane-washing water and the leaching of stillage. Cane-washing water in Brazil passes either through closed circuits (which were found in 54 of 60 mills surveyed) [22] or through aeration lagoons, and thus is no longer dumped directly on the ground. Stillage is now plowed back into the fields as a fertilizer. Years

---

3. 12.4 tonnes per hectare per year, compared with 38.1 for beans, 26.7 for peanuts, 25.1 for rice, 20.1 for soybeans, 12 for corn, and 33.9 for cassava [21].

of research on groundwater contamination show[4] that using stillage as a fertilizer does not contribute to groundwater contamination if its application is controlled [23] by limiting the volume of stillage per hectare. Stillage is valued as a sugarcane fertilizer because of its potassium levels; 100 cubic meters (m^3) of stillage per hectare provides 125 kilograms of K_2O that would otherwise cost $75.00 per hectare. Nonetheless, the volume of stillage that can be applied to a field varies from place to place; in regions with near-surface groundwater, for example, much less can be applied safely.

The impact of cane burning on air pollution

Despite widespread concern, sugarcane is burned in almost all countries where it is produced, including the United States and Brazil. Preharvest burning (dry leaves are burned) is intended to promote pest control and lower harvesting costs; it is generally carried out just a few hours before harvesting. Postharvest burning (tops and remaining green leaves), which involves smaller amounts of material, eliminates trashy residues and so expedites plowing and replanting.

In Brazil, as in other countries, air-pollution concerns are exacerbated by the number of sugarcane fields located near urban areas. Although an analysis of cane burning in Hawaii pursuant to the U.S. Clean Air Act of 1970 [25] shows that no health problems to date can be attributed to cane burning (investigated were herbicide and pesticide residues in smoke, airborne biogenic silica fibers, and the incidence of asthma), such studies must continue. Even if there are no significant health effects, cane-burning produces large amounts of smoke, and so is viewed as a nuisance.

The problem can be alleviated by selecting optimal burn times according to the time of day and the direction of prevailing winds [26, 27]; such strategies may soon be mandated in some regions of Brazil. An alternative approach is to harvest these sugarcane residues (dry and green leaves, and tops) for energy generation. Indeed, developing energy markets for these residues is probably the best way to advance the technologies that are needed to harvest green cane and to collect these residues more efficiently, thus eliminating the need to burn.

The impact of alcohol engines on air pollution

The introduction of gasohol has had an immediate impact on Brazil's air quality. By making it possible to eliminate tetraethyl lead in gasoline, airborne lead levels were reduced in São Paulo by 80 percent between 1978 and 1983. Emissions also dropped dramatically with the introduction of the neat-ethanol car (see table 6). Yet pollution-control measures in Brazil lag significantly behind those of the United States; one of the reasons, of course, is that in Brazil there are notably fewer cars per unit area. Therefore emissions data are based on cars that lack both electronic fuel-injection systems and catalytic converters. Nonetheless, ethanol cars have low emissions of carbon monoxide, oxides

---

4. Levels < 400 m^3 per hectare of stillage are safe from the contamination standpoint, except for protected areas (water sources); however, a limit of 200 m^3 per hectare has been recommended to avoid problems for the sugarcane plant. Those results were established for São Paulo.

Table 6: Light-vehicle emissions in Brazil[a]

Vehicle	CO	Pollutant grams per kilometer		Aldehyde
		HC	NO$_x$	
Gasoline, before 1980	54	4.7	1.2	0.05
Gasoline, 1986 (gasohol)	22	2.0	1.9	0.02
Ethanol, 1986	16	1.6	1.8	0.06

a. Data are provided by Companhia de Tecnologia de Saneamento Ambiental (CETESB) in São Paulo [24].

of nitrogen and hydrocarbons, and so have had a noticeable impact on urban air quality in regions where the vehicular mix is roughly divided between ethanol and gasohol cars. In short, given the same level of emission control technology, gasohol engines are superior to gasoline engines in abating air pollution.

To what extent might air quality be improved by adopting advanced emissions-control systems? It is possible [7] that nitric oxide emissions, which are lower in ethanol than in gasohol engines, could be controlled without catalysts or exhaust gas recirculation systems. A few studies show [6] there is no difference between gasoline and ethanol in terms of non-methane hydrocarbons; the increased acetaldehyde emissions should lead to increased peroxyacetyl nitrate formation.

The relation between ethanol engines and the formation of ground-level ozone is not well understood, although preliminary studies [6] indicate that ethanol and methanol are roughly equivalent on this score and superior to gasoline. But neat-ethanol vehicles have significantly reduced toxic (carcinogenic) emissions: unlike gasoline vehicles they emit almost no benzene and no 1,3 butadiene or polycyclic organic matter. Formaldehyde emissions, however, which are thought to be carcinogenic, are higher for ethanol engines and thus of some concern.

Global warming

Ethanol produced from sugarcane may be a potent force in slowing global warming brought on by the buildup of carbon dioxide in the atmosphere. To begin with, the production process uses only a small amount of external energy relative to the agro-industrial system as a whole. With presently deployed technology, bagasse provides 100 percent of the thermal energy and 92 percent of the electricity needed to process sugarcane and produce ethanol. Thus, only 8 percent of the electricity must be generated from outside sources. Details are provided in the section on "Net energy analysis".

Second, and most important, the carbon dioxide that is released during various stages (from the burning of crop residues to the fermentation of cane juice to the burning of bagasse and finally to the actual combustion of ethanol) is entirely recycled. Because sugarcane is grown renewably (that is, new crops are planted to replace those that are harvested) the carbon dioxide released to the atmosphere during the above stages is removed (fixed) by the photosynthetic activity of the new crops. Thus, the only net increase in atmospheric carbon dioxide attributable to ethanol production and use comes from fossil fuels that are used, say, to produce fertilizer in agricultural operations, to manufacture heavy equipment, and to run the machines needed for building construction.

If all input energy used by buildings and machinery were derived from fossil fuel, then the ratio of energy in ethanol to energy in input fossil fuel would be 5.9 (or 8.2 in the best cases) (see table 1). If one assumes an equivalence of 1.2 liters of ethanol per liter of gasoline, then the production and use of 1 liter of ethanol, as a replacement for gasoline, avoids the emission of 0.54 to 0.57 kilograms of carbon as carbon dioxide (a nearly 90 percent reduction over gasoline).

Substituting surplus bagasse for fossil fuels will add significantly to this figure; to some extent, this is already happening. A much larger impact would be possible, however, by promoting the greater use of cane residues (see "Net energy analysis"). Thus, ethanol from sugarcane is a truly renewable energy option, largely because bagasse provides so much industrial energy.

The situation is quite different for ethanol from corn [6], where the balance is much less interesting. For an equivalence of 1.2 liters of ethanol per liter of gasoline, corn ethanol reduces the amount of carbon dioxide emitted from gasoline by only 19 percent—using present techniques and procedures [9].

SOCIAL ISSUES

Brazil's Pro-álcool program raises two important questions. First, what long-term impact will the program have on the number and quality of jobs? And second, as the industry continues to expand, will sugarcane growers find themselves competing with food growers for use of the land? Before addressing those questions, an important point must be made: Brazil's economy is thought to have saved $17.5 billion in hard currency (based on official estimates first made in 1987 and updated in 1990), made possible in part by the country's increased production of ethanol, which in turn reduced the need for oil imports.

Employment and job quality
The impact of ethanol production on employment has been extensively analyzed. So far, ethanol production has generated some 700,000 jobs in Brazil, with a relatively low index of seasonal work [1, 28]. Data from 1983 [28] indicate that a typical autonomous distillery with a capacity for 120,000 liters of ethanol a day in a 167-day harvesting season requires 455 men per year in the Center–South region (1,775 men per year in the Northeast) at the agricultural end of production, and 150 men per year at the industrial end.

The cost of creating a single job in the ethanol agro-industry varies from $12,000 to $22,000 [2], a range that reflects working capital requirements and whether or not the number of actual men per year or number of jobs (some part-time) are included in the calculations. In Brazil, job generation in most other industries requires larger investments: for industrial projects in the Northeast from 1972 to 1982, the average job required an investment of $40,000. Petrochemical projects in that region cost as much as 20 times more per job [29].

Job quality must be evaluated relative to other employment sectors. Sugarcane laborers in São Paulo receive higher wages on average than 80 percent of the labor force employed in other agricultural sectors, including cattle raising, forestry, and fishing. Their incomes are also higher than 50 percent of the labor force in the service sector and 40 percent of those in industry. Social conditions are reported to be improving in the Northeast region [30]. Special legislation has mandated that 1 percent of the net sugarcane price and 2 percent of the net ethanol price be channeled into medical, dental, pharmaceutical, sanitary, and educational activities for sugarcane workers.

Although Brazilian workers have low incomes relative to workers in developed countries, there has been a significant increase in agricultural mechanization, particularly in São Paulo's Center–South region. There are two main reasons for this trend: social changes such as urban migration have reduced the number of workers available to the sugarcane industry and mechanization has proved cheaper than hand labor.

Thus, the delicate balance between mechanization and the number and quality of new jobs created by the industry is likely to remain a key issue for several years. One consequence is that the low cost of Brazilian sugarcane, initially made possible by a large underpaid work force, will probably drop further as a result of increased mechanization and industrial automation, provided that a shift to fewer but more highly paid jobs is deemed socially acceptable.

Land use

Sugarcane grown for ethanol production currently occupies only 4.2 percent of the land area devoted to primary food crops (see table 7). This area represents only 0.6 percent of the area registered for economic use, or 0.3 percent of Brazilian territory. Such low agricultural use holds true even in the state of São Paulo, where the largest tracts of sugarcane (and the highest use of cultivable land) are found. São Paulo has 15 million hectares of cultivable land, of which only 50 percent is currently used for agricultural purposes [1]. Thus, competition for land for food, export crops, and energy crops is not significant. This may not be true, of course, in other countries.

In addition, crop rotation in sugarcane areas has led to an increase in food production: beans and peanuts are sometimes rotated with sugarcane (see "Promoting new products). Furthermore, byproducts of ethanol production, such as hydrolyzed bagasse and dry yeast, are used in cattle, chicken, and pork feed. In

Table 7: Area devoted to the cultivation of primary food crops in Brazil in 1988[a]

Crop	Cultivated area *1,000 hectares*	Share *percent*
Corn	13,182	24.0
Soybean	10,524	19.2
Beans	5,905	10.7
Rice	5,961	10.8
Sugarcane	4,117[b]	7.5[b]
Wheat	3,480	6.3
Coffee	2,957	5.4
Cassava	1,757	3.2
Other	4,510	8.2
Total	**54,951**	**100.0**

a. Data obtained from Fundação Instituto Brasilia de Geografia e Estatística (FIBGE), 1989 [31].

b. Approximately half is harvested for sugar; ethanol requires about 2.3×10^6 hectares, or 4.1 percent of the total area.

fact, the potential for producing food in conjunction with sugarcane appears to be larger than expected and should be explored further.

PROSPECTS FOR FURTHER COST REDUCTIONS

Technological advances

Despite the advances that have been made over the past decade, technology development and transfer to Brazilian alcohol and sugar factories has been hampered by the low average technical level that prevailed during the mid 1970s; other drawbacks include differences in the scale of production, which might vary from 2,000 to 32,000 tons of cane a day, and in technology level (overall conversion efficiencies from raw cane to refined sugar might differ by 20 percent) [32], among the nearly 400 industrial units.

Although differences in conversion efficiencies, equipment productivity, and management have begun to diminish, they are still very high, which accounts for the wide range in industrial production costs (see "Economic analysis"). The same is true for sugarcane, where developments in crop strains, use of fertilizers, and management of agricultural operations have not been widely implemented. Indeed, implementing these developments should be the first step towards cost reduction. A second step might be the development of strategies to promote spin-off products for both agricultural and industrial markets.

Table 8: Potential for reducing costs of ethanol production[a]

Sector	Cost reductions[b] percent
Sugarcane production (agriculture)	
Variety selection and handling	9.8
Lime application	1.6
Liquid fertilizers	0.7
Stillage use	1.0
Weed removal	2.1
Transportation	0.5
Operations planning	3.4
Ethanol production (industry)	
Milling	1.3
Fermentation	3.3
Distillation	0.3
Energy	1.5[c]
Total	**23.1**[d]

a. Data were provided by Copersucar [16, 1] and reflect cost reductions based on existing technologies, some of which are already in use. In all cases the potential can be realized in a few years.

b. For each item, "cost reduction" corresponds to the ratio between net benefit (gains minus associated costs, including costs of downstream processing), and the total cost of production and storage of ethanol [16].

c. The goal is to reach self-sufficiency with respect to electricity and to produce 18 percent surplus bagasse, which can then be sold as fuel.

d. The total does not correspond to the sum of each item because some items are interrelated.

Processing improvements

Data provided by Copersucar associated mills [33, 16], based on 700,000 hectares of sugarcane and nearly 50 firms, suggest that ethanol costs can be significantly reduced (see table 8). An overall cost reduction of about 23 percent could be achieved within the next few years simply by adopting available technologies, some of which are already in use. Thus, it is likely that the average rate of cost reduction (4 percent annually over the past decade) could be maintained for several years.

Promoting new products

A number of commercial byproducts of sugarcane have already been marketed, including dry yeast and hydrolyzed bagasse for animal feed, amylic alcohols for fuel oil, and cogenerated electricity from bagasse [22]. Two additional options that need consideration are crop rotation and energy cogeneration.

Soybeans and peanuts can both be rotated with sugarcane; it is estimated that in the Center–South region of Brazil, 17 percent of the total cane area is renewed every year with 18-month cane, and 60 percent of this area (about 10 percent of the total) can be used for crop rotation. For example, in a sample plot of 700,000 hectares, 4 percent is rotated every year and planted with soybeans and peanuts.

If all the land that is potentially available for rotation (10 percent of the total) were used in this manner, then ethanol costs could be cut by an additional 1.7 percent. Broken down, 1.0 percent of the savings comes from the food crop; 0.3 percent comes from the increased productivity of the sugarcane obtained by rotation and 0.4 percent comes from the reduced need to treat the new sugarcane crop with herbicides and fertilizer. It should be mentioned that simultaneous intercropping does not apply to sugarcane, mainly because the technique requires selective herbicides and irrigation, although some day a suitable crop that can be grown along with cane with little intervention may be found.

Energy cogeneration

The role of biomass in energy cogeneration has been well studied [34] in Brazil. On average, about 280 kilograms of bagasse (which contains 50 percent moisture) are produced per tonne of cane, which is equivalent to 2.1 gigajoules of energy per tonne; 90 percent of the bagasse is burned to produce steam (450 to 500 kilograms of steam can be generated from 1 tonne of cane), which in turn can be used to cogenerate electricity and mechanical power for the mill drivers. A fraction (10 to 15 percent) of the electricity needed to run the mill is purchased from the local power company. Average steam generation pressures are low (2 megapascals) so that with currently deployed technologies, efficiencies are small. This means that the potential for cogeneration is largely unexplored.

In 1987, 60 industries were assessed for cogeneration [35], including the costs of building appropriate facilities. In 1989, costs were reevaluated [36] in the context of cogeneration, that is, limiting steam production to process needs. Under those conditions, the average costs of cogeneration were estimated to be U.S. $0.05 per kilowatt hour.

Bagasse was assumed to cost 50 percent of the energy equivalent of fuel oil (based on the prevailing market price of oil in São Paulo). A more detailed study currently under way [37] will evaluate the use of bagasse in condensing-extraction steam turbines (CEST electric generating systems) and off-season electricity production with recovered cane tops and leaves. So far the results indicate that for pressures ranging from 2 to 8 megapascals, and for process steam run at 350 kilograms of steam per tonne of cane, surplus electric energy values of 20 to 65 kilowatt-hours (kWh) per tonne of cane could be achieved with bagasse.[5] Moreover, the energy surplus could be doubled simply by using cane tops and leaves for power generation during the off-season. Although final cost esti-

---

5. Total generated power of 77 kWh per tonne of cane (8 megapascals) for bagasse availability of 252 kilograms per tonne of cane (5 percent reserved for start-up in rainy periods) and condensing pressure of 3 kilopascals. Exported energy can be up to 74 kWh per tonne of cane with better condensing conditions and bagasse storage. For higher fiber sugarcane (Brazilian Northeast; 300 kilograms per tonne of cane of bagasse) and lower condensing pressure (1 kilopascal, if cooler water is available at the end of the season) generated power could be above 100 kWh per tonne of cane.

mates are not yet available, they are expected to be about U.S. $0.05 per kilowatt hour.

The generation, transmission, and distribution of electricity in Brazil is state-owned or controlled, and privately cogenerated power must be sold to the state-controlled utilities. But because buying is not compulsory and the purchase price is not determined by the avoided cost of increasing power generation, utilities have few incentives to buy cogenerated power. The system is likely to change soon, however, with the government promising to introduce a special policy for cogeneration. The policy is likely to mandate long-term (10 year) contracts, with prices based on avoided cost.

Eletrobrás, the federal electric utility, estimates that the avoided cost of generation and transmission of electricity is about U.S. $0.05 (1989 values) per kilowatt hour (based on hydropower), with higher prices likely. So it seems that CEST generators could be competitive with more traditional sources of power, even with credits equivalent to the current market value of bagasse. Costs associated with collecting and transporting cane tops and leaves will be determining parameters for the overall costs of producing electricity, and both are site-dependent.

Clearly the potential for cogeneration justifies the efforts being made to overcome technical and institutional obstacles. In addition, the prospects for advanced technologies such as bagasse gasification-gas turbines (see chapter 17: *Advanced Gasification-based Power Generation*) are highly promising; such systems would significantly increase the potential for cogeneration and thus could lead to significant cost reduction. If, for example, a net income of U.S. $0.01 per kilowatt hour could be obtained, then a CEST unit producing 100 kWh per tonne of cane (in the form of bagasse and residues) would reduce the cost of ethanol by 6 percent.

Fuel-oil substitutes

Despite their importance, energy-conservation systems and processes have not been introduced in Brazilian sugar mills in large part because bagasse was widely viewed as worthless only 10 years ago. But the situation is likely to change dramatically as recognition of bagasse's high benefit-to-cost ratio forces its widespread use.

Other uses for bagasse include its incorporation in cattle feed, pulp, paper, and fiber boards. While those markets are likely to expand, in the near future the largest new market for bagasse will be as a fuel-oil substitute in industrial applications. In São Paulo the market price for bagasse, with a moisture content of 50 percent, has been set at 50 percent of fuel-oil prices (for an energy-equivalent amount). At those prices, the average net income for a sugar mill would be $6 per tonne of bagasse (or $0.80 per gigajoule).[6] The market, however, is site-dependent. It is estimated [37] that as much as 45 percent of the bagasse can

---

6. Average price paid is $1.46 per gigajoule; the difference is due to transportation costs and taxes. Estimates indicate 1.5×10^6 tonnes per year sold.

be saved with conventional technologies for energy efficiency; an even larger surplus is possible [38]. If 40 percent surplus bagasse is produced, for example, and all is distributed to local markets (a reasonable assumption in São Paulo), the net income from surplus bagasse alone would equal from 3 to 4 percent of the total cost of producing ethanol. The potential market value of cane residue is larger still. Another revenue-generating option is to process residues on-site to produce higher value fuels such as charcoal, or to generate electricity at the sugar mill itself and so reduce transportation costs.

The possibility of producing ethanol by subjecting bagasse to enzymatic or acid hydrolysis has also been considered (see chapter 21). If bagasse retains its value at the sugar mill of $0.80 per gigajoule, and if the costs verified in Thailand [13] for residue collection and transportation ($1.68 per gigajoule) can be met in Brazil, then a large amount of cheap cellulosic material would be available at the mill. Consequently, ethanol might be produced off-season, using almost all existing equipment: fermentation vats, boilers, and the distillery. In order for ethanol production to grow, however, the program must continue to develop at its present pace for at least the next several years.

Many studies on the feasibility of obtaining methane from stillage have been undertaken, and the data from some commercial units [37] indicate that for upflow anaerobic sludge-blanket (UASB) reactors with mesophilic bacteria (those that prefer temperatures near 35°C), as much as 0.32 gigajoules of methane can be generated per tonne of cane. Since potassium, applied as a fertilizer, is the main component in stillage, the effluent from the digester will have the same value as stillage for fertilizer–irrigation purposes. Because the effluent has a high biological oxygen demand, it must either be treated or used for irrigation. The costs of producing methane have been estimated at $3.2 per gigajoule (assuming that biogas is 60 percent methane and 40 percent carbon dioxide).

CONCLUSION

Brazil's Pro-álcool program was originally instituted for economic and social reasons. By promoting the production and use of ethanol, the country could address several needs at once: it could reduce fuel imports (largely because of balance-of-payment problems), create jobs in a depressed economy, and exploit idle capacity in the sugar agro-industry (which was hurt by the low price of sugar in the world market). In addition, the Pro-álcool program was an opportunity to reduce air pollution in metropolitan areas. Those objectives have been attained.

In some areas the program has exceeded its original scope; one unexpected bonus has been the accelerated pace of technological development, both in agriculture and industry, which has led to major cost reductions. Three overlapping technological phases can be identified: increased productivity, followed by greater conversion efficiencies and, most recently, adoption of integrated agro-industrial management techniques. Creating a "technological environment" for the agro-industry was critical in Brazil, especially in São Paulo, where it has served as a model for other agriculturally based industries such as orange growing and beef processing. Thus, the sugarcane-to-ethanol industry has introduced high-level production techniques to the agricultural sector, allowing the latter to become efficiently integrated with the industrial sector.

Critics might argue that the cost of producing ethanol from sugarcane is artificially low in Brazil because labor is cheap. While that may have been the case during the program's early years, today increasing mechanization has lowered production costs. Technology is likely to reduce costs still further by creating more opportunities for coproduction, exemplified by the increase in food production made possible by crop rotation and by the rise in energy production made possible by building cogeneration facilities.

The Pro-álcool program has also exceeded all expectations with respect to the environment. As significant as ethanol's role in reducing urban air pollution, is its potential to abate global warming by displacing fossil fuels. In addition, the development and implementation of coproduction facilities will greatly enhance ethanol's position as a successful clean-energy program. For all these reasons, ethanol can be expected to compete directly with oil derivatives in the near future.

WORKS CITED

1. Borges, J.M.M. 1990
   The Brazilian alcohol program: foundations, results, perspectives, *Energy Sources* 12:451–461.

2. Borges, J.M.M. and Caracciolo, E.B. 1988
   Liquid fuels for transportation purposes: a case study—Brazil, *Proceedings of the 14th Congress of the World Energy Conference, technical section 4.2a*, Montreal.

3. Comissão Nacional de Energia. 1987
   Avaliação do programa nacional do álcool, Ministéria das Minas e Energia, Brasilia.

4. Ministério das Minas e Energia. 1988
   Balanco energetico nacional, 1988, Brasilia.

5. Petróleo Brasileiro S.A.(Petrobrás). 1989
   Produção e importação de petróleo, *Jornal do Brasil*, 31 December.

6. Environmental Protection Agency. 1990
   Analysis of the economic and environmental effects of ethanol as an automotive fuel, special report, Office of Mobile Sources.

7. Gallopoulos, N.E. 1985
   Alcohol for use in motor fuels and motor fuel components, GMR-4933, General Motors Research Laboratory, Warren, Michigan.

8. Macedo, I.C. and Nogueira, L.A.H. 1985
   Balanco do energia na produção de cana-de-açúcar e álcool nas usinas cooperadas, *Boletim técnico Copersucar* 31:22–28.

9. Pimentel, D. 1990
   Ethanol fuels: energy security, economics and the environment, internal report, Cornell University (College Agriculture and Life Sciences), September.

10. Ripoli, T.C. and Molina, W.F. 1990
    Equivalente energético do palhiço da cana-de-açúcar—Congresso Brasiliero de Engenharia Agrícola, 19- Piracicaba, São Paulo.

11. De Beer, A.G. 1989
    The agricultural consequences of harvesting sugarcane containing various amounts of tops and trash, *Proceedings of the South African Sugar Association Experimental Station*, June.

12. Kinoshita, C.M. 1988
    "Composition and processing of burned and unburned sugarcane in Hawaii," *International Sugar Journal* 90(1070):34–37.

13. Winrock International. 1990
    Costs and performance of equipment for baling sugarcane tops and leaves, draft report.

14. Alexander. 1985
    The energy cane alternative, Elsevier, Amsterdam.

15. World Bank. 1984
    Economic aspects of the alcohol program, preliminary document.

16. Copersucar. 1989
    Pro-álcool: fundamentos e perspectivas, 13–20, Copersucar, São Paulo, May.

17. Comissão Estadual de Energia de São Paulo. 1985
    Investigação sobre custos do álcool e da cana-de-açúcar: síntese preliminar, Report on seminar, June.

18. Centro de Tecnologia Copersucar. 1991
    Relatório Final do PCTS, safra 90/91, Piracicaba, São Paulo.

19. Stumpf, U.E. 1979
    O álcool como combustível de motores, *Petro e Química*, 62–67, December.

20. Monaco, L.C. 1982
    The use of ethanol as gasoline extender and octane booster, Ministério da Industria e Comércio, Brasilia.

21. Lombardi Neto, F. 1989
    Simpósio sobre terraceamento agrícola, Fundação Cargill, São Paulo.

22. Centro de Tecnologia Copersucar. 1990
    IV Seminário de Tecnologia Industrial, CTC, Piracicaba, São Paulo, November.

23. Guichet, J., Matioli, C., and Donzelli, J.L. 1988
    Efeitos da vinhaça nos solos, águas lixiviadas e influência nas águas subterrâneas, internal report, CTAG-CTC, Copersucar, Piracicaba, São Paulo.

24. Branco, G.M. 1985
    Fuel alcohol and air pollution, *Copersucar international symposium on sugar and alcohol*, Ediserv, São Paulo, 437:46.

25. Whalen, S.A. 1989
    Cane burning: environmental and health impacts, paper 669, Hawaiian Sugar Planters Association Experimental Station.

26. Rozeff, N. 1989
    Sugarcane burning in south Texas, *International Sugar Journal* 83(970).

27. Sakuma, J. 1989
    Automated meteorological network for cane burning management, Hawaiian Sugar Planters Association Experimental Station Journal Series.

28. COALBRA. 1983
    O impacto da produção de álcool de cana-de-açúcar e de madeira na geração de empregos, Ministério da Industria e Comércio, Brasilia.

29. Geller, H.S. 1984
    Ethanol from sugarcane in Brazil: an investigation on some critical issues, Companhia Energético de São Paulo, São Paulo.

30. Secretaria de Tecnologia Industrial (STI). 1984.
    Previsão e análise tecnológica do Pro-álcool 1:364; 2:443, Brasilia.

31. Fundação Instituto Brasilia de Geografia e Estatística (FIBGE). 1989
    Levantamento sistemático de produção agrícola, R. Janeiro.

32. Macedo, I.C. 1989
    Sugar and alcohol production technology: advances at CTC in the 1980–1990 period, *International Sugar Journal* 91(1090):192–194.

33. Centro de Tecnologia Copersucar. 1990
    Análise econômica dos custos e benefícios das tecnologias do setor, Relatório Interno, Piracicaba, São Paulo.

34. Lorenz, K., Leal, M.R.L.V., and Macedo, I.C. 1989
    Energy generation and use in Brazilian sugar and alcohol factories, *Proceedings of the 20th Congress* 1:409–414, International Society of Sugar Cane Technologists, São Paulo.

35. Centro de Tecnologia Copersucar. 1986
    Análise preliminar da cogeração de energia nas unidades cooperadas, Piracicaba, São Paulo.

36. Campos, R. 1990
    O potencial e a viabilidade econômica da geração excedentes de energia a partir do bagaço de cana-de-açúcar, *Revista Brasilia de Energia* 1(3):68–78.

37. Centro de Tecnologia Copersucar. 1991
    Relatório técnico interno: convênio Eletrobrás-Copersucar, alternativas de cogeração, Piracicaba, São Paulo.

38. Ogden, J. and Fulmer, M. 1990
    Assessment of new technologies for co-production of alcohol, sugar and electricity from sugarcane, PU/CEES report number 250, Princeton University Center for Energy and Environmental Studies, Princeton, New Jersey.

21
ETHANOL AND METHANOL FROM CELLULOSIC BIOMASS

CHARLES E. WYMAN
RICHARD L. BAIN
NORMAN D. HINMAN
DON J. STEVENS

Cellulosic biomass includes agricultural and forestry wastes, municipal solid waste, and energy crops. Enough ethanol or methanol could be made from cellulosic biomass in countries such as the United States to replace all gasoline, thereby reducing strategic vulnerability and lowering trade deficits for imports. Direct alcohol blends and gasoline containing ethers of ethanol or methanol decrease emissions of carbon monoxide, and neat alcohols reduce smog. In addition, producing alcohol fuels from biomass that is grown sustainably does not contribute to the accumulation of carbon dioxide (CO_2) in the atmosphere. Significant progress has been made over the past few years in the technologies for converting biomass to ethanol or methanol. The simultaneous saccharification and fermentation (SSF) process is favored for producing ethanol from cellulose, because of its low cost potential. Technology has also been developed for converting hemicellulose into ethanol. Burning the remaining fraction—predominantly lignin—can provide enough heat and electricity for the conversion process and generate extra electricity for export. Developments in conversion technology have reduced the projected selling price of ethanol from about U.S.$45 per gigajoule ($0.95 per liter) ten years ago to only about $13 per gigajoule ($0.28 per liter) today.[1] For methanol production, improved gasification technology has been developed, and more economical syngas cleanup methods are available. The projected cost of methanol has been reduced from about $16 per gigajoule ($0.27 per liter) to less than $15 per gigajoule ($0.25 per liter) at present. Technical opportunities have been identified that could reduce the costs of ethanol and methanol produced from cellulosic biomass to levels competitive with gasoline ($0.21 per liter) derived from oil at $25 per barrel.

INTRODUCTION

Transportation issues
As underscored by political volatility in the Middle East, oil represents the most vulnerable component of the industrialized world's energy supply. For example, about 50 percent of all petroleum in the United States is imported, and imports have risen markedly over the past few years [1]. Because the Organization of Pe-

---

1. Pretax cost of production (COP), assuming a 12 percent discount rate.

troleum Exporting Countries (OPEC) controls about 75 percent of the world's oil reserves, petroleum imports will probably continue to rise in most countries unless alternatives are developed. Transportation fuels are almost totally derived from petroleum and are particularly vulnerable to disruptions. Moreover, imported petroleum accounted for about 40 percent of the balance-of-payments deficit for the United States in 1989 [2].

Much of the smog and carbon monoxide pollution in major cities is caused by automobiles [3]. In addition, global climate change may result from the accumulation of carbon dioxide (CO_2) in the atmosphere from burning petroleum and other fossil fuels [4].

Alcohols from cellulosic biomass

Ethanol and methanol are liquid fuels that can be readily substituted for gasoline in the transportation sector. Replacing gasoline with alcohol fuels produced from renewable sources of cellulosic biomass can improve energy security, reduce the balance-of-payments deficit, decrease urban air pollution, and reduce the atmospheric buildup of CO_2 [5–7]. However, it is necessary to reduce the cost of ethanol and methanol to the point that they can compete with gasoline without tax incentives, so that the benefits of these fuels can be realized.

Biomass types

Through photosynthesis, plants convert CO_2 and water into simple sugars. In plants such as sugarcane, solar energy is stored directly as the chemical energy of these sugars. In crops such as corn, the sugars are converted into starch. These sugars are also joined together to form the carbohydrate polymers cellulose and hemicellulose, which, together with lignin, provide structural support for the plant. In this way, the intermittent energy of the sun is captured in a solid material that can be burned directly to release the stored energy as heat, or thermally processed or biochemically transformed to produce liquid or gaseous fuels. Cellulosic biomass is the main constituent of most forms of plant matter and is much less expensive than cornstarch or sugar because it has no food value and costs less to produce.

Biomass availability

In the United States, researchers estimate that about 77 million hectares of land could be used to produce energy crops for the production of alcohol fuels. At an average productivity of 20 tonnes per hectare per year, about 1.5 billion tonnes of cellulosic biomass could be supplied annually [8]. When underused wood, agricultural residues, and municipal solid waste are included as well, about 2.3 billion dry tonnes of cellulosic biomass per year could be available at prices from $20 to $70 per dry tonne [5, 6]. These prices are equivalent to about $1.10 to $3.70 per gigajoule; in contrast, oil at $25 per barrel costs about $4.00 per gigajoule. This quantity of feedstock can generate more than 1 trillion liters of ethanol and methanol annually, more than enough to meet the total current U.S.

gasoline market of about 425 billion liters per year. Although these values are subject to significant uncertainty, they indicate that the resource base of renewable feedstocks is substantial while the cost is reasonable. Thus, cellulosic biomass is a promising feedstock for fuel alcohol production.

Fundamentals of ethanol production

Ethanol can be produced by biologically catalyzed reactions. In much the same way that sugars are fermented into beverage ethanol by various organisms including yeast and bacteria, sugars can be extracted from sugar crops, such as sugarcane, and fermented into ethanol. For starch crops such as corn, starch is first broken down to simple glucose sugars by acids or enzymes known as amylases. Acids or cellulase enzymes similarly catalyze the breakdown of cellulose into glucose, which can then be fermented to ethanol. The hemicellulose fraction of biomass breaks down into various sugars such as xylose in the presence of acids or enzymes known as xylanases; conventional organisms cannot ferment many of the sugars derived from hemicellulose into ethanol with reasonable yields. However, new technology has been developed to convert hemicellulose to ethanol.

Biological processing offers a number of advantages for converting biomass into biofuels. First, the enzymes used in bioprocessing are typically capable of catalyzing only one reaction, and so formation of unwanted degradation products and by-products is avoided. Additionally, biological transformations occur at near-ambient pressures and temperatures, so that the cost of containment is modest. Furthermore, materials not targeted for conversion can pass through the process unchanged and be used for other applications. Finally, biotechnology and bioprocessing are new and evolving areas with a demonstrated ability to dramatically alter a process and improve economics. Thus, former hurdles to developing cost-effective technologies for producing ethanol from cellulosic biomass may well be overcome.

Fundamentals of methanol production

Methanol is manufactured primarily by thermal processes, which occur at rapid rates. Methanol can be produced from fossil fuels such as natural gas, petroleum naphthas, and coal, and from biomass resources such as woody and herbaceous plants. In general, methanol production consists of three groups of chemical unit operations: 1) synthesis gas (syngas) generation, 2) syngas upgrading, and 3) methanol synthesis and purification. In the case of natural gas and naphtha, syngas generation consists of converting methane and light hydrocarbons to carbon monoxide (CO) and hydrogen (H_2) via steam reforming. In the case of coal and biomass, the solid feed is converted to syngas using gasification. For natural gas and naphtha systems, syngas upgrading consists primarily of CO_2 removal. For biomass and coal systems, the primary synthesis gas is either reformed or shifted to produce a syngas with low methane content and a proper H_2 to CO ratio. Carbon dioxide and sulfur compounds are removed before methanol synthesis. Commercial methanol synthesis operations involve reacting CO, H_2, and steam over a copper–zinc oxide catalyst in the presence of a small amount of CO_2 at temperatures of about 500 to 570 K and pressures of about 5.2 to 10.3 megapascals. The methanol synthesis reaction is equilibrium-controlled and excess reactants must be recycled to obtain economic yields. Thermal conversion methods can deal with a wide variety of biomass feedstocks, and methanol technologies are remarkably product specific. Improved gasification and gas conditioning technologies offer the potential to reduce methanol production costs.

ALCOHOLS AS FUELS

Alcohols may emerge as excellent alternative transportation fuels. They can be used in internal combustion engines as blends with gasoline, as neat fuels by themselves, or as oxygenated derivatives that are added to gasoline. Alcohols can also be employed in fuel cells.

Direct blends of ethanol and methanol with gasoline

Alcohol blends enable gasoline engines to run lean and reduce carbon monoxide emissions by 10 to 30 percent [9–12]. Alcohol also increases the octane of the gasoline with which it is blended. At a 10 percent blend with gasoline, ethanol increases the Reid vapor pressure[2] by about 3.4 kilopascals. Increasing vapor pressure increases evaporative emissions, while the addition of the oxygenate reduces tail pipe emissions of unburned hydrocarbons. Estimating the overall effect of ethanol blends on smog formation is complex, but modeling studies generally predict lower overall emissions with ethanol blends. Because methanol increases the Reid vapor pressure by about 10 kilopascals, methanol blends are not widely accepted in the United States, but are routinely used on a cosolvent basis in Europe at levels of up to 3 percent in gasoline [13]. In 1988 the worldwide use of methanol as a gasoline additive was 0.141 million tonnes [15].

Ethyl tertiary butyl ether (ETBE) and methyl tertiary butyl ether (MTBE) blends with gasoline

Alcohols are not fungible with gasoline: they do not ship and handle like other gasoline components. Therefore, they cannot be handled like conventional gasoline in common pipelines and tanks and are not swapped like normal gasoline [16]. This drawback can be overcome if ethanol and methanol are converted to fungible ether blend stocks. Ethanol or methanol can be reacted with isobutylene to form ethyl tertiary butyl ether (ETBE) or methyl tertiary butyl ether (MTBE), respectively, for blending with gasoline [17, 18]. Ethanol, methanol, ETBE, and

---

2. Reid vapor pressure is the pressure of a fuel mixture in a closed vessel with an air volume four times the volume of the liquid and heated to 38°C. Reid vapor pressure indicates the vapor-lock tendency of a fuel as well as explosion and evaporation hazards: ASTM D-323 [14]. Higher vapor pressure can improve cold starting of an engine but also increases emissions of unburned hydrocarbons that contribute to the formation of smog.

MTBE each contain one atom of oxygen per molecule, and addition of these alcohols or ethers to gasoline reduces CO emissions. However, because one molecule of the alcohol plus one molecule of isobutylene are needed to form one molecule of ether with no loss of oxygen, greater quantities of the ethers must be used to achieve the same blended oxygen content and equivalent alcohol use. For example, a 22 percent blend of ETBE and gasoline results in the same oxygen content and equivalent ethanol use as the direct addition of 10 percent ethanol to gasoline. MTBE is vapor-pressure neutral when blended with gasoline, while ETBE lowers the Reid vapor pressure of gasoline, thereby decreasing the release of smog-forming compounds.

MTBE is the major ether used by the refining industry because of its availability, pricing, fungibility, and its wide acceptance by consumers and the oil and automotive industries. MTBE is already an established gasoline component and has been in use since 1979 when it was approved by the U.S. Environmental Protection Agency (EPA) for unleaded gasoline [16]. Worldwide demand for MTBE was forecast at 2.8 million tonnes in 1990 and is expected to rise to about 8.5 million tonnes by the year 2000 [19]. Although ETBE is not yet a commercial product, substantial interest is mounting in its use [18].

Neat ethanol and methanol

Ethanol or methanol can be employed directly as a neat (close to 100 percent) fuel or hydrated ethanol (95 percent ethanol with 5 percent water), as it is in Brazil [5]. Using hydrated ethanol eliminates the cost of removing the final 5 percent water while providing better performance. Neat ethanol and methanol have many fuel properties that are desirable (see table 1). They provide superior efficiency and performance to gasoline in properly optimized engines because they require lower stoichiometric air/fuel ratios, have higher latent heats of vaporization, provide higher octane values, and have a lower flame temperature. Thus, ethanol and methanol are often preferred to gasoline for high performance automobile races such as the Indianapolis 500. The fact that the majority of new cars in Brazil run on neat ethanol clearly shows that it is a suitable fuel.

Currently, mixtures of 85 percent methanol and 15 percent gasoline (known as M85) or 85 percent ethanol and 15 percent gasoline (E85) are often preferred over pure alcohols for automotive use. The addition of gasoline increases the vapor pressure of the fuel enough to facilitate cold starting. Further engine development is needed to cold start engines with pure ethanol or methanol, particularly during winter months in colder climates.

Air quality problems associated with gasoline use in major cities throughout the world have reduced interest in gasoline options, such as the production of gasoline from methanol, in favor of new fuel and engine options. One of the important considerations favoring the use of neat alcohols is the air emission benefits they offer. Carbon monoxide emissions are similar for the combustion of neat ethanol, methanol, or gasoline in spark ignition engines. However, evaporative emissions during fueling and from the fuel system itself are less for the alcohols than for gasoline due to the lower vapor pressure of the alcohols. In addition, several types of hydrocarbons and their partial combustion products escape in the engine exhaust of gasoline-powered vehicles. Although some of these organic compounds are toxic, a larger concern is the photochemical reaction of these materials in the atmosphere that increases local ozone concentrations, more commonly referred to as smog. On the other hand, the primary products from the

Table 1: Properties of methanol, ethanol, MTBE, ETBE, isooctane, and unleaded regular gasoline

Property	Methanol	Ethanol	MTBE	ETBE	Isooctane	Gasoline
Formula	CH_3OH	C_2H_5OH	$(CH_3)_3COCH_3$	$(CH_3)_3COC_2H_5$	C_8H_{18}	$C_4 - C_{12}$
Molecular weight	32.04	46.07	88.15	102.18	114	
Density kg/m^3 @ 298 K	790	790	740	750	690	720–780
Air/fuel stoichiometric ratio						
Mole basis	7.14	14.29	35.71	42.86	59.5	
Mass basis	6.48	9.02	11.69	12.10	15.1	
Higher heating value *MJ/kg*	19.92	26.78	35.27	36.03	44.42	41.8–44.0
Lower heating value *MJ per liter*	15.74	21.16	26.10	27.02	30.65	31.4–33.0
RON	106	106		118	100	91–93
MON	92	89		102	100	82–84
(RON + MON)/2	99	98		110	100	88
Blending RON	135	114–141[a]	118	117–120[b]		
Blending MON	105	86–97[a]	101	101–104[b]		
(Blending RON + MON)/2	120	115	110	111	111	
Atmospheric boiling point *K*	337.8	351.6	328.6	344.8		
Heat of vaporization *MJ/kg*	1.1	0.84	0.34		0.41	
Flash point *K*	280	285	245			
Ignition point *K*	737	697	733			
Reid vapor pressure *kilopascals*						
Pure component		15.85		30.3		
Blending	214+	82.7–186	55.1	20.7–34.5		55.1–103.4
Water solubility *weight percent*						
Fuel in water	100	100	4.3	2	negligible	negligible
Water in fuel	100	100	1.4	0.6	negligible	negligible
Water azeotrope, (atm b.p.), *K*	(none)	351.4	325.4			
Water in azeotrope *weight percent*		4.4	3.2			

a. 10 percent blends.
b. Assumed 12.7 percent blend.

exhaust of an alcohol-fueled engine are unburned alcohol and aldehydes. The alcohols themselves have a lower photochemical reactivity to form ozone than the hydrocarbons associated with gasoline use, but the aldehydes (primarily formaldehyde for methanol combustion and acetaldehyde for ethanol use) are highly reactive. Nonetheless, combustion of ethanol or methanol is expected to contribute less to ozone formation than conventional gasoline when used in vehicles with equivalent emission controls.

Engines designed to run on alcohol emit less oxides of nitrogen (NO_x) because the alcohols burn at lower temperatures than gasoline, and NO_x emissions drop with decreasing temperature. However, as the engine compression ratio is increased to improve the efficiency of alcohol-fueled engine performance, NO_x emissions increase as well, negating much if not all of the NO_x benefit. Excessive NO_x is not a common problem in most cities, but it does contribute to smog formation. The benefits to be gained from the reductions in NO_x and hydrocarbon levels actually vary from city to city. Nevertheless, the consensus is that engines fueled with neat alcohols (including M85 and E85) will improve air quality, although the degree of improvement predicted is quite uncertain for existing atmospheric models [20].

Another plus for alcohols is their water solubility, which makes spills less environmentally threatening. In addition, ethanol has a relatively low toxicity, and both alcohols are readily biodegradable. The energy density of both alcohols is lower than that of gasoline. However, because of their higher octane, higher heat of vaporization, and other favorable properties, dedicated ethanol or methanol engines could achieve higher efficiencies than gasoline engines, thereby compensating for their lower volumetric energy content to some extent [21, 22].

Alcohol fuel cells

Alcohols could be used in fuel cells, devices in which fuel is electrochemically reacted with an oxidant. The fuel and oxidant are separated by an electrolyte that will readily transport ions but not electrons, and the electrons move through an external circuit as the reaction between the ions of fuel and oxidant occurs. Thus, chemical energy is converted into direct current electricity without first burning the fuel to produce heat, much like in a battery, except that the fuel and oxidant are supplied externally. The electricity produced can be used to power an electric motor for a vehicle.

Fuel cells offer a number of advantages compared with internal combustion engines. First, their emissions are several orders of magnitude less than for internal combustion engines, even when the latter are equipped with catalytic converters. Second, the efficiency of fuel use is at least two times greater for fuel cells than for gasoline-fueled, spark-ignited internal combustion engines and one and one half times that for diesel-fueled, compression-ignited engines, which reduces onboard fuel storage requirements. For example, methanol has an energy density deliverable to the wheels of 1,900 Wh per kilogram in a fuel cell versus 900 Wh per kilogram for gasoline in an internal combustion engine. For comparison, a lead acid battery has an energy density of less than 40 Wh per kilogram, while advanced batteries are expected to have energy densities ranging from 100 to 200 Wh per kilogram [23]. Third, fuel cells are far quieter than internal combustion engines, producing only minor noise associated with the electrical control systems for conversion of direct to alternating current. Like battery-powered cars, fuel cell cars would have electric drive trains, but fuel-cell cars using methanol or ethanol do not require the long charge times characteristic of a battery but are refueled by simply refilling the tank, as with gasoline cars. Compared with internal combustion engine systems, fuel-cell systems would occupy more space and cost more, but the technology could be improved to overcome many, if not all, such problems [23].

Hydrogen can be used directly in fuel cells at high efficiency and with zero pollutant emissions. Moreover, hydrogen can be produced from a variety of renewable energy sources, including electrolytic hydrogen produced from wind or photovoltaic sources and biomass-derived hydrogen (see chapter 22: *Solar Hydrogen*), the latter using the same gasification technology as for making methanol from biomass [24]. However, the low volumetric energy density of hydrogen makes on-board storage difficult, and the lack of a gaseous fuel infrastructure complicates the process of introducing hydrogen as a fuel [23, 25].

Because they are liquids, methanol and ethanol could be introduced and stored much more easily than hydrogen, and methanol in particular has been extensively studied for fuel-cell applications. However, the alcohols are not particularly reactive with known catalysts in aqueous solutions, and large amounts of noble metal catalysts are required for fuel cells that use alcohol fuels directly. Because of the formation of carbon dioxide during oxidation of alcohols, fuel cells employing carbon dioxide–rejecting electrolytes must be used. Direct charging of fuel cells with methanol results in catalyst poisoning by partial oxidation products and unreacted alcohol [26, 27].

The problems posed by fuel cells fueled directly with alcohol can be overcome by reforming the alcohol over solid catalysts to form hydrogen and carbon dioxide. With steam reforming, the alcohol serves as a medium to facilitate the handling and storage of hydrogen fuel [25, 28]. Steam reforming results in a higher fuel-cell efficiency than for fuel cells fueled with alcohol directly. However, reformers start up and change output slowly [25].

To date, most attention has focused on methanol reformers, since methanol is relatively easy to reform, requiring modest temperatures (200 to 250°C) and low-cost (e.g., copper–zinc or copper–chromium) catalysts. In contrast, ethanol reformers have to operate at 500 to 600°C and employ more costly nickel catalysts.

For fuel cells that operate at sufficiently high temperatures, waste heat from the fuel cell could provide the heat for reforming. The phosphoric acid fuel cell (PAFC), which operates at about 200°C, is a good candidate for use with methanol, while the solid oxide fuel cell (SOFC), which operates at a temperature of 1,000°C, would be a good candidate fuel cell for ethanol.

The PAFC is a commercially ready technology, with a demonstrated long life and simple water management requirements. Its disadvantages are its high platinum loading, a slow start-up (about 15 minutes to reach 200°C), the sensitivity of its stack and component structures to shock and vibration, and its low power per unit weight and volume. Its low power density makes it an inappropriate candidate for automotive applications, although it might be useful in other transport areas, such as buses, trucks, trains, or ships. Although the SOFC would have high power density and no noble metal catalyst requirements, this technology is at a relatively early stage in its development.

At present, the most promising fuel cell for automotive applications is the solid polymer electrolyte fuel cell (SPEFC). It offers high power density, fast start-up times, ruggedness, and the potential for low cost in mass production. While SPEFCs have required high platinum catalyst loadings (which limits the potential for cost reduction), recent advances have reduced platinum loadings in laboratory fuel cells to low levels [29]. The SPEFC requires a good water management system to prevent the polymer membrane from drying out, and its intolerance of carbon monoxide limits its inherent compatibility for use with reformed alcohol fuels, but there are various strategies for meeting these challenges [25, 29–31].

Because of the low operating temperature of the SPEFC (about 80°C), some of the alcohol must be burned to provide the heat to operate the reformer. For this reason, and also because of the lower partial pressure of the hydrogen at the fuel cell's anode, a SPEFC operated on alcohol fuel would be less efficient than one operated on hydrogen. Despite this drawback, as well as the fact that hydrogen derived from biomass would probably be less costly delivered to the consumer than methanol from biomass, it is likely that on a life-cycle cost basis a SPEFC car operated on methanol from biomass would be less than one operated on biomass-derived hydrogen, if the hydrogen is stored on-board as a compressed gas, owing to the high cost of gaseous fuel storage (see chapter 22). Moreover, if the cost targets for SPEFCs are reached, a biomass-based, methanol-fueled SPEFC car could well be competitive with a gasoline-fueled internal combustion engine (ICE) car if oil sells for $25 a barrel (see chapter 22).

At present, the U.S. Department of Energy is supporting two fuel-cell projects: the Georgetown Bus Project, which uses a PAFC operated on reformed methanol [32, 33], and a project with the General Motors Company, which is to deliver a methanol-fueled SPEFC prototype automobile within five years [34].

Although fuel cells must be markedly improved before they can widely replace internal combustion engines in motor vehicles, their significantly improved emissions characteristics and high efficiency make them highly desirable.

The value of ethanol and methanol

For blending with gasoline, the following relationship expresses the additional price that a blending company could afford to pay for a fuel additive:

$$(P + \Delta p) f + P(1 - f) = P + V \tag{1}$$

in which P is the price of the base gasoline to which the agent is added, Δp is the amount over the base price the blender would be willing to pay for the fuel additive, V is the increase in value of the fuel with the additive, and f is the volume fraction of additive used. This expression is on a volumetric basis and neglects any change in volume when the two components are mixed (generally small) or any loss in energy content (also small) [35]. Based on this expression, the additional price Δp that can be paid for the fuel additive is:

$$\Delta p = \frac{V}{f} \qquad (2)$$

In the case of ethanol, the 10 percent blend commonly used in the United States (known as gasohol) increases the octane level of regular unleaded gasoline to above that of midgrade. If midgrade is worth $0.013 per liter more than regular (V in the above expressions), then according to the second equation, ethanol is worth $0.13 per liter more than the base price of the gasoline to which it is added. For a base gasoline price of $0.18 per liter, the blender would be willing to pay up to $0.31 per liter for ethanol.

Currently, ethanol sells for about $0.33 per liter in the United States. Thus, an additional federal excise tax exemption of $0.013 per liter of 10 percent gasoline blend is used to encourage ethanol use, which translates into $0.13 per liter of ethanol. When added to the $0.31 per liter value of ethanol for blending, the total price one could afford to pay becomes $0.44 per liter, well above the $0.33 per liter selling price. Yet, ethanol is not widely blended with gasoline except in states with additional incentive programs.

This contradiction is partly due to the poor compatibility (fungibility) between ethanol and existing gasoline supplies, which causes problems with storage, transport in pipelines, and the effect of water on fuel properties. In addition, ethanol is typically blended with regular unleaded gasoline, and the vaporization characteristics and other fuel properties are different from typical gasoline [12]. Although the base gasoline formulation could be changed to compensate for these effects, the small companies that typically blend ethanol with gasoline are not equipped to make such adjustments, and the fuel that results can cause vapor lock and other problems in some engines. Consequently, ethanol blends are often viewed unfavorably by the public and must often be sold at lower prices than competing products. Another factor influencing the price differential is that ethanol is typically blended at the distributor's site, and the relationships above do not include the cost of blending equipment and additional storage vessels for the distributor.

For other fuel additives, such as MTBE, the picture is more complex because adjustments are made in the base gasoline composition. To simplify this analysis, if we consider the blending of MTBE with gasoline to have the same Δp as ethanol, then a blender would be willing to pay $0.31 per liter for MTBE. Currently [36], MTBE sells for about $0.25 per liter. In addition, MTBE is compatible with the existing gasoline infrastructure for storage and transportation, can be manufactured and blended at the refinery site, does not change fuel properties noticeably, and has little negative publicity [37]. Consequently, MTBE is widely used in the United States without special tax considerations and is the basis for "reformulated" gasolines being marketed to reduce smog and carbon monoxide.

The price A that a consumer would be willing to pay for a neat alcohol fuel is given by:

$$A = \eta e P \qquad\qquad (3)$$

where η is the relative efficiency of the alcohol as a fuel compared with the base fuel, e is the ratio of volumetric energy content of the alcohol fuel to that of the base fuel, and P is the price of the base fuel. The energy ratio e for methanol is about 0.50 and for ethanol the value is about 0.67 compared to gasoline (see table 1). Thus, if we assume no efficiency increase for the alcohols relative to gasoline, methanol must sell for only half the price of gasoline, and ethanol would have to sell for two-thirds the price of gasoline. Furthermore, ethanol is worth about 33 percent more than methanol. Yet, experience with the alcohols suggests that a 20 percent gain in engine efficiency can be obtained relative to gasoline in a well-designed engine [22, 38]. This improved efficiency is due to the greater octane, higher heat of vaporization, and other properties of ethanol and methanol. In this case, the price the consumer would be willing to pay for neat ethanol is 80 percent of that of gasoline; neat methanol is now worth 60 percent of the price of gasoline. Recent prices of gasoline, ethanol, and methanol [36] give a methanol-to-gasoline price ratio of about 80 percent and an ethanol-to-gasoline price ratio of 160 percent with natural gas and corn as the feedstocks, respectively. Thus, customers would prefer gasoline to either alcohol, if price is the only factor in their decision, and methanol from natural gas is now closer to being competitive with gasoline than ethanol derived from corn. For that reason, methanol is preferred over ethanol as a substitute for gasoline in programs such as the one being undertaken in California. Historically, however, prices exhibit substantial seasonal variation. In September 1990, U. S. Gulf Coast methanol prices (FOB barge) were $0.09 per liter; in January 1991, they were $0.16 per liter [39, 40]. Likewise, ethanol prices vary considerably both with season and with changes in gasoline prices. For example, ethanol prices in the United States fluctuated from about $0.40 per liter in September 1990 to $0.32 per liter in January 1991 [41, 42].

Currently, several major U.S. automobile manufacturers, such as Ford and General Motors, sell internal combustion engines that can use gasoline, methanol, ethanol, or any mixture of the three. The so-called flexible-fueled vehicles use fuel sensors to measure the composition of the fuel, and an on-board computer to adjust the air/fuel ratio and timing to ensure proper performance. However, some engine settings, such as the compression ratio, are not readily adjusted during operation and must be set to accommodate the generally poorer characteristics (such as lower octane) of gasoline compared with the alcohols. As a result, the alcohols achieve little, if any, performance advantage in these engines. On the other hand, since ethanol and methanol perform about the same, inflexible settings in a dedicated alcohol engine could be optimized to achieve better efficiencies with either alcohol than with gasoline, and the fuel line sensor could be employed to adjust the air/fuel ratio and timing to accommodate either ethanol or methanol or mixtures of the two. Such engines are not much more expensive than conventional ethanol or methanol engines, and widespread use of this technology would allow motorists to readily switch between ethanol and methanol, based on price and regional availability. Thus, it appears that the market infrastructure could allow introduction of both fuels and accommodate differences in regional availability or preference for the two alcohols.

If alcohol fuel cells are successfully commercialized, the value of alcohols relative to gasoline in internal combustion engines might increase, owing to the higher overall efficiency of fuel cell vehicles. However, this value is determined not only by relative efficiencies but also by relative system capital and operating costs.

ETHANOL PRODUCTION

Ethanol has experienced several periods of strong demand over the years. It became important during World War I and World War II because petroleum products were scarce. In the 1930s, a blend of ethanol and gasoline was sold in several U.S. midwestern states, but because it could not compete with inexpensive domestic petroleum, its use soon ended. As energy prices rose in the 1970s, interest in ethanol as a transportation fuel was revived in the United States and Brazil. Although ethanol from sugar and corn can be produced commercially, the inexpensive conversion of cellulosic biomass to ethanol is still under development. Production of ethanol from sugar crops in Brazil is discussed in chapter 20: *The Brazilian Fuel-alcohol Program* and will not be considered further here.

Biomass composition

Carbohydrates, including sugars, are among the most abundant constituents of plants and animals and serve many vital functions. They provide energy and also form supporting tissues of plants and some animals. Carbohydrates are classified as mono-, di-, tri-, tetra-, and polysaccharides, depending on the number of sugar molecules that form them. Practically all natural monosaccharides contain five or six carbon atoms, known as pentoses and hexoses, respectively.

The disaccharide known as cane sugar, or sucrose, can be broken down into the six-carbon sugars glucose and fructose by hydrolysis. Sugarcane and other plants contain about 10 to 15 percent sucrose. About 70 percent of corn seed is the polysaccharide known as starch (see figure 1), which is a mixture of straight-chained and branched polymers of glucose with molecular weights ranging up to 1 to 2 million. Hydrolysis of starch by acids or enzymes known as amylases forms glucose sugar.

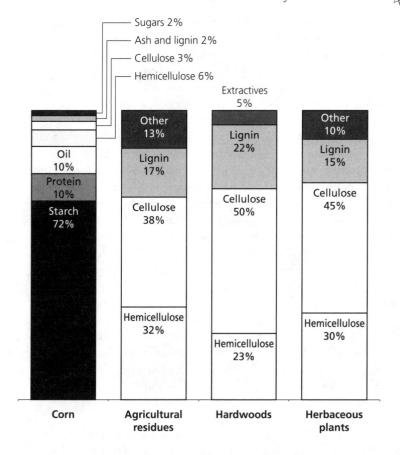

FIGURE 1: *Starch crops, such as corn, are made up of about 70 percent starch. The remaining fraction is primarily protein, oil, cellulose, hemicellulose, and lignin. Cellulosic biomass consists of cellulose, hemicellulose, lignin, and some extractives as shown here for representative examples of agricultural residues (corn cobs), hardwood, and herbaceous plants.*

Cellulosic biomass is actually a complex mixture of carbohydrate polymers known as cellulose and hemicellulose, plus lignin and a small amount of other compounds known as extractives. Examples include agricultural and forestry residues, municipal solid waste, herbaceous and woody plants, and underused standing forests. Cellulose is generally the largest fraction, representing about 40 to 50 percent of the material by weight (see figure 1); the hemicellulose portion represents 20 to 40 percent of the material. The remaining fraction is predominantly lignin with a lesser amount of extractives. The cellulose fraction is composed of glucose molecules bonded together in long chains that form a crystalline structure. The hemicellulose portion is made of long chains of different sugars and lacks a crystalline structure. For hardwoods, the predominant component of hemicellulose is xylose. Softwoods are generally not considered a viable feedstock for the dedicated production of energy in the near term because competition from the paper industry and other markets makes softwoods too expensive. In addition, hardwoods are more amenable to the short-rotation production methods that offer the potential for the low costs vital for large-scale energy applications (see chapter 14: *Biomass for Energy: Supply Prospects*).

Ethanol from starch crops

Two processes are used to produce ethanol from corn and other starch crops in the United States: dry milling and wet milling. In dry milling, the feed material is ground mechanically and cooked in water to gelatinize the starch. Enzymes are then added to break down the starch to form glucose, which yeast ferments to ethanol. The fermentation broth passes through a series of distillation columns to recover a 95 percent ethanol/5 percent water mixture, which is then passed through additional distillation columns or molecular sieves to recover pure (close to 100 percent) ethanol. The solids from fermentation are recovered, dried, and sold as a cattle feed called distillers' dried grains and solubles (DDGS), which contain about 27 percent protein. In this process, about 440 to 460 liters of ethanol, 380 kilograms of DDGS, and 340 kilograms of carbon dioxide are produced per dry tonne of corn.

In wet milling, the insoluble protein, oil, fiber, and some solids are removed from the corn first, with only a slurry of starch fed to the ethanol production step. The enzymatic breakdown of starch, fermentation of glucose, and recovery of ethanol parallel those of the dry milling operation, but only the enzyme, unconverted starch, and yeast are left for recovery as a solid material following the wet milling process. About 37 kilograms of corn oil per dry tonne of corn are refined for human consumption. About 70 kilograms of a 60 percent protein product known as corn gluten meal are recovered per tonne of corn and sold for poultry feed. Also recovered prior to ethanol fermentation are 275 kilograms per tonne of corn of a 21 percent protein product used in cattle feed called corn gluten feed. During fermentation, about 440 liters of ethanol and 330 kilograms of carbon dioxide are produced per tonne of corn.

In 1987, fuel ethanol production from starch crops (primarily corn) provided gasohol equivalent to 8 percent of the U.S. gasoline market, up from less than 1 percent in 1981. Today about 7 million tonnes of corn are now used each year to provide more than 3 billion liters of anhydrous ethanol, (the equivalent of 0.06 exajoules), for 10 percent blends with gasoline [9]. There are about 50 fuel-ethanol manufacturing facilities in the United States using corn and other grains as feedstocks for ethanol production with about two-thirds of the ethanol production coming from wet milling and the remainder from dry milling operations [9]. However, ethanol from corn sells for between $0.29 per liter and $0.41 per liter and the price of corn at about $110 per tonne is too high to produce ethanol at prices competitive with gasoline at today's wholesale prices of $0.15 to $0.20 per liter, even with substantial coproduct credits [5]. Thus, relaxation of gasoline taxes is employed to encourage use of ethanol blends in the United States. The U.S. Department of Agriculture (USDA) estimates that an additional 15 to 19 bil-

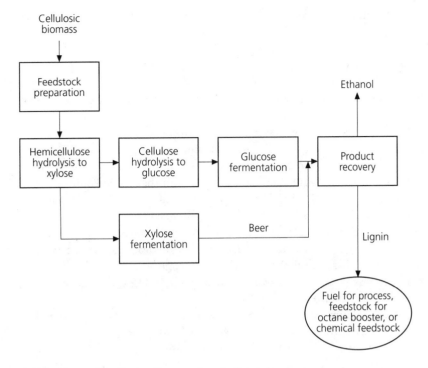

FIGURE 2: *Process flow diagram for conversion of cellulosic biomass to ethanol.*

lion liters of ethanol could be produced from 34 million tonnes of corn in the United States without jeopardizing agricultural resources and raising grain prices [43]. Thus, corn provides a valuable source of ethanol in the short run, but the increase in grain prices and the decrease in coproduct prices with increased production could hinder major displacement of gasoline by ethanol produced from corn.

Ethanol from cellulosic biomass

Various cellulosic feedstocks can be produced at costs much lower than the cost of growing corn and seemingly would be attractive for fuel production. However, these materials have evolved a natural resistance to decomposition that has ensured the survival of plant life. Thus, despite the low cost of the biomass substrate, the cost of biological conversion, until recently, has been too high to seriously consider economic application of cellulosic materials for biofuels production, but this situation is changing.

For production of ethanol, the cellulosic feedstock is first pretreated to reduce its size and open up the structure to facilitate conversion, as shown in figure 2. The cellulose fraction is hydrolyzed by acids or enzymes to produce glucose, which is subsequently fermented to ethanol. The soluble xylose sugars derived from hemicellulose are also fermented to ethanol, and the lignin fraction, which cannot be fermented into ethanol, can be used as fuel for the rest of the process, converted into octane boosters, or used as a feedstock for the production of chemicals.

Acid-catalyzed processes

Several dilute acid hydrolysis pilot plants were constructed in Germany and the United States during wartime to produce ethanol as a petroleum substitute [44], but the economics were too unfavorable for continued postwar operation. Dilute acid–catalyzed processes are currently operated in the former Soviet Union for converting cellulosic biomass into ethanol and single-cell protein. Thus, acid-catalyzed processes provide a near-term technology for production of fuel-grade ethanol from cellulosic biomass, but the low yields of sugars from cellulose and hemicellulose (about 50 to 60 percent of the theoretical maximum) typical of dilute acid systems make them unable to compete with existing fuel options in a free market economy [45, 46]. Concentrated sulfuric or halogen acids achieve high yields (essentially 100 percent of theoretical). However, because low-cost acids (such as sulfuric) must be used in large amounts while more potent halogen acids are relatively expensive, recycling of acid by efficient, low-cost recovery operations is essential to achieve economic operation [47,48]. Unfortunately, the acids must also be recovered at a cost substantially lower than that of producing them from raw materials, which is a difficult requirement.

Enzymatic hydrolysis technologies

Enzymatic hydrolysis emerged from U.S. Army research during World War II aimed at finding ways to overcome microbial attack on the canvas (cellulosic) webbing and tents of soldiers stationed in the tropics. These studies in turn led to research on the possibility of promoting the decomposition of cellulose by a fungus responsible for the breakdown of cotton, now named *Trichoderma reesei*, in hopes of generating glucose syrups for application to food and then fuel ethanol production [49].

Enzyme-catalyzed processes offer several key advantages. They achieve high yields under mild conditions with relatively low amounts of catalyst. Moreover, enzymes are biodegradable and thus environmentally benign. Although the cost of ethanol produced from enzyme- and acid-catalyzed processes may be comparable at present, enzyme-catalyzed processes have tremendous potential for technology improvements that could bring the cost of ethanol down to levels competitive with those for petroleum-based fuels. Nonetheless, considerable improvement is required to achieve economic application of this technology.

Enzymatic processing steps

The following major steps are involved in converting cellulosic biomass into ethanol based on the application of enzymatic hydrolysis technology (see figure 3).

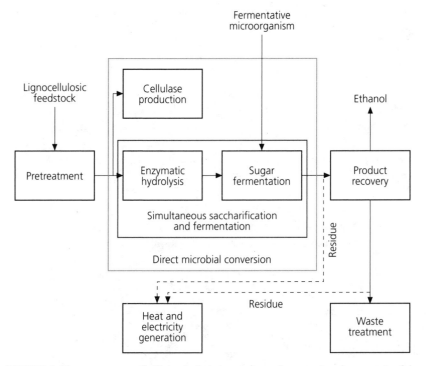

FIGURE 3: *The separate steps of cellulose hydrolysis and glucose fermentation characteristic of the separate hydrolysis and fermentation (SHF) process are combined in the simultaneous saccharification and fermentation (SSF) configuration. Direct microbial conversion (DMC) takes this consolidation one step further by using organisms that can produce cellulase as well as ferment sugars, thereby eliminating the separate enzyme production step required for both SHF and SSF.*

Pretreatment of cellulosic biomass

Cellulosic biomass is naturally resistant to enzymatic attack. A pretreatment step is required to overcome this resistance if the enzyme-catalyzed hydrolysis process is to proceed at a reasonable rate with the high yields vital to economic viability. The pretreatment step must facilitate conversion of both the cellulose and hemicellulose fractions of biomass into ethanol while minimizing the degradation of these fractions into compounds that cannot be fermented into ethanol. Several options have been considered for biomass pretreatment, including steam explosion, acid-catalyzed steam explosion, sulfur dioxide–catalyzed hydrolysis, treatment with organic solvents, base addition, and dilute acid. At this time, the dilute acid option appears to have the best near-term economic potential [50]. In this process, about 0.5 percent sulfuric acid is added to the milled feedstock, and the mixture is heated to around 140 to 160°C for 5 to 20 minutes. Under these conditions, most of the hemicellulose is hydrolyzed to form xylose, which is removed in solution, leaving a porous material of primarily cellulose and lignin that is more accessible to enzymatic attack. The primary reaction, hydrolysis of hemicellulose to xylose, can be represented as:

$$(C_5H_8O_4)_n + nH_2O \xrightarrow{\text{acid}} nC_5H_{10}O_5$$

The xylose sugars released by this process can be fermented into ethanol. Pretreating various agricultural residues, short-rotation hardwoods, and herbaceous energy crops with dilute acids has consistently shown that the yield of glucose from the cellulose left in the solid correlates well with the degree of hemicellulose removal [51]. Pretreatment options that remove the lignin fraction may also be effective if an inexpensive system can be devised.

Cellulose hydrolysis and fermentation
In the cellulose hydrolysis step, cellulase enzymes catalyze the hydrolysis of cellulose to glucose sugars by the following reaction:

$$\underset{\text{cellulose}}{(C_6H_{10}O_5)_n} + \underset{\text{water}}{nH_2O} \xrightarrow{\text{cellulase}} \underset{\text{glucose}}{nC_6H_{12}O_6}$$

Yeast can then ferment the glucose into ethanol:

$$\underset{\text{glucose}}{C_6H_{12}O_6} \xrightarrow{\text{yeast}} \underset{\text{ethanol}}{2C_2H_5OH} + \underset{\text{carbon dioxide}}{2CO_2}$$

Although almost half the weight of the glucose is lost to carbon dioxide, which has no heating value, about 96 percent of the heat of combustion of the cellulose is preserved in the ethanol product [5]. Thus, hydrolysis and fermentation efficiently convert the energy of a solid substrate into a more useful liquid form. A number of yeasts such as *Saccharomyces cerevisiae*, as well as the bacterium *Zymomonas mobilis,* are quite efficient ethanol formers with 92 to 95 percent or more of the glucose going to form ethanol by these reactions.

Although the individual steps for converting cellulosic biomass into liquid fuels can be conveniently isolated (see figure 2), these steps can be combined in various ways to minimize the overall conversion cost. The front-running integrated microbiology-based process configurations are described below.

Separate hydrolysis and fermentation (SHF). The SHF process uses distinct process steps for enzyme production, cellulose hydrolysis, and glucose fermentation (see figure 3) [52, 53]. The primary advantage of this configuration is that enzyme production, cellulose hydrolysis, and sugar fermentation can be treated separately, thus minimizing the interactions between these steps. However, cellulase enzymes are inhibited by the accumulation of sugars, and considerable effort is still needed to overcome this end-product inhibition, which impedes attainment of reasonable ethanol concentrations at high rates and with high yields even at high enzyme loadings.

Simultaneous saccharification and fermentation (SSF). The sequence of steps for the SSF process is virtually the same as for SHF except that hydrolysis and fermentation are combined in one vessel (see figure 3) [54, 55]. The presence of yeast along with the enzymes minimizes accumulation of sugar in the vessel, and because the sugar produced during breakdown of the cellulose slows down the action of the cellulase enzymes, higher rates, yields, and concentrations of ethanol are possible for SSF than SHF at lower enzyme loadings. Additional benefits are that the number of fermentation vessels is cut in half, and that the presence of ethanol makes the mixture less vulnerable to invasion by unwanted microorganisms.

Direct microbial conversion (DMC). The DMC process combines enzyme production, cellulose hydrolysis, and sugar fermentation in one vessel [56–58]. In the most extensively tested configuration, two bacteria are employed to produce cellulase enzymes and ferment the sugars formed by the breakdown of cellulose and hemicellulose into ethanol. Unfortunately, the bacteria also produce a number of products in addition to ethanol, and yields are lower than for the SHF or SSF processes.

Of the alternatives, the SSF process has emerged as an especially promising route to low-cost fuel ethanol production within a reasonable time frame [50, 59].

Cellulase production

Several organisms, including bacteria and fungi, produce cellulase enzymes that can be used to hydrolyze cellulose into glucose [49, 50, 52, 53, 57, 58]. Currently, genetically altered strains of the fungus *Trichoderma reesei* are favored because of the relatively high yields, productivities, and activities of cellulase that are realized. The best performance is generally achieved in the fed batch mode of operation, in which cellulosic biomass is metered into the fermenter during the growth of the fungus and production of cellulase. Simple batch production of cellulase with addition of all ingredients at the beginning of the enzyme production cycle may be used with good results.

Hemicellulose conversion

The hemicellulose polymers in cellulosic biomass such as hardwood, agricultural residues, and herbaceous plants can be readily broken down during the pretreatment step to form xylose and other sugars. Several options have been considered for utilization of the sugars formed from hemicellulose.

In the presence of acid, the xylose can be reacted to form furfural, either during acid hydrolysis or after xylose recovery [46]. Furfural can be sold for use in foundry and other applications, but the market would be quickly saturated by the volume of furfural that would accompany large-scale applications of fuel ethanol [60]. Anaerobic bacteria can convert xylose to methane gas, but methane is less valuable than ethanol.

Another avenue is to ferment xylose into ethanol, using strains of yeast such as *Candida shehatae, Pichia stipitis,* and *Pachysolen tannophilus* [61–64]. However, these strains require small amounts of oxygen in the fermentation broth to ferment xylose and typically cannot achieve high ethanol yields or rates or tolerate high ethanol concentrations [65]. Other microorganisms, such as thermophilic bacteria and fungi, can anaerobically ferment xylose into ethanol [6, 66–70]. Although ethanol tolerance, yields, and selectivity are historically low for those choices, some new evidence suggests such conclusions were premature for some bacteria [71]. The common bacterium *Escherichia coli* has been genetically engineered to produce large quantities of xylose isomerase enzyme. This enzyme can convert xylose into an isomer called xylulose, which many yeast can ferment into ethanol under anaerobic conditions [72–75]. By employing the enzyme and yeast together in one vessel, ethanol yields from xylose of 70 percent of theoretical have been achieved, but the need to provide the isomerase enzyme and adjust for differences in pH optima between the yeast and enzyme complicates the technology. In another approach, the genes from *Zymomonas mobilis* have been spliced into *E. coli* enabling it to ferment xylose directly into ethanol with high yields [76, 77]. However, further evaluation of the procedure is needed. The latter two options are favored for ethanol production at this time.

For xylose fermentation, the overall reaction stoichiometry can be represented as:

$$3C_5H_{10}O_5 \xrightarrow{\text{yeast}} 5C_2H_5OH + 5CO_2$$

$$\underset{\text{glucose}}{\phantom{3C_5H_{10}O_5}} \qquad \underset{\text{ethanol}}{} \quad \underset{\text{carbon dioxide}}{}$$

Once again, most of the heat of combustion of the hemicellulose is preserved in the ethanol despite the significant loss of weight to carbon dioxide during fermentation; fermentation serves to concentrate the energy from the hemicellulose in a liquid energy carrier [5]. The primary challenge is to ensure that most of the xylose follows the pathway to ethanol production without a significant loss of yield to other by-products such as xylitol.

Ethanol recovery

During fermentation, a 3 to 12 percent solution of ethanol in water is produced, with the exact concentration determined by the substrate, yeast, enzyme, and process configuration. In addition, yeast, inert substances such as lignin, enzymes, and unreacted carbohydrates remain in the broth. In most commercial applications, the entire mixture is fed to a distillation (beer) column that concentrates the ethanol in the overhead product while allowing the solids and water to exit from the bottom. The enriched ethanol stream then passes to a second distillation (rectification) column for concentration to about a 95 percent by weight ethanol-in-water product known as an azeotrope.[3] To use ethanol as a hydrated fuel, this azeotrope mixture needs no further processing.

Because water has a low miscibility in gasoline, almost all of it must be removed from the ethanol that will be blended with gasoline. Water in the azeotrope must be removed by some method other than simple distillation. A third component such as benzene or cyclohexane can be added to break the azeotrope and allow purification of ethanol by distillation. Alternatively, molecular sieves can be used to preferentially adsorb the ethanol or water on a solid material such as corn grits. Membranes can also be used that are permeable to one of the components, typically water, while retaining the other by a technology called pervaporation. At present, distillation with a third component and molecular sieves are favored commercially.

Lignin utilization

The amount of the third largest fraction of cellulosic biomass, lignin, is close to that of hemicellulose; thus it is important to derive value from the lignin if ethanol is to be produced economically. Because lignin has a high energy content, it can be used as a process fuel [47, 65, 78]. The amount of lignin in most feedstocks is more than enough to supply all the heat and electricity required for the entire ethanol production process. In addition, the excess electricity or heat can be sold for additional revenue. Alternatively, the phenolic fraction from lignin can be reacted with alcohols to form methyl or ethyl aryl ethers, which are oxygenated octane boosters [79], although high product yields must be realized at low costs to provide a net income gain for the ethanol plant. Or a number of chemicals could be produced from lignin, including phenolic compounds, aromatics, dibasic acids, and olefins [80], which could augment the revenue for the ethanol plant, but the cost must be low enough and chemical yields high enough to ensure a net gain.

METHANOL PRODUCTION

Methanol is produced thermochemically via a two-stage process. The feedstock is first converted to a synthesis gas, composed primarily of H_2 and CO, and the intermediate syngas is then catalytically converted to methanol at elevated pressures. Current global demand for methanol is approximately 23 billion liters per year [81]. Methanol is used as a precursor in the synthesis of many other chemicals. For example, approximately 8.5 billion liters are used each year to make formaldehyde, 1.8 billion liters for ethanoic (acetic) acid, and 3.6 billion liters for MTBE.

---

3. An azeotrope is a liquid mixture whose composition is the same as that of the vapor in equilibrium with the liquid; separation of the components of the mixture cannot be carried out by distillation or other methods that rely on differences in liquid and vapor composition.

At present, methanol is produced primarily from natural gas and to a lesser extent from other hydrocarbons including propane, naphtha, and heavy oil. These technologies have been commercially available since the 1930s and have evolved into efficient, highly selective processes [82, 83]. But methanol can be produced from almost any carbon-containing resource, including biomass, the only methanol feedstock that is renewable. Although biomass-derived methanol is not produced commercially at present, recent advances in gasification technologies offer potential for the future.

Methanol from biomass
The production of methanol from biomass requires pretreatment of the feedstock, its conversion to syngas, cleanup of the syngas, and then conversion to methanol.

Pretreatment
Biomass must be dried and sized prior to methanol synthesis. Drying to a moisture content of 5 to 15 percent is accomplished using waste process heat, which may come from various unit operations in an integrated methanol production facility. For systems using an indirectly heated gasifier, waste heat for drying typically comes from hot flue gases produced during char combustion. For partial oxidation gasifier/reformer systems, waste heat typically comes from reformer furnace flue gases. The minimum heat required to dry biomass from 50 to 10 percent moisture, about 2 gigajoules per tonne of dry biomass, represents approximately 10 percent of the lower heating value of typical woody biomass, about 19 gigajoules per tonne.

Sizing depends on the specific gasifier technology used to produce synthesis gas. Because of the equipment and energy requirements for biomass size reduction, processes capable of using biomass with the least amount of size reduction or waste have an economic advantage in pretreatment costs over those requiring either fine feeds or very uniform particle size. Some gasifiers require very fine particles; for example, the Koppers–Totzek (K–T) entrained flow gasifier uses a minus-30 mesh (about 595 microns) feed, while many fixed-bed gasifier feeds cannot contain fines. Fluid-bed gasifiers have the greatest flexibility in feed particle size, although extra fine material may be blown from the bed before being gasified.

Synthesis gas production
Synthesis gas, or syngas, is produced from biomass through a gasification process that maximizes carbon monoxide and hydrogen while minimizing unwanted products, including methane. Gasification occurs in an atmosphere of steam and/or oxygen at moderately high temperatures (> 1,000 K) and short residence times (0.5 to 20 seconds). The gasification reactors are operated at pressures of 0.1 to 2.5 megapascals.

Gasification may include both partial oxidation and thermal pyrolysis of the feedstock according to the simplified reactions:

$$Biomass + O_2 \rightarrow CO + H_2 + heat \text{ (partial oxidation)}$$

$$Heat + biomass + steam \rightarrow CO + H_2 \text{ (pyrolysis)}$$

As shown, the pyrolysis step is endothermic and requires heat input, while the partial oxidation step is exothermic. In fact, both reactions occur during gasification. In oxidative systems, the reaction with oxygen provides heat to drive the pyrolytic reactions that break apart the solid biomass. In systems that are primarily pyrolytic, heat must be added from an outside source. The overall heat balance depends on the gasification system and feedstock selected.

The high reactivity of biomass allows processing options that are not available for coal. Not only is there a higher fraction of volatile material in the feedstock, but the resulting char is highly reactive, and so it is possible to use either partial oxidation or thermal pyrolysis as the primary conversion route. With coal, partial oxidation must be used to bring about the high temperatures needed to attain sufficient gasification rates.

Gasifiers using partial oxidation inject oxygen, which has been separated from air, into the reactor to provide heat for the gasification reaction. Operated at high temperature, these gasifiers can produce a syngas with low methane content but require the added cost of an oxygen separation facility. Steam is frequently added to improve char gasification. Oxygen-blown biomass gasifiers, including entrained-flow, fluidized-bed, and fixed-bed configurations, have successfully been demonstrated at scales of 5 to 100 tonnes of wood per day (tpd) [84–86]. Gasifiers designed specifically for wood feedstocks include the Institute of Gas Technology (IGT) and SynGas gasifiers in the United States, the Creusot-Loire facility in France, and the Biosyn gasifier in Canada. Work is either under way or planned for some of these gasifiers.

Even though the use of oxygen will produce a gas suitable for downstream synthesis gas operations, oxygen production is expensive and accounts for a large percentage of capital and operating costs. For example, oxygen costs $40 to $60 per tonne, and is typically used at the rate of 0.25 to 0.30 tonnes per tonne of biomass. This translates to a cost of $10 to $21 per tonne of biomass processed. There are, of course, economies-of-scale in oxygen production. The Union Carbide Company has estimated that a plant producing 2,000 tonnes per day of oxygen has a capital investment, in dollars per tonne of oxygen per day, of approximately two-thirds of that required for a 500 tonne per day facility [87]. In 1990, Chem Systems estimated the capital cost of an approximately 4,340 tpd oxygen facility to be about $23,500 per daily tonne [88].

Biomass can also be converted to syngas in indirectly heated gasifiers, which use pyrolytic reactions. These reactors use heat generated externally by combustion of part of the biomass to drive the pyrolysis and steam gasification reactions. The heat can be provided through the use of fire tubes in a fluidized bed or with multivessel concepts. Examples include the Battelle Columbus Laboratory (BCL) or the Manufacturing and Technology Conversion International (MTCI) gasifiers in the United States and the Studsvik gasifier in Sweden [89–91]. The fire tube designs may be limited to atmospheric pressure. Other designs should be capable of higher pressures but have not yet been demonstrated above atmospheric. The main advantage of these reactors is the elimination of a separate facility to provide purified oxygen. However, higher methane yields in the product gas will require a reforming step prior to methanol synthesis. Indirectly heated gasifiers could also potentially be operated using heat from solar collectors or other sources. Such alternative methods would increase the amount of gas that would be produced from a given supply of biomass. Although not currently cost effective, they might become attractive with improvements in solar-thermal conversion technology and/or if biomass feedstock costs rise.

Typical biomass gasifiers

Three conceptual gasifiers, which are representative of oxygen-blown and indirectly heated gasifiers, have been used to develop material and energy balances and process economics [92, 93]. The conceptual processes include syngas generation using low-pressure oxygen gasification (LPO), high-pressure oxygen gasification (HPO), and indirectly heated gasification (IND), followed by syngas upgrading and methanol synthesis (see figure 4). Although not shown, all processes require syngas compression prior to methanol synthesis.

The K–T gasifier is a low-pressure oxygen-blown gasifier developed originally for coal and is considered representative of commercially available LPO technology. Table 2 presents typical operating conditions and yields for the K–T gasifier. Also given in table 2 is an estimate of the order of magnitude increase in potential synthesis gas that can be realized through steam reforming of methane and higher hydrocarbons. This estimate is simplified because the shift reaction equilibrium and gas composition after reforming are not shown; the estimate is meant only to indicate that gasifier exit gas rate and composition cannot be used to compare the final methanol production rate.

The process generates a synthesis gas with an H_2/CO molar ratio that is less than 1 and low levels of methane and other light hydrocarbons. The processing downstream of the gasifier is typical of systems proposed for methanol from coal: 1) the syngas H_2/CO ratio is adjusted to match that required for methanol synthesis ($H_2/CO > 2$) in a shift conversion reactor; 2) acid gases (primarily CO_2 for biomass) are removed; and 3) methanol is synthesized and purified. The K–T gasifier requires a small-feed particle size, less than about 600 microns. Because comminution for biomass is energy and equipment intensive, the size requirement results in substantial feed preparation costs. Also, adjustment of synthesis gas composition by rejection of carbon as carbon dioxide results in lower yields compared with the other processes.

The IGT "Renugas" is a high-pressure oxygen-blown fluid-bed gasifier developed specifically for biomass. The gasifier has been operated at a 10 tpd pilot scale, at temperatures up to about 1,255 K, and at pressures up to about 2.38

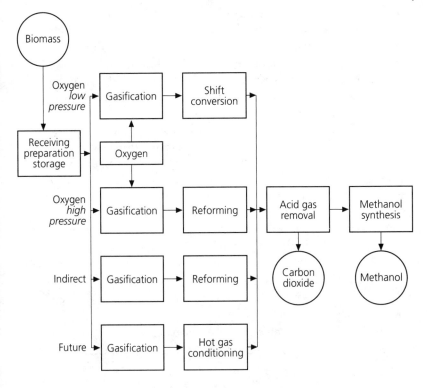

FIGURE 4: Thermochemical production of methanol from cellulosic biomass involves preparation of the feedstock, gasification through one of four primary technologies, removal of carbon dioxide, and synthesis of methanol.

megapascals [85]. Operation at high pressure produces a synthesis gas containing a high level of methane (see table 2). Methane acts as an inert diluent in methanol synthesis. Therefore, the synthesis gas must be reformed to reduce the concentration of methane before methanol is synthesized, in order to increase yields and improve economics. The Chem Systems study assumes that the IGT gasifier can be operated at pressures up to 4.44 megapascals and that no gas conditioning other than particulate removal is required before reforming. These assumptions will have to be confirmed during process scale-up. Operation of the gasifier at pressure eliminates costs and potential problems associated with the compression of syngas before reforming.

The BCL gasifier is a low-pressure, indirectly heated gasifier in which the product char is burned to heat sand, which in turn is mixed with fresh biomass to supply heat for gasification. Indirect gasification (IND) produces a synthesis gas with a low H_2/CO ratio, high levels of light hydrocarbons, and a low level of carbon dioxide. The high level of methane requires reforming prior to methanol synthesis. The synthesis gas is compressed prior to reforming in order to improve reforming economics. In the case study presented here, it is assumed that the raw synthesis gas is quenched, and that tars are recycled to extinction in the gasifier.

Alternatively, tars could be recycled to the combustor to provide process heat. A variation (IND2), in which catalytic hot-gas conditioning is used to both destroy tars and reform methane, gives the same overall conversion. Development of hot-gas conditioning in the United States is at the bench-scale stage and will need significant development; European hot-gas conditioning has not been directed toward elimination of downstream reforming. The BCL gasifier has been operated at the 12 tpd scale and at temperatures up to about 1,280 K [89].

A number of other systems have been developed at the pilot or demonstration scale (see table 3), but are not discussed in this chapter.

Syngas conditioning

Prior to methanol synthesis, the raw syngas must be cleaned and conditioned. Particulate removal is accomplished with cyclones, wet scrubbing, or high-temperature filters. For gasifiers that produce high yields of hydrocarbons, the unreacted methane and other organics must be reformed to generate additional CO and H_2. The reforming reaction is accomplished at 1,000 to 1,150 K, and catalysts, such as sulfided cobalt/molybdate, are used because of small amounts of sul-

Table 2: Gasifier comparison

	Units	K–T[a]	IGT[b]	BCL[c]
Temperature	*K*	1,255	1,255	1,255
Pressure	*megapascals*	0.1013	3.44	0.1013
Dry gas production	*Nm³ per tonne* [d]	1,347.5	1,065.8	1,027.2
Dry gas composition	*mol percent*			
H_2		36.2	30.9	30.6
CO		44.4	19.8	41.2
CO_2		19.1	36.2	10.9
CH_4		0.3	13.1	14.0
C_2		–	–	3.3
H_2/CO		0.82	1.56	0.74
Dry gas (normalized for *Nm3 per tonne*[e] CH_4 + decomposition)		1,360	1,485	1,510

a. Koppers - Totzek (K-T) gasifier [92].

b. Institute of Gas Technology (IGT) gasifier [85].

c. Battelle Columbus Laboratory (BCL) gasifier [89].

d. At 273.15 K and 0.10125 megapascals.

e. Example reaction: $CH_4(g) + H_2O(g) \rightarrow CO(g) + 3H_2(g)$ (steam reforming).

fur in the gas stream. In general, biomass is deficient in hydrogen for methanol production. The shift reaction

$$CO + H_2O \rightarrow CO_2 + H_2$$

will be used if necessary to obtain an H_2/CO ratio slightly greater than 2:1. The final step is acid-gas removal, where sulfides and most of the CO_2 are removed from the gas stream. The sulfur content (< 0.1 percent) of the syngas from most biomass is considerably lower than for coal, and a nonselective adsorption system can be used. Final sulfur removal is accomplished by adsorption on zinc oxide.

Methanol synthesis
Methanol is produced from clean syngas by catalytically recombining the carbon monoxide and hydrogen according to the reaction:

$$2H_2 + CO \xrightarrow{\text{catalyst}} CH_3OH + heat$$

If excess hydrogen is present, carbon dioxide in the syngas reacts with hydrogen to form additional methanol:

$$3H_2 + CO_2 \xrightarrow{\text{catalyst}} CH_3OH + H_2O$$

Table 3: Selected gasification processes and feed materials used

Process	Feed
American Thermogen (U.S.)	Municipal solid waste
Cruesot–Loire (France)	Wood
Davy McKee (U.K., U.S.)	Wood
HTW (Germany)	Peat, lignite
MINO (Studsvik, Sweden)	Wood
MTCI (U.S.)	Wood, agricultural wastes, black liquor
Omnifuel (Canada)	Wood
Pacific Northwest Laboratory (U.S.)	Wood
Pillard (France)	Vegetable wastes
Purox (U.S.)	Municipal solid waste
SynGas, Inc. (U.S.)	Wood, municipal solid waste
Twente (Netherlands)	Wood
University of Missouri-Rolla (U.S.)	Wood
Wellman Galusha	Wood

Methanol is not currently produced from biomass, but the production of clean syngas would lead to the use of systems very similar to those now used for natural gas. Methanol catalysts are highly selective. Less than 2,000 parts per million (ppm) of other products including ethanol and higher alcohols, dimethyl ether, and ketones are produced. Overall process efficiencies are discussed in the section "Methanol from cellulosic biomass, natural gas, and coal."

Methanol synthesis is accomplished at temperatures from 500 to 570 K and at pressures from 5.2 to 10.3 megapascals. Commercial methanol reactors use fixed catalyst beds but vary in the way that heat from the exothermic synthesis reaction is removed from the system. The Imperial Chemical Industries (ICI) axial flow reactor, for example, injects cold syngas at various points to moderate temperature. The Lurgi system uses a tubular reactor system surrounded by a pressurized water boiler. Other heat control designs are available or under development.

Approximately 20 to 25 percent of the syngas is converted to methanol on each pass through the reactor. Methanol is separated by condensation and the remaining syngas is recycled for additional conversion. Purging of the system is required to prevent buildup of inert materials, such as methane or excess CO_2. Because purging also removes CO and H_2, it is important to produce a high-quality syngas prior to methanol synthesis.

Methanol purification

The initial methanol product contains water and small quantities of other organic products. For high-purity chemical applications, the methanol is distilled in a two- or three-column system where the water and higher alcohol fractions are recovered separately. For fuel applications, a single stage distillation would reduce water content to less than 1 percent, and the higher alcohols would be blended back into the fuel.

Methanol from coal

The production of methanol from coal is similar in most respects to that from biomass. Initial processing involves washing and reduces the size of coal particles. Following front-end processing, the coal is gasified to syngas, the syngas is processed, and methanol is produced from the clean product. Again, the type of gasifier chosen will determine the size reduction needed.

Coal-gasification technology is commercially available. For example, the Lurgi gasifier is a fixed-bed reactor, the Winkler gasifier is a fluidized-bed reactor, and the Koppers–Totzek gasifier is an entrained-bed reactor. All three concepts partially oxidize the coal with oxygen separated from air. The Texaco gasifier [94] is a high-pressure, oxygen-blown, entrained-flow coal gasifier based on partial oxidation technology, which was originally commercialized in the late 1950s for converting hydrocarbon liquids and gases to carbon monoxide and hydrogen. The Texaco gasifier operates as a slagging, pressurized, downflow, entrained gasifier in which a coal/water slurry is pumped into the reactor along with oxygen.

Gasification takes place at temperatures in excess of 1,500 K, producing fixed gases such as carbon dioxide, carbon monoxide, and hydrogen, but no liquid hydrocarbons. The raw product gas contains some unburned carbon and molten ash, which must be removed before the gas is used. Texaco gasifier installations include the Tennessee Eastman plant in Kingsport, Tennessee, and the Cool Water plant in Daggett, California. The Texaco gasifier is the basis for coal-to-methanol economics presented later in this chapter.

In addition, a number of indirectly heated coal gasification systems have been developed. While coal is not reactive enough to be gasified indirectly, indirect gasification is feasible with two-stage processes in which coal is first converted to an intermediate product such as coke or char, which is then gasified. The historical "blue water gas" process in which coal is heated in the presence of air to produce an incandescent coke, which is then gasified by steam, is a classic example of indirect coal gasification. In recent times the COGAS process [95] exemplifies a staged pyrolysis-char gasification process.

The syngas from coal is cleaned using technologies similar to those for biomass. The ash content of coal is high and will require special attention, particularly in fixed-bed reactors. The syngas is also reformed to convert unreacted hydrocarbons into additional CO and H_2. The updraft Lurgi reactor, for example, produces significant quantities of partially reacted organics. Coal, being even more hydrogen deficient than biomass, will also require extensive shifting to obtain an acceptable H_2/CO ratio.

Higher sulfur concentrations in most coals require the use of selective adsorbent systems followed by sulfur recovery. CO_2 is also largely removed from the product stream. The resulting syngas is converted to methanol using the technology described above.

Methanol from natural gas

As indicated previously, more than 80 percent of current methanol is produced from natural gas in an efficient, highly selective process. Natural gas is usually processed to remove condensates, propane, and similar components. Sulfur is removed in the initial stages by passing the gas over metal-impregnated activated carbon or zinc oxide. Sulfur must be reduced to less than 0.5 ppm to prevent poisoning of the catalysts downstream.

Pretreated natural gas is then steam reformed to produce primarily CO and H_2 as follows:

$$CH_4 + H_2O \rightarrow CO + 3H_2$$
$$CO + H_2O \rightarrow CO_2 + H_2$$

Steam reforming has been practiced commercially since about 1930, and reactor units are available from several manufacturers.

The steam reformer consists of a furnace with an internal tube bundle. Part of the natural gas feedstock is burned to provide heat to the furnace. The natural gas/steam mixture is preheated to 800 to 870 K and passed through the tubes containing the catalyst, which is generally nickel on a ceramic support. The syngas exits the reactor at 1,170 to 1,270 K and usually contains 0.5 to 3.0 percent residual methane.

Reforming is typically performed at pressures of about 0.1 to 2 megapascals, although pressures up to about 4 megapascals can be used. Pressurization to intermediate levels leads to lower overall system costs, but also suppresses the reforming reaction thermodynamically. For this reason, the percent of unreacted methane will be greatest in the pressurized systems.

Steam reforming leads to a greater ratio of H_2/CO than is required for methanol synthesis. In contrast, biomass systems are hydrogen deficient. In recent natural gas facilities, steam reforming is combined with catalytic partial oxidation by adding purified oxygen from a separation plant. In other cases, CO_2 is purposely added to the reactor. In the methanol synthesis step, the catalyst reacts CO_2 with H_2 to form additional CO and H_2O by the reverse of the second reaction above, so that no additional processing of the syngas is necessary. Methanol is produced from the syngas as described previously.

ECONOMICS

Ethanol from starch crops

The cost of corn varies with the weather, fuel and fertilizer costs, agricultural policy, and other factors. For instance, between 1981 and 1988, the lowest average cost of number 2 yellow corn was $56.80 per dry tonne in 1986, while the highest cost was $164 per tonne in 1984 (see table 4). However, coproduct revenues also vary considerably from a low of $35.90 per tonne of corn in 1985 to a high of $76.90 per tonne in 1984 for the dry milling operation. For wet milling, the low coproduct revenue was $57.80 per tonne of corn produced in 1985, and the high was $86.70 per tonne in 1983. Values for both wet and dry milling coproduct revenues are shown in table 4, assuming no coproduct market for carbon dioxide.

Perhaps more important is the difference between the cost of corn and the revenue gained from the sale of coproducts produced from the corn, a quantity defined by many as the net cost of the corn. For dry milling operations, the highest net cost between 1981 and 1988 was in 1985, when corn cost $130 per tonne and generated only $35.90 per tonne in coproduct revenue; the lowest net cost for dry milling was in 1986, when corn cost $56.80 per tonne and coproduct revenues were $48.90 [96]. Wet milling operations have greater coproduct revenues and lower net corn costs. The lowest net cost was in 1986, when the cost was $56.80 per tonne, but revenues from corn oil, corn gluten meal, and corn gluten feed sales totaled $58.20. On the other hand, only two years earlier, the highest net corn cost occurred for wet milling, with corn costing $149 per tonne, producing only $62 of revenue in coproduct sales. By comparison, remember that ethanol is worth about $0.14 per liter as a neat fuel and about $0.31 per liter when blended with gasoline, depending on the prevailing price of gasoline. Thus, corn costs and coproduct prices for the dry and wet milling operations can quickly swing from unprofitable to profitable and back.

Table 5 presents the capital and operating costs for corn ethanol plants [9]. Capital costs and maintenance and personnel expenses vary with the size of the operation, with larger plants benefiting from economies-of-scale and more efficient use of personnel. In contrast, energy, chemical, enzyme, and yeast costs are more sensitive to the type of process and whether or not the enzymes and yeasts are purchased or produced on site. The costs in table 5 are for modern ethanol facilities built on bare ground (grassroots plants). The capital costs for wet milling

Table 4: Historical real prices of corn and coproducts for dry and wet milling processes[a]

	Year	Corn cost		Coproduct price[c]		Net cost[d]	
		$per tonne of dry corn[b]		*$per liter of ethanol produced*			
High corn cost							
Dry milling	1984	164.00	*0.358*	69.40	*0.152*	94.60	*0.206*
Wet milling	1984	164.00	*0.373*	80.20	*0.182*	83.80	*0.191*
Low corn cost							
Dry milling	1986	56.80	*0.124*	48.90	*0.107*	7.90	*0.017*
Wet milling	1986	56.80	*0.129*	58.20	*0.132*	−1.40	*−0.003*
High coproduct price							
Dry milling	1984	154.20	*0.337*	76.90	*0.168*	77.30	*0.169*
Wet milling	1983	158.00	*0.358*	86.70	*0.197*	71.30	*0.161*
Low coproduct price							
Dry milling	1985	130.00	*0.285*	35.90	*0.078*	94.10	*0.207*
Wet milling	1985	116.00	*0.265*	57.80	*0.131*	58.20	*0.134*
High net cost							
Dry milling	1985	130.00	*0.285*	35.90	*0.078*	94.10	*0.207*
Wet milling	1984	149.00	*0.338*	62.00	*0.141*	87.00	*0.197*
Low net cost							
Dry milling	1986	56.80	*0.124*	48.90	*0.107*	7.90	*0.017*
Wet milling	1986	56.80	*0.129*	58.20	*0.132*	−1.40	*−0.003*

a. Data from [96] without credit for sales of carbon dioxide for 1981 to 1988.

b. *Multiply values shown by 0.02151 to obtain price in $per bushel.*

c. Coproduct price is revenue generated by coproduct sales per tonne of corn processed.

d. Net cost is defined as the cost of the corn less the revenue generated from the sales of coproducts produced per dry tonne of corn.

operations tend to be near the high end of the range, whereas those for dry milling plants are near the low end. Capital costs are lower if the process is constructed on an existing site, perhaps tying into an existing wet milling operation that already produces other starch products.

Table 6 presents the cost of producing ethanol from corn by both wet and dry milling processes. The wet mill was assumed to be a large, efficient process with a capital cost of 1989 $0.65 per liter for a grassroots plant, and operating costs at the low end of the ranges shown in table 4. Because taxes were not included in this analysis, annual insurance costs from table 5 were taken as 0.5 percent of capital costs. In the first row of table 6, a low net corn cost corresponding to the values shown for 1986 in table 4 was used. Ethanol produced by this scenario would cost $7.76 per gigajoule for a 6 percent rate of return discounted over a 20 year operating life and would cost $9.90 per gigajoule for a 12 percent rate of return over the same period. For the highest net cost of corn values for 1984 in table 4, the cost of ethanol production rises to $17.27 per gigajoule at a 6 percent rate of return and $19.48 per gigajoule at a 12 percent rate of return (these values are presented in the second and sixth rows in table 6). Such tremendous swings in ethanol production costs, varying with corn costs and coproduct prices, show

Table 5: Ranges in production and capital costs for production of ethanol from corn[a]

| | Plant size | | | |
| | 10^3 liters/year | | (10^3 gallons/year) | |
	< 150	(< 40)	150–950	(40–250)
	$ per annual liter		*($ per annual gallon)*	
Capital investment	0.85	*(3.20)*	0.55–0.70	*(2.10–2.70)*
	$ per liter produced		*($ per gallon produced)*	
Fuel				
Coal	0.026–0.37	*(0.10–0.14)*	0.026–0.37	*(0.10–0.14)*
Gas	0.084	*(0.32)*	0.084	*(0.32)*
Electricity	0.010–0.013	*(0.04–0.05)*	0.010–0.013	*(0.04–0.05)*
Chemicals, enzymes, yeasts	0.010–0.032	*(0.04–0.12)*	0.010–0.032	*(0.04–0.12)*
Maintenance	0.042	*(0.16)*	0.029	*(0.11)*
Personnel	0.050	*(0.19)*	0.018	*(0.07)*
Taxes and insurance	2 percent of capital			

a. From [9] for a modern plant built on an undeveloped site (grass roots); transformed to 1989 dollars

how sensitive ethanol prices are to these factors, even for a low-cost, efficient operation. Furthermore, the U.S. Department of Agriculture (USDA) projects that the cost of corn will rise and coproduct prices will drop as ethanol production increases by four to five times current levels [43].

Also shown in table 6 are the costs of ethanol for a small dry milling operation for both low and high net corn cost scenarios from table 4. Capital costs were taken from the entries for smaller plants in table 5 and from the upper portion of the range of costs, as appropriate. It is apparent that the cost of ethanol is about 50 to 60 percent higher for dry milling of corn than for wet milling. All smaller plants in the United States are dry milling operations, mostly smaller in size than the example chosen in table 6.

Ethanol from cellulosic biomass

The preliminary economics of ethanol production from cellulosic biomass presented here are based on a 1990 study performed by Chem Systems for the National Renewable Energy Laboratory (NREL) [97]. Chem Systems developed cost estimates for SSF technology, for two biomass feed rates: 1,745 and 9,090 tonnes plant were comparable in size to the second Chem Systems case. Other improve-

Table 6: Pretax cost of production (COP) for ethanol from corn[a]
1989 U.S.$ per gigajoule (LHV) of ethanol

Process	Capital	Feed	O&M	Coproducts	Ethanol
COP at 6 percent discount rate					
Wet mill *265 million liters of ethanol per year*					
low net corn cost	3.139	6.086	4.771	6.236	7.760
high net corn cost	3.221	15.914	4.771	6.635	17.271
Dry mill *76 million liters of ethanol per year*					
low net corn cost	4.013	5.852	7.433	5.037	12.262
high net corn cost	4.076	13.431	7.433	3.694	21.247
COP at 12 percent discount rate					
Wet mill *265 million liters of ethanol per year*					
low net corn cost	5.274	6.086	4.771	6.236	9.895
high net corn cost	5.434	15.914	4.771	6.635	19.484
Dry mill *76 million liters of ethanol per year*					
low net corn cost	6.741	5.852	7.433	5.037	14.990
high net corn cost	6.864	13.431	7.433	3.694	24.035

a. Ethanol yields are 440 liters per tonne of corn for a wet mill process and 458 liters per tonne of corn for a dry mill process.

Table 7: Capital and operating cost estimates for producing ethanol from cellulosic materials

	Reference case[a]	Larger scale[a]	Improved technology[b]
Plant size, feed rate *dry tonnes per day (tpd)*	1,745	9,090	2,727
Product rate *Millions of liters per year (denatured, hydrated)*	219	1,096	507
Feed price *1989 U.S. $per GJ (LHV)*	2.45	2.45	2.00
Capital cost *millions of 1989 U.S. $*			
Feed handling	6.91	27.59	10.99
Prehydrolysis	22.54	89.60	36.35
Xylose fermentation	6.02	24.05	3.80
Cellulase production	2.60	10.39	1.37
SSF fermentation	21.60	86.26	10.58
Ethanol purification	3.73	10.71	5.18
Offsite tankage	3.04	7.40	9.09
Environmental systems	3.98	11.42	3.12
Utilities/auxiliaries	57.60	165.34	49.91
Erected plant cost	128.04	432.75	130.40
Owner's costs, fees, and profit	12.81	43.29	
Start-up	6.40	21.62	6.52
Total capital investment	147.24	497.66	136.92
Working capital	9.70	37.54	7.24
Operating costs *millions of 1989 U.S. $per year*			
Variable costs			
Feedstock	26.88	134.39	34.34
Catalyst and chemicals	9.25	46.24	14.83
By-product credits	0.40	2.00	0.63
Utilities	(3.10)	(15.52)	(3.72)
Fixed costs			
Labor	1.59	3.18	1.42
Maintenance	4.40	14.87	3.93
General overhead	3.77	11.31	3.27
Direct overhead	0.72	1.43	0.64
Insurance	0.74	2.49	0.68
Total operating cost	44.65	200.39	56.02

a. [97]. Because Chem Systems estimated all costs in 1987 U.S. dollars, the values were translated to the 1989 U.S. dollar values shown by application of the Nelson–Farrar index [99].

b. [98].

per day (see table 7). Hardwood costing $46 per tonne was received as 2.5 centimeter chips with a 50 percent by weight moisture content and milled to 3 millimeter size by disc refiners. The milled wood was pretreated with dilute sulfuric acid, and a recombinant *E. coli* strain fermented the xylose removed during pretreatment into ethanol. The fungus *Trichoderma reesei* produced the cellulase enzyme in a simple batch process for the SSF process step. Continuous processing was used for cellulase conversion to ethanol. Conventional distillation removed the ethanol from the fermentation broth to produce a neat fuel containing 5 percent water. Gasoline was added as a denaturant to give a product composition of 90.3 percent ethanol, 4.7 percent water, and 5 percent gasoline by weight. Lignin recovered from the bottom of the distillation unit was burned as boiler fuel to provide the heat and electricity for the process, with excess electricity sold. Based on the best estimates of current technical performance for each of these process steps, Chem Systems derived capital and operating costs for the process and carried out sensitivity analyses to identify opportunities for improving the technology. Substantial yield losses, slow reaction rates, large power requirements, and other problems were identified. Using this information, NREL developed preliminary economics for ethanol production from cellulosic biomass based on improved technology that increases yields, speeds fermentation rates, improves energy efficiencies, and addresses other important problems [98].

The improved technology case results in cost reductions from enhanced ethanol yields, rates, and efficiencies judged to be attainable through research. These changes are not exhaustive and do not include radical departures in technology. The costs shown in the improved case were derived from Chem Systems values, vendor quotes, and application of Aspen/SP simulation software [100], where necessary. Mature technology representative of an *n*th plant was assumed throughout.

Capital costs ranged from $498 million for the 9,090 tpd feed rate to $147 million for the 1,745 tpd rate for current technology. For the improved-technology case, the capital cost drops to $137 million for a feed rate of 2,727 tpd.

A cash flow analysis for the Chem Systems reference case with a feed rate of 1,745 tpd is presented in table 8 to estimate the unit cost of ethanol production; the cash flow analysis does not include taxes or depreciation. Similar cash flows were projected for the other two cases described in table 7, and the cost of ethanol production is summarized in table 9 for all three cases. Subtotals are given for unit capital, feedstock, and operating and maintenance costs. At a 12 percent discount rate, the projected price of ethanol ranges from $15.36 per gigajoule for the Chem Systems reference case with a feed rate of 1,745 tpd to $12.59 per gigajoule for a feed rate of 9,090 tpd. At a 6 percent discount rate, costs drop to $13.12 per gigajoule and $11.06 per gigajoule, respectively. With the improved technology, ethanol costs are projected to be only $7.54 per gigajoule for a 12 percent discount rate and $6.65 per gigajoule for a 6 percent discount rate with a feed rate of 2,727 tpd. Costs with improved technology could be reduced further if the

Table 8: Cash flow (excluding taxes) for ethanol production from cellulosic feedstocks
millions of 1989 U.S. $

Year	Ethanol revenue	Electricity revenue	Capital	Working capital	Feed cost	Fixed op costs	Variable op costs	Total costs	Pretax cash flow	Cumulative pretax cash flow
1			44.172					44.172	(44.172)	(44.172)
2			73.620					73.620	(73.620)	(117.792)
3			29.448					29.448	(29.448)	(147.240)
4	35.764	1.860		9.700	16.128	6.732	5.790	38.350	(0.726)	(147.966)
5	47.686	2.480			21.504	8.976	7.720	38.200	11.966	(136.000)
6	59.607	3.100			26.880	11.220	9.650	47.750	14.957	(121.043)
7	59.607	3.100			26.880	11.220	9.650	47.750	14.957	(106.086)
8	59.607	3.100			26.880	11.220	9.650	47.750	14.957	(91.129)
9	59.607	3.100			26.880	11.220	9.650	47.750	14.957	(76.171)
10	59.607	3.100			26.880	11.220	9.650	47.750	14.957	(61.214)
11	59.607	3.100			26.880	11.220	9.650	47.750	14.957	(46.257)
12	59.607	3.100			26.880	11.220	9.650	47.750	14.957	(31.300)
13	59.607	3.100			26.880	11.220	9.650	47.750	14.957	(16.343)
14	59.607	3.100			26.880	11.220	9.650	47.750	14.957	(1.386)
15	59.607	3.100			26.880	11.220	9.650	47.750	14.957	13.571
16	59.607	3.100			26.880	11.220	9.650	47.750	14.957	28.529
17	59.607	3.100			26.880	11.220	9.650	47.750	14.957	43.486
18	59.607	3.100			26.880	11.220	9.650	47.750	14.957	58.443
19	59.607	3.100			26.880	11.220	9.650	47.750	14.957	73.400
20	59.607	3.100			26.880	11.220	9.650	47.750	14.957	88.357
21	59.607	3.100			26.880	11.220	9.650	47.750	14.957	103.314
22	59.607	3.100			26.880	11.220	9.650	47.750	14.957	118.271
23	59.607	3.100		(9,700)	26.880	11.220	9.650	38.050	24.657	142.929

Table 9: Pretax production cost of ethanol from cellulosic biomass
1989 U.S.$ per gigajoule (LHV) of ethanol

Process	Capital	Feed	O&M	Electricity	Ethanol
COP at 6 percent discount rate					
Reference	3.292	5.915	4.593	0.682	13.117
219 million liters of ethanol per year[a]					
Larger scale	2.237	5.915	3.588	0.683	11.057
1,096 million liters of ethanol per year[a]					
Improved	1.314	3.268	2.417	0.354	6.646
507 million liters of ethanol per year[a]					
COP at 12 percent discount rate					
Reference	5.538	5.915	4.593	0.682	15.364
219 million liters of ethanol per year[a]					
Larger scale	3.767	5.915	3.588	0.683	12.587
1,096 million liters of ethanol per year[a]					
Improved	2.208	3.268	2.417	0.354	7.540
507 million liters of ethanol per year[a]					

a. Denatured hydrated ethanol.

ments in the process steps or application of other enzymatic options offer alternative routes for reducing costs.

Through these engineering studies, several technological opportunities have been identified for reducing the cost of ethanol to about $7 per gigajoule ($0.15 per liter) or lower. First, high yields of ethanol must be achieved with low enzyme costs, realized with improved enzymes, better enzyme production or more efficient pretreatment. High yields of ethanol from hemicellulose are important with high xylose yields from pretreatment, low-cost media for xylose fer-

Table 8: notes

Process: Chem Systems reference case	Basis: constant 1989 U.S. $
Capital cost: 147.24 million $	No escalation
Ethanol COP: $13.44 per gigajoule	Three-year construction, 20 year operating life
Production rate:	Investment schedule: year 1, 30 percent
219.189 × 10⁶ liters per year	
(denatured hydrated product)	year 2, 50 percent
4.54 × 10⁶ gigajoules per year	year 3, 20 percent
Feed cost: 26.88 million $ per year	Production schedule: year 4, 60 percent of nameplate
Discount rate: 6.00 percent	year 5, 80 percent of nameplate

mentations, and prolonged microbial stability. The economics of production would also benefit by increasing the ethanol concentration to 6 to 8 percent, although advanced distillation apparently can recover lower ethanol concentrations at a reasonable cost [101, 102]. Reducing the enzyme processing times from the current five to seven days to three days for complete fermentation would also achieve significant capital cost reductions through improvements in pretreatment or cellulase technology. The cost of ethanol production could also be reduced by use of low-power mixing devices. Improved fermenter designs can still improve process economics by reducing product accumulation and inhibition of yeast and enzymes and minimization of by-product formation [71, 103]. Although considerable progress has been made in this direction, full-process integration is a key step to establishing the performance of the entire SSF process when operated on actual feedstocks in the actual process sequence to establish interactions among steps.

The feedstock cost for present-day technology was $46 per dry tonne ($2.45 per gigajoule), and the improved technology was based on reducing the cost to about $37 per tonne ($2.00 per gigajoule). The former price is based on prices typical of wood-fired boilers and other operations. The future price is the goal of the Biomass Production Program of the U.S. Department of Energy for production of biomass on energy plantations. A key goal of this technology is to increase biomass productivities to about 20 tonnes per hectare per year.

The current projected selling price of $11.06 to $15.36 per gigajoule ($0.23 to $0.32 per liter) is for engineering designs based on data produced at the bench scale and information on commercial corn ethanol processes. Because significant portions of the process are similar to the production of ethanol from starch crops such as corn, there is good evidence that major sections of cellulosic conversion should perform as predicted from bench-scale experience. In addition, pilot plants for converting biomass to ethanol have been successfully operated in the past, although the level of development was inadequate to achieve economic viability. Nevertheless, before commercialization, the technology must be demonstrated at a large enough scale to gain experience with key items of equipment and gather accurate material and energy balance data. In any such scale-up, problems are to be expected with the selection of equipment, although the largest challenge is the handling of the viscous solids suspensions that must be processed. However, cellulosic biomass slurries are successfully used in the pulp and paper and other industries, so these problems should not be insurmountable. The process should be scaled-up as soon as possible to determine its performance and define areas where R&D is required to ensure that the process will operate as designed. Larger scale operation is particularly needed to verify energy requirements for such operations as size reduction and mixing, as well as to establish the effects of full material balance integration and mixing hydrodynamics.

Waiting for such a pilot plant until all the research targets are met would result in delays in final commercialization of the technology due to unanticipated operational problems and other issues measurable only at a larger scale. In addition, the pilot plant would provide an opportunity to prove the technology with low-cost feedstocks (such as cellulosic waste streams) once confidence is gained in the operation of the process. Technological improvements can then be integrated into the pilot unit as they become ready for commercial application. Currently, NREL, with funding from the Biofuels Systems Division of the U.S. Department of Energy, is undertaking operation of a pilot plant for ethanol production from cellulosic biomass.

The size and cost of scale-up units are dictated by certainty in process performance, the risk one is willing to accept, the type and cost of the feedstock, and the time frame desired to reach commercial applications. The smallest possible pilot plant [probably in the range of 1 tpd of feedstock] should be built first to provide the data needed so that larger plants can be built with confidence. Following successful operation of this unit, a large demonstration process at the scale of about 50 dry tpd might be built, followed by a commercial unit at the scale of approximately 2,000 dry tpd.

Methanol from biomass

In this section, preliminary economics for methanol production from biomass are presented. The costs of production are presented for a commercial coal gasifier that is adapted to biomass, and for gasifiers currently being developed in the United States specifically for biomass. In a Chem Systems study [92] carried out for NREL, preliminary economics were developed for two systems: one based on the Koppers–Totzek gasifier and one on the IGT "Renugas" gasifier. In addition, preliminary economics have been developed by NREL for the BCL gasifier [93].

Capital and operating cost estimates for the four systems (LPO, HPO, IND, and IND2) are shown in table 10. Costs for the indirect systems, IND and IND2, were based on gasifier costs presented separately [105], in conjunction with reformer and methanol synthesis costs based on Chem Systems estimates. Feed costs are for 2.5 centimeter chips with 50 percent moisture content delivered at the plant gate. Yields and energy balances were estimated using the Aspen/SP [100] process simulator. Plant size was for bimass input of about 1,814 dtpsd, and reported in table 10 in terms of tonnes per stream day of methanol product. An *n*th plant estimate for a 9,090 dtpsd biomass plant is presented for IND2. The cost for the large plant was estimated from the smaller 1,814 dtpsd IND2 plant using a 0.7 scaling factor and parallel trains for the gasifier and methanol synthesis. There is no "learning curve" factor used to reflect increased reliability, efficiency, or process improvements. For the 1,814 dtpsd plants, costs range from $158.19 million for the IND2 plant to $321.73 million for the LPO plant.

Table 11 presents a typical cash-flow analysis carried out to determine the cost of production (COP) of methanol (pre-tax basis) from biomass using the IND gasifier. Table 12 presents summary COP values for methanol production. Values (1989 U.S. $per gigajoule of methanol) are presented at 6 and 12 percent discount rates and are divided into capital, feed and operating and maintenance costs. At a 12 percent discount rate the COP ranges from $7.65 per gigajoule of

Table 10: Capital and operating costs for producing methanol[a]

	Biomass LPO[b]	Biomass HPO[b]	Biomass IND[c]	Biomass IND2[c]	Biomass IND2[c]	Natural gas[d,f]	Coal[d]
Plant size *tpsd*[e] *methanol*	790	920	1,110	1,110	5,550	2,500	5,000
Feed price *1989 U.S. $per GJ (LHV)*	2.45	2.45	2.45	2.45	2.00	1.90	1.42
Capital cost *millions of 1989 U.S. $*							
Feed handling, preparation	37.95	16.95	20.34	20.34	65.89		73.55
Oxygen plant	49.46	42.96					122.59
Gasification	96.69	29.10	7.93	7.93	25.69		129.95
Gas conditioning	10.77	0.75	1.67	12.79	41.42		
Compression	0.75	8.10	25.72	25.58	82.98	23.77	122.59
Shift reaction/heat exchange							122.59
Reforming		31.66	41.32			80.44	
Acid gas removal/cooling	10.98	13.75	11.64	11.64	26.90		46.59
Sulfur recovery							39.23
Methanol synthesis/purification	21.21	23.99	32.23	32.23	109.76	55.57	57.62
Utilities/auxiliaries	56.92	41.79	35.18	27.63	107.74	68.57	226.79
Overhead, E&C, if listed separately							286.62
Erected plant cost	284.73	209.04	176.03	138.14	460.36	228.55	1,105.53
Owner's costs, fees, and profit	24.87	20.90	14.67	11.51	46.04	110.55	110.55
Catalysts and chemicals						6.40	32.00
Land	2.13	2.13	2.13	2.13	13.50	5.33	10.66
Start-up	6.40	6.40	6.40	6.40	24.85	4.26	32.00

Table 10: (cont.)

	Biomass LPO[b]	Biomass HPO[b]	Biomass IND[c]	Biomass IND2[c]	Biomass IND2[c]	Natural gas[d,f]	Coal[d]
Total capital investment	321.73	238.48	199.22	158.19	544.77	244.54	1,290.76
Working capital	15.40	13.00	15.00	12.00	60.00	12.05	53.10
Operating costs *millions of 1989 U.S. $ per year*							
Variable costs							
Feedstock	28.50	28.10	28.00	28.00	113.30	55.56	101.12
Catalyst and chemicals	0.51	1.78	2.67	2.67	12.09	2.95	11.77
By-product credits							(8.89)
Utilities	2.12	3.25	5.52	5.52	25.01	0.00	4.11
Fixed costs							
Labor	0.92	0.92	1.09	1.09	5.10	0.82	4.42
Maintenance	7.67	5.69	4.74	4.03	14.10	9.97	26.84
General overhead	5.61	4.31	3.85	3.28	9.12	7.10	20.55
Direct overhead	0.42	0.42	0.49	0.49	2.65	0.42	2.07
Insurance	1.61	1.19	0.98	0.82	2.31	1.26	6.45
Total operating cost	47.36	45.66	47.34	45.90	183.68	78.08	168.44

a. LPO = low-pressure oxygen gasifier; HPO = high-pressure oxygen gasifier; IND = low-pressure indirect gasifier; IND2 = low-pressure indirect gasifier with hot-gas conditioning.

b. Costs for the LPO and HPO systems were taken from Chem Systems [92] as 1987 U.S. dollars and updated to 1989 U.S. dollars using the Nelson–Farrar index [99]. The Nelson–Farrar index was used for methanol because most plant unit operations (for example, reforming and methanol synthesis) are considered to be refining operations.

c. [93].

d. [104].

e. Tonnes per stream day.

f. Costs for a 2,500 tpd methanol plant located on the U.S. Gulf Coast have been converted to 1989 dollars using the Nelson–Farrar index.

Table 11: Cash flow (excluding taxes) for methanol production
millions of 1989 U.S. $

Year	Methanol revenue	By-product credit	Capital	Working capital	Feed cost	Fixed op costs	Variable op costs	Total costs	Pretax cash flow	Cumulative pretax cash flow
1			59.766					59.766	(59.766)	(59.766)
2			99.610					99.610	(99.610)	(159.376)
3			39.844					39.844	(39.844)	(199.220)
4	40.949	0.000		15.000	16.800	11.150	4.914	47.864	(6.915)	(206.315)
5	54.598	0.000			22.400	11.150	6.552	40.102	14.496	(191.639)
6	68.248	0.000			28.000	11.150	8.190	47.340	20.908	(170.731)
7	68.248	0.000			28.000	11.150	8.190	47.340	20.908	(149.824)
8	68.248	0.000			28.000	11.150	8.190	47.340	20.908	(128.916)
9	68.248	0.000			28.000	11.150	8.190	47.340	20.908	(108.008)
10	68.248	0.000			28.000	11.150	8.190	47.340	20.908	(87.101)
11	68.248	0.000			28.000	11.150	8.190	47.340	20.908	(66.193)
12	68.248	0.000			28.000	11.150	8.190	47.340	20.908	(45.285)
13	68.248	0.000			28.000	11.150	8.190	47.340	20.908	(24.377)
14	68.248	0.000			28.000	11.150	8.190	47.340	20.908	(3.470)
15	68.248	0.000			28.000	11.150	8.190	47.340	20.908	17.438
16	68.248	0.000			28.000	11.150	8.190	47.340	20.908	38.346
17	68.248	0.000			28.000	11.150	8.190	47.340	20.908	59.254
18	68.248	0.000			28.000	11.150	8.190	47.340	20.908	80.161
19	68.248	0.000			28.000	11.150	8.190	47.340	20.908	101.069
20	68.248	0.000			28.000	11.150	8.190	47.340	20.908	121.977
21	68.248	0.000			28.000	11.150	8.190	47.340	20.908	142.885
22	68.248	0.000			28.000	11.150	8.190	47.340	20.908	163.792
23	68.248	0.000		(15.000)	28.000	11.150	8.190	32.340	35.908	199.700

methanol for the large IND2 system to $19.60 per gigajoule of methanol for the LPO case.

Methanol from natural gas

More than 80 percent of the methanol commercially manufactured today is produced by natural gas reforming followed by methanol synthesis. The economics of methanol production from natural gas have been estimated, based upon a Chem Systems report published in 1989 [104], for comparison with the biomass cases. Capital and operating costs are shown in table 10, and cost of production in table 12. The COP is $6.24 per gigajoule of methanol at a 6 percent discount rate and $7.27 per gigajoule of methanol at a 12 percent discount rate. These COP values are lower than COPs for the smaller-scale biomass systems, and a little higher than a conceptual *n*th plant biomass system COP. Large-scale, 4,000 to 5,000 tpd, biomass-to-methanol plants have the potential to compete with world-scale natural-gas–based plants in areas that have natural gas costs comparable to those in the United States.

The impact of plant size on COP for the processes studied is shown in figure 5. Although the natural gas COP is comparable to the COP of a biomass IND2 system at half the scale of operation, no learning curve improvements have been included in the biomass plant estimates.

Process economics includes a natural gas feed cost of $1.90 per gigajoule ($2.00 per million Btu). This feed cost is considered representative of present-day natural gas costs in the United States and is used in most U.S. evaluation studies. Process economics is greatly influenced by feed costs, since natural gas represents approximately 44 percent of the COP at the above cost. Recent projections [106] of natural gas prices in the United States show the wellhead price increasing from $1.73 per gigajoule in 1990 to $3.58 per gigajoule in 2000, and to $5.77 per gigajoule in 2010. Much of the world's future methanol production will be located in remote locations having large amounts of natural gas and small domestic markets. Areas such as the Middle East, South America, and Southeast Asia are considered significant methanol production areas [19]. Low fuel costs in these areas are partially offset by increased transportation costs for their delivery

---

Table 11: notes

Process: biomass-indirect	Basis: constant 1989 U.S. $
Capital cost: 199.22 million $	No escalation
Methanol COP: $9.25 per gigajoule	Three-year investment, 15 year operating life
Production rate: 1,110 tpsd	Investment schedule: year 1, 30 percent
370,000 tonnes per year	year 2, 50 percent
7.37 × 10^6 gigajoules per year	year 3, 20 percent
Feed cost: 28 million $per year	Production schedule: year 4, 60 percent of nameplate
Discount rate: 6.00 percent	year 5, 80 percent of nameplate
	years 6–18, 100 percent of nameplate

Table 12: Pretax production cost of methanol
1989 U.S.$ per gigajoule of methanol

Process	Capital	Feed	O&M	Total
COP at 6 percent discount rate				
Biomass: LPO[a]	6.35	5.43	3.56	15.34
790 tpsd of methanol				
Biomass: HPO[b]	4.07	4.59	2.85	11.51
920 tpsd of methanol				
Biomass: IND[c]	2.85	3.79	2.61	9.25
1,110 tpsd of methanol				
Biomass: IND2[d]	2.27	3.79	2.42	8.48
1,110 tpsd of methanol				
Biomass: IND2	1.59	3.07	1.90	6.56
5,550 tpsd of methanol				
Natural gas	1.60	3.34	1.30	6.24
2,500 tpsd of methanol				
Coal	4.01	3.04	2.00	9.05
5,000 tpsd of methanol				
COP at 12 percent discount rate				
Biomass: LPO	10.62	5.43	3.55	19.60
790 tpsd of methanol				
Biomass: HPO	6.81	4.59	2.84	14.24
920 tpsd of methanol				
Biomass: IND	4.78	3.79	2.60	11.17
1,110 tpsd of methanol				
Biomass:IND2	3.81	3.79	2.41	10.01
1,110 tpsd of methanol				
Biomass: IND2	2.68	3.07	1.90	7.65
5,550 tpsd of methanol				
Natural gas	2.68	3.34	1.30	7.32
2,500 tpsd of methanol				
Coal	6.70	3.04	1.99	11.73
5,000 tpsd of methanol				

a. LPO = low-pressure oxygen gasifier.
b. HPO = high-pressure oxygen gasifier.
c. IND = low-pressure indirect gasifier.
d. IND2 = low-pressure indirect gasifier with hot-gas conditioning.

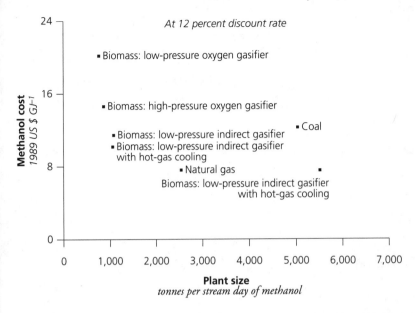

At 12 percent discount rate

FIGURE 5: *Methanol cost versus plant size.*

to major markets. For example, the fuel-plus-transportation cost estimates in table 13 were projected as part of methanol costs.

Capital and operating costs also vary somewhat, about ±15 percent, depending on the location, but the major variables for methanol production from natural gas are fuel and transportation costs. Methanol produced from natural gas in remote regions can often be delivered to areas such as the U.S. Gulf Coast at very competitive prices.

Methanol from coal

Systems are also being proposed for methanol production from coal. Comparative costs for a second-generation coal gasification–based system using a commercial Texaco gasifier have been developed by Chem Systems [88]. Second-generation coal gasification assumes operation at high pressure. Capital and operating costs are shown in table 10, and the cost of production is shown in table 12. The COP for a coal-based system at 5,000 tpd of methanol is $9.53 per gigajoule of methanol at a 12 percent discount rate; this is comparable to an indirect biomass system at the 1,110 tpd methanol size, despite the assumption that biomass feedstock is 65 percent more costly than coal. Because of added costs, coal systems will have to operate at a larger scale than biomass-based systems to obtain comparable economic returns. For example, for comparable 5,000 tpd methanol plants, the coal-based system has acid gas/sulfur recovery costs of $86 million, whereas the biomass system has acid gas removal costs of $27 million.

Table 13: Fuel-plus-transportation cost estimates[a]
$per gigajoule of methanol

Location of production	Natural gas	Transportation	Natural gas plus transportation
Venezuela	0.26	1.00	1.26
Saudi Arabia	0.26	1.50	1.76
Southeast Asia	1.90	1.75	3.65

a. FOB U.S. Gulf Coast, 1991 U.S. $ [103].

ENERGY BALANCES AND CARBON DIOXIDE

Ethanol from biomass

Energy use for ethanol production can be measured by subtracting energy inputs from the energy produced, and dividing this difference by the energy content of the feedstock:

$$\text{Energy Ratio} = \frac{\text{Energy in products} - \text{External energy inputs}}{\text{Energy in feedstock}}$$

See table 14 for the energy balance information and energy ratios for ethanol produced from corn and cellulosic biomass according to this definition.

For corn-to-ethanol plants in the United States, steam is generally provided by the combustion of coal. Steam usage is reported to be from 11.1 to 16.7 megajoules per liter of ethanol produced, with the lower values typical of modern plants [9, 109]. Electrical power requirements are around 0.95 megajoules per liter (1.0 kWh per gallon) produced [109]. These values are for medium- to large-size plants, with capacities of approximately 113 million liters per year (30 million gallons per year) or more.

For corn, only the starch is considered in the feedstock energy content in table 14 because the remaining portion of the kernel is sold as animal and human food and not used as fuel. On this basis, the energy ratio for ethanol from corn is about 27 percent based on modern processing technology. If the total energy content of the corn kernel is included, efficiencies drop to about 70 percent of those shown in table 14. Adding the amount of energy required to plant, fertilize, cultivate, and harvest the corn lowers the energy ratio further to between −13 and 1 percent because about one-third to more than half as much fossil energy is required in fuel and fertilizer to produce corn as can be provided by the ethanol produced [22].

The first column in table 14 assigns all of the energy to ethanol production and does not deduct energy for drying the feed coproducts resulting from ethanol production from corn. Several approaches have been employed in the past to account for the use of energy to produce coproducts, such as assigning energy based on the value of the coproducts [110] or on the energy required to produce the equivalent amount of protein from soybeans [107]. The latter approach is taken here because it considers the alternative to produce the same protein content an-

Table 14: Ethanol production energy ratio[a]

Process	Corn ethanol only	Corn energy to coproducts[b]	Cellulosic biomass reference case	Cellulosic biomass improved technology
Ethanol *liters per tonne*	440–458	440–458	338[c]	497[c]
Ethanol LHV *megajoules per liter*	21.2–111	21.2	21.2	21.2
Feedstock LHV *gigajoules per tonne*	16.2[d]	16.2[d]	18.87	18.87
Energy ratio for ethanol *percent*	26.5–27.5[e]	36.8–38.3[e]	38.0	55.8
Electricity *gigajoules per tonne feed*	–	–	0.658	0.365
Energy ratio for ethanol plus electricity *@ 3.6 MJ per kWh*	–	–	41.5	57.8
Energy ratio for ethanol plus electricity *@ 10.8 MJ per kWh*	–	–	48.4	61.6
Energy ratio including energy inputs for biomass production	−13.3 – +0.6	−2.5 – +11.4	41.6–43.5	54.8–56.7

a. Defined as energy produced minus fossil energy inputs divided by biomass energy content.

b. Energy assigned to coproducts based on energy required to produce soybeans and energy value of fuel oils @ 2.8 megajoules per liter ethanol [107].

c. Ethanol only after subtracting water and denaturant (gasoline) from product.

d. [78].

e. Corn energy ratios are based on the energy content of starch only. Other components of the corn kernel are sold as animal and human food. Average starch content is 73.4 percent [108].

imal feed. For this case the energy ratio increases to about 37 percent for conversion only, as shown by the second column in table 14. Including the energy required to grow and harvest the corn gives an energy ratio of –2 to 11 percent.

When ethanol is produced from corn, the total energy input from all fossil energy sources is approximately equal to the energy content of the ethanol produced, and the net amount of carbon released from fossil sources approximates that of gasoline [110, 107]. However, about 1 tonne of corn stover[4] is produced per tonne of corn harvested [108]. Burning about 40 percent of the stovers could displace all the coal used to produce ethanol via a modern, efficient process. The corn stover requirements could be reduced below that level by careful heat integration of the process. Except in locations where most of the residues are left on the field to control erosion, burning corn residues to provide process heat would thus be a viable alternative to coal. Because coal requirements are equivalent to about two-thirds of the lower heating value of the ethanol produced, this change would have a significant impact on the carbon dioxide buildup in the atmosphere attributable to ethanol from corn. Yet the input of fossil fuels in the form of liquid fuels and fertilizers for producing corn (one-third to well over one-half the lower heating value of the poduced ethanol) cannot be so easily reduced.

In the current conceptual design of a plant for enzyme-based cellulosic biomass conversion to ethanol [97], steam is provided by the combustion of lignin and waste organics, and power requirements are provided by cogeneration. The estimated total steam energy consumed is 9.2 megajoules per liter. The estimated total electricity produced for a plant size of 219 million liters per year of denatured fuel is 36.1 megawatts-electric (MW_e), of which 22.7 MW_e or 2.9 megajoules per liter (3.14 kWh per gallon) is consumed by the plant. The remaining 13.4 megawatts are sold [97].

For production of ethanol from cellulosic biomass with current technology, the estimated energy ratio is 38 percent when ethanol is the only product considered. Additional energy output results from the use of cellulosic biomass: electricity from the excess steam generated by burning the lignin and other unused substrate. Including electricity as an energy output increases the energy ratio to almost 42 percent. If electricity is counted at three times its energy content, to reflect the fuel needed at a central-station thermal-electric power plant to produce this much electricity, the ratio increases to more than 48 percent. The equivalent of only about 13 to 18 percent of the ethanol energy content is required to grow cellulosic biomass [22]. Including this energy requirement decreases the energy ratio for cellulosic biomass less than for corn, to 42 to 43 percent for the reference case.

For the improved technology case, the energy ratio for ethanol production is 55.8 percent when no credit is taken for by-product electricity. Projected advances in several areas account for the improved energy yield. One is a higher carbohydrate content in the feedstock (77.2 versus 70.2 percent in the base case). In

---

4. Stover is the field residue from growing corn.

addition, higher yields of ethanol are expected as a result of improvements in enzymes and fermentative organisms and retention of these biocatalysts to reduce loss of yield to biocatalyst production. Reductions in xylose and cellulose fermentation times, as well as cellulase production requirements and times, result in further decreases in power for mixing and other process steps, since substantially fewer vessels are needed. Projected increases in ethanol concentrations in the fermentations reduce the energy requirements for distillation as well. Overall, the result is a higher ethanol yield obtained with less process energy. Thus, although there is a slight drop in electricity exported from the plant, the overall ethanol plus electricity energy ratio is 57.8 percent, or 61.7 percent if the thermal equivalent of the electricity is included. When the energy needed for growth of cellulosic biomass is included, the energy ratio is still about 55 percent for the improved case. If process heat integration is improved, these ratios could be increased.

Production of ethanol from cellulosic biomass realizes two benefits regarding potential global climate change. First, burning the lignin that is left after production of ethanol provides enough energy to run the entire conversion process and leaves a surplus for export as heat or electricity, as discussed above [97]. Thus, all of the energy indicated in table 14 is derived from the lignin. And second, modest energy inputs are needed to grow and harvest cellulosic biomass (about 13 to 18 percent of the lower heating value of ethanol). The result is that about five times more energy is produced in the form of ethanol and heat than is required in fossil fuel inputs for feedstock growth, harvesting, conversion, and the distribution system [22]. If the modest energy requirements for feedstock growth and harvesting were also provided by renewable fuels, net carbon dioxide emissions to the atmosphere would be reduced further.

Methanol from cellulosic biomass, natural gas, and coal

Overall process energy balances are shown in table 15 for the methanol production processes described above for cellulosic biomass, natural gas, and coal, based on published reports for biomass [92] and for fossil fuels [104]. In addition, the process energy balance for indirect gasification has been estimated [93].

Overall process energy ratios (defined as above for ethanol) for biomass systems range from 40 to 53 percent. These efficiencies compare with about 60 percent for a natural gas-based system and 51 percent for a coal-based system. Carbon conversion efficiencies reflect the relative hydrogen/carbon ratios of the individual feedstocks and range from a low of 33 percent for a LPO-biomass based system to 79 percent for a natural gas-based system. Process yields reflect the carbon/hydrogen contents of the feeds and process efficiency, and range from 0.64 tonnes of feed per tonne of product for a natural gas-based system to 2.30 tonnes of feed per tonne of product for a LPO-biomass based system.

PERSPECTIVES ON PRODUCTION OF ALCOHOLS FROM BIOMASS

With potential improvements, production costs for ethanol and methanol from biomass are projected to be lower than those for methanol from coal. Furthermore, the investments required for plants that would produce alcohols from renewable feedstocks would be only about one-third to one-tenth of those for coal–methanol plants. A reduction in capital requirements of this magnitude

Table15: Methanol production energy balances

Process	Biomass: LPO[a]	Biomass: HPO[a]	Biomass: IND[b]	Natural gas[c]	Coal[c]
Methanol *tonnes per day*	790	920	1,110	2,500	5,000
Tonnes feed per tonne of methanol	2.30	1.97	1.63	0.64	1.76
Feed LHV *gigajoules per tonne*	18.87	18.87	18.87	52.25	24.75
External power required[d] *gigajoules per tonne feed*	0.84	1.29	1.94	e	e
Methanol LHV[f] *gigajoules per tonne*	19.92	19.92	19.92	19.92	19.92
Energy ratio *percent*	40.3	50.8	53.4	59.6	50.6
Carbon use *percent*	32.6	38.0	45.9	79.2	ND

a. [92].

b. [93].

c. [104].

d. External power reported assuming 33 percent electrical generation efficiency.

e. No external power required.

f. LHV methanol liquid.

could facilitate financing for biomass plants and allow smaller firms to commercialize the technology.

Because the costs of producing ethanol and methanol from biomass are projected to be quite similar if the technology for each can be improved to meet the goals identified, factors other than price may dictate the choice between alcohols. As mentioned before, engines with fuel-line sensors could be designed to accommodate both fuels.

One advantage offered by methanol is that much of the technology needed for biomass-derived methanol can be readily adapted from technology already developed for making methanol from natural gas and coal. Moreover, methanol production technology is much less sensitive to feedstock composition than ethanol technology and may thus be preferred in places where good ethanol feedstocks are not readily available. Methanol might also be favored because it is easier to reform than ethanol and because reforming technology is more advanced for methanol than for ethanol, making methanol a better candidate for fuel cell applications, at least in the near term. However, the fact that methanol can be produced from coal at a cost comparable to that for biomass-derived methanol poses the risk that coal might one day be the feedstock of choice for methanol production if land-use constraints should drive the cost of biomass feedstocks out of the competitive range. Such an outcome would be undesirable from the perspective of global climate change.

Ethanol may be preferred because of its low toxicity, although a denaturant would undoubtedly have to be added to prevent people from drinking it. Ethanol production also employs predominantly natural materials such as the biomass itself, proteins (enzymes), and yeast, and selectivity to target products is very high. Thus, the environmental impacts associated with its manufacture should be very low. Because wet cellulosic waste streams are well suited for conversion into ethanol, the need to dispose of such wastes could provide a near-term niche for ethanol production technology. Furthermore, ethanol production may well continue to benefit from the ongoing advances in biotechnology to realize even greater price reductions than presented herein.

CONCLUSIONS

As alternatives to petroleum-based fuels, alcohols from biomass can be blended with gasoline or used as neat fuels. The use of alcohols produces less air pollution than gasoline and helps solve local air quality problems. If alcohol-fuel cells can be successfully developed, it will be possible to achieve much lower emissions than for internal combustion engines. Moreover, the greater efficiencies of fuel cells would reduce the fuel use and associated land needed to grow biomass. When made from biomass that is subsequently regrown, alcohols contribute no net carbon dioxide to the atmosphere during their use cycle. This characteristic makes them particularly well suited to a world in which carbon dioxide emissions are limited. With new technologies currently being developed, both ethanol and methanol from biomass have the potential to be cost-competitive with gasoline.

Both acid- and enzyme-catalyzed reactions have been evaluated for conversion of cellulosic biomass into ethanol. Research has been focused on enzymatic hydrolysis technology because of its potential to achieve high yields under mild conditions. In particular, the ssf process is favored for ethanol production from the major cellulose fraction of the feedstock because of its low cost potential. Technology has also been developed for converting the second largest fraction, hemicellulose, into ethanol; the remaining lignin can be burned as boiler fuel to power the conversion process and generate extra electricity for export. Together, developments in conversion technology have reduced the selling price of ethanol from about $45 per gigajoule ($0.95 per liter) ten years ago to only about $13 per gigajoule($0.28 per liter) (at 12 percent discount rate) today. Additional technical targets have been identified to render ethanol competitive with gasoline produced from $25 per barrel oil, if an aggressive research and development program is followed.

Methanol can be produced from cellulosic biomass in the near-term with oxygen-blown pressurized fluidized-bed gasifiers and natural gas-derived reformer and methanol synthesis technology. The cost for this option is estimated to be $14.66 per gigajoule (12 percent discount rate) at 920 tonnes per day. Technical targets, including indirect gasification to eliminate the oxygen plant investment and to increase yields of syngas, coupled with new gas-conditioning technology, have been identified. Achieving these targets would make biomass derived methanol competitive with gasoline derived from $25 per barrel crude oil.

WORKS CITED

1. U.S. Department of Energy. 1990
   Energy Information Administration, *Annual Energy Review 1989*, Government Printing Office, Washington DC.

2. Wald, M.L. 1990
   Greater reliance on foreign oil feared as U.S. output tumbles, *New York Times*, 18 January.

3. U.S. Environmental Protection Agency. 1989
   EPA Lists Places Failing to Meet Ozone or Carbon Monoxide Standards, Office of Public Affairs, Washington DC.

4. Intergovernmental Panel on Climate Change. 1990
   J.T. Houghton, G.J. Jenkins, and J.J. Ephraums, eds., *Climate change—the IPCC scientific assessment*, Cambridge University Press, Cambridge, United Kingdom.

5. Wyman, C.E. and Hinman, N.D. 1990
   Ethanol: fundamentals of production from renewable feedstocks and use as a transportation fuel, *Appl. Biochem. Biotechnol.* 24/25:735.

6. Lynd, L.R. 1989
   Production of ethanol from lignocellulose using thermophilic bacteria: critical evaluation and review, *Adv. in Biochem. Eng./Biotechnol.* 38:1.

7. Lynd, L.R. 1990
Large-scale fuel ethanol from lignocellulose: potential, economics, and research priorities, *Appl. Biochem. Biotechnol.* 24/25:695.

8. Interlaboratory Report. 1990
The potential of renewable energy: an interlaboratory white paper, SERI/TP-260-3674, National Renewable Energy Laboratory (formerly Solar Energy Research Institute), Golden, Colorado.

9. U.S. Department of Agriculture. 1987
Fuel ethanol cost-effectiveness study, National Advisory Panel on Cost-Effectiveness of Fuel Ethanol Production, Washington DC.

10. Livo, K.B. and Gallagher, J. 1989
Environmental influence of oxygenates, presented at the AIChE national meeting, San Francisco, California.

11. Johnson, L. 1989
Vehicle performance and air quality issues of 10% ethanol and blends, presented at the 24th Intersociety Energy Conversion Engineering Conference, Washington DC.

12. Watson, E.B. 1988
Alternative fuels for cars and light trucks: an introduction to oxygenated fuels, American Society of Mechanical Engineers, state government technical brief, Washington DC.

13. Prezelj, M. 1987
Pool octanes via oxygenates, *Hydrocarbon processing,* Gulf Publishing, Houston, Texas.

14. Gary, J.H. and Handwerk, G.E. 1984
Petroleum refining technology and economics, second edition, Marcell Dekker, New York.

15. Veralin, C.H. 1989
Methanol growth is equivalent to one worldscale plant a year, *Hydrocarbon processing,* Gulf Publishing, Houston, Texas.

16. Ludlow, W.I. 1989
MTBE—a practical private sector route to clean fuels, presented at the World Methanol Conference, Houston, Texas.

17. Pahl, R.H. 1988
Motor fuel and automotive technology development providing cleaner air for all Americans, testimony before Joint Congressional Briefing, Washington DC.

18. Anderson, E. 1988
Ethyl tert-butyl ether shows promise as octane enhancer, *Chem. Eng. News* 65(43):11.

19. Crocco, J.R. 1989
Putting the future of methanol in proper perspective, presented at the World Methanol Conference, Houston, Texas.

20. Sperling, D. 1988
New transportation fuels: a strategic approach to technological change, University of California Press, Berkeley, California.

21. Gray, C.L. and Alson, J.A. 1989
The case for methanol, *Scientific American* 261(5):108.

22. Lynd, L.R., Cushman, J.H., Nichols, R.J., and Wyman, C.E. 1991
Fuel ethanol from cellulosic biomass, *Science* 251:1318.

23. Appleby, A.I. and Foulkes, F.R. 1989
Fuel cell handbook, Van Nostrand Reinhold, New York, 177–190, 236–238.

24. Larson, E.D. and Katofsky, R.E. 1992
Production of hydrogen and methanol via biomass gasification, in *Advances in Thermochemical Biomass Conversion,* Elsevier Applied Science, London (forthcoming).

25. Lemons, R.A. 1990
Fuel cells for transportation, *Journal of Power Sources* 29:251.

26. Srinivasan, S. 1987
Fuel cells for extraterrestrial and terrestrial applications, report LA-UR-87-2393, Los Alamos National Laboratory, Los Alamos, New Mexico.

27. Fritts, S.D. and Sen, R.K. 1988
Assessment of methanol electro-oxidation for direct methanol–air fuel cells, report PNL-6077, Battelle Pacific Northwest Laboratory, Richland, Washington.

28. Kumar, R., Ahmed, S., Krumpelt, M., and Myles, K.M. 1990
Methanol reformers for fuel cell powered vehicles: some design considerations, fuel cell seminar, Phoenix, Arizona, November 26–28.

29. Wilson, M.S., Springer, T.E., Zawodzinski, T.A., and Gottesfeld, S. 1991
Recent achievements in polymer electrolyte fuel cell (PEFC) research at Los Alamos National Laboratory, presented at the 26th IECEC, Boston, August,.

30. Srinivasan, S., Enayetullah, M.A., Somasundaram, S., Swan, D.H., Manko, D., Koch, H., and Appleby, A.J. 1989
Recent advances in solid polymer electrolyte fuel cell technology with low platinum loading electrodes, in W.D. Jackson, ed., *Proceedings of the 24th Intersociety Energy Conversion Engineering Conference,* volume 3.

31. Prater, K. 1990
The renaissance of the solid polymer fuel cell, *Journal of Power Sources* 29:239–250.

32. Kevala, R.J. 1990
Development of a liquid-cooled phosphoric acid fuel cell/battery power plant for transit bus applications, in P.A. Nelson, W.W. Schertz, and R.H. Till, eds., *Proceedings of the 25th Intersociety Energy Conversion Engineering Conference,* volume 3, 297–300, American Institute of Chemical Engineers, New York.

33. Romano, S. 1990
The DOE/DOT fuel cell bus program and its applications to transit missions, in P.A. Nelson, W.W. Schertz, and R.H. Till, eds., *Proceedings of the 25th Intersociety Energy Conversion Engineering Conference,* volume 3, 293–296, American Institute of Chemical Engineers, New York.

34. Prater, K. 1991
Testimony on behalf of Ballard Power Systems, Inc., at the hearing on S. 1269, Renewable Hydrogen Energy Research and Development Act of 1991, United States Senate, Committee on Energy and Natural Resources, Subcommittee on Energy Research and Development, 25 June.

35. Tosh, J.D., Stulsas, A.F., Buckingham, J.P., Russel, J.A., and Cuellar, J.P. Jr. 1985
Project for reliability fleet testing of alcohol/gasoline blends, final report DOE/CE/50004-1, U.S. Department of Energy, Washington DC.

36. *New Fuels Report.* 1991
Inside Washington Publishers, Washington DC, 22 April.

37. Piel, W.J. 1988
MTBE—the refiner's key to future gasoline production, National Petroleum Refiners' Association annual meeting, San Antonio, Texas.

38. Grayson, M., ed. 1984
Encyclopedia of chemical technology: alcohol fuels to toxicology, (supplement), John Wiley, New York.

39. *Chemical Marketing Reporter.* 1990
Schnell Publishing Co., New York, 2 September.

40. *Chemical Marketing Reporter.* 1991
Schnell Publishing Co., New York, 28 January.

41. *New Fuels Report.* 1990
Inside Washington Publishers, Washington DC, 2 September.

42. *New Fuels Report.* 1991
Inside Washington Publishers, Washington DC, 28 January.

43. U.S. Department of Agriculture. 1989
Ethanol's role in clean air, USDA Backgrounder Series, U.S. Department of Agriculture, Washington DC.

44. Wenzl, H.F. 1970
The chemical technology of wood, Academic Press, New York.

45. Wright, J.D. 1983
High temperature acid hydrolysis of cellulose for alcohol fuel production, SERI/TR-231-1714, National Renewable Energy Laboratory (formerly Solar Energy Research Institute), Golden, Colorado.

46. Wright, J.D. 1988
Ethanol from lignocellulose: an overview, *Energy Progress* 8(2):71.

47. Wright, J.D. and Power, A.J. 1985
Concentrated acid hydrolysis processes for alcohol fuel production, *Biotechnol. Bioeng. Symp.* 15:511.

48. Wright, J.D., Power, A.J., and Bergeron, P.W. 1985
Evaluation of concentrated halogen acid hydrolysis processes for alcohol fuel production. SERI/TR-232-2386, National Renewable Energy Laboratory (formerly Solar Energy Research Institute), Golden, Colorado.

49. Reese, E.T. 1976
History of the cellulase program at the U.S. Army Natick Development Center, *Biotechnol. Bioeng. Symp.* 6:9–20.

50. Wright, J.D. 1988
Ethanol from biomass by enzymatic hydrolysis, *Chem. Eng. Prog.* 84(8):62–74.

51. Torget, R., Werdene, P., Himmel, M., and Grohmann, K. 1990
Dilute acid pretreatment of short rotation woody and herbaceous crops, *Appl. Biochem. Biotechnol.* 24/25:115.

52. Mandels, M.L., Hontz, L., and Nystrom, J. 1974
Enzymatic hydrolysis of waste cellulose, *Biotechnol. Bioeng.* 16:1471.

53. Wilke, C.R., Yang, R.D., and Stockar, U.V. 1976
Preliminary cost analysis for enzymatic hydrolysis of newsprint, *Biotechnol. Bioeng. Symp. Ser.* 6:155.

54. Gauss, W.F., Suzuki, S., and Takagi, M. 1976
Manufacture of alcohol from cellulosic materials using plural ferments, U.S. patent 3,990,944, November 9.

55. Takagi, M., Abe, S., Suzuki, S., Evert, G.H., and Yata, N. 1977
A method of production of alcohol directly from yeast, *Proc. Bioconv. Symp.* Delhi: IIT.

56. Veldhuis, M.K., Christensen, L.M., and Fulmer, E.I. 1936
Production of ethanol by thermophilic fermentation of cellulose, *Ind. Eng. Chem.* 28:430.

57. Ng, T.K., Weimer, P.J., and Zeikus, J.G. 1977
Cellulolytic and physiological properties of *Clostridum thermocellum,* *Arch. Microbiol.* 114:1.

58. Cooney, C.L., Wang, D.I.C., Wang, S.D., Gordon, J., and Jiminez, M. 1978
    Simultaneous cellulose hydrolysis and ethanol production by cellulolytic anaerobic bacterium, *Biotechnol. Bioeng. Symp. Ser.* 8:103.

59. Wright, J.D., Wyman, C.E., and Grohmann, K. 1988
    Simultaneous saccharification and fermentation of lignocellulose: process evaluation, *Appl. Biochem. Biotechnol.* 18:75.

60. Gaines, L.L. and Karpuk, M. 1987
    Fermentation of lignocellulosic feedstocks: product markets and values, in D.L. Klass, ed., *Proceedings of the Energy from Biomass and Wastes X. Conference*, 1395–1416, Institute of Gas Technology, Chicago, Illinois.

61. Jeffries, T.W., Fardy, J.H., and Lightfoot, E.N. 1985
    Biotechnol. Bioeng. 27(2):171–176.

62. Gong, C.S., McCracken, L.D., and Tsao, G.T. 1981
    Direct fermentation of D-xylose to ethanol by a xylose-fermenting yeast mutant, Candida SP XT 217, *Biotechnol. Lett.* 3(5):245–250.

63. Beck, M.J., Johnson, R.D., and Baker, C.S. 1990
    Ethanol production from glucose/xylose mixes by incorporating microbes in selected fermentation schemes, *Appl. Biochem. Biotechnol.* 24/25:415.

64. Jeffries, T.W. 1989
    Comparison of alternatives for the fermentation of pentoses to ethanol by yeasts, in M.Z. Lowenstein, ed., *Energy applications of biomass*, 231, Elsevier Applied Science, New York.

65. Hinman, N.D., Wright, J.D., Hoagland, W., and Wyman, C.E. 1989
    Xylose fermentation: an economic analysis, *Appl. Biochem. Biotechnol.* 20/21:391.

66. Buchert, J., Pols, J., and Poutanen, K. 1989
    The use of steamed hemicellulose as substrate in microbial conversions, *Appl. Biochem. Biotechnol.* 20/21:309.

67. Antonopoulos, A.A. and Wene, E.G. 1987
    Fusarium strain development and selection for enhancement of ethanol production, *FY 1987 biochemical conversion program annual review*, C-51, National Renewable Energy Laboratory (formerly Solar Energy Research Institute), Golden, Colorado.

68. Asther, M. and Khan, A.W. 1984
    Conversion of cellobiose and xylose to ethanol by immobilized growing cells of *Clostridium saccharolyticum* on charcoal support, *Biotechnol. Lett.* 6(12):809.

69. Alexander, J.K., Connors, R., and Yamamato, N. 1981
    Production of liquid fuels from cellulose by combined saccharification—fermentation for cocultivation of *Clostridia*, in M. Moo-Young, ed., *Advances in Biotechnology, Vol. II*, Pergamon Press, New York.

70. Slapack, G.E., Russell, I., and Stewart, G.G. 1987
    Thermophilic microbes in ethanol production, CRC Press, Boca Raton, Florida.

71. Lynd, L.R., Ahn, H.J., Anderson, G., Hill, P.; Kersey, D.S., and Klapatch, T. 1991
    Thermophilic ethanol production: investigation of ethanol yield and tolerance in continuous culture, *Appl. Biochem. Biotechnol.* 28/29:549.

72. Jeffries, T.W. 1981
    Fermentation of xylulose to ethanol using xylose isomerase and yeast, *Biotechnol. Bioeng. Symp.* 11:315.

73. Chaing, L.C., Hsiao, H.Y., Ueng, P.P., Chen, L.F., and Tsao, G.T. 1981
    Ethanol production from xylose by enzymatic isomerization and yeast fermentation, *Biotechnol. Bioeng. Symp.* 11:263.

74. Lastick, S.M., Mohagheghi, A., Tucker, M.P. and Grohmann, K. 1990
Simultaneous fermentation and isomerization of xylose to ethanol at high xylose concentrations, *Appl. Biochem. Biotechnol.* 24/25:431.

75. Tewari, Y.B., Steckler, D.K., and Goldberg, R.N. 1985
Biophys. Chem. 22:181–185.

76. Ingram, L.O. and Conway, T. 1988
Expression of different levels of ethanologenic enzymes from *Zynomonas mobilis* in recombinant strains of *Escherichia coli, Appl. Environ. Microbiol.* 54(2):397.

77. Ingram, L.O., Conway, T., Clark, D.P., Sewell, G.W., and Preston, J.F. 1987
Genetic engineering of ethanol production in *Escherichia coli, Appl. Environ. Microbiol.* 53(10):2420.

78. Domalski, E.S., Jobe, T.L. Jr., and Milne, T.A. 1987
Thermodynamic data for biomass materials and waste components, American Society of Mechanical Engineers, New York.

79. Johnson, D.K., Chum, H.L., Anzick, P., and Baldwin, R.M. 1990
Preparation of a lignin-derived pasting oil, *Appl. Biochem. Biotechnol.* 24/25:31.

80. Busche, R.M. 1985
The business of biomass, *Biotech. Prog.* 1(3):165.

81. Philip, R.J. 1986
Methanol production from biomass, NRCC 27143, 29–32, National Research Council of Canada, Ottawa, August.

82. Herman, R.G. 1986
Catalytic conversions of synthesis gas and alcohols to chemicals, Plenum Press, New York.

83. Swedish Motor Fuel Technology Co. 1986
Alcohols and alcohol blends as motor fuels, volumes I and II, prepared for the International Energy Agency, Swedish National Board for Technical Development Information No. 580, Trosa, Sweden.

84. Reed, T.B., Levie, B., and Graboski, M.S. 1988
Fundamentals, development and scale-up of the air-oxygen stratified downdraft gasifier, Pacific Northwest Laboratory, Richland, Washington.

85. Evans, R.J., Knight, R.A., Onischak, M., and Babu, S.P. 1988
Development of biomass gasification to produce substitute fuels, report PNL-6518, prepared by the Institute of Gas Technology, Chicago, Illinois, for the Pacific Northwest Laboratory, Richland, Washington.

86. Gravel, G., Hoareau, R., Pelletier, J., Guerard J., Chornet, E., Bergougnou, M.A., Chamberland, A., Grace, J.R., Overend, R.P., Dolenko, A., Hayes, D., Levesque, Y., and Carbonneau, R. 1987
Gasification project: energy from biomass, *Sixth Annual Canadian Bioenergy Seminar,* Elsevier Applied Sciences, New York.

87. Drnevich, R.F., Ecelbarger, E.J., and Portzer, J.W. 1981
Industrial oxygen plants—a technology overview for users of coal gasification-combined cycle systems, EPRI report AP-1674, Union Carbide Corporation, Tonawanda, New York, for the Electric Power Research Institute, Palo Alto, California.

88. Chem Systems. 1990
Optimization of electricity-methanol coproduction: configurations of integrated-gasification-combined cycle/once-through methanol, EPRI report GS-6869, Chem Systems, Tarrytown, New York, for the Electric Power Research Institute, Palo Alto, California.

89. Feldmann, H.F., Paisley, M.A., Appelbaum, H.R., and Taylor, D.R. 1988
Conversion of forest residues to a methane-rich gas in a high-throughput gasifier, PNL report PNL-6570, prepared by Battelle Columbus Division, Columbus, Ohio, for the Pacific Northwest Laboratory, Richland, Washington.

90. Manufacturing and Technology Conversion International, Inc. 1990
Testing of an advanced thermochemical conversion reactor system, PNL-7245, Pacific Northwest Laboratory, Richland, Washington.

91. Rensfelt, E. and Ekstrom, C. 1989
Fuel from municipal waste in an integrated circulating fluid-bed gasification/gas cleaning process, in D.L. Klass, ed., *Energy from biomass and wastes XI,* Institute of Gas Technology, Chicago, Illinois.

92. Chem Systems. 1990
Assessment of cost of production of methanol from biomass, report DOE/PE-0097P, Chem Systems, Tarrytown, New York, December.

93. Bain, R.L. 1991
Methanol from biomass: assessment of production costs, presented at the November 1989 Hawaii Natural Energy Institute Renewable Transportation Alternatives Workshop, Honolulu, Hawaii.

94. Simbeck, D.R., Dickinson, R.L., and Oliver, E.D. 1983
Coal gasification systems: a guide to status, applications, and economics, EPRI report AP-3109, Synthetic Fuels Associates, Mountain View, California, for the Electric Power Research Institute, Palo Alto, California.

95. Probstein, R.F. and Hicks, R.E. 1982
Synthetic fuels, McGraw-Hill Book Company, New York.

96. Lewis, S.M. and Grimes, W.M. 1988
Economic time series analysis of the fuel alcohol industry, Finnsugar Bioproducts, Inc. (Finnsugar is now part of Genencor International), Schaumburg, Illinois.

97. Chem Systems. 1990
Technical and economic evaluation: wood to ethanol process, Chem Systems, Tarrytown, New York, prepared for the Solar Energy Research Institute (now the National Renewable Energy Laboratory), August.

98. Hinman, N.D., Schell, D.J., Riley, C.J., Bergeron, P.W., and Walter, P.J. 1990
Preliminary estimates of the cost of ethanol production for SSF technology, Ethanol from Biomass Annual Review Meeting, Lincoln, Nebraska, 12–13 September.

99. *Oil & Gas Journal.* 1990
Nelson-Farrar Cost Indices, 61, Pennwell, Tulsa, Oklahoma, 3 September.

100. Simulation Sciences. 1990
Aspen/SP users manual, Simulation Sciences, Denver, Colorado.

101. Lynd, L.R. and Grethlein, H.E. 1984
IHOSR/extractive distillation for ethanol separation, *Chem. Eng. Prog.* 80(11):59.

102. Lynd, L.R. and Grethlein, H.E. 1986
Distillation with intermediate heat pumps and optimal sidestream return, *AIChE J.* 32(8):1347.

103. Maiorella, B.L., Blanch, H.W., and Wilke, C.R. 1984
Biotechnol. Bioeng. 26:1003.

104. Chem Systems. 1989
Assessment of costs and benefits of flexible and alternative fuel use in the U.S. transportation sector, technical report three: methanol production and transportation costs, DOE/PE-0093, Chem Systems, Tarrytown, New York.

105. Wan, E. I. and Fraser, M.D. 1989
    Economic assessment of advanced biomass gasification systems, Science Applications International Corporation, McLean, Virginia.

106. Energy Information Administration. 1990
    Annual energy outlook 1990 with projections to 2010, report DOE/EIA-0383(90), Energy Information Administration, U.S. Department of Energy, Washington DC.

107. Ho, S.P. 1989
    Global impact of ethanol versus gasoline, presented at the 1989 National Conference on Clean Air Issues and America's Motor Fuel Business, Washington DC.

108. Benson, G.O. and Pearce, R.B. 1987
    Corn perspective and culture, in S.A. Watson, and P. Ramstad, eds., *Corn: chemistry and technology*, American Association of Cereal Chemists, St. Paul, Minnesota.

109. McInnis, J.R. 1986
    Economics of fuel ethanol production 1986, presented at the Fourth Annual National Conference on Alcohol Fuels, Kansas City, Kansas.

110. Segal, M. 1989
    Ethanol fuel and global warming, *CRS report for Congress*, Library of Congress, Washington DC.

22
SOLAR HYDROGEN

JOAN M. OGDEN
JOACHIM NITSCH

Although many renewable energy sources will probably be used first for the direct production of electricity, the potential for renewable electricity is limited by the intermittent character of solar radiation and wind energy and by the difficulty of using electricity for applications such as transportation. The role of renewables in the global energy economy could be greatly extended, if they could be converted to energy carriers that are easily stored and transported and that could serve fluid-fuel markets not readily served by electricity. Hydrogen is such an energy carrier.

Hydrogen is a high quality, low polluting fuel that can be used with high efficiency for transportation, heating and power generation. Hydrogen could be utilized in markets where it is difficult to use electricity. In particular, hydrogen could play an important role in emerging markets for "zero emissions vehicles." Hydrogen fuel-cell vehicles would have advantages over electric battery vehicles because onboard hydrogen storage systems would be less bulky, heavy, and expensive than electric batteries and could be refueled in only a few minutes (versus several hours for recharging batteries).

Hydrogen can be produced and used on a global scale with greatly reduced greenhouse gas emissions and little local pollution. Hydrogen can be produced from a variety of widely available renewable resources, using technologies such as electrolysis of water (powered by wind, solar, or hydro electricity), or gasification of renewably grown biomass. In short, hydrogen can make the vast potential of renewable energy sources widely available as substitutes for fossil fuels.

PROSPECTS FOR SOLAR HYDROGEN SYSTEMS

The first uses of many renewable energy sources (e.g., hydropower, biomass, wind, solar-thermal electric, and solar photovoltaics) are likely to involve the direct production and use of electricity. However, the potential for utilizing renewables for electricity is limited by the intermittent character of solar radiation and wind energy and by the difficulty of using electricity for some applications. The role of renewables in the global energy economy could be greatly extended if they could be converted to energy carriers that are easily stored and transported and

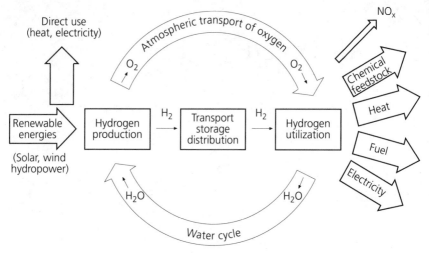

FIGURE 1: *The solar hydrogen cycle.*

could serve fluid-fuel markets are not easily served by electricity. There are a number of reasons why hydrogen could emerge as an attractive and potentially widespread energy carrier in the future.

To begin with, hydrogen has the potential to be the least polluting fuel. When hydrogen is burned, the main product is water (H_2O) (see figure 1). The only pollutants resulting from hydrogen combustion in air are oxides of nitrogen (NOx). With catalytic heaters (which operate at a lower temperature than combustion systems), NOx emissions would be reduced to negligible levels; with fuel cells NOx would be eliminated entirely. Among the various energy carriers that can be derived from renewable sources, only hydrogen and electricity could completely eliminate harmful emissions at the point of use. (Some other options such as methanol derived from biomass and used in fuel cells would come close to meeting this goal.) If hydrogen is made from renewable resources, essentially no greenhouse gases or other pollutants would be generated in energy production or use. Indeed, solar hydrogen is one of the few long-term energy options that could meet the world's growing energy needs without contributing to the greenhouse effect, to local air pollution, or to acid precipitation.

In addition, hydrogen can be produced efficiently from a variety of widely available renewable sources, using such methods as water electrolysis powered by solar electricity (wind, hydropower, solar-thermal electric, solar photovoltaics) and gasification of renewably grown biomass. The diversity of primary energy sources would facilitate hydrogen's role as a "universal" energy carrier in the future. The potential for hydrogen production from solar and wind resources is huge (a tiny fraction of available solar energy alone could meet the world's energy needs), and biomass could also be a significant source of hydrogen. Solar hydrogen could be produced in most areas of the world and thus would be much less geographically constrained than present energy supplies.

The introduction of hydrogen as an energy carrier requires no technological breakthroughs. Electrolysis is a commercially available technology, which is now being optimized for use with intermittent power sources such as wind and solar. Methods for producing hydrogen thermochemically from biomass could be commercialized during the next decade by adapting biomass gasifiers being developed for methanol production. Large-scale hydrogen storage and transmission technologies would be similar to those used in the chemical industry today. It is technically feasible for hydrogen to replace oil and natural gas in virtually all their present end-uses, which have already been demonstrated in a number of experimental hydrogen vehicles, heaters, and power generation systems.

Furthermore, hydrogen can be used very efficiently: in catalytic heaters with close to 100 percent efficiency, in fuel-cell vehicles with about three times the efficiency of comparable gasoline vehicles, and in stationary power applications at 80 percent efficiency in converting fuel energy into cogenerated electricity and heat. Because renewable energy supplies tend to be resource and/or capital intensive, it is critically important to use renewable energy as efficiently as possible.

Because large-scale hydrogen storage and long distance transmission systems would cost less than those for electricity, long distance transport of solar energy might be more attractive with hydrogen than with electricity.

Finally, hydrogen could supply end-uses not easily served by electricity. For transportation applications, onboard hydrogen storage systems are less expensive, bulky, and heavy than electric batteries. Moreover, hydrogen storage systems (compressed gas cylinders) can be refueled in about three minutes, as compared with several hours needed to recharge batteries. Unlike electric-battery vehicles where the ultimate market would be limited by the vehicle's range and refueling time, hydrogen fuel-cell vehicles could be general purpose "zero emission vehicles" suitable for long trips, as well as commuter use. Thus, hydrogen fuel-cell vehicles could potentially play much larger roles in transportation than battery-powered electric vehicles.

The use of electrolytic hydrogen on a large scale would take place after solar-electric technologies have been introduced and used extensively, so that costs have been reduced substantially. Similarly, biomass hydrogen would be produced after biomass gasifier and plantation technologies have been developed for electricity production. The development of a solar hydrogen energy system is not a strategy that would compete with a solar electric generation system. Rather, hydrogen would make the vast potentials of solar radiation and wind energy more widely accessible. With hydrogen it would be possible to gradually transform the present worldwide energy infrastructure based on fossil fuels to one based on solar-derived energy carriers.

SOLAR HYDROGEN PRODUCTION TECHNOLOGY

Performance and cost data are given for current technology, in addition to projections for what might be achieved in the near term (in the 1990s) and long term

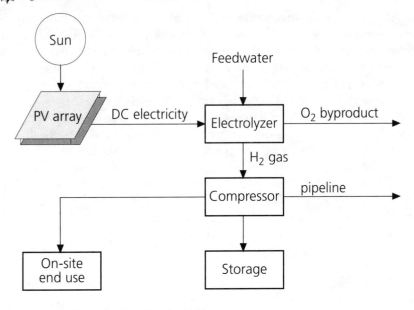

FIGURE 2: A solar photovoltaic electrolytic-hydrogen system.

(post 2000), with mature technologies widely employed so that economies-of-scale are fully exploited.

Hydrogen from solar powered electrolysis

Solar-electrolytic hydrogen systems

In solar-electrolysis systems, renewable electricity (which has been generated from wind, hydro-power, solar-thermal electric, or solar photovoltaic sources) is used to produce hydrogen and oxygen via water electrolysis. The hydrogen can be compressed for onsite use or storage or transmitted via pipeline to remote users (see figure 2).

Electrolysis technology

In electrolysis, water is split into its constituent elements hydrogen and oxygen by passing direct electric current through an electrolyte [1–8]. Today, commercially available electrolyzers are based on alkaline water electrolysis, although two other types are under development (see below). Alkaline systems employ an aqueous electrolyte (30 percent potassium hydroxide, KOH) and asbestos diaphragms to separate the anode and cathode spatially and to avoid the remixing of hydrogen and oxygen. Most commercially available electrolyzers have a bipolar configuration, where the individual cells are linked electrically and geometrically in series. Thus each metal wall separating one cell from the next functions in a bipolar fashion, supporting a cathode on one side and an anode on the other. The other commercially available technology, where electrolysis cells are connected in parallel, is called unipolar.

Electrolysis is a well established technology, but for cost reasons is used mainly in small sized plants where a reliable onsite source of high purity hydrogen is needed for chemical processes (for example, in metals reduction and semiconductor, pharmaceutical or food production). At present, there are only a few large-scale [5 to 25 megawatt] plants, with a total capacity of about 150,000 normal cubic meters (Nm^3) of hydrogen (H_2) per hour [1,900 gigajoules H_2 per hour or 530 megawatts H_2 on a higher heating value basis]. These large plants are sited near sources of cheap hydropower to produce hydrogen for ammonia synthesis. Due to a limited market, plant technology has advanced slowly.

Unipolar electrolyzers are currently less expensive than bipolar electrolyzers, but the latter have a larger development potential with respect to improved efficiency. Also, bipolar electrolyzers can operate under pressure [up to 3 megapascals absolute (MPa)], reducing the energy and capital requirements associated with hydrogen compression.

Achieving higher efficiency and lower cost with advanced alkaline water electrolysis will require the following:

▼ Changing the cell configuration and geometry with the goal of reducing the cell resistance by a factor of 3 to 10 ("zero gap" cell design) thereby reducing the Ohmic voltage losses.

▼ Developing new and inexpensive electrocatalyst materials able to reduce the sum of anodic and cathodic overvoltages to about 0.3 volts or less.

▼ Developing new diaphragm materials that are superior to conventional asbestos cloth.

Progress in these areas will allow considerably higher current densities [greater than 400 milliamperes per square centimeter (mA/cm^2)] which will make it possible to reduce the size of the electrolyzer, thus reducing costs. Moreover, reducing Ohmic voltage losses and electrode overvoltages could improve electrolysis efficiencies from their present values of 70 to 75 percent to perhaps 85 to 90 percent. (Efficiency is defined as hydrogen energy output divided by electricity input, based on the higher heating value of hydrogen.)

For solar hydrogen, special attention should be given to the dynamic behavior of large electrolysis plants because of the intermittent character of wind and solar electricity [1, 7, 9]. Characteristic data are given in table 1 for alkaline bipolar and unipolar electrolyzers for present and future technologies (the latter would be achievable after about one decade of further development, assuming increased production capacity) [1–3].

At present, only advanced alkaline electrolyzers offer the potential of efficient and economic production of solar hydrogen in large quantities. However, two other types of electrolyzers are in earlier stages of development. The **proton exchange membrane (PEM) electrolyzer** uses solid membrane electrolytes made of an acid resin [1, 8]. Its efficiency would be comparable to that of an advanced alkaline electrolyzer, about 80 to 90 percent. At present, platinum catalysts are required for stable operation, although less expensive catalyst materials are being sought. Due to constraints in membrane design, this type of electrolyzer seems better suited to the small electrolyzer market (up to 100 kilowatts of power) than for units in the multi-megawatt range. PEM electrolyzers offer the potential for low costs in mass production, if inexpensive membrane and catalyst materials could be developed. It has been suggested that combination PEM electrolyzer/ fuel cells could be designed, so that the same device could produce hydrogen and oxygen from electricity or electricity from hydrogen and oxygen. Such a device could be useful, for example, in fuel cell vehicles, which could produce hydrogen from electricity at night and then utilize the hydrogen for transport fuel.

The second advanced technology is **high-temperature steam electrolysis** [1]. Here part of the energy needed to split water is supplied as heat rather than electricity. Because the free enthalpy of water decreases with increasing temperature, only a very low voltage (1.2 to 1.3 volts) is needed to split water vapor at 900 to 1,000°C (compared with 1.7 to 2.0 volts at ambient temperature). Although high temperature heat is required, the electrical power input required for high temperature steam electrolysis is about 25 to 30 percent less than for advanced alkaline water electrolysis. (The total efficiency of hydrogen production counting both heat and electrical input would be about 90 to 95 percent; the hydrogen energy output would be about 1.1 to 1.2 times the electrical input.) The electrodes in a steam electrolyzer are electrically conductive oxide ceramics. The cathode is a zirconium dioxide cermet, while the anode is a thin layer of an electronically conductive mixed oxide. However, the high operating temperature creates various unsolved material and fabrication problems. Only small-scale laboratory cells have been tested up to now. One of the most important problems to be solved is the reduction of electrical cell resistance by scaling down the thickness of the cathode zirconium dioxide ceramic to 20 to 30 microns. Because of these difficulties, commercial use of high temperature steam electrolysis for hydrogen generation is still far away.

Sources of solar electricity

Present costs and cost projections for various solar electric technologies are summarized in table 2. (See tables 14 to 17 for more detailed cost and performance data.)[1]

In the near term, off-peak hydropower, which costs from 1 to 4 cents per kilowatt hour (kWh), may offer the lowest electricity costs [10, 11].

Wind power is expected to achieve low costs also. With current technology,

---

1.   All costs are given in average 1989 U.S. dollars. [Costs derived from German studies assume that 1.8 DM (1989) = 1 U.S. dollar (1989).] Levelized production costs for electricity and hydrogen are calculated in constant 1989 U.S. dollars for discount rates of 6 percent and 12 percent. For electric and hydrogen plants, the annual insurance cost is assumed to be 0.5 percent of the installed capital cost, and property taxes are neglected. All hydrogen costs and efficiencies are based on the higher heating value of hydrogen. The electricity cost is the levelized cost of intermittent electricity at the plant site, with no storage, transmission, or distribution costs included.

electricity from wind produced at a site with a hub height annual average wind power density of 500 watts per square meter (m^2) would cost perhaps 7 cents per kilowatt-hour alternating current (kWhAC) (10 cents per kWhAC), assuming a 6 percent (12 percent) discount rate (see chapter 3: *Wind Energy: Technology and Economics*).[2] In the near term (1990s) production costs are expected to drop to 4.3 cents (6.5 cents) per kWh and in the long term costs may reach 2.7 cents (3.9 cents) per kWh.

With current technology, solar thermal electric systems located in sunny regions such as the southwestern U.S. could produce electricity at a cost of 9 to 13 cents (13 to 18 cents) per kWh. By the end of the 1990s, the cost of electricity is projected to drop to 8 to 13 cents (11 to 18 cents) per kWh, and in the long term to 4.6 to 6.5 cents (6.7 to 9.4 cents) per kWh.

Solar photovoltaic direct current (DC) electricity produced in an area with high insolation would cost 12 to 30 cents (20 to 49 cents) per kWh today. [For photovoltaic (PV) electrolysis, the DC power produced by the PV array could be used directly. For grid power, PV electricity would have to be converted to alternating current (AC).] In the long term, however, electricity from thin film PV systems could drop to 2 to 4 cents per kilowatt-hour DC (kWhDC) (3 to 6 cents per kWhDC) or about 3 to 5 cents (4 to 7 cents) per kWhAC.

For a more detailed description of the solar electric technologies considered here, see chapters 3 to 10 (on wind power and solar electric technologies).

Cost of electrolytic hydrogen production

The cost of producing hydrogen from various renewable sources is projected to decrease markedly over the next 10 to 20 years[3] (see table 3).

Over the next decade, hydroelectricity would be the least expensive source for electrolytic hydrogen. If electricity were available for one to four cents per kWh during nighttime hours, the cost of hydrogen would be about $9 to $19 ($11 to $21) per gigajoule. Indeed, hydropower is already used for electrolytic hydrogen production (for chemical processes) at a few sites around the world, where low cost excess power is available.

Wind power will probably be the least costly source of intermittent solar electricity by the late 1990s. Around the year 2000, electrolytic hydrogen produced at an excellent wind site (with annual average wind power density of 630 watts per m^2) would cost $11 ($16) per gigajoule.[4] At a good wind site (with an-

---

2.   Throughout this chapter, energy costs are presented for a 6 percent discount rate, followed in parentheses by costs based on a 12 percent discount rate.

3.   In table 3, costs are given for 10 MW$_e$ wind and PV electrolysis plants (see tables 18 and 19), because there is little scale economy for these systems at sizes larger than about 10 MW. A 10 MW PV or wind system would produce about 180 gigajoules (14,000 Nm^3) of hydrogen per day, enough to fuel a fleet of about 1,000 fuel-cell automobiles.

4.   A region with Class 5 to 6 wind resource, according to the U.S. Wind Atlas classification scheme (see chapter 4), would have a wind power density of about 630 watts per m^2 at hub height of 50 meters. (A region with Class 3 wind resource would have an average wind power density of about 350 watts per m^2 at a hub height of 50 meters.

nual average wind power density of 350 watts per m²) hydrogen costs would be $17 ($25) per gigajoule.

In the long term, electrolytic hydrogen produced from solar-thermal electricity would cost about $18 to $25 ($27 to $36) per gigajoule.

While solar PV hydrogen is likely to be more expensive than most other sources of renewable hydrogen for the next decade or so, it is a potentially important long-term supply option because its production would be far less constrained by resource availability, land or water requirements than any other renewable resource. By the early part of the next century, solar PV hydrogen produced in a sunny area such as the southwestern United States (with an annual average insolation of 270 watts per m²) would cost about $10 to $16 ($15 to $25) per gigajoule. Reaching high efficiency is a key condition for reducing the cost of PV hydrogen (see figure 3, where production cost of PV hydrogen is plotted as a function of the PV module efficiency and cost). Ranges of PV hydrogen costs based on near-term and long-term projections are indicated (see tables 14, 15).

Hydrogen from biomass gasification

Biomass gasification technology
Hydrogen can also be produced by thermochemical gasification of biomass feedstocks, such as forest and agricultural residues, urban refuse, or wood chips from biomass plantations. The feedstock is introduced into a gasifier at high temperature, where it breaks down to form a gas (see chapter 17: *Advanced Gasification-based Biomass Power Generation*). The gasifier output consists mainly of hydrogen, carbon monoxide and methane. Steam is then added to this mixture to convert the methane into carbon monoxide and hydrogen, via a process known as steam reforming of methane. Then at a lower temperature, steam reacting with carbon monoxide gives rise to a final gaseous mixture consisting mainly of hydrogen and carbon dioxide. The carbon dioxide is then removed, leaving hydrogen.

Biomass gasifiers have been demonstrated at the laboratory and pilot-plant scale. Several biomass gasifiers, which are being developed for methanol production (see chapter 21: *Ethanol and Methanol from Cellulosic Biomass*), are also suitable for hydrogen production.[5] All the equipment for converting the gasifier output to hydrogen (methane reformers, shift reactors, carbon dioxide- (CO_2)- removal technology, and pressure-swing-adsorption technology for hydrogen purification) are well-established commercial technologies in the chemical process industries [12, 13].

Cost of biomass hydrogen production
In table 20, cost and performance projections are given for a biomass gasifier plant processing 1,650 dry tonnes per day [13], based on the Battelle Columbus

---

5.   Indirectly heated biomass gasifiers are particularly interesting because they do not require oxygen plants, which are very capital-intensive at the modest scales needed for biomass conversion facilities [13].

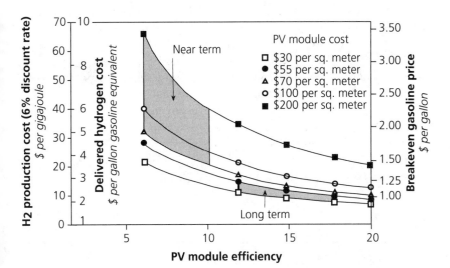

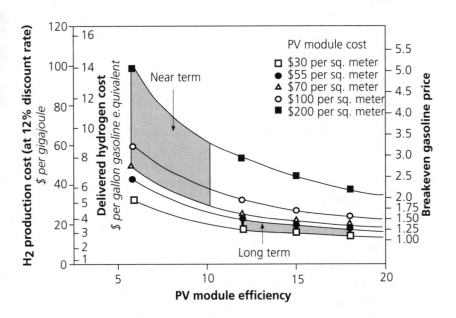

FIGURE 3: PV hydrogen production cost as a function of PV module efficiency and PV module cost.

Laboratories (BCL) gasifier. Hydrogen [at a pressure of 6.8 MPa [1,000 pounds per square inch (psia)] could be produced in this plant at a thermal efficiency of 70 percent and cost of $5.9 to $8.5 ($6.8 to $9.4) per gigajoule, assuming that the cost of the biomass feedstock cost is $2 to $4 per gigajoule. A plant this size could produce about 21,000 gigajoules of hydrogen per day (1.6 million Nm³ per day), enough fuel for a fleet of perhaps 300,000 fuel-cell passenger cars. Biomass hydrogen produced in smaller plants would be considerably more expensive because biomass gasifiers would exhibit strong scale-economies up to about this size.

In contrast, the cost of methanol from biomass from a plant of the same size would be $7.9 to $10.9 ($9.4 to $12.4) per gigajoule, again assuming that the cost of the biomass feedstock is $2 to $4 per gigajoule [13]. Costs are higher for methanol than for hydrogen primarily because the equipment is more expensive and because the conversion efficiency of biomass to methanol is lower (62 versus 70 percent).

At large production scales, biomass gasification could be the least costly source of renewable hydrogen, although hydrogen from solar electrolysis would probably be less expensive at small scale (see table 3).

LONG-DISTANCE TRANSPORT OF HYDROGEN

Hydrogen can be stored, transmitted, distributed, and delivered to consumers using gas transport technologies that are similar to those for natural gas. Alternatively, hydrogen can be liquefied and delivered by tanker or truck. (Both methods are employed by the chemical industry today.) In cases where energy demand is far from hydrogen production, long distance transport of hydrogen might be necessary.

Gaseous hydrogen

With present pipeline technology large quantities of natural gas (10 to 20 GW energy flow) can be transmitted over distances of several thousand kilometers. Existing pipelines typically have a diameter of 1.4 meters and a pipeline pressure of 8 MPa. By 2000, diameters of 1.6 to 1.8 meters and pressures of 10 to 12 MPa will be available. Typically, the gas is compressed using turbo-compressors driven by gas turbines fueled by gas from the pipeline. If electricity is available, electric motors could be used as well [1, 3].

Hydrogen pipelines would be similar to those for natural ga, although cost minimizing calculations for gas transport show that hydrogen pipelines will have larger diameters and fewer compressor stations.

It has been suggested that during a transition period, it may be possible to use existing natural gas pipelines for hydrogen. Transporting hydrogen through natural gas pipelines raises several concerns. First, to avoid hydrogen embrittlement, it would be necessary to check a particular pipeline to make sure that the steels were compatible with hydrogen service.[6] Moreover, equipment such as seals, meters, and compressors would have to be replaced with parts designed for hydrogen.[7]

---

6. Although hydrogen embrittlement can be a severe problem at the high temperatures and pressures found in oil refineries and other chemical process plants, it would be much less so at conditions typical of pipeline systems (ambient temperature and 6 to 8 MPa). Many of the steels and welds commonly used in natural gas pipelines could be used with hydrogen pipelines as well.

7. Hydrogen diffuses more easily than natural gas, and the flow rate through an orifice is almost three times faster. Thus, tighter seals and different sized metering equipment would be required.

In addition, because hydrogen has only about one-third the energy density of natural gas, its compression power requirements are greater. Moreover, because hydrogen is lighter than methane, it will flow through a pipeline about 2.5 times faster. As a result, a pipeline designed for natural gas can transport only about 80 percent as much energy when carrying hydrogen at the same pressure. Overall, the cost of transmission per unit energy would be about 50 percent higher for hydrogen than for natural gas.

An existing natural gas pipeline that might one day be used for delivering hydrogen to Europe runs from the Algerian gas fields at Hassi R'Mel to Sicily. Hydrogen produced electrolytically from solar-electric sources in the deserts of North Africa could be fed into the pipeline, which has been designed for an annual energy delivery of 12 billion Nm^3 of natural gas, an amount equivalent to about 5 percent of Germany's annual energy consumption.

To maintain constant hydrogen input into the pipeline, solar or wind hydrogen systems would need a large-scale hydrogen storage system. Appropriate large-scale storage options include rock or salt caverns, depleted gas wells, or aquifers.

The levelized cost of transport (with new equipment) from North Africa to Central Europe (a distance of 3,300 kilometers) would be about $4 to $5 ($7 to $9) per gigajoule, including compression (assuming pressurized electrolysis), underground storage, and pipeline transmission over 3,300 kilometers (with a segment of undersea pipeline) (see tables 12 and 21). For a 1,600 kilometer overland large-scale [16 gigawatt (GW)] pipeline in the United States, the levelized cost of compression (assuming electrolysis at atmospheric pressure), storage and pipeline transmission would be about $2 ($3) per gigajoule (see table 10), which is roughly 10 percent of the total delivered hydrogen cost [2].

Liquid hydrogen

Batch transport of liquid hydrogen (LH_2) in tankers is an alternative to pipeline transport of gaseous hydrogen (GH_2). The main attraction of LH_2 as an energy carrier is its high volumetric energy density (see table 22). However, liquefaction of hydrogen is very energy intensive. For present day plants, electricity requirements are from 10 to 11 kilowatt-hours of electricity (kWh_e) per kilogram of LH_2. In combination with a pressurized electrolyzer, specific electricity consumption for liquefaction can be reduced to 9 kWh_e per kilogram LH_2 (0.23 kWh_e per kWh H_2, HHV) in a 100 MW plant with post-2000 technology (see table 23).

Costs could be cut further by reducing storage losses (to 0.03 percent per day from today's 0.1 percent), by reducing filling and transfer losses (to 8 percent from today's 10 to 15 percent) and by lowering the cost of large LH_2-tankers.

Total energy consumption for liquefaction, storage, and tanker transport over a distance of 5,700 kilometers within a 20-day transport cycle is 34 percent of the hydrogen energy flow (HHV); thus 66 percent of the energy in the originally generated solar electricity is available at the tanker terminal in the user country.

The levelized cost of transporting liquid hydrogen from North Africa to Central Europe would be about \$9 to \$11 (\$14 to 16) per gigajoule, including liquefaction, storage, and tanker ship transport (see table 13). The high cost would be justified only for particular applications, such as aviation, where LH_2 is needed because of its high energy density relative to other hydrogen storage methods. Moreover, if long distance sea transport of hydrogen were required, liquid hydrogen tanker transport would be preferable because of the difficulty and cost of building an undersea pipeline.

Strategies for long distance transport of hydrogen
In those areas where energy demand is high and potential local renewable hydrogen resources are inadequate to meet local demands, gaseous pipeline transmission of hydrogen produced in remote regions would be less expensive than transporting liquid hydrogen by tanker, except for those cases involving long distance ocean transport. However, in many regions of the world, renewable hydrogen could be produced locally, obviating the need for long-distance transport.

HYDROGEN END-USE SYSTEMS

Hydrogen could replace fossil fuels in virtually all their present uses. Experimental hydrogen vehicles and heating and power generation systems already exist. Of these, transportation markets are likely to be the most important because of the growing demand for clean transportation fuels.

Hydrogen for residential heating
Although hydrogen might seem to be an exotic choice for a residential heating fuel, hydrogen-rich gases have been used for home heating and cooking for more than 100 years. "Town gas" (a mixture of approximately half hydrogen and half carbon monoxide that can be derived from coal, wastes, or wood) was piped into millions of urban homes in the United States before natural gas became widely available[8] and is still used in parts of Europe, South America, and Asia.

Researchers in the United States, Europe, and Japan have developed experimental home heating systems and appliances for use with pure hydrogen [14–21]. Hydrogen flame burners for these applications would resemble those in today's natural gas heating systems and appliances. The only important difference would be the size of the burner openings controlling the gas flow; to deliver the same amount of energy, hydrogen would have to flow three times as fast as natural gas. As with natural gas, hydrogen appliances with an open flame would be vented to disperse the NOx. The efficiency and cost of hydrogen home furnaces,

---

8.   In the United States, natural gas pipelines were not extended from the Gulf states to the northeast until the late 1940s, and town gas was still used as late as the early 1950s in some northeastern cities.

water heaters, and stoves would be essentially the same as those of natural gas systems.

Hydrogen appliances based on catalytic combustion have also been developed. In a catalytic heater the fuel gas combines with the oxygen in air in the presence of a catalyst at a relatively low temperature (200 to 400°C), creating a radiant glow. Because of the low operating temperature, very little NOx is emitted, so that the combustion product (which is almost pure water vapor) can be discharged directly into the heated space, thus giving an efficiency of almost 100 percent. Inexpensive catalytic heaters have been commercialized for natural gas. Similar systems could be used with hydrogen, although the installation costs would be less with hydrogen because vents would not be needed.

Alternatively, fuel cells can be used for the combined production of electricity and heat, providing heat for hot water and for space heating and air conditioning (e.g., using heat-driven absorption air conditioners) at scales ranging from individual dwellings to large-scale district heating [22]. Compared to cogeneration systems based on internal combustion engines, fuel cells would provide a larger fraction of useful energy in the form of electricity, which is more valuable than heat.

Prospects for residential cogeneration using proton exchange membrane fuel cells operated with natural gas (which is reformed onsite to hydrogen) are now being investigated by Rolls Royce with the Southern California Gas Company [23]. Under certain conditions, fuel-cell residential cogeneration systems fueled with hydrogen could become economically attractive to electric utilities as dispersed power generation systems (see appendix A).

Hydrogen for transportation

Concerns about urban air quality and energy supply security have led to increased interest in low-polluting alternatives to petroleum-based transportation fuels. Hydrogen is particularly attractive as a transport fuel because it produces few harmful emissions, can be used very efficiently, and can be made from a variety of feedstocks.

Two types of hydrogen-fueled vehicles are being developed: internal-combustion engine vehicles (ICEVs) and fuel-cell vehicles (FCVs). In prototype hydrogen ICEVs, the only pollutants produced are oxides of nitrogen (NOx), which can be controlled to very low levels with such techniques as water injection, lean operation, and exhaust-gas recirculation. (With engines optimized for hydrogen fuel, it is likely that NOx emissions could be made much lower than those from a comparable gasoline powered engine.) And in hydrogen FCVs which are now under development, NOx emissions would be eliminated entirely.

The technical feasibility of using hydrogen in transportation has been demonstrated in experimental automobiles, buses, trucks, and airplanes (see figure 4). In Germany, BMW and Daimler-Benz have ongoing programs in hydrogen vehicle research. Daimler-Benz, for example, has operated a fleet of five hydrogen vans and five dual fuel hydrogen/gasoline cars in Berlin, accumulating more than

FIGURE 4: Hydrogen automobile.

250,000 kilometers of experience over a five-year period [24]. Hydrogen vehicles have also been built in Japan, the former Soviet Union, Romania, the United States, Canada, Italy, New Zealand, and Australia [25–28]. A liquid hydrogen-fueled jet airplane was recently tested in the then Soviet Union [29].

Fuel storage is perhaps the greatest technical challenge affecting hydrogen's future as a transport fuel. Unlike liquid fuels such as gasoline, diesel fuel, or methanol, hydrogen is not easily stored at ambient pressure and temperature, so special onboard storage systems and dispensing systems are needed. Several methods have been tested successfully in experimental vehicles. Hydrogen has been stored as a compressed gas at 16 to 68 MPa (2,400 to 10,000 psia) in steel, aluminum or composite cylinders; as a liquid at $-253°C$ in cryogenic dewars; or as a metal hydride (a compound that absorbs hydrogen under pressure and releases it when heat is applied). Additional methods are being investigated, e.g., storage in liquid organic hydrides such as methylcyclohexane [30] or in refrigerated activated carbon systems at $-150°C$ [31].

Onboard hydrogen storage systems are much heavier and bulkier than those for liquid fuels such as gasoline or methanol, although they are lighter and more compact than electric batteries (see table 25). Storing sufficient fuel for a reasonable travelling range is therefore a major challenge. This suggests that hydrogen vehicles should be designed to be highly energy efficient in order to minimize the weight and volume needed for fuel storage.

The difficulty of storing hydrogen could be avoided altogether by producing hydrogen onboard the vehicle, either by reforming methanol [12, 32] or by steam oxidation of iron [33] (see appendix B). Both these approaches are being actively investigated for vehicle and cogeneration applications and should lead to substantial reductions in the storage system cost.

To date, most experimental hydrogen automobiles and buses have been equipped with modified internal combustion engines, which were originally designed for gasoline. Some investigators report that hydrogen automobiles are from 15 to 100 percent more energy-efficient than comparable gasoline-powered cars, primarily because hydrogen engines can run with leaner fuel mixtures (e.g., a lower fuel to air ratio) than gasoline engines [27, 28]. Hydrogen could also be used in high pressure direct injection or stratified charge engines, which might be up to 50 percent more efficient than current spark-ignited gasoline engines.

Hydrogen might find an early niche market as a pollution reducing additive to compressed natural gas for vehicles. Preliminary evidence suggests that adding from 5 to 15 percent hydrogen (by volume) to methane (producing a blend called "hythane") substantially reduces carbon monoxide emissions. The introduction of hythane could also help build experience with compressed hydrogen gas as a component of transport fuel [34].

Fuel-cell vehicles could be the innovation needed to make hydrogen a major option for transportation. Not only would NOx emissions be eliminated, but the efficiency would be perhaps three times that of a gasoline internal combustion engine vehicle [35–39], easing onboard storage requirements.

In a hydrogen fuel-cell vehicle, a hydrogen-air fuel cell powers an electric motor, which is coupled to an electric drive train similar to those used in battery-powered electric vehicles (BPEVs). Hydrogen fuel is stored directly (as compressed hydrogen gas or in a hydride), or in the form of methanol, which is reformed onboard to produce hydrogen.[9] In some designs, peak power demands are met by a small battery that supplements the fuel cell. Recently it has been proposed that iron could be stored onboard and steam oxidized to produce hydrogen plus iron oxide (see appendix B).

Various types of fuel cells have been considered for transportation applications (see table 26). At present the most attractive candidate for automobiles is the proton-exchange-membrane (PEM) fuel cell, which offers high power density, quick start-up time, modest operating temperatures (about $100°C$), and the potential to reach low costs in mass production. A major challenge in the development of a PEM fuel-cell vehicle is reducing the cost of the fuel-cell membrane via mass production. In addition, some engineering system integration design work remains to be done.

---

9. Although a methanol fuel-cell vehicle would be somewhat less efficient than a hydrogen fuel cell vehicle and would have the added complexity, weight, and cost of an onboard reformer, methanol storage would be much simpler and less expensive than other hydrogen storage methods. Methanol can be reformed at modest temperatures ($200°C$) using relatively simple devices suitable for vehicle use. Reforming other carbon-based fuels (methane or ethanol) for use in PEM fuel cells requires considerably higher temperatures and is much more technically challenging. However, with advanced fuel cells that operate at higher temperature (such as the solid oxide fuel cell), it might one day be feasible to reform a variety of fuels onboard.

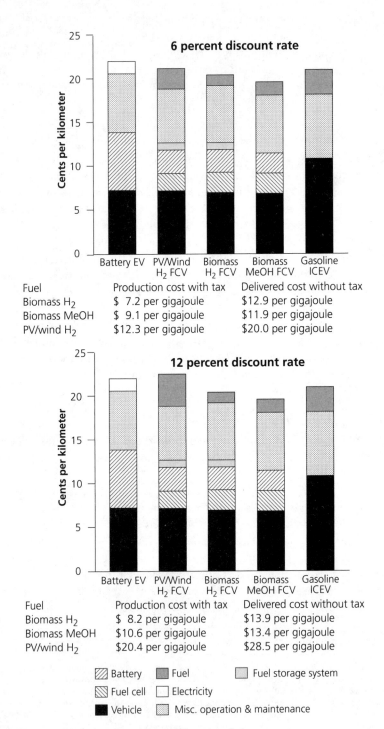

6 percent discount rate

Fuel	Production cost with tax	Delivered cost without tax
Biomass H_2	$ 7.2 per gigajoule	$12.9 per gigajoule
Biomass MeOH	$ 9.1 per gigajoule	$11.9 per gigajoule
PV/wind H_2	$12.3 per gigajoule	$20.0 per gigajoule

12 percent discount rate

Fuel	Production cost with tax	Delivered cost without tax
Biomass H_2	$ 8.2 per gigajoule	$13.9 per gigajoule
Biomass MeOH	$10.6 per gigajoule	$13.4 per gigajoule
PV/wind H_2	$20.4 per gigajoule	$28.5 per gigajoule

Battery Fuel Fuel storage system
Fuel cell Electricity
Vehicle Misc. operation & maintenance

FIGURE 5: Lifecycle cost of transportation for various fuels.

Recently, several companies including Ballard Power Systems of Canada, Elenco of Belgium, and Energy Partners of the United States have begun developing experimental hydrogen fuel-cell vehicles. The growing interest in fuel-cell vehicles has been spurred both by advances in electric-vehicle and fuel-cell technologies and by the expected demand for non-polluting or "zero emission vehicles".[10] The only technologies that could rigorously meet these standards are battery-powered electric vehicles and fuel cell vehicles run on hydrogen.[11]

Hydrogen fuel-cell vehicles could offer a number of advantages over other candidates for zero (or near-zero) emission vehicles:

▼ Because hydrogen storage is less heavy and bulky than electric batteries, the range of a hydrogen fuel-cell car would be longer than for a battery-powered electric car. [12]

▼ In addition, hydrogen could be stored in high pressure cylinders that could be refueled in a few minutes, in contrast to electric batteries that require several hours to recharge.

▼ Biomass is the only renewable energy source of methanol. (In some parts of the world, conditions are not conducive to biomass production, and eventually, a point will be reached when further expansion of worldwide biomass production for fuels will be limited by land and water constraints.) In contrast, hydrogen can be produced from a variety of renewable resources.

Recent studies have compared the performance and economics of hydrogen fuel-cell automobiles to other options [gasoline internal combustion engine vehicles, methanol fuel-cell vehicles and battery-powered electric vehicles, based on post-2000 projections for fuel-cell, battery, solar-electric and biomass technologies [32, 40] (see appendix C).

Surprisingly, these studies show that fuel-cell vehicles fueled with hydrogen from wind or pv could have lifecycle costs comparable to those of gasoline or battery-powered cars (see figure 5, table 29). If hydrogen or methanol were derived from biomass, costs would be even lower. Even though the initial cost of a fuel-cell car would be significantly higher than that of a gasoline car ($9,000 to

10.   In the United States, for example, the California Air Resources Board has mandated that zero emission vehicles (ZEVs) must be phased in starting in 1998. By 2003, 10 percent of the new cars sold in California, or several hundred thousand cars per year, must be ZEVs. In 1991, several northeastern states voted to adopt the California plan. Urban air pollution is a problem in many other countries, as well, some of which are considering similar measures.

11.   Although methanol FCVs would not be ZEVs because the reformer would emit small quantities of CO and NOx, emissions would be only about 1 percent of those from a typical gasoline-powered vehicle, or very close to zero.

12.   Because a hydrogen fuel-cell vehicle would be so much more efficient than a comparable gasoline internal combustion engine vehicle, the amount of onboard storage required would be relatively small. Recent studies by DeLuchi et al. [12, 32, 40] has shown that it should be possible to achieve a 400 kilometer travelling range in a 4 to 5 passenger hydrogen FCV, using compressed gas cylinders at 8,000 psia (55 MPa).

$13,000) (see table 27), and the delivered fuel cost for hydrogen or methanol would be several times that of gasoline (see table 28), the overall lifecycle costs for fuel-cell vehicles (see figure 5, tables 29 and 30) are about the same as those for gasoline or electric-battery vehicles (or slightly lower for biomass fuels). The reasons why the costs are comparable are as follows:

▼ Hydrogen can be used in a fuel-cell vehicle about three times as efficiently as gasoline in an internal combustion engine vehicle, so that the fuel cost per kilometer is less than for gasoline.

▼ The lifetime of the fuel-cell vehicle is about 50 percent longer than that of the gasoline vehicle, so that the contribution of the vehicle to the lifecycle cost is only slightly higher than for gasoline.[13]

▼ Maintenance costs are lower for fuel-cell cars than for gasoline cars, largely because there are many fewer moving parts.

But if hydrogen fuel-cell vehicles are to compete economically in the emerging (ZEV) market, they will have to be developed, tested, and commercialized on a large enough scale to significantly reduce fuel-cell costs. The first prototype hydrogen FCVs could be developed over the next few years. Testing small fleets would be an important step toward gaining experience with the vehicles and their refueling technology and with consumer acceptance. Thereafter, modest sized fleets of several hundred to several thousand vehicles that would be centrally refueled might be introduced. With a commitment from industry, it is possible that such fleets could be ready in about 10 years, in response to the ZEV markets in California and in the northeastern United States.[14] If hydrogen FCVs were successful in fleet service, the hydrogen distribution network might be expanded to general consumers. During the early decades of the next century, hydrogen FCVs might come to capture a large share of the ZEV market, say several hundred thousand vehicles per year or more.

POTENTIAL CONTRIBUTIONS OF RENEWABLE HYDROGEN TO WORLD ENERGY SUPPLY

The potential contributions of various renewable sources of hydrogen to future energy supply depend not only on the resource base, but on the land area and water required for hydrogen production (see tables 4 to 6 and table 31).

---

13. Experience with battery-powered electric vehicles in fleet service suggests that the fuel-cell vehicle lifetime would be about 50 percent longer than that of an internal combustion engine vehicle. If the fuel-cell vehicle's lifetime was only 10 percent longer than that of a gasoline ICEV, the lifecycle cost would be slightly higher than that of a gasoline ICEV (see appendix C and table 30).

14. Hydrogen fuel-cell vehicles would benefit from the development of battery-powered electric vehicles (the drive train would be essentially the same for BPEVs and FCVs), compressed natural-gas vehicles (the gaseous onboard storage system and filling station equipment would be similar) and methanol fuel-cell vehicles (even though the methanol vehicle would have a reformer and a different storage system, the fuel cell would be the same).

Although the global potential for electrolytic hydrogen from hydropower could be significant—about 56 exajoules, assuming that all technically usable hydropower were dedicated to hydrogen production—environmental considerations will limit hydropower development for all purposes to a fraction of this potential. Moreover, hydropower resources are geographically limited to good sites (many of which are already developed) and require much more land and water than wind or PV electrolysis. Thus, the global contribution of hydro-power to a hydrogen energy system would be relatively small. Still, because of its low cost, off-peak hydropower at existing sites might offer an opportunity to help launch electrolytic hydrogen as an energy carrier.

The global potential hydrogen production from biomass is substantial (see table 4). Assuming that 10 percent of the total land area presently committed to forest, woodland, and cropland is developed as biomass plantations, and that an average productivity of 15 dry tonnes of biomass per hectare per year could be realized, about 113 exajoules of hydrogen could be produced per year, an amount approximately equivalent to global oil use today. If all the degraded lands in developing countries suitable for reforestation [nearly 8 million square kilometers (km^2)] could be developed for biomass hydrogen, then about 159 exajoules could be produced. The hydrogen production potential from excess cropland within the European Economic Community (some 15 million hectares) would be about 3.0 exajoules, and from 30 million hectares of idle cropland in the United States, some 5.9 exajoules could be produced. Nevertheless, land use considerations are likely to be a key issue for biomass hydrogen development. Therefore, introduction of efficient end-use technologies such as fuel-cell vehicles, which would reduce the amount of land needed, are crucial for large-scale biomass development.

Wind power is a large and widespread resource, that would be a less land intensive source of hydrogen than biomass (see tables 5 and 6). Moreover, wind farms could perhaps serve dual functions as power producers and agricultural or grazing sites. In practice, only a fraction of the global electrolytic hydrogen potential of almost 1,200 exajoules per year (see table 4) could be developed because of rugged terrain and competing uses for land. Even with restrictions, however, the wind hydrogen potential would be huge. To supply an amount of hydrogen equivalent to current fossil fuel use (300 exajoules) would require only 6 percent of the world's land area. To match the current demand for oil in the United States (34 exajoules), 14 percent of the country's land area would be needed (see table 6). These areas are somewhat more than half the land areas which would be required to produce such quantities of hydrogen from biomass.

Although PV hydrogen would be more expensive than hydrogen from biomass, it could be produced wherever there is adequate insolation. Moreover, PV land requirements would be much lower than for any other option, about 1/30th those for biomass. Sufficient PV hydrogen to meet the world's foreseeable energy needs could be produced on a few percent of the earth's desert area. For example, southern Spain has more than 20,000 km^2 of unused arid zones, which could

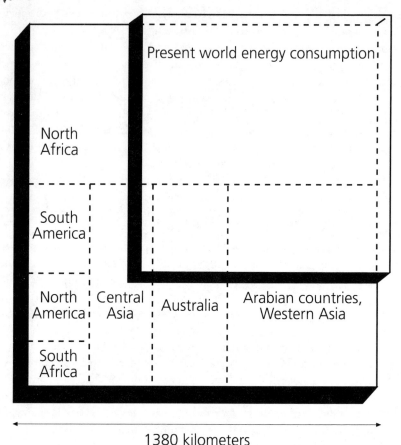

FIGURE 6: *Land areas in various continents that are appropriate for PV hydrogen production.*

produce 8 exajoules of hydrogen per year, about one-third of the estimated potential use in Western Europe. And in the United States, an amount of hydrogen equivalent in energy to current oil consumption (34 exajoules) could be produced on 0.7 percent of the U.S. land area (or 9 percent of the U.S. desert area). In order to produce solar hydrogen in an amount equal to the *present* global consumption of fossil fuels (300 exajoules), about 700,000 km^2 (or 2 percent of the world's desert area) would be needed. Under very strict criteria, about 1.9 million km^2 worldwide can be classified as excellent sites for the installation of large PV hydrogen plants (see figure 6). The total corresponds to 5 percent of the global desert area or 1.3 percent of the global land area [1]. Other possible sites in desert-like arid zones (in addition to the areas listed in figure 6) are to be found in North America and southern Europe.

One concern, however, is the need for water during electrolysis. Although desert sites are ideal for producing PV hydrogen, such sites have limited water resources. Fortunately, the water requirement for electrolytic hydrogen production is modest— amounting to about 63 liters per gigajoule of hydrogen, if the cooling water for the electrolysis plant is recycled. Typically, a small amount of the annual rainfall in the area covered by a solar hydrogen plant would be sufficient to supply feed water for electrolysis. For example, the annual water consumption of a pv-hydrogen plant corresponds to 2.7 centimeters of rain per year over an area equal to the plant size, which amounts to only 14 percent of the annual rainfall in El Paso, one of the most arid places in the United States [2]. In contrast, achieving a biomass productivity of 15 dry tonnes per hectare per year would require rainfall of from 75 to 150 centimeters per year (see chapter 14: *Biomass for Energy: Supply Prospects*). Alternatively, pv powered could be transmitted to electrolyzers installed at the coast, where the water for electrolysis could be provided via the desalination of sea water. About 1 to 2 percent of the hydrogen energy produced would be needed for the desalination of sea water. The current worldwide desalination capacity of 3.6×10^9 Nm^3 water per year would be sufficient for a production of 56 exajoules H$_2$, corresponding to about half the energy equivalent of today's global oil consumption [1, 3]. Thus, even in the most arid deserts, the water supply for electrolytic hydrogen would not be a severe problem.

Unlike fossil fuels, which are unevenly distributed throughout the world, renewable hydrogen can be generated almost anywhere. Using one or more indigenous renewable resources, large quantities of hydrogen could be produced in most parts of the world.

HYDROGEN SAFETY

Despite the public perception that hydrogen is more dangerous than such commonly used fuels as natural gas or gasoline, this concern is not borne out by hydrogen's physical properties or by the many years of industrial and residential experience with hydrogen. Studies of the relative safety of hydrogen, methane, and gasoline have concluded that no one fuel is inherently safer than the others in every respect, but that all three fuels can be and have been used safely [41, 42] (see table 7). Concerns that are often mentioned with hydrogen are:

▼ **Risk of fire or explosion is greater with hydrogen than with other fuels.**
One worry is that hydrogen can burn over a much wider range of concentrations in air (4 to 75 percent by volume) than either natural gas or gasoline. However, it is the lower flammability limit (minimum flammable concentration of fuel in air) that is most relevant to safety, if the gas were to build up gradually through a leak, as might occur in residential or indoor industrial settings. Because sources of ignition (sparks, open flames, hot surfaces) are often present in these environments, a fire would be likely when a fuel reached a flammable concentration. The lower flammability limit for hydrogen is 4 percent, which is only slightly lower than that for natural gas (5 percent) and significantly higher than that for gasoline (1 percent). To reach an explosive mixture requires much higher concentrations for hydrogen (18 to 59 percent) than for gasoline (1 to 3 percent) or methane (6 to 14 percent). The dangers associated with wide flammability limits usually arise

in industrial situations in connection with hydrogen handling equipment. To prevent the occurrence of flammable or explosive mixtures, hydrogen pipelines and other hydrogen handling equipment are routinely purged with an inert gas such as nitrogen during startup, before hydrogen is introduced into the system.

Hydrogen's low ignition energy also raises concern. Although the ignition energy for hydrogen is much lower than that for methane or gasoline, all three fuels have such low ignition energies that sparks from static electricity or weak thermal sources (such as open flames or hot surfaces) would ignite them. Similar precautions would be needed to avoid build-up of dangerous concentrations with any of these fuels.

Leakage presents another worry. Hydrogen would have a volumetric leak rate from a damaged seal or cracked weld about three times that of methane, but less than that of gasoline vapors. However, because it is much lighter than air, hydrogen disperses very quickly from a leak, making it difficult to form a flammable mixture. Hydrogen or methane leaks outdoors (from automotive tanks, storage or pipelines) should pose little problem, because these fuels are lighter than air and will disperse before a hazardous mixture can form. In contrast, gasoline vapors are heavier than air and can linger near a leak, even outdoors, causing a longer lasting danger of explosion or fire. For indoor use, adequate ventilation through roof vents would assure that buoyant hydrogen would not reach dangerous levels. As with natural gas, odorants would probably be added to hydrogen to aid in detecting leaks. In addition, hydrogen, unlike gasoline, is non-toxic.

▼ **Hydrogen would be more dangerous than natural gas or gasoline in a fire or an explosion.**

Unlike gasoline, hydrogen is odorless and burns with a very hot, almost invisible flame, which is difficult to detect and extinguish. The visibility problem could be overcome, by adding a colorant to make the flames more visible. Moreover, hydrogen fires tend to radiate less heat and are shorter lived than comparable gasoline fires. In an explosion, the amount of damage depends on the location. Hydrogen flame fronts move faster than those of other fuels; thus, in an enclosed space damage would be more, while in an open space there would be less damage than with methane or gasoline.

▼ **There is no safe way to store hydrogen.**

Questions have been raised about the safety of hydrogen storage. Yet hydrogen has great flexibility and can be stored in the form of hydrides, compressed gas, or liquid hydrogen. All three methods have been used safely in industry and utility applications. In motor vehicles, high-pressure compressed gas cylinders similar to those used with natural gas would probably be used. These cylinders have been shown to be very safe, even at high pressures.

▼ **Hydrogen embrittlement of metals can cause dangerous cracks and leaks.**

Under certain conditions, hydrogen can diffuse into metals (including some

steels), which tends to "embrittle" them, making them susceptible over time to cracks. At the high operating temperatures and pressures of hydrogen handling equipment used in oil refineries, hydrogen embrittlement can occur. At the lower temperatures found in most hydrogen energy systems, embrittlement should not be a serious problem.

The key point is that all chemical fuels including natural gas, gasoline, and hydrogen are potentially dangerous and must be handled properly. When appropriate precautions are taken, all three fuels can be produced, stored, and transmitted safely.

DESIGN OF SOLAR HYDROGEN ENERGY SYSTEMS: TWO CASE STUDIES

In imagining a transition from present energy carriers toward widespread use of hydrogen energy, it is important to provide a vision of how renewable hydrogen energy systems might evolve. We have chosen two potentially important markets, the United States and Europe, as case studies. These two regions offer interesting contrasts.

Solar hydrogen energy systems for the United States

A detailed description of renewable hydrogen energy systems for the United States offers insights into the role hydrogen might play in a country with high energy use and abundant resources for hydrogen production. Although hydrogen can be used in a wide range of markets, the focus here is on road transportation, which is likely to be the largest market for hydrogen in the United States. Not only does renewable hydrogen have the potential to help ameliorate environmental problems, but at the same time it could help relieve the country's dependence on petroleum for the transport sector.

Energy use for transportation (and the need for imported oil) in the United States could be reduced significantly, if hydrogen fuel-cell vehicles were to displace gasoline internal-combustion engine vehicles.[15]

Consider light-duty vehicles—automobiles and light trucks. In 1990, the U.S. Department of Energy projected that fuel use by light-duty vehicles will increase from 12.4 exajoules in 1990 to 14.4 exajoules in 2010 [43]. Yet, if highly efficient hydrogen fuel-cell vehicles were to replace gasoline vehicles, the energy use for light duty vehicles would drop 3-fold, to 4.8 exajoules per year.

---

15. Liquid fuel requirements in the United States in general and for light-duty vehicles in particular, are extraordinarily high by global standards. Demands are projected to rise even higher, putting increasing pressure on the world oil market unless there are major departures from present trends. In its 1990 analysis for the National Energy Strategy, the U.S. Department of Energy projected that total U.S. oil requirements in 2010 will be 44.6 exajoules per year (up from 35.9 exajoules in 1990) or 157 gigajoules per capita, and that oil import requirements will be 30.9 exajoules per year at that time (up from 17.0 exajoules in 1990). For comparison, total oil consumption in all developing countries was 32.5 exajoules or 8.9 gigajoules per capita in 1985 [43].

At this level of demand, hydrogen derived from any of the major renewable options could displace a significant amount of gasoline. For example, 4.8 exajoules of hydrogen could be provided by biomass grown on 23 million hectares of excess cropland,[16] which is equivalent to 70 percent of the amount of idle cropland in the United States in 1990, or less than 40 percent of the amount of idle cropland projected for the year 2000 [44], or 2.5 percent of the entire land area of the United States. Alternatively, obtaining the same amount of hydrogen (4.8 exajoules per year) from wind energy, would require 13 million hectares of land or 1.4 percent of the U.S. land area. Providing 4.8 exajoules of hydrogen annually from photovoltaic sources would require less than 0.8 million hectares of land area, (less than 0.1 percent of U.S. land area).

Meeting the fuel requirements for all light-duty vehicles in the U.S. from only one of these sources could require the development of hydrogen pipeline systems to transport the hydrogen from where it is produced to major energy demand centers elsewhere. However, long distance pipeline systems probably would not be necessary in the United States, where many centers of population are close to good solar, wind, or biomass resources. Substantial quantities of hydrogen could be produced locally, almost anywhere in the United States (see table 8).[17]

Forty-three states have locally significant statewide renewable-hydrogen energy resources and can produce more than 0.1 exajoules of hydrogen per year (see table 8): one tenth of an exajoule per year could supply enough hydrogen for several million fuel cell vehicles. In most states, PV hydrogen resources are larger than other renewable hydrogen resources, with the exception of a few states in the Great Plains, including Iowa, Kansas, Minnesota, Montana, North Dakota, South Dakota, Oklahoma, and Wyoming, where wind resources are dominant. Several Midwestern and Plains states have considerable biomass potential, and hydrogen from off-peak hydroelectricity could be important in Washington state.

Estimates show that if hydrogen fuel-cell vehicles entirely replace gasoline internal-combustion engine vehicles, enough solar hydrogen to fuel those vehicles could be produced from local resources, except for a few small, densely populated states in the northeast (see table 8).

Because the United States has such a diversity of renewable hydrogen supplies, several approaches for introducing hydrogen as an energy carrier can be si-

---

16. Biomass is assumed to be grown at an average productivity of 15 tonnes per hectare per year.

17. Off-peak hydropower potential assumes that existing installed hydroelectric capacity could be used 25 percent of the time for hydrogen production. Undeveloped hydropower is assumed to be dedicated to hydrogen production. The estimate for biomass hydrogen is based on plantations on the 13 million hectares of cropland idled under the U.S. Department of Agriculture's Conservation Reserve Program. [The potential biomass hydrogen resource could be larger if biomass residues or municipal solid wastes were also exploited or if the 17 million hectares of additional cropland idled under the Acreage Reduction Program were developed; in addition the amount of excess agricultural land is likely to increase in the future [44].] The potential for producing hydrogen from wind power is based on a scenario where the wind energy potential is fully developed and dedicated to hydrogen production, subject to constraints whereby all urban and environmentally sensitive lands, 50 percent of forests, 30 percent of agricultural land and 10 percent of range lands are excluded from development [45]. The PV resource is estimated assuming 1 percent of each state land area is used for PV hydrogen production.

multaneously pursued. An evolutionary strategy can be presented for producing hydrogen for fuel-cell vehicles based on the prospective economics of hydrogen from wind (see table 18), PV (see table 19), and biomass (see table 20) in the early part of the next century. The strategy calls for an ambitious development schedule, which would be driven by a growing demand for zero emission vehicles.

It should be noted that the various technologies involved are characterized by different economies-of-scale. On the one hand, PV, wind systems, and electrolyzers are highly modular technologies; PV or wind-powered electrolyzers would have no economies-of-scale above electric power levels of 10 MW_p, which corresponds to a daily hydrogen output of about 180 gigajoules (enough for a fleet of about 1,000 cars). On the other hand, hydrogen compression, storage, and transmission systems do exhibit economies-of-scale up to larger sizes. There are also economies-of-scale involved in producing hydrogen from biomass, up to production levels of about 20,000 gigajoules per day.

Markets for hydrogen transportation fuel might evolve as follows:

▼ Development of prototype vehicles for small test fleets of 5 to 50 vehicles, in the 1990s.

▼ Development of hydrogen production facilities for 1,000-vehicle centrally refueled fleets, from 2000 to 2010.

▼ Development of city-scale hydrogen production facilities, each serving some 300,000 vehicles, from 2010 to 2020.

▼ Development of large-scale hydrogen and transport facilities for serving markets remote from production centers, beyond 2020.

Hydrogen prototype vehicle development
During the 1990s, prototype hydrogen fuel-cell vehicles will be built and tested, first singly, and then in micro-fleets of five to 50 cars. Since the number of vehicles involved will be small, fuel could be obtained from existing industrial hydrogen market sources in the form of truck-delivered pressurized hydrogen.

Hydrogen production for fleets
Once hydrogen fuel-cell vehicles have been successfully demonstrated in test fleets, emphasis could shift, during the first decade of the 21st century to centrally refueled fleets of about 1,000 vehicles. At this scale, onsite hydrogen production systems would be less expensive than delivered hydrogen.

If fleet vehicles are driven 48,000 kilometers (30,000 miles) per year and have gasoline-equivalent fuel economies of 3.9 liters per 100 kilometers [60 miles per gallon (mpg)], hydrogen production capacities of about 180 gigajoules per day would be required. At this scale, the cost of producing electrolytic hydrogen from off-peak hydropower, wind, or PV sources, would be comparable to the cost of producing hydrogen via steam-reforming of natural gas, the least costly conventional technology (see table 9). In terms of land requirements, a PV hydrogen system would need from 12 to 19 hectares, depending on the system efficiency; a wind hydrogen system would need from 47 to 160 hectares, depending on wind-turbine spacing and the intensity of the wind resource. Hydrogen would be compressed and stored in steel pressure cylinders similar to those used in the chemical industry today. Hydrogen fuel-cell vehicles with gaseous storage systems could be refueled directly from these storage tanks.

Although hydrogen derived from biomass would probably be too costly to compete with electrolysis at this small scale, hydrogen could be produced at lower cost either from biomass or from steam-reformed natural gas at larger scales (see table 3). Accordingly, one might consider supplying hydrogen from one large centralized steam-reforming or biomass gasifier plant capable of supplying 10 to 100 cities, each having a fleet of 1,000 vehicles. However, having to build numerous small pipelines to supply those cities would make this option more costly than small, decentralized electrolytic-hydrogen production systems.

One major attraction of electrolysis is that production systems can be built with only modest front-end investments, measured in tens of millions, rather than hundreds of millions or billions of dollars that are typical for most conventional synfuels plants (see table 10).[18]

The delivered cost of PV hydrogen produced in the southwestern United States (including compression, storage, distribution, and filling station costs) would be $15.9 to $22.1 per gigajoule (equivalent in energy terms to $2.3 to $3.2 per gallon of gasoline), assuming a 6 percent discount rate (see table 9). (With a 12 percent discount rate, the delivered costs would be $22 to 32 per gigajoule.) Wind hydrogen produced in an area with good resources (wind class rating of 3, see table 18) would cost about $22.3 per gigajoule ($3.2 per gallon gasoline equivalent) delivered; in an area with excellent resources (wind class rating of 5 to 6), the delivered cost would be about $16.9 per gigajoule.

Another option for small-scale hydrogen production would be electrolysis powered by off-peak hydroelectricity. Although the capital cost of an externally powered electrolyzer with compression, storage, and filling station system would be lower than for PV or wind electrolysis ($5.9 million), the delivered hydrogen cost would be about the same: $17 to $25 per gigajoule, for off-peak hydropower costing 2 to 4 cents per kilowatt-hour (see table 9). (If hydropower costs were as low as 1 cent per kilowatt-hour, the delivered hydrogen cost would be $13.5 per gigajoule).

Hydrogen production at city scale
During the next stage—from 2010 to 2020—city-wide markets for hydrogen

---

18. The capital investment required for a 10 MW$_p$ PV-hydrogen production system (including equipment for hydrogen compression from 0.1 to 24 MPa and storage at 24 MPa in pressurized cylinders), would be $9 to $15 million, depending on the cost and efficiency of the PV modules (see table 9). The majority of the cost is due to the PV array ($5 to $11 million) and the electrolyzer ($2.3 million). Similarly, a 10 MW$_p$ wind hydrogen system would cost about $13.1 million ($7.5 million for the wind turbines, $3.7 million for the electrolyzer, and $1.4 million for compression and storage). The cost of the filling station equipment needed to serve a 1,000 car fleet would add perhaps $0.5 million [32].

fuel-cell vehicles could be developed, with a single supply system serving perhaps 300,000 such vehicles. Hydrogen would be piped short distances from production sites to the local distribution network and then to individual filling stations. City-scale production facilities might be located near large urban centers in the southwest for PV, near cities in the Great Plains for wind, or near biomass plantations on reclaimed idled cropland in the Midwest.

If each vehicle is driven an average of 16,000 kilometers (10,000 miles) per year and has a fuel economy of 3.9 liters per 100 kilometer (60 mpg), 300,000 vehicles would need an installed PV or wind capacity of 750 MW_p for hydrogen production; a PV system would require 9 to 14 km^2, a wind system 35 to 120 km^2. At this scale, underground gaseous hydrogen storage in rock caverns would become viable, reducing storage costs (see table 10).

The capital investment needed for a PV hydrogen system would be about $0.6 to $1.0 billion or about $1 billion for a wind system. An additional $65 million would be needed to build some 130 refilling stations. Local distribution systems would have to be built or adapted, and the total cost of local distribution and filling station expenses would add about $5.7 per gigajoule to the cost of hydrogen, for a total delivered cost of about $17.6 to $24.1 per gigajoule ($2.5 to $3.5 per gallon of gasoline equivalent).

In those regions of the country where there is adequate land and water, hydrogen could be derived from biomass. A biomass gasifier hydrogen plant processing 1,650 dry tonnes of biomass per day could provide enough hydrogen to support 300,000 vehicles. To do so would require about 410 km^2 of biomass plantation area. At this scale, hydrogen production based on biomass would be highly competitive with hydrogen from conventional sources: steam reforming of natural gas and coal gasification (see table 3). Certainly, the production facilities would be much less capital-intensive than electrolytic systems—requiring only $0.14 billion for a plant of this scale (see table 10). Moreover, the delivered cost of biomass hydrogen would be about 40 percent less than the cost for PV or wind hydrogen—some $11.6 to $14.0 per gigajoule ($1.52 to $1.83 per gallon of gasoline equivalent), assuming a 6 percent discount rate. (If a 12 percent discount rate were used, delivered costs would be $12.5 to $15.2 per gigajoule for biomass hydrogen, which is about half that of PV or wind hydrogen.)

Hydrogen production for remote markets
Eventually, as city-wide hydrogen systems become well-established, the scale of hydrogen production may expand in regions that are especially well-endowed with renewable resources (such as the great plains or the southwest) for export to those states (such as the northeast) where local hydrogen resources are inadequate.

In areas where long-distance transport is needed, production should be expanded to levels at which the scale economies of pipeline transmission are fully exploited.[19] A pipeline carrying 0.5 exajoules per year could support 30 million fuel-cell passenger vehicles (or 10 million fleet vehicles).

In the southwest, where the annual insolation averages 271 watts per m^2, a 75 GW$_p$ PV system covering a total land area of 900 to 1,400 km^2 could supply hydrogen at a peak output rate of 64 GW. If three-fourths of the daily output were compressed for storage and one-fourth was fed directly into the pipeline, a 16 GW pipeline could transport a continuous flow of hydrogen (from the electrolyzer during the day and from storage at night). The required capital investment would be $40 to $80 billion for the PV arrays, $17 billion for electrolyzers, $1 billion for storage in gas wells (the most economical storage option at this scale), $1 billion for compressors and $1.9 billion for a 1,600 kilometer pipeline.

With such a system, PV hydrogen could be produced for $10 to $16 per gigajoule. Compression and storage would add about $1.4 per gigajoule, and pipeline transmission about $0.3 per gigajoule. Adding local distribution costs of $0.5 per gigajoule and filling station costs of $5.2 per gigajoule, the delivered cost would be $17.3 to $23.5 per gigajoule (equivalent to $2.5 to $3.4 per gallon of gasoline), at a 6 percent discount rate.

Cost comparisons

The delivered costs for PV and wind hydrogen do not vary much over the scales considered here (see table 10), because scale economies in hydrogen storage and compression at large scale are balanced by increased transmission and distribution costs. (This is not true of hydrogen from biomass, coal, or natural gas, where hydrogen-production scale economies are more pronounced.)

Although the delivered costs of renewable hydrogen would generally be higher than gasoline prices (exclusive of taxes) within the time frame of this analysis, hydrogen fuel-cell vehicles would often be able to compete on a lifecycle cost basis with gasoline internal-combustion engine vehicles, as noted earlier, even if the environmental benefits of hydrogen fuel-cell vehicles are not taken into account.[20]

Solar hydrogen systems for Europe and North Africa

The Mediterranean perspective

Europe is a region of high energy consumption because of its high population density and intensive industrial and consumer activities. Presently, Western Europe (including Yugoslavia and Turkey, but excluding Eastern Europe) consumes 15 percent of the world's energy but has only 8 percent of the world's population. At the same time, the continent is comparatively poor in energy resources and heavily dependent on energy imports.

---

19.   Pipelines have economies-of-scale up to hydrogen flow rates of about 16 GW (HHV) or 0.5 exajoule per year (equivalent to 2.5 percent of current U.S. oil use for transportation or about 10 percent of projected use for light duty fuel-cell vehicles).

20.   This is shown in figure 3, where the breakeven gasoline price for PV hydrogen is plotted as a function of PV module efficiency and cost. In the near term, the breakeven gasoline price would be about $2.5 to $4.0 per gallon ($3.0 to $5.0 per gallon), and in the long term $1.00 to $1.25 per gallon ($1.25 to $1.70 per gallon) assuming a 6 percent (12 percent) discount rate for hydrogen production. This contrasts with the delivered cost of PV hydrogen which would be $4 to $10 ($6 to $14) per gallon in the near term and $2.3 to $3.5 ($3.2 to $4.6) per gallon in the long term.

Europe is not as well endowed with renewable energy resources as the United States. Except for the Mediterranean region, insolation is relatively poor throughout the continent. Hydroelectric resources, which are not yet developed, will probably be limited by environmental constraints, and the exploitation of wind resources will probably be limited to a fraction of the potential, mainly because of land use constraints (see chapter 4: *Wind Energy: Resources, Systems, and Regional Strategies*). Although considerable excess agricultural land will be available for biomass production (see chapter 14), the land areas are likely to be much less than in the United States, where cropland area per capita is nearly three times as large as in Europe. Thus, as the world shifts to renewable energy, Europe will have to depend on renewable imports for some of its energy needs.

Most North African countries have excellent solar resources. In the long term, many of these countries could become exporters of solar energy-carriers to European markets because their solar resources far exceed local requirements. Like today's oil-rich countries, they could finance their further development by selling solar hydrogen, but without the fear of exhausting their energy resources. The development of a large-scale solar energy system would involve joint ventures between European and African countries.

Development of a solar-based energy supply system may also include land areas in southern Europe (e.g., southern Spain). But these areas would serve only an interim function because of limited area available for large solar-hydrogen systems. Most of these areas would be needed to produce solar electricity and heat for the country's own demands and thus would play no role in the export market.

Reference solar hydrogen systems
To assess hydrogen's role in the development of solar-energy supply systems, some large solar-electrolytic hydrogen systems have been defined, based on solar-photovoltaic and solar-thermal electric power plants (see table 11). (Both these technologies will contribute considerably to direct solar electricity generation before hydrogen production will be of importance.)

Two types of advanced (post-2000) PV plants are considered (see table 32):

▼ An advanced (post-2000) polycrystalline technology (with a module efficiency of 16 percent and module cost $100 per m^2), assuming mass production in the 100 MW-per-year range for each factory. The balance-of-system costs are derived from detailed design studies by the German PV industry using a size of 15 MW (AC) for the smallest unit, including a DC/AC-converter and transformer to 20 kV, [3];

▼ An advanced (post-2000) thin-film technology (see chapters 9: *Amorphous Silicon Photovoltaic Systems* and 10: *Polycrystalline Thin-film Photovoltaics*).

The solar-thermal plant (see table 32) is based on solar tower technology. Costs are derived from different design studies mainly by German companies and institutes [e.g., for the PHOEBUS project (see chapter 5: *Solar Thermal-electric Technology*)] and represent the post-2000 status of commercially available plants in the 100 to 200 MW size with rigid heliostats [46].[21]

In contrast to the U.S. case study, the connection between the PV plant and the electrolyzer is made on the AC current level. Although direct coupling of the PV plant and electrolyzer seems to be more efficient and cheaper, it probably will not be used in very large systems. In PV plants, the DC voltage has to be restricted to a maximum of 500 to 1,000 volts, mainly because of short-circuit strength of modules and strings. Coupling large systems on DC level would therefore lead to both high currents and high losses because of the long distance the currents would travel to the electrolyzer. Moreover, using the AC level allows for a flexible combination of solar plants and electrolyzers on different sites using the AC electric grid as a link. Since the European case study focuses on very large-scale (multi-gigawatt) systems, producing hydrogen for export, AC coupling was chosen.

To explore the prospects for solar hydrogen in Europe and North Africa, detailed assessments were carried out for:

▼ *North Africa.* The Tinrhert Plateau, which lies directly north of the East Algerian crude oil area near In Amenas, offers 21,000 km^2 of absolutely smooth and flat lime plains called "hammadas." PV hydrogen could be transmitted to Central Europe via a 3,300 kilometer gas pipeline or it could be liquefied and transported by tankers a distance of 5,700 kilometers (see tables 21 and 23).

▼ *Southern Spain.* From Andalusia, PV hydrogen could be transmitted to Central Europe via a 2,000 kilometer pipeline (see table 21).

Each solar hydrogen plant is designed to generate 200 terawatt-hours of electricity per year, an amount equivalent to 50 percent of the gross electricity production in 1988 of the Federal Republic of Germany. This large plant size was selected mainly because of the economics of a transcontinental pipeline, which should be designed for some 10 GW(th) of energy flow.[22] Depending on the location and energy source, the annual rated utilization of the installations is 1,800 hours per year (PV Spain), 2,100 hours per year (PV Africa) and 3,600 hours per year (solar thermal with 4 to 5 hours of daily thermal storage), leading to the figures of rated power shown in table 11.

Energy losses in converting solar electricity into gaseous hydrogen (GH$_2$) and liquid hydrogen (LH$_2$) and transmitting them to Central Europe are 19 per-

---

21. The cost reduction potential of plastic membrane heliostats was not considered in this example. Also, solar thermal plants based on parabolic linear concentrators (trough collectors with SEGS technology) offer the same potential and, moreover, have the advantage of being commercially available today.

22. The natural gas pipeline from Algeria to Sicily is designed for 13.5 GW(th) energy flow. The plant size has no influence on the specific cost of the other system elements because solar hydrogen systems are modular and show no economy of scale beyond about 10 MW$_e$ (100 MW for solar tower). The first solar hydrogen plants could be much smaller without any increase in hydrogen cost, assuming hydrogen is distributed in a local gas network.

cent (GH$_2$ from Spain), 22 percent (GH$_2$ from North Africa), and 34 percent (LH$_2$ from North Africa). Storage and transmission account for about 10 percent of total losses. Most losses occur during electrolysis and liquefaction.

Land requirements are considerably higher for solar-thermal plants than for PV, because of the low land-use factor of 0.25 for heliostats, which have to track the sun's position.

Despite higher transmission losses, the production of hydrogen in North Africa is more favorable, reducing the land requirements to 90 percent of those required for siting in Spain. Producing the same amount of hydrogen in Central Europe would require about twice the land area (including transmission losses of the other options) and would cause a correspondingly higher hydrogen cost. This shows the importance of joint development of large solar-energy systems by European and North African countries.

Costs of solar hydrogen

Solar-electric generating systems dominate the capital cost of solar hydrogen plants, requiring 78 to 88 percent of the total investments. Total investments for PV hydrogen plants in North Africa range from $134 billion (PV II) to $192 billion (PV I) (see tables 12 and 13). (These costs are 11 to 15 percent lower than for equivalent installations in southern Spain.) Liquefaction adds about $40 billion.

The capital investments should be compared with the contributions to the energy consumption of the consumer country (see also figure of merit). Measured against the consumption of the FRG in 1989, the amounts of hydrogen delivered by **one** large installation as described here are 22 percent of natural gas consumption in the case of GH$_2$ supply and 20 percent of fuel consumption in the case of LH$_2$ supply. Costs of **gaseous hydrogen** delivered to Central Europe range from $22.0 to $33.1 per gigajoule for hydrogen from North Africa and from $23.3 to $37.2 per gigajoule for hydrogen from southern Spain, assuming a 6 percent discount rate (see table 12). The large cost difference between the PV I and PV II systems underscores the importance of low-cost electricity to the overall economics.

Storage and transmission accounts for 12 percent (in Spain) to 17 percent (in North Africa) of the cost of imported hydrogen.[23] Importing GH$_2$ from North Africa is roughly 15 percent less expensive than importing it from southern Spain. Production of hydrogen in Germany would result in a cost increase of 40 to 50 percent compared with imported hydrogen from North Africa [4]. The lower cost of hydrogen produced in North Africa highlights the importance of hydrogen trade and international cooperation in hydrogen energy development.

Liquefying hydrogen requires additional costs of about 50 to 60 percent compared with GH$_2$ production, resulting in LH$_2$ costs in Central Europe of $36.1 to $50.2 per gigajoule. Investments in liquefaction and transmission equipment would be five times the investment of storing and transporting GH$_2$

---

23. The assumed low cost underground storage of hydrogen supposes availability in the vicinity of the electrolyzer facility of either exhausted gas fields (e.g., in North Africa) or suitable geological formations where underground storage of hydrogen is possible.

from North Africa, if the electricity consumption of the liquefaction plant, as well as filling losses during transmission, are included (see table 13).

TOWARD RENEWABLE HYDROGEN ENERGY

If performance and cost goals are met, hydrogen from renewable sources could begin to offer an environmentally benign and economically acceptable alternative to fossil fuels early in the 21st century and eventually could provide a substantial portion of the world's energy needs. But such a scenario cannot be realized without continued research and development programs and new public policy initiatives to address environmental and energy supply security issues.

Development of intermittent renewable electric technologies
The development of low-cost solar-electric technologies is a prerequisite for hydrogen derived electrolytically from wind, PV, or solar-thermal electric sources. Also, engineering designs for central station electric generation would be applicable to large solar hydrogen systems. Thus, research and development on and market development initiatives for intermittent renewable electric technologies warrant high priority among the efforts needed to promote renewable hydrogen energy.

Development of electrolysis for intermittent power sources
Alkaline electrolysis is considered to be a mature commercial technology, and several large (greater than 10 MW) electrolyzers have been built at sites where low-cost hydroelectric power is available. However, these systems have not been optimized for use with diurnally varying or intermittent wind or solar power sources. At present, there is limited experience with PV-powered electrolyzers and none with wind or solar-thermal electric powered electrolyzers. While there have been no undue problems with PV-powered electrolysis experiments, the long-term behavior of solar-powered electrolyzers is not well known. Development of electrolyzers for use with solar electric sources must be pursued if issues of optimum design and electrode lifetime under intermittent operation are to be addressed.

Development of biomass hydrogen technologies
Because hydrogen produced from biomass is likely to be the least costly of the renewable hydrogen options, a range of efforts relating to the production of biofuels is needed. Such an approach should include research and development on biomass residue recovery and processing, the growing of biomass on dedicated plantations, and relevant conversion technologies.

The only component for producing hydrogen thermochemically from biomass that is not commercially available is a biomass gasifier. Coal gasifiers are commercially available, and modest efforts to develop biomass gasifiers were launched over the past two decades, aimed mainly at making methanol. Essentially the same gasification technology can be used for making hydrogen as well.

High priority should be given in research and development (R&D) programs to such technologies—with special emphasis placed on indirectly heated biomass gasifiers.

Biofuels production and marketing are likely to be carried out most efficiently by firms that differ in key respects from today's energy companies—perhaps combining features of existing oil, energy utility, chemical process, forest product, and agricultural industry firms. Policies to promote the development of a biofuels industrial infrastructure are also needed.

Development of hydrogen end-use technologies

In addition, research and development on hydrogen end-use technologies must be expanded. Technologies for hydrogen use in transportation (e.g., light-duty vehicles, buses, and trucks), fuel cells, and hydrogen-heating and cogeneration systems should receive high priority. Because hydrogen from renewable sources is likely to cost more than fluid fossil-fuels, emphasis should be given to highly energy-efficient technologies.

The fuel cell is perhaps the single most important end-use technology for hydrogen and should be emphasized accordingly. Because fuel-cell unit costs must be especially low for transportation markets, fuel-cell development and marketing strategies are needed that offer the potential for reducing costs by capturing economies of mass production. Fuel-cells for cogeneration applications in buildings (such as domestic water heater/cogeneration systems) are likely to reach cost-competitiveness before fuel cells for transportation, so the development and marketing strategies for the two sectors should be coordinated.

Demonstrations of hydrogen technology

A wide range of demonstration projects is needed to gain field experience with and to promote investor interest in the diverse set of technologies associated with the production and use of hydrogen as a fuel. Such wide-ranging demonstration efforts need not be especially costly: most will have costs that are modest compared with those needed to demonstrate more conventional technologies. In contrast to nuclear fission breeder reactors or fossil synfuel technologies that typically involve investments on the order of hundreds of millions of dollars, most hydrogen demonstration projects would costs hundreds of thousands of dollars, or at most tens of millions of dollars.

Some PV hydrogen demonstrations have already been carried out or are underway. Several small (one to ten kilowatt-sized) experimental PV-hydrogen systems have been built in the United States and Germany [47, 48–51, 9], and a number of other solar hydrogen demonstration projects are in the early stages of development [52]. In a joint venture, Germany and Saudi Arabia are building a 350 kW$_p$ PV electrolysis system in Saudi Arabia [16]. The largest PV hydrogen demonstration project planned is the 500 kW$_p$ Solar Wasserstoff project in Germany. In addition, the German government is studying the possibility of linking solar hydrogen systems in North Africa or southern Spain to Central Europe via pipeline or tanker transport of liquid hydrogen [52]. It has also been proposed that Canadian excess hydropower be used to produce hydrogen electrolytically and that the produced hydrogen be shipped to Europe via tanker [53] in the form of liquid hydrogen or methylcyclohexane for use as a transport fuel. Recently the South Coast Air Quality Management District and Electrolyzer, Inc. have begun a small solar PV hydrogen demonstration project, using the fuel for cars [54]. Although the great plains in the United States would be an ideal site for a wind hydrogen demonstration, none is planned at present.

Among possible new demonstration initiatives, proton exchange membrane fuel-cell vehicles and biomass gasifiers for hydrogen production warrant high priority. Both offer enormous potential, yet only recently have they been considered seriously for hydrogen applications, although both are sufficiently advanced technologically to be demonstrated. Projects now underway for building experimental hydrogen fuel-cell vehicles [35–39] should be expanded in number and in scope. They should, for example, include the demonstration of various ancillary technologies for dispensing hydrogen and storing both hydrogen and electricity. There are several candidate biomass gasifiers that have operated at pilot scale that could be scaled up for commercial demonstration with biomass fuels [13].

Complete hydrogen systems should be tested, even though most of the components (e.g., solar-electric generators, electrolyzers, and end-use devices) have already been demonstrated separately, to gain experience with hydrogen system integration and thereby help advance the commercialization of such systems by private energy companies, utilities, chemical companies, and/or automobile manufacturers.

Encouraging hydrogen market development

Commercializing any new energy technology is difficult, but hydrogen technology is likely to be especially so. First, the transportation sector, which is the most promising market, is dominated by liquid fuels that are extraordinarily easy to handle. Shifting to a gaseous fuel like natural gas or hydrogen means that fuel delivery and distribution would be more costly than for gasoline. Hydrogen is especially inconvenient for transportation; having just one-third the heating value of natural gas, it is more difficult to store. The challenge is made more daunting because in most instances a shift to hydrogen will require substantial changes in the production, delivery, and end-use technologies involved. And finally, even though renewable hydrogen (used in fuel cells) could become cost-competitive on a lifecycle cost basis with conventional technology, the cost advantage will be small. Thus, it is not likely that free market forces acting alone will lead to the replacement of internal-combustion-engine vehicles by fuel-cell vehicles.

Instead, strong market interventions are probably necessary if hydrogen is to play a major role in the global energy economy. Because the benefits of a cleaner environment and more secure energy supply are the major drivers for hydrogen energy, it is appropriate to shape public policies so as to hasten a transition to hydrogen energy. Both tax and regulatory policies have roles to play.

An oil tax or tariff, for example, might be levied to reduce a country's energy security risks. A carbon tax could help abate global warming. Other pollutant emissions might be taxed as well.

While taxes would be helpful, strict environmental regulations would be more powerful drivers of change. There are two reasons for this judgment. First, the cost of fuel constitutes a relatively small fraction of the overall cost of transportation, so that large taxes might be needed to promote a shift to such radically new technologies as hydrogen fuel-cell vehicles. Second, hydrogen technologies will often provide the most promising approaches for meeting tough environmental goals, such as those posed by requirements for ultra-low emission or zero-emission vehicles.

The challenge in promoting a shift to hydrogen energy is to frame policies sufficiently broadly (i.e., by setting performance criteria but not specifying what the technology should be) so as to foster a stream of innovations that will make hydrogen technology broadly attractive to consumers.

International cooperation

Although renewable hydrogen resources are more widely available than fossil fuel resources, some regions are much better endowed than others. Low-cost hydrogen could be produced in these regions at levels far in excess of local needs for export to distant markets. Accordingly, international joint ventures involving technology transfer, cooperative development, and international energy trade are likely to be important in a world where solar hydrogen is a major energy carrier. Such joint ventures could benefit all participants: the importing country will have access to less costly sources of hydrogen, and the exporter could use the sale of renewable hydrogen to help finance economic development, without the risk of resource depletion facing oil and natural gas exporters.

In addition, international cooperation is needed to adequately address the global environmental and energy security risks that are associated with fossil production and use. Renewable hydrogen could play an important role in reducing such risks.

Appendix A
PEM fuel cells for residential cogeneration

Proton exchange membrane (PEM) fuel cells, which are compact and modular, could potentially achieve low costs at sizes of less than 10 kilowatts. Fueled directly with hydrogen, PEM fuel cells based on near-term technology could convert 40 percent of the heating value of hydrogen to electricity and 45 percent to heat in the form of hot water at 70 to 100° C. These attributes make PEM fuel cells possible candidates for residential cogeneration applications. Here we consider the economics of PEM fuel cells for residential cogeneration from the perspective of an electric utility, comparing various dispersed and central station options for providing peak power (see table 24).

If the fuel cell cogeneration system is sized to meet just the domestic hot water requirements (a heat load that varies little throughout the year) it should be feasible to use essentially all the heat produced. For typical hot water loads in the United States (18.8 gigajoules per year), the average rate of electricity production from a fuel cell hot water cogeneration system would be about 0.5 kW$_e$ per household, and the annual electricity production per household would be about 4,600 kilowatt-hours per year (see table 24)—about half the average amount of electricity consumed per household in the U.S. If such systems had about one day's storage capacity for hot water, an electric utility could own and "dispatch" these electricity generators so that they would provide electricity when the utility needs it the most. Such systems would produce electricity and heat primarily during the utility's peak demand period. Electricity not consumed onsite would be exported to the utility grid and the hot water produced would be stored for use throughout the day.

This possibility is investigated elsewhere in this book (see chapter 23: *Utility Strategies for Using Renewables*) for a utility where the peak demand occurs in the afternoon. It is shown there that hydrogen fuel-cell water heaters operated on average at about 30 percent of the installed capacity (e.g., about 1.7 kW$_e$ of installed capacity per household) could economically displace large (100 MW$_e$) advanced combined-cycle power plants on the utility system, even though the delivered price of hydrogen to residential users may be 2.5 times the price of natural gas paid by utilities and the unit capital cost of the hot water heater cogeneration is likely to be perhaps 60 percent higher than that for the combined-cycle plant. When these systems operate on average at less than 39 percent of rated capacity (corresponding to a capacity factor CF $\leq$ 0.39), the net cost of electricity generation would be less for the fuel-cell system (see table 24).

The fuel-cell system is economically attractive for several reasons:

▼ The "net effective efficiency" of power generation (taking credit for the hot water that would otherwise have to be provided by burning extra fuel) is high— about 130 percent for a fuel cell cogeneration system that displaces a 65 percent efficient natural gas water heater.[24]

▼ The cost for the natural gas water heater which does not have to be purchased can be taken as a credit against the cost of the cogeneration system.

▼ These cogeneration systems offer the utility "distributed benefits" as deferral of investments in transmission and distribution, reduced transmission and distribution losses, increased power system reliability, etc. (see chapter 23).

---

24. For each unit of electricity produced, $1/0.4 = 2.5$ units of fuel are required. However, $0.45 \cdot 2.5 = 1.125$ units of heat are generated as a by-product. If this heat were instead provided by a natural gas water heater, some $1.125/0.65 = 1.731$ units of fuel would be required. Thus, effectively, only $2.5 - 1.731 = 0.77$ units of fuel are required to produce 1 unit of electricity, corresponding to an efficiency of $100 \cdot (1/0.77) = 130$ percent.

Appendix B
Small-scale production of hydrogen vial steam-oxidation of iron as an alternative to hydrogen storage

Hydrogen can be produced from renewable resources or steam reforming of natural gas or coal gasification. Hydrogen can also be produced by reacting steam with iron via the steam-oxidation reaction:

$$3\,Fe + 4\,H_2O \rightarrow Fe_3O_4 + 4\,H_2$$

In fact, industrial hydrogen was once produced using this method, before inexpensive natural gas made steam reforming a more attractive option. The steam oxidation reaction occurs at temperatures of $25\,°C$ to $900\,°C$, although the reaction rate is so slow below $500\,°C$, that a catalyst is needed.

A New Jersey company, H-Power Corp., recently launched a program for producing hydrogen at small scale by reacting sponge iron[25] with steam (or water) at 80 to 200°C, in the presence of a catalyst[26]. If this system were coupled with a fuel cell, it would be possible to provide steam for the hydrogen production reaction from the fuel cell exhaust and waste heat, even with a low temperature PEM fuel cell (see table 33).[27]

If small-scale systems for producing hydrogen from sponge iron and low temperature steam (or hot water) could be realized at low cost, the implications for hydrogen energy systems could be profound. The difficulties and expense of storing and transporting gaseous or liquid hydrogen could be avoided because free hydrogen would no longer be needed. Instead iron could be stored and transported and hydrogen produced as required. When the iron "fuel" is depleted (fully oxidized), the resulting iron oxide (Fe_3O_4) powder would be exchanged for fresh sponge iron. Reducing Fe_3O_4 back to sponge iron could be accomplished in direct reduction plants using a variety of reducing agents, [including hydrogen and synthesis gas (a gaseous mixture consisting largely of CO and H_2)], which could be produced from renewable resources or from reformed natural gas.

The Southern California Gas Company in conjunction with H-Power Corp. and Rolls Royce Company are investigating residential cogeneration systems coupling the H-Power steam oxidation system to a PEM fuel cell.

Here we examine a potential long term application of the H-Power system,

---

25. Today sponge iron is produced by direct reduction of iron ore (e.g., without smelting) by reacting a reducing gas such as hydrogen or CO or reducing solid with the ore (largely iron oxides) at 500 to 700 C. The resulting "sponge iron" product is about 90 to 94 percent iron by weight (83 to 89 percent metallic iron), with 1 to 3 percent carbon, 3 to 6 percent oxides, and traces of sulfur, phosphorus and other metals, depending on the composition of the iron ore and reducing gases. Sponge iron is a relatively inexpensive material that is sometimes used to provide iron for making steel in electric arc furnaces. It costs about $110 to $125 per tonne at the factory (compared with about $1,000 per tonne for structural steel).

26. J. Werth, H-Power Corp, U.S. Patent Application, April 1992.

27. The inputs to this system would be iron and air and the outputs, electricity and iron oxide. Hydrogen would be created as an intermediate step and consumed in the fuel cell.

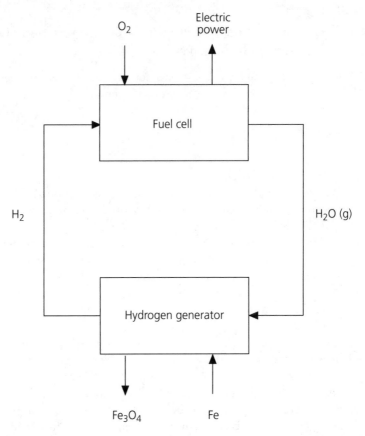

FIGURE B.1: Power system for "iron-powered" fuel cell vehicle.

a PEM fuel cell vehicle with an onboard iron hydrogen production system. Although a number of questions remain to be answered about practicality and cost of such a vehicle, several simple back-of-the-envelope calculations reveal the intriguing potential of this approach.

Onboard hydrogen production versus hydrogen storage

Systems for storing hydrogen onboard a fuel cell vehicle (in the form of compressed hydrogen gas, liquid hydrogen, and metal hydrides) are compared with systems for producing hydrogen onboard a vehicle (via reformation of methanol and steam oxidation of iron) (see figure B.1 and table 33). The H-Power system would produce about 4.8 percent hydrogen by weight of iron stored; whereas reformed methanol would produce 19 percent hydrogen by weight of methanol. By contrast, metal hydrides would store about 1.5 percent hydrogen by weight. For a given vehicle range, the weight and volume of a sponge iron hydrogen production system might be similar to that of compressed hydrogen gas storage at 8,000 psia (55 MPa). However, the cost would probably be much less [$500 versus $4,000 [32] for a system with a 250 mile (402 kilometer) range], and high pressure would not be required for refueling.

A "tankful" of sponge iron

The amount of iron needed for a fuel-cell vehicle with a 250 mile (402 kilometer) range—would be only 15 liters, less than half that for gasoline (see table 34). However, the iron would weigh 63 kilograms, more than twice as much as gasoline. After oxidation the iron oxide would weigh 88 kilograms and take up 33 liters. These figures do not include the weight and volume of the container holding the fuels and associated hardware. This would be small for gasoline but could be significant for sponge iron.

The price of sponge iron at the factory varies from $110 to 125 per tonne. Assuming that sponge iron could be delivered to the consumer for about $150 per tonne, a tankful would cost about $9.5. By comparison, a tankful of gasoline at $1.21 per gallon would cost $11.7.

Energy balance for producing sponge iron

The energy required to produce sponge iron from iron oxide (Fe_2O_3) (using the Midrex process) is about 11.1 gigajoules of natural gas per tonne of sponge iron. Assuming that 85 percent of the sponge iron by weight is metallic iron, this is equivalent to 13.0 gigajoules per tonne of metallic iron. For Fe_3O_4, the heat of reaction per kilogram of Fe produced is about 7 percent less than for Fe_2O_3. Thus the energy required for reducing Fe_3O_4 is assumed to be 7 percent less than for Fe_2O_3 or 12.1 gigajoules per tonne of metallic iron.

The hydrogen energy produced by steam oxidation of iron is 6.83 gigajoules per tonne of metallic iron. Thus the efficiency of producing hydrogen from natural gas via the intermediary of sponge iron is about 6.83/12.1 = 56 percent.

Resource requirements for U.S. light-duty vehicles

Sponge iron could be reduced with a variety of reductant gases, including pure hydrogen (which could be produced via wind or pv electrolysis) or with synthesis gas (which could be produced via biomass gasification).

Since the sponge iron used in a fuel-cell car will be recycled, the iron resource required per car is small, about 63 kilograms. However, if the car is driven 10,000 miles (16,000 kilometers) per year, this iron would be recycled 40 times per year, so that the recycling rate would be 2,500 kilograms per year per car. If the entire U.S. light duty fleet were based on this technology, the annual recycling rate would be about 500 million tonnes per year. This is approximately equal to seven times the global direct reduction capacity for producing sponge iron from iron ore (iron oxide) and ten times the pig iron production in the United States. Moving toward "iron fuel cell vehicles" would require a vast scaleup of sponge iron processing capacity, even though the primary iron resource requirements would be small.

One source of reducing gases for sponge iron production from iron oxide is biomass. If the overall energy efficiency of producing hydrogen in this way is 56 percent, (the same as for natural gas), then 5.9 exajoules per year of biomass would be needed. The biomass would require 20 million hectares of plantation land, assuming a productivity of 15 dry tonnes per hectare per year. This is about two-thirds of the idle cropland in the United States in 1990. If PV hydrogen were used to reduce iron oxide, the land requirements would be substantially less.

Unresolved questions

These back-of-the-envelope calculations suggest that onboard steam oxidation might be an interesting alternative to hydrogen storage for fuel cell vehicles. Not only do the resources to produce sponge iron from renewables exist, but the cost for a sponge iron based hydrogen production system could be considerably less than for hydrogen storage. Nevertheless, a number of questions would have to be addressed before the "iron-powered fuel-cell vehicle" could become a reality.

▼ What is the best catalyst? How much catalyst would be needed, and what would the added cost be? Would the reaction rates with the best catalyst be fast enough to make this system practical for vehicle applications? Although several catalysts have been identified, the optimum material would have to be determined from analysis and experimentation.

▼ By what process would the catalyst be removed from the spent fuel (iron oxide) and recycled into the fresh fuel (sponge iron)? This would depend on the catalyst chosen.

▼ What purity of sponge iron would be needed for fuel cell operation? Would impurities in sponge iron affect the operation of a PEM fuel cell? Since PEM fuel cells are sensitive to some contaminants, it would be important to assure that any potentially harmful impurities cannot be carried over into the fuel cell by the hydrogen.

▼ How would sponge iron be stored on a vehicle? What would be the cost and lifetime of an iron storage system?

▼ How would the vehicle be refueled?

▼ Could iron oxide be reduced cost effectively at a filling station or would transport of the iron oxide to a much larger central direct reduction plant be required?

H-Power researchers envision the storage system as several cylinders filled with sponge iron powder. When the sponge iron is fully oxidized, the cylinders could be removed, so that the spent fuel could be reduced to iron, then replaced in the car. Alternatively, the spent fuel could be "blown" out of the cylinders using pneumatic pressure and fresh fuel blown in. H-Power calculations indicate that small-scale iron reduction using reformed natural gas could be economically attractive, but these issues have not been investigated for renewable systems.

Appendix C
Economics of hydrogen fuel-cell automobiles

The performance and lifecycle cost of transportation are estimated for hydrogen fuel-cell automobiles as compared with other alternative vehicles. Table 27 presents estimated costs and characteristics of several alternative fueled vehicles that give similar performance and range (400 kilometers) [32, 40]:

▼ A gasoline-powered internal-combustion engine vehicle (ICEV) based on year 2000 projections for a compact car similar to a Ford Taurus.

▼ A methanol PEM fuel cell vehicle.

▼ A hydrogen PEM fuel cell vehicle with compressed gas storage at 55 MPa (8,000 psia).

▼ A battery powered electric vehicle (BPEV) based on a bipolar lithium alloy/iron disulfide battery.

Comparing the various vehicles we see that:

▼ The fuel-cell and battery-powered electric vehicles initially cost from $9,000 to $13,000 more than the gasoline vehicle[28].

▼ Fuel-cell vehicles have an energy efficiency about 2.5 to 3 times greater than that of a gasoline ICEV.

▼ The fuel-cell and battery-electric vehicles have a 50 percent longer lifetime than the gasoline vehicle and would have lower maintenance costs.

The delivered cost of fuel for these vehicles (including production costs, compression, storage, local distribution and filling station costs, but excluding taxes) is shown in table 28. In all cases:

▼ The delivered price of gasoline is $1.21 per gallon ($0.32 per liter or $9.2 per gigajoule), excluding taxes, which is the expected cost of reformulated gasoline for a crude oil price of $23 per barrel.

▼ Electricity is assumed to cost seven cents per kilowatt-hour, based on the projected average price of residential electricity in the United States for 2000.

For hydrogen and methanol, two production costs are given corresponding to 6 percent and 12 percent real discount rates.

▼ The delivered cost of methanol from biomass is $11.9 ($13.4) per gigajoule, assuming a 6 percent (12 percent) discount rate.

▼ Solar or wind hydrogen is estimated to cost $20.0 ($28.4) per gigajoule de-

---

28.   Compared with the gasoline vehicle, the methanol FCV costs about $9,000 more, primarily because of the added cost of the fuel cell and reformer; the hydrogen FCV with compressed gas storage costs about $13,000 more because of the cost of the fuel cell and the compressed gas storage tanks.

livered and biomass hydrogen $12.9 ($13.9) per gigajoule delivered.

While the initial cost of the vehicle and the fuel costs in dollars per gigajoule are higher for fuel-cell vehicles than for gasoline vehicles, hydrogen FCVs and methanol FCVs can compete with gasoline ICEVs or BPEVs on a **lifecycle cost** basis (see tables 29 and 30 and figure 5).[29] In all cases, biomass hydrogen and methanol fuel-cell vehicles would have a slightly lower life cycle cost than gasoline vehicles or battery-powered electric vehicles. Even with more expensive solar or wind electrolytic hydrogen, the lifecycle cost of transport with FCVs would be, at most, only slightly higher than that of gasoline ICEVs.

There are several reasons why FCVs compare so well to gasoline ICEVs:

▼ The energy efficiency of a FCV is about 2.5 to 3 times that of a gasoline powered ICEV. Even though hydrogen, methanol, and electricity are more expensive than gasoline on a dollar per gigajoule basis, the fuel cost per kilometer is substantially less (see table 29).

▼ Fuel-cell and battery-electric vehicles would have longer lifetimes than a gasoline ICEV. This means that the contribution of the initial cost to the total lifecycle cost is only about 10 percent higher for FCVs than for the gasoline ICEV, even though the initial cost of the FCV is about 60 to 75 percent higher.[30]

▼ Maintenance costs would be lower for fuel-cell vehicles than for gasoline ICEVs.

Savings relating to fuel, vehicle life, and maintenance approximately offset the higher initial costs of the fuel cell and hydrogen storage components, so that the overall lifecycle cost is similar for the FCV and gasoline ICEV.

The "breakeven price" of gasoline,(e.g., the price that would make the gasoline vehicle lifecycle cost equal to that of the alternative vehicles), can be estimated for each technology (see table 29). For example, if the fuel production cost discount rate is 6 percent for solar or wind electrolytic hydrogen, the breakeven gasoline price without taxes is $1.11 per gallon of gasoline ($0.33 per liter); for the biomass hydrogen case it is about $0.80 per gallon of gasoline $0.21 per liter), for the biomass methanol case, about $0.60 per gallon of gasoline ($0.16 per liter), and for the BPEV, $1.42 per gallon ($0.38 per liter).]

Although comparisons of fuel costs tend to dominate discussions of alternative transportation fuels, fuel costs are only a small fraction (typically about 1/8 or less) of the total cost of owning and operating a car (see figure 5). And the fuel cost per kilometer is lower for hydrogen, methanol, or electricity than for gasoline

---

29.   In figure 5 (top), the production cost of hydrogen and methanol are estimated based on a 6 percent discount rate; in figure 5 (bottom) a 12 percent discount rate is used.

30.   Table 30 shows the effect of vehicle lifetime on the lifecycle cost. Although experience with fleets of commercial, battery-powered electric vehicles indicates that vehicle life would be about 50 percent longer, this might not be achieved. Even if the fuel cell vehicle is assumed to have a lifetime only 10 percent longer than the ICEV, the total lifecycle cost would be only 10 percent higher for the FCVs than for the gasoline ICEV.

because of the high efficiency of electric vehicles as compared with ICEVs. In many important respects hydrogen resembles electricity more than gasoline because it is a high quality form of energy which can be used very efficiently and cleanly and can be made from a wide range of sources. Because of this difference in energy quality, direct comparisons of hydrogen and gasoline costs are less meaningful than comparing fuel costs per kilometer or total lifecycle costs.[31]

---

31. This is particularly important for PV hydrogen, the most widely available and least constrained but most expensive source of renewable hydrogen. In figure 3, the delivered cost of hydrogen transportation fuel (in dollars per gallon of gasoline equivalent energy) and the breakeven gasoline price are plotted as a function of PV module efficiency and cost. From this graph, it is possible to estimate what PV module efficiencies and costs would be required for PV hydrogen to compete with gasoline. If the delivered cost of fuel is taken as the sole figure of merit, we see that even in the long term (assuming PV module efficiencies of 12 to 18 percent, and costs of $30 to $55 per square meter), PV hydrogen would cost $2.5 to $3.6 ($3.1 to $4.9) per gallon of gasoline equivalent (GGE) assuming a 6 percent (12 percent) real discount rate. In the near term (with PV module efficiencies of 6 to 10 percent and costs of $70 to $200 per square meter), delivered hydrogen costs would be $4 to $10 ($6 to $14) per GGE. Based on the delivered fuel cost alone, it would be easy to conclude that PV hydrogen could never compete as a transport fuel. However, if we account for lifecycle costs by using the breakeven gasoline price as an indicator, near term estimates for PV would give breakeven gasoline prices of $1.6 to $3.5 ($2.0 to $5.0) per gallon, and long term estimates $0.9 to $1.26 ($1.2 to $1.7) per gallon. Clearly, the PV module efficiency and cost requirements to make PV hydrogen compete with gasoline would be much less stringent, if the total lifecycle cost is considered.

Table 1: Advanced alkaline electrolyzers

Electrolyzer type		Bipolar[a]		Unipolar[b]	
		Present	Future	Present	Future
Rated power	MW_e	10	100	10	100
Pressure	MPa	3	3	0.1	0.1
Temperature	°C	90	160	70	70
Type of diaphragm		Asbestos	$CaTiO_3$–Cermet	Asbestos	Synthetic
Rated current density	mA/cm^2	200	450	134	250
Maximum operating current density	mA/cm^2	267	600	168	333
Rated voltage	volts	1.86	1.7	1.9	1.74
Efficiency at rated current density					
HHV	percent	73	90	73	90
LHV	percent	62	76	62	76
Efficiency of rectifier	percent	96	98	96	98
Feed water	liters/GJ H_2 HHV	63	63	63	63
Cooling water	m^3/GJ H_2 HHV	2.5	2.5	2.5	2.5
Capital costs (including rectifier, building)	$/kWAC	600	330	600	400
Capital costs for DC plant	$/kWDC			474	274
Annual O&M costs (percent of capital costs, including feed and cooling water costs and regeneration of KOH)		4	4	2	2
Lifetime in years		20	20	20	20

a. Estimates for bipolar electrolyzers are from [3];1989 U.S. dollars, are calculated from 1.8 DM = $1 (1989).

b. Estimates for unipolar technology are from [4–7].

Table 2: Current and projected costs for solar electricity
¢ per kWh

Technology	1991	near term	post 2000
INTERMITTENT ELECTRICITY[a]			
Discount rate = 6 percent			
Wind[b] *630 W/m²*	–	3.7	2.3
500 W/m²	6.9	4.3	2.7
350 W/m²	9.5	5.8	3.5
Solar thermal electric (southwest U.S.)	9–13	8–13	4.6–6.5
Solar PV (southwest U.S.)	12–30	6–13	2–4 (DC)
			3–5 (AC)
Discount rate = 12 percent			
Wind[b] *630 W/m²*	–	5.4	3.3
500 W/m²	9.9	6.5	3.9
350 W/m²	14	8.8	5.2
Solar thermal electric (southwest U.S.)	13–18	11–19	6.7–9.4
Solar PV (southwest U.S.)	20–49	10–21	3–6 (DC)
			4–7 (AC)
Hydropower *Off peak*	1–4	1–4	1–4

a. The levelized production cost of intermittent electricity is given at the generation site assuming no storage. See tables 14-17 for details.

b. The annual average wind power density at hub height is indicated in italics.

Table 3: Current and projected production costs of hydrogen[a]
$ per GJ

	1991	Near term	post 2000
Discount rate = 6 percent			
RENEWABLE SOURCES			
Electrolytic hydrogen (for plants producing 14,160 Nm^3 (180 GJ) H$_2$ per day) from:[b]			
Solar pv (southwest U.S.)	45–101	24–47	10–16
Wind 630 W/m^2			11
500 W/m^2	31	21	15
350 W/m^2	44	32	17
Solar thermal (southwest U.S.)	37–50	31–52	18–25
Off peak hydroelectricity[c]	9–19	9–19	9–19
Hydrogen from biomass gasification[d]			
Large plant *17,360 GJ H$_2$/day*			5.9– 8.5
FOSSIL SOURCES			
Hydrogen from steam reforming of natural gas[e]			
Large plant *36,170 GJ H$_2$ per day*	5– 7	5– 7	7–10
Small plant *180 GJ H$_2$ per day*	11–14	11–14	14–16
Hydrogen from coal gasification[f]			
Large plant *36,170 GJ H$_2$ per day*	8	8	8
Small plant *9,041 GJ H$_2$ per day*	13	13	13
Discount rate = 12 percent			
RENEWABLE SOURCES			
Electrolytic hydrogen (for plants producing 180 GJ H$_2$ per day) from:[b]			
Solar pv (southwest U.S.)	73–167	37–76	15–25
Wind 630 W/m^2			17
500 W/m^2	45	30	21
350 W/m^2	65	46	25
Solar thermal (southwest U.S.)	52–69	45–73	27–36
Off peak hydroelectricity[c]	11–21	11–21	11–21
Hydrogen from biomass gasification[d]			
Large plant *17,360 GJ H$_2$ per day*			6.8– 9.4
FOSSIL SOURCES			
Hydrogen from steam reforming of natural gas[e]			
Large plant *36,170 GJ H$_2$ per day*	5– 8	5– 8	8–10
Small plant *180 GJ H$_2$ per day*	12–15	12–15	15–18
Hydrogen from coal gasification[f]			
Large plant *36,170 GJ H$_2$ per day*	9	9	9
Small plant *9,041 GJ H$_2$ per day*	14	14	14

a. Levelized hydrogen production costs are given.

b. A hydrogen plant producing 180 gigajoules per day could provide enough energy to fuel about 1,000 fuel cell fleet vehicles, each traveling 48,000 km per year.

c. Assuming that off-peak hydroelectricity at existing sites costs 1 to 4 cents per kWh.

d. Assuming that the biomass feedstock costs $2 to $4 per gigajoule.

e. Assuming that natural gas costs $2 to $4 per gigajoule in the 1990s and $4 to $6 per gigajoule beyond the year 2000, which is the range projected for the year 2000 for industrial and commercial customers.

f. Costs for hydrogen from coal gasification are based the steam–iron process [55], assuming coal costs $1.78 per gigajoule, which is the projected cost for the year 2000.

Table 4: Potential resources for renewable hydrogen production

| | Technically usable hydro-power | Electrolytic hydrogen from | | Biomass H$_2$[a] |
| | | Total wind potential | PV on 1% land area | |
	EJ H$_2$ per yr	EJ H$_2$ per yr	EJ H$_2$ per yr	EJ H$_2$ per yr
Africa	9.1	257	128	18
Asia	15.5	68	103	21
Australia	1.1	75	47	5
North America	9.1	308	94	17
South and Central America	11.0	122	77	24
Europe and the former USSR	10.6	366	130	24
World	56.3	1,196	579	113

a. produced on 10% of forests, woodlands, croplands

Table 5: Land and water requirements for renewable hydrogen production

	Land requirements		Water requirements
	$ha\ per\ MW_e$ peak	$m^2/(GJ/yr)$	liters/GJ HHV
Electrolytic hydrogen from:			
PV[a]	1.3	1.89	63
Solar thermal electric[b]	4.0	5.71	63
Wind[c]	4.7–16	6.3–33	63
Hydroelectric[d]	16–900	11–500	$\gg$63
Biomass Hydrogen[e]	–	50	37,000–74,000

Table 6: Land needed ($10^6\ km^2$) to produce hydrogen equivalent in energy to:

		Present demand			Projected world nonelectric fuel demand[f] (IPCC)	
Source	US light duty vehicles powered by fuel cells (4.8 EJ)	US oil (34 EJ)	World oil (115 EJ)	World fossil fuel (300 EJ)	2025 (286 EJ)	2050 (289 EJ)
PV	0.008	0.079	0.268	0.700	0.667	0.674
Wind	0.13	0.87	2.9	7.7	7.3	7.4
Biomass	0.23	2.2	7.6	19.8	18.9	19.0

a. It is estimated that a fixed, flat plate PV system is used, with array spacing so that half the land area is covered by arrays. The efficiency of the PV array is assumed to be 15 percent, the DC electrolyzer is taken to be 80 percent, based on the higher heating value of hydrogen, and the coupling efficiency between the PV array and the electrolyzer is taken to be 96 percent. Annual energy production is given for a southwestern U.S. location with average annual insolation of 271 watts per m^2. Water requirments are for electrolyzer feedwater.

b. Land use is estimated for a parabolic trough system, assuming that the efficiency (percentage of the solar energy falling on the collector area that is converted to electricity) is 10 percent, and that 1/4 of the land area is covered by collectors. (Land use per MW would be similar for central receiver or dish systems.) An electrolyzer with AC efficiency of 79 percent is used, and the coupling efficiency of the solar thermal electric plant and the electrolyzer is assumed to be 96 percent. Annual energy production is given for a southwestern U.S. location with average annual insolation of 271 watts per m^2. Water requirements are for electrolyzer feedwater only. If wet cooling towers were used for cooling the steam turbine condensors, there would be substantial water losses. The steam turbine would also consume some water during operation.

c. It is assumed that an array of 33 meter diameter 340 kW wind turbines is used. For areas with a unidirectional or bidirectional wind resource (as in some mountain passes), the wind turbine spacing could be 1.5 diameters in the direction perpendicular to the prevailing wind and 10 diameters in the direction parallel to the prevailing wind [56], without interference losses. In this case, the land use would be 4.7 hectares per megawatt of electric power. For areas with more variable wind direction (such as the Great Plains), the spacing would be 5 diameters by 10 diameters, with a land use of 16 hectares per MWe. An electrolyzer with AC efficiency of 79 percent is used. Coupling efficiency between the wind turbine and the electrolyzer is assumed to be 96 percent. The wind turbine capacity factor is assumed to be 26 percent, corresponding to a Class 4 site, with hub height of 50 meters. Water requirements are for electrolyzer feedwater.

d. Land use for hydroelectric power varies greatly depending on the location. The range shown is for large projects in various countries (see Chapter 2: *Hydropower and its constraints*.) Water requirements are for electrolyzer feedwater only. Evaporative losses at the reservoir would probably be much greater than feedwater consumption, depending on the site.

e. It is assumed that biomass productivity of 15 dry tonnes per hectare/year is achieved, and that the biomass has a higher heating value of 19.38 gigajoules per dry tonne. The energy conversion efficiency of biomass to hydrogen via gasification in a Battelle Columbus Laboratory gasifier is assumed to be 70.0 percent. Water use is based on a rainfall of 75 to 150 centimeters per year needed to achieve a biomass productivity of 15 dry tonnes per hectare (see chapter 14).

f. Projections are from the IPCC accelerated policy scenario (see chapter 1).

Table 7: Safety–related properties of hydrogen, methane, and gasoline[a]

		Hydrogen	Methane	Gasoline
Flammability limits in air	*percent volume*	4.0–75.0	5.3–15.0	1.0–7.6
Detonability limits in air	*percent volume*	18.3–59.0	6.3–13.5	1.1–3.3
Minimum energy for ignition in air	*mJ*	0.02	0.29	0.24
Diffusion velocity in air	*m s^{-1}*	2.0	0.51	0.17
Buoyant velocity in air	*m s^{-1}*	1.2–9.0	0.8–6.0	nonbuoyant
Leak rate in air (relative to methane)		2.8	1.0	1.7–3.6
Toxicity		nontoxic	nontoxic	toxic in concentrations > 500 ppm

Safety aspect	Relevant physical properties	Implications
Fire/explosion hazard		
Leak rate	Density, absolute viscosity	Hydrogen has a leak rate 2.8 x that of methane, 0.6–7.3 x that of gasoline
Ignitability	Minimum ignition energy, flammability/detonability limits, buoyant velocity, diffusion velocity	All three fuels ignite easily, hazard persists longest with gasoline, then methane and hydrogen
Physiological hazard	Toxicity	Gasoline toxic, hydrogen and methane nontoxic

a. Adapted from [41, 42]

Table 8: Potential for solar hydrogen production in the United States

	Hydroelectric[a]		Biomass[b]	Wind[c]	Solar PV[d]	Ratio of total H_2 potential to state energy use for transport with FCVs[e]
	off-peak	undeveloped				
Alabama	0.018	0.005	0.042	0	**0.810**	6.41
Arizona	0.015	0.021	0	0.029	**2.414**	21.4
Arkansas	0.007	0.012	0.018	0.064	**0.831**	10.9
California	0.056	0.096	0.015	**0.171**	**2.770**	3.5
Colorado	0.003	0.021	**0.159**	**1.396**	**2.020**	35
Connecticut	0.001	0.002	0	0.015	0.060	0.8
Delaware	0	0.002	0.0001	0.006	0.024	1.3
Florida	0.0003	0.0004	0.010	0	**0.816**	2.0
Georgia	0.013	0.010	0.054	0.003	**0.823**	3.7
Idaho	0.014	0.066	0.064	**0.212**	**1.315**	51
Illinois	0.0002	0.003	0.054	**0.177**	**0.79**	3.1
Indiana	0.0006	0.0007	0	0	**0.478**	2.7
Iowa	0.0008	0.004	**0.161**	**1.600**	**0.794**	28
Kansas	0.00001	0.001	**0.233**	**3.106**	**1.450**	55
Kentucky	0.005	0.011	0.035	0	**0.562**	4.8
Louisiana	0	0.004	0.012	0	**0.671**	4.9
Maine	0.004	0.025	0.004	**0.163**	**0.385**	12.8
Maryland	0.003	0.002	0.001	0.009	**0.131**	1.0
Massachusetts	0.002	0.003	0	0.073	0.097	1.0
Michigan	0.003	0.005	0.016	**0.189**	**0.757**	3.3
Minnesota	0.001	0.003	**0.150**	**1.907**	**1.057**	22
Mississippi	0	0.001	0.060	0	**0.712**	9.2
Missouri	0.003	0.008	**0.122**	0.151	**1.039**	7.2
Montana	0.014	0.032	**0.222**	**2.961**	**2.190**	182
Nebraska	0.002	0.003	**0.111**	**2.520**	**1.223**	73
Nevada	0.005	0.0004	0.0003	**0.145**	**2.240**	50
New Hampshire	0.003	0.005	0	0.012	**0.104**	0.3
New Jersey	0.0001	0.0006	0.00007	0.029	0.093	0.5
New Mexico	0.0004	0.0006	0.039	**1.263**	**2.473**	69
New York	0.026	0.016	0.005	**0.180**	**0.546**	2.1
North Carolina	0.012	0.013	0.012	0.020	**0.693**	3.3
North Dakota	0.003	0.004	**0.256**	**3.512**	**1.002**	195
Ohio	0.0008	0.002	0.020	0.012	**0.509**	1.6
Oklahoma	0.005	0.007	0.093	**2.105**	**1.217**	31
Oregon	0.036	0.036	0.042	**0.125**	**1.441**	19
Pennsylvania	0.005	0.020	0.008	**0.131**	**0.557**	2.2
Rhode Island	0.00005	0.0001	0	0.003	0.013	0.6
South Carolina	0.008	0.010	0.021	0.003	**0.455**	4.0
South Dakota	0.010	0.005	**0.169**	**2.99**	**1.212**	168
Tennessee	0.014	0.004	0.025	0.006	**0.584**	3.9
Texas	0.004	0.020	**0.319**	**3.454**	**4.412**	13.8
Utah	0.001	0.015	0.019	0.070	**1.600**	33
Vermont	0.002	0.004	0.00001	0.014	**0.107**	6.4
Virginia	0.005	0.011	0.007	0.035	**0.563**	3.0
Washington	**0.142**	0.070	0.080	0.096	**0.825**	8.9
West Virginia	0.001	0.017	0.00006	0.015	**0.299**	5.6
Wisconsin	0.003	0.004	0.049	**0.163**	**0.675**	6.3
Wyoming	0.002	0.015	0.021	**2.168**	**1.719**	184
Total	**0.455**	**0.660**	**2.76**	**31.30**	**47.56**	**10.40**

Important local resources are in bold.

a. Locally significant resources (defined here as 0.1 exajoules per year or more) are highlighted in boldface type.

b. The off-peak hydroelectric potential for hydrogen production is estimated assuming that power equal to the installed capacity could be available 25 percent of the time for off-peak hydrogen production. It is assumed that all the undeveloped hydropower devoted to hydrogen production (existing and undeveloped) is from the Federal Energy Regulatory Commission.

c. The biomass potential is based on lands held in the Conservation Reserve Program, which could be reforested with biomass assuming that biomass productivity of 15 dry tonnes per hectare per year is achieved and that the biomass has a higher heating value of 19.38 gigajoules per dry tonne. The higher heating value efficiency of converting biomass to hydrogen via gasification would be 70 percent. An additional amount of idled cropland would be available for biomass plantation development. Other sources such as residues and urban waste are not taken into account. They might add about 6 to 8 exajoules nationally, if they were available [57].

d. The wind energy available in each state is estimated for Class 3 and higher, assuming that 100 percent of urban and environmentally sensitive land, 50 percent of forest land and 30 percent of agricultural land are excluded [45]. An AC electrolyzer efficiency of 79 percent is assumed. The hydrogen produced by class is:

Wind Class	Hydrogen	Land use	Percent contiguous U.S. land area
	EJ per yr	km^2	
Class 3	15.2	579,449	7.5
Class 4	14.0	415,117	5.4
Class 5	0.9	27,944	0.4
Class 6	0.9	17,203	0.2
Class 7	0.2	273	0.003
Total wind class 3 to 7	**31.2**	**1,041,842**	**13.6**

e. Here the total renewable hydrogen potential in each state is compared to the energy which would be used for light vehicles in that state, based on projections for 2010 driving levels, if gasoline light duty vehicles were replaced by fuel cell vehicles with three times greater efficiency.

Table 9: Delivered cost of hydrogen from small plants c.2000

| | Electrolysis[a] 10 MW$_p$ | | | Steam reforming of natural gas | |
	PV	Wind	Hydro-power		
Land used by power system *hectares*	12–19	47–160			
Energy delivered *GJ per year*	66,000	76,000	84,000	64,000	
Vehicles fueled[b]	1,020	1,180	1,290	990	
Capital costs *1989$ millions*					
Power system	5.2–10.8	7.5	–	2.1	Reformer plant
Electrolyzer	2.3	3.7	4.0	–	
Compressor	0.43	0.41	0.41	0.11	Compressor
Storage	0.73	0.97	0.97	0.75	Storage
Filling station	0.5	0.5	0.5	0.5	Filling stat.
Total	**9.2–14.8**	**13.1**	**5.9**	**3.5**	**Total**
Contributions to hydrogen cost *$ per GJ*					
Power system	6.0–12.2	11.0	3.4–13.8	5.0	Plant cap.
Electrolyzer	3.9	5.5	5.2	2.3	O&M
Compression	2.2	2.0	0.9–1.8	0.8	Compress.
Hydrogen storage	1.2	1.4	1.1	1.4	H$_2$ storage
Filling station	2.5	2.5	2.5	2.5	Filling stat.
				5.4–8.1	Nat. gas[c]

Cost of hydrogen to consumer at filling station

$/GJ	15.9–22.1	22.3	13.5–24.6	18.4–21.1	
$/gal gasoline	2.29–3.17	3.23	1.97–3.56	2.66–3.05	

Breakeven gasoline price[d]

$/gal gasoline	0.90–1.18	1.18	0.79–1.29	1.01–1.13	

a. Costs and performance for pv and wind electrolysis systems are taken from tables 18 and 19. Costs of the electrolyzer for the hydropower system are taken from table 1. A discount rate of 6 percent is assumed for fuel production costs.

b. Fuel cell fleet vehicles with an efficiency equivalent to 60 miles per gallon gasoline, driven 48,000 kilometers per year.

c. For natural gas price of $4 to 6 per gigajoule.

d. The breakeven price of gasoline without taxes (see appendix C).

Table 10: Delivered cost of solar hydrogen based on post-2000 projections[a]

	Demonstration 10 MWp		City supply 750 MWp			Solar export 75 GWp
	PV	Wind	PV	Wind	Biomass	PV
	hectares		*km²*			*km²*
Land used	12–19	47–160	9–14	35–120	410	900–1,400
Energy delivered	0.066	0.076	5	7	7	500
PJ per year						
Vehicles fueled	~1,000[b]		~300,000[c]			~30 million
Capital costs $1989	*millions*		*billions*			*billions*
Power system	5–11	7.5	0.4–0.8	0.6	0.14	40–80
Electrolyzer	2.3	3.7	0.17	0.28		17
Compressor	0.43	0.41	0.02	0.02		1.0
Storage	0.73	0.97	0.04	0.04		1.1
Pipeline	–	–	–	–		1.9
Filling station	0.5	0.5				
Total	9.2–14.8	13.1	0.6–1	1.0	0.14	60–100
AT 6 PERCENT DISCOUNT RATE						
Contributions to hydrogen cost			*$ per gigajoule*			
Power system	6.0–12.2	11.0	6.0–12.2	11.0	5.9–8.5	6.0–12.2
Electrolyzer	3.9	5.5	3.9	5.5		3.9
Compression	2.2	2.0	1.2	1.2		1.2
Hydrogen storage	1.2	1.4	0.8	0.8		0.2
Pipeline (1,000 mi)	–	–	–	–		0.3
Local distribution	–	–	0.5	0.5	0.5	0.5
Filling station	2.5	2.5	5.2	5.2	5.2	5.2
Cost of hydrogen to consumer at filling station						
$ per gigajoule	15.9–22.1	22.3	17.6–23.8	24.1	11.6–14.0	17.3–23.5
$ per gal. gasoline	2.29–3.17	3.23	2.54–3.44	3.49	1.52–1.83	0.96–1.24
Breakeven gasoline price[d]						
$ per gallon gasoline			0.98–1.25	1.26	0.71–0.82	0.96–1.24

Table 10: (cont.)

	Demonstration 10 MWp		City supply 750 MWp			Solar export 75 GWp
	PV	**Wind**	**PV**	**Wind**	**Biomass**	**PV**
AT 12 PERCENT DISCOUNT RATE						
Contributions to hydrogen cost			*$ per gigajoule*			
Power system	9.8–20	15.8	9.8–20	15.8	6.8–9.4	9.8–20
Electrolyzer	5.5	7.7	5.5	7.7		5.5
Compression	2.5	2.3	1.4	1.4		5.5
Hydrogen storage	1.8	2.0	1.1	1.1		0.3
Pipeline (1,000 mi)	–	–	–	–		0.5
Local distribution	–	–	0.5	0.5	0.5	0.5
Filling station	2.5	2.5	5.2	5.2	5.2	5.2
Cost of hydrogen to consumer at filling station						
$ per gigajoule	22.0–32.3	30.4	23.4–33.6	31.7	12.5–15.2	23.0–33.9
$ per gal. gasoline	3.19–4.67	4.39	3.39–4.86	4.58	1.64–1.99	3.32–4.91
Breakeven gasoline price			*$ per gallon gasoline*			
	1.17–1.63	1.54	1.23–1.69	1.60	0.75–0.87	1.22–1.70

a. Performance for biomass gasifier, pv, and wind electrolysis systems are taken from tables 18, 19 and 20.

b. For passenger vehicles with efficiency equivalent to 60 mpg gasoline, driven 48,000 kilometers per year.

c. For passenger vehicles with efficiency equivalent to 60 mpg gasoline, driven 16,000 kilometers per year.

d. Breakeven price of gasoline without taxes (see appendix C).

Table 11: Main technical data for solar hydrogen production systems

Gaseous hydrogen

Power plant type		Southern Spain		North Africa	
		PV	ST	PV	ST
Electricity generation	GW_e	111	56	95	56
Hydrogen production	$GW\,H_2$	74.9[a]	50.1	64.3[a]	50.1
Underground storage	$GW\text{-}hours$	355	178	355	178
Transmission energy flow	$GW\,H_2$	29.3	29.6	29.3	29.6
Energy flow at pipeline outlet	$GW\,H_2$	26.9	27.1	25.9	26.2
Delivered hydrogen	*PJ/year*	583	587	562	566
Mean annual efficiency	*percent*				
Electric generation		13.0	16.1	12.0	18.0
H_2 production/storage		88.6	89.3	88.6	89.3
H_2 transmission		91.5	91.5	88.5	88.5
Total		10.5	13.2	9.4	14.2
Land area required	km^2	1,485	2,630	1,330	1,790
Land use factor		0.5	0.25	0.5	0.25

Liquid hydrogen (LH$_2$)

Power plant type		North Africa	
		PV	ST
Electricity generation	GW_e	95	56
Hydrogen production	$GW\,H_2$	53.3[a]	41.5
Liquefaction	$GW\,H_2$	53.3	41.5
Storage and transmission	$GW\text{-}hours$	8,790	8,790
Energy flow at LH$_2$ terminal	$GW\,H_2$	16.0	16.0
Delivered hydrogen	*PJ/year*	477	477
Mean annual efficiency	*percent*		
Electric generation		12	18
LH$_2$ production		74.6	74.6
LH$_2$ storage/transmission		89	89
LH$_2$/electricity		66.3	66.3
Total		7.94	12.0
Land area required	km^2	1,330	1,790
Land use factor		0.5	0.25

a. Rated power of electrolyzer is 85 percent of pv power.

Table 12: Main economic data of hydrogen production systems

Gaseous hydrogen from 200 TWh per year solar electricity

Location:	Southern Spain			North Africa		
Type of plant	PV I	PV II	ST	PV I	PV II	ST
Investments 10^9 $						
Power plant	185.0	117.2	184.0	158.7	100.5	147.0
Electrolysis	27.5	27.5	18.4	23.5	23.5	18.4
Gas storage and transmission	5.3	5.3	5.3	10.0	10.0	10.0
Total system	**217.8**	**150.0**	**207.7**	**192.2**	**134.0**	**175.4**
Figure of merit $/gigajoule/year	372	256	353	339	236	308
Specific energy cost (discount rate 6 percent) $ per GJ						
Electricity	26.1	13.3	26.1	20.3	11.4	20.8
Electrolysis	8.3	6.9	6.9	7.5	6.1	6.4
Storage and transmission	4.2	3.1	4.2	5.3	4.2	5.3
GH_2 in producer country[a]	32.2	20.2	33.0	27.8	17.5	27.2
GH_2 in central Europe[a]	36.4	23.3	37.2	33.1	21.9	32.5
Specific energy cost (discount rate 12 percent) $ per GJ						
Electricity	36.9	21.7	39.2	31.7	18.6	31.1
Electrolysis	12.2	10.0	9.4	10.3	8.6	8.6
Storage and transmission	6.4	4.4	6.4	8.6	6.7	8.1
GH_2 in producer country	49.1	31.8	48.6	42.0	27.2	39.7
GH_2 in central Europe	55.3	36.4	55.0	50.6	33.9	47.8

a. Rated power of electrolyzer is 85 percent of pv power.

Table 13: Liquid hydrogen (LH$_2$) from 200 TWh per year solar electricity (plant location is North Africa only)

Type of plant	PV I	PV II	ST
Investments 10^9 $			
Power plant	158.7	100.5	147.0
Electrolysis	19.8	19.8	15.4
Liquefaction	27.5	27.5	21.4
Storage and transmission	29.6	29.6	29.6
Total system	**235.6**	**177.4**	**213.4**
Figure of merit $ per gigajoule per year	492	369	444
Specific energy cost (discount rate 6 percent) $/GJ			
Electricity	20.3	11.4	20.8
Electrolysis	7.5	6.1	6.4
Liquefaction	11.6	9.7	10.0
Storage and transmission	10.6	8.9	10.3
LH$_2$ in producer country	39.4	27.2	37.2
LH$_2$ in central Europe	50.0	36.1	47.5
Specific energy cost (discount rate 12 percent) $/GJ			
Electricity	31.7	18.6	31.1
Electrolysis	10.3	8.6	8.6
Liquefaction	16.9	13.9	14.8
Storage and transmission	16.1	13.9	15.8
LH$_2$ in producer country	58.9	41.1	54.5
LH$_2$ in central Europe	75.0	55.0	70.3

Table 14: Cost and performance of solar photovoltaic modules

Solar PV technology	PV module efficiency *percent*			PV module manufacturing cost $\$ m^{-2}$		
	1990	**near term**	**post 2000**	**1990**	**near term**	**post 2000**
Flat plate modules						
Thin films						
Amorphous silicon[a]	6	8–10	12–18	100	70	30–55
CuInSe$_2$[b]	10	10	15	200	75–200	45
CdTe[b]	8	10	15	200	75–200	45
Thin film silicon[b]			16			50
Polycrystalline[b]	13		17	250–400		170–340
Crystalline[b]	15		20	500–800		200–400
Concentrator modules[c]	20	25	35	300-700	200	150

a. From chapter 9: *Amorphous Silicon Photovoltaic Systems.*

b. From chapter 10: *Polycrystalline Thin-Film Photovoltaics.*

c. From chapter 8: *Photovoltaic Concentrator Technology.*

Table 15: Cost and performance of solar photovoltaic systems[a,b]

		1990	near term	post 2000
Balance of system costs	$ m^{-2}			
Fixed flat plate		50–80	37–55	37
One-axis tracking			75	75
Two-axis tracking			125	100
Balance-of-system-efficiency[c]	*percent*	85	89	89
System lifetime	*years*	30	30	30
Annual O&M costs				
Fixed flat plate	$ m^{-2}/year	1.2	0.5	0.5
One- or two-axis tracking	*cents/kWh*	1	1	1
Indirect costs	*percent of capital cost*	33	25	25

Total installed system cost $ per kW_p

	1990	near term	post 2000
Flat plate systems			
Thin films	3,900–4,400	1,500–3,500	500–1,100
Polycrystalline	3,600–5,800		1,700–3,200
Crystalline	5,700–9,200		1,700–3,200
Concentrator systems[d] (two-axis tracking)	4,300–7,400	1,900	1,100

Cost of electricity[e] *cents per kWh DC*
Discount rate = 6 percent

	1990	near term	post 2000
Flat plate			
Thin films	14–15	5.6–11	1.8–3.8
Polycrystalline	12–19		5.7–10.6
Crystalline	19–30		5.7–10.6
Concentrator (two-axis tracking)	14–23	7.2	4.7

Discount rate = 12 percent

	1990	near term	post 2000
Flat plate			
Thin films	24–25	8.9–19	2.8–6.2
Polycrystalline	20–32		9.4–17.5
Crystalline	32–57		9.4–17.5
Concentrator (two-axis tracking)	23–38	11.4	7.2

a. PV system costs, except long term balance-of-system costs, from fixed flat plate systems are from K. Zweibel and A. Barnett, "Polycrystalline Thin Film Photovoltaics," chapter 10.

b. Long term balance of system costs ($37 m^{-2}) are from R. Matlin, Advanced Photovoltaic Systems, Princeton, New Jersey [58].

c. The balance of system efficiency is equal to dc system efficiency divided by module efficiency and accounts for system losses.

d. Levelized cost of DC electricity (in cents per kWhDC) is in southwestern United States, with average annual insolation of 271 watts per m^2. Estimates for concentrators are from Boes and Luque, chapter 8.

e. If AC power were produced instead of DC power, the power conditioning equipment would add an extra $150/kW. The balance of system efficiency for an AC system would be 85 percent rather than 89 percent because of energy losses in the inverter which is assumed to be 96 percent efficient. The cost of power would be about 0.6 cents per kWh greater than the costs shown here.

Table 16: Cost and performance of solar thermal electric technologies[a,b]

		1990	near term	post 2000
Parabolic trough systems				
Capital cost	$/kW	2,800–3,500	2,400–3,000	2,000–2,400
Peak capacity	MW_e	80	80	160
Annual energy efficiency solar mode	percent	13–17	13–17	13–17
Method for enhanced load matching			Natural-gas firing	
Fraction of kWh from gas	percent	25	25	25
Solar capacity factor	percent	22–25	18–26	22–27
O&M cost	cents/kWh	1.8–2.5	1.6–2.4	1.3–2.0
System lifetime	years	30	30	30
AC electricity cost[c]	cents per kWhAC			
discount rate = 6 percent		9.3–13	7.9–13.5	6.5–9.9
discount rate = 12 percent		13–18	11–19	8.7–13
Central receiver systems				
Capital cost	$/kW	3,000–4,000	2,225–3,000	2,900–3,500
Peak capacity	MW_e	100	200	200
Annual energy efficiency solar mode	percent	8–15	10–16	10–16
Method for enhanced load matching			Thermal storage	
Solar capacity factor	percent	25–40	30–40	55–63
O&M cost	cents/kWh	1.3–1.9	0.8–1.6	0.5–0.8
System lifetime	years	30	30	30
AC electricity cost	cents per kWhAC			
discount rate = 6 percent		8–16	5.8–10	4.6–6.5
discount rate = 12 percent		11–23	8.3–14	6.7–9.4
Parabolic dish systems				
Capital cost	$/kW	3,000–5,000	2,000–3,500	1,250–2,000
Peak capacity	MW_e	3	30	300
Annual energy efficiency solar mode	percent	16–24	18–26	20–28
Method for enhanced load matching			Solar only	
Solar capacity factor	percent	16–22	20–26	22–28
O&M cost	cents/kWh	2.5–5.0	2.0–3.0	1.5–2.5
System lifetime	years	30	30	30
AC electricity cost[c]	cents per kWhAC			
discount rate = 6 percent		15–33	8.8–1.9	5.5–11
discount rate = 12 percent		21–47	12–27	7.5–15

a.   Adapted from chapter 5.
b.   Levelized cost of electricity is computed for southwestern U.S. location.

Table 17: Cost and performance of wind power technologies

		1990[a]	near term[b]	post 2000[c]
Total installed cost	$ kW$_p^{-1}$	1,100	1,000	750
Turbine output	kW	100	340	1,000
Turbine diameter	meters	17.5	33	52
Availability	percent	90	95	95
Annual O&M costs including retrofits	¢/kWh	1.5	1.1	0.6
Rent on land	¢/kWh	0.3	0.3	0.3
Lifetime	years	25	30	30
Annual average capacity factor[d]				
Wind power density = 350 W m^{-2}		0.132	0.202	0.273
500 W m^{-2}		0.205	0.288	0.390
630 W m^{-2}		–	0.362	0.490
AC electricity cost	cents per kWh			
discount rate = 6 percent				
Wind power density = 350 W m^{-2}		9.5	5.8	3.5
500 W m^{-2}		6.9	4.3	2.7
630 W m^{-2}		–	3.7	2.3
discount rate = 12 percent				
Wind power density = 350 W m^{-2}		14	8.8	5.2
500 W m^{-2}		9.9	6.5	3.9
630 W m^{-2}		–	5.4	3.3

a. Cost and performance estimates are for U.S. Windpower 100 kW models (see chapter 3 and (9,10)).

b. Costs are estimated for mid 1990s wind turbine technology based on the U.S. Windpower 33 meter diameter variable speed drive model (see chapter 3 and [59]).

c. Costs and performance projections for advanced wind turbines are from projections for post 2000 technology by the Solar Energy Research Institute (see chapter 3 and [60, 57]).

d. The annual average capacity factor is given for three levels of average wind power density (350, 500, and 630 W m^{-2} of swept rotor area), measured at the rotor hub height. With present wind turbines, the hub height is typically 30 meters, and the average wind power density is 350 W m^{-2} in Class 4 wind regions and about 500 W m^{-2} in Class 5 to 6 wind regions. With near term technology, it should be possible to extend the height to 50 meters. In this case, the average power density would be 350 W m^{-2} for a Class 3 region, and about 500 W m^{-2} for a Class 4 to 5 region. Class 3 and 4 resources are widely found throughout the world (see chapter 4). In areas such as the U.S. Great Plains, the wind energy density at 50 meters is often much higher at night than during the day. Based on measured values at Bushland, Texas, the annual average energy density would be about 630 W m^{-2} at a Class 4 site with a nighttime "jet." The average wind power density in a Class 5 to 6 region would be about 600 watts m^{-2}. Class 6 regions, which are not as widely distributed geographically as Class 3 or Class 4 regions, would offer a wind power density of about 700 W m^{-2} at 50 meter hub height.

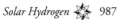

Table 18: Post 2000 wind electrolytic hydrogen system parameters

---

Horizontal axis wind turbine[a]

Turbine capacity	1,000 kW
Turbine diameter	52 meters
Hub height	50 meters
Total installed system cost	$750/kW peak
Annual O&M cost	$0.005/kWh AC
Land rent	$0.003/kWh AC
System lifetime	30 years
System availability	95 percent
Turbine spacing/turbine diameter	5×10
Hectares per MW_e	16
Efficiency of coupling to electrolyzer[b]	94 percent

Atmospheric pressure unipolar electrolyzer[c]

Rated voltage	1.74 volts
Rated current density	250 mA cm^{-2}
Maximum operating current density	333 mA cm^{-2}
Efficiency at maximum operating voltage	85 percent
Rectifier cost	$130/kW AC in
Rectifier efficiency	96 percent
Installed AC plant capital cost @ maximum operating current density	$371/kW AC in
Electrolyzer annual O&M cost	2 percent of capital cost
Electrolyzer lifetime	20 years
Total wind/electrolysis system capital cost	$1,461–1,592/kW H$_2$ out

Wind resource[d]		**Class 5**	**Class 4**	**Class 3**
Annual average wind power density (power per unit of area swept by turbine)	$W\,m^{-2}$	630	450	350
Annual average capacity factor		0.49	0.35	0.27
Levelized cost of DC electricity	$/kWh			
Discount rate = 6 percent		0.023	0.029	0.035
Discount rate = 12 percent		0.033	0.042	0.052
Levelized cost of wind hydrogen	$/GJ			
Discount rate = 6 percent		11.3	14.6	16.7
Discount rate = 12 percent		16.2	21.1	25.0

a. Costs and performance for wind systems are from chapter 3.

b. It is assumed that the wind system produces AC power, which is then rectified to DC for use in electrolysis. AC losses in transmission from the wind tower to the electrolyzer are assumed to be 6 percent. See Winter and Nitsch [1].

c. Electrolyzer operating characteristics and costs are based on currently available unipolar technology. It is assumed that the rectifier is sized for maximum current density [5–7]. The maximum current density is taken to be 1.25 times the rated current density [1, 9].

d. The annual average capacity factor is given for three cases, corresponding to sites with typical Class 3, Class 4, and Class 5/6 wind resources (350, 450, and 630 W m^{-2}, respectively). A Class 4 site with a nighttime "jet" would also have a wind power density of 630 W m^{-2}.

Table 19: Post 2000 pv electrolytic hydrogen system parameters

---

Thin film PV modules, tilted, fixed flat-plate array (> 10 MW$_p$)^a

pv module efficiency	12–18 percent
pv module manufacturing cost	$30–55 m^{-2}
Area-related balance of system cost	$37 m^{-2}
Balance of system efficiency	89 percent
pv system efficiency	10.7–16.0 percent
pv annual O&M cost	$0.5 m^{-2}/year
pv system lifetime	30 years
pv system indirect cost factor	25 percent
pv system capital cost	$522–1,077/kW dc
Efficiency of coupling to electrolyzer^b	93 percent (direct connection)
Cost of coupling to electrolyzer	negligible

Atmospheric pressure unipolar electrolyzer^c

Rated voltage	1.74 volts
Rated current density	250 mA cm^{-2}
Maximum operating current density	333 mA cm^{-2}
Efficiency at maximum operating voltage	85 percent
Installed dc plant capital cost @ maximum operating current density	$231/kW dc in
Electrolyzer annual O&M cost	2 percent of capital cost
Electrolyzer lifetime	20 years

Cost and performance of PV hydrogen system

System efficiency (H$_2$ HHV)/insolation	8.4–12.7 percent
Total capital cost	$954–$1,654/kW H$_2$ out

Solar resource

Annual average insolation^d	271 W m^{-2}
Land area required in SW U.S.	
10.7 percent efficient pv system	1.87 ha MW$_e^{-1}$
16.0 percent efficient pv system	1.25 ha MW$_e^{-1}$

Energy costs

Module efficiency	18 percent	12 percent
Module manufacturing cost	$30 m^{-2}	$55 m^{-2}

Levelized cost of DC electricity *$/kWh*		
Discount rate = 6 percent	0.018	0.038
Discount rate = 12 percent	0.030	0.061

Levelized cost of PV hydrogen *$/GJ*		
Discount rate = 6 percent	10.0	16.2
Discount rate = 12 percent	15.3	25.5

a.   Projected efficiencies and manufacturing costs for thin-film pv modules are from Carlson and Wagner (chapter 9) and

Zweibel and Barnett (chapter 10). Area-related balance of system costs are based on conceptual designs for large fixed, flat plate arrays are from [58] and [2]. Balance of system efficiency for a DC system is derived from USDOE estimates [62]. Operation and maintenance costs are projections based on field experience from EPRI [61] and SMUD [20]. Indirect costs of 25 percent are assumed based on Sandia experience with fixed, flat plate arrays [61, 62]. PV system lifetime of 30 years is taken from USDOE year 2000 goals [62].

b.   PV/electrolyzer coupling efficiencies are based on small experimental systems [47, 48–51, 9].

c.   Electrolyzer operating characteristics and costs are based on currently available unipolar technology. It is assumed that no rectifier is needed [48, 9]. The maximum current density is taken to be 1.25 times the rated current density [1, 9].

d.   Average annual insolation is given for the southwestern United States.

Table 20: Production of hydrogen from biomass[a]

Dry tonnes biomass per day		1,650
Biomass energy input	GJ/hour	1,382
External electricity input	MW_e	18.2
Thermal conversion efficiency		70%
GJ–hydrogen out/GJ–energy in (biomass + electricity)		
Plant lifetime	years	25
Plant capacity factor	percent	90
Total investment cost	10^6 $	137
Working capital	10^6 $	10.1
Land	10^6 $	2.05
Cost of biomass	$/GJ	2–4
Variable operating costs excluding biomass	10^6 $/year	9.24
Biomass costs	10^6 $/year	21.8–43.7
Fixed operating cost	10^6 $/year	7.20

Levelized costs	$/GJ
Discount rate = 6 percent	
Capital	1.43
Labor, maintenance, chemicals	1.15
Purchased electricity	0.79
Biomass	2.57–5.15
Total	**5.94–8.51**
Discount rate = 12 percent	
Capital	2.27
Labor, maintenance, chemicals	1.15
Purchased electricity	0.79
Biomass	2.57–5.15
Total	**6.78–9.35**

a. Based on the Battelle Columbus gasification technology. E. Larson et al., 1992 [13].

Table 21: Gaseous hydrogen: compression, storage, and transmission[a]

Pipeline diameter	1.7 meters	
Pipeline pressure	10 MegaPascals	
Hydrogen flow rate	29.6 gigawatts H_2 (HHV) = 8.3 million Nm^3/hour	
Capacity factor (daily storage assumed)	6,040 hr/year	
Useable storage volume	355 GWh H_2	
Storage losses (compressor drive)	2.9%	
Transmission distance (to Central Europe from)	2,000 km Southern Spain	3,300 km North Africa
Number of compressor stations[b]	4	8
Mechanical compressor power (3 megapascal electrolyzer pressure assumed)	720 MW_M	975 MW_M
Transmission efficiency if H_2-fueled gas turbines are used for compressor drive	91.5 percent	88.5 percent
Pipeline capital cost	$1,900/m	
Compressor station capital cost	$1,100 kW_M^{-1}	
Underground storage[c]	$2.4/kWh H_2	
Total specific capital cost	$177/kW H_2	$338/kW H_2[d]
Annual O&M cost	1.5 percent of capital cost	
Lifetime	30 years	

a. [3].

b. Optimized number with respect to pipeline diameter and pressure cost.

c. Existing underground storage volume assumed (e.g., depleted gas field).

d. Includes underwater pipeline of 300 kilometers length which has five times higher capital cost than overland pipeline.

Table 22: Energy density of fuels[a]

	Volumetric energy density MJ/liter		Mass energy density MJ/kg		Mass density kg/m³
	HHV	LHV	HHV	LHV	
Gasoline	34.5	32.3	45.9	43.0	720
LPG (100 psia)	25.5	23.6	50.0	46.3	510
Methane gas (2,400 psia)	6.16	5.51	55.5	50.0	111
Liquefied methane (−161°C)	23.9	21.5	43.4	39.1	550
Methanol	18.1	15.9	22.7	19.9	791
Ethanol	24.0	21.6	30.2	27.2	790
Hydrogen gas (2,400 psia)	2.09	1.76	142.4	120.2	14.7
(8,000 psia)	6.97	5.87	142.4	120.2	49
Liquefied hydrogen (−253°C)	9.95	8.4	142.4	120.2	71
Metal hydride	5.8	4.9	2.13	1.80	2,706
Lead-acid battery	0.42		0.144		2,900
Advanced battery	0.90		0.324		2,778

a. Adapted from [2]. Densities include the fuel only, not the container.

Table 23: Liquid hydrogen: liquefaction, storage, and transmission[a]

Capacity of liquefaction plant	118 megawatts H_2 (HHV)
Electricity consumption	0.23 kWh$_e$/kWh H_2
Capacity factor (adapted to electricity production from pv)[b]	2,800 kWh H_2/yr/kW H_2
Total efficiency (electrolysis included) LH_2/electricity	74.6 percent (HHV)
Capital cost of LH_2 plant	$515/kW H_2 (HHV)
Annual O&M cost (including cost for cooling water)	6 percent of capital cost
Lifetime	30 years
Unit size of LH_2 tanker	26,000 m^3 LH_2 (= 0.26 PJ LH_2)
Unit size of storage vessel	27,500 m^3 LH_2
Transmission distance	5,700 kilometers
Overall transmission efficiency at 17 transport cycles/year including storage and filling losses	89 percent[c]
Mean annual energy flow (8,300 hours per year):	0.149 gigawatts H_2 (HHV)
Capital cost of total system related to mean annual energy flow[c]	$1,646/kW H_2 (HHV)
Annual O&M cost	1.5 percent of capital cost
Lifetime	30 years

a. Reference technologies from [3] for mature technology.

b. pv: 2,100 hr per year, electrolysis and LH_2-plant: 2,800 hr per year at rated power.

c. Storage losses 0.03 percent per day; filling losses 8 percent.

Table 24: Hydrogen water heater cogeneration versus natural gas-fired advanced gas turbine combined cycle system[a,b]

	Net Electricity Production Cost cents per kWh	
	Hydrogen WH cogeneration[a,b]	Natural gas combined cycle[c]
Capital	1.23/CF	0.55/CF
Capital credit for natural gas water heater that is not needed	– 0.93[d]	–
Distributed benefits	– 0.91/CF[e]	–
Fuel	9.91	3.14
Fixed op. and maintenance cost	–[f]	0.17/CF
Variable op. and maintenance cost	–[f]	0.20
Credit for natural gas not needed for water heating	– 4.61[g]	–
Net electricity generation cost	4.37 + 0.32/CF	3.34 + 0.72/CF

a. It is assumed that the cogeneration system is sized to meet the hot water demand, some 18.8 gigajoules per year or 595 Wth per household in the U.S. [63].

b. With near-term technology, a hydrogen-fired proton exchange membrane fuel cell could convert 40 percent of the fuel energy (HHV basis) to electricity and 45 percent to heat, so that the average electrical output would be $(40/45) \cdot 595 = 529$ We and the annual electricity production would be 4,637 kilowatt-hours per year. It is assumed that the fuel cell system costs $1,000 per kWe installed, the discount rate is 6 percent, the annual insurance charge is 0.5 percent of the installed cost, the system life is 15 years, and the installed capacity is 0.529/CF kWe, where CF is the annual average capacity factor.
  Assuming the hydrogen is derived from biomass at $2.43 per gigajoule (see R. Williams and E. Larson, chapter 17: *Advanced Gasification-Based Biomass Power Generation*), the production cost would be $6.53 per gigajoule. To this should be added a distribution cost to large users of $0.5 per gigajoule plus local distribution charges. It is assumed that 80 percent as much energy will flow through a hydrogen pipeline as through a natural gas pipeline. For natural gas the difference between the gas prices for utility and residential users in the U.S. is $3.18 per gigajoule. For hydrogen it is assumed to be $3.18/0.80 = $3.98 per gigajoule. Thus the total delivered cost of hydrogen is $11 per gigajoule and the annual fuel cost would be:

(18.8 gigajoules/year) · ($11/gigajoule)/0.45 = $459.5

c. A 100 MWe advanced natural gas-fired combined cycle power plants with an efficiency of 50 percent (HHV basis) is expected to have an installed cost of $618 per kWe, a variable O&M cost of 0.2 cents per kilowatt-hour, and a fixed O&M cost of $15 per kilowatt-year (see chapter 23). It is assumed that the discount rate is 6 percent , the annual insurance charge is 0.5 percent of the installed cost, and the system life is 30 years. The natural gas price is assumed to increase from $4 per gigajoule in 2000 to $5 per gigajoule in 2030, corresponding to a levelized gas price of $4.36 per gigajoule. Here CF is the annual average capacity factor.

d. High efficiency natural gas water heaters are 65 percent efficient and typically cost $400. This cost is annualized, assuming a 6 percent discount rate, an annual insurance charge of 0.5 percent of the installed cost, and a system life of 15 years.

e. A conservative estimate of distributed benefits for distributed power systems is $80 per kilowatt-year (see chapter 23).

f. It is assumed that the operation and maintenance costs for fuel cell cogeneration systems are the same as for the natural gas water heaters they replace, so that net incremental operation and maintenance costs are zero.

g. The base case assumptions about natural gas prices for electric utilities in this book are that the price will rise from $4 per gigajoule in 2000 to $5 per gigajoule in 2030; this implies that the levelized price for the period 2000-2015 would be $4.21 per gigajoule, assuming a 6 percent discount rate. Also, the difference in natural gas prices for residential and utility users in the U.S. is $3.18 per gigajoule. Thus the levelized natural gas price to residential users would be $7.39 per gigajoule. Thus, for a 65 percent efficient natural gas water heater, the annual natural gas savings would be:

(18.8 gigajoules/year) · ($7.39/gigajoule)/0.65 = $213.7

Table 25: Onboard energy storage systems for automobiles[a]

Storage system	Installed energy density[b]		Container cost[c]	Refuel time[d]	Filling station markup[e]
	MJ/liter	MJ/kg	$	min	$/GJ
Hydrogen storage systems					
Carbon wrapped aluminum cylinder 55 MPa (8,000 psia)	3.4	7.0	4,000	2–3	5
Liquid hydrogen Dewar (–253 °C)	5.0	15.0	1,000–2,000	>5	11
FeTi metal hydride	2–4	1–2	3,300–5,500	20–30	3
Cryoadsorption (Carbon at –150 °C)	2.1	6.3	2,000–4,000	5	4–5
Thermocooled pressure vessel	2.5	8.0	4,000+	5+	5–8+
Organic liquid hydride	0.5	1.0	?	6–10	?
H–Power system (proprietary)	5.8	5.0	500?	2–3	?
Storage systems for other automotive fuels					
Gasoline	32.4	34.0	20	2–3	0.6
Methanol	15.9	14.9	20	2–3	1.2
Batteries:					
Lead–acid	0.187	0.115	$100/kWh	60–360	0.5[f]
Nickel–iron	0.407	0.18	$100/kWh	60–360	0.5[f]
Sodium–sulfur	0.432	0.378	$100/kWh	60–360	0.5[f]
Bipolar lithium alloy	0.925	0.511	15,000	60–360	0.5[f]

a. Adapted from [32]. It is assumed in all cases that enough energy is stored to travel 400 kilometers.

b. Energy density is based on the weight and volume of the full container plus auxiliaries.

c. Cost to the original equipment (automotive) manufacturer.

d. Time to deliver fuel only.

e. The mark-up (the cost in excess of the hydrogen cost to the filling station operator) is needed to cover the full cost of owning and operation the refueling station.

f. The cost of the home recharging station for the electric battery car does not include the price of electricity.

Table 26: Characteristics of fuel cells for transportation[a]

Type of fuel cell	Status[b] (1991)	Specific power[c]		Operating temp	Contam. by	Startup time
		kW/kg	*kW/liters*	*°C*		*min*
Phosphoric acid	CA	0.12	0.16	150–250		300
Alkaline	CA	1.49	1.47	65–220	CO, CO_2	120–720
Proton exchange membrane	D	1.33	1.20	25–120	CO	5
Molten carbonate	D	–	0.7	650		500
Monolithic solid oxide	L	8.3	4.0	700–1,000		100

a. Adapted from [32].

b. CA = commercially available, D = development of prototypes, L = laboratory.

c. Specific power includes only the fuel cell stack but no auxiliaries.

Table 27: Cost and performance of alternatively fueled vehicles[a]

		Gasoline ICEV	MeOH FCV[b]	Hydrogen FCV[b]	Battery EV[c]
Power to wheels	*kW*	98.4	74	72	81
Gasoline-equivalent fuel economy	*liters/100 km*	9.08	3.59	3.06	2.18
	mpg	25.9	64.8	76.5	108
Curb weight	*10³ kg*	1.37	1.27	1.24	1.44
Initial price	*10³ $*	17.3	26.3	30.4	29.6
Vehicle life	*10³ km*	193	289	289	289
Annual maintenance cost	*$*	516	416	401	358

a. Adapted from [32].

b. Fuel cell vehicles are assumed to use proton exchange membrane fuel cells. In the methanol vehicle, onboard reforming is used to produce hydrogen for the fuel cell. Compressed hydrogen is stored at 55 MPa (8,000 psia).

c. The battery considered for the battery-powered electric vehicle is a promising advanced battery—the bipolar lithium alloy/iron disulfide battery.

Table 28: Production and consumer costs for alternative renewable fuels[a]

	Alternative sources of hydrogen[b]		Methanol from biomass
	Wind/PV	**Biomass**	
		$ per gigajoule	
AT 6 PERCENT DISCOUNT RATE			
Production cost	12.3[c]	7.2[d]	9.1[d]
Compression	1.2[e]	– [e]	–
Storage	0.8[f]	– [f]	–
Local distribution	0.5	0.5	–
Refueling station	5.2	5.2	2.8[g]
Cost to consumer	20.0	12.9	11.9
AT 12 PERCENT DISCOUNT RATE			
Production cost	20.4[c]	8.2[d]	10.6[d]
Compression	1.4[e]	– [e]	–
Storage	1.1[f]	– [f]	–
Local distribution	0.5	0.5	–
Refueling station	5.2	5.2	2.8[g]
Cost to consumer	28.5	13.9	13.4

a. Adapted from [32].

b. See tables 18, 19, 20, and references [12, 13, 32].

c. This hydrogen production cost is a mid-range value for wind and photovoltaic (PV) power sources expected in the 2000+ time period, for the assumed discount rate. In the case of wind power, it is estimated that hydrogen can be produced at a cost in the range $16.2 to $25.0 per gigajoule, for projected AC wind electricity costs in the range $0.033 to $0.052 per kilowatt-hour. In the case of photovoltaic power, it is estimated that hydrogen can be produced at a cost of $15.3 to $25.5 per gigajoule, for DC PV electricity costs in the range $0.030 to $0.061 per kilowatt-hour. In both cases it is assumed that hydrogen is produced at atmospheric pressure using unipolar electrolyzers.

d. Both methanol and hydrogen can be produced from biomass via thermochemical gasification [13]. The present analysis is based on the use of biomass feedstocks costing $3 per gigajoule. The gasifier involved is the Battelle Columbus Laboratory (BCL) gasifier, which is under development and not yet commercially available. In the gasification process, the biomass is heated and converted into a gaseous mixture consisting mainly of methane, carbon monoxide, and hydrogen. This gaseous fuel mixture can then be converted into methanol or hydrogen, using well-established industrial technologies. The production of methanol based on the BCL gasifier is described in chapter 20: *The Brazilian Fuel-Alcohol Program*. The costs for methanol and hydrogen produced using the BCL gasifier are from [13].

e. For the present analysis it is assumed that hydrogen produced from intermittent renewable sources is generated using electrolyzers operated at atmospheric pressure, so that the hydrogen must be compressed before it enters the pipeline. The output of the hydrogen-from-biomass plant would be hydrogen pressurized to 7.5 MPa (1.098 psia), so that in this case no further pressurization is needed before the hydrogen is put into the pipeline.

f. Hydrogen storage is needed with intermittent renewable sources in order to keep the distribution lines operating at near capacity. In the case of hydrogen derived from biomass these storage costs would be small and are neglected here, because the biomass hydrogen plant would be producing hydrogen continuously.

g. The cost for local distribution plus the cost for refueling operations is estimated to be $0.050 per liter ($0.19 per gallon of methanol), or $2.77 per gigajoule [32].

Table 29: Lifecycle cost of transportation[a]
¢ per kilometer

Battery EV	PV/wind H₂ FCV	Biomass H₂ FCV	Biomass MeOH FCV	Gasoline ICEV	Cost component
		Vehicle type			Cost component
1.47					Purchased electricity
7.09	6.72	6.72	6.73	11.17	Vehicle (excluding fuel cell, battery, H₂ storage)
6.71	2.67	2.67	2.52		Battery
	0.83	0.83	0.02		Fuel storage system (compressed H₂ gas)
	2.25	2.25	2.60		Fuel cell system
0.05					Home recharging sta.
	2.15	1.39	1.50	2.89	Fuel for vehicle (excluding fuel taxes)[b]
1.70	1.90	1.90	1.97	2.89	Maintenance
4.95	4.77	4.77	4.72	4.56	Misc. other costs
21.97	21.29	20.53	20.06	21.51	Total cost *¢ per km*
					Breakeven gasoline price [c]
1.42	1.11	0.80	0.60	–	*$ per gallon*
0.38	0.33	0.21	0.16	–	*$ per liter*

If the discount rate for fuel production were 12 percent (see figure 5)[d]

	22.19	20.63	20.25		Total cost *cents/km*
	1.49	0.84	0.68		Breakeven price *$ per gallon*
	0.39	0.22	0.18		*$ per liter*

a. Adapted from [32]. In all cases the consumer cost of gasoline (without tax) is assumed to be $1.21 per gallon ($0.32 per liter) and the cost of electricity is assumed to be 7 cents per kilowatt-hour. Results are shown in figure 5 (top), assuming a fuel production discount rate of 6 percent and in figure 5 (bottom) for 16 percent discount rate.

b. Delivered fuel costs (excluding taxes) for hydrogen and methanol are given in table 28 for various cases.

c. The "breakeven price of gasoline" is the price (excluding tax) that would make the lifecycle cost of the gasoline vehicle equivalent to that of the alternative vehicle.

d. The lifecycle cost of transportation is given assuming a fuel production discount rate of 12 percent.

Table 30: Lifecycle costs for alternative motor vehicle systems
cents per kilometer

	Alternative fuel cell vehicle/fuel combinations[a] for vehicle lifetimes[b]					Gasoline ICEV
	50% longer than for ICEVs			10% longer than for ICEVs		
	MeOH	Biomass H_2	pv/wind H_2	Biomass H_2	pv/wind H_2	
Vehicle	6.73	6.72	6.72	8.43	8.43	11.17
Fuel Cell	2.60	2.25	2.25	2.80	2.80	–
Battery	2.52	2.67	2.67	2.58	2.58	–
Fuel storage	0.02	0.83	0.83	0.83	0.83	–
Misc. O&M	6.69	6.67	6.67	7.24	7.24	7.45
Fuel	1.69	1.49	3.05	1.49	3.05	2.89
Total	20.25	20.63	22.19	23.37	24.93	21.51
Breakeven gasoline price[c]						
$ per liter	0.179	0.221	0.393	0.527	0.700	
$ per gallon	0.68	0.84	1.49	1.99	2.65	

a. The costs shown here for gasoline internal combustion engine cars, for methanol fuel cell cars, and for hydrogen fuel cell cars are the same as those shown in table 27.

b. An important parameter in the comparison of alternative technologies is the lifetime of the fuel cell vehicle. There is considerable evidence from experience with electric battery-powered vehicles suggesting that vehicles with electric drive trains will last considerably longer than vehicles with mechanical drive trains. In the analysis reported here [32] the base case estimate is that vehicle life will be increased 50 percent. If instead the fuel cell vehicle lifetime were only 10 percent longer than the ICEV, the lifecycle cost would rise by 2.7 cents per kilometer, and the breakeven gasoline price would be $1.15 per gallon (30 cents per liter) higher.

c. This is the gasoline price (without retail taxes) at which a car powered by a fuel cell would compete on a lifecycle cost basis with the gasoline-fueled car having an internal combustion engine.

Table 31: Potential for hydrogen production from hydro, biomass, and wind

POTENTIAL FOR HYDROGEN PRODUCTION FROM HYDROELECTRICITY[a]

	Theoretical potential electricity *TWh/year*	Technically usable electricity *TWh/year*	Technically usable potential hydrogen *EJ H$_2$/year HHV*
Africa	10,118	3,140	9.1
North America	6,150	3,120	9.1
South and Central America	5,670	3,780	11.0
Asia	16,486	5,340	15.5
Australasia	1,500	390	1.1
Europe	4,360	1,430	4.2
USSR	3,940	2,190	6.4
World	44,280	19,390	56.3

POTENTIAL FOR HYDROGEN FROM BIOMASS[b]

	Total land area *10^6 ha*	Cropland, forest, and woodland *10^6 ha*	Energy production on 10 percent of cropland, forest, and woodland — Biomass *EJ/year*	Energy production on 10 percent of cropland, forest, and woodland — Hydrogen *EJ/year*
Africa	2,966	886	26	18
North America	1,839	827	25	17
South and Central America	2,050	1,173	34	24
Asia	2,679	1,013	30	21
Europe	473	295	9	6
USSR	2,227	1,160	34	24
Oceania	843	207	6	5
World	13,081	5,560	162	113

POTENTIAL FOR HYDROGEN PRODUCTION FROM WIND RESOURCES[c]

	Land area *10^6 km^2*	Class 3 Percent land area	Class 3 H$_2$ *EJ/year*	Class 4 Percent land area	Class 4 H$_2$ *EJ/year*	Class 5–7 Percent land area	Class 5–7 H$_2$ *EJ/year*	Total class 3–7 Percent land area	Total class 3–7 H$_2$ *EJ/year*
Africa	29.7	12	116	11	131	1	10	24	257
Asia	24.4	6	48	2	18	1	10	9	68
Australia	10.6	8	26	4	16	5	26	17	75
North America	21.9	12	79	8	68	15	161	35	308
South America	17.8	8	43	5	33	5	46	18	122
Europe + USSR	30.3	12	110	9	107	10	149	31	366
World	137.7	10	423	7	372	6	401	23	1,196

a. The annual hydroelectricity production potential was estimated in chapter 3. (Source: World energy Conference, 1980.) The hydrogen production was estimated assuming an ac electrolysis efficiency of 79 percent.

b. The figure for total land areas in croplands, forests, and woodland was based on estimates in U.S. Environmental Protection Agency, "Policy Options for Stabilizing Global Climate," Report to Congress Technical Appendices, 21P-2003.3, December 1990. The biomass production was calculated assuming a productivity of 15 dry tonnes biomass per hectare per year and a higher heating value for biomass of 19.38 gigajoules per dry tonne. Hydrogen production was estimated assuming that gasification has an energy efficiency of 70 percent.

c. The theoretical potential for producing hydrogen from wind electricity is estimated for various regions of the world. The percentage of the total regional land area in each wind class and the hydrogen production from this area are given. It is assumed that the turbine spacing is 5 × 10 rotor diameters. The land area which could be practically developed for wind energy production would be significantly less than these estimates because of the difficulty in siting on rugged terrain, competing land uses (urban, agriculture, forest), and environmental considerations (exclusion of wilderness lands). In the United States, for example, if all urban and environmentally sensitive lands, 50 percent of forest lands, 30 percent of agricultural lands, and 10 percent of range lands are excluded, about 2/3 of the theoretically available wind resources could be developed. If all forest and agricultural land is excluded, only about 1/4 of the total wind resources could be developed [45].

Table 32: Solar electricity generation in Spain (SP) and North Africa (NA)

Type of plant		PV I[a]	PV II[b]	ST[c]
Module efficiency	*percent*	16	15	–
Mean annual efficiency, AC	*percent*	13	12.4	16.1[d]/18.0[e]
Module/heliostat cost	$ m^{-2}	100	50	110
Area related BOS cost	$ m^{-2}	60	37	170/218 $ m^{-2}
Power related BOS cost	$ kW_{AC}^{-1}	400	150	
Indirect cost factor		–[f]	1.25	–[f]
Total investment	$ m^{-2}	217	130	280/328
	$ kW_{AC}^{-1}	1,670	1,060	3,290/2,625
Annual O&M cost *percent of investment*		1.5	0.4	2.5

Electricity production cost[g] *¢/kWh*

	SP	NA	SP	NA	SP	NA
Discount rate 6 percent	8.6	7.3	4.8	4.1	9.4	7.5
Discount rate 12 percent	13.3	11.4	7.8	6.7	14.1	11.2

a. Advanced polycrystalline technology from [3]; 1 1989 $ = 1.8 DM.
b. Advanced thin film technology; modified from Ogden/Williams [2].
c. Advanced solar-tower technology from [1]: 1 $ (1989) = 1.8 DM.
d. SP = southern Spain with global horizontal insolation of 1,950 kWh m^{-2}/year.
e. NA = North Africa with global horizontal insolation of 2,300 kWh m^{-2}/year.
f. Included in other cost figures.
g. Lifetime 30 years; annual insurance rate 0.5 percent of capital cost.

Table 33: Onboard hydrogen storage systems versus hydrogen production systems for fuel cell automobiles[a]

	Installed energy density[b]		Container cost[c]	Refuel time[d]	Filling station mark-up[e]
	MJ/liter	*MJ/kg*	*dollars*	*minutes*	*$/GJ*
Hydrogen storage systems					
Carbon wrapped Al cylinder 55MPa (8,000 psia)	3.4	7.0	4,000	2–3	5
Liquid hydrogen dewar (–253°C)	5.0	15.0	1,000–2,000	>5	11
FeTi metal hydride	2–4	1–2	3,300–5,500	20–30	3
Hydrogen production systems					
Methanol reforming	19.2	17.2	20	2–3	1.2
Fe steam oxidation	5.8	5.0	500?	2–3	?

a. Adapted from [32]. It is assumed in all cases that enough energy is stored to travel 400 kilometers.

b. For hydrogen storage systems the energy density is based on the weight and volume of the full container plus auxiliaries. For hydrogen production systems, the energy density is based on the amount of hydrogen produced per unit of iron or methanol stored. The estimated weight of the container plus auxiliaries is included but not the reformer.

c. Cost to the original equipment (automotive) manufacturer.

d. Time to deliver fuel only.

e. The markup (the cost in excess of the hydrogen cost to the filling station operator) is needed to cover the full cost of owning and operating the refueling station.

Table 34: Operating characteristics for alternative motor vehicles

	Gasoline internal combustion engine vehicle	Iron powered fuel cell vehicle	
Gasoline equivalent fuel economy			
miles per gallon	25.9	76.5	
liters per 100 km	9.08	3.05	
Volume of fuel needed for 250 mile (402 km) range		Fe in	Fe_3O_4
gallons	9.7	4.1	8.7
liters	36.5	15.3	32.8
Mass of fuel needed for 250 mile (402 km) range			
pounds	57.9	140	193
kilograms	26.3	63.4	87.7

WORKS CITED

1. Winter, C.-J. and Nitsch, J., eds. 1988
   Hydrogen as an energy carrie, Springer-Verlag, Berlin, New York, 377 pp.

2. Ogden, J. M. and Williams, R.H. 1989
   Solar hydrogen: moving beyond fossil fuels, World Resources Institute, Washington DC, 123 pp.

3. Nitsch, J., Klaiss, H., and Meyer, J. et al. 1990
   Bedingungen und Folgen von Aufbaustrategien fur eine solare Wasserstoffwirtschaft (Conditions and consequences of a development strategy for a solar hydrogen economy), Study for the Enquete-Commission of the German Parliament on Technikfolgenabschaftzung und -bewertung (technology assessment and evaluation), Bonn, Germany, April.

4. Stuart, A.K. 1991
   A perspective on electrolysis, in *Proceedings: transition strategy for hydrogen as an energy carrier, first annual hydrogen meeting,* Electric Power Research Institute, EPRI Report No. GS-7248, March.

5. Hammerli, M. 1984
   When will electrolytic hydrogen become competitive? *International Journal of Hydrogen Energy* 9:25–51.

6. Leroy, R. L. and Stuart, A. K. 1978
   Unipolar water electrolysers: a competitive technology, in *Hydrogen Energy System, Proceedings of the 2nd World Hydrogen Energy Conference,* Zurich.

7. Pirani, S. N. and Stuart, A. T. B. 1991
   Testing and evaluation of advanced water electrolysis equipment and components, *Proceedings of the 2nd International Energy Agency Hydrogen Production Workshop,* Julich, Germany, September 4–6.

8. Stucki, S. 1991
   Operation of membrel electrolyzers under varying load, *Proceedings of the 2nd International Energy Agency Hydrogen Production Workshop,* Julich, Germany, September 4–6.

9. Steeb, H., Brinner, A., Bubmann, H., and Seeger, W. 1990
   Operation experience of 10 kW pv-electrolysis system in different power matching modes, in *Hydrogen Energy Progress VIII, Proceedings of the 8th World Hydrogen Energy Conference,* Vol. 2: 691, Honolulu, Hawaii.

10. Miller, R. 1992
    Oil, Chemical and Atomic Workers Union, private communications.

11. Stuart, A.T.B. 1992
    Electrolyzer Corporation, Ontario, Canada, private communications.

12. DeLuchi, M.A., Larson, E.D., and Williams, R.H. 1991
    Biomass methanol and hydrogen for transportation, Princeton University Center for Energy and Environmental Studies Report, October.

13. Larson, E.D. and Katofsky, R. 1992
    Production of hydrogen and methanol from biomass, in *Advances in thermochemical biomass conversion,* Interlaken, Switzerland, May 11–15.

14. Haruta, M., Souma, Y., and Sano, H. 1982
    Catalytic combustion of hydrogen-II, *International Journal of Hydrogen Energy* 7:729–736.

15. Haruta, M., Souma, Y., and Sano, H. 1982
    Catalytic combustion of hydrogen-III, *International Journal of Hydrogen Energy* 7:737–740.

16. Haruta, M., Souma, Y., and Sano, H. 1982
    Catalytic combustion of hydrogen-IV, *International Journal of Hydrogen Energy* 7:801–807.

17. Ledjeff K. 1990
    New hydrogen appliances, in *Hydrogen energy progress VIII* ,Vol. 3, 1429–1443, Plenum Press, New York.

18. Mercea, J., Grecu, E., and Fodor, T. 1981
    Heating of a testing room by hydrogen fueled catalytic heater, *International Journal of Hydrogen Energy* 6:389–395.

19. Pangborn, J. B. 1980
    Catalytic combustion of hydrogen in model appliances, *Proceedings of the 15th Intersociety Conference on Energy Conversion,* Seattle, Washington.

20. Sharer, J. C. and Pangborn, J. B. 1974
    Utilization of hydrogen as an appliance fuel, in T.N. Veziroglu, ed., *Hydrogen Energy: part B,* Plenum Press, New York.

21. Billings, R.E. 1988
    Hydrogen appliances, presented at the 8th World Hydrogen Energy Conference, Moscow, USSR.

22. Williams, R. H. 1992
    The potential for reducing CO_2 emissions with modern energy technology: an illustrative scenario for the power sector in China, *Science and Global Security,* Vol. 3, Nos. 1–2:1–42.

23. Wills, J. 1992
    Rolls Royce, private communications.

24. Feucht, K., Holzel, G., and Hurwich, W. 1988
    Perspectives of mobile hydrogen transportation, *Hydrogen Energy Progress VII,* Vol. 3, 1963–1974, Plenum Press, New York.

25. Buchner, H. 1984
    Hydrogen use-transportation fuel, *International Journal of Hydrogen Energy* 9:501–515.

26. Buchner, H. and Povel, R. 1982
    The Daimler-Benz hydride vehicle project, *International Journal of Hydrogen Energy* 7:259–266.

27. DeLuchi, M. A. 1989
    Hydrogen vehicles: an evaluation of fuel storage, performance, safety, environmental impacts and cost, *International Journal of Hydrogen Energy* 14:81–130.

28. Furuhama, S. 1988
    Hydrogen engine systems for land vehicles, *Hydrogen Energy Progress VII,* Vol. 3, 1841–1854, Plenum Press, New York.

29. Prosenstko, K. 1988
    Liquid hydrogen in air transportation, presented at the 8th World Hydrogen Energy Conference, Moscow, USSR.

30. Schucan, T. 1991
    Paul Scherrer Institute, Switzerland, private communications.

31. Amankawah, K. A. G., Noh, J. S., and Schwarz, J. A. 1990
    Hydrogen storage on superactivated carbon at refrigeration temperatures, *International Journal of Hydrogen Energy* 14:437–447.

32. DeLuchi, M. A. and Ogden, J. M. 1992
    Solar hydrogen transportation fuels, PU/CEES Report, Center for Energy and Environmental Studies, Princeton University, Princeton, New Jersey, February.

33. Maceda, J. 1992

H-Power Corporation, New Jersey, private communications.

34. Foute, S.1991
Hythane, presented at the Second Annual National Hydrogen Association Meeting, Arlington, Virginia, March 13–15.

35. Appleby, A. J. 1991
Alkaline fuel cells for transportation, presented at the Fuel Cells for Transportation TOPTEC workshop sponsored by the Society of Automotive Engineers, Washington DC.

36. Romano, S. 1990
The DOE/DOT fuel cell bus program and its application to transit missions, *Proceeding of the 25th Intersociety Energy Conversion Engineering Conference*, Vol. 3, 293–296, American Institute of Chemical Engineers, New York.

37. Kumar, R., Krumpelt, M., and Mistra, B. 1989
Fuel cells for vehicle propulsion applications: a thermodynamic systems analysis, *Proceeding of the 24th Intersociety Energy Conversion Engineering Conference*, Vol. 3, 297–300, American Institute of Chemical Engineers, New York.

38. Lemmons, R. A. 1990
Fuel cells for transportation, *Journal of Power Sources* 29: 251–264.

39. Prater, K. 1990
The renaissance of the solid polymer fuel cell, *Journal of Power Sources* 29: 239–250.

40. DeLuchi, M.A. 1992
Assessment of fuel cell vehicles, PU/CEES Report Center for Energy and Environmental Studies, Princeton University, Princeton, New Jersey (forthcoming).

41. Hord, J. 1978
Is hydrogen a safe fuel? *International Journal of Hydrogen Energy* 3:157–176.

42. Brewer, G. D. 1978
Some environmental and safety aspects of using hydrogen as a fuel, *International Journal of Hydrogen Energy* 3:461–474.

43. US Department of Engery. 1991/1992
National energy strategy, technical annex 2: integrated analysis supporting the national energy strategy: methodology, assumptions and results, DOE/S-0086P.

44. US Department of Agriculture. 1990
The second RCA appraisal: soil, water and related resources on non-federal land in the US, miscellaneous publication #1482, May.

45. Elliott, D.L., Wendell, L.L., and Glower, G.L. 1990
U.S. areal wind resource estimates considering environmental and land-use exclusions, AWEA Windpower Conference, Washington DC, September.

46. Winter, C.-J. and Nitsch, J., eds. 1988
Hydrogen as an energy carrier, Springer-Verlag, Berlin, New York, 377 pp.

47. Hug, W., Divisek, J., Mergel, J., Seeger, W., and Steeb, H. 1990
High efficiency advanced alkaline electrolyzer for solar operation, in *Hydrogen Energy Progress VIII, Proceedings of the 8th World Hydrogen Energy Conference*, Vol. 2:681–690, Honolulu, Hawaii.

48. Metz, P. D. and Piraino, M. 1985
Brookhaven National Laboratory Report BNL-51940; Leigh, R. W., Metz, P. D., and Michalek, K. 1983, Brookhaven National Laboratory Report BNL-34081.

49. Metz, P. D. 1985
Technoeconomic analysis of PV hydrogen systems, Chapter 3.0, Brookhaven National Laboratory Report BNL-B199SPE(3).

50. Hammerli, M. 1990
personal communications.

51. Lehmann, P. 1990
Presentation at the 8th World Hydrogen Energy Conference, Honolulu, Hawaii.

52. Winter, C.-J. 1990
Hydrogen and solar energy—avoiding a lost moment in the history of energy, *Hydrogen Energy Progress VIII*, Vol.1, 3–47, Plenum Press, New York.

53. Wurster, R. and Malo, A. 1990
The Euro-Quebec hydro-hydrogen pilot project, *Hydrogen Energy Progress VIII*, Vol.1, 59–?, Plenum Press, New York.

54. Lloyd, A. 1991
South Coast Air Quality Management District, private communications.

55. Gregory, D. P., Tsaros, C. L., Arora, J. L., and Nevrekar, P. 1980
The economics of hydrogen production, American Chemical Society Report, 0-8412-0522-1/80/47-116-003.

56. Smith, D. 1991
Pacific Gas and Electric Company, private communications.

57. Solar Energy Research Institute 1990
The potential of renewable energy, an interlaboratory white paper, Appendix F, Report No. SERI/TP-260-3674, March.

58. Matlin, R. 1990
Advanced Photovoltaic Systems, Princeton, New Jersey, private communications.

59. Lucas, E. J., McNerney, G. M., DeMeo, E. A., and Steeley, W. J. 1990
The EPRI-utility US Windpower advanced wind turbine program—status and plans, Windpower '90 Conference.

60. Hock, S. M., Thresher, R. W., and Cohen, J. M. 1990
Performance and cost projections for advanced wind turbines, presented at the American Society of Mechanical Engineers, Winter Annual Meeting, Dallas, Texas, November 25.

61. Noel, G. T., Carmichael, D. C., Smith, R. W., and Broehl, J. H. 1985
Optimization and modularity studies for large-size, flat-panel array fields, *IEEE PV Specialists Conference,* Las Vegas, Nevada.

62. Zweibel, K. 1990
Harnessing solar power: the photovoltaics challenge, Plenum Press, New York, 412 pp.

63. Electric Power Research Institute. 1987
Technical assessment guide, Vol. 2: electricity end use; Part 1: residential electricity use—1987, September.

23
UTILITY STRATEGIES FOR USING RENEWABLES

HENRY KELLY
CARL J. WEINBERG

Electric utilities will play a key role in determining the contribution renewables will make in electric markets. These institutions are in a unique position to ensure that electric services are delivered at the lowest cost and with minimal impact on the environment. Utility investment in renewables will be considered in the context of investment portfolios that will include a range of new technologies for generating electricity as well as new technologies for increasing the efficiency of energy use, for electric transmission and distribution, and for storing electric energy. New generating equipment is likely to be smaller in scale, cleaner, and closer to demand sites than typical contemporary equipment. Intelligent networks will dispatch generation and manage demands with increasing efficiency.

Utilities flexible enough to take full advantage of the new investment opportunities will be substantially different from existing utilities. Their investment portfolios will be much more diverse and they will pay much more attention to all the production and service activities that contribute to the cost and quality of energy services delivered to their customers. Changes of this sort, however, depend on a change in the basic social contract connecting the utility industry to the public it serves.

If public policy encourages utilities to invest in ways that minimize energy service costs and environmental costs, renewable electric generation could play a major role. Renewable equipment likely to be available early in the 21st century could provide more than 80% of the energy inputs of modern utilities at prices no higher than those charged for electricity today. The appropriate mix of hydroelectric, biomass, and intermittent (wind, photovoltaics, solar thermal) systems depends on local resources. Intermittent resources should be able to meet at least one third of the utility's energy needs without adverse effects on costs. The value of each type of renewable technology is strongly affected by its scale, location and the characteristics of other generating equipment operating in the region. Advanced natural-gas turbine systems, for example, make a good match with renewables.

It is possible that advanced, fossil-powered equipment capable of very high efficiencies may be able to produce electricity at a slightly lower cost than advanced renewable equipment (both could be producing at prices below current electricity costs), but renewable equipment may still be preferred because of its environmental benefits.

ELECTRIC UTILITIES IN THE 21st CENTURY

Electricity markets are critical if renewables are to play a significant role in meeting world energy needs. Growing electric demands are likely to dominate energy demand growth worldwide in spite of electricity's relatively high cost. Electricity makes it possible to apply power only where it is needed and when it is needed and proves to be an ideal energy source for modern factory, commercial, and household equipment.

Electricity can be produced from renewable resources in many different ways. Hydroelectric power already provides about 20 percent of the world's electricity. Biomass can be used as a direct substitute for coal in conventional and advanced generators. The power of the wind can be tapped using wind turbines. The sun's energy can be used to heat fluids for turbines in solar thermal systems or converted directly to electricity using photovoltaic equipment. The diversity of options means that most parts of the world have a renewable resource that can be used to generate electricity at competitive prices. But diversity frustrates a simple assessment of value. Techniques used to evaluate traditional utility investments do not work well when applied to renewable equipment whose value depends heavily on details of local conditions (local wind or sunlight resources, the hourly pattern of electric demand in the region, and the characteristics of other equipment operating in the utility system).

Existing evaluation techniques, however, are also inadequate for evaluating many other areas open for utility investment. Future utility portfolios will include investments in thermal generating equipment with characteristics substantially different from those in place today, electric storage systems, advanced transmission and distribution systems, advanced control systems, as well as investments in renewable electric generation equipment (see figure 1). Utilities are also likely to invest in equipment and services that can increase the efficiency with which their customers use electricity [1].

Utilities will, therefore, not simply ask whether renewable electric technology can compete with a particular type of conventional generating plant. They will ask whether investment in renewables make sense as a part of an efficient electric production and delivery system. The central challenge is finding a mix of investments that minimizes overall cost of meeting the ultimate demands of electric customers: a reliable source of lighting, space conditioning, mechanical power, and other end uses that meets environmental goals.

Assessing the performance of an electric system that combines equipment with widely varying operating characteristics is a complex analytical task. Unlike most manufacturing, electricity is typically produced only when it is needed. Utilities must constantly adjust their production rates to meet demands which vary from hour to hour and from season to season. Since it is expensive to maintain even a small inventory of electricity in the form of "electric storage," utilities practice an extreme form of "just in time" inventory control.

In a dynamic electric system, the performance of each piece of equipment affects the economics of other devices in the system in complex ways. A balanced analysis must consider such things as: (1) how is overall reliability of electric service affected by different kinds of investment (renewables with intermittent output affect reliability but reliability also depends on the details of transmission and distribution systems and other generating equipment); (2) what is the optimum size for individual generating plants; and (3) where should generating equipment and storage systems be located (there are clear advantages in locating generating equipment close to consumption sites since delivering electricity to customers can cost as much or more than generating electricity).

These are not new issues. But they are receiving renewed attention as utilities struggle to understand the merits of new technologies and to cope with increasing pressure to reduce costs.

The Investment Portfolio

Traditional systems

The basic features of a traditional utility system include centralized fossil, hydro-electric, and nuclear generating equipment as well as the transmission and distribution systems needed to move power to customers, convert it to voltages appropriate for different customers, and ensure power quality (frequency and voltage stability and reliability) (see figure 1).

Traditional electric generation is divided into three classes: baseload plants designed to meet the portion of utility loads that do not vary greatly during the day; peaking plants designed to respond quickly to provide power during periods of highest demand, and load following plants designed to meet routine daily load fluctuations. Hydroelectric plants are used primarily for baseload power but the output of these plants can be adjusted to meet fluctuating loads when reservoirs are adequate and environmental, agricultural, and other demands can be satisfied.

Typically, baseload plants are comparatively expensive to build but have low operating costs because they use low cost coal or nuclear fuels. Since they have high fixed costs, these baseload plants must be operated many hours a year to be economical. Economies of scale in conventional baseload steam-plant designs mean that large (600 to 1,000 megawatts (MW) or millions of watts) plants produce the cheapest electricity. Peaking plants in traditional systems have relatively high operating costs because they use natural gas or oil, but are comparatively inexpensive to build and thus can be economical even if they operate only a few hours a year. Load following plants have characteristics that fall between baseload and peaking units. The least cost mix of baseload, intermediate, and peaking plants can be estimated using a procedure described in Box A.

New investment opportunities

Utilities today must assemble investment portfolios (see figure 1) whose merits cannot be easily assessed using standard methods. Some renewable electric generators have unique characteristics that complicate investment decisions. While

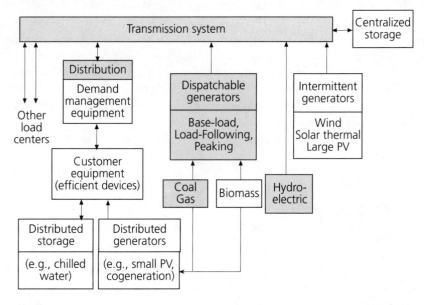

FIGURE 1: Delivering electric energy services. This figure outlines all the equipment that contributes to the cost of meeting the energy service needs of utility customers: lighting, space conditioning, mechanical power, etc. The list includes generation, transmission, distribution, controls, storage, and devices which put electricity to work (light bulbs, refrigerators, machine tools). Traditional utilities invested only in centrally located, fossil-fired generating equipment, hydroelectric equipment, transmission, and distribution. But technical advances have greatly increased the range of investment opportunities for utility and nonutility customers alike. The challenge becomes one of identifying the investment portfolio that provides the highest quality service at the lowest possible costs.

biomass power plants, like fossil fuel power plants, are "dispatchable", i.e., their output is under the control of the utility, intermittent renewable systems, such as wind, photovoltaic, and solar thermal units, are not. But the characteristics of other potential utility investments (advanced fossil generators, storage, transmission and distribution, customer end-use equipment) are also very different from conventional utility investments. New analytical techniques would be needed even if renewables were not available.

While biomass generation equipment would typically be much smaller than conventional baseload plants (biomass units would typically be 100 MW or less), biomass generators can be dispatched like conventional fossil-powered equipment. Other renewable systems have more complex operating characteristics. The output of intermittent renewable electric systems depends on the availability of sunlight or wind. The value of power produced in this way is sensitive to local conditions. Photovoltaic and solar thermal devices located in southern regions where electricity demand is highest during the day can reduce expensive peak generation as well as peak loads on transmission and distribution (T&D) systems.

Hydroelectric systems with large reservoirs can be dispatched to follow fluctuating electric demands and can help offset variations introduced by intermittent electric generators operating in the same system. Concern for the environment, and the needs of agriculture and recreational activities, limit the variation possible. Some hydroelectric systems (such as "run-of-the-river units) cannot be dispatched.

New generating equipment using fossil fuels, however, also require new ways of thinking about utility investments. The new generation of small gas-turbines, for example, can come close to competing with baseload coal plants. The value of small-scale generators increases if a market can be found for the heat left over after electric power is generated, if the equipment is located where it can reduce the investment in transmission and distribution equipment, or if generation systems located close to consumption sites can contribute to system reliability.

Unfortunately, it is particularly difficult to compare investments in T&D with investments in generation. Transmission and distribution have too often been considered as an "overhead" by utility planners —an inflexible fixed percentage of generation investment. Pressure to minimize costs is, however, forcing many utilities to improve the way transmission and distribution investment and reliability analysis is integrated with other utility investment planning. But much work remains.

Wise investment in transmission and distribution is clearly important. This equipment represents 40 percent of total utility capital investment in the United States [2] and a much higher fraction of all new capital investment in many regions. Investments are lower in congested areas (like northern Europe) and greater in sparsely settled areas. Transmission and distribution systems can also be costly because of energy losses. The percent of electricity generated that is lost in transmission and distribution systems averages 6.2 percent in the United States[3]. And customer power outages are at least as likely to result from faults in transmission and distribution as they are from generation. The cost and reliability of transmission is strongly affected by the location and size of individual generators and storage systems.

Computer networks, telephone switching, and many other large systems operating in modern businesses are evolving away from hierarchical models toward more flexible systems having many multiply-connected nodes, each operating with comparative independence.

Utility systems are likely to follow a similar path as they attempt to manage a complex set of demand management and generating equipment distributed through their service areas. At present, most utility grids are operated essentially like an irrigation system— delivering a commodity from a large reservoir to many customer sites. But future utilities are more likely to resemble computer networks, with many sources, many consumers, continuous reevaluation of delivery priorities, and continuous management of faults. All customers and producers will be able to communicate freely through this system to signal changed priorities and costs. A modern system would, in effect, intermediate a constantly shifting market for electricity producers and consumers.

Box A
Selecting the Optimum Mix of Dispatchable Plants

A standard strategy for estimating the least-cost mix of generating equipment begins with data about the utility's loads throughout a year (figure A.1 shows data for a typical week). The loads are then sorted by size to create a "load duration curve" (see figure A.2). The load duration curve can be interpreted as the number of hours each year that a utility's load exceeds an indicated level.

The mix of plants that minimizes a utility's production costs can be estimated using the load duration curves and the fixed and variable costs of generating equipment. The challenge is to determine the number of plants of each type that can meet the system peak while minimizing annual costs.

The basic method is to compute the cost of operating each type of plant as a function of hours of annual operation (the straight lines plotted in figure A.2). The least expensive set of plants (assuming no operational constraints) is determined by the points where the production cost lines intersect. For example, the intercooled LM6000 simple-cycle (SC) gas peaking plant provides the lowest cost power for loads that last less than 614 hours per year (point "A" on the curve). The CIG/ISTIG coal plant provides the lowest cost power for all loads that occur more than 4,654 hours per year (point "B" on the curve). The intercooled LM6000 combined cycle (CC) provides power at intermediate points. The load duration curve can be used to convert these operating times into a mix of plants that would minimize production costs. Systems with sharp peaks will best be served by systems with comparatively large numbers of simple cycle machines while systems with flat load duration curves will have a large fraction of baseload generating plants. Once a set of plants is selected, they are dispatched in order of their variable operating costs.

This method provides a good initial estimate of the optimum mix of plants, but the initial estimate must be adjusted to reflect a variety of practical considerations. The most obvious is that plants must be purchased in discrete sizes. It is not possible to install the precise capacity computed from the line intersections. The mismatch between the capacity recommended by the calculation and the capacity actually purchased is greatest if the size of individual plants is large (e.g., baseload coal plants).

Actual investment decisions also include a variety of other operational factors (see Box E). And additional plants are purchased to ensure system reliability.

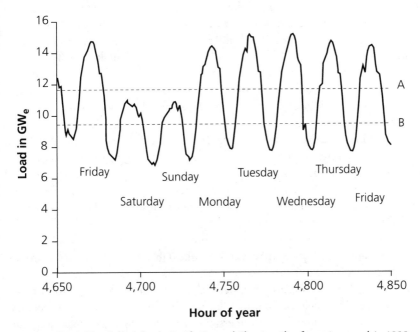

FIGURE A.1: *Hourly loads for the Pacific Gas and Electric utility for a winter week in 1989. Notice the comparatively small peaks occurring during the weekends days.*

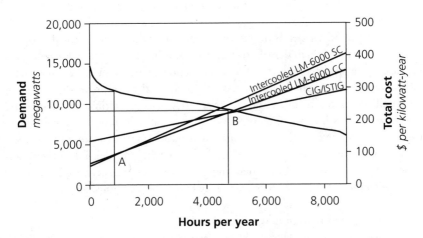

FIGURE A.2: *A load duration curve for PG&E in 1989 constructed from a year's worth of data of the type illustrated in figure A.1. The utility needs to provide a 15 GW$_e$ for at least one hour (the point on the far left of the curve). Loads exceeded 6 GW$_e$ for all 8,760 hours of the year (the point at the far right of the curve). Loads exceeded 10 GW$_e$ for 3,700 hours out of the year. The straight lines represent the annual cost of operating different kinds of generating equipment as a function of the number of hours each year they operate. At zero operating hours, costs represent only the capital carrying charges of the plant.*

New Business Structures

Regulatory changes have changed the investment climate for electric utilities. The changes increase the risks utilities take in any new investment, open up new investment opportunities for utilities, and permit much greater competition in generation. The competitive pressure has already led to dramatic changes in the investment patterns of many U.S. utilities. Utility investment in consumer efficiency measures is over 6 percent of utility gross revenues in many large U.S. utilities and several are investing more than 2 percent. The Sacramento Municipal Utility spent 6.4 percent of revenues on demand side management programs, Seattle City Light 6.2 percent, New England Electric 4.9 percent and Wisconsin Electric 4.8 percent [4]. Relaxation of regulations barring entry into electric generation markets created large markets for industrial co-generation systems and other independent power producers. Independent producers—or non-utility owners of electric generation equipment—are responsible for about 15.6 percent of California's total electric generating capacity [5].

The new competitive pressures have forced utilities to behave much more like modern manufacturing enterprises and less like stable state monopolies. Automobile producers, producers of consumer electronics, and many other manufacturing firms have been forced by competition to pay much more attention to the needs of individual clients. They have learned that their survival in a competitive world depends on an ability to understand what it is their customers want and react quickly. Often what customers are looking for is quality. Utilities are quickly learning the same lesson. They have come to understand that customers are not necessarily interested in low-cost kilowatt hours, but in low-cost, high-quality energy services.

Quality clearly depends in part on the nature of the services provided by the utility and on the reliability of service. But techniques used to understand how consumers value different levels of reliability, and how different utility investment decisions affect reliability, are inadequate. The need to improve these tools is made all the more urgent by the fact that renewable generating technology and the other new utility investment opportunities can affect system reliability in complex ways (see Box B). Improved analytical tools for understanding the operation of dynamic utility systems are only a part of the solution. Like other manufacturing enterprises, utilities operating in competitive markets must be sensitive to a wide range of subtle factors that shape consumer decisions.

The role of renewables in new utility systems

The remainder of this chapter will be devoted to evaluating a variety of investment portfolios built from the elements illustrated in figure 1. The analysis concentrates on a single simple measure of a portfolio's value: what is the cost of meeting the demand for electric services in the utility's region during a year assuming that the utility receives a fixed rate of return on its capital investments? This measure forces attention to the complex interactions between different kinds of utility equipment. Analytical methods used to estimate costs are discussed in the final sections of this chapter [6]. The basic ground rules are outlined in Box C.

While much work remains to be done, and results must be tested in a large number of different climate and resource regions, the analysis resulted in the following conclusions for a utility operating in the first quarter of the 21st century:

▼ Utility portfolios that rely on intermittent renewable sources for 30 percent of their electricity can serve loads at lower costs than utilities using typical new equipment and at costs about 5 percent higher than those achievable using advanced coal and gas production equipment now in development.

▼ The cost of the renewable portfolio just described would be unchanged if biomass is substituted for coal. This system would rely on renewables for over 90 percent of its energy.

▼ A system meeting 50 percent of its loads from intermittent renewable resources would cost 10 percent more than a system using intermittents for 30 percent of its loads.

▼ The three conclusions just described assume a 6 percent real discount rate (typical of regulated utilities in industrialized nations) and equipment costs reasonable for the first quarter of the next century. If a 12 percent real rate of return is used (or if all capital costs are increased by two thirds), the cost would be approximately 20 percent more.

▼ Small generating equipment that can be located close to demand centers can reduce transmission and distribution costs and thereby reduce delivered costs. These credits can reduce the gap between cost of a utility with 30 percent of its power from intermittents and an advanced fossil-based utility to well below 5 percent (assuming a 6 percent discount rate).

▼ Advanced turbine systems using natural gas as a fuel provide an excellent complement to renewables-intensive utility portfolios. They can operate efficiently while following loads and they can be added quickly. Their low cost makes them an excellent way to buy time to determine the best investment strategy in the future.

▼ Hydroelectric sites with large reservoirs also provide an excellent match for intermittent renewable technologies. The output of hydroelectric systems can be adjusted to fill in gaps in the output of intermittent equipment.

▼ Electric storage equipment is not needed to achieve the high levels of penetration of renewables just described. In fact, the value of storage to a utility is decreased, and not increased, if photovoltaic or solar-thermal equipment is added to utility systems where peaks occur during daylight hours.

It is important to recognize that these cost estimates are made under the assumption that utilities are free to optimize their investments to minimize costs based on new investments. No attempt was made to develop a detailed schedule of plant additions and retirements. Utilities in most industrialized nations are likely to experience slow or even declining demand for electricity given the attractive options for investing in energy efficiency. As a result, new investments will be made primarily to replace obsolete equipment or equipment forced into obsolescence by new regulations. There are few occasions when any new plant—renewable or otherwise—can offer costs so low that a justification can be found for replacing equipment which has not reached the end of its useful life. In such cases the full cost of new generating equipment must compete with the operating costs of the existing units.

Countries in the process of building utility systems may be in a better position to exploit new technologies than countries with large investments in old equipment. Industrialized nations with large sunk costs may be tempted by investments that make near-term sense but that will not lead to a system well matched for future markets. Commitment to large generating plants having 30-year lifetimes can create a self-fulfilling prophecy by making investments in new technologies uneconomical for a generation to come.

ECONOMIES OF SCALE

The range of new energy production technologies just described undermine longstanding assumptions about the value of scale economies—both the value of scale in the size of individual generating units and the value of scale in the size of transmission and distribution systems that connect different customers and generators [7]. The issues of scale fall roughly into the categories illustrated in conceptual terms in figure 2.

Generating costs
Generating costs are affected by scale in two primary ways: (1) larger generators may cost less per unit of capacity and have higher efficiencies, and (2) the total capacity installed in a region depends on the diversity of the load created by interconnections.

To understand this second point consider a case where 100 homes operate with no distribution system connecting them. Each house would need to have a generator large enough to meet its own peak load and possibly a backup unit to maintain reliability. The total capacity would be much larger than the generating capacity that would be required if the homes were connected with a distribution grid. Since it is unlikely that everyone's refrigerator would turn on simultaneously, the peak of the combined load is lower than the sum of the peak loads of individual units. But the value of adding additional homes to a grid declines as the system becomes larger and the combined load approaches average daily patterns. On average, all homes will have their air conditioners running during peak cooling hours and will have their lights on during the evening. These underlying load variations will not be reduced by adding more units. While there are no universal rules in such matters, the value of diversity declines rapidly in residential areas

Box B
Reliability of Electrical Service

Growing competitive pressures have stimulated electric utilities to pay new attention to consumer interests in the quality of their product. Particular attention is being paid to the reliability of electric service. Many customers place such a high premium on reliability that they are willing to invest in emergency battery or generation systems on their own premises. Hospitals, telephone companies, computer-intensive service firms, and other businesses routinely invest in such systems. Since the equipment is seldom, if ever, used, the cost of electricity is extremely high. This price provides a crude measure of the minimum value they place on reliability.

It is much more difficult to measure the value other types of customers place on reliability. Surveys conducted in recent years suggest that the value placed on reliability depends on the nature of the customer, the season, the time of day, and the duration of the power outage. Residential customers would be willing to pay $4.5 to $5.5 per kWh to avoid a one hour loss of power in the winter and $5 to $6.6 per kWh to avoid a one hour loss in summer. The total amount customers would be willing to pay to avoid a loss of power increases with the length of the power loss but the price per kWh of outage declines somewhat for longer outages.

It is not surprising that business customers place higher value on reliability than residential customers. Business customers face major losses from power failures. Employee time is wasted, equipment stands idle, customers may be lost, and serious safety problems or equipment damage can occur. Industrial outage costs depended strongly on time of day. Surveys suggest that a one hour outage at 8 am is valued at $16 per kWh but only $6 per kWh at 6 pm. Outages lasting 3 to 12 hours were valued at $3 to $6 per kWh.

Utilities have traditionally used a somewhat arbitrary standard of reliability for their generating system. They attempt to ensure that their equipment failure will not lead to more than one hour of lost power in ten years. There is a long-standing concern that this standard may not be economical. At least one major utility has taken the position that the standard of reliability should be based on the real value customers place on reliability. They argue that utility investments should minimize customer costs—including the cost of outages. This argument has forced a number of long-hidden assumptions about utility planning into the open and raises a number of unresolved issues. For example, should the standard of reliability reflect the needs of an average customer or the needs of customers unusually sensitive to power outages?

The cost of increasing reliability near the utility's traditional design point (1 day in 10 years) may be $10 to $20 per kWh if the utility uses conventional generating equipment and lower if advanced technologies are used. The new equipment would also increase reliability because unit capacities are smaller (see figure 3). Utility systems which incorporate large amounts of intermittents may be slightly more sensitive to reliability criteria but the effect appears to be very small.

The high cost of meeting standard reliability criteria suggests that investments in efficiency, storage, and generation on or near customer premises and improved design of distribution systems should be considered. It must also be recognized that in most systems, failures in distribution and transmission are more frequent than generation failures. Indeed, failures in the distribution system alone exceed the one day in 10 years (2.4 hours per year) standard (see table below).

Average outage rates due to failures in the distribution system for the PG &E utility system (*in hours per year*)

	Residential	Small Light & Industrial	Power
1990	3.9	4.25	2.5
1991	2.55	2.83	1.98

Source: Pacific Gas and Electric Company, data prepared for General Rate Case Application, 4 March 1992.

Box C
Ground Rules for Portfolio Analysis

Utility portfolios are assessed by estimating the cost of operating a hypothetical utility during the first quarter of the 21st century. Hourly utility loads, wind, and sunlight data were taken from the PG&E service territory in 1989. The following assumptions were used in the analysis:

▼ The utility is free to invest in biomass and conventional generating equipment that minimize production costs (real utilities, of course, would need to consider the sunk costs of existing equipment).

▼ Coal prices remain fixed at $2.1 per gigajoule and natural gas prices rise from $4 per gigajoule in the initial year to $5 per gigajoule after 30 years. These prices are lower than those used in most long-term forecasts (the U.S. Energy Information Administration, for example, suggests that natural gas prices may approach $8 per gigajoule during this period), because of an assumption that rapid introduction of renewable fuels would increase the diversity of world fuel supplies thus increasing competition and stabilizing prices.

▼ The utility is financed assuming both a 6 percent real cost of capital (typical of regulated utilities) and a 12 percent real cost of capital (typical of non-regulated owners). A 0.5 percent annual insurance is also assumed. No taxes are used in computing fixed charge rates. (Electricity from capital-intensive equipment like renewables is more expensive if high discount rates are used.)

▼ All cost-effective investments in energy efficiency and demand management have already been made. in many cases such investments would be less expensive than any investment in new generating equipment. The goal of the present analysis is to minimize the cost of meeting the remaining load.

▼ No accounting is made for environmental costs except for the fact that the cost of meeting existing (1992) U.S. environmental regulations is implicit in the price of conventional generating equipment.

▼ No energy storage equipment is used except for energy stored in hydroelectric reservoirs and in biomass fuels.

▼ No attempt was made to optimize the mix of wind, solar-thermal, and photovoltaic equipment used by the utility.

▼ Any intermittent renewable output which exceeds the load is wasted. This is not a realistic assumption since a market would undoubtedly be found for this inexpensive energy—however unreliable. Surplus energy could be sold to neighboring utilities or sold for pumping, desalination, or producing other storable products.

▼ Hydroelectric equipment is dispatched to reduce utility peaks as much as possible within specified constraints on reservoir size and maximum filling and discharge rates (see appendix).

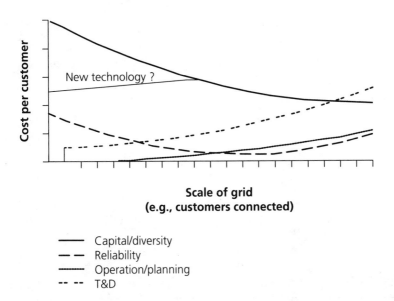

FIGURE 2: *Delivered energy costs as a function of scale. The cost of utility system scale requires considering a number of factors which depend in complicated ways on the size of the system. These include (i) economies of scale in the cost of the generating equipment itself, (ii) the greater diversity achieved by combining different load types (which changes the ratio of peak loads to average loads and thereby affects the total capacity which must be installed to serve the area), (iii) transmission and distribution costs, (iv) reliability, and (v) planning and operational considerations.*

after about 100 homes are connected [8]. Residential units have a comparatively uneven pattern of electric demand. Table 1 suggests that residential distribution systems have "load factors" around 0.41 (a load factor is the ratio of average demand to peak demand). Diversity can still be gained, however, by connecting residential areas with commercial or industrial customers. Industrial customers typically have load factors greater than those of residential customers and may have peaks that occur at times of the year and times of day that are significantly different from residential peaks.

The cost of generation is high if the "scale" of a system is very small (i.e., the situation where each home has its own generator) both because of generation economies and lack of diversity (see figure 2). But new generation and load-management technologies could reduce the cost penalty associated with small systems. Small generators are modular and can be mass produced using highly productive modern manufacturing methods. Larger units typically require large construction projects in the field where productivity gains are much more difficult.

Reliability

Reliability can be increased, rather than decreased, by using a large number of small generators instead of a small number of large ones. Reserves must be main-

Table 1: Load factors by customer class[a]

	Bulk transmission	Area transmission	Primary distribution
Residential	0.55	0.44	0.41
Agricultural	0.43	0.37	0.37
Small L&P	0.41	0.43	0.43
Medium L&P	0.50	0.49	0.48
Large (0.5 to 1 megawatt)	0.54	0.52	0.51
Large (> 1 megawatt)	0.66	0.59	0.78
Total[b]	0.54	0.47	0.47

a. [2].
b. Total includes streetlights.

tained to cover unexpected failure of the largest operating plant. The total installed generating capacity required to meet the "one day in 10 years" reliability criteria is higher if large plants are substituted for smaller plants with identical failure rates (see figure 3). Moreover, systems consisting of a large number of small generators scattered throughout a region can be designed to fail far more gracefully than systems heavily dependent on a few critical pieces of equipment. Large, system-wide blackouts should be easier to avoid with distributed systems [7].

Transmission and distribution costs
Transmission and distribution costs clearly increase as the scale and complexity of a system grows. Customers located more than 0.5 kilometers from a utility distribution system [10] may be best served with an independent generation system not connected to the grid (although it may be owned and operated by a utility).

Renewable generators add a new dimension to the scale-economies of transmission. Distributed photovoltaics may reduce T&D costs, but the system may also benefit if grids connect wide areas and increase the diversity of renewable supplies. Grids can move intermittent energy from areas with temporary over-supply to regions lacking power because of wind or sunlight conditions.

Planning and operational costs
Large generating plants can incur high planning and operational costs. In the past these costs were offset by the fact that the larger generating equipment produced much less expensive power. But planning and operational costs must be given much more careful consideration as the price advantage of the larger equipment declines [11].

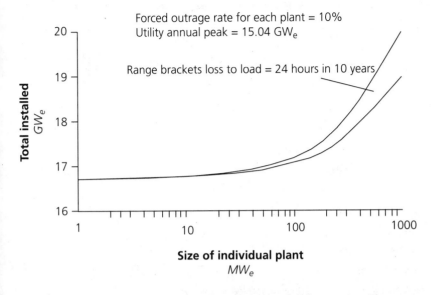

FIGURE 3: Capacity needed to meet a one-day-in-10 years reliability criterion. The capacity needed to maintain a given level of reliability declines as size of individual units declines. The figure shows the capacity needed to meet 1989 PG&E demands for plants with a 10 percent forced outage rate, assuming that all plants in the system are the same size. This obviously artificial assumption illustrates the underlying point: a real utility would use small plants to achieve reliability. The curve approaches the theoretical limit: peak demand divided by 0.9. The range indicates that capacity can be more precisely adjusted to meet a reliability standard if small units are installed than if units are installed in comparatively large sizes.

Table 2: Marginal customer costs[a,b]

	Installation costs (including overheads) $	Annual O&M costs $ per year
Residential	250–800	60
Small light and power	1,100–1,700	90
Medium light and power[c]	4,300–9,700	670
Small industrial (.5–1 megawatt)	14,700–21,200	9,000
Large industrial (> 1 megawatt)	34,600–40,300	9,000

a.   [2].

b.   Costs of connecting individual customers (meters, lines from distribution lines to individual residences or buildings, etc.) not included in the transmission and distribution costs cited in the text.

c.   Secondary costs only (see source for details).

Table 3: Operational benefits of energy storage[a,b]

	$ per kilowatt
Dynamic operational benefits	0–400
Operating and spinning reserve	0–120
Unit commitment	0–100
Frequency regulation	0–200
Thermal unit minimum loading	0–50
Black start capability	0–20
Voltage regulation	0–50
Transmission and distribution benefits for dispersed siting	
Transmission capital costs	0–300
Transmission loss	0–50
Corporate planning benefits	
Modularity and short construction lead-time	0–200

a.   [25].

b.   The value of electric storage technologies to an electric utility *in addition* to the value resulting from moving loads from peaks to off-peak periods. A storage system would be preferred if its cost per kilowatt is lower than the sum of operational benefits and peak-shaving benefits.

Table 3 summarizes estimates of the "dynamic benefits" of energy storage prepared by several U.S. utility companies. The wide range of estimates (zero to several hundred dollars per kilowatt) results both from (1) extreme sensitivity to the specific characteristics of each utility's loads and generating options, and (2) use of a variety of analytical tools and economic assumptions. Although much remains to be learned in this area, several general results are becoming clear.

Smaller plants can be added quickly as they are needed and even disassembled and moved if loads decline, but the economics of larger plants depend heavily on long-term forecasts. Many U.S. utilities paid dearly for construction programs justified by mistakenly high forecasts of electric demand growth. The costs of overbuilding can be offset to some extent if the utility is able to sell excess capacity to neighboring utilities [12].

New efficiency and renewable technologies can change both rate of growth of electric demand and the hourly pattern of electricity demand. A utility must consider whether its demand will increase by simply scaling up the pattern of loads experienced in previous years or whether new demand equipment, new load management, and penetration of intermittent renewable equipment will change demand patterns in ways that can have a major effect on the optimum mix of plants [13, 14]. Analysis presented later in this chapter suggests that utilities using

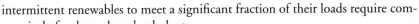

intermittent renewables to meet a significant fraction of their loads require comparatively few large, base-load plants.

If plants can only be added in very large increments, it is much more difficult for utilities to optimize their systems (see Box A). If, for example, the least-cost generating mix calls for 2,300 megawatts of coal capacity, but coal plants only come in 1,000 megawatt sizes, a compromise must be made—either too much or too little coal capacity must be purchased.

Small, distributed storage and generation units also lead to a number of operational advantages (see Box D and table 3). These dynamic considerations are difficult to detect without hourly production-simulation models [15].

METHODS FOR ASSESSING NEW INVESTMENT PORTFOLIOS

Ground rules

The issues just reviewed demonstrate the complexity of designing a utility investment portfolio. The basic rules are changing. Critical information about resources are lacking, and the preferred portfolios are likely to be different in every region. While much work remains to be done, it is possible to show that renewable electric systems can be an important part of investment strategies in many regions.

·A number of indices of merit can be computed for the individual technologies. But the most important indicator is the value of the equipment when integrated in a plausible utility system. The analysis presented here is designed to estimate the total cost of meeting yearly demand for electricity from a variety of different portfolio mixes in a way that provides a rough measure of many of the most important operational issues described in the previous section. It does this using the assumptions summarized in Box C.

The procedures used to calculate utility costs are described in more detail in the appendix, but in broad terms the analysis proceeds using the following steps:

▼ Determine the demand for electricity at each hour of the year.

▼ Compute the amount of intermittent electric generation available from wind, solar-thermal, and photovoltaic plants at each hour of the year.

▼ Subtract the intermittent electricity available from the electric demand in each hour (discarding electricity generated from intermittents in excess of the amount needed during the hour).

▼ Dispatch available hydroelectric equipment to meet demand peaks within the constraints imposed on operation of the hydroelectric equipment. The output of the hydroelectric system is subtracted from the demand in each hour.

▼ Design and operate a set of fossil and biomass equipment that minimizes the cost of meeting the remaining hourly demand.

Generating equipment characteristics
Before attempting to compare the merits of different investment portfolios, it is important to understand the characteristics of the complex range of equipment that may be a part of utility portfolios in coming years.

Intermittents
Unlike all conventional generating technologies, electric production depending on the wind or sunlight cannot be controlled by utility dispatchers. Unless storage is available, the power must be used when it is available. But the problems created by intermittent renewable electric production are not qualitatively different from the problems already faced by utilities when they dispatch plants to meet demand for electricity that fluctuates during the day [16]. Demand for utility electricity can more than double during the day and seasonal variations can be much higher.

While utilities have little direct experience with predicting wind or solar inputs, a significant fraction of the seasonal and daily variations in utility loads result from sunlight and weather conditions. Utilities with significant amounts of wind are finding that wind forecasts can provide good guidance for anticipating the need for conventional generating equipment.

While utilities in regions like northern Europe, Canada, and the northern parts of the United States have peaks during the winter, most of the world's electric growth is occurring in sunny regions where peaks occur in the summer. It is likely that improvements in the efficiency of building heating systems and possibly even climate change will increase the number of areas where utilities experience summer peaks.

Intermittent electric production could increase the variability of loads if it is not correlated with demand for electricity. It could actually decrease the variation if the renewable energy is available at periods of peak utility demand. This is clearly the case with photovoltaics used in areas where peak demands are associated with air-conditioning loads, which are typically highest during the summer days when solar energy is available.

In any case, large amounts of renewable capacity would have to be added before the fluctuations in demand created by intermittent equipment begin to approach those of demand variation. In fact, if the wind generation is small compared to total system loads, and completely uncorrelated with demand, the intermittents can be treated as a completely reliable plant with a capacity equal to the average annual output [17].

At higher levels of penetration, more care must be taken to consider the impact of intermittents on system costs and reliability. The effect clearly depends on the nature of the load and the intermittent resource. The advantage of combining wind sites with transmission systems is immediately apparent by examining load duration curves (see figure 4). The number of hours when no wind is available to the system is greatly reduced when several sites are combined (see also chapter 4: *Wind Energy: Resources, Systems, and Regional Strategies*).

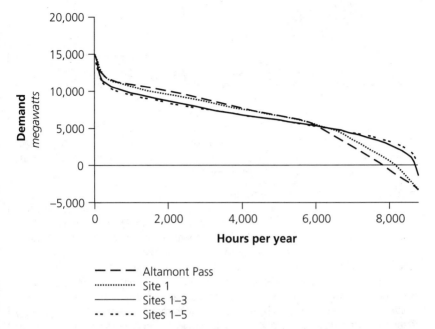

FIGURE 4: *Load duration curves for wind generation (wind energy supplies 30 percent of the utility's loads). These load duration curves represent the utility loads reduced by the output of hypothetical wind systems. This was done by subtracting the output of the wind equipment from the utility's load at each hour of the year. The net load was then sorted to produce the load duration curve. The first wind system shown is the actual output of the Altamont Pass wind farm in 1989 scaled to provide 30 percent of the load. The other two wind curves represent the output of advanced wind systems operating with random wind patterns. The wind statistics were generated using a Rayleigh distribution with average wind speed = 8 mps. Wind speeds were converted to electric output using data for the U.S. Wind Power Variable Speed Wind Turbine, 1991.*

The correlation between wind and loads is very sensitive to the site. In the Netherlands, for example, 10 years of data indicate no correlation with load. But correlations can be quite good for sites like the Solano Pass region in California (see chapter 3: *Wind Energy, Technology and Economics*). The Altamont Pass data (see figure 4) are an intermediate case.

The statistics of intermittent energy also depend on the number of production sites that are combined to serve loads. The output of wind machines located close to each other is likely to be highly correlated [19] (see table 5); if one wind machine is producing no power it is very likely that adjoining wind machines will also be producing no power. It is much less likely that two geographically distant sites will both have no wind and even less likely that no wind will be blowing at any of three or more sites (see table 4). While the output of a 2 megawatt wind farm located at a single site (Sacramento) would have output below 200 kilowatts 44 percent of the time, a system combining output from four sites would have

Box D
Operational Issues and Scale Economies

The cost of meeting the constantly changing electricity needs of a utility's customers obviously depends on the cost, efficiency, and fuel costs of each plant in a utility's system. But costs also reflect a variety of operational details whose significance cannot be recognized without a detailed review of how the system is operated to satisfy changing loads throughout the year. These operational factors include:

▼ **Load regulation.** Because several hours and a considerable amount of energy are needed to start large power plants, there is an incentive to keep plants operating once they are started. Indeed, fluctuating loads are often met by operating the plants at some fraction of their design capacity. But there are also incentives to avoid operating a plant at partial loads. A typical coal plant operating at half of its capacity is 16 percent less efficient than it is at full capacity. In cases where a utility's load begins to decline but is likely to increase again within a few hours, it may be preferable to keep them in a "warm" state in which they produce no output but are not shut down. Smaller plants, in contrast, can be started quickly to meet fluctuating loads since they require little or no start-up energy. In the past, smaller plants were so much more costly to operate than larger units that it was often better to meet fluctuating loads by adjusting the output of large plants rather than use the smaller, less efficient generators. But with new, efficient gas turbines this may no longer be the case.

▼ **Operating & spinning reserve.** A utility must maintain reserves to ensure that load can be met should a large plant go off-line unexpectedly or the load change in ways not anticipated. With large systems, generators can be operated at partial loads so that their capacity can be increased rapidly if necessary. But, as suggested above, this decreases operating efficiency.

▼ **Thermal unit minimum loading.** Generators cannot be operated below a fixed fraction of their design capacity. This means that larger units must be shut down when loads fall below a specified threshold and restarted when demand increases.

▼ **Other thermal unit operating constraints.** The rules governing the starting and stopping of large steam plants can be comparatively rigid. Carefully prescribed rules govern minimum allowed down times, minimum up times and the rates at which units can be started or stopped without thermal overloads. Each start/stop cycle leads to stresses which increase maintenance costs and reduce equipment lifetimes. The power level of nuclear plants can be adjusted over wide ranges but it is expensive and time consuming to stop them. Restarting a nuclear plant can take weeks.

The range of costs associated with these constraints is summarized in table 3. The details of methods used to calculate the costs, and assumptions about the operational characteristics of equipment, are described in the appendix (see also Box E).

Table 4: Probability that wind power is less than indicated (2 megawatt rated)[a]

	Case 1[b]	Case 2[c]	Case 3[d]	Case 4[e]
1,000	0.84	0.71	0.44	0.40
800	0.77	0.58	0.33	0.29
600	0.67	0.44	0.24	0.20
400	0.57	0.31	0.16	0.12
200	0.44	0.20	0.10	0.07

a. [37].
b. Sacramento, California.
c. Four sites: Bakersfield, Stockton, Sacramento, and San Francisco, California.
d. PG&E territory (most of northern California).
e. All California sites.

Table 5: Spatial cross correlation of wind speed for California sites[a]

	BFL[b]	SAC[c]	SFO[d]	SCK[e]	RBL[f]	SDB[g]
BFL	0.35	0.40	0.37	0.32	0.17
SAC		0.30	0.49	0.29	0.01
SFO			0.44	0.23	0.19
SCK				0.30	0.08
RBL					0.07
SDB					

a. 1= Perfect correlation; –1 = inverse correlation.
b. Bakersfield.
c. Sacramento.
d. San Francisco.
e. Stockton.
f. Red Bluff.
g. Sandberg.

outputs below 200 kilowatts only 20 percent of the time. If a large number of sites from all over the state of California are combined, output falls below 200 kilowatts only 7 percent of the time (see chapter 4).

Even greater diversity can be realized by combining several types of renewable electric generators (photovoltaics, wind, and solar thermal equipment) in the utility grid. The residual loads that must be met by dispatchable generating equipment are much more uniform for systems that provide 30 percent of the system's

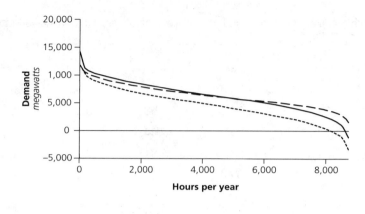

FIGURE 5: *Load duration curves for different mixes of intermittent generation (intermittents meet 30 or 50 percent of utility loads). These curves were prepared using methods similar to those described for figure 4. Photovoltaic and solar-thermal data outputs were developed from sunlight data collected in the PG&E service area in 1989, the year represented by the load data.*

electricity with a combination of wind, photovoltaics, and solar-thermal than they would be if 30 percent of the power were provided by wind alone.

The value of intermittent renewables declines as their fraction of total loads increases (see figure 6). The value of the photovoltaic equipment increases rapidly if relatively small amounts are added to the utility system. This is because photovoltaic equipment is highly correlated with system peaks in utilities with summer peaks; the photovoltaic equipment replaces peaking requirements. The value of tracking photovoltaic equipment is somewhat higher than fixed systems since trackers are able to provide more power late in the afternoon, when peaks typically occur. The value of the tracking photovoltaic systems also provides a rough guide to the value of solar-thermal equipment.

After 10 to 15 percent of the utility's power is provided by photovoltaic equipment, however, the value increases less rapidly. At this level of penetration photovoltaic units have a significant impact on a utility's peak. They may even shift peaks to evening hours. When this happens they must compete with load following and base-load equipment, which operate at lower costs than peaking plants.

The value of wind machinery increases more slowly than photovoltaic equipment since wind is not necessarily correlated with peak loads. While the value of wind energy increases rather steadily, however, the cost of wind can increase as the best wind sites are all occupied. Expanded wind use will force investors to use poorer wind sites.

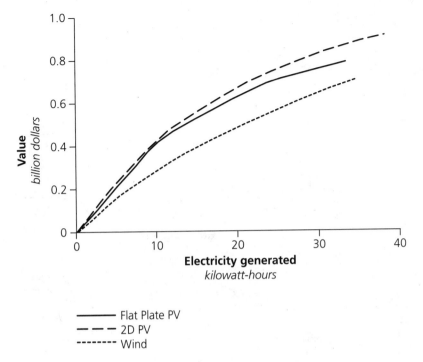

FIGURE 6: The value of intermittent electric generation. These curves represent the amount by which a utility's costs are reduced if different amounts of intermittent electrical equipment is added. Each point is computed by estimating the cost of meeting a utility load which has been reduced by the output of the intermittent equipment in each hour of the year. The optimum mix of conventional generating equipment is computed and utility operations are simulated each hour of the year using methods discussed later. The conventional system assumes use of advanced gas turbines, combined cycles, and coal gasification equipment (see table 7) (see appendix for details).

Hydroelectric power

About 20 percent of the world's electricity is generated from hydroelectric resources (see chapter 2: *Hydropower and Its Constraints*). These resources not only provide an excellent source of comparatively inexpensive power, but the output of hydroelectric systems can often be adjusted quickly to meet fluctuating demands. This dispatching feature can help offset increased load variability created by adding intermittent outputs to a utility system.

It is difficult to generalize about the capabilities of hydroelectric systems since costs and performance depend heavily on local conditions. Systems called "run of the river" have no reservoirs. Their output depends directly on water flows which may change with seasons. Systems with reservoirs large enough to store water for weeks or months can vary their output over large ranges to meet fluctuating loads. Other systems represent many intermediate conditions.

A variety of factors limit a utility's ability to change the output of hydroelectric plants to follow loads even when it is technically possible to do so. Water management must reflect the needs of irrigation systems. Minimum flows must be

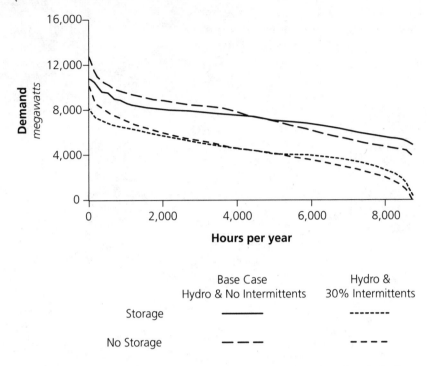

FIGURE 7: *The impact of hydroelectric equipment. These curves show how hydroelectric equipment can affect the loads which must be met with other kinds of equipment. The hydroelectric system is assumed to operate with an average output of 2 gigawatts during the year but has an ability to adjust output between 1 and 3 gigawatts to follow loads. The reservoir is assumed adequate to hold 60 hours of average output. In the hourly model used to generate this data, it is assumed that reservoir levels have no seasonal swings. At the end of each week the reservoir level returns to the level experienced at the beginning of the week. The system is dispatched by determining a threshold for each week that is consistent with the weekly refill constraint. When utility loads exceed this threshold, the hydroelectric output is allowed to go above the average level within allowed constraints. The reservoir is recharged whenever loads fall below the weekly average (see appendix for details).*

maintained for fish runs. Maximum flows must be limited to prevent shore erosion; sudden changes in flow rates can cause particularly serious damage to stream beds. Reservoir capacity must be maintained within limits to ensure that water levels do not change in ways that damage shore ecology or interfere with recreational activities. In many areas, complex regulatory procedures make it difficult to find a way to optimize recreational, agricultural, and energy needs. Even with these constraints, however, most hydroelectric equipment can vary power over wide ranges.

Hydroelectric power is an economically attractive source of power both because of its low operating costs and because the equipment can be dispatched to reduce peaks (see figure 7).

Dispatchable plants

Plants using fossil, biomass, or nuclear fuels can be dispatched to meet fluctuating energy demands within a complex set of operational constraints. In traditional systems, large coal or nuclear plants provided the lowest cost power and the core of a utility's cost-minimizing strategy was finding ways to provide the largest possible fraction of electricity from such units. Power that could not be provided by such generators was met with peaking units that were much less efficient. The operational problems associated with using large plants discussed earlier were more than offset by their enormous cost advantages. But new dispatchable equipment that can be built in comparatively small sizes and that can meet fluctuating loads at comparatively low costs has challenged conventional thinking in this area.

Selecting the optimum plant mix

The set of dispatchable plants that minimize production costs can be determined using methods described in Box A. Additional plants must be purchased to ensure adequate system reliability. (Methods for doing this are described in the appendix. Also see Box B.)

Hourly dispatching

Once a set of equipment has been purchased that ensures adequate reliability, production costs can be minimized by dispatching equipment in the order of ascending variable operating costs. This means, for example, that gas peaking plants are operated only when all available coal plants are already in use. An accurate assessment of costs requires an hourly accounting which keeps track of when individual plants are started and stopped and accounts for the need to ensure that loads can be met if a plant fails or loads increase unexpectedly. Strategies for dispatching large plants that require many hours to start and stop can be quite complex [16]. Hourly analysis is also needed to ensure that both fuel costs and emissions associated from plant dispatching are estimated accurately.

The analysis used in this chapter is based on a simplified model of actual dispatch operations using assumptions listed in Box E. Greater precision would require a level of analysis so complex that it would not be appropriate for a broad assessment of investment alternatives.

One controversial point is the extent to which intermittent units add to requirements for reserves. There would be no need for reserves if the output of intermittents could be predicted with precision. While it might be difficult to predict the output of a single wind or solar site, forecasts of the output of systems scattered over a wide region can be quite reliable. More operational experience is needed before utility dispatchers can perfect methods for using such forecasts.

The analysis employed here uses the simple assumption that variations in intermittent generation can be treated like variations in demand. Utilities equipped with large amounts of responsive gas peaking capabilities and dispatchable hydroelectric power—as was the case in all of the advanced systems examined in this analysis—are less sensitive to rapid fluctuations in intermittent output than systems forced to follow loads primarily by adjusting the output of comparatively large thermal plants. The value of intermittent equipment is lower if the utility system relies heavily on relatively inflexible thermal generating equipment that adjusts to load fluctuations by operating at partial loads [16].

Transmission and distribution

Large transmission and distribution systems are justified for four primary reasons: (i) to aggregate loads to take advantage of large-scale generation (presumed to produce power at lower cost), (ii) to create diversity in load so that generators do not need to follow the sharp variation in demand for individual customers, (iii) to increase reliability by connecting many different generating systems, and (iv) to connect remote generating sites to customers. The logic of existing investment strategies is being undercut by a recognition that: (i) scale economies in generation are declining, (ii) most utilities are far larger than the minimum size needed to saturate load diversity, and (iii) decentralized generation can increase system reliability.

The unique features of renewable generators underline the importance of reviewing transmission and distribution (T&D) investment strategies. Small fossil-fired cogenerators and photovoltaic units can reduce transmission and distribution requirements. But remote hydroelectric, solar-thermal and wind sites can increase transmission needs.

Distributed systems

Utility transmission and distribution costs can be lowered if efficiency improvements or demand management reduce a regions's peak demand, or if generation equipment can be located close to the customer's site. Individual distribution areas must be examined since many costs can only be reduced if the peaks experienced at local substations are reduced. Individual distribution areas do not necessarily experience peak demands when overall utility loads are at a peak. Residential regions may peak during the evening hours, while business peaks occur during the afternoon. But peak management in substation regions is also important since the load factor of individual substations is typically far lower than that of the utility as a whole (see table 1).

Investments in distributed equipment can lead to savings in a variety of areas. They can reduce the size of transmission conductors and the size of transformer substations serving local distribution grids. They can also reduce operating costs associated with fluctuating loads within an individual distribution area [20]. The effect is reinforced in utility systems whose peaks occur during hot summer months. The carrying capacity of typical aluminum conductors may be 11 to 14 percent below average during periods of peak demand [21].

The complexity of transmission and distribution systems makes it difficult to estimate the cost implications of different generation strategies. New equipment typically costs far more than existing equipment because of growing land costs, increasing concern about the appearance of overhead lines, and concern about electromagnetic fields. Distribution systems in new urban areas are now typically built underground—a process that greatly adds to costs.

Fortunately, a recent study conducted by a large utility provides a good contemporary estimate of average costs [21, 22, 23]. The study examined investments actually planned in transmission and distribution for the next decade and compares these investments with expected growth in load. Planned investments were developed for each distribution region. The marginal costs of meeting load growth attributed to new T&D costs were calculated in three steps: (i) estimate the present value of planned T&D investments, (ii) estimate the present value of postponing the planned T&D investments by one year, and (iii) divide the difference between the results of step (i) and step (ii) by the predicted load growth.

The results varied widely by region but averaged $6.7 per kilowatt-year for bulk transmission, $13 per kilowatt-year for area transmission, and $45 per kilowatt-year for primary distribution systems.[1] A distributed generation system that reduced a region's peak by one kilowatt would, therefore, have a value of about $65 per kilowatt-year for the system as a whole. A distributed system can also avoid other transmission-related costs. Maintaining control over voltage requires complex switching operations at transformer stations. This is typically done by changing transformer taps, or switching in static capacitors and synchronous condensers.[2] A typical 12-kilovolt transformer substation may need to switch tap positions 5,000 times a year for an average cost of about $1 per tap change. A recent estimate of a distributed photovoltaic system suggests that the system would reduce voltage control costs by about $4 to $5 per kilowatt-year [22].

Distributed systems can also contribute to system reliability. But other factors must be considered. Photovoltaic systems distributed through a utility system can increase rather than decrease the reliability of power actually delivered to a customer's site. Distributed photovoltaic systems are "intermittent" in that dispatchers have no control over their output, but photovoltaic output is highly correlated with air-conditioning peak loads. The reliability gains result primarily from the fact that utilities would have more options for routing power around a fault in the distribution system [21]. In many cases a customer must be disconnected from the grid while faults are repaired, even though a physical connection could be established between the customer and the transmission system. Figure 8 (top) illustrates such a situation. The customers in the lower part of the distribution system shown in the diagram would have to be cut off if a fault occurred at the point illustrated. The distribution system's voltage would drop dangerously if an attempt

---

1. See [23]. The estimates were prepared using 1993 dollars and a fixed charge rate of 9.1%. These costs were converted into 1989 dollars and the fixed charge rate used for this analysis 7.8%. The costs of were computed using the following method: (1) actual planned investments in new T&D systems were estimated for an area, (2) the present value of the investment stream was calculated, (3) the present value of the investment stream postponed for one year was calculated, (4) the difference between the two present values was divided by the projected growth in demand for the region, and (5) averages were computed. The methods are based largely on [24].

2. See [25]. These costs could presumably be reduced considerably if solid-state devices could be developed to operate over the full range of power and voltage conditions that occur at the station.

Reliablity benefits from distributed generation

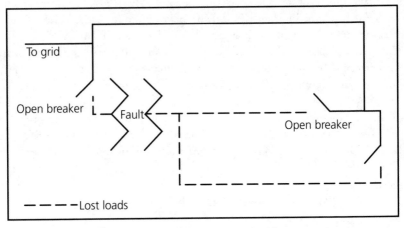

A. Loads lost with conventional grid

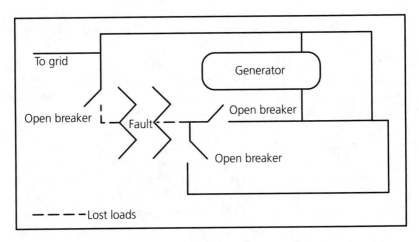

B. Loads lost with distributed generator support

FIGURE 8: Distributed generators can increase the reliability of local electric service. Many utility customers lose power when a distribution line is broken even though there is a physical connection between their premises and the utility grid. They are cut off by utility service crews because serving them would mean routing power through lengthy bypass routes using distribution lines which are already fully loaded serving existing customers. A typical problem is illustrated here (top). A fault in the distribution system requires the utility to open breakers at the points indicated while repairs are underway, thereby cutting-off customers in the lower loop. If a distributed generator is available at a strategic point in the distribution system, however, power can be maintained in a much larger area while repairs are underway (bottom).

Box E
Hourly Dispatching

Detailed analysis of utility operations can be complex and time consuming. The portfolio analysis used in this chapter balances the need to consider a variety of important operational issues with the need compare a wide range of possible investment strategies. Uncertainties about the cost and performance of advanced generating equipment limit the precision achievable with any method.

The method used for hourly dispatching involves the following steps. (See the appendix for details about the method. Table A in the appendix details assumptions made about start times, part load efficiencies, and other factors important for hourly dispatch calculations).

▼ Plants are closed for planned maintenance during the year and unexpected failures may force them to shut down (a "forced outage" condition). The analysis accounts for this down time by assuming that a plant's effective operating capacity is equal to its actual capacity reduced by the probability that it will be available. In other words, if the utility operates 10 gas peaking plants with 50 megawatt capacities and an average availability of 90 percent, the hourly analysis dispatches 10 plants with 45 megawatt capacity. (The forced outage rate is also used to calculate the reliability of the generating system).

▼ The hourly calculations track the time and energy needed to "cold start" a unit that has been shut down and "warm start" a unit already in standby condition thus ensuring that units are available when they are needed.

▼ Equipment must operate at partial loads when demand is not a precise increment of the available capacity and often must operate at a fraction of their full output capability in order to provide reserves. The analysis keeps track of the decline in efficiency that results from operating at part load.

▼ The analysis assumes that plants are maintained in a warm, standby condition rather than shutting down entirely when loads are expected to increase in the near future. The analysis assumes that plants are not turned off if the unit has to be restarted during a period shorter than its cold start time.

▼ Plants are not allowed to operate below a minimum level (specified in table A). This would create difficulties for utility systems with so much intermittent generating capacity that the net load placed on the conventional equipment is very small or even zero during some periods. Conventional units, such as large base-load plants, may need to be kept in operation at minimum output levels during such periods and some electric output is wasted (in practical situations, some market is likely to be found).

were made to serve them through existing lines. Generators located at strategic sites close to consumers (see figure 8, bottom) would allow the utility to keep the lower part of the grid energized while the fault was isolated and repaired. Automated fault location and switching equipment would greatly facilitate the process of determining which breakers to open. Equipment designed to reduce customer loads to minimum levels (e.g., by temporarily cutting off hot water heaters or refrigerators) would provide even more options for minimizing service disruptions.

Unfortunately there is no general rule for computing the reliability gains possible. Reliability gains are highly sensitive to the details of individual distribution systems. One careful study has been conducted for a substation in central California. Five years of data on actual power disruptions were examined and estimates made of the number of customers whose power could have been restored quickly if distributed generation had been available. The analysis suggested that a distributed generator would have a value of $200 per year for each kilowatt of distributed power installed based on estimates of the value of lost service.[3] The system examined may have had a comparatively weak distribution system and much more work is needed to determine whether such high values could be realized by distributed systems operating in other areas.

When the value of distributed generation is estimated in the analysis presented in this chapter, a credit of $80 per kilowatt-year is assumed. This includes the $56 per kilowatt-year for avoided T&D capital costs, $4 per kilowatt-year for avoided T&D maintenance, and a $20 per kilowatt-year credit for reliability. The credit for reliability assumed is conservative because detailed estimates are available for only a single site. The value of the distributed credit would presumably decline as large amounts of distributed generation enter a utility's system. The value of the distributed credit also depends on the reliability of the distributed system. The highest credits would apply to highly reliable systems such as rooftop PV.

While it is beyond the scope of the current analysis it is important to recognize that T&D costs can also be reduced through investments in efficient electric use and demand-side management automation. Modern communication and control systems have opened many opportunities for dynamic management of complex electric networks. These systems can serve a variety of purposes (see Box F). Moreover, utility customers placing a particularly high value on reliability often have large, expensive, electric battery systems as an insurance against power failures. For example, virtually all telephone systems, computer facilities, and hospitals have electric storage systems as well as backup generators for use in the event of a power loss. These systems are costly and seldom used. It is likely that a careful analysis of the reliability needs of different customer classes, and the options for

---

3.   The credit for transmission and distribution have been multiplied by 0.85 to reflect the availability of the photovoltaic system at system peaks [22]. The reliability credit is taken to be 10 percent of the amount calculated for the site described earlier. The numbers used have been rounded to avoid a false sense of precision.

locating utility and non-utility generating and storage equipment, could achieve improved reliability for all classes of customers at reduced costs.

Remote intermittent generation

There are, of course, circumstances where renewable equipment will increase rather than decrease demand for investments in bulk transmission. Many wind, solar-thermal, geothermal, and hydroelectric sites are located at some distance from load centers. Transmission is required to exploit good resources and inter-connections offer significant operational advantages. Hydroelectric resources can be dispatched to minimize system peaks. And the combination of diverse intermittent sites can provide more reliable power to a grid. It is unlikely that no direct solar or wind power is available on a system integrated by a transmission system, even though the probability of zero production at any given site is large.

Long-range transmission costs for wind systems, exclusive of rights-of-way are estimated to be about $0.036 per kilowatt-year per kilometer [26]. Transmission lines for a wind system 100 kilometers from a grid would add $3.6 per kilowatt-year to system costs. Actual costs would depend on the amount of energy delivered per kilowatt (measured by the load factor of the wind site). Transmission costs for wind systems would be high compared to those of coal or nuclear equipment since the load factor of wind systems (typically 20 to 30 percent) is comparatively low. The optimum way to load transmission lines in systems employing large amounts of intermittent electric generation remains largely unexamined.

Storage

An ability to store electricity would have a number of operational benefits for an electric utility—whether or not the utility employs significant amounts of renewable electricity[27, 28]. Storage systems allow utilities to operate their least expensive plants and their most expensive plants fewer hours each year. Low-cost plants are used to charge storage during periods of slack demand and energy is taken from storage during peak demand periods when costs are high (see figure 9). Storage can also offer a number of operational benefits (see table 5).

Storage is, of course, inherent in conventional energy systems. Hydroelectric reservoirs are able to store water until it is needed to meet load. Fossil-fuel systems store energy inexpensively in inventories of fluids or solids. There are also many ways to store electric energy close to where it will be consumed. Industries have the option of accumulating inventories of electric-intensive products during periods of low demand. Commercial heating and cooling systems can store energy inexpensively in the form of heated or chilled water or ice. At least 30 utilities have experimented with cool-water storage, offering payments ranging from $115 to $550 per kilowatt of peak demand reduced. A two-million square-foot office building in Dallas, Texas, for example, has used such a system for several years [29]. Utility customers seldom find it profitable to invest in storage since utilities have not been able to communicate costs effectively. Increased consumer invest-

Table 6: Electric storage technologies[a]

	Power -related cost	Energy -related cost	Hours of storage	Total cost
	$ per kilowatt	$ per kilowatt-hour		$ per kilowatt
COMPRESSED AIR				
Small module (25–50 MW)	575	5	10	625
Large module (110–220 MW)	415	1	10	425
PUMPED HYDRO				
Conventional (500–1,500 MW)	1,000	10	10	110
Underground (2,000 MW)	1,040	45	10	1,490
BATTERY				
Mid-term target (10 MW)	125	170[b]	3	635
Advanced target (10 MW)	125	100	3	425
SUPERCONDUCTING MAGNET TARGET (1,000 MW)	150	275	3	975

a. [27].
b. Replace after 15 years at a cost of $85 per kilowatt-hour for 250 cycles per year.

ment could be expected if electric rates reflected hourly utility costs and if a system for communicating these costs were available.

Strategies for converting electricity into some other form (e.g., chemical energy in batteries or mechanical energy in the form of flywheels) has proven much more difficult and expensive (see table 6). They include:

▼ *Pumped hydroelectric systems.* Energy can be stored in a two-reservoir hydroelectric system by pumping water from the lower to upper reservoir during periods of slack demand. Up to 75 percent of the electricity used to charge the storage system can be recovered [28]. Pumped hydro facilities are by far the most widely used electric storage systems today. The United States has 37 pumped hydroelectric plants. The systems are typically very large (1,000 to 2,000 megawatts) and therefore lack many of the operational advantages of storage systems that can be located at strategic points in a utility grid [27].

▼ *Storage batteries.* Battery storage is expensive but offers the enormous advantages of speed and modularity [30]. Batteries can be located at convenient sites and can be started in 20 milliseconds. West Berlin has operated a 52 gigajoule lead-acid battery since 1987 that can deliver 8.5 megawatts over 5 hours and can deliver 17 megawatts at some penalty in energy. There is also a 1 megawatt battery storage system under test in Osaka, Japan, and a10 megawatt system in Chino,

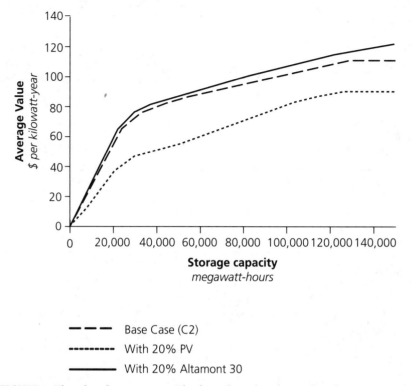

FIGURE 9: The value of energy storage. This figure shows the average value of energy storage for utilities with different generation mixes. In all cases, the marginal value of additional storage declines sharply after 20 gigawatt-hours of installed capacity. The cases include a "base case" using advanced generating equipment, and systems combining advanced generating equipment with wind and photovoltaics. Introduction of photovoltaics decreases the value of storage because it reduces the system peaks.

California. The United States also has a two megawatt, 22 gigajoule zinc-chloride experimental facility.

▼ *Mechanical storage.* Energy can be stored in mechanical form by spinning flywheels or by compressing gas that is later expanded over turbines [31]. A 290 megawatt compressed gas storage built in Huntorf, Germany has operated well since 1981 and a large test is also underway in the United States. These may be particularly valuable when used with coal or biomass gasification and gas turbines where both hot water and humid hot air can be stored [32].

▼ *Thermal storage.* Energy in solar-thermal plants can be stored in the form of heated fluids (see chapter 5: *Solar-thermal Electric Technology*). The value of energy storage to a utility system shown in figure 9 is much the same as the value of the hydroelectric systems described earlier. It is important to notice that the value of storage depends on the nature of the loads it is meant to optimize. When 20 percent of a utility's load is produced from photovoltaic equipment, for example,

Box F
Distribution Automation Equipment

Advanced communications and control technologies can improve the management of utility generating equipment and offer a unique opportunity to manage distributed generators (some of which may not be owned by utilities), transmission and distribution equipment, and even electric equipment operating in homes and businesses. Electric meters equipped with microprocessors and connected to a utility-managed communication system provide instant information about customer's current rates of consumption and historic data that can be used for billing and evaluation. Communication equipment operating through radio, cable, microwave, and optical fiber give utilities dispatchers an ability to monitor the status of lines, transformers, and other equipment.

The new technologies provide an efficient way to communicate seasonal, weekly, and even hourly changes in the cost of producing electricity so that customers can be encouraged to (1) purchase energy efficient products, distributed generating devices, and storage equipment most likely to reduce costs over the entire system, and (2) operate their equipment in ways that minimize system costs. Electricity is typically most expensive during periods of peak demands and cheapest during the night when base-load equipment may be operating at partial capacity (in the future power may be cheapest when the wind is blowing or sunlight is most intense).

▼ The new equipment allows more flexibility in agreements utilities reach with customers to automatically change thermostat settings or shut off water heaters or refrigerators for brief periods during periods when power is most expensive. With good controls and widespread participation, the impact on each customer would be minimal. Most systems allow the customer to override the automatic utility control if they are willing to pay a price.

▼ New equipment can allow more careful management of multiple load centers within a facility—multiple residences in an apartment complex or complex sets of production equipment in a manufacturing facility.

▼ The equipment allows utilities to dispatch generation and storage equipment located on the customer's premises to minimize overall system costs

▼ Advanced sensors and communication equipment can report faults in the distribution system (possibly even faults on a customer's premises) and automatically disable lines in a way that minimizes load losses. (This is now done manually by trial and error). The communication systems can also shut off non-essential loads on a customer's premises during emergencies so that the smallest possible number of customers would be completely without power.

▼ The new systems can automatically dispatch a large number capacitor banks and other voltage-management devices distributed throughout a utility network.

▼ They can provide automatic billing and bill monitoring for both customers and the utility.

▼ The complex flow of energy through transmission and distribution systems can be understood and managed through advanced computation and control systems. These controls can reduce energy lost in unwanted current flows in the transmission systems and can improve system stability—particularly during emergencies.

system peaks are reduced, and storage is worth less to the utility than it would be if no photovoltaics were used. When wind generators are added to the system, peaks increase and storage becomes more valuable.

PORTFOLIO COSTS

The remainder of this chapter will describe methods for measuring the value of the different renewable electric technologies, making different assumptions about the way electric utilities may evolve during the next few decades. No utility can be considered "typical" but a representative system was selected for an analysis that included a realistic examination of utility dispatching problems. The system chosen, the PG&E utility in northern California in 1989, has a pattern of electric demand representative of many growing utility systems around the world. This system experiences peak demands during the summer largely because of air-conditioning loads. Because this analysis focuses on renewable supplies, no attempt was made to compare investments in energy efficiency and load management with investments in generation.

It was assumed that the system examined had average access to hydroelectric power (20 percent of world electricity results from hydroelectric power), and access to wind and solar resources likely to be available in most parts of the world. A variety of hypothetical investment portfolios were examined. None represents equipment actually installed in the PG&E utility (42 percent of California's electricity comes from hydroelectric power, geothermal, biomass, wind, and solar resources [33]). Cost and performance estimates of renewables are largely based on analysis detailed in other chapters of this volume (see Boxes C and D). The estimates used in the present analysis are conservative in that equipment with lower costs or better performance than those assumed in the present analysis is clearly possible (see individual chapters for details). (Detailed assumptions about the cost of equipment are shown in table A in the appendix and are summarized in table 7).

Reference system costs
Three types of fossil-powered equipment were evaluated:

▼ Conventional generating equipment (C0) typical of equipment now operating in utility systems,

▼ Best new generating equipment (C1) intended to represent the best generating equipment being installed today, and

▼ Advanced generating equipment (C2) which represent technologically advanced equipment likely to be available to utilities by the end of the century.

The new generating equipment (C2) clearly offers significant cost advantages over the current generation of fossil equipment (C0 and C1) (see figures 10 and 11). Costs, of course, depend on the cost of capital available to a utility. In this

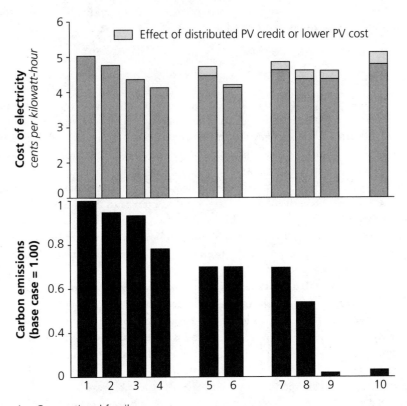

1 = Conventional fossil
2 = Best new fossil
3 = Advanced fossil
4 = Advanced fossil with 21% hydro
5 = Advanced fossil with 21% hydro & 10% PV
6 = Advanced fossil with 10% mixed intermittents
7 = Advanced fossil with 30% mixed intermittents (3 wind sites)
8 = Advanced fossil with 21% hydro & 30% mixed intermittents (3 wind sites)
9 = Advanced biomass & gas, 21% hydro & 30% mixed intermittents (3 wind sites
10 = Advanced biomass & gas, 21% hydro & 50% mixed intermittents (3 wind sites

FIGURE 10a: This figure shows the average cost of meeting the annual electricity demands of the PG&E utility in northern California in 1989 (top) and the CO_2 emissions as a fraction of the emissions from the reference case (bottom) for 10 different portfolios of generating equipment. The upper segment of bars 4 through 10 represents the magnitude of the transmission and distribution savings that can be realized by distributed photovoltaic systems. This distributed credit was valued at $80 per kilowatt-year for the installed photovoltaic capacity and includes credits for transmission and distribution capital costs, distribution losses, and 10 percent of the reliability credit computed for the one case in which a detailed reliability study has been conducted (see text for details). The upper bar segment can also represent the effect on overall electricity cost of lowering the cost of installed photovoltaic capacity from $1,800 per kilowatt to $900 per kilowatt if no distributed credit is applied.

Table 7: Cost and performance of utility generating equipment[a]

Plant name	Capacity	Overnight cost	Forced outage	Efficiency at full load[b]
Source	MW_e	$ per kW_e	% time	%
COMMERICAL (C0)				
Simple cycle	80	387	3.5	27.9
Combined cycle	120	503	6	42.7
Sub-critical steam	1,000	1,290	6	33.9
BEST NEW (C1)				
Simple cycle	140	355	3.5	29.7
Combined cycle	210	473	6	45.4
Coal/gasifier/CC	800	1,275	10	38.2
ADVANCED (C2)				
LM6000 Intercool SC	80	440	2	40.9
LM6000 Intercool CC	100	600	4	49.9
Coal gasifier/ISTIG	100	1,090	4	42.1
HYDRO	4,000	1,800	1	c
WIND (post 2000)	100	800	2	d
SOLAR THERMAL (post 2000)	200	1,625	1	e
PV (flat plate post 2000)	10	900 - 1,800	0.5	f

a.   Hourly calculations require many other assumptions about the dynamic performance of this equipment. Assumptions made about start-up times and energy use, part-load heat rates, data sources and other factors are explained in detail in table A in the appendix.

b.   These efficiencies apply at 100 percent loads and refer to lower heating values.

c.   The characteristics of the reservoirs vary with applications discussed in the text. The cost applies to the peak output. In typical applications shown the calculations (including those summarized in figures 10 and 11), it is assumed that the system operates with a 50 percent capacity factor and that output can be adjusted from 25 to 100 percent of the peak within limits imposed by the reservoir size (assumed to be 60 hours at average output).

d.   Wind output varies with location. When Altamont is specified, actual wind data is from the Altamont Pass field in California in 1989. Outputs scaled to reflect the performance of the U.S. Wind Power Equipment now under test. When random winds are used, output is computed from a Rayleigh wind distribution. It is assumed that sites are available with average wind speeds of 8 mps until wind represents 5 percent of the system's load. An average wind speed of 7.5 mps is then used until capacity reaches 10 percent of the utility's load and any additional wind energy needed is supplied from regions with wind speed averaging 7.25 mps. All speeds are for 50 meter hub heights. Wind electricity is computed from wind speeds using the performance characteristics of the U.S. Wind Power Equipment. Wind Class 5 has average winds at 50 meters which range from 7.5 to 8 mps. Wind Class 4 areas average 7.25 mps.

e.   Solar thermal output is based on hourly measurements taken in the PG&E service territory in 1989. They assume two axis tracking and no storage or on-site generation.

f.   Photovoltaic output was based on hourly measurements taken in the PG&E service territory (Carissa Plains) in 1989. The system was a flat plate tilted at the latitude angle. Credits for distributed systems are estimated to be $80 per kilowatt-year based on $65 per kilowatt-year for transmission and distribution capital credits, $5 per kilowatt-year for T&D O&M credits, and $10 per kilowatt-year for reliability credits. Since detailed studies of this credit have been made in only one utility region, it is conservatively estimated that the credits would be 10 percent of the credit computed in that region (see text for details). PV costs were $1,800 per kw. This is a conservative estimate of mid-term costs. Longer term costs could be nearly half this (see chapters 6–10). The range of PV costs computed with and without the $80 per kilowatt-year distributed credit can also be read to reflect a range of PV costs with the lower end of the range showing optimistic assumptions about future photovoltaic prices.

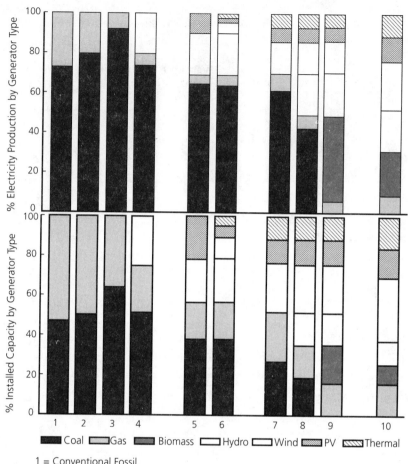

1 = Conventional Fossil
2 = Best New Fossil
3 = Advanced Fossil
4 = Advanced Fossil with 21% Hydro
5 = Advanced Fossil with 21% Hydro & 10% PV
6 = Advanced Fossil with 10% Mixed Intermittents
7 = Advanced Fossil with 30% Mixed Intermittents (3 Wind Sites)
8 = Advanced Fossil with 21% Hydro & 30% Mixed Intermittents (3 Wind Sites)
9 = Advanced Biomass & Gas, 21% Hydro & 30% Mixed Intermittents (3 Wind Sites)
10 = Advanced Biomass & Gas, 21% Hydro & 50% Mixed Intermittents (3 Wind Sites)

FIGURE 10b: This figure shows the energy produced (top) and installed capacity (bottom) of each generator type for the portfolios examined. Generator types with low capacity factors (such as gas peaking plants) produce comparatively little energy per unit of installed capacity.

analysis it is assumed that regulated utilities use a 6 percent real discount rate and private investors 12 percent. Higher capital costs disadvantage high-capital cost equipment (including renewables). Doubling the discount rate has about the same effect as holding capital costs constant and increasing the cost of equipment by about two thirds. None of the costs displayed include the effect of taxes or the cost of externalities.

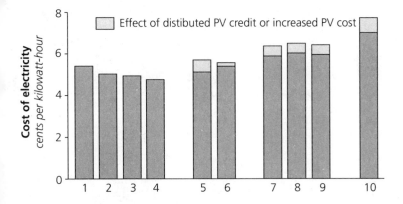

1 = Conventional fossil
2 = Best new fossil
3 = Advanced fossil
4 = Advanced fossil with 21% hydro
5 = Advanced fossil with 21% hydro & 10% PV
6 = Advanced fossil with 10% mixed intermittents
7 = Advanced fossil with 30% mixed intermittents (3 wind sites)
8 = Advanced fossil with 21% hydro & 30% mixed intermittents (3 wind sites)
9 = Advanced biomass & gas, 21% hydro & 30% mixed intermittents (3 wind sites)
10 = Advanced biomass & gas, 21% hydro & 50% mixed intermittents (3 wind sites)

FIGURE 11: Comparing utility investment portfolios (12 percent discount rate). This figure uses the same assumptions as figure 10 except that the utility discount rate is assumed to be 12 percent (real). This has roughly the same effect on costs as keeping the discount rate at 6 percent and increasing all capital costs by two thirds.

Systems with intermittents

The cost of portfolios containing intermittent renewables is sensitive to the equipment selected, the fraction of the energy supplied by intermittents, and the cost of capital (see figures 10 and 11). The analysis examined utility generation systems in which 10, 30 and 50 percent of demand is met with intermittents. No attempt was made to develop a mix of intermittents that minimized production costs.

Perhaps the most striking result of the analysis is that, with only one exception, all of the renewable scenarios considered can meet the system's load at a cost lower than the cost of operating the system with typical new coal and gas equipment. And many of the renewable cases provide 90 to 95 percent of all electricity from renewables. Portfolios making more than 10 percent use of intermittents had difficulty competing with advanced coal and natural gas technologies (technologies likely to be available by the end of the decade). The cost of meeting loads ranges from 4.5 to 5 cents per kilowatt-hour for systems in which mixtures of wind, photovoltaics, and solar-thermal provide 30 percent of the load. But even the extreme case examined (50 percent intermittents) was only 1.2 cents per kilowatt-hour more expensive than the lowest cost conventional system using advanced technology. This translates into a selling price that is about 15 percent higher than the average delivered electric price today.

The high-renewables cases produce only 2 to 4 percent of the carbon of the conventional fossil system and 3 to 6 percent of the carbon produced by the advanced fossil system. These enormous reductions in carbon production are all achieved at cost premiums below 1.2 cents per kWh. For comparison, average electric prices today are 6.3 cents per kilowatt-hour and residential prices 7.5 cents per kilowatt-hour (these prices include transmission, distribution, and management costs as well as the generation costs displayed in figure 10a).

Biomass

Biomass systems could compete directly with advanced coal systems (see figures 10a and 11)—particularly if advanced integrated biomass gasifier technologies are available. Biomass could economically be substituted for the coal in any of the cases examined.

Hydroelectricity

The advantages of buffering demand on a utility system using hydroelectric equipment is apparent in both conventional and renewable-intensive investment portfolios. Hydroelectric power is particularly attractive when used to buffer the large fluctuations that result from high penetration of intermittent equipment. For example, when the system is operated to provide 30 percent power from solar and wind systems, system costs average 4.5 cents per kilowatt-hour. When hydroelectric power is added, costs fall to 4.3 cents per kilowatt-hour. The cost difference would be much greater if today's equipment (C0 Technology) provided the peaking power. The advanced gas-turbine systems assumed in this analysis, however, can be operated to meet peaks at costs not strikingly greater than those of base-load plants, thereby reducing the contribution hydroelectric equipment can make by reducing costs associated with peaking.

Distributed systems

The costs shown reflect the operational advantages of modular generating equipment. They do not include the cost of transmission and distribution systems, customer hookups, or miscellaneous service costs such as billing and meter reading. The previous discussion suggested that these costs can be 1.5 cents per kilowatt-hour (assuming a 50 percent capacity factor).

Figure 10 shows the value of distributed systems which reduce transmission and distribution costs by giving distributed systems a capital credit equal to the T&D savings. Given these credits, much larger penetration of renewable systems can be justified.

Distributed credits would also apply to decentralized fuel cells and other distributed generation equipment. A hydrogen-powered fuel cell might be an attractive element in a utility production network if it can be operated as a cogenerator and dispatched by the utility (see chapter 22: *Solar Hydrogen*). For the assumed costs, the fuel-cell cogeneration systems would be dispatched as a substitute for advanced combined cycle systems. The costs of such fuel-cell systems could be justified only if credit were given for reduced transmission and distribution costs.

Environmental costs

The investment portfolios have very different impacts on production of carbon dioxide (see figure 10a) and other pollutants. But the renewable-intensive systems claim no credit for these benefits under the accounting systems used in the analysis presented here (e.g., in figures 10 and 11). Renewable portfolios are therefore systematically undervalued in conventional accounting. While there is disagreement about the magnitude of their environmental benefits, the value is plainly greater than zero.

Over half the states in the United States either already include environmental costs in their evaluation of proposed new utility generation investment or are considering such rules [34]. For the most part, the costs computed do not include greenhouse gases such as carbon dioxide. While methods vary, costs of several cents per kilowatt-hour are under consideration—far greater than the 1.2 cents per kilowatt-hour needed to make even the largest renewable penetration considered here competitive.

Care must be taken to avoid counting emission costs twice. The cost of complying with existing environmental regulations is already included in the capital and fuel costs of fossil-burning power plants. The question remaining is whether additional reductions in emissions are worth additional investments.

Increasing the cost of fossil fuels by adding externality taxes would change the optimum mix of generating equipment, as well as change the way equipment is dispatched (see figure 12). A carbon tax would increase the cost of electricity from coal and, to a much lesser extent, natural gas-fired equipment. A carbon tax also increases the cost penalty associated with repeatedly starting and stopping equipment, since fuel is needed each time a plant is started. This increases the cost penalty associated with intermittent renewables that increase the variability of loads placed on thermal plants.

If a $30 per ton carbon tax were imposed, and advanced gas turbine equipment (C2 assumptions) were available, utility costs would be lowest if only gas generation were used. A carbon tax of about $100 per ton would allow a utility system using intermittents for 30 percent of its energy to compete with a system using advanced generating equipment (see figure 12). A similar tax would allow a system that obtains 50 percent of its energy from intermittents and employs advanced biomass generation, to compete with utilities using advanced generating equipment.

Considerable caution must be exercised in interpreting these results since small changes in cost assumptions lead to significant changes in the level of taxes that equalizes average production costs. The calculation also places no value on the externality costs of emissions other than carbon dioxide production. Some utility jurisdictions in the United States are already imposing taxes for non-carbon

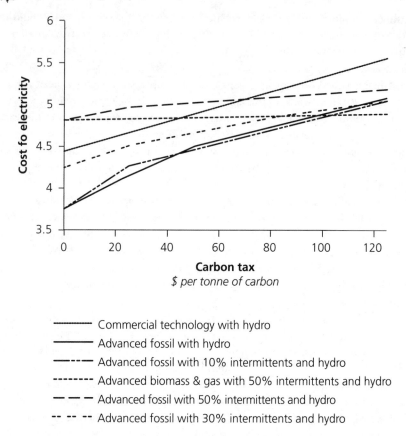

FIGURE 12: *The impact of a carbon tax. The imposition of a carbon tax would affect utility operations by changing the mix of fossil-burning plants that minimize production costs. A tax would increase the slopes of the production cost curves shown in figure A2. The slope of coal-burning plants would be affected most strongly since coal burning produces 25 kgC per GJ of fuel. Gas burning generates 13.5 kgC per GJ. As a result, fewer coal plants would be chosen in a cost-minimizing mix that includes a carbon tax. The dispatch of plants also must change to reflect the difference in production costs. The figure demonstrates the impact of a carbon tax on a utility employing equipment of a type typically in use today including hydroelectric contributions (C0H), and a utility using advanced conventional equipment. It also shows the cost of production for systems using intermittents for 10, 30, and 50 percent of a utility's electricity production.*

emissions that would have a larger effect on system cost comparisons than the carbon taxes suggested in figure 12.

NEW REGULATORY PARADIGMS

The analysis just presented demonstrates that renewables provide a unique tool for achieving ambitious environmental goals while keeping the cost of energy services low and the quality of services high. But the regulatory environment governing utility decisions in most regions has evolved around an old paradigm of utility technology and is likely to frustrate the emergence of utility portfolios that make optimum use of renewables. New approaches to public policy are needed to encourage the transition to utility businesses better able to meet the cost, quality, flexibility, and environmental requirements of future electricity markets.

Based on the assumptions of scale economies, the strategy of utility regulation has typically involved protection against the risk of large capital investments [35]—although few of these programs worked to the advantage of renewable generators. Other utility regulations, however, allow utilities to pass fuel price increases directly to customers. This clearly reduces the incentive to create an investment portfolio designed to provide protection against the risk of future price increases and leads to under investment in renewable generation [36]. Decisions made without accounting for the environmental benefits of renewables will clearly also lead to under investment in renewables. And regulations which prevent utilities from making balanced investments in all technologies relevant to minimizing the cost of electric service (including energy efficiency, demand management, centralized and distributed generation, storage, transmission and distribution) can slow the emergence of balanced investment portfolios.

Many areas have already recognized the need for fundamental change in utility policy and revised regulations to encourage balanced long-range planning, but much remains to be done. Each nation will react to these changed circumstances in different ways. Some will be try to protect conventional strategies while others will move more actively to encourage investment in non-traditional areas. But the underlying pressure for change seems clear. Utilities will be encouraged to consider a much more diversified set of products and services than they have in the past. Competition will be encouraged, but regulated utilities will be needed to set prices and establish other rules that allow investors to find the most profitable use of energy-related capital. Given the right incentives, utilities can be a powerful force for encouraging investment in electric production and delivery systems that make optimum use of renewable electric technologies—technologies that represent key tools for keeping electric costs low while meeting ambitious environmental goals.

ACKNOWLEDGMENTS

The authors thank Gregory Terzian of Princeton University for his assistance with the calculations undertaken for this analysis and his insights on the results and Tom Hoff for providing renewable resource data.

Appendix
COMPUTING THE COST OF PROVIDING ELECTRIC SERVICE[4]

The economic analysis is based on a highly simplified dispatching model. This model computes the levelized cost of providing service based on (1) the financing available to the utility, (2) forecasted fuel prices, and (3) the cost and performance of different kinds of generation systems and storage systems.

The calculations distinguish between "dispatchable" equipment whose output is under control of the utility (biomass, fossil and nuclear plants, and geothermal facilities), and "intermittent" equipment, wind, solar-thermal, and photovoltaic. Hydroelectric equipment is treated separately.

Basic financial calculations

Capital costs
The utility discount rate (d) is the weighted cost of capital:

$$d = (1 - Ta) \; rb \cdot fb + rp \cdot fp + re \cdot fe$$

where Ta is the marginal tax rate (set at zero in this analysis–see footnote 2 in the appendix for chapter 1 at the back of the book) and d was set at 6 and 12 percent (see Box C).

The utility fixed cost rate is computed from the discount rate assuming that a sinking fund is used to accumulate capital for repayment at the end of the plant's life:

$$FIX = PT + IN + \mathrm{CRF}(L,d) \cdot [1 - Ta/(Ld \cdot \mathrm{CRF}(Ld,d)]/(1 - Ta)]$$

where $\mathrm{CRF}(L,d)$ = the capital recovery factor for an annuity of period L paying interest d, IN is the insurance rate (set at .005 in this analysis), and PT is the property tax rate (set at zero in this analysis).

The cost of a plant is computed on the basis of its construction schedule as follows:

$$C = \sum_{i=0}^{i=5} (\text{funds expended in year } t) \cdot (1 + d)^{t}$$

The funds expended in year t are equal to the difference in investment in place at the beginning and end of the year. This is specified by the percent of cost expended in year prior to plant start-up.

---

4.   A list of variables appears at the end of this appendix.

The annual fixed cost of a plant type j per kilowatt per year (CAj) can then be computed from its capital cost per kilowatt (Cj) and its fixed charge rate ($FIXj$) as follows:

$$CAj = Cj \cdot FIXj$$

The fixed cost of all equipment of this type can be found by multiplying CA by the total installed capacity (in kilowatts).

Fuel costs

The levelized cost of fuel is computed as follows:

$$FP = \sum_{t=0}^{L} \frac{F(t) + EX}{(1+d)^t}$$

where $EX(i)$ is the total externality cost of fuel-type i in \$/MMBTU:

$$EX = \sum_{i=0}^{5} EX(i) \cdot EP(i)$$

Operational issues

The cost of meeting load is calculated using the following sequence:

1. The hourly utility load is read from a file.
2. The output of intermittent producers is subtracted from the load, producing a net load to be met by hydro and dispatchable equipment.
3. The hydroelectric system is operated together with storage in a way that minimizes utility peaks.
4. The dispatchable plants are operated to meet the load at minimum cost.

Intermittents

Wind and Direct Solar. The intermittent output is provided by an hourly file specifying the fraction of the intermittent peak load provided at any hour. The total intermittent power produced in hour t is called $I(t)$. The output is simply subtracted from the utility load at each hour of the year.

Hydroelectric. Hydroelectric energy is provided using the following highly simplified algorithm for the hourly output of hydroelectric facilities:

$$H(t) = Hav + Hs \cdot cos\left(2\pi(t/8{,}760 - P/12)\right)$$

where Hav = the average hydroelectric output (megawatts) and Hs is the annual swing above and below the average (megawatts) and P is the month of the peak. The annual swings depend on availability of rainfall and other factors. Hs is set at zero for the calculations presented in figures 10 and 11.

The analysis assumes that the output can be varied by operators to go above and below $H(t)$ by an amount determined by the capacity of the generators at the dam sites and many other operational restrictions such as recreational uses of streams, fish migration, agricultural needs, or flood control. The amount of peaking power available, and the time for which it is available, depends on the capacity of installed turbines and the reservoir capacity. The output of run of the river hydro, for example, cannot be adjusted to meet fluctuating loads. These facilities are run continuously.

The amount that can be produced from a reservoir system also depends on the timing with which water is moved from reservoir to reservoir in large river systems. Many systems have bottlenecks. For example, a downstream dam may be able to meet high loads for short periods but its reservoir may not have enough capacity to operate at high levels for long periods. Upstream facilities may not be able to resupply the lower dam fast enough. If, however, it were known that a large peak would need to be met, large amounts could be "parked" in the downstream reservoir in anticipation of the peak.

It is assumed that the hydroelectric facilities can produce an amount Hu above the output $H(t)$ to meet peaks and that the output can fall below $H(t)$ by Hd. The reservoir size is specified as HR.

This provides a significant capability for reducing peaks. The hydroelectric plant is dispatched as follows:
1. The average hourly output $H(t)$ is subtracted from the load.
2. The dispatch is conducted week by week. For each week the average remaining load for the week (WKAV) and the weekly peak (WKPK) are calculated. The hydroelectric system is then dispatched with the rule that hydroelectric peaking is used to meet all loads which exceed a specified threshold T (constrained by the maximum increase Hu) and that hydroelectric output is reduced whenever the load is below the weekly average (constrained by the maximum reduction Hd). T is set initially at the weekly peak. The threshold is then lowered until one of the following two criteria is not met:
    (i)   The weekly peak is not reduced by lowering the threshold, or
    (ii) The energy in the reservoir at the end of the week is below the energy in the reservoir at the beginning of the week.

When the threshold for the week is determined, the load is reduced and increased appropriately for each hour of the week.

Storage
Storage is dispatched using an approach identical to that of the peaking hydroelectric facilities. The maximum charge and discharge rates (equivalent to Hu and Hd) and the total storage capacity are specified. The only difference is that the

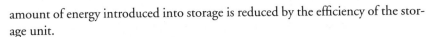

amount of energy introduced into storage is reduced by the efficiency of the storage unit.

Determining the number of dispatchable plants

Dispatchable generating equipment must be used to meet the load remaining after intermittents have been removed and storage has been used to minimize peaks. The optimum mix of plants can be determined by minimizing the total cost of meeting the annual load. This means minimizing the following:

$$\text{Cost} = \sum_j CA_j + T_j \cdot V_j$$

where CA_j is the annual fixed cost per kilowatt of plant type j (capital costs plus fixed operating costs), V_j is the variable cost per kilowatt of operating plant type j (levelized fuel costs plus variable operating costs), and T_j is the number of hours per year plant j operates. The sum of all capacity installed must equal the total peak demand on the system.

Using the method of Lagrange multipliers, the following rules can be shown to minimize production costs. A graphic solution is shown in figure 8. This figure plots:

1. The load duration curve, or the total number of hours the utility load exceeds a given capacity. The peak capacity P is exceeded for 0 hours while the load exceeds capacity 0 for 8,760 hours (the total number of hours in the year).

2. The total cost of operating plants of type 1 and 2 as a function of the number of hours each operates during the year. In this construction, plant type 2 is least expensive if it is operated more than T_x hours per year, and plant type 1 is least expensive if operated less than T_x hours per year. This means that costs are minimized if $P-X$ kw of plant type 1 is purchased and X kw of plant type 2 is purchased. Obviously real purchasing decisions involve more complex problems—one of which is that plants cannot be purchased in arbitrary capacity sizes. A typical new coal plant is 800 megawatts, for example.

Using the method suggested above, the following rules were used to compute optimum recommended equipment sizes:

1. Find the plant type with the lowest annual fixed cost (smallest CA_j—CA1 in the figure). This plant is last in the dispatch order; plants of this type are used only after all other available plants are in operation.

2. Compute the cost of operating plant j for T hours:

Table A: Characteristics of alternative generating technologies

Plant name	Fuel type	Capacity	Overnight cost	Percent completed by year before start-up[k]					Plant life
				1	2	3	4	5	
		MW_e	$ per kW_e						yrs
COMMERCIAL (C0)									
Simple cycle (EPRI)[a]	gas	80	367	0	0	0	0	0	30
Combined cycle (EPRI)[a]	gas	120	503	50	0	0	0	0	30
Sub-crit. steam[a]	coal	1,000	1290	80	60	40	20	0	30
BEST NEW (C1)									
Simple cycle[a]	gas	140	355	0	0	0	0	0	30
Combined cycle[a]	gas	210	473	50	0	0	0	0	30
Coal gasifier/CC[a]	coal	800	1,275	75	50	25	0	0	30
ADVANCED (C2)									
LM6000 Intercool CC[b]	gas	100	600	60	0	0	0	0	30
LM6000 Intercool SC[b]	gas	80	440	0	0	0	0	0	25
Coal gasifier/ISTIG[c]	coal	110	1,090	50	0	0	0	0	30
Biomass gasifier/ISTIG[c]	biomass	110	950	50	0	0	0	0	30
Hydro[d]	hydro	4,000	1,800	66	33	0	0	0	50
Wind (post-2000)[e]	[f]	100	800	0	0	0	0	0	30
Solar-thermal (post-2000)[g]	[h]	200	1,625	0	0	0	0	0	30
PV (flat plate post-2000)[i]	[j]	10	900–1,800	0	0	0	0	0	30

a.   [38].

b.   From General Electric Corporation and Bechtel Corporation sources.

c.   See chapter 17.

d.   [26].

e.   See chapter 3.

f.   Output from a site with a Rayleigh wind speed distribution and an average wind speed of 8 meters per second for up to 5 percent of total system load, an average wind speed of 7.5 meters per second for up to an additional 5 percent of total system load, and an average wind speed of 7.25 meters per second for all wind output in excess of 10 percent of total system load.

g.   See chapter 5.

h.   Output calculated using insolation data from Carissa Plains in northern California in 1989.

i.   See chapter 6.

j.   Simulated photovoltaic output based on energy available to a flat-plate from sunlight data taken in the PG&E service area (Carissa Plains) in 1989.

k   Percent of plant construction completed by year before start-up, measured in value terms by a plant's overnight construction cost. For example, 50 percent of the value of a commercial combined cycle plant is in place one year before plant start-up.

Table A: (cont.)

Plant name	Fixed O&M	Variable O&M	Planned outage	Forced outage	Cold starts Time	Cold starts Energy	Warm starts Time	Warm starts Energy
	$ per kW-year	*$ per kW-hour*	*% of hours*	*% of hours*	*hours*	*MMBTU*	*hours*	*MMBTU*
COMMERCIAL (C0)								
Simple cycle (EPRI)[a]	0.7	0.0077	4.2	3.5	0	0	0	0
Combined cycle (EPRI)[a]	4	0.0039	4	6	4	1,000	0	0
Sub-crit. steam[a]	23	0.0059	9	6	6	8,000	4	4,000
BEST NEW (C1)								
Simple cycle[a]	0.7	0.0074	4.2	3.5	0	0	0	0
Combined cycle[a]	4	0.0037	4	6	3	840	0	0
Coal gasifier/CC[a]	34	0.0036	4	10	20	15,360	4	1,600
ADVANCED (C2)								
LM6000 Intercool CC[b]	15	0.0020	0	4	2	100	0	20
LM6000 Intercool SC[b]	22	0.0020	2	2	0	0	0	0
Coal gasifier/ISTIG[c]	48	0.0010	4	4	20	4,800	4	1,000
Biomass gasifier/ISTIG[c]	34	0.0009	4	4	20	4,800	4	1,000
Hydro[d]	10	0.0000	1	1				
Wind (post-2000)[e]	0	0.0090	2	2				
Solar-thermal (post-2000)[f]	0	0.0010	0	1				
PV (flat plate post-2000)[g]	0	0.0015	0	0.5				

a. [38].
b. From General Electric Corporation and Bechtel Corporation sources.
c. See chapter 17.
d. [26].
e. See chapter 3.
f. See chapter 5.
g. See chapter 6.

Table A: (cont.)

Plant name	Heat rates BTU per kilowatt-hour				Minimum load
	100%	**75%**	**50%**	**25%**	
COMMERCIAL (C0)					
Simple cycle (EPRI)[a]	12,200	13,420	15,140	21,700	25
Combined cycle (EPRI)[a]	7,990	8,380	9,420	12,610	30
Sub-crit. steam[a]	10,056	10,233	10,880	12,313	50
BEST NEW (C1)					
Simple cycle[a]	11,500	12,480	14,270	20,450	25
Combined cycle[a]	7,514	7,885	8,660	11,860	25
Coal gasifier/CC[a]	8,930	9,540	11,090	15,570	25
ADVANCED (C2)					
LM6000 Intercool CC[b]	6,832	7,141	8,029	12,300	40
LM6000 Intercool SC[b]	8,334	8,667	14,605		40
Coal gasifier/ISTIG[c]	8,104	8,334	9,466		50
Biomass gasifier/ISTIG[c]	7,952	8,178	9,289	9,289	50
Hydro[d]					25
Wind (post-2000)[e]					
Solar-thermal (post-2000)[f]					
PV (flat plate post-2000)[g]					

a.   [38].
b.   From General Electric Corporation and Bechtel Corporation sources.
c.   See chapter 17.
d.   [26].
e.   See chapter 3.
f.   See chapter 5.
g.   See chapter 6.

$$CAj + T\,Vj$$

Find the intersection of this cost function with the cost functions of all other plants. The costs of plant type j and plant type k are equal when both operate Tjk hours where $Tjk = (CAj - CAk)/(Vk - Vj)$. If a plant must be operated more than Tjk hours, plant type k is preferred. If a plant must be operated less than Tjk hours, plant type j is preferred.

3. Find the plant type for which Tjk is the smallest. This plant type is next to last in the dispatch order.

4. Compute the Tkm intersection points for all remaining plants and find the plant with the smallest Tkm. This is next in the dispatch order.

5. Continue this process until all plant types have been examined or all Tkm are greater than 8,760.

The above procedure specifies the capacity of each plant type that will minimize the cost of meeting load. The "recommended number of plants" specified in the program for plant type j (RNj) is the capacity estimated using the previous procedure (call it CCj) adjusted by the plant's availability and the unit size of the plant (Sj):

$$RNj = CCj/[(1 - \text{forced outage rate}) \cdot (1 - \text{planned outage rate}) \cdot Sj]$$

Loss of load probability

The capacity installed must provide enough spare capacity to allow plants to be taken off-line for maintenance and to provide reserves to cover plants that fail. A precise estimate of necessary reserve requirements requires a maintenance schedule for each plant. As plants seldom fail completely, typically performance is degraded for limited periods. These, and other complex issues that must be considered in practical systems are beyond the scope of this simple analysis and a series of simplifying approximations will be made [43]. Since there is a small but finite chance that a large number of plants will fail simultaneously, reserve margin specifications must be stated in statistical terms. Utilities typically try to ensure that the probability of failing to meet load does not exceed one day in 10 years or an average of 2.4 hours per year. This "loss of load probability" can be computed using the following methods. For simplicity, it is assumed that all plants are specified in units of 10 megawatts. Let $S(n)$ be the probability that n units (or $n \cdot 10$ megawatts of capacity) is available in a given hour. The loss of load probability for an hour when the load is n units is therefore given by $1 - S(n)$. The probability depends on the load in the hour. The sum of the loss of load probabilities for each hour is the annual loss of load probability.

The probability of failing to meet load is orders of magnitude higher when the demand approaches peak demand levels than during periods of average loads. It is assumed that maintenance is scheduled during periods when load is likely to be low. As a result, the loss of load analysis presented here considers only forced outage rate and ignores planned outages for scheduled maintenance. This is a good approximation since the bulk of the annual loss of load probability occurs during the highest 100 load hours.

The probability distribution $S(n)$ is computed as follows:

1. Let N_j be the number of plants of type j. Then the probability that n plants will be available is given by the binomial distribution using the forced outage rate Fo:

$$P(n, N_j) = \frac{N_j!}{n! \, (N_j - n)} \cdot (Fo)^{N_j - n} \cdot (1 - Fo)^n$$

2. These individual binomial distributions can be combined to produce a net probability $S(n)$ that n units of capacity will be available. The peak capacity available will be $\Sigma c_j \cdot N_j$ where c_j the size of each plant of type j (measured in units of 10 megawatts). $S(0)$ is the probability that no plants are available and $S(P)$ is the probability that all are available.

3. Compute $A(n)$ = the probability that n or fewer units will be available:

$$A(n) = \sum_{i=0}^{i=n} S(i)$$

The loss of load probability for an hour in which the load is p units is therefore given by $A(p\text{-}1)$.

4. The loss of load probability for the year is computed by calculating $A(p\text{-}1)$ for the load at each hour of the year and summing.

The loss of load computed in this way is specified in the program in hours per year. It should be close to 2.4 hours.

Dispatching

Once the hourly load has been computed by subtracting intermittents, hydro, and adjusting for storage, the program uses the dispatchable plants specified to minimize costs. It does this by using the dispatchable equipment in order of variable operating costs—that is, an attempt is made to meet the load using plants with the lowest variable costs. If the load cannot be met, plants with the next highest variable costs are used. The sequence continues until all available plants are used.

Reserves

Before the dispatch is begun, however, an estimate must be made of the amount of reserve that must be provided in each hour. Reserve must be provided to react quickly to the following kinds of events:

▼ Failure of an operating plant.

▼ Unexpectedly rapid increase in the load.

▼ Rapid decline in wind or direct solar output (clouds).

Rapid response to such situations can be achieved with the following kinds of equipment:

▼ Hydroelectric plants that have surplus capacity for short periods (most reservoirs have such capacity).

▼ Peaking units with short start times (15 minutes or less).

▼ Thermal plants operated at partial capacity so that they can be brought quickly to higher levels of output or kept warm so that they can be started quickly. Most units can be operated at 40 percent capacity to provide needed quick-reserve capacity.

The last of these options, commonly called "spinning reserve" is used when hydro or peaking units are not available. But the strategy carries a significant penalty since plants operating at part loads are 20 percent or more less efficient than plants operating at full capacity.

The program uses the following algorithm to specify the spinning reserve capacity needed in at any given time:

$$R(t) = \alpha \cdot \text{load}(t) + \beta I(t) + B - Hu - Pe$$

where α is the load uncertainty (2 percent is used throughout the analysis), β is the drop in intermittent output covered by reserve, B is the size of the largest plant in the system, Hu is the amount of surge hydroelectric capacity available, and Pe is the total installed peaking capacity (defined to be plants with "cold start" capabilities less than 30 minutes). $R(t)$ is never less than zero.

The maximum drop in intermittent output can be computed from the information available in $I(t)$ and the program user can simply specify the β computed in this way. However, since most declines in renewable output are predictable, it is not obvious that all of this computed drop needs to be covered by reserve [44]. There would be ample time to plan for the expected decline. There is, however, no general rule about the accuracy of intermittent forecasts; the program user is asked to specify the fraction of the maximum β that should be used in the analysis.

Hourly dispatch

The final calculation requires determining the fuel used by each class of thermal unit. The system must be dispatched to meet real loads and to ensure adequate spinning reserve. The calculations use the number of plants of each type specified but assume that the capacity is reduced by outage rates. This means that if plant type j has a specified capacity of Sj, the dispatch calculations assume that plants of this type have a capacity equal to $Sj/[(1 -$ forced outage rate$) \cdot (1 -$ planned outage rate$)]$.

Meeting variable loads means that plants must be constantly turned on and off. Most plants require many hours to start; start-up must be begin well before output is required. During this start-up period steam pressure is raised in boilers and turbines are gradually brought up to operating temperatures. A considerable amount of fuel can be used in the process.

Typically a "cold start-up" and a "warm start-up" period are specified. Cold start means starting from a situation where the plant is taken off-line for an extended period for maintenance or other purposes. Units can be maintained in a "warm start" condition for considerable periods if necessary using comparatively little fuel. In cases where two combined cycle units are in the same building, the waste heat from one unit can be used to keep a second unit warm with no additional fuel input.

The program divides all units into four types:

1. Peakers: units with cold start times less than 30 minutes.

2. Fast starters: units with warm start times less than 30 minutes. When these units are called upon for spinning reserve, they are maintained in a "warm start" condition in which they consume fuel at one percent of the full power rate.

3. Slow starters: units with warm start times more than 30 minutes but less than 24 hours. When these units are called upon for spinning reserve, they are maintained at part load (the minimum fraction is specified as an input).

4. Very slow starters: units with cold start times more than 24 hours (typically nuclear plants). These units are kept running at all times. They can be operated at partial load (down to a specified minimum) but are turned off only for maintenance. They may produce power even when it is not needed.

The dispatch calculation for each hour proceeds as follows:

1. The reserve is calculated.

2. The first plant in the dispatch order is used to meet as much of the load and reserve capacity as possible. The fuel used is computed based on the part-load heat rates specified (the program uses a linear interpolation between specified points of part-load capacity).

3. The number of plants of the type that are on-line at the specified hour, running(t) are compared with the number of plants of this type on-line in previous hours. If the number has increased from the previous hour [running(t) > running ($t-1$)], the number of hours prior to t in which the number running has been less than running(t) is calculated. If this number is greater than the warm start time, a warm start is counted. And if this number is greater than the cold start time, a cold start is counted. The fuel required for these cold and warm starts is computed at the end of the annual run.

4. If the first type of plant in the dispatch order cannot meet the load and reserve, all plants of this type are operated at peak capacity and an attempt is made to meet the load and provide reserve margin using the next type of plant.

Steps 3 and 4 are repeated until the load is met.

This sequence allows an analysis of the fuel used by each plant type and the number of warm and cold starts. This in turn is used to compute the total amount, and cost, of fuel used during the year.

When loads fall but are expected to increase in the near future, it is often preferable to maintain these plants at partial loads or even in a warm, standby condition rather than shutting them down entirely. It is assumed that plants are not turned off if the unit would have to be restarted during a period of time shorter than its cold start time.

Plants are not allowed to operate below a minimum level even if this means that some energy is wasted.

Variables

α	=	uncertainty in load covered by reserve
$A(p)$	=	probability that the system will not be able to meet a load of $p \cdot 10$ megawatts.
ß	=	maximum expected drop in intermittent output (fraction of intermittent output in the hour specified)
B	=	capacity of largest dispatchable unit selected
C	=	total capital costs of equipment
CA_j	=	fixed operating cost of plant type j ($/kw/year)
c_j	=	capacity of plant type j in units of 10 megawatts
d	=	utility discount rate
EX	=	externality cost of fuel ($/MMBTU)
$EX(i)$	=	externality cost of emission type i ($/kg)

$EP(i)$ = emission of product i per MMBTU of fuel used (kg/MMBTU)

FIX = utility fixed charge rate

Fo = forced outage rate (fraction)

FP = levelized fuel price

$F(t)$ = fuel price in year t

FT = Federal tax rate

fb = fraction of utility capital derived from bonds

fe = fraction of utility capital derived from equity

fp = fraction of utility capital derived from preferred bonds

$H(t)$ = output of hydroelectric plant at hour t

Hav = average annual output of hydroelectric energy (megawatts)

Hd = the amount below $H(t)$ that the hydroelectric plant can produce (megawatts)

HR = reservoir size

Hs = seasonal swing above and below average (megawatts)

Hu = hydroelectric peaking capability = amount above $H(t)$ that the hydro electric plant can produce (megawatts)

$I(t)$ = total intermittent output in hour t

IN = insurance rate

L = expected plant life

Ld = depreciation life

Nj = number of plants of type j

Pe = total installed peaking capacity (defined in text)

$P(n,N)$ = probability that n out of N units will be available

PT = property tax

r_b = rate of return on bonds

r_e = rate of return expected on equity

r_p = rate of return expected on preferred bonds

$RE(t)$ = reserve margin provided in hour t

$S(j)$ = probability that j*10megawatts of capacity will be available

Sj = size of plant type i (kw)

ST = state tax rate

Ta = effective tax rate = FT(1-ST)+ST

Tij = intersection point of cost of service curves (hours)

Vj = variable operating costs of plant type j ($/kw/hour)

WORKS CITED

1. Lonnroth, M. 1989
   The coming reformation of the electric utility industry, in T.B.Johansson, B.Bodlund, and R.H.Williams, eds. *Electricity: efficient end-use and new generation technologies and their planning implications,* Lund University Press, Lund, Sweden.

2. Pacific Gas and Electric Company. 1991
   1993 test year, Application No 91-11-036, 2–9.

3. Financial Statistics of Selected Electric Utilities. 1988
   EIA:1990

4. Financial Statistics of Selected Electric Utilities. 1988
   EIA:1990, table 29, 42.

5. Moskowitz, D. and Nadel, S., and Geller, H. 1991
   Increasing the efficiency of electricity production and use: barriers and strategies, American Council for an Energy Efficient Economy, Energy Study Issues Paper #5, Washington, D.C.

6. See also M. J. Grubb, "Value of Variable sources on power systems" IEEE proceedings - C, vo. 138, No.2, March 1991 pp. 149-165; W.C. Turkenburg, K. Blok, E.A. Alsema, and A.J.N. VanWijk, The value of storage facilities in a renewable electricity system, in W. Palz and F.C. Treble, eds., *Proceedings of the 6th EC Photovoltaic Solar Energy Conference* (Reidel Publishing Co.,Dordrecht NL,1985) 337–342; W.C. Turkenburg, On the potential and implementation of wind energy, in *Wind Energy; Technology and Implementation, Proceedings of the Amsterdam EWEC '91,* Part II (Elsevier Publishers: Amsterdam NL,1992); D.R.Smith and R. Dracker, wind and solar energy potential in Northern California, Wind Energy Technology and Implementation (Amsterdam, WEEC 1991); F.J.L. Van Julle, P.T. Smulders, and J.B. Dragt, eds.(Elsevier Science Publishers, B.V.)1991, 347; T. Flaim, T. Conidine, T. Winholderm, and M. Edesses, Economic assessments of intermittent grid-connected solar technologies: a review of methods (Solar Energy Research Institute: Golden, CO, 1981); S. Hock and T. Flaim, Wind energy systems for electric utilities: a synthesis of value studies, 3rd annual meeting of the American Solar Energy Society (Minneapolis, MN, 1983); P.L. Surman and J.F.Walker, System integration of new and renewable energy sources: A UK view, *Proceedings of the European New Energies Congress,* vol. 1 (October 24-28, 1988, Saarbrucken, Germany); G.Botta, A. Invernizzi, S. Panichelli, and L. Salvadori, EC wind power penetration study, *Proceedings of the European New Energies Congress,* Vol. 1 (October 24-28, 1988, Saarbrucken, Germany); G. T. Chinery and J.M. Wood, Estimating the value of photovoltaic generation to electric power systems (IEEE 0160-8371/85/0000-13471985); N. Halberg, Power production simulation for wind power assisted systems (Dutch Electricity Generating Board); E. O'Dwyer and A. Harpur, Simulation of wind power generation (Electricity Supply Board, Dublin Ireland, n.d.).

7. Kahn, E. 1979
   The compatibility of wind and solar technology with conventional energy systems, *Ann. Rev. Energy* 4:313–52.

8. Willis H.L. and Northcote-Green, J.E. 1983
   Spatial electric load forecasting: a tutorial review, *Proceedings of the IEEE,* Vol. 71, No. 2, February.

9. Whyle, M.D., Farris, W.E., Hugo, R.V., Maliszewski, R.M., and Mallard, S.A. 1977
   Load management: its impact on system planning and operation, Phase II, Edison Electric Institute, cited in Kahn (1979), op. cit.

10. Bigger, J.E., Kern, E.C., and Russel, M.C. 1991
    Cost-effective photovoltaic applications for electric utilities, presented at 22nd IEEE Photovoltaic Specialists Conference, Las Vegas, Nevada, October 7–11.

11. Electric Power Research Institute. 1986
Dynamic operating benefits of energy storage, EPRI AP-4875.

12. Hirst, E. 1990
Benefits and costs of flexibility—short-lead-time power plants, long range planning, Vol. 23, No. 5, 106–115.

13. Akbari, H., Eto, J., Turiel, I., Heinemeier, K., Lebot, B., Nordman, B., and Rainer, L. 1989
Integrated estimation of commercial sector end-use load shapes and energy use intensities, LBL 27512.

14. Electric Power Research Institute. 1979
Integrated analysis of loadshapes and energy storage, EPRI EA-970.

15. Kahn, E.P., Marnay, C., and Berman, D. 1991
Evaluating dispatchability features in competitive bidding, IEEE/PES International Joint Power Generation Conference and Exposition, San Diego, California, October 6–9.

16. Grubb, M.J. 1991
Value of variable sources on power systems, IEEE Proceedings C, Vol. 138, no.2, March 149–165.

17. Grubb, M.J. 1991
The integration of renewable electricity sources, *Energy Policy*, June.

18. Farmer, E.E. et al. 1980
Economic and operational implications of a complex of wind generators on a power system, IEEE Proceedings A., Vol. 127, June.

19. Verbruggen, A. 1985
Decision model of electricity generation planning with an application for Belgium, Vol. 7, No. 4, Electrical Power and Energy Systems.

20. Ruger, G.M. and Manzoini, G. 1990
Utility planning and operational implications of photovoltaic power systems, presented at IEA/ENEL Photovoltaic Systems for Utility Applications Conference, Taormina, Sicily, December.

21. Pacific Gas and Electric Company. 1991
Benefits of distributed generation in Pacific Gas and Electric's T&D system: a case study of photovoltaics serving Kerman substation, GM663024-8, August.

22. Shugar, S.D. 1990
Photovoltaics in the distribution system, presented at the 21st IEEE Photovoltaic Specialists Conference, Kissimmee, Florida, May 22.

23. Pacific Gas and Electric Company.
1993 test year, marginal costs, Pacific Gas and Electric -1, 4–19.

24. Orens, R. 1989
Area-specific marginal costing for electric utilities: a case study of transmission and distribution costs, dissertation submitted to the Department of Civil Engineering, Stanford University, for the Doctor of Philosophy, September.

25. Electric Power Research Institute. 1987
DYNASTORE—A computer model for quantifying dynamic energy storage benefits, EPRI AP-550, December.

26. Northwest Power Planning Council. 1991
1991 Northwest Conservation and Electric Power Plan, Vol. II, Part II, 625.

27. Electric Power Research Institute. 1989
Emerging strategies for energy storage, *EPRI Journal*, Vol.14, No. 5, 14–15.

28. Cultu, N. 1989
    Energy storage systems in operation, in B. Kilkis and S. Kakc, eds., *Energy storage*, 551–574, Kluwer, Dordrecht.

29. Electric Power Research Institute,n.d.
    Commercial Cool Storage, EPRI Energy Management and Utilization Division.

30. Electric Power Research Institute. 1987
    Benefits of battery storage as spinning reserve: quantitative analysis, EPRI AP-5327.

31. Jense, J. and Sorensen, B. 1984
    Fundamentals of Energy Storage, Wiley, New York.

32. Nakhamkin, M., Patel, M. Swensen, E. Cohn, A., and Louks, B. 1991
    Application of air saturation to integrated coal gasification/CAES power plants, presented at the International Power Generation Conference, San Diego, California, October 6–10.

33. Smith, D.R. and Dracker, R. 1991
    Wind and solar energy potential in Northern California, in F.J.L. Van Julle, P.T. Smulders, and J.B. Dragt, eds., *Wind Energy Technology and Implementation*, 347, Amsterdam, WEEC '91, Elsevier Science Publishers, B.V.

34. Pace University Center for Environmental Legal Studies. 1990
    Environmental Costs of Electricity, The Oceana Group, New York.

35. Averch, H. and Johnson, L.
    "Behavior of the firm under Regulatory Constraint" *Am. Economic Review* 52:1052–69.

36. Electric Power Research Institute. 1990
    Strategic assessment of storage plants, economic studies, January (EPRI GS 6646).

37. Kahn, E. 1979
    "The reliability of distributed wind generators", Electric Power Systems Research, pp 1-14, Elsevier Sequoia, Lausanne.

38. Electric Power Research Institute. 1989
    Technology Assessment Guide, EPRI P-6587-L

Appendix to Chapter 1
A RENEWABLES-INTENSIVE GLOBAL ENERGY SCENARIO

THOMAS B. JOHANSSON
HENRY KELLY
AMULYA K.N. REDDY
ROBERT H. WILLIAMS

The tables at the end of this appendix describe the renewables-intensive global energy scenario (RIGES) presented in chapter 1: *Renewable Fuels and Electricity for a Growing World Economy.* The RIGES is designed to help understand the outlook for renewable energy in a global context. Results are shown for 1985 (base year), 2025, and 2050.

The energy supply mix shown was assembled to satisfy the energy demands for one of the global energy scenarios constructed by the Response Strategies Working Group (RSWG) of the Intergovernmental Panel on Climate Change [1],[1] using renewable energy technologies that proved to be the most attractive, based on the technical assessments in this book[2]. (No independent analysis was carried out in the present study on future energy demand.)

The RSWG constructed energy demand/supply scenarios for nine world regions. For the present study these were expanded into the 11 world regions shown in table A. The two additional regions were obtained by separating the RSWG region OECD Europe/Canada into OECD Europe and Canada, and by separating the RSWG region OECD Pacific into Japan and Australia/New Zealand. This disaggregation helps highlight the extent of interregional energy trade in a renewables-intensive energy future.

---

1. Under the auspices of the World Meteorological Organization and the United Nations Environment Program, the Intergovernmental Panel on Climate Change (IPCC) established three working groups to assess the scientific information on climate change, to assess the environmental and socioeconomic impacts of climate change, and to formulate response strategies.

2. Energy production costs presented in this book were calculated for 6 and 12 percent real (inflation-corrected) discount rates, with taxes (corporate income and property) neglected. Taxes were ignored in part because the intent is to highlight costs to society; taxes are transfer payments, not true costs. In addition, the international perspective of the book dictates that taxes not be taken into account, because tax codes differ markedly from one region to another.

THE RSWG ENERGY DEMAND SCENARIOS

The population estimates used by the RSWG and adopted for the present analysis were taken from the World Bank [2] and were not varied among the RSWG scenarios. World population increases from 4.87 billion in 1985 to 8.19 billion in 2025 and to 9.53 billion in 2050, with the developing country share increasing from 75 percent in 1985 to 82 percent in 2025 and to 85 percent in 2050.

The RSWG constructed both high and low economic growth variants of its scenarios. For the present study only the high economic growth variants are considered, as the spirit of the analysis is to understand the prospects for renewables in a world where future living standards are much higher than at present. Gross Domestic Product (GDP) grows at a global average rate of 3.5 percent per year, 1985–2025, and 3.0 percent per year, 2025–2050, in the high economic growth variants of the RSWG scenarios. However, there are substantial differences among regions—ranging from 2.7 percent per year, 1985–2025, and 2.0 percent per year, 2025–2050, for the United States, Canada, and OECD Europe, to 5.2 percent per year, 1985–2025, and 4.5 percent per year, 2025–2050, for Centrally Planned Asia.

The sets of scenarios developed by the RSWG are indicated in table B. The resulting demand levels for electricity and for the direct use[3] of solid, liquid, and gaseous fuels are shown in tables C and D. Only commercial fuels are considered in these scenarios.

The RSWG scenarios involve high and low energy demand growth scenarios for each economic growth variant, reflecting both the variance in historical trends and the range of views as to what is feasible in the future. The high energy demand growth trajectory is embodied in the 2030 High Emissions Scenario (called the "IPCC High Emissions Scenario" throughout this discussion). This scenario is characterized by its authors as involving "moderate energy efficiency" improvements. The global average energy intensity (the ratio of primary energy consumption to GDP) declines at an average rate of 1.15 to 1.45 percent per year in the period 1985–2050, for the high economic growth variant of this scenario. All the other IPCC scenarios are characterized by its authors as emphasizing "high energy efficiency." The global average energy intensity declines at an average rate of 2.0 to 2.1 percent per year in this period for the high economic growth variants of these scenarios.

As the pursuit of cost-effective opportunities for improving energy efficiency is necessary to achieve sustained economic growth and the analysis in this book indicates that prior emphasis on energy efficiency would facilitate a shift to renewables, one of the high efficiency variants of the RSWG scenarios was adopted as the energy demand projection for the RIGES. Specifically, the electricity consumption and direct fuel use projections of the Accelerated Policies (AP) Scenario are adopted here as the demand projection for the RIGES.

---

3. The "direct use" of fuels refers to the use of fuels by final consumers and does not include the use of fuels for power generation or the processing losses associated with producing synthetic fuels.

THE ENERGY SUPPLY SCENARIO

Assuming the demand levels from the AP Scenario of the RSWG, an energy supply scenario was constructed. In what follows, the details of this energy supply scenario are described—with respect to roles for conventional energy, biomass production, renewables-intensive strategies for electricity generation, fluid fuels production from biomass, and hydrogen production from intermittent renewable sources.

Future energy prices

While detailed energy price projections were not made for the RSWG, adoption of the set of technologies chosen for the scenario would give rise to future energy prices that are much lower than those of most other long-term energy forecasts. For example, in its 1991 National Energy Strategy report, the United States Department of Energy projects that between 1989 and 2030 the world oil price will more than double, natural gas prices for electric utilities in the U.S. will increase 3½-fold, and retail electricity prices will increase some 10 percent [3]. In contrast, it is expected that if the RIGES were realized, the world oil price in 2030 would not be much different from at present, while the price of gas paid by utilities would perhaps double, and electricity prices would decline somewhat. This finding is a consequence of the prospect that renewables will be able to compete with conventional energy in a wide range of circumstances—a competition that will bring downward pressure on energy prices.

The prospect of a stable world oil price is a consequence of emphasis given in the RIGES to biofuels in transportation and the expectation that in the post-2000 period biofuels may well be able to compete with oil in transportation at near the present world oil price. Specifically, the studies in this book indicate that there are at least two alternative technological paths[4] that could lead to a world oil price not much higher than at present:

▼ With estimated costs for producing ethanol from biomass via improved enzymatic hydrolysis technology (a 40 percent cost reduction relative to what could be achieved with present technology), this ethanol could compete in internal combustion engine vehicle applications with gasoline derived from oil at $23 per

---

4. The costs presented here for the three alternative options are based on biomass feedstock prices of $3 per gigajoule (HHV basis) and an evaluation of fuel production costs assuming a 12 percent real discount rate with taxes neglected.

barrel.[5] The key uncertainty is whether the targeted production cost reduction can be achieved.

▼ At the estimated costs for producing methanol and hydrogen from biomass via indirectly heated biomass gasifiers, fuel cell cars using these fuels could compete with gasoline internal combustion engine cars at costs well below the present world oil price.[6] The key uncertainty is whether the estimated 50 percent longer vehicle life will be achieved in practice.[7]

While there are technological uncertainties associated with each of these alternative transportation options, the probability of realizing at least one of these options or a comparably attractive alternative[8] over the next one to two decades is high.

The high likelihood of achieving at least one of these goals justifies on purely economic grounds a substantial effort to commercialize the technologies involved and build up quickly the productive fuel capacity that would be required to stabilize the world oil price at levels below those usually projected. The RIGES is constructed in such a way that this might be achieved.[9] For example, global oil production is 76 exajoules in the RIGES in 2025, compared to 117 exajoules in the RSWG's Accelerated Policies (AP) Scenario (the supply scenario that is the closest in spirit to the RIGES), while production in the Middle East is 27 exajoules in the RIGES, compared to 52 exajoules for the AP supply scenario, de-

---

5. The ethanol production cost with advanced technology is estimated to be $0.21 per liter (see chapter 21: *Ethanol and Methanol from Cellulosic Biomass*) corresponding to a cost at the retail level of $0.26 per liter. Since ethanol has a heating value just 0.67 times that for gasoline and it can be used in internal combustion engines at about 20 percent greater efficiency, this ethanol cost is equivalent to a gasoline price of $0.26/(1.2 × 0.67) = $0.32 per liter.

To convert a gasoline price into an equivalent crude oil price, it is assumed that crude oil is converted at 90 percent efficiency into gasoline, that ordinary refining costs $0.0660 per liter, that producing reformulated gasoline adds an incremental refining cost of $0.0396 per liter, that distribution and storage cost $0.0317 per liter, and that the gasoline station adds $0.0185 per liter (private communication from M. DeLuchi, March 1992). Thus the gasoline price P_{gas} ($ per liter) is related to the crude oil price P_{crude} ($ per barrel) by:

$$P_{gas} = (6.99 \times 10^{-3}) \times P_{crude} + 0.156 \text{ dollars per liter.}$$

Thus a gasoline price of $0.32 per liter corresponds to a crude oil price of $23 per barrel.

6. The break-even gasoline price for fuel cell cars operated on methanol (hydrogen) is $0.18 ($0.22) per liter (see endnote for figure 11, chapter 1), corresponding to a crude oil price of $3 ($9) per barrel.

7. If the methanol fuel cell car lasts only 10 percent longer than the internal-combustion-engine car, the break-even gasoline price would be $0.51 per liter ($1.92 per gallon), corresponding to a crude oil price of $50 per barrel.

8. Advanced hydrogen storage technologies or alternative hydrogen carrier technologies (e.g., the "iron-powered fuel cell car discussed briefly in the main text) or alternatives to batteries (which were estimated in the present analysis to contribute more to the cost that the fuel cells themselves--see endnote for figure 11, chapter 1) for providing peaking power (e.g., advanced low-cost capacitors) could bring down both the initial purchase price of fuel-cell vehicles and the cost of the purchased fuel. Such possibilities might enable fuel-cell vehicles to compete even if they were to last no longer than gasoline-powered internal-combustion-engine vehicles.

9. For specificity, the RIGES emphasizes the methanol fuel-cell car as an alternative to the gasoline-fueled internal-combustion-engine car, but a stabilization of the world oil price could also be achieved if ethanol used in internal-combustion-engine vehicles were emphasized.

spite the same level of demand for direct liquid fuel use in both scenarios.

The lower world oil price that would result from building up this capacity for liquid biofuels production would help stabilize other energy prices as well. For example, it is assumed in the RIGES analysis that the price paid for natural gas by typical gas utilities increases from present levels (in the range $2.2 to $3.1 per gigajoule in 1989 in Western European and U.S. markets) to about $4 per gigajoule in 2000 and $5 per gigajoule in 2030. This rise is much less than is typically projected for this period but is consistent with both the expected oil price stabilization and the major expansions of natural gas use projected for the RIGES (see table F).

The relative stability of the gas price and deemphasis of coal in the RIGES should lead to stable coal prices in the RIGES. It is assumed that utility coal prices remain constant at $2 per gigajoule in the period 2025 to 2050.

The prospects are auspicious that 1) there will be improvements in fossil fuel-based power-generating technologies, and 2) that renewables will be able to compete with typical existing fossil fuel technologies and come very close to competing with advanced fossil fuel technologies (see chapter 1). Therefore, emphasis on such technologies in the RIGES, in the context of the general expectations about fuel prices in the RIGES as discussed above, is likely to lead to stable or slightly lower prices for electricity in the future.

Conventional energy supplies in the RIGES

The point of departure for the RIGES supply analysis is the set of demands for solid, liquid, and gaseous fuels, and electricity for the AP Scenario of the RSWG. The first task in constructing the supply scenario is the development of a set of algorithms for conventional energy supplies—coal, oil, natural gas, and nuclear power.

Because the environmental problems posed by the use of coal are more intense than for other fossil fuels, attention was focused on identifying alternatives to coal. The direct use of coal and coal used for power generation are treated differently in the analysis. Biomass is the only solid fuel alternative to the direct use of coal, but there are various alternatives to coal that can be considered for power generation.

In the AP scenario, the direct use of solid fuels declines from 36 exajoules per year in 1985 to 25 exajoules per year in 2050 in regions outside Centrally Planned Asia (CPA), while in CPA the direct use of solid fuels increases from 16 exajoules in 1985 to 31 exajoules in 2050. At the same time, the direct use of liquid and gaseous fuels increases from 165 exajoules in 1985 to 221 exajoules in 2050 in regions outside CPA, while their use varies from 4 exajoules in 1985 to 12 exajoules in 2050 in CPA.

Because of the declining importance of solid fuels used directly in regions outside CPA in the AP scenario and because potential biomass supplies are limited, the use of biomass is restricted to providing either electricity or synthetic liquid and gaseous fuels outside CPA. In these regions, only coal is used in the

RIGES to meet the demand for solid fuel used directly. In CPA, where future demand for solid fuels used directly is much greater than for liquid and gaseous fuels and is projected to grow substantially in the AP Scenario, biomass provides one-third of the solid fuel used directly by the year 2050.

Where the cost of renewable power generation is expected to be about the same as for coal power, the renewables-intensive mix was preferred in constructing the RIGES. Biomass power generation in particular can be competitive with coal power under a wide range of circumstances (see chapter 17: *Advanced Gasification-based Power Generation*). In the RIGES the use of coal for power generation is phased out by the middle of the next century in regions where biomass-based power generation is a major option. The share of coal in power generation declines at the global level, from 40 percent in 1985 to 11 percent in 2025 and to 6 percent in 2050.

For Centrally Planned Asia and South and East Asia, where the prospects for biomass power generation are more limited than in most other regions, coal-based power generation actually increases six-fold and three-fold, respectively, between now and 2050 (see table Q). It is assumed that the expanding market for coal power in these regions provides a favorable environment for innovation, such that coal-integrated gasifier/gas turbine (CIG/GT) power cycles with efficiencies of 42 percent become the norm by 2025 and coal-integrated gasifier/fuel cell (CIG/FC) technologies with efficiencies of 57 percent become the norm by 2050 (see table N).

The role of oil is constrained in the RIGES by the limited remaining resources of conventional oil outside the Middle East. Unconventional oil resources were not considered in light of the expectation that in the RIGES there would not be a substantial rise in the world oil price. In the RIGES, oil production in all regions outside the Middle East was assumed to decline exponentially at a rate such that in the year 2050 one third of the estimated ultimately recoverable conventional resources remain available for future exploitation (see table E). With this assumption, oil production outside the Middle East declines from 103 exajoules in 1985 to 49 exajoules in 2025 and to 31 exajoules in 2050.

Despite an 11 to 12 percent increase in liquid fuel requirements in the period 2025–2050 relative to 1985 (see table D) and sharp declines in oil production outside the Middle East (see table E), the scenario does not involve corresponding sharp increases in oil production in the Middle East. Instead, oil exports by the Middle East decline slightly from the 1985 level of 18 exajoules to 16 exajoules by 2025 and then rise to 21 exajoules by the year 2050 (see table Gb). By the year 2050, the Middle East contains about 58 percent of the world's remaining oil resources, compared to about 45 percent today (see table E). The relatively constant level of oil exports from the Middle East in this period is a consequence of expected competition from methanol derived from biomass fuels in transportation markets, as discussed below.

As in the case of oil, only conventional natural gas resources were considered in constructing the RIGES. Estimates of remaining conventional gas resources are comparable to those for oil; yet, at the global level natural gas is produced at just half the rate at which oil is produced (compare tables E and F). Thus the prospects are auspicious for major expansions in gas production in many parts of the world. As the cleanest fossil fuel and often the most economically attractive, natural gas is the fossil fuel emphasized in the RIGES where competitive renewable energy supplies are limited and where gas supplies are adequate. Natural gas production algorithms vary by region, but production schedules were constrained such that by the year 2050 at least one third of the estimated ultimately recoverable conventional resources remain available for future exploitation in each region (see table F). For the world as a whole, about half of the current gas resources will remain in the year 2050 (see table F).

Under the assumptions about production outside the Middle East, natural gas exports from the Middle East increase from near zero in 1985 to 10 exajoules in 2025 and to 21 exajoules in 2050. Exports of oil plus natural gas from the Middle East increase sharply, from 18 exajoules in 1985 to 27 exajoules in 2025 and to 42 exajoules in 2050 (see tables Ga and Gb). Because of projected rapid growth in energy demand in the Middle East, growth in oil plus gas production there increases from 27 exajoules in 1985 to 46 exajoules in 2025 and to 67 exajoules in 2050 (see table H).

In most regions nuclear power is limited in the period 2025–2050 to what can be produced with capacity already in place and under construction. Globally, the share of electricity provided by nuclear power declines from 15 percent in 1985 to 9 percent in 2025 and 6 percent in 2050 (see table R). It is assumed that the economics of coal and nuclear power are about the same, so that the logic that leads to deemphasizing coal applies to nuclear power as well.

Biomass supplies for energy

In the RIGES most biomass is used for electricity generation and for fluid fuels production (especially methanol and hydrogen produced thermochemically). The biomass resources considered are biomass recoverable from existing forests, agricultural and industrial biomass residues, and biomass grown on dedicated plantations. Potential biomass supplies were estimated on a regional basis.

In considering wood resources, it is useful to distinguish between three supply sources: virgin forests, forests that are presently managed for roundwood production, and plantations established on already deforested or previously unforested lands. For the RIGES, only the latter two categories are considered for biomass energy purposes.

Biomass supplies (other than residues) from forests presently managed for roundwood production in the period 2025 to 2050 are assumed to be limited to three-fourths of the level of roundwood harvested in 1985 for fuelwood and charcoal applications, about 10 exajoules per year, 5 to 7 percent of total biomass supplies (see table M). About five-sixths of this comes from forests in developing countries (see table I). It is assumed that in the period 2025 to 2050 wood from the natural forests is no longer used for traditional fuelwood and charcoal applications but is instead used to produce electricity or modern fuels.

Such a limited role for biomass from forests presently managed for fuelwood production is assumed for the scenario because of concerns that harvesting forests poses environmental problems, including loss of biological diversity. The assumed level of harvesting is lower than at present to help ensure that the harvesting is done sustainably.

The biomass residues considered are residues of the forest product industries (logging residues plus mill residues), sugarcane residues, dung, residues of cereals production, and urban refuse.

Because of environmental concerns about overuse of the natural forests, it is assumed that roundwood production from forests increases only slowly—in proportion to total population and thus much more slowly than the economy. It is further assumed that one-half of logging residues and three-fourths of mill residues are recoverable. Under these assumptions the total residue supply from forests in the period 2025 to 2050 amounts to about 14 exajoules per year (see table I)—some 7 to 9 percent of total biomass supplies (see table M).

It is also assumed that sugarcane production increases only as fast as population, and that all of the bagasse (the residue left after crushing the cane) plus two-thirds of the cane tops and leaves are recoverable for energy. When sugarcane processing factories are made energy efficient, only a tiny fraction of this residue energy is needed to run the factories, so that there are major opportunities for producing energy for other purposes (see chapter 17). Total recoverable cane residues, nearly all of which are in developing countries, amount to 10 exajoules in 2025 and 12 exajoules in 2050 (see table K).

Estimates of future dung production rates are based on RSWG projections for the production of meat and dairy products by region, assuming fixed dung production coefficients for different animals. It is assumed that one-fourth of the produced dung is recoverable for energy conversion. Dung supplies for energy amount to 17 exajoules in 2025 and 25 exajoules in 2050 (see table K).

Future cereals production levels by region are assumed to be those projected by the RSWG. It is assumed that residues are produced at a rate of 1.3 tonnes per tonne of harvested grain (the global average for 1983) and that one-fourth of these residues are recoverable for energy purposes. Recoverable cereals residues amount to 11 exajoules in 2025 and 13 exajoules in 2050.

The use of urban refuse for energy is considered only for industrialized countries. It is assumed that the production rate per capita is constant and that three-fourths of the produced urban refuse is recoverable for energy. This biomass supply amounts to about 3 exajoules per year in the period 2025 to 2050 (see table K).

It is assumed that in industrialized countries, biomass plantations are located primarily on excess agricultural lands. For developing countries, it is assumed that such plantations are located primarily on deforested or otherwise degraded lands that are not needed for food production. Considerable amounts of land on low-lying mountains that are already targeted for reforestation in China could be mitted to large-scale biomass plantations for energy purposes. The largest amounts of lands in developing countries suitable for biomass plantations and not needed for food production are located in sub-Saharan Africa and in Latin America.

The total amount of land required for biomass plantations in the RIGES amounts to about 370 million hectares by 2025 and to about 430 million hectares by 2050; the amount of energy produced would be 80 exajoules in 2025 and 128 exajoules in 2050 (see table L)—55 percent and 62 percent of total biomass supplies, respectively (see table M).

Total primary biomass supplies amount to 145 exajoules in 2025 and 206 exajoules in 2050 (see table M). For comparison, total primary global energy consumption was 323 exajoules in 1985.

A renewables-intensive strategy for electricity

In the power sector, intermittent renewable electric technologies (wind, photovoltaic, and solar thermal power), hydropower, natural gas, and biomass power generation are emphasized in constructing the RIGES.

The future use of hydropower in the scenario is based on estimates of the economic potential by region from a study carried out for the World Energy Conference [4], with these potentials reduced to take into account growing environmental concerns about many hydropower projects. Globally there is less than a doubling of hydroelectric power generation by 2025 and a 2½-fold increase by 2050; the percentage of total electricity provided by hydropower at the global level declines from the 1985 level of 20 percent to 17 percent in 2025 and to 15 percent in 2050 (see table O), at which time hydropower production equals about one-half of the estimated economic potential [10] and less than one-third of the technically exploitable potential (see chapter 2: *Hydropower and Its Constraints*).

It is assumed that oil is phased out of power generation, but natural gas use for power generation is expanded using advanced gas turbine-based power cycles. The share of total electricity provided by natural gas increases at the global level from 12 percent in 1985 to 23 percent by 2025 and to 25 percent by 2050 (see table P)—essentially the same as the 23 percent share of power generation accounted for by oil plus natural gas in 1985.

This emphasis on natural gas is a result of several considerations. First, new gas turbine power-generating technologies characterized by high thermodynamic efficiency (see table N) and low unit capital cost are making it feasible for natural gas-based power generation to compete in a wide range of circumstances with conventional fossil fuel-based steam-electric generating technologies. Second, many of these new power-generating technologies are characterized by very low levels of air pollution, without the use of costly air pollution control technologies. Third, the modular nature of these technologies makes it feasible to add new generating capacity quickly and thereby avoid overinvesting in new generating capacity in the now commonplace situations where future electrical demand is highly uncertain. And finally, the capacity to adjust output quickly and low unit capital cost make this natural gas-based generating option an attractive complement to intermittent renewable technologies on the utility grid.

In generating electricity from biomass, emphasis is given to energy-efficient conversion processes based largely on similar technology being developed for coal. It is assumed that advanced biomass-integrated gasifier/gas turbine (BIG/GT) power cycles with efficiencies of 43 percent become the norm by 2025 and biomass-integrated gasifier/fuel cell (BIG/FC) technologies with efficiencies of 57 percent become the norm by 2050 (see table N).

Biomass provides 17 to 18 percent of global power generation in the period 2025 to 2050 (see table S). Particularly important for developing countries is the cogeneration of electricity and steam from sugarcane residues (bagasse and sugarcane tops and leaves) at factories that process cane to produce either sugar or ethanol. This "cane power" accounts for 40 to 49 percent of all biomass-based power generation in developing countries (see table S). The rest of the electricity from biomass would be provided in stand-alone power plants using as fuel various other biomass residues and biomass grown on plantations.

The analysis of this book indicates that a high level of penetration of intermittent renewables can be accommodated on most utility systems, with little electrical storage requirements, if a substantial fraction of the rest of the generating capacity on the system has the flexibility to vary output in response to changes in the output of the intermittent renewable system (see chapter 23: *Utility Strategies for Using Renewables*). This flexibility is provided by natural gas-fired gas turbines and hydropower. Intermittent renewables provide 22 percent of electricity requirements in 2025 and 30 percent in 2050, although there are substantial variations among regions (see table T). The lowest level of penetration is 10 percent in 2025 and 2050 for Africa and Latin America, largely because these regions are expected to have abundant low-cost electricity from biomass sources. Likewise, relatively modest contributions of 14 percent in 2025 and 18 percent in 2050 are projected for OECD Europe, owing to expected severe land-use constraints on wind power and to the low insolation levels in much of northern Europe.

Biomass-derived fluid fuels

The analysis of this book shows that the least costly liquid fuels that can be produced from renewable sources are ethanol and methanol derived from biomass, and that the prospects are auspicious that all these options could become competitive with oil in transportation options, as noted above (see chapters 17, 20: *The Brazilian Fuel-alcohol Program*, 21, and 22: *Solar Hydrogen*).

It is assumed that liquid fuel requirements not provided by oil are met in part by alcohol fuels produced from biomass—some ethanol from cane, for sugarcane-producing countries, plus methanol produced thermochemically from estimated available domestic biomass sources.

While both ethanol derived from cellulosic biomass feedstocks via enzymatic hydrolysis and methanol derived via thermochemical gasification are promising liquid fuels, methanol was emphasized in the scenario, because it is especially well-suited for use with fuel-cell vehicles, which are assumed to become the propulsion technology of choice for road transportation in the period 2025 to 2050. For much the same reason, hydrogen produced using the same thermochemical gasification technology is emphasized as the biomass-derived gaseous fuel of choice in the scenario. It is assumed that methanol and hydrogen are produced at efficiencies of 62 to 63 percent and 70 to 72 percent, respectively (see table N). These efficiency levels are achievable with indirectly heated biomass gasifier technology that could be commercialized over the next decade.

Biomass-derived methanol is produced in the scenario at a level of 45 exajoules in 2025 and 61 exajoules by 2050 (see table U), representing 37 percent and 50 percent of global liquid fuel demand, respectively. This very high level of methanol use is a result of the finding that methanol produced from biomass and used in fuel cells for transportation would be competitive at about the present world oil price with gasoline used in internal-combustion-engine vehicles (see chapter 22).

This finding provides the basis for the assumption made in constructing the scenario that in interregional liquid fuel trade there will be equal shares of imports coming from oil and methanol derived from biomass feedstocks. This implies that methanol traded interregionally amounts to 17 exajoules in 2025 and 21 exajoules in 2050 (see table Gb). Most of the methanol involved in interregional commerce would be produced in sub-Saharan Africa and Latin America, where there are large amounts of land not needed for food production that are suitable for growing biomass (see table L).

Biomass-derived hydrogen is produced at a level of 16 exajoules in 2025 and 25 exajoules by 2050 (see table U), representing 12 percent and 20 percent of total gaseous fuel demand, respectively. Even though the economics of biomass-derived hydrogen would be about as favorable as for biomass-derived methanol in fuel-cell car applications (see chapter 22), it is expected that the use of hydrogen will grow more slowly than the use of methanol, because of the present lack of infrastructure for the use of gaseous fuels in transportation. Moreover, the overall more favorable outlook for natural gas compared to oil in regions outside the Middle East will make it relatively more difficult to introduce hydrogen as an alternative gaseous fuel than it will be to introduce an alternative liquid fuel.

It is assumed that all recoverable dung resources—one fourth of the dung generated (see table K)—are used to produce biogas via anaerobic digestion, and that this biogas can be produced at an efficiency of 57 percent (90 percent of the stoichiometric rate)—a value achievable with modern biogas conversion technologies (see chapter 19: *Anaerobic Digestion for Energy Production and Environmental Protection*). The resulting level of biogas production is 10 exajoules in 2025 and 14 exajoules in 2050—of which the developing country share is 83 percent in 2025 and 88 percent in 2050.

Electrolytic hydrogen

Hydrogen is an extraordinarily clean fuel that is especially attractive when used in fuel cell vehicles. While land and water supply constraints limit the production of hydrogen derived from biomass in some regions today and will ultimately limit the global level of hydrogen production from biomass, no such constraints will limit the production of hydrogen from wind and photovoltaic sources (see chapter 22).

Even though the delivered price of hydrogen from wind and photovoltaic sources would be about twice that for hydrogen from biomass,[10] fuel-cell cars operated on electrolytic hydrogen would become competitive with gasoline internal-combustion-engine cars at a gasoline price of $0.39 per liter ($1.5 per gallon), corresponding to a crude oil price of $34 per barrel (see endnote for figure 11, chapter 1, and footnote 4, above), if, as projected, these cars last 50 percent longer than cars with gasoline-internal-combustion engines. This economic prospect provides that basis for the projection that electrolytic hydrogen production would reach 10 exajoules in 2025 and 14 exajoules in 2050.

Because deserts in North Africa and the Middle East are especially good sites for producing hydrogen from photovoltaic sources, it is assumed that hydrogen exports from North Africa to Europe amount to 1.3 exajoules in 2025 and 2.1 exajoules in 2050, while exports from the Middle East to former Centrally Planned Europe and South and East Asia amount to 2.2 exajoules in 2025 and 3.3 exajoules in 2050 (see table Ga). While these production levels for export are modest [amounting, for the Middle East, to only 5 percent of total fluid fuel production in the period 2025 to 2050 (see table H)], they represent the beginnings of what could become major sources of income for these now oil-producing regions in the long term, as oil resources there approach depletion.

If concerns about greenhouse warming become a major driver for energy policy, one way to reduce emissions further than what could be achieved with the RIGES would be to increase the production of wind and photovoltaic hydrogen as a fossil fuel substitute. In particular, strategies for displacing coal used for direct heat applications with hydrogen might be explored. This use of coal accounts for carbon dioxide emissions amounting to 1.7 GtC in 2025 and 1.1 GtC in 2050 [nearly four-fifths of which would be in developing countries (see table W)], accounting for one-third and one-fourth of total CO_2 emissions for the RIGES in those years, respectively (see table X). Such possibilities were not considered in the present study.

---

10. On an energy content basis, the gasoline-equivalent "pump price" of hydrogen from wind or photovoltaic sources would be about $1 per liter ($3.75 per gallon)—see endnote for figure 11, chapter 1.

COMPARISON OF THE RIGES AND THE AP SCENARIO OF THE RSWG

The Renewables-Intensive Global Energy Scenario constructed here differs in some important respects from the energy supply scenario constructed by the RSWG for its Accelerated Policies Scenario.

▼ The analysis in this book shows that a strong emphasis on natural gas in power generation using advanced gas turbine power-generating cycles would be a desirable complement to the use of intermittent renewable electric technologies. Yet, despite the global abundance of natural gas (see table F), the AP scenario de-emphasizes the role of natural gas in the global energy economy—with global natural gas consumption falling from 65 exajoules in 1985 to 60 exajoules in 2025 and to 42 exajoules in 2050. In contrast, natural gas use expands in the RIGES, to 92 exajoules in 2025 and to 107 exajoules in 2050 (see table F).

▼ Both the IPCC AP scenario and the RIGES emphasize biomass, but the amounts are somewhat different: the overall level of biomass supplies amount to 135 exajoules in 2025 and 245 exajoules in 2050 in the AP scenario, compared to 145 exajoules in 2025 and 206 exajoules in 2050 for the RIGES.

▼ The thrust of the AP scenario is to emphasize roles for non-fossil fuel energy technologies beginning in the first half of the 21st century—nuclear as well as solar technologies. In the AP scenario of the RSWG, nuclear power doubles by 2025 and increases five-fold by 2050. In contrast, nuclear power increases only 30 percent, 1985–2025, and is constant thereafter in the RIGES.

▼ Global CO_2 emissions from fossil fuel burning amount to 5.5 GtC in 2025 and 3.5 GtC in 2050 in the AP Scenario of the RSWG. In the RIGES, the emissions are 5.0 GtC in 2025 and 4.2 GtC in 2050 (see table X). If biomass consumption levels were as high in 2050 in the RIGES scenario as in the AP scenario and if this biomass were substituted for coal on a one-to-one energy equivalence basis, CO_2 emissions in 2050 would be reduced in the RIGES by an additional 1.0 GtC, to 3.2 GtC.

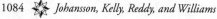

Table A: World regions for RIGES analysis

Africa

Latin America

South and East Asia

Centrally Planned Asia

Japan

Australia/New Zealand

United States

Canada

OECD Europe

Former Centrally Planned Europe

Middle East

Table B: Characteristics of IPCC global energy scenarios

2030 High Emissions: moderate rate of energy efficiency improvement, fossil fuel-intensive energy supply with strong emphasis on coal, with the concentration of total greenhouse gases (in CO_2-equivalent units) in the atmosphere reaching double the pre-industrial level by 2030.

2060 Low Emissions: high rate of energy efficiency improvement, emphasis on natural gas, with the concentration of total greenhouse gases (in CO_2-equivalent units) in the atmosphere reaching double the pre-industrial level by 2060.

Control Policies: high rate of energy efficiency improvement, emphasis on non-fossil energy technologies (solar and nuclear) in the latter half of the 21st century, with the concentration of total greenhouse gases (in CO_2-equivalent units) in the atmosphere reaching double the pre-industrial level by 2090, with stabilization of atmospheric concentrations thereafter.

Accelerated Policies: high rate of energy efficiency improvement, emphasis on non-fossil energy technologies (solar and nuclear) in the first half of the 21st century, with the concentration of total greenhouse gases (in CO_2-equivalent units) kept below a doubling throughout the next century.

Table C: IPCC global electricity consumption scenarios
TWh per year

	2030 High emissions	2060 Low emissions
1985	8,432	8,432
2000	14,833	12,500
2025	29,472	19,694
2050	49,861	30,583

	Control policies	Accelerated policies
1985	8,432	8,432
2000	12,444	12,139
2025	19,889	19,861
2050	31,639	30,750

Table D: IPCC global direct fuel use scenarios
exajoules per year

	Solids	Liquids	Gases	Total fuel	Solids	Liquids	Gases	Total fuel
	2030 High emissions				**2060 Low emissions**			
1985	52.4	114.9	51.7	219.0	52.4	114.9	51.7	219.0
2000	52.9	124.6	53.0	230.5	42.3	113.9	48.2	204.4
2025	101.2	151.6	78.5	331.3	64.1	109.2	65.7	239.0
2050	147.7	214.0	91.7	453.4	84.9	118.1	78.7	281.7
	Control policies				**Accelerated policies**			
1985	52.4	114.9	51.7	219.0	52.4	114.9	51.7	219.0
2000	42.8	114.2	45.7	202.7	59.2	108.7	48.9	216.8
2025	63.7	109.1	68.0	240.8	71.0	127.9	86.8	285.7
2050	82.8	119.8	86.2	288.8	56.1	128.5	104.5	289.1

Table E: Oil production schedule for the RIGES

	Oil production[a] exajoules per year				**Resources** exajoules	
	1985	**1989**	**2025**	**2050**	**12.31.88**[b]	**12.31.50**[a]
Africa	12.32	13.64	6.25	3.63	692	231
Latin America	15.31	15.93	9.28	6.38	953	318
S & E Asia	6.08	6.84	2.39	1.15	292	97
CP Asia	5.67	6.25	4.93	4.18	470	157
Japan	0.03	–	–	–	–	–
Australia/NZ	1.29	1.15	0.47	0.25	54	18
United States	22.60	19.29	9.19	5.49	1,005	335
Canada	3.85	4.13	2.94	2.32	287	96
OECD Europe	8.74	9.16	2.45	0.98	335	112
Former CP Europe	27.38	27.66	11.46	6.21	1,314	438
Middle East	24.31	37.78	27.10	33.99	4,468	2,504
Total	**127.58**	**141.83**	**76.46**	**64.58**	**9,870**	**4,306**
Total (excl. ME)	**103.27**	**104.05**	**49.36**	**30.59**	**5,402**	**1,802**

a. It is assumed that for regions outside the Middle East, oil production $P(t)$ declines exponentially, such that by the end of 2050 the remaining oil resources $R(2050)$ are one-third of the resources $R(1988)$ at the end of 1988:

$$P(t) = P(1989)e^{-a(t - 1989)}$$
$$P(1989) \cdot [1 - e^{-61a}]/a = (2/3) \cdot R(1988).$$

The oil production schedule for the Middle East is determined by oil import requirements for the other regions.

b. Ultimately recoverable resources as of 31 December 1988. Ultimately recoverable resources consist of identified reserves plus estimated recoverable undiscovered resources. The estimates for all regions but the United States were made by the U.S. Geological Survey [5]. For the United States, the ultimately recoverable resource estimates were developed by the U.S. Dept. of Energy [6].

Table F: Natural gas production schedule for the RIGES

	1985	Gas production[a] exajoules per year 1989	2025	2050	Resources exajoules 12.31.88[b]	12.31.50[a]
Africa	2.08	2.53	4.45	8.12	706	422
Latin America	2.91	3.17	8.83	8.83	665	222
S & E Asia	3.14	3.96	7.49	7.49	599	200
CP Asia	0.51	0.58	4.61	4.61	315	105
Japan	0.09	–	–	–	–	–
Australia/NZ	0.67	0.78	1.07	1.09	155	94
United States	17.96	18.83	15.38	13.36	1,457	486
Canada	3.19	3.79	3.12	2.80	502	297
OECD Europe	7.10	7.56	4.36	2.97	449	150
Former CP Europe	24.56	29.98	25.40	25.40	3,319	1,659
Middle East	2.64	3.82	18.45	33.19	2,578	1,559
Total	**64.85**	**75.00**	**93.16**	**107.86**	**10,745**	**5,194**
Total (excl. ME)	**62.21**	**71.18**	**74.71**	**74.67**	**8,167**	**3,635**

a. The production schedules vary by region, but in no region are remaining natural gas resources less than one-third of the resources remaining as of 31 December 1988.

b. Ultimately recoverable resources as of 31 December 1988. Ultimately recoverable resources consist of identified reserves plus estimated recoverable undiscovered resources. The estimates for all regions but the United States were made by the U.S. Geological Survey [5]. For the United States, the ultimately recoverable resource estimates were developed by the U.S. Dept. of Energy [6].

Table Ga: Imports of gaseous fuels by region for the RIGES[a]
exajoules per year

	1985		2025		2050	
	Natural gas	**Hydrogen**	**Natural gas**	**Hydrogen**	**Natural gas**	**Hydrogen**
Africa	− 0.89	−	− 1.34	− 1.34	− 2.07	− 2.07
Latin America	−	−	−	−	−	−
S & E Asia	− 1.41	−	3.42	0.50	9.79	1.00
CP Asia	−	−	2.90	−	6.64	−
Japan	1.54	−	0.67	−	0.75	−
Australia/NZ	−	−	−	−	−	−
United States	1.23	−	0.82	−	−	−
Canada	− 0.95	−	− 0.95	−	− 0.95	−
OECD Europe	1.81	−	1.34	1.34	2.07	2.07
Former CP Europe	− 1.13	−	3.41	1.70	4.61	2.30
Middle East	− 0.12	−	−10.27	− 2.20	−20.84	− 3.30
Total	**0.08**	**−**	**0.00**	**0.00**	**0.00**	**0.00**

Table Gb: Imports of liquid fuels by region for the RIGES[a]
exajoules per year

	1985		2025		2050	
	Oil	**Methanol**	**Oil**	**Methanol**	**Oil**	**Methanol**
Africa	− 8.31	−	−	−10.07	−	−14.57
Latin America	− 5.23	−	−	− 5.21	−	− 5.21
S & E Asia	2.00	−	3.43	3.42	9.80	9.79
CP Asia	− 1.45	−	0.51	0.51	0.85	0.84
Japan	8.95	−	2.78	2.79	2.32	2.32
Australia/NZ	0.08	−	0.13	0.13	0.11	0.11
United States	9.68	−	2.60	2.60	1.42	1.43
Canada	− 0.71	−	− 0.65	− 1.65	− 0.68	− 1.67
OECD Europe	15.87	−	5.68	5.67	4.02	4.01
Former CP Europe	− 3.03	−	1.82	1.81	2.95	2.95
Middle East	−18.23	−	−16.30	−	−20.79	−
Total	**− 0.38**	**−**	**0.00**	**0.00**	**0.00**	**0.00**

a. Summaries from the regional scenarios.

Table H: Fluid fuels production in the Middle East for the RIGES[a]
exajoules per year

	1985	1989	2025	2050
Oil	24.31	37.78	27.10	33.99
Natural gas	2.64	3.82	18.45	33.19
Hydrogen	–	–	2.20	3.30
Total	**26.95**	**41.60**	**47.75**	**70.48**

a. Summary from the Middle East regional scenario for the Renewables-intensive Global Energy Scenario.

Table I: Biomass energy supplies from forests for the RIGES
exajoules per year

	Residues[a]						Total	
	Industrial[b]		Fuel-wood[c]	Subtotal		Fuel-wood[d]		
	2025	2050	2025/ 2050	2025	2050	2025/ 2050	2025	2050
Africa	0.72	0.98	0.78	1.50	1.76	2.43	3.93	4.19
Latin America	0.87	1.01	0.51	1.38	1.52	1.59	2.97	3.11
S & E Asia	0.95	1.12	1.00	1.95	2.12	3.13	5.08	5.25
CP Asia	0.78	0.84	0.39	1.17	1.23	1.21	2.38	2.44
Japan	0.20	0.19	–	0.20	0.19	–	0.20	0.19
Australia/NZ	0.15	0.15	0.00	0.15	0.15	0.02	0.17	0.17
United States	2.18	2.14	0.20	2.38	2.34	0.61	2.99	2.95
Canada	0.96	0.95	0.01	0.97	0.96	0.04	1.01	1.00
OECD Europe	1.28	1.27	0.10	1.38	1.37	0.31	1.69	1.68
Former CP Europe	2.18	2.28	0.19	2.37	2.47	0.58	2.95	3.05
Middle East	0.18	0.23	0.00	0.18	0.23	0.02	0.20	0.25
Total	**10.45**	**11.16**	**13.63**	**14.34**	**14.34**	**9.94**	**23.57**	**24.28**

a. Residues are associated with both industrial roundwood production and roundwood produced for fuelwood and charcoal. Roundwood production for both purposes in 1985 is presented in table J. It is assumed that the density of roundwood is 0.4 dry tonnes per m^3, and that the heating value of wood is 20 gigajoules per dry tonne, or 8 gigajoules per m^3.

b. It is assumed that 50 percent of the forest residues and 75 percent of the mill residues associated with industrial roundwood production are recoverable, and that industrial roundwood production increases with population. It is assumed that in all regions, residues are related to industrial roundwood production using coefficients derived for the United States. In the United States, industrial roundwood production in the late 1970s amounted to 61 percent of felled timber and the final forest products amounted (in energy terms) to 55 percent of industrial roundwood [7]. Thus total recoverable residues are

$$0.50 \cdot (1 - 0.61)/0.61 + 0.75 \cdot 0.45 = 0.65 \cdot (\text{industrial roundwood production}).$$

c. As for industrial roundwood it is assumed that 50 percent of forest residues are recoverable and that residues are related to roundwood production for fuelwood and charcoal using coefficients derived for industrial roundwood production in the United States (note b). Thus recoverable residues amount to

$$0.50 \cdot (1 - 0.61)/0.61 = 0.32 \cdot (\text{roundwood produced for fuelwood and charcoal}).$$

It is assumed that 75 percent of the 1985 production level of roundwood for fuelwood and charcoal is available for energy in the period 2025–2050.

d. It is assumed that during 2025–2050 fuelwood and charcoal production for applications such as cooking has been largely phased out and that 75 percent of the roundwood currently produced for such purposes (assumed to to be less than the 1985 level to help ensure that yields are sustainable; at present there is overharvesting in some areas) is available for the modern energy uses that characterize the Renewables-intensive Global Energy Scenario.

Table J: Roundwood production in 1985[a]
million m³

	Industrial roundwood	Fuelwood and charcoal
Africa	54	405
Latin America	94	265
South & East Asia	102	522
Centrally Planned Asia	98	202
Japan	33	–
Australia/New Zealand	25	3
United States	347	102
Canada	165	6
OECD Europe	219	52
Former Centrally Planned Europe	350	97
Middle East	14	3
Total	**1,501**	**1,657**

a. [8].

Table K: Biomass supplies from residues other than forest residues for the RIGES
exajoules per year

	Sugarcane[a]		Dung[b]		Cereals[c]		Urban refuse[d]	
	2025	**2050**	**2025**	**2050**	**2025**	**2050**	**2025**	**2050**
Africa	1.17	1.58	3.46	5.19	0.68	0.85	–	–
Latin America	5.33	6.19	3.23	4.18	0.98	1.70	–	–
South & East Asia	3.20	3.79	6.12	11.53	2.34	2.98	–	–
CP Asia	0.32	0.35	1.23	1.39	1.13	1.19	–	–
Japan	–	–	–	–	0.16	0.23	0.53	0.53
Australia/NZ	0.18	0.18	0.48	0.62	0.27	0.38	0.063	0.062
United States	–	–	0.62	0.41	1.72	1.81	1.14	1.12
Canada	–	–	–	–	0.35	0.35	0.11	0.11
OECD Europe	–	–	0.76	0.80	1.41	1.41	1.30	1.28
Former CP Europe	–	–	1.10	1.14	1.81	2.07	–	–
Middle East	–	–	–	–	–	–	–	–
Total	**10.20**	**12.09**	**17.00**	**25.26**	**10.85**	**12.97**	**3.14**	**3.10**

a. Sugarcane residues consist of 150 dry kilograms of bagasse (2.85 gigajoules) plus 279 dry kilograms of barbojo (5.30 gigajoules) (see chapter 17) per tonne of cane (tc). It is assumed that all the bagasse plus two thirds of the barbojo is recovered for energy and that cane production increases in proportion to the population [9]. In China, however, cane residues are often used for papermaking. Accordingly, it is assumed that residues from only half of sugarcane production are available for energy in Centrally Planned Asia.

b. Based on dung production coefficients and dung heating values for different animals[10] and livestock inventories [9]. It is assumed that 25 percent of the produced dung is recoverable. For all regions except South and East Asia and OECD Europe, dung residues are assumed to increase in proportion to meat production, as projected by the IPCC Response Strategies Working Group [11]. For South and East Asia and for OECD Europe dung residues are assumed to increase in proportion to the production of dairy products, as projected by the IPCC Response Strategies Working Group [11].

c. It is assumed that cereal residues are produced at a rate of 1.3 tonnes per tonne of grain, the global average for 1983[12]; and that one fourth of these residues, with a heating value of 12 gigajoules per tonne, are recoverable. Cereals production levels are assumed to be the projections of the IPCC Response Strategies Working Group [11].

d. It is assumed that 75 percent of the energy in urban refuse can be recovered and that the waste generation rate per capita is constant over time. For the U.S., based on an annual generation rate of 330 kilograms with a heating value of 15.9 megajoules per kilogram—parameters for the U.S. with a 50 percent recycling rate [13]; the same values are assumed to be applicable for Canada. For other OECD countries, based on the average generation rate in 1985 of 300 kilograms per capita per year for 12 countries that accounted for 80 percent of Europe's population [14], assuming the average heating value for municipal solid waste in the United States without recycling (12.7 megajoules per kilogram).

Table L: Biomass energy supplies from plantations for the RIGES

	Land in plantations million hectares		Energy crop production exajoules per year	
	2025	**2050**	**2025**	**2050**
Africa	95[a]	106[a]	18.9[b]	31.8[c]
Latin America	161[d]	165[d]	32.3[b]	49.6[c]
South & East Asia	–	–	–	–
CP Asia	25[e]	50[e]	5.0[b]	15.0[c]
Japan	–	–	–	–
Australia/NZ	–	–	–	–
United States	32[f]	32[f]	9.6[c]	9.6[c]
Canada	6	6	1.2[b]	1.2[b]
OECD Europe	30	30	9.0[c]	9.0[c]
Former CP Europe	20	40	4.0[b]	12.0[c]
Middle East	–	–	–	–
Total	**369**	**429**	**79.9**	**128.2**

a. For developing countries it is assumed that mainly degraded lands are used for biomass plantations. Grainger [15] estimates that there are 842 million hectares of degraded lands in Africa and that all of the degraded lands involving logged forests (39 million hectares), humid tropics forest fallow (59 million hectares), and deforested watershed (3 million hectares) are suitable for reforestation—totalling 101 million hectares. In addition, he estimates that one fifth of desertified drylands globally are potentially available for reforestation. If this fraction is applicable to Africa as well, an additional 148 million hectares could be reforested.

b. For an average productivity of 10 dry tonnes per hectare per year @ 20 gigajoules per tonne. For an average productivity of 10 dry tonnes per hectare per year @ 20 gigajoules per tonne.

c. For an average productivity of 15 dry tonnes per hectare per year.

d. Grainger [15] estimates that there are 318 million hectares of degraded lands in Latin America and that all of the degraded lands involving logged forests (44 million hectares), humid tropics forest fallow (85 million hectares), and deforested watershed (27 million hectares) are suitable for reforestation—totalling 156 million hectares. Up to one fifth of desertified drylands (32 million hectares) might also be suitable for reforestation.

e. Plantations in China would be primarily on deforested land on low-lying mountains that are already targeted for reforestation. In the early 1980s China announced reforestation goals to raise its forest cover to 20 percent of its land area by 2000, with an eventual goal of 30 percent [16]. These targets correspond to increments of 52 million hectares and 145 million hectares, respectively, over the 1983–1985 level of forest cover.

f. In industrialized countries the most likely sites for biomass plantations would be excess agricultural lands. For the United States at present this amounts to 32 million hectares [the amount of land currently held out of agricultural production for the Acreage Reduction Program (18 million hectares) and the Conservation Reserve Program (14 million hectares)].

Table M: Total biomass supplies for energy for the RIGES
exajoules per year of primary energy

	Forests[a]	Residues[b]		Plantations[c]		Total	
	2025/ 2050	2025	2050	2025	2050	2025	2050
Africa	2.43	6.81	9.38	18.94	31.81	28.18	43.62
Latin America	1.59	10.92	13.59	32.30	49.60	44.81	64.78
S & E Asia	3.13	13.61	20.42	–	–	16.74	23.55
CP Asia	1.21	3.85	4.16	5.00	15.00	10.06	20.37
Japan	–	0.89	0.95	–	–	0.89	0.95
Australia/NZ	0.02	1.14	1.39	–	–	1.16	1.41
United States	0.61	5.86	5.68	9.60	9.60	16.07	15.89
Canada	0.04	1.43	1.42	1.20	1.20	2.67	2.66
OECD Europe	0.31	4.85	4.86	9.00	9.00	14.16	14.17
Former CP Europe	0.58	5.28	5.68	4.00	12.00	9.86	18.26
Middle East	0.02	0.18	0.23	–	–	0.20	0.25
Total	**9.94**	**54.82**	**67.76**	**80.04**	**128.21**	**144.80**	**205.91**

a. Roundwood only, from table I; residues are not included.

b. Forest and forest product residues from table I, plus sugarcane residues, dung, cereals residues, and urban refuse from table K.

c. Plantation energy from table L.

Table N: Average energy efficiencies for conversion technologies
involved in the RIGES[a]
percent

	2025	2050
Electricity generation		
Natural gas-fired combined cycles or variants thereof	50 [b]	55 [b]
Methanol-fired chemically recuperated gas turbines	55 [c]	60 [c]
Coal-fired plants in regions where coal-based power generation is declining	35	–
Coal-fired plants in regions where coal-based power generation is expanding	42.1 [d]	57 [d]
Biomass-fired power plants		
Gas-turbine systems for cogeneration in the sugarcane industrial applications	40 [e]	40 [e]
Gasification-based power systems for stand-alone applications	42.9 [f]	57 [f]
Biomass fuels production		
Methanol production from biomass	61.9 [g]	62.9 [g]
Hydrogen production from biomass	69.7 [h]	71.5 [h]

a. All fuel heating values and energy conversion efficiencies in this study are expressed in terms of the higher heating value (HHV) of the fuel. In many parts of the world, fuel heating values and efficiencies are based on the use of lower heating values. To express the results presented here in terms of lower heating values (LHVs), one needs to know the ratio of lower to higher heating values for the fuels involved:

	LHV/HHV
Coals	
Illinois No. 6	0.965
East Texas Lignite	0.960
Oils	
No. 6 fuel oil	0.946
No. 2 fuel oil	0.939
Ethanol	0.901
Methanol	0.878
Ethane (a minor constituent of natural gas)	0.915
Methane (the primary constituent in natural gas)	0.901
Hydrogen	0.842

Thus, for example, a combined cycle operated on natural gas with an efficiency of 50 percent on a HHV basis would have an efficiency of 50/0.9 = 55.6 percent on a LHV basis.

b. An efficiency of 50 percent (HHV basis, or 55.5 percent, LHV basis) will be achieved for commercial systems in the 1990s. Substantial further gains are likely both because of expected continuing improvements in basic gas turbine technology (e.g., turbine inlet temperatures for state-of-the-art gas turbines are increasing at a rate of 20°C per year) and because a variety of innovative gas turbine cycles will be developed [17].

c. Heat recovery by steam reforming of fuel is a more efficient way of recovering gas turbine exhaust heat than recovery in a steam boiler [18]. Methanol is especially easy to reform with the heat from the exhaust of a simple cycle gas turbine: the efficiency of a methanol-fueled system based on mid-1980s gas turbine technology (with a turbine inlet temperature of 1,200°C) is 47 percent, HHV basis or 53.5 percent, LHV basis [19].

d. For coal integrated gasifier/intercooled steam-injected gas turbine technology in 2025 and coal integrated gasifier/advanced molten carbonate fuel cell technology in 2050 (see chapter 17).

e. For biomass integrated gasifier/intercooled steam-injected gas turbine technology operated in the cogeneration mode, with all fuel consumption charged to power generaton—i.e., no credit is taken for the byproduct steam and electricity used to run the sugar factory or ethanol distillery (see chapter 17).

f. For biomass integrated gasifier/intercooled steam-injected gas turbine (BIG/ISTIG) technology in 2025 and biomass integrated gasifier/advanced molten carbonate fuel cell (BIG/FC) technology in 2050 (see chapter 17).

g. Using the Battelle Columbus Laboratory indirectly heated biomass gasifier, gasification of 1 gigajoule of biomass yields 0.661 gigajoule of methanol but requires 12.29 kWh of purchased electricity [20]. Here it is assumed that in 2025 the purchased electricity is generated in a BIG/ISTIG plant with an efficiency of 42.9 percent, so that the overall energy efficiency (biomass to methanol) is 61.9 percent, and that in 2050 the purchased electricity is generated in a BIG/FC plant with an efficiency of 57 percent, so that the overall energy efficiency (biomass to methanol) is 62.9 percent.

h. Using the Battelle Columbus Laboratory indirectly heated biomass gasifier, gasification of 1 gigajoule of biomass yields 0.777 gigajoules of hydrogen but requires 17.75 kWh of purchased electricity [20]. Here it is assumed that in 2025 the purchased electricity is generated in a BIG/ISTIG plant with an efficiency of 42.9 percent, so that the overall energy efficiency (biomass to hydrogen) is 69.7 percent, and that in 2050 the purchased electricity is generated in a BIG/FC plant with an efficiency of 57 percent, so that the overall energy efficiency (biomass to hydrogen) is 71.5 percent.

Table O: Hydroelectricity generation by region, for the RIGES

	TWh per year			As fraction of electricity		
	1985	2025	2050	1985	2025	2050
Africa	45.3	261	522	0.19	0.23	0.25
Latin America	307.9	505	702	0.64	0.25	0.22
South & East Asia	101.7	413	500	0.23	0.14	0.08
CP Asia	120.1	650	911	0.26	0.17	0.14
Japan	76.5	77	77	0.12	0.12	0.10
Australia/NZ	32.0	40	40	0.23	0.31	0.24
United States	262.7	335	345	0.11	0.11	0.10
Canada	294.0	368	368	0.73	0.80	0.69
OECD Europe	398.0	510	510	0.22	0.24	0.21
Former CP Europe	232.4	600	800	0.12	0.15	0.15
Middle East	9.5	10	10	0.06	0.01	0.01
Total	**1,880**	**3,769**	**4,785**	**0.20**	**0.18**	**0.15**

Table P: Natural gas-based power generation by region, for the RIGES

	TWh per year			As fraction of electricity		
	1985	2025	2050	1985	2025	2050
Africa	36.4	306	633	0.16	0.28	0.30
Latin America	82.0	206	691	0.17	0.10	0.21
South & East Asia	30.8	558	1,886	0.07	0.20	0.30
CP Asia	5.8	946	1,579	0.01	0.25	0.25
Japan	120.8	58	83	0.19	0.09	0.10
Australia/NZ	16.1	35	45	0.12	0.27	0.27
United States	289.0	498	424	0.12	0.16	0.13
Canada	6.6	37	34	0.02	0.08	0.06
OECD Europe	103.5	211	404	0.06	0.10	0.17
Former CP Europe	327.6	1,257	1,299	0.16	0.32	0.24
Middle East	80.4	636	1,092	0.52	0.64	0.64
Total	**1,099**	**4,748**	**8,170**	**0.12**	**0.23**	**0.25**

Table Q: Coal-based power generation by region, for the RIGES

| | TWh per year | | | As fraction of electricity | | |
	1985	2025	2050	1985	2025	2050
Africa	116.5	–	–	0.50	–	–
Latin America	16.4	–	–	0.03	–	–
S & E Asia	185.1	778	1,177	0.41	0.28	0.19
CP Asia	267.3	739	855	0.58	0.20	0.14
Japan	94.4	–	–	0.15	–	–
Australia/NZ	82.7	–	–	0.60	–	–
United States	1,387.8	125	–	0.56	0.04	–
Canada	77.0	–	–	0.19	–	–
OECD Europe	593.6	148	–	0.32	0.07	–
Former CP Europe	961.2	481	–	0.48	0.12	–
Middle East	–	–	–	–	–	–
Total	**3,782**	**2,271**	**2,032**	**0.40**	**0.11**	**0.06**

Table R: Nuclear power generation by region, for the RIGES

| | TWh per year | | | As fraction of electricity | | |
	1985	2025	2050	1985	2025	2050
Africa	5.5	20	20	0.02	0.02	0.01
Latin America	8.4	33	33	0.02	0.02	0.01
S & E Asia	48.5	104	104	0.11	0.04	0.02
CP Asia	–	89	89	–	0.02	0.01
Japan	150.3	300	300	0.24	0.48	0.38
Australia/NZ	–	–	–	–	–	–
United States	376.2	565	565	0.15	0.19	0.17
Canada	58.6	84	84	0.15	0.18	0.16
OECD Europe	538.7	539	539	0.29	0.26	0.22
Former CP Europe	213.1	100	100	0.11	0.03	0.02
Middle East	–	–	–	–	–	–
Total	**1,399**	**1,834**	**1,834**	**0.15**	**0.09**	**0.06**

Table S: Electricity from biomass by region, for the RIGES

	Sugarcane		Other		Total		As fraction of electricity	
	2025	2050	2025	2050	2025	2050	2025	2050
Africa	129	174	284	519	413	693	0.37	0.33
Latin America	589	683	510	811	1,099	1,494	0.54	0.46
S & E Asia	354	419	–	–	354	419	0.13	0.07
CP Asia	36	39	378	632	414	671	0.11	0.11
Japan	–	–	57	82	57	82	0.09	0.10
Australia/NZ	20	20	–	–	20	20	0.15	0.12
United States	–	–	456	504	456	504	0.15	0.15
Canada	–	–	–	–	–	–	–	–
OECD Europe	–	–	399	529	399	529	0.19	0.22
Former CP Europe	–	–	522	1,355	522	1,355	0.13	0.25
Middle East	–	–	–	–	–	–	–	–
Total	**1,128**	**1,335**	**2,606**	**4,432**	**3,734**	**5,767**	**0.18**	**0.18**

Table T: Intermittent renewable electricity use by region, for the RIGES

	TWh per year		As fraction of electricity	
	2025	2050	2025	2050
Africa	111	208	0.10	0.10
Latin America	201	325	0.10	0.10
South & East Asia	736	2,200	0.25	0.35
CP Asia	946	2,211	0.25	0.35
Japan	126	240	0.20	0.30
Australia/NZ	33	58	0.25	0.35
United States	760	1,177	0.25	0.35
Canada	110	221	0.18	0.31
OECD Europe	293	429	0.14	0.18
Former CP Europe	987	1,913	0.25	0.35
Middle East	348	594	0.35	0.35
Total	**4,651**	**9,576**	**0.22**	**0.30**

Table U: Fluid fuels from biomass by region, for the RIGES
exajoules per year

| | Methanol | | | | Hydrogen | | | |
| | Production | | Consumption | | Production | | Consumption | |
	2025	2050	2025	2050	2025	2050	2025	2050
Africa	13.43	21.53	3.36	6.96	–	–	–	–
Latin America	17.36	22.30	12.15	17.09	3.32	10.61	3.32	10.61
S & E Asia	–	–	3.42	9.79	5.17	5.88	5.17	5.88
CP Asia	1.50	2.97	2.01	3.81	–	–	–	–
Japan	0.55	0.60	3.34	2.92	–	–	–	–
Australia/NZ	0.31	0.38	0.44	0.49	–	–	–	–
United States	4.81	5.16	7.41	6.59	2.71	2.93	2.71	2.93
Canada	1.65	1.67	–	–	–	–	–	–
OECD Europe	3.11	3.15	8.78	7.16	3.50	3.59	3.50	3.59
Former CP Europe	1.81	3.59	3.62	6.54	1.02	2.04	1.02	2.04
Middle East	–	–	–	–	–	–	–	–
Total	**44.53**	**61.35**	**44.53**	**61.35**	**15.72**	**25.05**	**15.72**	**25.05**

Table V: Hydrogen from intermittent renewable energy sources by region,
for the RIGES
exajoules per year

| | Production | | Consumption | |
	2025	2050	2025	2050
Africa	1.34	2.07	–	–
Latin America	–	–	–	–
South & East Asia	0.50	1.00	1.00	2.00
CP Asia	0.50	1.00	0.50	1.00
Japan	0.66	0.74	0.66	0.74
Australia/NZ	–	–	–	–
United States	2.13	2.96	2.13	2.96
Canada	0.75	0.75	0.75	0.75
OECD Europe	–	–	1.34	2.07
Former CP Europe	1.70	2.30	3.40	4.60
Middle East	2.20	3.30	–	–
Total	**9.78**	**14.12**	**9.78**	**14.12**

Table W: Coal consumption sources by region, for the RIGES
exajoules per year

	Direct use			Electricity			Total		
	1985	2025	2050	1985	2025	2050	1985	2025	2050
Africa	1.80	3.90	3.70	1.60	–	–	3.40	3.90	3.70
Latin America	0.73	1.60	1.20	0.18	–	–	0.91	1.60	1.20
S & EAsia	3.16	9.40	9.60	2.22	6.65	7.43	5.38	16.05	17.03
CP Asia	16.46	37.89	21.29	3.25	6.32	5.40	19.71	44.21	26.69
Japan	2.27	1.71	1.14	0.91	–	–	3.18	1.71	1.14
Australia/NZ	0.32	0.39	0.26	1.04	–	–	1.36	0.39	0.26
United States	3.14	2.10	1.20	15.43	1.29	–	18.57	3.39	1.20
Canada	0.27	0.41	0.29	0.86	–	–	1.13	0.41	`0.29
OECD Europe	4.64	3.39	2.41	6.74	1.52	–	11.38	4.91	2.41
Former CP Europe	14.85	7.30	5.10	10.06	4.95	–	24.91	12.25	5.10
Middle East	0.12	–	–	–	–	–	0.12	–	–
Total	**47.76**	**68.09**	**46.19**	**42.29**	**20.73**	**12.83**	**90.05**	**88.82**	**59.02**

Table X: Carbon dioxide emissions from fossil fuel use by region, for the RIGES

	MtC per year			Emissions relative to 1985		
	1985	2025	2050	1985	2025	2050
Africa	179	261	245	1.00	1.46	1.36
Latin America	258	340	274	1.00	1.32	1.06
S & E Asia	315	659	873	1.00	2.09	2.77
CP Asia	582	1,313	917	1.00	2.26	1.58
Japan	277	106	84	1.00	0.38	0.30
Australia/NZ	72	36	28	1.00	0.50	0.39
United States	1,353	533	345	1.00	0.39	0.25
Canada	125	84	64	1.00	0.67	0.51
OECD Europe	902	358	226	1.00	0.40	0.25
Former CP Europe	1,414	954	711	1.00	0.67	0.50
Middle East	156	321	424	1.00	2.06	2.72
Total	**5,633**	**4,965**	**4,191**	**1.00**	**0.88**	**0.74**

REGIONAL SCENARIO TABLES

In what follows, the details of the renewables-intensive global energy scenario (RIGES) are presented in tabular form, for each of the 11 regions indicated in table A. For each region, the following tables are presented to characterize the scenario:

Table 1. Growth schedule for per capita GDP and population, assuming IPCC growth rates

Table 2. Baseline data for energy, 1985

Table 3. IPCC electricity consumption scenarios (TWh per year)

Table 4. IPCC direct fuel use scenarios (exajoules per year)

Table 5. Electricity supply scenario (TWh per year)

Table 6. Fuel supply scenario (exajoules per year)

The starting point for the construction of the RIGES are the IPCC Accelerated Policies Scenario projections (high economic growth variant) for electricity consumption and the direct use of fuels (tables 3 and 4).

In these tables, the transmission and distribution losses account for the differences between the electricity generation levels listed in tables 2 and 5 and the electricity consumption levels listed in table 3. The fuel supplies listed in table 6 are the sum of fuels for direct use (table 4) and fuels required for power generation (table 5).

The first sets of tables are a global summary, an industrialized country summary, and a developing country summary.

RENEWABLES-INTENSIVE ENERGY SCENARIO—GLOBAL TOTALS

Table 1: Growth schedule for per capita GDP and population for the world, assuming IPCC growth rates

	GDP per capita *1989 U.S.$*	Population *millions*
1985	3,232	4,874
2000	4,428	6,172
2025	7,676	8,185
2050	13,636	9,521

Table 2: Baseline world energy data (commercial fuels only), 1985

	Electricity *TWh per year*	Primary energy[a] *EJ per year*	Direct fuel use *EJ per year*	Fuel for electricity *EJ per year*
Coal	3,782.0	90.0	47.7	42.4
Other solids	21.4	4.8	4.7	0.1
Oil	1,042.0	127.2	115.8	11.4
Natural gas	1,098.9	64.9	52.8	12.1
Hydro	1,880.1	20.5	–	–
Nuclear	1,399.3	15.3	–	–
Geothermal	15.0	0.1	–	–
Totals	**9,239**	**323**	**211**	**66**

a. The primary energy equivalent of hydropower, nuclear power, and geothermal power is taken to be a fossil fuel-equivalent heat rate of 10.91 megajoules per kWh (33 percent efficiency) here and in all regional tables.

Table 3: IPCC global electricity consumption scenarios
TWh per year

	High emissions	Accelerated policies
1985	8,432	8,432
2000	14,833	12,139
2025	29,472	19,861
2050	49,861	30,750

Table 4: IPCC global direct fuel use scenarios (commercial fuels only)
EJ per year

	High emissions				Accelerated policies			
	Solids	Liquids	Gases	Total fuel	Solids	Liquids	Gases	Total fuel
1985	52.5	115.8	52.8	221.1	52.5	115.8	52.8	221.1
2000	52.9	124.6	53.0	230.5	59.2	108.7	48.9	216.8
2025	101.2	151.6	78.5	331.3	71.0	127.9	86.8	285.7
2050	147.7	214.0	91.7	453.4	56.1	128.5	104.5	289.1

Table 5: Global electricity supply scenario
TWh per year

	1985	2025	2050
Coal	3,782	2,271	2,032
Oil	1,042	–	–
Natural gas	1,099	4,748	8,170
Nuclear	1,399	1,834	1,834
Hydro	1,880	3,769	4,785
Intermittent renewables	–	4,651	9,576
Sugarcane residues	–	1,128	1,335
Biomass (stand-alone plants)	–	2,386	4,084
Methanol	–	220	348
Geothermal	15	197	212
Other	21	–	–
Totals	**9,239**	**21,204**	**32,376**

Table 6: Global fuel supply scenario (commercial fuels only)
EJ per year

	1985	2025	2050
Solid fuels			
Coal	90.0	88.8	59.0
Biomass and other	5.0	33.1	47.8
Subtotal	*95.1*	*121.9*	*106.8*
Liquid fuels			
Oil	127.6	76.5	64.3
MeOH from biomass	–	44.5	61.5
EthOH from sugarcane	–	0.9	1.0
Subtotal	*127.6*	*121.9*	*126.8*
Gaseous fuels			
Natural gas	64.9	93.2	107.9
H_2 **from biomass**	–	15.7	25.1
Biogas (dung)	–	9.7	14.1
Biogas (distilleries)	–	0.2	0.2
H_2 **from intermittent** renewables	–	9.8	14.4
Subtotal	*64.9*	*128.6*	*161.7*
CO_2 emissions in *MtC*	5,633	4,965	4,191
[Index, 1985 = 100]	[100.]	[88.1]	[74.4]
CO_2 emissions per capita in *tC*	1.158	0.607	0.440
[Index, 1985 = 100]	[100.]	[52.4]	[38.0]

RENEWABLES-INTENSIVE ENERGY SCENARIO—TOTALS FOR INDUSTRIALIZED COUNTRIES

Table 1: Growth schedule for per capita GDP and population for industrialized countries, assuming IPCC growth rates

	GDP per capita *1989 U.S.$*	Population *millions*
1985	10,354	1,225
2000	15,825	1,333
2025	30,693	1,433
2050	55,267	1,443

Table 2: Baseline energy data for industrialized countries (commercial fuels only), 1985

	Electricity *TWh per year*	Primary energy[a] *EJ per year*	Direct fuel use *EJ per year*	Fuel for electricity *EJ per year*
Coal	3,197.0	60.5	25.5	35.0
Other solids	21.4	4.8	4.7	0.1
Oil	736.7	94.7	87.0	7.7
Natural gas	863.5	56.1	46.7	9.3
Hydro	1,295.6	14.1	–	–
Nuclear	1,336.9	14.6	–	–
Geothermal	15.0	0.1	–	–
Totals	**7,446**	**245**	**164**	**52**

a. The primary energy equivalent of hydropower, nuclear power, and geothermal power is taken to be a fossil fuel-equivalent heat rate of 10.91 megajoules per kWh (33 percent efficiency) here and in all regional tables.

Table 3: IPCC electricity consumption scenarios for industrialized countries
TWh per year

	High emissions	Accelerated policies
1985	6,892	6,892
2000	11,528	8,833
2025	18,861	9,806
2050	24,833	12,111

Table 4: IPCC direct fuel use scenarios (commercial fuels only) for industrialized countries
EJ per year

	High emissions				Accelerated policies			
	Solids	Liquids	Gases	Total fuel	Solids	Liquids	Gases	Total fuel
1985	30.3	87.0	46.7	164.0	30.3	87.0	46.7	164.0
2000	24.3	85.5	45.0	154.8	28.1	74.1	42.1	144.3
2025	31.2	78.7	56.2	166.1	15.3	62.1	56.7	134.1
2050	34.8	91.8	58.4	185.0	10.4	48.6	58.5	117.5

Table 5: Electricity supply scenario for industrialized countries
TWh per year

	1985	2025	2050
Coal	3,197	754	–
Oil	737	–	–
Natural gas	864	2,096	2,289
Nuclear	1,337	1,588	1,588
Hydro	1,296	1,930	2,140
Intermittent renewables	–	2,309	4,038
Sugarcane residues	–	20	20
Biomass (stand-alone plants)	–	1,377	2,388
Methanol	–	57	82
Geothermal	15	197	212
Other	21	–	–
Totals	**7,463**	**10,328**	**12,757**

Table 6: Fuel supply scenario (commercial fuels only) for industrialized countries
EJ per year

	1985	2025	2050
Solid fuels			
Coal	60.5	23.0	10.4
Biomass and other	5.0	11.7	15.3
Subtotal	*65.5*	*34.7*	*25.2*
Liquid fuels			
Domestic oil	63.9	26.5	15.3
Imported oil	30.8	12.4	10.1
MeOH from biomass	–	12.2	14.6
Imported MeOH	–	11.4	9.2
Subtotal	*94.7*	*62.5*	*49.2*
Gaseous fuels			
Domestic natural gas	53.6	49.3	45.6
Imported natural gas	2.5	5.3	6.5
H_2 from biomass	–	7.2	8.6
Biogas (dung)	–	1.7	1.7
H_2 from intermittent renewables	–	5.3	6.8
Imported H_2	–	3.0	4.4
Subtotal	*56.1*	*71.8*	*73.6*
CO_2 emissions in *MtC*	4,143	2,071	1,458
[Index, 1985 = 100]	[100.]	[50.0]	[35.2]
CO_2 emissions per capita in *tC*	3.382	1.450	1.010
[Index, 1985 = 100]	[100.]	[42.9]	[29.9]

RENEWABLES-INTENSIVE ENERGY SCENARIO—TOTALS FOR DEVELOPING COUNTRIES

Table 1: Growth schedule for per capita GDP and population for developing countries, assuming IPCC growth rates

	GDP per capita *1989 U.S. $*	Population *millions*
1985	842	3,649
2000	1,289	4,839
2025	2,792	6,752
2050	6,198	8,078

Table 2: Baseline energy data for developing countries (commercial fuels only), 1985

	Electricity *TWh per year*	Primary energy[a] *EJ per year*	Direct fuel use *EJ per year*	Fuel for electricity *EJ per year*
Coal	585.3	29.5	22.1	7.4
Other solids	–	–	–	–
Oil	305.3	32.5	28.8	3.7
Natural gas	235.4	8.9	6.1	2.8
Hydro	584.5	6.4	–	–
Nuclear	62.4	0.7	–	–
Geothermal	–	–	–	–
Totals	**1,773**	**78**	**57**	**14**

a. The primary energy equivalent of hydropower, nuclear power, and geothermal power is taken to be a fossil fuel-equivalent heat rate of 10.91 megajoules per kWh (33 percent efficiency) here and in all regional tables.

Table 3: IPCC electricity consumption scenarios for developing countries
TWh per year

	High emissions	Accelerated policies
1985	1,540	1,540
2000	3,306	3,306
2025	10,611	10,056
2050	25,028	18,639

Table 4: IPCC direct fuel use scenarios (commercial fuels only) for developing countries
EJ per year

	High emissions				Accelerated policies			
	Solids	Liquids	Gases	Total fuel	Solids	Liquids	Gases	Total fuel
1985	22.2	28.8	6.1	57.1	22.2	28.8	6.1	57.1
2000	28.6	39.1	8.0	75.7	31.1	34.6	6.8	72.5
2025	70.0	72.9	22.3	165.2	55.7	65.8	30.1	151.6
2050	112.9	122.2	33.3	268.4	45.7	79.9	46.0	171.6

Table 5: Electricity supply scenario for developing countries
TWh per year

	1985	2025	2050
Coal	585	1,517	2,032
Oil	305	–	–
Natural gas	235	2,652	5,881
Nuclear	62	246	246
Hydro	584	1,839	2,645
Intermittent renewables	–	2,342	5,538
Sugarcane residues	–	1,108	1,315
Biomass (stand-alone plants)	–	1,009	1,696
Methanol	–	163	266
Totals	**1,771**	**10,876**	**19,619**

Table 6: Fuel supply scenario (commercial fuels only) for developing countries
EJ per year

	1985	2025	2050
Solid fuels			
Coal	29.5	65.8	48.6
Biomass and other	–	21.4	32.5
Subtotal	*29.5*	*87.2*	*81.1*
Liquid fuels			
Domestic oil	63.7	50.0	49.3
Imported oil	– 31.2	– 12.4	– 10.4
MeOH from biomass	–	32.3	46.8
Imported MeOH	–	– 11.4	– 9.1
EthOH from sugarcane	–	0.9	1.0
Subtotal	*32.5*	*59.4*	*77.6*
Gaseous fuels			
Domestic natural gas	11.3	43.8	62.2
Imported natural gas	– 2.4	– 5.3	– 6.5
H_2 from biomass	–	8.5	16.5
Biogas (dung)	–	8.0	12.7
Biogas (distilleries)	–	0.2	0.2
H_2 from intermittent renewables	–	4.5	7.4
Imported H_2	–	– 3.0	– 4.4
Subtotal	*8.9*	*56.7*	*88.1*
CO_2 emissions in *MtC*	1,491	2,897	2,732
[Index, 1985 = 100]	[100.]	[194.3]	[183.2]
CO_2 emissions per capita in *tC*	0.410	0.430	0.340
[Index, 1985 = 100]	[100.]	[104.9]	[82.9]

RENEWABLES-INTENSIVE ENERGY SCENARIO FOR AFRICA

Table 1: Growth schedule for per capita GDP and population for Africa, assuming IPCC growth rates

	GDP per capita *1989 U.S.$*	Population *millions*
1985	809	579
2000	1,023	872
2025	1,587	1,498
2050	2,773	2,026

Table 2: Baseline data for energy in Africa, 1985

	Electricity[a] *TWh per year*	Primary energy[a] *EJ per year*	Direct fuel use[b] *EJ per year*	Fuel for electricity[c] *EJ per year*
Coal	116.5	3.40	1.80	1.60
Oil	29.1	4.01	3.61	0.40
Natural gas	36.4	1.19	0.69	0.50
Hydro	45.3	0.49	–	–
Nuclear	5.5	0.06	–	–
Totals	**232.5**	**9.15**	**6.10**	**2.50**

a. Data from the [21].

b. Commercial fuels only.

c. The average efficiency of power generation is assumed to be 26 percent. From the IPCC report [11] the ratio of delivered electricity to fuels input is 0.222. Thus T&D losses in 1985 were 100·(1 − 22.2/26) = 15 percent.

Table 3: IPCC electricity consumption scenarios for Africa
TWh per year

	High emissions	Accelerated policies
1985	198	198
2000	417	389
2025	1,055	1,028
2050	2,333	1,972

Table 4: IPCC direct fuel use scenarios (commercial fuels only) for Africa
EJ per year

	High emissions				Accelerated policies			
	Solids	Liquids	Gases	Total fuel	Solids	Liquids	Gases	Total fuel
1985	1.80	3.61	0.69	6.10	1.80	3.61	0.69	6.10
2000	2.50	4.70	0.80	8.00	2.10	4.40	0.70	7.20
2025	6.80	10.40	1.50	18.70	3.90	9.30	2.90	16.10
2050	12.30	16.10	2.30	30.70	3.70	10.10	4.90	18.70

Table 5: Electricity supply scenario for Africa
TWh per year

	1985	2025	2050
Hydro	45.3	261 [a]	522 [a]
Coal	116.5	– [b]	– [b]
Oil	29.1	– [b]	– [b]
Nuclear	5.5	20 [c]	20 [c]
MeOH	–	62 [d]	104 [d]
Intermittent renewables	–	111 [e]	208 [e]
Sugarcane	–	129 [f]	174 [f]
Biomass	–	222 [g]	415 [g]
Natural gas	36.4	306 [h]	633 [h]
Totals	**232.5**	**1,111** [i]	**2,076** [i]

a. It has been estimated that the economic potential for hydropower in Africa is 1,569 TWh per year [4]. Here it is assumed that one sixth of this potential is developed by 2025 and one third by 2050.

b. It is assumed that oil and coal power are phased out.

c. As of 31 December 1989 some 1,840 MW$_e$ of nuclear power was operating in South Africa and another 1,800 MW$_e$ was under construction in Egypt. It is assumed that this capacity operates at an average capacity factor of 62 percent and that there is no further expansion of nuclear generating capacity.

d. It is assumed that methanol used in chemically recuperated gas turbines accounts for 5 percent of electricity in 2025 and 2050, with efficiencies of 55 percent in 2025 and 60 percent in 2050. The corresponding methanol fuel requirements are 0.41 exajoules in 2025 and 0.62 exajoules in 2050.

e. It is assumed that intermittent renewables (PV, solar thermal, and wind) account for 10 percent of electricity in 2025 and 2050.

f. It is assumed that sugarcane production increases in proportion to the population from 73.3 million tonnes in 1985, to 190 million tonnes in 2025, and to 256 million tonnes in 2050, and that by-product electricity is produced from cane residues at a rate of 720 kWh per tonne of cane (tables K and N) using BIG/ISTIG technology or the equivalent (see chapter 17).

g. It is assumed that 20 percent of electricity is provided by biomass in stand-alone plants at 42.9 percent efficiency (BIG/ISTIG or equivalent) in 2025 and 57 percent efficiency (BIG/FC or equivalent) in 2050. The corresponding biomass requirements are 1.86 exajoules in 2025 and 2.62 exajoules in 2050.

h. It is assumed that natural gas used in advanced gas turbine cycle accounts for electricity requirements not provided by other sources, with efficiencies of 50 percent in 2025 and 55 percent in 2050. The corresponding natural gas requirements are 2.20 exajoules in 2025 and 4.14 exajoules in 2050.

i. From table 3, assuming 7.5 percent T&D losses in 2025, 5 percent losses in 2050.

Table 6: Fuel supply scenario for Africa
EJ per year

	1985	2025	2050
Solid fuels			
Coal	3.40	3.90[a]	3.70[a]
Cane residues for power	–	1.17[b]	1.58[b]
Other biomass for power	–	1.86[c]	2.62[c]
Subtotal	*3.40*	*6.93*	*7.90*
Liquid fuels			
Domestic oil	12.32	6.25[d]	3.63[d]
EthOH from sugarcane	–	0.10[e]	0.13[e]
MeOH from biomass	–	13.43[f]	21.53[f]
Imported oil	– 8.31	–	–
Imported MeOH	–	– 10.07[f]	– 14.57[f]
Subtotal	*4.01*	*9.71[g]*	*10.72[g]*
Gaseous fuels			
Domestic natural gas	2.08	4.45[h]	8.12[h]
Biogas at alcohol distilleries	–	0.02[i]	0.03[i]
Biogas from dung	–	1.97[j]	2.96[j]
Hydrogen from intermittent renewables	–	1.34[k]	2.07[k]
Imported hydrogen	–	– 1.34[k]	– 2.07[k]
Imported natural gas	– 0.89	– 1.34[l]	– 2.07[l]
Subtotal	*1.19*	*5.10[m]*	*9.04[m]*
CO_2 **emissions**[n] in *MtC*	179	261	243
CO_2 **emissions per capita** in *tC*	0.309	0.174	0.120

a. Direct coal use (table 4) plus coal for power generation (tables 2 and 5).

b. See table K and note f, table 5.

c. See note g, table 5.

d. See table E.

e. It is assumed that one third of sugarcane production (63.3 million tonnes per year in 2025 and 85.3 million tonnes per year in 2050) is dedicated to alcohol production, at a rate of 70 liters (1.60 gigajoules) per tonne of cane @ (22.8 megajoules per liter for 96 percent hydrated ethanol).

f. It is assumed that methanol is produced from cellulosic materials via thermochemical gasification to provide for (i) domestic liquid fuel needs not met by petroleum or ethanol from sugar cane plus (ii) export requirements. Large amounts of degraded lands could be available for biomass plantations in sub-Saharan Africa (see table L), making it possible for Africa to produce large quantities of methanol for export markets. It is assumed that Africa accounts for 58% of the world trade in methanol in 2025 and 66% in 2050 (see table Gb). Assuming conversion efficiencies of 61.9% in 2025 and 62.9% in 2050 (see table N), the biomass inputs required for methanol production are 20.81 exajoules in 2025 and 32.80 exajoules in 2050. The required biomass would be provided by a mix of forest biomass, cereals residues, and biomass from plantations (see tables I, K, and L).

g. Direct liquid fuel consumption for the IPCC Accelerated Policies Scenario (table 4) plus the methanol required for power generation (0.41 exajoules in 2025 and 0.62 exajoules in 2050—see table 5, note d).

h. It is assumed that African natural gas provides gaseous fuels export requirements (see note l) plus African gaseous fuel demand in excess of what can be provided by biogas (see notes i and j). According to the U.S. Geological Survey [5], remaining conventional supplies of natural gas (identified reserves plus estimated recoverable undiscovered resources) in Africa amount to 706 exajoules, as of the end of 1988. With the natural gas production schedule assumed here more than 3/5 of these gas resources would remain in 2050 (see table F).

i. It is assumed that biogas is recovered from stillage at cane alcohol distilleries at a rate of 0.33 gigajoules per tonne of cane.

j. It is assumed that biogas is produced from recoverable dung residues (see table K) at 90 percent of the stoichiometric rate (see chapter 19), corresponding to a 57 percent energy conversion efficiency.

k. While much land is available for photovoltaic hydrogen production in southern Europe, the cost of hydrogen delivered from production sites in the Sahara would be less costly (see chapter 22). Thus it is assumed that PV hydrogen demand in OECD Europe (half of gaseous fuel import requirements) is met by imports from North Africa.

l. It is assumed that natural gas from North Africa accounts for half of OECD European gas import requirements.

m. Direct gaseous fuel consumption for the IPCC Accelerated Policies Scenario (table 4) plus the natural gas required for power generation (2.20 exajoules in 2025 and 4.14 exajoules in 2050—see table 5, note h).

n. Assuming CO_2 emission rates of 25 MtC per exajoule for coal, 19.5 MtC per exajoule for petroleum, and 13.5 MtC per exajoule for natural gas.

RENEWABLES-INTENSIVE ENERGY SCENARIO FOR LATIN AMERICA

Table 1: Growth schedule for per capita GDP and population for Latin America, assuming IPCC growth rates

	GDP per capita *1989 U.S.$*	Population *millions*
1985	2,283	402
2000	3,450	530
2025	7,152	715
2050	15,298	829

Table 2: Baseline data for energy in Latin America, 1985

	Electricity[a] *TWh per year*	Primary energy[a] *EJ per year*	Direct fuel use[b] *EJ per year*	Fuel for electricity[c] *EJ per year*
Coal	16.4	0.91	0.73	0.18
Oil	65.6	10.07	9.33	0.74
Natural gas	82.0	2.91	1.99	0.92
Hydro	307.9	3.35	–	–
Nuclear	8.4	0.09	–	–
Totals	**480.3**	**17.33**	**12.05**	**1.84**

a. Data from[21].

b. Commercial fuels only.

c. It is assumed that the average efficiency of thermal power generation is 32 percent and that T&D losses amount to 10 percent.

Table 3: IPCC electricity consumption scenarios, Latin America
TWh per year

	High emissions	Accelerated policies
1985	432	432
2000	778	722
2025	2,194	1,891
2050	4,944	3,083

Table 4: IPCC direct fuel use scenarios (commercial fuels only) for Latin America
EJ per year

	High emissions				Accelerated policies			
	Solids	Liquids	Gases	Total fuel	Solids	Liquids	Gases	Total fuel
1985	0.73	9.33	1.99	12.05	0.73	9.33	1.99	12.05
2000	1.50	14.40	3.50	19.40	1.10	12.60	3.20	16.90
2025	5.20	26.00	10.10	41.30	1.60	21.20	12.60	35.40
2050	7.50	39.70	14.00	61.20	1.20	23.00	17.40	41.60

Table 5: Electricity supply scenario for Latin America
TWh per year

	1985	2025	2050
Hydro	307.9	505 [a]	702 [a]
Oil	65.6	– [b]	– [b]
Coal	16.4	– [b]	– [b]
Nuclear	8.4	33 [c]	33 [c]
Methanol	–	101 [d]	162 [d]
Intermittent renewables	–	201 [e]	325 [e]
Sugarcane	–	589 [f]	683 [f]
Biomass	–	409 [g]	649 [g]
Natural gas	82.0	206 [h]	691 [h]
Totals	**480.3**	**2,044 [i]**	**3,245 [i]**

a. It is assumed that, because of environmental constraints, hydropower is developed by 2050 to a level that is only one third of the economic potential, some 2105 TWh per year [4].

b. It is assumed that oil and coal are phased out.

c. As of 1989 there was 2,245 MW$_e$ of nuclear generating capacity in Latin America and an additional 3,804 MW$_e$ under construction. It is assumed that all this capacity is built and is operated at an average capacity factor of 62 percent and that no more nuclear capacity is added.

d. It is assumed that methanol used in chemically recuperated gas turbines accounts for 5 percent of electricity in 2025 and 2050, with efficiencies of 55 percent in 2025 and 60 percent in 2050. The corresponding methanol fuel requirements are 0.66 exajoules in 2025 and 0.97 exajoules in 2050.

e. It is assumed that intermittent renewables (PV, solar thermal, and wind) account for 10 percent of electricity in 2025 and 2050.

f. It is assumed that sugar cane production increases in proportion to the population, from 460 million tonnes in 1985, to 818 million tonnes in 2025, and to 949 million tonnes in 2050, and that byproduct electricity is produced from cane residues at a rate of 720 kWh per tonne of cane (tables K and N), using BIG/ISTIG technology or the equivalent (see chapter 17).

g. It assumed that 20 percent of electricity is provided by biomass in stand-alone plants @ 42.9 percent efficiency (BIG/ISTIG or equivalent) and 57 percent efficiency in 2050 (BIG/FC or equivalent) in 2050. The corresponding biomass requirements are 3.43 exajoules in 2025 and 4.10 exajoules in 2050.

h. It is assumed that natural gas used in advanced gas turbine cycle accounts for 5 percent of electricity in 2025 and 2050, with efficiencies of 50 percent in 2025 and 55 percent in 2050. The corresponding natural gas requirements are 1.48 exajoules in 2025 and 4.52 exajoules in 2050.

i. From table 3, assuming 7.5 percent T&D losses in 2025, 5 percent losses in 2050.

Table 6: Fuel supply scenario for Latin America
EJ per year

	1985	2025	2050
Solid fuels			
Coal	0.91	1.60[a]	1.20[a]
Cane residues for power	–	5.33[b]	6.1˙9[b]
Other biomass for power	–	3.43[c]	4.10[c]
Subtotal	*0.91*	*10.36*	*11.49*
Liquid fuels			
Domestic oil	15.30	9.28[d]	6.38[d]
EthOH from sugarcane	–	0.44[e]	0.50[e]
MeOH from biomass	–	17.36[f]	22.30[f]
Imported oil	– 5.23	–	–
Imported MeOH	–	– 5.21[f]	– 5.21[f]
Subtotal	*10.07*	*21.87[g]*	*23.97[g]*
Gaseous fuels			
Domestic natural gas	2.91	8.83[h]	8.83[h]
Biogas at alcohol distilleries	–	0.09[i]	0.10[i]
Biogas from dung	–	1.84[j]	2.38[j]
H_2 from biomass	–	3.32[k]	10.61[k]
Subtotal	*2.91*	*14.08[l]*	*21.92[l]*
CO_2 emissions[m] in *MtC*	258	340	274
CO_2 emissions per capita in *tC*	0.64	0.48	0.33

a. Direct coal use (see table 4) plus coal for power generation (see table 5).

b. See table K and note f, table 5.

c. See note g, table 5.

d. See table E.

e. It is assumed that one third of sugarcane production (273 million tonnes per year in 2025 and 316 million tonnes per year in 2050) is dedicated to alcohol production, at a rate of 70 liters (1.60 gigajoules) per tonne of cane.

f. It is assumed that methanol is produced from cellulosic materials via thermochemical gasification to provide for (i) domestic liquid fuel needs not met by petroleum or ethanol from sugar cane plus (ii) export requirements. Large amounts of degraded lands could be available for biomass plantations in Latin America (see table L), making it possible for Latin America to produce large quantities of methanol for export markets. It is assumed that Latin America accounts for 32 percent of the world trade in methanol in 2025 and 25 percent in 2050 (see table Gb). Assuming conversion efficiencies of 61.9 percent in 2025 and 62.9 percent in 2050 (see table N) the biomass inputs required for methanol production are 28.04 exajoules in 2025 and 35.45 exajoules in 2050. The required biomass would be provided by a mix of forest biomass, cereals residues, and biomass from plantations (see tables I, K, and L).

g. Direct liquid fuel consumption for the IPCC Accelerated Policies Scenario (table 4) plus the methanol required for power generation (0.66 exajoules in 2025 and 0.97 exajoules in 2050—note d, table 5).

h. Natural gas production is assumed to rise linearly until 2025 and then remain constant, with the level chosen such that one third of estimated gas resources remain by 2050 (see table F).

i. It is assumed that biogas is recovered from stillage at cane alcohol distilleries at a rate of 0.33 gigajoules per tonne of cane.

j. It is assumed that biogas is produced from recoverable dung residues (see table K) at 90 percent of the stoichiometric rate (see chapter 19), corresponding to a 57 percent energy conversion efficiency.

k. It is assumed that gaseous fuel requirements in excess of what can be provided by domestic natural gas and biogas are provided by hydrogen derived from biomass. Hydrogen can be produced from biomass at a rate of 0.771 exajoules per exajoule of biomass feedstock, along with 7.07 TWh of byproduct electricity per exajoule of produced hydrogen [20]. Thus some 6.63 exajoules of biomass is needed for hydrogen production in 2025 and 12.57 exajoules in 2050.

l. Direct gaseous fuel consumption for the IPCC Accelerated Policies Scenario (table 4) plus the natural gas required for power generation (1.48 exajoules in 2025 and 4.52 exajoules in 2050—note h, table 5).

m. Assuming CO_2 emission rates of 25 MtC per exajoule for coal, 19.5 MtC per exajoule for petroleum, and 13.5 MtC per exajoule for natural gas.

RENEWABLES-INTENSIVE ENERGY SCENARIO FOR SOUTH AND EAST ASIA

Table 1: Growth schedule for per capita GDP and population for South and East Asia, assuming IPCC growth rates

	GDP per capita 1989 U.S.$	Population millions
1985	581	1,417
2000	962	1,858
2025	2,276	2,534
2050	5,512	2,998

Table 2: Baseline data for energy in South and East Asia, 1985

	Electricity[a] TWh per year	Primary energy[a] EJ per year	Direct fuel use[b] EJ per year	Fuel for electricity[c] EJ per year
Coal	185.1	5.38	3.16	2.22
Oil	82.3	8.08	7.10	0.98
Natural gas	30.8	1.73	1.36	0.37
Hydro	101.7	1.11	–	–
Nuclear	48.5	0.58	–	–
Totals	**448.4**	**16.88**	**11.62**	**3.57**

a. Data from [21].

b. Commercial fuels only.

c. The IPCC report estimates that the ratio of electricity consumed to fuel input at power stations is 26.1 percent. Here it is assumed that the average efficiency of power generation is 30 percent. Thus, T&D losses amount to 13 percent.

Table 3: IPCC electricity consumption scenarios, South and East Asia
TWh per year

	High emissions	Accelerated policies
1985	390	390
2000	778	806
2025	2,778	2,722
2050	7,194	5,972

Table 4: IPCC direct fuel use scenarios (commercial fuels only) for South and East Asia
EJ per year

	High emissions				Accelerated policies			
	Solids	Liquids	Gases	Total fuel	Solids	Liquids	Gases	Total fuel
1985	3.16	7.10	1.36	11.62	3.16	7.10	1.36	11.62
2000	5.90	8.90	1.90	16.70	5.30	9.20	1.90	16.40
2025	17.30	13.40	4.60	35.30	9.40	17.00	9.10	35.50
2050	29.90	26.20	8.10	64.20	9.60	24.70	15.80	50.10

Table 5: Electricity supply scenario for South and East Asia
TWh per year

	1985	2025	2050
Hydro	101.7	413[a]	500[a]
Oil	82.3	– [b]	– [b]
Nuclear	48.5	104[c]	104[c]
Natural gas	30.8	558[d]	1,886[d]
Sugarcane	–	354[e]	419[e]
Intermittent renewables	–	736[f]	2,200[f]
Coal	185.1	778[g]	1,177[g]
Totals	**448.4**	**2,943**[h]	**6,286**[h]

a. It has been estimated that the economic potential for hydropower in South and East Asia is 803 TWh per year [4]. It is assumed that by 2050 five eights of this potential is developed.

b. It is assumed that oil power is phased out.

c. As of 1989 there was 13,643 MW$_e$ of nuclear generating capacity in South and East Asia and an additional 5,590 MW$_e$ under construction. It is assumed that all this capacity is built and is operated at an average capacity factor of 62 percent and that no more nuclear capacity is added.

d. It is assumed that natural gas used in advanced gas turbine cycles accounts for 20 percent of electricity (with an efficiency of 50 percent) in 2025 and 30 percent in 2050 (with an efficiency of 55 percent). The corresponding natural gas requirements are 4.02 exajoules in 2025 and 12.34 exajoules in 2050.

e. It is assumed that sugar cane production increases in proportion to the population, from 275 million tonnes in 1985 to 492 million tonnes in 2025 and to 582 million tonnes in 2050, and that byproduct electricity is produced from cane residues at a rate of 720 kWh per tonne of cane (tables K and N), using BIG/ISTIG technology or the equivalent (see chapter 17).

f. It is assumed that intermittent renewables (PV, solar thermal, and wind) account for 25 percent of electricity in 2025 and 35 percent in 2050.

g. It is assumed that coal accounts for electricity not provided by other sources, with an average efficiency of 42.1 percent (CIG/ISTIG technology) in 2025 and 57 percent (advanced CIG/FC technology) by 2050. The corresponding coal requirements are 6.65 exajoules in 2025 and 7.43 exajoules in 2050.

h. From table 3, assuming 7.5 percent T&D losses in 2025, 5 percent losses in 2050.

Table 6: Fuel supply scenario for South and East Asia
EJ per year

	1985	2025	2050
Solid fuels			
Coal	5.38	16.05[a]	17.03[a]
Cane residues for power	–	3.20[b]	3.79[b]
Subtotal	*5.38*	*19.25*	*20.82*
Fluid fuels			
Domestic oil	6.08	2.39[c]	1.15[c]
Domestic natural gas	3.14	7.49[d]	7.49[d]
Ethanol from sugarcane	–	0.26[e]	0.31[e]
H$_2$ **from biomass**	–	5.17[f]	5.88[f]
Biogas at alcohol distilleries	–	0.05[g]	0.06[g]
Biogas from dung	–	3.49[h]	6.57[h]
H$_2$ **from intermittent** renewables	–	0.50[i]	1.00[i]
Imported H$_2$	–	0.50[i]	1.00[i]
Imported oil	2.00	3.43[j]	9.80[j]
Imported alcohols	–	3.42[j]	9.79[j]
Imported natural gas	– 1.41	3.42[j]	9.79[j]
Subtotal	*9.81*	*30.12[k]*	*52.84[k]*
CO$_2$ **emissions**[l] in *MtC*	315	659	873
CO$_2$ **emissions per capita** in *tC*	0.22	0.26	0.29

a. Direct coal use (see table 4) plus coal for power generation (see table 5).

b. See table K and note e, table 5.

c. See table E.

d. Domestic natural gas production is assumed to rise at a linear rate until 2025 and then remain constant. The level is chosen such that one third of estimated gas resources remain by 2050 (see table F).

e. It is assumed that one third of sugarcane production (164 million tonnes per year in 2025 and 194 million tonnes per year in 2050) is dedicated to alcohol production, at a rate of 70 liters (1.60 gigajoules) per tonne of cane (@ 22.8 megajoules per liter for 96 percent hydrated ethanol).

f. It is assumed that biomass supplies other than cane residues and dung (7.42 exajoules in 2025 and 8.23 exajoules in 2050—see tables I and K) are used to produce hydrogen. It is assumed that hydrogen is produced at efficiencies of 69.7 percent in 2025 and 71.5 percent in 2050 (see table N).

g. It is assumed that biogas is recovered from stillage at cane alcohol distilleries at a rate of 0.33 gigajoules per tonne of cane.

h. It is assumed that biogas is produced from recoverable dung residues (see table K) at 90 percent of the stoichiometric rate (see chapter 19), corresponding to a 57 percent energy conversion efficiency.

i. It is assumed that hydrogen from intermittent sources (wind, PV) provides 1.0 exajoules of energy in 2025 and 2.0 exajoules in 2050—half of which is imported from the Middle East. Hydrogen produced in the Middle East and delivered to markets in South and East Asia could often be cheaper than domestically produced hydrogen, because of the good insolation in the Middle East (see chapter 22).

j. It is assumed that fluid fuel requirements in excess of what can be provided by domestic sources is provided by equal amounts of oil, natural gas, and alcohol imports.

k. Direct liquid plus gaseous fuel consumption for the IPCC Accelerated Policies Scenario (table 4).

l. Assuming CO$_2$ emission rates of 25 MtC per exajoule for coal, 19.5 MtC per exajoule for petroleum, and 13.5 MtC per exajoule for natural gas.

RENEWABLES-INTENSIVE ENERGY SCENARIO FOR CENTRALLY PLANNED ASIA

Table 1: GDP and population for Centrally Planned Asia, for IPCC growth rates

	GDP per capita 1989 U.S.$	Population millions
1985	337.1	1,140
2000	609.1	1,408
2025	1,681.3	1,728
2050	4,679.3	1,866

Table 2: Baseline data for energy in Centrally Planned Asia, 1985

	Electricity[a] TWh per year	Primary energy[a] EJ per year	Direct fuel use[b] EJ per year	Fuel for electricity[c] EJ per year
Coal	267.3	19.71	16.46	3.25
Oil	63.9	4.22	3.43	0.79
Natural gas	5.8	0.51	0.43	0.08
Hydro	120.1	1.31	–	–
Nuclear	–	–	–	–
Totals	**457.1**	**25.75**	**20.32**	**4.12**

a. Data from [21].

b. Commercial fuels only.

c. The average heat rate for power generation in 1985 was 11.66 megajoules per kWh (LHV) [22], which becomes 12.15 megajoules per kWh (HHV) for coal (29.6 percent), 12.40 megajoules per kWh (HHV) for oil (29.0 percent), and 12.96 megajoules per kWh (HHV) for natural gas (27.8 percent). It is assumed that T&D losses are 15 percent.

Table 3: IPCC electricity consumption scenarios, Centrally Planned Asia
TWh per year

	High emissions	Accelerated policies
1985	389	389
2000	944	1,111
2025	3,306	3,500
2050	8,000	6,000

Table 4: IPCC direct fuel use scenarios (commercial fuels only) for Centrally Planned Asia
EJ per year

	High emissions				Accelerated policies			
	Solids	Liquids	Gases	Total fuel	Solids	Liquids	Gases	Total fuel
1985	16.46	3.43	0.43	20.32	16.46	3.43	0.43	20.32
2000	18.70	4.10	0.30	23.10	22.60	3.30	0.30	26.20
2025	40.70	7.70	0.50	48.90	40.80	7.50	1.90	50.20
2050	63.10	16.00	0.80	79.90	31.20	8.90	2.70	42.80

Table 5: Electricity supply scenario for Centrally Planned Asia
TWh per year

	1985	2025	2050
Hydro	120.1	650[a]	911[a]
Nuclear	–	89[b]	89[b]
Oil	63.9	–[c]	–[c]
Sugarcane	–	36[d]	39[d]
Biomass	–	378[e]	632[e]
Intermittent renewables	–	946[f]	2,211[f]
Natural gas	5.8	946[g]	1,579[g]
Coal	267.3	739[h]	855[h]
Totals	**457.1**	**3,784[i]**	**6,316[i]**

a. The economic potential for hydropower in Centrally Planned Asia has been estimated to be 1,546 TWh per year [4]. It is assumed that 3/5 of this potential is developed by 2050.

b. Some 3.9 GW_e of nuclear power is planned for the year 2000. It is assumed that this increases fourfold by 2025 and then levels off.

c. It is assumed that oil-based power generation is phased out.

d. In China, sugarcane residues are often used for making paper. Accordingly, it is assumed that only half of sugarcane production (50 and 54 million tonnes in 2025 and 2050, respectively) is associated with byproduct electricity production, at a rate of 720 kWh/tc in excess of onsite needs, using BIG/ISTIG technology or the equivalent (see chapter 17).

e. It is assumed that 10 percent of electricity is provided by biomass in stand-alone power plants @ 42.9 percent efficiency (BIG/ISTIG or equivalent) in 2025 and 57 percent efficiency (BIG/FC or equivalent) in 2050. The corresponding biomass requirements are 3.18 exajoules in 2025 and 4.00 exajoules in 2050.

f. It is assumed that intermittent renewables provide 25 percent of power generation in 2025 and 35 percent in 2050.

g. It is assumed that 25 percent of electricity requirements are provided by advanced gas turbine cycles fired with natural gas (at efficiencies of 50 percent and 55 percent in 2025 and 2050, respectively). Thus natural gas requirements for power generation are 6.81 exajoules in 2025 and 10.34 exajoules in 2050.

h. It is assumed that coal provides electricity not provided by other sources, at efficiencies of 42.1 percent (CIG/ISTIG or equivalent) in 2025 and 57 percent (CIG/FC or the equivalent) in 2050. The coal requirements for power generation are thus 6.32 exajoules in 2025 and 5.40 exajoules in 2050.

i. From table 3, assuming 7.5 percent T&D losses in 2025 and 5 percent losses in 2050.

Table 6: Fuel supply scenario for Centrally Planned Asia
EJ per year

	1985	2025	2050
Solid fuels			
Coal for direct fuel	16.46	37.89[a]	21.29[a]
Coal for power	3.25	6.32[b]	5.40[b]
Biomass for direct fuel	–	2.91[a]	9.91[a]
Cane residues for power	–	0.32[c]	0.35[c]
Other biomass for power	–	3.18[d]	4.00[d]
Subtotal	*19.71*	*50.62[a]*	*40.95[a]*
Liquid fuels			
Domestic oil	5.67	4.93[f]	4.18[f]
EthOH from sugarcane	–	0.05[g]	0.06[g]
MeOH from biomass	–	1.50[h]	2.97[h]
Oil imports	– 1.45	0.51[i]	0.85[i]
MeOH imports	–	0.51[i]	0.84[i]
Subtotal	*4.22*	*7.50[e]*	*8.90[e]*
Gaseous fuels			
Domestic natural gas	0.51	4.61[k]	4.61[k]
Biogas	–	0.70[l]	0.79[l]
H_2 **from intermittent renewables**	–	0.50[m]	1.00[m]
Imported natural gas	–	2.90[n]	6.64[n]
Subtotal	*0.51*	*8.71[j]*	*13.04[j]*
CO_2 emissions[o] in *MtC*	582	1,313	917
CO_2 emissions per capita in *tC*	0.51	0.76	0.49

a. Direct solid fuel consumption for the IPCC Accelerated Policies Scenario (see table 4), with the biomass share constrained by total estimated biomass supplies (see tables I, K, L, and M) and biomass requirements for electricity and fluid fuels production (see note e, table 5, and note h, below).

b. See note h, table 5.

c. See note e, table 5 and table K.

d. See note e, table 5.

e. Direct liquid fuel consumption for the IPCC Accelerated Policies Scenario (see table 4).

f. See table E.

g. It is assumed that one third of sugar cane (33 million tonnes in 2025 and 36 million tonnes in 2050) is used to produce alcohol, at a rate of 70 liters (1.60 gigajoules) per tonne of cane (@ 22.8 megajoules per liter for 96 percent hydrated ethanol).

h. It is assumed that MeOH from biomass accounts for one fifth of liquid fuel requirements in 2025 and one third in 2050. Assuming conversion efficiencies of 61.9 percent in 2025 and 62.9 percent in 2050 (see table N) the biomass inputs required for methanol production are 2.42 exajoules in 2025 and 4.72 exajoules in 2050. The required biomass would be provided by a mix of forest biomass, cereals residues, and biomass from plantations (see tables I, K, and L).

i. It is assumed that liquid fuel requirements not met by domestic oil, ethanol, and methanol are provided by imports of equal quantities of oil and biomass-derived methanol.

j. Direct gaseous fuel consumption for the IPCC Accelerated Policies Scenario (see table 4) plus natural gas requirements for power generation (see note g, table 5).

k. It is assumed that gas production rises linearly until 2025 to a level that is maintained until 2050, with the level chosen such that in 2050 one third of the assumed gas resources (315 exajoules) remain (see table F). This may be an underestimate of natural gas resources in China. According to China's national oil and gas resources assessment, completed in 1987, total natural gas resources on land and on the continental shelf are about 1,300 exajoules [22].

l. It is assumed that biogas is produced from recoverable dung residues (see table K) at 90 percent of the stoichiometric rate (see chapter 19), corresponding to a 57 percent energy conversion efficiency.

m. It is assumed that electrolytic hydrogen derived from intermittent sources—derived mainly from wind power in especially windy areas of China—provides 0.5 exajoules of energy in 2025 and 1.0 exajoules in 2050.

n. It is assumed that gas requirements not provided by domestic natural gas, biogas, and electrolytic hydrogen are provided by imported gas (probably Siberian gas from the former Soviet Union).

o. Assuming CO_2 emission rates of 25 MtC per exajoule for coal, 19.5 MtC per exajoule for petroleum, and 13.5 MtC per exajoule for natural gas.

RENEWABLES-INTENSIVE ENERGY SCENARIO FOR JAPAN

Table 1: Growth schedule for per capita GDP and population for Japan, assuming IPCC growth rates

	GDP per capita *1989 U.S.$*	Population *millions*
1985	12,506	121
2000	19,095	132
2025	33,695	139
2050	55,614	138

Table 2: Baseline data for energy in Japan, 1985

	Electricity[a] *TWh per year*	Primary energy[a] *EJ per year*	Direct fuel use *EJ per year*	Fuel for electricity[b] *EJ per year*
Coal	94.4	3.18	2.27	0.91
Oil	183.2	8.98	7.23	1.75
Natural gas	120.8	1.63	0.41	1.22
Hydro	76.5	0.85	–	–
Nuclear	150.3	1.66	–	–
Geothermal	1.4	0.02	–	–
Totals	**626.6**	**16.32**	**9.91**	**3.88**

a. International Energy Agency data [23]. IEA data (in LHV) converted to HHV by dividing by 0.96 for coal, 0.94 for oil, and 0.90 for natural gas.

b. T&D losses averaged 4.584 percent of net generation.

Table 3: IPCC electricity consumption projections for Japan
TWh per year

	High emissions	Accelerated policies
1985	597.9	597.9
2000	737	599
2025	1,082	599
2050	1,497	760

Table 4: IPCC secondary commercial fuel use projections for Japan
EJ per year

	High emissions				**Accelerated policies**			
	Solids	Liquids	Gases	Total fuel	Solids	Liquids	Gases	Total fuel
1985	2.27	7.23	0.41	9.91	2.27	7.23	0.41	9.91
2000	1.96	7.10	0.45	9.51	2.37	6.68	0.59	9.64
2025	2.37	5.75	0.45	8.57	1.71	5.75	0.91	8.37
2050	2.77	7.27	0.41	10.45	1.14	4.73	0.95	6.82

Table 5: Electricity supply scenario for Japan
TWh per year

	1985	2025	2050
Hydro	76.5	77 [a]	77 [a]
Nuclear	150.3	300 [b]	300 [b]
Geothermal	1.4	12 [c]	18 [c]
Coal	94.4	– [d]	– [d]
Oil	183.2	– [d]	– [d]
Intermittent renewables	–	126 [e]	240 [e]
Natural gas	120.8	58 [f]	83 [f]
MeOH	–	57 [f]	82 [f]
Totals	**626.6**	**630 [g]**	**800 [g]**

a. The economic potential for hydropower in Japan has been estimated to be 67 TWh per year [4]. As this level is already exceeded, it is assumed that there is no further expansion of hydropower in Japan.

b. It is assumed that the nuclear generating capacity under construction is completed but that no more nuclear power is built.

c. In 1985 geothermal capacity totaled 215 MW_e; it is expected to increase to 457 MW_e (see chapter 13: *Geothermal Energy*). Here it is assumed that geothermal capacity increases further, to 2,000 MW_e by 2025 and to 3,000 MW_e by 2050 and that this capacity is operated at an average capacity factor of 67 percent.

d. It is assumed that coal- and oil-based power generation are phased out.

e. It is assumed that intermittent renewables (mainly PV) provide 20 percent of power generation in 2025 and 30 percent in 2050.

f. It is assumed that remaining electricity requirements are provided by advanced gas turbine cycles, with equal shares from natural gas (at efficiencies of 50 percent and 55 percent in 2025 and 2050, respectively) and by methanol in chemically recuperated gas turbine cycles (at efficiencies of 55 percent and 60 percent in 2025 and 2050, respectively). Thus natural gas requirements for power generation are 0.42 exajoules in 2025 and 0.54 exajoules in 2050, while methanol fuel requirements are 0.37 exajoules in 2025 and 0.49 exajoules in 2050.

g. From table 3, assuming 5 percent T&D losses.

Table 6: Fuels consumption scenario for Japan
EJ per year

	1985	2025	2050
Solid fuels			
Coal for direct use	2.27	1.71[a]	1.14[a]
Coal for power	0.91	– [b]	– [b]
Subtotal	*3.18*	*1.71*	*1.14*
Liquid fuels			
Domestic oil	0.03	–	–
MeOH from biomass	–	0.55[c]	0.60[c]
Imported MeOH	–	2.79[d]	2.32[d]
Imported oil	8.95	2.78[d]	2.32[d]
Subtotal	*8.98*	*6.12[e]*	*5.24[e]*
Gaseous fuels			
Domestic natural gas	0.09	–	–
H_2 from intermittent renewables	–	0.66[f]	0.74[f]
Imported natural gas	1.54	0.67[f]	0.75[f]
Subtotal	*1.63*	*1.33[g]*	*1.49[g]*
CO_2 emissions[h] in *MtC*	277	106	84
CO_2 emissions per capita in *tC*	2.29	0.76	0.61

a. Direct solid fuel consumption for the IPCC Accelerated Policies Scenario (table 4) plus coal for power generation (table 5).

b. See table 2 and note d, table 5.

c. It is assumed that biomass supplies (a mixture of forest biomass, cereals residues, and urban refuse—see tables I and K) are used to produce methanol at conversion efficiencies of 61.5 percent in 2025 and 62.9 percent in 2050 (see table N).

d. It is assumed that liquid fuel imports are a 50/50 mix of oil and methanol (the latter derived from biomass).

e. Direct liquid fuel consumption for the IPCC Accelerated Policies Scenario (table 4) plus methanol requirements for power generation, some 0.37 exajoules in 2025 and 0.49 exajoules in 2050 (see note f, table 5).

f. It is assumed that gaseous fuel requirements are provided by a 50/50 mix of imported natural gas and hydrogen produced domestically from intermittent renewable electric sources (mainly PV).

g. Direct gaseous fuel consumption for the IPCC Accelerated Policies Scenario (table 4) plus natural gas requirements for power generation, some 0.42 exajoules in 2025 and 0.54 exajoules in 2050 (see note f, table 5).

h. Assuming CO_2 emission rates of 25 MtC per exajoule for coal, 19.5 MtC per exajoule for petroleum, and 13.5 MtC per exajoule for natural gas.

RENEWABLES-INTENSIVE ENERGY SCENARIO FOR AUSTRALIA/NEW ZEALAND

Table 1: GDP and population for Australia/New Zealand, for IPCC growth rates

	GDP per capita 1989 U.S.$	Population millions
1985	10,817	19.1
2000	16,516	20.9
2025	29,144	22.0
2050	48,103	21.8

Table 2: Baseline data for energy in Australia/New Zealand, 1985[a,b]

	Electricity TWh per year	Primary energy EJ per year	Direct fuel use EJ per year	Fuel for electricity EJ per year
Coal	82.70	1.361	0.323	1.038
Other solids	0.42	0.195	0.189	0.006
Oil	4.37	1.373	1.331	0.041
Natural gas	16.07	0.668	0.495	0.173
Hydro	31.98	0.349	–	–
Nuclear	–	–	–	–
Geothermal	1.12	0.012	–	–
Totals	**136.66**	**3.958**	**2.338**	**1.258**

a. International Energy Agency data [23]. IEA data (in LHV) converted to HHV by dividing by 0.96 for coal, 0.95 for other solids, 0.94 for oil, and 0.90 for natural gas.

b. T&D losses averaged 9.694 percent of net generation.

Table 3: IPCC electricity consumption projections for Australia/New Zealand
TWh per year

	High emissions	Accelerated policies
1985	123.4	123.4
2000	152	124
2025	224	124
2050	309	157

Table 4: IPCC secondary fuel use projections for Australia/New Zealand
EJ per year

	High emissions				Accelerated policies			
	Solids	Liquids	Gases	Total fuel	Solids	Liquids	Gases	Total fuel
1985	0.512	1.331	0.495	2.338	0.512	1.331	0.495	2.338
2000	0.440	1.300	0.550	2.290	0.630	1.220	0.710	2.560
2025	0.530	1.050	0.550	2.130	0.390	1.050	1.090	2.530
2050	0.630	1.330	0.490	2.450	0.260	0.870	1.150	2.280

Table 5: Electricity supply scenario for Australia/New Zealand
TWh per year

	1985	2025	2050
Hydro	32.0	40[b]	40[b]
Nuclear	–	–	–
Geothermal	1.1	2[c]	2[c]
Coal	82.7	–[d]	–[d]
Other solids	0.4	–[d]	–[d]
Oil	4.4	–[d]	–[d]
Sugarcane	–	20[e]	20[e]
Intermittent renewables	–	33[f]	58[f]
Natural gas	16.1	35[g]	45[g]
Totals	**136.7**	**130**[a]	**165**[a]

a. From table 3, assuming 5 percent T&D losses.

b. The economic potential for hydropower in Australia/New Zealand has been estimated to be 44 TWh per year [4]. Here it is assumed that 90 percent of this potential is developed.

c. Installed geothermal capacity in 1985 was 167 MW$_e$, and some 342 MW$_e$ is projected for 1995 (see chapter 13). The latter level is assumed for the years 2025 and 2050.

d. It is assumed that coal- and oil-based power generation are phased out.

e. It is assumed that cane production increases in proportion to the population, from 24.4 million tonnes in 1985 to 27.8 million tonnes in 2025 and to 28.1 million tonnes in 2050, and that byproduct electricity is produced from cane residues at a rate of 720 kWh per tonne of cane (tables K and N) using BIG/ISTIG technology or equivalent (see chapter 17).

f. It is assumed that intermittent renewables provide 25 percent of power generation in 2025 and 35 percent in 2050.

g. It is assumed that remaining electricity requirements are provided by advanced gas turbines cycles fired with natural gas (at efficiencies of 50 percent and 55 percent in 2025 and 2050, respectively). Thus natural gas requirements for power generation are 0.25 exajoules in 2025 and 0.29 exajoules in 2050.

Table 6: Fuels consumption scenario for Australia/New Zealand
EJ per year

	1985	2025	2050
Solid fuels			
Coal	1.361	0.390[a]	0.260[a]
Other solids	0.195	–	–
Cane residues for power	–	0.180[b]	0.180[b]
Subtotal	*1.556*	*0.570*	*0.440*
Liquid fuels			
Domestic oil	1.293	0.470[d]	0.250[d]
EthOH from sugarcane	–	0.015[e]	0.015[e]
MeOH from biomass	–	0.313[f]	0.383[f]
Imported MeOH	–	0.126[g]	0.111[g]
Imported oil	0.080	0.126[g]	0.111[g]
Subtotal	*1.373*	*1.050[c]*	*0.870[c]*
Gaseous fuels			
Biogas	–	0.271[i]	0.353[i]
Domestic natural gas	0.668	1.069[j]	1.087[j]
Subtotal	*0.668*	*1.340[h]*	*1.440[h]*
CO_2 emissions[k] in *MtC*	72	36	28
CO_2 emissions per capita in *tC*	3.78	1.63	1.29

a. Direct solid fuel consumption for the IPCC Accelerated Policies Scenario (table 4) plus coal for power generation (table 5).

b. See table K and note e, table 5.

c. Direct liquid fuel consumption for the IPCC Accelerated Policies Scenario (table 4).

d. See table E.

e. It is assumed that one third of sugarcane production (9.4 million tonnes per year in 2025 and 2050) is dedicated to alcohol production, at a rate of 70 liters (1.60 gigajoules) per tonne of cane (@ 22.8 megajoules per liter for 96 percent hydrated ethanol).

f. It is assumed that biomass supplies (other than dung and cane residues—tables I and K) are used to produce MeOH, at conversion efficiencies of 61.9 percent in 2025 and 62.9 percent in 2050 (table N).

g. It is assumed that the remaining requirements for liquid fuels are made up of equal contributions from imported oil and imported MeOH (the latter produced from biomass).

h. Direct gaseous fuel consumption for the IPCC Accelerated Policies Scenario (table 4) plus natural gas requirements for power generation, some 0.25 exajoules in 2025 and 0.29 exajoules in 2050.

i. It is assumed that biogas is produced from recoverable dung residues (see table K) at 90 percent of the stoichiometric rate (see chapter 19), corresponding to a 57 percent energy conversion efficiency.

j. It is assumed that gaseous fuel requirements not provided by biogas are provided by natural gas. With the indicated gas production schedule, in the year 2050 about three fifth of gas resources remain to be exploited (see table F).

k. Assuming CO_2 emission rates of 25 MtC per exajoule for coal, 19.5 MtC per exajoule for petroleum, and 13.5 MtC per exajoule for natural gas.

RENEWABLES-INTENSIVE ENERGY SCENARIO FOR THE UNITED STATES

Table 1: Growth schedule for per capita GDP and population for the U.S., assuming IPCC growth rates

	GDP per capita *1989 U.S.$*	Population *millions*
1985	18,287	239
2000	25,503	267
2025	43,681	289
2050	72,670	285

Table 2: Baseline data for energy in the U.S., 1985[a,b]

	Electricity *TWh per year*	Primary energy *EJ per year*	Direct fuel use *EJ per year*	Fuel for electricity *EJ per year*
Coal	1,387.8	18.57	3.14	15.43
Other solids	1.4	2.95	2.93	0.02
Oil	99.2	32.28	31.16	1.12
Natural gas	289.0	19.19	15.88	3.31
Hydro	262.7	2.87	–	–
Nuclear	376.2	4.10	–	–
Geothermal	9.2	0.10	–	–
Imports	40.9	0.45	–	–
Totals	**2,466.4**	**80.51**	**53.11**	**19.88**

a. International Energy Agency data [23]. IEA data (in LHV) converted to HHV by dividing by 0.96 for coal, 0.95 for other solids, 0.94 for oil, and 0.90 for natural gas.

b. T&D losses averaged 7.506 percent of net generation.

Table 3: IPCC electricity consumption scenarios for the U.S.
TWh per year

	High emissions	Accelerated policies
1985	2,281	2,281
2000	3,444	2,722
2025	5,556	2,889
2050	6,500	3,194

Table 4: IPCC direct fuel use scenarios for the U.S.
EJ per year

	High Emissions				Accelerated policies			
	Solids	Liquids	Gases	Total fuel	Solids	Liquids	Gases	Total fuel
1985	6.07	31.16	15.88	53.11	6.07	31.16	15.88	53.11
2000	5.8	32.1	16.0	53.9	4.3	24.7	13.3	42.3
2025	8.1	25.2	17.9	51.2	2.1	19.2	17.8	39.1
2050	9.0	26.8	16.7	52.5	1.2	13.5	16.7	31.4

Table 5: Electricity supply scenario for the U.S.
TWh per year

	1985	2025	2050
Hydro	262.7	335[b]	345[b]
Nuclear	376.2	565[c]	565[c]
Coal	1,387.8	125[d]	–[d]
Other solid fuel	1.4	–	–
Oil	99.2	–[e]	–[e]
Geothermal	9.2	171[f]	180[f]
Biomass	–	456[g]	504[g]
Intermittent renewables	–	760[h]	1,177[h]
Imports	40.9	131[i]	167[i]
Natural gas	289.0	498[j]	424[j]
Totals	**2,466.4**	**3,041[a]**	**3,362[a]**

a. From table 3, assuming 5 percent T&D losses.

b. The economic potential for hydropower in the U.S. has been estimated to be 638 TWh per year [4]. However, owing to environmental constraints, it is assumed here that only about half of this potential is developed before 2050. The projections to the years 2025 and 2050 are the same as in the U.S. National Energy Strategy for the years 2025 and 2030, respectively [3] the increase over present hydro output is due mainly to efficiency improvements at existing sites.

c. As of 1989 nuclear installed capacity in the U.S. was 98 GW_e, and capacity will increase to 104 GW_e when plants under construction are completed. It is assumed that capacity remains at this level in the period 2025–2050 and that this capacity is operated at 62 percent average capacity factor.

d. It is assumed that by 2025 the only coal-fired power plants operating are the 19 GW_e of capacity announced as of 12/31/89 for utility addition in the period 1990–2010 [24]. These plants are assumed to have an efficiency of 35 percent and to operate at 75 percent capacity factor. It is assumed that coal power is phased out by 2050.

e. It is assumed that oil-based power generation is phased out.

f. The geothermal projections to the years 2025 and 2050 made here are the same as in the U.S. National Energy Strategy for the years 2025 and 2030, respectively [3].

g. It is assumed that biomass provides 15 percent of electricity 2025–2050. For 2025 it is assumed this biomass is used at an average efficiency of 42.9 percent (BIG/ISTIG or the equivalent); for 2050, at an average efficiency of 57 percent (BIG/FC or the equivalent); the corresponding biomass requirements are 3.83 exajoules in 2025 and 3.18 exajoules in 2050.

h. It is assumed that intermittent renewables account for 25 percent of generation in 2025 and 35 percent in 2050. For comparison, in the U.S. Department of Energy's Interlaboratory White Paper on Renewable Energy [25], intermittent renewable electricity production is projected to increase to the following levels (in TWh per year) for the Business-As-Usual (BAU) and Research, Development, and Demonstration Intensification (RD&D) scenarios:

	2010		2020		2030	
	BAU	RD&D	BAU	RD&D	BAU	RD&D
Wind	98	221	201	547	318	1,026
Solar thermal-electric	28	97	114	300	292	870
Photovoltaic	15	65	72	184	275	641
Total	**141**	**383**	**387**	**1,031**	**885**	**2,537**

The projection for intermittent renewables in 2025 in the present analysis is only about 20 percent higher than in the BAU scenario and much lower than for the RD&D Scenario. Because of the emphasis given to efficient use of electricity in the IPCC Accelerated Policies Scenario constructed here, less intermittent renewables development is needed.

i. With large potential wind and hydroelectric resources and having only 1/10 the population of the United States, Canada has the potential to export large quantities of renewable electricity to the United States. These levels of imports correspond to about 1/4 of total Canadian electricity production. The projected increase in imports is consistent with ongoing trends. The U.S. Department of Energy projects that Canadian imports will increase to 68.2 TWh per year by 2010 [24].

j. It is assumed that electricity requirements not provided by other means are provided by advanced natural gas-based power generating systems, with an average efficiency of 50 percent in 2025 and 55 percent in 2050. The corresponding natural gas requirements are 3.59 exajoules in 2025 and 2.78 exajoules in 2050.

Table 6: Fuel supply scenario for the U.S.
EJ per year

	1985	2025	2050
Solid fuels			
Coal	3.14	2.10[a]	1.20[a]
Other solids, direct use	2.93	–	– [d]
Coal for power	15.43	1.29[b]	– [d]
Other solids, power	0.02	–	– [d]
Biomass for power	–	3.83[c]	3.18[c]
Subtotal	*21.52*	*7.22*	*4.38*
Liquid fuels			
Domestic oil	22.60	9.19[e]	5.49[e]
MeOH from biomass	–	4.81[f]	5.16[f]
Imported oil	9.68	2.60[g]	1.42[g]
Imported MeOH	–	2.60[g]	1.43[g]
Subtotal	*32.28*	*19.20[d]*	*13.50[d]*
Gaseous fuels			
Domestic natural gas	17.96	15.38[i]	13.36[i]
Imported natural gas	1.23	0.82[j]	– [j]
Biogas from dung	–	0.35[k]	0.23[k]
H_2 from biomass	–	2.71[l]	2.93[l]
H_2 from intermittent renewables	–	2.13[m]	2.96[m]
Subtotal	*19.19*	*21.39[h]*	*19.48[h]*
CO_2 emissions[n] in *MtC*	1,353	533	345
CO_2 emissions per capita in *tC*	5.65	1.85	1.21

a. Direct solid fuel consumption for the IPCC Accelerated Policies Scenario (table 4).

b. See note d, table 5.

c. See note g, table 5/

d. Direct liquid fuel consumption for the IPCC Accelerated Policies Scenario (table 4).

e. See table E. The projection of oil production for the year 2025 is roughly midway between the two alternative projections made by the U.S. Department of Energy in support of the National Energy Strategy: 8.65 exajoules per year for its projection for the Current Policy Base Case Scenario and 12.00 exajoules per year for its National Energy Strategy Scenario [3].

f. It is assumed that MeOH is produced from two thirds of the biomass supplies other than dung (see tables I, K, and L) in excess of what is needed for power generation (see note g, table 5), at conversion efficiencies of 61.9 percent in 2025 and 62.9 percent in 2050 (see table N).

g. It is assumed that liquid fuel requirements not otherwise met by domestic sources are provided by equal shares of oil and MeOH imports—the latter derived from biomass.

h. Direct gaseous fuel consumption for the IPCC Accelerated Policies Scenario (table 4) plus the fuel required for natural gas-fired power generation some 3.59 exajoules in 2025 and 2.78 exajoules in 2050 (see table 5, note j).

i. It is assumed that U.S. natural gas production declines at a linear rate such that by 2050 one third of estimated ultimately recoverable resources remain (see table F). The projection of gas production for the year 2025 using this model is lower than the 20.2–20.7 exajoules per year projections made by the U.S. Departmentof Energy in support of the National Energy Strategy [3].

j. It is assumed that gas imports decline one third by 2025 and are phased out by 2050.

k. It is assumed that biogas is produced from recoverable dung residues (see table K) at 90 percent of the stoichiometric rate (see chapter 19), corresponding to a 57 percent energy conversion efficiency.

l. It is assumed that H_2 is produced from one third of biomass supplies other than dung (see tables I, K, and L) in excess of what is needed for power generation (see note g, table 5), at conversion efficiencies of 69.7 percent in 2025 and 71.5 percent in 2050 (see table N).

m. It is assumed that hydrogen produced electrolytically from PV and wind sources provides gaseous fuel requirements not met by other means.

n. Assuming CO_2 emission rates of 25 MtC per exajoule for coal, 19.5 MtC per exajoule for petroleum, and 13.5 MtC per exajoule for natural gas.

RENEWABLES-INTENSIVE ENERGY SCENARIO FOR CANADA

Table 1: Growth schedule for per capita GDP and population for Canada, assuming IPCC growth rates

	GDP per capita 1989 U.S.$	Population millions
1985	15,564	25.38
2000	22,717	27.09
2025	40,021	28.51
2050	66,346	28.21

Table 2: Baseline data for energy in Canada, 1985[a,b]

	Electricity TWh per year	Primary energy EJ per year	Direct fuel use EJ per year	Fuel for electricity EJ per year
Coal	77.0	1.13	0.27	0.86
Other solids	1.6	0.46	0.44	0.02
Oil	6.6	3.14	3.07	0.07
Natural gas	6.6	2.24	2.17	0.07
Hydro	294.0	3.21	–	–
Nuclear	58.6	0.64	–	–
Exports	– 43.8	– 0.48	–	–
Totals	**400.6**	**10.34**	**5.95**	**1.02**

a. International Energy Agency data [27]. IEA data (in LHV) converted to HHV by dividing by 0.96 for coal, 0.95 for other solids, 0.94 for oil, and 0.90 for natural gas.

b. T&D losses averaged 7.825 percent of net generation.

Table 3: IPCC electricity consumption scenarios for Canada
TWh per year

	High emissions	Accelerated policies
1985	369.3	369.3
2000	518	398
2025	897	438
2050	1,101	504

Table 4: IPCC direct fuel use scenarios for Canada
EJ per year

	High emissions				Accelerated policies			
	Solids	Liquids	Gases	Total fuel	Solids	Liquids	Gases	Total fuel
1985	0.71	3.07	2.17	5.95	0.71	3.07	2.17	5.95
2000	0.67	3.01	1.62	5.30	0.72	2.83	2.08	5.63
2025	0.88	2.55	1.88	5.31	0.41	2.29	2.65	5.35
2050	0.94	2.72	1.75	5.41	0.29	1.64	2.38	4.31

Table 5: Electricity supply scenario for Canada
TWh per year

	1985	2025	2050
Hydro	294.0	368[b]	368[b]
Wind	–	110[c]	221[c]
Nuclear	58.6	84[d]	84[d]
Coal	77.0	– [e]	–
Other solids	1.6	–	–
Oil	6.6	– [e]	– [e]
Export to the U.S.	– 43.8	– 138[f]	– 176[f]
Natural gas	6.6	37[g]	34[g]
Totals	400.6	**461[a]**	**531[a]**

a. From table 3, assuming 5 percent T&D losses.

b. The economic potential for hydropower in Canada has been estimated to be 610 TWh per year [10]. It is assumed here that only three fifths of this potential is exploited because of environmental constraints.

c. For Canada, the wind-electric potential, even with severe exclusions, would be many times total present electricity demand [30]. Thus, the potential for electricity generation will be constrained not so much by resource or environmental considerations but rather by the extent to which wind electricity can be accommodated on the electrical grid. For Canada this can be large, owing to the large contribution to the electrical supply from hydropower. Here it is assumed that the level of wind power development grows to 30 percent of that for hydropower by 2025 and to 60 percent that for hydro by 2050.

d. As of 1989 installed nuclear capacity was 11,872 MW$_e$ and an additional 3,524 MW$_e$ was under construction. It is assumed that in the period 2025–2050 this much capacity is operating, at an average capacity factor of 62 percent.

e. It is assumed that electricity from coal and oil sources is phased out by 2025.

f. It is assumed that electricity exported to the U.S. increases to three tenth of total production by 2025 and to one third of total production by 2050. The projected increase in exports is consistent with ongoing trends. The U.S. Department of Energy projects that these imports will increase to 68.2 TWh per year by 2010 [28].

g. It is assumed that electricity requirements not provided by other means are provided by advanced natural gas–based power-generating systems with an average efficiency of 50 percent in 2025 and 55 percent in 2050. The corresponding natural gas requirements are 0.27 exajoules in 2025 and 0.22 exajoules in 2050.

Table 6: Fuel supply scenario for Canada
EJ per year

		1985	2025	2050
Solid fuels				
Coal		1.13	0.41	0.29
Other solids		0.46	–	–
Subtotal		*1.59*	*0.41[a]*	*0.29[a]*
Liquid fuels				
Domestic oil		3.85	2.94[c]	2.32[c]
MeOH from biomass		–	1.65[d]	1.67[d]
Imported MeOH		–	– 1.65[d]	– 1.67[d]
Imported oil		– 0.71	– 0.65[e]	– 0.68[e]
Subtotal		*3.14*	*2.29[b]*	*1.64[b]*
Gaseous fuels				
Domestic natural gas		3.19	3.12[g]	2.80[g]
H_2 from intermittent renewables		–	0.75[h]	0.75[h]
Imported natural gas		– 0.95	– 0.95[i]	– 0.95[i]
Subtotal		*2.24*	*2.92[f]*	*2.60[f]*
CO_2 emissions[j]	MtC	125	84	64
CO_2 emissions per capita	tC	4.94	2.95	2.28

a. Direct solid fuel consumption for the IPCC Accelerated Policies Scenario (table 4).

b. Direct liquid fuel consumption for the IPCC Accelerated Policies Scenario (table 4).

c. See table E.

d. It is assumed that biomass supplies (see table M) are used for MeOH production, at conversion efficiencies of 61.9 percent in 2025 and 62.9 percent in 2050 (see table N), and that all the produced MeOH is exported.

e. It is assumed that oil produced in excess of domestic needs is exported.

f. Direct gaseous fuel consumption for the IPCC Accelerated Policies Scenario (table 4) plus the fuel required for natural gas-fired power generation (some 0.27 exajoules in 2025 and 0.22 exajoules in 2050—see table 5, note g).

g. With this natural gas production schedule 60 percent of gas resources would remain in 2050 (see table F).

h. This hydrogen would be produced primarily from wind resources.

i. Gas exports are assumed to be constant.

j. Assuming CO_2 emission rates of 25 MtC per exajoule for coal, 19.5 MtC per exajoule for petroleum, and 13.5 MtC per exajoule for natural gas.

RENEWABLES-INTENSIVE ENERGY SCENARIO FOR OECD-EUROPE

Table 1: Growth schedule for per capita GDP and population for OECD-Europe, assuming IPCC growth rates

	GDP per capita *1989 U.S.$*	Population *millions*
1985	8,333	404.3
2000	12,163	431.6
2025	21,428	454.2
2050	35,521	449.4

Table 2: Baseline data for energy in OECD-Europe, 1985[a,b]

	Electricity *TWh per year*	Primary energy *EJ per year*	Direct fuel use *EJ per year*	Fuel for electricity *EJ per year*
Coal	593.6	11.38	4.64	6.74
Other solids	18.0	1.42	1.20	0.22
Oil	172.8	24.61	22.70	1.91
Natural gas	103.5	8.91	7.76	1.15
Hydro	398.0	4.34	–	–
Nuclear	538.7	5.87	–	–
Geothermal	3.2	0.04	–	–
Totals	**1,827.8**	**56.57**	**36.30**	**10.02**

a. International Energy Agency data [27]. IEA data (in LHV) converted to HHV by dividing by 0.96 for coal, 0.95 for other solids, 0.94 for oil, and 0.90 for natural gas.

b. T&D losses averaged 7.609 percent of net generation in 1985.

Table 3: IPCC electricity consumption scenarios for OECD-Europe
TWh per year

	High emissions	Accelerated policies
1985	1,689	1,689
2000	2,371	1,824
2025	4,103	2,006
2050	5,038	2,302

Table 4: IPCC direct fuel use scenarios for OECD-Europe
EJ per year

	High emissions				Accelerated policies			
	Solids	Liquids	Gases	Total fuel	Solids	Liquids	Gases	Total fuel
1985	5.84	22.70	7.76	36.30	5.84	22.70	7.76	36.30
2000	5.43	22.29	5.78	33.50	5.88	20.97	7.42	34.27
2025	7.22	18.85	6.72	32.79	3.39	16.91	9.45	29.75
2050	7.76	20.08	6.25	34.09	2.41	12.16	8.52	23.09

Table 5: Electricity supply scenario for OECD-Europe
TWh per year

	1985	2025	2050
Hydro	398.0	510[b]	510[b]
Wind	–	187[c]	187[c]
Geothermal	3.2	12[d]	12[d]
Direct solar	–	106[e]	242[e]
Nuclear	538.7	539[f]	539[f]
Coal	593.6	148[g]	–[g]
Other solid fuel	18.0	–	–
Oil	172.8	–[h]	–[h]
Biomass	–	399[i]	529[i]
Natural gas	103.5	211[j]	404[j]
Totals	**1,827.8**	**2,112**[a]	**2,423**[a]

a. From table 3, assuming 5 percent T&D losses.

b. The economic potential for hydropower in OECD-Europe has been estimated to be 696 TWh per year [10]. Also, the level projected for the year 2010 by the CEC is 623 TWh per year [31]. A much lower level of development is assumed here, owing to environmental constraints.

c. The practical wind energy potential for Europe has been estimated to be 573 TWh per year with moderate land-use exclusions and 187 TWh per year with severe land-use exclusions (see chapter 4). The latter value is assumed here.

d. Geothermal capacity in 1985 was 545 MW_e and is expected to increase to 1,064 MW_e by 1995 (see chapter 13). It is assumed that the capacity doubles again by 2025–2050 and operates at a 67 percent average capacity factor.

e. Direct solar (photovoltaic and solar thermal-electric power) is assumed to be 5 percent of total generation in 2025 and 10 percent in 2050.

f. Nuclear power is assumed to be constant.

g. It is assumed that coal-based power generation is reduced to 25 percent of the 1985 level by 2025 and entirely phased out by 2050. Assuming an efficiency of 35 percent, the coal requirements for power generation are 1.52 exajoules in 2025.

h. It is assumed that oil-based power generation is phased out by 2025.

i. It is assumed that one fourth of the biomass supplies other than dung (tables K and M) are used for power generation in stand-alone plants @ 42.9 percent efficiency (BIG/ISTIG or equivalent) in 2025 and 57 percent efficiency in 2050 (BIG/FC or equivalent) (see table N).

i. It is assumed that natural gas-fired advanced gas turbine cycles account for electricity requirements not provided by other means. The average efficiency of power generation with natural gas is assumed to be 50 percent in 2025 and 55 percent in 2050. Thus natural gas fuel requirements for power generation are 1.52 exajoules in 2025 and 2.64 exajoules in 2050.

Table 6: Fuel supply scenario for OECD-Europe
EJ per year

	1985	2025	2050
Solid fuels			
Coal for direct use	4.64	3.39[a]	2.41[a]
Coal for power	6.74	1.52[b]	–
Other solids, direct use	1.20	–	–
Other solids, power	0.22	–	–
Biomass for power	–	3.35[c]	3.34[c]
Subtotal	*12.80*	*8.26*	*5.75*
Liquid fuels			
Domestic oil	8.74	2.45[e]	0.98[e]
MeOH from biomass	–	3.11[f]	3.15[f]
Imported oil	15.87	5.68[g]	4.02[g]
Imported MeOH	–	5.67[g]	4.01[g]
Subtotal	*24.61*	*16.91[d]*	*12.16[d]*
Gaseous fuels			
Domestic natural gas	7.10	4.36[i]	2.97[i]
Biogas	–	0.43[j]	0.46[j]
H_2 from biomass	–	3.50[k]	3.59[k]
Imported H_2	–	1.34[l]	2.07[l]
Imported natural gas	1.81	1.34[l]	2.07[l]
Subtotal	*8.91*	*10.97[h]*	*11.16[h]*
CO_2 emissions[l] *MtC*	902	358	226
CO_2 emissions per capita *tC*	2.23	0.79	0.50

a. Direct solid fuel consumption for the IPCC Accelerated Policies Scenario (table 4).

b. See note g, table 5.

c. See note i, table 5.

d. Direct liquid fuel consumption for the IPCC Accelerated Policies Scenario (table 4).

e. See table E.

f. It is assumed that MeOH is produced from three eighth of the biomass supplies other than dung (see tables K and M) at efficiencies of 61.9 percent in 2025 and 62.9 percent in 2050 (see table N).

g. It is assumed that liquid fuel requirements not provided by domestic sources are provided by equal quantities (in energy terms) of oil imports and MeOH imports, with the latter derived from biomass.

h. Direct gaseous fuel consumption for the IPCC Accelerated Policies Scenario (table 4) plus the fuel required for natural gas-fired power generation, some 1.52 exajoules in 2025 and 2.64 exajoules in 2050 (table 5, note j).

i. It is assumed that natural gas production declines at an exponential rate such that in 2050, one third of estimated natural gas resources remain to be exploited (see table F).

j. It is assumed that biogas is produced from recoverable dung residues (see table K) at 90 percent of the stoichiometric rate (see chapter 19), corresponding to a 57 percent energy conversion efficiency.

k. It is assumed that hydrogen is produced from three eighth of the biomass supplies other than dung (see tables K and M) at efficiencies of 69.7 percent in 2025 and 71.5 percent in 2050 (see table N).

l. It is assumed that gas requirements not provided by other sources are provided by imports of equal quantities of natural gas and hydrogen produced electrolytically from photovoltaic sources in North Africa.

l. Assuming CO_2 emission rates of 25 MtC per exajoule for coal, 19.5 MtC per exajoule for petroleum, and 13.5 MtC per exajoule for natural gas.

RENEWABLES-INTENSIVE ENERGY SCENARIO FOR FORMER CENTRALLY PLANNED EUROPE

Table 1: GDP and population for former Centrally Planned Europe, for IPCC growth rates

	GDP per capita *U.S.$ per capita*	Population *millions*
1985	6,794	416
2000	12,222	454
2025	30,304	500
2050	62,387	521

Table 2: Baseline data for energy in former Centrally Planned Europe, 1985[a,b]

	Electricity *TWh per year*	Primary energy *EJ per year*	Direct fuel use *EJ per year*	Fuel for electricity *EJ per year*
Coal	961.2	24.91	14.85	10.06
Oil	270.5	24.35	21.52	2.83
Natural gas	327.6	23.43	20.00	3.43
Hydro	232.4	2.54	–	–
Nuclear	213.1	2.32	–	–
Totals	**2,004.8**	**77.55**	**56.37**	**16.32**

a. Data from the Energy Information Administration (1991).

b. The IPCC estimates that the ratio of delivered electricity to fuel inputs is 31.4 percent. In 1985 T&D losses amounted to 8.66 percent of generation in the Soviet Union [32], a value assumed here for all of former Centrally Planned Europe. Thus the average efficiency of thermal power generation was 34.4 percent in 1985.

Table 3: IPCC electricity consumption scenarios, former Centrally Planned Europe *TWh per year*

	High emissions	Accelerated policies
1985	1,831	1,831
2000	4,306	3,167
2025	7,000	3,750
2050	10,389	5,194

Table 4: IPCC direct fuel use scenarios (commercial fuels only) for former Centrally Planned Europe *EJ per year*

	High emissions				Accelerated policies			
	Solids	Liquids	Gases	Total fuel	Solids	Liquids	Gases	Total fuel
1985	14.85	21.52	20.00	56.37	14.85	21.52	20.00	56.37
2000	10.00	19.70	20.60	50.30	14.30	17.70	18.00	50.00
2025	12.10	25.30	28.70	66.10	7.30	16.90	24.80	49.00

Table 4: IPCC direct fuel use scenarios (commercial fuels only)
for former Centrally Planned Europe
EJ per year

	High emissions				Accelerated policies			
	Solids	Liquids	Gases	Total fuel	Solids	Liquids	Gases	Total fuel
2050	13.70	33.60	32.80	80.10	5.10	15.70	28.80	49.60

Table 5: Electricity supply scenario for former Centrally Planned Europe
TWh per year

	1985	2025	2050
Hydro	232.4	600[b]	800[b]
Nuclear	213.1	100[c]	100[c]
Oil	270.5	–[d]	–[d]
Coal	961.2	481[e]	–[e]
Biomass	–	522[f]	1,355[f]
Intermittent renewables	–	987[g]	1,913[g]
Natural gas	327.6	1,257[h]	1,299[h]
Total	**2,004.8**	**3,947[a]**	**5,467[a]**

a. From table 3, assuming 5 percent T&D losses.

b. The economic potential for hydropower in former Centrally Planned Europe has been estimated to be 1,199 TWh per year [10]. It is assumed that two thirds of this potential is developed by 2050.

c. It is assumed that nuclear power is reduced in half.

d. It is assumed that oil-based power generation is phased out.

e. It is assumed that coal-based electricity generation is reduced 50 percent by 2025 and phased out by 2050. Power plants operating in 2025 are assumed to have an efficiency of 35 percent, so that coal requirements for power generation amount to 4.95 exajoules in 2025.

f. It is assumed that one half of biomass supplies other than dung (see tables K and M) are used for power generation in stand-alone plants @ 42.9 percent (BIG/ISTIG or equivalent) in 2025 and 57 percent (BIG/FC or the equivalent) in 2050 (see table N).

g. It is assumed that intermittent renewables provide 25 percent of power generation in 2025 and 35 percent in 2050.

h. It is assumed that electricity requirements provided by other sources are produced by advanced gas turbines cycles fired with natural gas (at efficiencies of 50 percent and 55 percent in 2025 and 2050, respectively). Thus natural gas requirements for power generation are 9.05 exajoules in 2025 and 8.50 exajoules in 2050.

Table 6: Fuel supply scenario for former Centrally Planned Europe
EJ per year

	1985	2025	2050
Solid fuels			
Coal for direct use	14.85	7.30[a]	5.10[a]
Coal for power	10.06	4.95[b]	–
Biomass for power	–	4.38[c]	8.56[c]
Subtotal	*24.91*	*16.63*	*13.66*
Liquid fuels			
Domestic oil	27.38	11.46[e]	6.21[e]
MeOH from biomass	–	1.81[f]	3.59[F]
Oil imports	– 3.03	1.82[g]	2.95[g]
MeOH imports	–	1.81[g]	2.95[g]
Subtotal	*24.35*	*16.90[d]*	*15.70[d]*
Gaseous fuels			
Domestic natural gas	24.56	25.40[i]	25.40[i]
Biogas	–	0.62[j]	0.65[j]
H_2 **from biomass**	–	1.02[k]	2.04[k]
H_2 **from intermittent** renewables	–	1.70[l]	2.30[L]
Imported H_2	–	1.70[l]	2.30[L]
Imported natural gas	– 1.13	3.41[l]	4.61[l]
Subtotal	*23.43*	*33.85[h]*	*37.30[h]*
CO_2 **emissions**[k] *MtC*	1,414	954	711
CO_2 **emissions per capita** *tC*	3.40	1.91	1.37

a. Direct solid fuel consumption for the IPCC Accelerated Policies Scenario (table 4).

b. See note e, table 5.

c. See note f, table 5.

d. Direct liquid fuel consumption for the IPCC Accelerated Policies Scenario (table 4).

e. See table E.

f. It is assumed that one third of biomass supplies other than dung (see tables K and M) are used for MeOH production at efficiencies of 61.9 percent in 2025 and 62.9 percent in 2050 (see table N).

g. It is assumed that liquid fuel requirements not met by domestic sources are provided by equal quantities (on an energy basis) of imported oil and imported MeOH—the latter derived from biomass.

h. Direct gaseous fuel consumption for the IPCC Accelerated Policies Scenario (table 4) plus natural gas requirements for power generation (see table 5, note h).

i. It is assumed that gas production rises linearly until 2025 to a level that is maintained until 2050, with the level chosen such that in 2050 one half of the assumed gas resources remain to be exploited (table F).

j. It is assumed that biogas is produced from recoverable dung residues (see table K) at 90 percent of the stoichiometric rate (see chapter 19), corresponding to a 57 percent energy conversion efficiency.

k. It is assumed that hydrogen is produced from one sixth of the biomass supplies other than dung (see tables K and M) at efficiencies of 69.7 percent in 2025 and 71.5 percent in 2050 (see table N).

l. It is assumed that gas requirements not provided by domestic natural gas, biogas, and biomass-derived H_2 are provided by equal amounts of electrolytic hydrogen from intermittent sources and imported natural gas, with half of the electrolytic hydrogen imported from the Middle East (from photovoltaic sources) and half produced domestically (from wind and photovoltaic sources).

m. Assuming CO_2 emission rates of 25 MtC per exajoule for coal, 19.5 MtC per exajoule for petroleum, and 13.5 MtC per exajoule for natural gas.

RENEWABLES-INTENSIVE ENERGY SCENARIO FOR THE MIDDLE EAST

Table 1: GDP and population for the Middle East, for IPCC growth rates

	GDP per capita *1989 U.S.$*	**Population** *millions*
1985	4,306	111
2000	5,108	171
2025	9,705	277
2050	18,129	359

Table 2: Baseline data for energy in the Middle East, 1985[a,b]

	Electricity *TWh per year*	**Primary energy** *EJ per year*	**Direct fuel use** *EJ per year*	**Fuel for electricity** *EJ per year*
Coal	–	0.12	0.12	–
Oil	64.4	6.08	5.33	0.75
Natural gas	80.4	2.52	1.58	0.94
Hydro	9.5	0.10	–	–
Nuclear	–	–	–	–
Totals	**154.3**	**8.82**	**7.03**	**1.69**

a. Data from the [25].

b. The IPCC estimates that the ratio of delivered electricity to fuel inputs is 26.3 percent. It is assumed that in 1985 T&D losses amounted to 15 percent of generation. Thus the average efficiency of thermal power generation was 30.9 percent in 1985.

Table 3: IPCC electricity consumption projections for the Middle East
TWh per year

	High emissions	**Accelerated policies**
1985	131	131
2000	389	278
2025	1,278	944
2050	2,556	1,611

Table 4: IPCC secondary fuel use projections for the Middle East
EJ per year

	High emissions				**Accelerated policies**			
	Solids	**Liquids**	**Gases**	**Total fuel**	**Solids**	**Liquids**	**Gases**	**Total fuel**
1985	0.12	5.33	1.58	7.03	0.12	5.33	1.58	7.03
2000	–	7.00	1.50	8.50	–	5.10	0.60	5.70
2025	–	15.40	5.60	21.00	–	10.80	3.60	14.40
2050	–	24.20	8.10	32.30	–	13.20	5.20	18.40

Table 5: Electricity supply scenario for the Middle East
TWh per year

	1985	2025	2050
Hydro	9.5	10	10
Oil	64.4	– [b]	– [b]
Intermittent renewables	–	348 [c]	594 [c]
Natural gas	80.4	636 [d]	1,092 [d]
Totals	**154.3**	**994** [a]	**1,696** [a]

a. From table 3, assuming 5 percent T&D losses.

b. It is assumed that oil-based power generation is phased out.

c. It is assumed that intermittent renewables provide 35 percent of power generation.

d. It is assumed that electricity requirements not provided by renewables are provided by advanced gas turbine cycles fired with natural gas (at efficiencies of 50 percent and 55 percent in 2025 and 2050, respectively). Thus natural gas requirements for power generation are 4.58 exajoules in 2025 and 7.15 exajoules in 2050.

Table 6: Fuels consumption scenario for the Middle East
EJ per year

	1985	2025	2050
Solid fuels			
Coal for power	0.12	– [a]	– [a]
Subtotal	*0.12*	*–*	*–*
Liquid fuels			
Domestic oil	24.31	27.10 [d]	33.99 [d]
Oil imports	– 18.23	– 16.30 [c]	– 20.79 [c]
Subtotal	*6.08*	*10.80* [b]	*13.20* [b]
Gaseous fuels			
Domestic natural gas	2.64	18.45 [g]	33.19 [g]
Imported natural gas	– 0.12	– 10.27 [f]	– 20.84 [f]
H_2 from intermittent renewables	–	2.20 [h]	3.30 [h]
Imported H_2	–	– 2.20 [h]	– 3.30 [h]
Subtotal	*2.52*	*8.18* [e]	*12.35* [e]
Total exports	**18.35**	**28.77**	**44.93**
CO_2 emissions [i] *MtC*	156	321	424
CO_2 emissions per capita *tC*	1.40	1.16	1.18

a. Direct solid fuel consumption for the IPCC Accelerated Policies Scenario (table 4).

b. Direct liquid fuel consumption for the IPCC Accelerated Policies Scenario (table 4).

c. Net oil import requirements of other world regions.

d. With this oil production schedule more than half the oil resources are still to be exploited after 2050 (table E.).

e. Direct gaseous fuel consumption for the IPCC Accelerated Policies Scenario (table 4) plus natural gas requirements for power generation, some 4.58 exajoules in 2025 and 7.15 exajoules in 2050 (see note d, table 5).

f. Net natural gas import requirements of other world regions.

g. With this natural gas production schedule three fifth of the gas resources are still to be exploited after 2050 (table F).

h. Hydrogen is produced electrolytically from photovoltaic sources in the deserts of the Middle East for export to former Centrally Planned Europe and to South and East Asia.

i. Assuming CO_2 emission rates of 25 MtC per exajoule for coal, 19.5 MtC per exajoule for petroleum, and 13.5 MtC per exajoule for natural gas.

WORKS CITED

1. Intergovernmental Panel on Climate Change (World Meteorological Organization/United Nations Environment Program). 1991
   Climate change: The IPCC response strategies, Island Press, Washington, DC.

2. Zachariah, K.C. and Vu, M.T. (World Bank). 1988
   World population projections, 1987–1988 Edition, Johns Hopkins University Press, Baltimore, Maryland.

3. U.S. Department of Energy. 1992
   Integrated analysis supporting the national energy strategy: methodology, assumptions, and results, DOE/S-0086P.

4. Strong, E.L. 1978
   Hydraulic resources, in *Renewable energy resources,* the full reports to the Conservation Commission of the World Energy Conference, IPC Science and Technology Press, London.

5. Masters, C.D., Root, D.H., and Attanasi, E.D. 1990
   World oil and gas resources—future production realities, *Annual Review of Energy,* Vol. 15, 23–31.

6. Energy Information Administration. 1990
   The domestic oil and gas recoverable resource base: supporting analysis for the national energy strategy, U.S. Department of Energy, SR/NES/90-05, Washington, DC.

7. Office of Technology Assessment of the United States Congress. 1980
   Energy from biological processes, Washington, DC.

8. Food and Agriculture Organization of the United Nations. 1986
   1985 yearbook of forest products, 1974–1985, Rome.

9. Food and Agriculture Organization of the United Nations. 1986
   FAO production yearbook 1985, Vol. 39, Rome.

10. Taylor, T.B., Taylor, R.B., and Weiss, S. 1982
    Worldwide data related to potentials for widescale use of renewable energy, PU/CEES Report No. 132, Center for Energy and Environmental Studies, Princeton University, Princeton, New Jersey.

11. Response Strategies Working Group of the Intergovernmental Panel on Climate Change. 1990
    Emissions scenarios, Appendix of the Expert Group on Emissions Scenarios, U.S. Environmental Protection Agency, Washington, DC, April.

12. Strehler, A. and Stutzle, W. 1987
    Biomass residues, in D.O. Hall and R.P. Overend, eds., *Biomass: regenerable energy,* John Wiley & Sons, Chichester.

13. Beyea, J., Cook, J., Hall, D., Socolow, R., and Williams, R. 1992
    Toward ecological guidelines for large-scale biomass energy development, report of a workshop for engineers, ecologists, and policy makers, May 1991, National Audubon Society, New York, New York.

14. World Resources Institute. 1990
    World resources 1990–91, Oxford University Press, New York.

15. Grainger, A. 1988
    Estimating areas of degraded tropical lands requiring replenishment of forest cover, *International Treecrops Journal,* Vol. 5, 1–2.

16. Food and Agriculture Organization of the United Nations. 1982
    Forestry in China, FAO Forestry Paper 35, Rome.

17. Williams, R.H. and Larson, E.D. 1989
Expanding roles for gas turbines in power generation, in T.B. Johansson, B. Bodlund, and R.H. Williams, eds., *Electricity: efficient end-use and new generation technologies and their planning implications,* 503–553, Lund University Press, Lund, Sweden.

18. Lloyd, A. 1991
Thermodynamics of chemically recuperated gas turbines, PU/CEES Report No. 256, Center for Energy and Environmental Studies, Princeton University, Princeton, New Jersey, January.

19. Klaeyle, S., Laurent, R., and Nandjee, F. 1987
New cycles for methanol-fueled gas turbines, Paper 87-GT-175, presented at the ASME Gas Turbine Conference and Exhibition, Anaheim, California, May 31–June 4.

20. Larson, E.D. and Katofsky, R.E. 1992
Production of hydrogen and methanol via biomass gassification, in *Advances in thermochemical biomass conversion,* Elsevier Applied Science, London (forthcoming).

21. Energy Information Administration. 1991
1989 International energy annual, DOE/EIA- 0219(89), Washington, DC.

22. Ministry of Energy, People's Republic of China. 1990
Energy in China, Beijing.

23. International Energy Agency. 1991
Energy balances of OECD countries 1980–1989, Paris.

24. Energy Information Administration. 1991
Annual outlook for U.S. electric power 1991, with projections through 2010, U.S. Department of Energy, DOE/EIA-0474(91), Washington, DC.

25. Idaho National Engineering Laboratory, Los Alamos National Laboratory, Oak Ridge National Laboratory, Sandia National Laboratories, and the Solar Energy Research Institute. 1990
The potential of renewable energy, an interlaboratory white paper, prepared for the Office of Policy, Planning, and Analysis of the U.S. Department of Energy, SERI/TP-260-3674, Washington, DC.

26. Thompson, G. 1981
The prospects for wind and wave power in North America, PU/CEES Report No. 117, Center for Energy and Environmental Studies, Princeton University, Princeton, New Jersey.

27. Directorate General for Energy. 1990
Energy for a new century: the European perspective, Commission of the European Communities, Energy in Europe, Brussels, Belgium.

28. U.S. Central Intelligence Agency. 1990
Soviet energy data resource handbook, Washington, DC.

INDEX

Advanced gasification-based biomass power generation, 620, 729–85
aeroderivative turbines, 736
BIG/GT, *see* Biomass-integrated gasifier/gas turbine (BIG/GT) technologies
coal-integrated gasifier/gas turbine (CIG/GT) technologies, 743–46
cogeneration, 735, 736, 766–70
efficient gas-turbine cycle options, 738–43
  air bottoming cycle, 741, 742
  combined cycle, 740–41
  intercooled steam-injected gas turbine (ISTIG), 739, 744
  prospects for improvements, 741–43
  steam-injected gas turbine (STIG), 738–39, 744
environmental issues, 761–64
gas-turbine cycle technology, 734–38
generating capacity in the United States, 729–30
hydrogen from, 932, 943, 949, 950, 951
overview, 729–36
public policy issues, 778–80
steam-turbine cycle technology, 731–34, 735
Advanced Photovoltaic Systems (APS), 409, 420, 422
Advanco Corporation, 267, 268, 283
Africa, wind energy potential in, 194–95
Agence Française pour la Maîtrise de l'Energie (French Solar Energy Agency), 246
Agriculture:
  biomass plantations versus land for, 637–40
  future agricultural policy, 45, 46, 55–59
Ahlstrom, 754, 758
Air bottoming cycle, 741, 742
Air pollution from biomass fuels, 761–62
  from burning sugarcane, 852
  from cooking, 656, 685, 686, 690
  ethanol cars and, 852–53
Alabama Power, 505
Alcohol fuels, 37, 671, 681, 868–76
  alcohol fuel cells, 871–73
  ethanol, *see* Ethanol
  methanol, *see* Methanol
Ametek, 455, 456
Amorphous silicon photovoltaic systems, 299, 308–11, 403–35, 443, 953
  advantages of, 308–309
  alloying, 411, 415–417, 418
  charge transport and defects, 414–15
  competing thin-film materials, 309
  cost of electricity from, 325–28
  doping, 415–16
  future prospects for, 432–33
  historical perspective on, 404–11
    approaches to higher efficiency and better stability, 404–405

    discovery of amorphous silicon cells, 404
    efficiency of amorphous silicon cells, 407–409
    field experience, 410–11
    manufacturing advances, 409–10
    multijunction cells, 405–407
    light-induced degradation, 416–18
  manufacturing technology, 418–32
    methods of depositing amorphous silicon, 418–19
    module performance, 424–25
    projected manufacturing costs, 427–32
    reliability and stability, 425–27
    solar cell structures, 420–22
    state of the art, 422–24
  multijunction cells, 405–407, 412–13
  optical properties, 411–14
    light trapping, 413–14
    optical absorption spectrum, 411–13
  overview, 403–404
  production costs, 309–11
Anaerobic digestion, 817–36
  carbon dioxide production, 818, 819, 834–35
  conclusions, 835
  economics of, 835, 836
  environmental factors affecting, 819–21
  methane production, 818, 819, 834, 860
  overview, 817–18
  principles of anaerobic wastewater treatment, 821–24
    anaerobic filter, 823
    contact process, 823
    fluidized-bed systems, 823–24
    schematic diagram, 822
    upflow anaerobic sludge blanket (UASB) reactor, 824–25, 860
  role in energy generation, 828–35, 836
  upflow anaerobic sludge blanket (UASB) technology, 824–28, 860
    applicability and present use of, 826
    design of anaerobic treatment systems, 826–27
    economics of, 835, 836
    future developments, 827–28
    operation of high-rate anaerobic wastewater-treatment reactors, 824–26
Animal dung, *see* Dung
Annualized life-cycle cost (ALC), 656
Aparasi, R. R., 237
Aquatic ecosystems, dams and, 104–105
ARCO Solar, 247, 250, 446, 448, 449, 450, 451, 488, 505
Arizona Public Service Company, 250, 487, 490, 504, 505, 506
Array interference, 152, 158, 171–72
Asia, wind energy potential in, 195–96
ASTRA ole, 658, 659, 677, 678, 679, 680, 683, 687, 688

Astropower Company, 305–306,
Aswan High Dam, 104
Austin, Texas, 487, 504, 506
Australia:
   photovoltaic systems in, 509
   wind energy potential in, 196
Austria, photovoltaic systems in, 485, 487,
   494–96
Automobiles, fuels for future, 33–41
   ethanol and, *see* Ethanol
   hydrogen power, *see* Solar hydrogen, for
      transportation
   methanol, *see* Methanol
   vehicle redesign and, 45

Back-pressure turbine, 560, 561, 565, 731,
   733
Bagasse, 607, 748, 750, 769, 828, 830,
   858–60
   anaerobic digestion of, 834
   energy derived from, 845–47, 858–59
Bain, Richard L., 865–916
Baldwin, Samuel, 655, 658–59
Ballard Power Systems, 941
Barber-Nickels Company, 264
Barbojo, 769
Barnett, Allen M., 437–81
Barr, C. G., 236
Battelle Columbus Laboratory (BCL)
   gasifier 888, 889–90, 932–33
Battery storage of electricity, 1042–43
Baum, V. A., 237
Bechtel Corporation, 250, 255, 262, 743
Bell Laboratories, 446
BIG/GT, *see* Biomass-integrated gasifier/gas
   turbine (BIG/GT) technologies
Binary–cycle plants, 565, 567–68, 569
Biogas, 612
   from anaerobic wastewater treatment
      systems, 817, 818
   for cooking, 673, 681
   as decentralized energy source, 787–88
   from dung, 612, 681–82
   production of, sources for, 667–68
   Pura village case study, *see* Pura village
      biogas electricity case study
   stoves, 667–68, 680–81
   supplies, 681–82
   from vinasse, 831, 833
Biogas plants, 791–94, 799
   fixed-drum-type, 791
   floating-drum, 791, 792
   long-term performance at Pura village,
      795–96
Biological diversity, 615
   biomass plantations and, 630–31, 764
   hydroelectric power plants and, 99–100
Biomass for energy:
   advanced gasification-based biomass
      generation, *see* Advanced
      gasification-based biomass generation
   anaerobic digestion, *see* Anaerobic
      digestion
   BIG/GT, *see* Biomass-integrated gasifier/

gas turbine (BIG/GT) technologies
   biogas, *see* Biogas
   Brazilian fuel-ethanol program, *see* Brazilian
      fuel-ethanol program
   from cellulosic biomass, *see* Cellulosic
      biomass as source of ethanol and
      methanol
   for cooking, *see* Cooking with biofuels
   in global energy scenario, 12–14, 18, 31
   agricultural policy, 55–59
   biomass-based methanol, 36
   biomass supply, 41–43
   hydrogen from, 932–34, 943, 948, 949,
      950, 951, 956, 990
   overview of future role of, 5
   plantations, *see* Biomass plantations
      costs of, 970–71
   potential of, 943, 1001
   supply prospects, *see* Biomass supply
   utility costs using, 1050
   wood gasifier, *see* Wood gasifiers, open-
      top
Biomass-integrated gasifier/gas turbine
      (BIG/GT) technologies, 746–61
   biomass residues as fuel for, 765–78
      kraft pulp, 771–78
      sugarcane, 765–71, 773–75
   cogeneration systems, 767–70
   commercialization prospects, 756–60
   environmental issues, 761–64
   fixed-bed gasifiers, 748–53
      hot-gas cleanup, 749
      tar, 751, 753
   fluidized-bed gasifiers, 752, 753–54
   future of, 760–61
   public policy issues, 778–80
   schematic representation of, 751
   system performance and cost, 754–56,
      757
Biomass plantations, 616–31
   agricultural policy and, 55–59
   biological diversity and, 630–31, 764
   economics of, 624–26
   energy costs of, 624
   erosion and, 629
   eucalyptus plantations, 618–23
      in Brazil, 619–23, 624–25, 634
      in Ethiopia, 618–19
   Jari estate, experiences with, 616, 618
   land area in, 616, 617
   pests and diseases, 628–29
   Philippine Dendrothermal Power
      Program, 618
   potential of, 633–44
      degraded lands, 58–59, 640–43
      food versus fuel, 637–40
      globally, 643–44
      land areas, 55–59, 636–44
      yields, 633–36
   short-rotation intensive-culture (SRIC),
      616
   site establishment, 627
   soil fertility, 628
   species selection, 628

strategies for sustaining high yields, 626–27
in temperate climates, 623
water pollution and, 630
Biomass residues:
BIG/GT applications using, 765–78
kraft pulp, 766, 771–72, 776–78
sugarcane, 765–71, 773–75
stoves, 666–67
supply of, *see* Biomass supply, from biomass residues
Biomass supply, 593–651
from biomass plantations, *see* Biomass plantations
from biomass residues, 607–14
burned for fuel, 607, 609
crop residues, 609–12
dung residues, 612
energy content of selected residues, 608
forest-product industry residues, 612–14
potential of, 631–44
conclusions, 644–45
conversion into modern energy carriers, 594–96
in developing countries, 594
from existing forests, 614–15
global consumption, 594, 595
as non-commercial fuel, 594
overview, 593–96
photosynthesis fundamentals, 598–607
biomass yields with adequate water and nutrients, 602–605
CO$_2$ buildup and biomass production, 605–607
estimating maximum efficiencies for field production of biomass, 599–601
water and nutrient requirements, 601–602
physical and chemical properties, 596–98
potential of, 631–44
biomass residues, 631–33
plantation biomass, 633–44
Biosyn gasifier, 887
Birds, windfarms and, 173
Black liquor, 772, 776, 778
Black & Veatch, 250
BMW, 937
Boeing Corporation, 389, 446, 447, 448, 449, 451
Boes, Eldon C., 361–401
Bonneville Power Authority, 79
BP Solar, 455–56
Brayton-cycle engines, 264, 734
Brazil:
biomass plantations in, 619–23, 624–25
fuel-ethanol program, *see* Brazilian fuel-ethanol program
hydroelectric plants in, 77–78, 89, 90, 91, 92, 94, 95, 96, 97, 104
Brazilian fuel-ethanol program, 37, 49–50, 828–31, 841–64

anaerobic digestion applied to, 828–35
cost reductions, prospects for, 856–60
crop rotation, 857–58
energy cogeneration, 858–59
fuel-oil substitutes, 859–60
processing improvements, 857
technical advances, 856
economic analysis, 848–51
environmental issues, 851–54
ethanol as fuel, 844–45
government incentives, 844
history and original objectives of, 841–42
net energy analysis, 845–47
bagasse and other residues, use of, 845–47
for sugarcane versus corn, 845
overview, 841
patterns of energy consumption in Brazil, 842–43
pollution from, 830–31
recent trends in, 844
social issues, 854–56
switch from gasoline to ethanol, 843–44
Briquettes, biomass, 666–67, 678, 681
British Petroleum, 455
Bulb turbine, 519–20, 524

Cadmium as toxin, 452, 458–62
Cadmium telluride (CdTe), 308, 309, 442, 443, 455–62
field stability of, 457–58
health and safety issues, 458–62
historical perspective, 455–56
Photon Energy and, 457
Caissons, 518–19, 523
California, 34
utility photovoltaic systems in, 503–504, 505–507, 508
wind energy in, 50–51, 121–22, 123, 132–33, 140, 149–51, 179, 181, 200, 203
California Energy Commission, 240, 254–55, 410–11, 503
California Public Utilities Commission, 250
Canada, wind energy potential in, 190–91
Canon-ECD plant, 422
Capacity credit of wind energy, 179, 180, 182
Capacity factor:
of amorphous silicon photovoltaic systems, 427
of wind turbines, 132–33
Carbon dioxide (CO$_2$), 1, 5, 6, 31, 32, 1051, 1052
from anaerobic digestion, 818, 819, 828, 834–35
burning of biofuels and emissions of, 687–88, 690, 762–63
from ethanol production:
from cellulosic biomass, 913
from corn, 854, 912
from sugarcane, 853–54

Carbon tax, 60–61, 1051–52
Carlson, David E., 403–35
Castel Gandolfo Colloquium, 60
Cavallo, Alfred J., 121–56
Cavanagh, James E., 513–47
Cellulosic biomass as source of ethanol and
    methanol, 865–923
  acid-catalyzed process, 880
  alcohols as fuels, qualities of, 868–76
  conclusions, 915–16
  economics of, 903–907
    of ethanol from starch crops, 894–97
    of ethanol from cellulosic biomass,
      897–903
    of methanol from biomass, 903–907
  energy balances, 910–13, 914
  enzymatic hydrolysis technologies,
      application of, 880–85
    cellulase production, 883
    direct microbial conversion (DMC),
      883
    ethanol recovery, 884–85
    hemicellulose conversion, 883–84
    lignin utilization, 885
    pretreatment of cellulosic biomass,
      881–82
    separate hydrolysis and fermentation
      (SHF), 882–83
    simultaneous saccharification and
      fermentation (SSF), 883
  ethanol production, 876–85
    biomass composition, 876–78
    from cellulosic biomass, 879–85, 897–
      903
    economics of, 894–903
    energy balances, 910–13
    fundamentals of, 867
    from starch crops, 878–79, 894–97
  methanol from biomass, 886–92
    economics of, 903–907
    energy balances, 913, 914
    fundamentals of, 867–68
    methanol purification, 892
    methanol synthesis, 891–92
    pretreatment, 886
    syngas conditioning, 890–91
    synthesis gas production, 867–68,
      886–88
    typical biomass gasifiers, 888–90
  overview, 865–68
  perspectives on, 914–15
  value of ethanol and methanol, 873–76
Center for Renewable Energy Sources, 492
Centers for Excellence on New and
    Renewable Energy Sources, 60
Central-receiver systems, see Solar-thermal
    electric technology, central-receiver
    systems
Central Research Institute of the Electric
    Power Industry (CRIEPI), 488, 493
Centre National de la Recherche
    Scientifique, 246
CESA-1, 238–39, 244

Charcoal:
  charcoal conversion in wood gasifier,
    703–704
  for cooking, 672–73, 675, 680 stoves,
    660–65, 680
Chem Systems, 897–901, 903, 907, 909
China:
  biogas programs in, 788, 793
  hydropower in, 86, 87, 96
  ocean energy devices, 535
  wind energy potential in, 195
  windmills in, 123
Chinese Academy of Meteorological
    Science, 195
Chronar, 420, 422
CIS, see Copper indium diselenide (CIS)
Clarke, John H., 513–47
Clean Air Act, 50, 852
Climate:
  global warming, see Global warming
  hydroelectric power plants' effect on, 97
Coal, 2, 3, 19, 23, 31, 912
  methanol from, 892–93, 908
    economics of, 909–10
    energy balances, 913, 914
Coal-integrated gasifier/gas turbine (CIG/
    GT) technologies, 743–46
Cogeneration of energy, 735, 736, 765–71,
    858–59, 1018
  PEM fuel cells for residential, 959–60,
    994
Combined cycle (CC) gas turbine/steam
    turbine, 740–41
Commission of the European Communities,
    243
Commonwealth Scientific and Industrial
    Research Organization, 605
Companhia Hidro Elétrica do São Francisco
    (CHESF), 759
Comsat Laboratories, 341
Concentrator systems, photovoltaic, 300,
    314–18, 361–401
  basic principles of, 362–73
    applications, 372–73
    economics of, 362–66
    novel cell-light coupling concepts,
      370–72
  cell efficiency and, 364–66
  component development, 387–90
  compound parabolic concentrator (CPC),
    369
  concentrating diffuse solar radiation,
    369–70
  concentrator cell development, 387–90
  concentrator modules, 377
  concentrator ratio, 314, 362, 365, 366
    limits to, 366–68
  conversion efficiency of solar cells, 387,
    388
  costs of, 390–97
    concentrator-cell costs, 392–94
    electricity costs, 325–28, 362, 365,
      366, 392, 395–97

high-efficiency concentrators, potential of, 395, 396
low-cost concentrator concepts, 395–97
lower array-field BOS costs, 394
manufacturing costs, 391–92
direct-beam radiation and, 373
early development of, 374–77
existing and planned systems, 378–80
Fresnel concentrator design concept, 363
Fresnel lenses, 363, 367, 375–76, 377, 378, 380–81, 383, 386
gallium arsenide cells, 368, 377, 380, 381, 387, 388–89
light-confining cavities, 371–72
linear concentrators, 364, 367, 369, 377, 380, 386
module costs, 362–63
module development, 380–86
    ENTECH linear Fresnel modules, 381–83
    Sandia experimental and baseline modules, 383–84, 385
    Sandia's Concept-90, 386
    Sandia's 100X, 380, 381
    SEA's 10X linear Fresnel module, 386
    Varian 1000X, 380, 381, 382, 384
module efficiency, 364–66, 381
modules, 316–18
multijunction cells, 377, 387, 389–90
overview, 361–62
passive cooling of cells, 377
point-focus concentrators, 364, 367, 369
point-focus Fresnel modules, 377, 378, 380
prismatic cover, 370–71, 383, 388
reflective parabolic-concentrator design concept, 364
silicon concentrator cells, 387, 388
solar resources and choice of, 318–320, 373, 374, 375
summary and conclusions, 397–98
tracking systems, 314–16, 318–19, 390
Condensing-extraction turbine, 731, 732, 766
Controlled cooking tests, 655
Cooking with biofuels, 653–97
air pollution and, 685–87, 690
biomass-residue stoves, 666–67
charcoal stoves, 660–65
climate change and, 687–88
comparative analysis of cooking systems, 672–81
    economics, 677–81
    energy requirements, 674–77
    energy use, 674
    user preference, 672–74
deforestation concerns, 684–85
equitable access to biomass resources, 684
ethanol, 671–72
evaluation of cooking systems, 653–57
    economics, 656

fuel availability, 653–4
fuel consumption, 655–56
smoke and air pollution, 656
social and cultural factors, 657
stove suitability, 654–55
health issues and, 685–86
with improved fuelwood stoves, 657–60
    insert stoves, 660, 661
    portable stoves, 658–60, 661
    site-built stoves, 658
overview, 653
policy guidelines, 689–91
producer gas-based cooking system, 668–69
resources, 681–84
three-stone open fire, 657, 676, 677
with traditional stoves, 657, 677
Copersucar Company, 848, 851
Copper indium diselenide (CIS), 308, 309, 407, 422, 424, 425, 443, 446–54
economics, 453–54
health and safety issues, 452
historical perspective, 446
indium supply, 453
technology, 447–52
Corn, ethanol derived from, 37–38, 845, 867, 876
as starch crop, 867, 876, 877, 878–79
economics of, 894–97
see also Cellulosic biomass as source of ethanol and methanol
Cotton, 610
Creusot-Loire gasifier, 887
Crop residues as biomass, 609–12
for cooking, 673
potential of, 611, 631–33
removal of, 610–11
Crude Oil Windfall Profits Act, 149
Crystalline and polycrystalline silicon solar cells, 299, 337–60
aluminum alloying, 342
back-surface field, 341
carrier lifetimes, 342
carrier recombination, 341
commercial terrestrial cells, 344–48
    manufacturing costs, 347–48
    metalization approaches, 344–47
    overview, 344
cost of components of, 304
    cell fabrication, 306
    encapsulation and module construction, 306–308
    prospects for, 307
    wafer production, 304–306
cost of electricity from, 325–28
Czochralski process, 348, 350–51, 393
dedrilic web approach, 352, 353
edge-defined film-fed growth (EFG) method, 352–54
efficiency limits, 355–56
electroless plating, 345
float-zone process, 351
history of cell performance, 338–44

laser-grooved cells, 346, 347, 348
light trapping, 342, 355
materials issues, 349–55
  crystalline-ingot technologies, 350–51
  polycrystalline-ingot technologies, 351–52
  polysilicon source material, 349
  self-supporting sheets, 352–54
  supported sheets, 355
  overview, 337–38
  PERL cell, 344, 347
  photolithography, 345
  screen-printing technique, 345
  spherical solar cells, 354
  summary, 356–57
  wafer cutting, 348, 350, 352
Crystalline thin-film silicon (THIN *x*-Si), 463–73
  cell design, 465–68
  economics, 471–72
  efficiency of, 463–64, 465, 469, 471, 473
  field stability, 471
  future of, 473
  health and safety issues, 471
  historical perspective of, 463–64
  technologies, 468–71
Cummins Engine Company, 275
Czochralski process, 348, 350–51, 393

Daimler-Benz, 937
Dams, 74
  architecture, 81
  design advances, 84–85
  large, 79–81
  retrofitting of, 81
  seismicity and, 97–99
  see also Hydroelectric power plants; Hydropower
Danish Utility Association, 134
Darrieus turbine, 135
Das, T. K., 680
Dasappa, S., 666, 699–728
Deforestation, biofuels for cooking and, 684–85
De Laquil, Pascal, III, 213–96
Demonstration of renewables, future agenda for, 46, 51–55
  links between laboratory and commercial markets, 52–55
  setting of priorities, 51–52
Denmark:
  ocean energy and, 535
  wind energy in, 51, 121–22, 134, 135, 140, 150–51, 173, 174, 175–76, 192–93, 200–201
Detroit Edison, 427
Deutsche Forschungsanstalt für Luft- und Raumfahrt (DLR), 261, 274
Developing countries:
  biomass as energy source in, 594
  hydropower in, 77
  photovoltaics in, 484, 507–509
Direct-absorption receiver (DAR), 255

Direct fuel use, 32–41
Direct-steam generation (DSG), 233–34
  gas turbine combined-cycle power plant with, 234–36
Diseases and biomass plantations, 628–29
Diver, Richard, 213–96
Dow Chemical, 743
Drag, 129, 130
Dry-steam condensing turbine with compressor for noncondensable gas removal, 560, 561
Dry steam turbine with upstream reboiler, 560, 561
Dung, 612
  biogas from, 612, 681–82, 788
    at Pura village, *see* Pura village biogas electricity case study
  as biomass resource, 612, 631
  for cooking, 673, 681–82, 788
  as fertilizer, 612, 788
Dunkerley, Joy, 674

Earthquakes, *see* Seismicity
East-West Center, 687
Economic Recovery Tax Act, 149
Egypt, wind energy potential in, 195
Eindhoven University of Technology, 662, 672
Electric Consumer Protection Act, 92
Electricité de France, 246
Electricity:
  from advanced gasification-based biomass generation, *see* Advanced gasification-based biomass generation
  from biogas, *see* Pura village biogas electricity case study
  from biomass fuels, *see* Biomass-integrated gasifier/gas turbine (BIG/GT) technologies
  for cooking, 672, 673, 675–76, 680
  costs of, *see specific renewable energy sources*
  from hydropower, *see* Hydropower
  from renewables, overview of, *see* Renewable fuels and electricity, overview of
  from solar energy, *see* Solar energy
  from wind energy, *see* Wind energy
Electric Power Research Institute (EPRI), 142, 246, 307, 380, 410–11, 734, 743, 760
  photovoltaic systems and, 488, 491, 492, 507
Electric utilities, *see* Utilities
Electrodeposition processes, 455–56, 462
Electrolysis, hydrogen from solar powered, 928–32, 949–50
  costs of, 931–32, 970–71
  electrolysis technology, 928–30
  potential of, 943–45
  sources of solar electricity, 930–31
Elenco Company, 941
Eletrobrás, 859

Elliott, D. L., 189, 198
Energy Institute of Moscow, 237
Energy markets, creation of rational, 46, 47–51
Energy Partners, 941
Energy Performance Systems, Inc., 733–34
Entech Company, 316, 381–83
Ente Nazionale per L'Energia Elettrica (ENEL), 485, 488, 493, 497–500 509
Environmental evaluation process for hydroelectric power plants, 106
Environmental impact statements, 106
Environmental issues, 62
    of anaerobic digestion, 830–31
    of biomass energy gasifiers, 720, 721, 761–64
    of ethanol production, 851–54
    of forest-industry residue removal, 612–14
    of geothermal energy, 580–82
    of hydropower, *see* Hydropower, environmental and social issues
    of ocean thermal energy conversion, 543–44
    of solar hydrogen, 926
    of tidal energy, 529–30
    utility investment portfolios and, 1051–52
    of wave energy, 537
Environmental Protection Agency, 452, 759, 849–50, 869
Erosion, 629, 851
Ethanol, 671–72, 673–74, 769, 770
    Brazilian programs, *see* Brazilian fuel-ethanol program
    from cellulosic biomass, *see* Cellulosic biomass as source of ethanol and methanol
    fundamentals of production of, 867
    in global energy scenario, 33, 42
    for transportation, 33, 35, 37–38
    land requirements for, 682, 683
    neat ethanol, 869–71
Ethiopia, eucalyptus plantations in, 618–19
Ethyl tertiary butyl ether (ETBE), 868–69, 870
Eucalyptus plantations, 618–23, 634
Eurelios, 237, 243
Europe:
    solar hydrogen energy systems for, 952–56
    tidal energy potential in, 517
    wind energy potential in, 191–94
Eutrophication, 102

Falling-sand receiver, 263–64
Federal Energy Regulatory Commission, 224
Fertilizer, 612, 628, 788, 793–94
    sludge fertilizer from biogas, 793–94
    from vinasse, 830, 831
Firor, Kay, 483–512

Fish ladders, 104
Fitzgerald, K. B., 674
Flachglas Solartechnik, 277
Flashed-steam-condensing turbines, 563, 564, 565
Flat-plate systems, *see* Photovoltaic technology, flat-plate systems
Float-zone process, 351, 393
Florida Solar Energy Center, 427
Food and Agricultural Organization (FAO), U.N., 594, 607, 612, 616, 639, 657, 778
Ford Motor Company, 875
Forest-product industry, biomass residues from, 612–14, 631
Forestry policy, future, 46, 55–59
Forests, biomass from, 612–15
Forest Service, U.S., 613–14, 624
Fossil fuels, 5
    carbon tax, 60–61, 1051–52
    in global energy scenario, 12, 14
    utility portfolio costs, 1045–51
France, 192, 542
Free-piston Stirling engine, 274, 283
Fresnel lenses, 363, 367, 375–76, 377
Fuel cells:
    alcohol, 871–73
    hydrogen, 872, 1050–51
        onboard hydrogen production, 939, 961–64, 995
        PEM fuel cells, 939, 959–60
    to power vehicles, 34–35, 937, 939–42, 947–49, 957, 965–67, 996–1000
Fuelwood for cooking, 673, 674, 675, 676, 689
    deforestation concerns, 684–85
    land requirements for, 682–83

Gallium arsenide cells, 368, 377, 380, 381, 387, 388–89
Garf, B. A., 237
Gasahol, *see* Alcohol fuels
Gas-cooled solar tower (GAST) technology, 256
Gasifiers:
    advanced gasification-based biomass generation, *see* Advanced gasification-based biomass generation
    BIG/GT, *see* Biomass=integrated gasifier/gas turbine (BIG/GT) technologies
    biomass, 887–90, 932, 956–57
    coal, 892–93
    open-top wood, *see* Wood gasifiers, open-top
Gasoline substitutes, 37, 671, 681
    ethanol, *see* Ethanol
    hydrogen fuel-cell vehicles (FCVs), 34–35, 937–38, 939–42, 947–49, 957, 965–67, 996–1000
    methanol, *see* Methanol
Gas-turbine cycle technology, 22–23, 734–38

biomass integrated gasifier and, *see*
Biomass-integrated gasifier/gas
turbine (BIG/GT) technologies
Gas-turbine/steam turbine combined-cycle
(CC), 740–41
Gedser Wind Turbine, 134
General Electric Company, 743, 745, 748,
759
General Motors Company, 873, 875
Geopressured geothermal systems, 570
Georgetown Bus Company, 873
Georgia Power Company, 265, 268
Geosynthesis, 84–85
Geothermal energy, 549–91
classification of geothermal systems,
552–54
conceptual models, 552
development of, factors influencing, 579–
82
environmental considerations, 580–82
legal constraints and institutional issues,
580
direct uses, 569
economics of, 573–79
exploration and drilling technologies,
556–59
fluid utilization, 559–69
direct uses, 569
flash cycles, 566–68
steam-dominated reservoirs, 559–65
water-dominated reservoirs, 565–69
future of, 584–89
in global energy scenario, 22
high-enthalpy (temperature) applications,
556
history of exploitation of, 554–55
low-enthalpy (temperature) applications,
556
new technologies, 570–73
overview, 549–54
present status of, 582–84, 585, 587
regions of concentration, 550–52
German Agency for Technical Cooperation,
86
Germany:
photovoltaic systems in, 485, 487, 493–
94, 496–97
wind energy in, 201, 202
Geyer, Michael, 213–96
Gifford, Roger, 605
Gill, Jas, 672
Global energy scenario, *see* Renewable fuels
and electricity, overview of, global
energy scenario
Global warming, biomass fuels and, 606,
687–88, 762
ethanol from sugarcane, 853–54
Glow-discharge deposition, 418–19, 420
Goldemberg, José, 841–64
Great Plains, wind energy of, 127, 151–52
Greece, 192
Green, Martin A., 337–60
Grubb, Michael J., 157–212

Hall, David O., 593–651
Hawaiian Electric Renewable Systems, 135
Health issues:
cooking with biomass fuels, 685–87
hydroelectric power plants and, 94, 105–
106
thin-film photovoltaics and, 452, 458–62
Heat-recovery steam generator, 735
Heaving floats, 531–32
Heliostats, 219–20, 236, 247–49
Solar One, 240–41
Hidroeléctrica Española, 502
High-temperature steam electrolysis, 930
Hildebrandt, Alvin, 237
Hinman, Norman D., 865–923
Hock, Susan M., 121–56
Hog fuel, 772, 776, 778
Hosahilli open-top gasifiers, 717–18
Hot dry rock (HDR) systems, 570–72
H-Power Corporation, 961, 964
Hunt, Arlon, 262
Hutter, Ulrich, 134
Hydraulic turbines, 74
design advances, 84
history of, 78–79
kinetic, 85
types of, 81–82
Hydroelectric Company of San Francisco,
620
Hydroelectric power plants, 81–87
environmental issues, *see* Hydropower,
environmental and social issues
financial considerations, *see* Hydropower,
economic and financial issues
hydroelectric complementation, 89–90
pumped-storage plants, 87, 1042
refurbishing, 82–84
small, mini, and micro, 85–87
technological advances, 84–85
transmission lines, 79, 84, 85
turbines, *see* Hydraulic turbines
Hydrogen, *see* Solar hydrogen
Hydrogen selenide, 452
Hydropower, 3, 73–120
in Brazil, 77–78, 89, 90, 92, 94, 95, 97,
104
in China, 87, 96
conclusion, 115
conversion to electricity, 74
costs of hydroelectricity, 93, 97, 109–13,
931–32
dams, *see* Dams
economic and financial issues, 107–15
costs of hydropower, 93, 97, 109–13
distinctive features of hydropower,
107–109
externalities, 113–14
financial constraints, 114–15
utility portfolio costs, 1050
in electrical systems, 88–92, 1033–34,
1042
environmental and social issues, 92–107
aquatic ecosystems, 104–105

construction's impact, 94
land inundation and reservoir filling,
  94–100
overview, 92–93
planning process and, 106–107
public health concerns, 94, 105–106
sedimentation, 103–104
water quality, 100–103
in global energy scenario, 14, 17–18, 31
global water availability and potential,
  75–78
  economic potential, 76–77
  technical potential, 76, 77
  undeveloped potential, 77
history of, 78–81
hydroelectric complementation, 89–90
hydrogen from, 930–31, 943, 950, 1001
overview, 73–74
power plants, *see* Hydroelectric power
  plants
pumped-storage plants, 87, 1042
storage reservoirs, 89
thermal complementation, 91–92
variation in river flow and, 88–92
Hythane, 939

Iannucci, Joseph J., 483–512
Imatran Voima Oy (IVO), 760
India:
  biogas electricity in, *see* Pura village
    biogas electricity case study
  biogas plants in, 667, 668
  Department of Nonconventional Energy
    Sources (DNES), 712, 715, 716, 717
  ocean energy and, 535, 536
  photovoltaic system in, 509
  wind energy potential in, 196
Indian Institute of Science, 666, 680
Indian Institute of Technology, 715
Indium, 453
Indonesia, 508–509
Inflatable weirs, 85
Insects, biomass plantations and, 628–29
Institute of Gas Technology, 759, 887
  Renugas system, 754, 759, 889
Intergovernmental Panel on Climate
  Change (IPCC), 11, 15, 32, 605, 639
Intermediate Technology Department
  Group, 667
Internal-film receiver (IFR), 256
International Development Research
  Centre, Canada, 684–85
International Energy Agency (IEA), 51
International Institute for Applied Systems
  Analysis (IIASA), 197
International organizations, future policy
  for, 46, 59–62
International Rice Research Institute, 60
International Solar Electric Technology
  (ISET), 451, 453
Ireland, 192
Italy, photovoltaic systems in, 484–85, 487,
  493, 494, 497–500

Jadavpur University, 680
James, L. W., 384
Japan:
  hot dry rock (HDR) technology in, 571
  ocean energy and, 535, 536, 542–43
  photovoltaic systems in, 486, 487, 490,
    494, 500–501, 509
Japan Cool Water Program, 743
Jayakumar, S., 787–815
Jet Propulsion Laboratory, 264, 344, 348
Johansson, Thomas B., 1–71, 1071–1141
Jones, Donald, 674

Karnataka State Council for Science and
  Technology, 716–17
Kearney, David, 213–96
Kelly, Henry, 1–71, 297–336, 1011–69,
  1071–1141
Kenyan jiko (charcoal stove), 662, 664,
  665
Kerosene for cooking, 672, 674, 675, 680,
  681, 689
Kesseli, James, 264
Kinematic Stirling engines, 274, 283
Kinetic turbines, 85
King Abdul Aziz City Center for Science
  and Technology, 268
Kitchen performance tests (KPTS), 655
Kodak, 455, 456
Kraft pulp, 771–78
K–T gasifier, 888

LaJet Energy Company, 268, 273, 284
Land:
  for biomass plantations, 636–44
  hydroelectric plants, disappropriation for,
    96–97
  photovoltaic technology and costs of,
    313
  requirements for wood-based fuels, 682
  sugarcane for ethanol and, 855–56
  wind energy and use of, 149, 172–73
Larson, Eric D., 729–85
Latin America, wind energy potential in,
  196–97
Lawrence Berkeley Laboratory, 262
Ledig, F. Thomas, 624
Lettinga, Gatze, 817–40
Lift, 129, 130
Light trapping, 342, 355, 465, 466–68
Liquid-phase epitaxial devices, 468
LOCUS plant, 177–78
Los Angeles Department of Water and
  Power, 238, 240
LPG for cooking, 672, 673, 674, 681
Ludwig, D. K., 616
Luque, Antonio, 361–401
Lurgi, 754
Luz International, Ltd., 21, 223, 230, 232,
  234, 277, 279
Lynette, Robert, 140, 147

Macedo, Isias C., 841–61
Magma systems, 572

Maize, *see* Corn, ethanol derived from
Manufacturing and Technology Conversion
  International gasifiers, 888
Market stimulation incentives for wind
  energy, 202–203
Matsushita Company, 455, 456
McDonnell Douglas, 250, 268, 270, 283,
  284
Mechanical Technology, Inc., 275
Methane, 687, 688–89, 690, 762
  from anaerobic digestion, 818, 819, 834,
    860
  see also Biogas
Methanol, 671
  from cellulosic biomass, *see* Cellulosic
    biomass as source of ethanol and
    methanol
  from coal, 892–93
    economics of, 909–10
    energy balances, 913, 914
  fundamentals of production of, 867–68
  in global energy scenario, 33, 42
    for transportation, 35–36, 38–40
  from natural gas, 893–94
    economics of, 907–909
    energy balances, 913, 914
  neat methanol, 869–71
  overview of future role of, 4, 5, 6
Methyl tertiary butyl ether (MTBE), 868–
  69, 870, 874–75
Mexico, 507
Meyers, Niels I., 157–212
Middle East:
  oil production, 14–15, 45, 865–66
  wind energy potential in, 196
Midway Laboratories, 384
Mobile Solar USA, 305
Molten-Salt Electric Experiment (MSEE),
  245, 246–47
Monaco, Lourival C., 841–64
Monosolar Company, 455–56
Monteith, John, 602
Moreira, José Roberto, 73–120
Mukunda, H. S., 666, 699–728
Multijunction cells, 377, 387, 389–90,
  405–407, 412–13, 421–22, 424–
  425

National Aeronautics and Space
  Administration, 271
  Lewis Research Center, 265, 274, 275
National certification programs for wind
  energy projects, 203
National Renewable Energy Laboratory, 38,
  700, 897, 903
Natural gas:
  for cooking, 672
  in global energy scenario, 15, 22–23, 31,
    36
  methanol from, 893–94
    economics of, 907–909
    energy balances, 913, 914
  overview of future role of, 5

Netherlands:
  ocean energy and, 542
  wind energy in the, 123, 176, 192, 201
Nevada Power Company, 268
New England Electric System, 487, 488,
  1018
  Gardner, Massachusetts project, 505
New York Power Authority, 314
New Zealand, 196
Niagara Mohawk Power Company, 142
Nimbkar Agricultural Research Institute,
  672
Nitrate-salt central receiver and storage
  technology, 249–56
Nitrous oxide (NO_x), 762, 871, 926, 936,
  937
Nitsch, Joachim, 925–1009
North Africa, solar hydrogen energy
  systems for, 952–56, 1003
Northern Research and Engineering
  Company, 264
Norway, ocean energy and, 535, 536, 542
Norwegian Agency for International
  Development, 86
Nuclear power, 23, 31

Oak Ridge National Laboratory, 624, 626,
  674
Ocean energy systems, 513–47
  in global energy scenario, 22
  ocean thermal energy conversion
    (OTEC), 538–44
    conclusions, 544
    development trends, 542–43
    economics and potential market for,
      543
    environmental effects of, 543–44
    principle of, 538
    the resource, 539
    technology of, 539–42
  overview, 513–15
  salinity gradients, 544–45
  tidal energy, *see* Tidal energy
  wave energy, *see* Wave energy
Ocean thermal energy conversion, *see*
  Ocean energy systems, ocean
  thermal energy conversion
Odeillo, 286
Office of Technology Assessment, 611, 614
Ogden, Joan M., 925–1009
Oil:
  in global energy scenario, 14–15, 45
  from the Middle East, 14–15, 45
Oil companies, 45
Oscillating water column, 534, 535

Pacific Gas & Electric (PG&E), 142, 151,
  323, 324, 410–11, 743, 1045
  photovoltaic systems and, 488, 490, 492,
    493, 503, 504, 507
Pacific Northwest Laboratories, 122, 189
Palmerini, Civis G., 549–91
Parabolic-dish systems, *see* Solar-

thermal electric technology, parabolic-dish systems
Parabolic-trough systems, *see* Solar-thermal electric technology, parabolic-trough systems
Particle Injection Receiver, 262–63
Passivated emitter, rear locally diffused (PERL) cell, 343, 344, 347
Pests, biomass plantations and, 628–29
Petroleum companies, 45
PHALK-Mont Soliel, 502
Philadelphia Electric, 487, 492
Philippine Dendrothermal Power Program, 618
PHOEBUS consortium, 257–61
Phosphoric acid fuel cell, 872–73
Photon Energy, Inc., 455, 456–57
Photosynthesis, *see* Biomass supply, photosynthesis fundamentals
Photoronics, 409
Photovoltaics, *see* Solar energy, photovoltaic technology
Photovoltaics for Utility Scale Applications project, *see* PVUSA project
Photovoltaic technology, 297–336
    amorphous silicon, *see* Amorphous silicon photovoltaic systems
    balance-of-system costs, 318, 325
    concentrator systems, *see* Concentrator systems, photovoltaic
    costs of photovoltaic electricity, 300–303, 325–28
    crystalline and polycrystalline, *see* Crystalline and polycrystalline silicon solar cells
    dye-sensitive films, 312
    flat-plate systems, 299, 301, 303–14
        amorphous silicon, *see* Amorphous silicon photovoltaic systems
        cost components of, 303, 304
        cost-reduction from, 299–300
        costs of electricity from, 315–18
        polycrystalline thin-film, *see* Polycrystalline thin-film photovoltaics
        single-crystal and polycrystalline materials, *see* Crystalline and polycrystalline silicon solar cells
        system costs related to area, 312–14
    module costs, 300–302
    overview, 297–98
    photovoltaic cells:
        efficiency of, 300, 301, 322–24, 332–34
        how they work, 328–32
    silicon spheres, 312
    system components of, 299
    system design issues, 318–25
        efficiency issues, 322–24
        operating costs, 322
        overhead costs, 321–22
        power conversion, 320–21
        rooftop units, 324–25

solar resource, 318–20
thin films, *see* Amorphous silicon photovoltaic systems; Polycrystalline thin-film photovoltaics
unique features of, 298
Pitching floats, 532
Plantations, biomass, *see* Biomass plantations
Plasma-enhanced chemical vapor deposition, 420
Plataforma de Solar Almería, 274, 286
Pollution:
    air, *see* Air pollution from biomass fuels
    solar hydrogen and, 926
    from vinasse, 830–31
Polycrystalline silicon solar cells, *see* Crystalline and polycrystalline silicon solar cells
Polycrystalline thin-film photovoltaics, 437–82, 953
    cadmium telluride (CdTe), *see* Cadmium telluride
    conclusions, 476
    copper-indium-diselenide (CIS), *see* Copper indium diselenide (CIS)
    cost analysis:
        long-term potential of photovoltaics and, 441, 473–76
        overview, 438–41
    crystalline thin-film silicon (THIN *x*- Si), *see* Crystalline thin-film silicon
    global sunlight utilized by, 437–38
    overview of, 437, 441–43
    system components, 438
    thin films, 441–46
        advantages of, 441
        manufacturing costs, 443–46
        manufacturing of, 442–43
Polytechnic University of Madrid, 372
Poole, Alan Douglas, 73–120
Poplar plantations, 623
Population resettlement, hydroelectric plant construction and, 96–97
Port Blair open-top gasifier, 718
Portugal, 536
Price, Roger, 513–47
Pro-álcool program, *see* Brazilian fuel-ethanol program
Producer-gas based cooking system, 668–69, 673, 681
Proton exchange membrane (PEM) electrolyzer, 929–30
Proton exchange membrane (PEM) fuel cells, 939, 959–60
Public health, *see* Health issues
Public policy issues for renewable energy future, 7–9
Public safety, *see* Safety
Public Utility Regulatory Policy Act (PURPA), 149, 200, 202, 224, 279, 486, 779
    biomass-electric generating capacity and, 729

Pumped-storage plants, 87, 1042
Pura village biogas electricity case study, 787–815
    administrative arrangements, 801–802
    advanced biogas plants of the future, 812–13
    biogas-diesel engines, use of, 788–89, 796
    conclusions, 813–14
    diagram of biogas plant system, 789
    economics of the biogas system, 805–13
        capital costs, 805
        comparison with central-station generation, 806–11
        disaggregation of costs, 808–809
        economic viability, 811–12
        equivalent generation-end costs, 809–11
    field performance, 799, 801, 802
    financial operations, 802–804
    institution-building, 801–802
    long-term performance of biogas plants, 795–96
    maintenance of the system, 799, 800
    operation of the system, 798, 800
    overview, 787–90
    subsystems, 790–98
        biogas-diesel engine for electricity generation, 796
        biogas plants, 791–94
        electrical illumination of homes, 796–97
        sandbed filtration system, 794–95
        water supply, 797–98
PVUSA project, 313, 378–80, 410–11, 427, 457
    photovoltaic systems and, 489, 490–91, 503–504, 505

Rajabapaiah, P., 787–815
Ramakrishna, Jamuna, 687
Rankine-cycle engines, 221, 264
Rayleigh distribution, 125, 126, 128, 165
RCA Laboratories, 404
Reddy, Amulya K. N., 1–71, 720, 787–815, 1071–1141
Reed, T. B., 700
Remote Area Power Systems, 509
Renewable fuels and electricity, overview of, 1–71
    agenda for action, 43–63
        agricultural and forestry policies, 45, 46, 55–59
        creation of rational energy markets, 46, 47–51
        environmental concerns, 62–63
        international programs, 46, 59–62
        outline of strategy for increased utilization of renewables, 44–46
        research, development, and demonstration, 46, 51–55
    benefits of renewables ignored in standard economic accounts, 4
    global energy scenario, 9–43, 1071–114

electricity generation strategies, 15–31
    energy demand, 10–12
    energy resources, 12–15
    strategies for fuels, 32–43
    major findings, 1–9
        key elements of renewables-intensive energy future, 3–7
        public policy issues, 7–9
    see also individual renewable fuels, e.g. Hydropower; Ocean energy systems; Wind energy
Renugas gasifier, 754, 759, 889
Research and development, future agenda for renewables, 46, 51–55
    links between laboratory and commercial markets, 52–55
    setting of priorities, 51–52
Research Triangle Institute, 307
Rheinbraun-Uhde HTW (High Temperature Winkler), 754
Rheinisch-Westfälisches Electrizitätswerk, 490, 496–97
Rice husks, cookstoves utilizing, 666
Rim-generator (STRAFLO) turbine, 520, 524
Risø National Laboratory, 135, 150
Rockwell International, 250
Roller compacted concrete (RCC), 84
Rolls Royce Company, 937, 961
Rooftop photovoltaic systems, 324–25, 493–94
Rosillo-Calle, Frank, 593–651
Rossby circulation, 158
Rotors in a wind turbine, 136–39
    movable tip brakes, 138
    stall controlled blades, 136, 137–38
    teetered, 136
    variable-pitch blades, 136–37
Roundwood, 612, 613
Rural electricity, biogas for, see Pura village biogas electricity case study

Sacramento Municipal Utility District (SMUD), 504, 1018
Safety:
    of gasifiers, 720
    of hydrogen, 945–47, 974
    of photovoltaics, 489–90
    of thin-film photovoltaics, 452, 458–62
    of wind turbines, 174
Salinity gradients, 544–45
Salt-density gradient, 221
Sanders Associates, 264
Sandia National Laboratories, 261, 264, 265, 271, 274, 275, 388, 391, 469
    photovoltaic concentrator systems, 380–86
San Diego Gas & Electric Company, 268–69, 488
Sangen, Ernst, 662
Sanyo Electric Co., 409, 410, 420
Sawdust, cookstoves utilizing, 666, 680
Schlaich Bergermann & Partner, 268, 270, 271, 272, 273, 274

Science Applications International
Corporation, 248
SEA Corporation, 386, 395, 397
Seattle City Light, 1018
Sedimentation, hydroelectric power plants
and, 103–104
SEGS, *see* Solar Energy Generating Stations
Seismicity, hydroelectric plants and, 97–99
Selenium, 452
Selenization, 449–50
Shell Corporation, 743
Shenandoah Solar Total Energy Project,
265, 266
Shrinivasa, U., 699–726
Siemens Company, 305, 409, 420
Siemens Solar Industries(SSI), 409, 420,
446, 449, 451, 452, 488, 505
Silane, 418, 420
Silicon concentrator cells, 387–88
Silicon film, 464
Silicon solar cells:
amorphous silicon, *see* Amorphous silicon
photovoltaic systems
crystalline, *see* Crystalline and
polycrystalline silicon solar cells
thin-film, *see* Crystalline thin-film silicon
Simmons, D. M., 194, 195, 196
SiO$_2$ coatings on silicon solar cells, 466–67
Small-particle receivers, 261–63
Small Solar Power Systems
Project/Central Receiver System
(SSPS/CRS), 245–46
Small Solar Power Systems
Project/Distributed Collector
System (SSPS/DCS), 223
Smith, Don R., 121–56
Social issues:
of Brazilian fuel-ethanol program, 854–56
of hydropower, *see* Hydropower,
environmental and social issues
Sodium-heat engines, 264
Soil fertility:
biomass plantations and, 628
sugarcane for ethanol and, 851
Solar Cells Inc., 455, 456
Solar Central-Receiver Utility Studies, 250–
52
Solar energy:
as complement to hydropower, 91–92
photovoltaic technology:
amorphous silicon, *see* Amorphous
silicon photovoltaic systems
concentrator systems, *see* Concentrator
systems, photovoltaic
crystalline and polycrystalline, *see*
Crystalline and polycrystalline silicon
solar cells
in global energy scenario, 14, 21–22
overview, *see* Photovoltaic technology
overview of future role of, 4
polycrystalline thin-film, *see*
Polycrystalline thin-film
photovoltaics
solar PV hydrogen, 931, 932, 933,

943–45, 949–50, 951, 952, 953,
955, 983–84, 988–89
utility field experience with, *see* Utility
field experience with photovoltaic
systems
in utility portfolios, 1028–33
solar hydrogen, *see* Solar hydrogen
-thermal electric technology, *see* Solar-
thermal electric technology
Solar Energy Generating Stations (SEGS),
223–33, 277, 279–82
future, 232–33
SEGS I, 223, 225
SEGS II, 224–25
SEGS III-SEGS VII, 225–31
SEGS VIII-SEGS IX, 231–32
Solar Energy Research Institute, 255, 700
Solarex Company, 310, 409, 411, 420, 422,
425
Solar hydrogen, 4, 925–1009
advantages of, 926–27
from biomass gasification, 932–34, 943,
949, 950, 951, 956–57, 990
costs of, 932–34, 970–71
potential of, 943, 1001
delivered cost of, 977–79
design of solar hydrogen energy systems,
947–56
for Europe and North Africa, 952–56
for the United States, 947–52
environmental advantages of, 926
future of, measures to ensure a, 956–59
demonstrations of hydrogen
technology, 957–58
development of biomass hydrogen
technologies, 956–57
development of electrolysis for
intermittent power sources, 956
development of hydrogen end-use
technologies, 957
development of intermittent renewable
electric technologies, 956
hydrogen market development, 958–
59
international cooperation, 959
in global energy scenario, 33, 36, 40–41
fuel-cell vehicles (FCVs), 34–35, 937–
38, 939–42, 947–49, 957, 965–67,
996–1000
overview, 925
potential contribution of, 942–45, 972–
76, 1001–1002
prospects for, 925–27
for residential heating, 936–37, 957,
959–60, 994
safety of, 945–47, 974
solar hydrogen cycle, 926
from solar powered electrolysis, 928–32,
949–50, 956, 968
costs of, 931–32, 970–71
electrolysis technology, 928–29
potential of, 943–45, 972–76
sources of solar electricity, 930–31,
969, 983–86

for transportation, 937–42, 957
fuel-cell vehicles (FCVs), 34–35, 937–
38, 939–42, 947–49, 957, 965–67,
996–1000
hydrogen storage systems, 938–39,
995, 1004
internal-combustion engine vehicles
(ICEVs), 937, 939
onboard production of hydrogen, 939,
961–64, 995, 1004–1005
transportation of, 934–36
gaseous hydrogen, 934–35, 981, 991
liquid hydrogen, 935–36, 982, 993
strategies for, 936
utilities and fuel-cell cogeneration
systems, 1050–51
Solar Kinetics, Inc., 270–71, 384
Solar One, 238–43
conversion project, 252–55
Solar 100, 250
Solarplant I, 268–69, 273
Solar ponds, see Solar-thermal electric
technology, solar ponds
Solar Power Engineering Company, 247
Solar Test and Research (STAR) Center,
504
Solar-thermal electric technology, 213–96
basis of, 214
capital costs, 217
central-receiver systems, 218, 219–20,
236–64, 277–78, 287–88
advanced-technology pilot plants, 244–
47
air-receiver and storage technology,
256–63, 284–85
cost projections, 282–83
falling-sand receiver, 263–64
heliostat development and status, 247–
49
history of, 236–37
nitrate-salt receiver and storage
technology, 249–56
water/steam pilot plants, 238–44
collector system, 215–16
cost and performance projections, 217,
277–85
advanced systems, 284–85
of central-receiver power plants, 282–
83
levelized energy cost projections, 279–
84
of parabolic-dish power plants,
283–84
of parabolic-trough plants, 279–82
technology attributes, 277–79
efficiency of, 216–17
fossil-fuel driven heat source, 216
in global energy scenario, 14, 20–21
hydrogen from, 931, 932, 953–54, 985
major types of, 214
market potential of, 214–15
operational and maintenance costs, 217
overview, 213–14

overview of future role of, 4
parabolic-dish systems, 218, 220–21,
264–77, 278–79, 288–89
collector technology, 269–71
cost projections, 283–84
design parameters, 272
engines driven by, 264–65, 271–75
pilot projects, 265–69
thermochemical energy transport,
276–77
parabolic-trough systems, 217–19, 236,
277–79 285–87
commercial projects, 223–33
cost projections, 279–82
direct-steam generation, 233–34
early pilot plants, 223
gas-turbine combined-cycle power
plant with direct steam generation,
234–36
history of, 222–23
thermal storage concepts for, 236
power conversion system, 215, 216
receiver system, 215, 216
solar ponds, 221–22, 289
storage of energy, 1043–45
summary, 285–89
test facilities, 286
transport-storage system, 215, 216
in utility investment portfolios, 1028–33
portfolio costs, 1049–50
Solar-Thermal Test Facility (STTF), 286
Solar Two, 254–55
Solar Wasserstoff project, 957
Solid oxide fuel cell, 872
Solid polymer electrolyte fuel cell, 873
South Coast Air Quality Management
District and Electroyzer, Inc., 958
Southern California Edison, 225, 238, 240,
250, 254, 268, 743
photovoltaic demonstration plants, 505–
507
Southern California Gas Company, 937
Southwest Technology Development
Institute, 492
Soviet Union (former), 77, 196
Sovonics, 410, 411, 421
Space, solar cells in, 338–41
Spain, 502, 954–55, 1003
Spanish Ministry of Industry and Energy,
244
Staebler-Wronski effect, 416, 417
Stall, 129, 132
Standard Telecommunications Laboratories,
404
Starch crops, ethanol from, 867, 877, 878–
79
economics of, 894–97
energy balances, 910–12
Steam-injected gas turbine (STIG), 738–39,
744
intercooled (ISTIG), 739, 744
Steam (Rankine) engines, 220, 264–65,
731–34

Stevens, Don J., 865–923
Stirling-cycle engines, 265, 267, 268, 270, 271–75, 278, 279, 283–84
Stirling Technology Company, 275
Stirling Thermal Motors, Inc., 274
Straw, 610, 611
Studsvik, 754
  gasifier, 888
Subsidies distorting energy markets, 47
Sugarcane, 42, 610, 611, 627, 631, 634, 681, 688, 748, 867
  anaerobic digestion of, 834–35
  anaerobic wastewater treatment of by-products of, 831
  BIG/GT application using residues, 765–71, 773–75
  Brazilian ethanol program, *see* Brazilian fuel-ethanol program
  direct production of energy from, 834–35
  see also Cellulosic biomass as source of ethanol and methanol
Sulaibyah Solar Power Station, 267
Sunpower, Inc., 375
Sweden:
  BIG/GT in, 758–59
  biomass plantations in, 623
  ocean energy and, 536, 542
Swiss Center for Appropriate Technology, 86
Switzerland, photovoltaic systems in, 486, 487, 488, 502
Sydkraft, 758
SynGas gasifiers, 887
Synoptic winds, 159
Synthesis gas (syngas):
  conditioning of, 890–91
  generation of, 867–68, 886–88

Taiwan, 543
Tampella Company, 759
Tapered channel device (Tapchan), 534
Tar, 702, 715, 720, 751–53
Tax policy, 48–49, 486
  carbon tax, 60–61, 1051–52
TCLP test, 452
Telecommunications interference, wind turbines causing, 174
Tellerium, 462–63
Tennessee Valley Authority, 79
Texaco, 743
  coal gasifier, 892–93
Themis, 239, 245, 246
Thermochemical-energy transport system, 276–77
Thin films, 308–311
  amorphous silicon, *see* Amorphous silicon photovoltaic systems
  polycrystalline, *see* Polycrystalline thin-film photovoltaics
Tidal energy, 515–30
  conclusions, 530

development trends and possibilities, 523–25
double regulation capabilities of turbines, 519, 520
ebb generation, 520, 521
economics and markets, 525–30
  estimation of economic potential, 526
  general considerations, 525–26
  national studies, 527–29
environmental effects of, 529–30
existing tidal energy schemes, 522–23
flood generation, 520
historically, 518
identified sites for development of, 517, 518
modes of operation, 520
origin of tides, 515
other barrage configurations, 521–22
overview, 513–15
potential, 515, 516, 517
pumping, 519, 520
tidal barrage design and construction, 518–20
tidal stream devices, 523
turbine generators, types of, 519–20, 524
two-way generation, 520, 521
Tobacco, 610
Toshiba Corporation, 275
Tracking systems, 314–16, 318–19, 438
Transportation, fuels for future, 33–42
  ethanol, *see* Brazilian fuel-ethanol program; Ethanol
  solar hydrogen, *see* Solar hydrogen, for transportation
  vehicle redesign and, 45

Umeme (stove), 662, 664
Union Carbide Company, 887
United Kingdom:
  ocean energy and, 536, 542
  wind energy in, 192, 193, 201–202
United Nations, 580
United Nations Conference on Environment and Development, 1992, (Rio de Janeiro), 615, 645
United Nations Development Program (UNDP), 54
United Nations Environmental Program (UNEP), 54
United Nations International Children Emergency Fund (UNICEF), 662, 664
United States:
  biomass plantations in, 623
  fuel-ethanol program, 37–38
  ocean energy and, 543
  photovoltaic systems in, 486, 487, 488, 503–507, 509
  solar hydrogen energy systems for, 947–52
  wind energy in, 187–90
U.S. Agency for International Development, 759

U.S. Department of Agriculture, 634, 638, 878–79
U.S. Department of Defense, 735–36
U.S. Department of Energy (DOE), 947
  Biofuels Systems Division of, 903
  biomass plantations and, 623
  fuel-cell projects and, 873
  gasifiers and, 745
    BIG/GT, 759
  solar energy and:
    cadmium telluride cells and, 456, 457
    central-receiver systems, 238, 246, 250, 254
    CIS development and, 446
    parabolic-dish systems, 265, 267
    parabolic-trough power plants, 222
    photovoltaics and, 410–11, 503
  Solar Energy Research Institute (SERI), 446, 448, 456, 457, 459
  wind energy and, 135, 141, 142, 175, 202
United States Wind Energy Program, 145
U.S. Windpower Inc., 140, 141, 142, 143, 150
United Stirling AB (USAB), 268, 270, 283, 284
Upflow anaerobic sludge blanket, *see* Anaerobic digestion, upflow anaerobic sludge blanket (USAB) technology
Utilities, 1011–69
  BIG/GT power generation and, 779
  competitive pressures on, 16, 1018
  computing cost of providing electric service, 1054–66
  economies of scale, 29–30, 1020–27, 1030
    generating costs, 1020–23
    planning and operational costs, 1024–27
    reliability, 1023–24, 1025
    transmission and distribution costs, 1024
  government policy and:
    buy-back regulations, 202
    national development targets, 203
    wind energy, 202, 203
  government regulation of, 8, 16, 1052–53
  investment portfolios for the future, 8, 23–31, 44–45, 1012–17
    case study, 23–27
    conclusions, 1019–20
    economies of scale, 29–30
    ground rules for portfolio analysis, 1022
    new investment opportunities, 1013–17
    reoptimizing the choice and use of thermal-electric equipment, 27–29
    the scenario, 30–31
    selecting the optimum mix of dispatchable plants, 1016
    traditional electric generation, 1013

  transmission and distribution, 1015
    methods for assessing new investment portfolios, 1027–45
      dispatchable plants, 1035
      ground rules, 1022, 1027
      hourly dispatching, 1035–36, 1039
      hydroelectric power, characteristics of, 1033–34
      intermittents, characteristics of, 1028–34
      load duration curves, 1032
      loads lost, 1037–40
      selection of optimal plant mix, 1035
      storage, 1041–45
      transmission and distribution, 1036–41, 1044
    overview of future role of, 6, 8, 1011
    photovoltaics and, *see* Utility field experience with photovoltaic systems
    portfolio costs, 1045–52
      biomass systems, 1050
      hydroelectricity, 1050
      distributed systems, 1050–51
      environmental costs, 1051–52
      reference system costs, 1045–49
      systems with intermittents, 1049–50
    reliability of electrical service, 1018, 1021, 1023–24, 1025, 1037–40
    sensitivity to consumer interests, 1018
Utility field experience with photovoltaic systems, 483–512
    distributed systems and reliability, 1037–40
    future of, 492–94
    lessons learned by, 489–92
      inverters, 490
      operation and maintenance, 491–92
      prediction of system power output, 490–91
      safety, 489–90
      scheduling and installation, 489
      system performance, 489
    national programs to encourage, 484–86
    overview, 483–84
    research, development, and demonstration, 487–88
    test facilities and demonstrations by country, 494–509
    utility investment portfolio costs, 1049–50
Utility Power Group, 411
Utility Studies, U.S., 282–83

Vacuum evaporation, 446, 447, 448
Vanguard I, 267–68, 338
Van Haandel, Adriaan C., 817–40
Vant-Hull, Lorin, 237
Varian Company, 380, 381, 384, 388, 389
Vattenfall, 759
Vermont Department of Public Service, 759
Vigotti, Roberto, 483–512
Vinasse, 828–34

Virginia Integrated Solar Test Arrays
(VISTA) facility, 504
Visser, Piet, 662
Volumetric receivers, 257

Wagner, Sigurd, 403–35
Wake effects, 133
Water boiling tests, 655, 657, 675, 676, 677
Water quality:
  biomass plantations and, 630
  ethanol production's effects on, 851–52
  hydroelectric power plants and, 100–103
Wave energy, 530–38
  areas suitable for extraction of, 531
  arrays of devices, 533–34
  attenuators, 533
  conclusions, 537–38
  determinants of, 530
  development trends, 534–36
  device size, 533
  economics of, 536–37
  environmental effects, 537
  large-scale offshore wave-energy devices,
    531–32
  nearshore wave-energy stations, 534
  point absorbers, 533
  shoreline devices, 534
  terminators, 532–33
Wave refraction, 530
Weibull distribution, 165
Weinberg, Carl J., 1011–69
Weizmann Institute of Science, 276–77,
  286
Wells air turbine, 534, 535
Westinghouse Electric Company, 135, 305
White Cliffs Solar Power Station, 266–67
Wild and Scenic Rivers Act, 92
Williams, Robert H., 1–71, 593–651, 729–
  85, 1071–1141
Willow plantations, 623
Wind atlases, 163–64, 187, 191–92, 194
Wind energy:
  costs, *see* Wind energy economics
  in global energy scenario, 14, 18–20
  integration studies and system-operating
    penalties at high power system
    penetrations, 205–208
  overview of future role of, 3–4
  potential, *see* Wind energy potential
  resources, *see* Wind energy resources
  technology, *see* Wind energy technology
Wind energy economics, 146–53, 930–31
  capital costs, 146–47
  conclusions, 153
  insurance costs, 149
  land rental costs, 149
  legislation affecting, 149–50, 200, 202
  operation and maintenance costs, 147–
    49
  other factors, 151–52
  policies for developing and deploying,
    199–204
    in California, 200

    in Denmark, 200–201
    market prospects, 203–204
    types of, 200, 202–203
  projected future costs, 152–53
  transmission costs, 151–52
  utilities with investment portfolios
    including wind energy, 1028–33
    portfolio cases, 1049–50
Wind energy potential:
  in Africa, 194–95
  in Asia, 195–96
  in Australasia, 196
  in Europe, 191–94
  globally, 197–99
  in Latin America, 196–97
  in North America, 187–91
  technical assumptions, 186–87
Wind energy resources, 157–76
  conclusions, 204–205
  constraints on energy extraction, 171–76
    array interference and siting densities,
      152, 171–72
    bird strikes, 173
    land requirements, 172–73
    noise, 173
    safety, 174
    siting constraints, 175–76
    telecommunication interference, 174
    visual impact and public acceptance,
      174–75
  general features of, 158
  global wind systems, 158–59
  height above sea level, 163–64
  overview, 157–58
  potential of, *see* Wind energy potential
  surface roughness and height above
    ground, 159–63
  surveys of, 202
  system integration, *see* Wind system
    integration
  terrains and, 163
  variations in wind energy, 165–70
    energy fluctuations, 167–69
    predictability, 170, 171
    windspeed distributions, 165–67
  see also Wind energy technology
Wind energy technology, 121–56
  aerodynamics of, 129–30
    drag, 129, 130
    lift, 129, 130
    pitch angle, 129, 130
    stall, 129, 131
  assumptions to estimate wind energy
    potential, 125–27
  Betz limit, 124
  conclusions, 153
  economics and, *see* Wind energy
    economics
  environmental advantages of, 129
  future developments, 142–46
  history of, 123–24, 133–36
  horizontal axis wind turbine (HAWT),
    133, 134, 135–36
  MOD-0 turbine, 135

operating experience, 140
overview, 121–24
power output, 130–33
  average, 132
  capacity factor, 132–33
  coefficient of performance, 131
  number of blades, 131
  stall control, 131–32
presently available wind turbines, 140, 141
variable-speed wind turbines, 140–42
vertical axis wind turbine (VAWT), 133, 134, 135
wake effects, 133
wind speed distributions, 127, 128, 165–67
wind hydrogen and, 930–31, 943, 949, 951, 952, 986–87, 1001–1002
wind power density, 124–25
wind speed, 125, 126–27, 158
wind turbine subsystems, 136–40
  drive trains, 139
  electrical systems, 139–40
  rotors, 136–39
  yaw-control systems, 139
Windfall Profits Act, 149
Wind frequency distributions, 127, 128, 165–67
Windmills, 123–24
Wind system integration, 176–85
  capacity credit, 179, 180, 182
  capacity remix value, 182
  fuel savings, 182
  large-scale systems:
    at high wind penetration, 181–85
    at low wind penetration, 178–81
  operational impacts and reserve, 181
  overview, 176
  small-scale systems, 176–78
Wind turbines:
  constraints on use of, *see* Wind energy resources, constraints on energy extraction
  economics of, *see* Wind energy economics
  system integration, *see* Wind system integration

technology of, *see* Wind energy technology
Wisconsin Electric, 1018
Wood, *see* Fuelwood; Roundwood
Wood gasifiers, open-top, 670, 699–728
  the blower, 708, 709, 710, 711
  the burner, 708, 709
  the chip-loading system, 713, 714
  versus closed-top gasifier, 700–705
  conclusions, 726
  cooling and cleaning systems, 706–709, 711
  demonstration projects, 716–18
    Hosahalli project, 717–18
    Port Blair sawmill project, 718
  economics of, 720–26
    maintenance costs, 721–22
    operation costs, 722–26
    payback period, 721, 723–26
  elements of, 705–706
  environmental considerations, 720, 721
  field trials of, 716–20
  instrumentation and control, 713–15
  overview, 699–700
  performance of gasifier-engine system, 715–16
  pyrolysis, 700, 705
  the reactor, 706, 707
  safety considerations, 720
  steps in conversion of biomass to gas, 703–704
  tar and particulates from, 702, 715, 720
  the wood-processing system, 711–13
Woods, Jeremy, 593–651
World Bank, 580, 672, 848
  Global Environment Facility, 54
  on hydropower, 96, 107–108, 111
World Energy Conference, 14, 196
World Meteorological Organization, 122, 194, 197
World Solar Challenge, 346–47, 389
Wyman, Charles E., 865–923

Zero emission vehicles (ZEV), 941, 942
Zweibel, Ken, 437–82

AUTHORS

Richard L. Bain
National Renewable Energy Laboratory
Golden, Colorado
United States

Allen M. Barnett
AstroPower, Inc.
Newark, Delaware
United States

Eldon C. Boes
National Renewable Energy Laboratory
Washington DC
United States

David E. Carlson
Solarex
Newtown, Pennsylvania
United States

Alfred J. Cavallo
Princeton University
Princeton, New Jersey
United States

James E. Cavanagh
Harwell Laboratory
Didcot, Oxfordshire
United Kingdom

John H. Clarke
Harwell Laboratory
Didcot, Oxfordshire
United Kingdom

S. Dasappa
Indian Institute of Science
Bangalore
India

Pascal De Laquil III
Bechtel Group, Inc.
San Francisco, California
United States

Richard Diver
Sandia National Laboratories
Albuquerque, New Mexico
United States

Gautam S. Dutt
Villa Ballester
Buenos Aires
Argentina

Kay Firor
Blue Mountain Energy
Cove, Oregon
United States

Michael Geyer
Flachglas Solartechnik
Cologne
Germany

José Goldemberg
Secretary for the Environment
Brasília
Brazil

Martin A. Green
University of New South Wales
Kensington
Australia

Michael J. Grubb
The Royal Institute of International Affairs
London
United Kingdom

David O. Hall
King's College
London
United Kingdom

Norman D. Hinman
National Renewable Energy Laboratory
Golden, Colorado
United States

Susan M. Hock
National Renewable Energy Laboratory
Golden, Colorado
United States

Joseph J. Iannucci
Pacific Gas and Electric Company
San Ramon, California
United States

S. Jayakumar
Indian Institute of Science
Bangalore
India

Thomas B. Johansson
University of Lund
Lund
Sweden

David Kearney
LUZ Development and Finance Corporation
Los Angeles, California
United States

Henry Kelly
Office of Technology Assessment
Washington DC
United States

Eric D. Larson
Princeton University
Princeton, New Jersey
United States

Gatze Lettinga
Agricultural University
Wageningen
The Netherlands

Antonio Luque
Solar Energy Institute
Polytechnic University of Madrid
Spain

Isaias C. Macedo
Copersucar Technical Center
Piracicaba, São Paulo
Brazil

Niels I. Meyer
The Technical University of Denmark
Lyngby
Denmark

Lourival C. Monaco
Office of Science and Technology
Brasília
Brazil

José Roberto Moreira
Office of Science and Technology
Brasília
Brazil

H. S. Mukunda
Indian Institute of Science
Bangalore
India

Joachim Nitsch
DLR (German Aerospace Research Establishment)
Stuttgart
Germany

Joan M. Ogden
Princeton University
Princeton, New Jersey
United States

Civis G. Palmerini
Italian Electricity Board
Pisa
Italy

Alan Douglas Poole
National Institute of Energy Efficiency
Rio de Janeiro
Brazil

Roger Price
Harwell Laboratory
Didcot, Oxfordshire
United Kingdom

P. Rajabapaiah
Indian Institute of Science
Bangalore
India

N. H. Ravindranath
Indian Institute of Science
Bangalore
India

Amulya K. N. Reddy
International Energy Initiative
Bangalore
India

Frank Rosillo-Calle
King's College
London
United Kingdom

U. Shrinivasa
Indian Institute of Science
Bangalore
India

Don R. Smith
Pacific Gas and Electric Company
San Ramon, California
United States

Don J. Stevens
Cascade Research, Inc.
Kennewick, Washington
United States

Adriaan C. van Haandel
Federal University of Paraiba
Campina Grande
Brazil

Roberto Vigotti
Italian Electricity Board
Pisa
Italy

Sigurd Wagner
Princeton University
Princeton, New Jersey
United States

Carl J. Weinberg
Pacific Gas and Electric Company
San Ramon, California
United States

Robert H. Williams
Princeton University
Princeton, New Jersey
United States

Jeremy Woods
King's College
London
United Kingdom

Charles E. Wyman
National Renewable Energy Laboratory
Golden, Colorado
United States

Ken Zweibel
National Renewable Energy Laboratory
Golden, Colorado
United States

ABOUT THE EDITORS

THOMAS B. JOHANSSON is professor of Energy Systems Analysis, Department of Environment and Energy Systems at the University of Lund in Sweden. He chairs the United Nations Solar Energy Group for Environment and Development, serves as a member of Sweden's Environmental Council, and is co-author with J. Goldemberg, A.K.N. Reddy, and R.H. Williams of *Energy for a Sustainable World*. He has a Ph.D. in nuclear physics from the University of Lund.

HENRY KELLY is senior associate at the Office of Technology Assessment (OTA) of the U.S. Congress.* He has served as special assistant to the Assistant Secretary for Conservation and Renewables at the U.S. Department of Energy, and as Assistant Director of the Solar Energy Research Institute in Colorado. He was project director and principal author of *A New Prosperity: Building a Sustainable Energy Future* and *Technology and the American Economic Transition*. He has a Ph.D. in physics from Harvard University.

AMULYA K.N. REDDY is founding director and president of the International Energy Initiative, a nonprofit North-South organization based in Bangalore, India, that promotes the efficient use and production of energy, including renewable sources of energy. He was formerly professor of the Department of Inorganic and Physical Chemistry and Chairman of the Department of Management Studies at the Indian Institute of Science at Bangalore. He has a Ph.D. in applied physical chemistry from London University.

ROBERT H. WILLIAMS is senior research scientist at the Center for Energy and Environmental Studies, Princeton University. He is recipient of the American Physical Society's Leo Szilard Award for Physics in the Public Interest and the U.S. Department of Energy's Sadi Carnot Award for his contributions to the field of energy efficiency. He has a Ph.D. in physics from the University of California, Berkeley.

LAURIE BURNHAM consults on energy and environmental issues; she was formerly associate editor and editor in charge of single-topic issues at *Scientific American*.

---

*The views expressed in this document are those of the author and not the Office of Technology Assessment or the OTA board.

Also Available from Island Press

Balancing on the Brink of Extinction: The Endangered Species Act and Lessons for the Future
Edited by Kathryn A. Kohm

Coastal Alert: Ecosystems, Energy, and Offshore Oil Drilling
By Dwight Holing

The Complete Guide to Environmental Careers
By The CEIP Fund

Crossing the Next Meridian: Land, Water, and the Future of the West
By Charles F. Wilkinson

The Energy-Environment Connection
Edited by Jack. M. Hollander

Farming in Nature's Image
By Judith Soule and Jon Piper

Ghost Bears: Exploring the Biodiversity Crisis
By R. Edward Grumbine

The Global Citizen
By Donella Meadows

Green at Work: Making Your Business Career Work for the Environment
By Susan Cohn

Holistic Resource Management
By Allan Savory

The Island Press Bibliography of Environmental Literature
By The Yale School of Forestry and Environmental Literature

Last Animals at the Zoo: How Mass Extinction Can Be Stopped
By Colin Tudge

Learning to Listen to the Land
Edited by Bill Willers

Lessons from Nature: Learning to Live Sustainably on the Earth
By Daniel D. Chiras

The Living Ocean: Understanding and Protecting Marine Biodiversity
By Boyce Thorne-Miller and John G. Catena

Making Things Happen
By Joan Wolfe

Nature Tourism: Managing for the Environment
Edited by Tensie Whelan

Not by Timber Alone
By Theodore Panayotou and Peter S. Ashton

Our Country, The Planet: Forging a Partnership for Survival
By Shridath Ramphal

Overtapped Oasis: Reform or Revolution for Western Water
By Marc Reisner and Sarah Bates

Population, Technology, and Lifestyle: The Transition to Sustainability
Edited by Robert Goodland, Herman E. Daly, and Salah El Serafy

Rain Forest in Your Kitchen: The Hidden Connection Between Extinction and Your Supermarket
By Martin Teitel

The Snake River: Window to the West
By Tim Palmer

Spirit of Place
By Frederick Turner

Taking Out the Trash: A No-Nonsense Guide to Recycling
By Jennifer Carless

Turning the Tide: Saving the Chesapeake Bay
By Tom Horton and William M. Eichbaum

Visions upon the Land: Man and Nature upon the Western Range
By Karl Hess, Jr.

The Wilderness Condition
Edited by Max Oelschlaeger

For a complete catalog of Island Press publications, please write: Island Press, Box 7, Covelo, CA 95428, or call: 1–800–828–1302